Building Energy Simulation

Building Energy Simulation

A Workbook Using DesignBuilder™

Second Edition

Vishal Garg
Jyotirmay Mathur
Aviruch Bhatia

CRC Press is an imprint of the
Taylor & Francis Group, an **informa** business

Second edition published 2020
by CRC Press
6000 Broken Sound Parkway NW, Suite 300, Boca Raton, FL 33487-2742

and by CRC Press
2 Park Square, Milton Park, Abingdon, Oxon, OX14 4RN

First edition published by CRC Press 2017

Second edition published by CRC Press 2020

CRC Press is an imprint of Taylor & Francis Group, LLC

ISBN13: 978-0-367-37470-9 (Hardback)
ISBN13: 978-0-367-37468-6 (pbk)
ISBN13: 978-0-429-35463-2 (eBook)

Visit the eResources: https://www.routledge.com/9780367374686

Dedication

This book is dedicated to our beloved teacher, the late Professor N. K. Bansal, who introduced the use of building energy simulation in India for energy-efficient building design.

Contents

Preface

Building Energy Simulation: A Workbook Using DesignBuilder is an outcome of a series of training programs conducted for participants with varied backgrounds. The authors experimented with various teaching techniques and arrived at the conclusion that the most effective method of imparting these training programs is through tutorials and step-by-step instructions along with graphical illustrations. This book is an updation of its first edition, released in 2017. Several new tutorials have been added, as was suggested to the authors through feedback received from book users across the world. The authors thankfully acknowledge these inputs.

The simulations in this workbook are performed using the DesignBuilder software for illustration purpose to help explain the aspects of a whole-building energy simulation process. This workbook adopts the 'learning by doing' principle to explain the fundamentals of building physics and building services and, in turn, help readers understand the concept of building energy performance. Based on participant feedback during the training programs, the authors decided to use EnergyPlus with DesignBuilder as the front end to explain the simulation process.

The book has been organized as follows:

- The first 11 chapters of this workbook cover various aspects of simulation, such as creating the building geometry, assigning material and equipment and analysing the results.
- Chapter 12 is new in this edition and explains the method of simulating renewable-energy systems fitted on buildings.
- Chapter 13 is also a new one that explains the method of carrying out life-cycle cost analysis of various design options, including sensitivity and uncertainty analysis.
- Chapter 14 explains simulation for the whole-building performance method of the ASHRAE 90.1 standard.
- Chapters 15 through 17, which are available online, provide exercises to simulate three different building projects.

The authors would highly appreciate any feedback or suggestions for further improving this workbook.

Acknowledgements

The authors thank all those who helped during the research, writing, review and editing process. Their help immensely contributed to making this workbook a reality. We would like to start by thanking all professionals, researchers and students from all over the globe for providing their feedback during the various building simulation training programs that we conducted during the past few years. This feedback helped us improve the building simulation teaching methodology and motivated us to create this workbook.

We thank late Professor N. K. Bansal, who not only introduced us to this subject of building science but also served as our role model in learning the art and science of the teaching process. We are also grateful to DesignBuilder Software, Ltd., Stroud, United Kingdom, for their technical support in developing this book and answering our queries during the writing process.

This workbook would not have been possible without all those reviewers who took the time to patiently go through the contents and provide their valuable feedback. We especially appreciate the contributions from Gaurav Chaudhary, Hema Rallapalli, Kopal Nihar, Nishesh Jain, Sraavani Gundepudi, Shivraj Dhaka and all the students from the International Institute of Information Technology (IIIT), Hyderabad, India, and the Malaviya National Institute of Technology (MNIT), Jaipur, India, who reviewed this workbook and provided feedback on the technical contents and accuracy.

Our special thanks go to Naresh Arthem for running the simulations for all the tutorials, capturing screen shots and closing the technical and editorial comments provided by various reviewers. We also thank Vijay Singh, Sahil Chilana, and Kuntal Chattopadhyay for rerunning all the simulations and for their thorough review of the manuscript.

Our thanks to all friends and family members for their support, and encouragement. Without their support this book would have not been possible.

Our sincere gratitude goes to the team at CRC Press/Taylor & Francis Group, especially Dr Gagandeep Singh for his trust in us and the numerous extensions to the timelines for delivering this workbook.

Authors

Vishal Garg is professor and head of the Center for IT in Building Science at the International Institute of Information Technology (IIIT), Hyderabad, India. His current research interests are in the areas of energy simulation and cool roofs. He teaches building automation and controls, energy simulation and lighting design and technology. He has conducted several national and international workshops on intelligent buildings, green buildings and energy simulation. He holds a BTech (Hons.) degree in civil engineering from MBM Engineering College, Jodhpur, India, and a PhD from the Indian Institute of Technology, Delhi, India. Dr. Garg is actively involved in the green building movement and in developing eTools for advancing energy efficiency in buildings and energy efficiency building code and its implementation. He was the founding president of the Indian chapter of the International Building Performance Simulation Association (IBPSA) and chaired the organizing committee of the International Conference for Building Simulation 2015 and the International Conference on Countermeasures to Urban Heat Islands (IC2UHI) in 2019. He is fellow of IBPSA and received the inaugural Arthur H. Rosenfeld Urban Cooling Achievement Award in 2018.

Jyotirmay Mathur is professor of mechanical engineering and the founding head of the Centre for Energy and Environment at Malaviya National Institute of Technology, Jaipur, India. He has done postgraduate work in energy studies at the Indian Institute of Technology, Delhi, India, and has received a doctorate in energy systems from the University of Essen, Germany. Dr. Mathur has published 80 research papers in refereed international journals and has presented more than 150 papers and talks at international seminars and conferences, besides writing five books. Dr. Mathur works in the field of energy modelling, codes and standards, energy conservation in buildings, passive cooling, adaptive thermal comfort and building integrated photovoltaic systems.

Aviruch Bhatia is assistant professor at the TERI School of Advanced Studies, New Delhi, India. He holds a PhD from the International Institute of Information Technology, Hyderabad, India, an MTech degree in energy engineering from the Malaviya National Institute of Technology, Jaipur, India, and MSc and MPhil degrees in physics from the University of Rajasthan, Jaipur, India. His areas of interest include building physics, calibrated energy simulation and fault detection and diagnostics in heating, ventilation and air conditioning systems. He has also worked for three years as an assistant manager at Sustainability Group of Spectral Consultant, Pvt. Ltd. (an AECOM company).

1 Getting Started with Energy Simulation

This chapter is designed to acquaint you with building energy simulation. It starts with the introduction of various components and requirements for simulation and then gradually progresses to the concept of simulation using a ready example file. It establishes the impact of key simulation inputs such as lighting power density (LPD), activity, setpoint, window-to-wall ratio (WWR), orientation and fresh air intake on energy consumption. There are 11 tutorials to help you navigate through the chapter. By the end of this chapter, you will be familiar with key input parameters, output visualization and the overall simulation process.

BUILDING ENERGY SIMULATION

Building energy simulation is performed using a computer to virtually represent a building design and perform physics-based calculations. The simulations can range from a building component to a cluster of buildings. For energy simulation, the building model, the usage pattern and the weather of the location are required to determine various outputs, such as peak loads, system sizing and energy consumption for any given period. This information can be used for estimating the utility bills and for evaluating cost-benefit analyses of various design strategies.

Some of the uses for energy simulation tools include the following:

Early design decisions. In the early design stage, decisions such as the orientation and layout of the building are taken. Energy simulation can help in evaluating various design strategies. However, a detailed simulation may not be possible because of the limited information available at this stage.

Component or material selection. Simulation helps in the decision-making process when selecting individual components of the building envelope or systems. It is quite commonly used to carry out cost-benefit analyses of various designs and components. Therefore, modelling at this stage needs to be performed with greater accuracy than the modelling for early design decisions.

Retrofitting decisions. For retrofits of existing buildings, energy simulation can help in selecting cost-effective solutions. For an accurate analysis, the simulation model should be calibrated using the measured performance data of the building.

WHAT IS NEEDED FOR ENERGY SIMULATION

Energy simulation of buildings can be performed using a systematic approach. A lot of data is required. It is recommended that you collect the required data before you start the modelling. The following basic information is required:

Location and weather file. Energy simulation tools need hourly data on ambient conditions (i.e. temperature, humidity, wind velocity, solar radiation, etc.) at the building location. This information is available in weather files. Simulation tools use these weather files to extract the hourly ambient conditions while carrying out the simulation. However, for some locations, the weather file may not be available. In such cases, the weather file of some other location with similar weather conditions can be used. In addition, weather files are available in different formats. Different simulation tools use different formats of weather files. Utilities are available on the web for converting a weather file from one format to another.

Building geometry. Building elevation and floor plans are required to create the geometric model of a building. Architectural drawings may have many details that might not be directly useful for energy simulation. It is useful to simplify the drawings based on thermal zoning into a single line drawing by removing unnecessary details.

Envelope components. It is necessary to have construction details, such as the thickness and thermophysical properties of materials used in each layer of the building envelope. Besides the opaque components, it is very important to know the properties of the window glass, frames and shading devices.

Building services. Information on various services such as heating, ventilation and air conditioning (HVAC) and lighting is required. This includes equipment capacities, energy efficiency, location and controls.

Use of the building. The hourly values of the following are required:

- Occupancy
- Lighting
- Equipment
- Thermostat setpoint
- HVAC operation

HOW SIMULATION SOFTWARE WORKS

The simulation program enables simultaneous interaction of the geometric model with outdoor conditions, occupancy and use of building systems to predict various loads arising in the building on an hourly basis. Basic laws of physics and energy balance equations are used for calculations. The energy consumption for the operation of systems corresponding to the heat and other loads is also calculated on the same time scale. Results of the processing are passed to the calculations of the next time slice and are also supplied to the output file. This process continues for the entire duration of the simulation, and the final output is seen as aggregated or on the same time slice for which calculations have been carried out. Most simulation tools are capable of simulating the energy flows through different building components on an hourly basis, including the transient effects of the envelope and systems.

The advantage of energy simulation over the classical method is that various effects of the thermophysical properties of the materials and the performance of the various systems under varying external and internal environmental conditions are considered in energy calculations. Most energy simulation tools do not require any special computing power because they can be run on commercially available desktop computers or laptops.

This chapter provides basic tutorials on creating simple geometry and analysing the impact of building orientation, WWR, internal loads and fresh air delivery, with a special focus on analysing building energy performance and system sizing. The tutorials are followed by exercises. There are many energy simulation tools available. For this book, we will be using DesignBuilder v6.1. The latest version of DesignBuilder can be downloaded from https://designbuilder.co.uk/download. Please note that there can be minor differences in the user interface and simulation results between the latest version of DesignBuilder and the version used for this book.

Tutorial 1.1

Opening and Simulating an Example File

GOAL

To evaluate the energy performance of a building model provided in an example file with DesignBuilder installation

WHAT ARE YOU GOING TO LEARN?

- How to simulate an example file
- How to view energy consumption results based on utility type or fuel, such as electricity or gas
- How to view the end-use energy consumption (i.e. lighting, equipment, fan, cooling, heating, pump and domestic hot water)
- How to view results on a daily, monthly and annual basis
- How to switch between metric (*Système international* [SI]) and English (inch-pound [IP]) measurement units
- How to calculate energy use intensity (EUI)

PROBLEM STATEMENT

In this tutorial, you are going to simulate an existing template, **Courtyard with VAV Example**, from the DesignBuilder library and simulate it for the climate of London. You will learn how to view the daily, monthly and annual energy consumptions in a graphical format. You will also learn how to view the annual fuel breakdown and fuel total.

SOLUTION

Step 1: Start DesignBuilder. The DesignBuilder main screen appears.

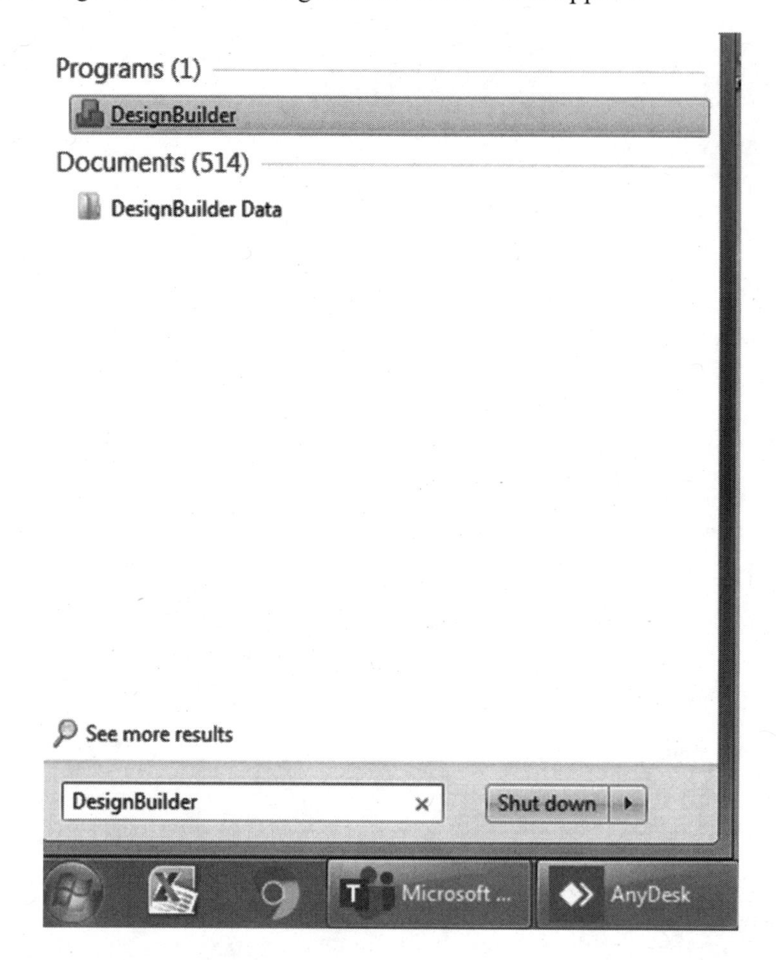

Step 2: Double-click Courtyard with VAV Example in the DesignBuilder templates under the **Recent Files** tab. A **New project** screen appears.

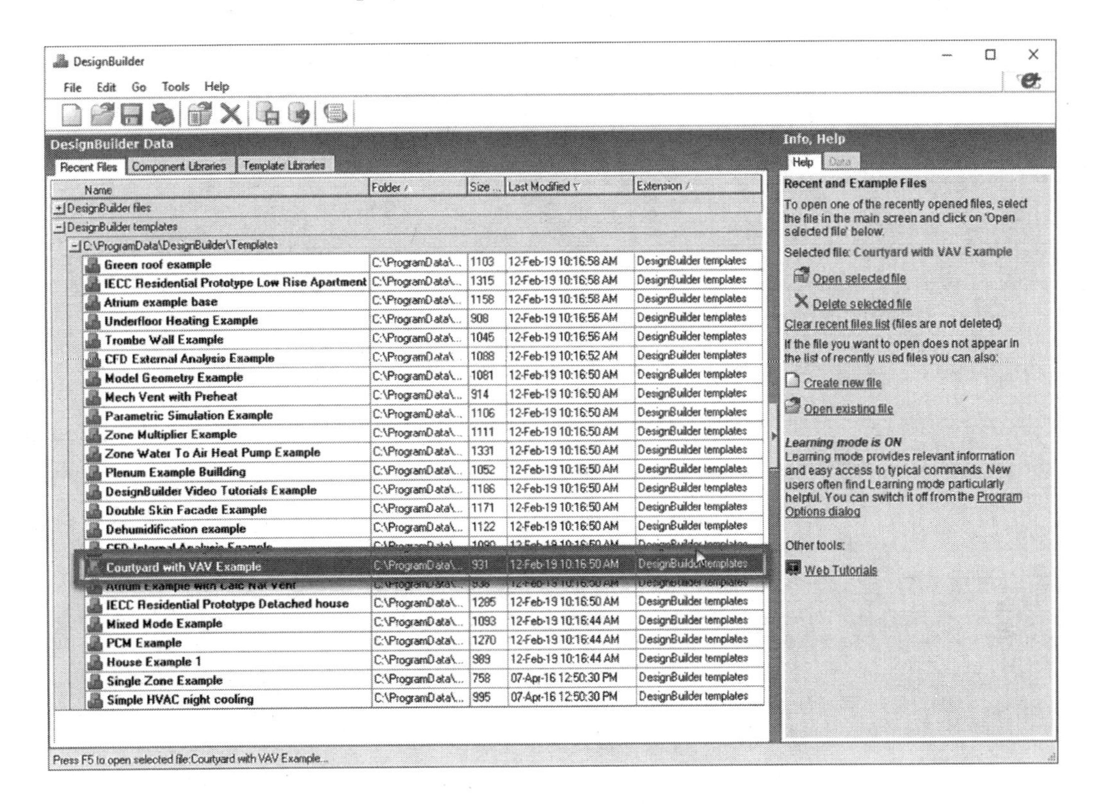

Courtyard with VAV Example is a DesignBuilder template file. A DesignBuilder template file helps to provide examples of various building typologies and systems such as geometry, HVAC system and passive strategies. Relevant data from the example files can be exported and imported into other DesignBuilder models.

Step 3: Click CA-SAN FRANCISCO INTL. Three dots (. . .) appear.

If you have installed DesignBuilder for the first time on your system, **CA-SAN FRANCISCO INTL** may appear as your location. If you have used DesignBuilder previously, you may get some other location based on your previous settings.

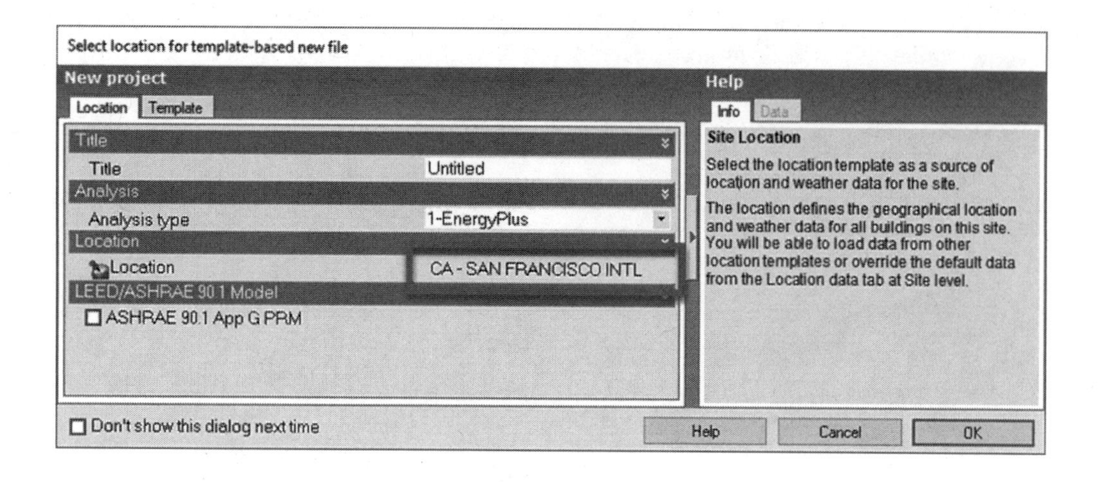

Step 4: Click the three dots (. . .). The **Select the location template** screen appears.

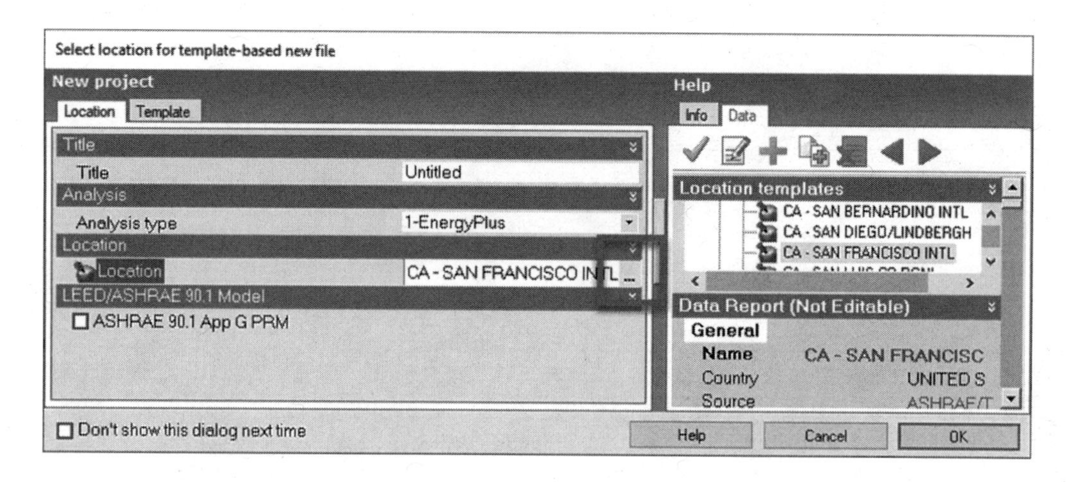

Step 5: **Select LONDON/GATWICK ARPT**. Click **OK**. The **Select location for template-based new file** screen appears.

LONDON/GATWICK ARPT is a weather file that comes along with the Design-Builder installation. For all other locations, an internet connection is needed to download the weather files.

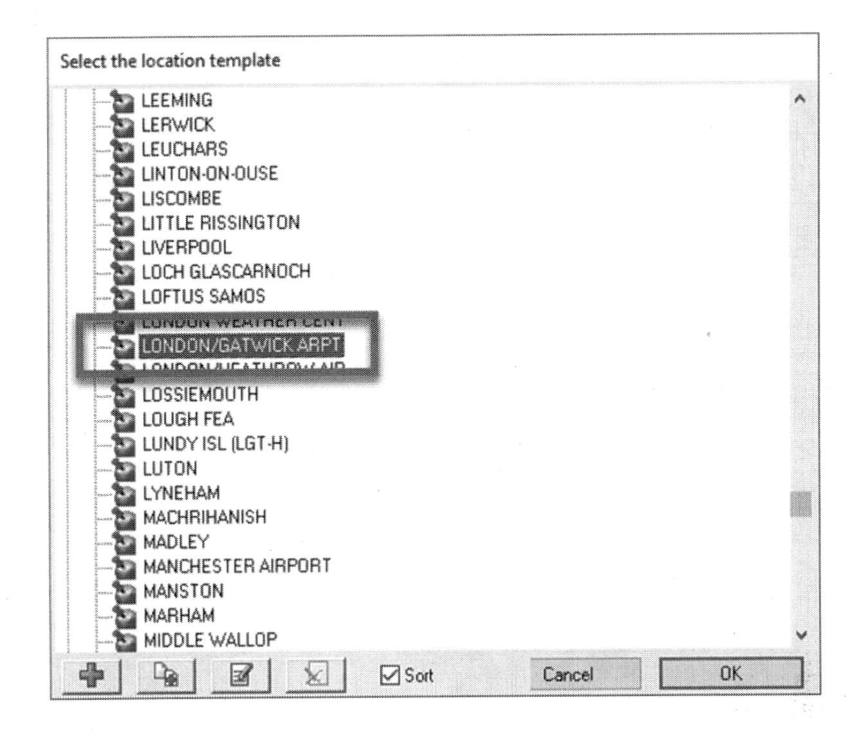

Step 6: Click **OK**.

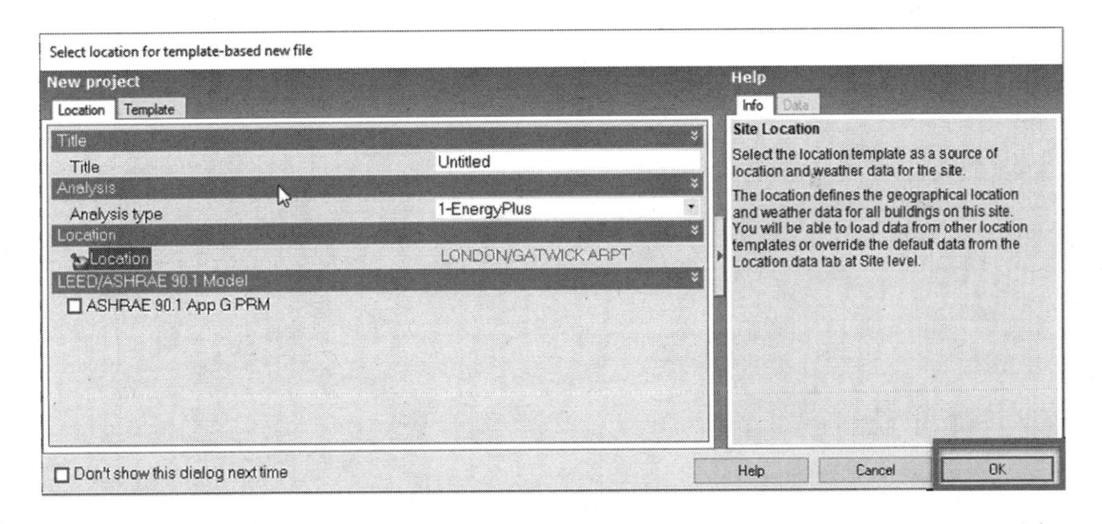

The building layout appears with the following tabs at the bottom of the central display screen:

- **Edit**
- **Visualise**
- **Heating design**
- **Cooling design**
- **Simulation**
- **CFD**
- **Daylighting**
- **Cost and carbon**

Step 7: Select the **Simulation** screen tab. Click **Update data**.

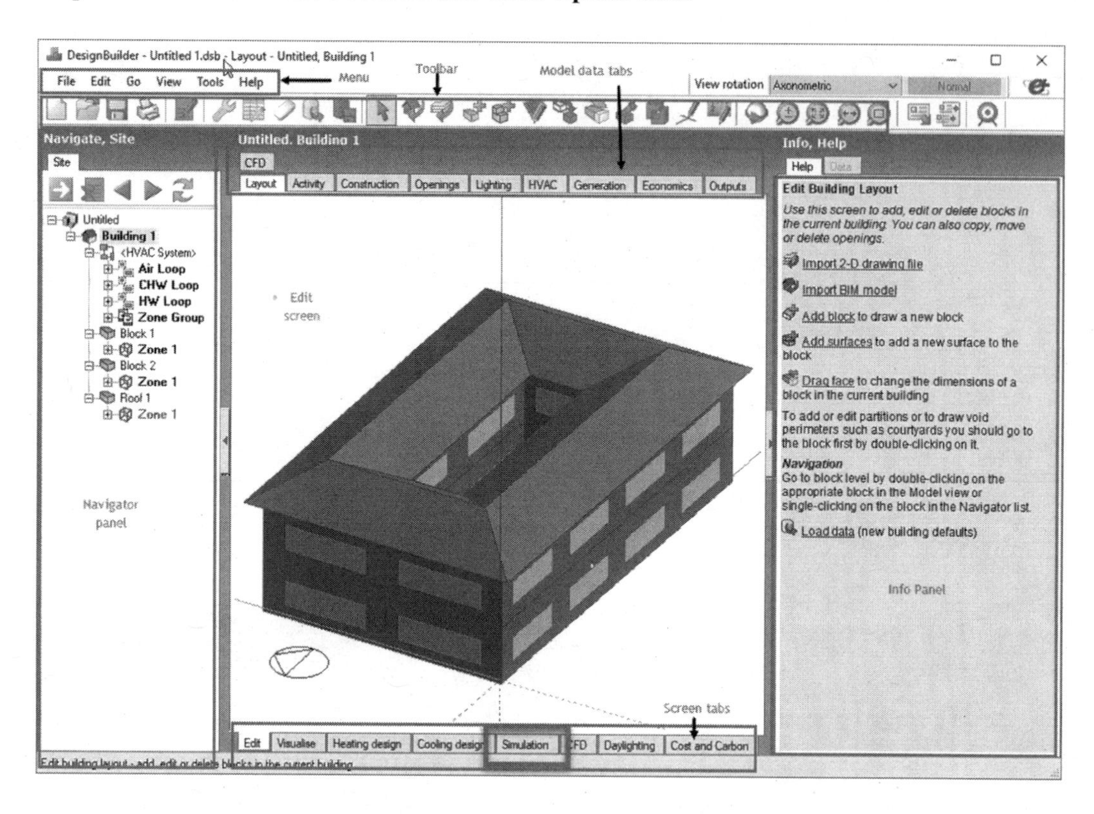

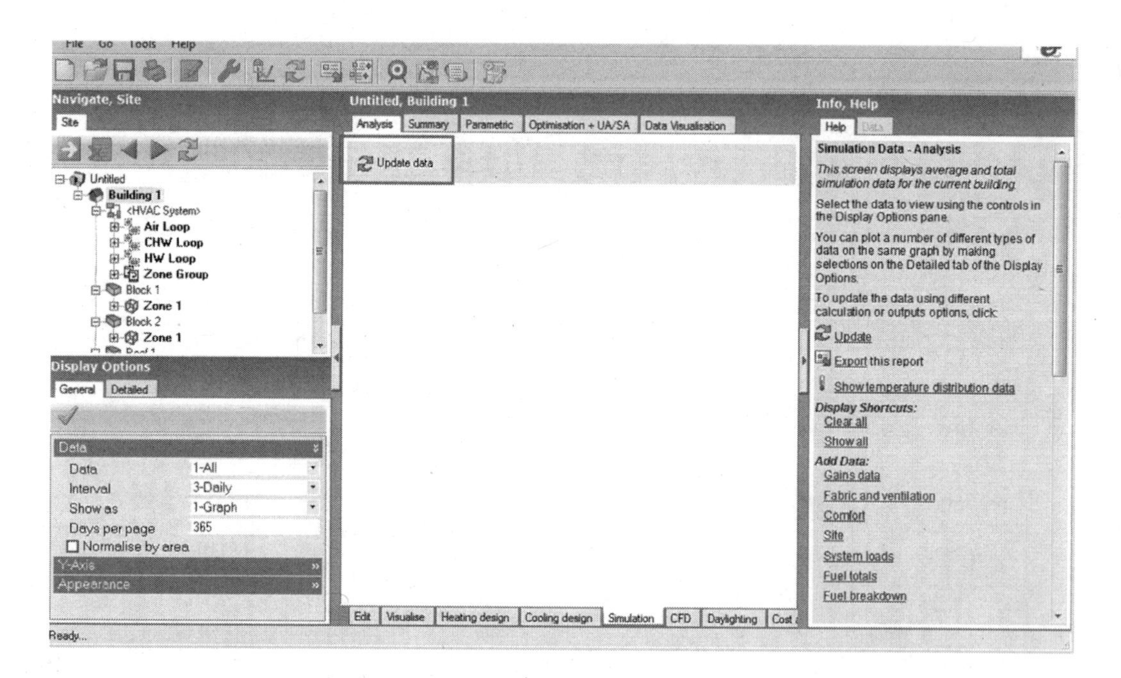

The **Edit Calculation Options** screen appears.

Step 8: Enter the **Simulation Period** start date as **1 Jan** and the end date as **31 Dec**. Click **OK** to start the simulation.

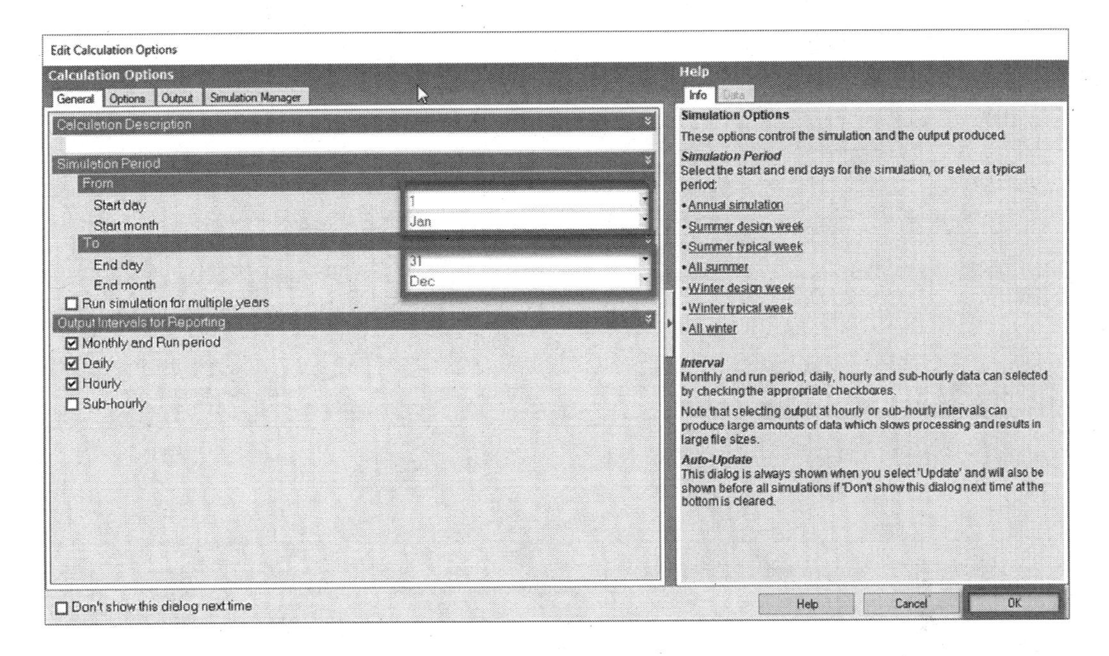

A screen appears showing the simulation progress. Depending on the configuration of the computer, it can take several minutes to complete the simulation.

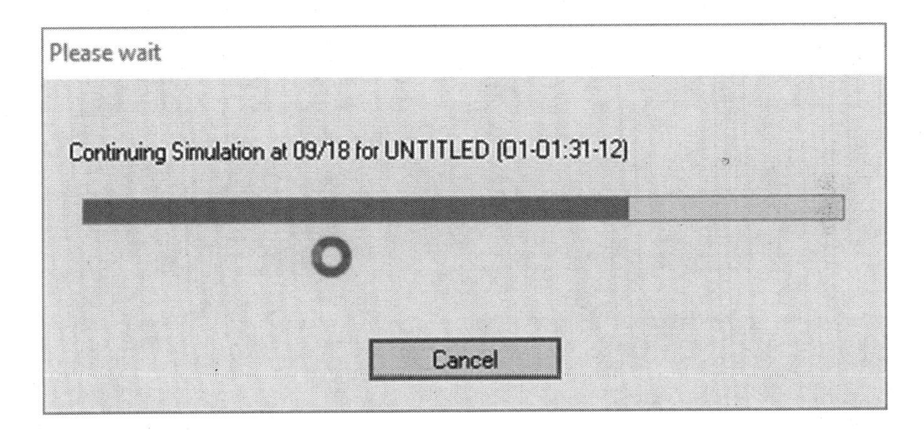

After the simulation is completed, results are displayed in the **Analysis** subtab of the **Simulation Screen** tab.

Step 9: Under **Display Options**, select **Fuel totals** from the **Data** drop-down list and **Run period** from the **Interval** drop-down list. A screen appears with a bar graph for annual fuel totals for electricity and gas.

Note: The results are either displayed in IP or SI units based on your configuration of DesignBuilder. Switching between SI and IP units is explained in Step 15.

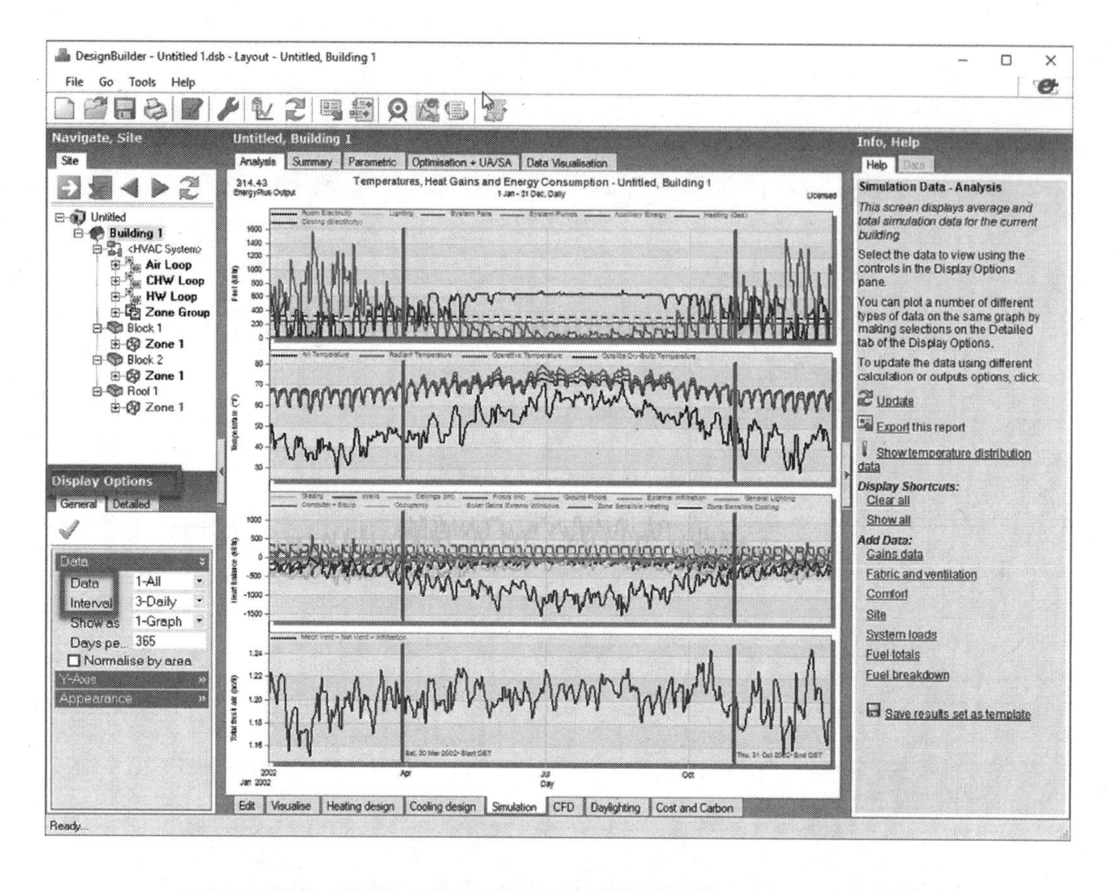
DesignBuilder - Untitled 1.dsb - Layout - Untitled, Building 1
File Go Tools Help
Navigate, Site
Site
Untitled
Building 1
<HVAC System>
Air Loop
CHW Loop
HW Loop
Zone Group
Block 1
Zone 1
Block 2
Zone 1
Roof 1
Zone 1
Display Options
General Detailed
Data
Data 1-All
Interval 3-Daily
Show as 1-Graph
Days pe... 365
Normalise by area
Y-Axis
Appearance
Untitled, Building 1
Analysis Summary Parametric Optimisation + UA/SA Data Visualisation
Temperatures, Heat Gains and Energy Consumption - Untitled, Building 1
Edit Visualise Heating design Cooling design Simulation CFD Daylighting Cost and Carbon
Info, Help
Help Data
Simulation Data - Analysis
This screen displays average and total simulation data for the current building.
Select the data to view using the controls in the Display Options pane.
You can plot a number of different types of data on the same graph by making selections on the Detailed tab of the Display Options.
To update the data using different calculation or outputs options, click:
Update
Export this report
Show temperature distribution data
Display Shortcuts:
Clear all
Show all
Add Data:
Gains data
Fabric and ventilation
Comfort
Site
System loads
Fuel totals
Fuel breakdown
Save results set as template
Ready...

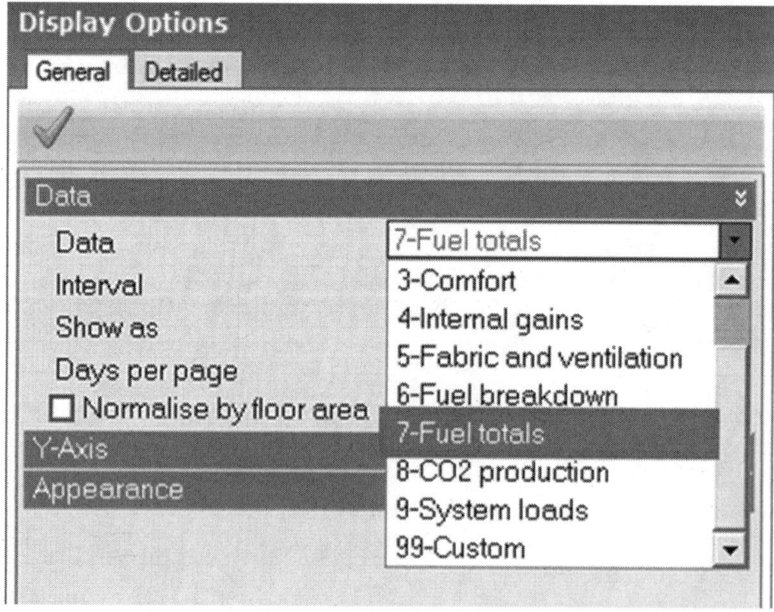
Display Options
General Detailed
Data
Data 7-Fuel totals
Interval
Show as
Days per page
Normalise by floor area
Y-Axis
Appearance
3-Comfort
4-Internal gains
5-Fabric and ventilation
6-Fuel breakdown
7-Fuel totals
8-CO2 production
9-System loads
99-Custom

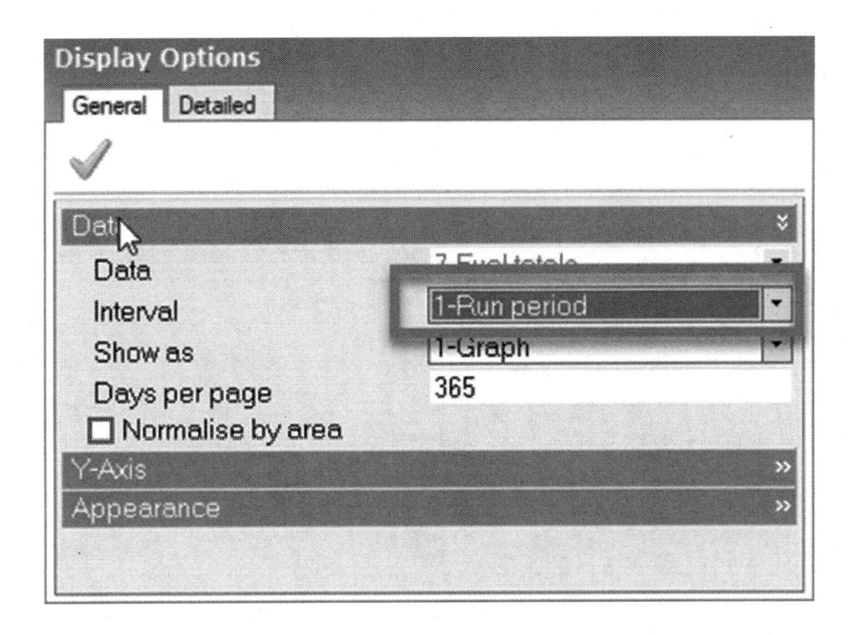

Step 10: Select the **Normalise by floor area** check box and **All floor area** from the **By** drop-down list to view the energy use intensity (EUI) of the building.

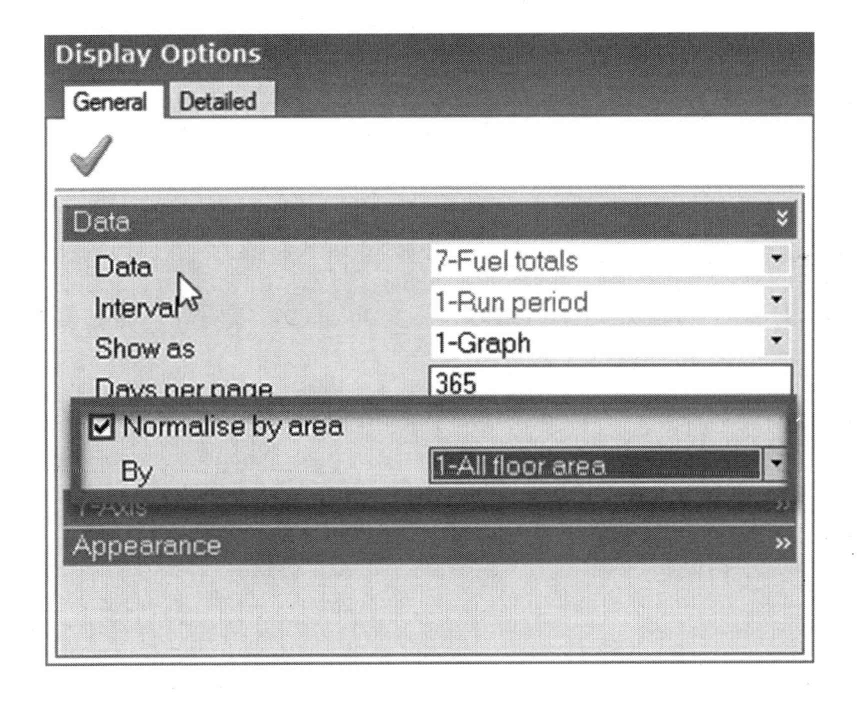

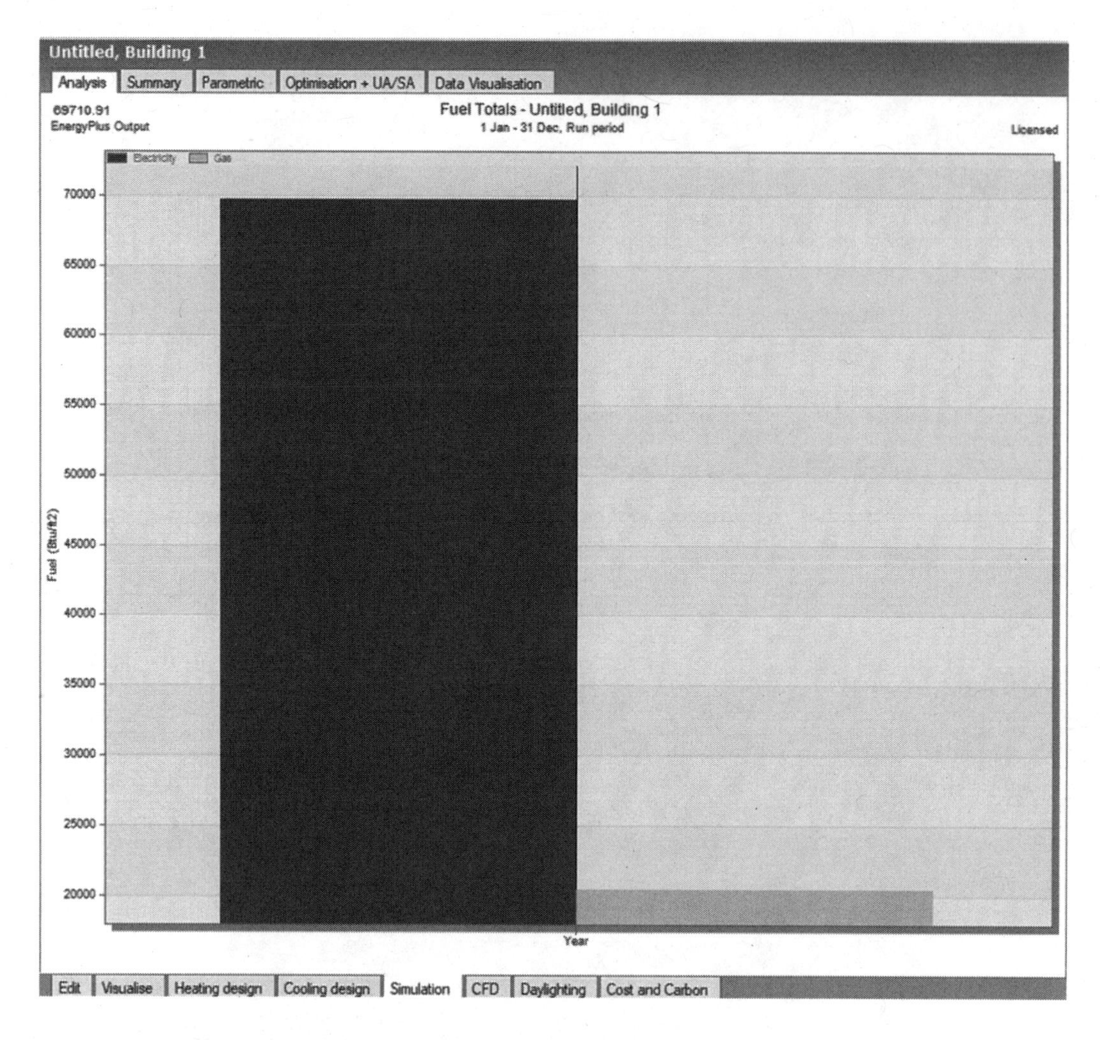

Units for the **Fuel totals** can be seen on the *y*-axis of the graph.

The term *energy use intensity* (EUI) refers to the energy consumption of the building per unit area per annum. The same term, in some countries, is also called the *energy performance index* (EPI). The EUI of a building is calculated by dividing the annual total energy consumption (all fuel types) of the building by its gross floor area. The unit of measurement is kWh/m^2 yr or kWh/ft^2 yr or Btu/ft^2 yr. A lower EUI means a better energy performance.

The *gross floor area* of a building is the total built-up area, which includes all conditioned and unconditioned enclosed spaces of the building.

Step 11: Change the **Interval from Run** period to **Monthly** to view monthly results. Refer to Step 9.

A DesignBuilder message box appears. Click **Yes**. Click **Update Data**. Click **OK** to perform a simulation to update the results.

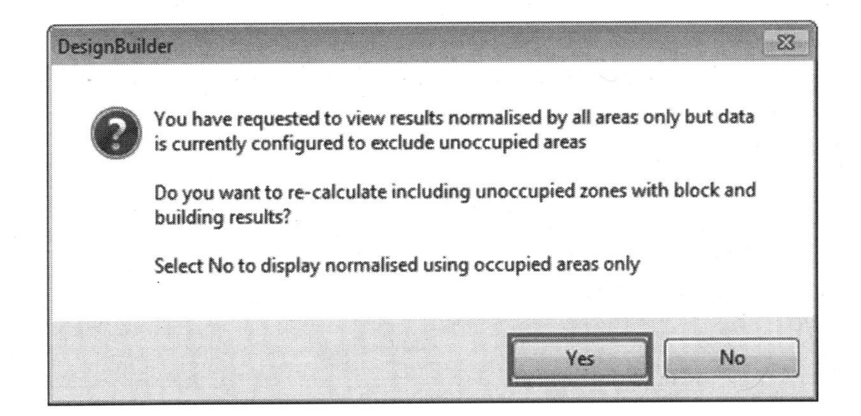

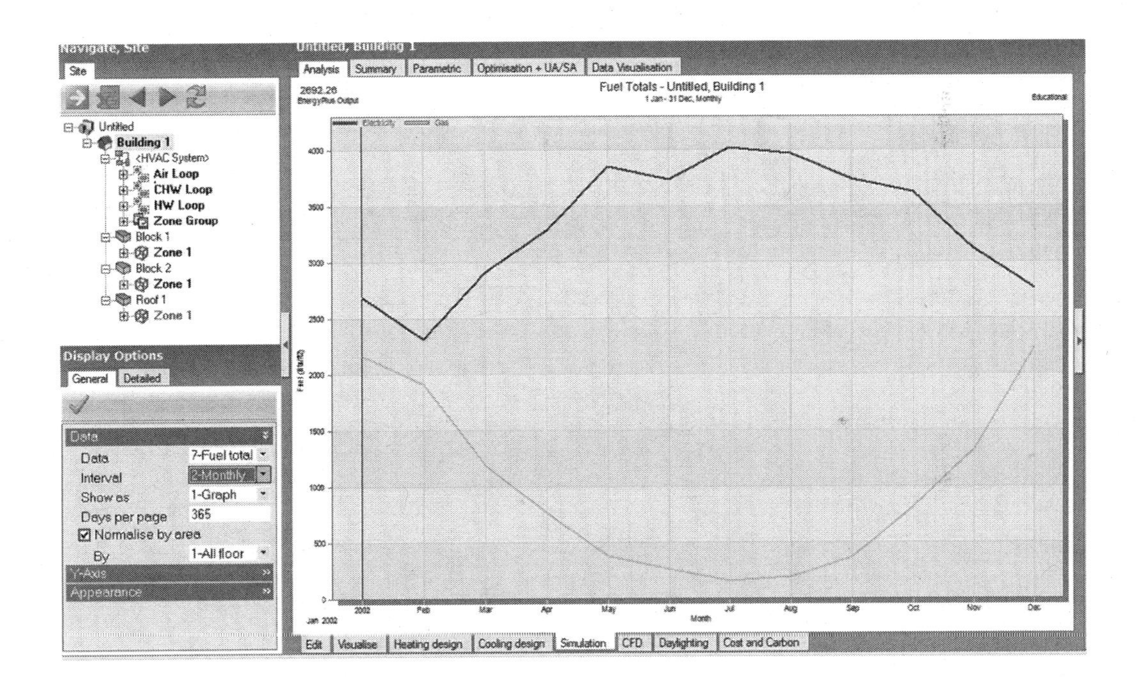

Step 12: Select **Grid** from the **Show as** drop-down list to view results in a grid format.

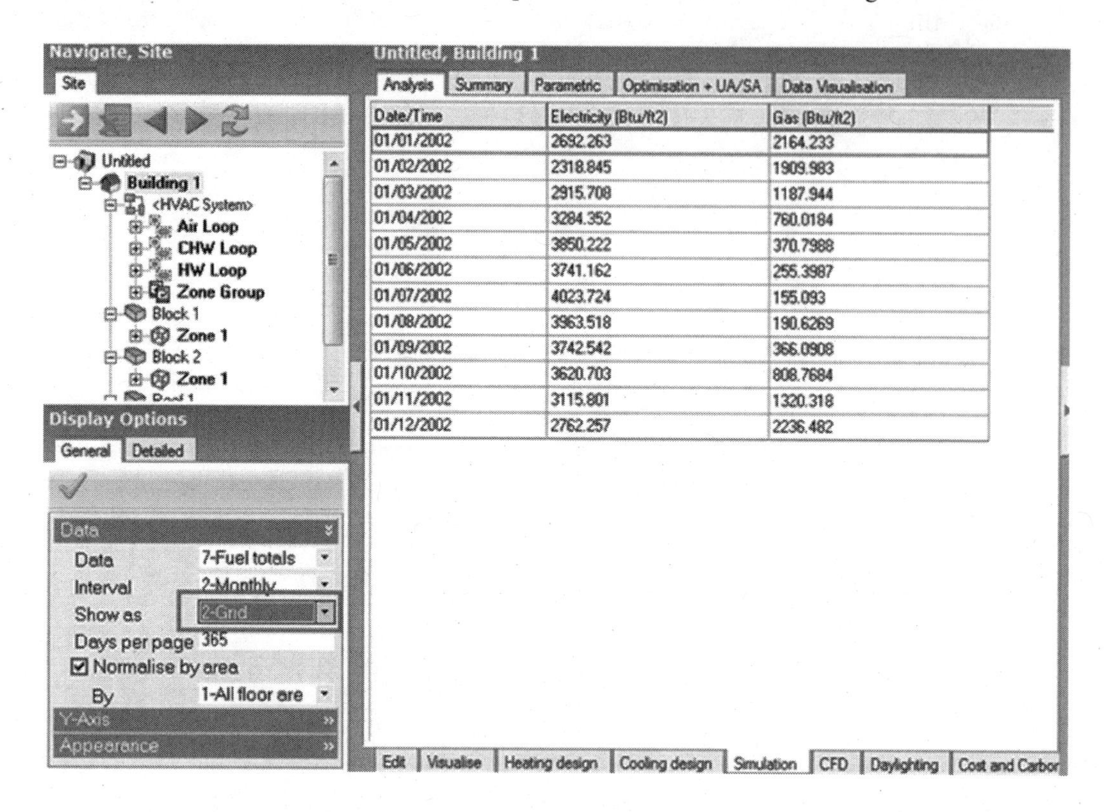

Step 13: Select **Fuel breakdown** from the **Data** drop-down list to view results for fuel breakdown.

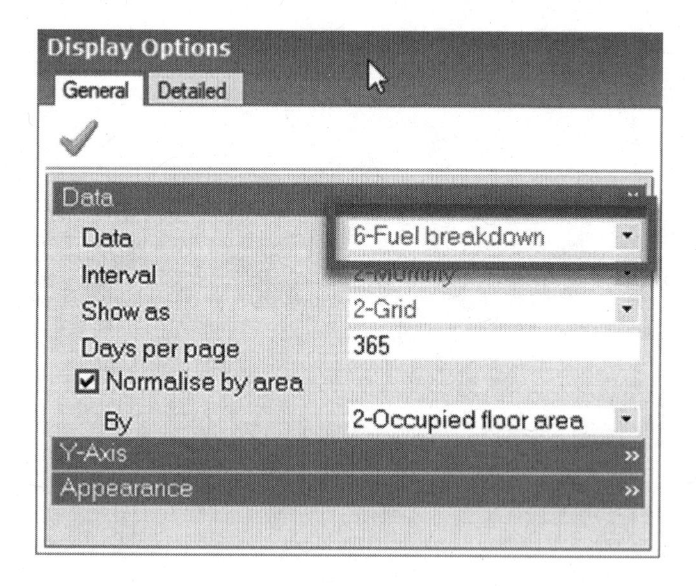

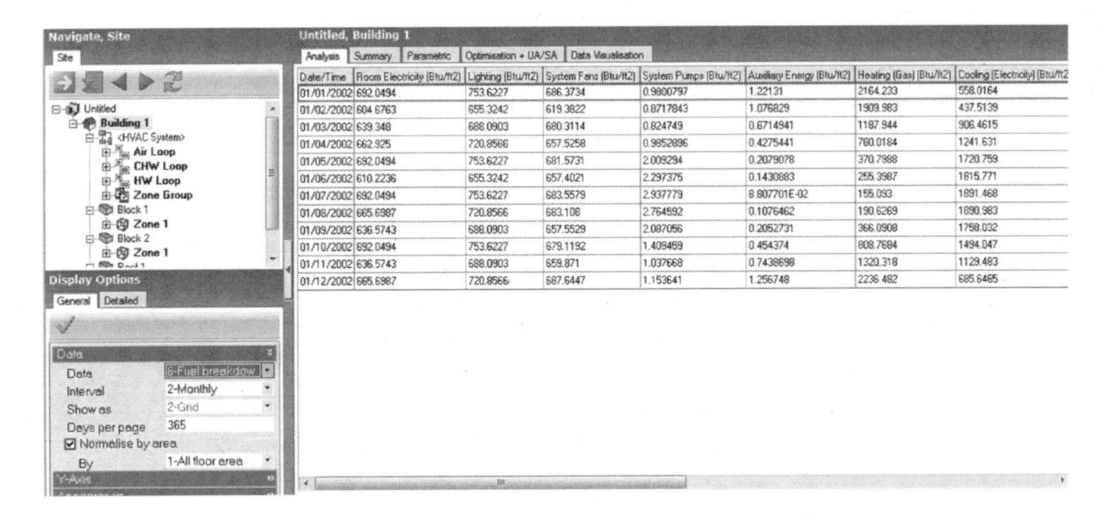

In all the preceding steps, the results are in SI or IP units based on the existing configuration of DesignBuilder in your system. To change the units, perform the following steps.

Step 14: From the **Tools** menu, select **Program options**. The **Edit Program Options** screen appears.

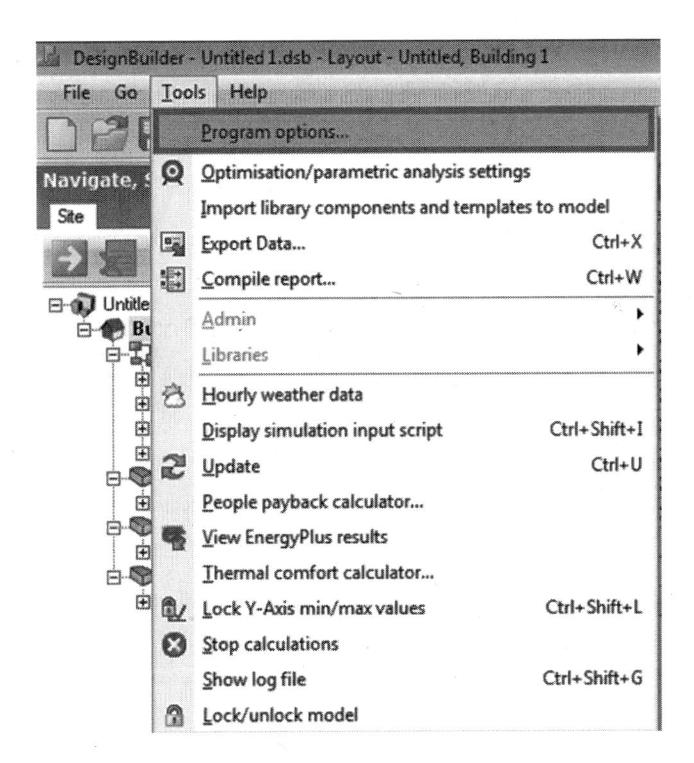

Step 15: Select the **International** tab, and select **Units** as per your requirement. Click **OK**. DesignBuilder updates all the values in the selected units. SI units will be used in this workbook.

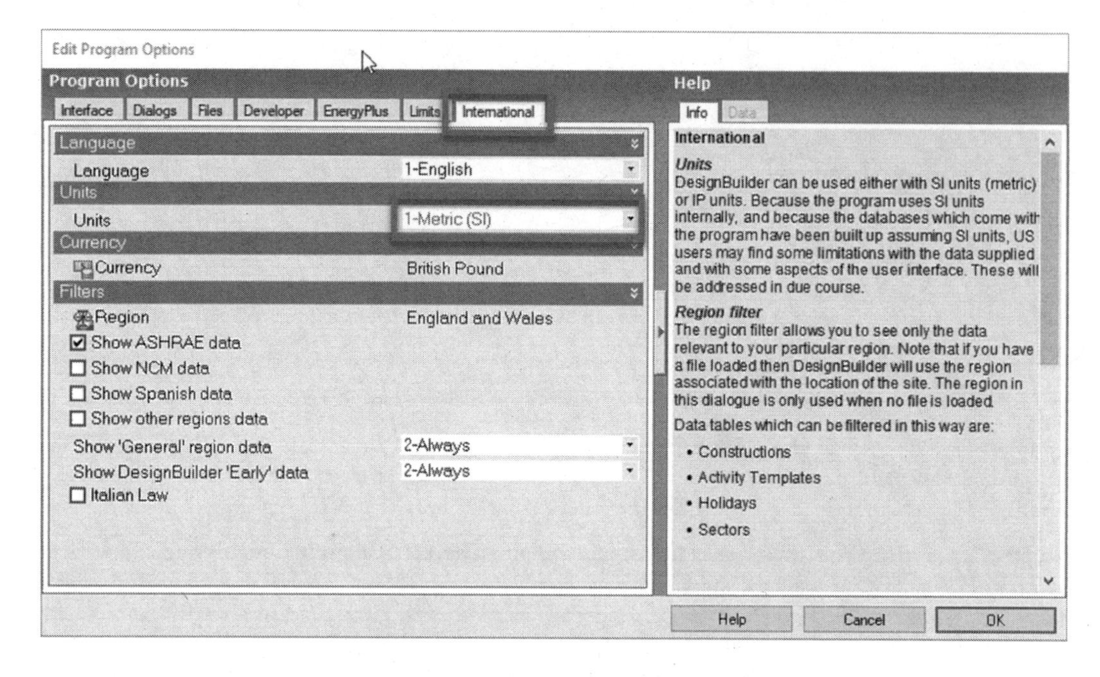

Step 16: From the **File** menu, select **Save** to save this model.

EXERCISE 1.1 COMPARE THE EUIS OF RESIDENTIAL AND OFFICE BUILDINGS

Courtyard with VAV Example is for an office space. Simulate a file with residential usage and compare the results.

> DesignBuilder can open only one file at a time.

Hint: Open the file **House example 1**. Ensure that the location is London Gatwick. Repeat the steps given in Tutorial 1.1 to find the EUI. Enter the EUI values in the last column of Table 1.1.

You can see that even with the same location, the EUIs of office buildings and residential buildings are different. This is mainly due to the difference in building usage, which includes difference in timing and duration of use, occupancy density, nature of equipment used and other factors. Office buildings usually have a higher EUI, as they have higher internal loads (see Table 1.1).

TABLE 1.1
Energy use intensities for office and residential buildings

	EUI (kWh/m^2)	
Energy use	**Office building**	**Residential building**
Electricity		
Gas		

Tutorial 1.2

Creating a Single-Zone Model

GOAL

To create a single-zone model and find heating and cooling capacities

WHAT ARE YOU GOING TO LEARN?

- how to draw, view and render a building model
- How to size runs for heating and cooling capacities
- How to perform an annual energy simulation

PROBLEM STATEMENT

Create a new file and draw a 40- × 20-m rectangular building of height 3.5 m with location as **LONDON/GATWICK ARPT**. Perform a cooling and heating sizing and annual energy simulation.

SOLUTION

Step 1: Click **File** and select **New file**. The **New file** screen appears.

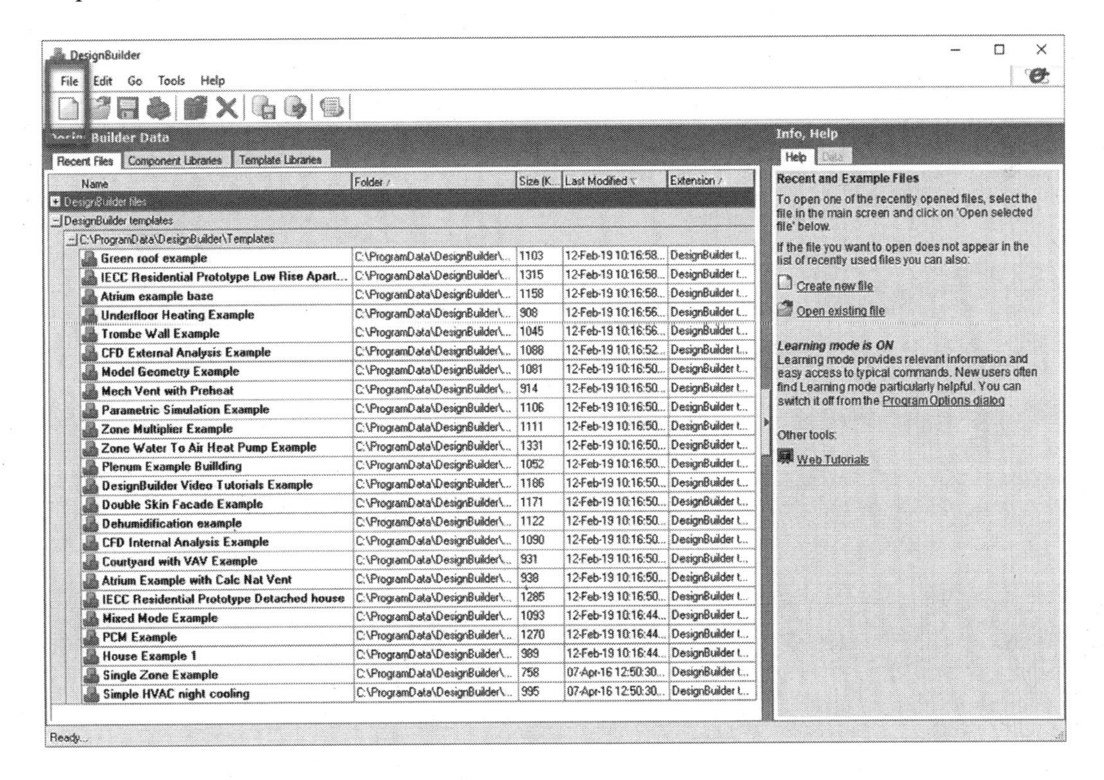

Step 2: Enter **First Building** in the **Title** text box and click **OK**. A blank building layout appears

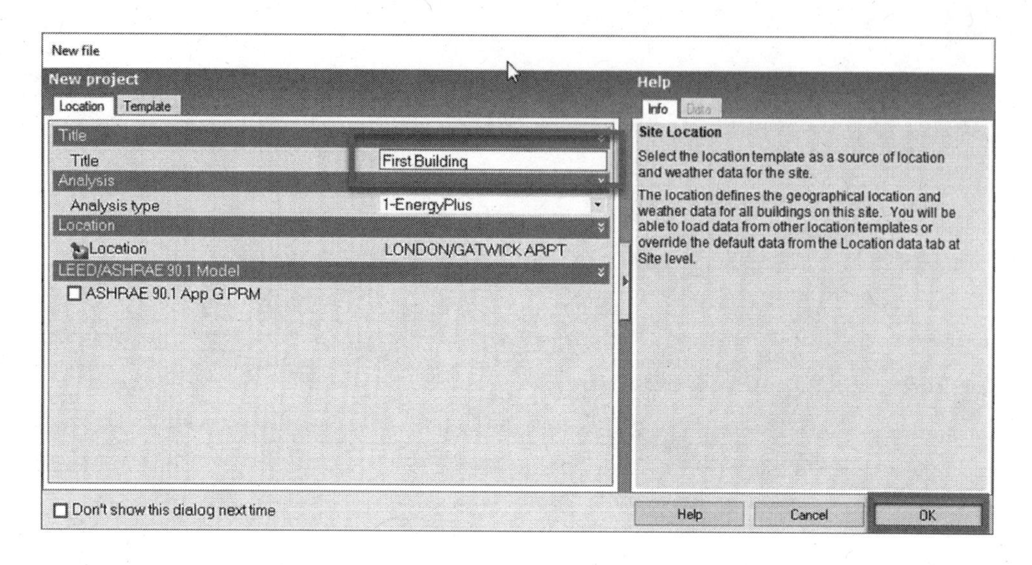

Step 3: Select **Plan** from the **View rotation** drop-down list.

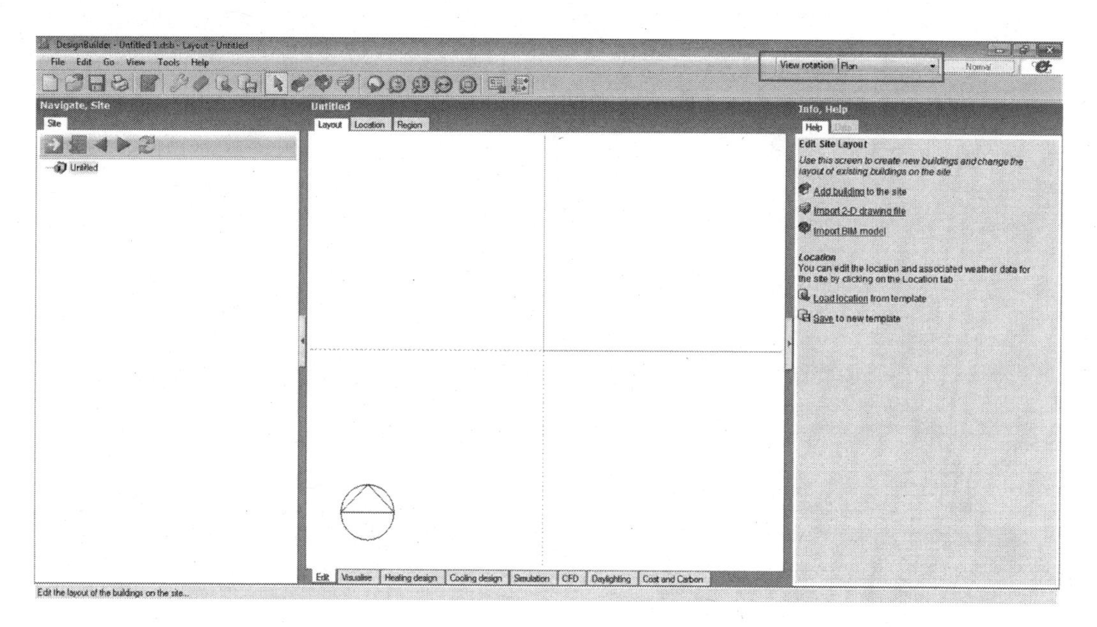

In the DesignBuilder display, the red line represents the x-axis, the green line the y-axis and the blue line the z-axis.

Step 4: Click the **Add new building** button. The **Add new building** screen appears.

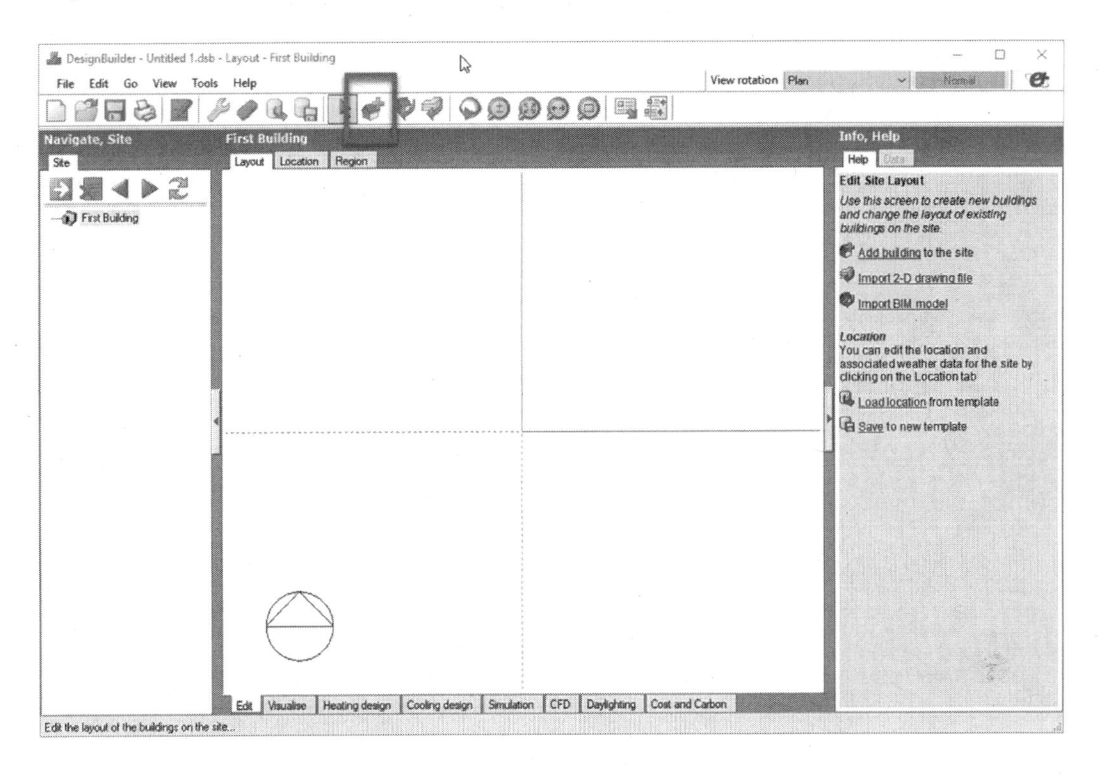

Step 5: Click **OK**. The screen appears with a cursor to draw a building.

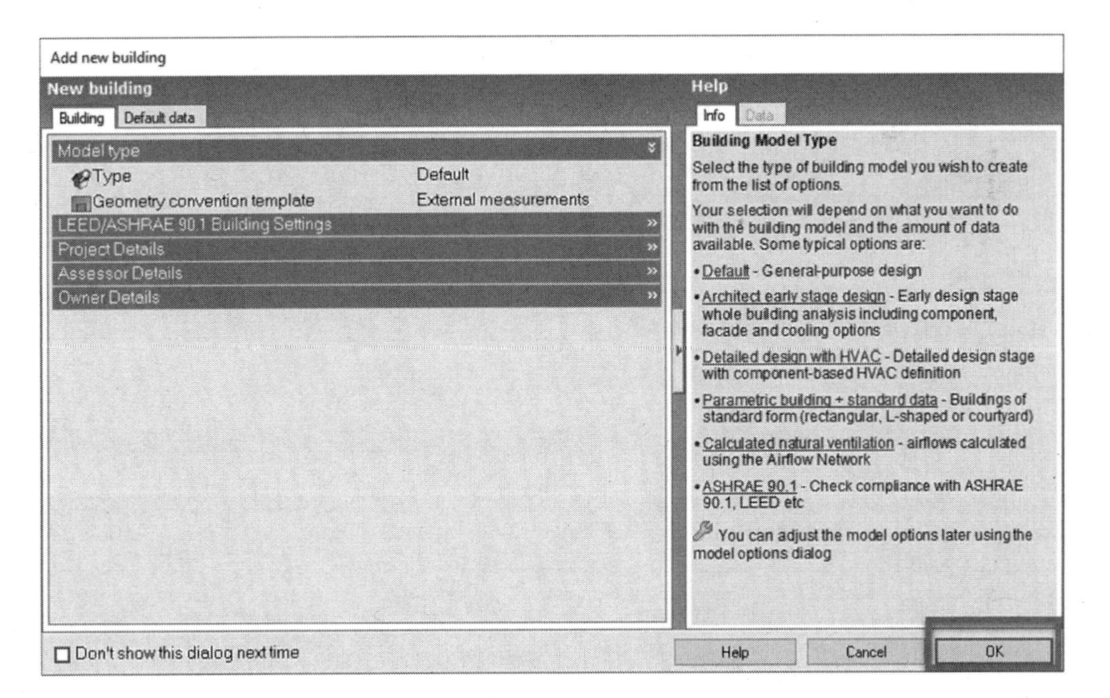

Step 6: Move the pointer to the **Origin**. Left-click, and move on the positive *x*-axis. The length of the segment is shown. (In case of any mistake, you can choose to cancel the drawn line by pressing the **ESC** key and then try again.)

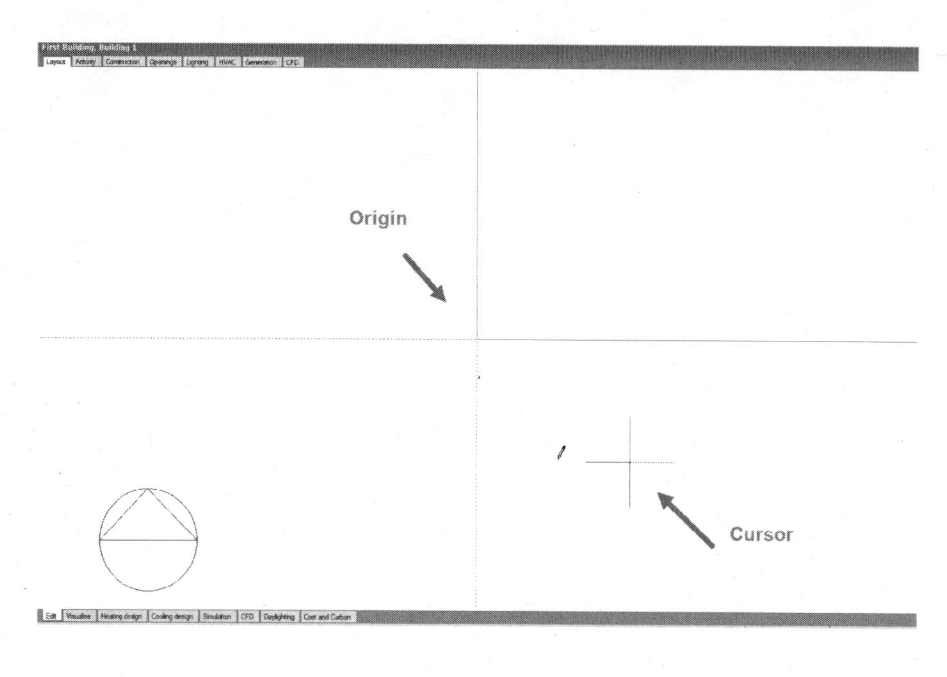

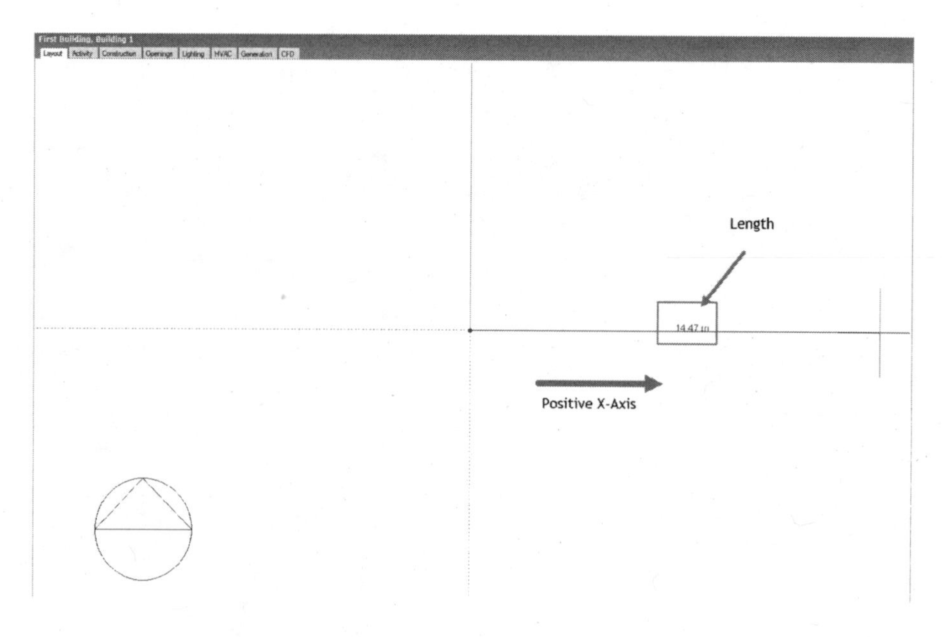

Step 7: Type **40**. You can type directly, and the text box for length will be highlighted. Fill in the value for the length in the **Drawing Options** section. The **Length** of the wall is set to **40** m. Ensure that the properties under the **Geometry** section are as follows:

- Block type: **Building Block**
- Form: **Extruded**
- Height: **3.5 m**

Press **Enter**.

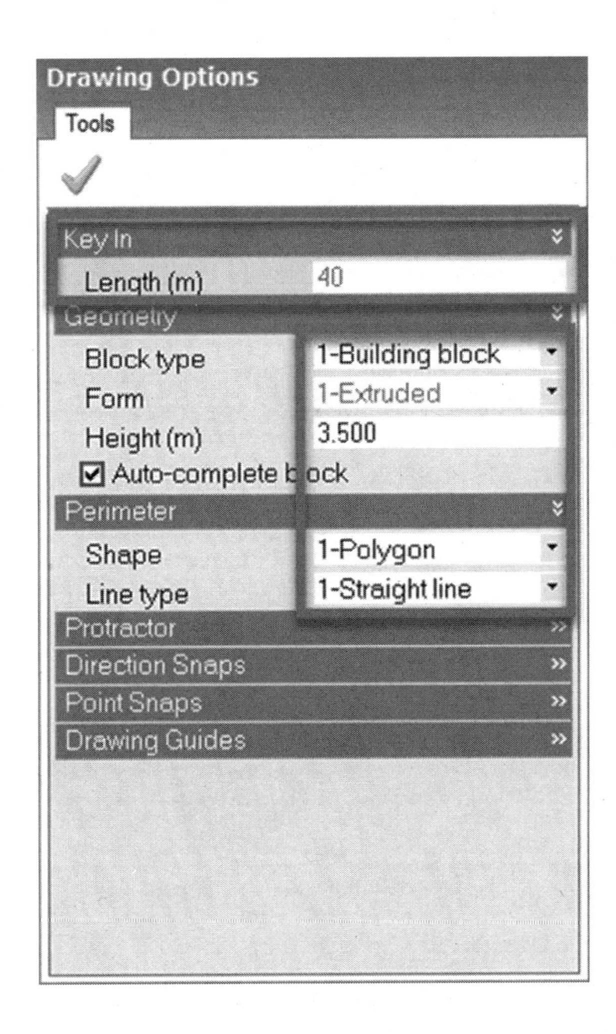

Step 8: Scroll the mouse wheel to zoom in or zoom out until you can see the complete line of 40 m in length. Move the pointer parallel to the positive *y*-axis. It snaps when a dashed vertical line appears. Type **20** to draw the second side of the rectangle. Press **Enter**.

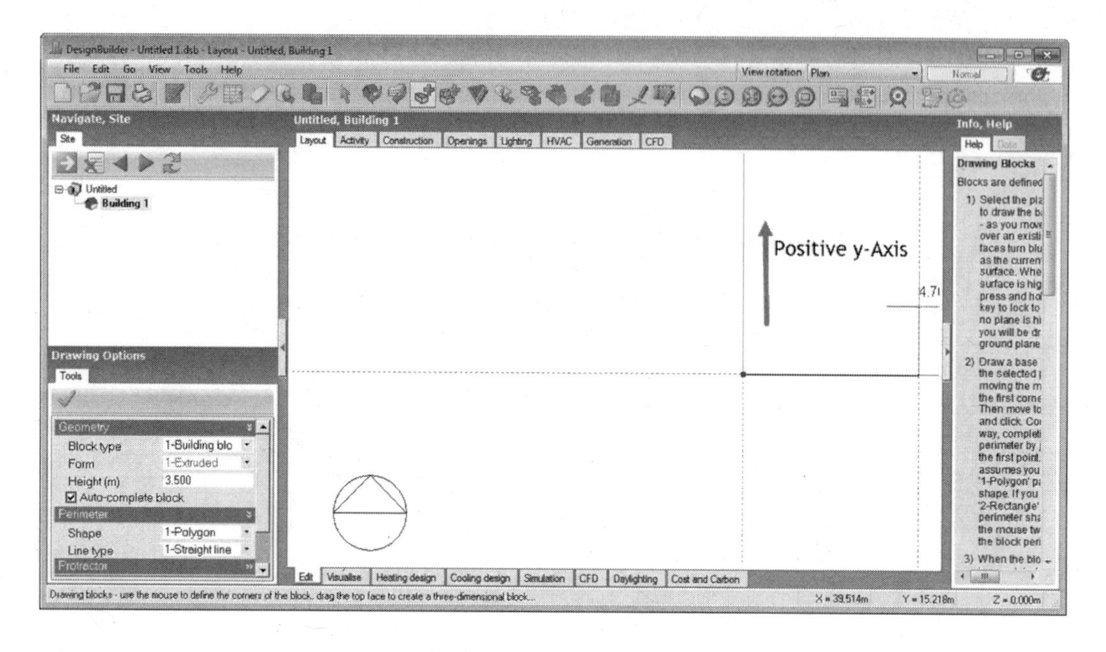

Step 9: Similarly draw other segments and complete the rectangular building block of dimensions 40 × 20 m. The following screen shows the completed block.

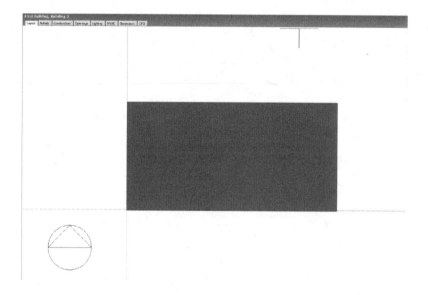

Step 10: Select the **Dynamic orbit** tab on the main menu bar. This shows the block in the orbit mode. Click and drag the mouse in the layout to view the block in the orbit mode.

DesignBuilder Snap

When drawing perimeter lines, block partition or surface opening direction snaps can be used to draw lines

- Parallel with the active drawing plane axes (axis snap),
- Parallel with existing elements (parallel snap) or
- Perpendicular to existing elements (normal snap).

Source: https://designbuilder.co.uk/helpv6.0/Content/Direction_snaps1.htm.

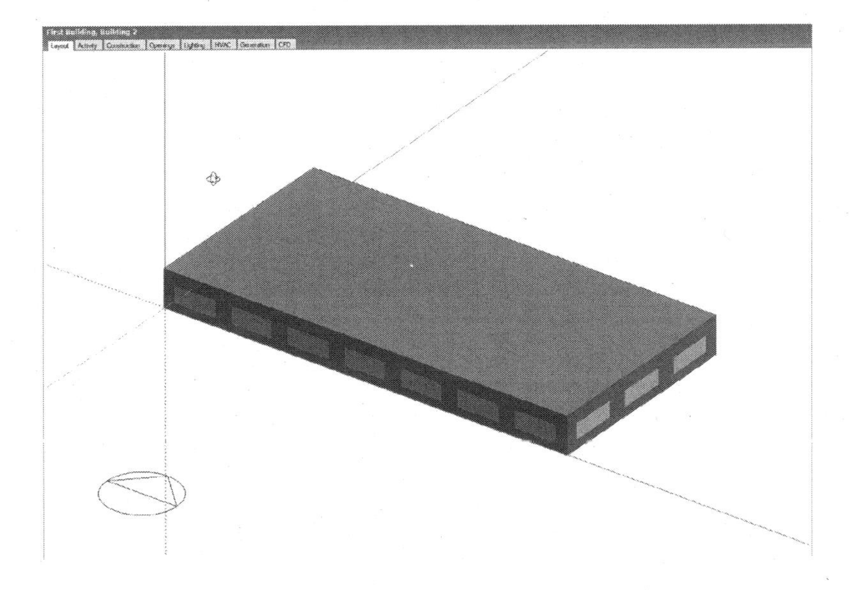

Now you will learn to run the system sizing.

Step 11: Select the **Heating design** screen tab at the bottom. Click **Update data**. The **Edit Calculation Options** screen appears.

Heating design calculations show the heating system size in the worst conditions (no internal loads and solar heat gains) for a winter design day. Steady-state simulations assume that the temperature across the envelope does not vary with time.

Step 12: Select **OK**. The screen appears with the average and total winter steady-state design data for the current building in graphical format.

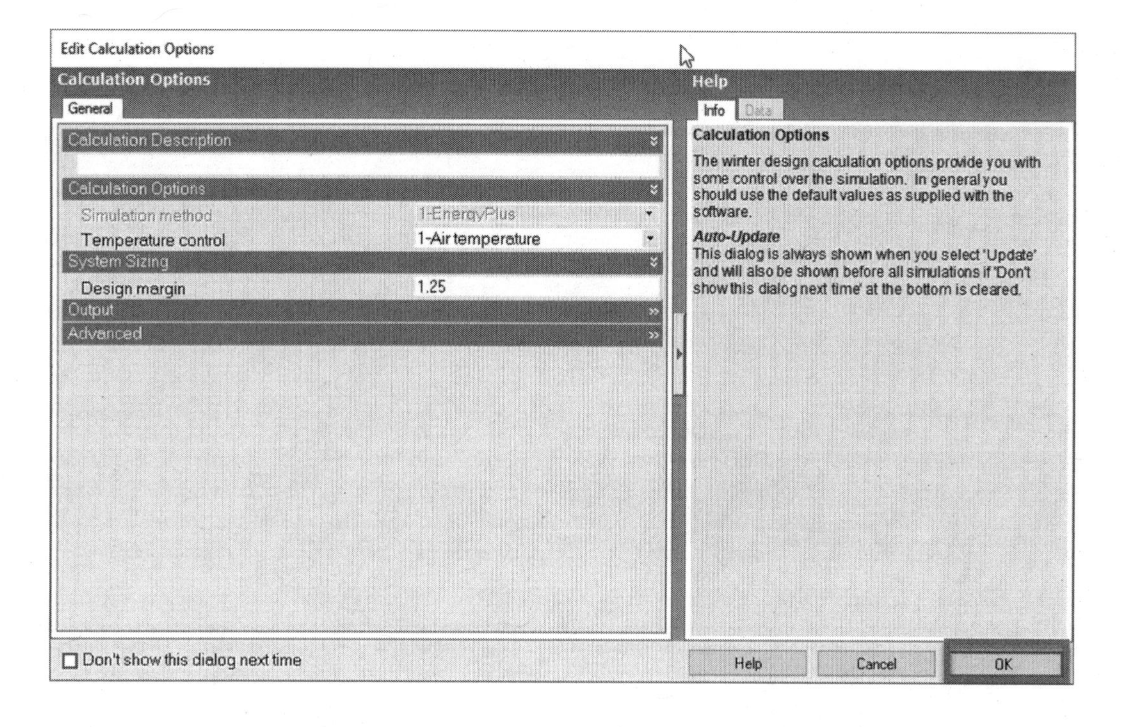

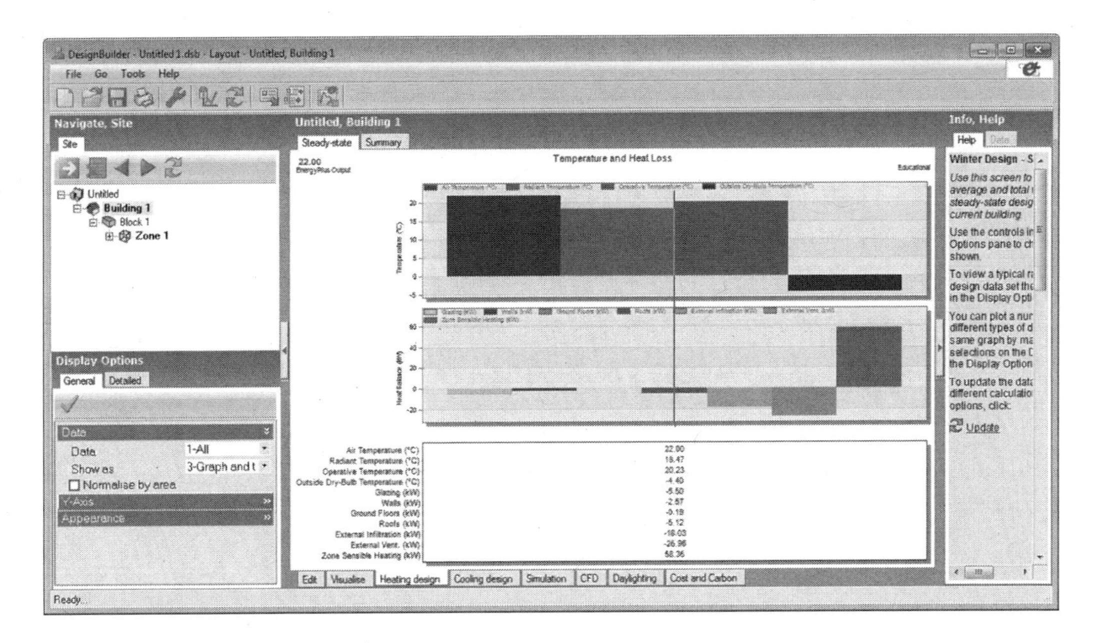

The graph in the preceding figure shows the heat balance to maintain an inside temperature of 22°C when the outdoor dry bulb temperature is −4.4°C. You can see that there is a heat flow from the inside to the outside (shown as negative numbers; the convention is that a positive value is used for a heat flow from the outside to the inside) through the envelope as a result of infiltration and ventilation. To balance this heat loss, the required zone sensible heating is 58.36 kW. The table in the preceding figure also shows the air temperature, radiant temperature and operative temperature, where the operative temperature is calculated as the average of the radiant and air temperatures.

Steady-state heat loss is the total heat loss from the building. In other words, it is the amount of heat needed to maintain the given indoor comfort temperature.

Step 13: Select the **Summary** tab. The heating design data summary appears.

First Building, Building 1

Steady-state | Summary

Zone	Comfort Temperature (°C)	Steady-State Heat Loss ...	Design Capacity (kW)	Design Capacity (W/m2)	Glazing Gains (...	Wall Gains (kW)	Floor Gains (kW)	Roof and Ceilin...	Ventilation Gain...	Infiltration Gain...
Building 1 Total Design Heating Capacity = 72.950 (kW)										
Block 1 Total Design Heating Capacity = 72.950 (kW)										
Zone 1	20.23	58.36	**72.95**	95.3319	-5.501	-2.566	-0.188	-5.118	-26.965	-18.028

Design capacity is the total heat loss multiplied by the sizing factor. Sizing factor is the safety factor considered in sizing an HVAC system. For example, heating systems are commonly oversized by 25%.

The design capacity of the building is 72.950 kW, which is calculated considering the sizing factor of 25% over the steady-state heat loss of 58.36 kW. The design capacity normalized by area is 95.55 W/m^2, which is calculated for the building floor area. In this case, the building dimensions are 40 × 20 m, and the building area is 800 m^2 (net floor area is 765.24 m^2).

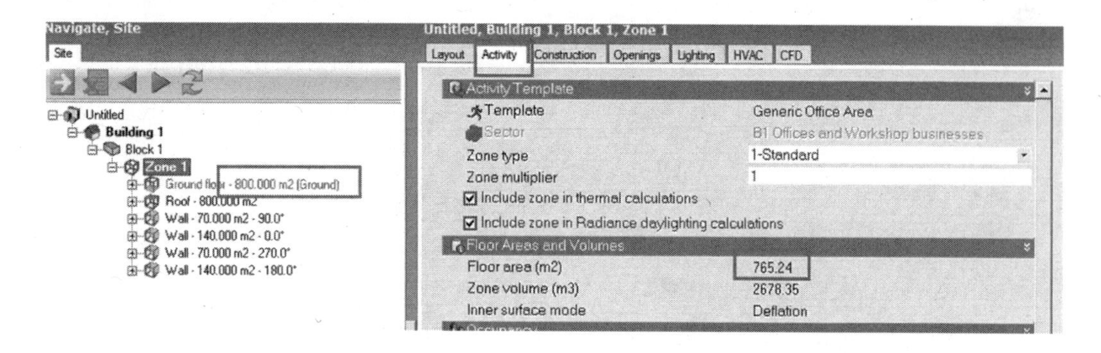

Step 14: Similarly select the **Cooling design** screen tab at the bottom. Click on **Update data.** The **Calculation Options – Building 1** screen appears. Click **OK**. The temperature, heat gains and other parameters appear in graphical format.

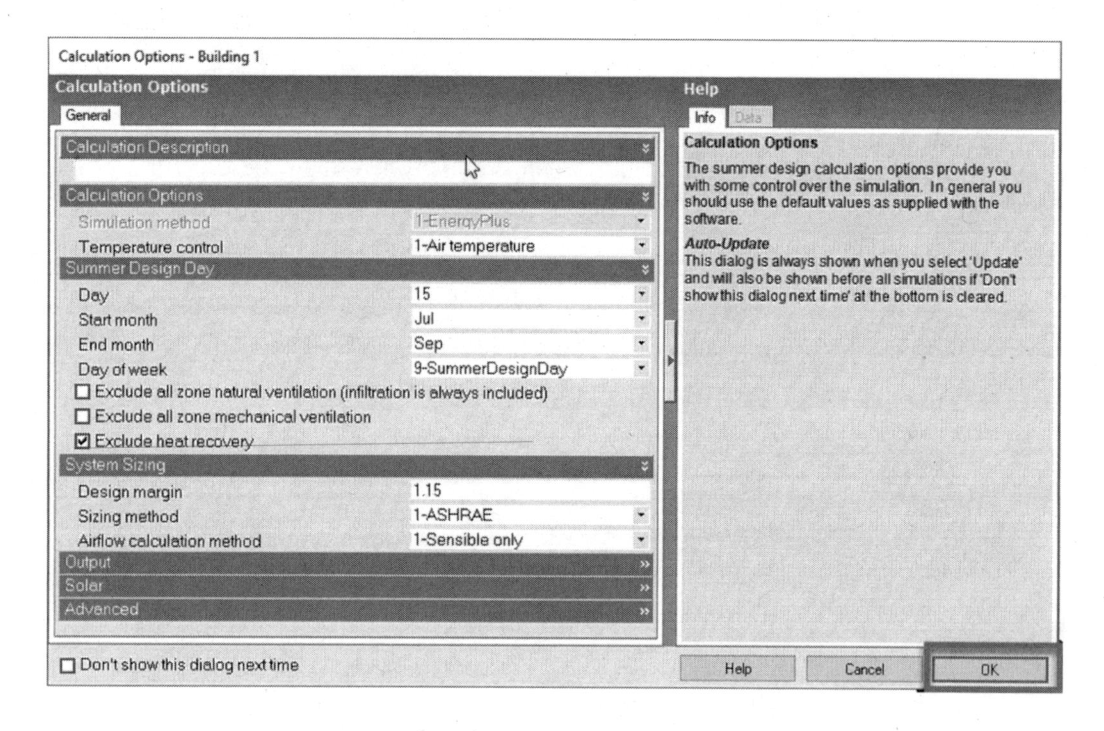

A cooling system is commonly oversized by 15%.

Step 15: Select the **Summary** tab. The screen appears with the cooling design data summary.

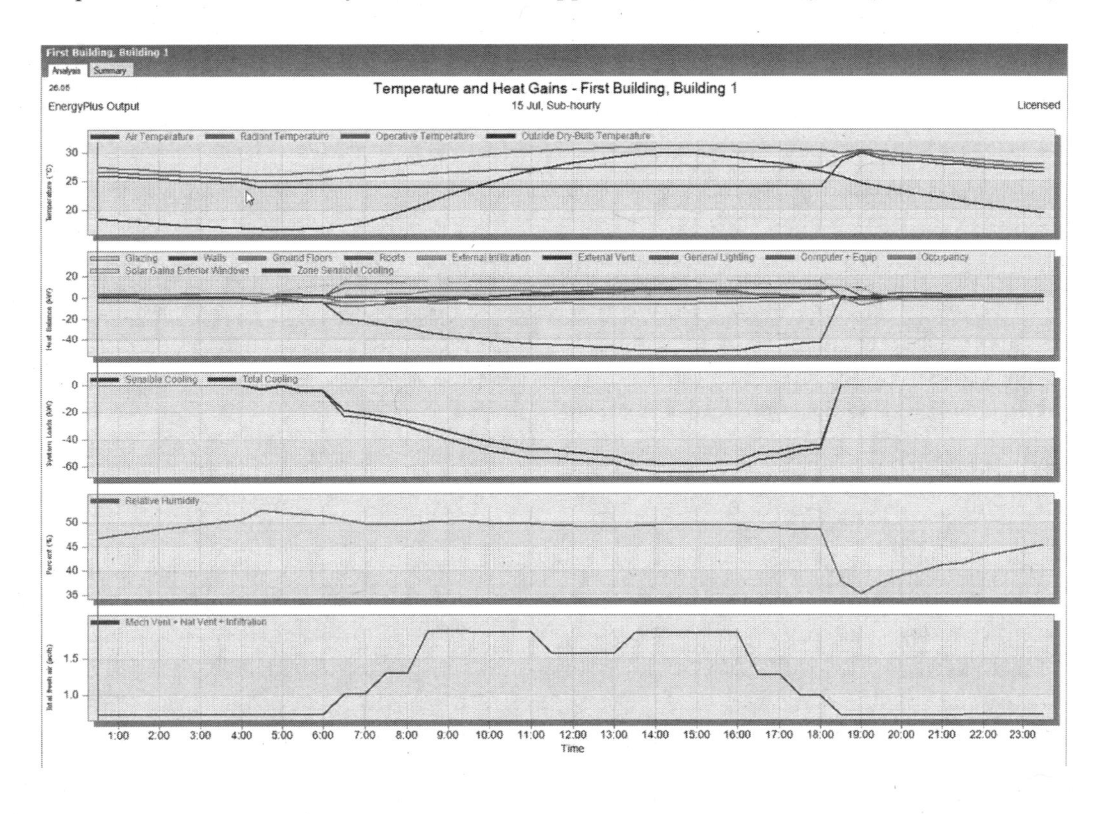

First Building, Building 1

Analysis | Summary

Zone	Design Capacity (kW)	Design Flow Rate (m3/s)	Total Cooling Load (kW)	Sensible (kW)	Latent (kW)	Air Temperature (°C)	Humidity (%)	Time of Max Cool...	Max Op Temp in Day (°C)	Floor Area (m2)
Building 1										
Block1:Zone1	79.36	4.7604	69.01	59.73	9.28	24.0	49.7	Aug 15:00	29.5	765.2
Totals	**79.36**	**4.7604**	**69.01**	**59.73**	**9.28**	**24.0**	**49.7**	**N/A**	**29.5**	**765.2**

The screen data are provided in tabular format in Table 1.2. The load calculated in the building is 79.36 kW. The cooling load is also described in tons of refrigeration (TR). For this building, it is 22.6 TR. Enter the cooling and heating capacities in Table 1.4.

TABLE 1.2
Data for the single-zone model

Zone	**Block 1: Zone 1**
Design capacity (kW)	79.36
Design flow rate (m^3/s)	4.76
Total cooling load (kW)	69.01
Sensible (kW)	59.73
Latent (kW)	9.28
Air temperature (°C)	24.0
Humidity (%)	49.7
Time of max cooling	August 15:00
Maximum operating temperature in day (°C)	29.5

(Continued)

TABLE 1.2 (Cont.)

Zone	Block 1: Zone 1
Floor area (m^2)	765.3
Volume (m^3)	2,678.3
Flow/floor area [litres (l)/s-m^2]	6.22
Design cooling load per floor area (W/m^2)	103.7
Outside dry bulb temperature at time of peak cooling load (°C)	30.5

A *ton of refrigeration* (TR) describes the heat-extraction capacity of refrigeration and air-conditioning equipment. It is defined as the rate of heat transfer that results in the melting of 1 short ton (907 kg) of pure ice at 0°C in 24 hours. A refrigeration ton is approximately equivalent to 3.5 kW of cooling effect.

Sensible vs. Latent Heat

Sensible heat is the energy required to change the temperature of a substance with no phase change. Sensible heat represents only the dry bulb temperature change. *Latent heat*, however, does not affect the temperature of a substance. Heat that causes a change of state with no change in temperature is called *latent heat*. A cooling system should be capable of removing both the sensible and latent heats from a building. Therefore, the total cooling capacity of a system will be (sensible heat load + latent heat load) × sizing factor.

In this tutorial, sensible heat is 59.73 kW and latent heat is 9.28 kW. The sum of both sensible and latent heats is 69.01. Therefore, the design capacity calculated by multiplying 69.01 by the sizing factor (1.15) is 79.36. You can divide this by 3.51 to convert it into the unit of tons of refrigeration: $\frac{79.36}{3.51} = 22.6\text{TR}$

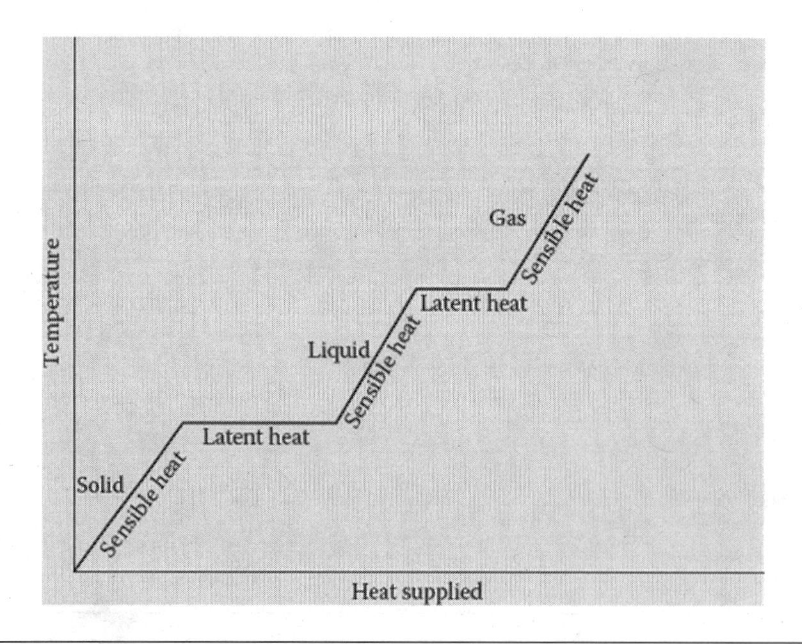

Step 16: Select the **Simulation** screen tab to perform the annual simulation. Select **Fuel breakdown** from the **Data** drop-down list, **Run period** from the **Interval** drop-down list and **Grid** from the **Show as** drop-down list.

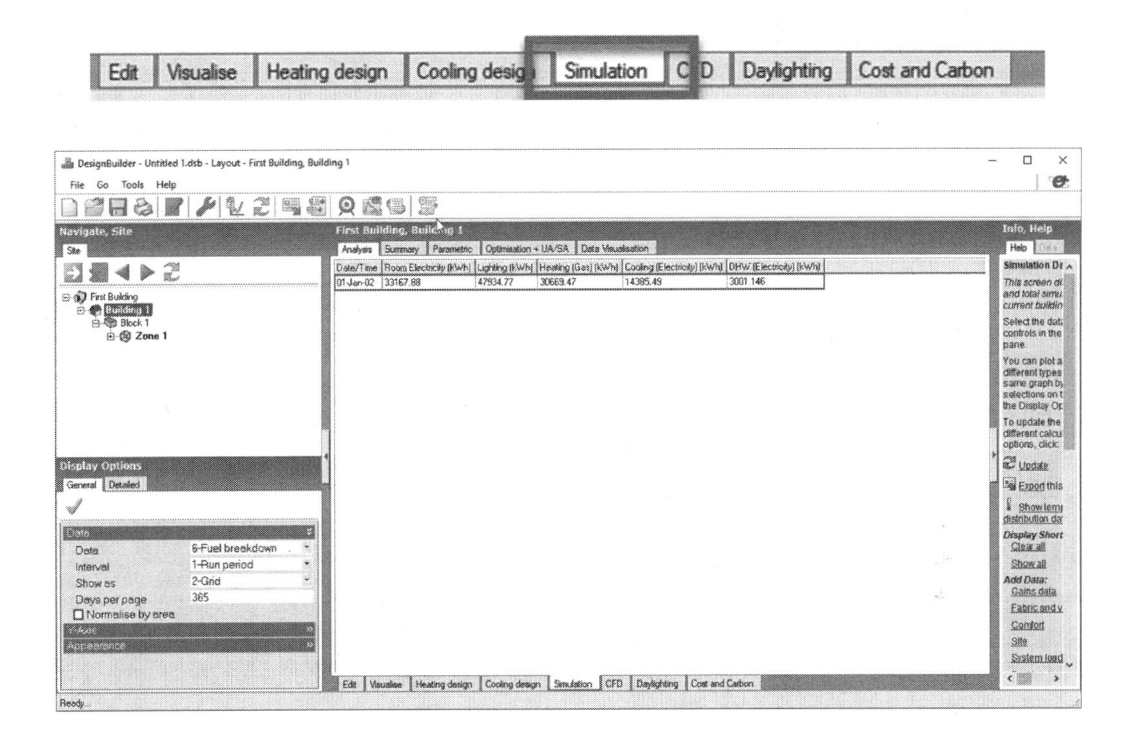

Step 17: Select the **Normalise by floor area** checkbox and **All floor area** from the **By** drop-down list to view the EUI of the building. Also select **Fuel totals** from the **Data** drop-down list. The DesignBuilder message screen appears. Click **Yes** to perform simulations for all floor area. Enter the electricity EUI value in Table 1.3.

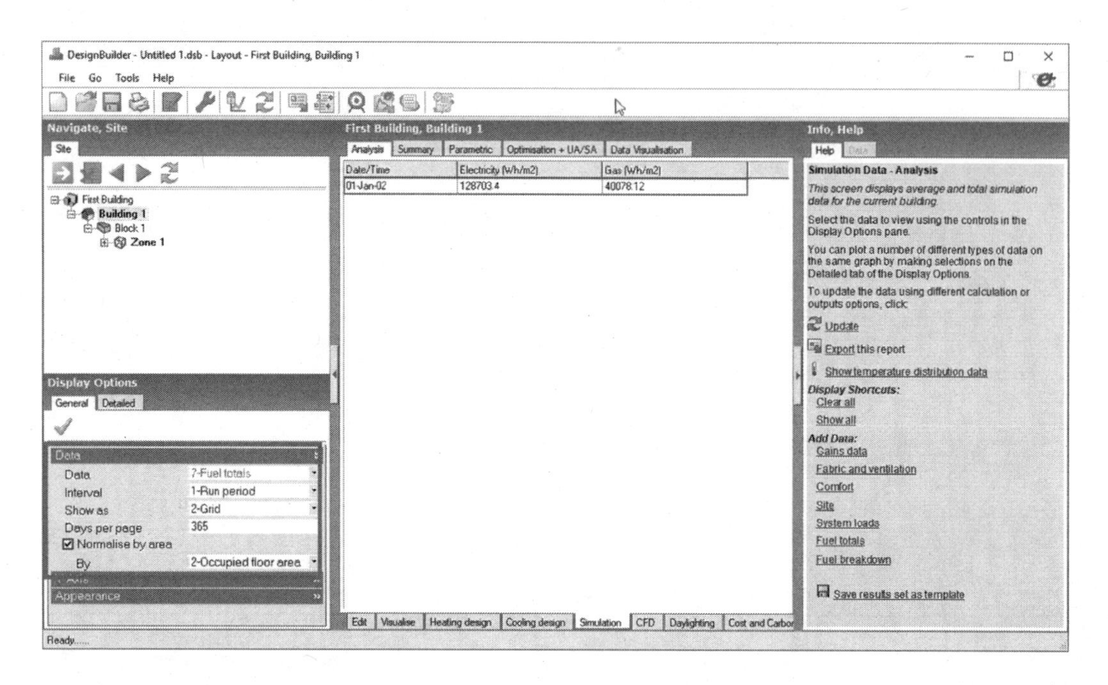

Step 18: Save the model as **First Building** on your desktop. You are going to use this model in forthcoming tutorials.

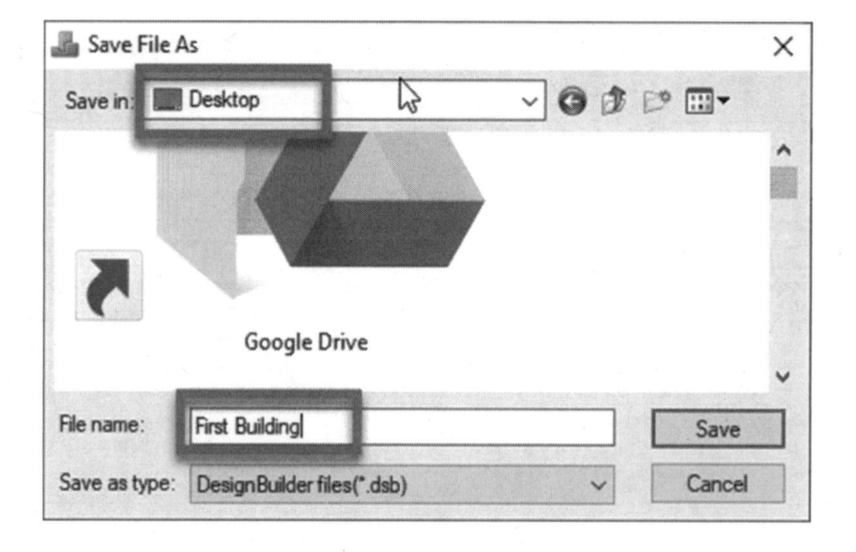

EXERCISE 1.2

Compare the EUI of a large building with that of a small building. Repeat the tutorial to create an 80- × 40-m rectangular building. Compare the EUI of the larger building with that of the smaller building (the 40- × 20-m building created in Tutorial 1.2), and enter the result in Table 1.3. Enter and compare the cooling and heating capacities of the larger building in Table 1.4. The method to calculate the EUI (kWh/m^2 yr) is explained in Tutorial 1.1.

TABLE 1.3
EUI for a large and small building

	EUI (kWh/m^2 year)	
Energy use	**Building with 40- × 20-m dimensions**	**Building with 80- × 4-m dimensions**
Electricity		
Gas		

TABLE 1.4
Cooling and heating capacity for a large and small building

Capacity	**Building with 40- × 20-m dimensions**	**Building with 80- × 40- m dimensions**
Cooling equipment (kW)		
Heating equipment (kW)		
Cooling per unit area (kW/m^2)		
Heating per unit area (kW/m^2)		

The following observations can be made as a result of Exercise 1.2:

1. The EUIs of both the buildings are different.
2. The cooling and heating equipment capacities are greater for the 80- × 40-m building than for the 40- × 20-m building. Larger buildings require higher-capacity equipment because of their greater envelope area, occupancy and internal loads.
3. Cooling/heating equipment capacities per unit area are also different. A large building has a lower capacity per unit area. The main reason behind such a difference is that although most of the internal loads, such as occupancy, lighting and equipment, vary linearly with area, the heat gain/loss through the building envelope does not follow the same trend because of the nonlinear variation in the exposed surface area versus the floor or carpet area.

Tutorial 1.3

Evaluating the Impact of Building Location and Orientation

GOAL

To evaluate the impact of building location and orientation on HVAC system sizing and building annual energy consumption

WHAT ARE YOU GOING TO LEARN?

- How to change the location of a building
- How to download a weather file
- How to change the orientation of a building
- How to analyse the impact of weather and orientation on a building's performance

PROBLEM STATEMENTS

1. In Tutorial 1.2, a 40- × 20-m rectangular building was simulated for London. Simulate the same building model for New Delhi/Palam, and compare the HVAC sizing, monthly energy consumption and annual energy consumption.
2. Rotate the building 90 degrees clockwise, and simulate it for New Delhi. Compare the results of the two models (without rotation and with 90 degrees of rotation).

SOLUTION

Step 1: Open the model saved in **Tutorial 1.2**.

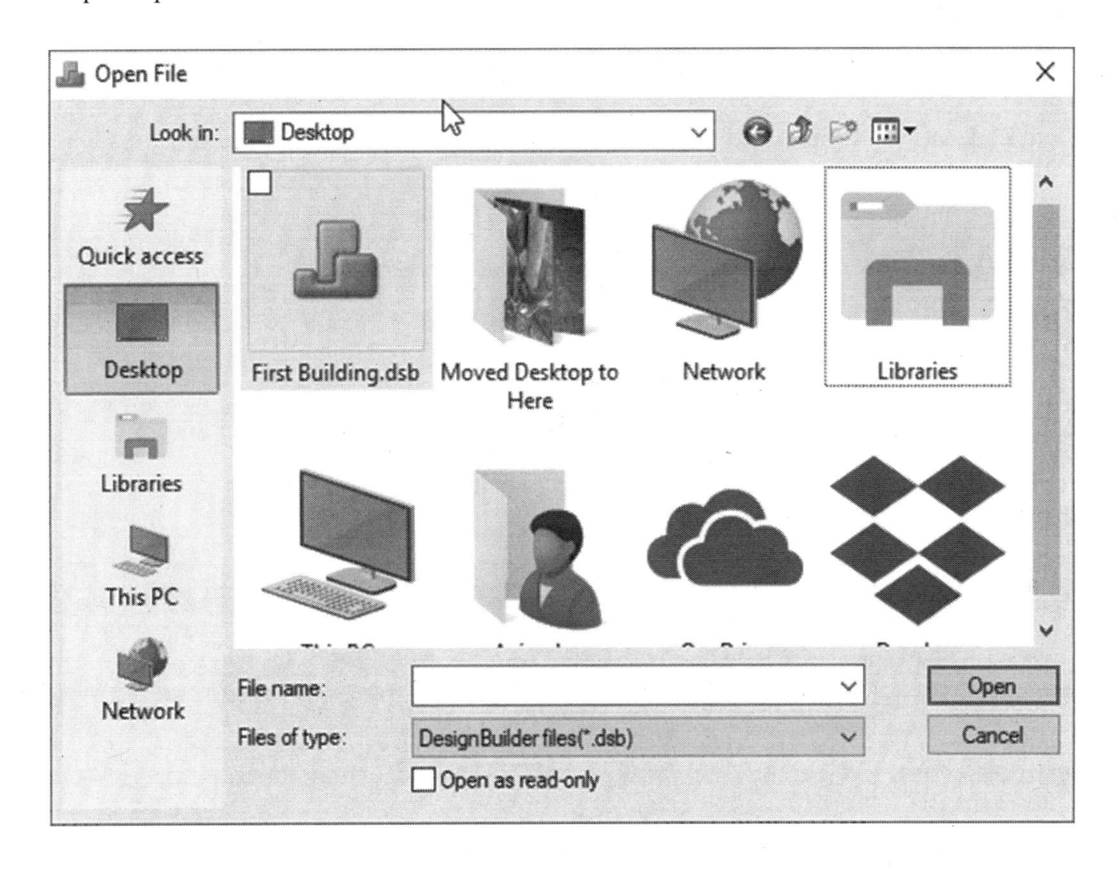

Step 2: Click **First Building**.

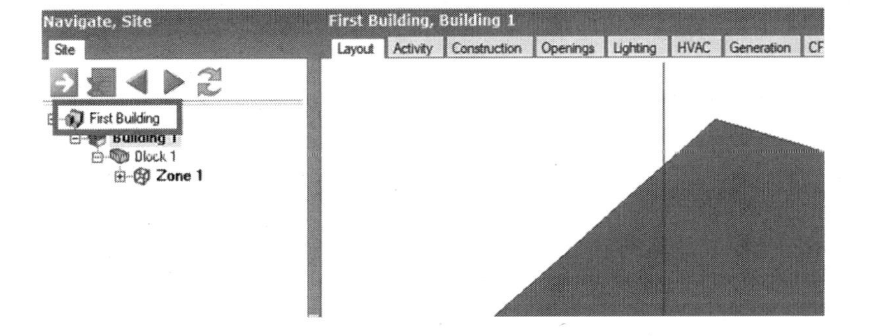

Step 3: Select the **Location** tab. The **Location Template** screen appears.

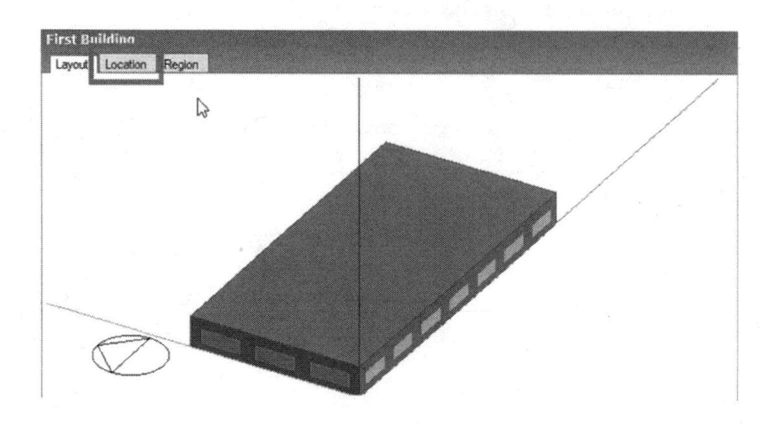

Step 4: Click **LONDON/GATWICK ARPT**. Three dots (...) appear. Click the three dots. The **Select the location template** screen appears.

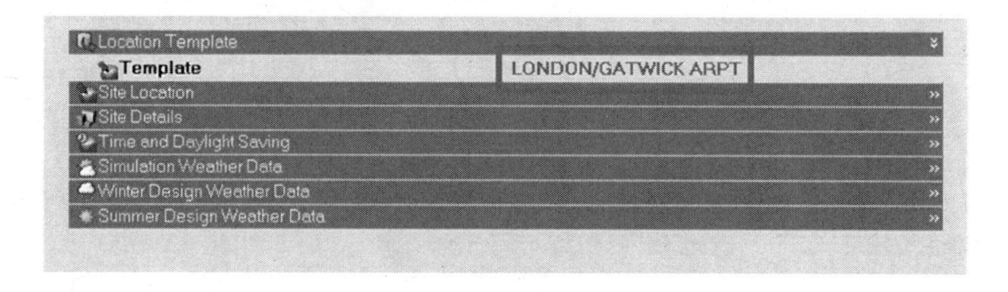

Step 5: Scroll to select **INDIA.** Double-click **INDIA,** and select **NEW DELHI/PALAM.** Click OK.

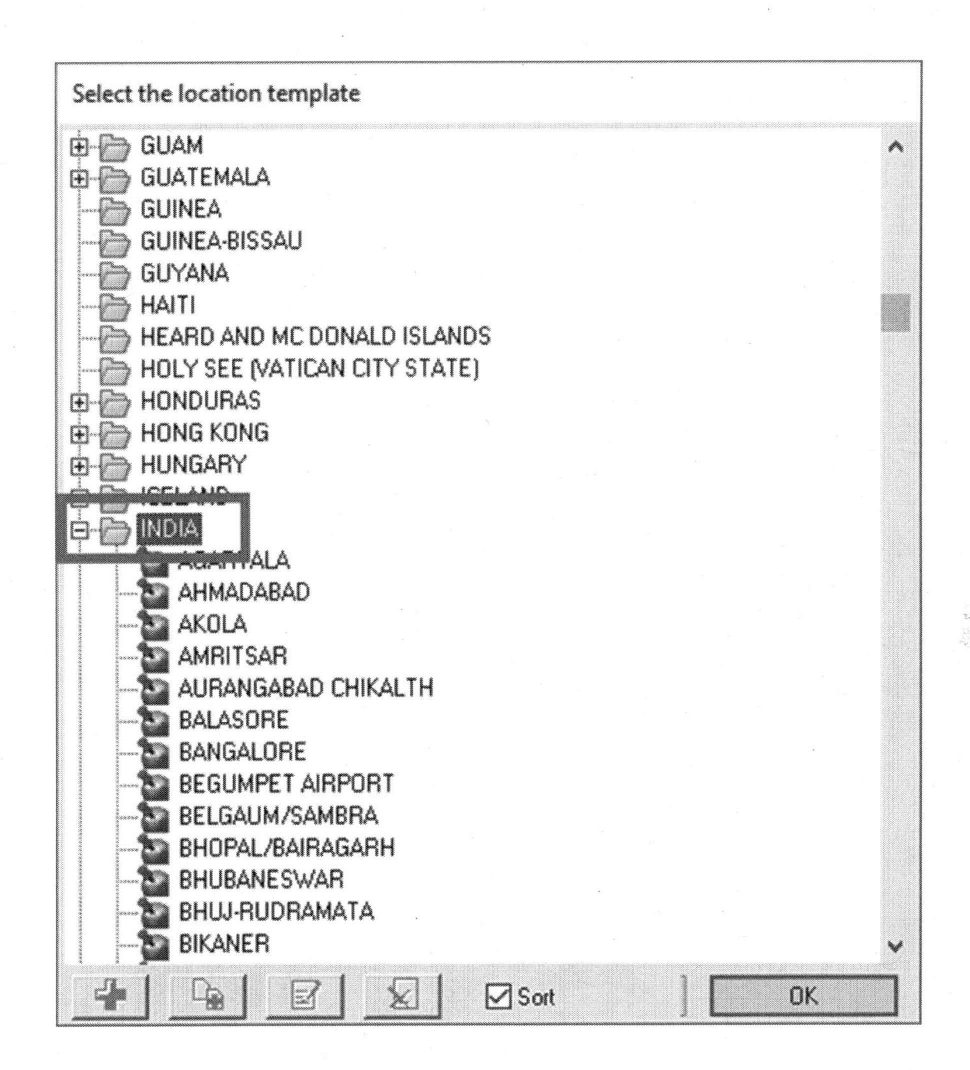

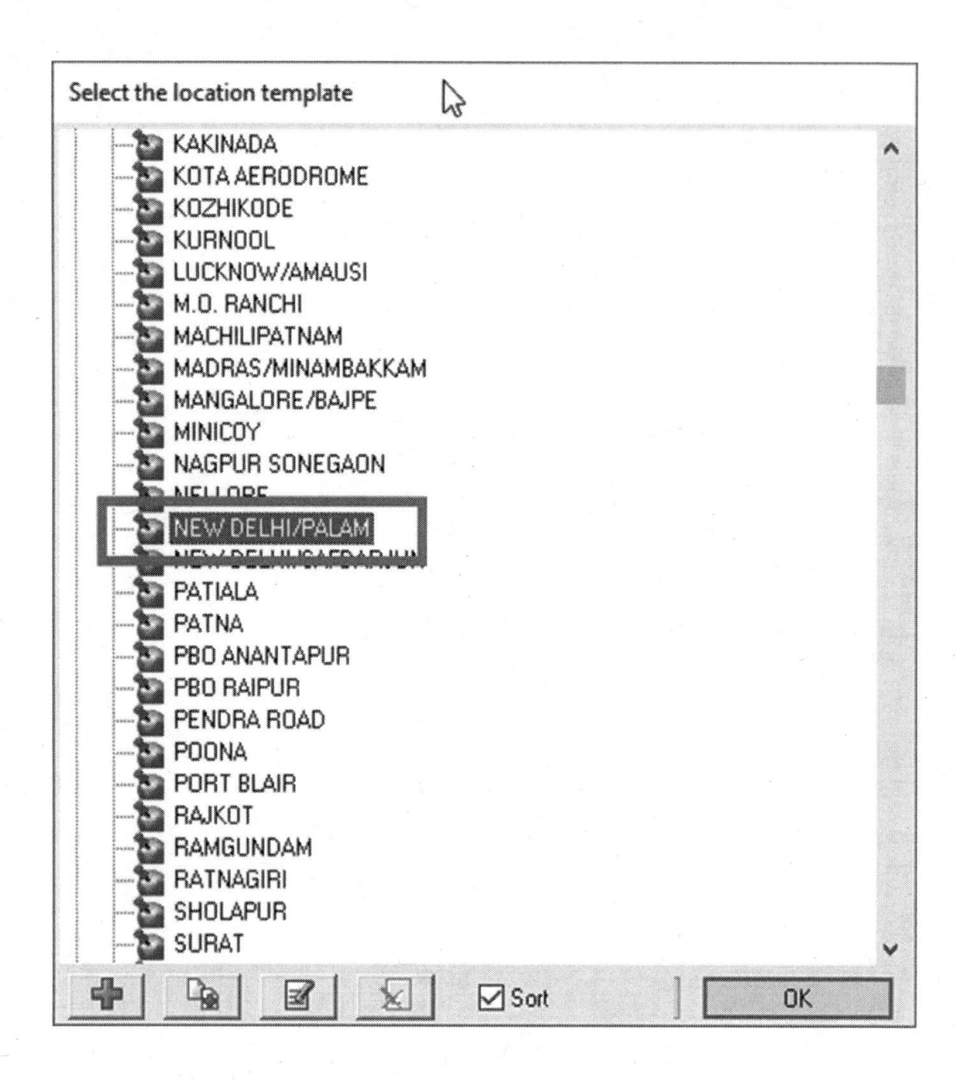

Step 6: Click **Building 1.**

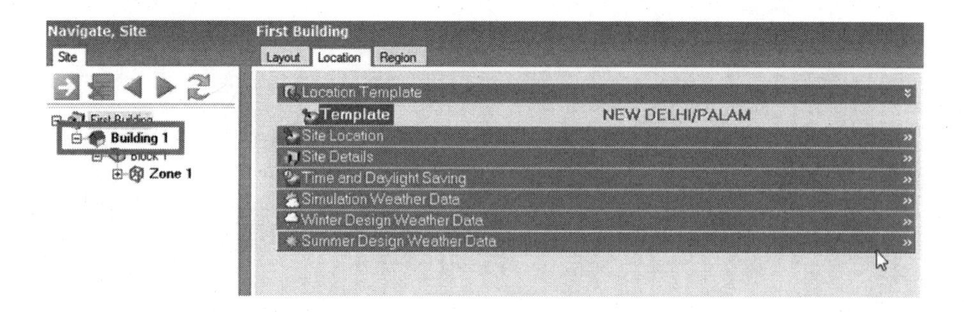

Step 7: Select the **Simulation tab.** Click on **Update data**. The **Edit Calculation Options** screen appears. Click **OK.** If your computer is connected to the internet, DesignBuilder directly downloads the weather file. If, for some reason, DesignBuilder is not able to download the weather file, you can manually download the weather file from https://energyplus.net/weather.

Place the downloaded weather file in the folder **C:\ProgramData\DesignBuilder\Weather Data**

DesignBuilder uses hourly weather files with the **.epw** extension. Weather files can be downloaded from the following link: https://energyplus.net/weather. Weather files that are downloaded directly from the website should be copied in the DesignBuilder weather data folder located at **C:\ProgramData\DesignBuilder\Weather Data**.

Step 8: Compare the simulation results for the London and New Delhi locations. Results for the London location can be obtained from Tutorial 1.2.

The impact of location on building energy consumption can be seen from the results reported in Table 1.5. You can see that there is no change in room electricity and lighting. This is because there were no change in internal loads; further, as there is no daylight-based control, the change in location does not change these values. From the results in Tables 1.5 and 1.6, the following observations can be made:

- The heating energy consumption is more in London than in New Delhi. This is because London has a colder climate than New Delhi.
- The reverse is true for the cooling energy consumption because New Delhi is warmer than London in summer.
- Similarly, there is a significant effect of location on the cooling and heating equipment capacities.

TABLE 1.5
Energy consumption with a change in building location

	Annual fuel breakdown data (kWh)	
Type of consumption	**London (kWh)**	**New Delhi (kWh)**
Room electricity	33,167.88	33,167.88
Lighting	47,934.77	47,934.77
Heating (gas)	30,669.47	1516.51
Cooling	14,385.49	102,428.10
Domestic hot water (DHW, electricity)	3,001.15	3,001.15

TABLE 1.6
Heating and cooling equipment capacity with a change in location

	London	New Delhi
Capacity of cooling equipment (kW)	79.36	119.35
Capacity of heating equipment (kW)	72.95	43.32

The following steps show how to change the orientation.

Step 9: Ensure that you are at the **First Building** level in the **Navigation** panel. Select the Location tab (refer to Steps 2 and 3 of this tutorial), and click **Site Details**. Enter **90** in the **Site Orientation** text box. Select the Layout tab. You can observe the change in the north arrow direction before and after changing the site orientation.

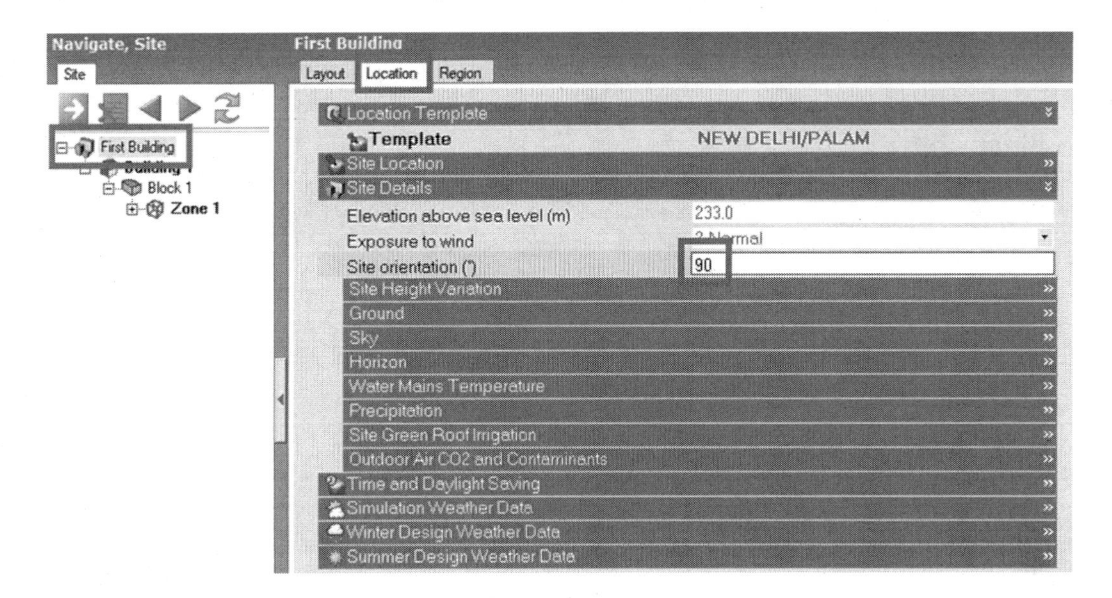

North is indicated by the direction of the north arrow in the sketch plan view. Working at the 'untitled' level (here **First Building**) of the tree means working at the site level. For example, if there are multiple buildings on a site that are to be modelled in DesignBuilder, after creating one building, return to the site level in the tree to start creating the next building. Therefore, all site-specific information such as location and orientation can be assigned to the model only when the site level is selected in the navigation tree.

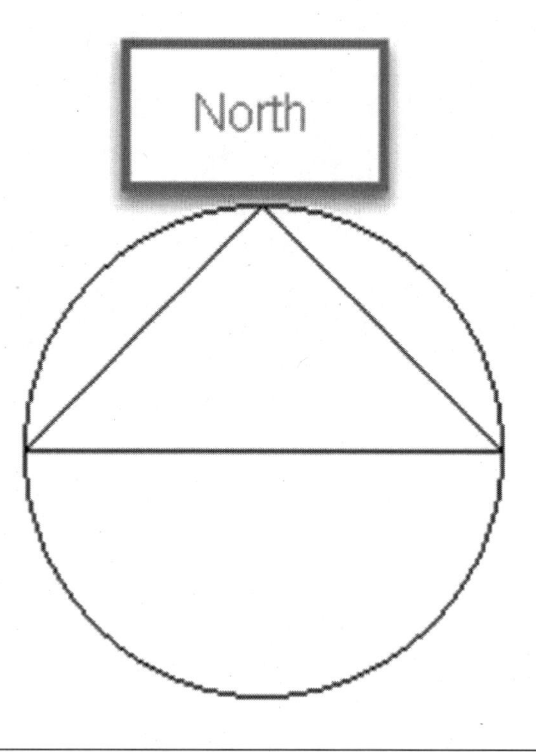

Site orientation (°) in DesignBuilder represents the alignment of the building(s) with respect to true north. However, because there could be multiple buildings on the site, this is called *site orientation* instead of *building orientation*.

Step 10: Select the **Simulation** screen tab to get the results.

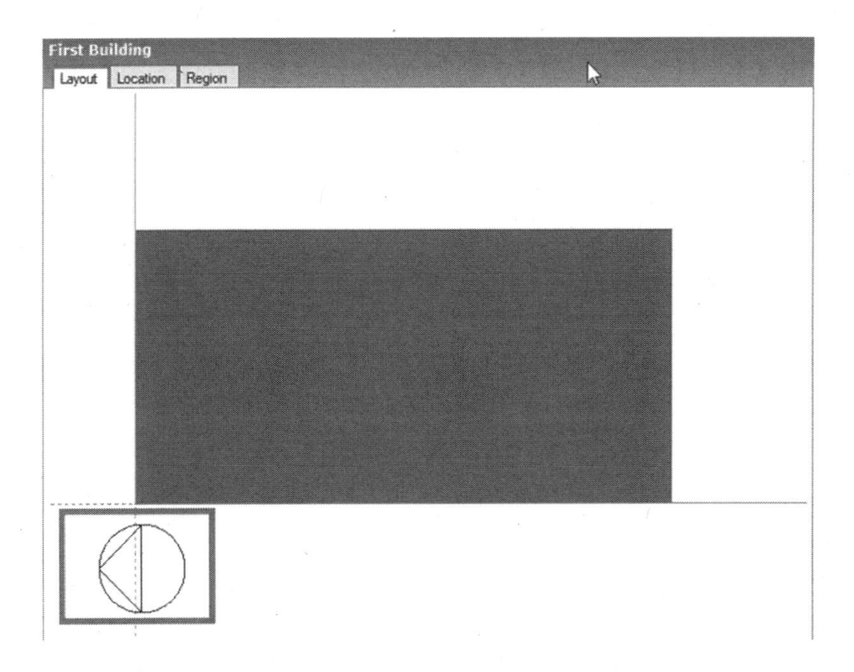

The impact of building orientation on building energy consumption is shown in the results recorded in Table 1.7.

Changing the orientation of a building changes the heating and cooling energy consumption as well as the size of the system. You may note that heating capacity is not affected by the change in orientation because of the fact that the peak requirement for heating occurs in non-sunshine hours. Hence, any orientation would require the same amount of heating (Table 1.8).

TABLE 1.7
Annual energy consumption with a change in building orientation

Type of consumption	Orientation 0 degrees (kWh)	Orientation 90 degrees (kWh)
Room electricity	33,167.88	33,167.88
Lighting	47,934.77	47,934.77
Heating (gas)	1,516.51	1,662.15
Cooling (electricity)	102,428.10	104,099.10
DHW (electricity)	3,001.146	3,001.146

TABLE 1.8
Heating and cooling equipment capacity with a change in building orientation

Type of equipment	Orientation 0 degrees (kW)	Orientation 90 degrees (kW)
Cooling	119.35	125.60
Heating	43.32	43.32

EXERCISE 1.3A

For New Delhi, run simulations for three more orientations: 45, 180 and 270 degrees. Compare the results (Tables 1.9 and 1.10).

TABLE 1.9
Energy consumption with a change in building orientation

	Annual fuel breakdown data (kWh)				
Type of consumption	**0 Degrees**	**45 Degrees**	**90 Degrees**	**180 Degrees**	**270 Degrees**
Room electricity					
Lighting					
Heating (gas)					
Cooling (electricity)					
DHW (electricity)					

TABLE 1.10
Heating and cooling equipment capacity with a change in building orientation

Type of equipment	0 Degrees	45 Degrees	90 Degrees	180 Degrees	270 Degrees
Heating capacity (kW)					
Cooling capacity (kW)					

Compare the energy consumption for 0 and 180 degrees. What do you observe? It can be seen that the results for 0 and 180 degrees are the same, and those for 90 and 270 degrees are the same. This is due to symmetry in the shape and window distribution of the building such that solar exposure for these two cases becomes similar. A comparison of the energy consumption for 0 and 90 degrees shows a difference in the cooling capacity as well as energy consumption owing to the change in solar exposure.

EXERCISE 1.3B

Change the weather locations of the model to New York and Singapore, and enter the values in Tables 1.11 and 1.12.

TABLE 1.11
Energy consumption with a change in building location

	Annual fuel breakdown data (kWh)	
Type of consumption	**New York**	**Singapore**
Room electricity		
Lighting		
Heating (gas)		
Cooling (electricity)		
DHW (electricity)		

TABLE 1.12
Heating and cooling equipment capacity with a change in building location

Type of equipment	New York (kW)	Singapore (kW)
Heating		
Cooling		

Observe and write why energy consumption and system sizing changed? It can be seen that the extent of variation is significantly less for a change in orientation versus a change in location. The change in energy consumption due to orientation may be more pronounced in buildings that have a higher aspect ratio (the ratio of longer and shorter sides) because of the change in solar exposure of the building that influences heat gain.

The change in energy consumption with location is largely due to the change in the harshness of the climatic conditions. This can be proportional to the ***cooling degree days*** (CDDs) for cooling energy consumption and ***heating degree days*** (HDDs) for heating energy consumption. Values for CDDs and HDDs can be found in weather files, design data books and references such as American Society of Heating, Refrigerating and Air-Conditioning Engineers (ASHRAE) handbooks.

HDDs and **CDDs** are common measures used to interpret the heating and cooling needs of a location. Degree days are the summation of the product of the difference in temperature (ΔT) between the average outdoor and the hypothetical average indoor temperatures and the number of days the outdoor temperature is above or below the hypothetical average indoor temperature.

The degree days are calculated in reference to a baseline line temperature that is commonly 18°C. Temperatures above 18°C need to be cooled, and temperatures below 18°C need to be heated.

Example: To calculate CDDs for two consecutive days:

Baseline temperature: 18 °C
Daily average outdoor dry bulb temperature (DBT) on day 1: 20°C
Daily average outdoor DBT on day 2: 17°C
CDD on day 1: $20 - 18 = 2$
CDD on day 2: $17 - 18 = -1$
Total CDD: 2 (sum for days that have positive values only)

The unit of measurement for CDDs and HDDs is degree days.

Tutorial 1.4

Evaluating the Impact of Opaque Envelope Components

GOAL

To understand the impact of the thermal properties of opaque envelope components (the external wall and roof) on HVAC system sizing and energy consumption

WHAT ARE YOU GOING TO LEARN?

- How to change wall and roof construction

PROBLEM STATEMENT

For the **Courtyard with VAV Example**, change the external wall from **Best practice Wall, Medium weight** to **Brickwork single leaf construction dense plaster,** and record its impact on the building energy performance for the London weather file.

SOLUTION

Step 1: Open the **Courtyard with VAV Example** model from the template.

DesignBuilder Data

Recent Files | Component Libraries | Template Libraries

Name	Folder	Size (KB)	Last Modified	Extension
IECC Residential Prototype Low Rise Apa...	C:\ProgramData\DesignBuilde...	1315	12-Feb-19 10:16:58 AM	DesignBuilder templates
Atrium example base	C:\ProgramData\DesignBuilde...	1158	12-Feb-19 10:16:58 AM	DesignBuilder templates
Underfloor Heating Example	C:\ProgramData\DesignBuilde...	908	12-Feb-19 10:16:56 AM	DesignBuilder templates
Trombe Wall Example	C:\ProgramData\DesignBuilde...	1045	12-Feb-19 10:16:56 AM	DesignBuilder templates
CFD External Analysis Example	C:\ProgramData\DesignBuilde...	1088	12-Feb-19 10:16:52 AM	DesignBuilder templates
Model Geometry Example	C:\ProgramData\DesignBuilde...	1081	12-Feb-19 10:16:50 AM	DesignBuilder templates
Mech Vent with Preheat	C:\ProgramData\DesignBuilde...	914	12-Feb-19 10:16:50 AM	DesignBuilder templates
Parametric Simulation Example	C:\ProgramData\DesignBuilde...	1106	12-Feb-19 10:16:50 AM	DesignBuilder templates
Zone Multiplier Example	C:\ProgramData\DesignBuilde...	1111	12-Feb-19 10:16:50 AM	DesignBuilder templates
Zone Water To Air Heat Pump Example	C:\ProgramData\DesignBuilde...	1331	12-Feb-19 10:16:50 AM	DesignBuilder templates
Plenum Example Buillding	C:\ProgramData\DesignBuilde...	1052	12-Feb-19 10:16:50 AM	DesignBuilder templates
DesignBuilder Video Tutorials Example	C:\ProgramData\DesignBuilde...	1186	12-Feb-19 10:16:50 AM	DesignBuilder templates
Double Skin Facade Example	C:\ProgramData\DesignBuilde...	1171	12-Feb-19 10:16:50 AM	DesignBuilder templates
Dehumidification example	C:\ProgramData\DesignBuilde...	1122	12-Feb-19 10:16:50 AM	DesignBuilder templates
CFD Internal Analysis Example	C:\ProgramData\DesignBuilde...	[illegible]	12-Feb-19 10:16:50 AM	DesignBuilder templates
Courtyard with VAV Example	C:\ProgramData\DesignBuilde...	931	12-Feb-19 10:16:50 AM	DesignBuilder templates
Atrium Example with Calc Nat Vent	C:\ProgramData\DesignBuilde...	[illegible]	12-Feb-19 10:16:50 AM	DesignBuilder templates
IECC Residential Prototype Detached ho...	C:\ProgramData\DesignBuilde...	1285	12-Feb-19 10:16:50 AM	DesignBuilder templates
Mixed Mode Example	C:\ProgramData\DesignBuilde...	1093	12-Feb-19 10:16:44 AM	DesignBuilder templates
PCM Example	C:\ProgramData\DesignBuilde...	1270	12-Feb-19 10:16:44 AM	DesignBuilder templates
House Example 1	C:\ProgramData\DesignBuilde...	989	12-Feb-19 10:16:44 AM	DesignBuilder templates
Single Zone Example	C:\ProgramData\DesignBuilde...	758	07-Apr-16 12:50:30 PM	DesignBuilder templates
Simple HVAC night cooling	C:\ProgramData\DesignBuilde...	995	07-Apr-16 12:50:30 PM	DesignBuilder templates

Step 2: Select the **Construction** tab.

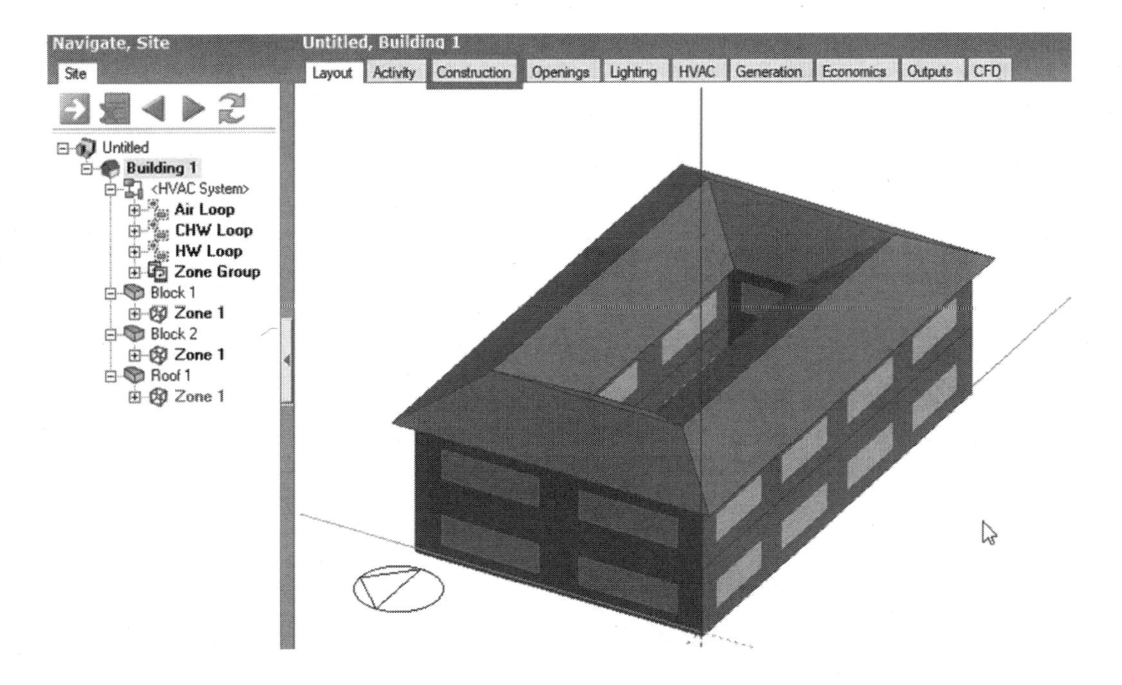

DesignBuilder uses a tree structure for data organization. The template is the *root*, and all the other fields are the *branches.* DesignBuilder templates are databases of typical generic data. Therefore, when editing the template at the root level, it changes at all branches, but if changes are made at the branch level, root and other branches will not change.

Objects such as external walls in the **Construction** template can also be edited individually without changing the complete root template. However, in this case, it is to be noted that this particular object will be decoupled from the root. This means that changing the root object does not impact the value of this object.

Step 3: Make sure that **Best practice Wall, Medium weight** is selected as External walls.

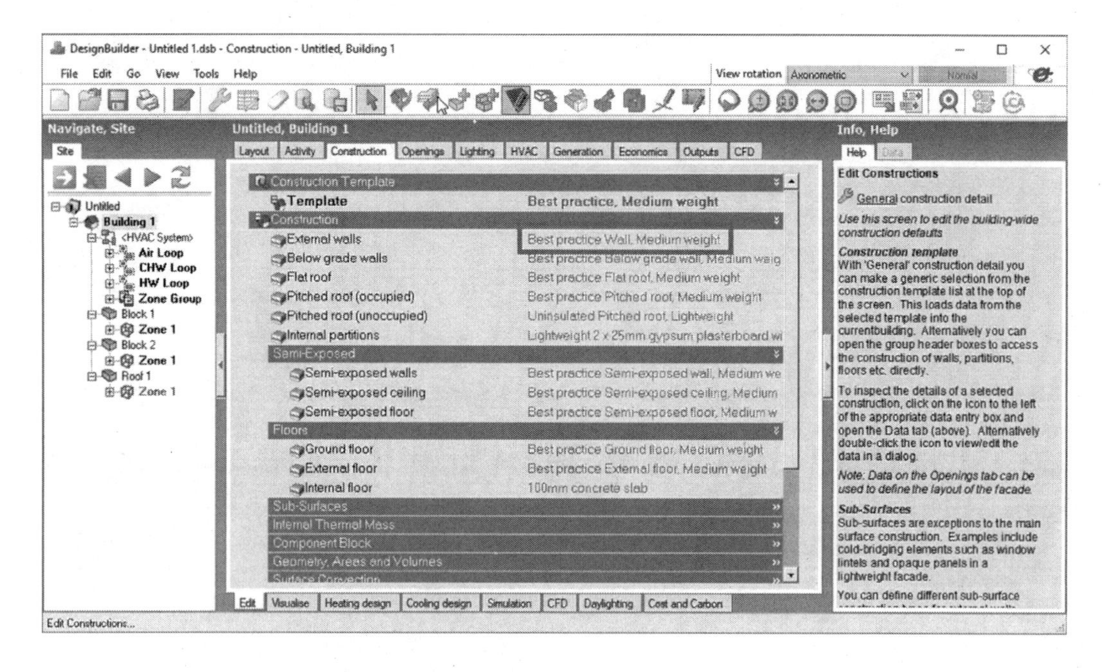

Step 4: Select the **Simulation** screen tab. Click on **Update data**. The **Edit Calculation Options** screen appears. Click the **Annual simulation** link to set **the** annual simulation period.

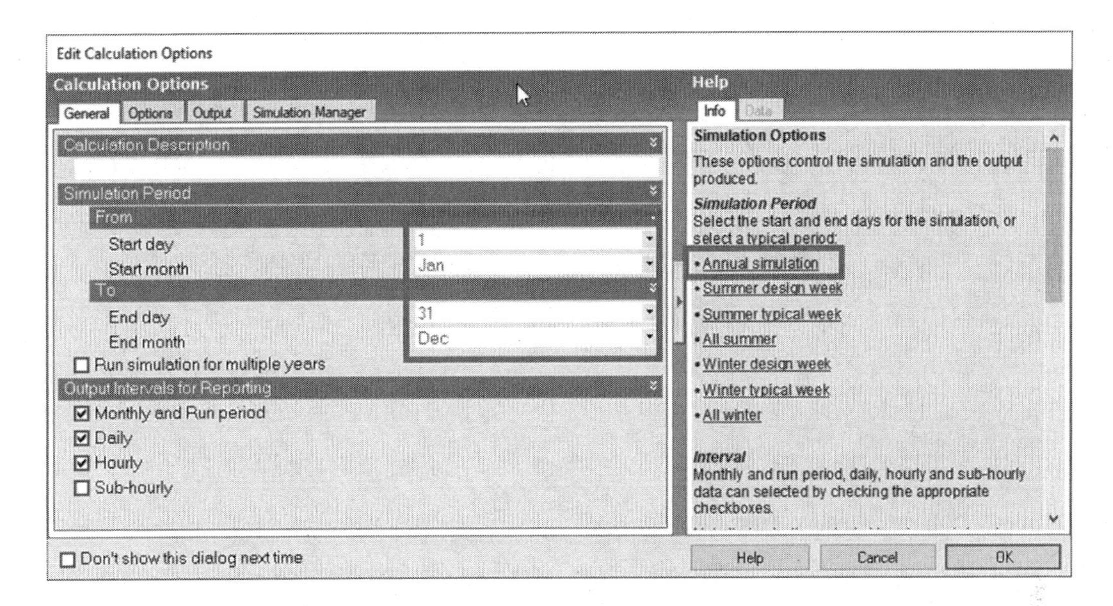

Step 5: Record the energy simulation results.

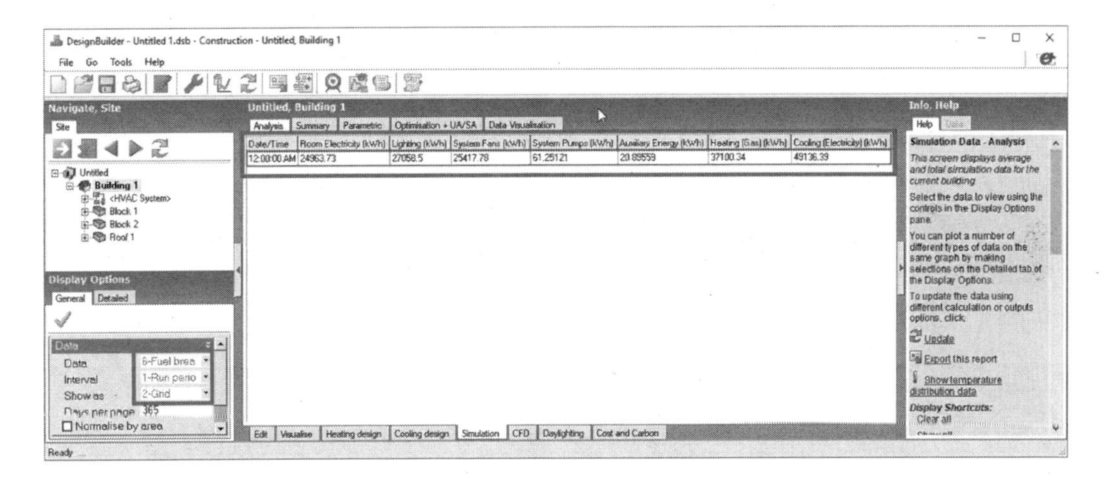

In the following steps, you are going to change the external wall construction.

Step 6: Click **the Edit** screen**. The Construction Template** screen appears.

Step 7: Click **External walls** under the **Construction** section. Three dots (...) appear. Click the three dots. The **Select the construction** screen appears.

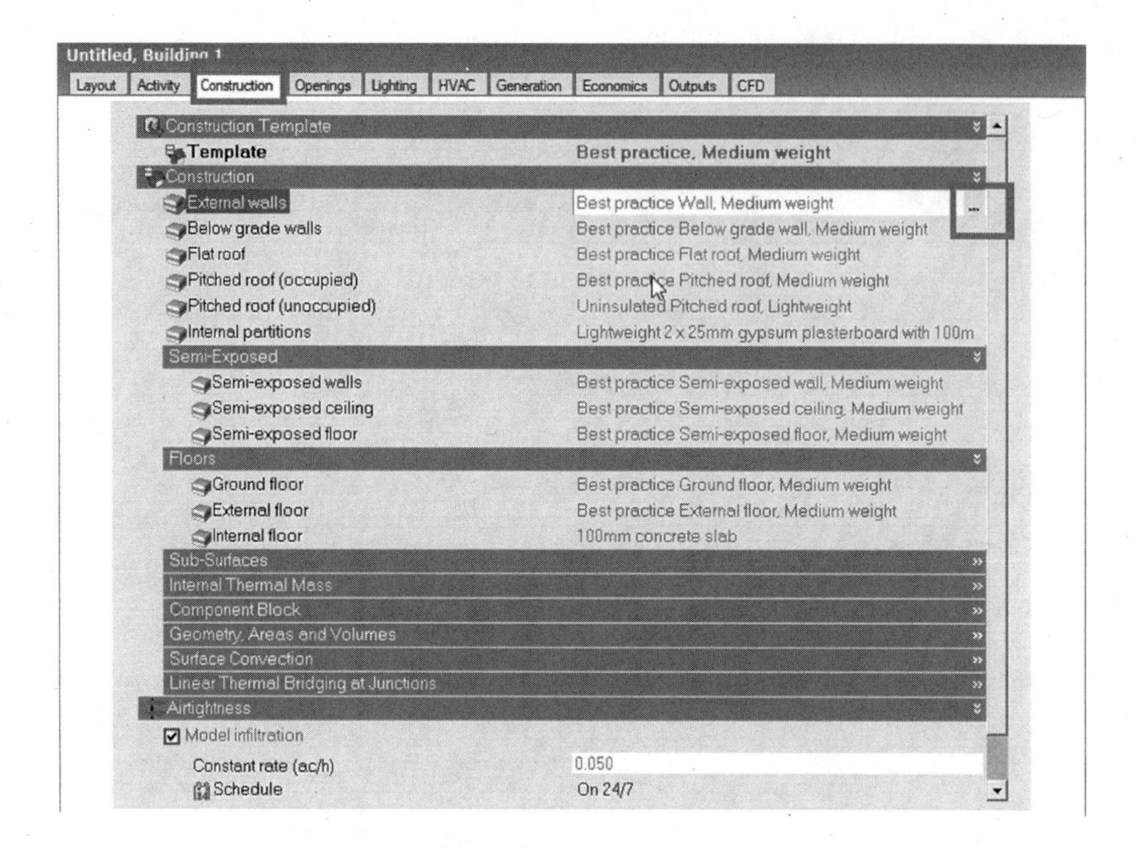

Step 8: Select **Brickwork single leaf construction dense plaster.**

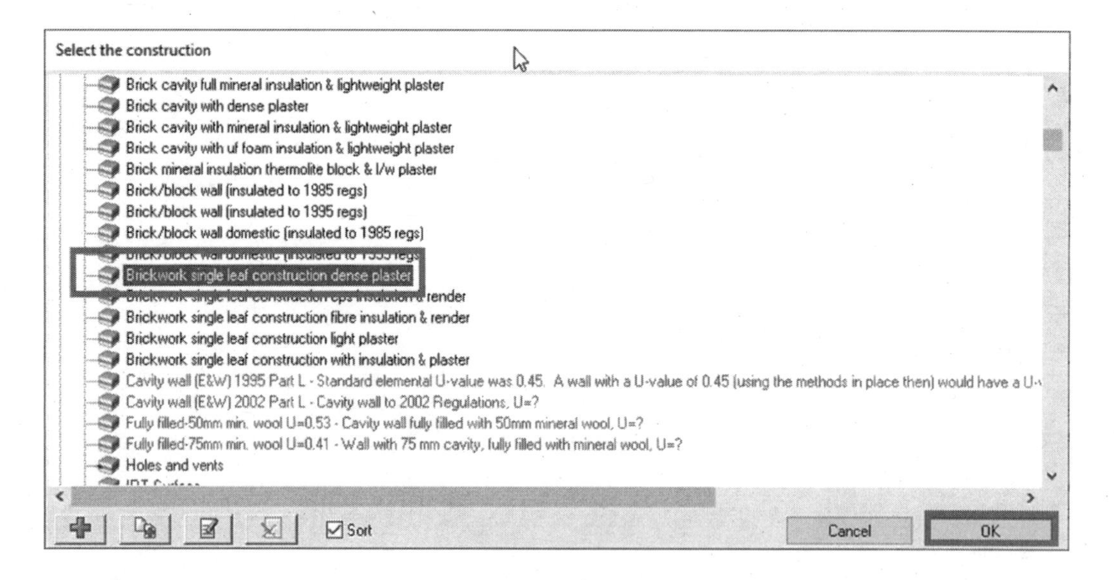

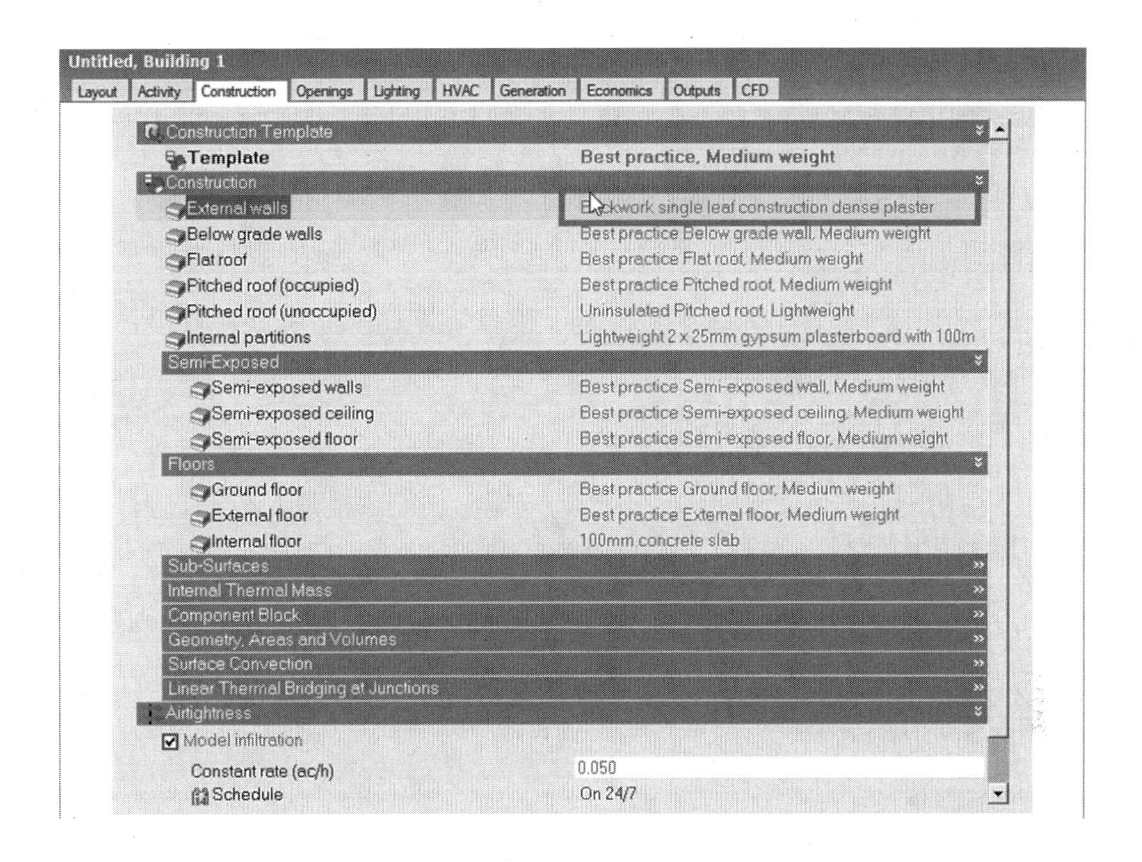

Step 9: Select the **Simulation** screen tab, and perform the annual energy simulation. View the energy simulation results in Tables 1.13 **and** 1.14 and compare. It can be seen that changing the wall impacts both equipment sizing and energy consumption.

TABLE 1.13
Energy consumption with a change in external wall construction

	Annual fuel breakdown data (kWh)	
Type of consumption	**Best practice Wall, Medium weight**	**Brickwork single leaf construction dense plaster**
Room electricity	24,964	24,964
Lighting	27,059	27,059
System fans	25,418	26,119
System pumps	61	68
Auxiliary energy	21	25
Heating (gas)	37,100	80,716
Cooling (electricity)	49,136	48,741

TABLE 1.14
Heating and cooling equipment capacity with a change in external wall construction

Type of capacity	Best practice Wall, Medium weight (kW)	Brickwork single leaf construction dense plaster (kW)
Cooling	70.37	72.62
Heating	55.87	93.07

EXERCISE 1.4

Open the model prepared in Tutorial 1.2. Assign flat roof construction as shown in Tables 1.15 and 1.16. Simulate and compare the results for London.

TABLE 1.15
Energy consumption with a change in roof construction

	Annual fuel breakdown data (kWh)	
Type of consumption	Roof, insulated entirely above deck, R-24 (4.2), U-0.040 (0.23)	Uninsulated flat roof, medium weight
Room electricity		
Lighting		
System fans		
System pumps		
Heating (electricity)		
Heating (gas)		
Cooling (electricity)		

TABLE 1.16
Heating and cooling sizing with a change in roof construction

	Heating and cooling sizing (kW)	
Type	Roof, insulated entirely above deck, R-24 (4.2), U-0.040 (0.23)	Uninsulated flat roof, medium weight
Heating		
Cooling		

You can follow similar steps for the external wall. Please refer to the following screenshots for help.

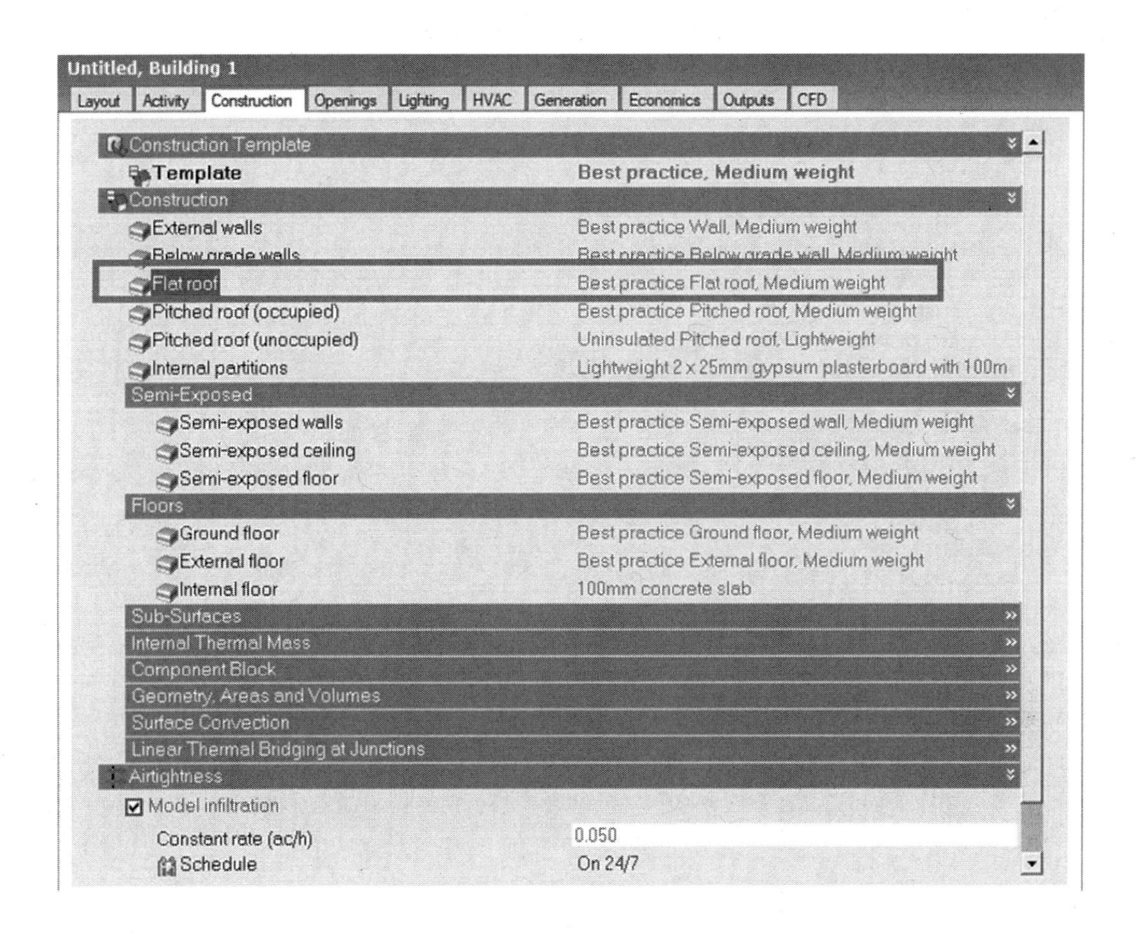
Untitled, Building 1
Layout
Activity
Construction
Openings
Lighting
HVAC
Generation
Economics
Outputs
CFD
Construction Template
Template
Best practice, Medium weight
Construction
External walls
Best practice Wall, Medium weight
Below grade walls
Best practice Below grade wall, Medium weight
Flat roof
Best practice Flat roof, Medium weight
Pitched roof (occupied)
Best practice Pitched roof, Medium weight
Pitched roof (unoccupied)
Uninsulated Pitched roof, Lightweight
Internal partitions
Lightweight 2 x 25mm gypsum plasterboard with 100m
Semi-Exposed
Semi-exposed walls
Best practice Semi-exposed wall, Medium weight
Semi-exposed ceiling
Best practice Semi-exposed ceiling, Medium weight
Semi-exposed floor
Best practice Semi-exposed floor, Medium weight
Floors
Ground floor
Best practice Ground floor, Medium weight
External floor
Best practice External floor, Medium weight
Internal floor
100mm concrete slab
Sub-Surfaces
Internal Thermal Mass
Component Block
Geometry, Areas and Volumes
Surface Convection
Linear Thermal Bridging at Junctions
Airtightness
Model infiltration
Constant rate (ac/h)
0.050
Schedule
On 24/7

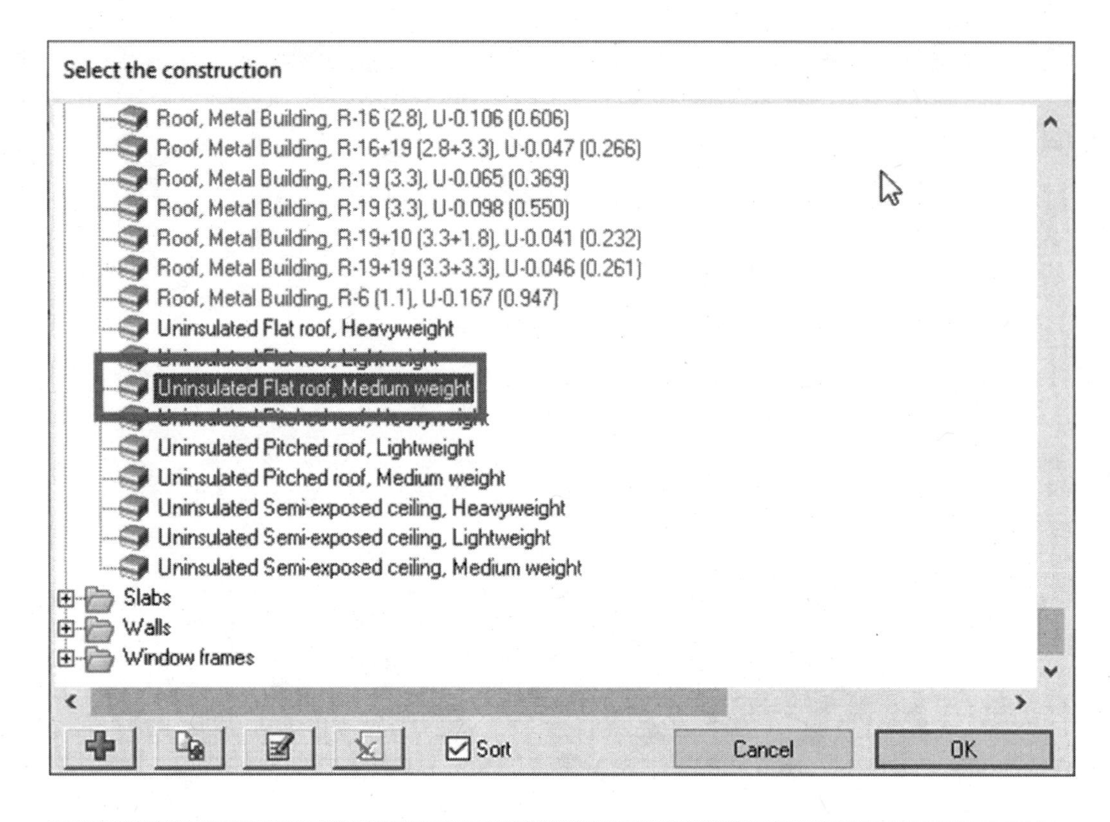
Select the construction
Roof, Metal Building, R-16 (2.8), U-0.106 (0.606)
Roof, Metal Building, R-16+19 (2.8+3.3), U-0.047 (0.266)
Roof, Metal Building, R-19 (3.3), U-0.065 (0.369)
Roof, Metal Building, R-19 (3.3), U-0.098 (0.550)
Roof, Metal Building, R-19+10 (3.3+1.8), U-0.041 (0.232)
Roof, Metal Building, R-19+19 (3.3+3.3), U-0.046 (0.261)
Roof, Metal Building, R-6 (1.1), U-0.167 (0.947)
Uninsulated Flat roof, Heavyweight
Uninsulated Flat roof, Medium weight
Uninsulated Pitched roof, Lightweight
Uninsulated Pitched roof, Medium weight
Uninsulated Semi-exposed ceiling, Heavyweight
Uninsulated Semi-exposed ceiling, Lightweight
Uninsulated Semi-exposed ceiling, Medium weight
Slabs
Walls
Window frames
Sort
Cancel
OK

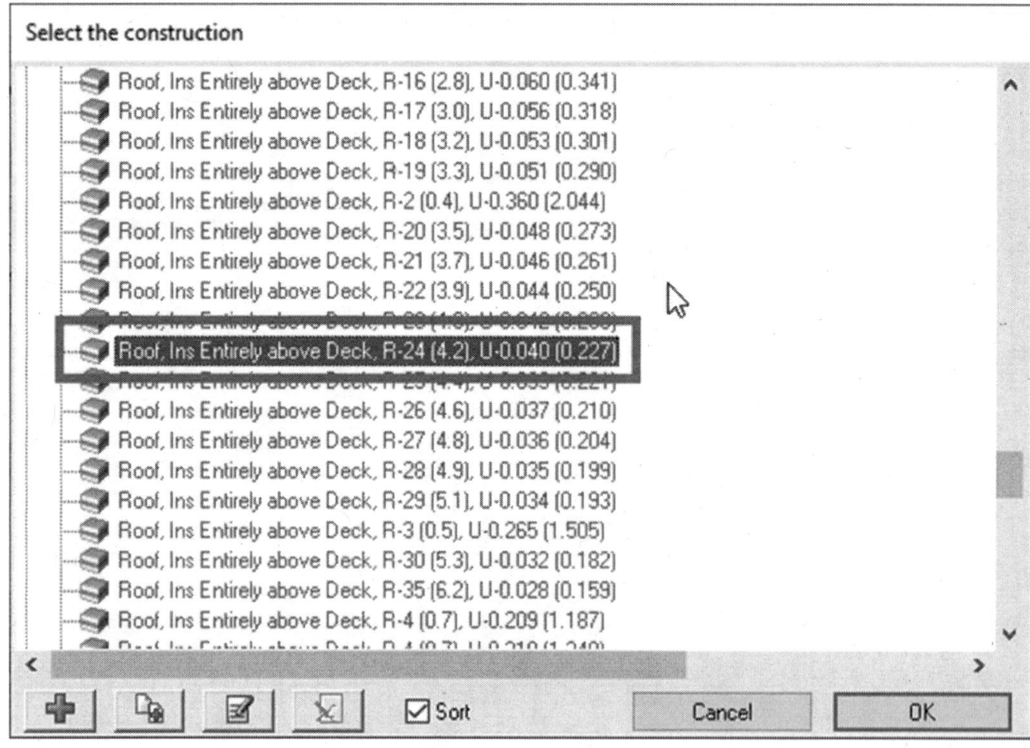
Select the construction
Roof, Ins Entirely above Deck, R-16 (2.8), U-0.060 (0.341)
Roof, Ins Entirely above Deck, R-17 (3.0), U-0.056 (0.318)
Roof, Ins Entirely above Deck, R-18 (3.2), U-0.053 (0.301)
Roof, Ins Entirely above Deck, R-19 (3.3), U-0.051 (0.290)
Roof, Ins Entirely above Deck, R-2 (0.4), U-0.360 (2.044)
Roof, Ins Entirely above Deck, R-20 (3.5), U-0.048 (0.273)
Roof, Ins Entirely above Deck, R-21 (3.7), U-0.046 (0.261)
Roof, Ins Entirely above Deck, R-22 (3.9), U-0.044 (0.250)
Roof, Ins Entirely above Deck, R-24 (4.2), U-0.040 (0.227)
Roof, Ins Entirely above Deck, R-26 (4.6), U-0.037 (0.210)
Roof, Ins Entirely above Deck, R-27 (4.8), U-0.036 (0.204)
Roof, Ins Entirely above Deck, R-28 (4.9), U-0.035 (0.199)
Roof, Ins Entirely above Deck, R-29 (5.1), U-0.034 (0.193)
Roof, Ins Entirely above Deck, R-3 (0.5), U-0.265 (1.505)
Roof, Ins Entirely above Deck, R-30 (5.3), U-0.032 (0.182)
Roof, Ins Entirely above Deck, R-35 (6.2), U-0.028 (0.159)
Roof, Ins Entirely above Deck, R-4 (0.7), U-0.209 (1.187)
Sort
Cancel
OK

Tutorial 1.5

Evaluating the Impact of WWR and Glass Type

GOAL

To evaluate the impact of glazing area and glazing properties on the energy performance of a building

In the field of building energy simulations, glazing area is most commonly quantified in terms of *window-wall ratio* (WWR). The WWR is the ratio of the total glazing area to the total external wall area in conditioned zones.

WHAT YOU ARE GOING TO LEARN?

- How to creating zones in a building
- How to change the WWR
- How to change glazing type

PROBLEM STATEMENT

Create a 60- × 40-m five-zone single-story building model for London. Take WWRs of 30% and 80% and evaluate the building energy performance. Compare cooling and heating equipment sizing of the north and south zones. Similarly, compare sizing in east and west zones. Evaluate the impact of glazing by changing the single-pane clear glazing (**Sgl Clr 3mm**) to high-performance, low solar heat gain coefficient (SHGC) dual-pane glazing (**Dbl Blue 6mm/13mm Air**).

SOLUTION

Step 1: Open a new project, and draw a building with a size of **60 × 40 m**.

Data Structure Hierarchy in DesignBuilder

Site→Building→Block→Zone→Surface→Opening

Assigning a template to the upper-level branch will result in the same template for all the sub-branches. For example, if a construction template named 'Framed Construction' is assigned to the building, all floors and zones will have the same template: 'Framed Construction'. When a particular sub-branch is edited, it is separated from the main branch, resulting in a need for a separate edit from the main branch.

For example, if a particular zone (such as zone 1) of the first floor of the building is selected and the construction template is changed from 'Framed Construction' to 'Mass Construction', and later the whole building is selected and the construction template is changed from 'Framed Construction' to 'Steel Construction', the construction template of zone 1 still retains 'Mass Construction'.

Source: https://designbuilder.co.uk/helpv6.0/#_Model_data_hierarchy_and_data_inheritance.htm.

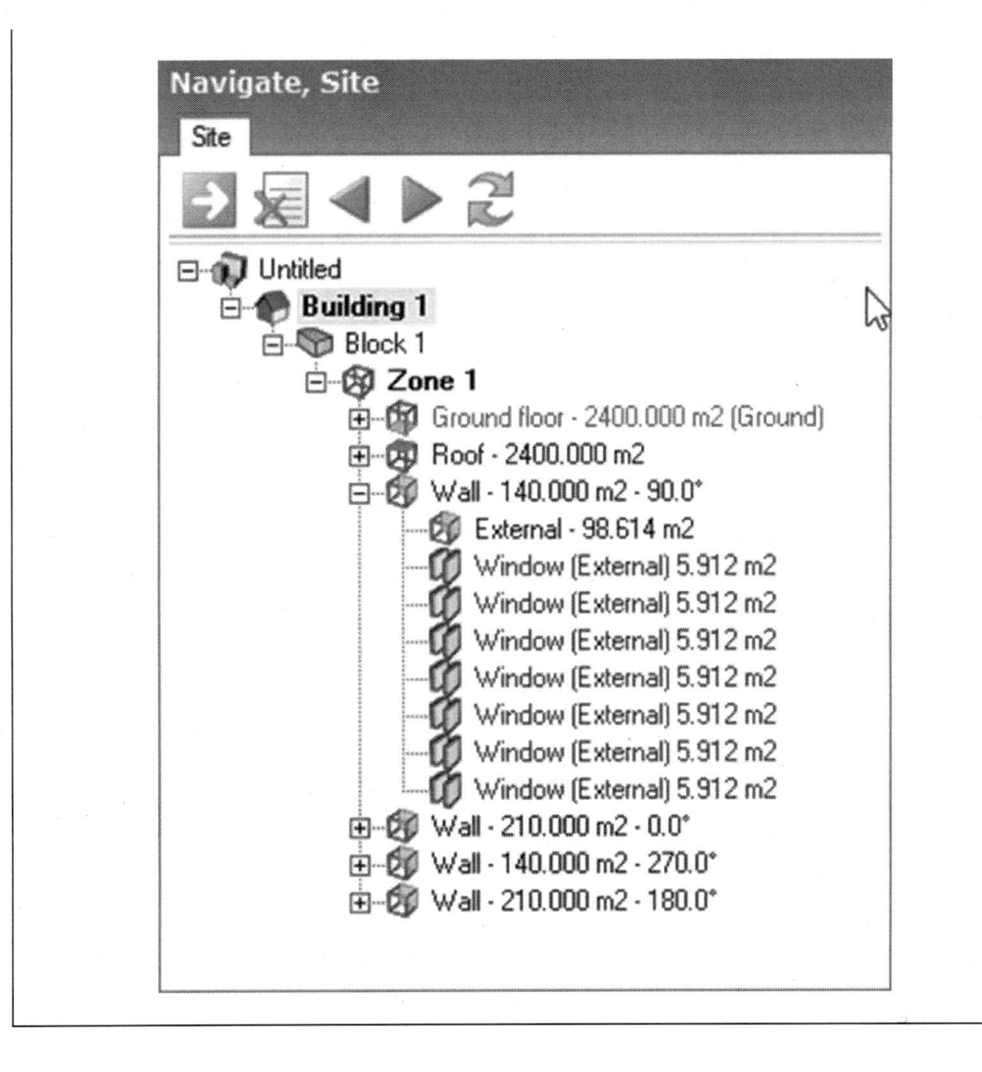

Step 2: Click **Block 1** in navigation tree. The display changes to **Zone 1**.

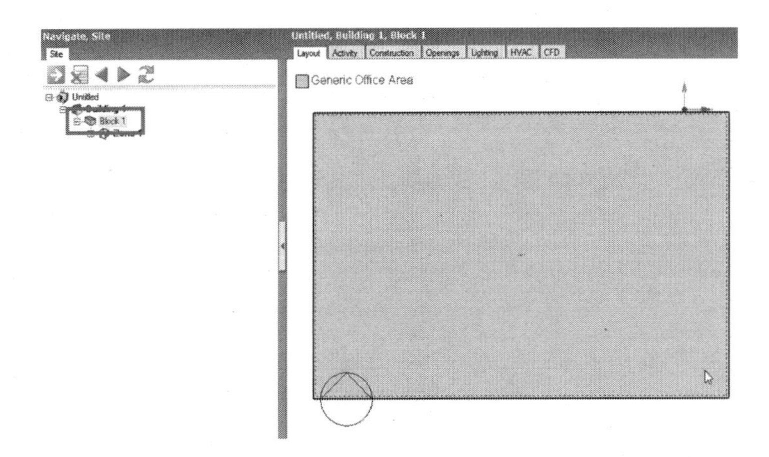

Step 3: Click the **Draw construction line** icon to mark lines on the zone. You need to draw an internal partition with an offset of 10 m.

Similar to other drafting programs, DesignBuilder provides some important drafting options such as snap points, gridlines and direction snap. Make sure to select the relevant options while creating the custom geometry.

Step 4: Place the pointer near the top-left corner. The pointer snaps to the corner (green square appears at the centre of the crosshairs). Left-click once the pointer has snapped, and then move the mouse in the right direction over the north wall of the building and type **10**. This draws a 10-m construction line from the top-left corner on the north wall.

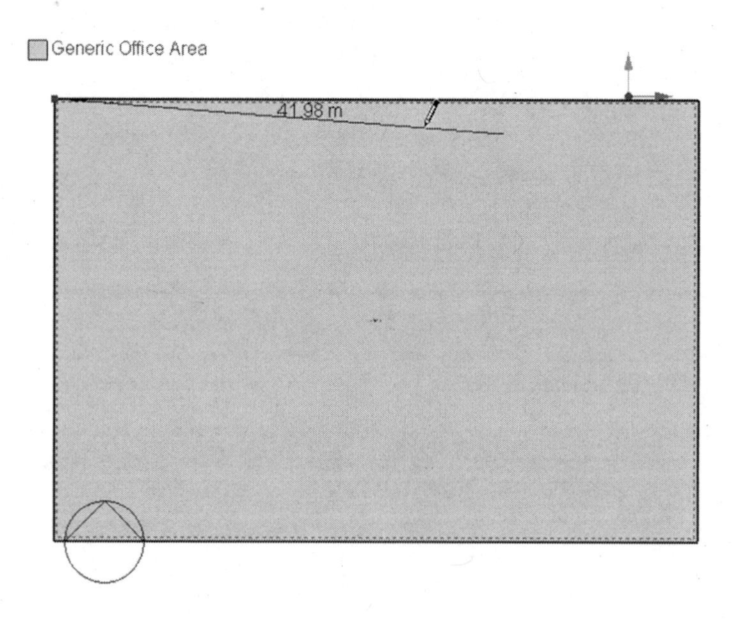

Step 5: Click the right end of the construction line. Now move the mouse vertically down parallel to the west wall of the building with an offset of 10 m (as achieved in the preceding step), and snap to the south wall (this time the crosshairs will change to the red square because it will be edge snapping) and click.

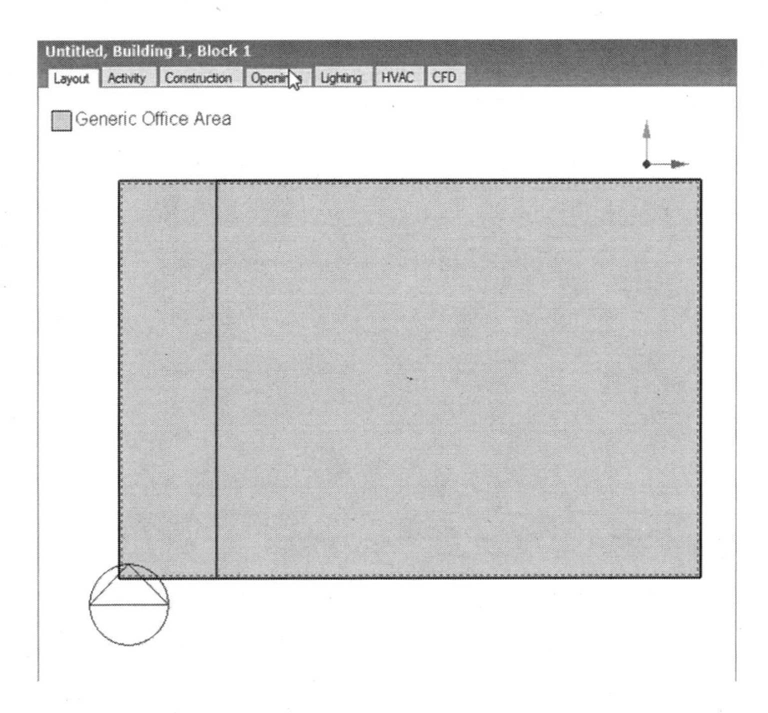

Step 6: Continue drawing all construction lines to mark core and perimeter zones.

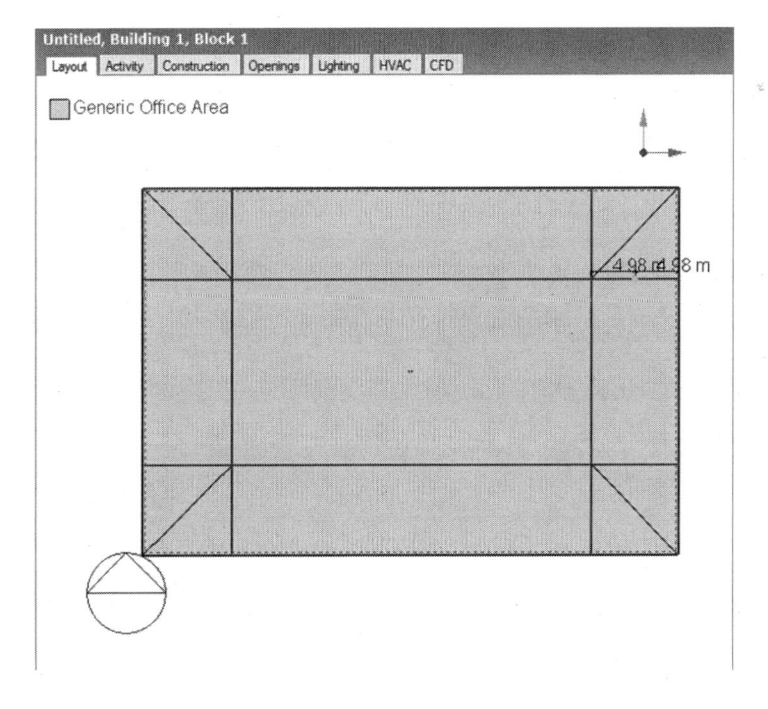

Step 7: Click the **Draw partition** icon. The screen appears with a cursor to draw the internal partitions.

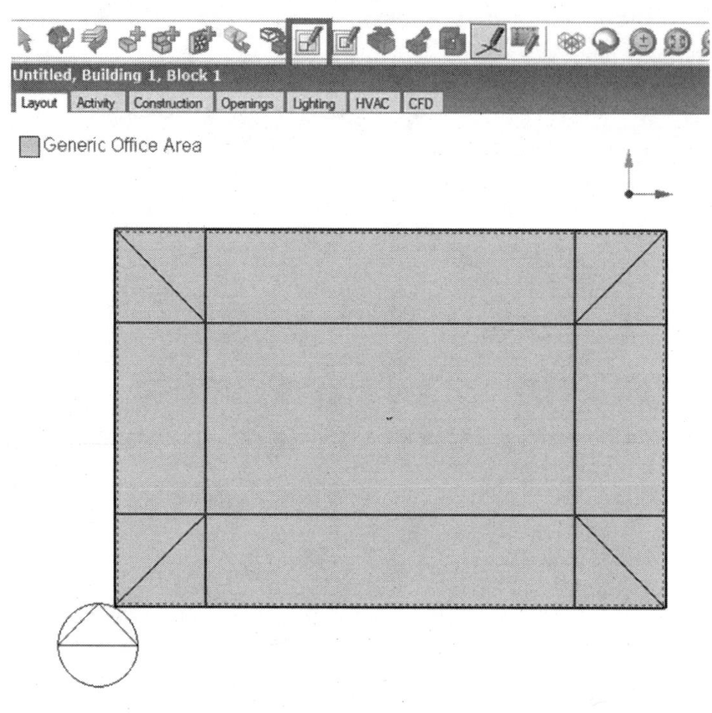

Step 8: Trace the construction line to draw a perimeter zone on the north.

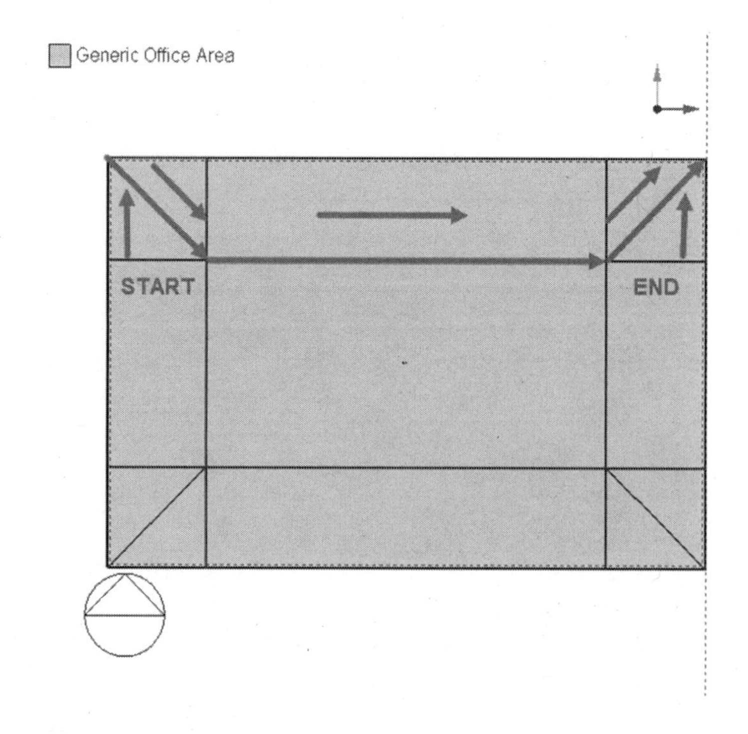

Step 9: Repeat Step 8 to draw the other three zones. This step results in a total of five zones, four perimeters and one core, as shown below with **Axonometric** view. (*Note:* There is no need to redraw the partition drawn in the preceding steps.)

> *Perimeter and core zoning* is common in building energy simulations, especially when the internal layout of the building that is being modelled is not designed. Perimeter and core zoning is also practiced while modelling large, open floor plans.

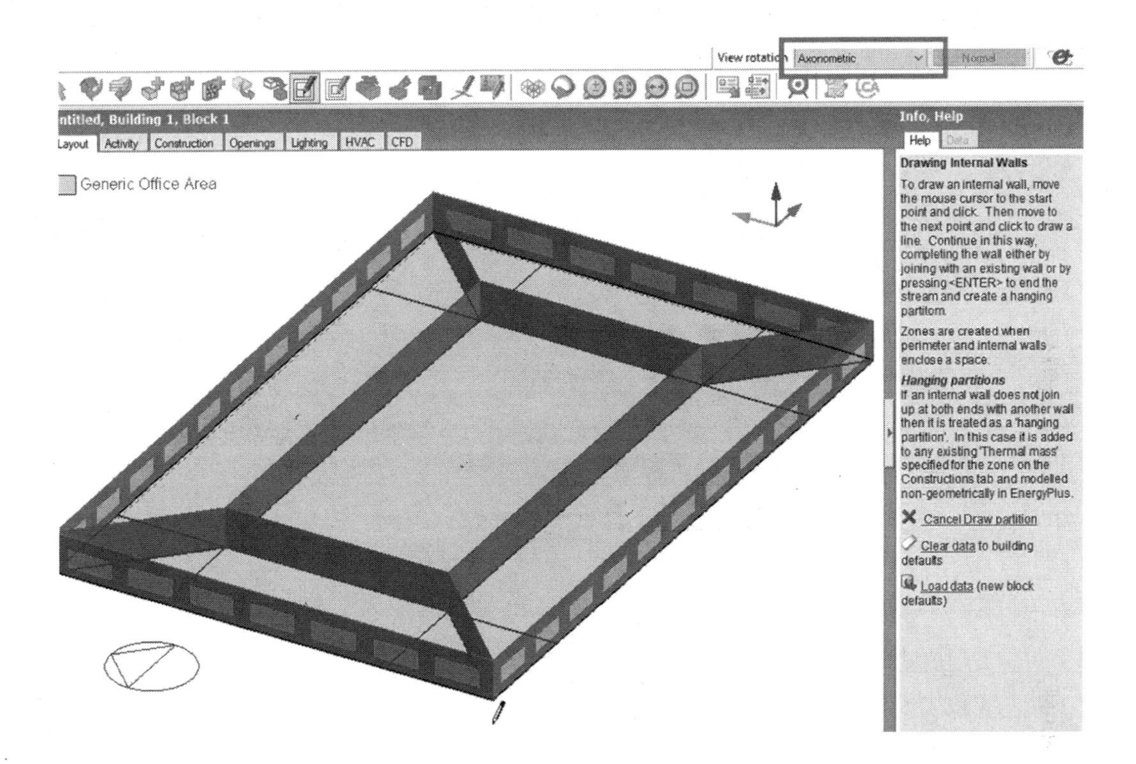

Step 10: Select and rename each zone based on its orientation. (For renaming, you need to single-click the zone name and wait for about 1 second and then click again. A text box appears to enter a new name.)

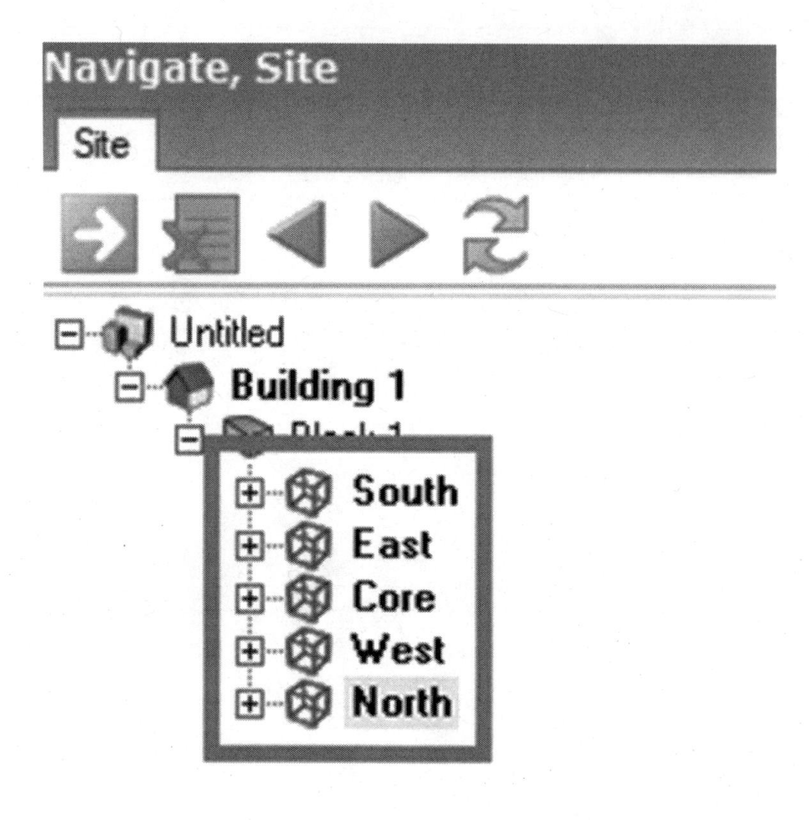

Step 11: Click **Building 1** in the navigation tree. Now you are going to set the WWR for the whole building.

Step 12: Select the **Openings** tab. The **Glazing Template** screen appears.

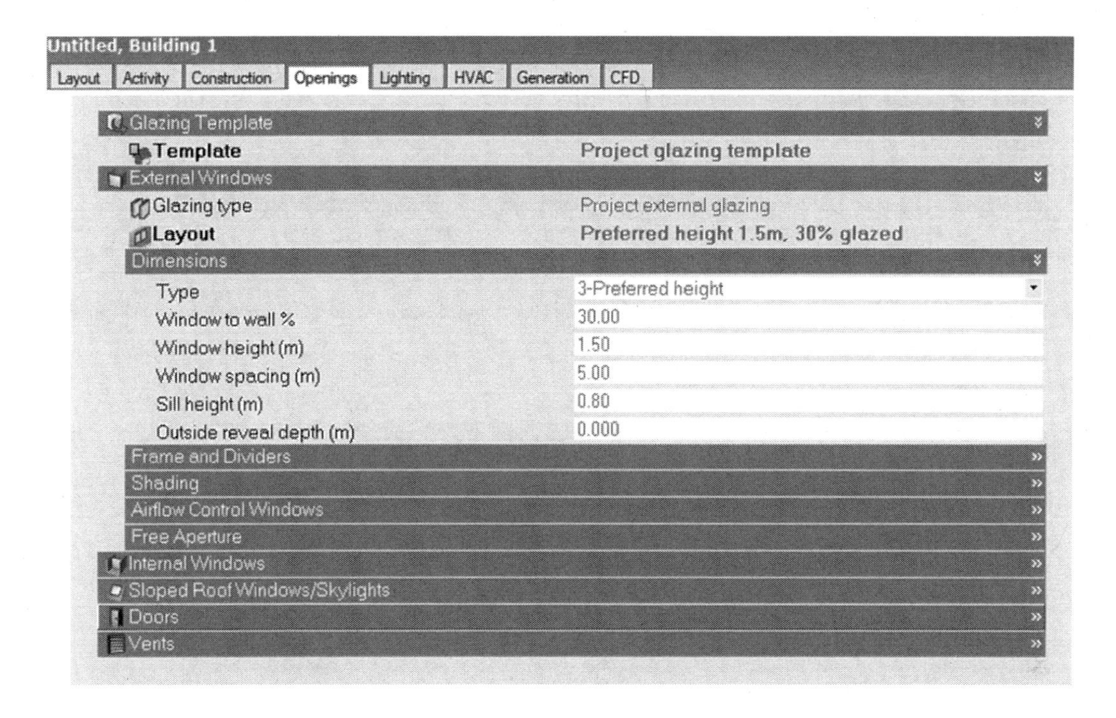

Step 13: Make sure that the window-to-wall percentage is **30**.

Dimensions	
Type	d height
Window to wall %	30.00
Window height (m)	[illegible]
Window spacing (m)	5.00
Sill height (m)	0.80
Outside reveal depth (m)	0.000

Step 14: Select the **Visualise** screen tab. Use the orbit tool for three-dimensional (3D) visuals.

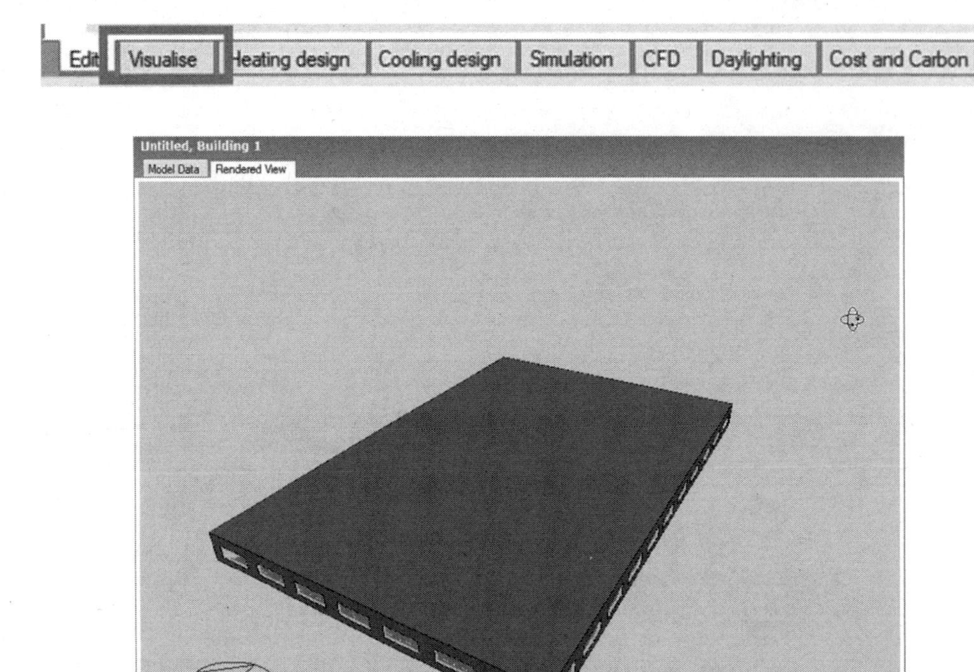

Step 15: Select the **Heating design** screen tab. Click on **Update data.** The **Calculation Options** screen appears.

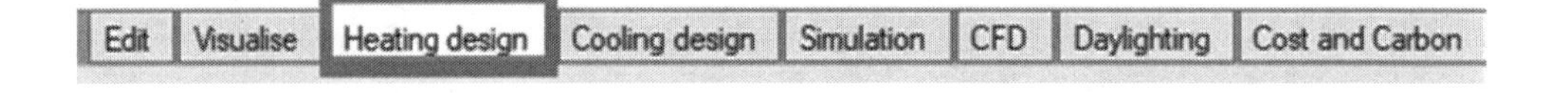

Step 16: Ensure that **Air temperature** is selected in the **Temperature control** drop-down list. Click **OK**.

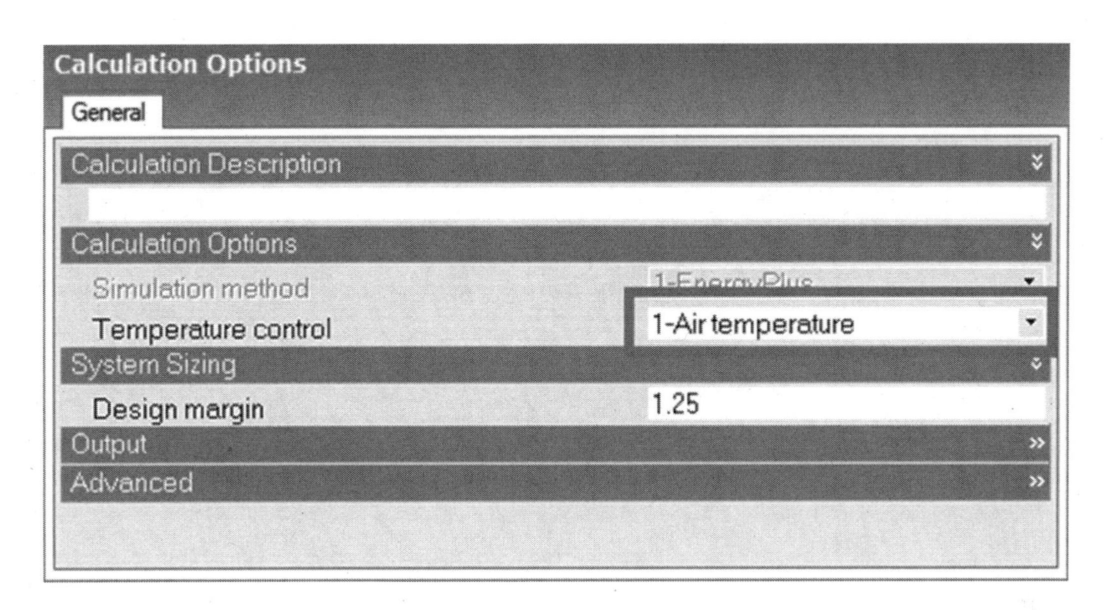

Air temperature control means controlling the mean air temperature of the zone to the assigned setpoint temperatures. Other control types mainly include operative temperature and adjusting zone radiant temperature control fraction. These control types are mainly used in advanced research and comfort analysis studies.

Step 17: Select the **Summary** tab. The results are displayed in a grid view. For each zone, the heating capacity is shown in the table.

Untitled, Building 1

Steady-state | Summary

Building 1 Total Design Heating Capacity = 207.580 (kW)

Block 1 Total Design Heating Capacity = 207.580 (kW)

Zone	Comfort Temp...	Steady-State ...	Design Capac...	Design Capac...	Glazing ...	Wall Gai...	Floor Ga...	Roof an...	Ventilati...	Infiltratio...
North	20.56	35.35	**44.19**	93.0288	-2.827	-1.171	-0.183	-3.246	-16.736	-11.190
East	20.55	21.37	**26.72**	94.3230	-1.871	-0.796	-0.109	-1.947	-9.981	-6.673
West	20.55	21.46	**26.83**	94.2702	-1.[illegible]72	-0.797	-0.109	-1.956	-10.028	-6.705
South	20.56	34.83	**43.54**	93.0707	-2.770	-1.181	-0.180	-3.200	-16.484	-11.021
Core	21.00	53.04	**66.30**	83.2626	0.000	-0.404	-0.437	-5.381	-28.061	-18.761

Step 18: Select the **Cooling design** screen tab. Click on **Update data.** The **Calculation Options** screen appears. Ensure that **Air temperature** is selected as temperature control. Click **OK**.

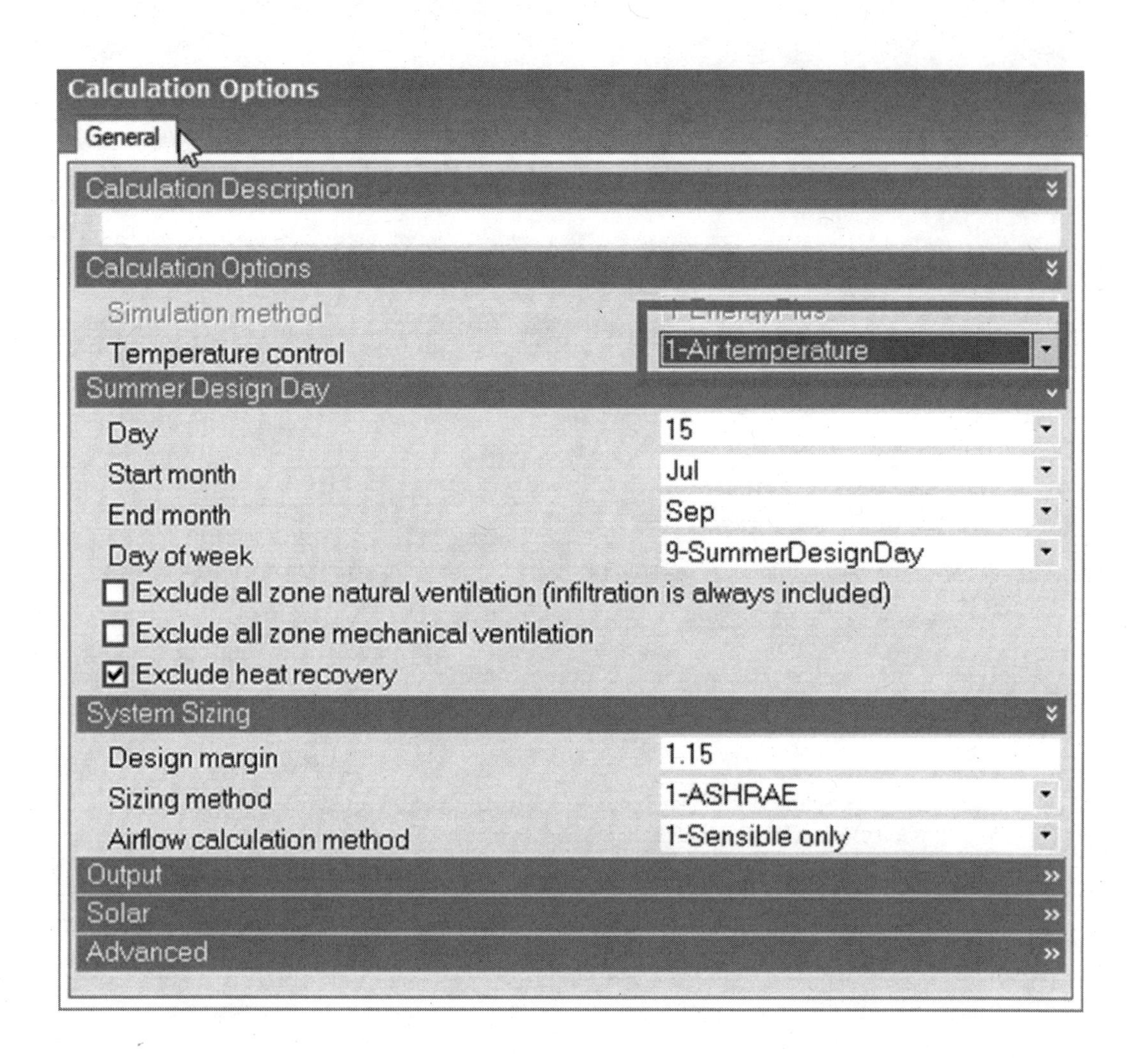

Step 19: Select the **Summary** tab to view the cooling system sizing results of each zone (Table 1.17).

TABLE 1.17
Cooling design data for 30% WWR

Zone	Design capacity (kW)	Design flow rate (m^3/s)	Total cooling load (kW)	Sensible (kW)	Latent (kW)	Air temperature (°C)	Humidity (%)	Time of max cooling	Max operating temperature in day (°C)
West	34.96	2.131	30.4	26.75	3.65	24	49.2	Aug 16:00	30.4
North	40.43	2.3681	35.16	29.71	5.44	24	50.4	Aug 15:00	28.2
East	27.59	1.6342	23.99	20.50	3.48	24	49.7	Aug 10:00	28.6
South	53.79	3.275	46.77	41.10	5.67	24	49.4	Aug 14:00	29.0
Core	63.06	3.66	54.83	45.93	8.9	24	50.7	Aug 15:00	28.2
Total	219.38	13.06	191.15	163.99	27.16	24	50.1	N/A	30.4

It can be seen that the peak cooling for the east zone is at 10:00 and for the west zone at 16:00. Further, note that the cooling capacity for the west zone is more than that for the east zone. Also, the design capacity for the south zone is more than that for the north zone. This is because London is located at 51.5 degrees north latitude, resulting in more solar radiation on the south facade.

Step 20: Run the energy simulation, and record the results for fuel breakdown. Save the model to use in forthcoming tutorials.

Step 21: Repeat the preceding steps, and set the WWR to **80**.

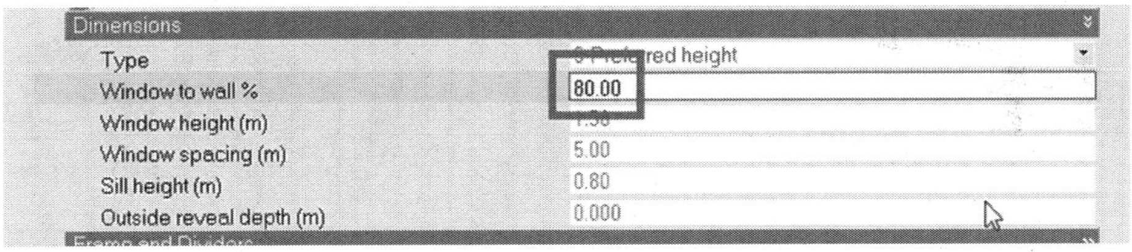

Step 22: Select the **Visualise** screen tab. Use the orbit tool for 3D visuals.

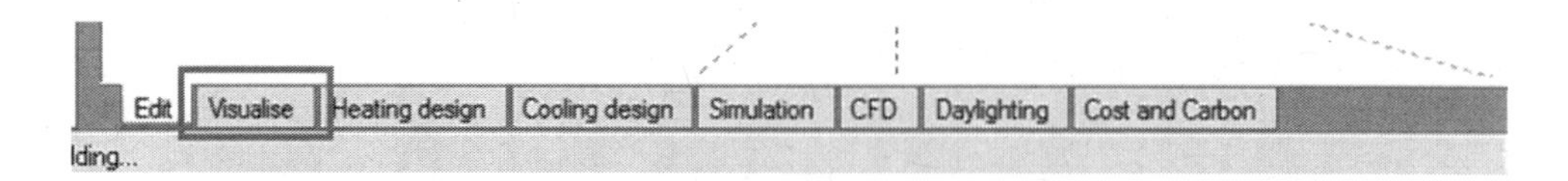

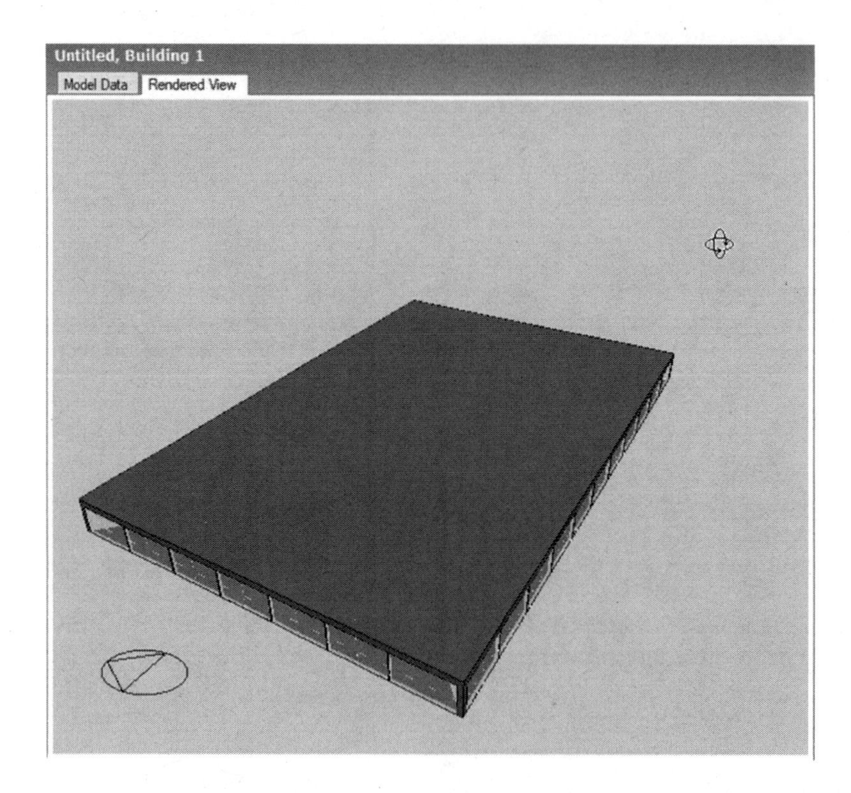

Step 23: Record the cooling design data and fuel breakdown (Tables 1.18 and 1.20).

Now compare the results for each zone to gain a better understanding of the impact of WWR on sizing. The comparison is shown in Table 1.19. Also compare the energy consumption of the two models (Table 1.20).

In this tutorial so far, you have learned to change WWR and seen its impact on sizing and energy consumption. Now you will see the impact of glazing by changing from a single-pane clear glazing (**Sgl Clr 3mm**) to a dual-pane, low SHGC high-performance glazing (**Dbl Blue 6mm/13mm Air**) with 30% WWR. Before going to the next step, make sure that the WWR is 30% and that you are at the building level in the navigation tree.

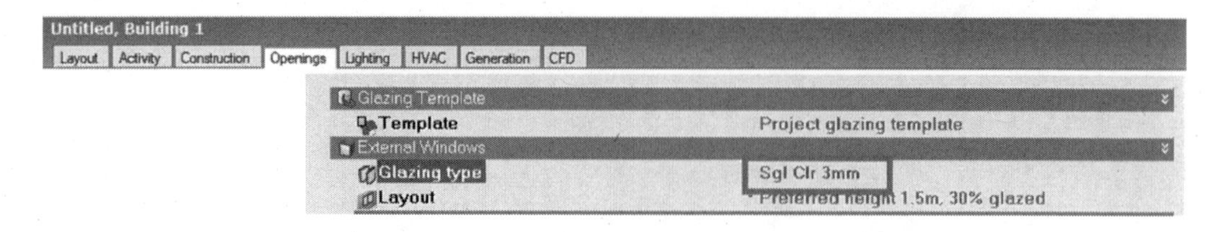

TABLE 1.18
Cooling design data for 80% WWR

Zone	Design capacity (kW)	Design flow rate (m^3/s)	Total cooling load (kW)	Sensible (kW)	Latent (kW)	Air temperature (°C)	Humidity (%)	Time of max cooling	Max operating temperature in day (°C)
West	58.79	3.76	51.12	47.12	4.00	24	48.3	Aug 16:00	34.7
North	46.85	2.80	40.74	35.06	5.68	24	49.9	Aug 15:00	29.0
East	49.31	3.11	42.88	39.03	3.85	24	49.9	Aug 09:30	30.5
South	85.33	5.45	74.20	68.35	5.84	24	48.4	Aug 13:30	30.7
Core	65.28	3.81	56.77	47.75	9.02	24	50.5	Aug 15:00	28.5
Total	305.57	18.91	265.71	237.32	28.39	24	49.4	N/A	34.7

TABLE 1.19
Heating and cooling sizing for 30% and 80% WWRs

	Heating sizing (kW)		Cooling sizing (kW)	
Zone	**WWR 30%**	**WWR 80%**	**WWR 30%**	**WWR 80%**
North	44.19	48.19	40.43	46.85
South	43.54	47.45	53.79	85.33
East	26.72	29.37	27.59	49.31
West	26.83	29.48	34.96	58.79
Core	66.30	66.81	63.06	65.28

TABLE 1.20
Energy consumption with 30% and 80% WWRs

	Annual fuel breakdown data	
Type	**WWR 30% (kWh)**	**WWR 80% (kWh)**
Room electricity	99,989.89	99,989.89
Lighting	144,507.10	144,507.10
Heating	89,866.72	91,160.98
Cooling	36,469.75	63,557.62
DHW	9,047.44	9,047.44

Step 24: Select the **Openings** tab, and choose **Sgl Clr 3mm** from the library, and run the annual simulation.

Step 25: Now select **Dbl Blue 6mm/13mm Air** from the library, and run the annual simulation.

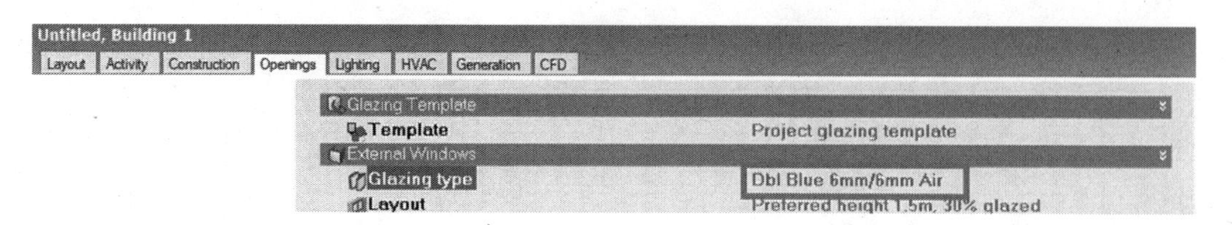

When you compare the results by changing the glazing from single to double glazing, notice that there is a change in the consumption of electricity for cooling and gas for heating (Table 1.21).

TABLE 1.21
Annual energy consumption with a change in glass type

	Annual fuel breakdown data	
Type	**Single glazing (Sgl Clr 3mm) (kWh)**	**Double glazing (Dbl Blue 6mm/13mm Air) (kWh)**
Room electricity	99,989.89	99,989.89
Lighting	144,507.10	144,507.10
Heating (gas)	110,262.70	103,146.60
Cooling	35,827.89	30,561.47
DHW	9,047.44	9,047.44

EXERCISE 1.5

Repeat this tutorial for Sydney, Australia.

a) Observe the effect when WWR is changed from 30% to 80% (Tables 1.22 and 1.23).

TABLE 1.22
Energy consumption with 30% and 80% WWRs

	Annual fuel breakdown data	
Type	**WWR (30%) (kWh)**	**WWR (80%) (kWh)**
Room electricity		
Lighting		
Heating (electricity)		
Heating (gas)		
Cooling (electricity)		

It can be seen that cooling sizing and cooling energy consumption increase with an increase in the WWR. This is because a larger glass area results in a larger solar gain. Further, as the U-value of the window is inferior to the U-value of the wall, a larger window result in a larger heat gain in summer. It is interesting to note that heating sizing and heating energy consumption also increase with an increase in WWR. Though an increased solar

TABLE 1.23
Heating and cooling sizing capacity with 30% and 80% WWRs

	Heating sizing (kW)		Cooling sizing (kW)	
Zone	**WWR 30%**	**WWR 80%**	**WWR 30%**	**WWR 80%**
North				
South				
East				
West				
Core				

heat gain through the glass tends to reduce the requirement for heating during the daytime, the heat loss through the glass due to conduction offsets this effect. At any given time, glass of only one or two orientations will allow solar radiation to enter, whereas glass of other orientations will lose more heat through conduction compared to a facade with a smaller WWR, assuming that the walls are more insulated than the glass. Further, during off-sunshine hours, the entire glazed area results in more heat loss if the U-value of the glass is inferior to that of the wall.

However, in subsequent chapters, it will be seen that this effect is combined with a reduction in lighting energy consumption if artificial light is simulated with a dimming feature.

b) Observe the effect with the change in glass type for 30% WWR (Tables 1.24 and 1.25).

TABLE 1.24
Energy consumption with a change in glass type

	Annual fuel breakdown data	
Type	**Single glazing (Sgl Clr 3mm) (kWh)**	**Double glazing (Dbl Blue 6mm/13mm Air) (kWh)**
Room electricity		
Lighting		
Heating (electricity)		
Heating (gas)		
Cooling (electricity)		

TABLE 1.25
Heating and cooling sizing capacity with a change in glass type

	Cooling and heating system sizing	
Type	**Single glazing (Sgl Clr 3mm) (kW)**	**Double glazing (Dbl Blue 6mm/13mm Air) (kW)**
Heating		
Cooling		

Tutorial 1.6

Evaluating the Impact of Occupancy Density

GOAL

To evaluate the impact of occupancy density on cooling and heating loads and the whole-building energy consumption.

WHAT ARE YOU GOING TO LEARN?

- How to change occupancy density

PROBLEM STATEMENT

Use the model created in Tutorial 1.5 (60- × 40-m building with five zones). Set the WWR to 30%, minimum fresh air (l/s-person) to 2.5, mechanical ventilation per area (l/s-m^2) to 0.3, and model infiltration [air changes (ac)/h] to 0.20. Run simulations with occupancy density (people/m^2) set to 0.07 and 0.10. Analyse the change in energy consumption. Use the **AZ-PHOENIX/SKY HARBOR, USA** weather file.

SOLUTION

Step 1: Select the **Activity** tab, and under the **Occupancy** section, set the occupancy density (people/m^2) to 0.111. Under the **Minimum Fresh Air** section, set Fresh air (l/s-person) to **2.5** and mechanical vent per area (l/s-m^2) to **0.3**.

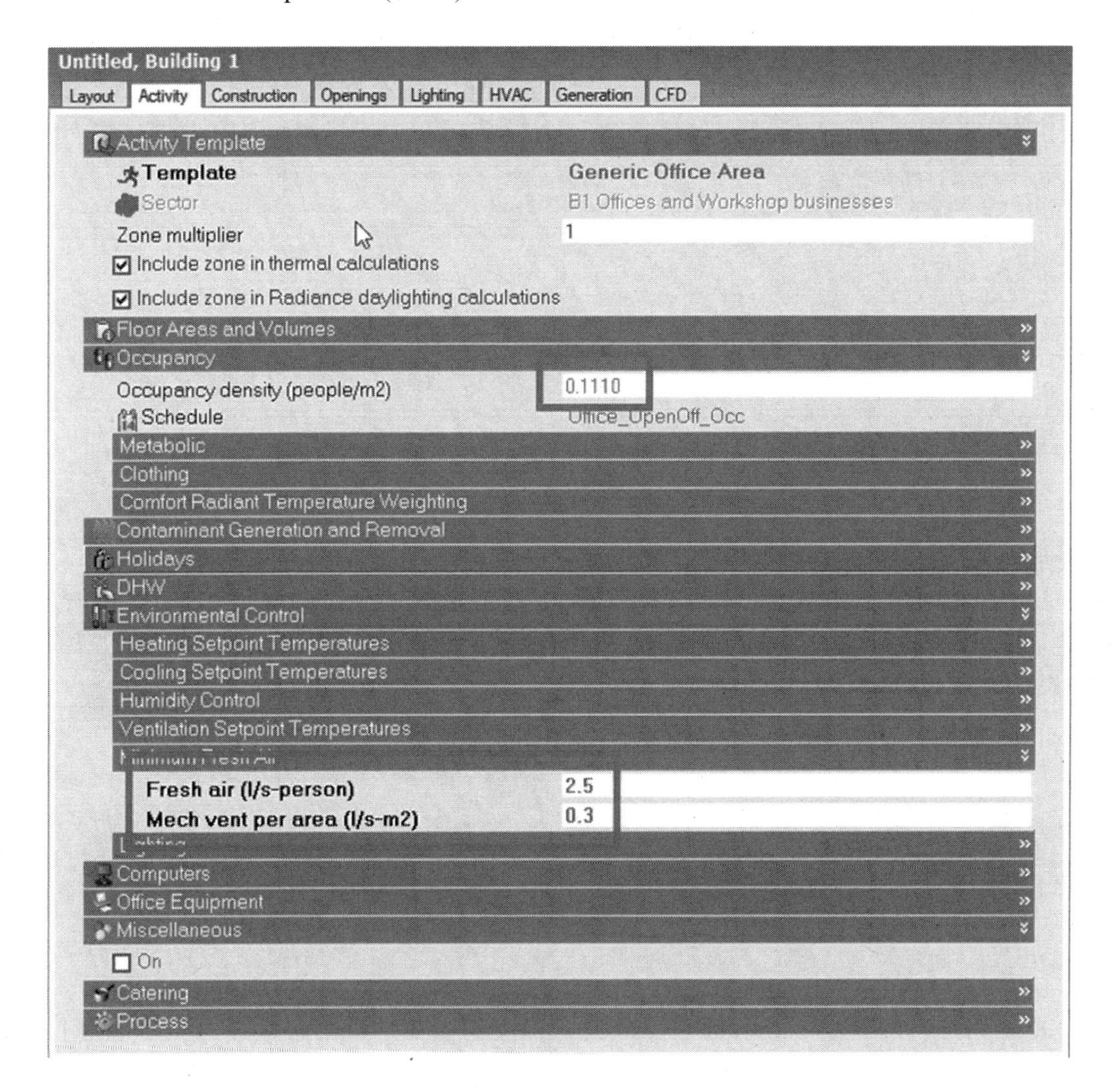

Step 2: Select the **Construction** tab. Under the **Airtightness** section, set **Model infiltration, Constant** rate as 0.200 ac/h.

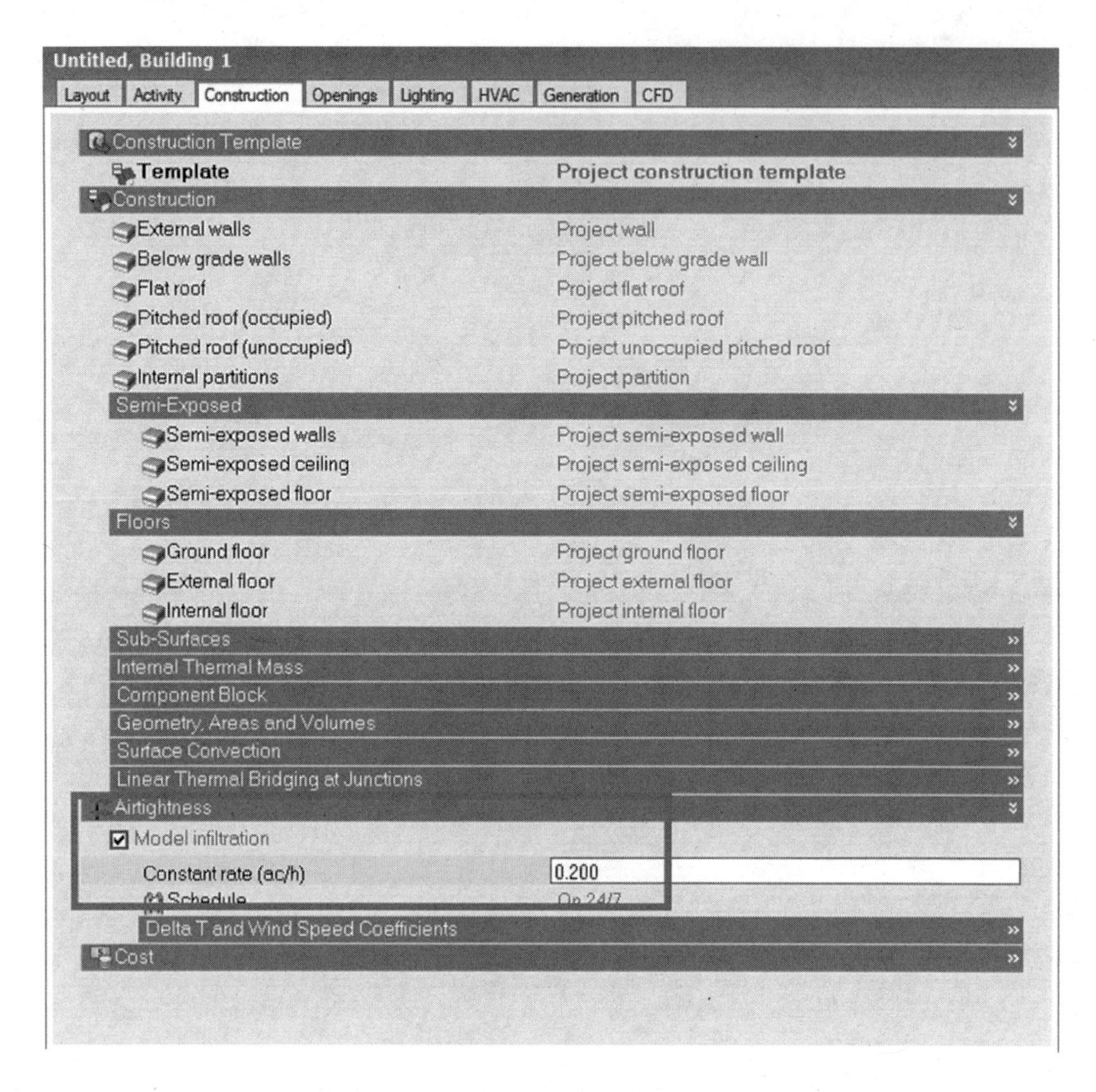

Step 3: Perform the annual simulation, and record the results.

Step 4: Set the occupancy density to **0.07** people/m^2. Perform the annual simulation, and record the results (Table 1.26).

TABLE 1.26
Annual energy consumption with a change in occupancy density

Type	0.111 people/m^2	0.07 people/m^2
Room electricity	99,989.90	99,989.90
Lighting	144,507.10	144,507.10
Heating (gas)	1,049.10	1,088.10
Cooling (electricity)	202,291.50	192,084.20
DHW (electricity)	9,047.40	9,047.40

Here you can observe that a change in occupancy density has an impact on the total energy consumption. The impact of occupancy density on energy consumption can be explained as follows:

- An increase in occupancy density increases cooling energy consumption because of the increased load (addition of sensible and latent loads from the occupants) in the zone air.
- An increase in occupancy density decreases the heating energy consumption because the heat added by the occupants to the indoor air helps to reduce the heating loads. However, this effect is not straightforward because a higher occupancy would require a higher fresh air intake that would, in turn, increase heating energy consumption. Further, for blowing more air into the space, the increase in fan power is also a factor that cannot be neglected.
- Occupancy density has no impact on lighting or equipment energy consumption unless these are directly related to the occupancy.

Two important parameters are used to show the occupancy of a zone: *occupancy density* and *schedule*. As seen in this tutorial, occupancy density is the maximum number of people in a zone. Schedule of occupancy defines when a zone is occupied or unoccupied and by how many people. *Metabolic activity* is the amount of heat given out by people, and it depends on the activity they engage in. For example, a person who is exercising gives out more heat than a person who is sleeping.

EXERCISE 1.6

For the same tutorial, observe the effect on cooling and heating equipment sizing (Table 1.27).

TABLE1.27
Heating and cooling sizing capacity with a change in occupancy density

Type	0.1 people/m^2	0.07 people/m^2
Cooling		
Heating		

It can be seen from the results that change in occupancy alters heating and cooling sizing as well.

Increasing occupancy density results in additional sensible and latent heat into the space, leading to higher cooling capacity and higher energy consumption. Higher occupancy density leads to higher fresh air and supply air requirements. Hence, system capacity needs to be increased.

Tutorial 1.7

Evaluating the Impact of Space Activity

GOAL

To understand the impact of space activity on cooling and heating loads and energy consumption

WHAT ARE YOU GOING TO LEARN?

- How to understand various activity types
- How to change space activity
- How to understand the impact of activity type on sizing and energy consumption

PROBLEM STATEMENT

Create a 20- × 15-m single-zone model. Set the **Activity Template** to **Office** and **Restaurant** and compare. Study the effect on sizing and energy consumption for London.

SOLUTION

Step 1: Create a 20- × 15-m single-zone model. You can refer to Tutorial 1.2 to create a single-zone model.

Step 2: Select the **Activity** tab. The **Activity Template** appears.

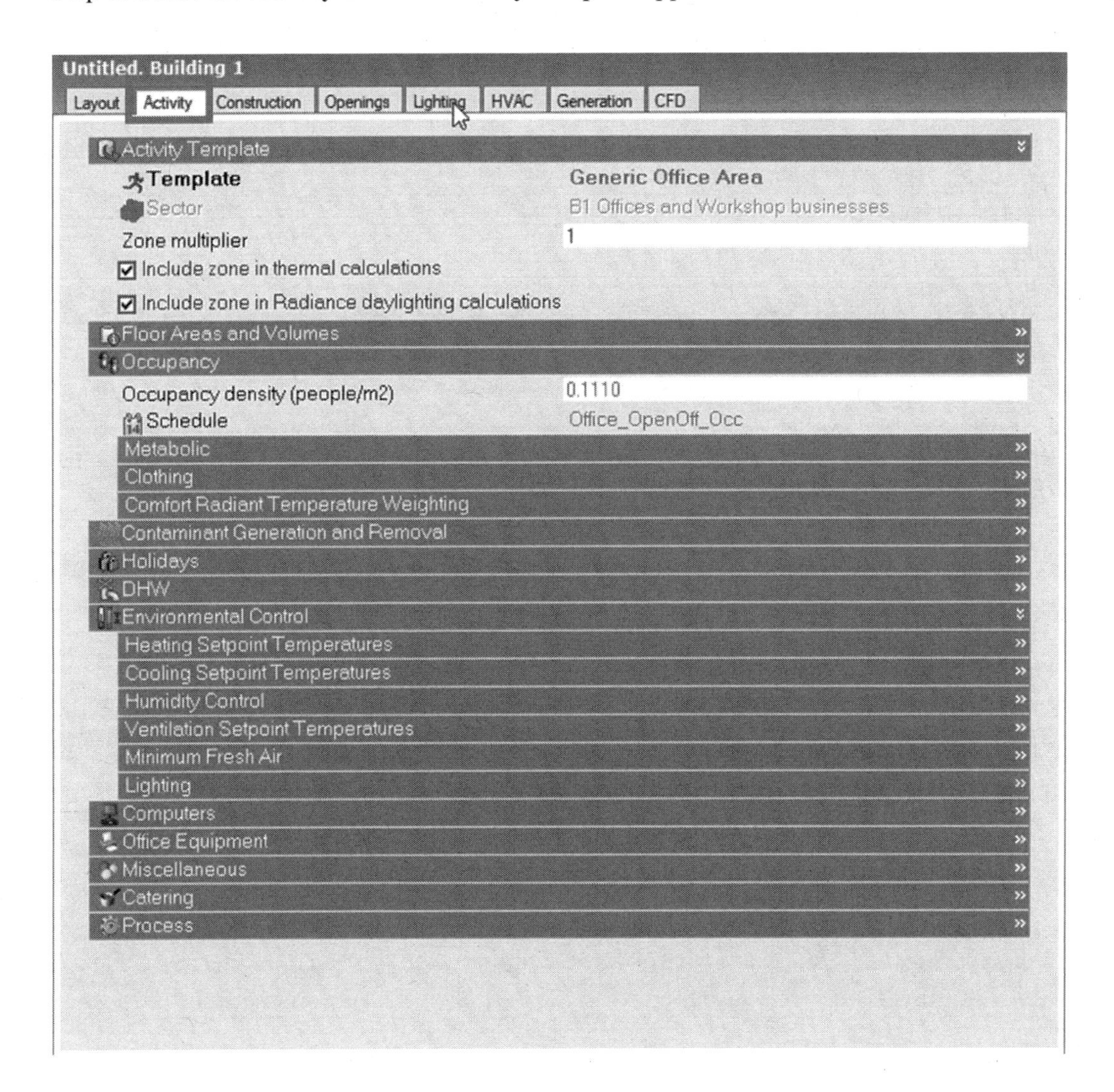

Step 3: Simulate the model, and record the results.

Step 4: On the **Activity Template**, select **Eating/drinking area**. This changes internal gains and schedules of the space.

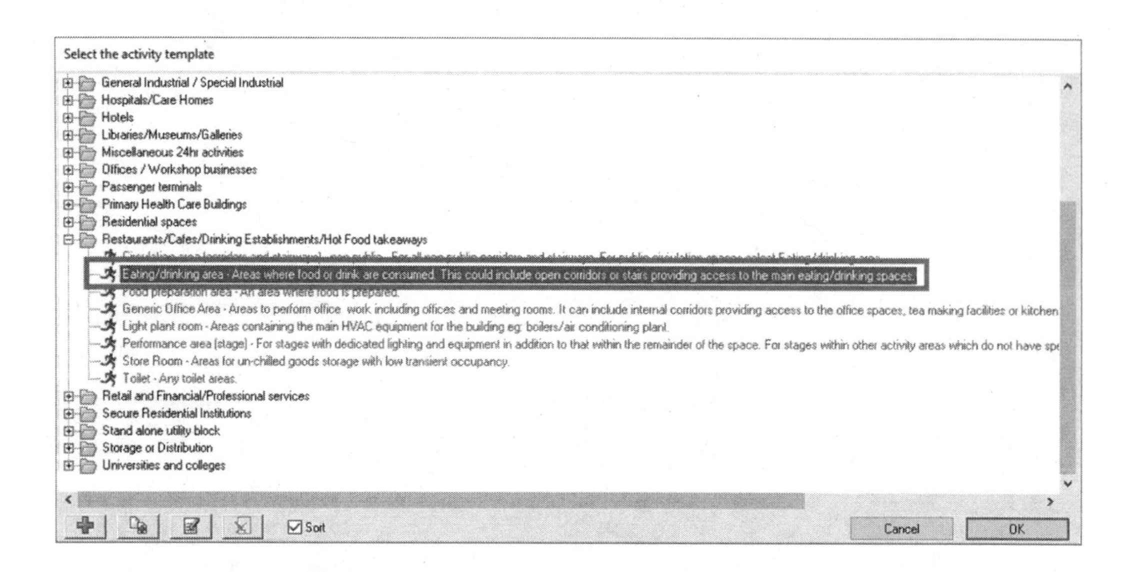

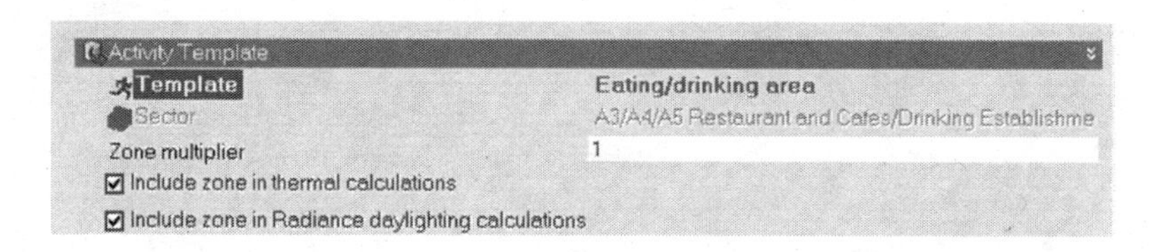

Step 5: Double-click on the **RestPub_EatDrink_Occ** schedule under the **Occupancy** tab. The **Edit schedule-RestPub_EatDrink_Occ** screen appears. (You can note the occupancy schedule for comparison with the other activity template options, as shown in Table 1.28.)

Schedules are used in DesignBuilder to define with respect to time the following:

- Occupancy
- Equipment, lighting and HVAC system operation
- Heating and cooling temperature setpoints
- Transparency of component blocks (usually seasonal)

Occupancy, equipment and lighting schedule are defined by a fraction (0–1). The maximum gain values (e.g. people/m^2) are multiplied by the values in the schedule to obtain the actual value to use at each time step in the simulation.

Source: https://designbuilder.co.uk/helpv6.0/Content/Schedules_-_EnergyPlus_Compact_Schedules.htm.

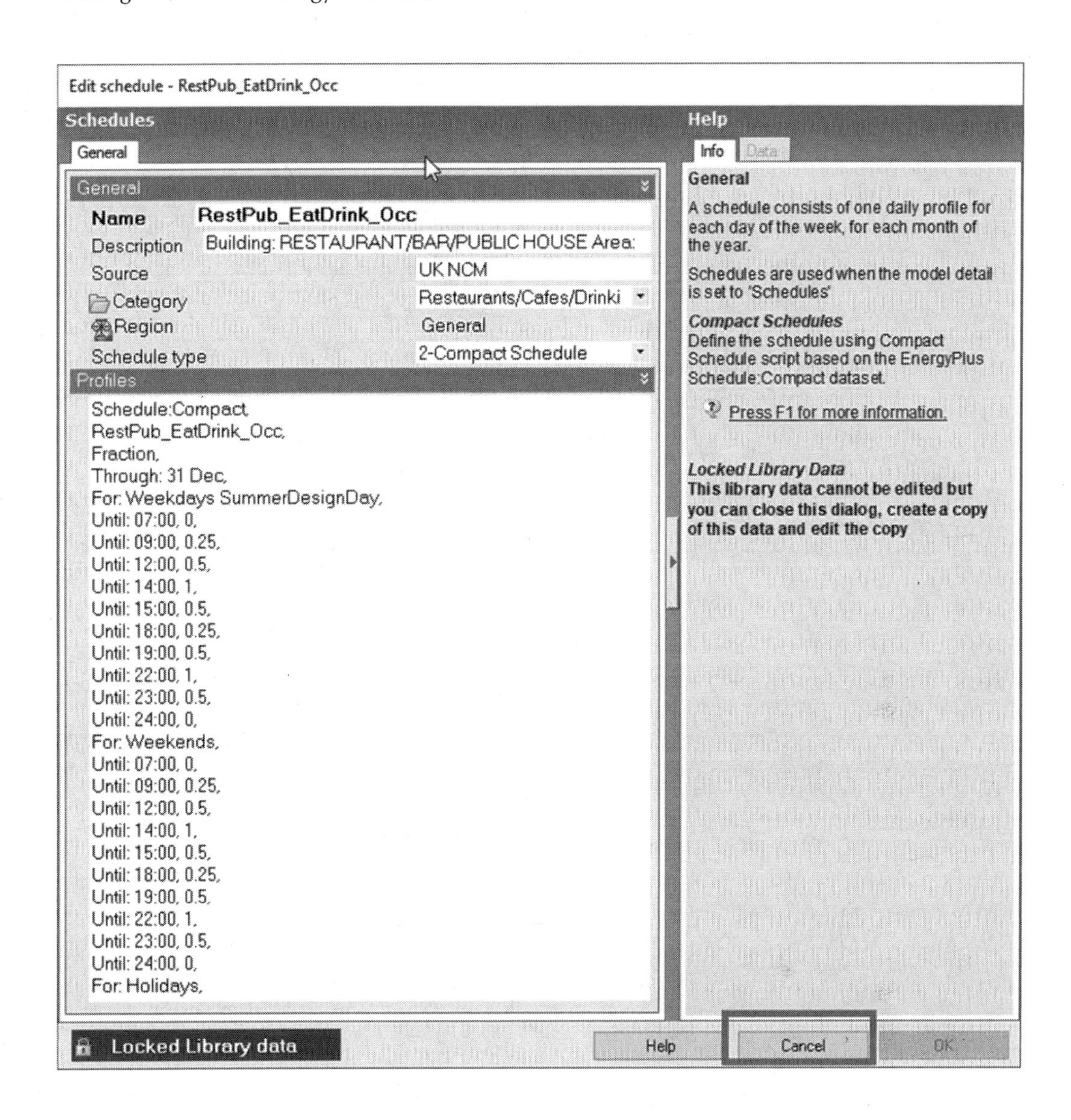

TABLE 1.28
Workday schedules for a generic office and eating/drinking area

Generic office area **Office_OpenOff_Occ**	**Eating/drinking area** **RestPub_EatDrink_Occ**
Schedule: Compact,	Schedule: Compact,
Office_OpenOff_Occ,	RestPub_FoodPrep_Occ,
Fraction,	Fraction,
Through: 31 Dec,	Through: 31 Dec,
For: Weekdays	For: Weekdays SummerDesignDay,

(Continued)

TABLE 1.28 (Cont.)

Generic office area Office_OpenOff_Occ	Eating/drinking area RestPub_EatDrink_Occ
SummerDesignDay,	Until: 06:00, 0,
Until: 07:00, 0,	Until: 07:00, 0.25,
Until: 08:00, 0.25,	Until: 08:00, 0.75,
Until: 09:00, 0.5,	Until: 14:00, 1,
Until: 12:00, 1,	Until: 15:00, 0.75,
Until: 14:00, 0.75,	Until: 17:00, 0.25,
Until: 17:00, 1,	Until: 18:00, 0.75,
Until: 18:00, 0.5,	Until: 22:00, 1,
Until: 19:00, 0.25,	Until: 23:00, 0.75,
Until: 24:00, 0,	Until: 24:00, 0.25,
For: Weekends,	For: Weekends,
Until: 24:00, 0,	Until: 06:00, 0,
For: Holidays,	Until: 07:00, 0.25,
Until: 24:00, 0,	Until: 08:00, 0.75,
For: WinterDesignDay	Until: 14:00, 1,
AllOtherDays,	Until: 15:00, 0.75,
Until: 24:00, 0;	Until: 17:00, 0.25,
	Until: 18:00, 0.75,
	Until: 22:00, 1,
	Until: 23:00, 0.75,
	Until: 24:00, 0.25,
	For: Holidays,
	Until: 06:00, 0,
	Until: 07:00, 0.25,
	Until: 08:00, 0.75,
	Until: 14:00, 1,
	Until: 15:00, 0.75,
	Until: 17:00, 0.25,
	Until: 18:00, 0.75,
	Until: 22:00, 1,
	Until: 23:00, 0.75,
	Until: 24:00, 0.25,
	For: WinterDesignDay AllOtherDays,
	Until: 24:00, 0;

Step 6: Click **Cancel**. The screen closes.

Step 7: Select the **Simulation** screen tab to get energy consumption results. Compare both cases (Table 1.29).

The results show the impact of activity type on energy consumption. From the model, you can determine the interior load parameters (Table 1.30). You can get these data from the **Activity** and **Lighting** tabs. You can correlate the change in energy consumption with the change in interior loads.

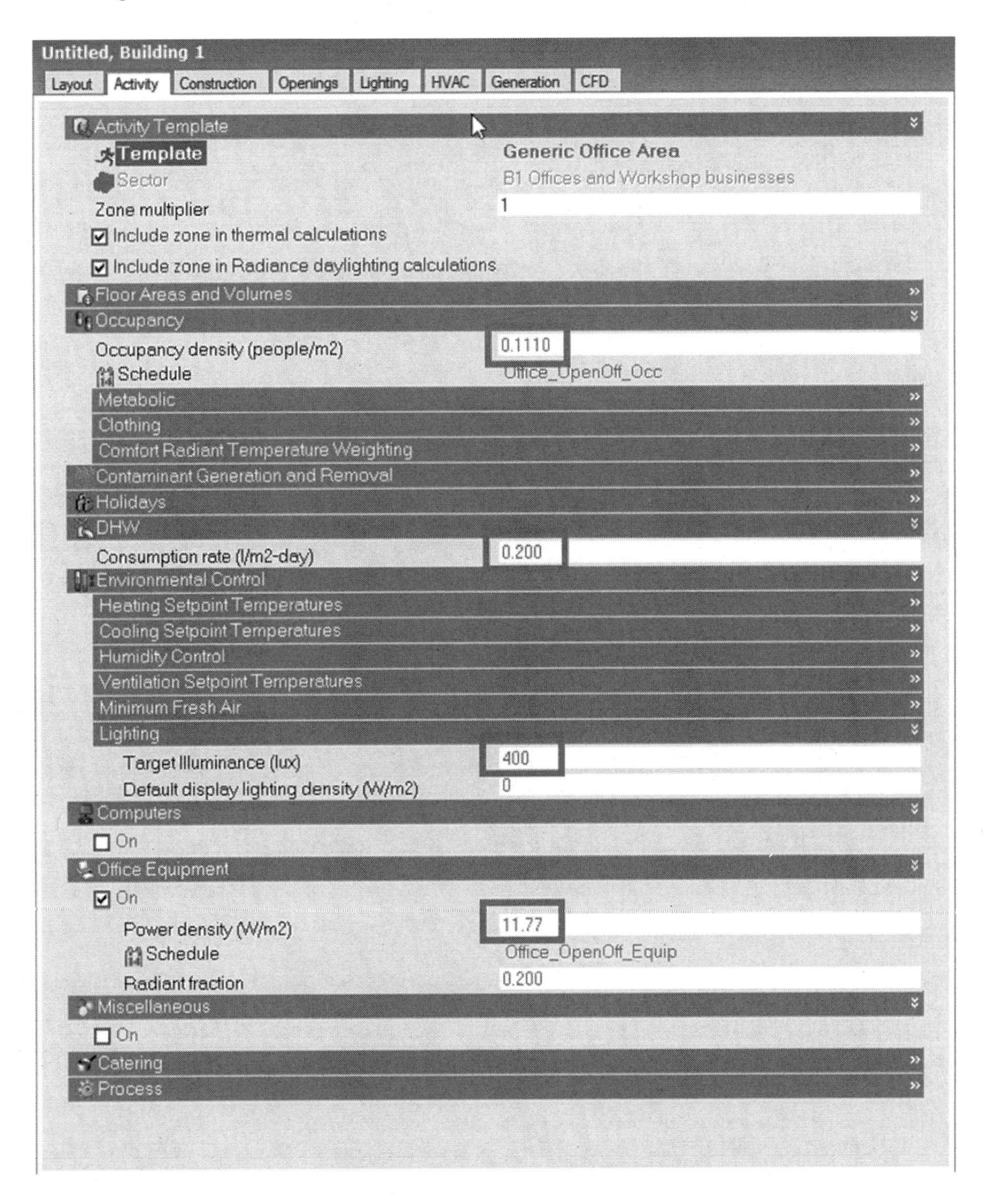

TABLE 1.29
Annual energy consumption with a change in space activity

Type	Generic office area (kWh)	Eating/drinking area (kWh)
Room electricity	12,130.28	32,222.99
Lighting	17,530.88	12,258.18
Heating (gas)	12,662.27	24,015.45
Cooling (electricity)	6,532.64	8,601.14
DHW (electricity)	1,097.59	43,669.16

TABLE 1.30
Internal load data for generic office and eating/drinking area

Interior load	Unit	Generic office area	Eating/drinking area
Occupancy	People/m^2	0.11	0.20
Target illuminance	lux	400	150
Interior light	W/m^2 per 100 lux	5	5
Equipment	W/m^2	11.77	18.88
DHW	l/s-day	0.20	5.69

EXERCISE 1.7

Compare the cooling and heating energy consumption for a generic office area and classroom (Table 1.31).

> Activity type: **Classroom** can be found under the **Universities and college** category.

TABLE 1.31
Annual energy consumption for a generic office and eating/drinking area

Type	Generic office area (kWh)	Class room area (kWh)
Room electricity		
Lighting		
Heating (gas)		
Cooling (electricity)		
DHW (electricity)		

From the model, you can find the change in the internal load parameter (Table 1.32).

TABLE 1.32
Internal load for a generic office and eating/drinking area

Interior load	Unit	Generic office area	Classroom area
Occupancy	People/m^2		
Target illuminance	lux		
Interior light	W/m^2 per 100 lux		
Equipment	W/m^2		
DHW	l/s-day		

Compare office and classroom schedules in Table 1.33.

TABLE 1.33
Workday schedules for a generic office and classroom area

Generic office area Office_OpenOff_Occ	Classroom area

Tutorial 1.8

Evaluating the Impact of Lighting and Equipment Power

GOAL

To understand the impact of lighting and equipment power density on HVAC system sizing and energy consumption

WHAT ARE YOU GOING TO LEARN?

- How to change lighting power density (LPD)
- How to change equipment power density (EPD)

PROBLEM STATEMENT

Compare the annual energy consumption when you set the LPD to 13.13 W/m^2 and to 10.0 W/m^2. Use the **Courtyard with VAV Example** template for London Gatwick Airport.

SOLUTION

Step 1: Select the **Lighting** tab, and under the **General Lighting** section, set the LPD shown as **Lighting energy (W/m^2)** to 13.13.

> *Lighting power density* (LPD) is the lighting power per unit area of a building, space or outdoor area expressed in W/m^2.

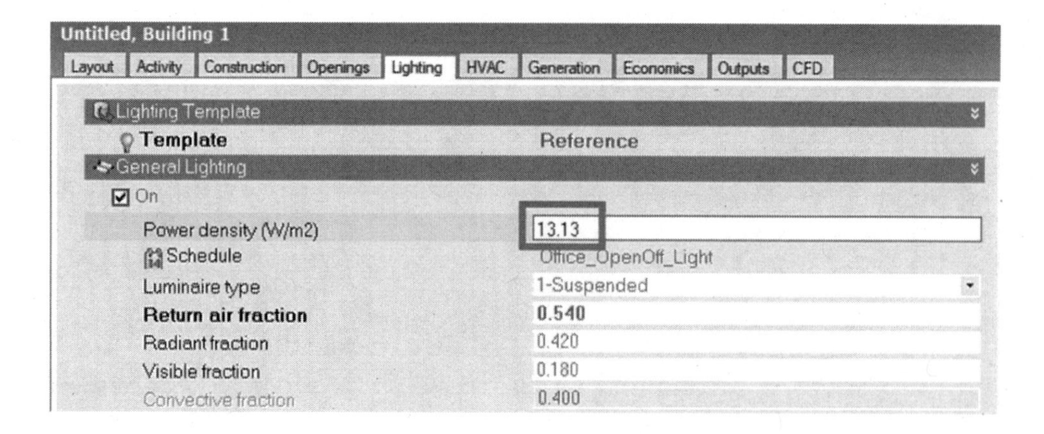

> ***Tip:*** Does your lighting energy have units as W/m^2 per 100 lux rather than W/m^2?
>
> If yes, change this to W/m^2 in the following way:
>
> go to Edit→Model Options Data→Data→Gain Data→Lighting Gains Units, and change the lighting gain units to W/m^2.

Step 2: Click the **Simulation** screen. Click on **Update data.** Click the **Annual simulation** link to perform annual simulation.

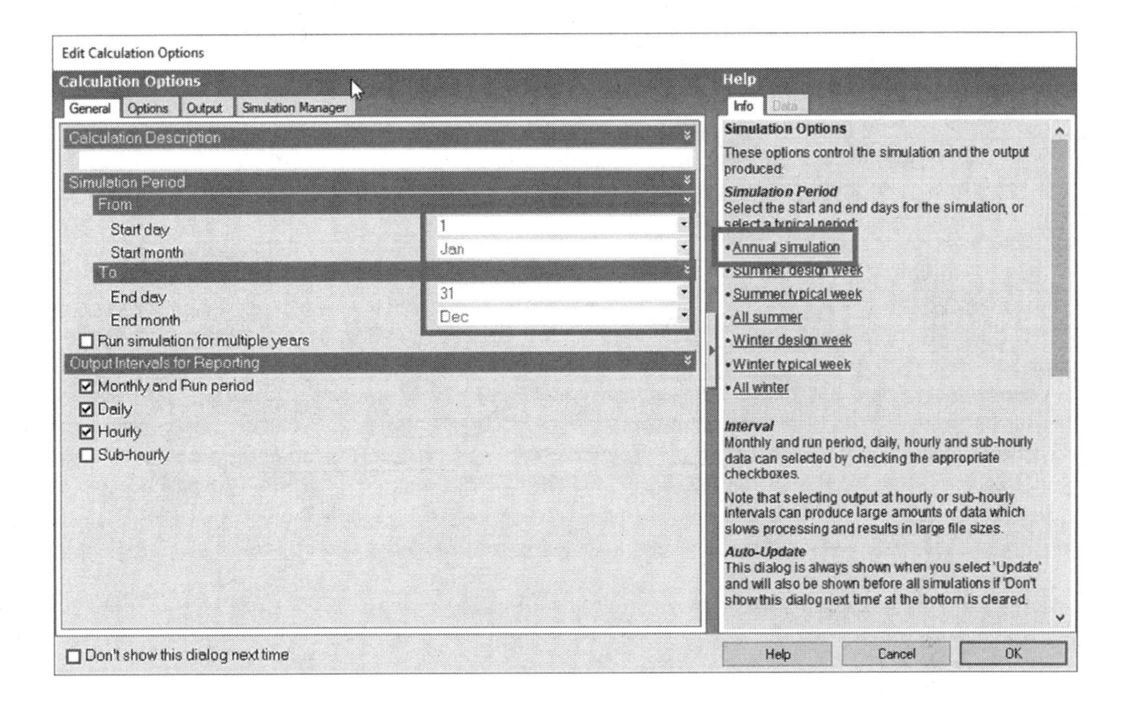

Step 3: Repeat the steps to set the LPD to 10.00. Perform annual energy simulation. Compare the end-use energy consumption of both cases (Table 1.34). From the end-use energy consumption shown in this table, it is clear that a reduction in LPD decreases the lighting energy consumption. Also note the decrease in cooling and fans and pump energy consumption.

TABLE 1.34
Annual fuel breakdown with a change in LPD

Type	LPD: 13.13 W/m^2 (kWh)	LPD: 10.0 W/m^2 (kWh)
Room electricity	24,963.70	24,963.70
Lighting	23,685.20	18,039.00
System fans	24,844.70	23,466.80
System pumps	59.20	54.60
Heating (gas)	38,374.00	43,004.70
Cooling (electricity)	47,318.20	42,804.20

EXERCISE 1.8

Set the equipment power density to 20 and 10 W/m^2, and study the impact on building energy consumption and sizing (Tables 1.35 and 1.36).

Equipment Power Density (EPD)

In DesignBuilder, under the **Activity** tab, you can set **Equipment gain** units. This option provides control over the method used to enter **Internal gains** such as computers, office equipment, miscellaneous, catering and process.

1. *Power density*, where zone equipment gains are entered as a power density, that is, as a power per zone floor area in W/m^2 or W/ft^2
2. *Absolute zone power*, where zone equipment gains are entered as a power value in watts (W). This option is usually used when detailed zone-by-zone survey or design data are available. Unless you need to enter specific known lighting equipment gains zone by zone, this option is best avoided, as it increases the amount of work involved because of the inheritance mechanism not working effectively for gains that are not normalised by floor area.

Source: https://designbuilder.co.uk/helpv6.0/Content/_Gains_data_model_detail.htm#Equipmen%20gain.

It can be clearly seen from the results that reducing the LPD and EPD not only reduces the energy consumption for lighting or equipment, respectively, but it also leads to a reduction in the cooling energy. This is because the energy consumed by lighting or equipment ultimately gets converted into heat and usually gets added into the space. This additional heat increases the energy consumption of the fans and pumps and also increases system sizing. In cases of heating energy consumption, the reverse effect is observed because lighting and equipment add additional heat to the space, thereby reducing energy requirements for heating.

TABLE 1.35
Energy consumption with a change in equipment power density

	Annual fuel breakdown data	
Type	**EPD (20 W/m^2) (kWh)**	**EPD (10 W/m^2) (kWh)**
Room electricity		
Lighting		
System fans		
System pumps		
Auxiliary energy		
Heating (gas)		
Cooling (electricity)		

TABLE 1.36
Heating and cooling sizing capacity with a change in equipment power density

Type	EPD (20 W/m^2) (kW)	EPD (10 W/m^2) (kW)
Heating		
Cooling		

Tutorial 1.9

Evaluating the Impact of Daylight Controls

GOAL

To evaluate the impact of daylight controls on energy consumption

WHAT ARE YOU GOING TO LEARN?

- How to specify daylight controls in a building model
- How to evaluate the impact of daylight controls on energy consumption

PROBLEM STATEMENT

Create a 60- × 40-m model with a core and four perimeter zones. Consider a perimeter depth of 5 m. (Refer Tutorial 1.5 for steps to create the model.) Set the WWR to 60%, and use **Sgl Grey 3mm** glass (VLT ~ 60%). Perform an annual energy simulation without daylight sensors, and then install daylight control in the north zone. Compare the lighting energy consumption with and without daylight control for the London Gatwick Airport location, United Kingdom.

> ***Tip:*** For single-pane glass with 60% visible light transmittance, select **Sgl Grey 3mm** with a light transmission value of 0.61 under the **Single** category of the glazing library.

SOLUTION

Step 1: Create a five-zone core and perimeter model of size 60 × 40 m. Set the WWR to 60%, and select **Sgl Grey 3mm** glass.

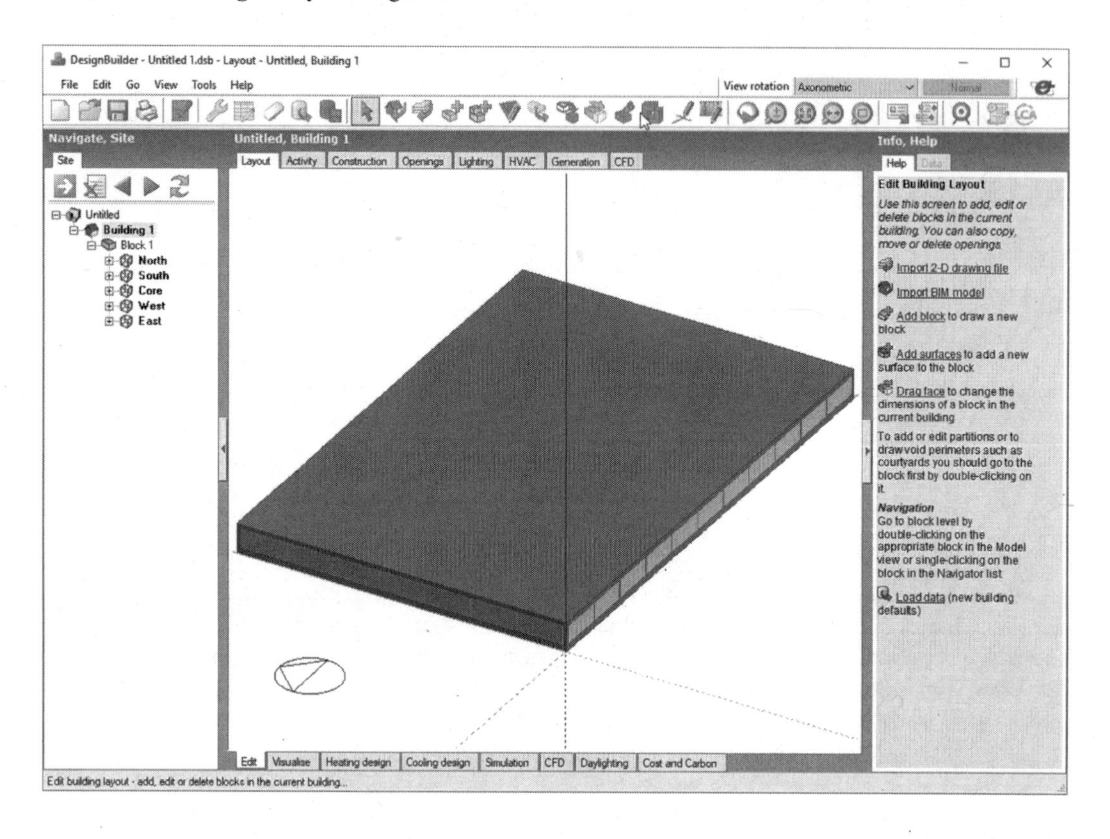

Step 2: Click the **Simulation** screen tab. Click on **Update data.** The **Edit Calculation Options** screen appears. Select the **Hourly** checkbox under **Output Intervals for Reporting**, and then click **OK**. After the simulation, hourly results appear under the **Analysis** tab.

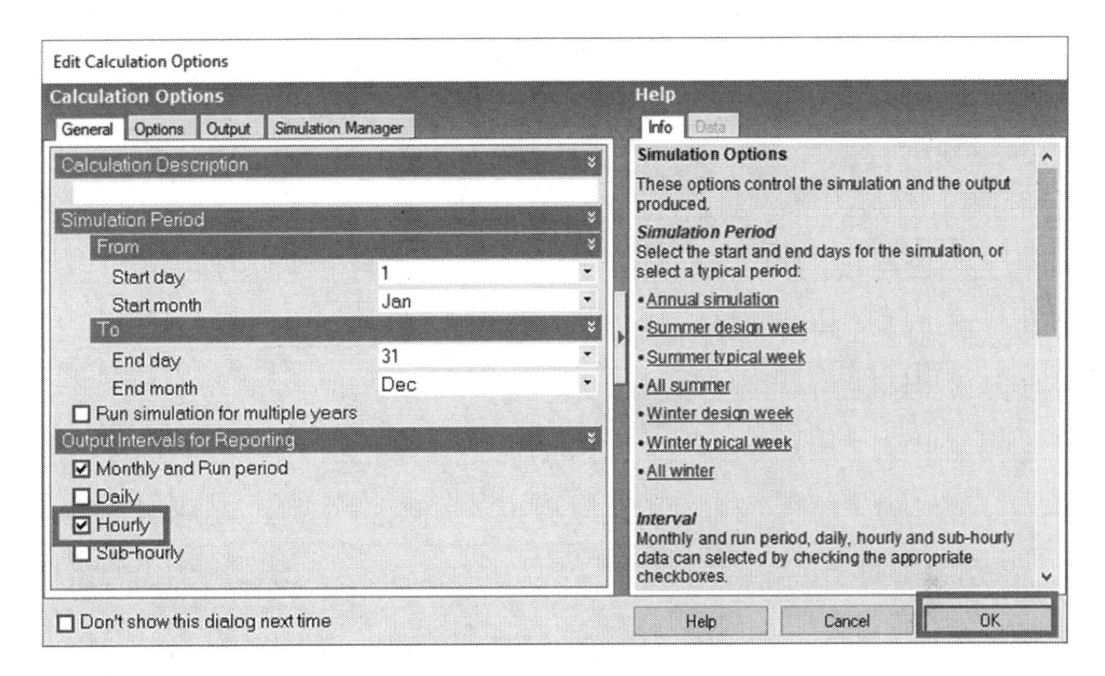

Step 3: Click the **Export data** icon to export the hourly energy consumption in the spreadsheet. (Ensure that **Fuel breakdown**, **Hourly interval** and **Grid** options are selected.)

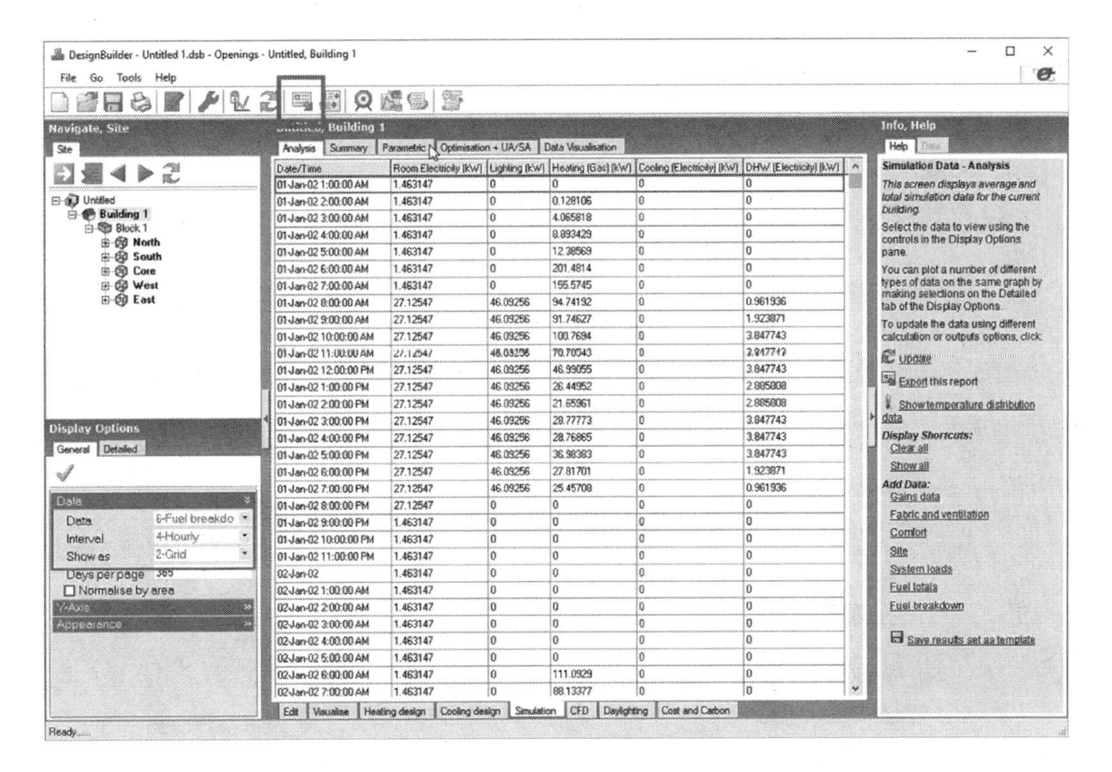

Step 4: Select **File** from the **Export to** drop-down list. Select **CSV spreadsheet** from the **Format** drop-down list. Click **OK**. The **Export CSV file** screen appears.

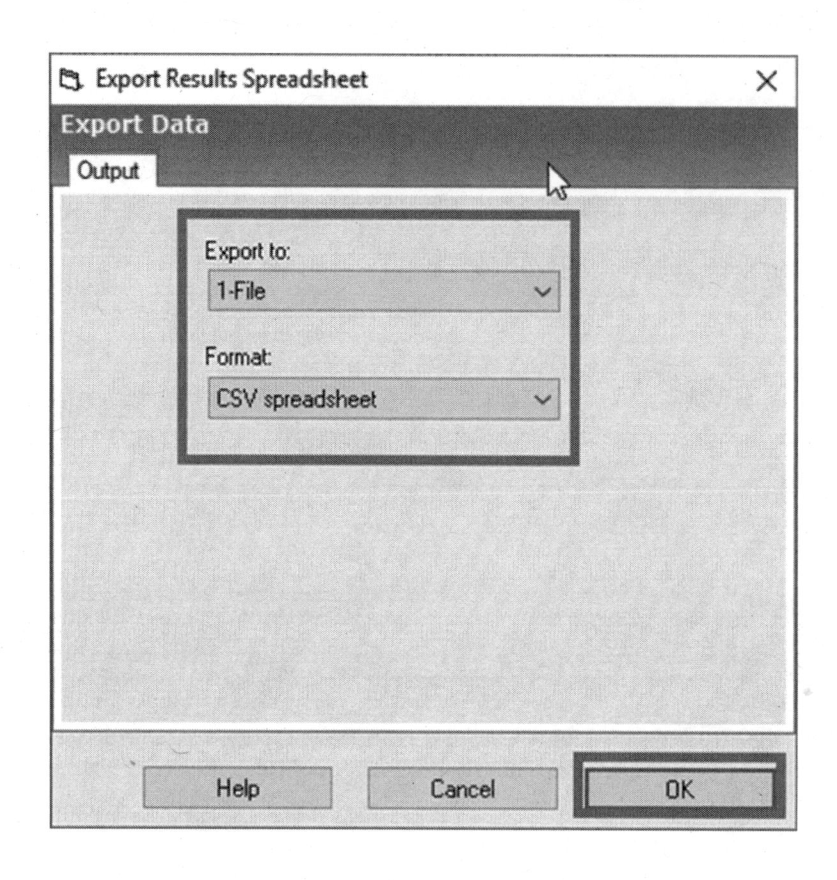

Step 5: Enter the file name **Without daylight sensor**, and save the file to **Desktop**.

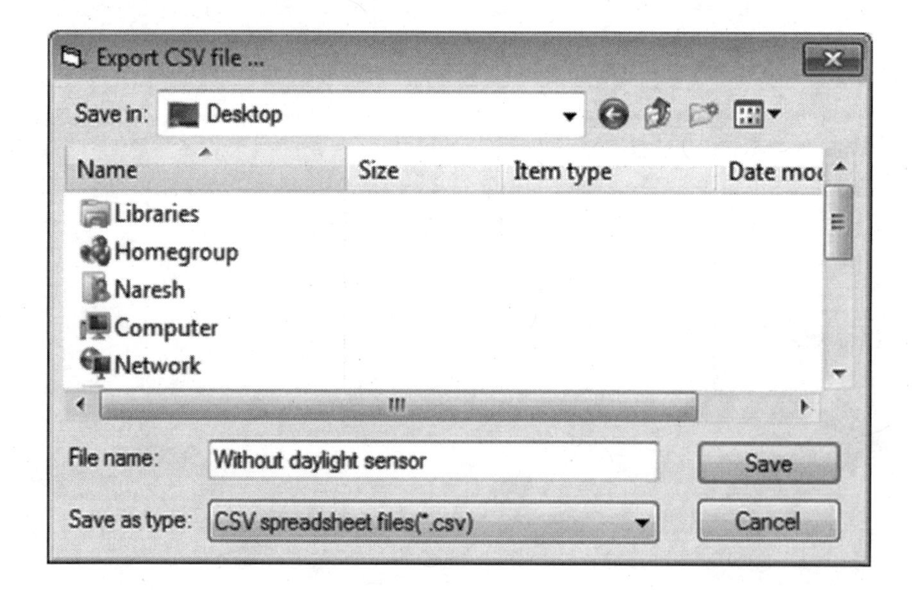

Step 6: Select **Run period** from the **Interval** drop-down list.

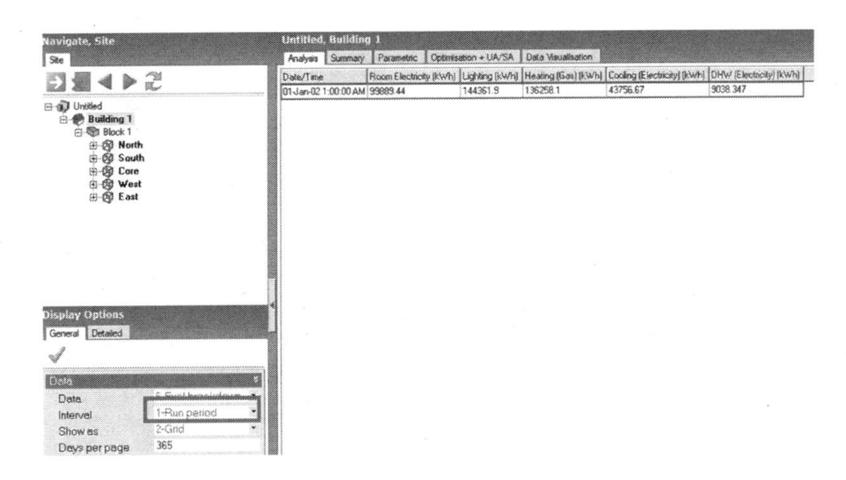

Record the simulation results. Save the model to use in other tutorials.

In the next step, you are going to install the daylight sensor in the north zone.

Step 7: Select the **Edit** screen tab. Click the **North** zone in the navigation tree. It shows the north zone in the **Layout** tab. Select the **Lighting** tab. The lighting properties are displayed.

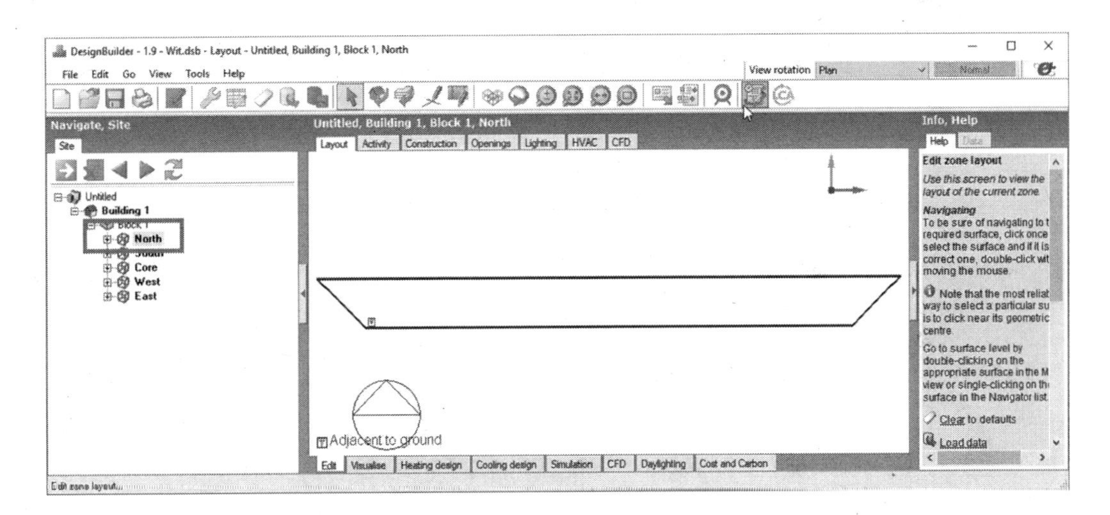

Step 8: Expand the **Lighting control** section. Select the **On** checkbox. Select **Linear** from the **Control type** drop-down list.

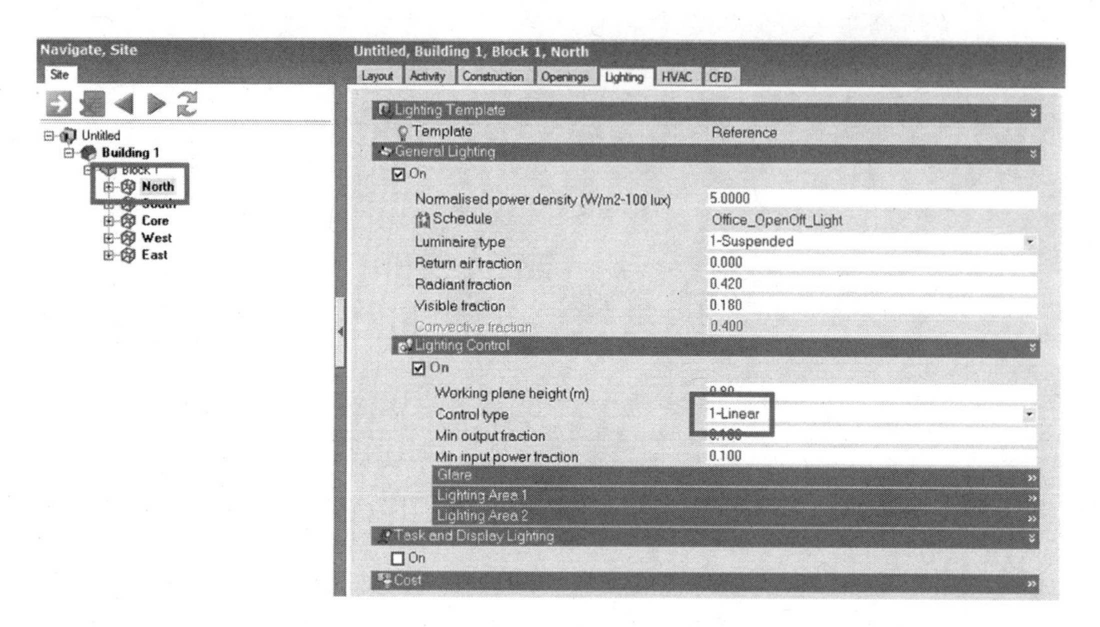

Different types of lighting control options that exist in DesignBuilder are as follows:

1. Linear
2. Linear/off
3. Stepped

Linear control, also called as *continuous control*, reduces the power input of the luminaire continuously, thereby decreasing the output light from the lamp until it reaches the minimum input and output fraction provided as a user input. This is possible with continuously dimmable fixtures and lamps. The decrease in input power (proportionally the lighting output) depends on the daylight illuminance requirement in the space. As the daylight illuminance increases, the input power decreases until it reaches the minimum input power and light output fraction and remains at the minimum specified ratio with a further increase in daylight illuminance.

Step 9: Select the **Layout** tab. The **Daylight Sensor** is displayed.

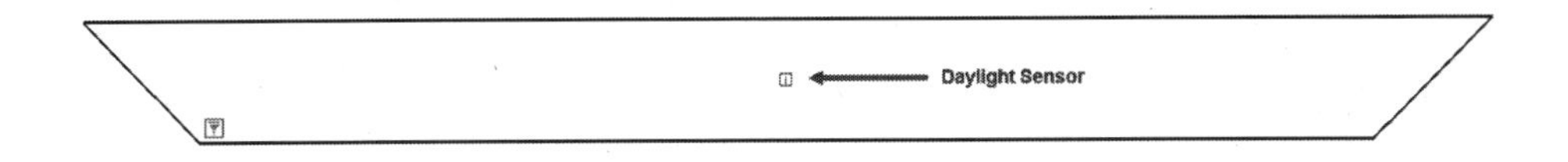

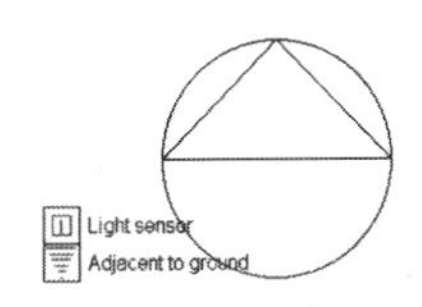

Step 10: Perform the simulation, and record the annual and hourly results of lighting energy consumption.

Step 11: Compare the results with and without lighting controls (Table 1.37).

Step 12: Draw the graph for hourly lighting energy consumption for 1 January with and without a daylight sensor by using the data exported using a spreadsheet program.

TABLE 1.37
Annual energy consumption with and without lighting control for the north zone

Type	Without lighting controls (kWh)	With lighting controls (kWh)
Room electricity	99,889.40	99,889.40
Lighting	144,361.90	132,794.80
Heating (gas)	136,258.10	141,686.60
Cooling (electricity)	43,756.60	41,783.80
DHW (electricity)	9,038.30	9,038.30

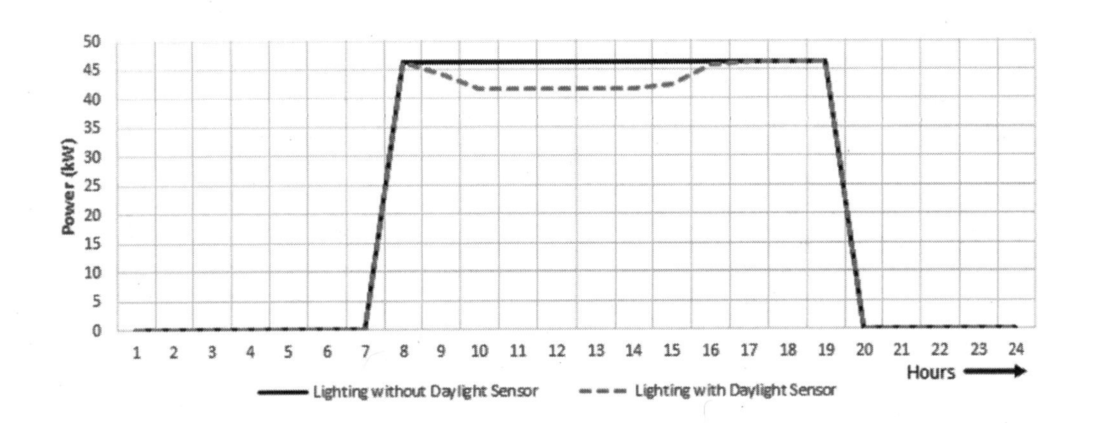

The preceding figure shows the hourly lighting energy consumption for 1 January. It can be observed that there is a reduction in energy consumption with the use of a daylight sensor and a dimmable luminaire.

The effect of daylight control is similar to that of reducing LPD, as discussed in the preceding exercise. The daylight control reduces the artificial lighting load whenever daylight is available in the space. The lighting loads are reduced to the extent that the combined lux levels of artificial light and daylight equal the setpoint lux level. This reduction in LPD reduces the lighting energy consumption. Different zones would have different daylight extents and timing of availability and hence would have different energy savings.

EXERCISE 1.9

Add daylight control in the south zone of the model, and study the effect on the energy consumption (Table 1.38).

It can be seen that the deployment of lighting control increases the heating energy consumption and decreases the cooling energy consumption.

TABLE 1.38
Annual energy consumption with and without lighting control for the south zone

Type	Without daylight controls (kWh)	With daylight controls (kWh)
Room electricity		
Lighting		
System fans		
System pumps		
Heating (electricity)		
Heating (gas)		
Cooling (electricity)		

Tutorial 1.10

Evaluating the Impact of Setpoint Temperature

GOAL

To evaluate the impact of setpoint temperature on sizing and energy consumption

WHAT ARE YOU GOING TO LEARN?

- How to change the setpoint for heating and cooling

PROBLEM STATEMENT

Open the model created in Tutorial 1.9 (60- × 40-m model with a core and four perimeter zones; consider a perimeter depth of 5 m) without a daylight sensor. Set the heating setpoint to 20 and 22°C. Simulate the model with the **PARIS-AEROPORT CHAR** weather file. Analyse the effect of setpoint on energy consumption and the HVAC system sizing.

SOLUTION

Step 1: Create a five-zone core and perimeter model of size 60 × 40 m.

Step 2: Select the **Activity** tab, and go to **Heating Setpoint Temperatures**. Enter **20** in **Heating (°C)** box.

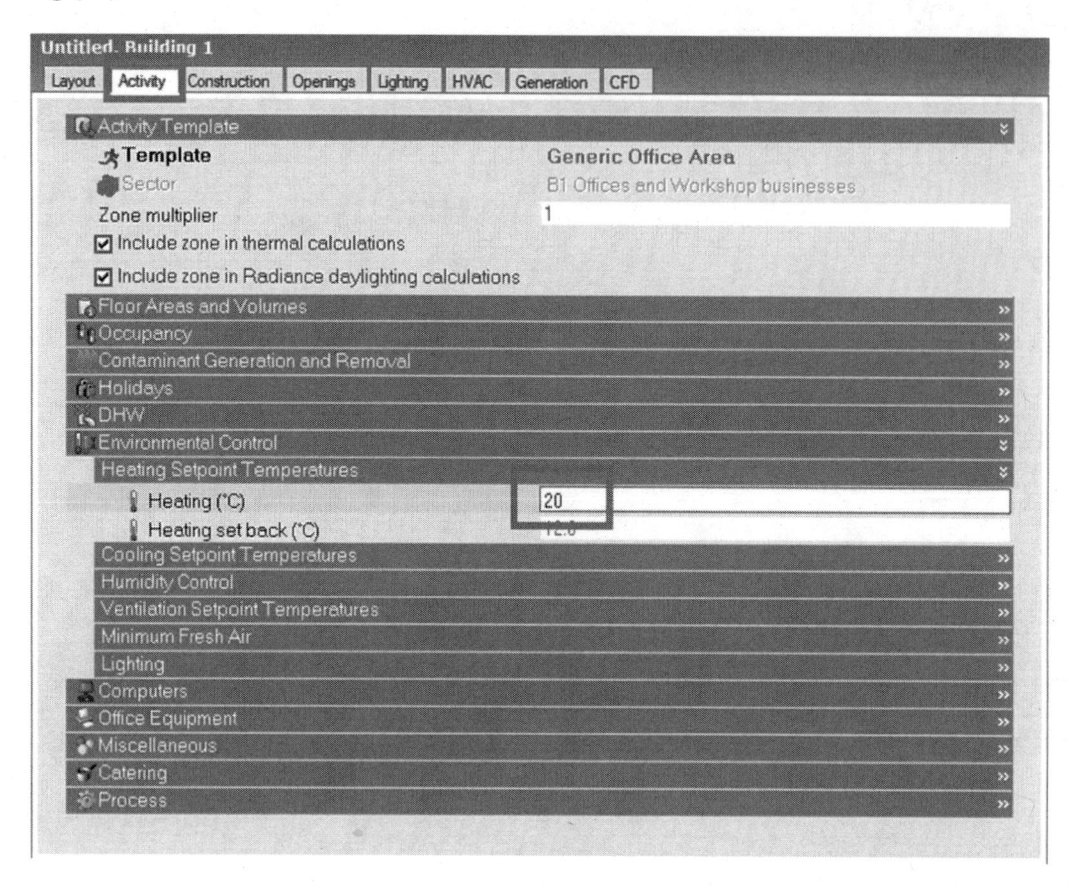

Step 3: Perform the annual simulation, and record all end-use energy consumption.

Step 4: Repeat the steps to set the heating temperature to 22°C. Simulate the model.

Step 5: Compare result for both cases (Tables 1.39 and 1.40). You can see that increasing the heating setpoints increases the heating energy consumption and heating sizing (Table 1.40).

TABLE 1.39
Energy consumption with a change in heating setpoint temperature

	Annual fuel breakdown data	
Type	**Heating setpoint temperature 20°C (kWh)**	**Heating setpoint temperature 22°C (kWh)**
Room electricity	99,889.40	99,889.40
Lighting	144,361.90	144,361.90
Heating (gas)	99,414.10	137,411.30
Cooling (electricity)	60,146.20	61,491.60
DHW (electricity)	9,038.30	9,038.30

TABLE 1.40
Heating sizing capacity with a change in heating setpoint temperature

	Heating sizing (kW)	
Type	**With heating setpoint temperature 20°C**	**With heating setpoint temperature 22°C**
Heating	244.61	250.15

EXERCISE 1.10

Repeat the tutorial for a change in the cooling setpoint from 24 to 25°C. Compare the results for cooling sizing and energy consumption. Set the weather location as Brisbane, Australia (Tables 1.41 and 1.42).

The following can be observed:

- Increasing the heating setpoint results in an increase in energy consumption because more heat must be added to the space to keep a higher temperature. This also results in a higher system capacity because the rate of heat addition at an elevated temperature is to be matched with the higher rate of heat loss. In the case of gas heaters using hot-water panels, the energy consumption of the pump also increases with the increase in setpoint.
- Similarly, a lower cooling setpoint demands the removal of more heat, thereby causing more energy consumption. This also results in a higher capacity of the cooling equipment and more pump and fan energy consumption.

TABLE 1.41
Energy consumption with a change in cooling setpoint temperature

	Annual fuel breakdown data	
Type	**With cooling setpoint 24°C (kWh)**	**With cooling setpoint 25°C (kWh)**
Room electricity		
Lighting		
System fans		
System pumps		
Heating (electricity)		
Heating (gas)		
Cooling (electricity)		

TABLE 1.42
Cooling sizing capacity with change in cooling setpoint temperature

Type	**With cooling setpoint 24°C (kW)**	**With cooling setpoint 25°C (kW)**
Cooling		

Tutorial 1.11

Evaluating the Impact of Fresh Air Supply Rate

GOAL

To evaluate the impact of the fresh air supply quantity on energy consumption

WHAT ARE YOU GOING TO LEARN?

- How to change the fresh air flow rate

PROBLEM STATEMENT

Create a single-zone 20- × 15-m model. Set the fresh air supply rate (l/s-person) to 5 and 7.5 and compare. Study the effect on energy consumption for the **AZ-PHOENIX DEER VALLEY, USA** location.

SOLUTION

Step 1: Create a single-zone 20- × 15-m model. Select the **Activity** tab.

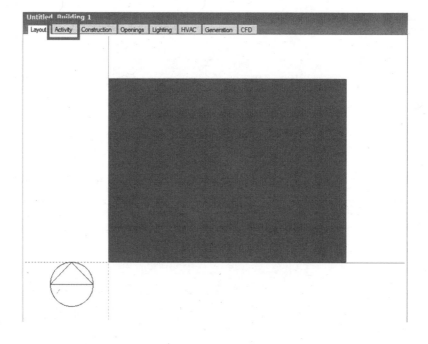

Step 2: Expand the **Minimum Fresh Air** section, and enter **5** in the **Fresh air (l/s-person)** section.

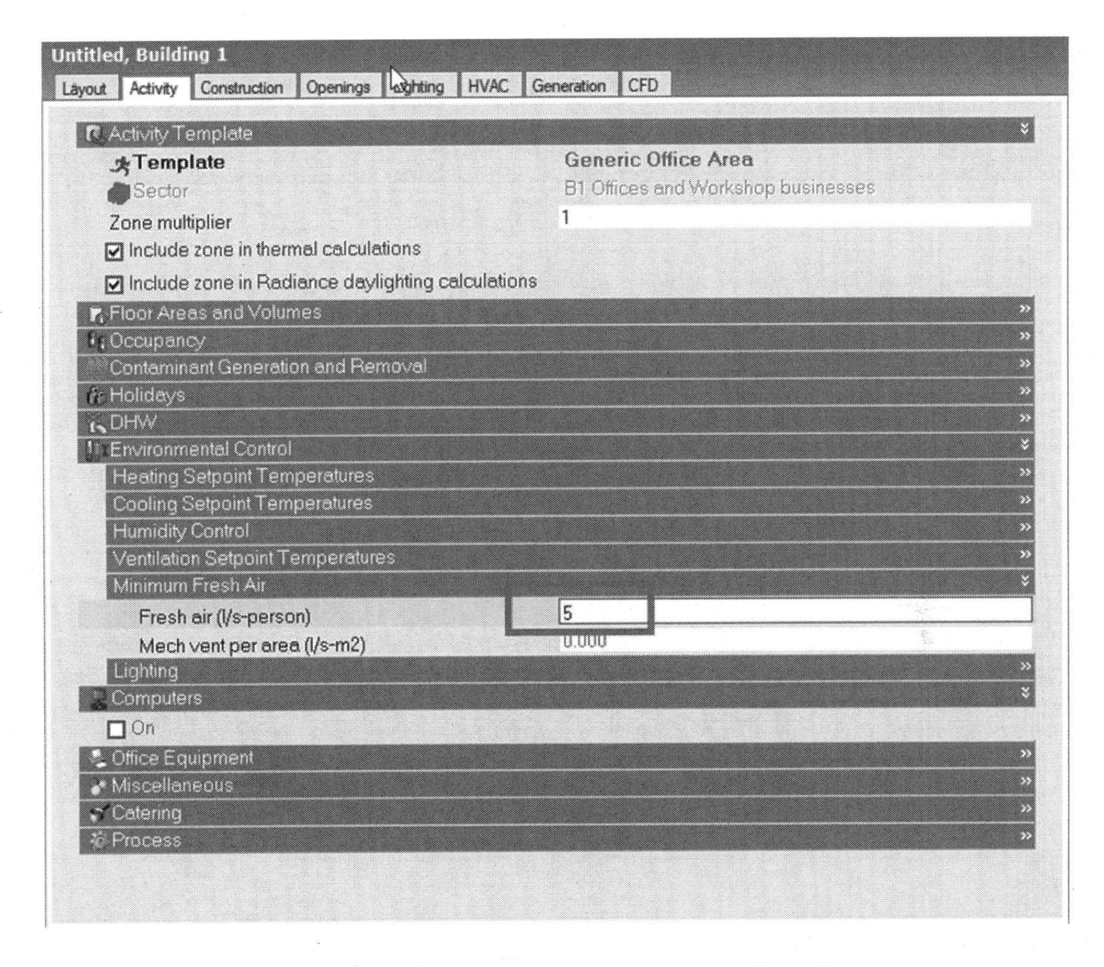

Step 3: Simulate the model, and record the results. Compare the end-use energy consumption results for the fresh air supply of 5 and 7.5 l/s-person (Table 1.43).

It can be seen from the table that with the increase in the fresh air rate, there is an increase in the cooling/heating energy consumption.

TABLE 1.43
Energy consumption with a change in fresh air supply quantity

	Annual fuel breakdown data	
Type	**Fresh air supply of 5 l/s-person (kWh)**	**Fresh air supply of 7.5 l/s-person (kWh)**
Room electricity	12,131	12,131
Lighting	17,530	17,530
Heating (gas)	1,633	1,672
Cooling (electricity)	31,036	31,396
DHW (electricity)	1,098	1,098

EXERCISE 1.11

Compare cooling and heating energy consumption for a change in fresh air supply volume from 5 to 9 l/s-person (Table 1.44).

An increase in the fresh air supply rate increases the energy consumption because of several factors:

- More fresh air brings in more sensible as well as latent load from outside. This, in turn, increases the load on the equipment, which increases the cooling/heating energy consumption.
- The energy consumption of the fan is proportional to the volumetric flow rate. With the increase in flow rate because of the additional energy required to blow more supply air (for heating/cooling), the fan energy consumption increases.
- More heating/cooling load and higher air flow rate result in the requirement of more water flow as a heat-adding/removing medium for a fixed change in temperature across the air handling unit because more heat is to be taken away per unit time. This requirement of a higher flow rate increases the pump rating and results in a higher energy consumption by the pump.

TABLE 1.44
Heating and cooling energy consumption by changing the fresh air supply quantity

	Cooling energy consumption, electricity (kWh)		Heating energy consumption, gas (kWh)	
Month	5 l/s-person	9 l/s-person	5 l/s-person	9 l/s-person
Jan				
Feb				
Mar				
Apr				
May				
Jun				
Jul				
Aug				
Sept				
Oct				
Nov				
Dec				

2 Geometry of Buildings

In this chapter, we discuss the geometrical aspects, such as thermal zoning, aspect ratio, floor multiplier and surface adjacency. Usually buildings have several rooms; however, from a modelling perspective, there may not be a requirement to model each room separately. If adjacent spaces have the same specifications, such as schedule, occupancy and cooling and heating temperature setpoint, you can combine these spaces and model the result as a single zone. By doing this, the complexity of the model and its simulation run time can be reduced without affecting the energy simulation results very much. Similarly, in cases of multi-storied buildings with typical floors, the model can be simplified by modelling only three floors: ground, typical intermediate and top. One of the important parameters that need to be considered while designing the building is the *aspect ratio*, which is the ratio of the floor length to width of the building. The aspect ratio affects the envelope area. An increase in the envelope area leads to higher heat gains/losses for the building. However, a higher *aspect ratio* helps in better distribution of daylight and more access to windows for occupants. The impact of these aspects on building performance is explained through four tutorials in this chapter.

Tutorial 2.1

Defining Thermal Zoning for a Building

GOAL

To evaluate the effect of architectural and thermal zoning on the end-use energy consumption and simulation run time

WHAT ARE YOU GOING TO LEARN?

- How to define thermal zoning
- How to add internal mass

PROBLEM STATEMENT

Create a multi-zone (15 zones, as shown in the following figure) building with a rectangular footprint of 100 × 50 m. Find its energy performance and simulation run time. Create another model by combining similar spaces into thermal zones, and compare the energy performance and run-time duration with those of the first model. Use the **FRANKFURT MAIN ARPT**, Germany, weather file.

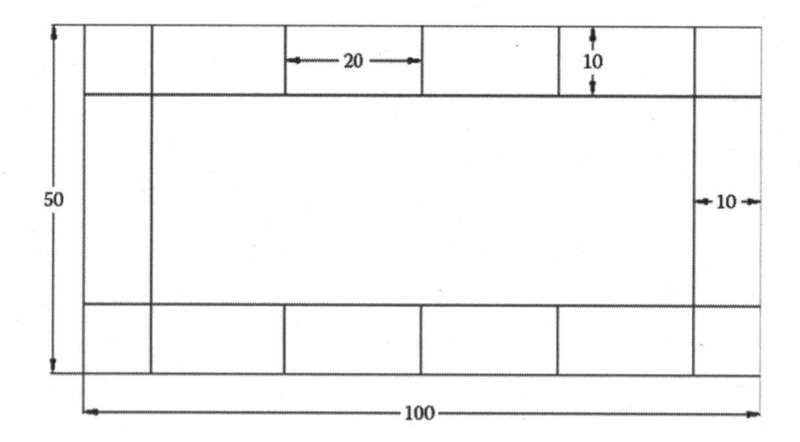

All dimensions are in metre.

SOLUTION

Step 1: Open a **New Project**, and create a **100- × 50-m** block with internal partitions. (Use construction lines to facilitate easy snapping while creating the partitions.)

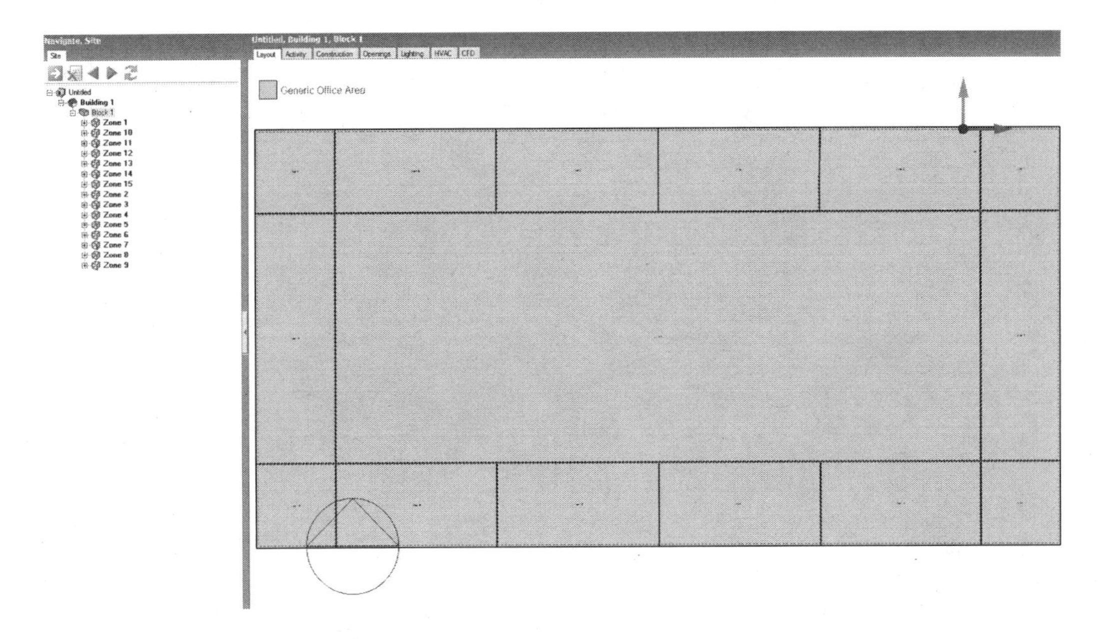

Step 2: Select the **Construction** tab, and select **Outer volume** in the Zone floor area calculation method drop-down list.

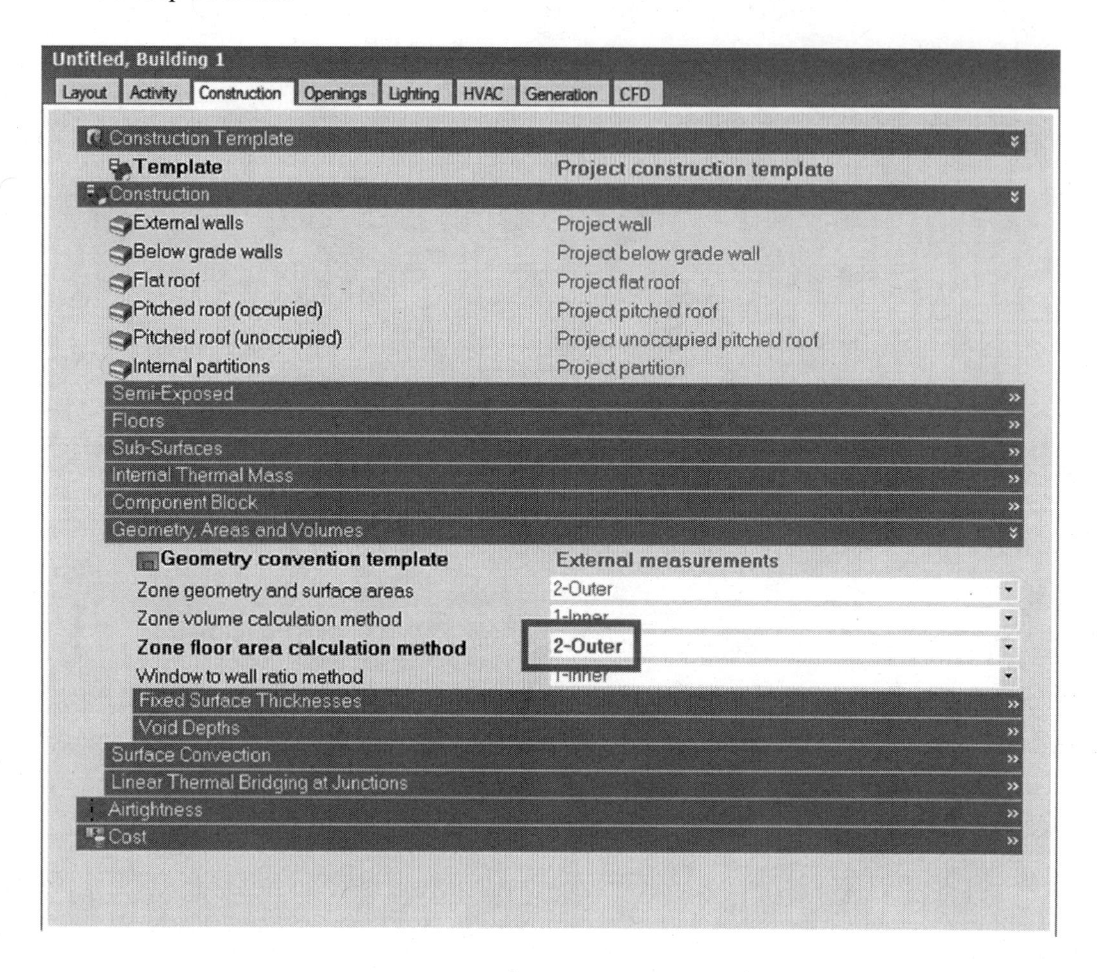

Zone Floor Area Calculation Method

This setting dictates whether internal or external measurements (include external wall thickness inside the zone boundary or not) are used to calculate the zone floor area. This is required to calculate per square-metre values such as occupancy and other internal gains as well as floor area values for general reporting.

Inner: Where the zone volume used in thermal calculations is derived from the zone inner geometry.
Outer: Where the zone volume used in thermal calculations is derived from the zone outer geometry.

Source: http://designbuilder.co.uk/helpv6.0/Content/GeometryAreasAndVolumes.htm.

Step 3: Perform hourly simulation for the whole year, and record the energy consumption and run time.

Recording a Run Time

After the simulation is complete, type the following path to open the eplusout.err file. You can use any text editor to open this file.

C:\Users\User\AppData\Local\DesignBuilder\EnergyPlus\eplusout.err

The actual folder name on a computer depends on the language setting and your Windows user name. In this path given, the Windows user name is 'User'. Alternatively, you can open the **EnergyPlus** folder from the **DesignBuilder** file option. At the end of the file, you can find **Elapsed Time**. You need to record the elapsed time.

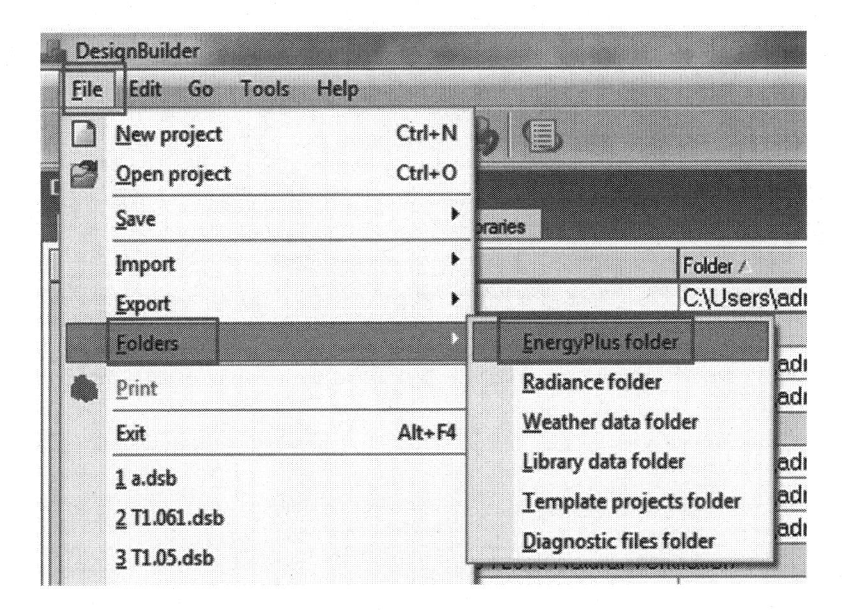

Source: www.designbuilder.co.uk/helpv6.0/Content/_DesignBuilder_files_location_and_extensions.htm.

Untitled, Building 1

Analysis | Summary | Parametric | Optimisation + UA/SA | Data Visualisation

Date/Time	Room Electricity (kWh)	Lighting (kWh)	Heating (Gas) (kWh)	Cooling (Electricity) (kWh)	DHW (Electricity) (kWh)
12:00:00 AM	216714.9	313200	237839.9	99471.71	19609.12

Step 4: Repeat Step 1 while combining zones for spaces with similar activity, schedule and setpoints. In this example, let us assume that the north zones (zones 7–10), as shown in the following figure, are similar and can be combined into a single thermal zone. Similarly, the south zones (zones 12–15) can be combined into a single thermal zone. Do note that zone numbers in your model might differ because zones are numbered in the sequence in which they were created.

You can select an internal partition and delete an internal partition by pressing the **Delete** key. (Please ensure that you are at the block level.)

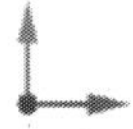

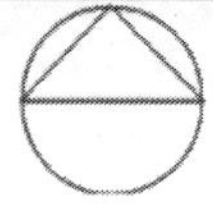

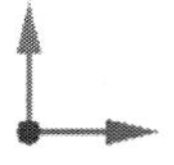

> A *thermal zone*, usually termed simply *zone*, is a virtual or real segment of a building that has a homogeneous enclosed volume of air. In a simple approach, each physical space can be treated as one zone. However, to simplify the modeller's work and to reduce the calculation time, areas having similar thermal and usage conditions such as occupancy, setpoint and solar exposure, and those which are serviced by common mechanical equipment, can be clubbed to create one zone.

Step 5: Perform annual simulation, and record the energy consumption and the run time. Compare the energy simulation results and the simulation run time (Tables 2.1 and 2.2).

TABLE 2.1
Annual fuel breakdown energy with architectural zoning and lumped thermal zones

Type	With architectural zoning (kWh)	With lumped thermal zones (kWh)
Room electricity	216,715	216,715
Lighting	313,200	313,200
Heating (gas)	238,755	238,281
Cooling (electricity)	99,454	99,166
Domestic hot water (DHW, electricity)	19,609	19,609

TABLE 2.2
Simulation run time with architectural zoning and lumped thermal zones

With architectural zoning	With lumped thermal zones
48.04 seconds	31.03 seconds

Simulation run times may vary from the values given in the preceding table based on your computer configuration.

As the results show, combining architectural zones to form thermal zones reduces the simulation run time. In this case, there is an approximately 35.4% decrease in the simulation run time. Note that because there is a difference in the models, there is a slight difference in the energy simulation results.

EXERCISE 2.1

Create thermal zoning for the plan shown in the following figure. Compare the simulation run time and energy consumption for the models with architectural and thermal zoning (Tables 2.3 and 2.4).

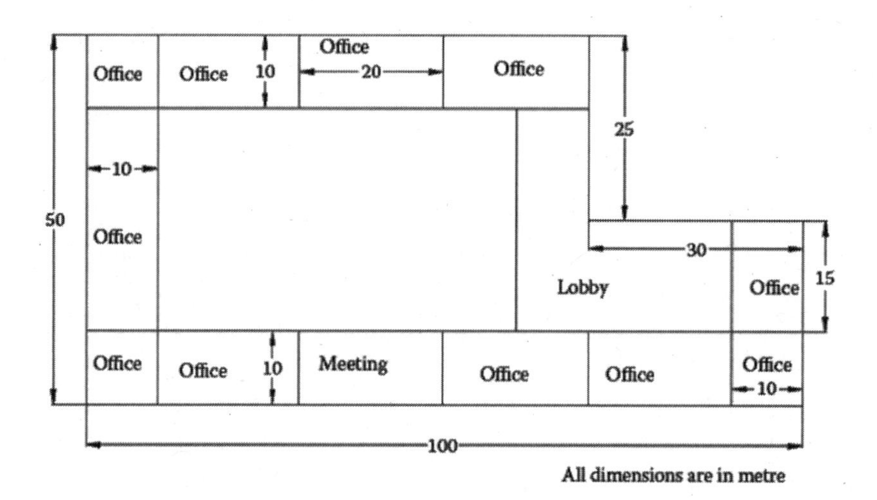

TABLE 2.3

Annual fuel breakdown energy with architectural zoning and lumped thermal zones

Type	With architectural zoning (kWh)	With lumped thermal zones (kWh)
Room electricity		
Lighting		
Heating (gas)		
Cooling (electricity)		
DHW (electricity)		

TABLE 2.4
Simulation run time with architectural zoning and lumped thermal zones

With architectural zoning (seconds)	With lumped thermal zones (seconds)

Tutorial 2.2

Evaluating the Effect of a Zone Multiplier

GOAL

To evaluate the effect of a zone multiplier on the energy consumption and simulation run time

WHAT ARE YOU GOING TO LEARN?

- How to use a zone multiplier

PROBLEM STATEMENT

Use the **Zone Multiplier Example** template file to evaluate the impact of a floor multiplier on a building energy consumption and simulation run time. This file contains two buildings with and without the floor multiplier. You need to select one building at a time and simulate for London Gatwick Airport, United Kingdom.

DesignBuilder has the concept of *zone multiplier*. The zone multiplier data allow you to reduce the size of your model in cases where there are similar zones by specifying that certain zones are repeated and so only need to be simulated once. A typical use is for multi-storey buildings with identical (or very similar) floors. The concept of zone multiplier when applied at the floor level multiplies all the zones on the given floor, effectively working as a floor multiplier. Hence, in a building where there are several identical floors, you can model one floor and use a zone multiplier on that floor.

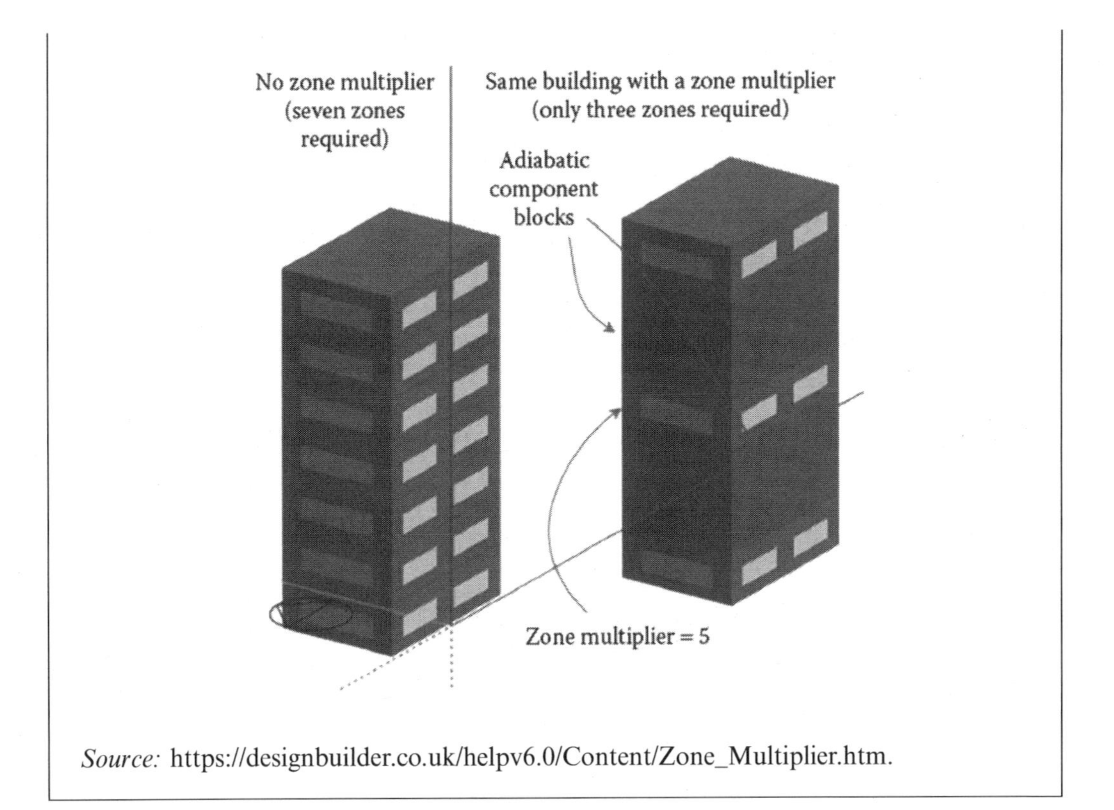

Source: https://designbuilder.co.uk/helpv6.0/Content/Zone_Multiplier.htm.

SOLUTION

Step 1: Open the **Zone Multiplier Example** template from DesignBuilder templates. The layout appears with two buildings.

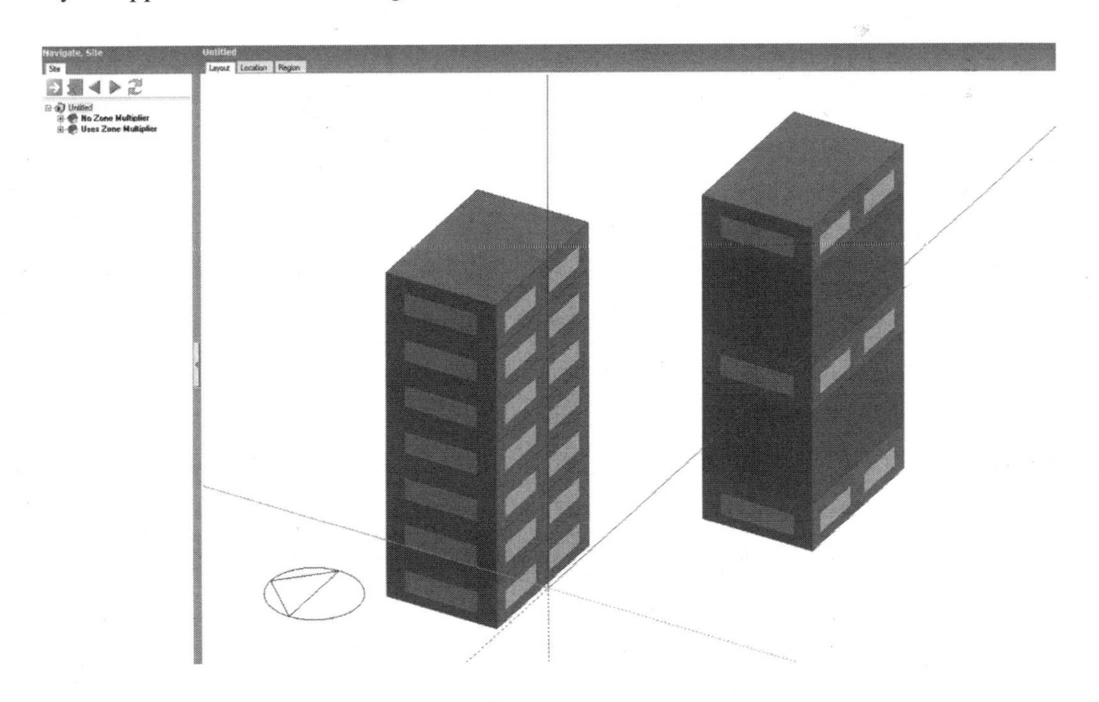

Step 2: Click the **No Zone Multiplier** building in the navigation tree. It selects the building that does not use a floor multiplier.

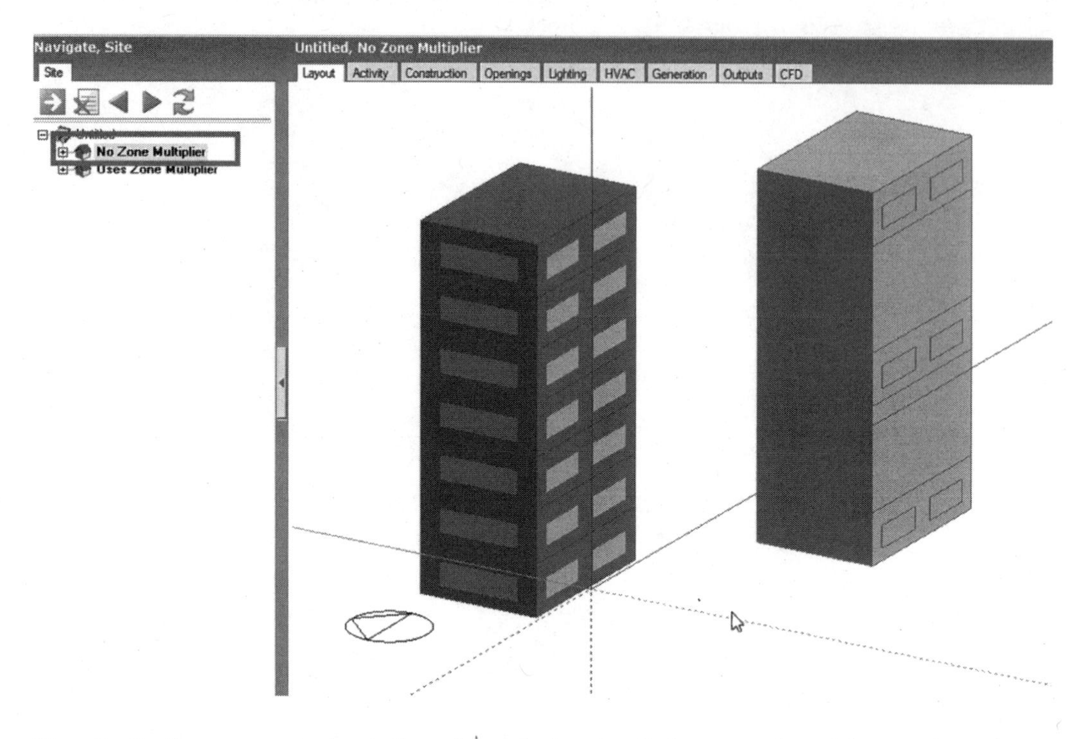

Step 3: Perform annual simulation. Record the simulation run time. The results appear. Record the energy simulation results.

Step 4: Select the **Edit** tab. Click the **Uses Zone Multiplier** building in the navigation tree. Expand the **Uses Zone Multiplier**.

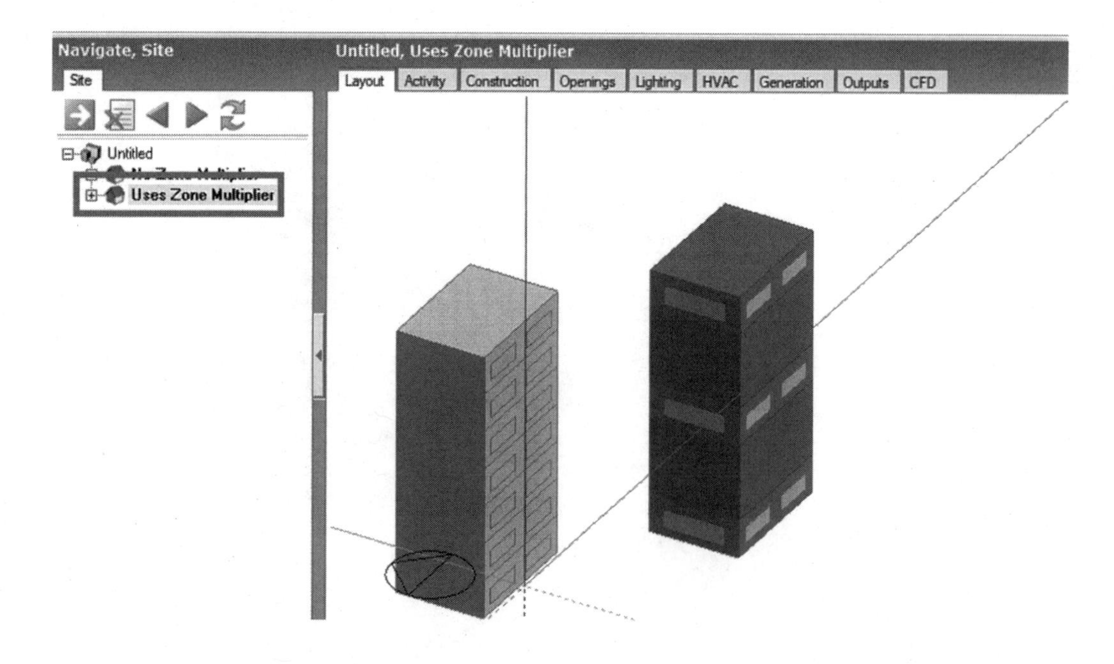

Step 5: Under **Mid**, click **Zone 1**.

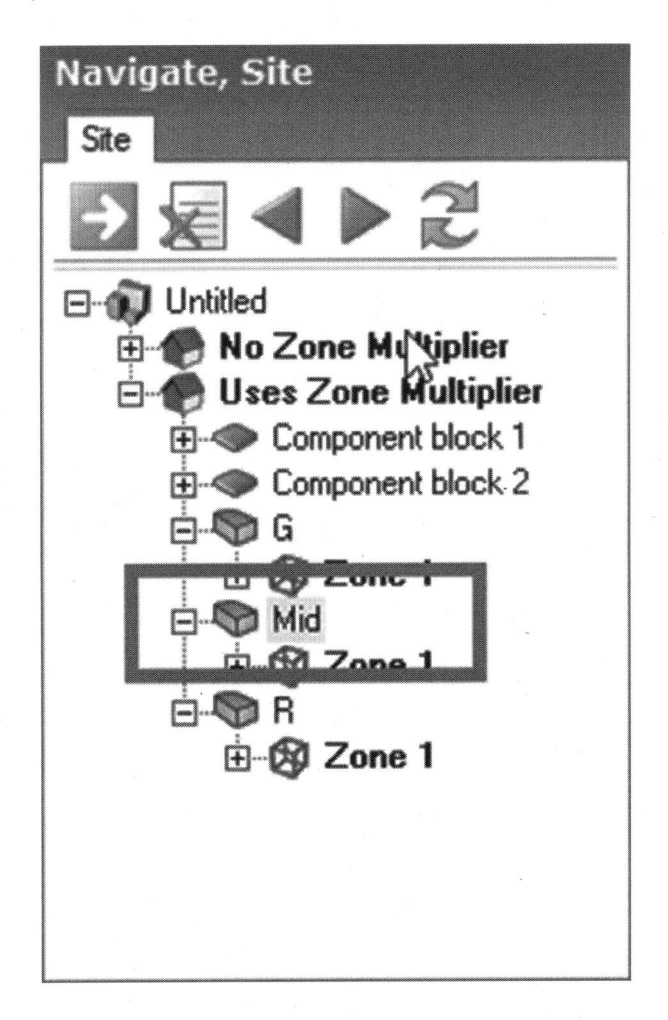

Step 6: Select the **Activity** tab. Because the middle five floors are typical floors, the **Zone Multiplier** for the whole floor is set to **5**.

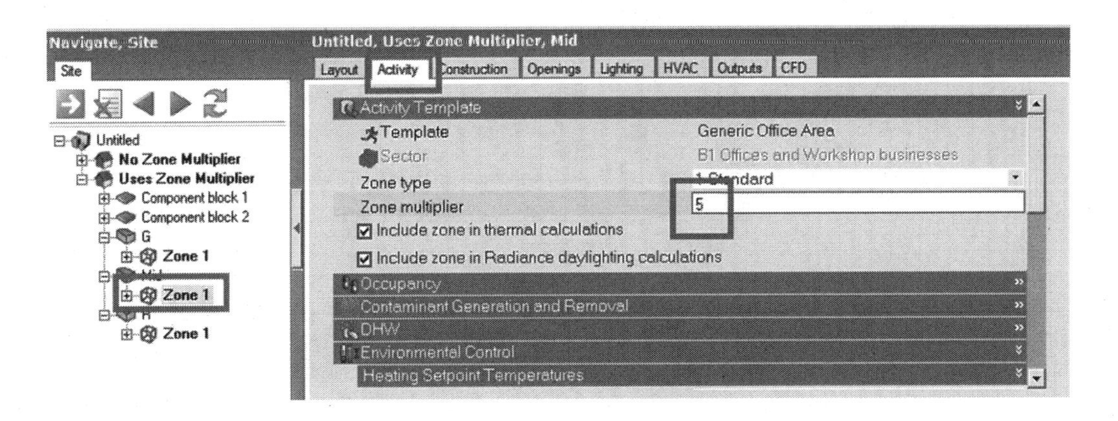

When using a floor multiplier, calculations are performed only for one floor and multiplied by the zone multiplier. This helps in reducing the simulation run time. Compare the results for the models with and without the zone multiplier (Table 2.5).

TABLE 2.5
Impact of a zone multiplier

	Annual fuel breakdown data	
Type	**Without floor multiplier (kWh)**	**With floor multiplier (kWh)**
Room electricity	29,238.68	29,238.68
Lighting	52,820.29	52,820.29
Heating (gas)	33,237.48	33,203.70
Cooling (electricity)	22,250.89	21,744.07
DHW (gas)	2998.37	2998.37
Simulation time	25.13 seconds	13.67 seconds

The results show that using a zone multiplier reduces the simulation run time. However, this has a slight impact on the energy consumption.

EXERCISE 2.2

Create a building with a total of 20 floors. The first 10 floors have a floor plate of dimensions 50 × 50 m. Floors 11 to 20 have a floor plate of dimensions 25 × 25 m. Each floor has a height of 3 m. All floors are centrally aligned. Perform an annual energy simulation of the building with and without a floor multiplier. Compare energy and run time of the simulation.

Tutorial 2.3

Evaluating the Impact of the Aspect Ratio

GOAL

To evaluate the impact of building aspect ratio on energy performance with and without daylight sensors

WHAT ARE YOU GOING TO LEARN?

- How to model a building with different aspect ratios but with the same floor area

PROBLEM STATEMENT

In this tutorial, you are going to analyse the impact of the aspect ratio of a building on the building's energy performance. You need to create models with different aspect ratios for floor plates of areas 64 m^2 and 625 m^2. Simulate various cases as given below, and compare their energy consumption for the London Gatwick weather location. Make sure that the WWR is 30%.

> For a rectangular building, *aspect ratio* is the ratio of the longest dimension of the building footprint to the narrowest dimension. An aspect ratio of 1.0 represents a square building footprint.

a. For a floor area of 64 m^2, Table 2.6 gives the length and breadth values for various aspect ratios

TABLE 2.6
Different aspect ratios for a 64 m^2 floor area

Serial no.	Length l (m)	Breadth b (m)	Aspect ratio l/b	Floor area $l \times b$ (m^2)	Facade area (window + wall) (m^2)	Facade area/floor area
1	8.00	8.00	1	64	112.00	1.75
2	11.31	5.65	2	64	118.79	1.86
3	13.85	4.62	3	64	129.33	2.02
4	16.00	4.00	4	64	140.00	2.19

b. For a floor area of 625 m^2, Table 2.7 gives the length and breadth values for various aspect ratios

TABLE 2.7
Different aspect ratios for a 625 m^2 floor area

Serial no.	Length l (m)	Breadth b (m)	Aspect ratio l/b	Area $l \times b$ (m^2)	Facade area (window + wall area) (m^2)	Facade area/floor area
1	25.0	25.0	1	625	350.00	0.56
2	43.3	14.4	3	625	404.15	0.65
3	55.9	11.2	5	625	469.57	0.75
4	66.1	9.4	7	625	529.15	0.85

SOLUTION

Step 1: Open a new project file.

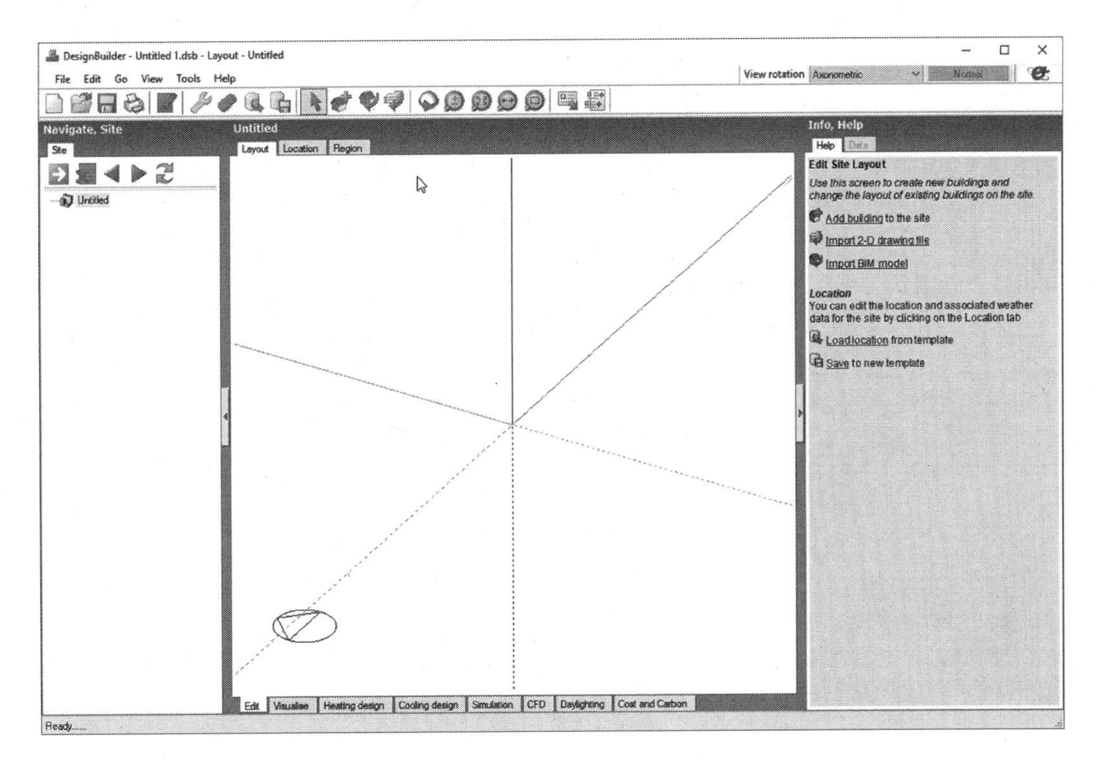

Step 2: Create an **8- × 8-m** building (aspect ratio 1).

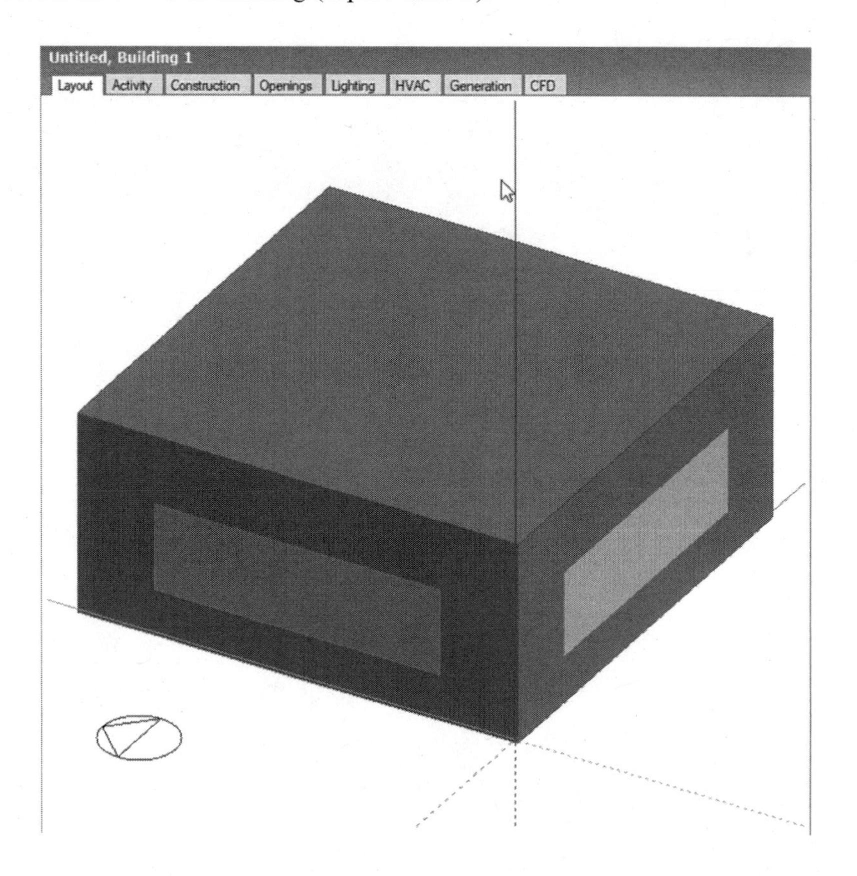

Step 3: Simulate the model and store the results normalized by area (Table 2.8).

TABLE 2.8
Energy simulation results for an 8- × 8-m model

Room electricity (Wh/m^2)	Lighting (Wh/m^2)	Heating (gas) (Wh/m^2)	Cooling (electricity) (Wh/m^2)	DHW (electricity) (Wh/m^2)
43,342.96	62,639.97	65,200.95	37,738.93	3921.82

Step 4: Repeat Steps 1 to 3 for aspect ratios 2, 3 and 4, respectively. Compare the energy simulation results for all cases (Table 2.9).

TABLE 2.9
Energy consumption with different aspect ratios (without a daylight sensor)

Aspect ratio	1 (8 × 8 m)	2 (11.3 × 5.7 m)	3 (13.9 × 4.6 m)	4 (16 × 4 m)
Room electricity (Wh/m^2)	43,342.96	43,342.96	43,342.96	43,342.96
Lighting (Wh/m^2)	62,639.97	62,639.97	62,639.97	62,639.97
Heating (gas) (Wh/m^2)	65,200.95	65,959.48	69,259.36	72,891.05
Cooling (electricity) (Wh/m^2)	37,738.93	38,256.89	41,034.55	43,458.84
DHW (electricity) (Wh/m^2)	3,921.82	3,921.82	3,921.82	3,921.82

The following figure shows the energy consumption with aspect ratio when daylight sensors are not modelled.

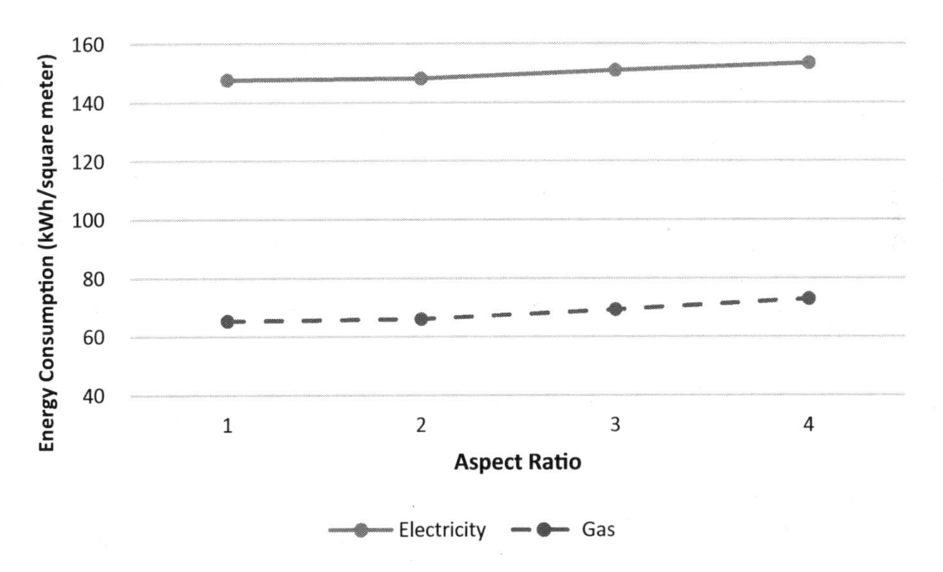

Step 5: Repeat all the preceding steps with a daylight sensor (Refer to Tutorial 1.9 for modelling of a daylight sensor) (Table 2.10).

TABLE 2.10
Energy consumption with varying aspect ratios (with a daylight sensor)

Aspect ratio	1 (8 × 8 m)	2 (11.3 × 5.7 m)	3 (13.9 × 4.6 m)	4 (16 × 4 m)
Room electricity (Wh/m^2)	43,342.96	43,342.96	43,342.96	43,342.96
Lighting (Wh/m^2)	16,058.78	16,068.64	15,861.8	14,721.81
Heating (gas) (Wh/m^2)	82,163.28	82,649.38	85,776.92	90,108.44
Cooling (electricity) (Wh/m^2)	26,791.65	27,290.38	29,933.03	32,244.44
DHW (electricity) (Wh/m^2)	3,921.82	3,921.82	3,921.82	3,921.82

The following figure shows energy consumption with varying aspect ratios when a daylight sensor is modelled.

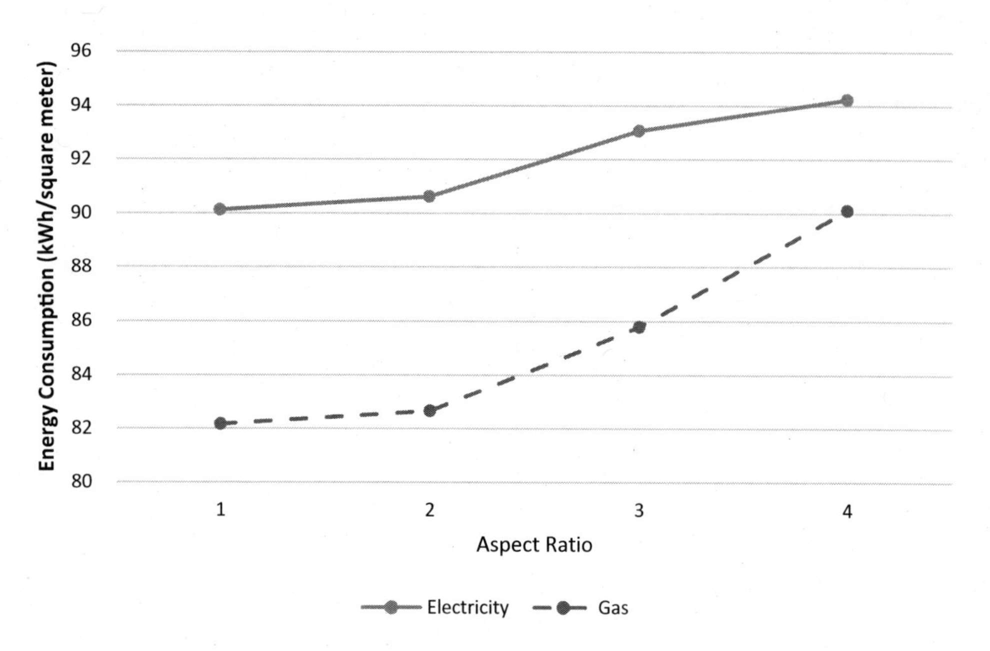

Step 6: Repeat all the preceding steps for a floor area of 625 m^2 (Tables 2.11 and 2.12). You need to draw the model with the help of plans as shown in the following figures with aspect ratios of 1, 3, 5 and 7. Perimeter depth for all plans is 600 mm.

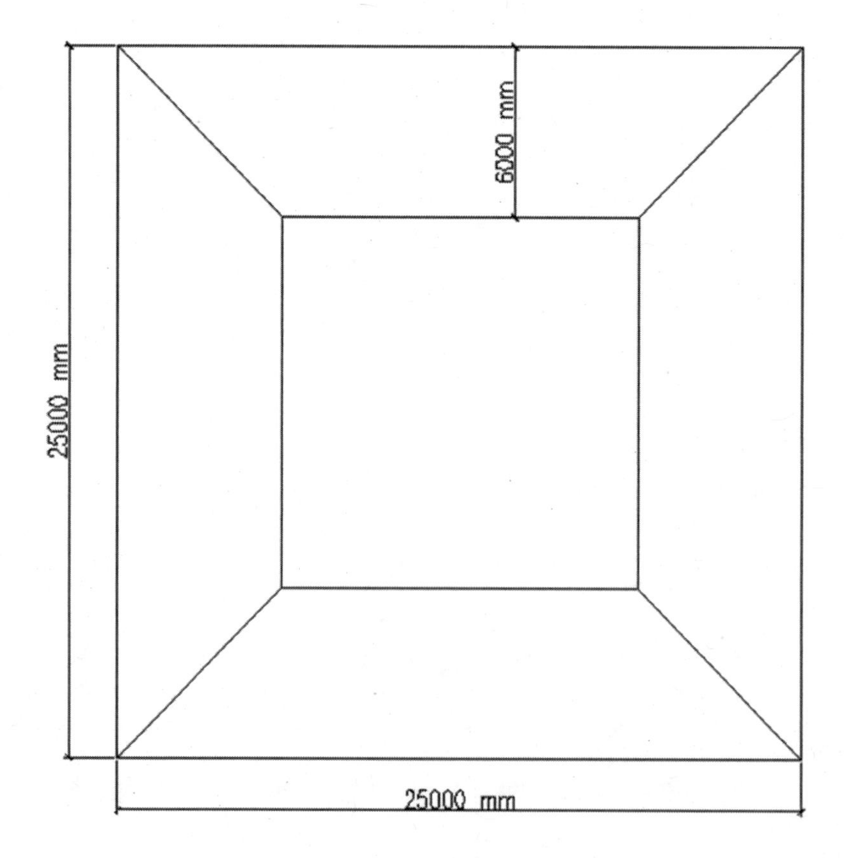

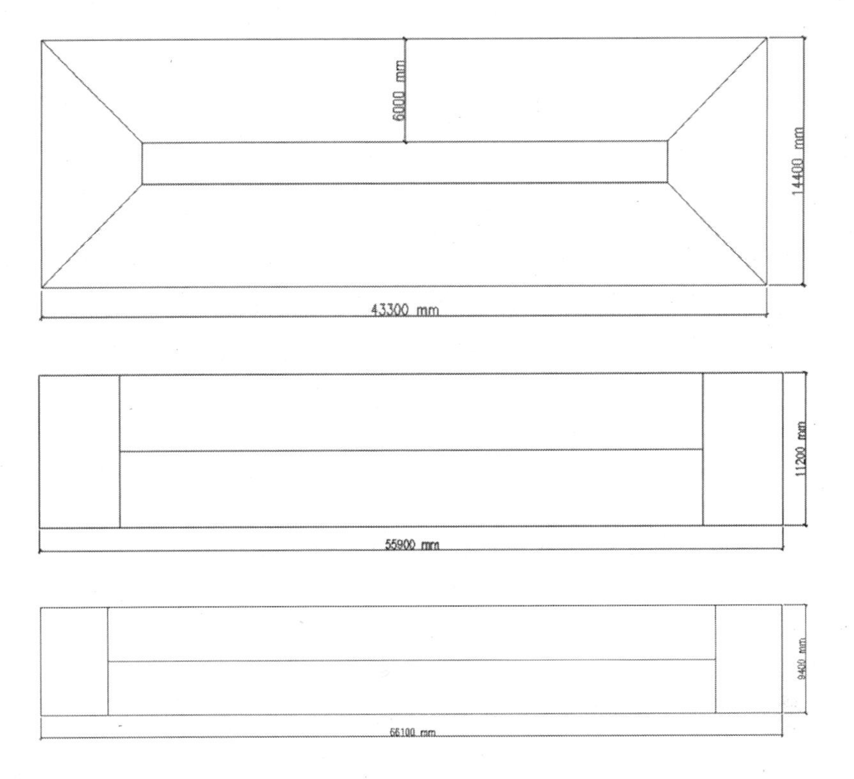

TABLE 2.11
Energy consumption for different aspect ratios (without a daylight sensor)

Aspect ratio	1 (25 ×25 m)	3 (43.3 × 14.4 m)	5 (55.9 × 11.2 m)	7 (66.1 × 9.4 m)
Room electricity (Wh/m^2)	43,342.98	43,342.98	43,342.98	43,342.98
Lighting (Wh/m^2)	62,640.00	62,640.00	62,640.00	62,640.00
Heating (gas) (Wh/m^2)	41,099.50	41,757.79	43,132.25	44,612.56
Cooling (electricity) (Wh/m^2)	19,559.21	20,505.61	21,906.63	23,400.11
DHW (electricity) (Wh/m^2)	3,921.82	3,921.82	3,921.82	3,921.83

TABLE 2.12
Energy consumption for different aspect ratios (with daylight sensor)

Aspect ratio	1 (25 ×25 m)	3 (43.3 × 14.4 m)	5 (55.9 × 11.2 m)	7 (66.1 × 9.4 m)
Room electricity (Wh/m^2)	43,342.98	43,342.98	43,342.98	43,342.98
Lighting (Wh/m^2)	25,782.53	23,210.38	20,955.56	19,514.44
Heating (gas) (Wh/m^2)	53,027.22	55,007.47	57,603.27	59,789.09
Cooling (electricity) (Wh/m^2)	10,333.91	10,888.03	11,947.66	13,180.83
DHW (electricity) (Wh/m^2)	3,921.82	3,921.82	3,921.82	3,921.83

The first of the following figures shows the energy consumption with aspect ratio when a daylight sensor is not modelled, and the second figure shows energy consumption with varying aspect ratios when a daylight sensor is modelled.

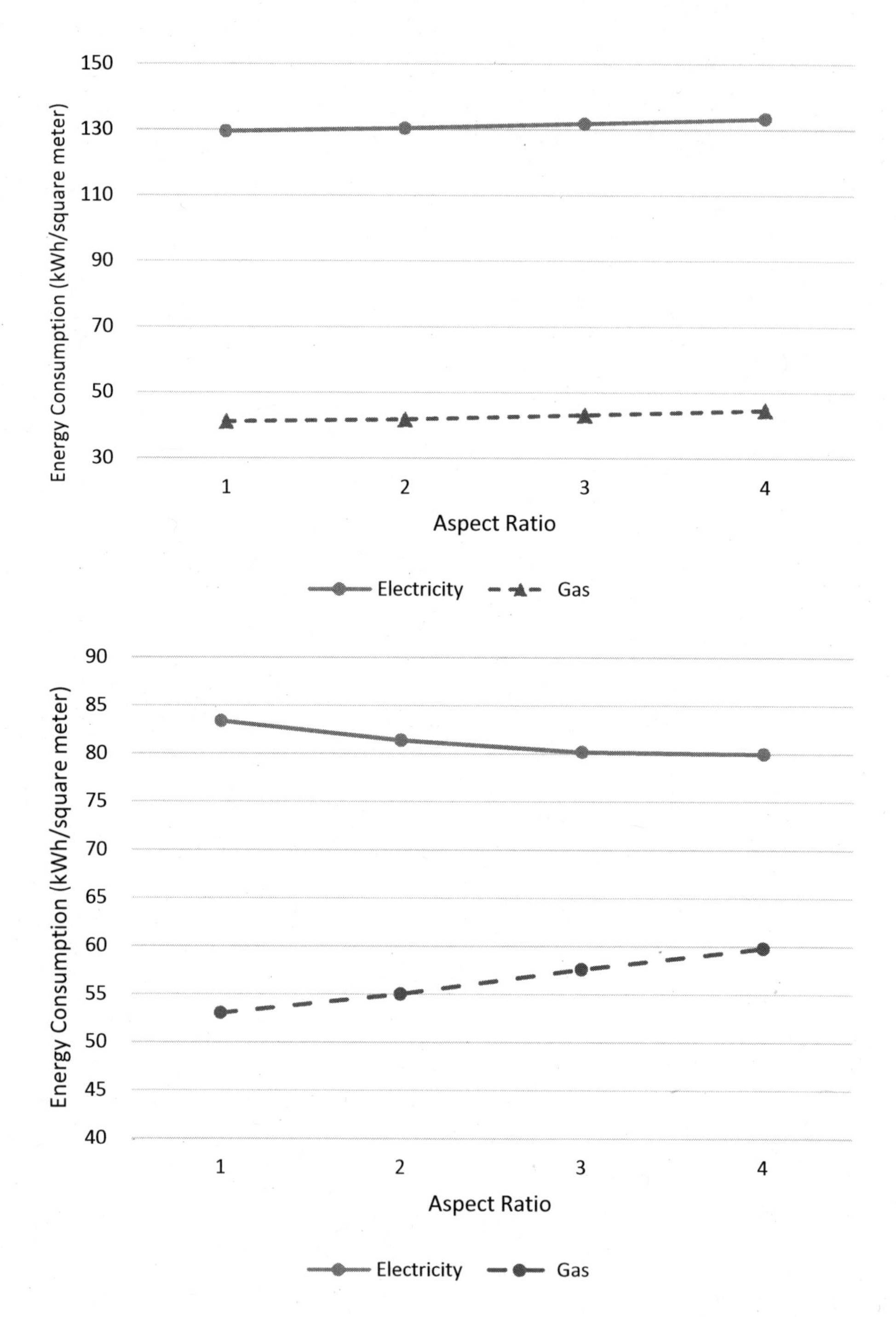

When a building is modelled with daylight sensors, with an increase in aspect ratio, there is a reduction in building electricity consumption. This reduction in electricity use is due to savings in interior lighting electricity consumption.

EXERCISE 2.3

a. Analyse the effect of the aspect ratio for all cases described in this tutorial for the 90-degree orientation.

b. Repeat the tutorial for a hot and dry climate (**UAE**), and observe the energy simulation results.

Tutorial 2.4

Evaluating the Impact of Adjacency of the Surface

GOAL

To evaluate the impact of the ground surface on energy performance

WHAT ARE YOU GOING TO LEARN?

- How to assign surface adjacency
- How to make a surface adiabatic

PROBLEM STATEMENT

In this tutorial, you are going to create a 50- × 25-m single-zone model. Assign ground floor construction to a 200-mm aerated concrete slab. Set the ground floor surface adjacency to **Auto** and as **Adiabatic**. Compare the effect of the surface adjacency on the energy consumption for **QC – Montreal/Mirabel INT'L A, Canada**.

SOLUTION

Step 1: Create a **50- × 25-m single zone** model. Expand **Zone 1** in the navigation tree.

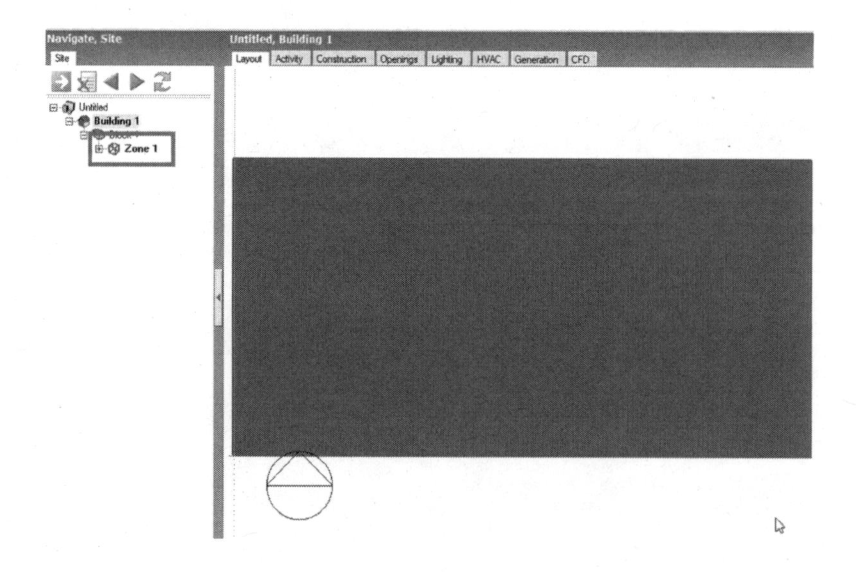

Step 2: Click **Ground Floor**.

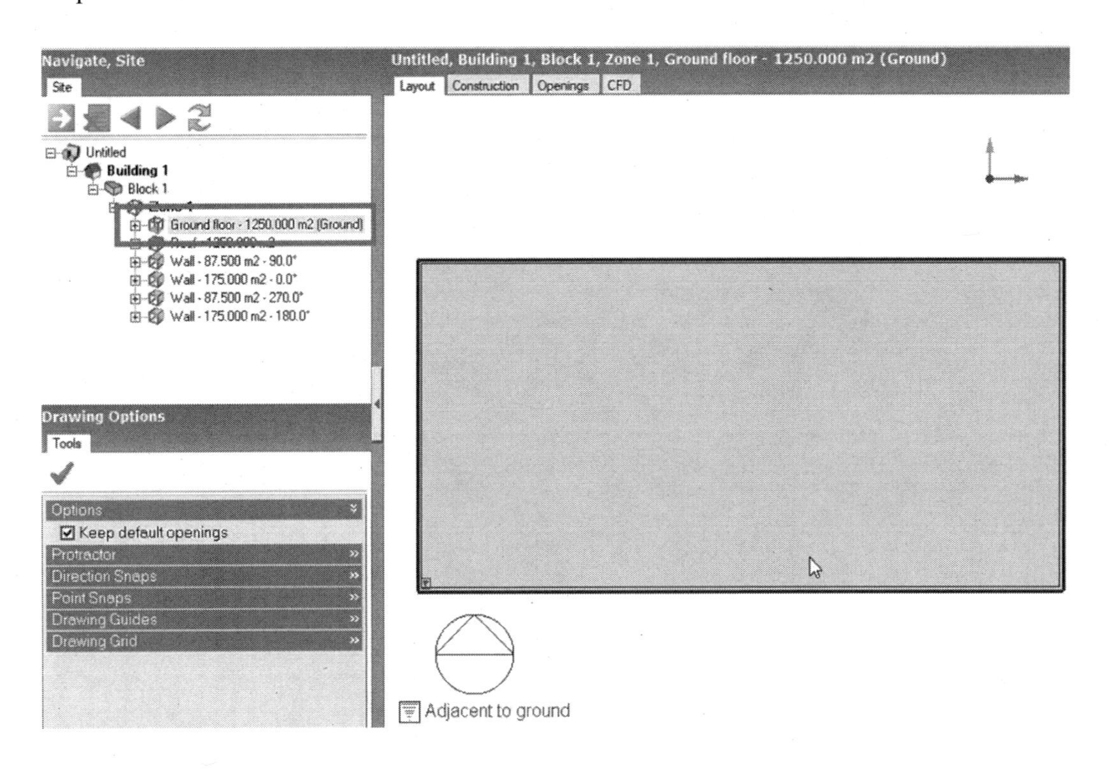

Step 3: Select the **Construction** tab.

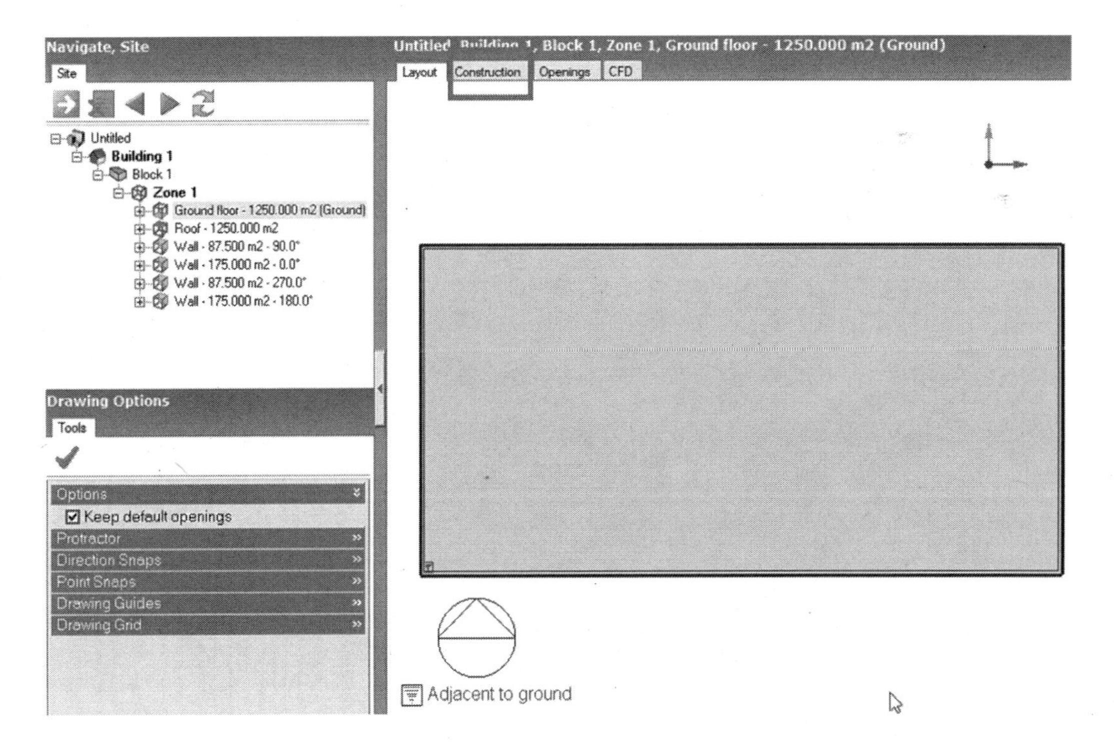

Step 4: Under the **Floors** section, click **Ground floor**. Three dots (...) appear. Click the dots. The **Select the Construction** screen appears.

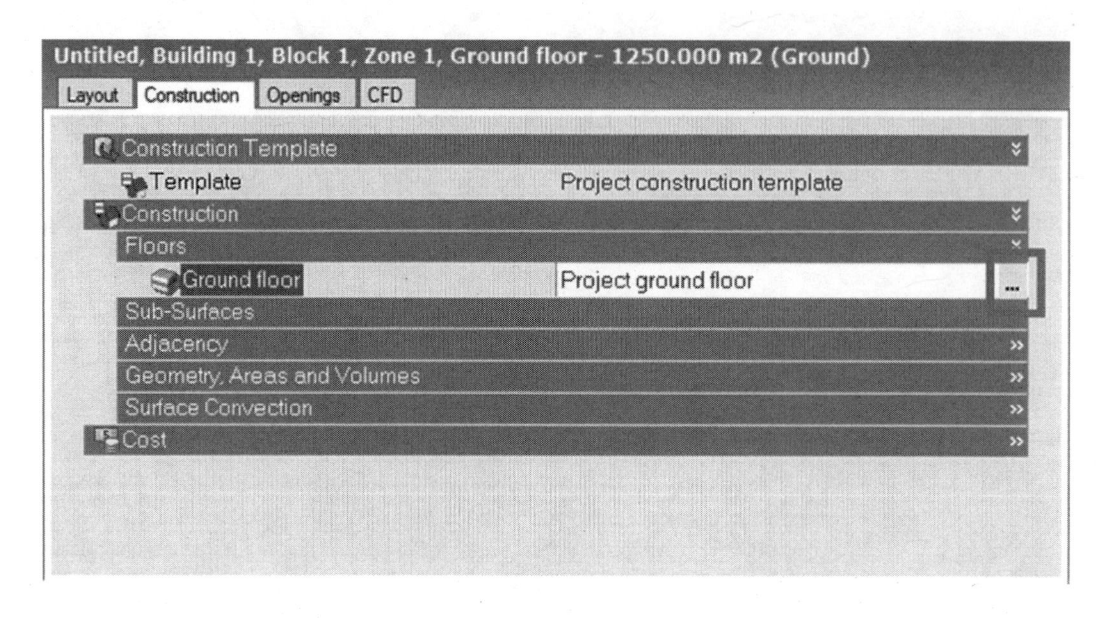

Step 5: Click the **Add new data** button to create a new construction.

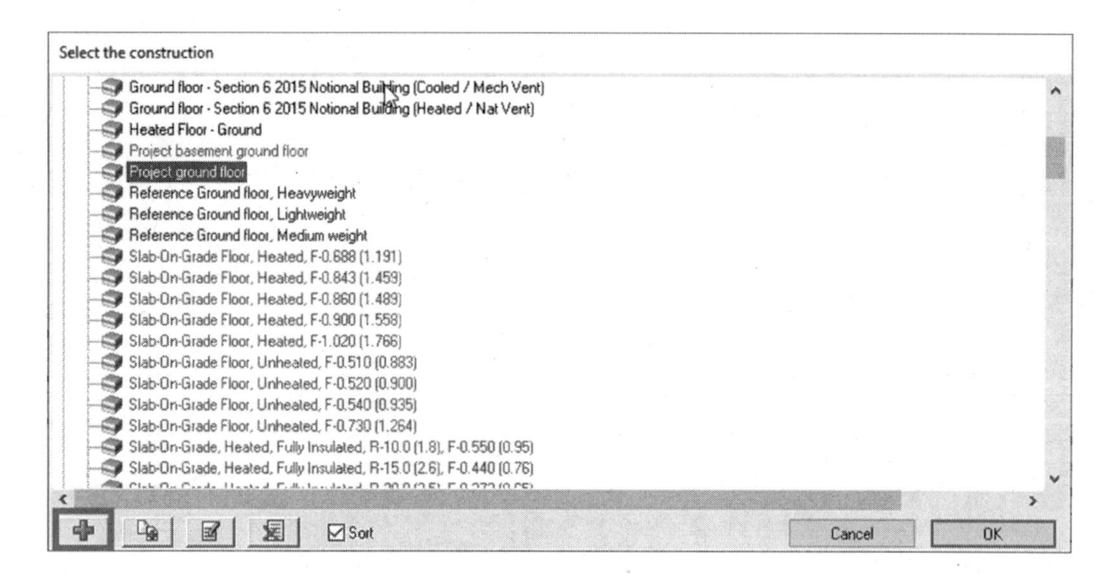

Step 6: Go to the **Layers** section, and make sure that **Areated Concrete Slab** is the material and that the thickness is **0.2000** m. Click **OK**.

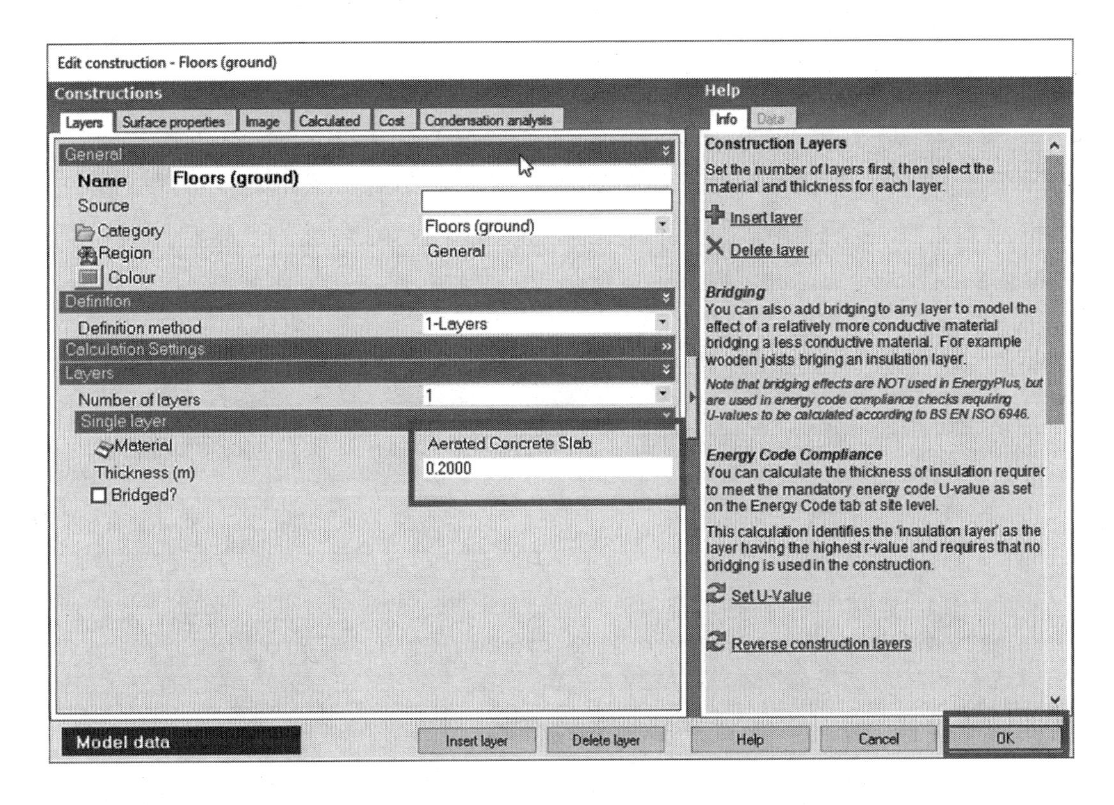

Step 7: Click the **Adjacency** section. It displays the adjacency property of the selected surface. Ensure that **Auto** is selected in the **Adjacency** drop-down list.

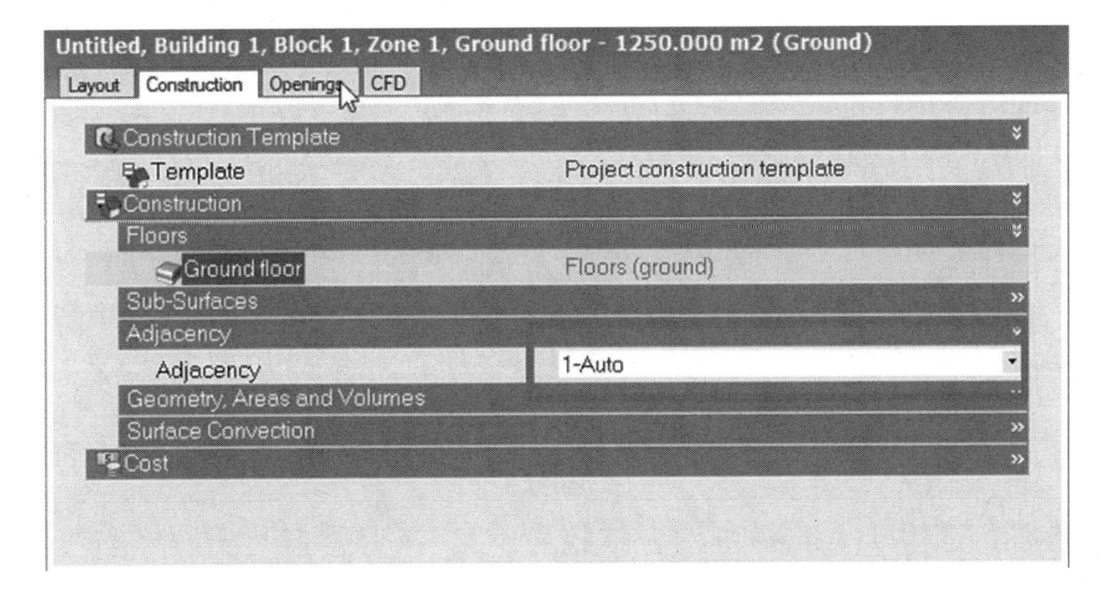

Step 8: Perform an annual energy simulation. Record the results for the end-use energy consumption.

Untitled, Building 1

Analysis | Summary | Parametric | Optimisation + UA/SA | Data Visualisation

Date/Time	Room Electricity (kWh)	Lighting (kWh)	Heating (Gas) (kWh)	Cooling (Electricity) (kWh)	DHW (Electricity) (kWh)
12:00:00 AM	52291.88	75573.1	139240.3	24968.78	4731.553

Step 9: Repeat Step 6 to select **Adiabatic** from the **Adjacency** drop-down list.

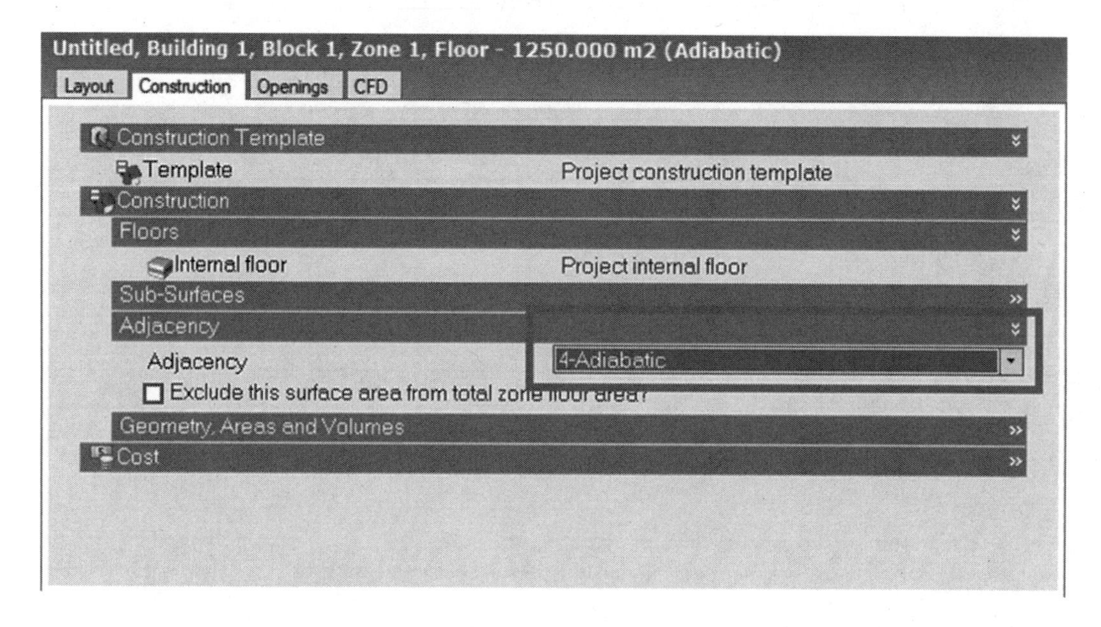

Step 10: Run the simulation to view the results. Compare the end-use energy consumption breakdown (Table 2.13). You can see that there is a change in the cooling and heating energy consumption when the ground floor surface property is set to adiabatic.

TABLE 2.13

Annual fuel breakdown data for building with ground floor adjacency set to adjacent to ground and adiabatic

Type	Adjacent to the ground (kWh)	Adiabatic (kWh)
Room electricity	52,291.80	52,291.80
Lighting	75,573.10	75,573.10
Heating (gas)	13,9240.30	13,5103.00
Cooling (electricity)	24,968.70	29,099.40
DHW (electricity)	4,731.50	4,731.50

The EnergyPlus weather file provides ground temperatures for undisturbed sites. However, you should not use the ground temperatures in the EnergyPlus weather file header because they are for undisturbed sites.

The temperature beneath a building is significantly affected by the building itself – the EnergyPlus documentation recommends using a ground temperature of 2°C below average internal temperatures for large commercial buildings (where the perimeter heat loss is relatively less important). Note that this temperature should be applied directly below the slab and should not include ground material; thus, if you use this approach to ground temperature definition, you should switch off the use of the ground construction at the site level.

EnergyPlus cannot model very thick constructions, so it is necessary to use less thick constructions (2 m or less) combined with some assumptions about temperatures at about half a meter below the floor.

Source: www.designbuilder.co.uk/helpv6.0/Content/Ground_Modelling.htm.

Many modellers prefer to define the ground temperature just below the slab and exclude the earth layers from the model. This has the advantage of simplicity and clarity and is the approach recommended by the EnergyPlus developers.

Note: The default ground temperatures provided in DesignBuilder assume that an earth layer is included in the constructions adjacent to the ground. If an earth layer is not included, then you should increase the default site ground temperatures to values closer to those typically found just below the ground slab.

Source: www.designbuilder.co.uk/helpv6.0/Content/Ground_Modelling.htm.

3 Material and Construction

The aim of this chapter is to explain how to create a model while defining materials and constructions and to evaluate their impact on the energy consumption of buildings. The chapter starts with a tutorial on evaluating the impact of thermal mass in the envelope by comparing the performance of lightweight and heavyweight external wall construction. Learners can also find a method for calculating the thickness of insulation on the roof or in external walls. The tutorials cover analysis of roof insulation location (overdeck or underdeck) and use of a cool roof and a radiant barrier. Materials are the backbone of building construction. Choosing materials for construction is most important for building as it involves a huge cost. There is need to carefully look at various properties such as thermal mass, conductivity, surface finish and so on.

Tutorial 3.1

Evaluating the Effect of Lightweight and Heavyweight Construction

GOAL

To evaluate the effect of thermal mass – the lightweight and heavyweight external wall construction in a building with night purge – on the thermal performance of the building

WHAT ARE YOU GOING TO LEARN?

- How to assign lightweight and heavyweight construction
- How to get zone temperatures

PROBLEM STATEMENT

In this tutorial, you are going to use a 50- × 25-m model and set external wall construction to lightweight and heavyweight. Determine the air temperature inside the zone for both the cases. You need to analyse it for the **DUBAI INTERNATIONAL, United Arab Emirates** weather location.

You are going to use the following construction for external walls:

- Uninsulated wall, lightweight (wood derivative – plywood 15 mm)
- Uninsulated wall, heavyweight (stone – granite, 450 mm)

SOLUTION

Step 1: Open a new project file. Create a **50-** × **25-m** building.

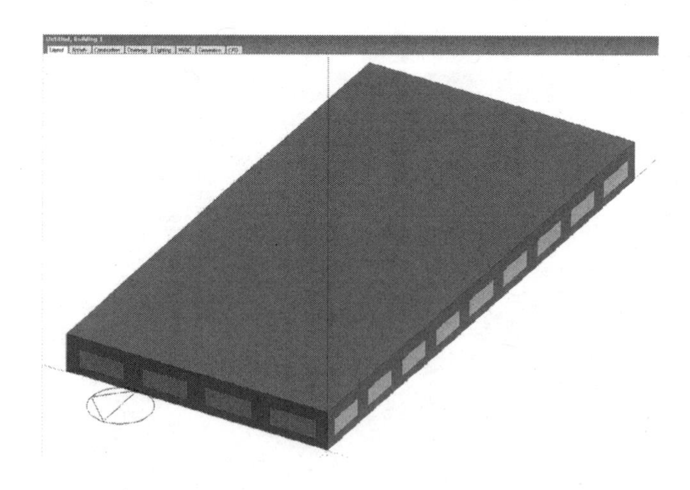

Step 2: Select the **Construction** tab.

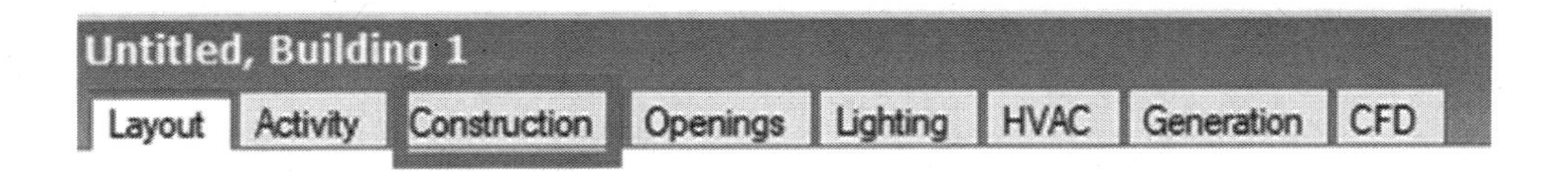

Step 3: In the **Construction** section, click **Project wall. External walls** gets highlighted. Click the **Add new item** icon. The **Edit construction – Wall** screen appears.

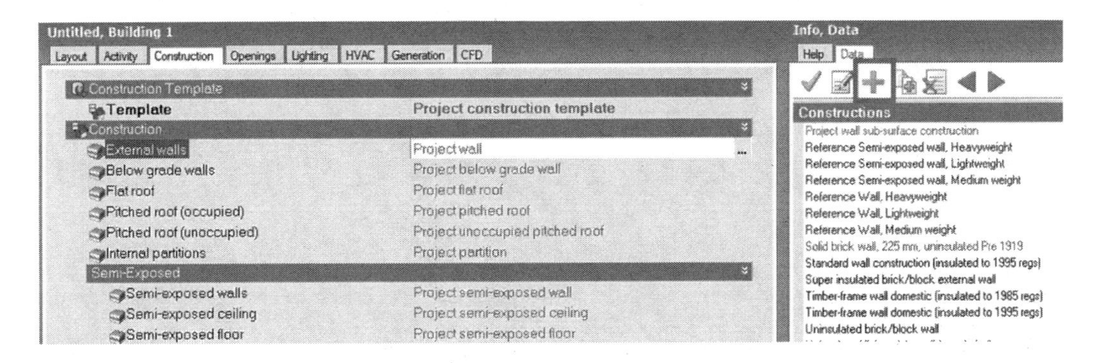

Step 4: Enter **Light weight** as the **Name**. Go to the **Layers** section, and select **Plywood** as the material (you can find it under the **Wood** branch). Set the thickness to **0.025**. Click **OK**. A message box appears.

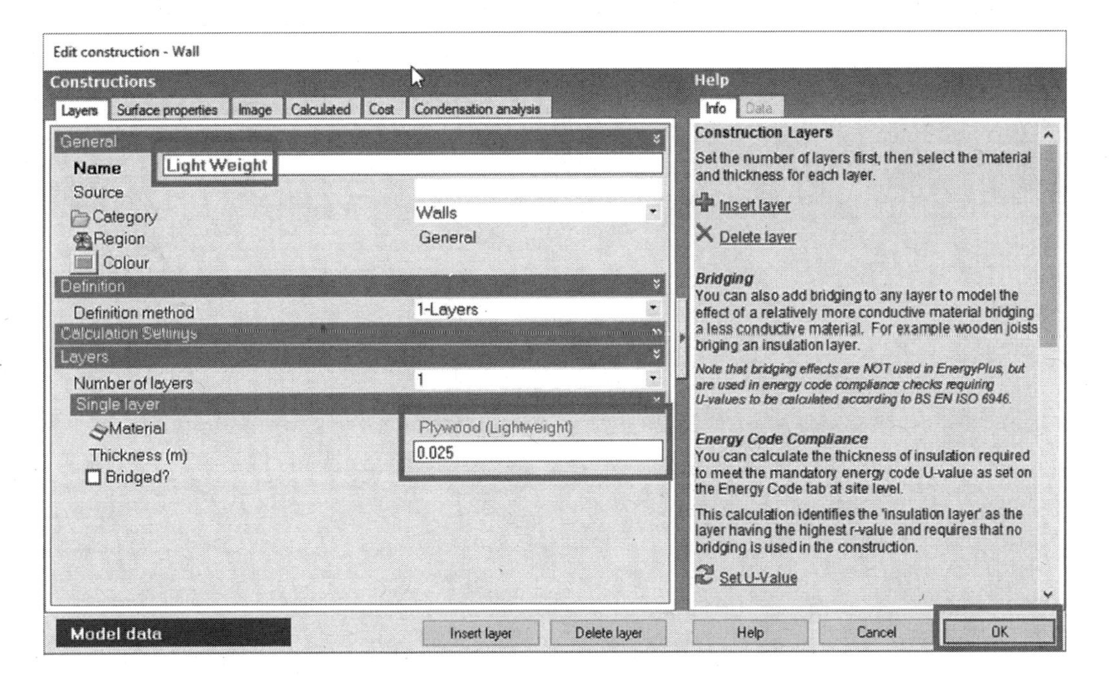

Step 5: Select **Light Weight** as the external wall.

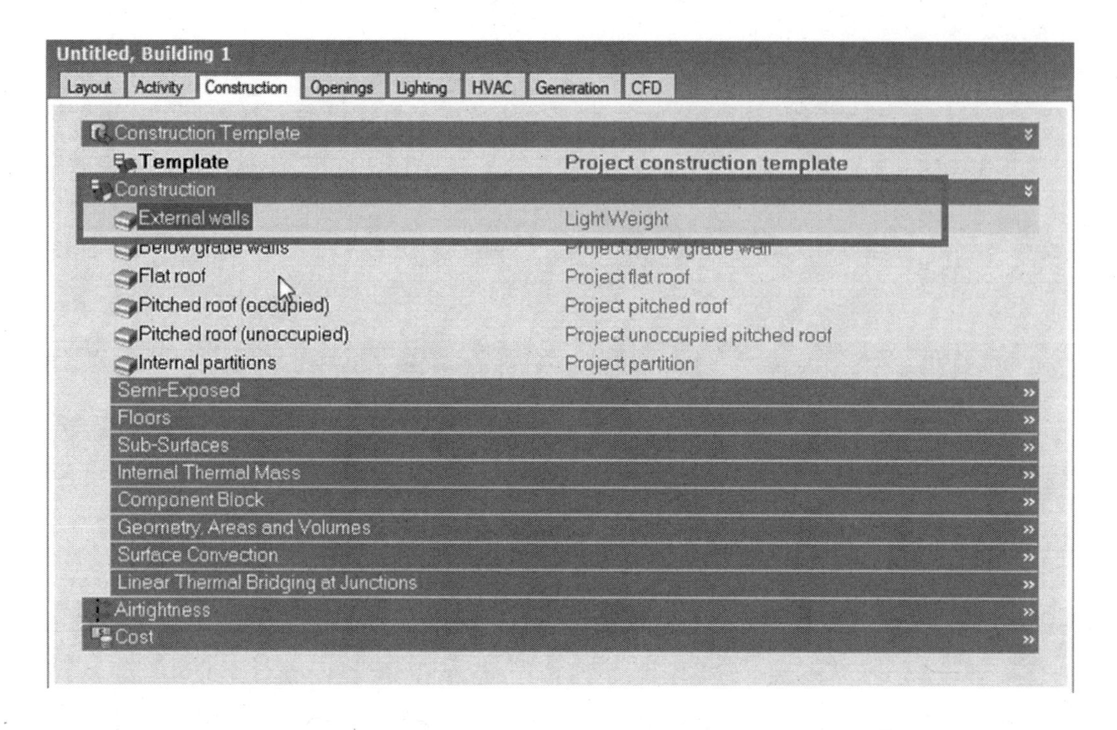

Step 6: Go to the **Airtightness** section and enter **3.000** in the **Constant rate (ac/h)** box. Click **On 24/7** in the **Schedule** field. Three dots appear (. . .). Click on the **dots**. The **Select the Schedule** screen appears.

Step 7: Click the **Create copy of highlighted item** icon. Make sure that the copy of the item is highlighted.

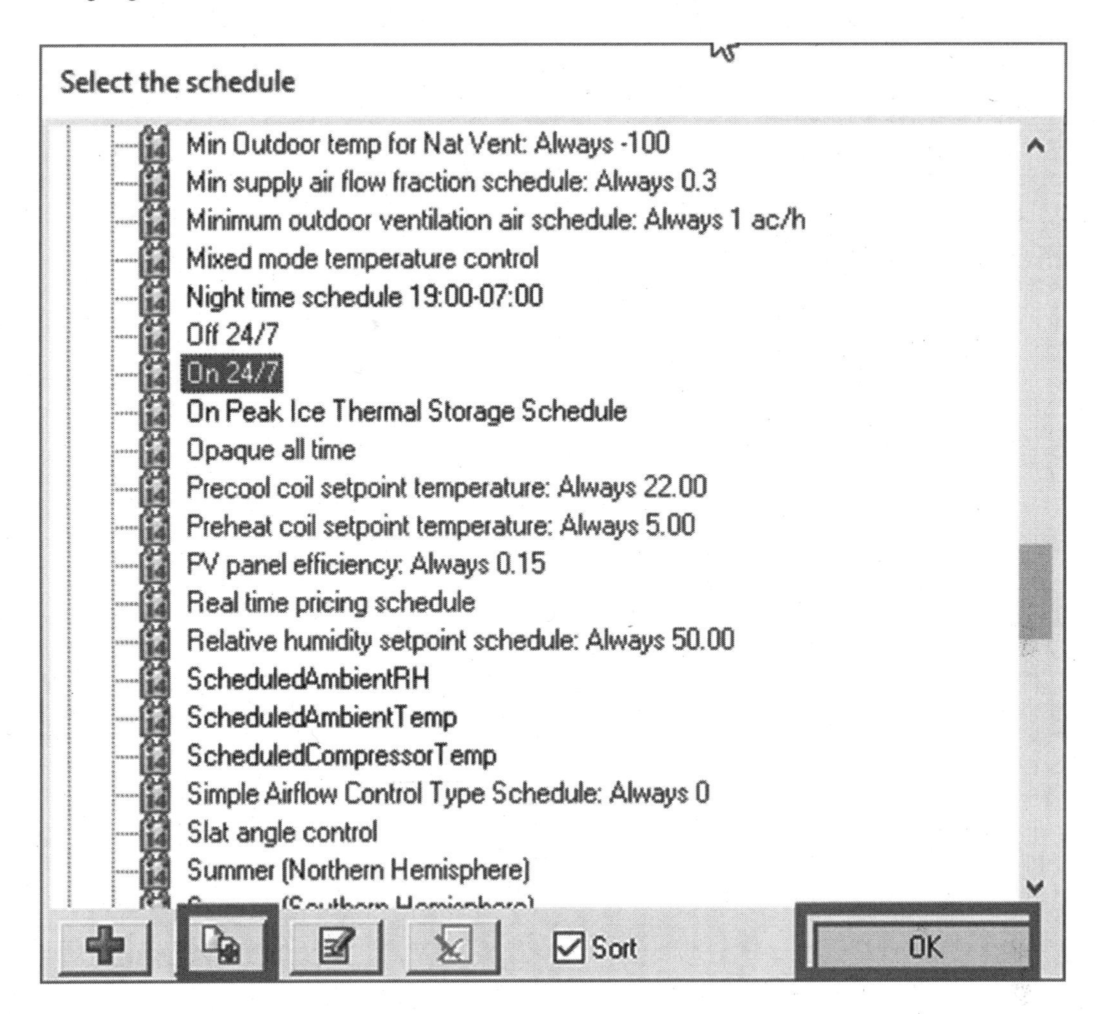

Step 8: Click the **Edit selected data** icon. The **Edit Schedule** screen appears.

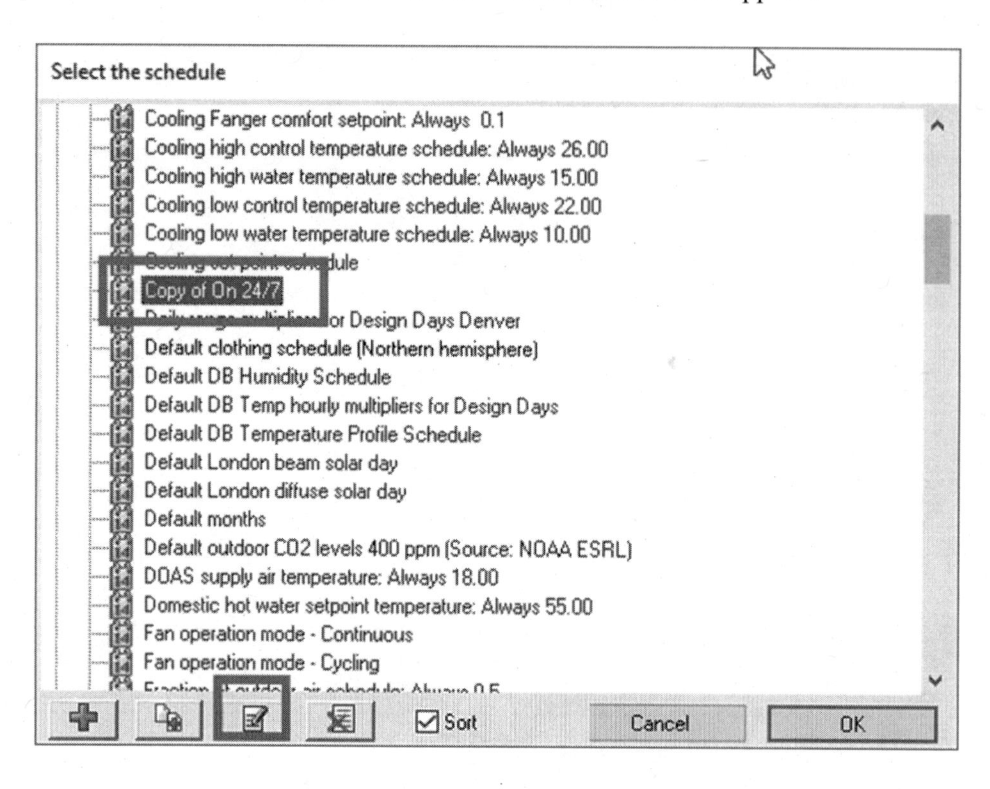

Step 9: In the **General** section, enter **Night Purge** in the **Name** box. Under the **Profiles** section, edit the schedule as shown below. Click **OK**.

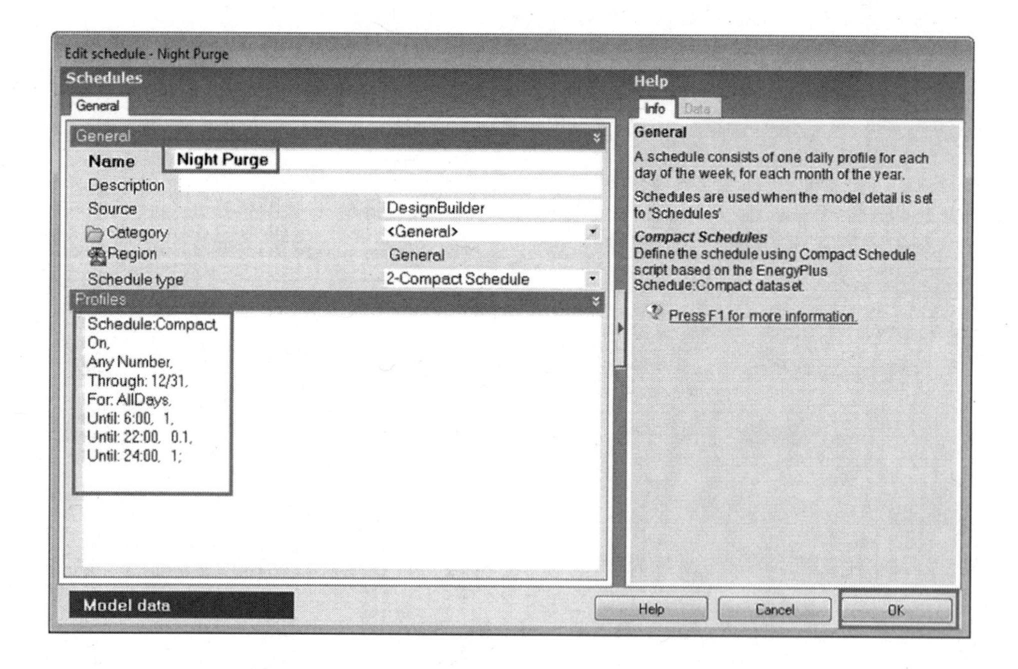

Step 10: Make sure that the **Night Purge** schedule is set.

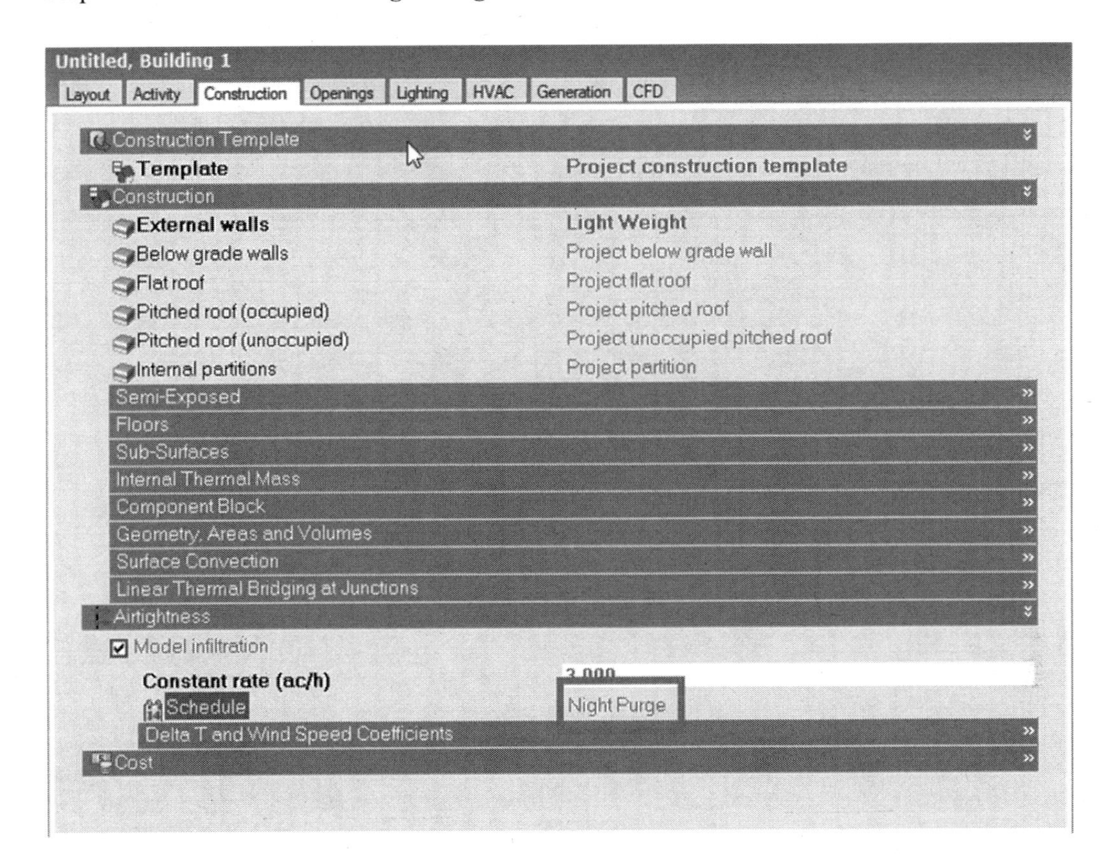

> *Night purge* is a technique used in conditioned buildings in which during unoccupied night hours, cool ambient air is passed through the building to flush out the heat released/accumulated in the building.

Step 11: Click **Flat roof element – 1250.000 m²** in the navigation tree, and select **Adiabatic** from the **Adjacency** drop-down list. Click **Building 1** in the navigation tree.

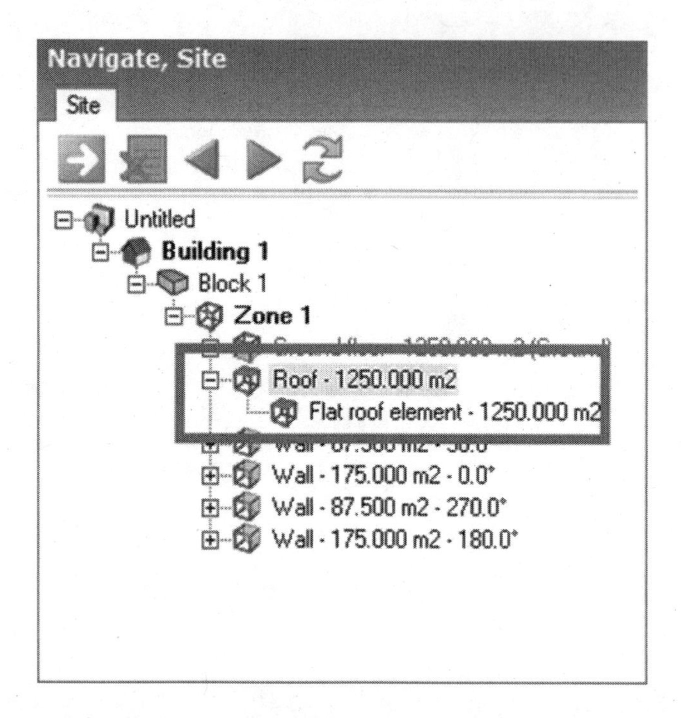

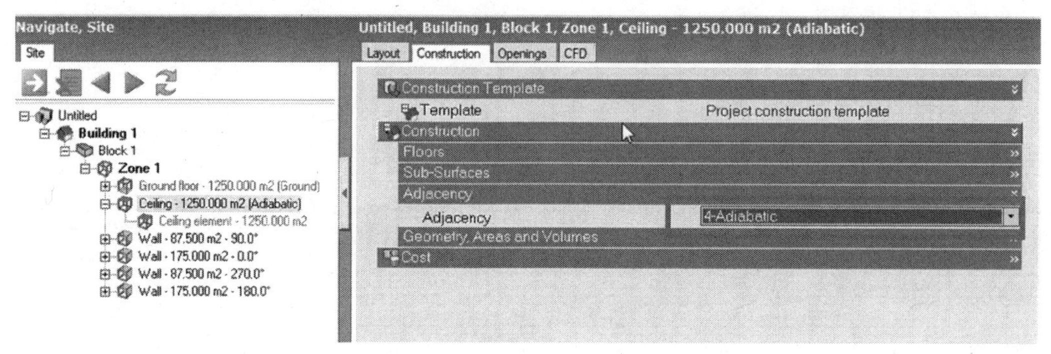

To understand the effect of the thermal mass of the external wall, we want to remove the heat gains and losses from the roof. To achieve this, the roof has been made adiabatic.

Adiabatic surface means that heat is not transferred across its external surface.

Source: https://designbuilder.co.uk/helpv6.0/#Adjacency.htm.

Step 12: Go to the **Building 1** level. Select the **Activity** tab, and select **Density (people/m^2)** as **0.06** and **Office Equipment Gain (W/m^2)** as **1.00**. (The internal load has been reduced so that the effect of the thermal mass is clearly visible. If the internal loads are higher, then the inside zone temperature is dominated by the internal load, and you cannot observe the effect of the thermal mass of the envelope.)

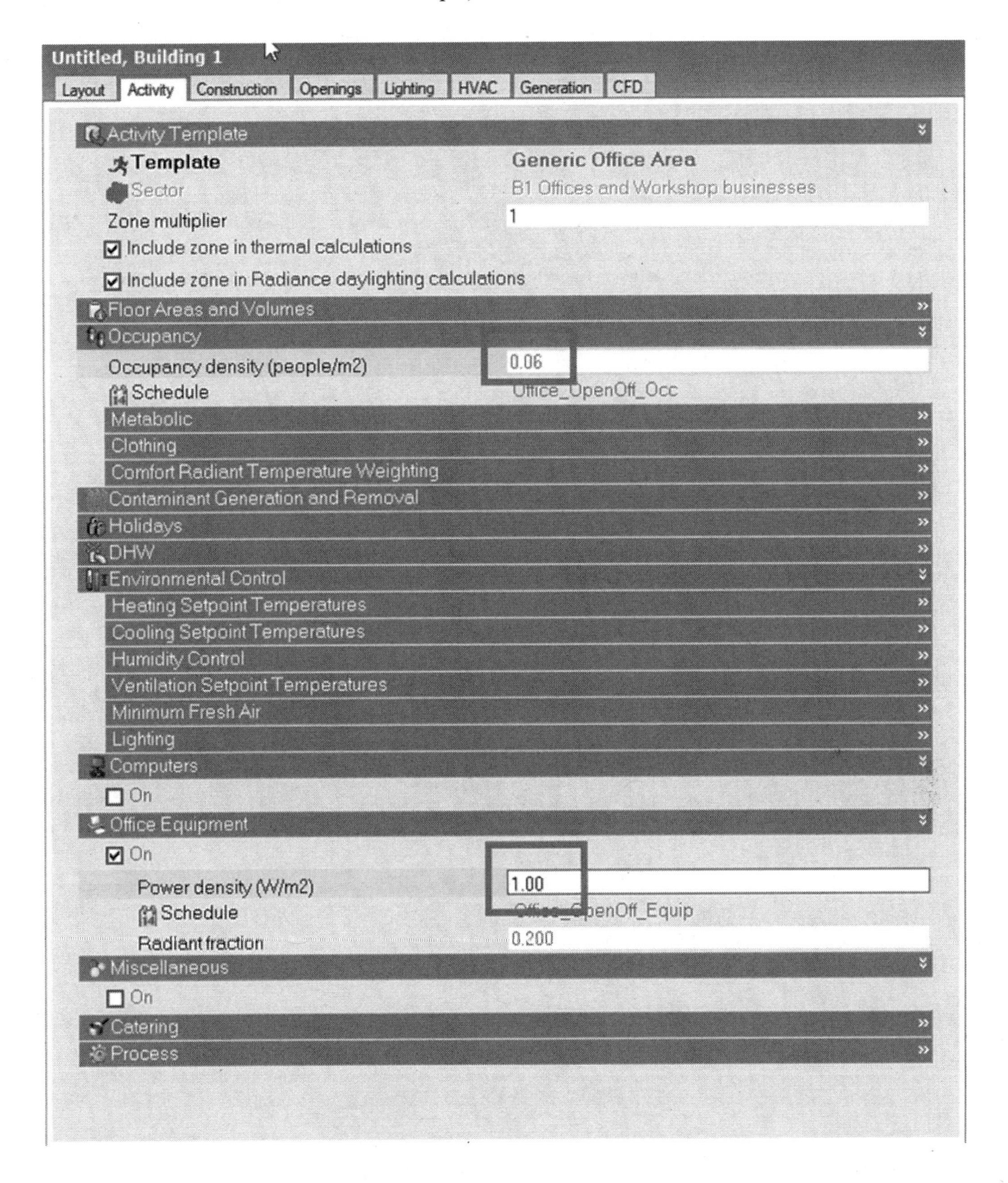

Step 13: Select the **Openings** tab, and ensure that **Preferred height** is selected and **window-to-wall percentage** is set as **0.00**.

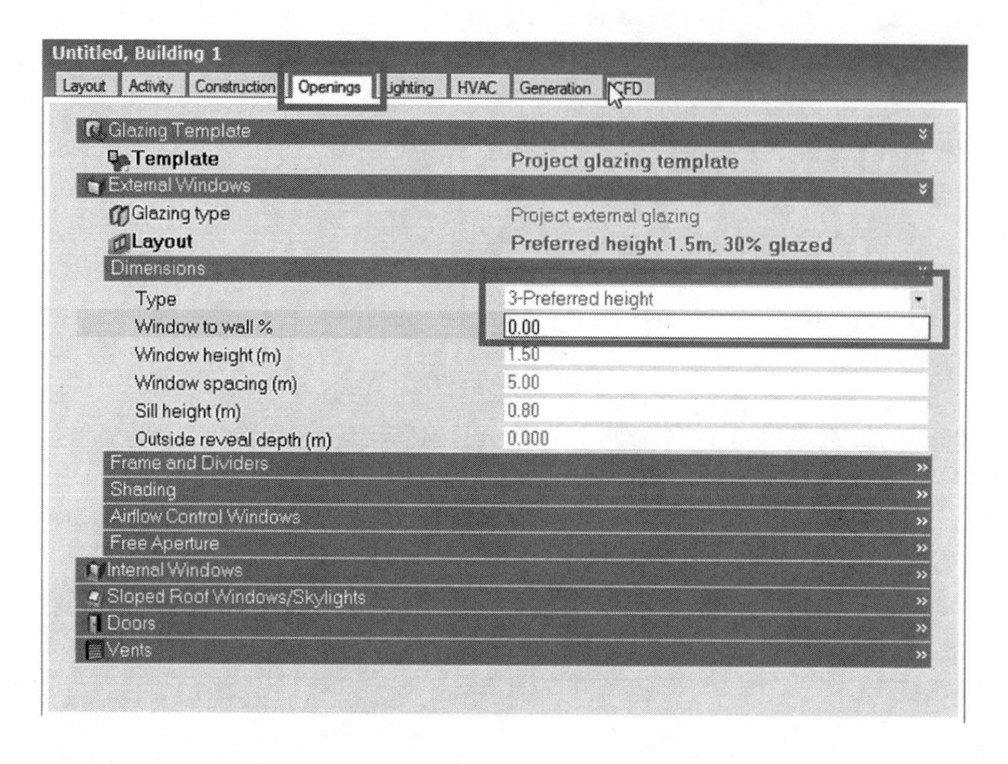

Step 14: Select the **Lighting** tab, and set the **Lighting energy (W/m^2-100 lux)** as **0.25**.

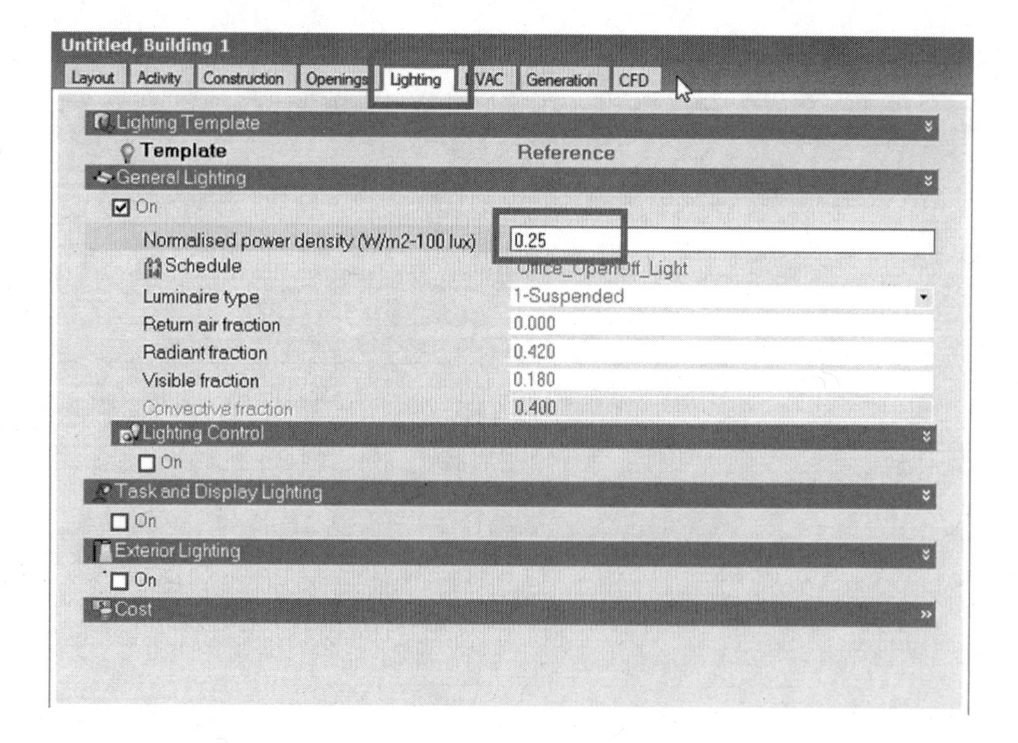

Step 15: Select the **HVAC** tab, and set the **template** as **<None>**. Also clear all checkboxes including **DHW** and **Natural Ventilation**.

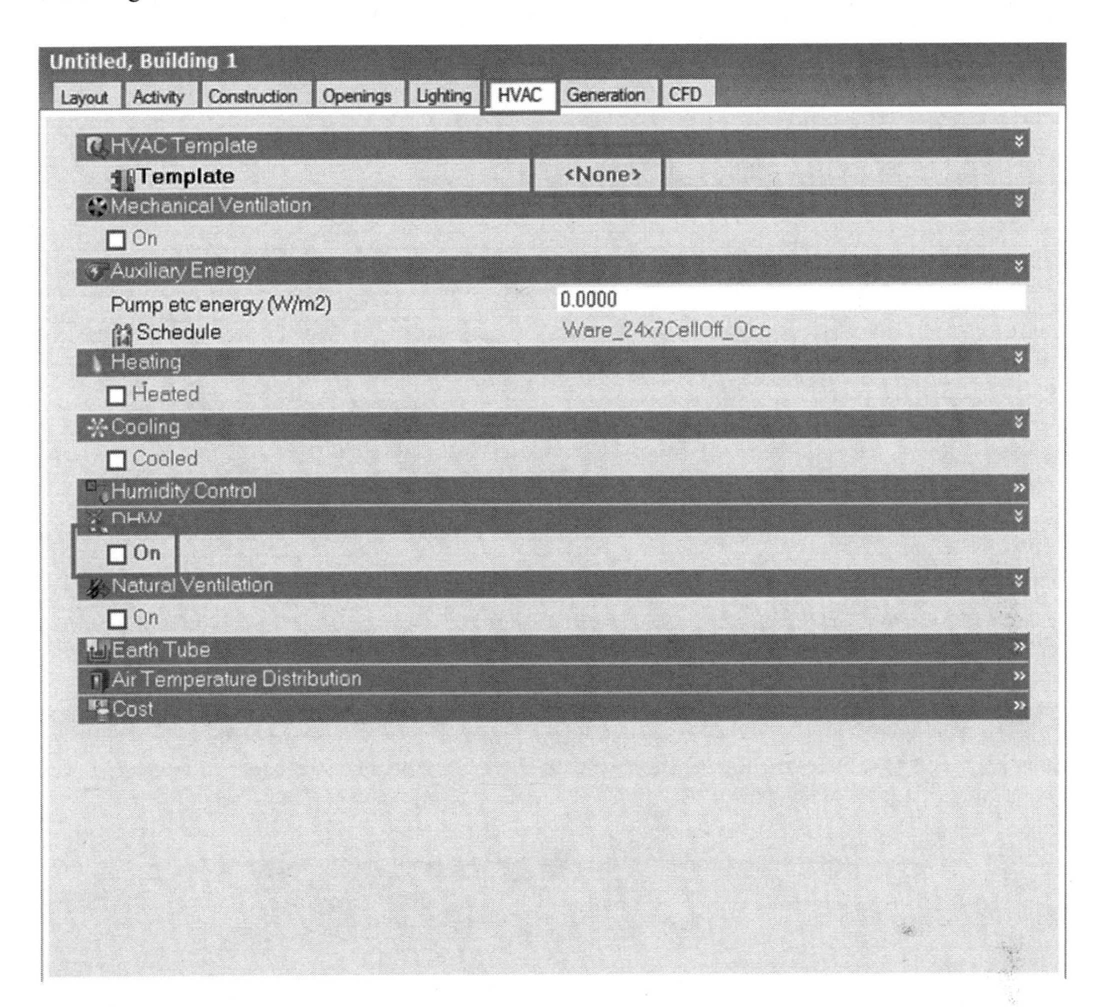

Step 16: Simulate the model for hourly interval reporting.

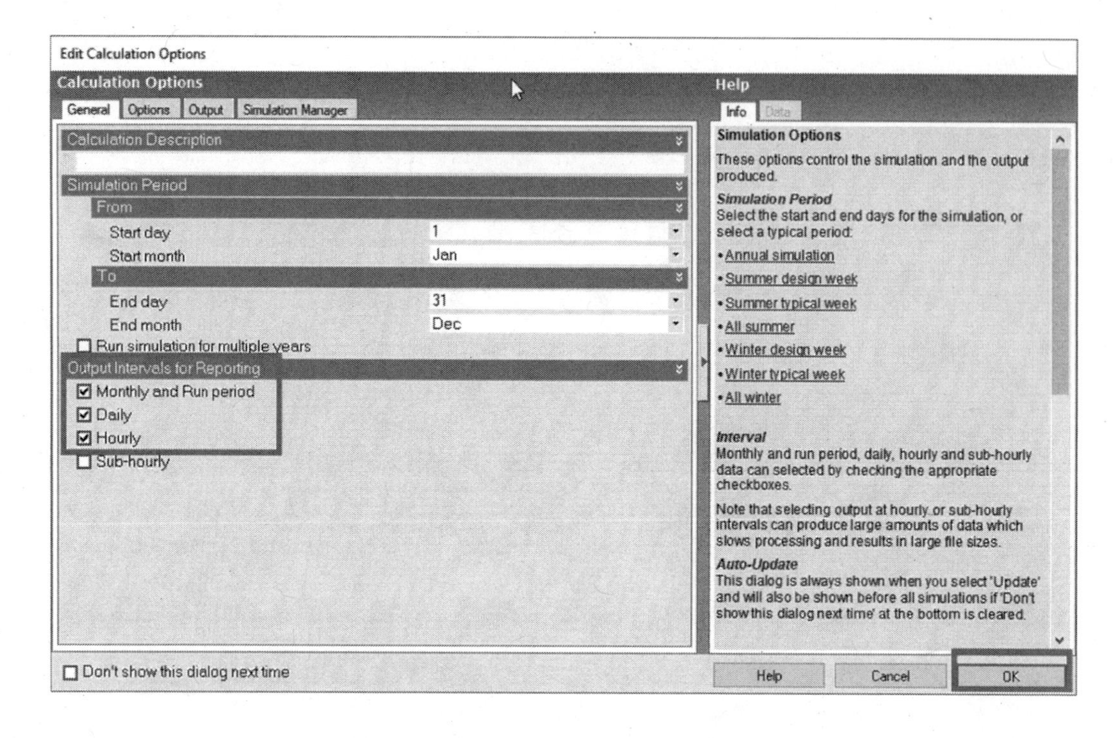

Step 17: Click **OK**. The results are displayed in the grid. You need to click on **Zone 1** to get the results at the zone level.

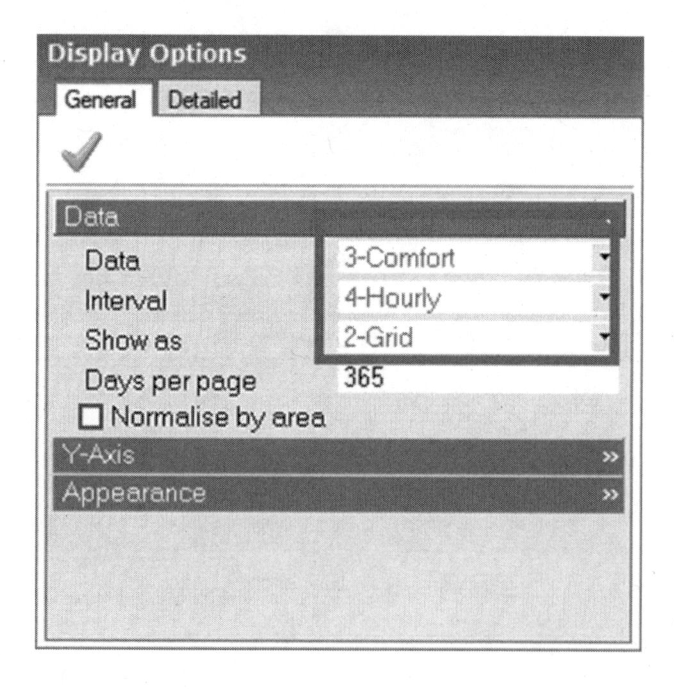

Untitled, Building 1

Analysis | Summary | Parametric | Optimisation + UA/SA | Data Visualisation

Date/Time	Air Temperature (°C)	Radiant Temperature (°C)	Operative Temperature (°C)	Outside Dry-Bulb Temperature (°C)	Relative Humidity (%)
01-Jan-02 1:00:00 AM	19.95737	22.60195	21.27966	16.75	67.63164
01-Jan-02 2:00:00 AM	19.18012	22.02603	20.60308	15.7	66.78936
01-Jan-02 3:00:00 AM	18.66659	21.57818	20.12239	15.225	68.14505
01-Jan-02 4:00:00 AM	18.22886	21.16847	19.69866	14.875	69.80441
01-Jan-02 5:00:00 AM	17.87271	20.79539	19.33405	14.5	71.07277
01-Jan-02 6:00:00 AM	17.55357	20.44208	18.99782	14.25	71.93153
01-Jan-02 7:00:00 AM	18.83224	20.23926	19.53575	13.825	66.30685
01-Jan-02 8:00:00 AM	19.69298	20.51493	20.10396	14.675	63.28962
01-Jan-02 9:00:00 AM	21.09112	21.41086	21.25099	16.8	59.69405
01-Jan-02 10:00:00 AM	23.24634	22.5772	22.91177	19.275	55.82804
01-Jan-02 11:00:00 AM	24.56033	23.5969	24.07861	22.225	55.40837
01-Jan-02 12:00:00 PM	25.55263	24.44096	24.9968	23.675	55.85431
01-Jan-02 1:00:00 PM	26.19285	25.0589	25.62588	24.575	56.92506
01-Jan-02 2:00:00 PM	26.72119	25.60469	26.16294	24.95	56.03151
01-Jan-02 3:00:00 PM	27.20889	26.0353	26.6221	25.75	54.77272
01-Jan-02 4:00:00 PM	27.55969	26.31138	26.93553	25.25	54.39156
01-Jan-02 5:00:00 PM	27.46161	26.24481	26.85321	24.25	56.24996
01-Jan-02 6:00:00 PM	26.6543	25.57448	26.11439	22.5	59.42007
01-Jan-02 7:00:00 PM	25.61065	24.87498	25.24282	21.4	63.07621
01-Jan-02 8:00:00 PM	24.68468	24.42342	24.55405	20.3	65.70653
01-Jan-02 9:00:00 PM	23.87904	24.07526	23.97715	20	67.56434
01-Jan-02 10:00:00 PM	23.21457	23.77371	23.49414	19.25	67.66742
01-Jan-02 11:00:00 PM	21.53504	23.39912	22.46708	19	69.85327
02-Jan-02	21.08634	22.99302	22.03968	19	69.28355
02-Jan-02 1:00:00 AM	20.26243	22.6403	21.45137	17.35	70.56285
02-Jan-02 2:00:00 AM	19.39335	22.13673	20.76504	16.125	69.91914
02-Jan-02 3:00:00 AM	18.68801	21.65969	20.17385	15.225	67.60561
02-Jan-02 4:00:00 AM	18.12729	21.20115	19.66422	14.55	68.30294
02-Jan-02 5:00:00 AM	17.69503	20.7783	19.23667	14.1	70.45697
02-Jan-02 6:00:00 AM	17.42261	20.4014	18.91201	14	70.7485
02-Jan-02 7:00:00 AM	18.70582	20.18133	19.44358	12.725	65.06431
02-Jan-02 8:00:00 AM	19.50633	20.38259	19.94446	13.575	62.313
02-Jan-02 9:00:00 AM	20.8487	21.13146	20.99008	15.5	58.92493
02-Jan-02 10:00:00 AM	22.63821	22.18693	22.41257	18.7	55.57981
02-Jan-02 11:00:00 AM	23.8565	23.144	23.50025	21.4	54.34824
02-Jan-02 12:00:00 PM	24.87431	23.78313	24.32872	22	54.63141
02-Jan-02 1:00:00 PM	25.14627	24.21216	24.67922	22.75	56.33355
02-Jan-02 2:00:00 PM	25.55248	24.61541	25.08395	23	56.13988
02-Jan-02 3:00:00 PM	26.05913	25.04504	25.55208	23.75	55.26085
02-Jan-02 4:00:00 PM	26.44112	25.35258	25.89685	23.175	55.13739
02-Jan-02 5:00:00 PM	26.38009	25.26748	25.82378	22.225	56.85746
02-Jan-02 6:00:00 PM	25.57104	24.69571	25.13338	21.25	59.3655
02-Jan-02 7:00:00 PM	24.60911	24.0734	24.34126	20.475	61.83529
02-Jan-02 8:00:00 PM	23.7159	23.67904	23.69747	20.075	63.94576
02-Jan-02 9:00:00 PM	22.9479	23.36686	23.15743	19.25	65.42835

Step 18: Click the **Export Data** icon. The **Export Results Spreadsheet** dialog box appears. You can save this results file on your desktop to retrieve it easily.

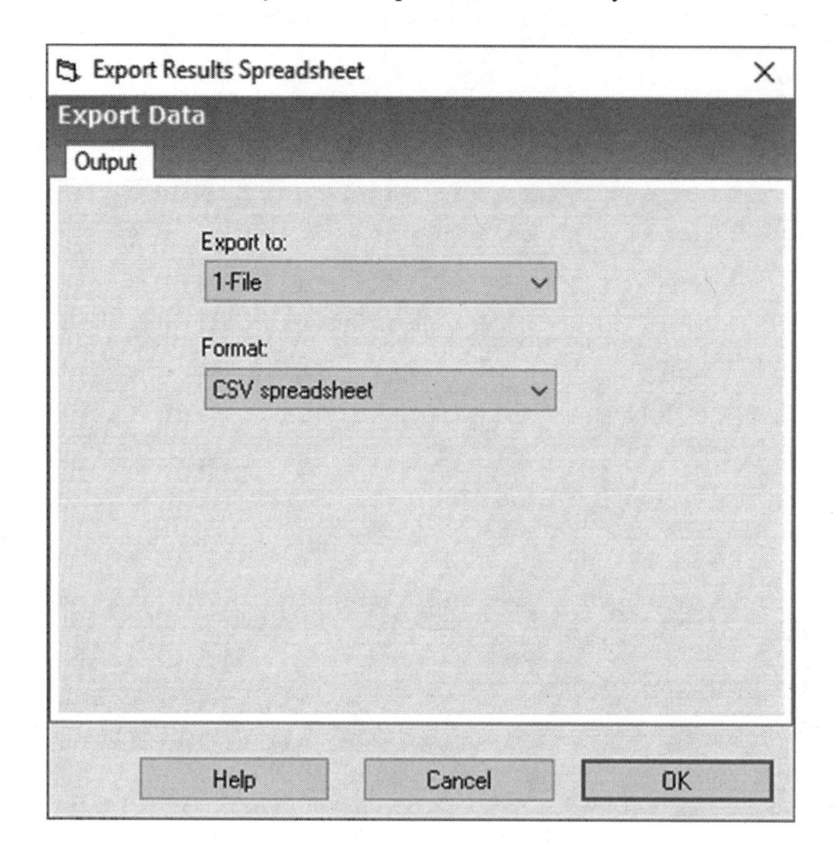

Step 19: Repeat the preceding steps to create a high-thermal-mass external wall.

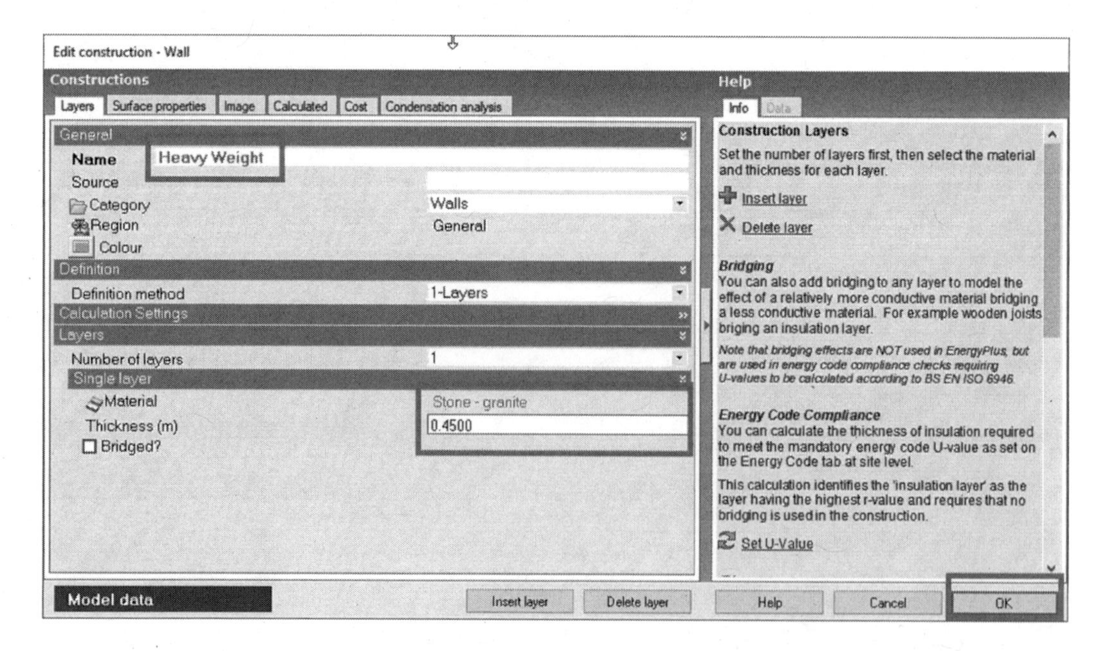

Step 20: Perform an hourly simulation, and record the results. Compare the indoor air temperature for both cases (5th and 6th of June) with the outside dry-bulb temperature.

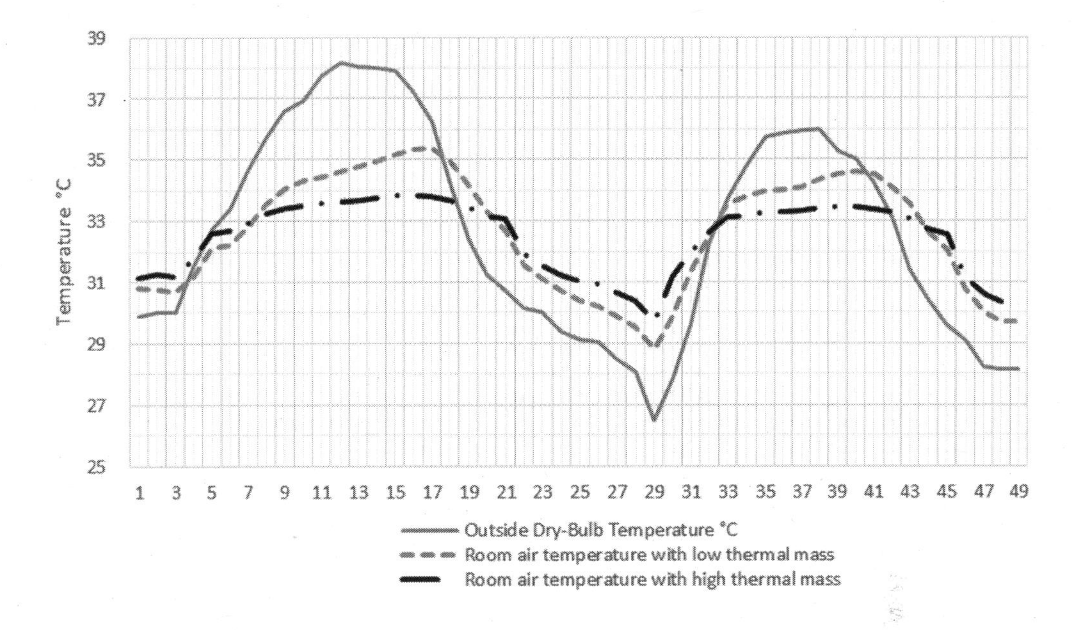

- As shown in the preceding figure, the temperature difference between the indoor air and outside air is higher for high-thermal-mass buildings. This is because building walls with a high thermal mass tend to store and increase the time taken for heat to transfer from the outside to the inside.
- The room air temperature in a high-thermal-mass building has less swing than the room air temperature in a low-thermal-mass building, which follows the outside temperature pattern.
- The occurrence of the highest room air temperature in a high-thermal-mass building is not at the same time when the outside air temperature peaks. Rather, it is shifted to a later time of day. This difference in time when the peaks are observed in room temperature and the outside air temperature is defined as the *thermal lag*.

EXERCISE 3.1

Repeat the tutorial with insulation on the lightweight external wall. Compare the results with and without insulation on the lightweight wall.

Tutorial 3.2

Evaluating the Impact of Roof Insulation

GOAL

To study the effect of roof insulation on building energy consumption

WHAT ARE YOU GOING TO LEARN?

- How to create roof construction
- How to set the *U*-value of a roof

PROBLEM STATEMENT

In this tutorial, you are going to use a 50- × 25-m model. Construct a roof with a 100-mm aerated concrete slab and glass fibre slab insulation of varying thicknesses. Achieve the *U*-values given in Table 3.1 by varying the insulation thickness. Find out the energy consumption for each thickness. Use the weather file for **WIEN/SCHWECHAT- FLUG, AUSTRIA**.

TABLE 3.1
***U*-Values and R-values of a roof**

Serial no.	*U*-value (W/m^2-K)	*R*-value (m^2-K/W)	S. No.	*U*-value (W/m^2-K)	*R*-value (m^2-K/W)
1	1	1	10	0.31	3.25
2	0.8	1.25	11	0.29	3.5
3	0.67	1.50	12	0.27	3.75
4	0.57	1.75	13	0.25	4.00
5	0.50	2.00	14	0.24	4.25
6	0.44	2.25	15	0.22	4.50
7	0.40	2.50	16	0.21	4.75
8	0.36	2.75	17	0.20	5.00
9	0.33	3.00			

***R*-Value or Thermal Resistance *R*.** The *R*-value of any section having one or more layers with parallel surfaces is an indication of the resistance offered by the section to heat flow. It is the reciprocal of thermal conductance. For a structure having plane parallel faces, the thermal resistance is equal to the thickness *L* of the structure divided by the thermal conductivity *k*:

$$R = \frac{L}{k} \tag{3.1}$$

The *R*-value of individual layers can be added to arrive at the total *R*-value of the section. It is also expressed as the ratio of the temperature difference across an insulator and the heat flux (the heat transfer per unit area per unit time).

Thermal transmittance (*U*-Factor) – Thermal Transmittance (*U*). The thermal transmission in the unit time through the unit area of the given section divided by the temperature difference between the fluids on either side of the building unit in steady-state conditions. It is also called the *U-value.* Its unit is W/m^2-K. It can be treated as a measure of the heat loss through the unit area of a building section such as a wall, floor or roof. A low *U*-value generally indicates high levels of insulation.

SOLUTION

Step 1: Open a new project and create a **50- × 25-m** building.

Step 2: Select the **Construction** tab.

Step 3: Go to the **Construction** section, click **Flat roof**, and then click the **Add new item** icon. The **Edit construction – Roof** screen appears.

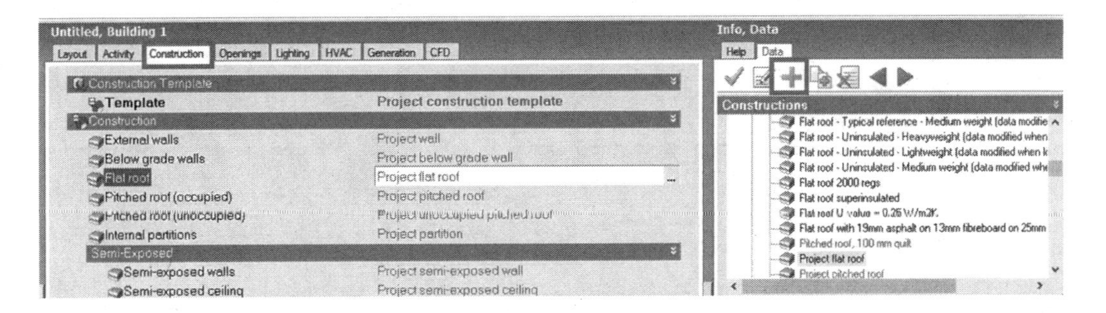

Step 4: Select **2** from the **Number of layers** drop-down list.

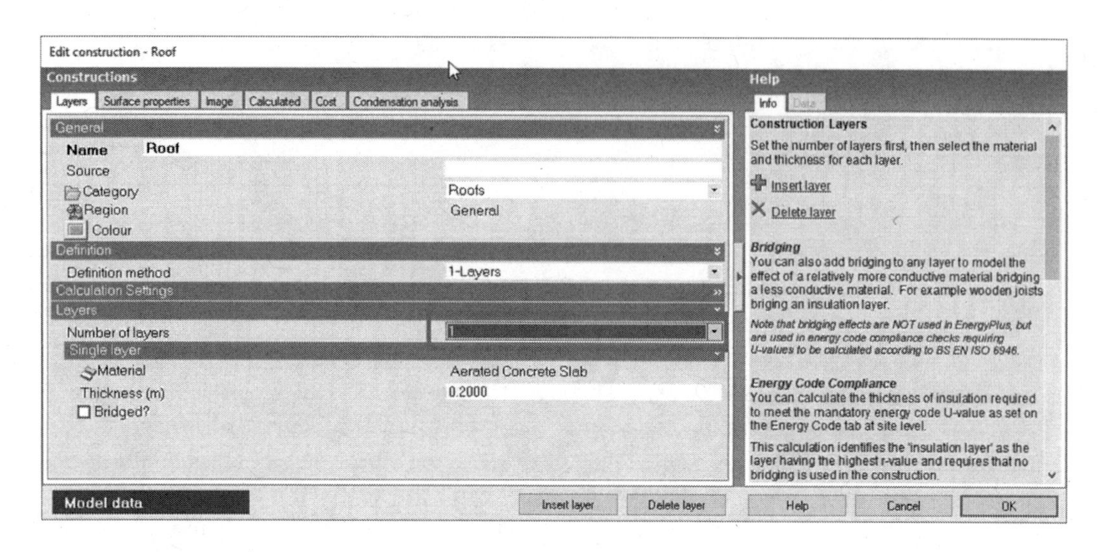

Step 5: Select **Aerated Concrete Slab** as the **Material,** and enter **0.1000** as the **Thickness (m)** in the **Outermost layer** section. Select **Glass Fibre Slab** from **Insulating materials**, and enter **0.2000** as the **Thickness (m)** for the innermost layer.

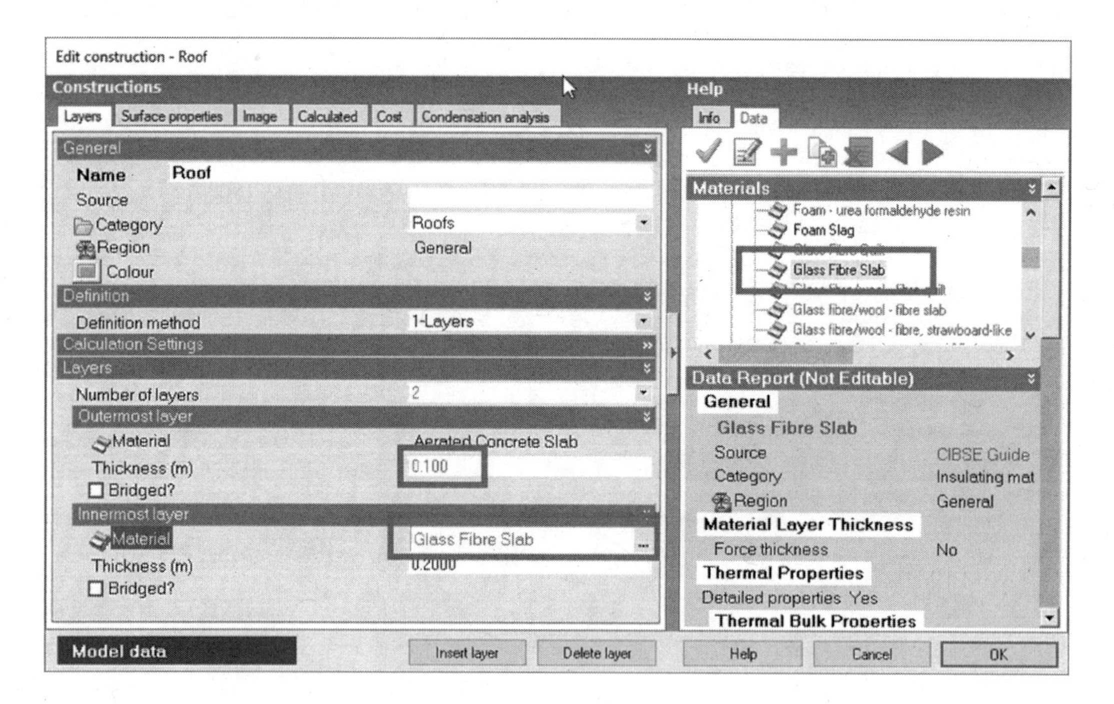

Step 6: Click anywhere in the blank space under the **Innermost layer** section to update the help section.

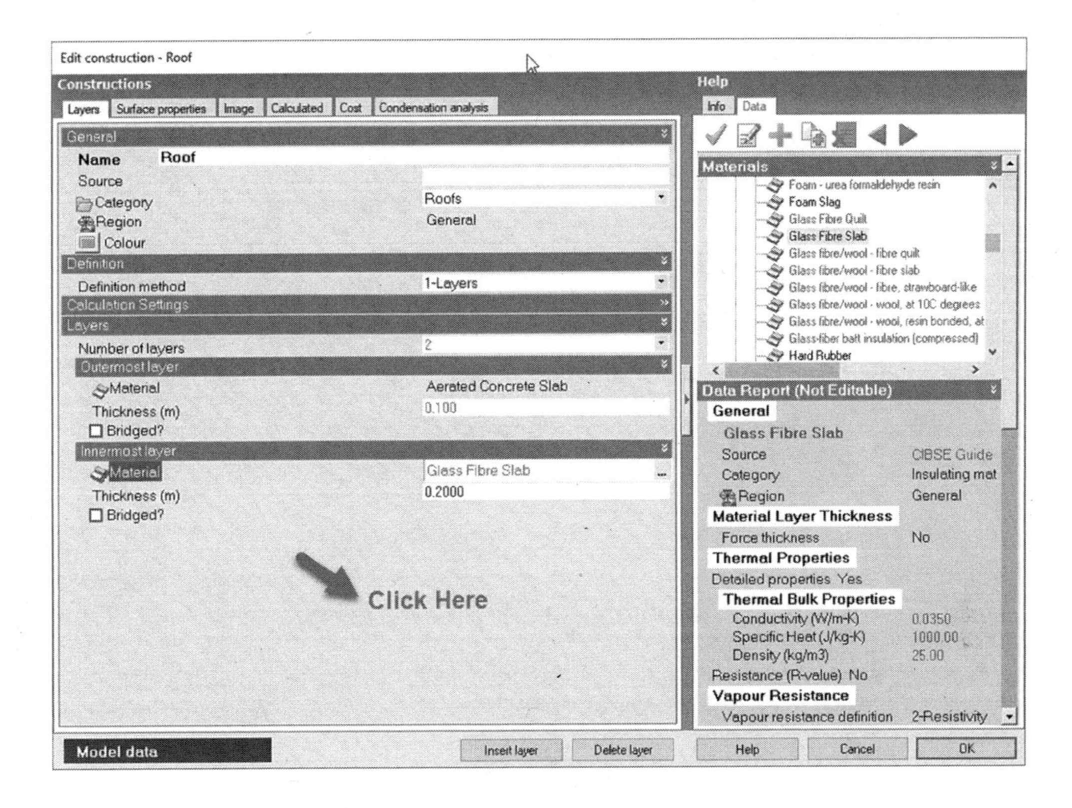

Step 7: Click the **Set U-value** link. The **Set Construction U-value** screen appears.

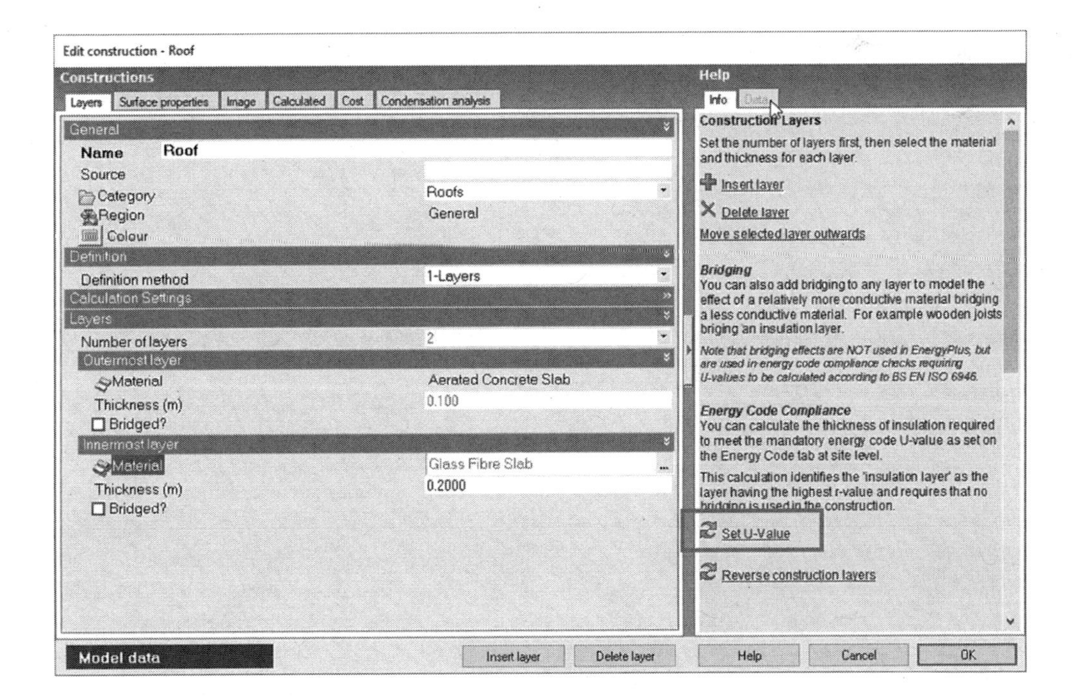

Step 8: Enter the **U-value (W/m^2-K)** as **1.0**. Click **OK**. A confirmation message appears with the updated insulation thickness.

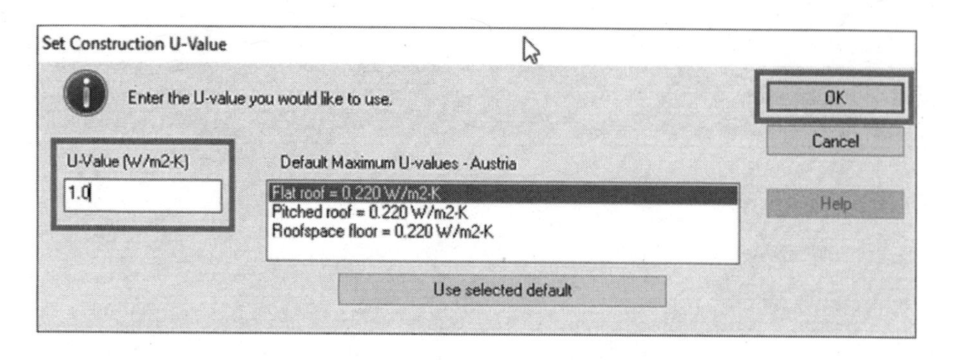

Step 9: Click **OK**. The insulation thickness is updated.

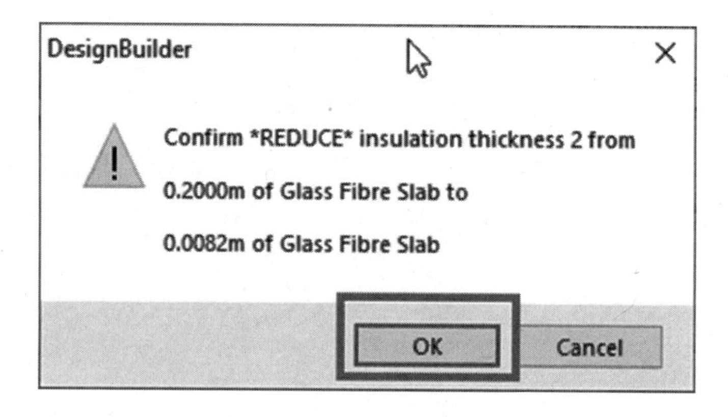

Step 10: Select the **Calculated** tab. The updated **U-value** of the construction appears.

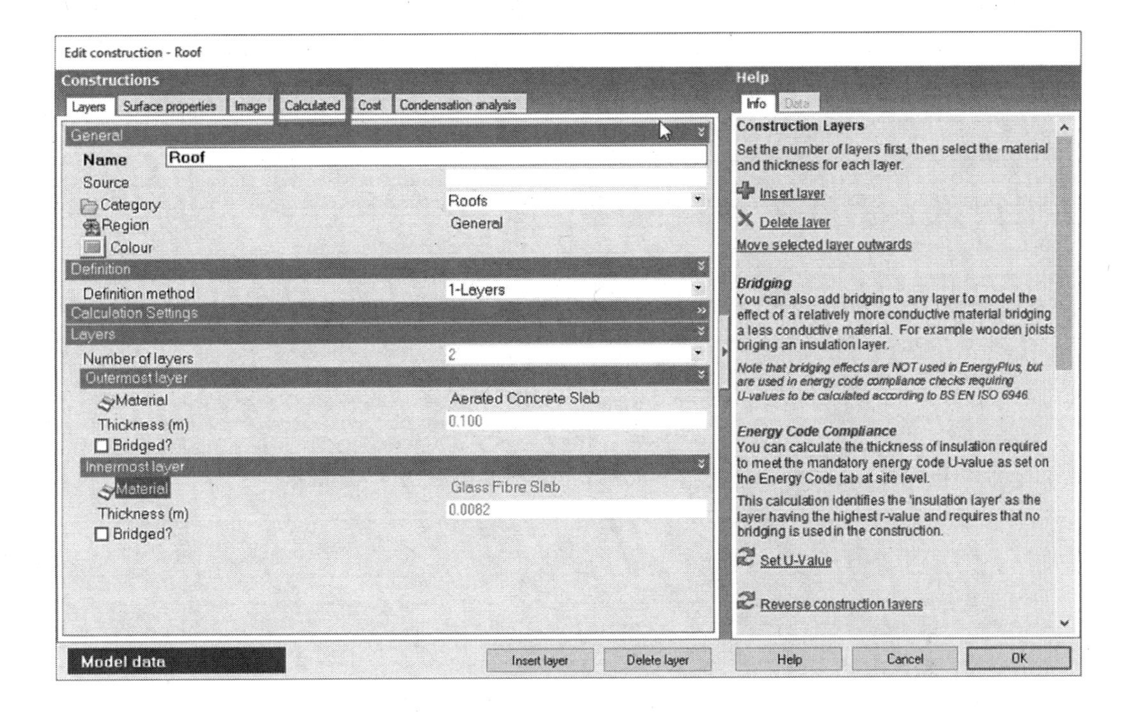

Step 11: Click **OK**.

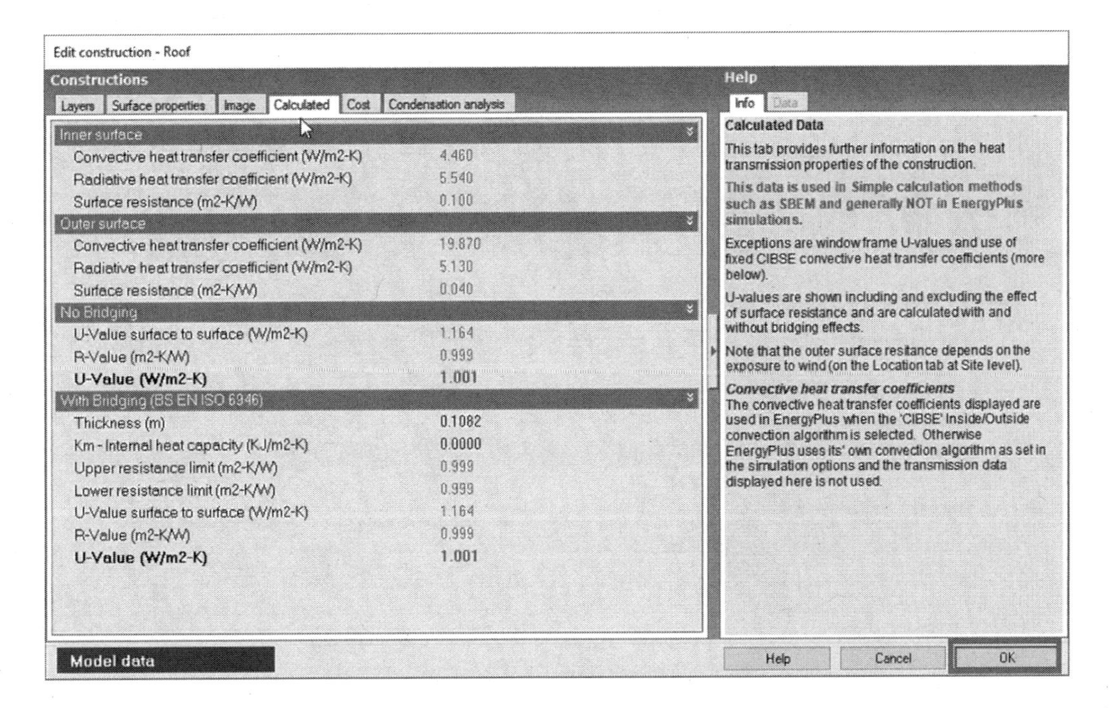

Step 12: Perform an annual energy simulation, and record the results.

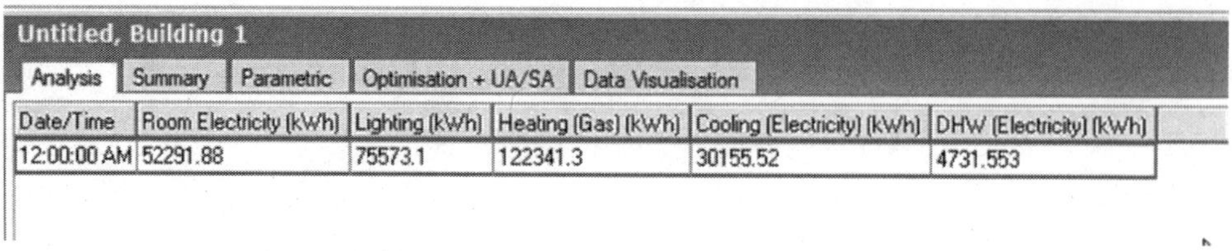

Untitled, Building 1

Analysis | Summary | Parametric | Optimisation + UA/SA | Data Visualisation

Date/Time	Room Electricity (kWh)	Lighting (kWh)	Heating (Gas) (kWh)	Cooling (Electricity) (kWh)	DHW (Electricity) (kWh)
12:00:00 AM	52291.88	75573.1	122341.3	30155.52	4731.553

Step 13: Repeat the preceding steps to set the *U*-values as given in Table 3.1. For each *U*-value, simulate and record the results. Compare the results for all simulations (Table 3.2).

TABLE 3.2

Heating and cooling energy consumption for different *U*-values

Serial no.	*U*-value (W/ m^2-K)	*R*-value (m^2-K/W)	Total heating consumption (gas) (kWh)	Total cooling consumption (electricity) (kWh)
1	1.00	1.00	122,341.30	30,155.52
2	0.80	1.25	108,081.90	30,445.38
3	0.67	1.50	98,859.23	30,782.29
4	0.57	1.75	91,779.54	31,121.71
5	0.50	2.00	86,764.41	31,407.00
6	0.44	2.25	82,456.66	31,681.44
7	0.40	2.50	79,626.97	31,884.26
8	0.36	2.75	76,730.62	32,096.06
9	0.33	3.00	74,553.87	32,270.18
10	0.31	3.25	73,122.60	32,387.88
11	0.29	3.50	71,667.70	32,512.29
12	0.27	3.75	70,224.38	32,634.77
13	0.25	4.00	68,803.48	32,761.72
14	0.24	4.25	68,075.55	32,827.73
15	0.22	4.50	66,650.73	32,960.28
16	0.21	4.75	65,932.87	33,025.67
17	0.20	5.00	65,225.09	33,091.81

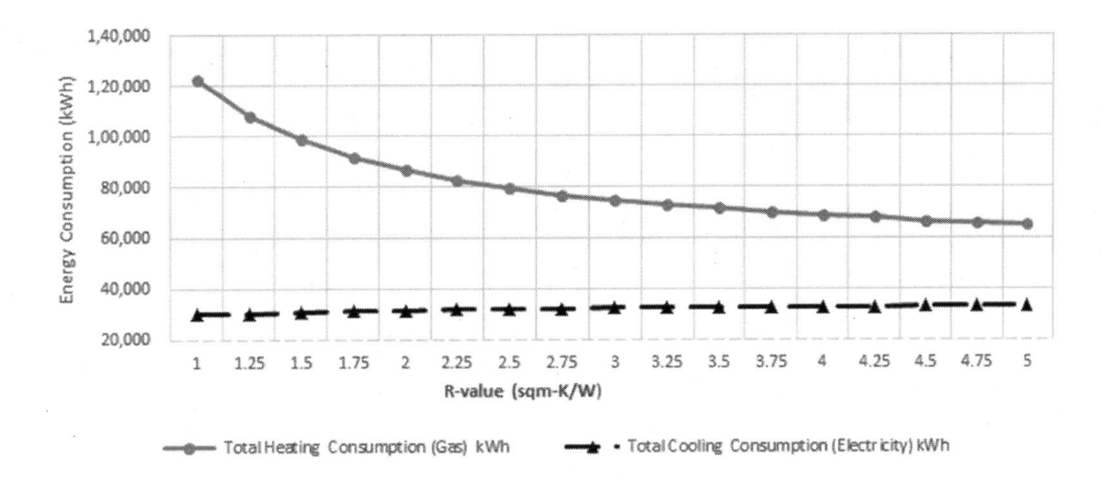

- In this scenario, the impact of insulation on the heating energy consumption is more than that on the cooling energy consumption.
- The law of diminishing returns can be seen here, as the heating energy consumption does not proportionally decrease with the increase in insulation thickness.

EXERCISE 3.2

Repeat the preceding steps with the Miami, Florida, location.

Tutorial 3.3

Evaluating the Impact of the Position of Roof Insulation

GOAL

To evaluate the impact of the position of roof insulation (overdeck and underdeck)

WHAT ARE YOU GOING TO LEARN?

- How to edit roof construction

PROBLEM STATEMENT

In this tutorial, you are going to use a 50- × 25-m model. Use the construction layers as given in Table 3.3. Find out the energy consumption for each variation. Use the weather data for **FRANKFURT MAIN ARPT, Germany** and **DUBAI INTERNATIONAL, United Arab Emirates**.

TABLE 3.3
Construction layers

Layer	Case I	Case II
Outermost	Concrete, medium density, 0.15 m	XPS extruded polystyrene – HFC blowing, 0.05 m
Innermost	XPS extruded polystyrene – HFC blowing, 0.05 m	Concrete, medium density, 0.15 m

In most locations, except for very high latitudes, the external surface of the roof is directly exposed to solar radiation for the longest duration compared with other surfaces of the building. The solar radiation, upon being absorbed by the external surface of the roof, turns into heat that subsequently gets transmitted into the rooms below through the roof slab. The slab, because of its thermal mass, also accumulates heat that continues to be transmitted even after sunset because of the temperature difference between the slab and room interiors. To avoid this transmission and accumulation of heat, insulation on the top surface is required in the form of *overdeck insulation*. A reduction in heat transmission via the use of insulation results in low energy consumption for operating cooling devices.

If insulation is provided on the inner surface of the roof, it is termed *underdeck insulation*. It results in decreasing energy consumption for cooling devices because it reduces the radiant heat from entering the rooms.

In locations requiring heating of buildings, the reverse is the approach. Insulation is provided on the inner side for reducing heat flow from the inside to the outside. As a concept, insulation should be put as early as possible in the path of a heat flow whether from outside to inside or from inside to outside.

SOLUTION

Step 1: Open a new project. Create a **50- × 25-m** building. Change the weather location to Frankfurt, Germany.

Step 2: Select the **Construction** tab.

Step 3: Go to the Construction section, click **Flat roof**, and then click the **Add new item** icon. The **Edit construction – Roof** screen appears. Add a roof with underdeck insulation: **0.15 m** of **Concrete, Medium density** (from **Concrete materials**) and **0.05 m** of **XPS Extruded Polystyrene – HFC Blowing** (from **Insulating materials**).

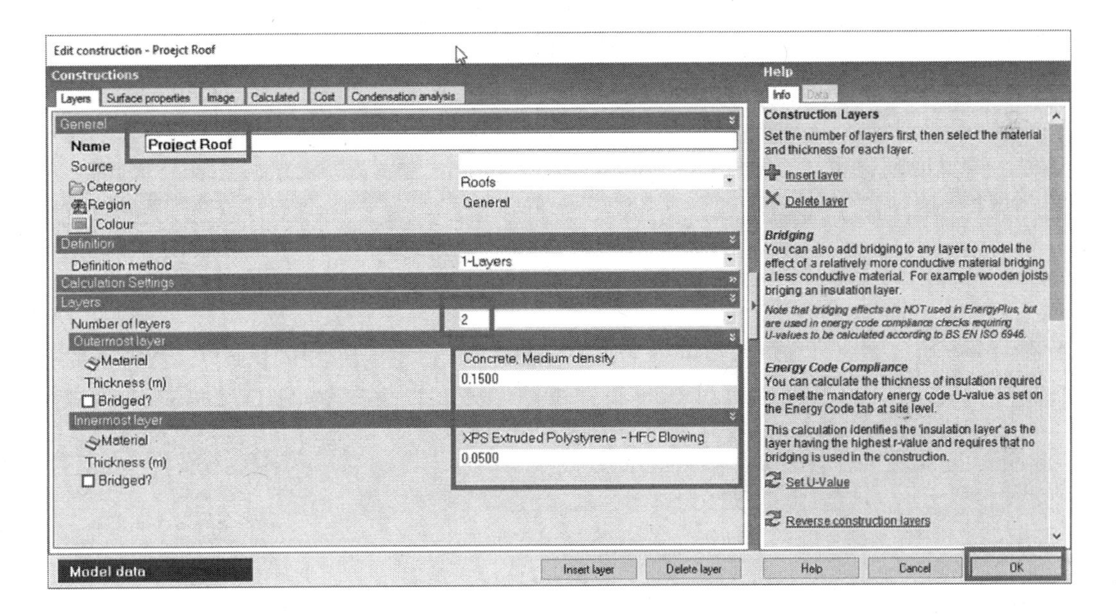

Step 4: Perform an annual simulation, and record the results.

Untitled, Building 1

Analysis | Summary | Parametric | Optimisation + UA/SA | Data Visualisation

Date/Time	Room Electricity (kWh)	Lighting (kWh)	Heating (Gas) (kWh)	Cooling (Electricity) (kWh)	DHW (Electricity) (kWh)
12:00:00 AM	52291.88	75573.1	77888.68	25347.96	4731.553

Step 5: Repeat the tutorial for overdeck insulation.

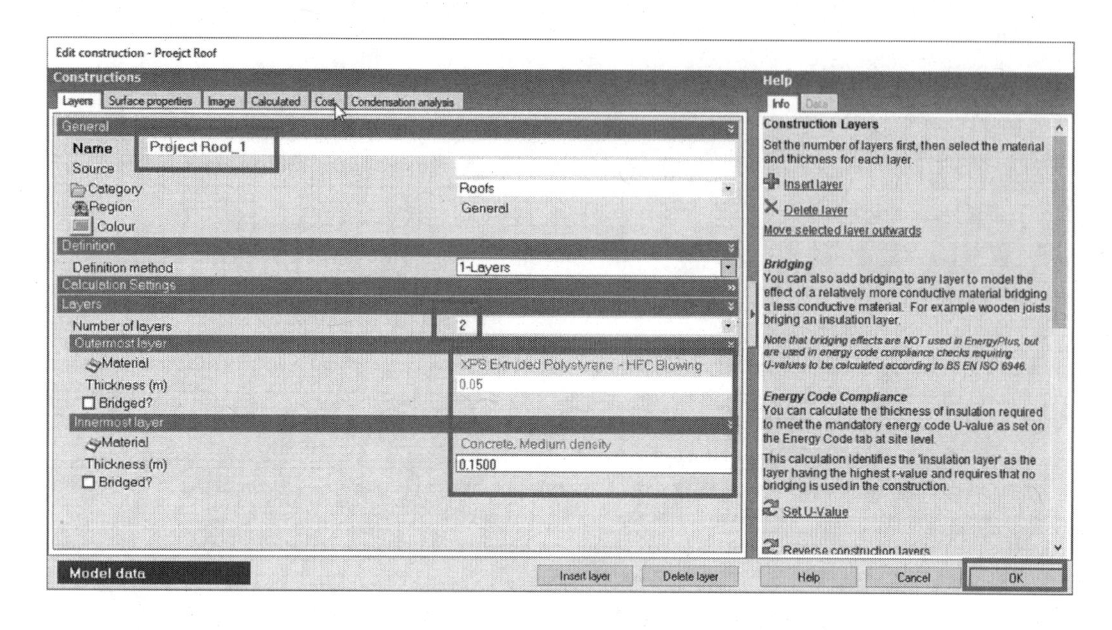

Step 6: Simulate the model and compare the results for the two cases (Table 3.4). Frankfurt requires predominantly heating and underdeck insulation, preventing the internal heat from being absorbed by the slab mass and thereby reducing the heating energy consumption.

TABLE 3.4

Annual fuel breakdown for underdeck and overdeck roof insulation for Frankfurt location

Type	With underdeck roof insulation (kWh)	With overdeck roof insulation (kWh)
Room electricity	52,291.88	52,291.88
Lighting	75,573.10	75,573.10
Heating (gas)	77,888.68	83,326.80
Cooling (electricity)	25,347.96	21,448.00
DHW (electricity)	4,731.55	4,731.55

Step 7: Change the weather location to **DUBAI INTERNATIONAL, United Arab Emirates** and compare the energy consumption for overdeck and underdeck insulation (Table 3.5). Cooling is the predominant requirement in Dubai. Overdeck insulation prevents the external heat from being absorbed by the slab mass, thereby reducing cooling energy consumption.

TABLE 3.5
Annual fuel breakdown for underdeck and overdeck roof insulation for Dubai location

Type	With underdeck roof insulation (kWh)	With overdeck roof insulation (kWh)
Room electricity	52,291.88	52,291.88
Lighting	75,573.10	75,573.10
Heating (gas)	656.09	70.50
Cooling (electricity)	171,933.20	169,859.30
DHW (electricity)	4,731.55	4,731.55

EXERCISE 3.3

Repeat this tutorial for the external wall. Use the weather data for **FRANKFURT MAIN ARPT, Germany** and **DUBAI INTERNATIONAL, United Arab Emirates** (Table 3.6).

TABLE 3.6
Construction layers for external walls

Layer	Case I	Case II
Outermost	Brickwork 230 mm	XPS extruded polystyrene – HFC blowing, 0.05 m
Innermost	XPS extruded polystyrene – HFC blowing, 0.05 m	Brickwork 230 mm

Tutorial 3.4

Evaluating the Impact of the Air Gap between Roof Layers

GOAL

To evaluate the effect of the air-gap thickness in roof construction on energy consumption

WHAT ARE YOU GOING TO LEARN?

- How to add the air gap between the roof layers
- How to change the air-gap thickness

PROBLEM STATEMENT

In this tutorial, you are going to use a 50- × 25-m model with a roof consisting of the following layers (starting with the outermost layer):

1. 0.01 m of cement/plaster/mortar-plaster
2. 0.15 m of concrete, medium density
3. Air gap with varying thicknesses (as given below)
4. 0.01 m of cement/plaster/mortar-plaster

Thickness of the air gap:

1. No air gap
2. Air gap of 15 mm (downwards)
3. Air gap of 17 mm (downwards)
4. Air gap of 25 mm (downwards)
5. Air gap of 50 mm (downwards)
6. Air gap of 100 mm (downwards)
7. Air gap of 300 mm (downwards)

Find out the energy consumption for **LONDON/GATWICK ARPT, United Kingdom**.

An air gap is different from the airspace layer. An *air gap* is enclosed on either side, whereas an *airspace* is a gap left between exterior finish layers and interior insulation layers – as commonly practiced in lightweight construction. The main role of the airspace layer is to act as a vapour and water drain/barrier compared with an air gap, which improves the overall insulation properties of the wall/roof section.

SOLUTION

Step 1: Open a new project. Create a **50- × 25-m** building.

Step 2: Select the **Construction** tab.

Step 3: Add a new roof with three layers. Select **Cement/plaster/mortar – cement plaster** from **Plaster** materials with thickness (m) as **0.0100, Concrete, Medium density** from **Concretes** materials with thickness (m) as **0.1500 m**, and select **Cement/plaster/mortar – cement plaster** from **Plaster** materials with thickness (m) as **0.0100 m**.

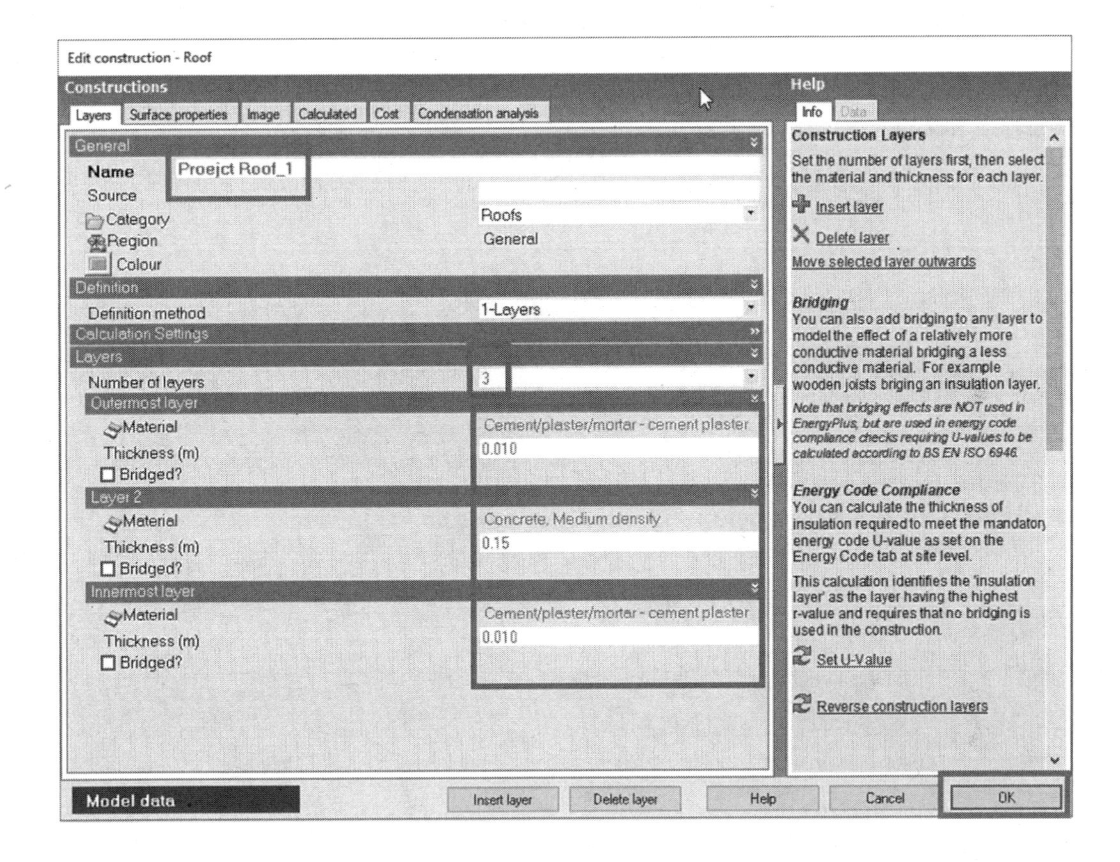

Step 4: Simulate the model, and record the results.

Untitled, Building 1

Analysis | Summary | Parametric | Optimisation + UA/SA | Data Visualisation

Date/Time	Room Electricity (kWh)	Lighting (kWh)	Heating (Gas) (kWh)	Cooling (Electricity) (kWh)	DHW (Electricity) (kWh)
12:00:00 AM	52291.88	75573.1	210243.3	10039.44	4731.553

Step 5: Again, select the **Construction** tab. Edit the roof construction.

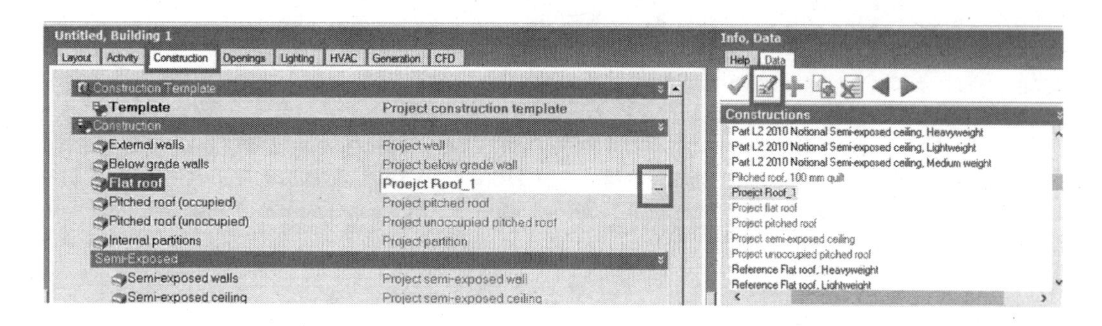

Step 6: Select the **Thickness** of the **Innermost layer** (we want to insert a new layer above this layer; hence, we have to select this layer). Click **Insert layer**. The **Edit Insert Material Layer** screen appears. Click **OK** to insert a new layer.

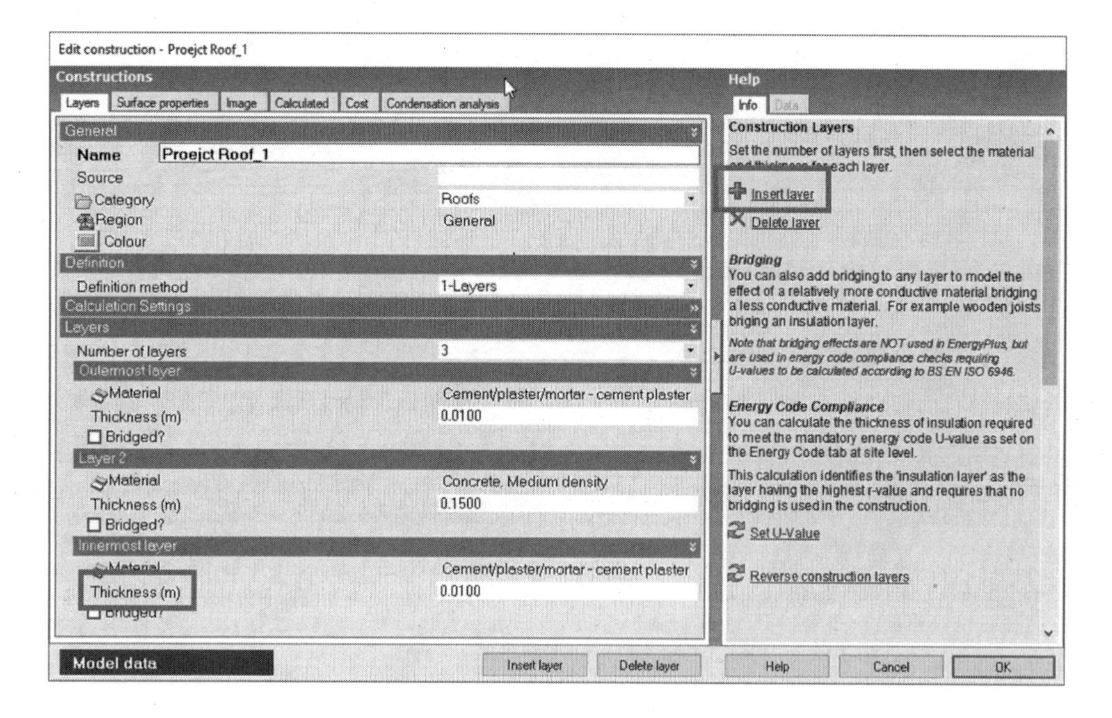

Step 7: Select **Air gap 15mm (downwards)** from **Gases** materials for the newly inserted layer (**Layer 3**), and enter **0.001** as thickness (m).

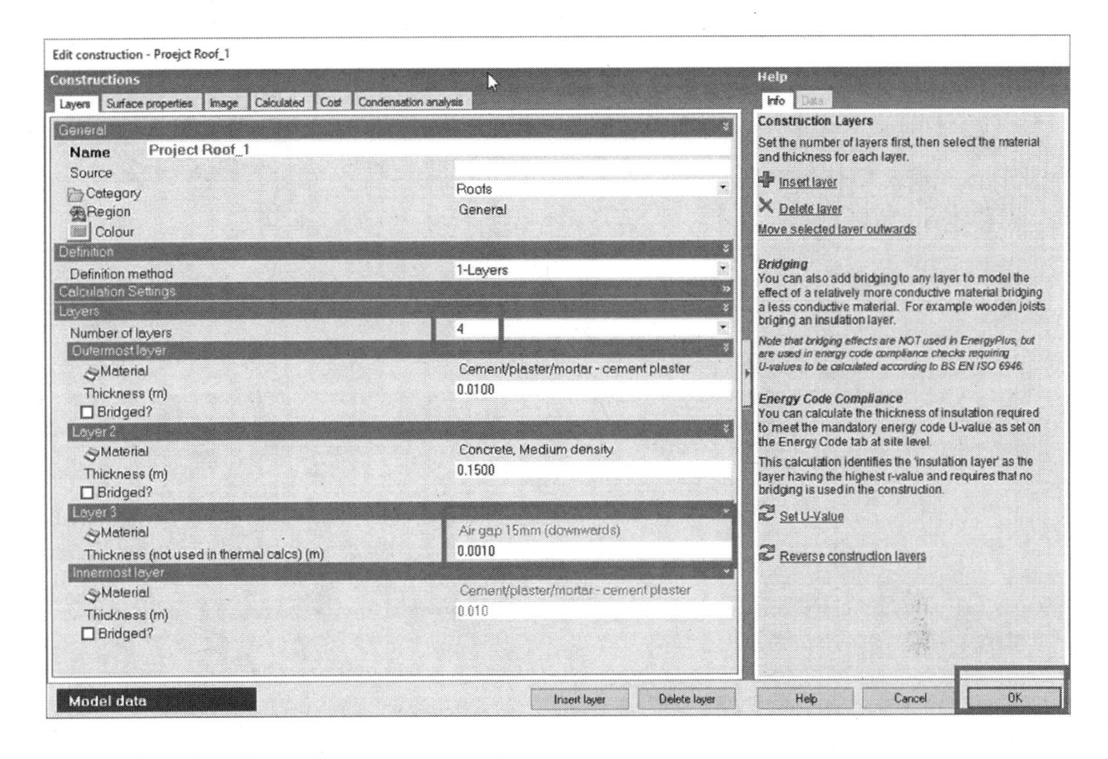

Step 8: Simulate the model, and record the results.

Untitled, Building 1

Analysis | Summary | Parametric | Optimisation + UA/SA | Data Visualisation

Date/Time	Room Electricity (kWh)	Lighting (kWh)	Heating (Gas) (kWh)	Cooling (Electricity) (kWh)	DHW (Electricity) (kWh)
12:00:00 AM	52291.88	75573.1	146450.2	11600.01	4731.553

Step 9: Repeat the preceding steps for all the air-gap thicknesses given in the **Problem Statement**. Compare the results for all simulations (Table 3.7).

TABLE 3.7

Comparison of annual fuel breakdown data for various air gaps

Annual fuel breakdown data

Gap	Room electricity (kWh)	Lighting (kWh)	Heating (gas) (kWh)	Cooling (electricity) (kWh)	DHW (electricity) (kWh)
No air gap	52,291.88	75,573.10	210,243.30	10,039.44	4,731.55
Air gap 15 mm	52,291.88	75,573.10	146,450.20	11,600.01	4,731.55
Air gap 17 mm	52,291.88	75,573.10	144,083.60	11,691.71	4,731.55
Air gap 25 mm	52,291.88	75,573.10	141,799.90	11,783.45	4,731.55
Air gap 50 mm	52,291.88	75,573.10	137,495.90	11,958.20	4,731.55
Air gap 100 mm	52,291.88	75,573.10	135,463.50	12,045.45	4,731.55
Air gap 300 mm	52,291.88	75,573.10	133,511.60	12,130.34	4,731.55

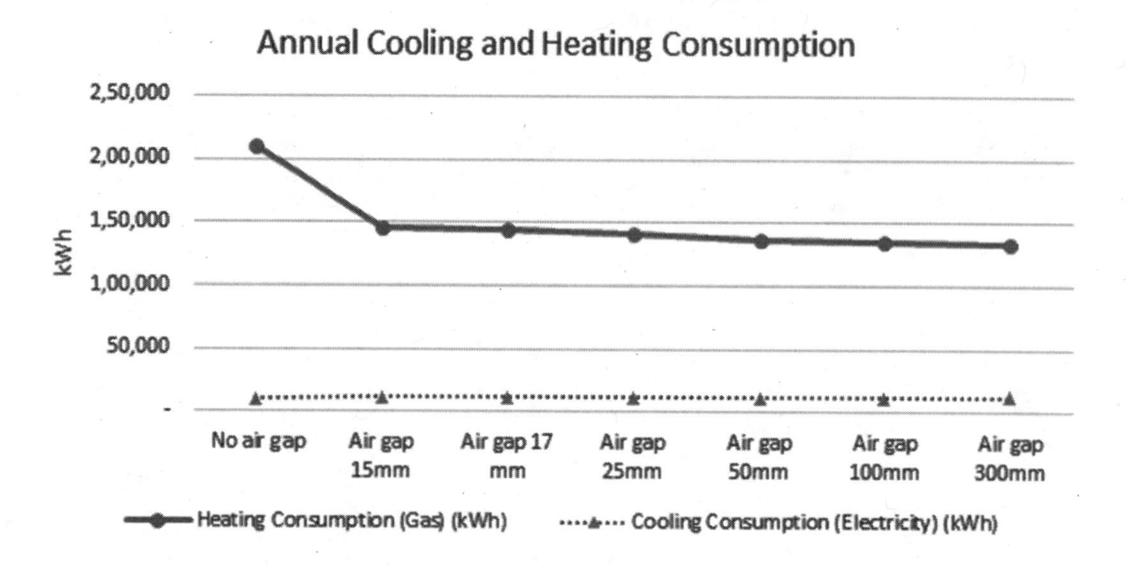

With proving air gap, there is a decrease in heating energy consumption, but for air gaps greater than 15 mm, energy consumption does not change. This situation indicates that an air gap provides insulation, but for larger air gaps, because of the convection currents, there is not much benefit.

Tutorial 3.5

Evaluating the Impact of Surface Reflectance

GOAL

To evaluate the effect of surface reflectance on the energy performance of a building

WHAT ARE YOU GOING TO LEARN?

- How to change surface reflectivity

PROBLEM STATEMENT

In this tutorial, you are going to use a 50- × 25-m model with a roof consisting of the following layers (starting with the outermost layer):

- 0.015 m of cement/plaster/mortar-plaster
- 0.150 m of concrete, medium density
- 0.015 m of cement/plaster/mortar-plaster

Vary the roof solar reflectivity on the outermost material from 0.9 to 0.1 in steps of 0.1. Determine the impact of roof solar reflectivity on the energy consumption for SINGAPORE/PAYA LEBA.

Surface absorptance is the propensity of the surface material to absorb radiation, and it is the opposite surface reflectance, which is the ability to reflect radiation. For opaque surfaces, surface absorptance and surface reflectance values are therefore ratios whose sum is always equal to 1.

Solar reflectivity = 1 – solar absorptance

SOLUTION

Step 1: Open a new project. Create a **50- × 25-m** building.

Step 2: Select the **Construction** tab.

Step 3: Add a new roof with three layers. Select **Cement/plaster/mortar – cement plaster** from **Plaster** materials with thickness (m) as **0.01**, **Concrete, Medium density** from **Concretes** materials with thickness (m) as **0.15** and **Cement/plaster/mortar-cement plaster** from **Plaster** materials with thickness (m) as **0.01**.

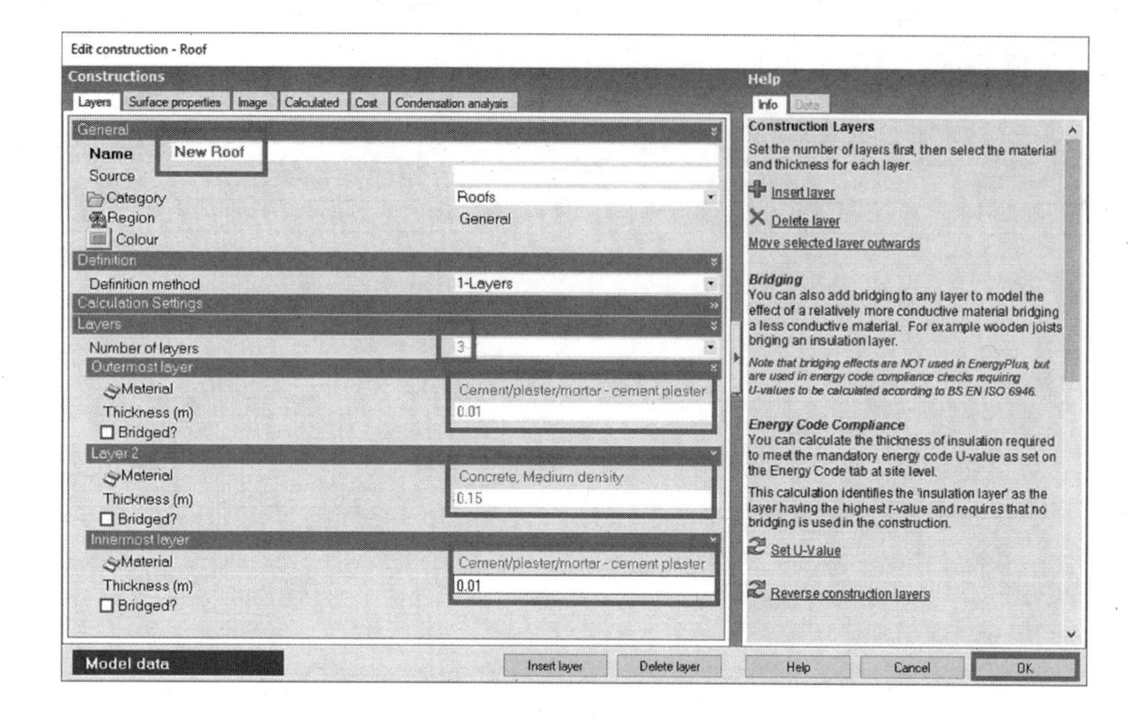

Step 4: Select **Cement/plaster/mortar – cement plaster** in the outermost layer, and then click the **Create copy of highlighted item** icon.

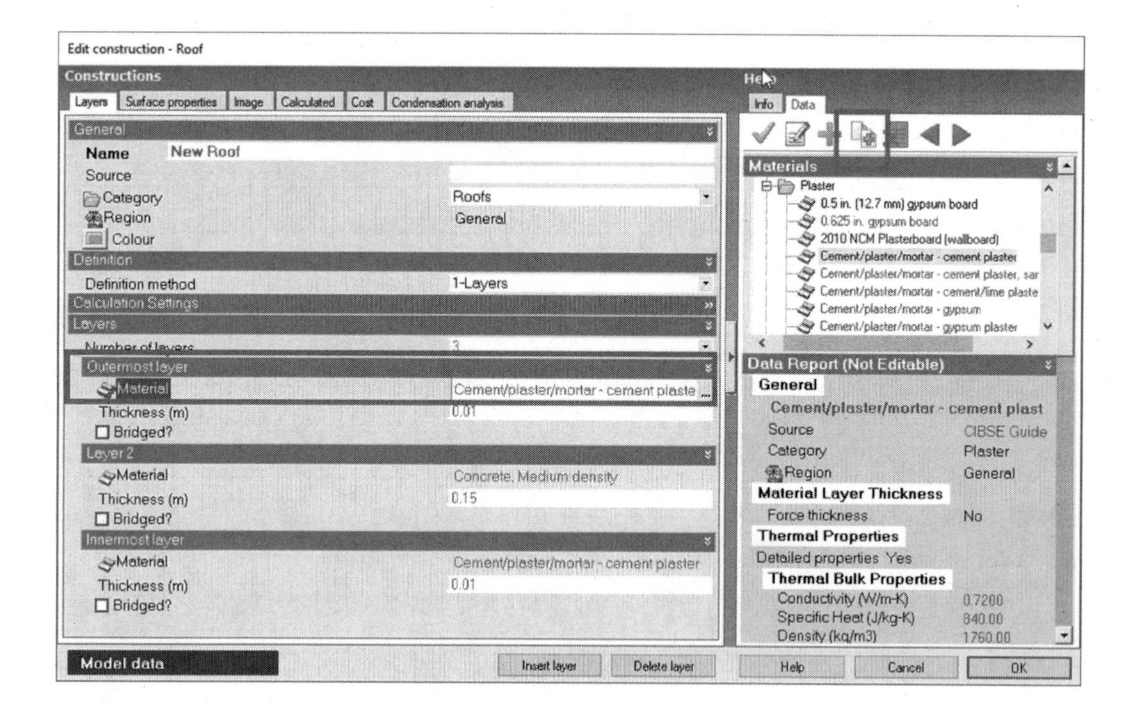

Step 5: Select the **Copy of Cement/plaster/mortar – cement plaster,** and then click the **Edit highlighted item** icon.

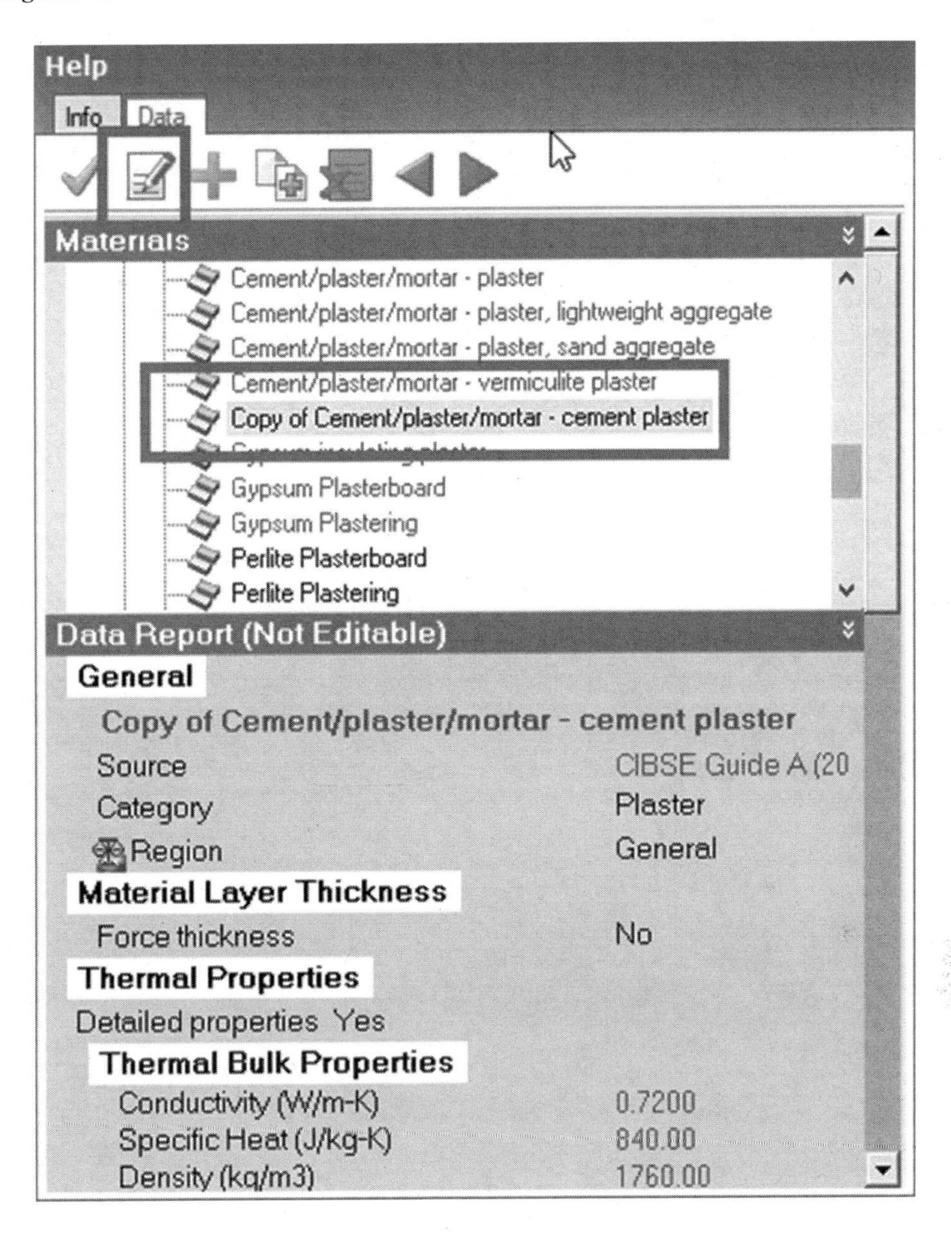

Step 6: Enter **Plaster – Surface Absorptance 0.9** in the **Name** box.

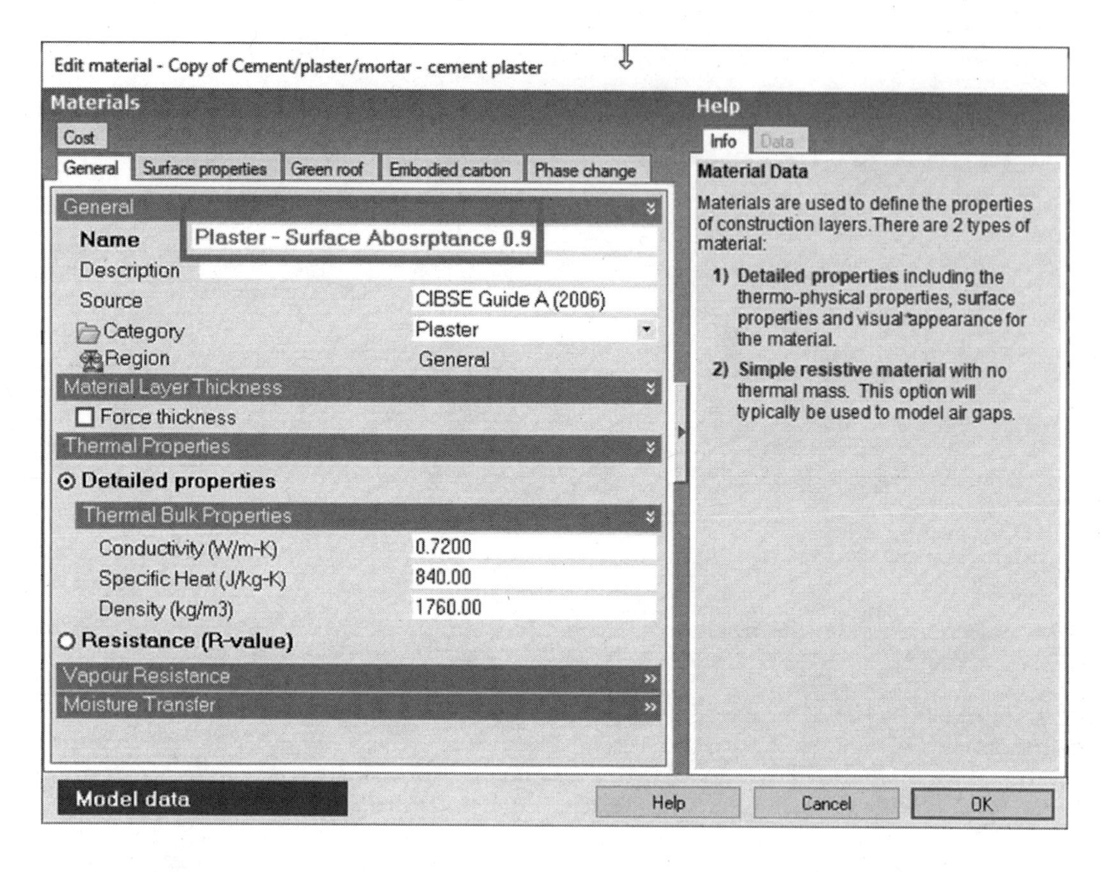

Step 7: Select the **Surface Properties** tab, and enter **Solar absorptance** as **0.900**.

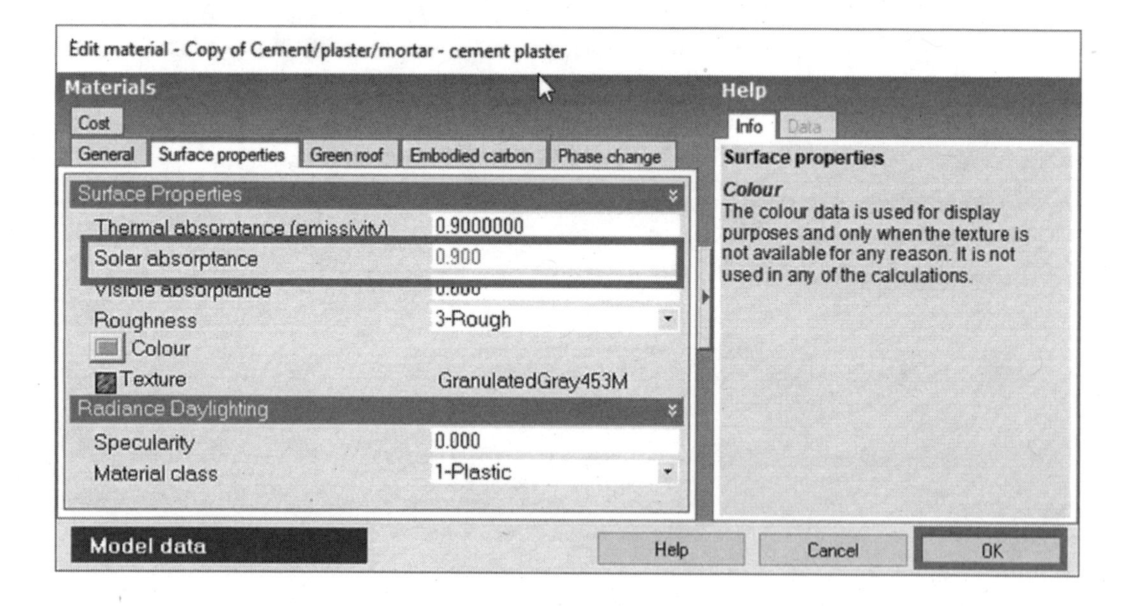

Step 8: Click the **Select this data** icon.

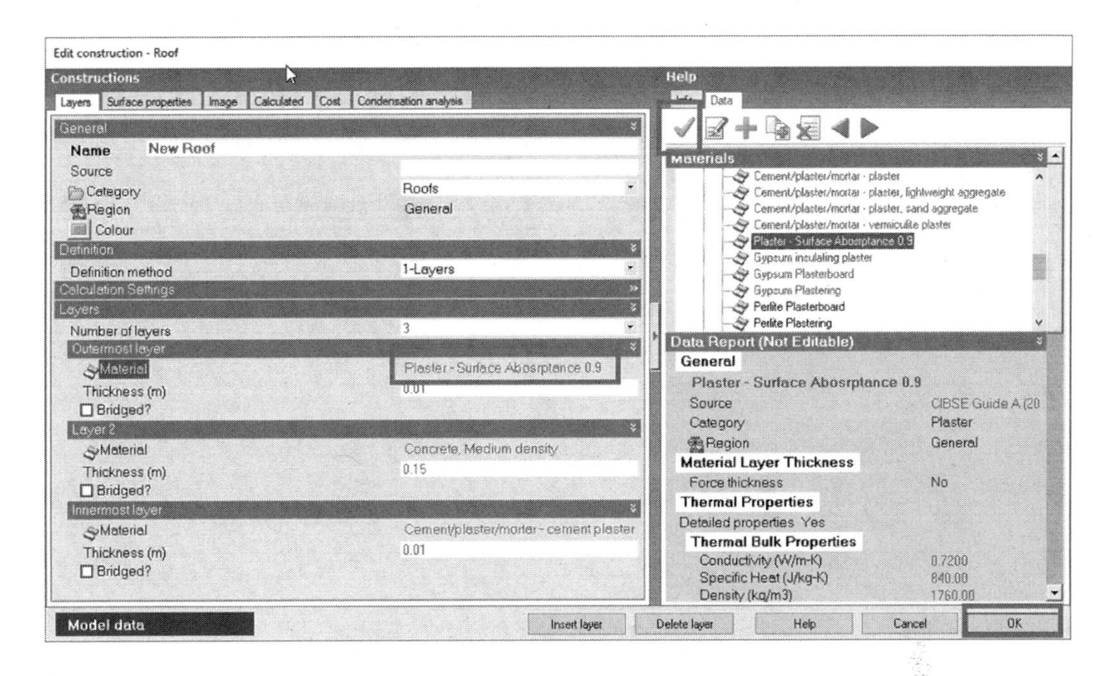

Step 9: Perform the annual simulation, and record the results.

Untitled, Building 1

Analysis | Summary | Parametric | Optimisation + UA/SA | Data Visualisation

Date/Time	Room Electricity (kWh)	Lighting (kWh)	Cooling (Electricity) (kWh)	DHW (Electricity) (kWh)
12:00:00 AM	52291.88	75573.1	251839.6	4731.553

Step 10: Repeat the preceding steps for all the values of solar absorptance given in the problem statement. Compare the results for all simulations (Table 3.8).

TABLE 3.8
Annual fuel breakdown data for different surface absorptances

Surface absorptance	Surface reflectance	Room electricity (kWh)	Lighting (kWh)	Cooling (electricity) (kWh)	DHW (electricity) (kWh)
0.9	0.1	52,291.88	75,573.10	251,839.60	4,731.55
0.8	0.2	52,291.88	75,573.10	242,780.20	4,731.55
0.7	0.3	52,291.88	75,573.10	233,636.10	4,731.55
0.6	0.4	52,291.88	75,573.10	224,349.60	4,731.55
0.5	0.5	52,291.88	75,573.10	214,943.60	4,731.55
0.4	0.6	52,291.88	75,573.10	205,524.40	4,731.55
0.3	0.7	52,291.88	75,573.10	195,968.10	4,731.55
0.2	0.8	52,291.88	75,573.10	186,375.80	4,731.55
0.1	0.9	52,291.88	75,573.10	176,912.90	4,731.55

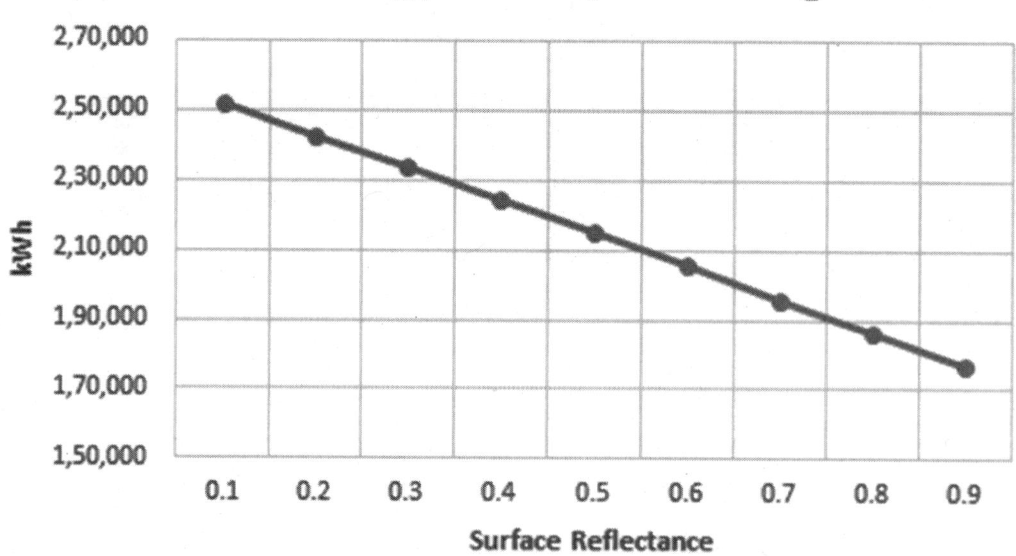

With an increase in surface reflectance, the amount of heat that makes it into a building through the roof decreases, thereby reducing the need for mechanical cooling to maintain a comfortable temperature inside the building. Hence, there is a reduction in cooling energy consumption with an increase in surface reflectance of roof.

EXERCISE 3.5

Repeat this tutorial for the **FRANKFURT MAIN ARPT, Germany** weather location (Table 3.9).

TABLE 3.9
Annual fuel breakdown data for different surface absorptance

Surface absorptance	Surface reflectance	Room electricity (kWh)	Lighting (kWh)	Heating (gas) (kWh)	Cooling (electricity) (kWh)	DHW (electricity) (kWh)
0.9	0.1					
0.8	0.2					
0.7	0.3					
0.6	0.4					
0.5	0.5					
0.4	0.6					
0.3	0.7					
0.2	0.8					
0.1	0.9					

Tutorial 3.6

Evaluating the Impact of Roof Underdeck Radiant Barrier

GOAL

To evaluate the impact of underdeck radiant barrier on the energy performance

WHAT ARE YOU GOING TO LEARN?

- How to change the emissivity of a material

PROBLEM STATEMENT

In this tutorial, you are going to use a 50- × 25-m model with a roof consisting of the following layers (starting with the outermost layer):

- 0.01 m of cement/plaster/mortar-plaster
- 0.15 m of concrete, medium density
- 0.01 m of cement/plaster/mortar-plaster

Vary the thermal emittance on the innermost material of the roof (ceiling) from 0.1 to 0.9 in steps of 0.1. Determine the impact of roof thermal emittance on the energy consumption for **CA-SAN FRANCISCO INTL, USA**.

Thermal Absorptance (Emissivity)

Thermal absorptance represents the fraction of incident long-wavelength radiation that is absorbed by the material. This parameter is used when calculating the long-wavelength radiant exchange between various surfaces and affects the surface heat balances (both inside and outside as appropriate). Values for this field must be between 0.0 and 1.0 (with 1.0 representing 'black body' conditions). *Radiant barrier* is the coating used under the roof deck to limit the heat transfer from the inside to the outside and the outside to the inside.

Source: www.designbuilder.co.uk/helpv6.0/Content/SurfaceProperties.htm.

SOLUTION

Step 1: Open a new project. Create a **50- × 25-m** building.

Step 2: Select the **Construction** tab.

Step 3: Add a new roof with three layers. Select **Cement/plaster/mortar – cement plaster** from **Plaster** materials with thickness (m) as **0.01**, **Concrete, Medium density** from **Concretes** materials with thickness (m) as **0.1500**, and **Cement/plaster/mortar – cement plaster** from **Plaster** materials with thickness (m) as **0.0100**.

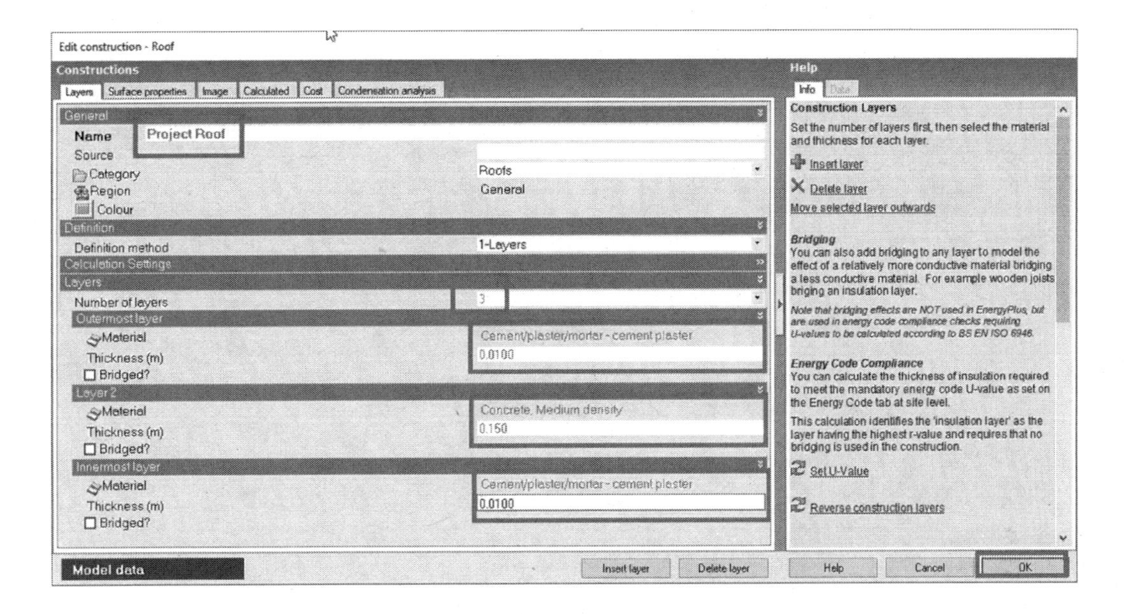

Step 4: Click **Cement/plaster/mortar – cement plaster** in the innermost layer. Create a copy and rename it as **New Cement/plaster/mortar – cement plaster**.

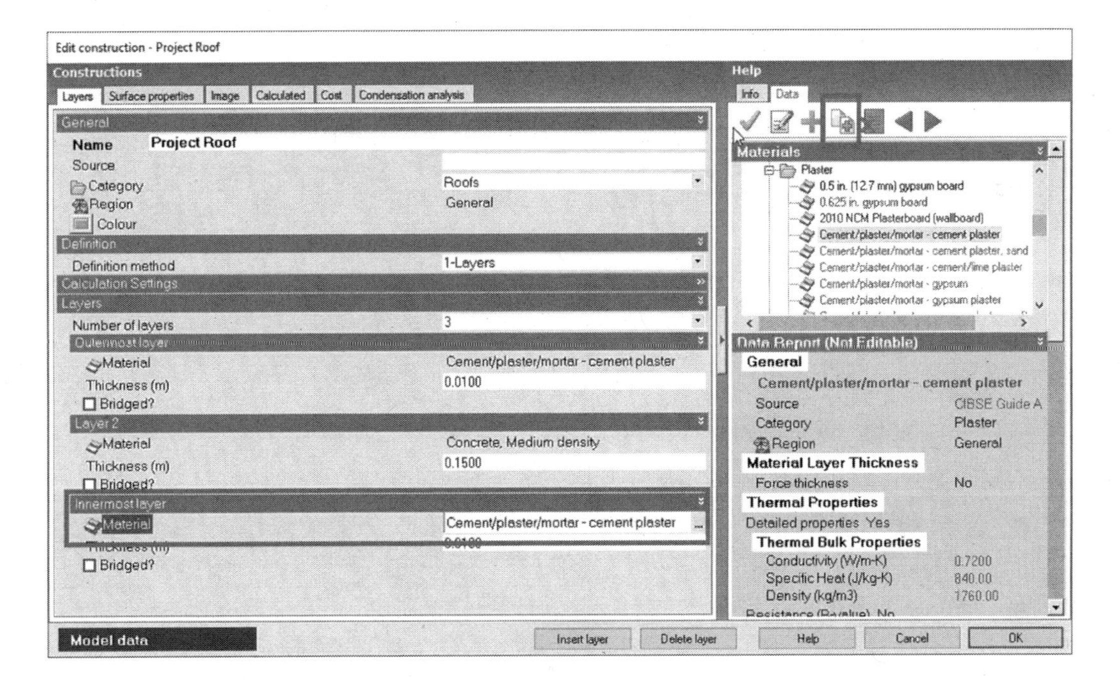

Step 5: Click the **Edit highlighted item** icon.

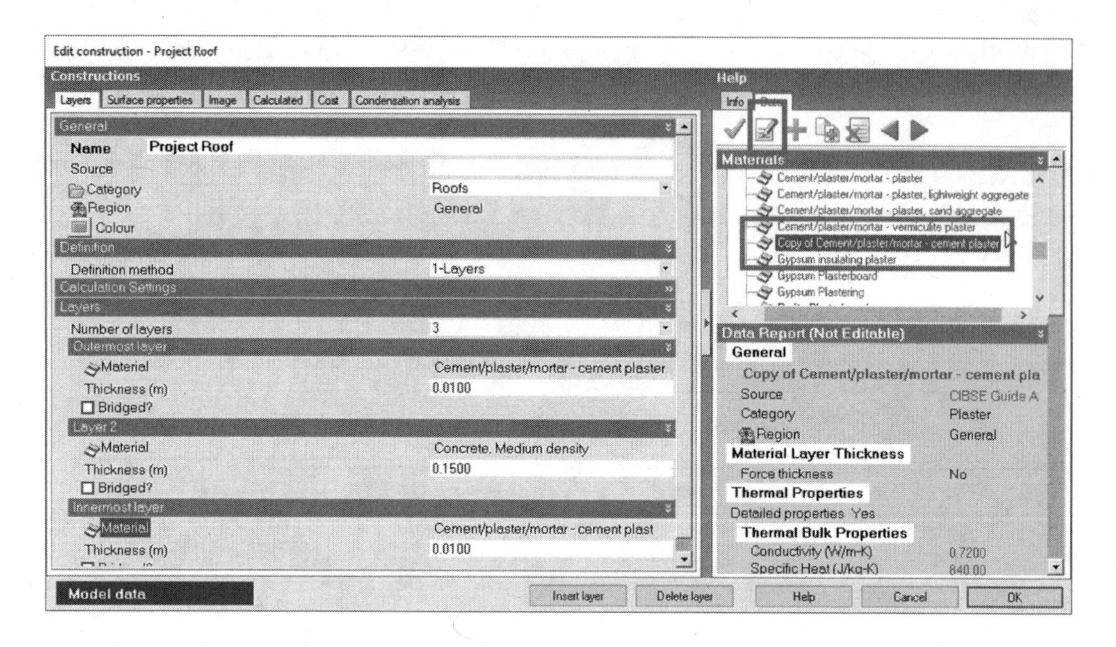

Step 6: Select the **Surface properties** tab, and enter **0.900** as the **Thermal absorptance (emissivity)**.

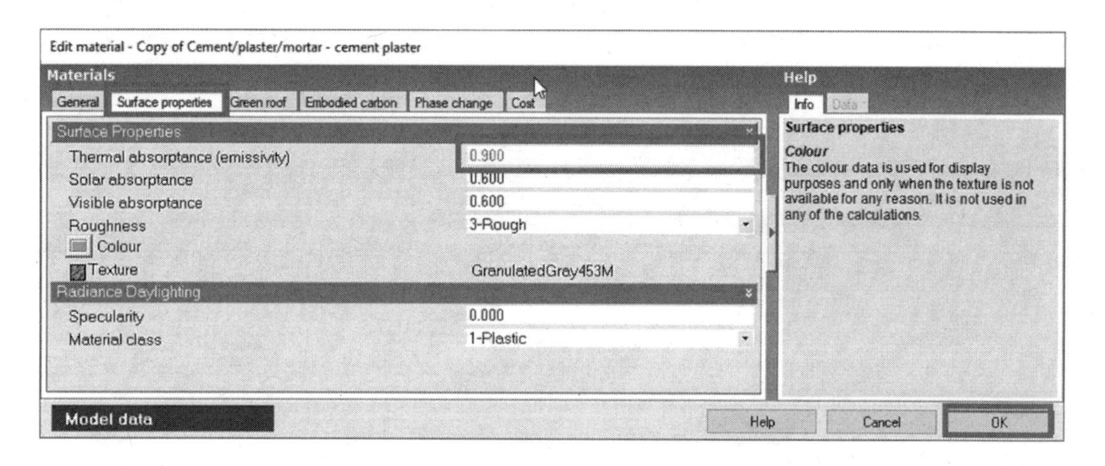

Step 7: Simulate the model, and record the results.

Untitled, Building 1

Analysis | Summary | Parametric | Optimisation + UA/SA | Data Visualisation

Date/Time	Room Electricity (kWh)	Lighting (kWh)	Heating (Gas) (kWh)	Cooling (Electricity) (kWh)	DHW (Electricity) (kWh)
12:00:00 AM	52291.88	75573.1	62947.03	15497.76	4731.553

Step 8: Repeat the preceding steps for all the values of thermal emissivity given in the problem statement. Compare the results for all simulations (Table 3.10).

TABLE 3.10
Comparison of annual fuel breakdown data for thermal absorptance

Thermal absorptance (emissivity)	Room electricity (kWh)	Lighting (kWh)	Heating (gas) (kWh)	Cooling (electricity) (kWh)	DHW (electricity)
0.9	52,291.88	75,573.10	62,947.03	15,497.76	4,731.55
0.8	52,291.88	75,573.10	62,668.83	15,520.37	4,731.55
0.7	52,291.88	75,573.10	62,342.30	15,560.60	4,731.55
0.6	52,291.88	75,573.10	61,936.19	15,625.79	4,731.55
0.5	52,291.88	75,573.10	61,468.10	15,727.79	4,731.55
0.4	52,291.88	75,573.10	60,841.12	15,888.41	4,731.55
0.3	52,291.88	75,573.10	60,063.18	16,135.11	4,731.55
0.2	52,291.88	75,573.10	58,981.66	16,516.90	4,731.55
0.1	52,291.88	75,573.10	57,433.11	17,137.45	4,731.55

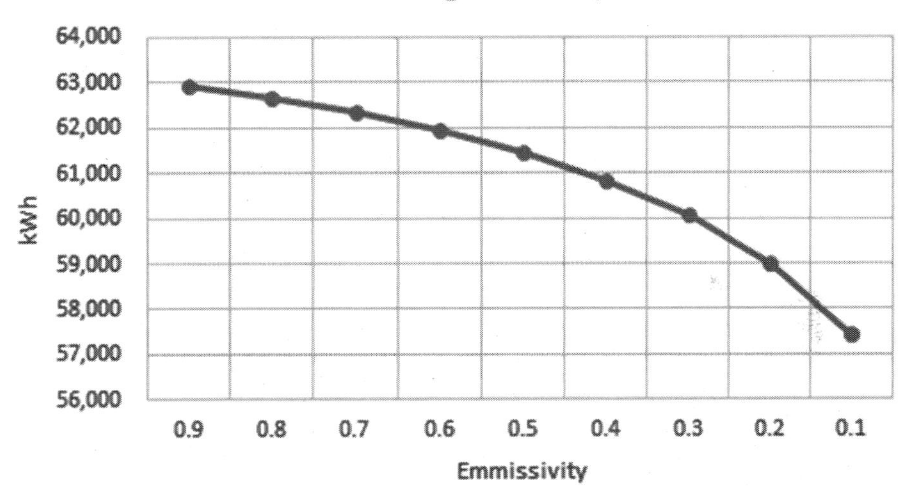

As radiant barriers reduce the radiant heat transfer, a decrease in thermal emittance reduces energy consumption.

EXERCISE 3.6

Repeat this tutorial for the **DUBAI INTERNATIONAL, United Arab Emirates** weather location (Table 3.11).

TABLE 3.11
Annual fuel breakdown data

Thermal absorptance (emissivity)	Room electricity (kWh)	Lighting (kWh)	Heating (gas) (kWh)	Cooling (electricity) (kWh)	DHW (electricity)
0.9					
0.8					
0.7					
0.6					
0.5					
0.4					
0.3					
0.2					
0.1					

Tutorial 3.7

Evaluating the Impact of a Green Roof

GOAL

To evaluate the impact of a green roof on energy consumption

WHAT ARE YOU GOING TO LEARN?

- How to create a green roof

PROBLEM STATEMENT

In this tutorial, you are going to use a 50- × 25-m model with a roof consisting of the following layers (starting with the outermost layer):

- 0.01 m of cement/plaster/mortar-plaster
- 0.15 m of concrete, medium density
- 0.01 m of cement/plaster/mortar-plaster

Add a green roof to the outermost layer of the roof. Determine the change in energy consumption with and without a green roof for **DUBAI INTERNATIONAL, United Arab Emirates**.

> A *green roof*, or *living roof*, is a roof of a building that is partially or completely covered with vegetation and a growing medium planted over a waterproofing membrane.
>
> *Source:* https://designbuilder.co.uk/helpv6.0/Content/GreenRoof.htm.

SOLUTION

Step 1: Open a new project. Create a **50- × 25-m** building.

Step 2: Select the **Construction** tab.

Step 3: Add a **New roof** with three layers. Select **Cement/plaster/mortar – cement plaster** from **Plaster** materials with thickness (m) as **0.0100**, **Concrete, Medium density** from **Concretes** materials with thickness (m) as **0.1500,** and **Cement/plaster/mortar – cement plaster** from **Plaster** materials with thickness (m) as **0.0100**.

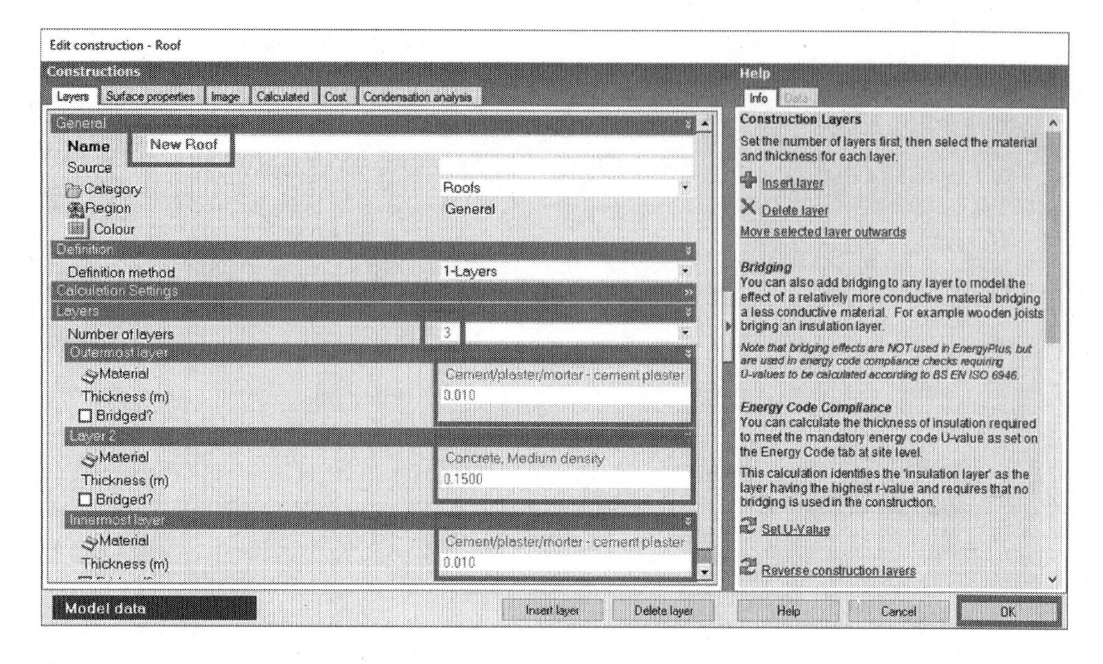

Step 4: Simulate the model, and record the results without a green roof. Now you are going to add a green roof.

Step 5: Add a **New roof** with four layers. Select **12 in. Soil at R-0.104/in.** from **Sands, Stones and Soil** materials with thickness (m) as **0.305**, **Ethylene propylene diene monomer (EPDM)** from **Rubber** materials with thickness (m) as **0.015**, **Concrete, Medium density** from **Concretes** materials with thickness (m) as **0.1500**, and **Cement/plaster/mortar – cement plaster** from **Plaster** materials with thickness (m) as **0.0100**.

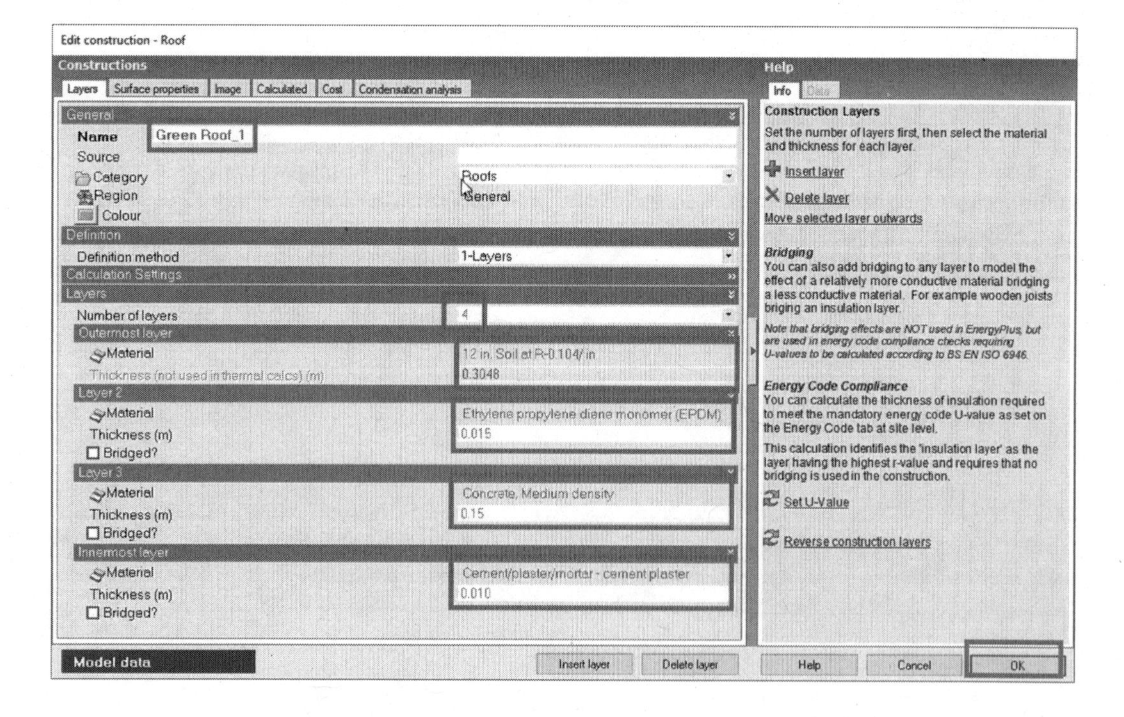

Step 6: Click **12 in. Soil at R-0.104/in.** In the outermost layer, it gets highlighted. Now create a copy.

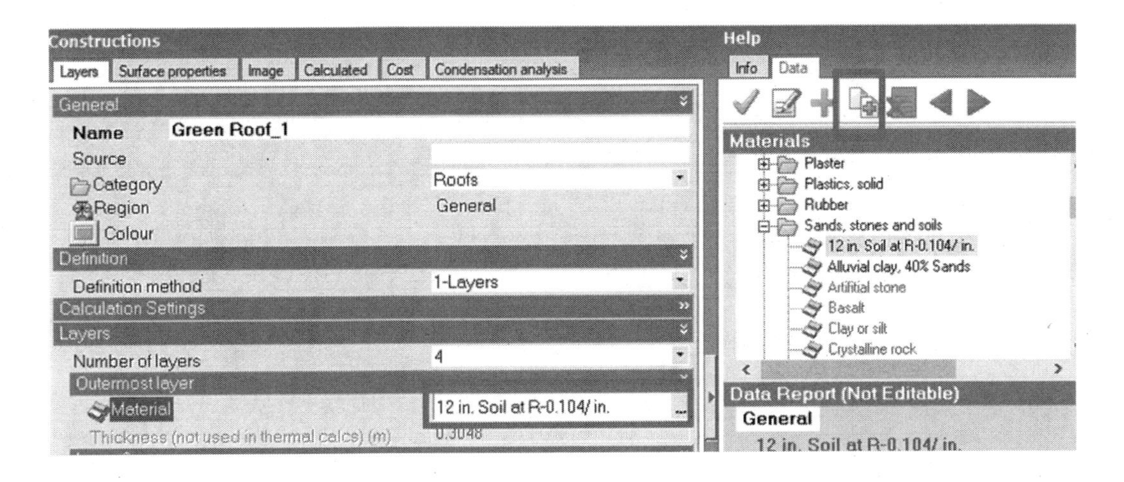

Step 7: Select the **Copy of 12 in. Soil at R-0.104/in.** Rename it **Green Roof Layer**.

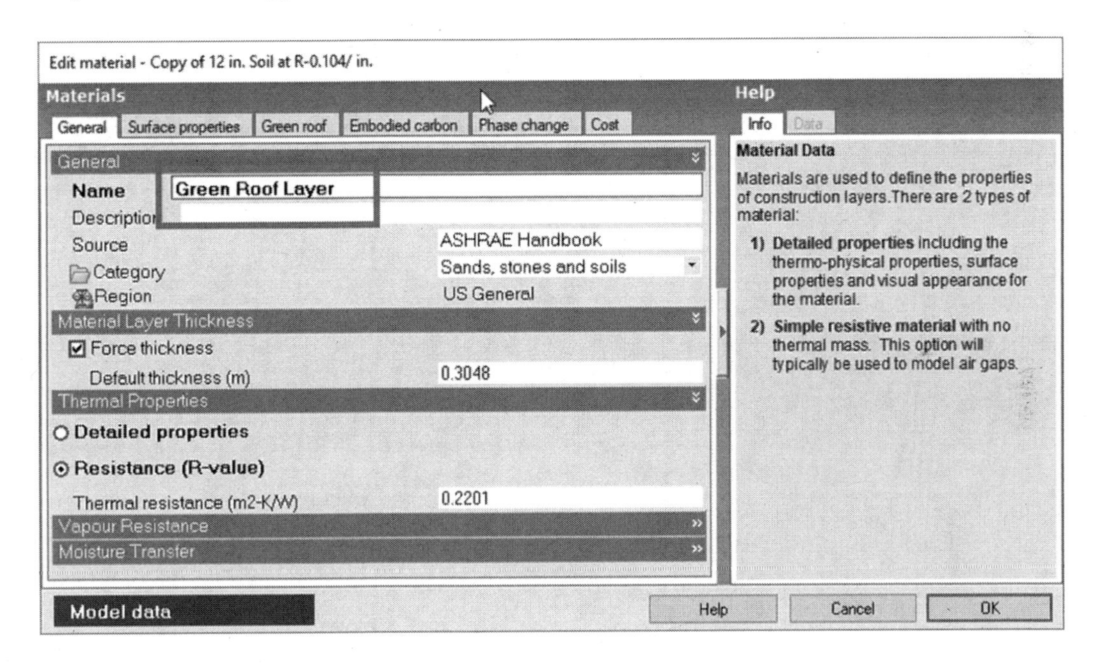

Step 8: Select the **Green roof** tab. Select the **Green roof** checkbox.

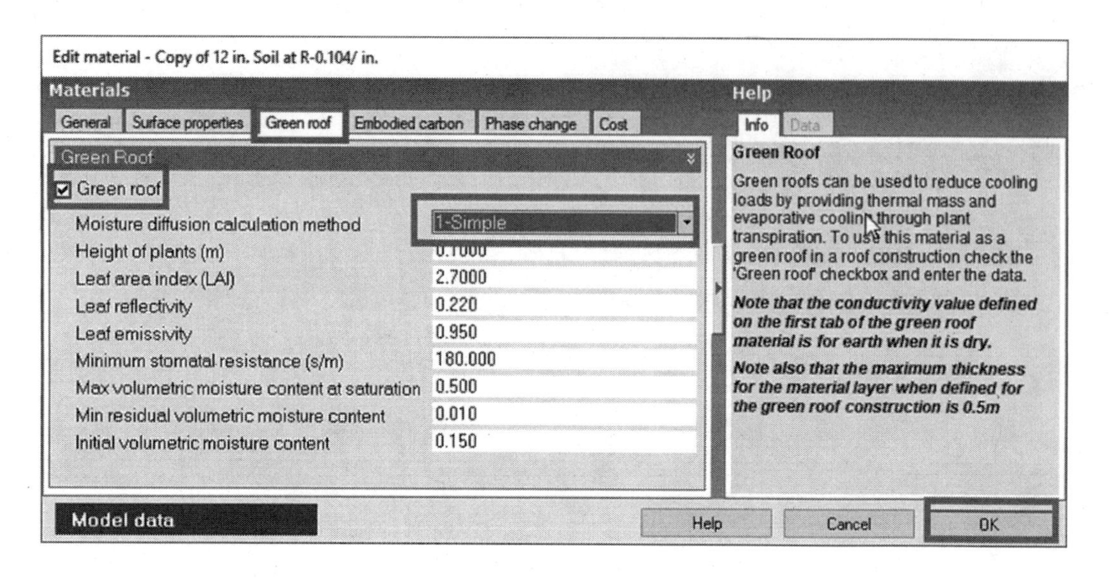

Step 9: Click **OK. Green Roof Layer** appears as the outermost layer.

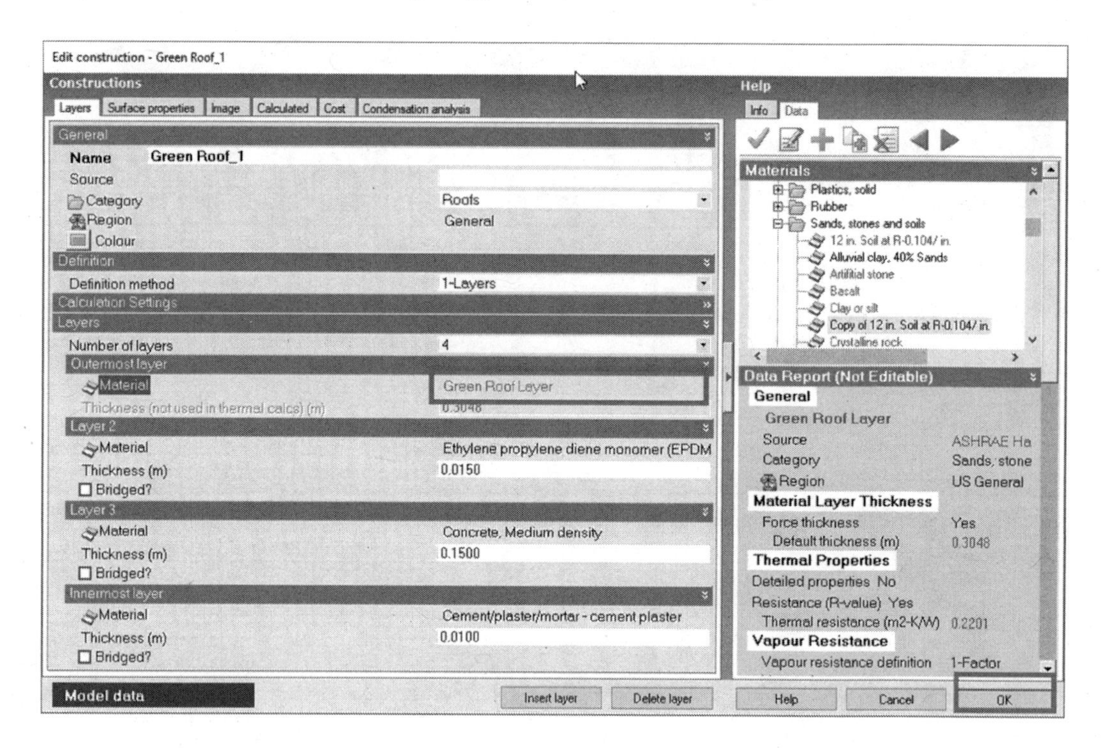

Step 10: Perform an annual simulation, and record the results. Compare the results for both simulations (Table 3.12).

TABLE 3.12
Annual fuel breakdown with and without a green roof

Type	Without a green roof (kWh)	With a green roof (kWh)
Room electricity	52,291.88	52,291.88
Lighting	75,573.10	75,573.10
Heating (gas)	2,505.46	473.02
Cooling (electricity)	196,164.10	188,460.60
DHW (electricity)	4,731.55	4,731.55

EXERCISE 3.7

Repeat this tutorial for the **CA – San Francisco, USA** weather file (Table 3.13).

TABLE 3.13
Annual fuel breakdown consumption with and without a green roof

	Annual fuel breakdown consumption (kWh)	
Type	Without a green roof	With a green roof
Room electricity		
Lighting		
Heating (gas)		
Cooling (electricity)		
DHW (electricity)		

Tutorial 3.8

Evaluating the Impact of Phase-Change Material in an External Wall

GOAL

To evaluate the effect of phase-change material (PCM) in external wall construction on cooling energy consumption and wall temperatures

WHAT ARE YOU GOING TO LEARN?

- How to assign PCM construction
- How to determine external and internal surface temperatures

PROBLEM STATEMENT

In this tutorial, you are going to use a five-zone 50- × 25-m model and set external wall construction to brick wall and then add a PCM. Then you will determine south facade external wall inside and outside face temperatures for both cases. Your analysis will be for the **JAIPUR/SANGANER, India** weather location, and you are going to use following construction for external walls:

- Brick (thickness 100 mm)
- Brick (thickness 100 mm) + PCM wall (thickness 25 mm)

> *Phase-change material* (PCM) works on two phases. By completing a transition between solid and liquid phases, PCMs can absorb large quantities of latent heat. The high latent heat storage capacity of these materials works to effectively increase thermal mass, which tends to moderate interior temperatures and improve comfort conditions. PCMs are frequently used to reduce the need for mechanical cooling and peak-load shifting and to improve solar energy use.
>
> *Source:* https://designbuilder.co.uk/helpv6.0/Content/Phase_Change.htm.

SOLUTION

Step 1: Open a new project file. Create **50- × 25-m** five-zone building with a 5-m perimeter depth.

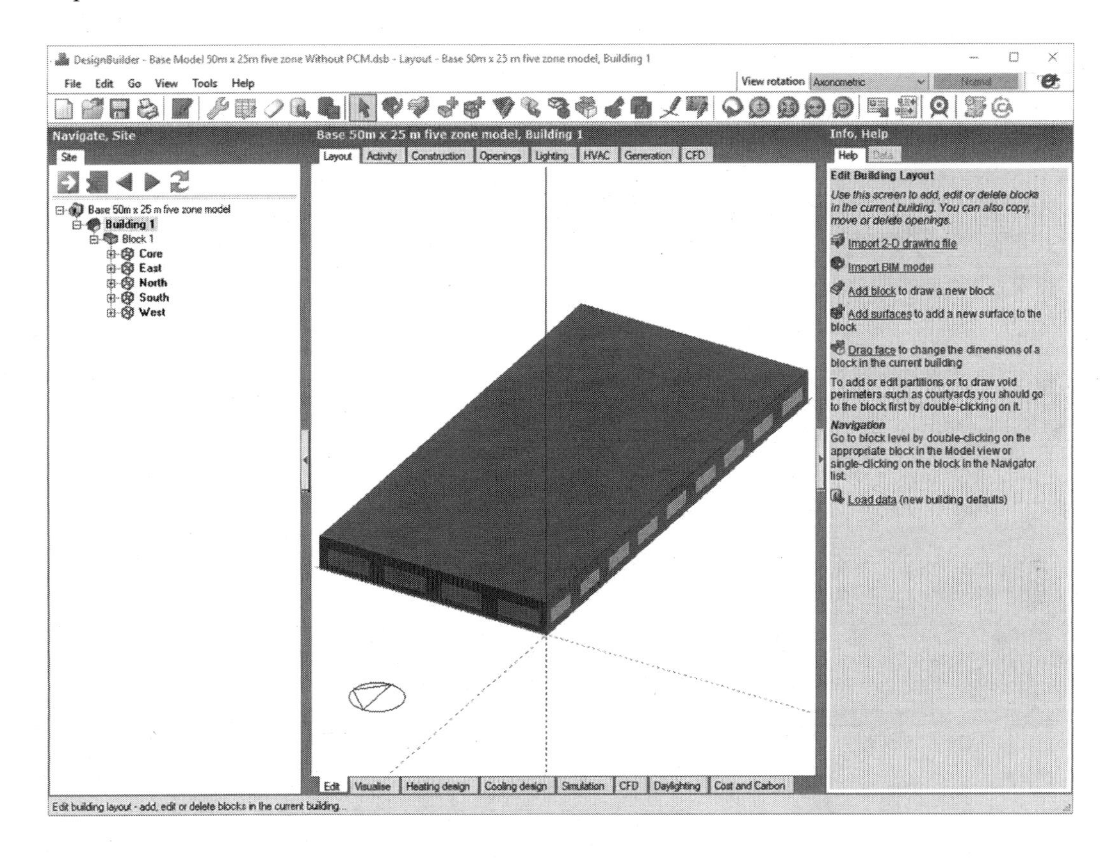

Step 2: Click the **Construction** tab.

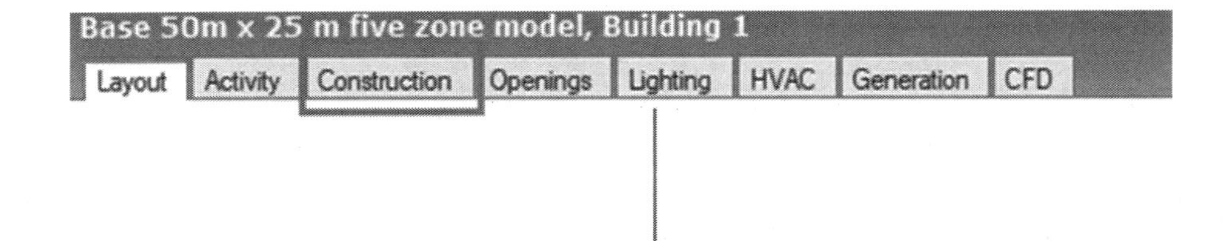

Step 3: In the **Construction** section, click **Project wall, External walls**. Click the **Add new item** icon. The **Edit construction-Wall** screen appears.

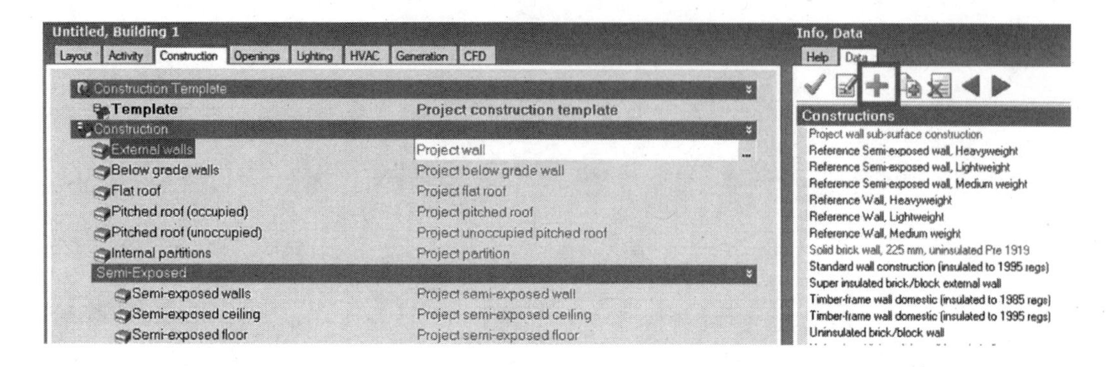

Step 4: Enter **Brick Wall** as the name. Go to the **Layers** section, and select **1** from the **Number of layers** drop-down list. Select **Brick** as material.

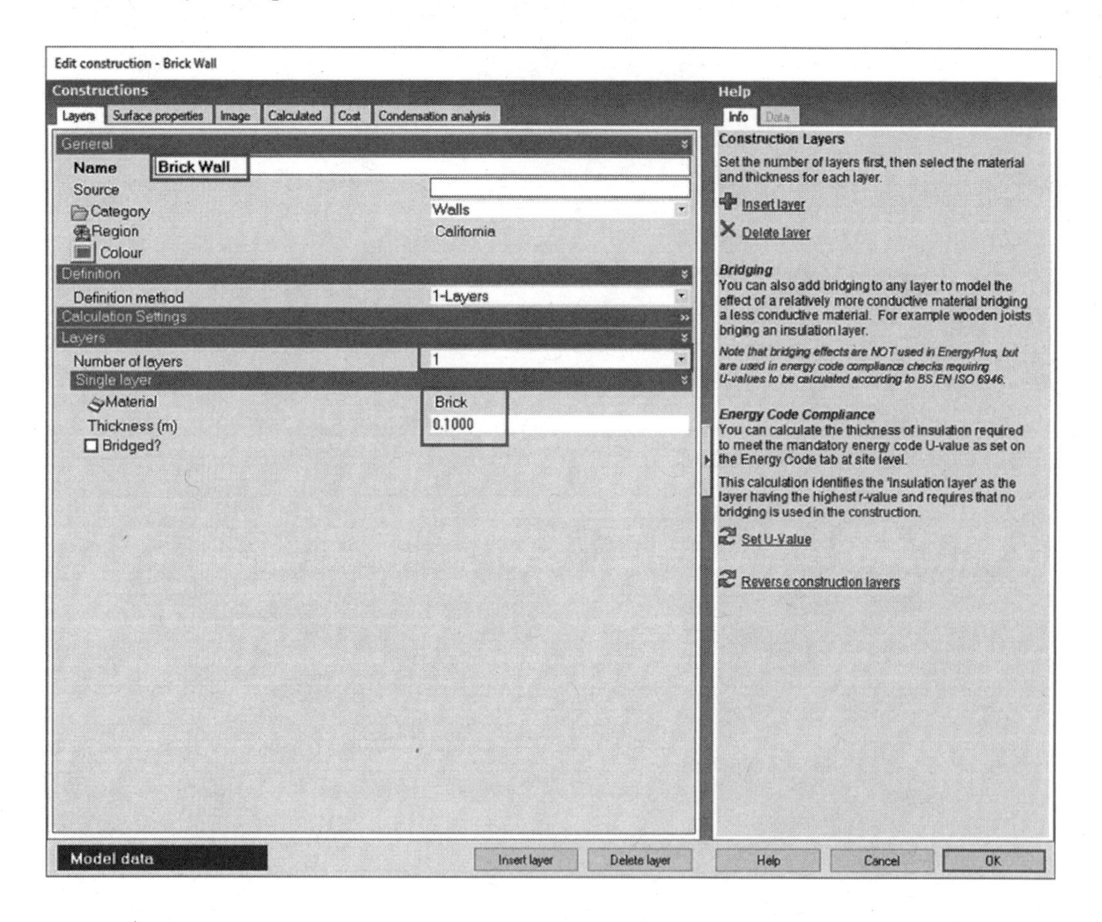

Step 5: Select **Brick Wall** as external wall.

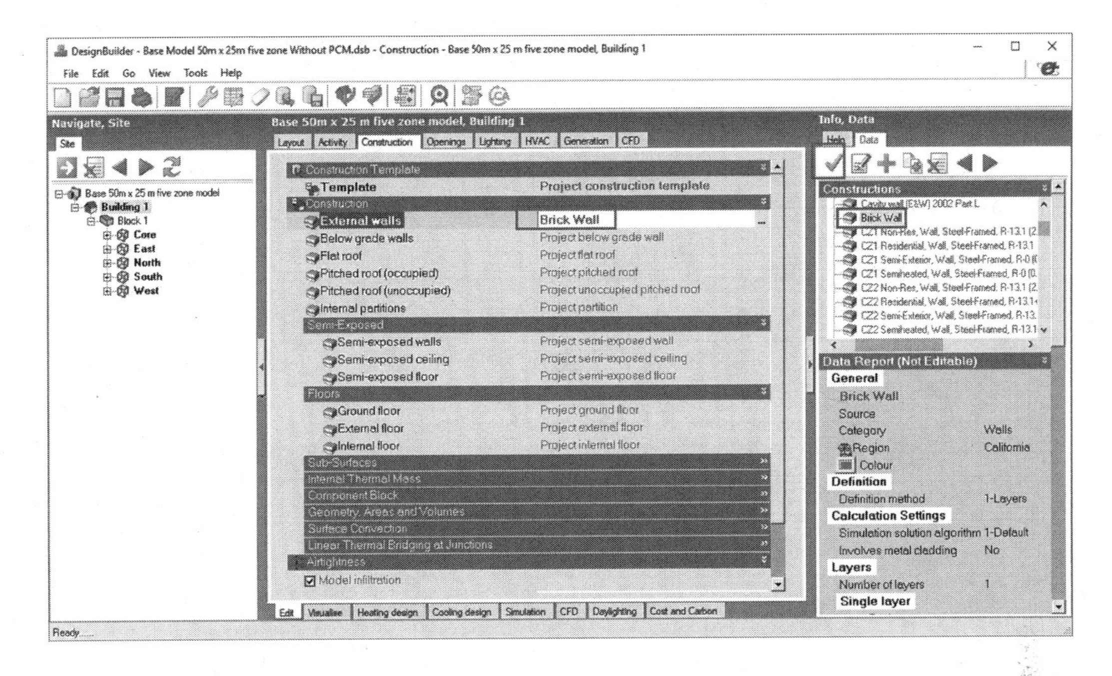

Step 6: Click the **Simulation** screen tab. Click **Update data.** The **Edit Calculation Options** screen appears.

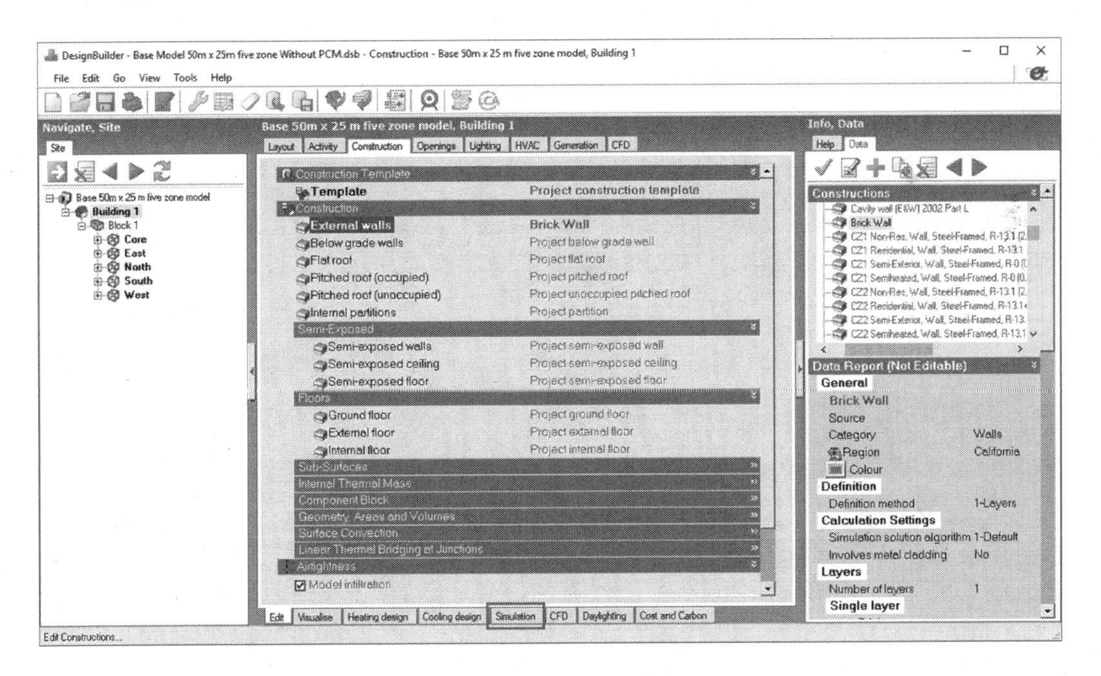

Step 7: Click the **All summer** link under **Simulation Options**. Select the **Hourly** checkbox.

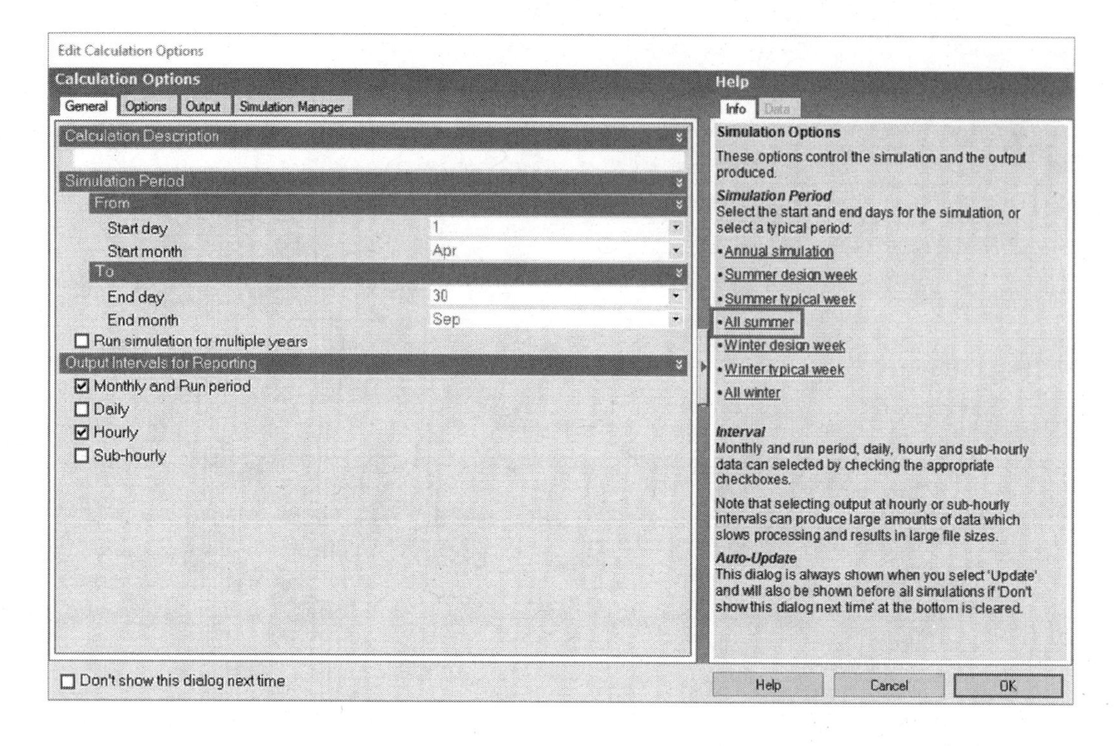

Step 8: Click the **Output** tab. Select the **Store surface output** checkbox. Click **OK**.

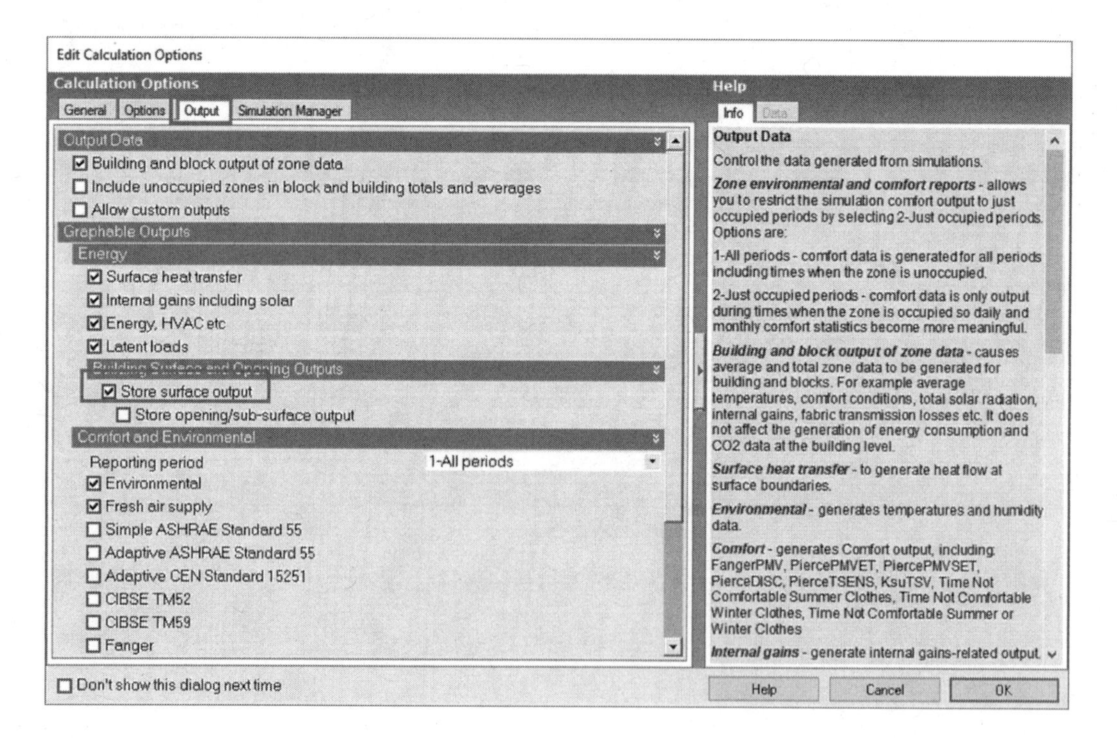

Step 9: Once the simulation is finished, record the energy consumption.

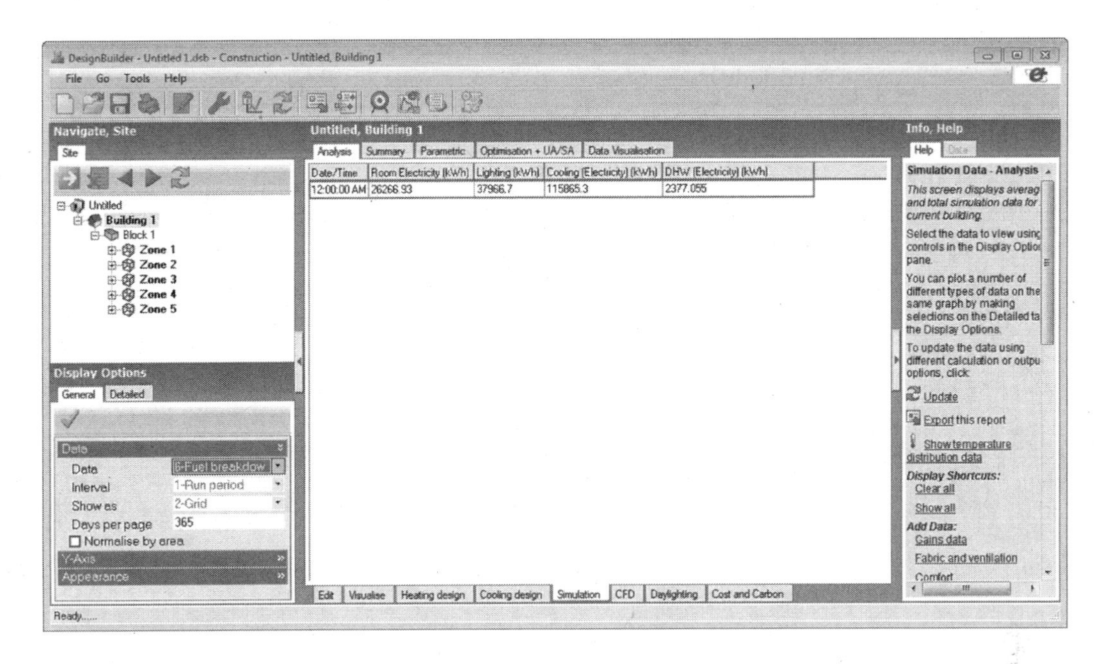

Step 10: Click the **View EnergyPlus results** icon. The **DesignBuilder Results Viewer** appears on the screen. If the **View EnergyPlus results** icon is not displayed, then you need to install the **Results Viewer** application.

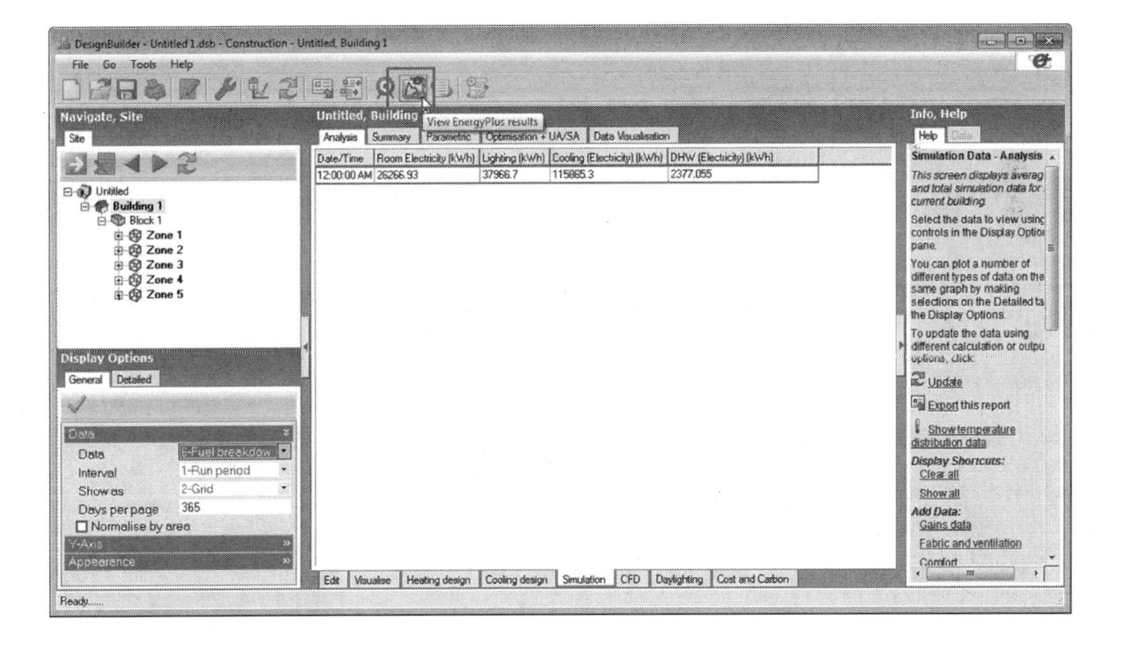

The *Results Viewer* is a separate application that can be used to view EnergyPlus results stored in one or more .eso files. It can be downloaded from the main **Downloads > Software** area of the DesignBuilder website (https://designbuilder.co.uk/download/release-software). When installed, the application allows you to view any results contained within EnergyPlus .eso and .htm files.

Source: https://designbuilder.co.uk/helpv6.0/Content/ResultsProcessor.htm.

Step 11: Click the **Hourly** tab.

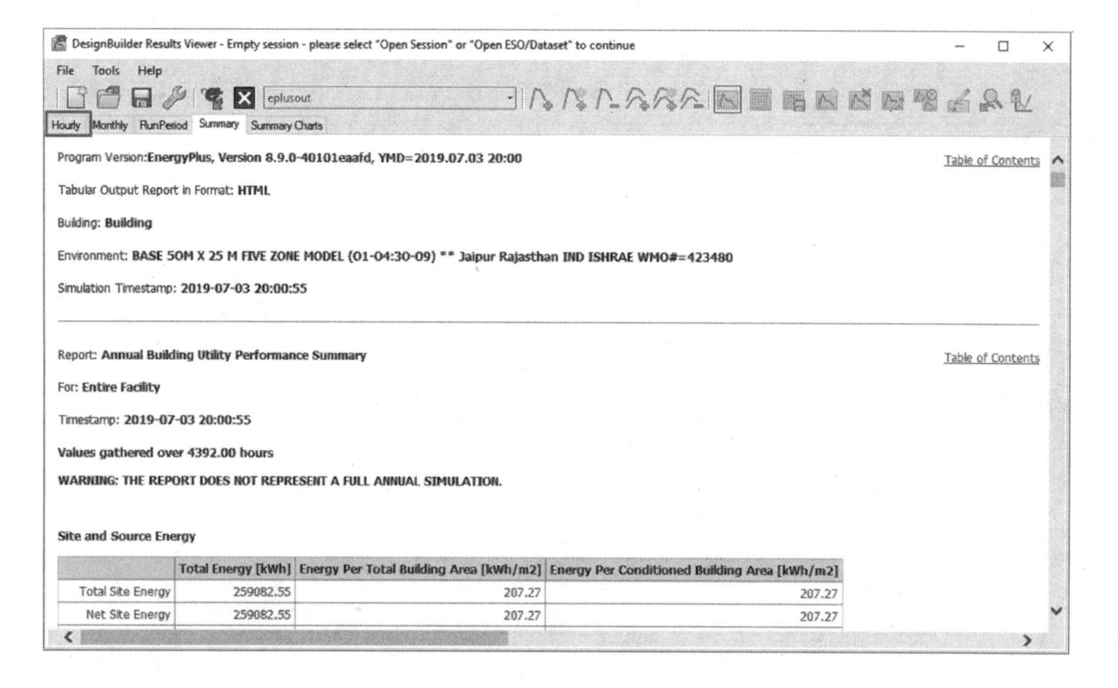

Step 12: Scroll down and click **Surface Inside Face Temperature** under **Report Type** and **Block1:SOUTH_WALL_5_0_0_0** under **Area**.

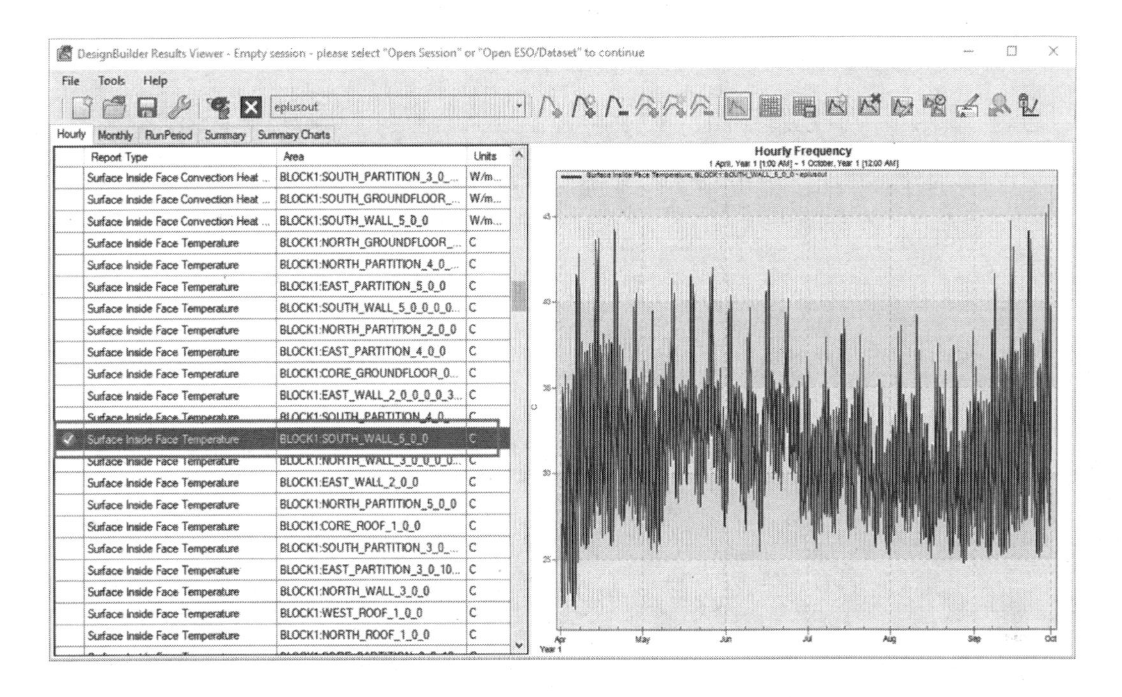

Step 13: Further scroll down and click **Surface Outside Face Temperature** under **Report Type** and **Block1:SOUTH_WALL_5_0_0_0** under **Area**.

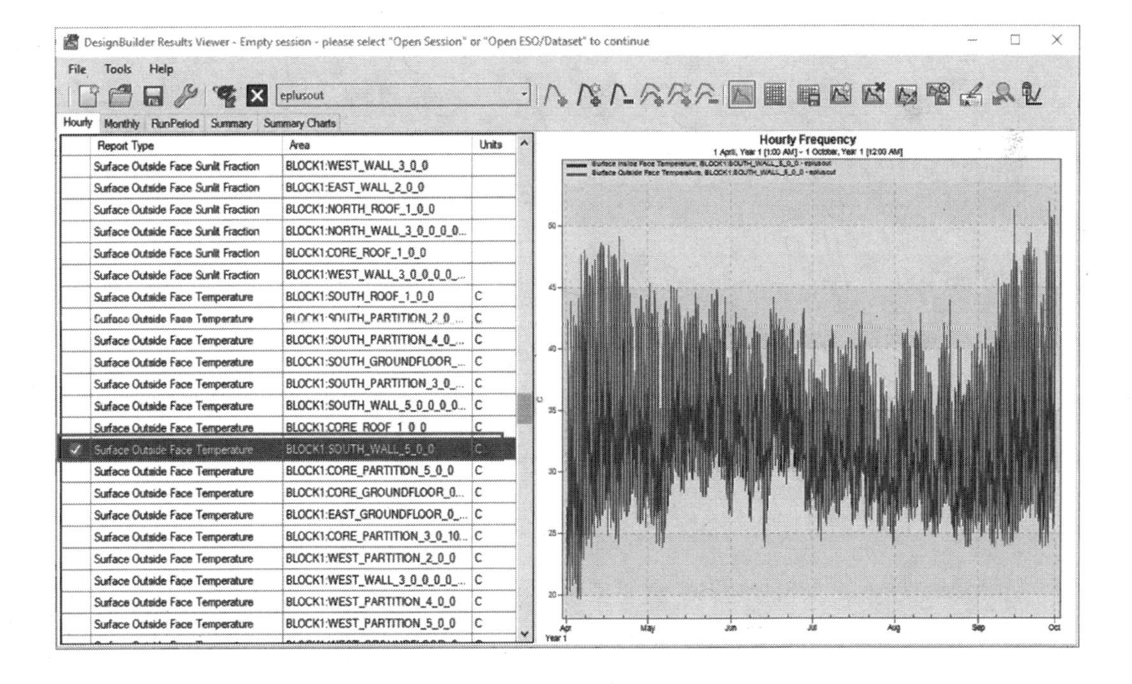

Step 14: Zoom the graph to see the data (from 27 to 31 May) for the following four days.

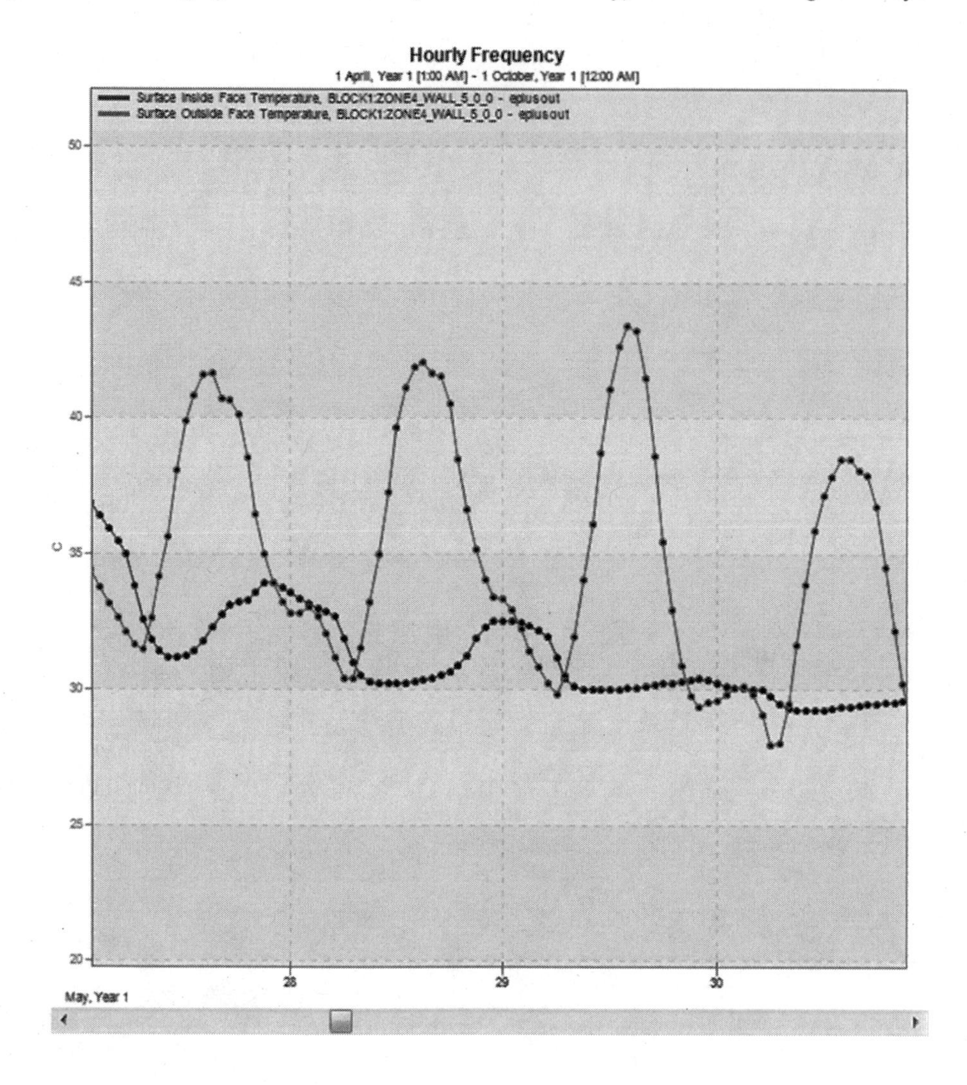

Step 15: Save as the model. Click on the **Edit** screen tab.

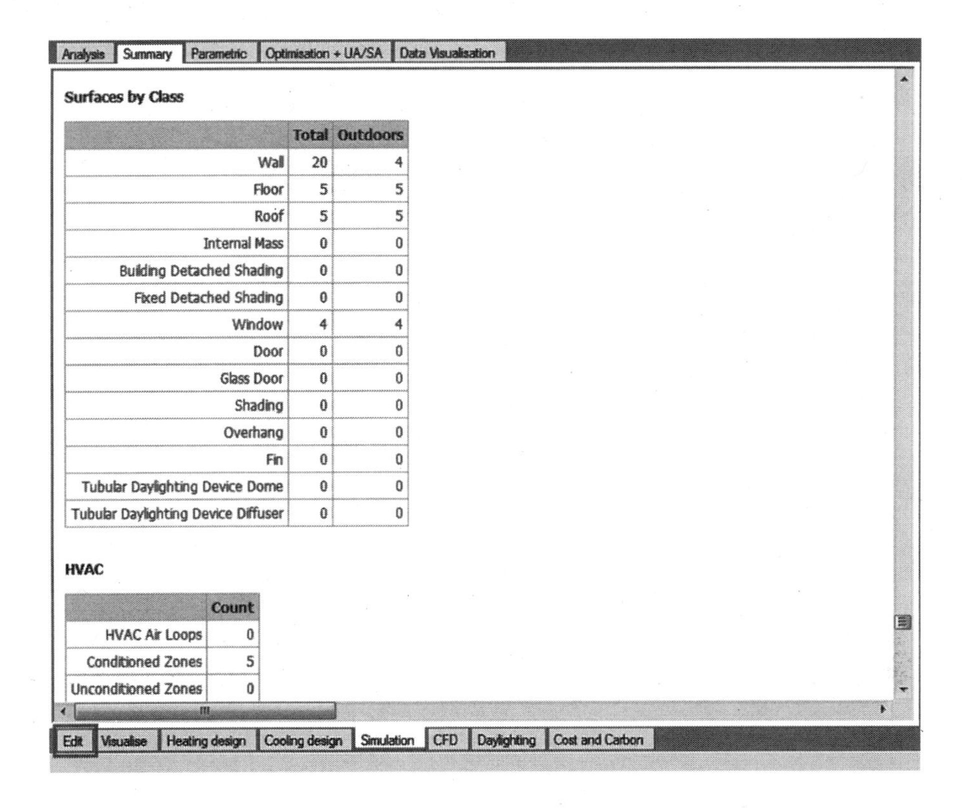
Analysis | Summary | Parametric | Optimisation + UA/SA | Data Visualisation

Surfaces by Class

	Total	Outdoors
Wall	20	4
Floor	5	5
Roof	5	5
Internal Mass	0	0
Building Detached Shading	0	0
Fixed Detached Shading	0	0
Window	4	4
Door	0	0
Glass Door	0	0
Shading	0	0
Overhang	0	0
Fin	0	0
Tubular Daylighting Device Dome	0	0
Tubular Daylighting Device Diffuser	0	0

HVAC

	Count
HVAC Air Loops	0
Conditioned Zones	5
Unconditioned Zones	0

Edit | Visualise | Heating design | Cooling design | Simulation | CFD | Daylighting | Cost and Carbon

Step 16: Click the **Construction** tab. The **Edit construction** screen appears.

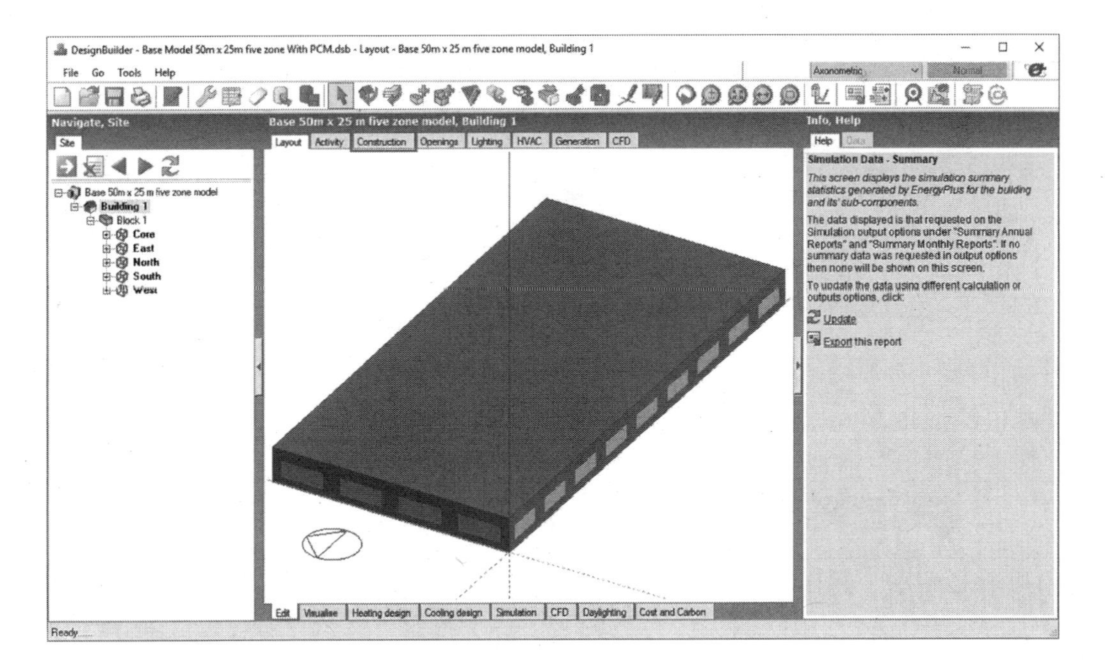

Step 17: Edit the **External wall construction – Brick Wall** (created in the preceding steps of this tutorial). Enter **PCM External Wall** in the **Name** text box. Select **2** from the **Number of layers** drop-down list.

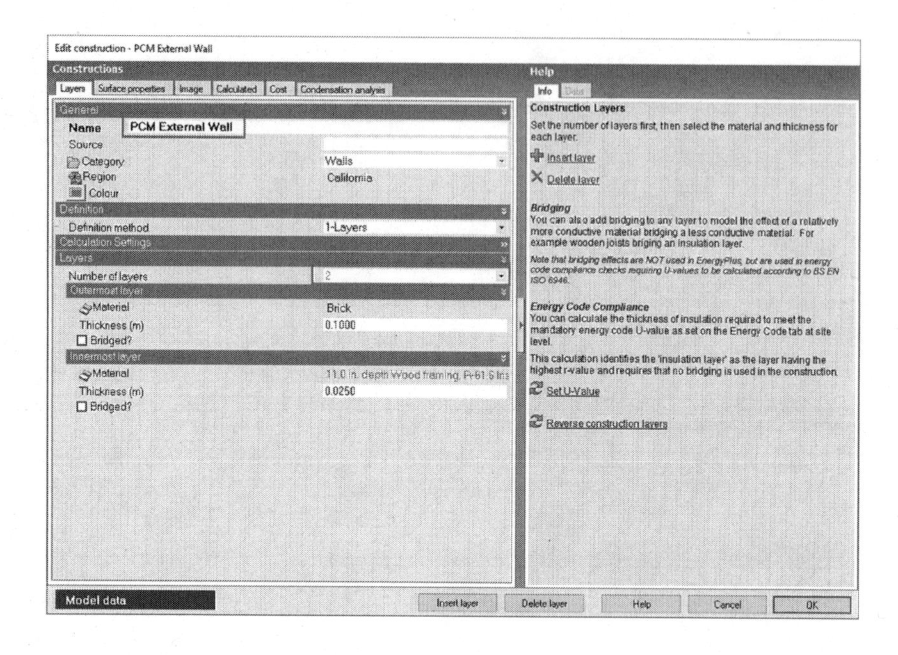

Step 18: Click **Material** under **Innermost layer**. Click on the three dots. Expand **Phase Change**.

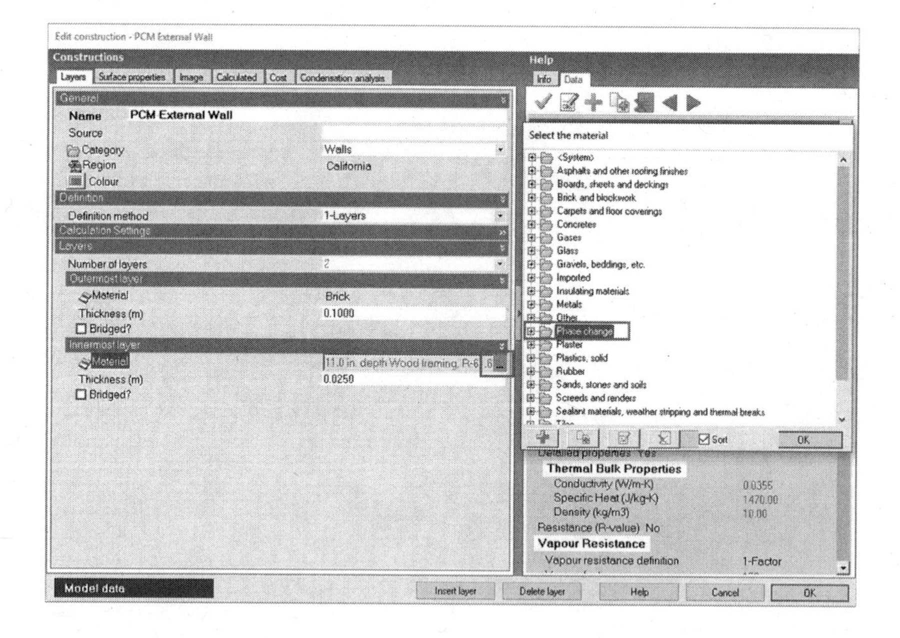

Step 19: Click **InfiniteRPCM29C**.

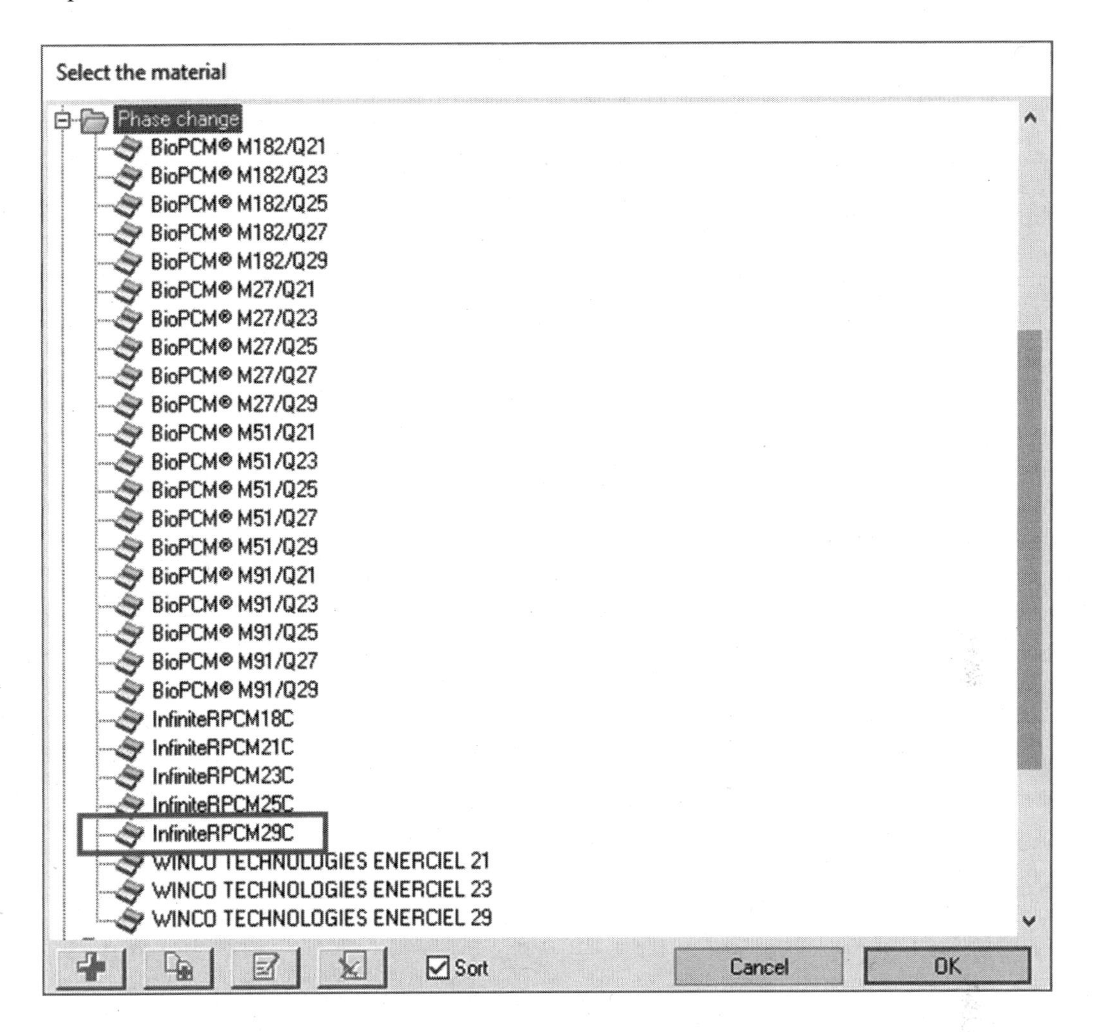

Step 20: Click the **Phase Change** tab to view properties of the PCM.

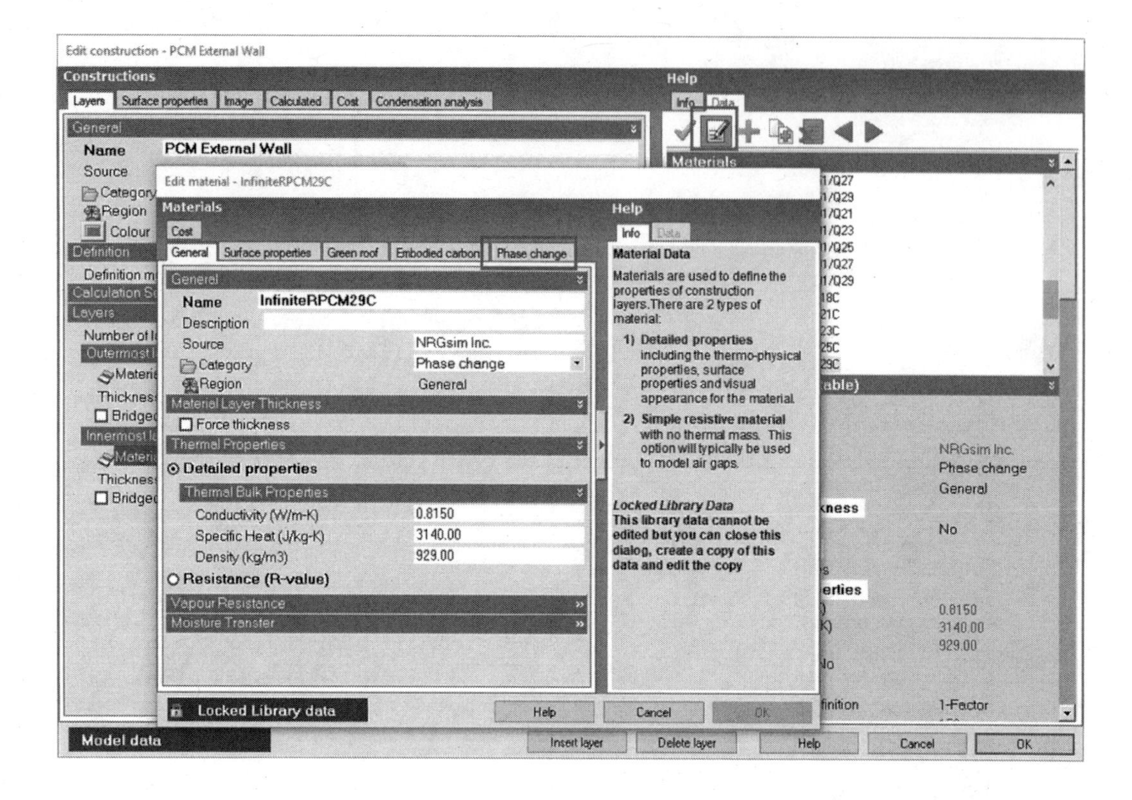

Step 21: Note the peak freezing and melting temperatures.

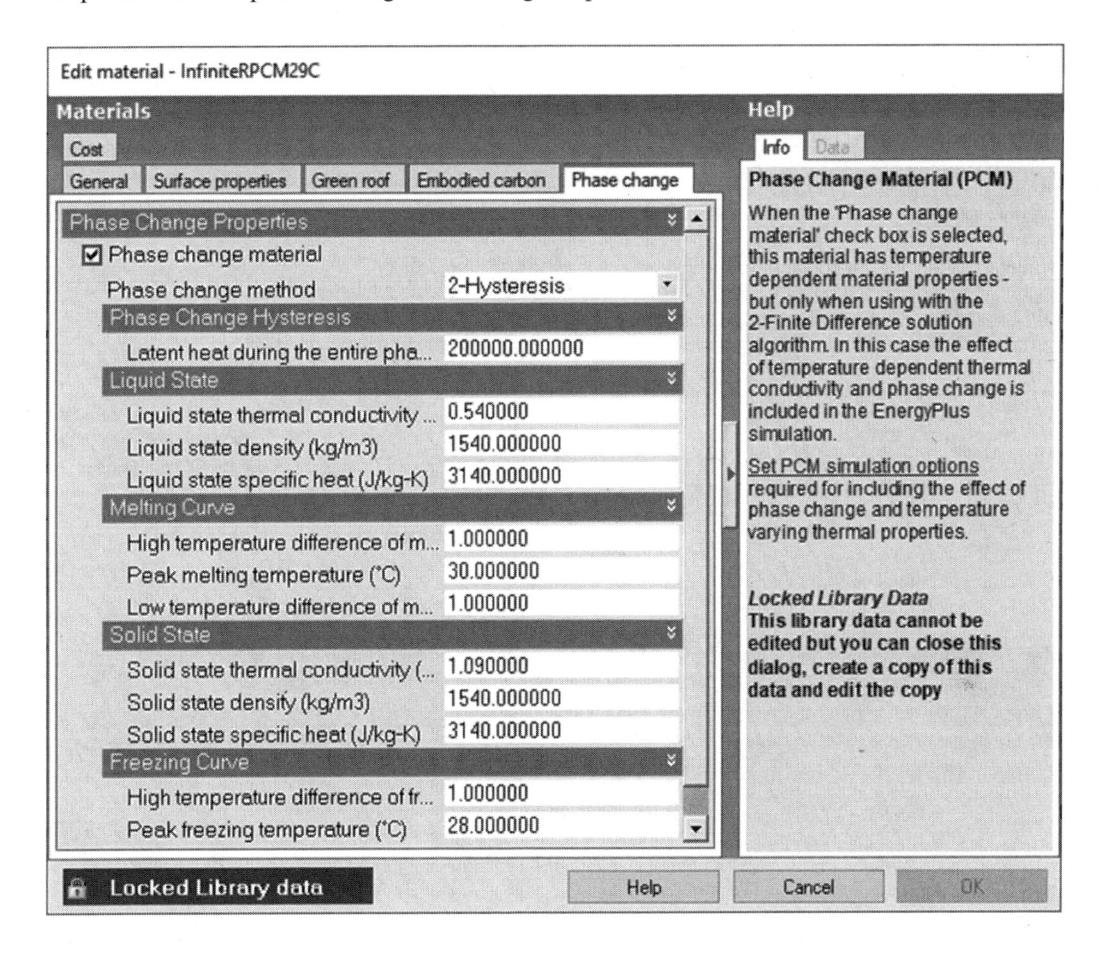

Step 22: Move the **PCM** layer to **Outermost** layer.

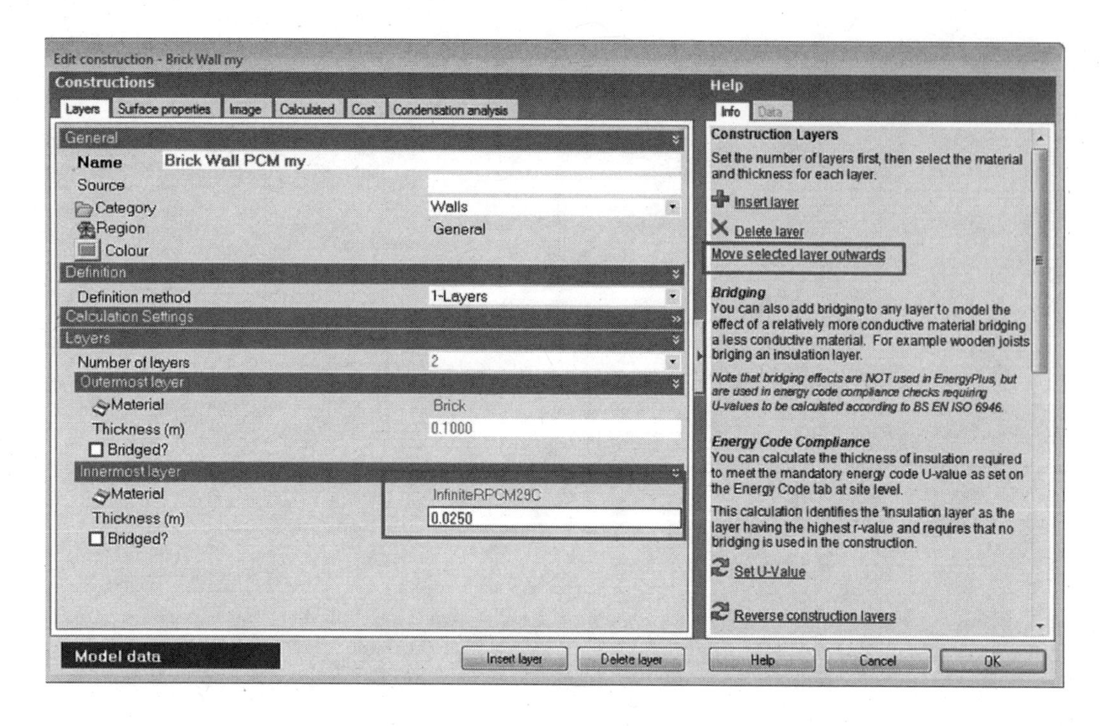

Step 23: Click the **Select item** button. Click the **Simulation** screen tab.

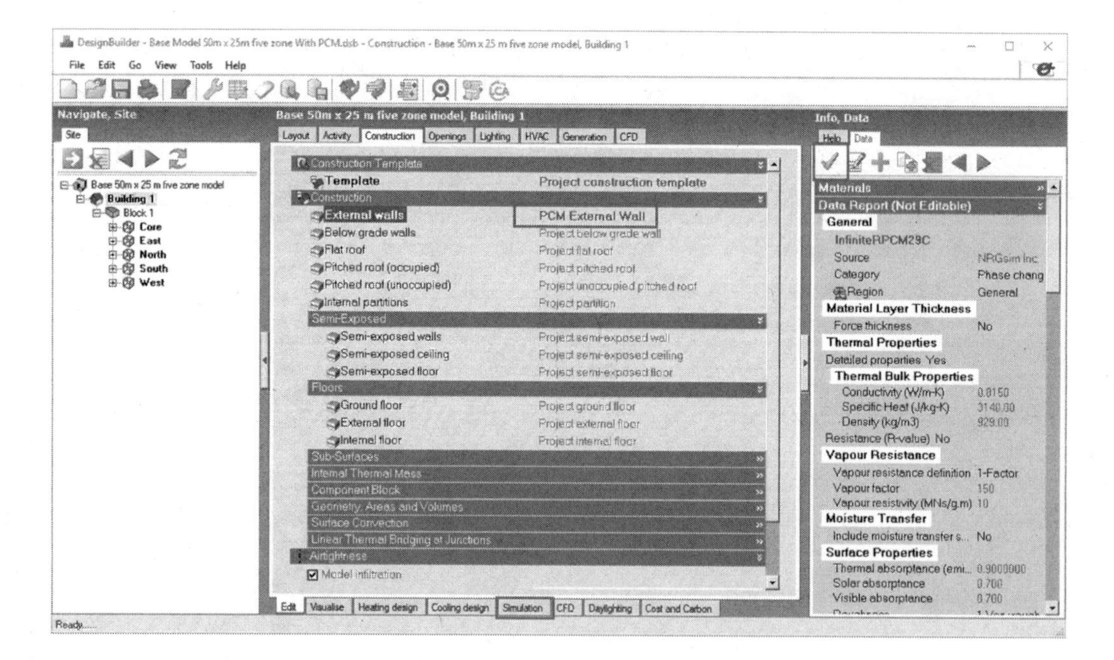

Step 24: Click **Update Calculation**. Click the **Options** tab. Select **12** from **Time steps per hour** drop-down list. Select **2-Finite Difference** from the **Solution algorithm** drop-down list.

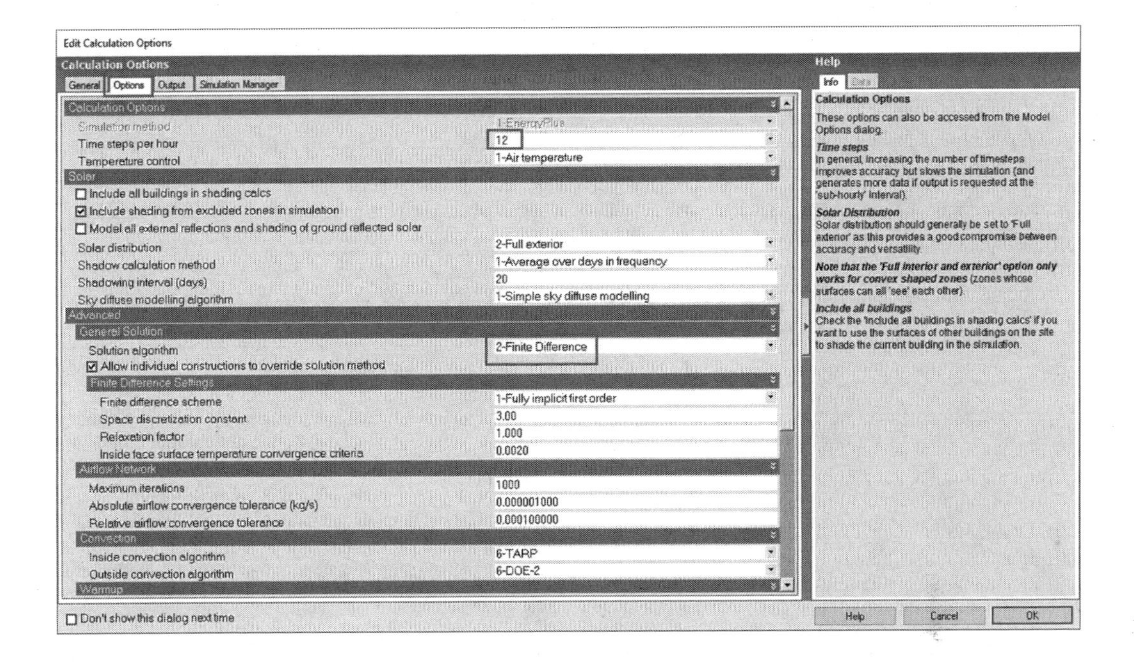

Solution Algorithm

The *finite difference solution algorithm* must be used to include the effect of material phase-change properties in simulations. If you use the conduction transfer function (CTF) algorithm, the material will behave as if its PCM option were not selected. Note also that the CTF solution algorithm is the only option available for heating and cooling design, so the phase-change effect of the material will not be seen in these calculations. An error message to this effect will be generated for PCM simulations where the CTF solution method is selected.

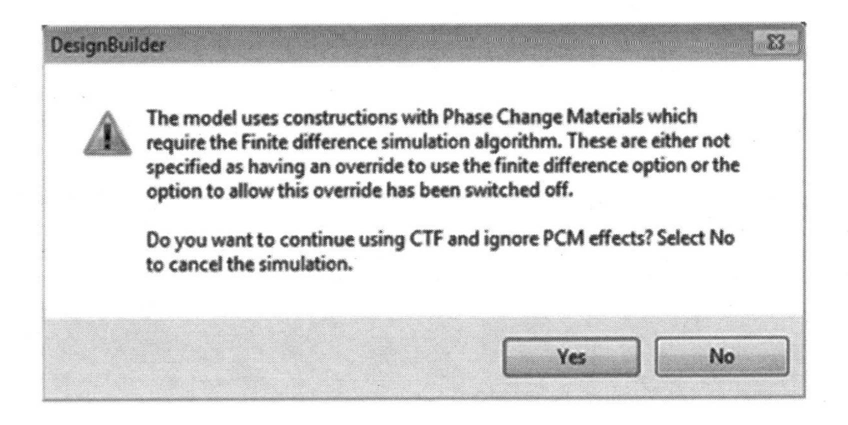

Recommended Simulation Parameters for PCM

Time step: 12 (higher values should increase accuracy at the expense of longer simulation times)
Solution algorithm: 2-Finite difference
Difference scheme: 1-Fully implicit first order
Space discretisation constant: 3
Relaxation factor: 1
Inside face surface temperature convergence criterion: 0.01 (lower values should increase accuracy at the expense of longer simulation times)

Source: https://designbuilder.co.uk/helpv6.0/Content/Phase_Change.htm.

Step 25: Simulate the model, and record the results.

Untitled, Building 1

Analysis | Summary | Parametric | Optimisation + UA/SA | Data Visualisation

Date/Time	Room Electricity (kWh)	Lighting (kWh)	Cooling (Electricity) (kWh)	DHW (Electricity) (kWh)
12:00:00 AM	26186.47	37850.41	109866.1	2369.774

Step 26: Note the inside and outside surface temperatures for the south wall.

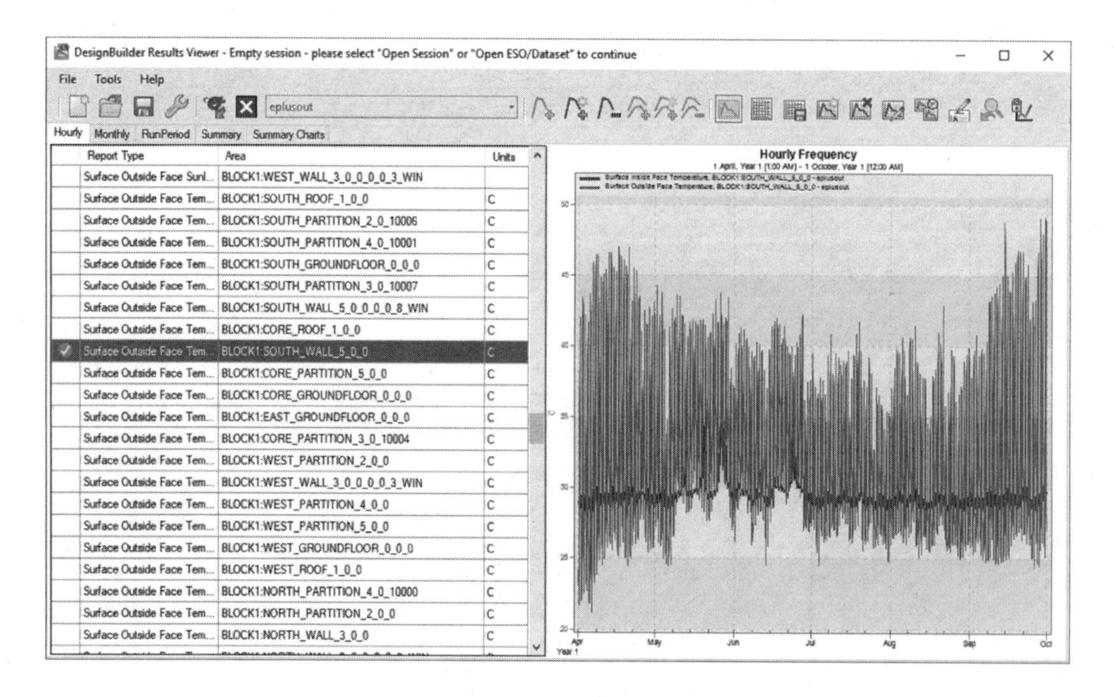

Step 27: Zoom the graph to see the data (from 27 to 31 May) for the following four days.

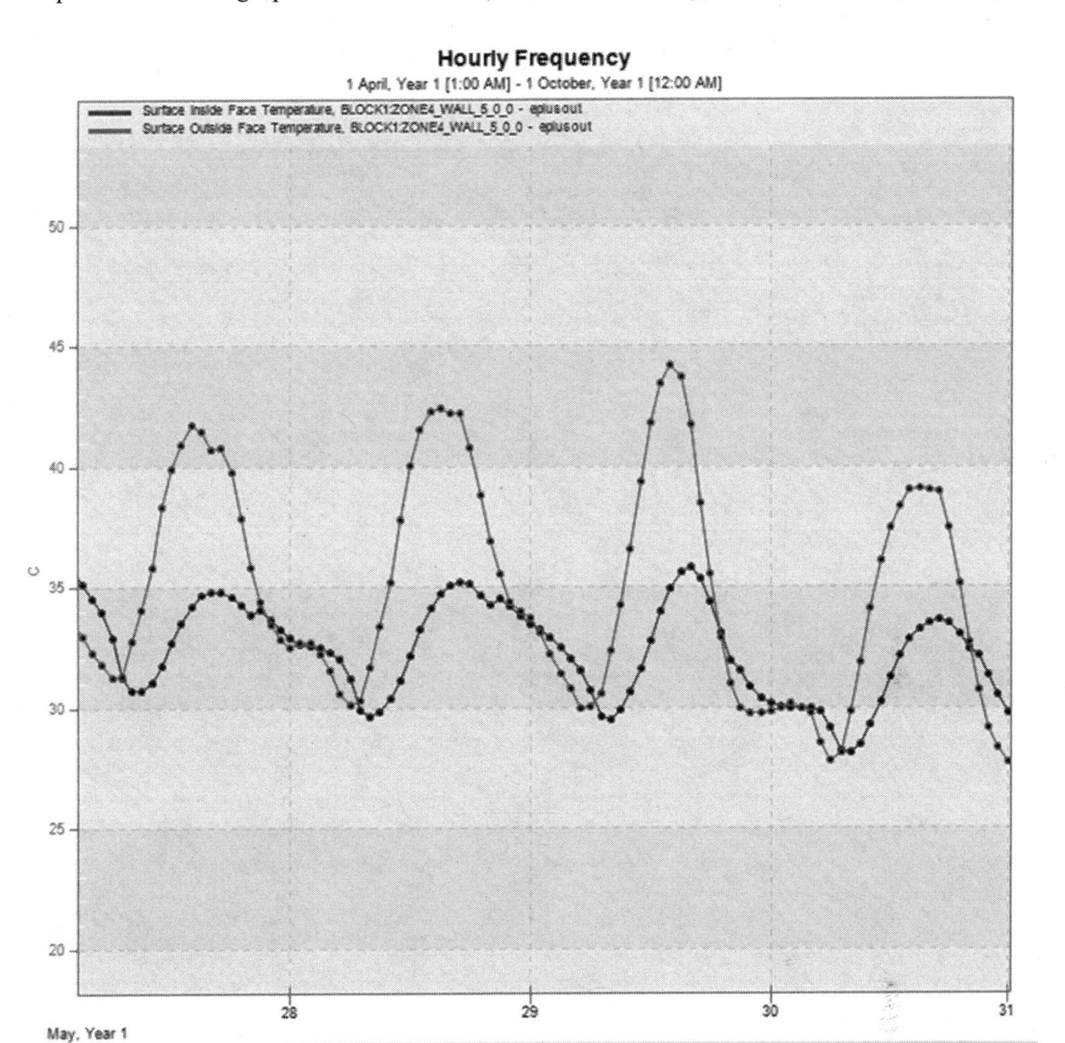

Step 28: Compare the results:

Brick wall construction

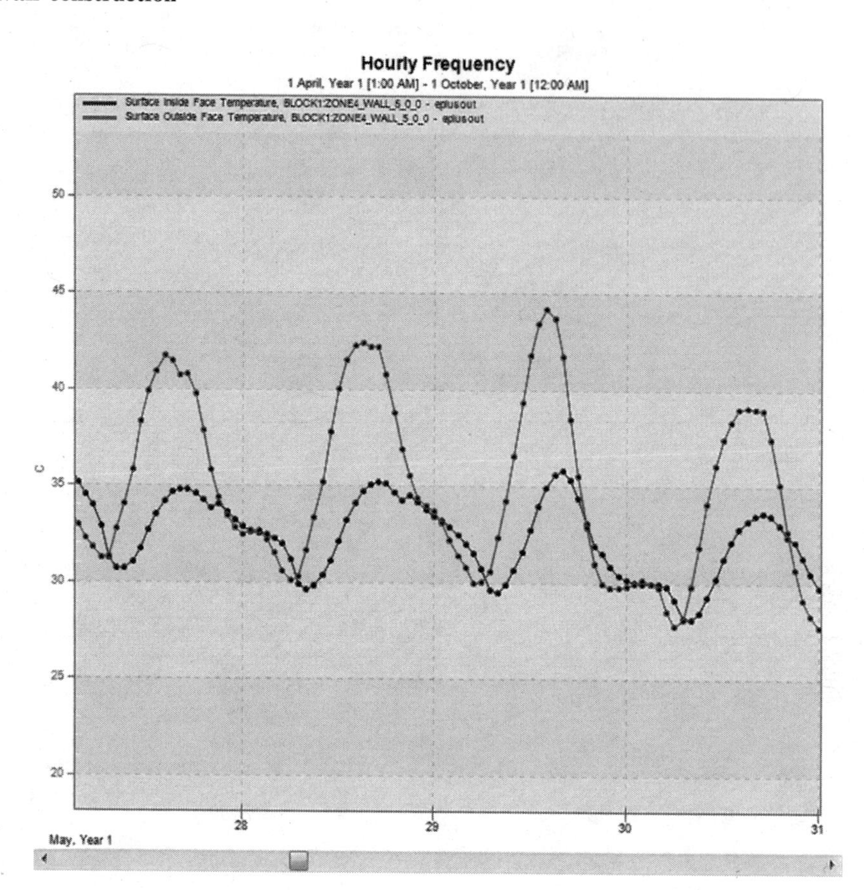

Brick with PCM wall construction

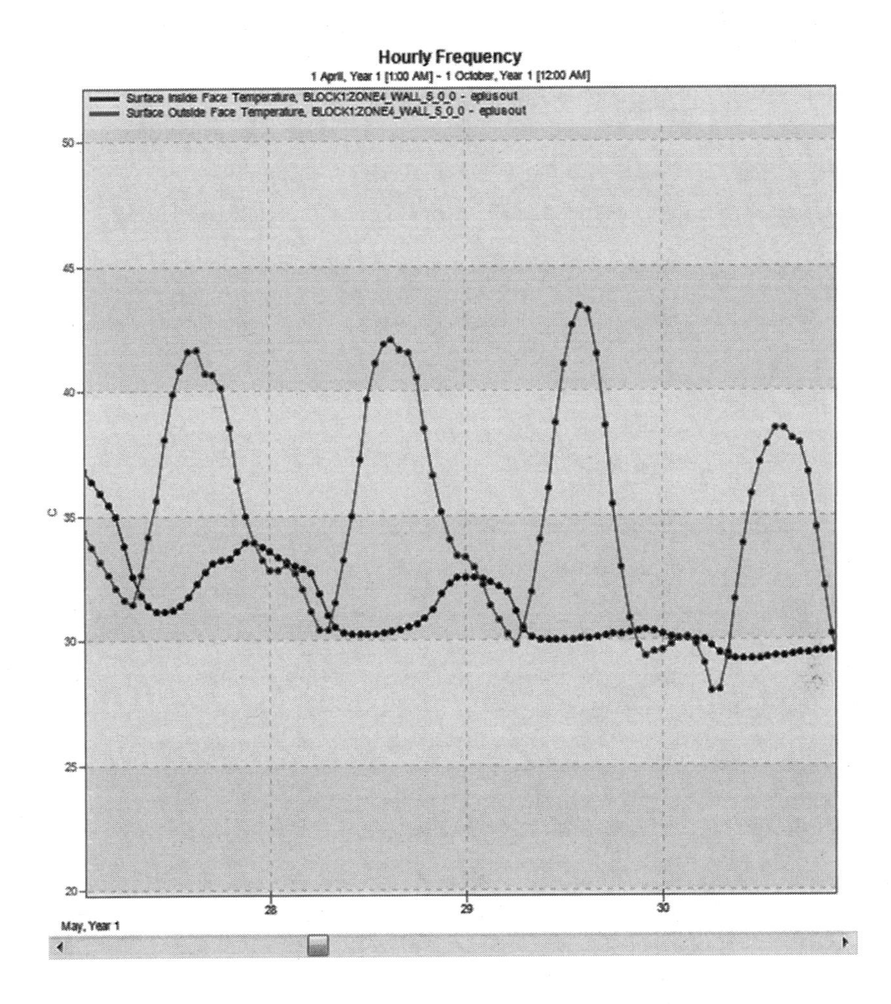

You can see that with the application of PCM, fluctuations in the inside temperature of the external wall construction have been reduced, and the cooling energy consumption also decreases, as shown in Table 3.14.

TABLE 3.14
Comparison of cooling energy consumption

External Wall Construction	Cooling Energy Consumption (kWh)
Brick wall	115,865
Brick wall + PCM	109,866

4 Openings and Shading

Openings are required in buildings to bring in daylight and fresh air and provide outdoor views. The energy for artificial electric light can be reduced if there is sufficient daylight in the space. However, daylight increases the heat gain in a space. If the climate is hot/warm, this results in an increase in the air-conditioning energy consumption. Hence, there is a need to optimize the window-to-wall ratio (WWR) to get the minimum energy consumption while getting sufficient daylight.

Choosing glass type is also important for buildings; building glass is specified by certain important properties, such as *U*-value, solar heat gain coefficient (SHGC) and visible light transmittance (VLT). Generally, the VLT-to-SHGC ratio is taken as an indicator of glass performance in cooling-dominated locations. The higher this ratio is, the better is the glass. Building shades can be used to cut the direct solar radiation in buildings and bring diffused daylight inside the perimeter space. Overhangs and fins are classified as fixed building shades. Operable shades can also be used to cut the direct radiation from windows.

In this chapter, through the four tutorials, you are going to learn how to analyse the impact of different glazing types, fixed shades and operable shades for a given climate. This can be useful in the analysis of different designs and approaches for reducing solar heat gain through windows.

Tutorial 4.1

Evaluating the Impact of Window-to-Wall Ratio and Glazing Type

GOAL

To evaluate the impact of the window-to-wall ratio and glazing type on energy consumption

> The *window-to-wall ratio* (WWR) is the ratio of the total glazing area to the total external wall area in conditioned zones.

WHAT ARE YOU GOING TO LEARN?

- How to set the WWR
- How to select glazing type

PROBLEM STATEMENT

In this tutorial, you are going to use a 50- × 25-m five-zone model with a 5-m perimeter depth. You are going to use the glass types and other configurations for the simulations, as given in Table 4.1. You will determine energy consumption for all cases for the **SINGAPORE/PAYA LEBA** weather location.

TABLE 4.1
Glass types and their properties

Tutorial part	Glass	Properties	Light-to-solar-gain ratio (*L/S*)	WWR	Daylight controls	Shade
I	Double glazed	Green 6mm/6mm Air, SHGC-0.49, VLT-0.66	1.35	0%–90% in steps of 10%	With and without	None
II	Single glazed	Sgl Clr 6mm. SHGC-0.81 and VLT-0.88	1.09	0%–90% in steps of 10%	With and without	None
III	ASHRAE 90.1 equivalent glass	U-6.81 W/m^2 K, SHGC-0.25 and VLT 0.53	2.12	0%–90% in steps of 10%	With and without	None and with a 0.5-m overhang on all windows

> The *light-to-solar gain* (*L/S*) *ratio* is the ratio between the VLT and SHGC. It provides a gauge of the relative efficiency of different glass or glazing types in transmitting daylight while blocking heat gain. The higher the number, the more the light is transmitted without adding excessive amounts of heat.

SOLUTION

Step 1: Create a **50- × 25-m** five-zone model with a 5-m perimeter depth.

PART I: WITH DBL GREEN 6MM/6MM AIR GLASS

Step 2: Click on the **Openings** tab, and select **Glazing type** as **Dbl Green 6mm/6mm Air**. Set WWR to 0.00%.

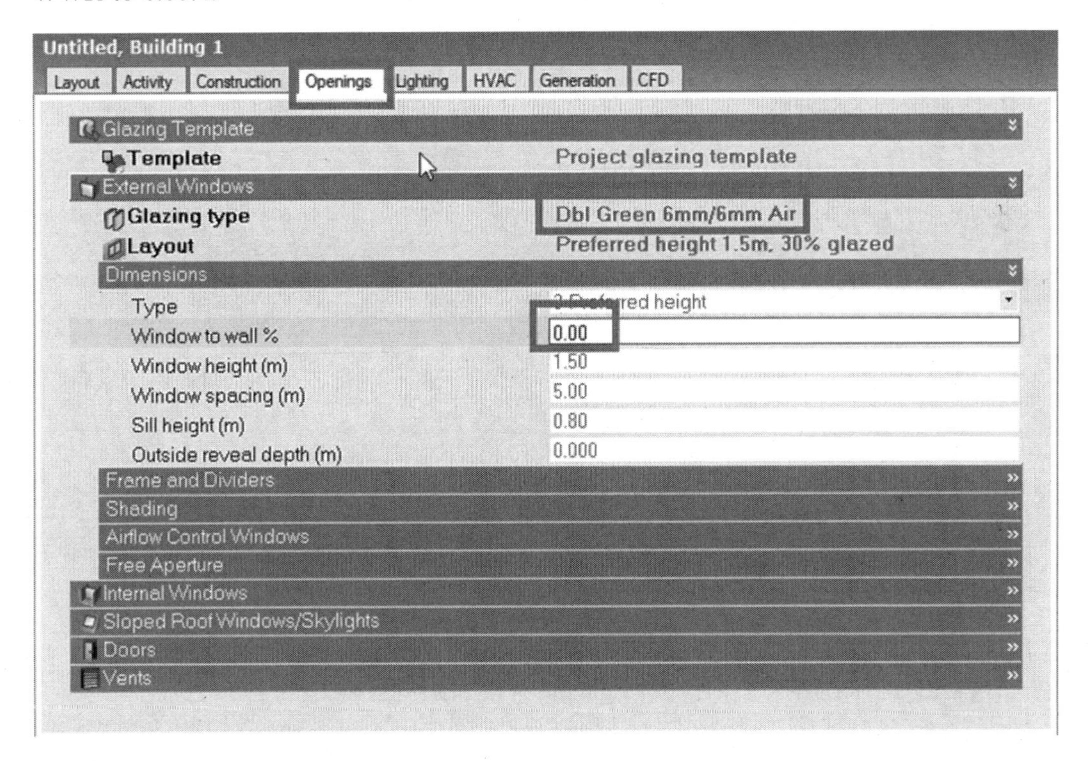

Step 3: Simulate the model, and record the results.

Step 4: Repeat the preceding steps to set the WWR (from 10% to 90% in steps of 10%) as given in the problem statement. Record the results for all WWRs without lighting control (Table 4.2).

TABLE 4.2
Annual energy consumption for a double-glazed window without daylight controls

Double glazing without daylight controls (Dbl Green 6mm/6mm Air)					
WWR (%)	**Room electricity (kWh)**	**Lighting (kWh)**	**Cooling (electricity) (kWh)**	**Domestic hot water (DHW, electricity) (kWh)**	**Annual consumption (electricity) (kWh)**
0	51,112.41	73,868.52	188,370.70	4,624.83	317,976.46
10	51,112.41	73,868.52	195,081.40	4,624.83	324,687.16
20	51,112.41	73,868.52	201,801.20	4,624.83	331,406.96
30	51,112.41	73,868.52	208,116.00	4,624.83	337,721.76
40	51,112.41	73,868.52	214,009.50	4,624.83	343,615.26
50	51,112.41	73,868.52	219,573.00	4,624.83	349,178.76
60	51,112.41	73,868.52	224,767.20	4,624.83	354,372.96
70	51,112.41	73,868.52	229,627.50	4,624.83	359,233.26
80	51,112.41	73,868.52	234,298.50	4,624.83	363,904.26
90	51,112.41	73,868.52	238,610.80	4,624.83	368,216.56

Step 5: Select the **Lighting** tab. In the **Lighting Control** section, select the **ON** checkbox.

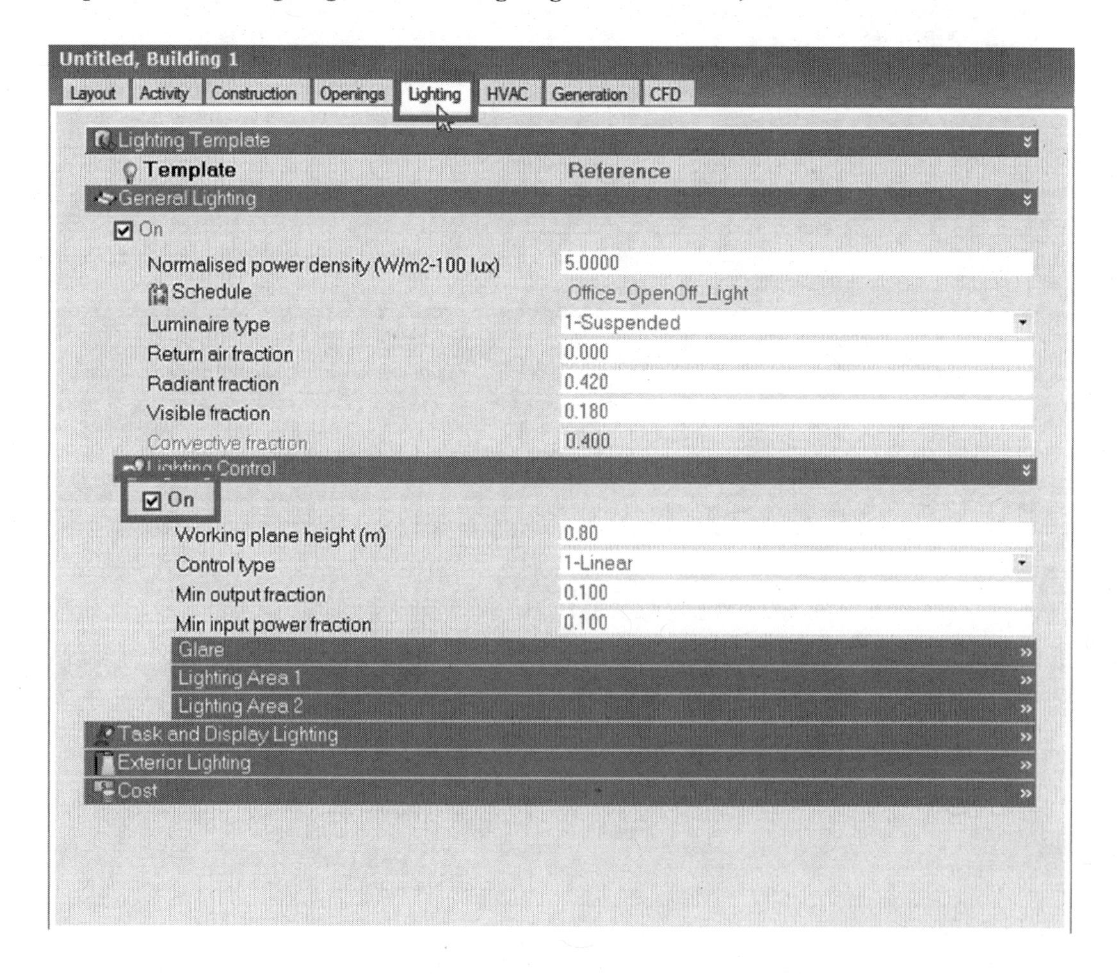

When you select the ON checkbox, you get the daylight sensor placed in all zones. You can see that the lighting energy consumption for the daylit perimeter zones will decrease.

Step 6: Simulate the model, and record the results. Record the results for all WWRs with lighting control (Table 4.3).

TABLE 4.3
Annual energy consumption for a double-glazed window with daylight controls

Double glazing with daylight controls (Dbl Green 6mm/6mm Air)					
WWR (%)	**Room electricity (kWh)**	**Lighting (kWh)**	**Cooling (electricity) (kWh)**	**DHW (electricity) (kWh)**	**Annual consumption (electricity) (kWh)**
0	51,112.41	73,868.52	188,370.70	4,624.83	317,976.46
10	51,112.41	49,144.95	180,210.30	4,624.83	285,092.49
20	51,112.41	45,076.97	185,136.90	4,624.83	285,951.11
30	51,112.41	43,982.02	191,277.60	4,624.83	290,996.86
40	51,112.41	43,422.12	197,293.10	4,624.83	296,452.46
50	51,112.41	42,904.71	202,983.50	4,624.83	301,625.45
60	51,112.41	42,539.73	208,329.30	4,624.83	306,606.27
70	51,112.41	42,288.48	213,379.50	4,624.83	311,405.22
80	51,112.41	42,517.13	218,468.20	4,624.83	316,722.57
90	51,112.41	42,270.16	222,877.30	4,624.83	320,884.70

Step 7: Compare the results with and without daylight controls (Table 4.4).

TABLE 4.4
Comparison of the total annual energy consumption for a double-glazed window with and without daylight controls

WWR (%)	Annual energy consumption without daylight controls (kWh)	Annual energy consumption with daylight controls (kWh)
0	317,976.46	317,976.46
10	324,687.16	285,092.49
20	331,406.96	285,951.11
30	337,721.76	290,996.86
40	343,615.26	296,452.46
50	349,178.76	301,625.45
60	354,372.96	306,606.27
70	359,233.26	311,405.22
80	363,904.26	316,722.57
90	368,216.56	320,884.70

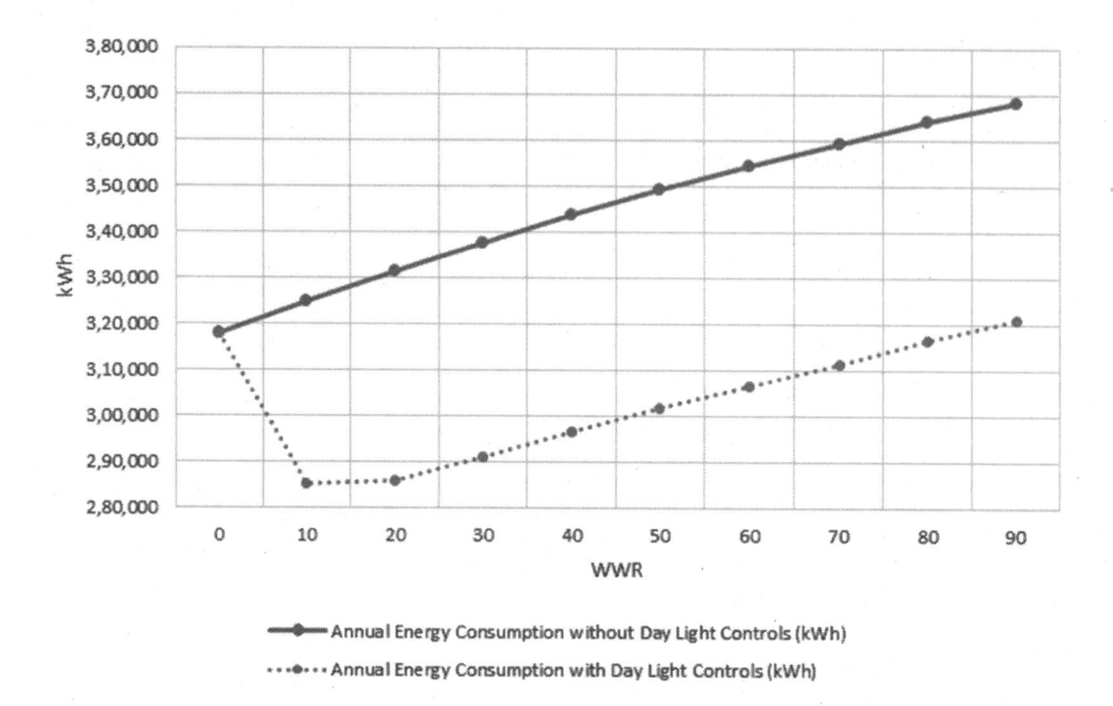

The results show that with double-glazed windows, the building consumes minimum energy at 20% WWR when daylight controls are installed in the building in all its daylit perimeter spaces.

Part II: With Sgl Clr 6mm Glass

Step 8: Repeat the preceding steps to get simulation results with **Sgl Clr 6mm** glass (Tables 4.5 through 4.7).

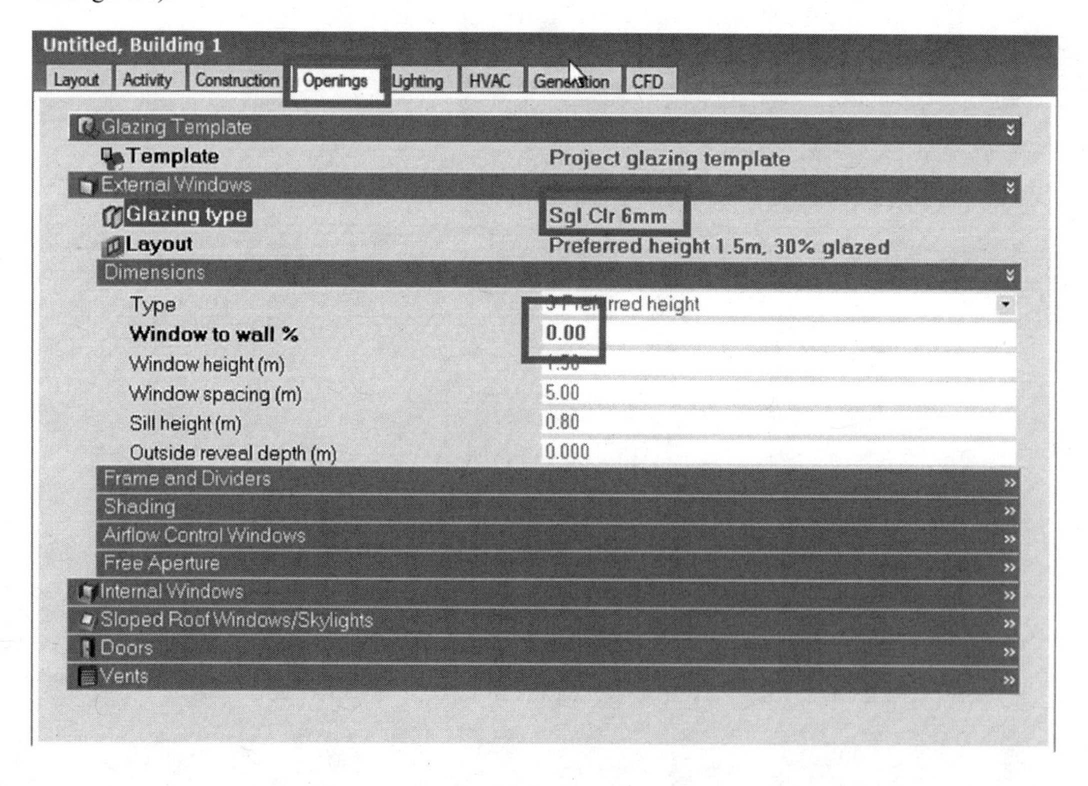

TABLE 4.5
Annual energy consumption for a single-glazed window without daylight controls

Single glazing without daylight controls (Sgl Clr 6mm)					
WWR (%)	Room electricity (kWh)	Lighting (kWh)	Cooling (electricity) (kWh)	DHW (electricity) (kWh)	Annual consumption (electricity) (kWh)
0	51,112.41	73,868.52	188,370.70	4,624.83	317,976.46
10	51,112.41	73,868.52	199,169.90	4,624.83	328,775.66
20	51,112.41	73,868.52	209,540.40	4,624.83	339,146.16
30	51,112.41	73,868.52	218,740.40	4,624.83	348,346.16
40	51,112.41	73,868.52	227,101.60	4,624.83	356,707.36
50	51,112.41	73,868.52	234,878.50	4,624.83	364,484.26
60	51,112.41	73,868.52	242,001.50	4,624.83	371,607.26
70	51,112.41	73,868.52	248,497.00	4,624.83	378,102.76
80	51,112.41	73,868.52	254,449.60	4,624.83	384,055.36
90	51,112.41	73,868.52	259,919.00	4,624.83	389,524.76

TABLE 4.6
Annual energy consumption for a single-glazed window with daylight controls

Single glazing with daylight controls (Sgl Clr 6mm)					
WWR (%)	Room electricity (kWh)	Lighting (kWh)	Cooling (electricity) (kWh)	DHW (electricity) (kWh)	Annual consumption (electricity) (kWh)
0	51,112.41	73,868.52	188,370.70	4,624.83	317,976.46
10	51,112.41	46,795.33	183,243.70	4,624.83	285,776.27
20	51,112.41	44,168.18	192,909.00	4,624.83	292,814.42
30	51,112.41	43,349.78	202,343.20	4,624.83	301,430.22
40	51,112.41	42,887.94	210,930.70	4,624.83	309,555.88
50	51,112.41	42,450.09	218,971.10	4,624.83	317,158.43
60	51,112.41	42,145.08	226,326.60	4,624.83	324,208.92
70	51,112.41	41,942.79	233,081.10	4,624.83	330,761.13
80	51,112.41	42,127.75	239,404.60	4,624.83	337,269.59
90	51,112.41	41,929.42	245,068.50	4,624.83	342,735.16

TABLE 4.7
Comparison of the total annual energy consumption for a single-glazed window with and without daylight controls

WWR (%)	Annual energy consumption without daylight controls (kWh)	Annual energy consumption with daylight controls (kWh)
0	317,976.46	317,976.46
10	328,775.66	285,776.27
20	339,146.16	292,814.42
30	348,346.16	301,430.22
40	356,707.36	309,555.88
50	364,484.26	317,158.43
60	371,607.26	324,208.92
70	378,102.76	330,761.13
80	384,055.36	337,269.59
90	389,524.76	342,735.16

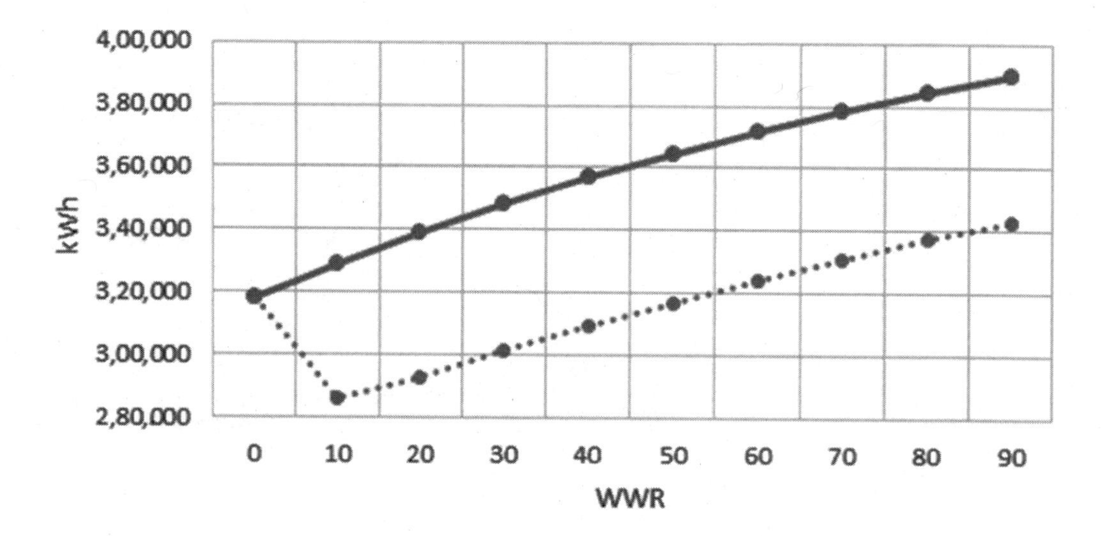

When there are no daylight controls, annual energy consumption increases linearly with increases in the WWR. With the use of daylight controls, first, energy decreases and then increases; minimum energy consumption is at 10% WWR.

Part III(a): With ASHRAE 90.1-2007 Equivalent Glass (U-1.20 (6.81), SHGC-0.25 and VLT 53%)

Step 9: Now refer to the preceding steps to get simulation results with **ASHRAE 90.1** equivalent glass, **Vertical glazing, 0%–40% of wall, U-1.20 (6.81) and SHGC-0.25** (Tables 4.8 and 4.9).

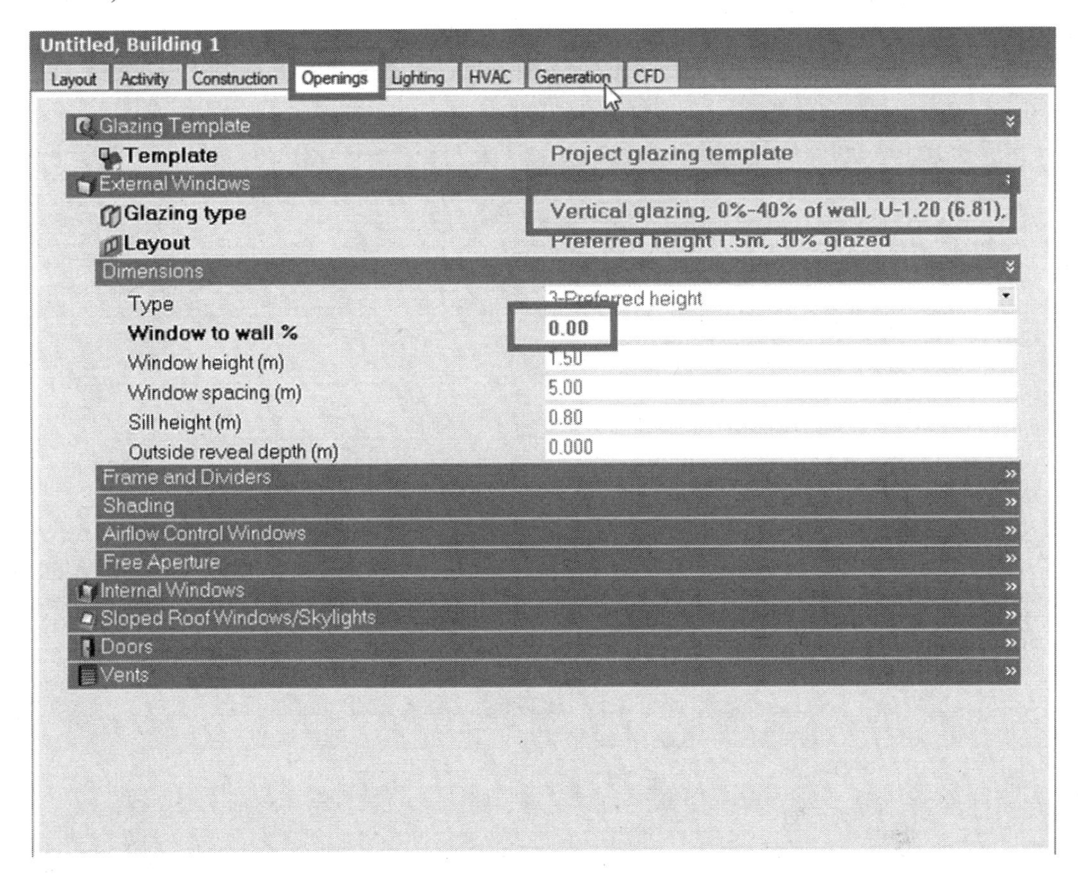

TABLE 4.8
Annual energy consumption with ASHRAE 90.1 equivalent glass without daylight controls

ASHRAE 90.1 Glass without daylight controls (vertical glazing, 0%-40% of wall, U-1.20 (6.81) and SHGC-0.25)					
WWR (%)	Room electricity (kWh)	Lighting (kWh)	Cooling (electricity) (kWh)	DHW (electricity) (kWh)	Annual consumption (electricity) (kWh)
0	51,112.41	73,868.52	188,370.70	4,624.83	317,976.46
10	51,112.41	73,868.52	191,684.00	4,624.83	321,289.76
20	51,112.41	73,868.52	195,014.00	4,624.83	324,619.76
30	51,112.41	73,868.52	198,166.10	4,624.83	327,771.86
40	51,112.41	73,868.52	201,121.50	4,624.83	330,727.26
50	51,112.41	73,868.52	203,840.10	4,624.83	333,445.86
60	51,112.41	73,868.52	206,318.40	4,624.83	335,924.16
70	51,112.41	73,868.52	208,638.70	4,624.83	338,244.46
80	51,112.41	73,868.52	210,965.00	4,624.83	340,570.76
90	51,112.41	73,868.52	212,987.90	4,624.83	342,593.66

TABLE 4.9
Annual energy consumption with ASHRAE 90.1 equivalent glass with daylight controls

Good glazing with daylight controls (vertical glazing, 0%–40% of wall, U-1.20 (6.81) and SHGC-0.25)

WWR (%)	Room electricity (kWh)	Lighting (kWh)	Cooling (electricity) (kWh)	DHW (electricity) (kWh)	Annual consumption (electricity) (kWh)
0	51,112.41	73,868.52	188,370.70	4,624.83	317,976.46
10	51,112.41	50,573.21	177,596.60	4,624.83	283,907.05
20	51,112.41	45,625.00	178,511.50	4,624.83	279,873.74
30	51,112.41	44,339.81	181,450.50	4,624.83	281,527.55
40	51,112.41	43,710.06	184,506.80	4,624.83	283,954.10
50	51,112.41	43,150.93	187,323.50	4,624.83	286,211.67
60	51,112.41	42,752.63	189,992.80	4,624.83	288,482.67
70	51,112.41	42,477.01	192,513.50	4,624.83	290,727.75
80	51,112.41	42,728.87	195,263.60	4,624.83	293,729.71
90	51,112.41	42,458.06	197,479.70	4,624.83	295,675.00

Step 10: Compare the results with and without daylight controls (Table 4.10).

TABLE 4.10
Comparison of simulation results of glass with and without daylight controls

WWR (%)	Annual energy consumption without daylight controls (kWh)	Annual energy consumption with daylight controls (kWh)
0	317,976.46	317,976.46
10	321,289.76	283,907.05
20	324,619.76	279,873.74
30	327,771.86	281,527.55
40	330,727.26	283,954.10
50	333,445.86	286,211.67
60	335,924.16	288,482.67
70	338,244.46	290,727.75
80	340,570.76	293,729.71
90	342,593.66	295,675.00

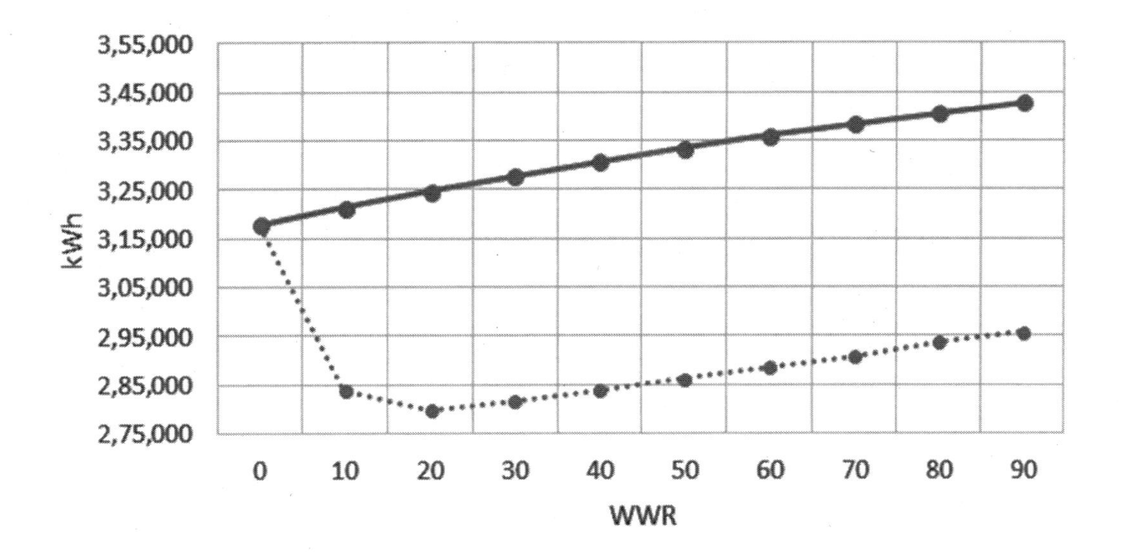

Step 11: Now compare the results for all simulations of double-glazed, single-glazed and ASHRAE 90.1 equivalent glass with lighting controls (Table 4.11).

TABLE 4.11
Comparison of the total annual energy consumption for all glass types with daylight controls

WWR (%)	Double glazing (L/S = 1.35) (kWh)	Single glazing (L/S = 1.09) (kWh)	ASHRAE 90.1 equivalent glazing (L/S = 2.12) (kWh)
0	317,976.46	317,976.46	317,976.46
10	285,092.49	285,776.27	283,907.05
20	285,951.11	292,814.42	279,873.74
30	290,996.86	301,430.22	281,527.55
40	296,452.46	309,555.88	283,954.10
50	301,625.45	317,158.43	286,211.67
60	306,606.27	324,208.92	288,482.67
70	311,405.22	330,761.13	290,727.75
80	316,722.57	337,269.59	293,729.71
90	320,884.70	342,735.16	295,675.00

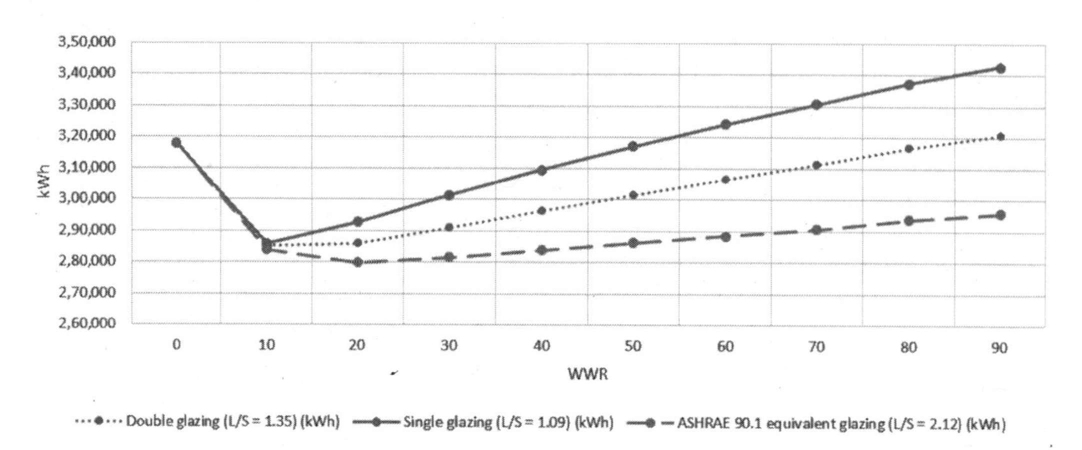

Part III(b): With ASHRAE 90.1-2007 Equivalent Glass (U-1.20 (6.81), SHGC-0.25 and VLT 53%) with the Shading of a 0.5-m Overhang

Step 12: Select the **Openings** tab, and check the **Local shading** checkbox under the **Shading** section. Select **0.5m Overhang** from the **Type** drop-down list (Table 4.12).

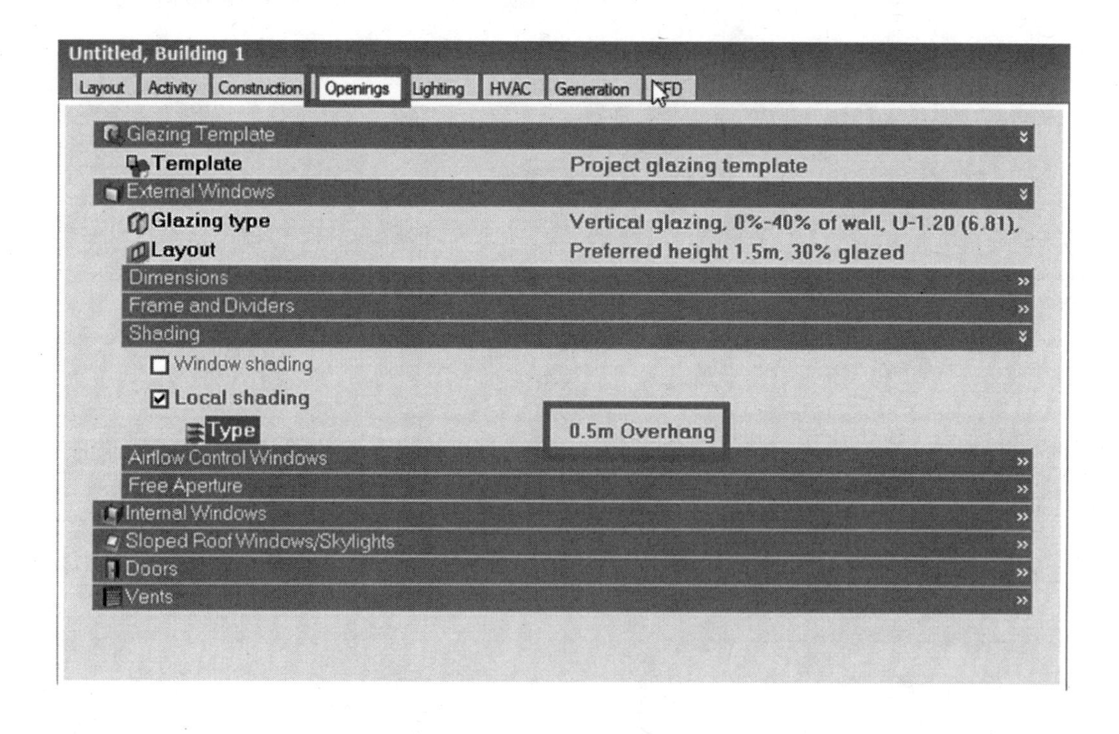

TABLE 4.12
Annual energy consumption with ASHRAE 90.1 equivalent glass with fixed shade and daylight controls

ASHRAE 90.1 equivalent glass with daylight controls (vertical glazing, 0%–40% of wall, U-1.20 (6.81) and SHGC-0.25)					
WWR (%)	Room electricity (kWh)	Lighting (kWh)	Cooling (electricity) (kWh)	DHW (electricity) (kWh)	Annual consumption (electricity) (kWh)
0	51,112.41	73,868.52	188,370.70	4,624.83	317,976.46
10	51,112.41	50,971.09	177,118.50	4,624.83	283,826.83
20	51,112.41	45,764.11	176,963.50	4,624.83	278,464.85
30	51,112.41	44,419.88	179,041.70	4,624.83	279,198.82
40	51,112.41	43,773.63	181,303.70	4,624.83	280,814.57
50	51,112.41	43,186.71	183,935.20	4,624.83	282,859.15
60	51,112.41	43,210.14	186,877.70	4,624.83	285,825.08
70	51,112.41	42,851.43	189,416.40	4,624.83	288,005.07
80	51,112.41	43,201.72	192,320.00	4,624.83	291,258.96
90	51,112.41	42,812.13	194,538.80	4,624.83	293,088.17

Compare the results for with and without shades for ASHRAE 90.1 equivalent glass and daylight controls (Table 4.13).

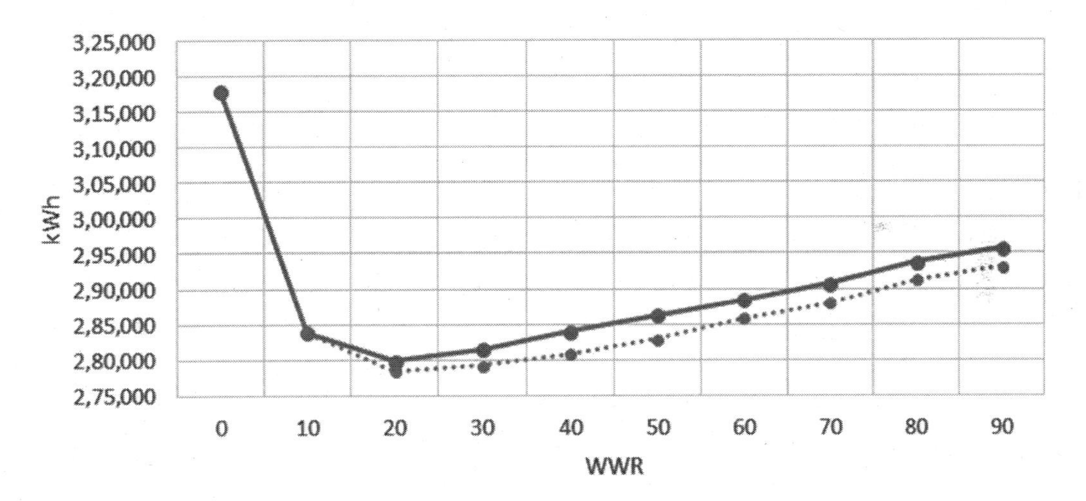

TABLE 4.13
Annual energy consumption results for ASHRAE 90.1 equivalent glass with shading simulated with and without daylight controls

WWR (%)	Annual energy consumption without daylight controls (kWh)	Annual energy consumption with daylight controls (kWh)
0	317,976.46	317,976.46
10	283,907.05	283,826.83
20	279,873.74	278,464.85
30	281,527.55	279,198.82
40	283,954.10	280,814.57
50	286,211.67	282,859.15
60	288,482.67	285,825.08
70	290,727.75	288,005.07
80	293,729.71	291,258.96
90	295,675.00	293,088.17

For warm/hot climates:

1. If a building is without daylight controls, then the energy consumption increases with an increase in WWR. With the increase in WWR, heat gains through glass increase in the building, and because of the absence of artificial lights dimming, the energy consumption increases.
2. If a building has daylight sensors, then artificial lights can be dimmed when sufficient daylight is available. With the increase in WWR, more daylight is available to the perimeter spaces. Daylight sensors help to reduce the artificial lighting load and offset the heat gains through the glass. However, after a point where the perimeter spaces are daylit, an increase in WWR does not save the artificial lighting energy as the lamps are fully dimmed. After this point with the increase in WWR, the heat ingress increases, thereby increasing the overall energy consumption.
3. Glass with a higher visible-light-to-solar-gain (*L*/*S*) ratio needs to be selected to get the maximum benefit from daylight.

EXERCISE 4.1

Repeat this tutorial for **WIEN/HOHE VARTE, Austria** weather location.

Tutorial 4.2

Evaluating the Impact of Overhangs and Fins

GOAL

To evaluate the impact of window overhangs and fins on energy performance

WHAT ARE YOU GOING TO LEARN?

- How to model overhangs and fins

PROBLEM STATEMENT

In this tutorial, you are going to use a 30- × 30-m model with a 5-m perimeter depth. Use square windows of 1.5 m. You need to select the **Split no fresh air** HVAC system for the simulation model. You also need to simulate the model with the following options:

1. No shades
2. Overhangs
3. Vertical fins

Use location DUBAI INTERNATIONAL, United Arab Emirates.

SOLUTION

Step 1: Open a new project, and create a **30- × 30-m** building with a 5-m perimeter depth.

Step 2: Select the **Openings** tab. Select the **Single glazing, clear, no shading** template. Select **Sgl Grey 6mm** in **Glazing** type.

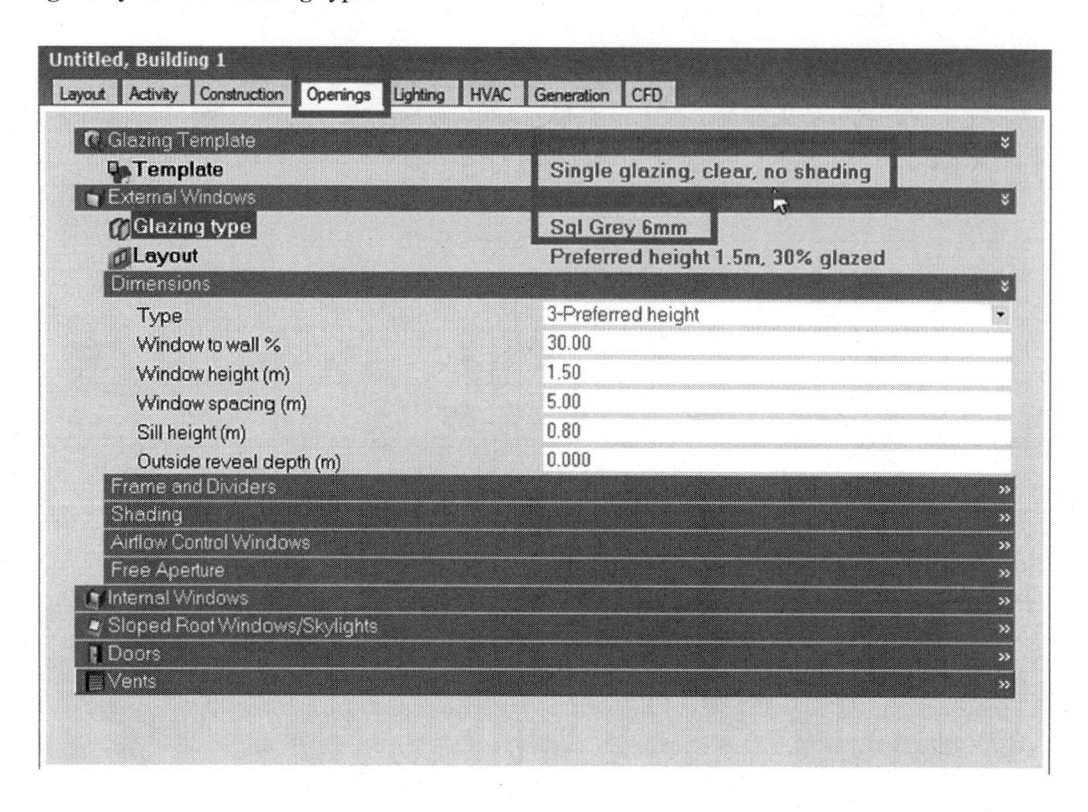

Step 3: Select **4-Fixed width** and **Height** from the **Type** drop-down list. Enter **1.5** in the **Window width (m)** and **Window height (m)** textbox. Enter **2.00** in the **Window spacing (m)** textbox.

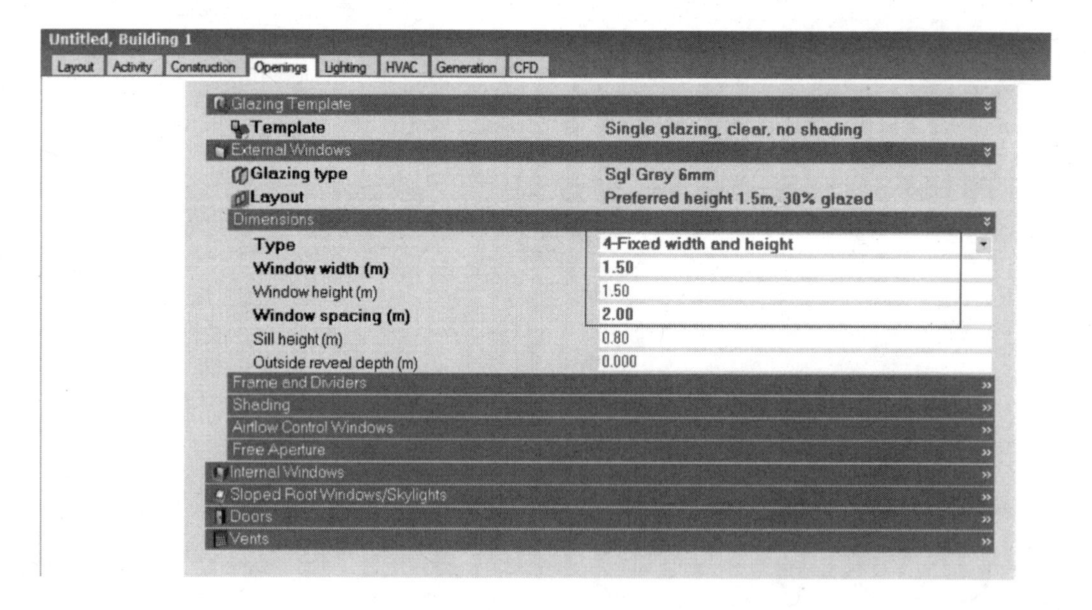

Step 4: Select the **HVAC** tab. Select the **Split no fresh air** template. Clear the **Heated** checkbox under the **Heating** section, and clear the **On** checkbox under **DHW**. Select the checkbox under **Cooling system**, and enter **3.0** as **CoP**.

When you use the **Split no fresh air** template, you can get the cooling energy consumption for all zones separately.

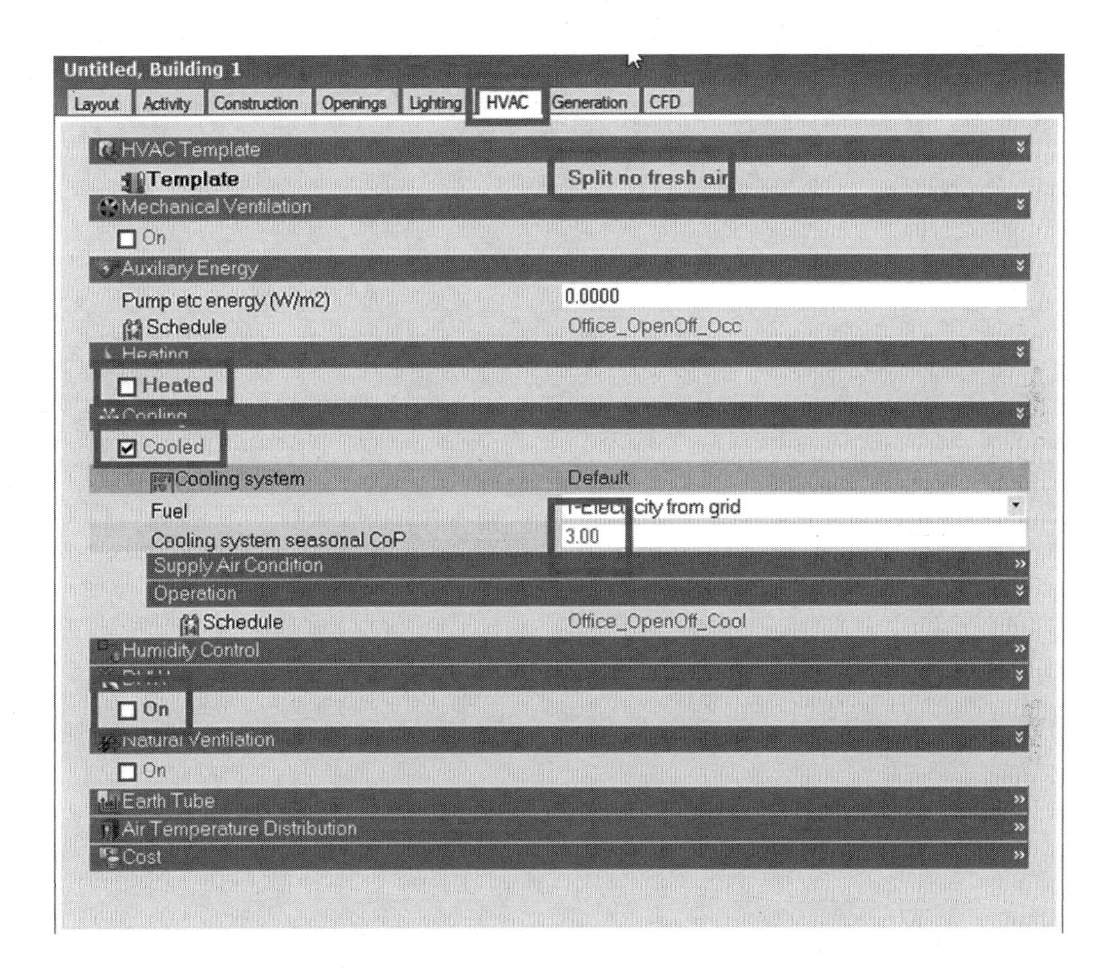

Step 5: Select the **Simulation** screen tab and then the **Output** tab. Expand **Summary Tables**, and select the **All Summary** checkbox in the **Summary Annual Reports** section.

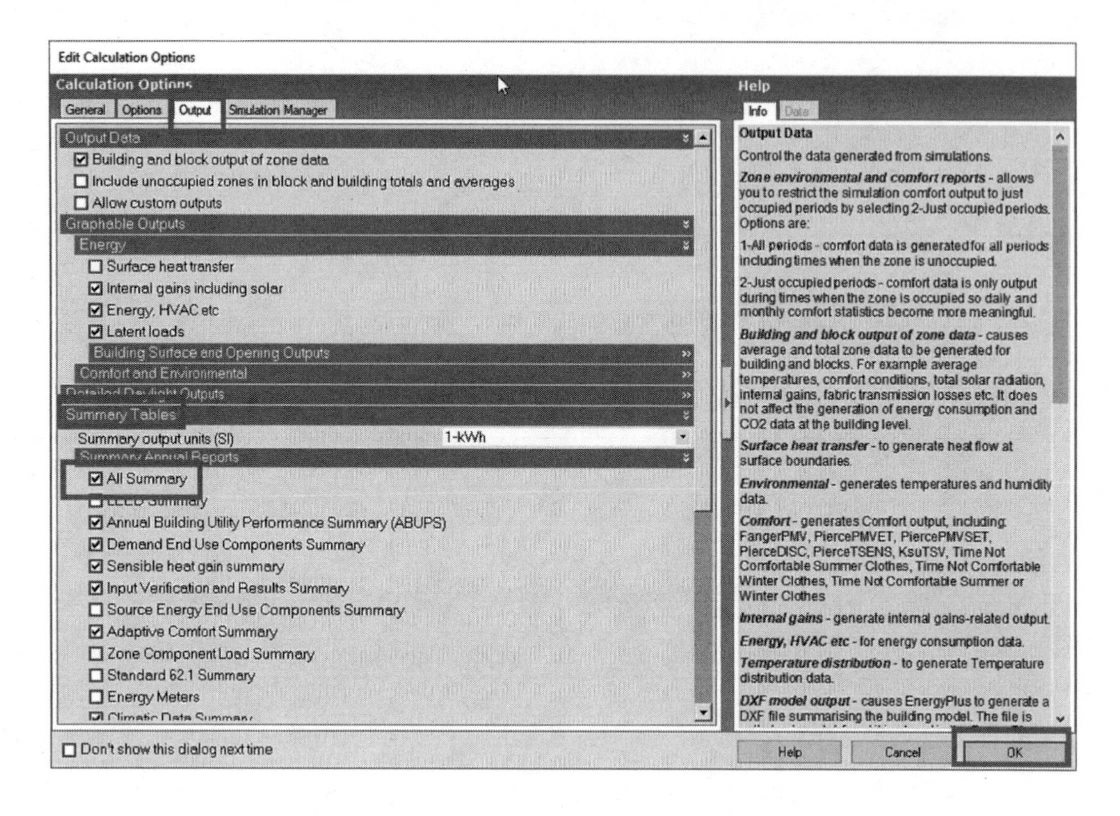

Step 6: Perform an annual simulation, and view the data in graphical format.

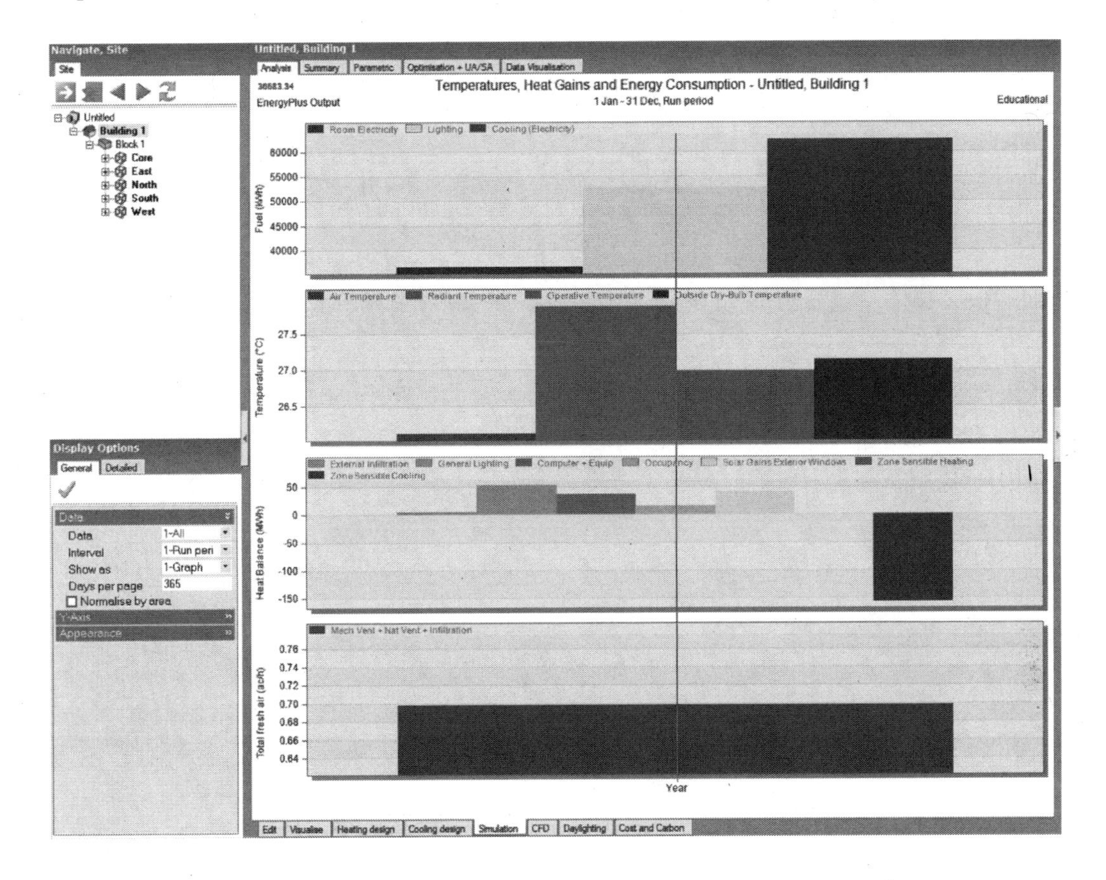

Step 7: Select the **Summary** tab.

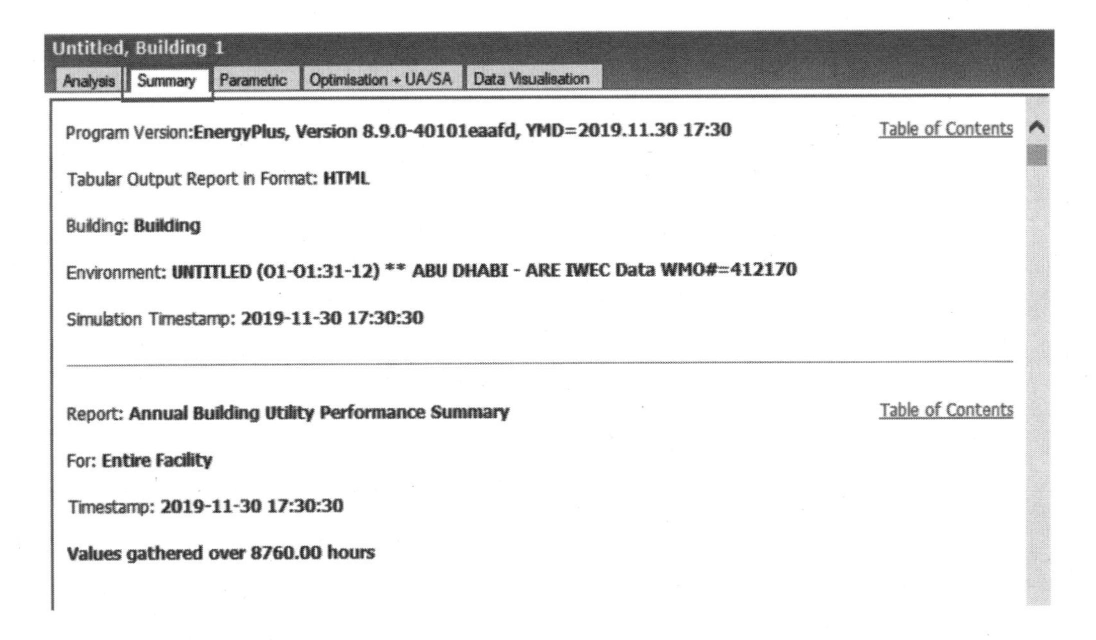

Step 8: Scroll down to read **Annual and Peak Values – Other**.

Untitled, Building 1

Analysis | Summary | Parametric | Optimisation + UA/SA | Data Visualisation

Annual and Peak Values - Other

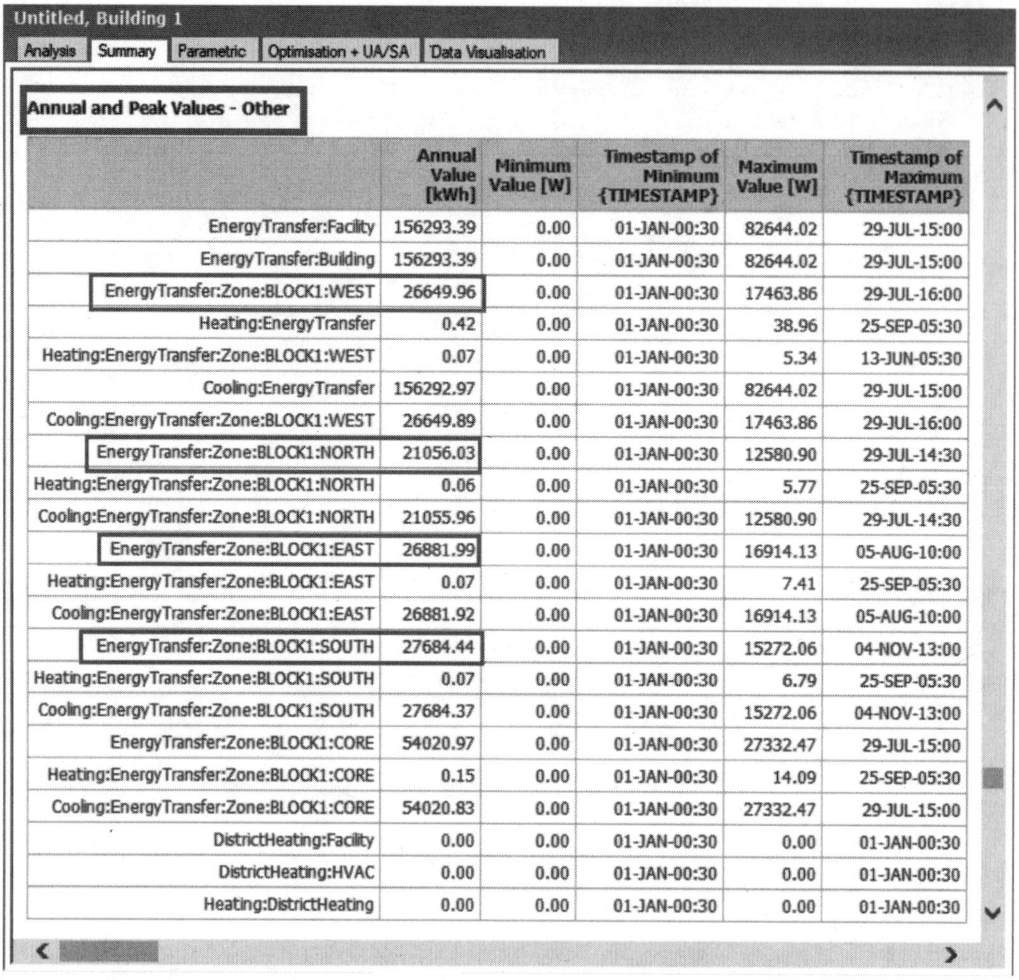

	Annual Value [kWh]	Minimum Value [W]	Timestamp of Minimum {TIMESTAMP}	Maximum Value [W]	Timestamp of Maximum {TIMESTAMP}
EnergyTransfer:Facility	156293.39	0.00	01-JAN-00:30	82644.02	29-JUL-15:00
EnergyTransfer:Building	156293.39	0.00	01-JAN-00:30	82644.02	29-JUL-15:00
EnergyTransfer:Zone:BLOCK1:WEST	26649.96	0.00	01-JAN-00:30	17463.86	29-JUL-16:00
Heating:EnergyTransfer	0.42	0.00	01-JAN-00:30	38.96	25-SEP-05:30
Heating:EnergyTransfer:Zone:BLOCK1:WEST	0.07	0.00	01-JAN-00:30	5.34	13-JUN-05:30
Cooling:EnergyTransfer	156292.97	0.00	01-JAN-00:30	82644.02	29-JUL-15:00
Cooling:EnergyTransfer:Zone:BLOCK1:WEST	26649.89	0.00	01-JAN-00:30	17463.86	29-JUL-16:00
EnergyTransfer:Zone:BLOCK1:NORTH	21056.03	0.00	01-JAN-00:30	12580.90	29-JUL-14:30
Heating:EnergyTransfer:Zone:BLOCK1:NORTH	0.06	0.00	01-JAN-00:30	5.77	25-SEP-05:30
Cooling:EnergyTransfer:Zone:BLOCK1:NORTH	21055.96	0.00	01-JAN-00:30	12580.90	29-JUL-14:30
EnergyTransfer:Zone:BLOCK1:EAST	26881.99	0.00	01-JAN-00:30	16914.13	05-AUG-10:00
Heating:EnergyTransfer:Zone:BLOCK1:EAST	0.07	0.00	01-JAN-00:30	7.41	25-SEP-05:30
Cooling:EnergyTransfer:Zone:BLOCK1:EAST	26881.92	0.00	01-JAN-00:30	16914.13	05-AUG-10:00
EnergyTransfer:Zone:BLOCK1:SOUTH	27684.44	0.00	01-JAN-00:30	15272.06	04-NOV-13:00
Heating:EnergyTransfer:Zone:BLOCK1:SOUTH	0.07	0.00	01-JAN-00:30	6.79	25-SEP-05:30
Cooling:EnergyTransfer:Zone:BLOCK1:SOUTH	27684.37	0.00	01-JAN-00:30	15272.06	04-NOV-13:00
EnergyTransfer:Zone:BLOCK1:CORE	54020.97	0.00	01-JAN-00:30	27332.47	29-JUL-15:00
Heating:EnergyTransfer:Zone:BLOCK1:CORE	0.15	0.00	01-JAN-00:30	14.09	25-SEP-05:30
Cooling:EnergyTransfer:Zone:BLOCK1:CORE	54020.83	0.00	01-JAN-00:30	27332.47	29-JUL-15:00
DistrictHeating:Facility	0.00	0.00	01-JAN-00:30	0.00	01-JAN-00:30
DistrictHeating:HVAC	0.00	0.00	01-JAN-00:30	0.00	01-JAN-00:30
Heating:DistrictHeating	0.00	0.00	01-JAN-00:30	0.00	01-JAN-00:30

Step 9: Record the annual value for the variable energy transfer zone (kWh) for all perimeter zones (Table 4.14).

TABLE 4.14
Annual values for the energy transfer for each zone with no shading

Zone	Annual value for energy transfer (kWh)
North	21,056.03
East	26,881.99
West	26,649.96
South	27,684.44

Save the simulation model for the next step. Now you are going to model window overhangs.

Step 10: Select the **Openings** tab.

Step 11: Click the **Shading** section. **Shading options** appears.

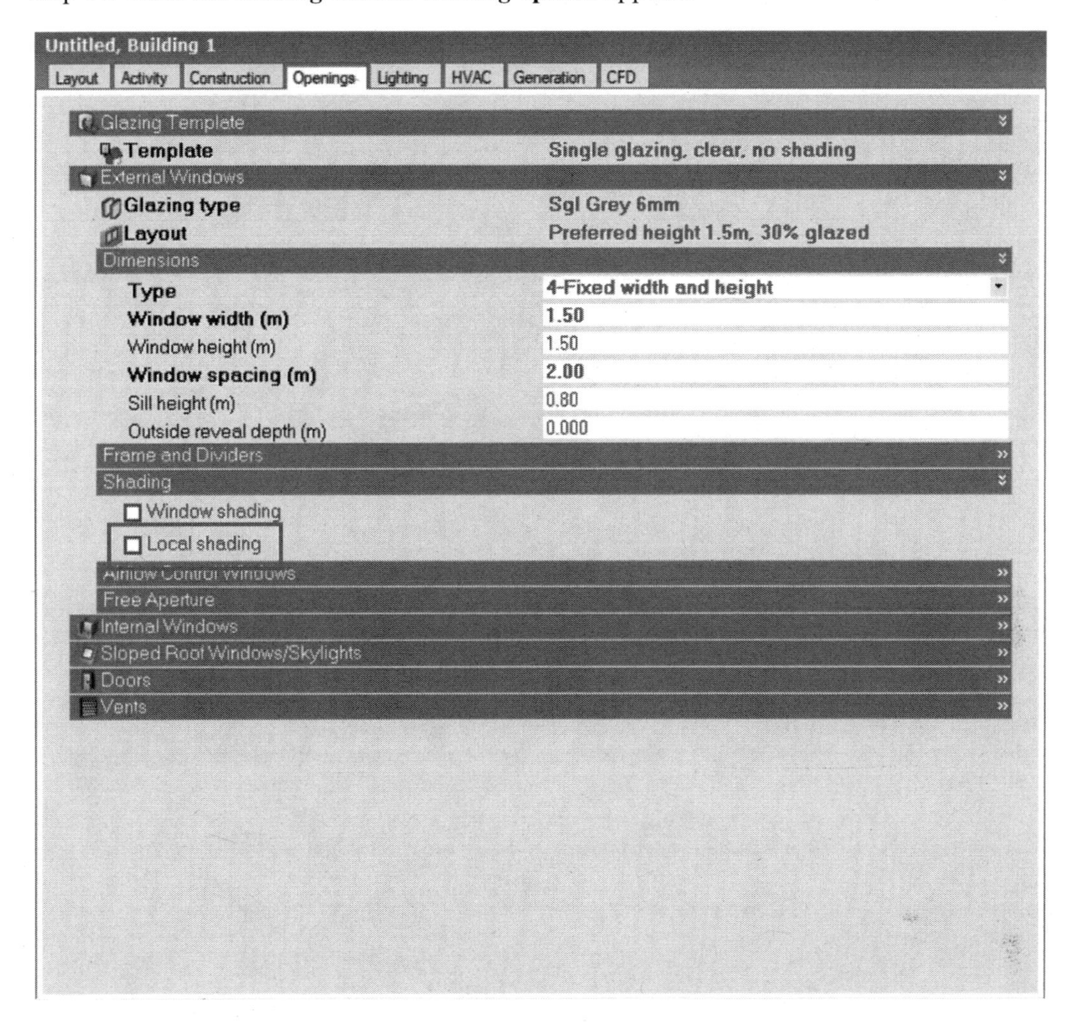

Step 12: Select the **Local shading** checkbox. It displays shading type. Select **0.5 m Overhang** in **Type**.

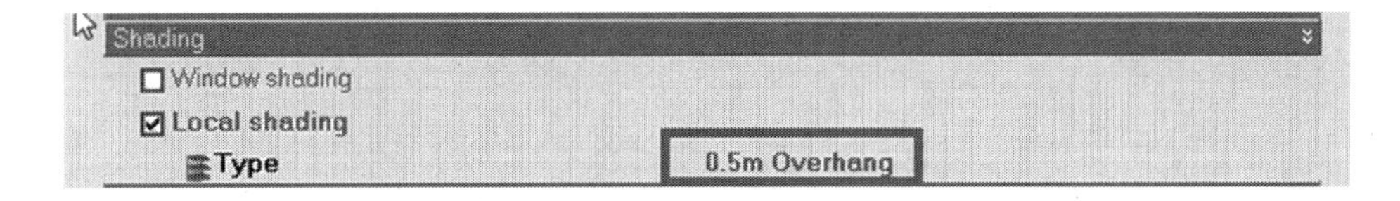

Step 13: Select the **Visualize** screen tab. Select the **Rendered View** tab. It displays the rendering of the building. Make sure that all windows are modelled with an overhang.

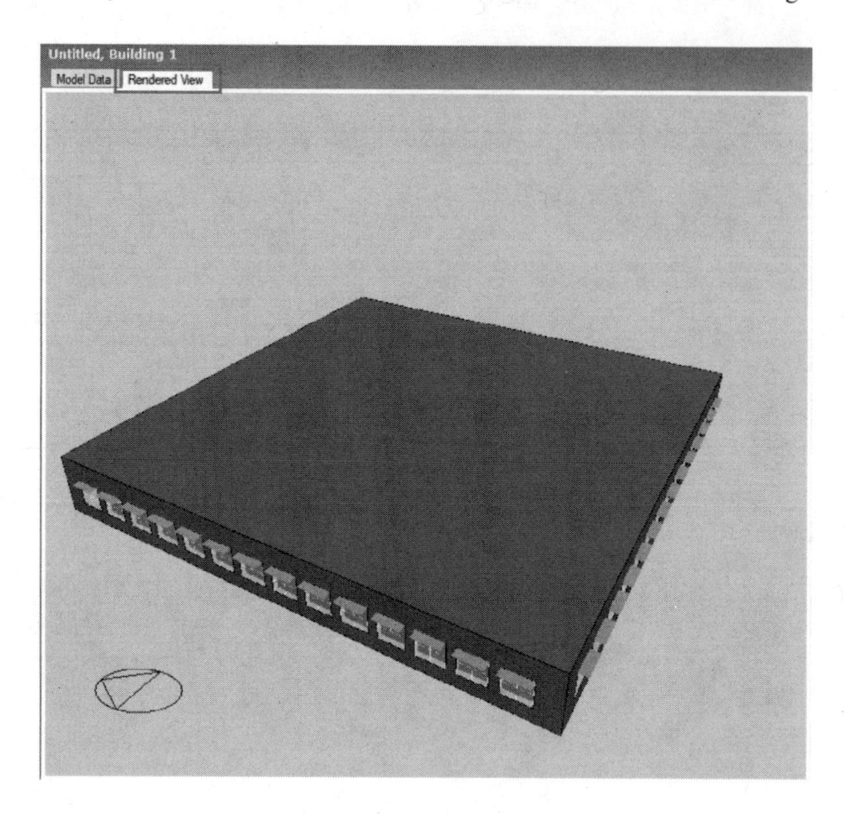

Step 14: Repeat the preceding steps to get simulation results (Table 4.15).

TABLE 4.15
Annual value for the energy transfer with overhangs

Zone	Annual value for energy transfer (kWh)
North	20,670.75
East	24,625.71
West	24,507.83
South	24,748.19

Save the simulation model with the name **DB_overhang**. In the following steps, you are going to model vertical fins.

Step 15: Open the saved simulation model.

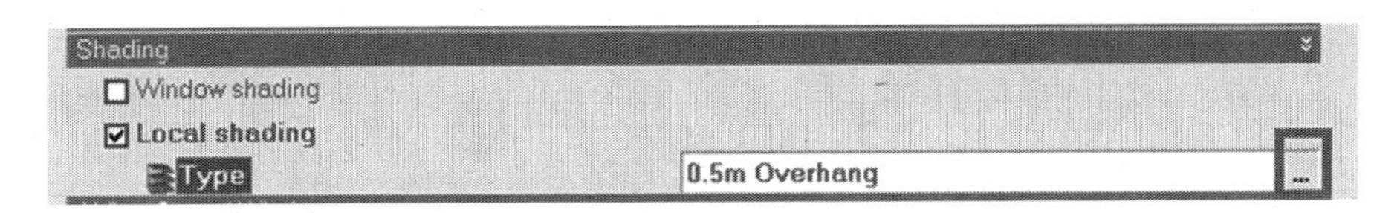

Step 16: Select **Overhang + sidefins (0.5m projection)**, and make a copy of the current selection and edit for changes.

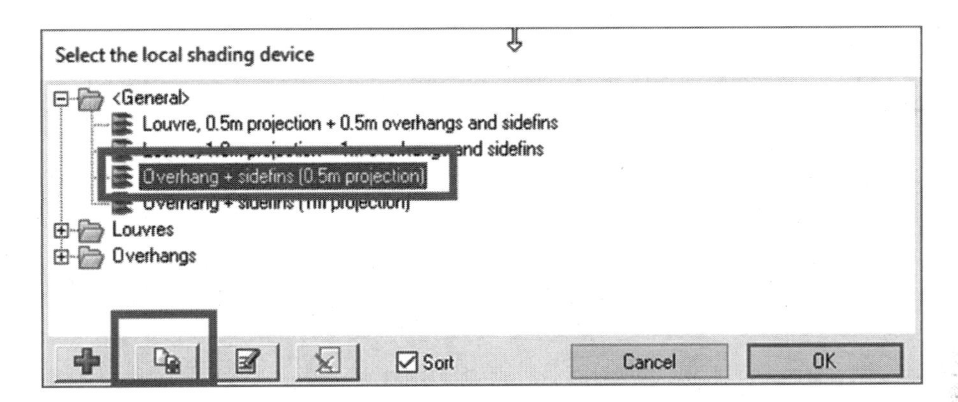

Step 17: Enter **0.5 m Fins** as the name.

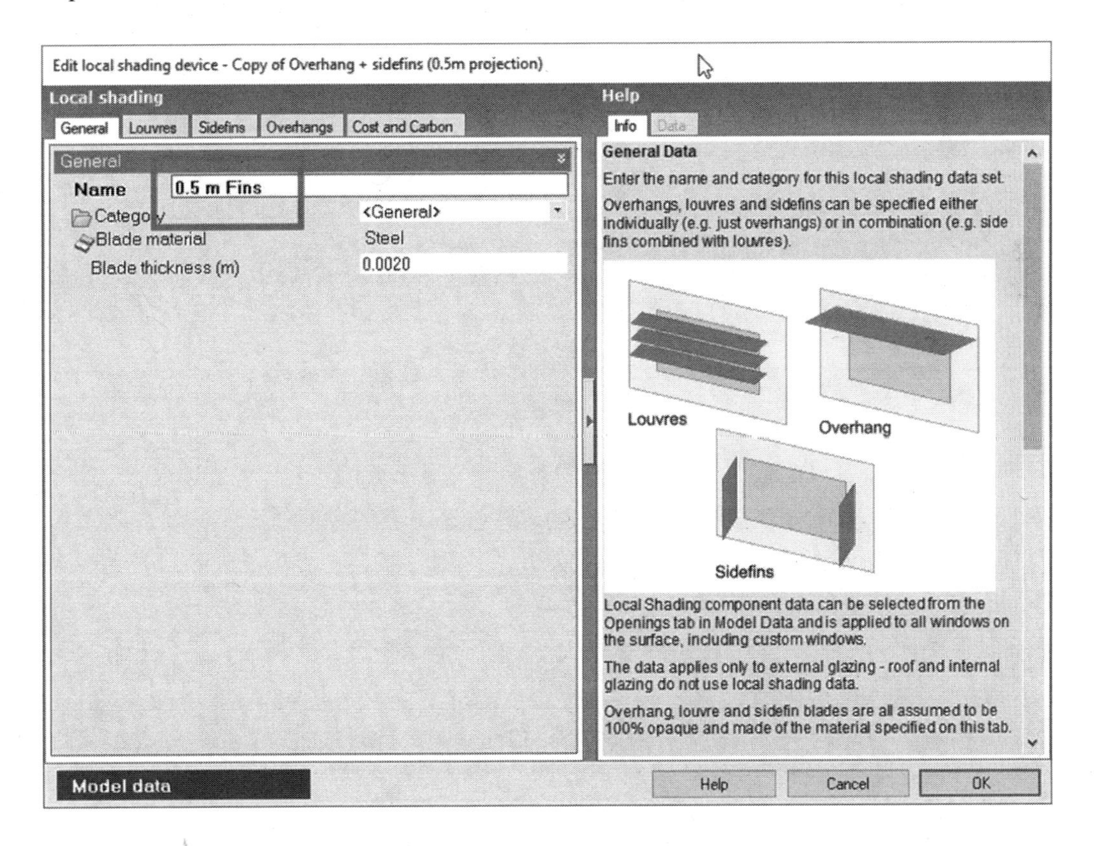

Step 18: Select the **Overhangs** tab.

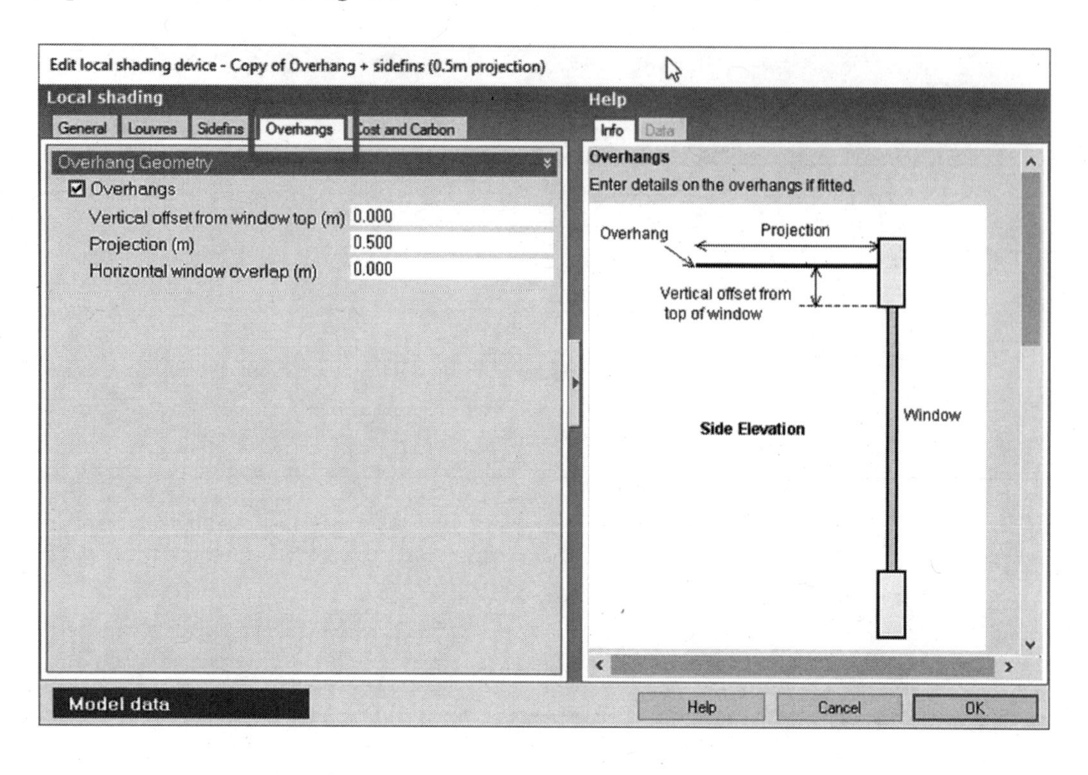

Step 19: Clear the **Overhangs** checkbox.

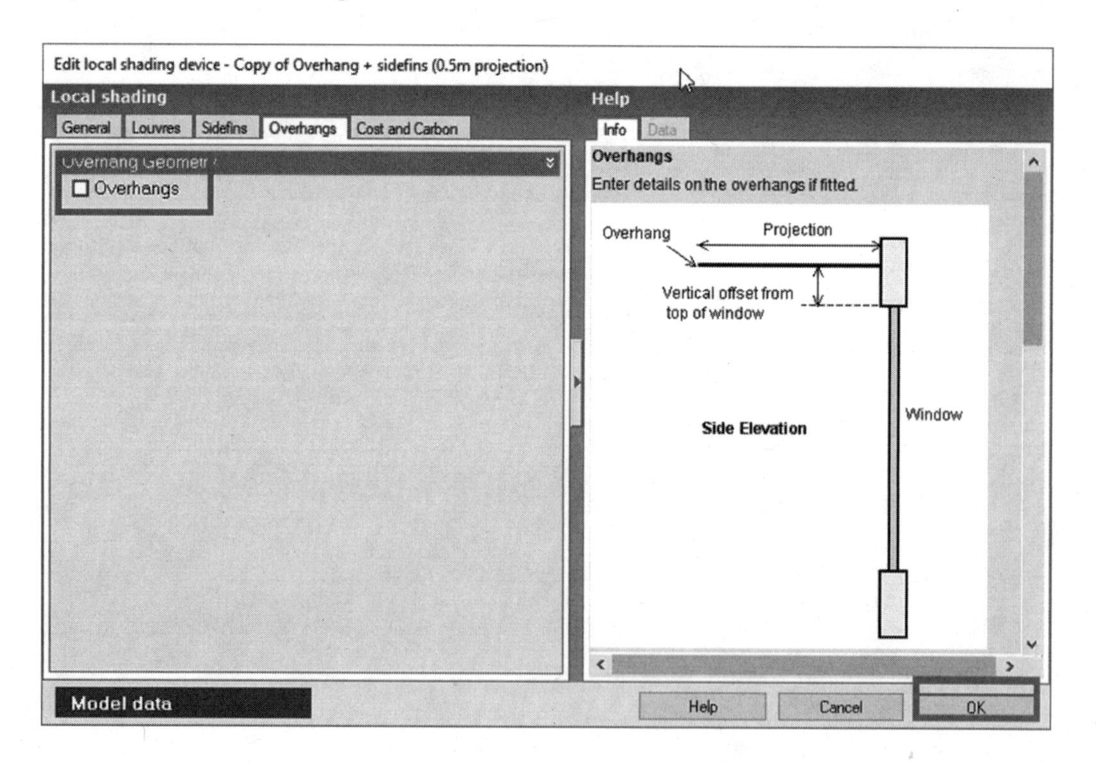

Step 20: Select the **Visualize screen** tab. Select the **Rendered View** tab. It displays the rendering of the building. Make sure that all windows are modelled with side fins.

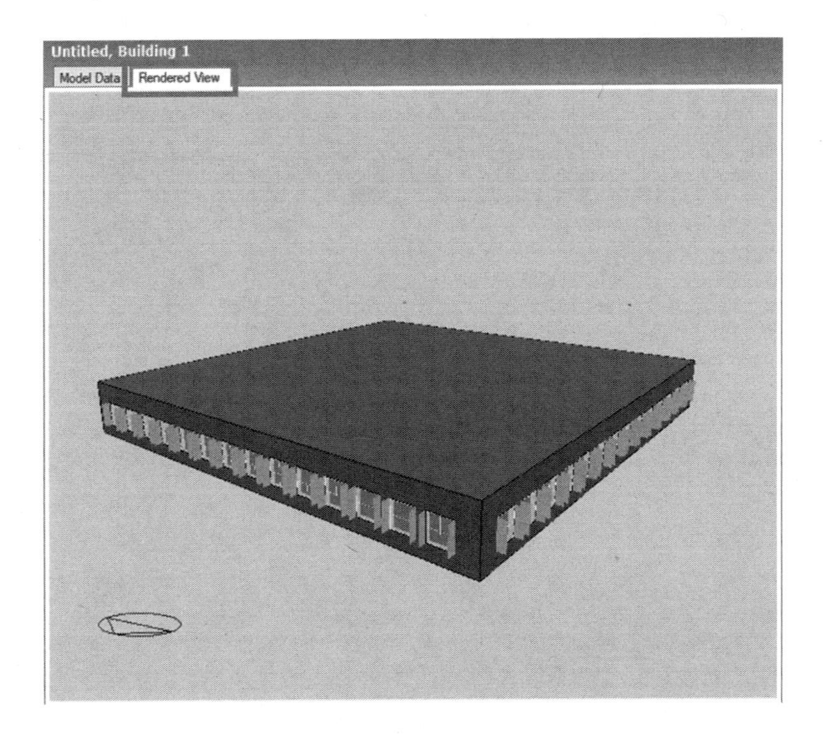

Step 21: Under the **Display options**, select the **Show sunpath** checkbox. Select the **Show shadows** checkbox. Under **Period**, enter **17:00 Time (dec hrs)** and **21** in the **Day** textbox. Select **Jun** in **Month** drop-down list.

Click the **Apply changes** button.

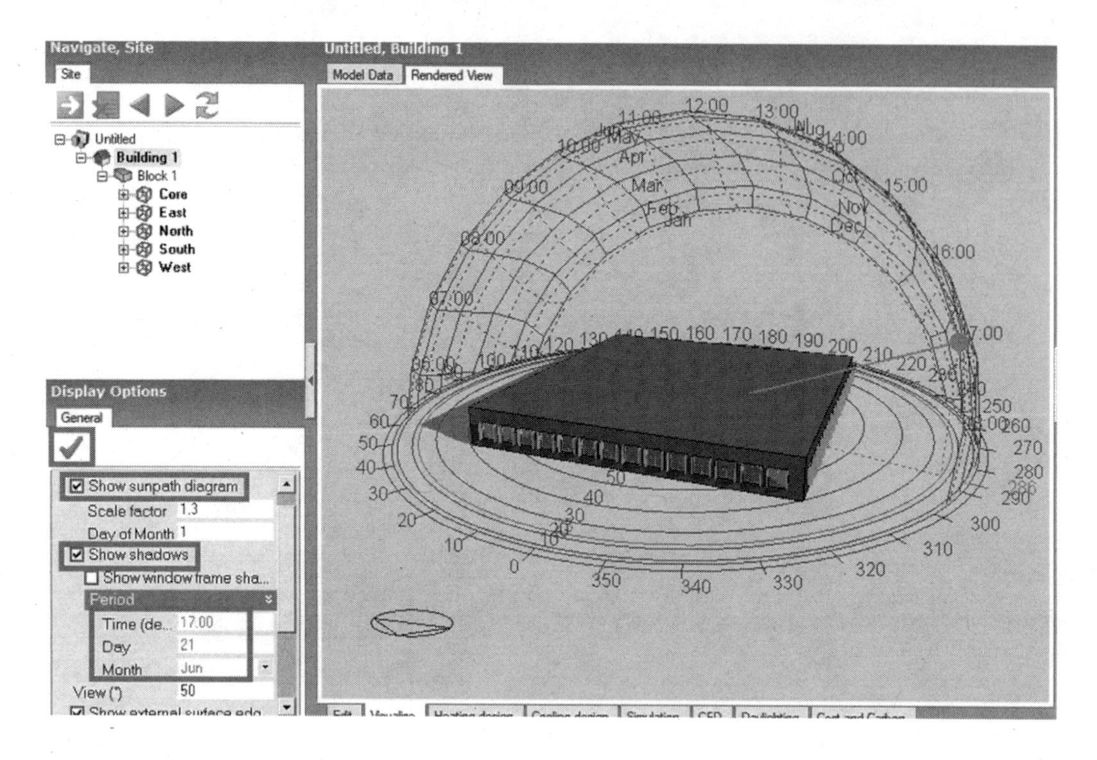

Step 21: Perform an annual simulation, and record the results (Table 4.16).

TABLE 4.16
Annual value of energy transfer for vertical fins

Zone	Annual value for energy transfer (kWh)
North	20,516.90
East	25,831.74
West	25,619.98
South	25,746.66

Compare the results for all cases (Table 4.17).

TABLE 4.17
Comparison of the annual cooling energy for all types of shades

	Annual cooling energy (kWh)		
Zone	**No shade**	**With overhangs**	**With fins**
North	21,056.03	20,670.75	20,516.90
East	26,881.99	24,625.71	25,831.74
West	26,649.96	24,507.83	25,619.98
South	27,684.44	24,748.19	25,746.66

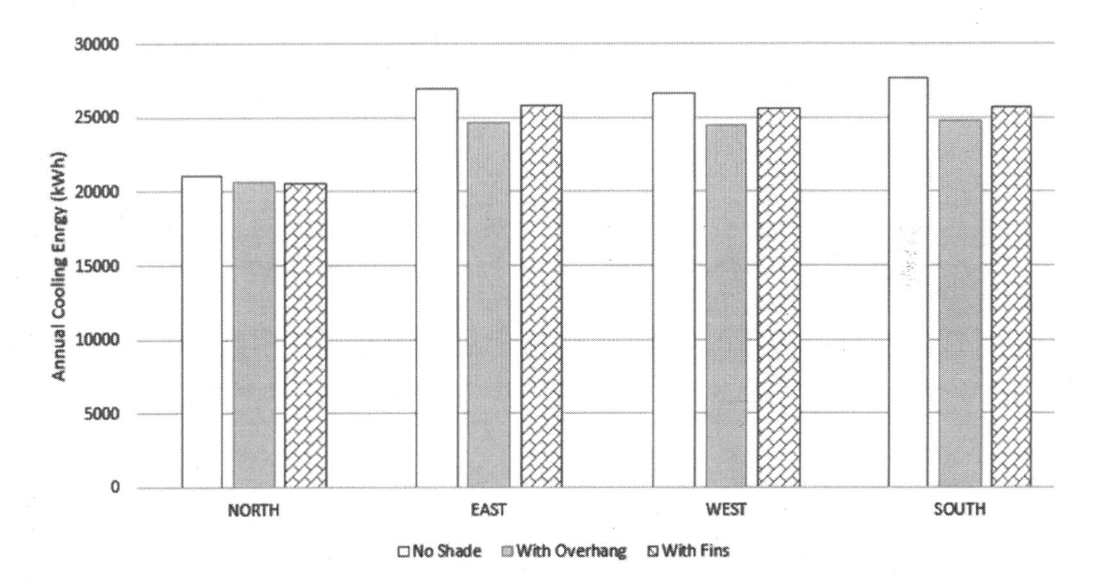

The analysis show that overhangs are effective in the south facade and that fins are more effective in the north facade windows.

EXERCISE 4.2

Repeat this tutorial for both overhangs and fins to observe the combined effect. Compare the energy consumption for all the perimeter zones.

Tutorial 4.3

Evaluating the Impact of Internal Operable Shades

GOAL

To evaluate the impact of window internal operable shades on energy consumption

WHAT ARE YOU GOING TO LEARN?

- How to model operable shades

PROBLEM STATEMENT

In this tutorial, you are going to use the **30- × 30-m** model with a 5-m perimeter depth. Determine the energy consumption and solar gain for all perimeter zones. You need to simulate the model with the following options:

1. Overhang with a 0.5-m depth
2. Overhang and internal operable shades with solar control

Use the **New Delhi/Palam, India** weather location.

SOLUTION

Step 1: Open a new project, and create a **30- × 30 m** model with a 5-m perimeter depth. Select the **Openings** tab, and change window height to **1.00 m**. Select the **Local shading** checkbox. Select **0.5 m** overhang in **Type**. Then select the **Simulation** screen tab. The **Edit Calculation Options** screen appears.

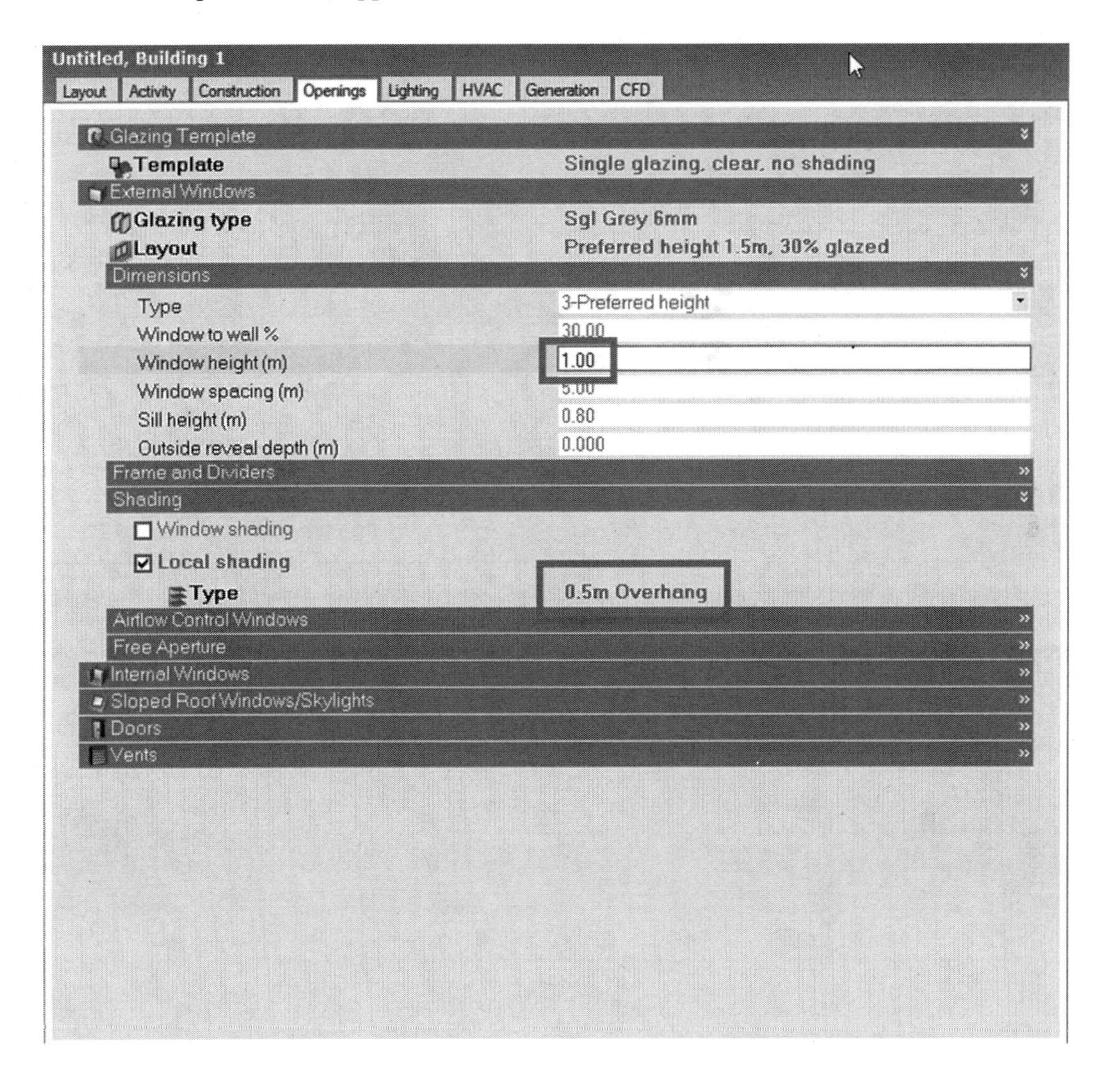

Step 2: Simulate the model, and select the **Sub-hourly** checkbox. Click **OK**. The results are displayed.

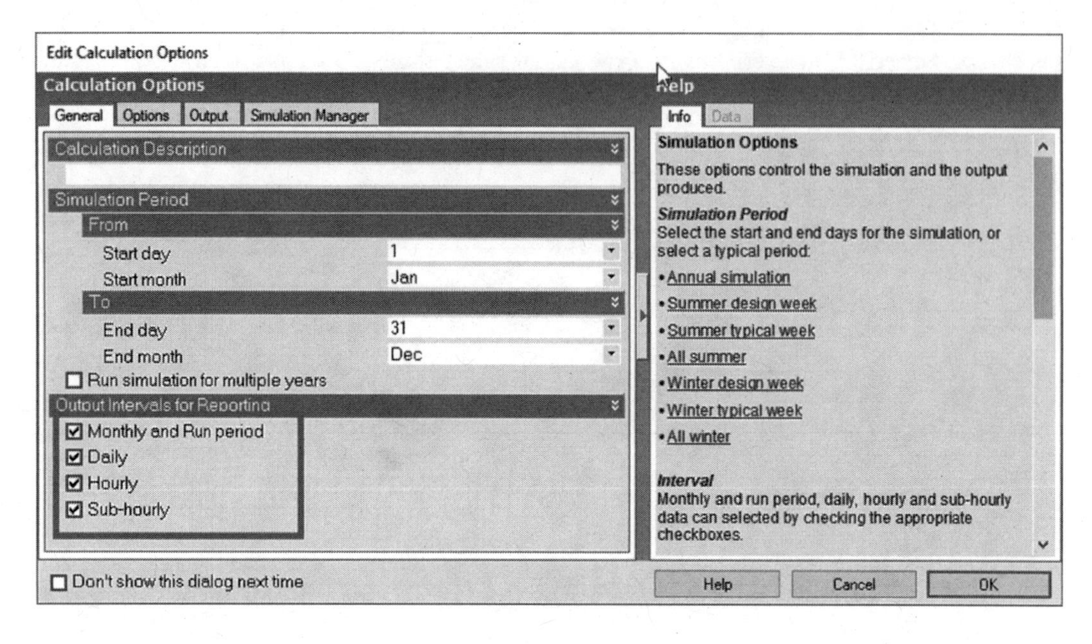

Step 3: Select **Internal gains** from the **Data** drop-down list. Click **East** in the navigation tree. It shows internal gains for the east zone.

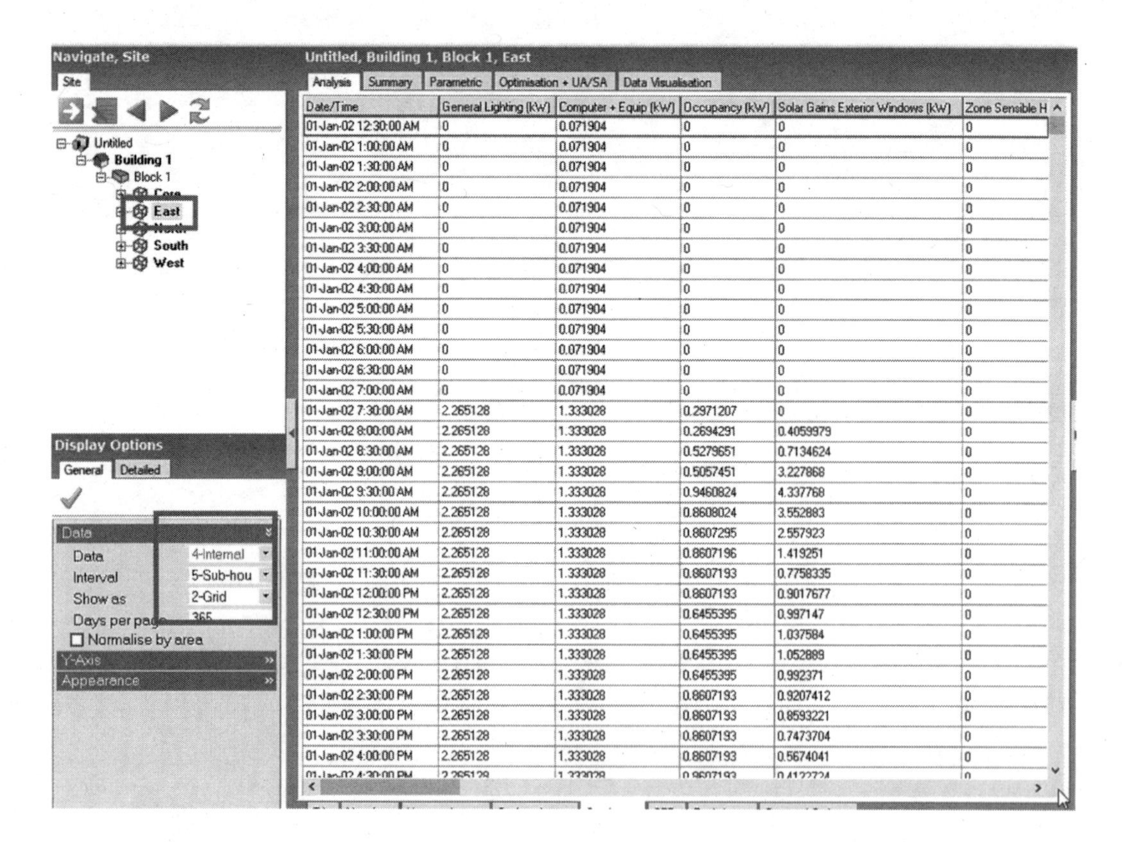

Step 4: Click the **Export data** button. The **Export Results Spreadsheet** screen appears.

Step 5: Select **File** from the **Export to** drop-down list and **CSV spreadsheet** from the **Format** drop-down list. Click **OK**.

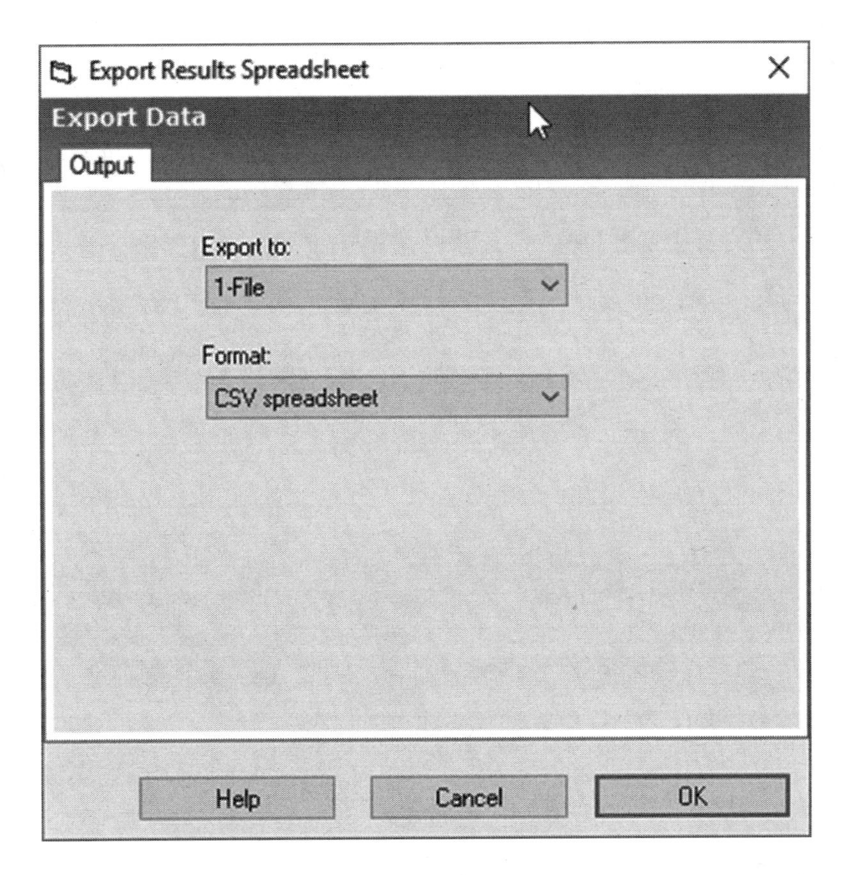

Step 6: Name the file **East Zone Results**, and save it for a comparison of results.

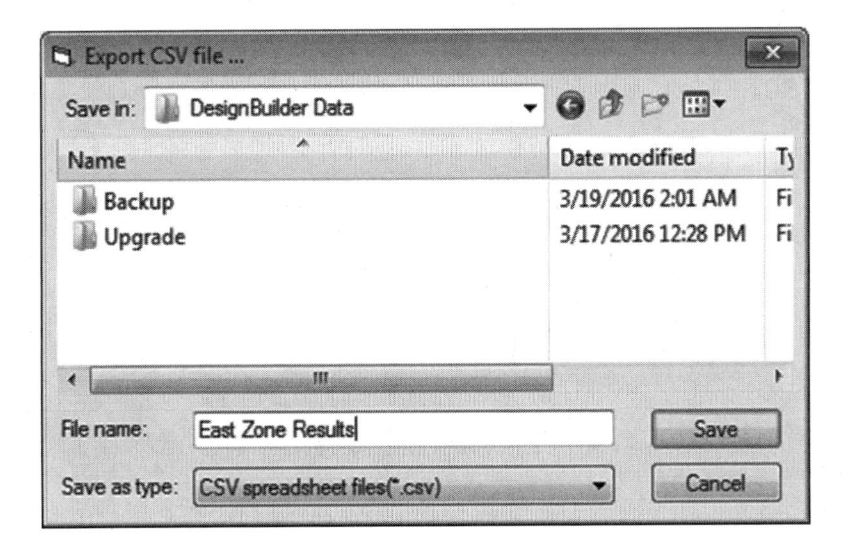

Similarly, repeat the preceding steps, and save the results files for all the perimeter zones. The following steps show how to model internal shades.

Step 7: Select the **Openings** tab. Select the **Window shading** checkbox under the **Shading** section. It displays shading type.

Step 8: Select **High reflectance – low transmittance shade** from the **Type** drop-down list. Select **4-Solar** from the **Control type** drop-down list. Set **250** as solar setpoint (W/m^2).

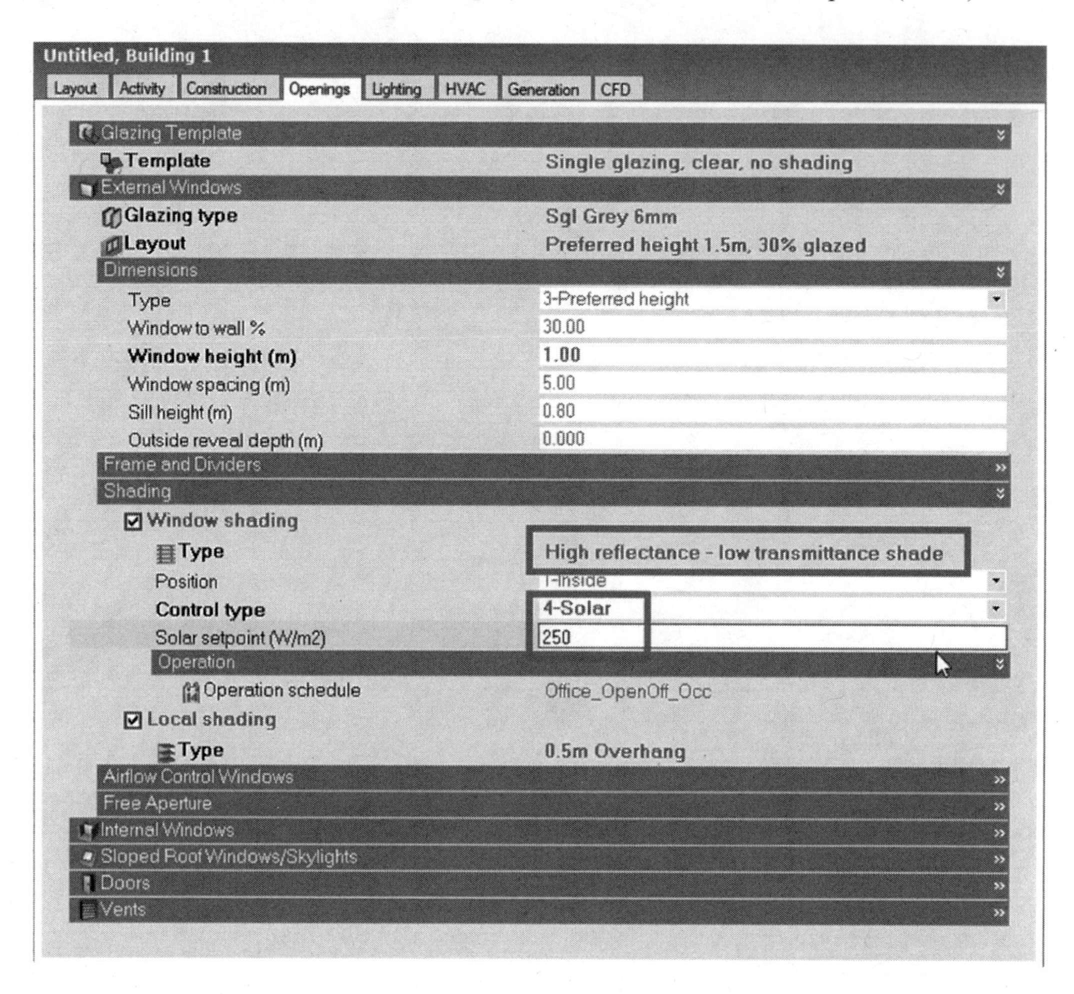

Step 9: Simulate the model, and record the results for all four zones. Compare the results for two cases in each orientation: only overhang and overhang with internal shade for solar gains (kW) for 4 April.

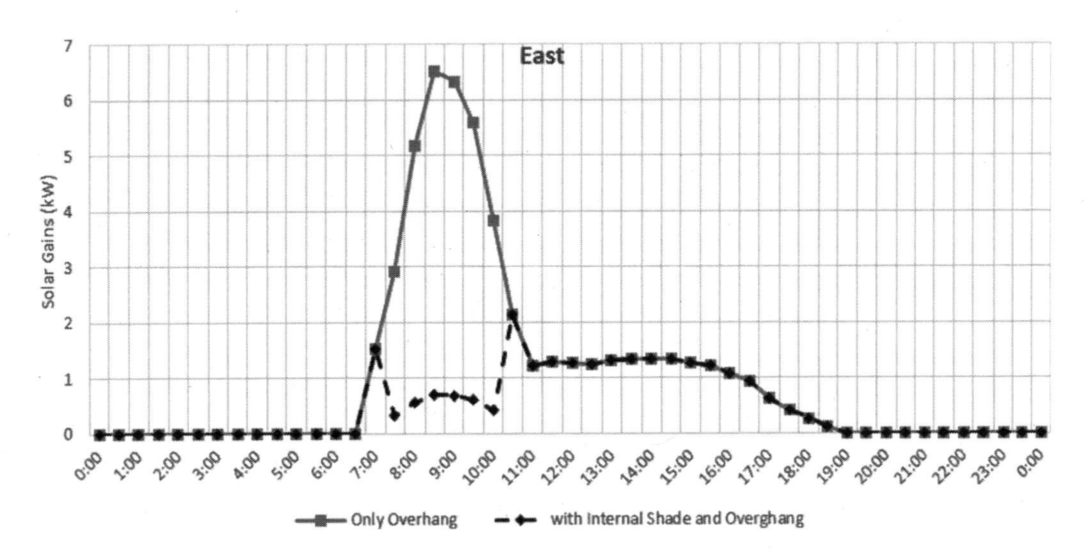

Solar control type operates the indoor shading devices based on the amount of solar radiation. In this example, the solar setpoint considered is 250 W/m^2; hence, whenever the solar radiation on the window is above this setpoint, the window shading will be move down to reduce the incoming heat through radiation. Indoor operable shades have other control types such as *schedules* and *illuminance levels.*

The preceding graph shows the profile of solar gains through the window on the east facade with overhang and overhang plus internal shades for a day (4 April); you can observe the reduction in solar gains due to internal shades between 7:00 and 10:00 h. The following diagram shows direct radiation coming from the east window at 7:00 h. Use of an internal shade can cut direct solar radiation.

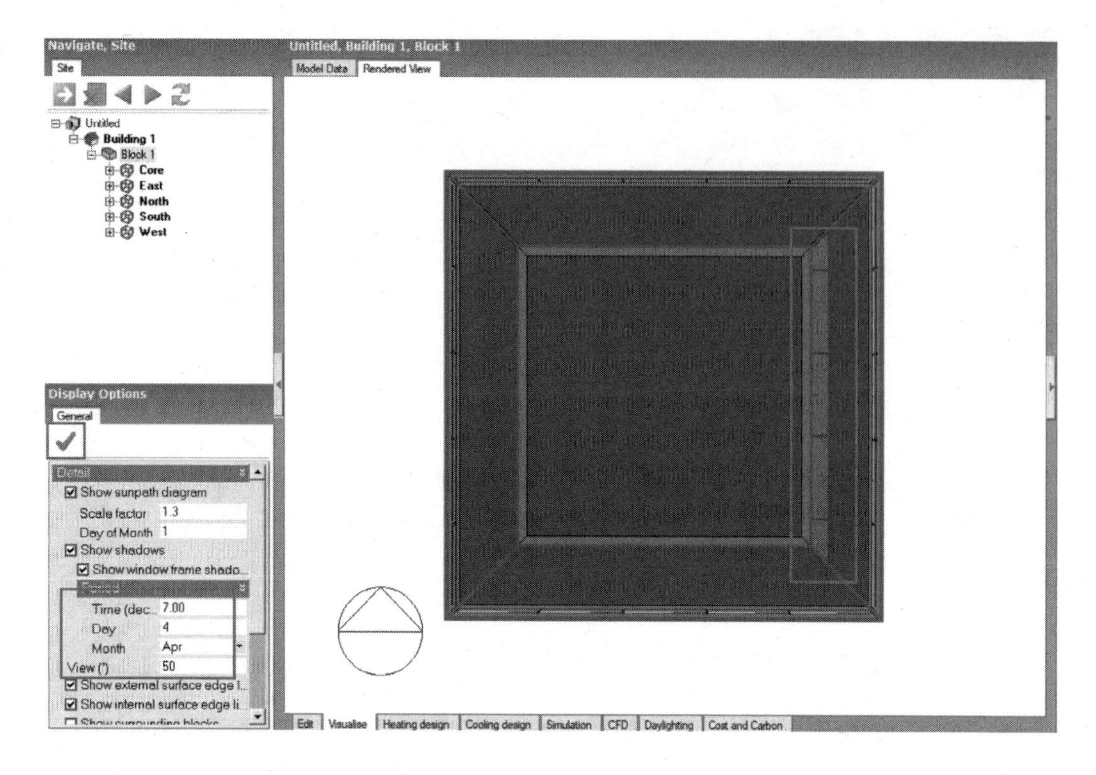

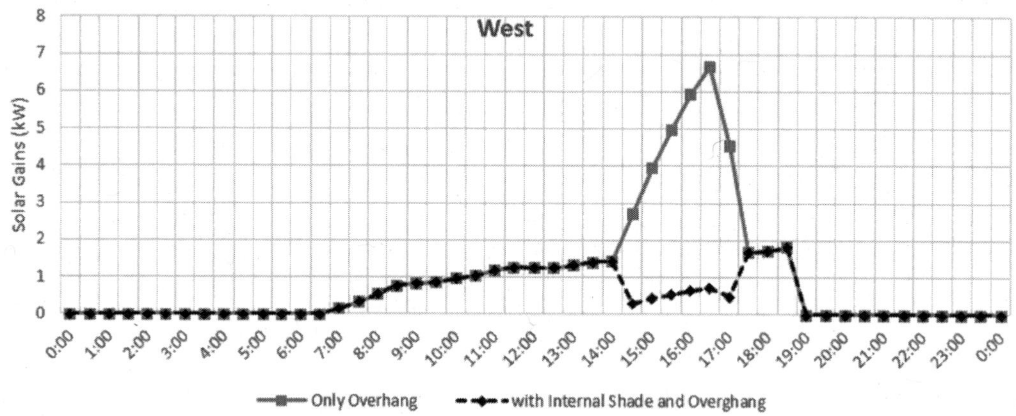

You can observe from the preceding graph that the internal shade is effective on the west window from 14:00 to 19:00 h on 4 April. It can also be seen from the following sunpath diagram that when there are no internal shades, west-facing windows get direct solar radiation at 16:00 h.

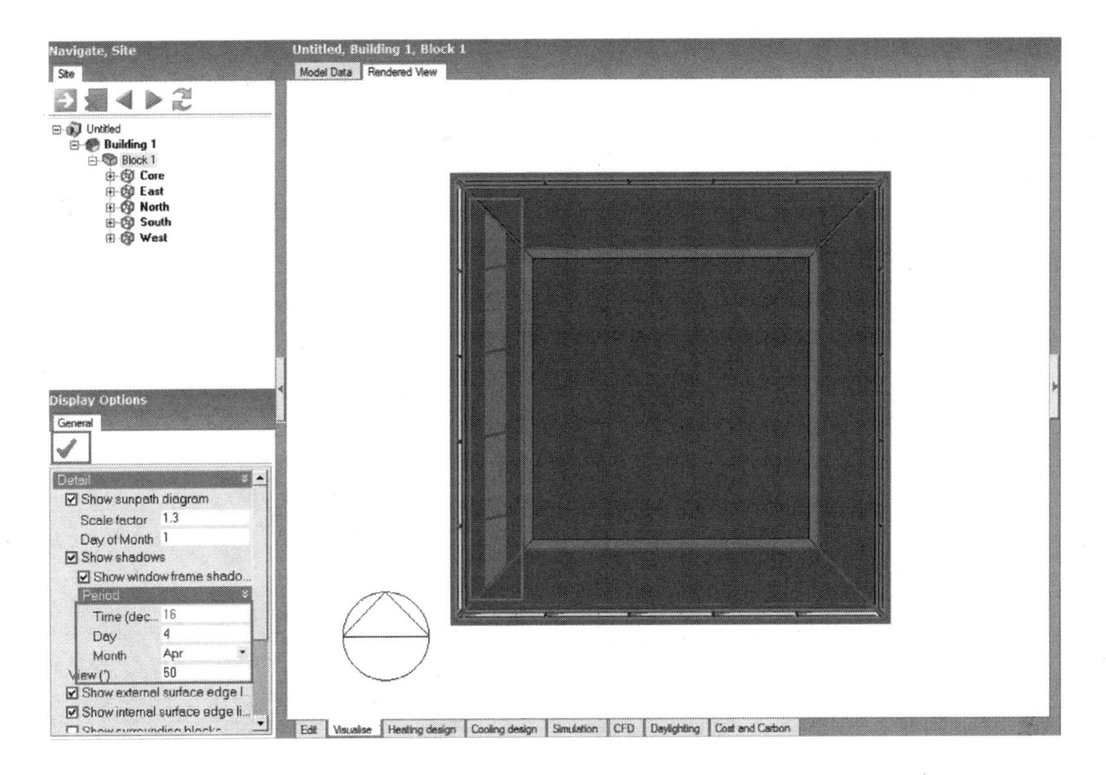

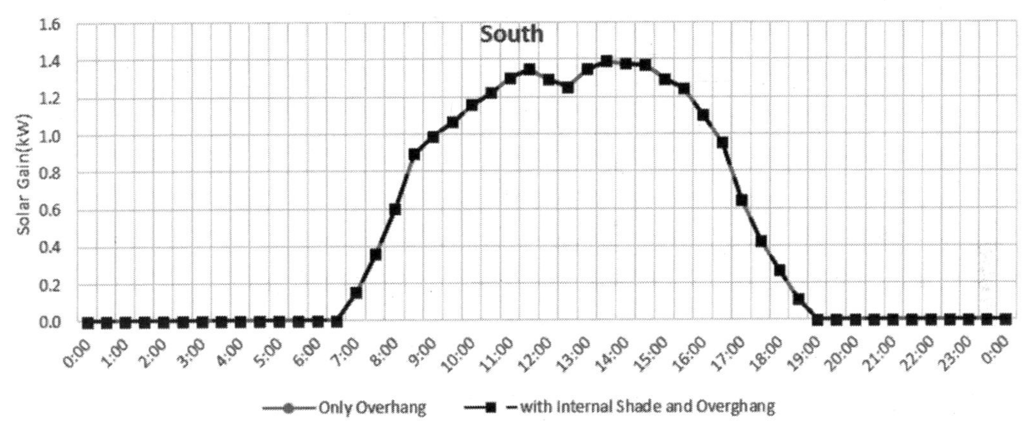

You can see from the preceding graph that internal shades are not required on the south window on 4 April for this building. It can also be seen from the following diagram that the south zone is not getting any direct glare with an overhang of 0.5 m. This is due to the fact that the altitude of the sun in daytime, except for sunrise and sunset hours, is high enough on 4 April to not cause glare through the windows.

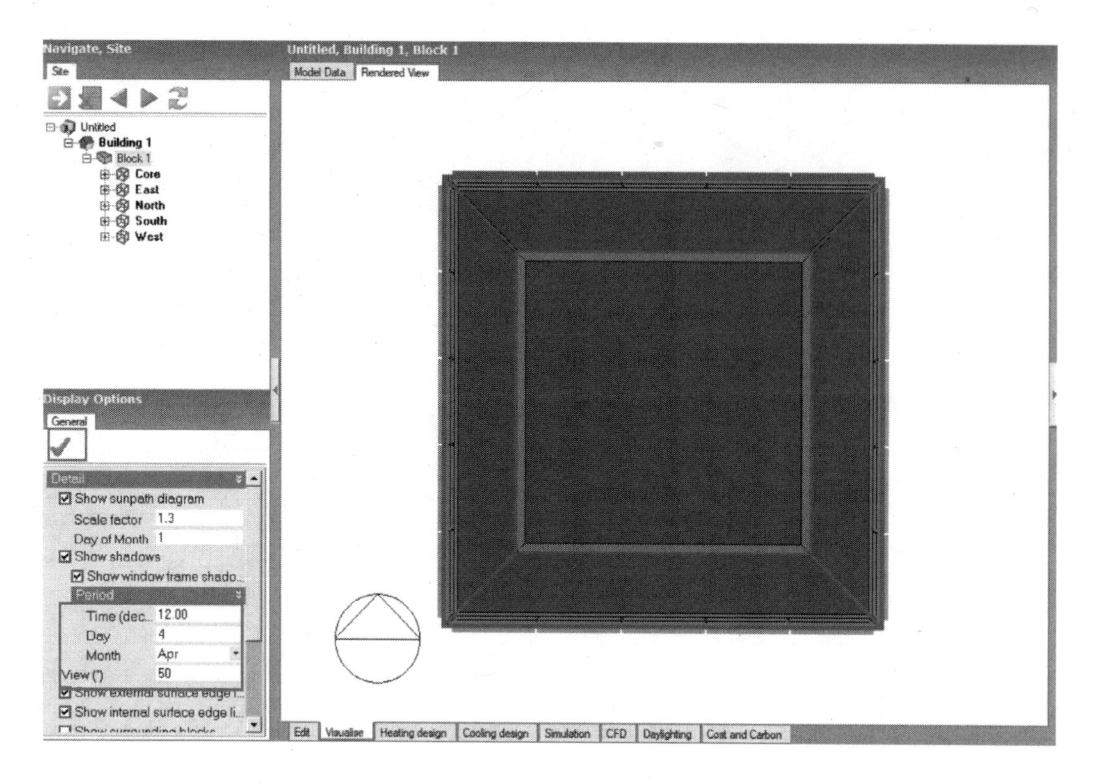

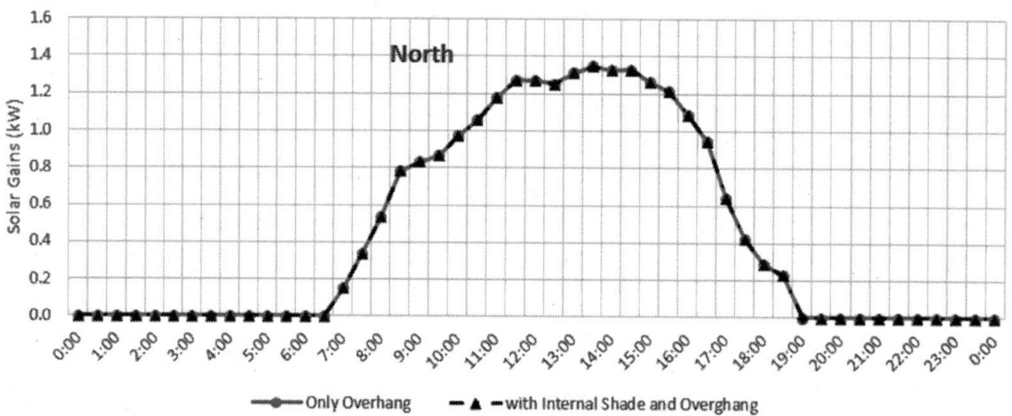

You can see from the preceding graph that the internal shade is not required on the north facade on 4 April. It can also be seen from the sunpath diagram that the window in the north facade is shaded on 4 April. Now plot the solar gains from the south facade on 6 February when the sun is at a lower altitude.

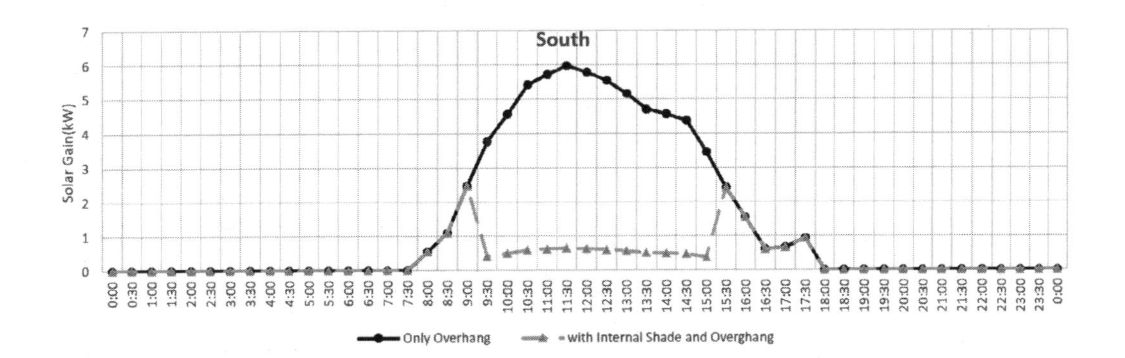
South
Solar Gain(kW)
Only Overhang
with Internal Shade and Overghang

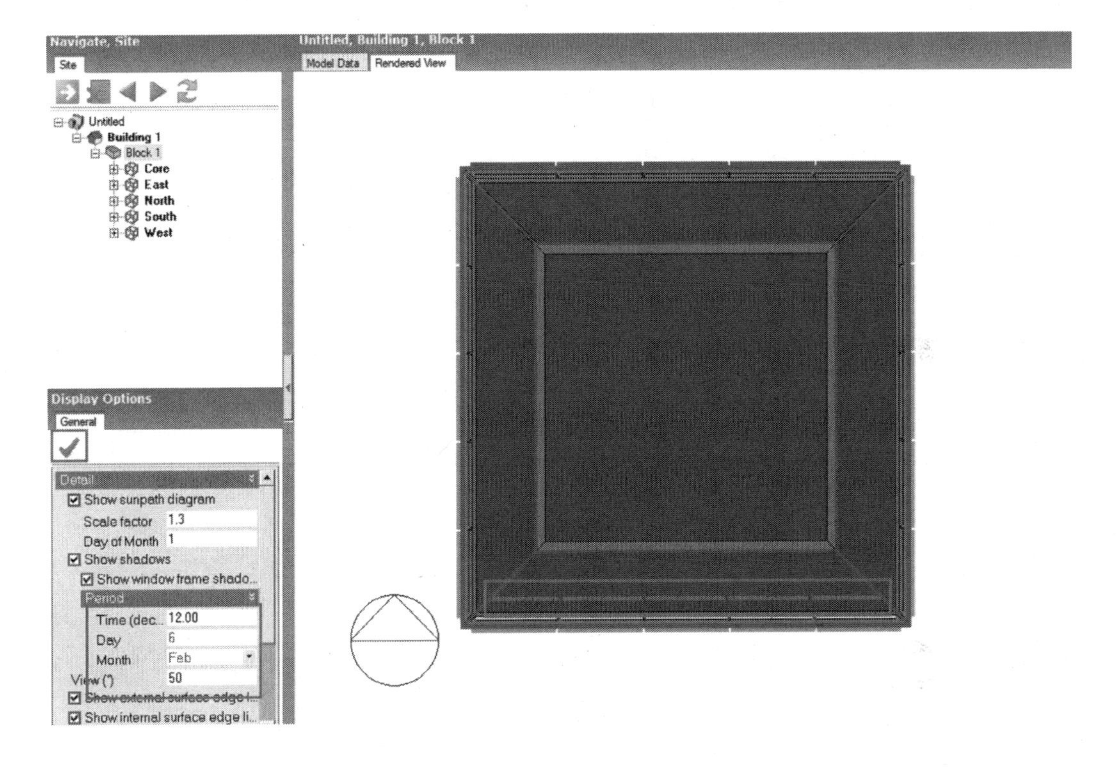
Navigate, Site
Site
Untitled
Building 1
Block 1
Core
East
North
South
West
Untitled, Building 1, Block 1
Model Data
Rendered View
Display Options
General
Detail
Show sunpath diagram
Scale factor 1.3
Day of Month 1
Show shadows
Show window frame shado...
Period
Time (dec... 12.00
Day 6
Month Feb
View (°) 50
Show external surface edge l...
Show internal surface edge li...

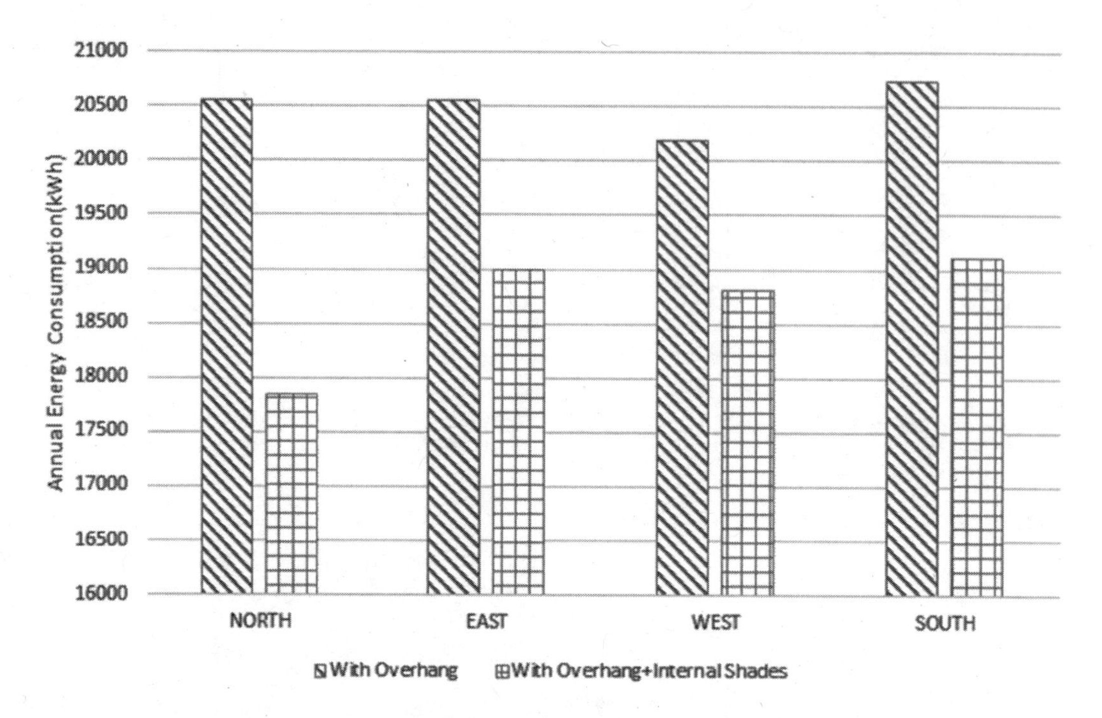

You can see that with internal shades, there is a reduction in solar gains from the south window between 12:00 and 16:00 h. You can also see from the preceding graph that internal shades are more effective for the east, west and south zones for the New Delhi climate.

The following two graphs give the direct normal radiation and global horizontal radiation on 4 April. You can see that the solar radiation curve is smooth, implying that the sky is clear.

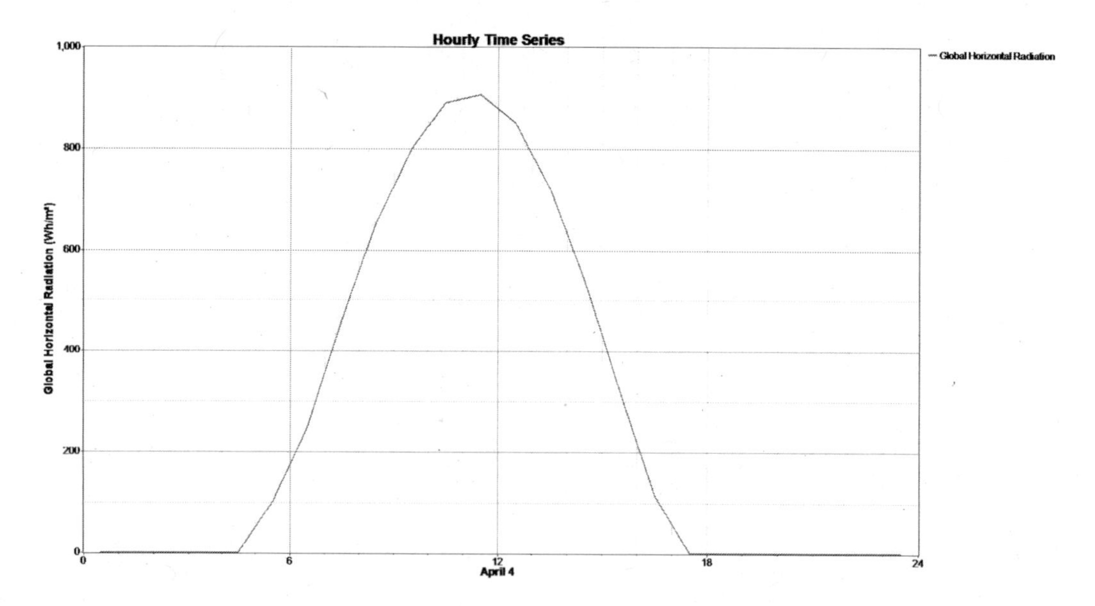

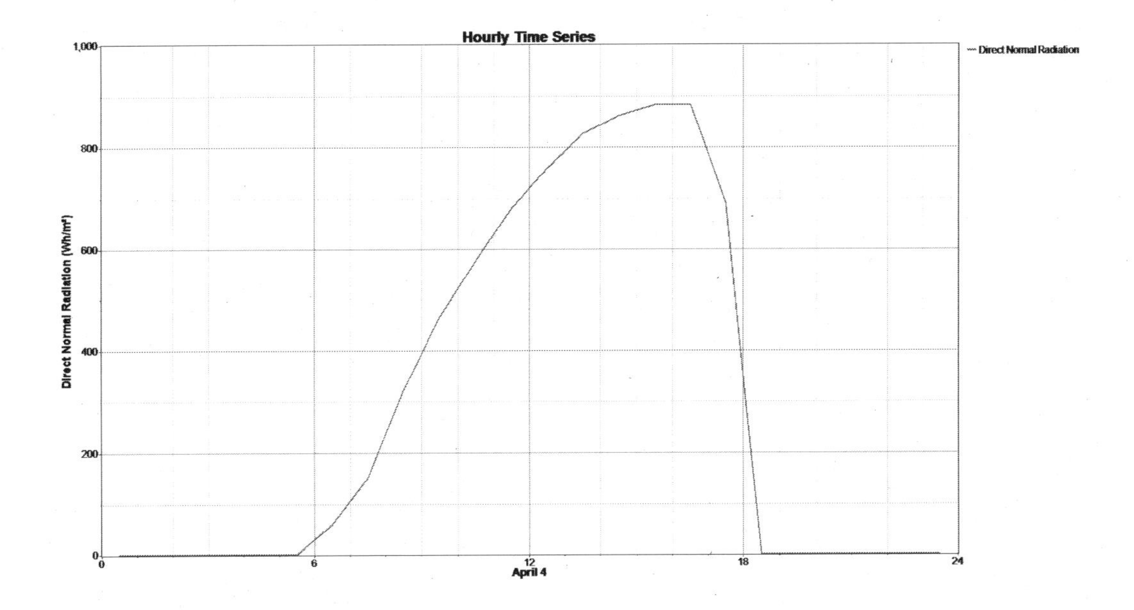

EXERCISE 4.3

Repeat this tutorial for operable shade–controlled **Outdoor air temp + Solar on window**. You can consider the outdoor temperature setpoint as 35°C for the **New Delhi/Palam, India** weather location.

Tutorial 4.4

Evaluating the Impact of Electrochromic Switchable Glazing on Windows' Solar Gains

GOAL

To evaluate the impact of electrochromic switchable glazing on windows' solar gains

WHAT ARE YOU GOING TO LEARN?

- How to model electrochromic glazing

PROBLEM STATEMENT

In this tutorial, you are going to use a five-zone 30- × 30-m model with a 5-m perimeter depth and WWR of 30%. You will use a **Split no fresh air** HVAC system. You need to simulate a model with the following options:

1. Single glazed (**Sgl Clear 6mm**)
2. Electrochromic reflective 6 mm (darkened state whenever the solar radiation incident on the outside of the window is greater than 120 W/m^2)

Compare solar gains from exterior windows in the south zone. Use weather location **Ganzhou, China**.

> *Electrochromic windows* are part of a new generation of technologies called *switchable glazing* or *smart windows*. Switchable glazing can change the light transmittance, transparency or shading of windows in response to an environmental signal such as sunlight, temperature or an electrical control. Electrochromic windows change from transparent to tinted by applying an electric current. Potential uses for electrochromic technology include daylighting control, glare control, solar heat control and fading protection in windows and skylights. By automatically controlling the amount of light and solar energy that can pass through the window, electrochromic windows can help save energy in buildings.
>
> *Source:* https://designbuilder.co.uk/helpv6.0/Content/Electrochromic_Glazing.htm.

SOLUTION

Step 1: Open a new project, and create a **30- × 30-m** building with five zones with a 5-m perimeter depth. Rename all zones. Click the **Openings** tab.

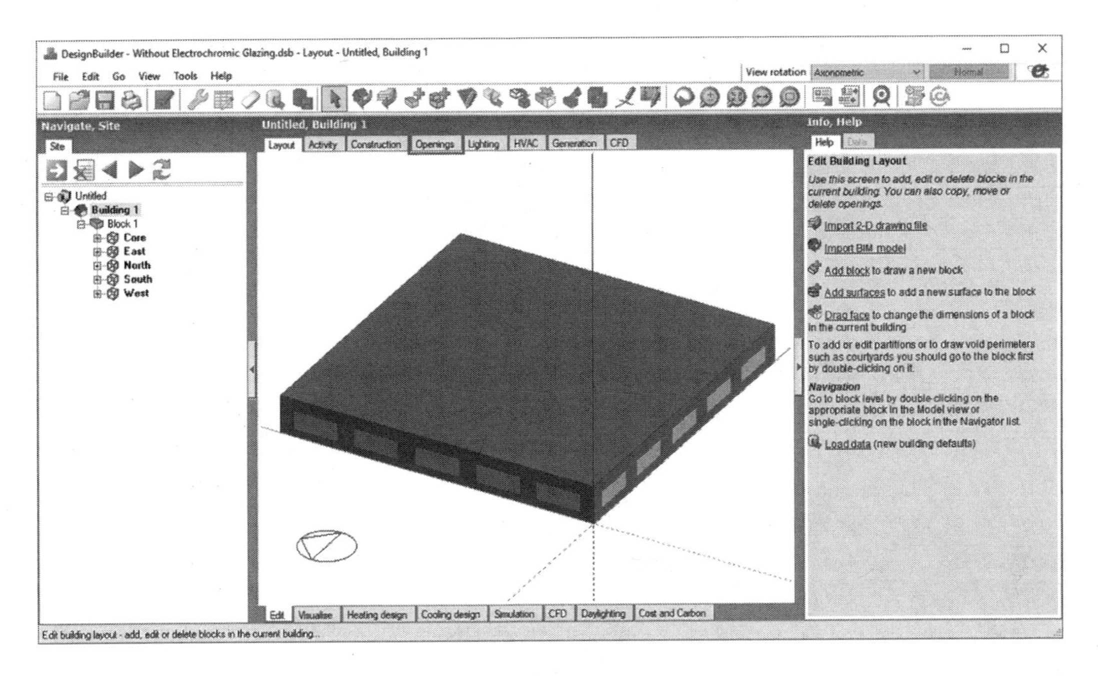

Step 2: Select the **Single glazing, clear, no shading** template. Select **Sgl Clr 6mm** in **Glazing** type.

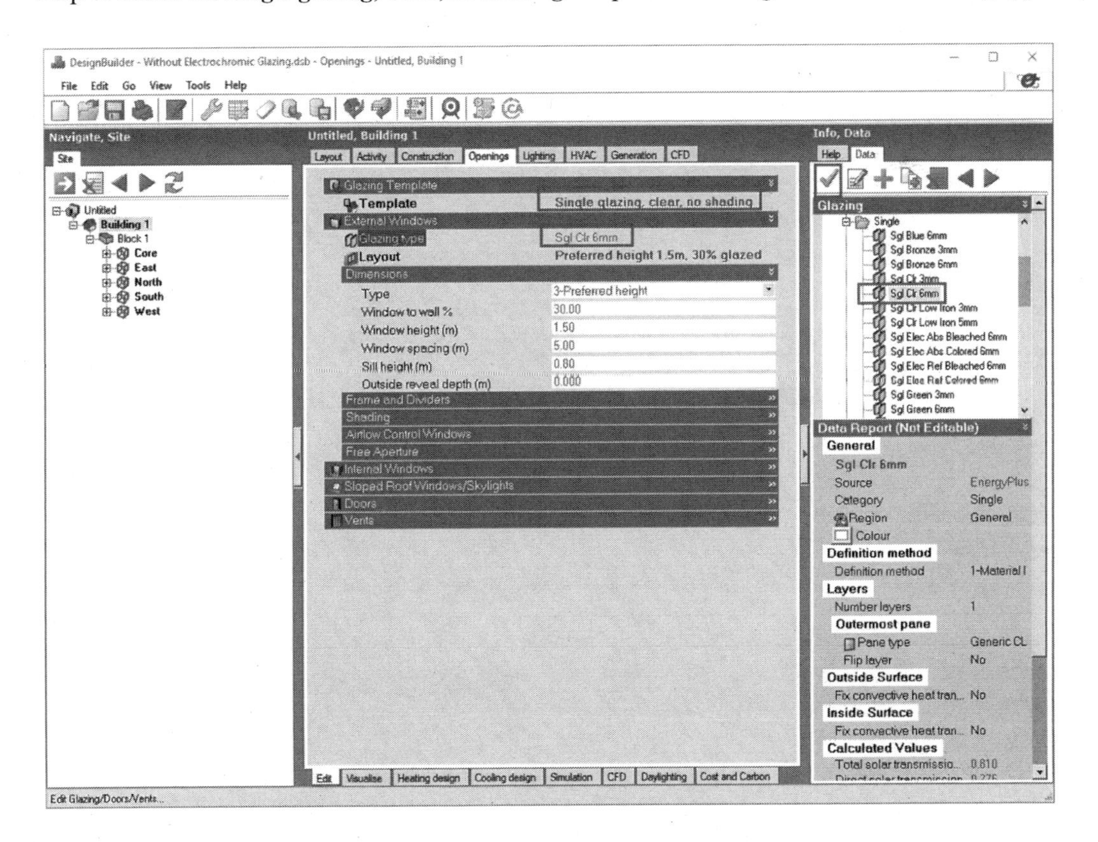

Step 3: Make sure that the **Window shading** checkbox is clear.

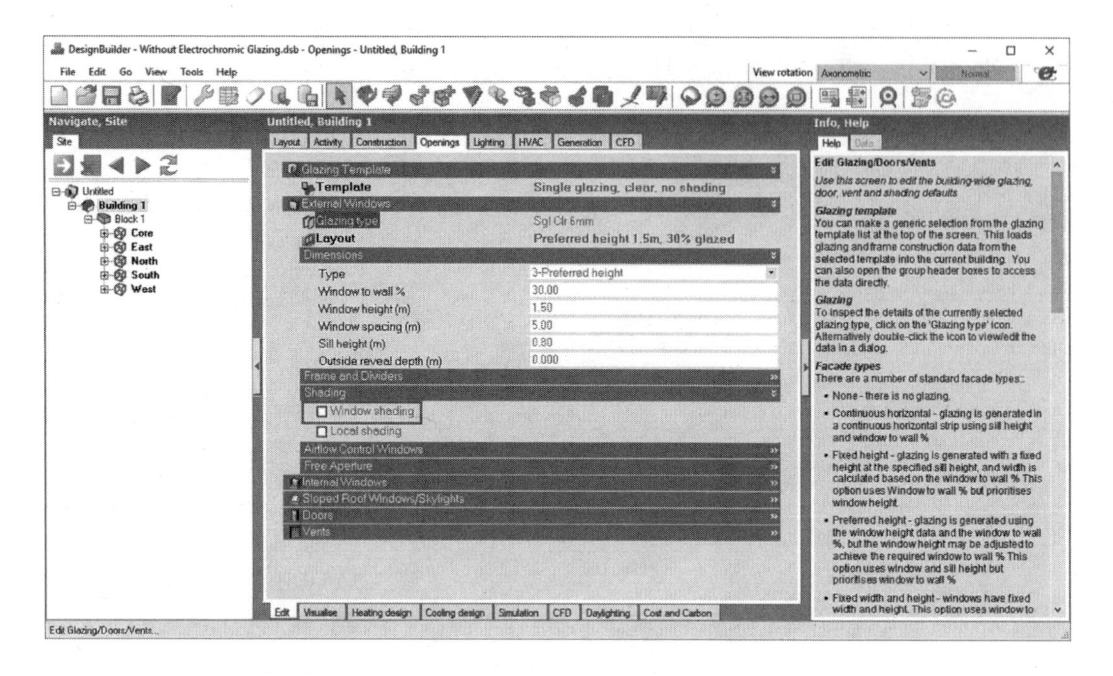

Step 4: Click the **HVAC** tab. Select the **Split no fresh air** template. Clear the **Heated** checkbox under the **Heating** section, and clear the **On** checkbox under **DHW**. Select the checkbox under **Cooling system**, and enter **3.0** as **CoP.**

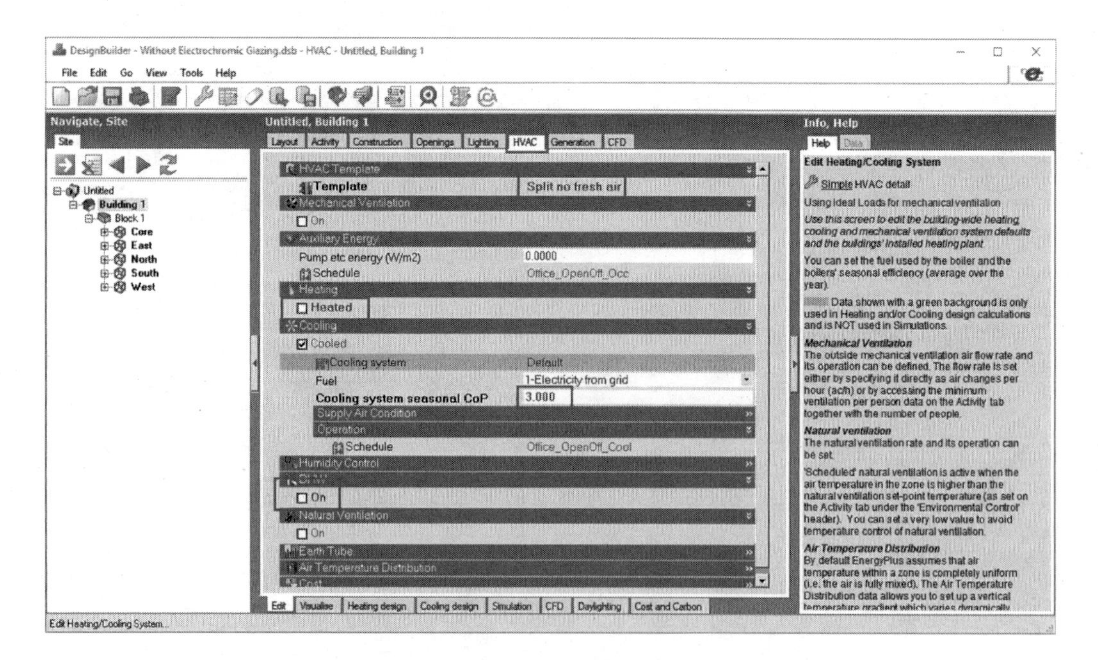

Step 5: Perform the simulation for **Summer design week**.

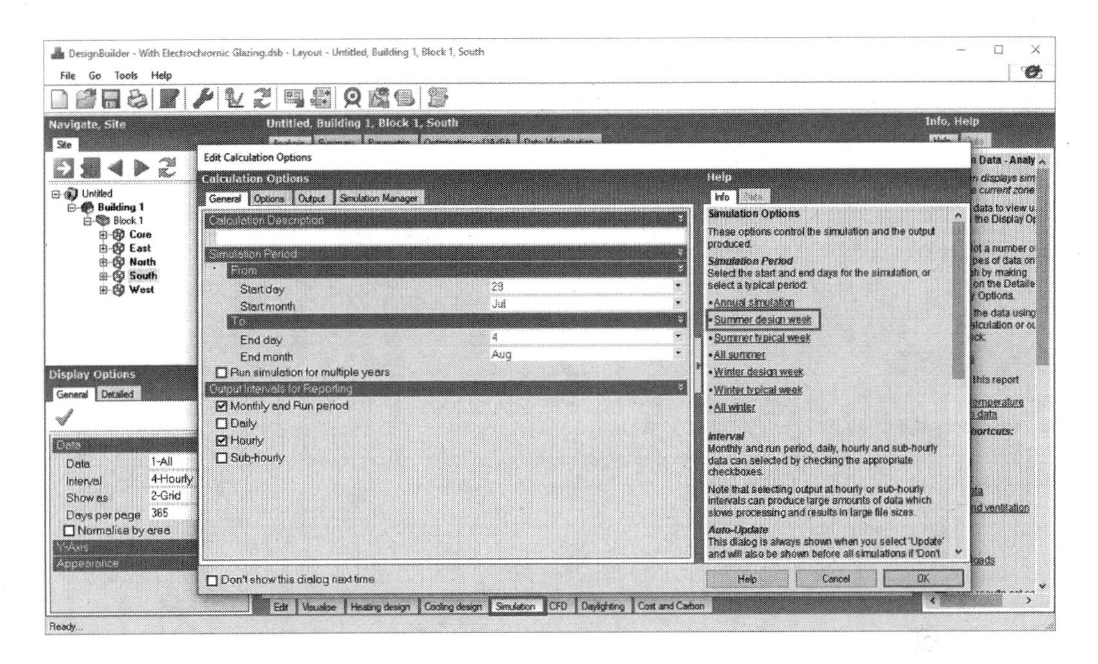

Step 6: Click the **South** zone, and then click the **Export data** icon.

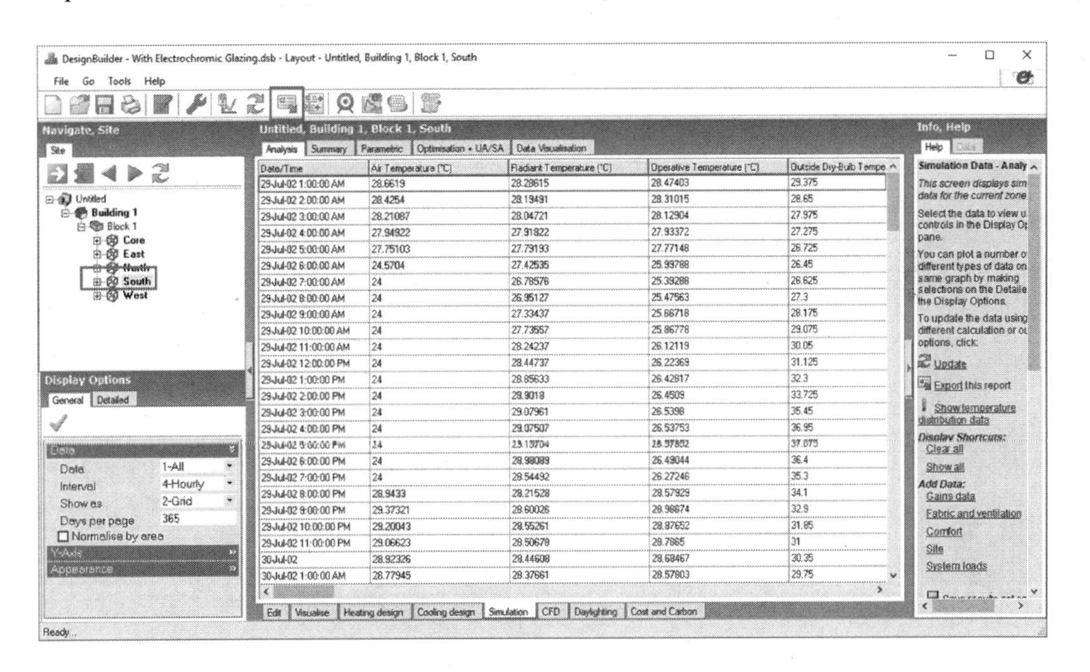

Date/Time	Air Temperature (°C)	Radiant Temperature (°C)	Operative Temperature (°C)	Outside Dry-Bulb Tempe
29-Jul-02 1:00:00 AM	28.6619	28.28615	28.47403	29.375
29-Jul-02 2:00:00 AM	28.4254	28.19491	28.31015	28.65
29-Jul-02 3:00:00 AM	28.21087	28.04721	28.12904	27.975
29-Jul-02 4:00:00 AM	27.94922	27.91822	27.93372	27.275
29-Jul-02 5:00:00 AM	27.75103	27.79193	27.77148	26.725
29-Jul-02 6:00:00 AM	24.5704	27.42535	25.99788	26.45
29-Jul-02 7:00:00 AM	24	26.78576	25.39288	26.625
29-Jul-02 8:00:00 AM	24	26.95127	25.47563	27.3
29-Jul-02 9:00:00 AM	24	27.33437	25.66718	28.175
29-Jul-02 10:00:00 AM	24	27.73557	25.86778	29.075
29-Jul-02 11:00:00 AM	24	28.24237	26.12119	30.05
29-Jul-02 12:00:00 PM	24	28.44737	26.22369	31.125
29-Jul-02 1:00:00 PM	24	28.85633	26.42817	32.3
29-Jul-02 2:00:00 PM	24	28.9018	26.4509	33.725
29-Jul-02 3:00:00 PM	24	29.07961	26.5398	35.45
29-Jul-02 4:00:00 PM	24	29.07507	26.53753	36.95
29-Jul-02 5:00:00 PM	24	29.15704	26.57852	37.075
29-Jul-02 6:00:00 PM	24	28.98089	26.49044	36.4
29-Jul-02 7:00:00 PM	24	28.54492	26.27246	35.3
29-Jul-02 8:00:00 PM	28.9433	28.21528	28.57929	34.1
29-Jul-02 9:00:00 PM	29.37321	28.60026	28.98674	32.9
29-Jul-02 10:00:00 PM	29.20043	28.55261	28.87652	31.85
29-Jul-02 11:00:00 PM	29.06623	28.50678	28.7865	31
30-Jul-02	28.92326	28.44608	28.68467	30.35
30-Jul-02 1:00:00 AM	28.77945	28.37661	28.57803	29.75

Step 7: Save the results to a spreadsheet program.

Step 8: Click the **File** menu and **Save file As.**

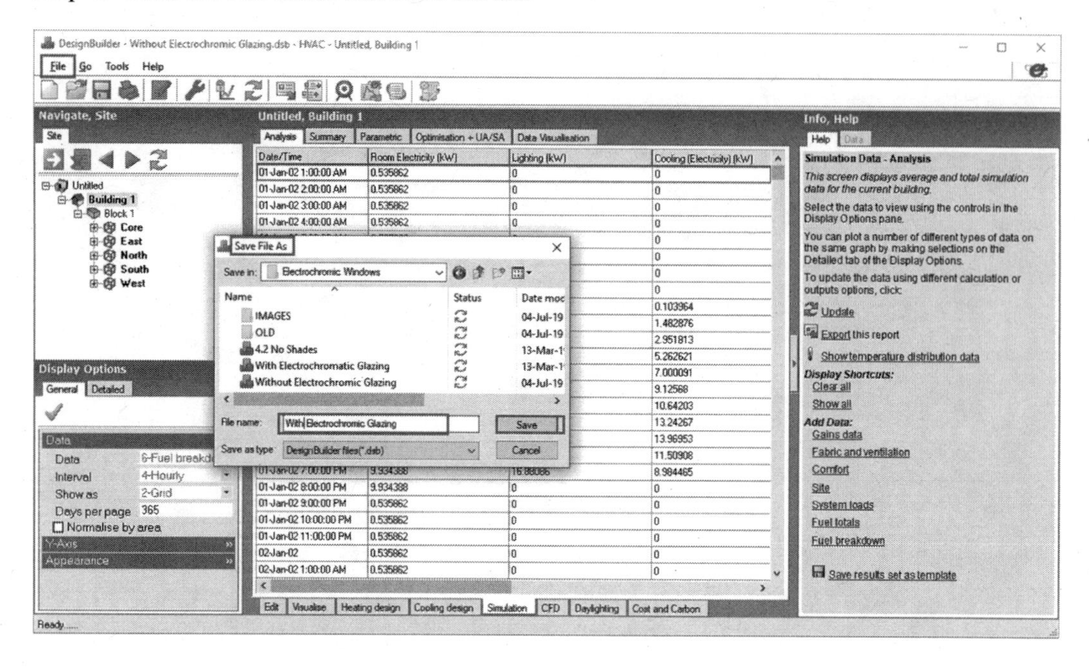

Step 9: Click the **Edit screen** tab.

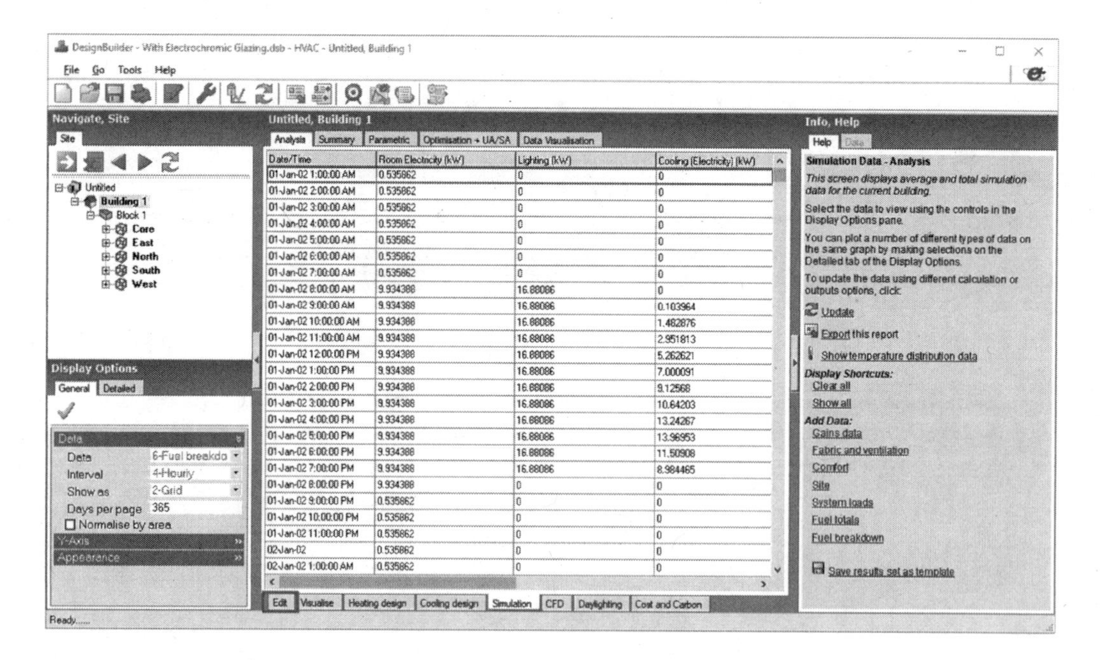

Step 10: Click the **Openings** tab.

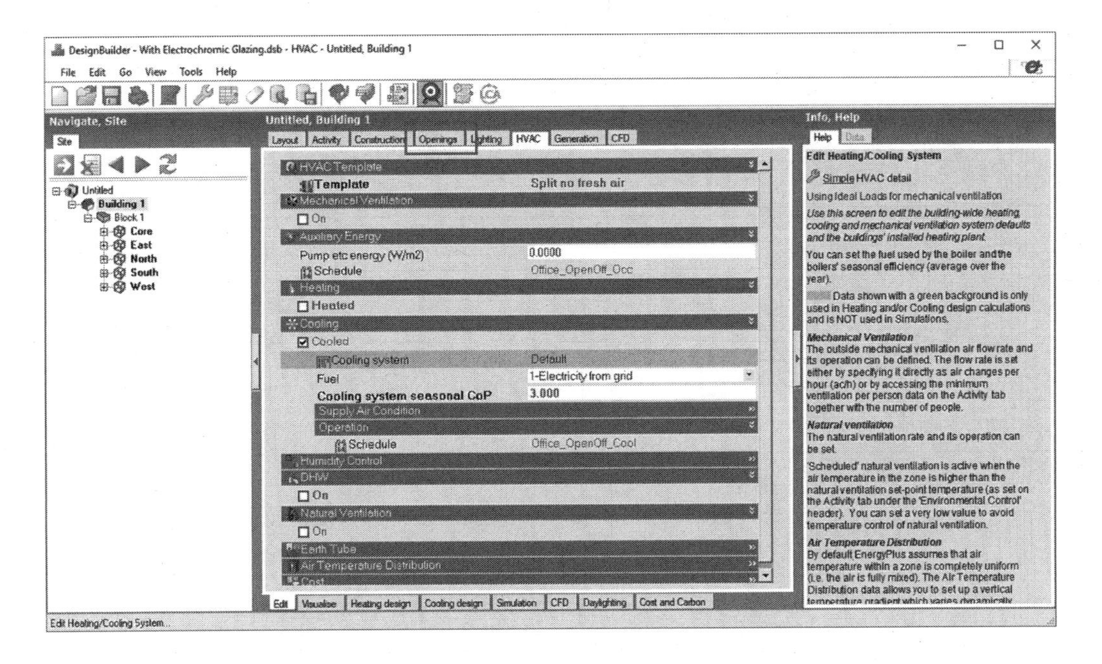

Step 11: Select the **Window shading** checkbox. Click on the three dots to change **Type**. The **Select the Window shade** screen appears.

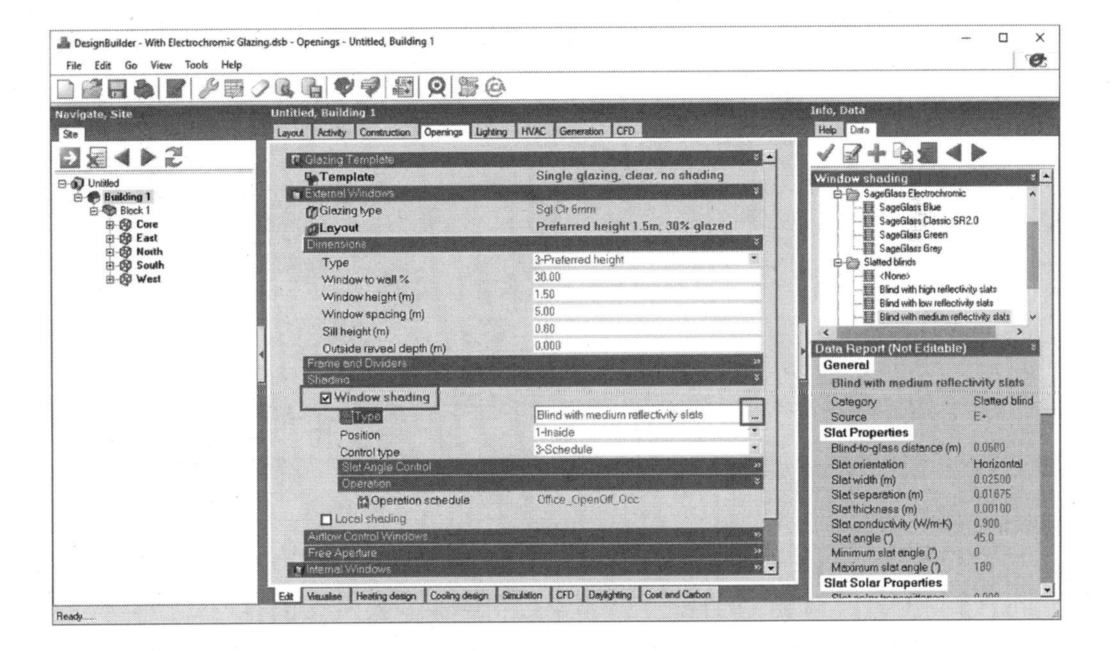

Step 12: Expand the **Electrochromic switchable** tree, and select **Electrochromic reflective 6mm**.

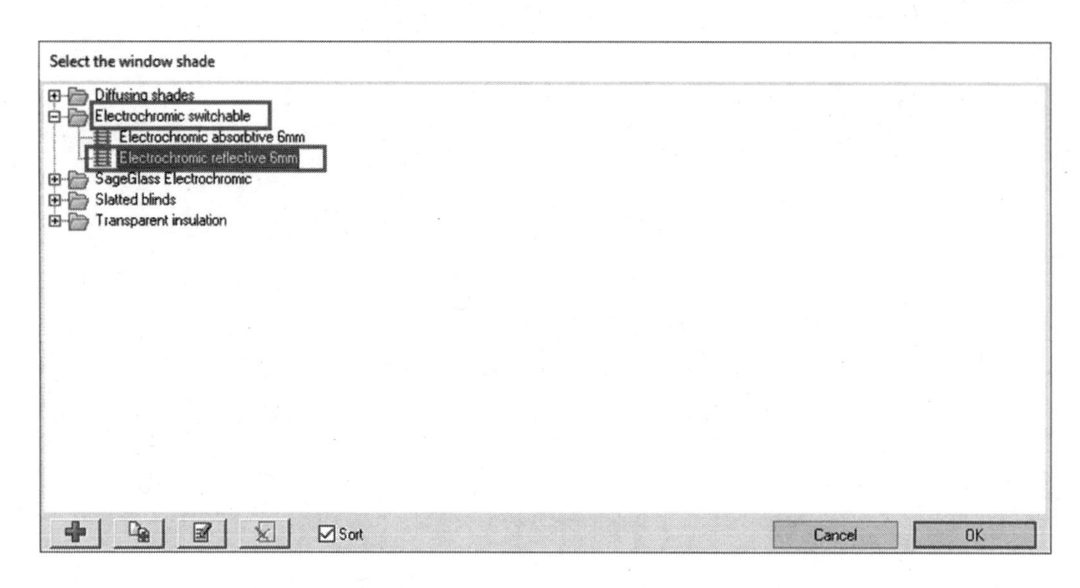

Step 13: Select **4-Switchable** from the **Position** drop-down list. Select **4-Solar** from the **Control type** drop-down list. Enter **120** in the **Solar setpoint(W/m^2)** textbox.

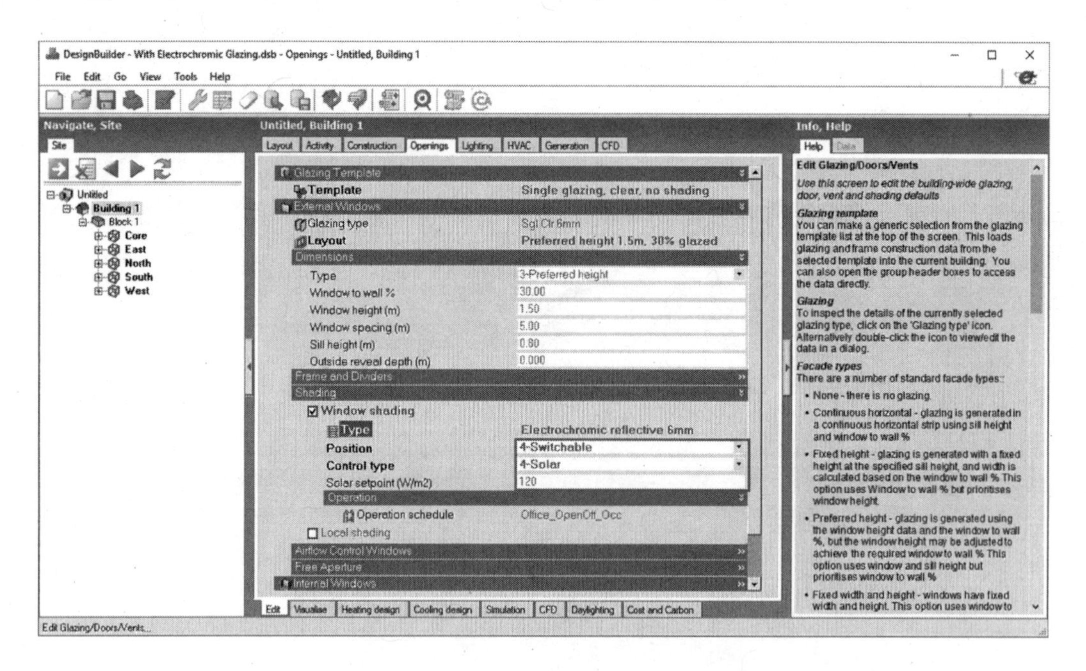

Step 14: Simulate the model, and record the results for the south zone.

Step 15: Compare hourly results for **Exterior Windows Solar Gain** (plot for 29 and 30 July).

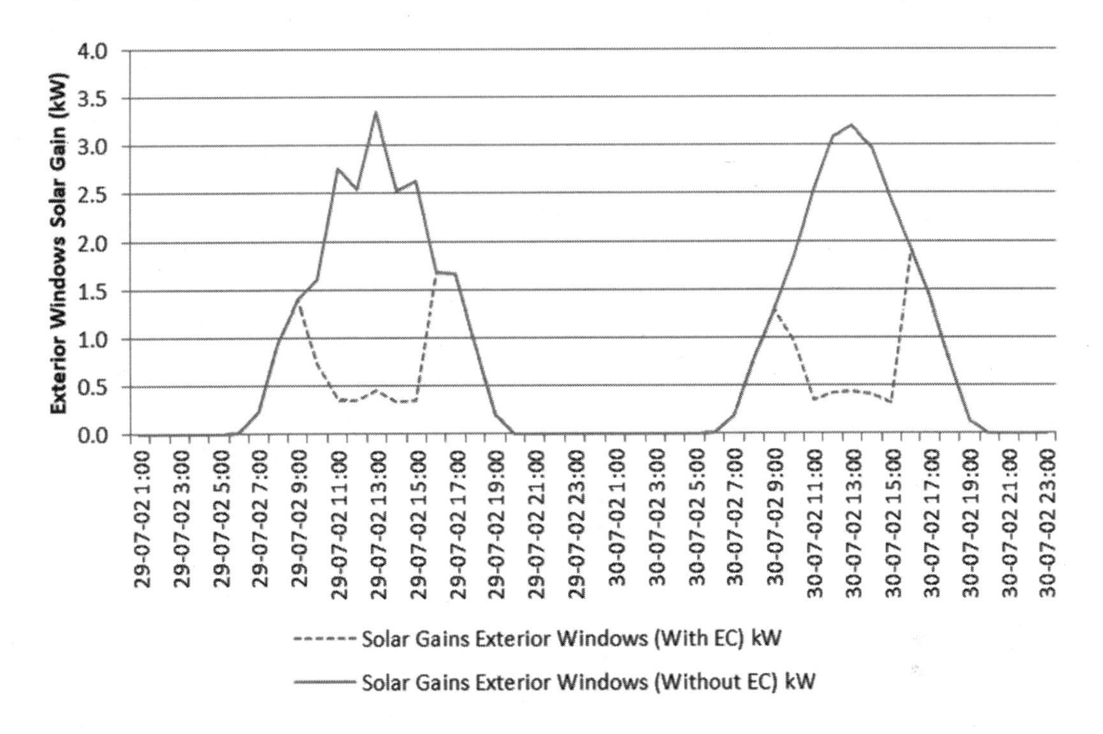

You can see that with electrochromic switchable glazing there is a reduction in solar gain from the glazing whenever the solar radiation incident on the outside of the window is greater than 120 W/m^2.

5 Lighting and Controls

As daylight varies throughout the day, it alone cannot provide targeted illuminance levels the whole day. Sensors can be used to measure the deficit in the illuminance levels and can control the artificial lighting to provide the balance lumens. The operating level of artificial lights in daylit areas can be varied to achieve energy savings. Energy simulation tools are capable of handling this phenomenon. This is explained via two tutorials in this chapter. The third tutorial explains daylight simulations.

Tutorial 5.1

Evaluating the Impact of Daylighting-Based Controls

GOAL

To evaluate the effect of daylighting-based controls on energy consumption

WHAT ARE YOU GOING TO LEARN?

- How to model daylight controls

PROBLEM STATEMENT

In this tutorial, you are going to use a 50- × 25-m five-zone model with a 5-m perimeter depth. You are going to evaluate the following lighting controls:

1. No lighting control
2. Linear control

Use the **WIEN/HOHE WARTE, Austria** weather location.

SOLUTION

Step 1: Open a new project, and create a **50- × 25-m** building with a **5-m** perimeter depth.

Step 2: Select the **Openings** tab, and select **Dbl Grey 6mm/6mm Air** as the **Glazing type** from the drop-down list.

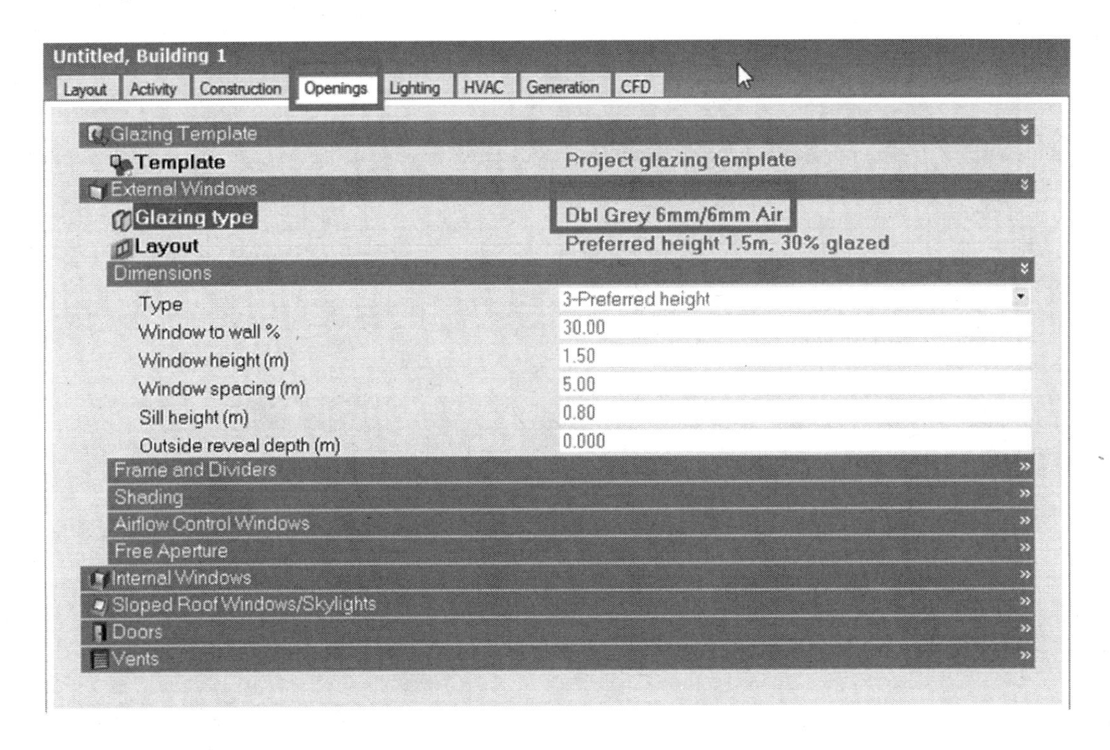

Step 3: Select the **Activity** tab. Select the **24 × 7 Generic Office Area** template from the **Miscellaneous 24hr activities** folder.

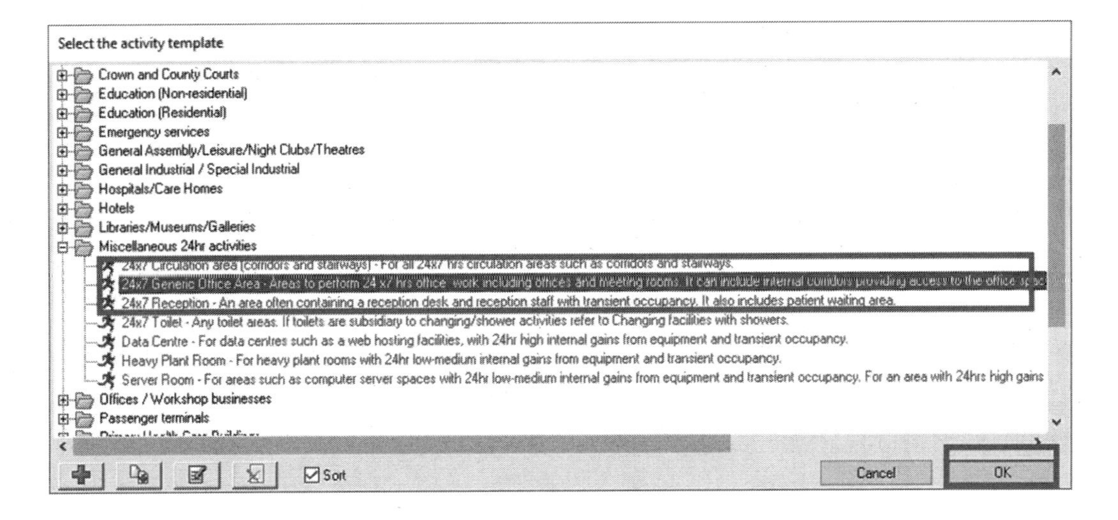

Step 4: Select the **Lighting** tab. Make sure to clear the **On** checkbox under **Lighting Control**.

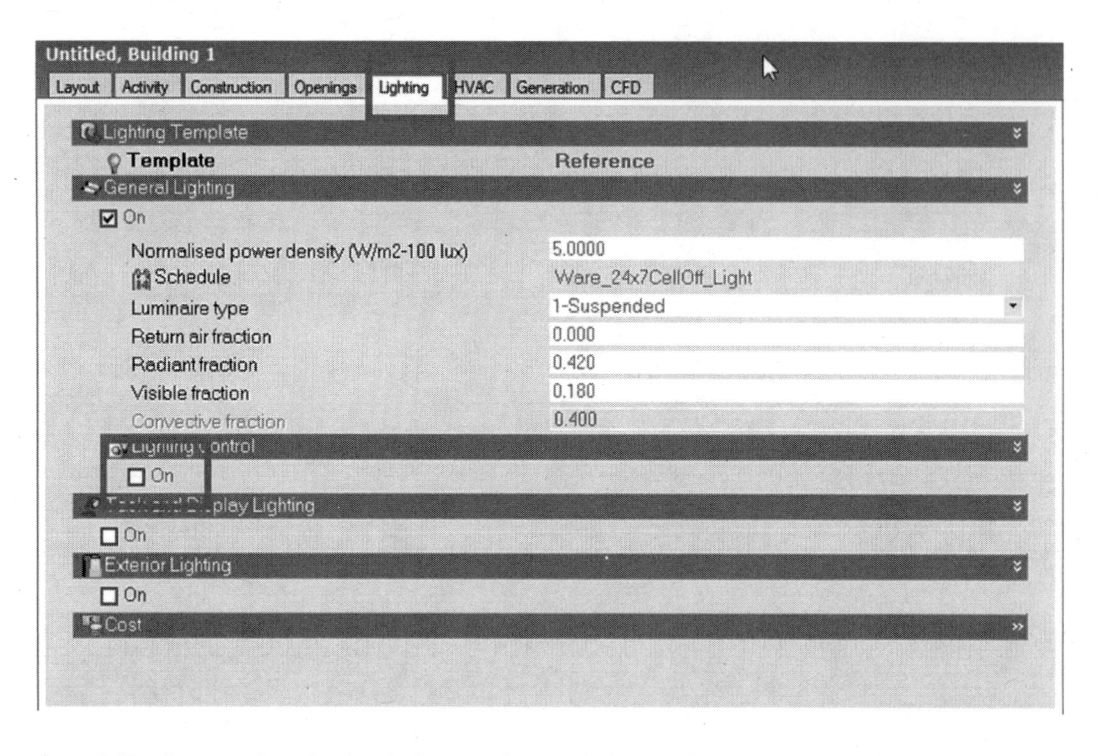

Step 5: Perform an hourly simulation, and record the results. For determining the results of each zone, go to the navigation tree on the left, select the zone, and select **Internal gains** from the **Data** drop-down list and **Hourly** from the **Interval** drop-down list.

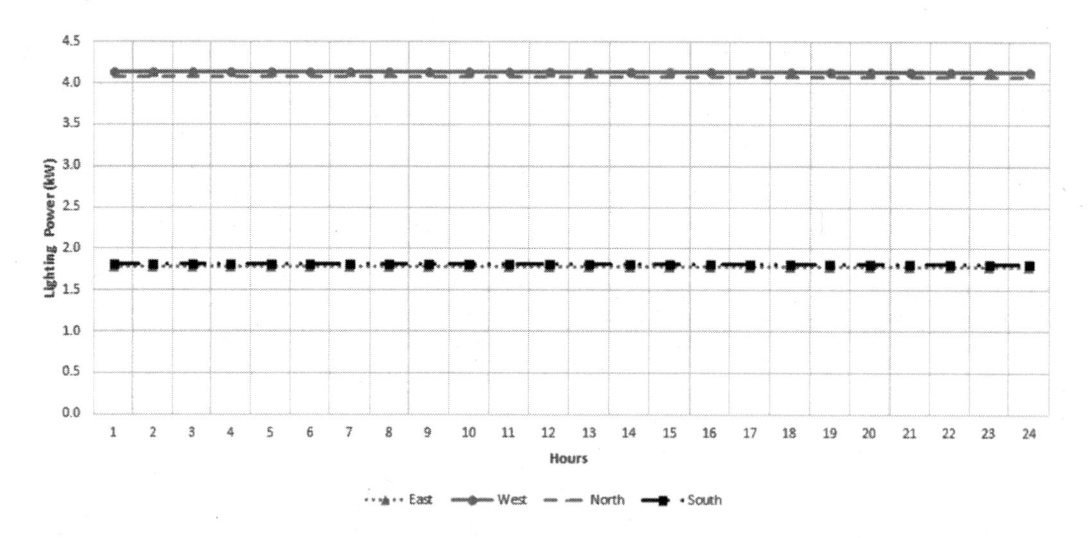

Step 6: Save the model using the **Save as** option. In the following steps, you are going to install daylight controls.

Step 7: Select the **Lighting** tab.

Step 8: Select the **On** checkbox under the **Lighting Control** section. Select **Linear** from the **Control type** drop-down list.

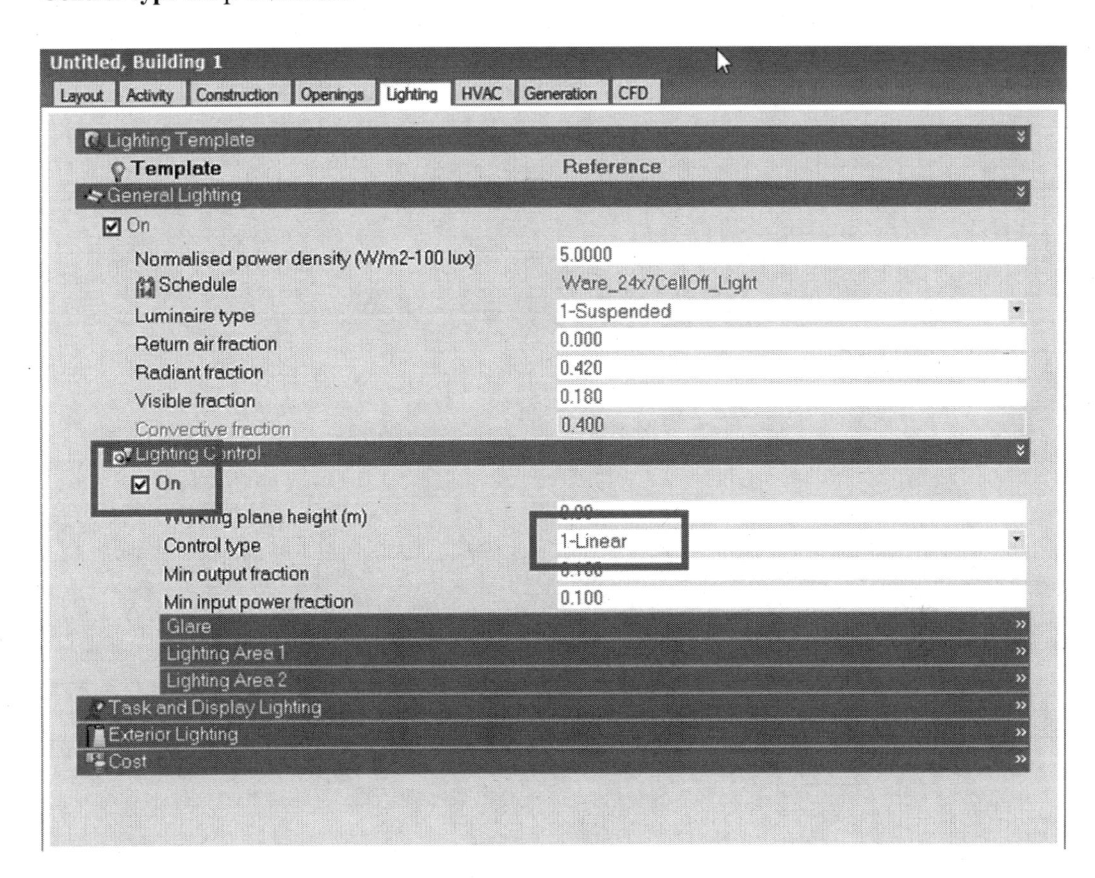

> Based on the control type, lighting controls dim or turn off the internal lighting when the assigned illuminance level is met. This reduces the lighting energy consumption as well as the internal heat gain due to artificial lighting. This reduction in heat gain decreases the cooling load and hence the cooling energy consumption.

Step 9: Select the **Hourly Simulation** tab, and record the results. For determining the results of each zone, go to the navigation tree on the left, select the zone, and select **Internal gains** from the **Data** drop-down list and **Hourly** from the **Interval** drop-down list.

Step 10: Compare the hourly lighting energy consumption for all options for 18 May. You can use a spreadsheet program to plot the comparative graphs.

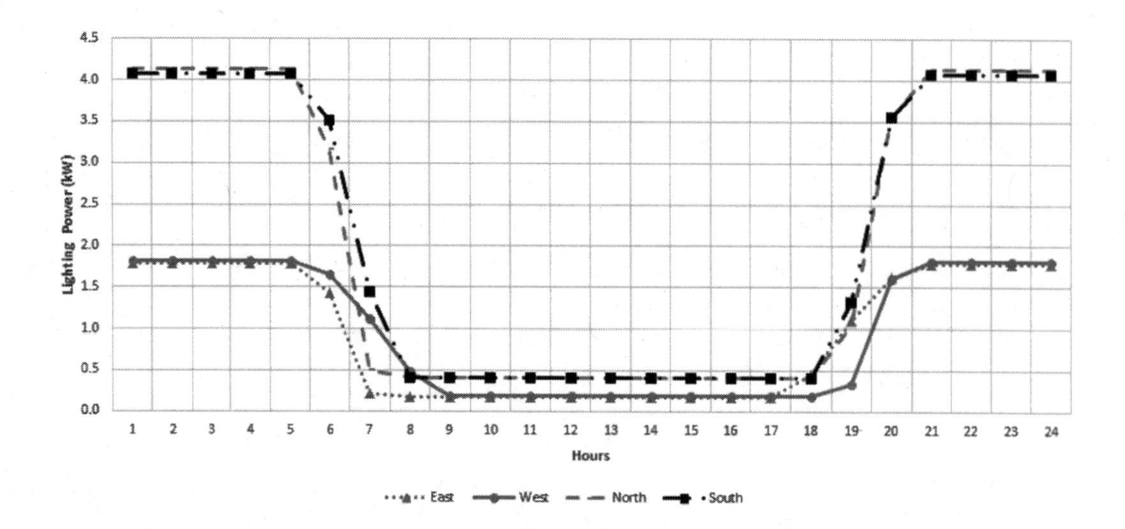

From this graph, it is clear that with the installation of controls, the lighting power consumption has decreased with the lighting being automatically switched off between 06:00 and 18:00 h in different zones based on daylight availability.

You can also get the solar radiation profile for 18 May by selecting **Site Data** from the **Data** drop-down list. Export the **Direct Normal Solar Radiation** plot using a spreadsheet program.

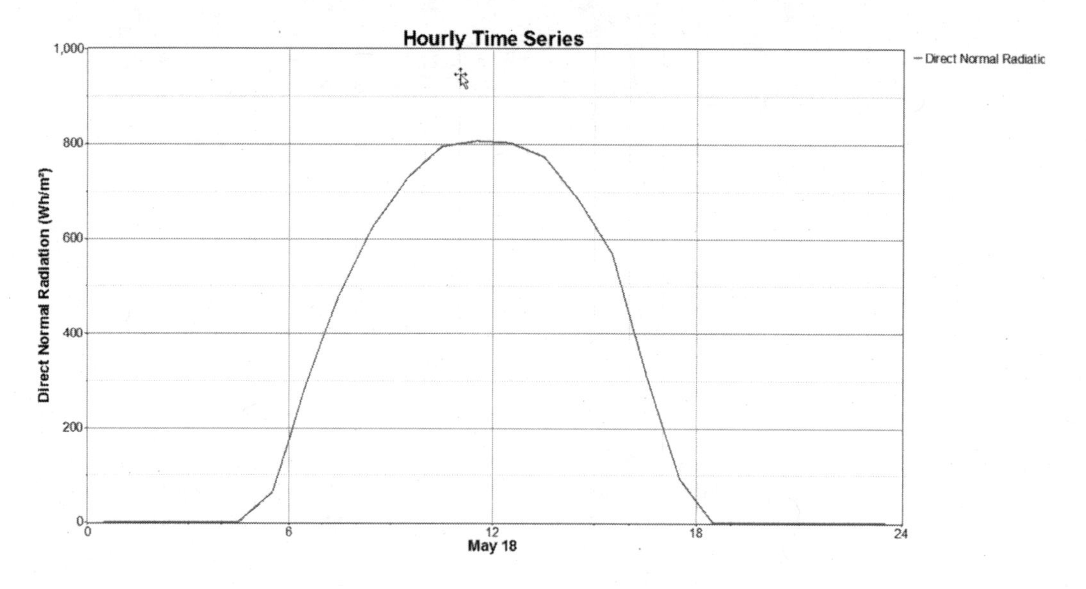

The date of 18 May is selected for the simulation. The solar radiation curve is smooth with no perturbations due to clouds.

EXERCISE 5.1

Repeat this tutorial and compare the energy consumption for stepped lighting controls.

Tutorial 5.2

Evaluating the Impact of Daylight Sensor Placement

GOAL

To evaluate the impact of daylight sensor positioning on energy consumption

WHAT ARE YOU GOING TO LEARN?

- How to define daylight sensors positioning

PROBLEM STATEMENT

In this tutorial, you are going to use a 10- × 10-m single-zone model with one window only on the south facade. Window area is 40% of gross south facade area. Determine the change in internal lighting gains in the zone with the use of a daylight sensor with **Linear/Off** control. Place daylight sensor(s) at the following locations:

- 1 m from the window
- 2 m from the window
- At 2 and 8 m from the window

Use the **New Delhi/Palam, India** weather location.

SOLUTION

Step 1: Open a new project, and create a **10- × 10-m** single-zone model. Press the **ESC** button to get out of the **Edit mode**.

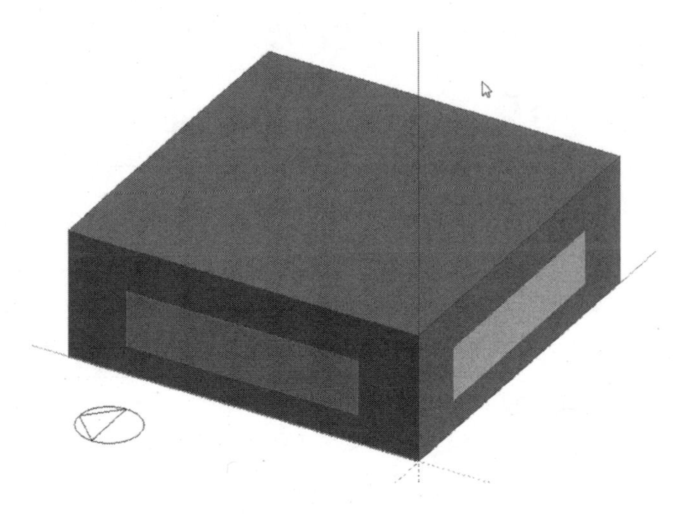

Step 2: Click on the west window; this highlights the window.

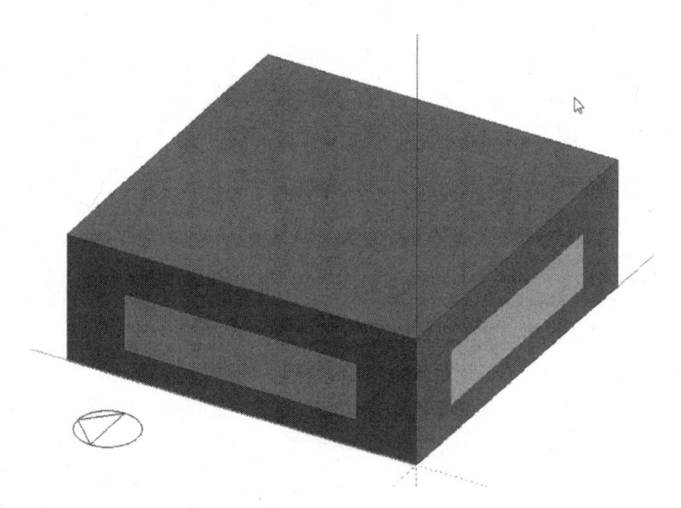

Step 3: Click **Delete selected object(s) (Del)**. Then click **Yes** on the message box. The window disappears. Similarly, you can delete windows on the north and east facades with the help of **Dynamic orbit**.

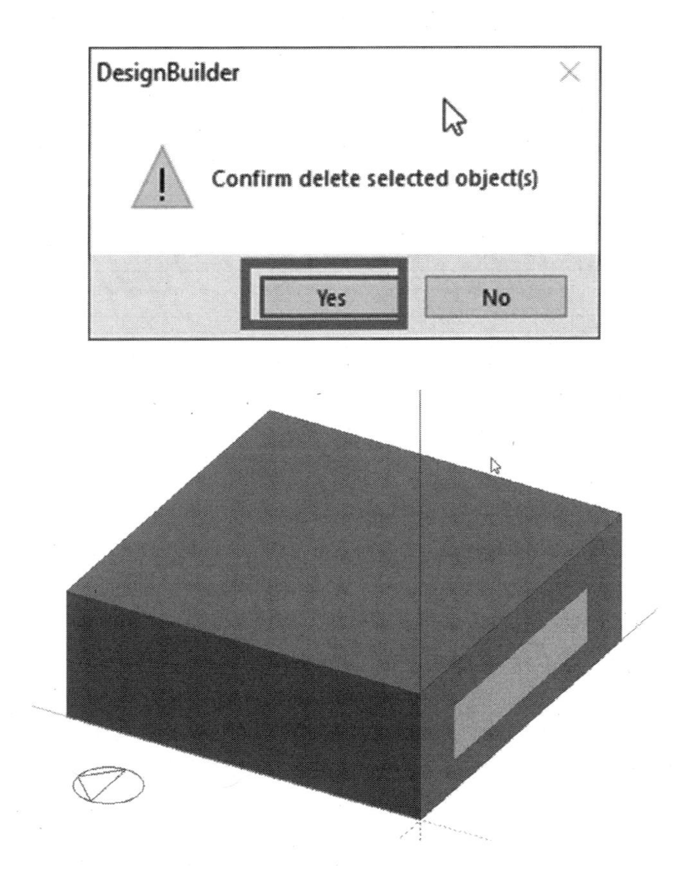

Step 4: Select the **Openings** tab, and select **Dbl Grey 6mm/6mm Air** from the glazing type. Enter **40.00** in the **Window to wall %** text box. You can cross-check the window area on the south wall in **Navigation** area.

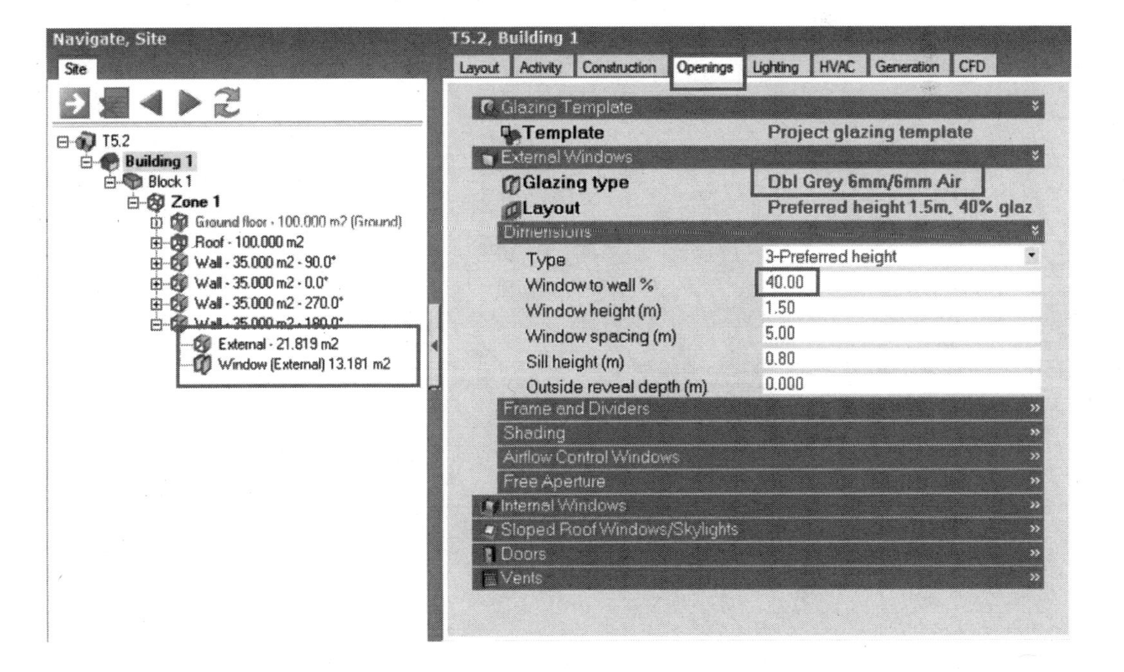

Step 5: Select the **Activity** tab. Select the **24 × 7 Generic Office Area** template from the **Miscellaneous 24hr activities** branch.

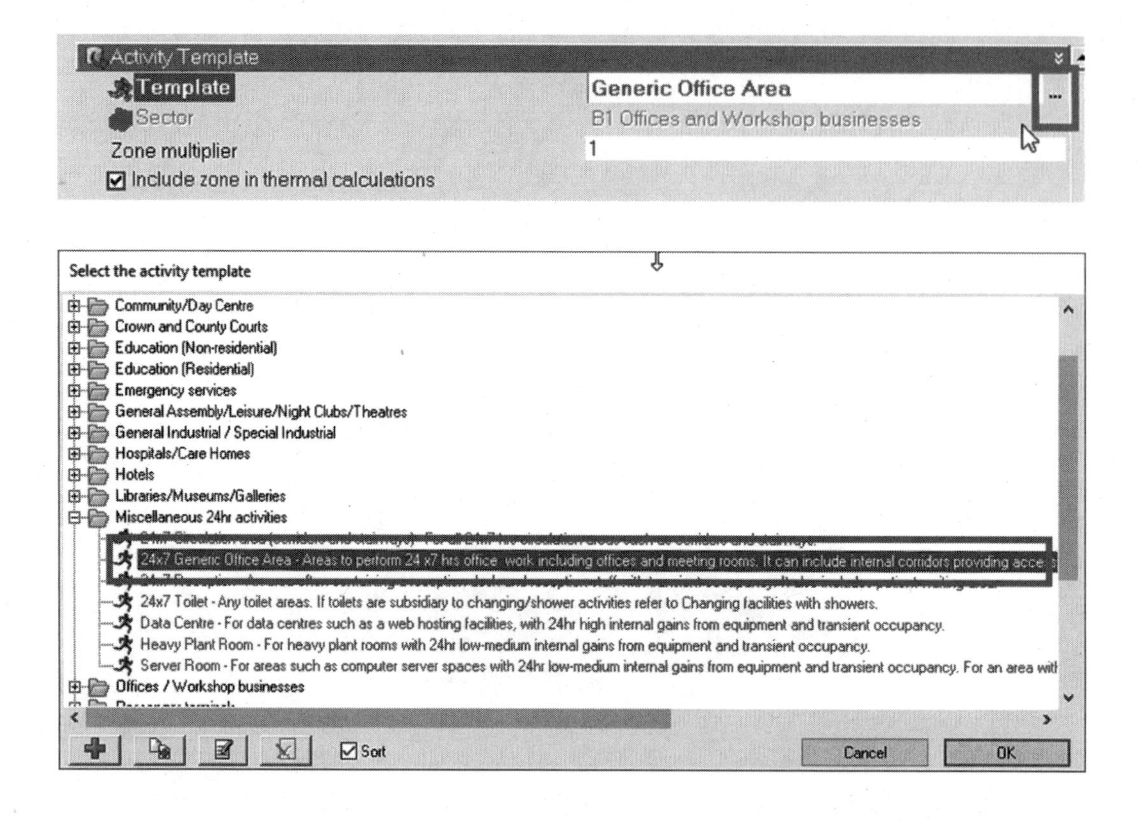

Step 6: Select the **Lighting** tab. Set the **Normalised power density (W/m^2-100 lux)** to **3**.

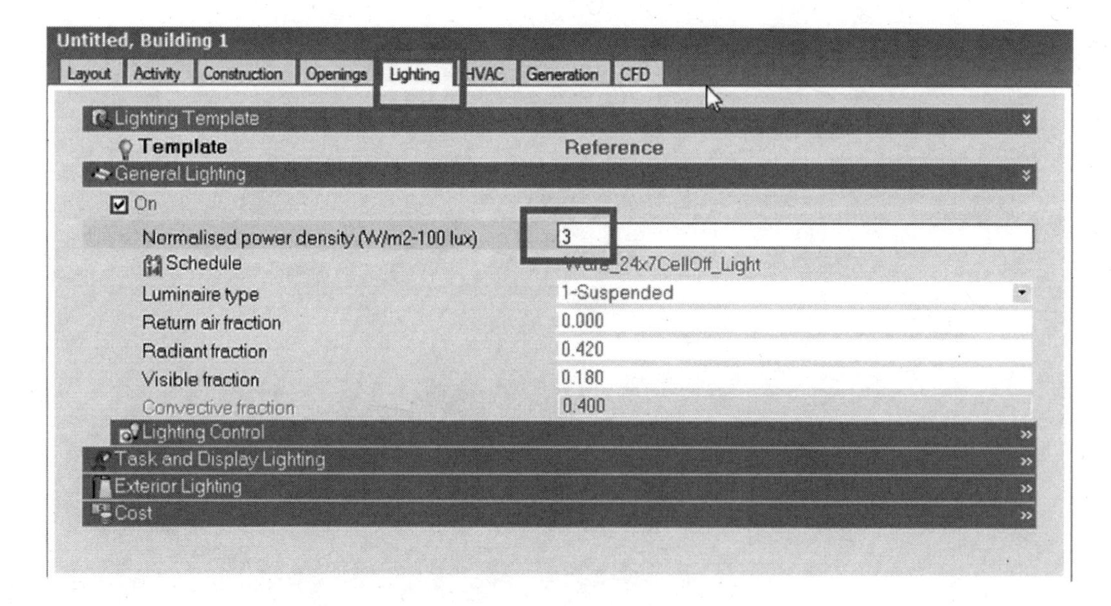

Step 7: Make sure to clear the **On** checkbox under the **Lighting Control** section.

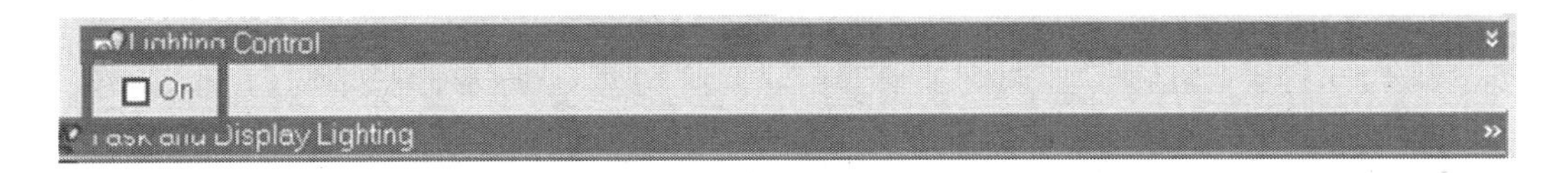

Step 8: Perform an hourly simulation, and record the results.

Untitled, Building 1

Analysis | Summary | Parametric | Optimisation + UA/SA | Data Visualisation

Date/Time	Room Electricity (kWh)	Lighting (kWh)	Heating (Gas) (kWh)	Cooling (Electricity) (kWh)	DHW (Electricity) (kWh)
12:00:00 AM	6800.086	9318.068	27.87282	21678.25	923.7079

Now you are going to enable daylight control in the model.

Step 9: Select the **Lighting** tab.

Step 10: Select the **On** checkbox under the **Lighting Control**, and select **Linear/off** in the **Control type**. Set **100** for **% Zone covered by Lighting Area 1**. Select the **Layout** tab.

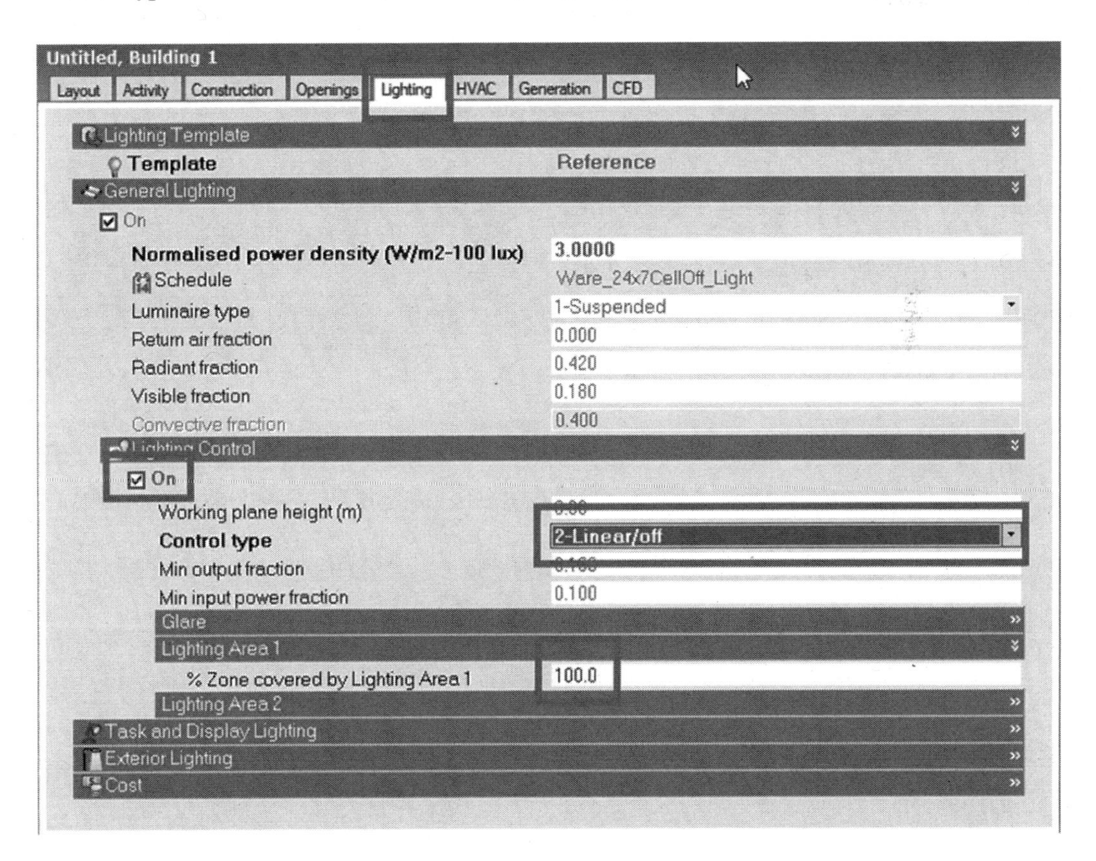

Step 11: Select the **Activity** tab, and make sure that the target illuminance (lux) is **400**.

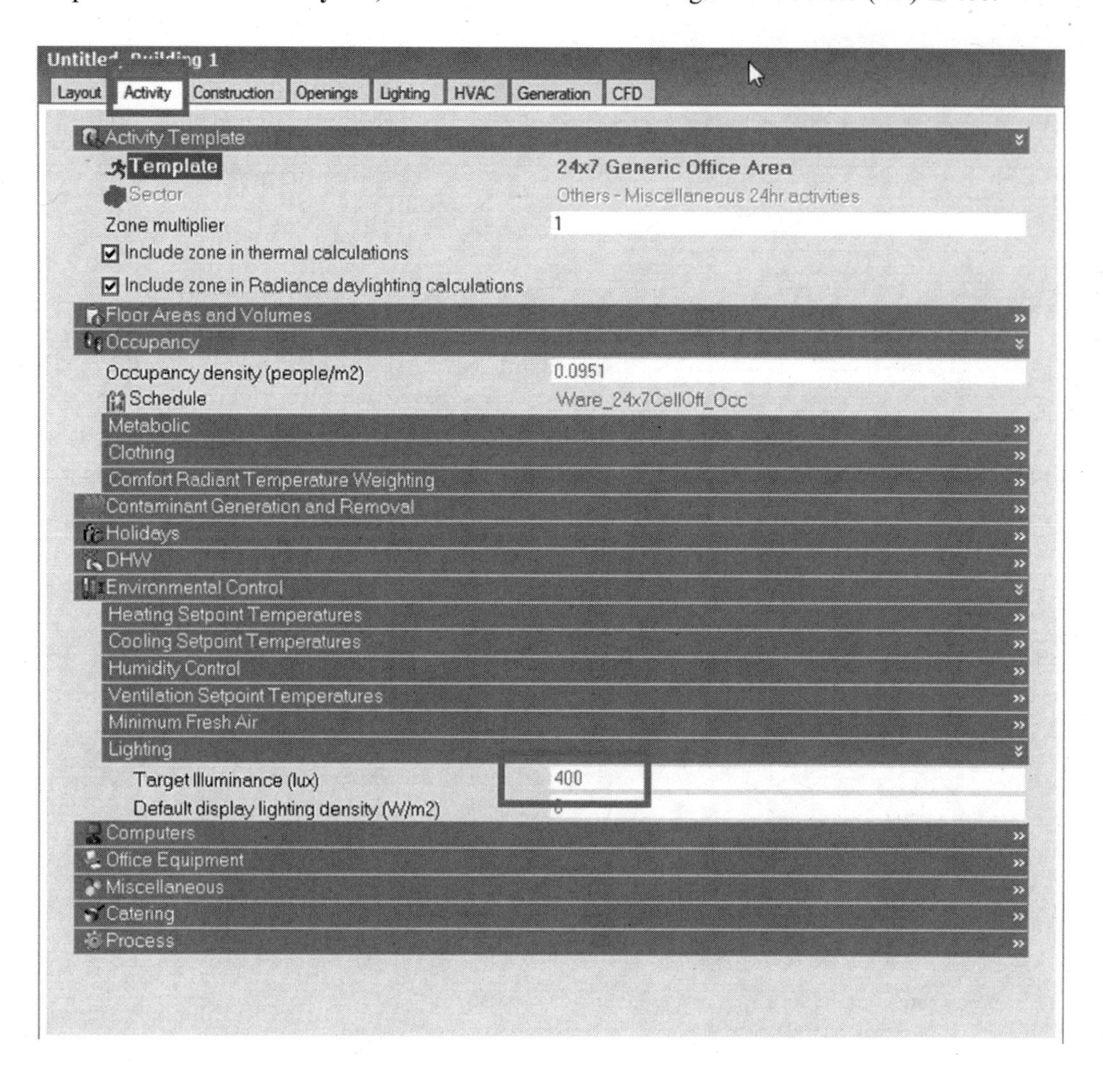

Step 12: Select **Zone 1** from the navigation tree. The daylight sensor can be seen in the layout.

Step 13: Select the **Sensor**, and click the **Move selected object** icon. Click the daylight sensor and place it 2 m away from the window with the help of construction lines.

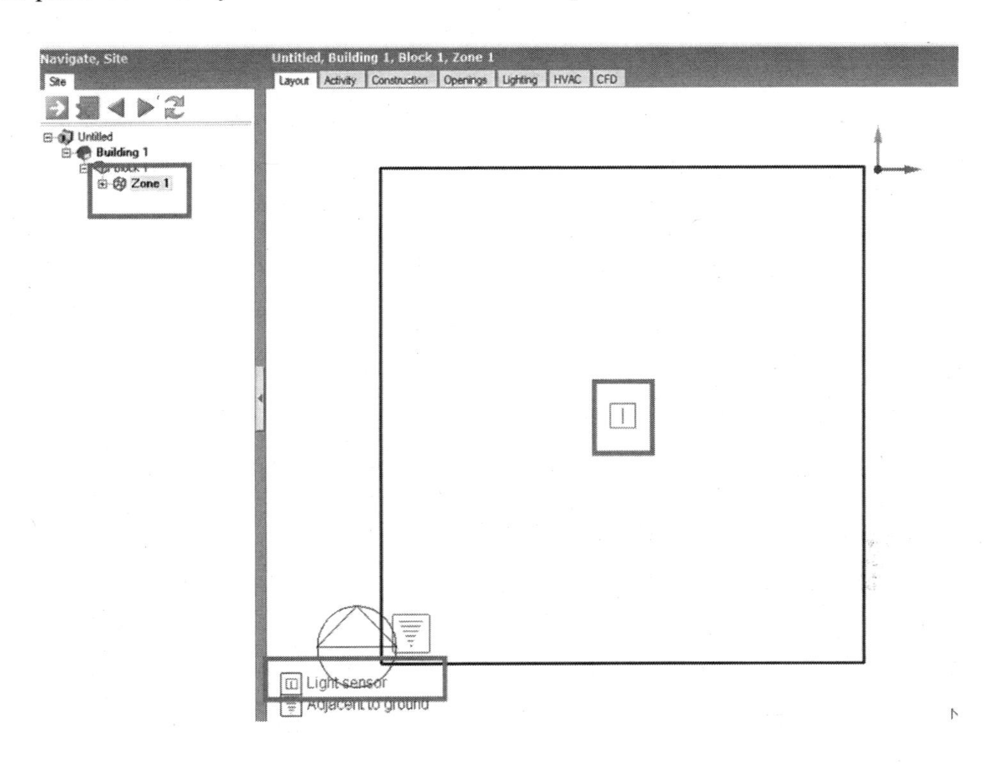

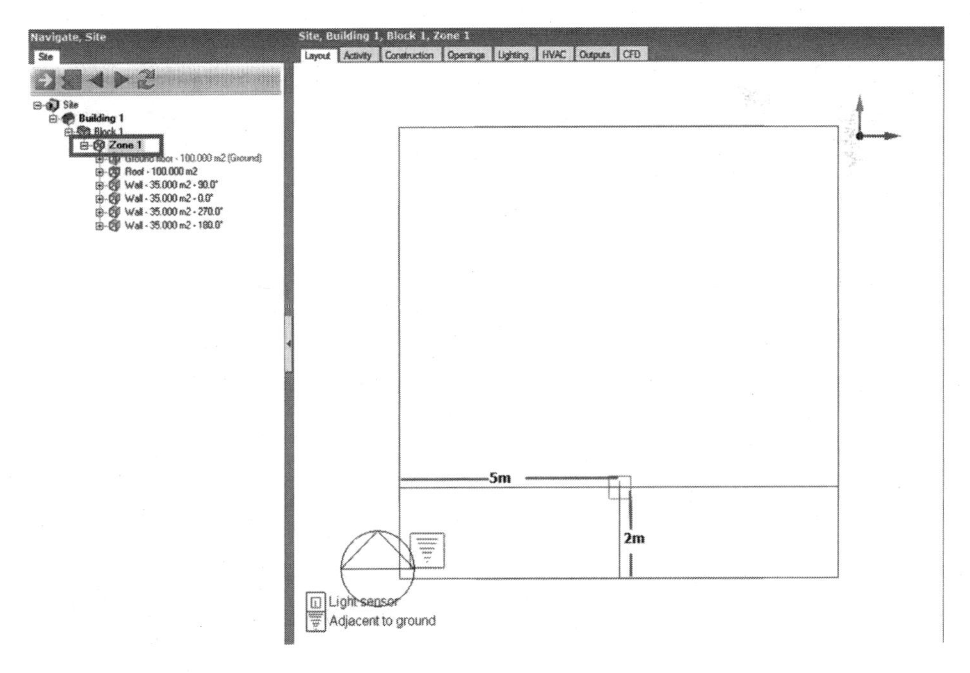

Step 14: Select the **Simulation** screen tab. The **Edit Calculation Options** screen appears. Select the **Output** tab, and click the **Detailed Daylight Outputs** section. Select the **Daylight map output** checkbox. Click **OK**. Perform an hourly simulation.

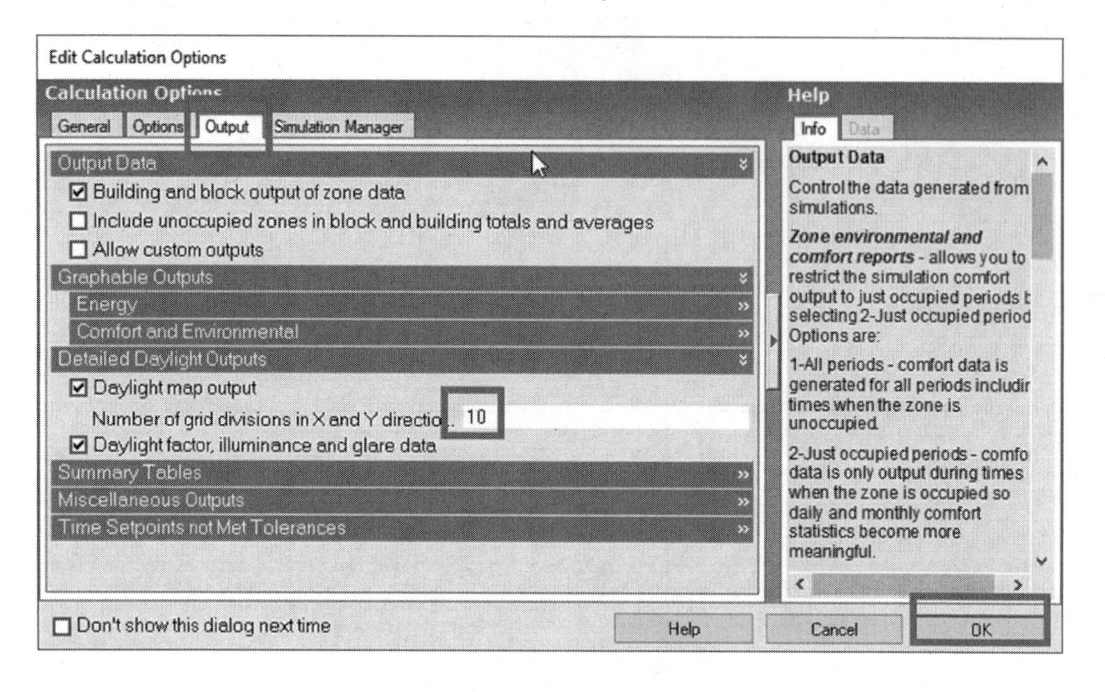

Step 15: Record the results. Plot hourly lighting data for 4 April.

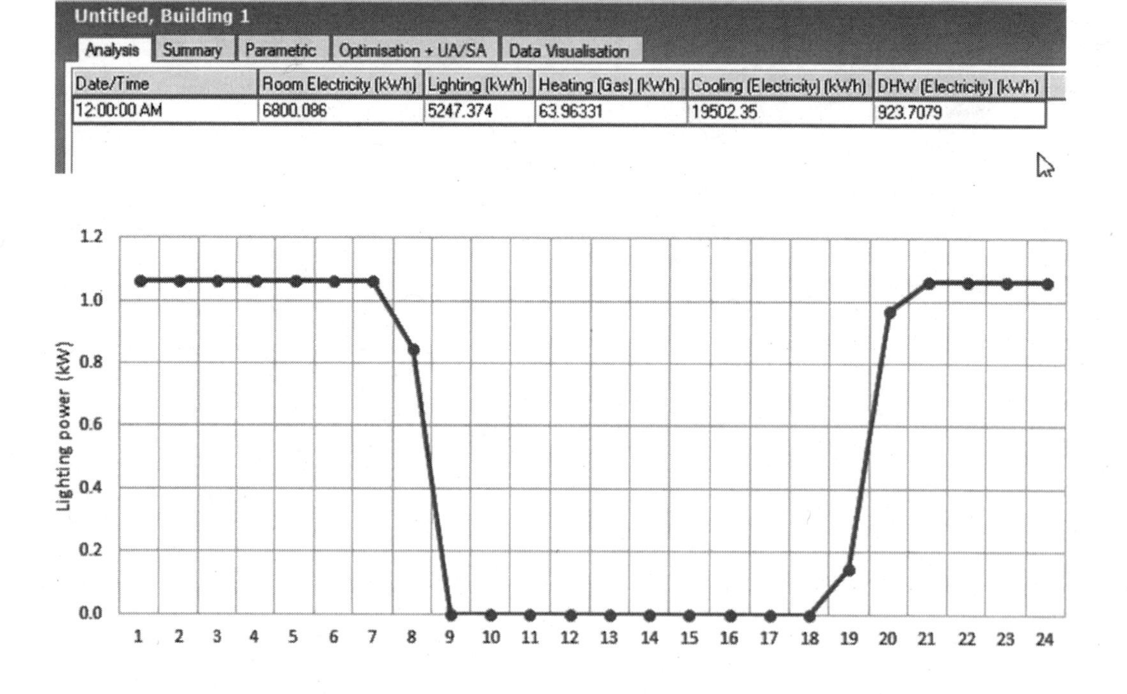

In the preceding graph, you can see that the lighting consumption is zero between 09:00 and 17:00 h. To get the lux level at this time, you need to get the illuminance map.

Step 16: Open the **eplusmap** file that exists in the **EnergyPlus** folder.

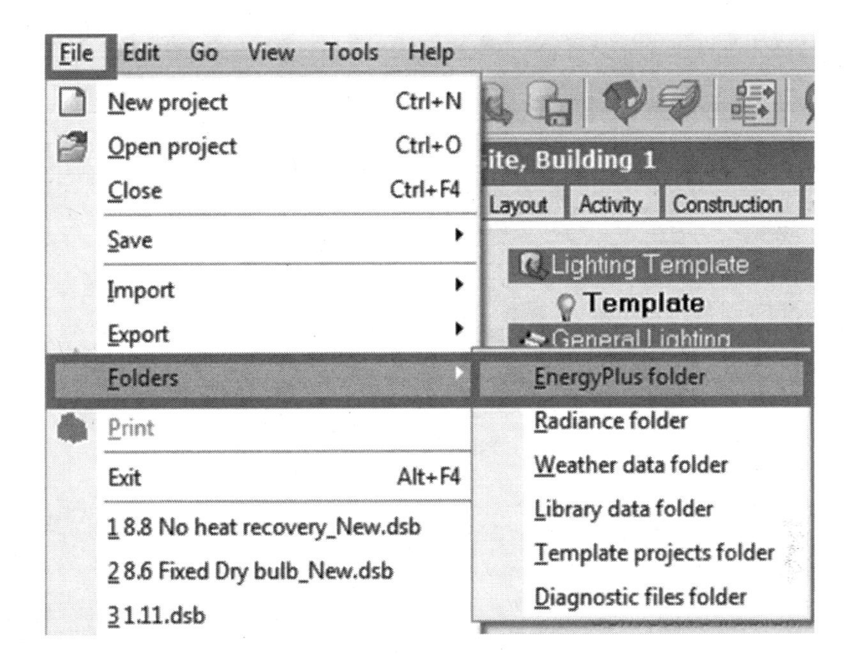

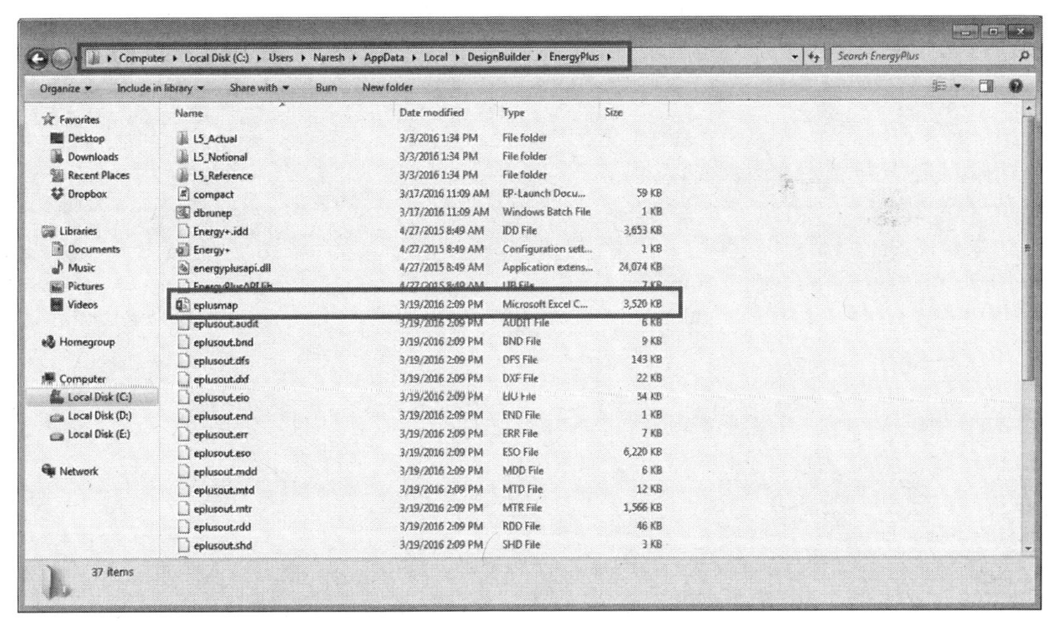

Step 17: Get the data for 11:00 h. Table 5.1 shows the illuminance map when the sensor is placed near the window. In the map, each cell reports the illuminance (in lux) at the location specified by the (X; Y) coordinates in the column and row headers. These are XY pairs separated by a semicolon for ease in importing into the spreadsheet. In the **eplusmap** file,

the Z coordinate of the map is shown in the title (the illuminance map is set in a plane), and the date and time are indicated in the upper-left cell of the map. You can see that at 11:00 h, all artificial lights are off because of sufficient illuminance near the sensor. This leads to low illuminance in the interiors of the space.

Daylight Map Output

EnergyPlus generates a file called **eplusmap.csv** when one or more zones have daylight control and the **Daylight map** simulation output option has been selected for those zones. This file contains a series of grids for each zone with daylight controls and has the daylight map output requested and for each hour of the day in the simulation.

Source: https://designbuilder.co.uk/helpv6.0/Content/Daylighting.htm.

TABLE 5.1
Illuminance map for 4 April: Reference Point 1 (RefPt1) = (4.98:2.00:0.80)

4/4 11:00	(0.18: 0.22)	(1.25: 0.22)	(2.31: 0.22)	(3.38: 0.22)	(4.45: 0.22)	(5.51: 0.22)	(6.58: 0.22)	(7.65: 0.22)	(8.71: 0.22)	(9.78: 0.22)
(0.18; 0.22)	306	20,414	20,417	20,417	20,418	20,421	20,425	20,428	20,416	285
(0.18; 1.29)	422	564	604	611	613	612	608	589	495	340
(0.18; 2.35)	352	395	417	426	428	427	420	401	363	318
(0.18; 3.42)	320	338	349	355	357	355	349	338	321	303
(0.18; 4.49)	304	312	318	321	322	321	317	311	302	294
(0.18; 5.55)	294	299	302	304	304	303	301	297	293	288
(0.18; 6.62)	289	292	294	295	295	294	293	290	288	285
(0.18; 7.69)	285	287	288	289	289	288	287	286	284	282
(0.18; 8.75)	282	283	284	285	285	284	284	283	281	280
(0.18; 9.82)	280	281	282	282	282	282	281	281	280	279

You can also get the solar radiation profile on 4 April by selecting **Site Data** from the **Data** drop-down list. Export the **Direct Normal Solar Radiation** and plot using a spreadsheet program.

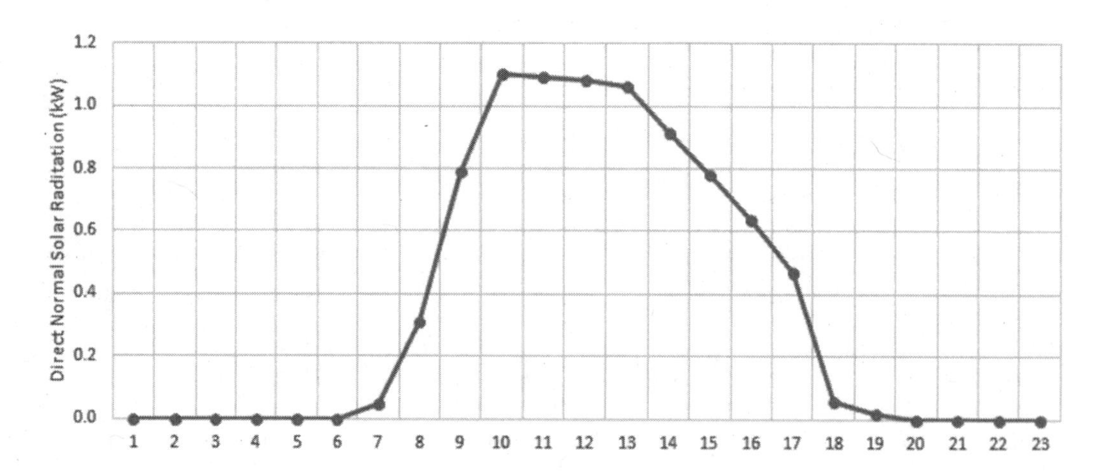

Step 18: Save the model using the **Save as** option. Now you are going to change the location of the daylight sensor to 8 m from the south facade window.

Step 19: Move the daylight sensor 8 m away from the window towards the north (as explained earlier).

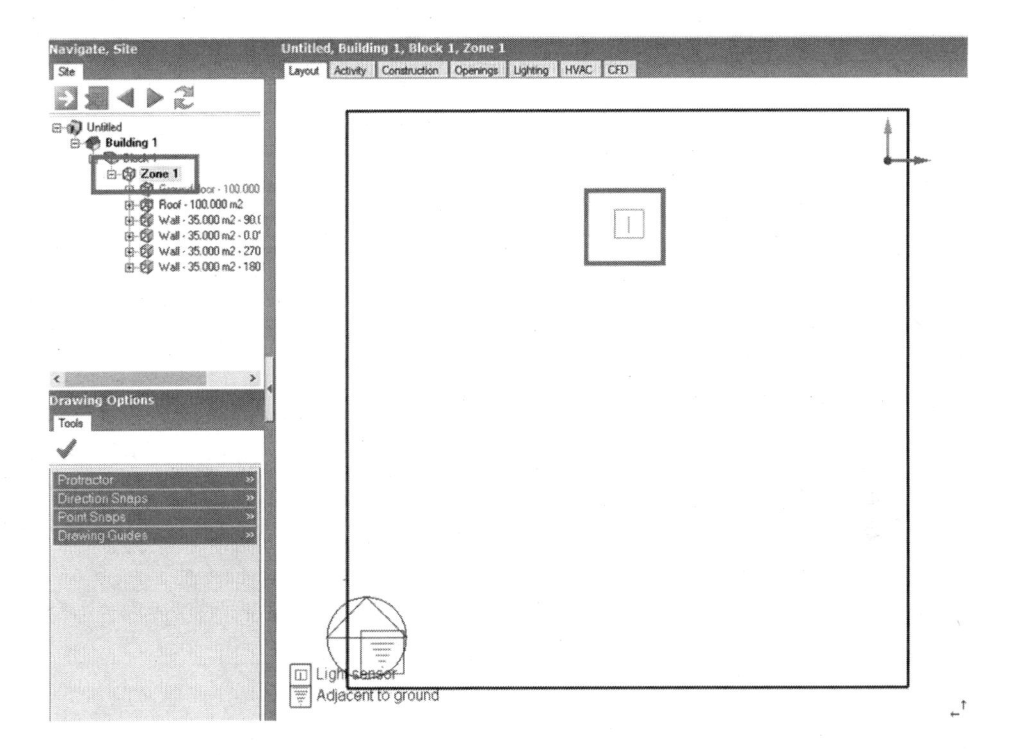

Step 20: Perform an hourly simulation. Record the annual results. Also record hourly lighting data for 4 April.

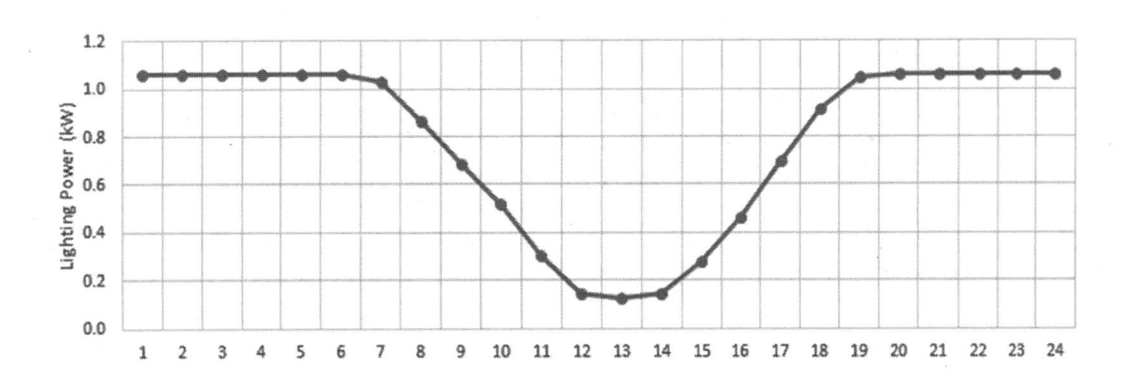

Untitled, Building 1

Analysis | Summary | Parametric | Optimisation + UA/SA | Data Visualisation

Date/Time	Room Electricity (kWh)	Lighting (kWh)	Heating (Gas) (kWh)	Cooling (Electricity) (kWh)	DHW (Electricity) (kWh)
01-Jan-02 1:00:00 AM	6800.086	6911.67	44.58714	20409.95	923.7079

The preceding figure shows the hourly lighting load for 4 April (with the daylight sensor 8 m from the window). In this graph, you can see that lighting energy consumption between 08:00 and 17:30 h is higher than when the sensor is placed near the window. When the sensor is moved away from the window, it is in a darker portion of the room and is triggered when more daylight enters the room. This increase in daylight might take some time, thereby delaying the time when the lights are switched off. This increases the lighting energy consumption but ensures that sufficient light is available even in the interior areas of the room when the lights are switched off. However, this might lead to higher illuminance levels near the window, resulting in visual discomfort. So there is a need to place two sensors at different positions to get both energy savings and visual comfort.

Step 21: Save the model using the **Save as** option. Now you are going to install one more daylight sensor.

Step 22: Select the **Lighting** tab, and select the **Second lighting area** checkbox. Enter **50.0** as the % **Zone covered by Lighting Area 1** and % **Zone covered by Lighting Area 2**. Make sure that the **Target Illuminance (lux)** is set to **400**.

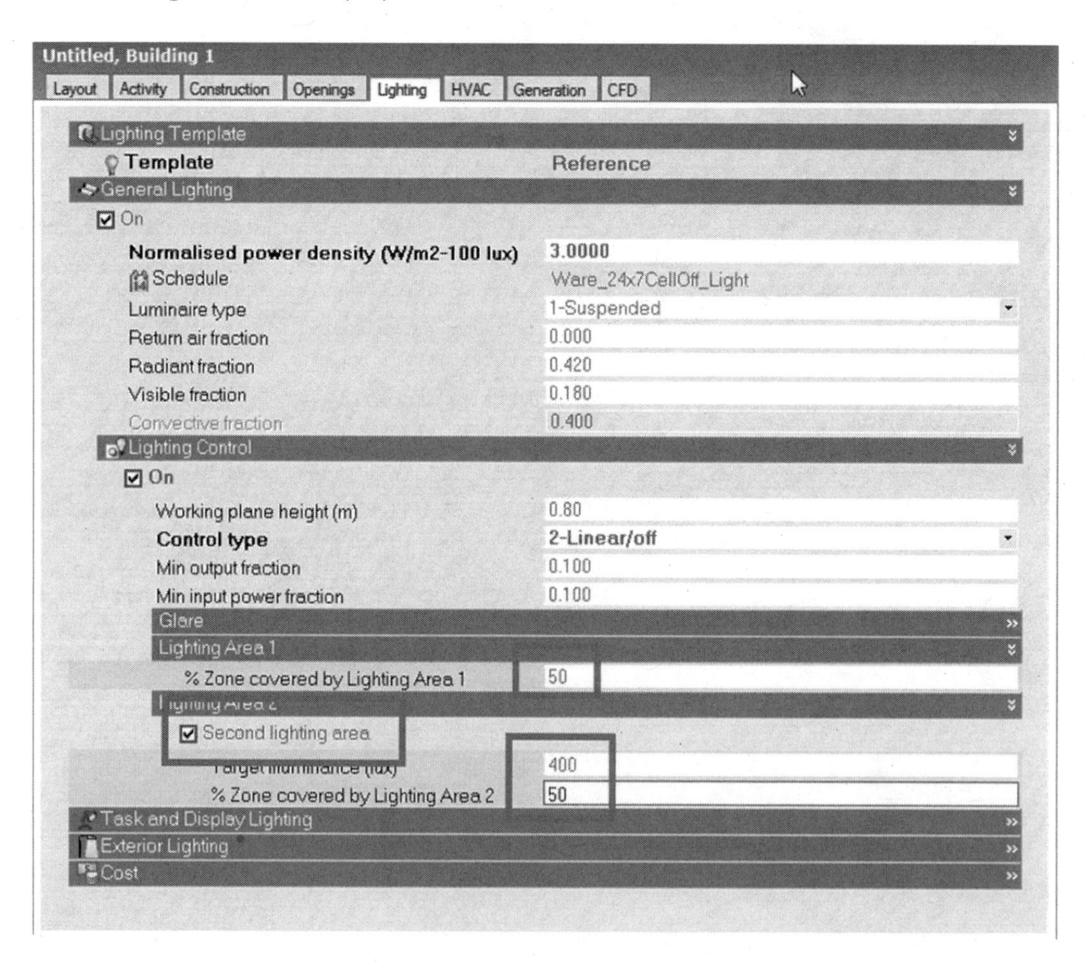

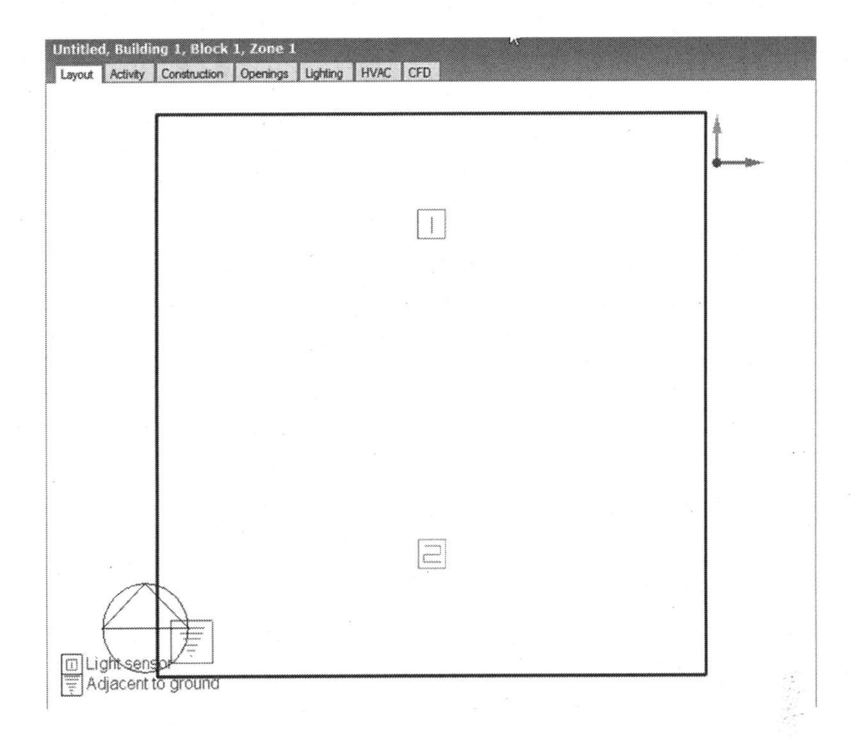

Step 23: Perform an hourly energy simulation, and record the results.

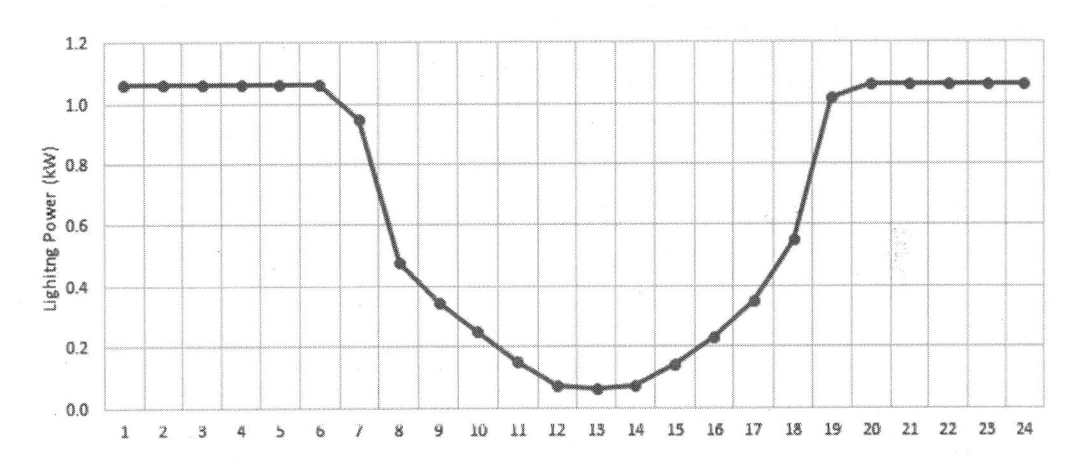

In this figure, you can see the lighting consumption between 08:00 and 17:30 h. The energy consumption is less than when the sensor is placed far from the window, as artificial lights are switched off near the window when there is sufficient daylight. Get the data for 11:00 h for illuminance levels (Table 5.2). Now compare the annual energy consumption for all three cases (Table 5.3).

It can be observed that the placement of the sensor affects energy consumption. This is due to the fact that while using a single sensor, the controller assumes the same illuminance level in the entire zone as is found on the sensor. With this approach, when the sensor is placed close to the window, the model calculates the requirement for artificial light against a higher daylight level than when there is a lower daylight level when the sensor is placed

TABLE 5.2
Illuminance map for 4 April: RefPt1 = (4.98:8.06:0.80), RefPt2 = (4.98:2.19:0.80)

4/4 11:00	(0.18: 0.22)	(1.25: 0.22)	(2.31: 0.22)	(3.38: 0.22)	(4.45: 0.22)	(5.51: 0.22)	(6.58: 0.22)	(7.65: 0.22)	(8.71: 0.22)	(9.78: 0.22)
(0.18; 0.22)	306	20,414	20,417	20,417	20,418	20,421	20,425	20,428	20,416	285
(0.18; 1.29)	422	564	604	611	613	612	608	589	495	340
(0.18; 2.35)	352	395	417	426	428	427	420	401	363	318
(0.18; 3.42)	320	338	349	355	357	355	349	338	321	303
(0.18; 4.49)	304	312	318	321	322	321	317	311	302	294
(0.18; 5.55)	294	299	302	304	304	303	301	297	293	288
(0.18; 6.62)	289	292	294	295	295	294	293	290	288	285
(0.18; 7.69)	285	287	288	289	289	288	287	286	284	282
(0.18; 8.75)	282	283	284	285	285	284	284	283	281	280
(0.18; 9.82)	280	281	282	282	282	282	281	281	280	279

TABLE 5.3
Annual lighting energy consumption with different sensor placements

Sensor placement	Annual lighting energy consumption (kWh)
No sensor	9,318.07
Near the window	5,247.37
Far from the window	6,911.67
With two sensors	6,103.38

far away. With two sensors, the space is divided into two zones independently controlled, and hence the energy consumption is between the two cases, as discussed earlier. In practice, however, even when only one sensor is used, different fixtures can be calibrated to adjust tot different daylight levels at various depths.

Tutorial 5.3

Evaluating the Impact of Window External Shades and WWR on Daylight Performance

GOAL

To evaluate the impact of window external shades and WWR on daylight performance

WHAT ARE YOU GOING TO LEARN?

- How to perform simulations for daylight assessment

PROBLEM STATEMENT

In this tutorial, you are going to use a 10- x 10-m single-zone model. Model the building with the **Glazing** template **Double glazing; clear no shading (Dbl Clr 6 mm/13 mm Air)**. Take an occupancy schedule as 9:00 to 17:00 h. You need to simulate three models with the options shown in Table 5.4.

Find the preceding grade-floor areas that meet or exceed the illuminance level between 100 and 2,000 lux for at least 90% of the potential daylit time in a year. Consider a work-plane height of 0.75 m above the finished floor. Use the **New Delhi Palam, India** weather location.

> *Useful daylight illuminance* (UDI) is defined as the annual occurrence of daylight between 100 and 2,000 lux on a work plane. The degree to which UDI is not achieved because illuminances exceed the upper limit is indicative of the potential for occupant discomfort.
>
> *Source:* Nabil, A., and Mardaljevic, J. (2005). Useful daylight illuminance: a new paradigm for assessing daylight in buildings. *Lighting Research & Technology* 37(1):41–57. Available at https://doi.org/10.1191/1365782805li128oa.

TABLE 5.4
Window-to-wall ratio (WWR) and shading variations

Serial no.	Glass	WWR (%)	Overhang and fins
1	Dbl Clr 6 mm/13 mm Air	30	No shade
2	Dbl Clr 6 mm/13 mm Air	30	Overhang (1 m) and fins (1 m)
3	Dbl Clr 6 mm/13 mm Air	25	Overhang (1 m) and fins (1 m)

SOLUTION

Step 1: Open a new project, and create a single-zone **10- × 10-m** building. Click the **Activity** tab.

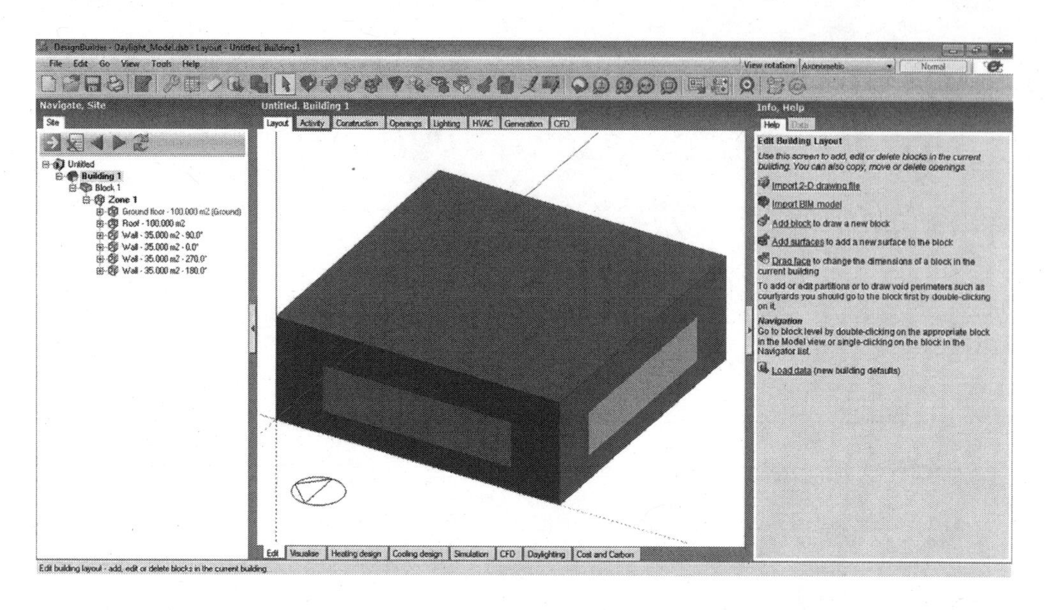

Step 2: Select **Schedule** under **the Occupancy** tab. Copy and edit the existing schedule.

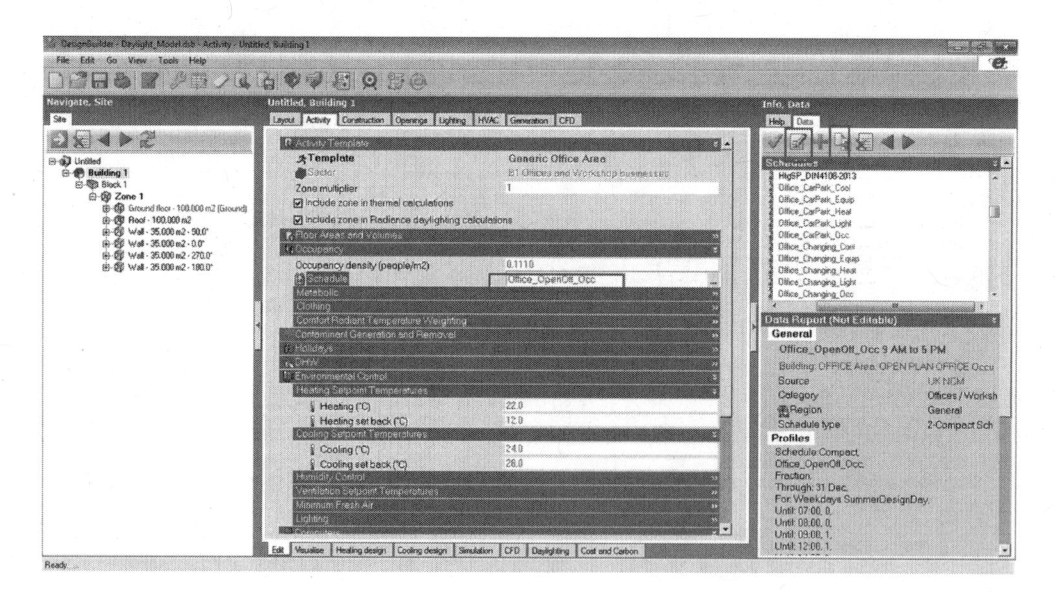

Step 3: Type **Office_OpenOff_Occ 9 AM to 5 PM** in the **Name** tex tbox. Set **1** from 09:00 to 17:00 h and the remaining to **0**.

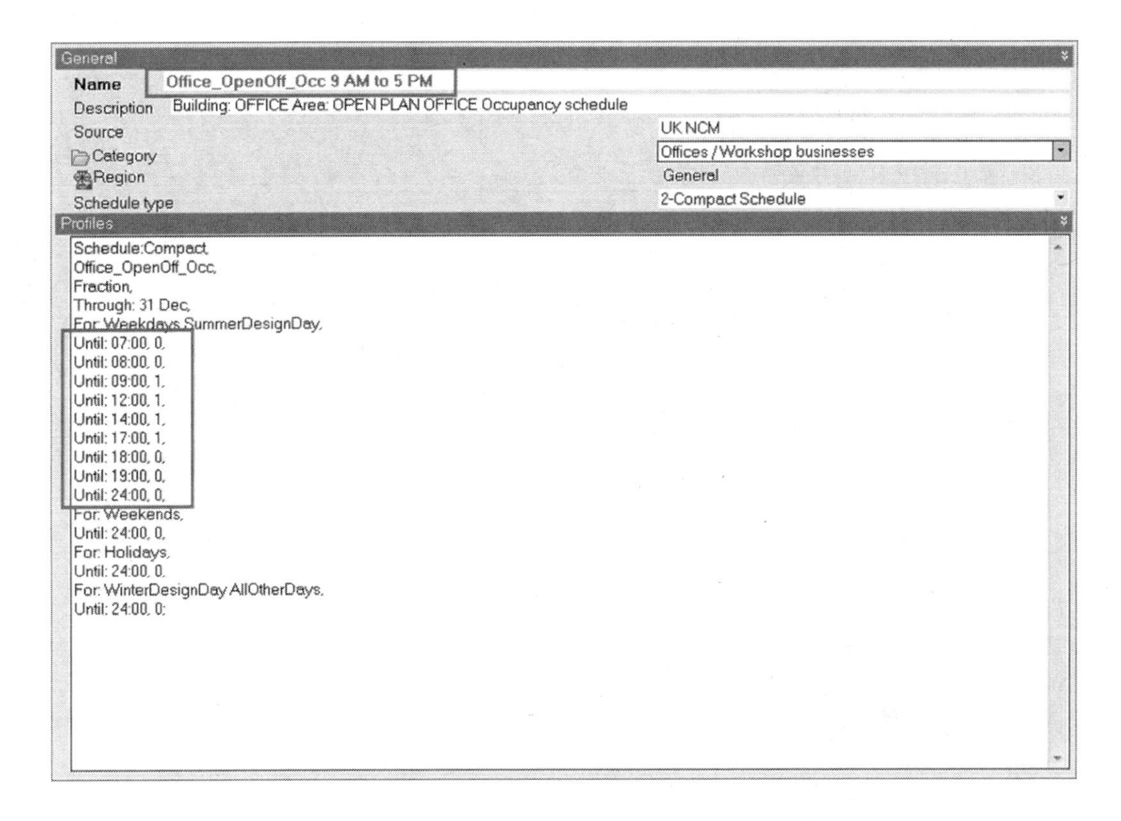

Step 4: Apply the **Office_OpenOff_Occ 9 AM to 5 PM** schedule.

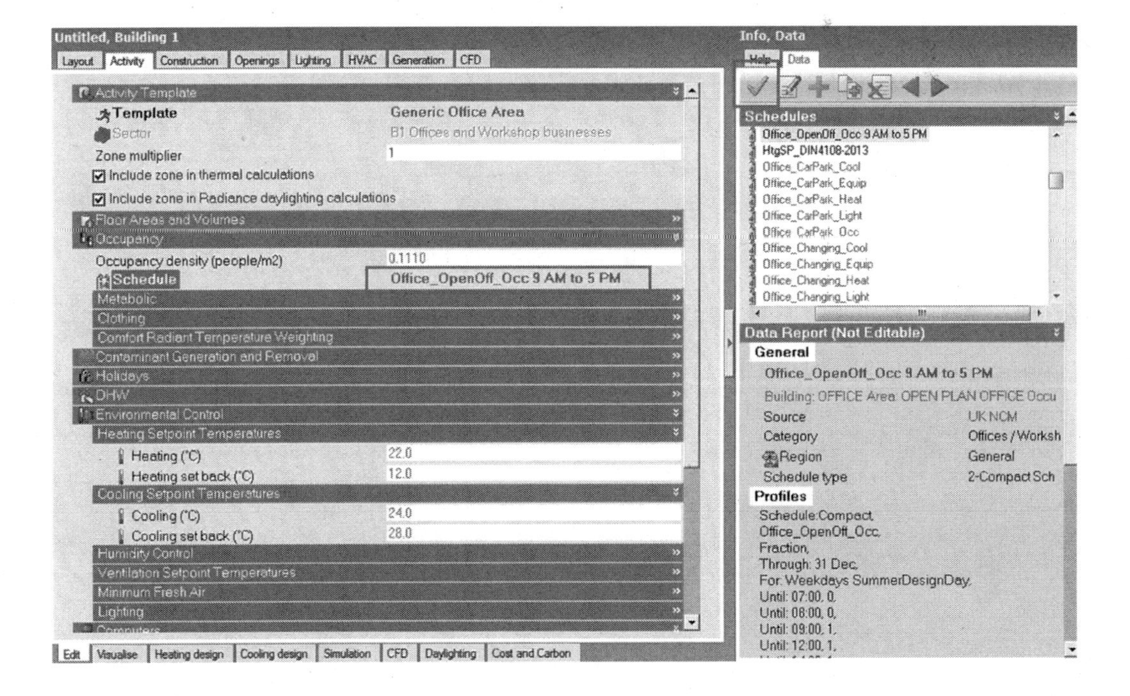

Step 5: Click **Openings** tab.

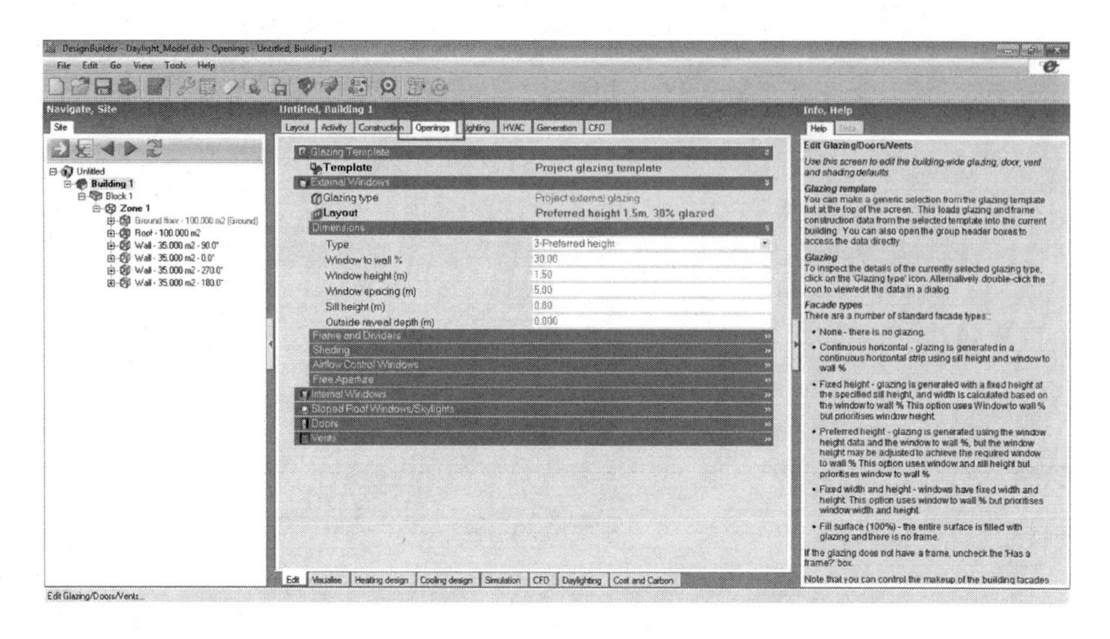

Step 6: Select the **Double glazing, clear, no shading** template.

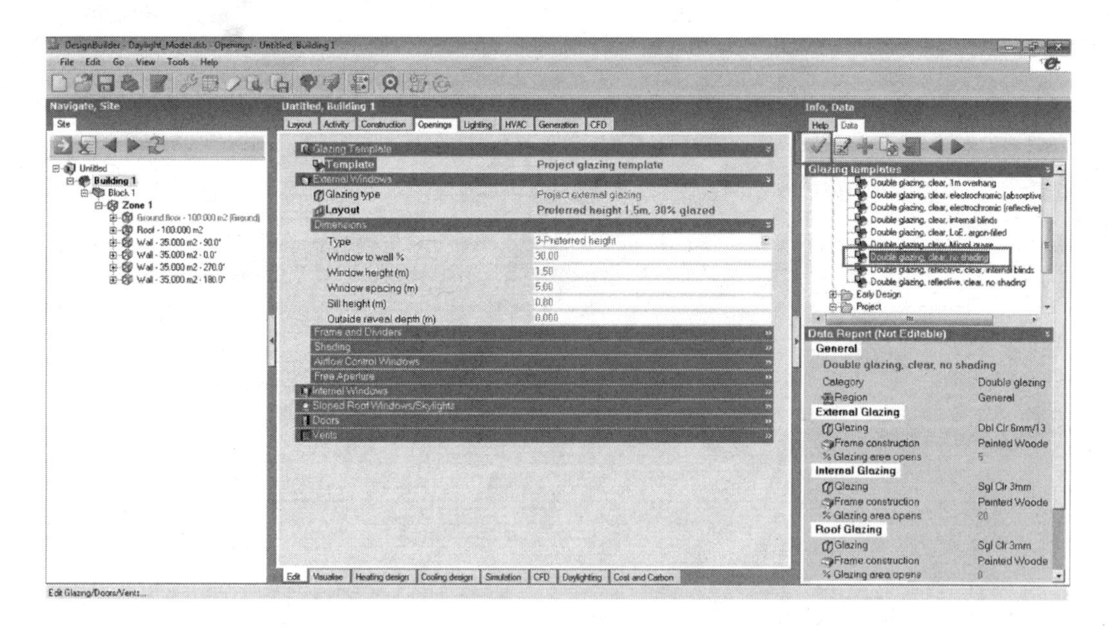

Step 7: Make sure that the **Glazing** type is **Dbl Clr 6mm/13mm Air** and that the WWR is set to **30%.**

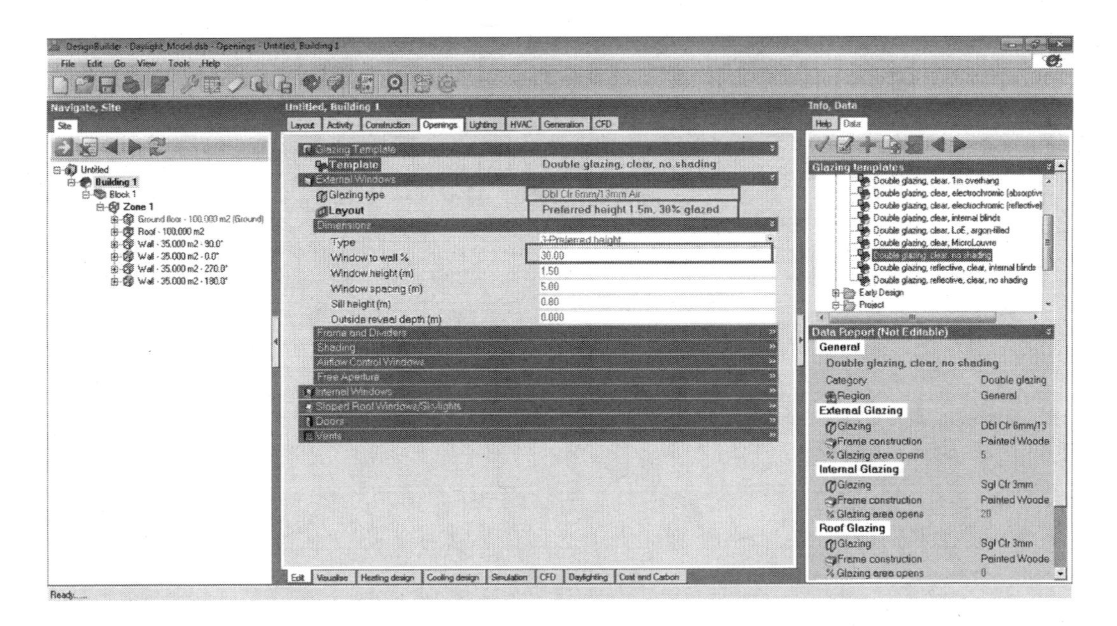

Step 8: Click the **Daylighting** screen tab.

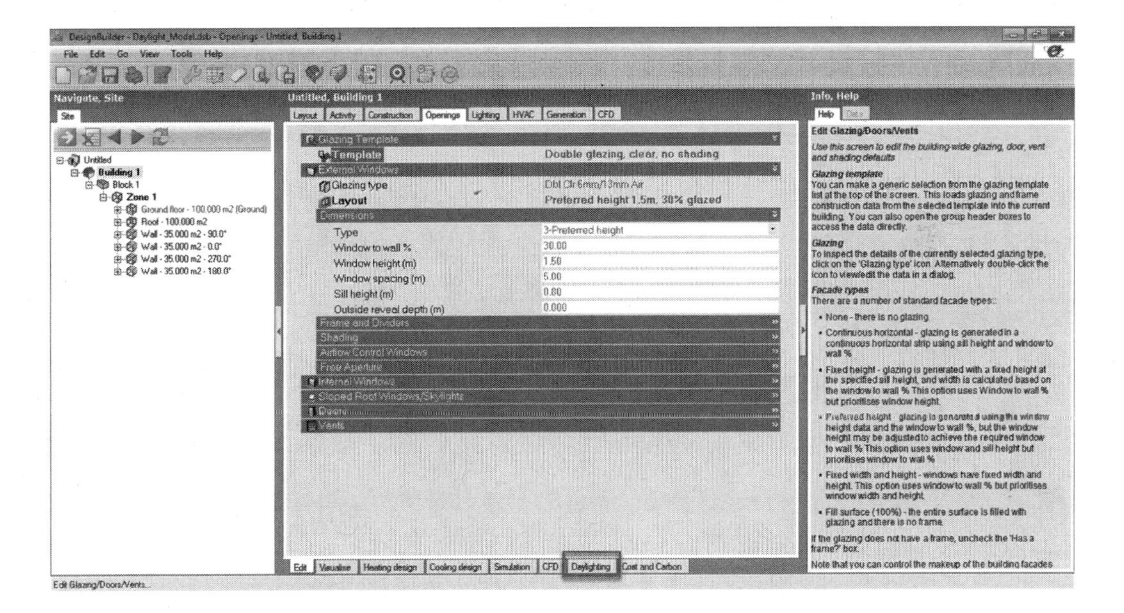

Step 9: Click the **Annual daylighting** tab.

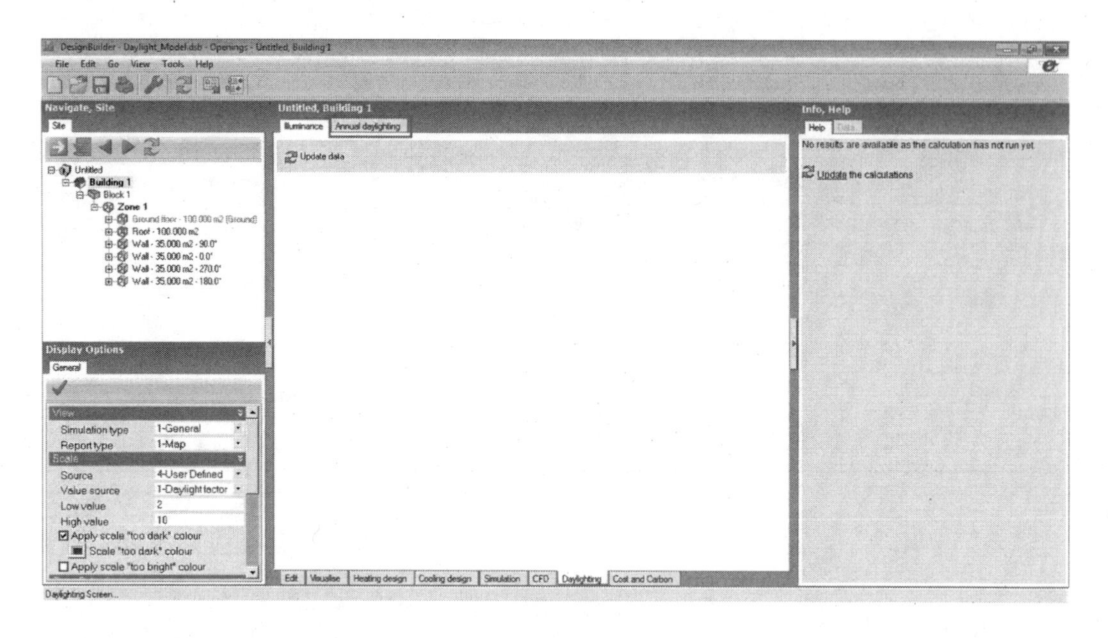

Step 10: Click **Update data**. The **Edit Calculation Option** screen appears.

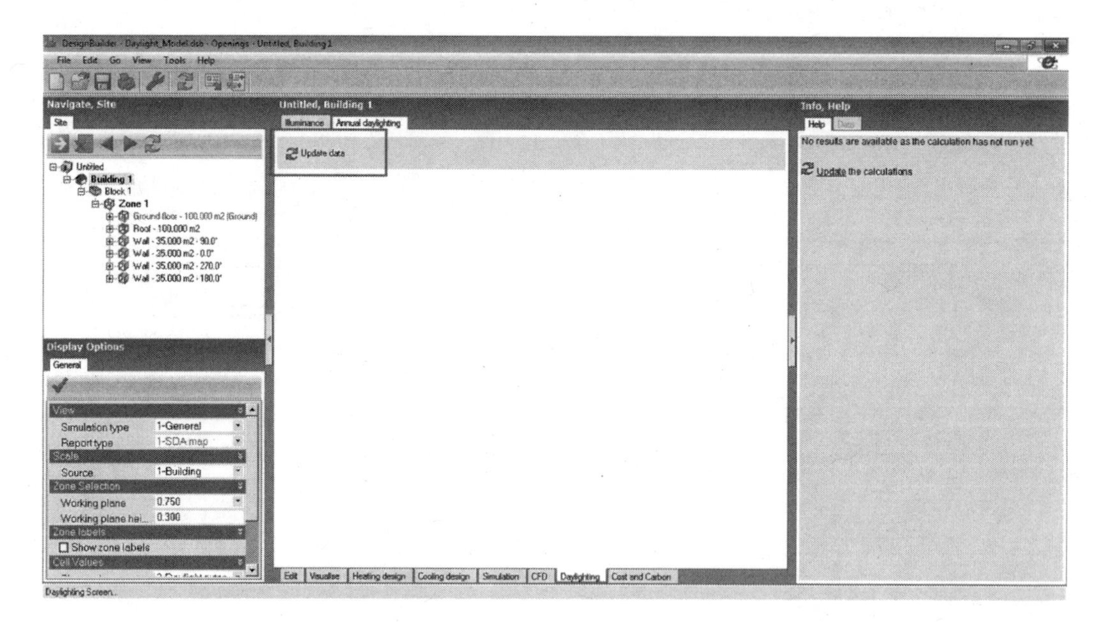

Step 11: Expand the **Output and Thresholds** section. Enter **100** in the **UDI lower illuminance threshold (lux)** text box and **2000** in the **UDI upper illuminance threshold (lux)** text box. Click **OK** to perform the simulations.

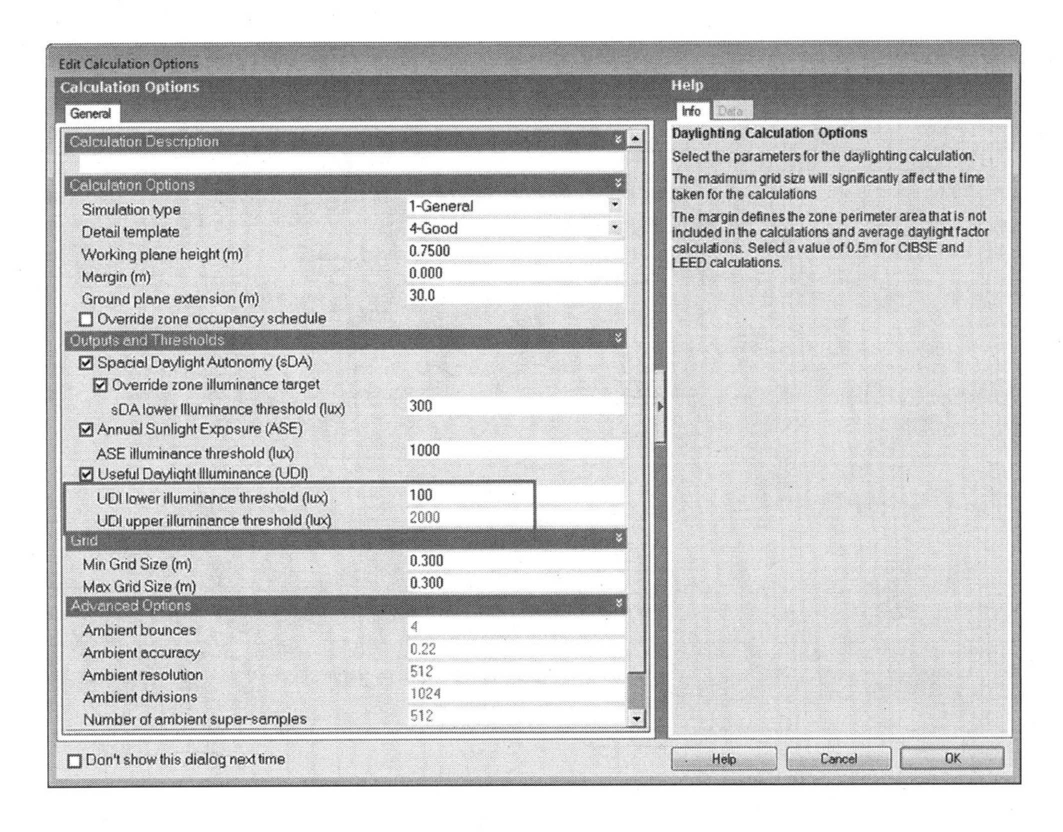

Step 12: Select **4-Grid** from the **Report type** drop-down list. Enter **90** in the **UDI percentage annual hours** text box. Record the results. Note the value of **UDI Area in Range (%)**.

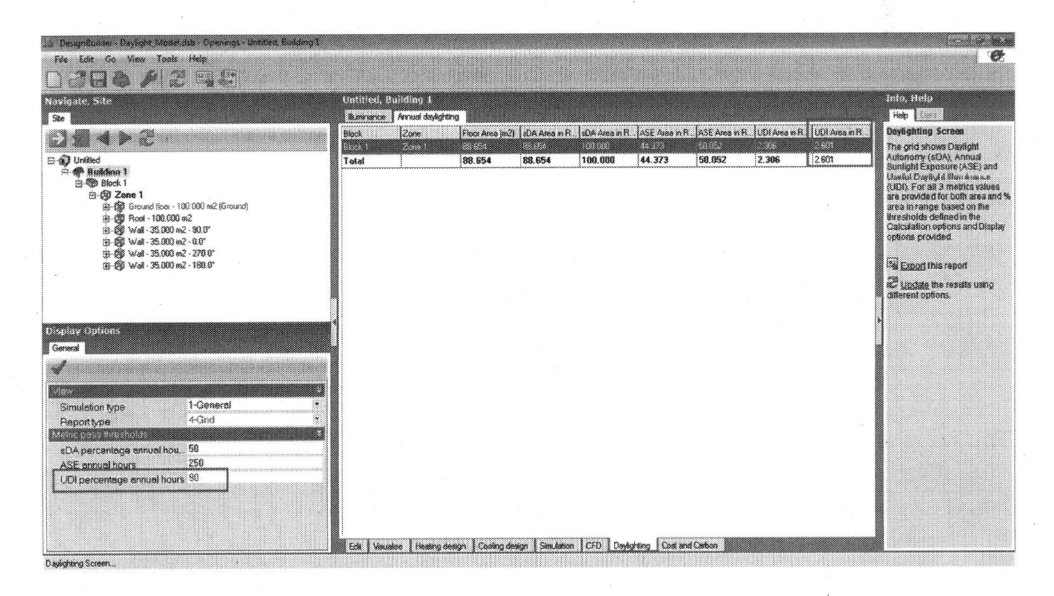

Step 13: Select the **3-UDI map** from the **Report type** drop-down list.

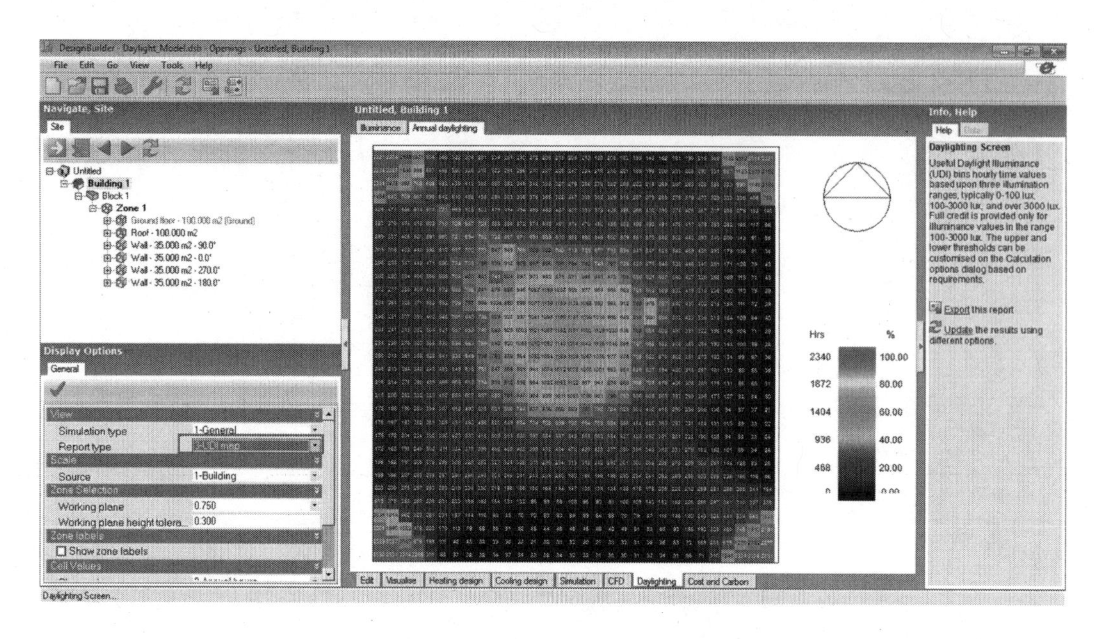

You can see that for most of the area, especially near windows, the UDI percentage is low because of excess daylight available near the windows. To increase the percentage of UDI area in range, you can provide overhangs and fins.

Step 14: Click the **Edit** tab.

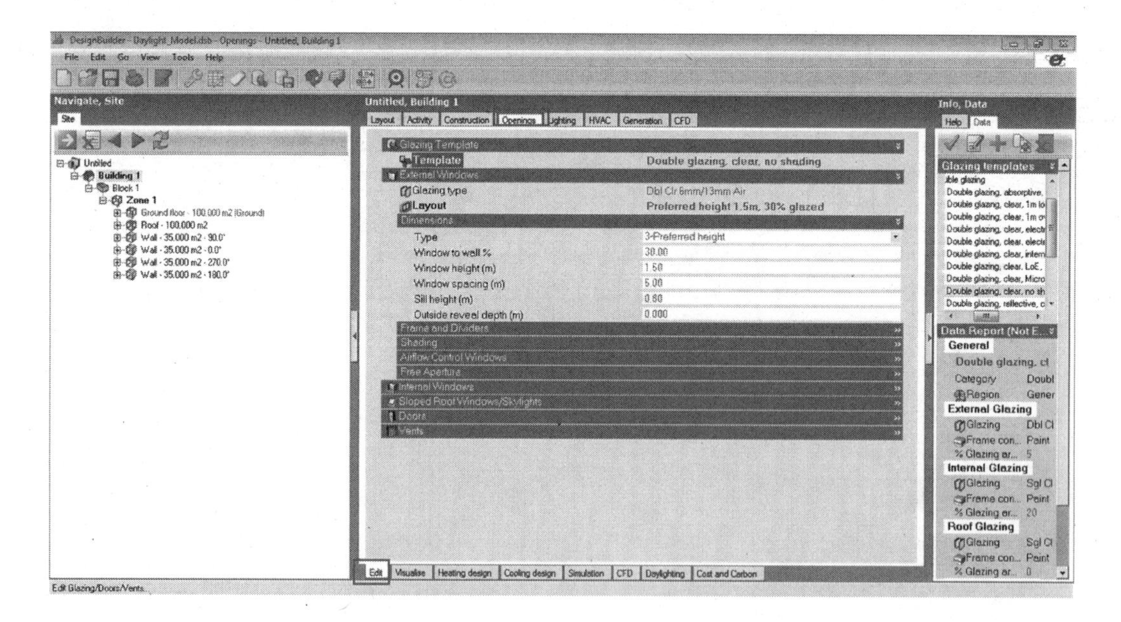

Step 15: Select the **Local shading** checkbox. Select **Overhang + sidefins (1m projection)** for **Type**.

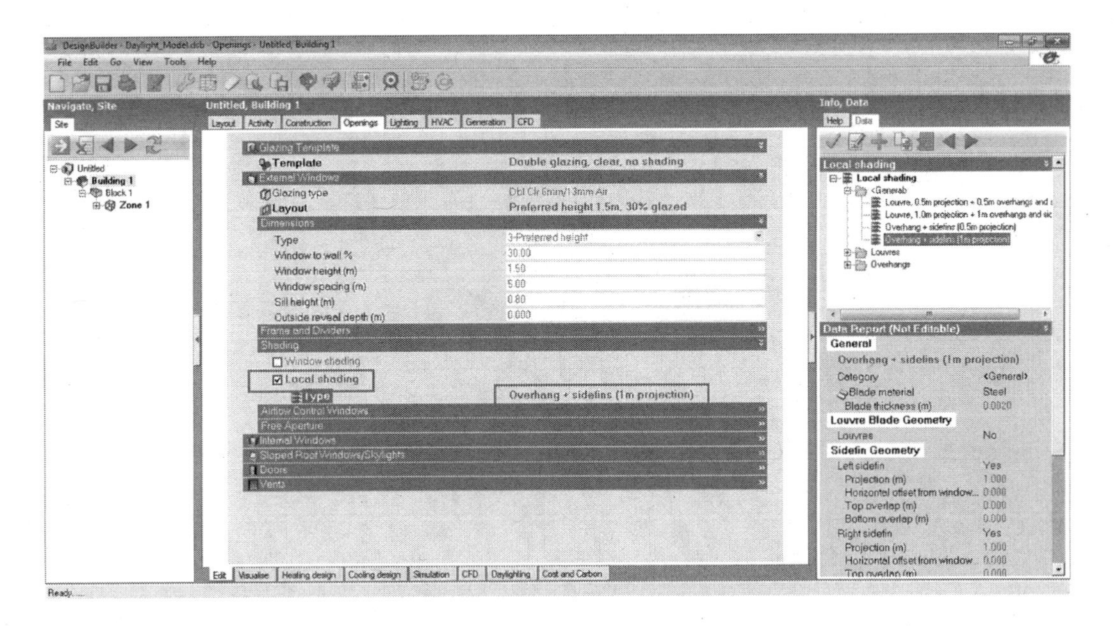

Step 16: Simulate the model for daylighting analysis, and record the UDI area.

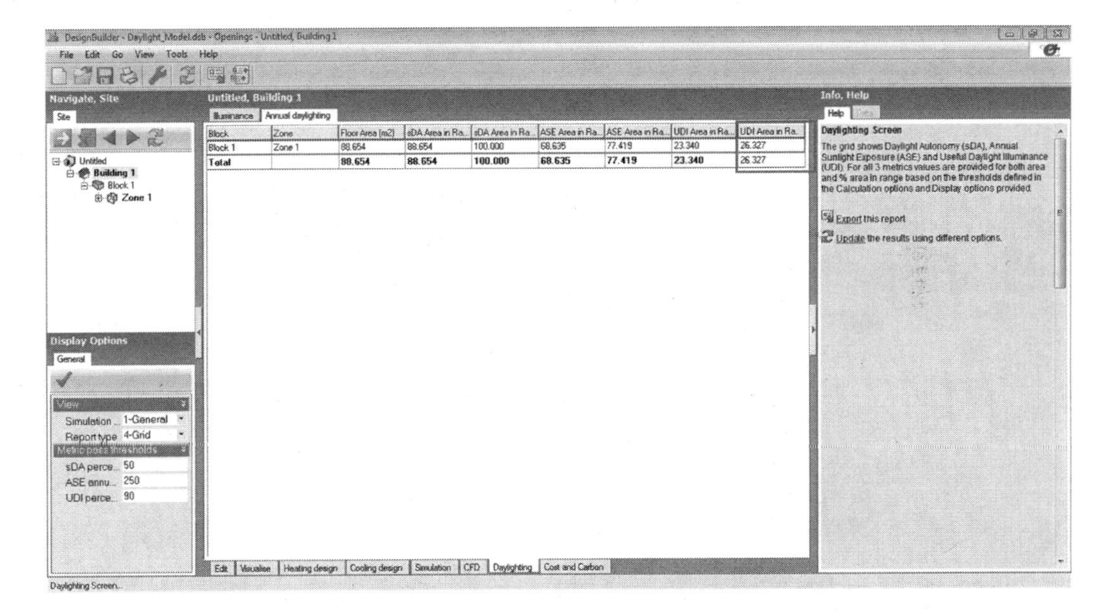

You can see that now 26.27% of the floor area is under the required UDI range.
Step 17: Select **3-UDI map** from the **Report type** drop-down list.

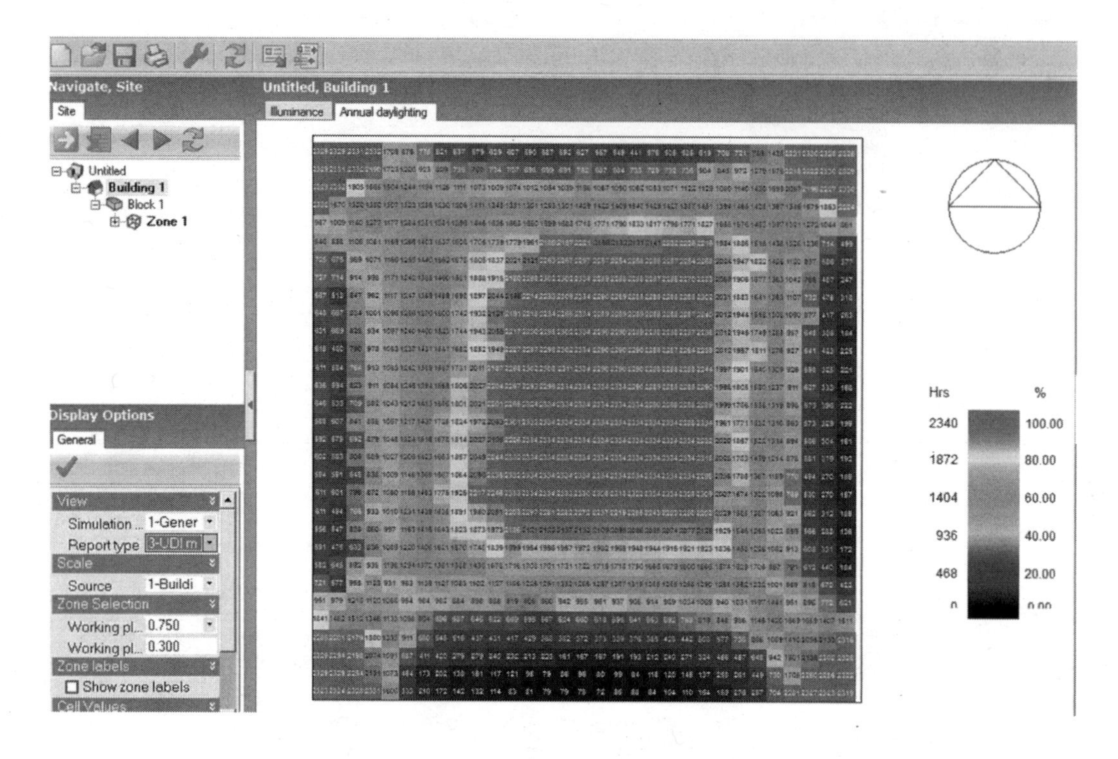

Step 18: Click the **Edit** tab. Click the **Openings** tab. Enter **25.00** in the **Window to wall %** text box.

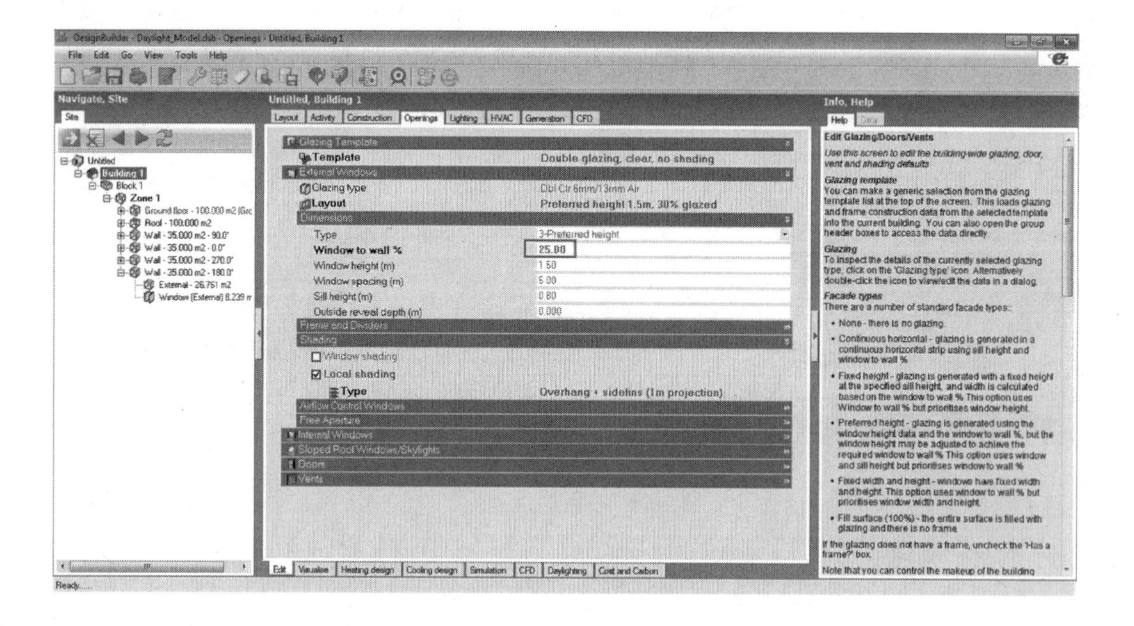

Step 19: Simulate the model for daylighting analysis, and record the results.

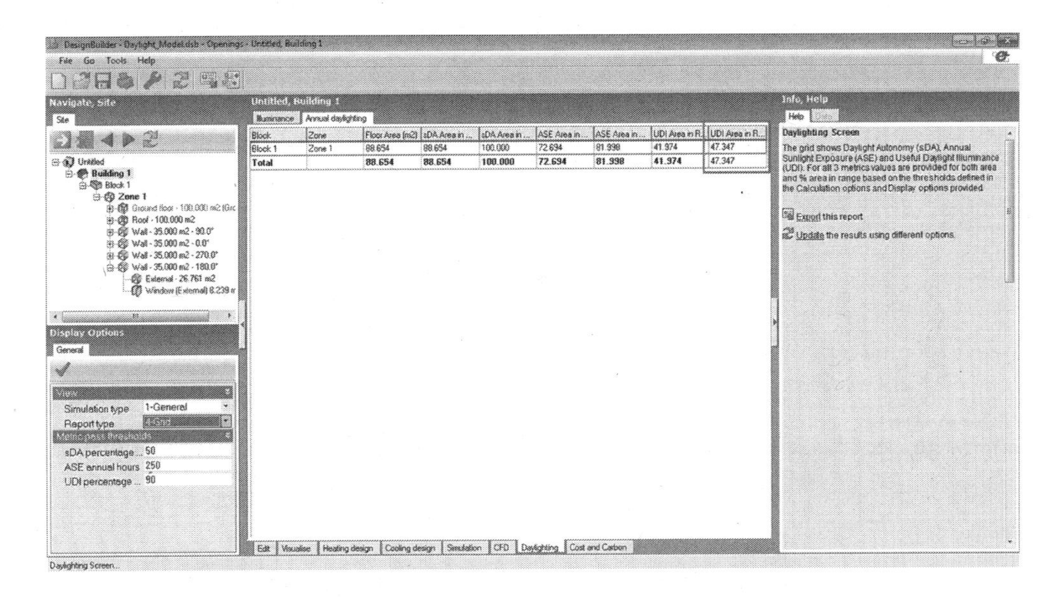

Step 20: Select **3-UDI map** from the **Report type** drop-down list.

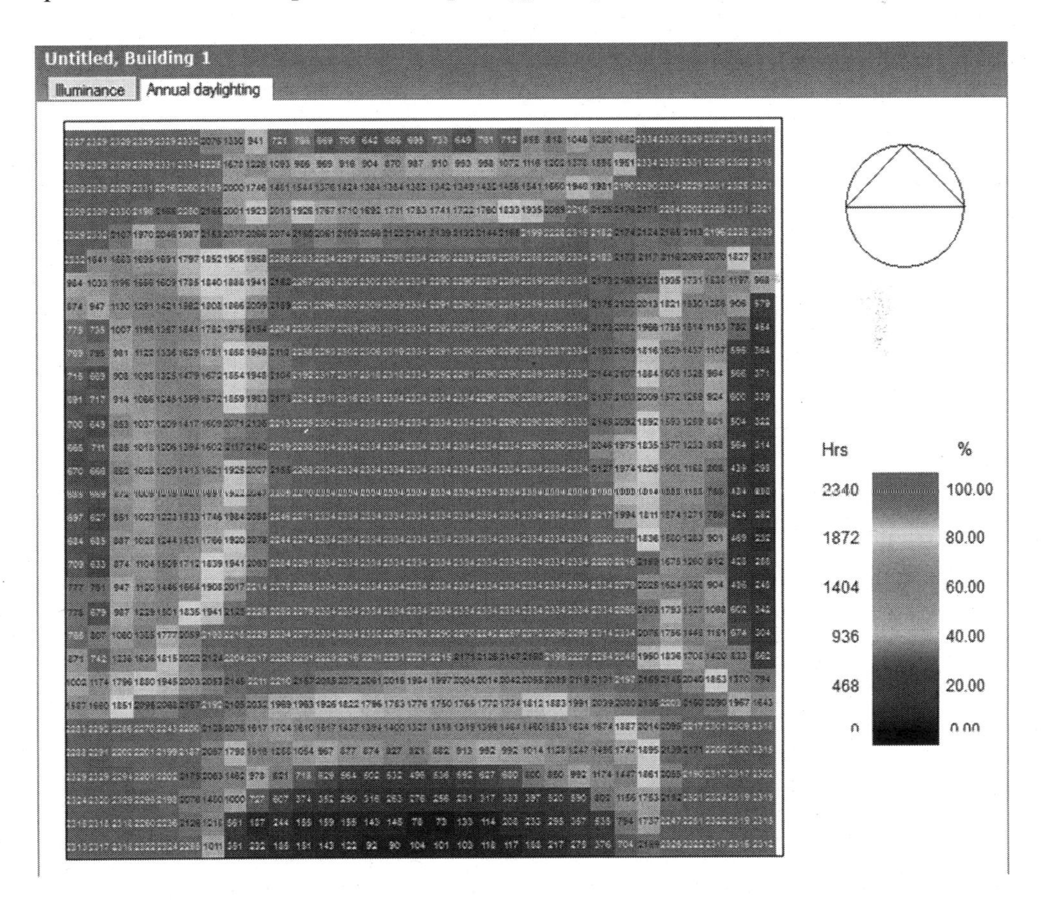

TABLE 5.5
Floor area meeting UDI requirements (%) for different combinations of WWR and shading

Serial no.	Glass	WWR (%)	Overhang and fins	Floor area meeting UDI requirements (%)
1	Dbl Clr 6 mm/13 mm Air	30	No Shade	2.601
2	Dbl Clr 6 mm/13 mm Air	30	Overhang (1 m) and Fins (1 m)	26.32
3	Dbl Clr 6 mm/13 mm Air	25	Overhang (1 m) and Fins (1 m)	47.34

You can see that now 47.34% the above-grade floor area meets the illuminance level between 100 and 2,000 lux for at least 90% of the potential daylit time in a year.

Table 5.5 shows the floor area meeting UDI requirements (%) for different combinations of WWR and shading. It can be seen that excess daylight can have a negative impact on UDI. In this tutorial, we are able to increase the UDI by providing shading and reducing the WWR.

6 Heating and Cooling Design

This chapter explains how to size and model heating, ventilation and air-conditioning (HVAC) systems. Of the three tutorials, one explains the effect of HVAC operating criteria on energy consumption. Often thermostats of HVAC systems are operated by sensing the air temperature, and the same criterion is also used for the evaluation of thermal comfort hours through simulation. This tutorial shows the difference in alternative approaches using the case of 'operative temperature', that is, a combination of air temperature and mean radiant temperature. The second tutorial explains the method of sizing HVAC systems, and the third tutorial covers the effect of using different algorithms for performing HVAC calculations in the simulation.

Tutorial 6.1

Evaluating the Impact of Temperature Control Types

GOAL

To evaluate the impact of temperature control type – air temperature and operative temperature – on HVAC equipment sizing and energy consumption

WHAT ARE YOU GOING TO LEARN?

- How to change temperature setpoint control type and evaluate its impact

PROBLEM STATEMENT

In this tutorial, you are going to use a 50- × 25-m model with a 5-m perimeter depth with the following specifications:

- Number of floors: **G+1**
- Window-to-wall ratio (WWR): **40%**
- Glass type: **Dbl Blue 6mm/13mm Air** (*U*-value: 2.70 W/m^2–K; SHGC: 0.48; VLT 0.50)
- Roof construction: **Roof, Ins Entirely above Deck, R-50(8.8), U-0.020(0.114)**
- Activity template: **24 × 7 Generic Office Area**

Find the cooling equipment sizing and energy consumption for the ground floor with the following two temperature controls:

1. Air temperature (AT)
2. Operative temperature (OT)

Use the **New Delhi/Palam, India** weather location.

SOLUTION

Step 1: Open a new project, and create a **50- × 25-m** building with a **5-m** perimeter depth, and select the template as **24 × 7 Generic Office Area**.

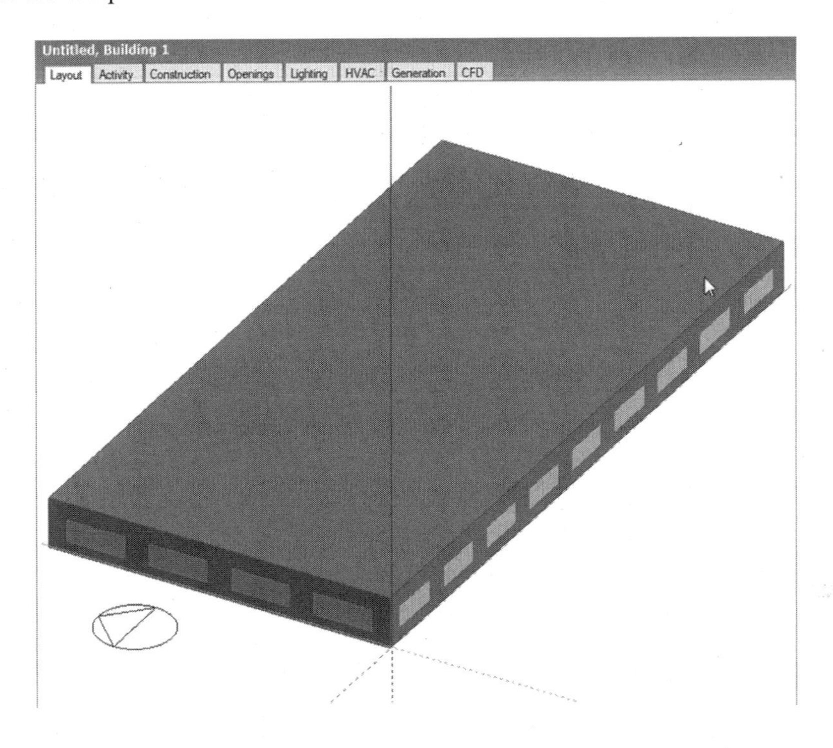

Step 2: Go to the **Building level**, and select **Building** on the **Edit** screen.

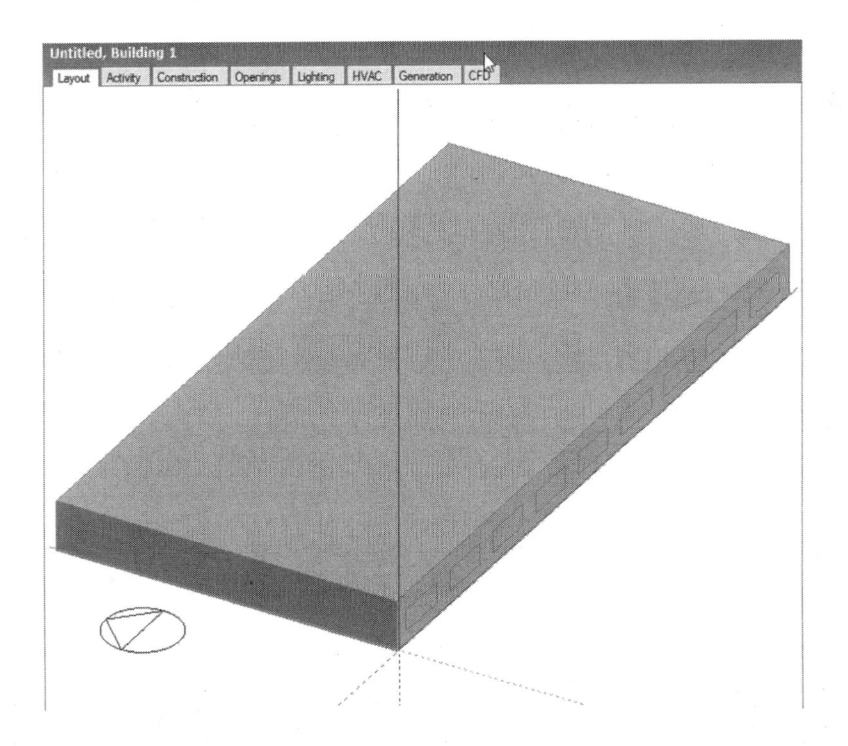

Step 3: Click **Clone selected object(s)**. Click the origin of the floor, and move the cursor to the top of the floor to paste the cloned floor.

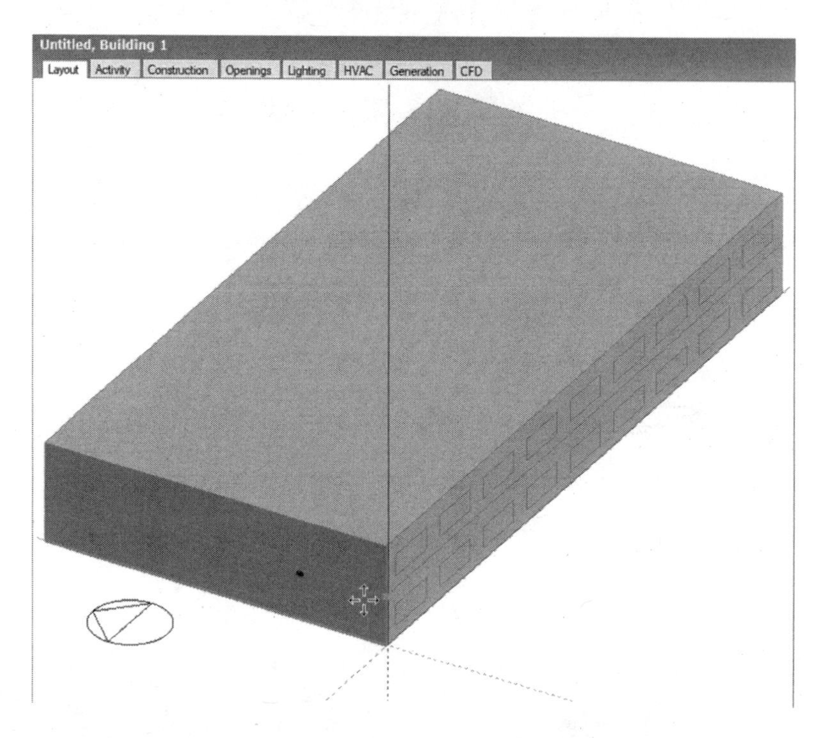

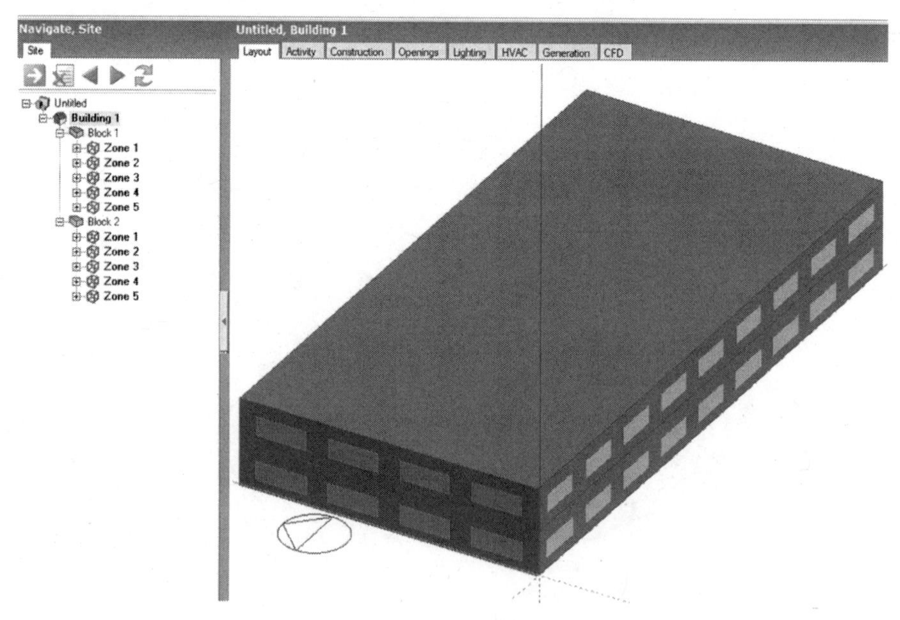

Step 4: Select the **Openings** tab. Select **Dbl Blue 6mm/13mm** Air as the **Glazing type**, and set the layout as **Preferred height 1.5m, 40% glazed**.

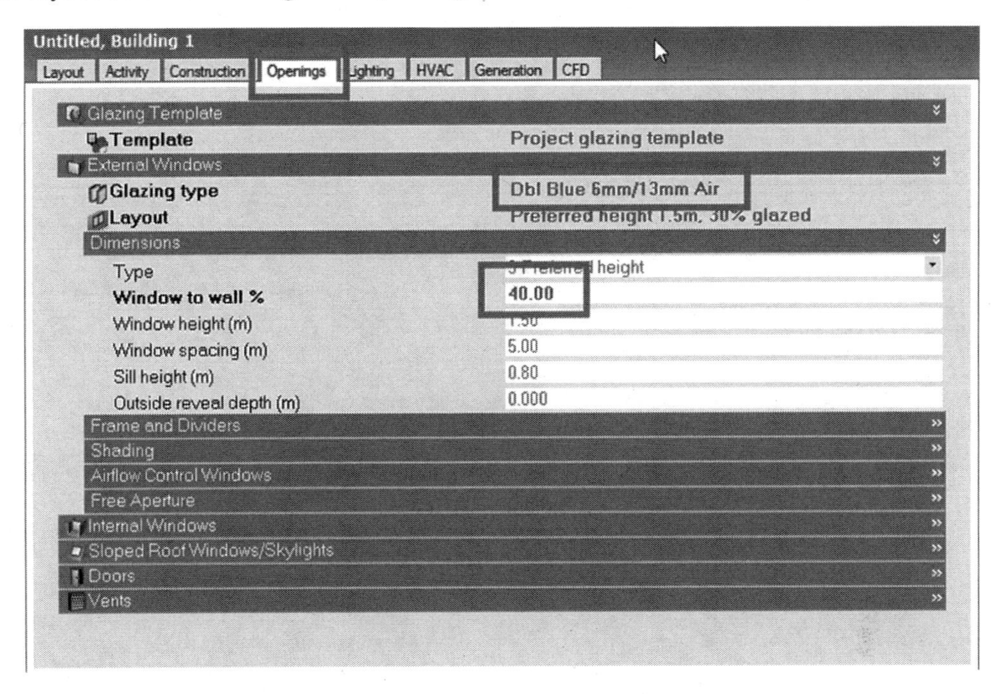

Step 5: Select the **Construction** tab, and select **Roof, Ins Entirely above Deck, R-50 (8.8), U-0.020 (0.114)** as the **Flat roof** type.

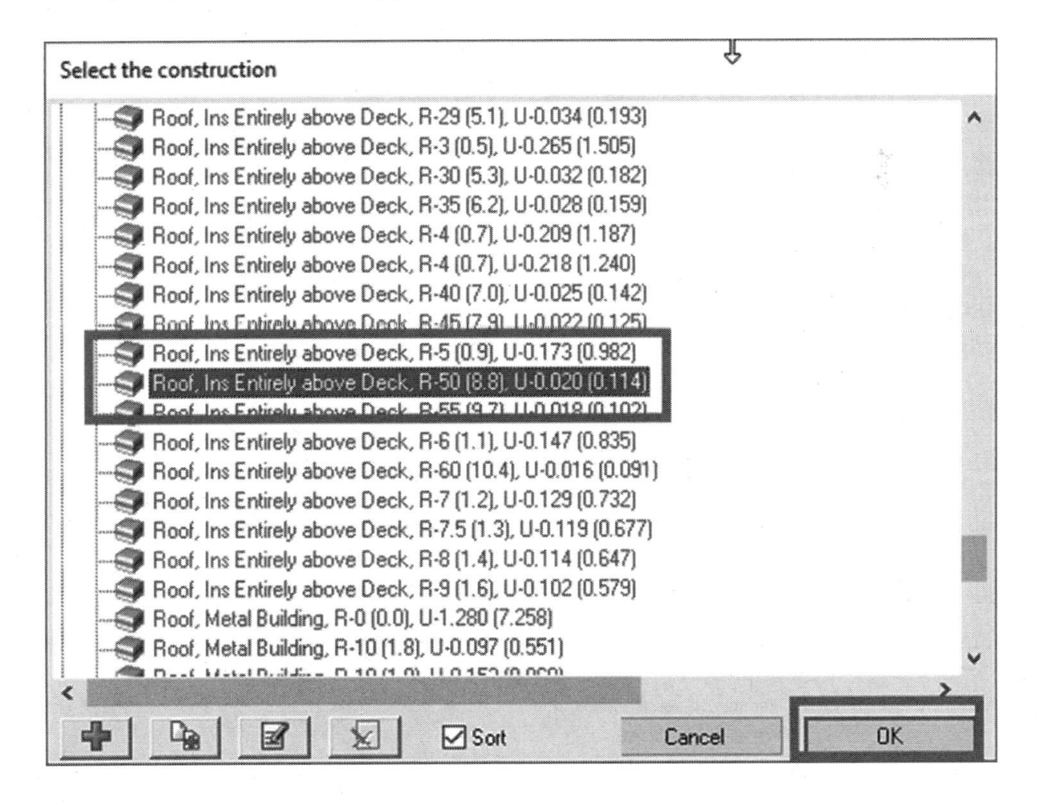

Step 6: Select the **Cooling design** screen tab. Click **Update data**. The **Calculation Options** screen appears.

Step 7: Select **Air temperature** from the **Temperature control** drop-down list. Click **OK**. The results are displayed in the **Analysis** tab.

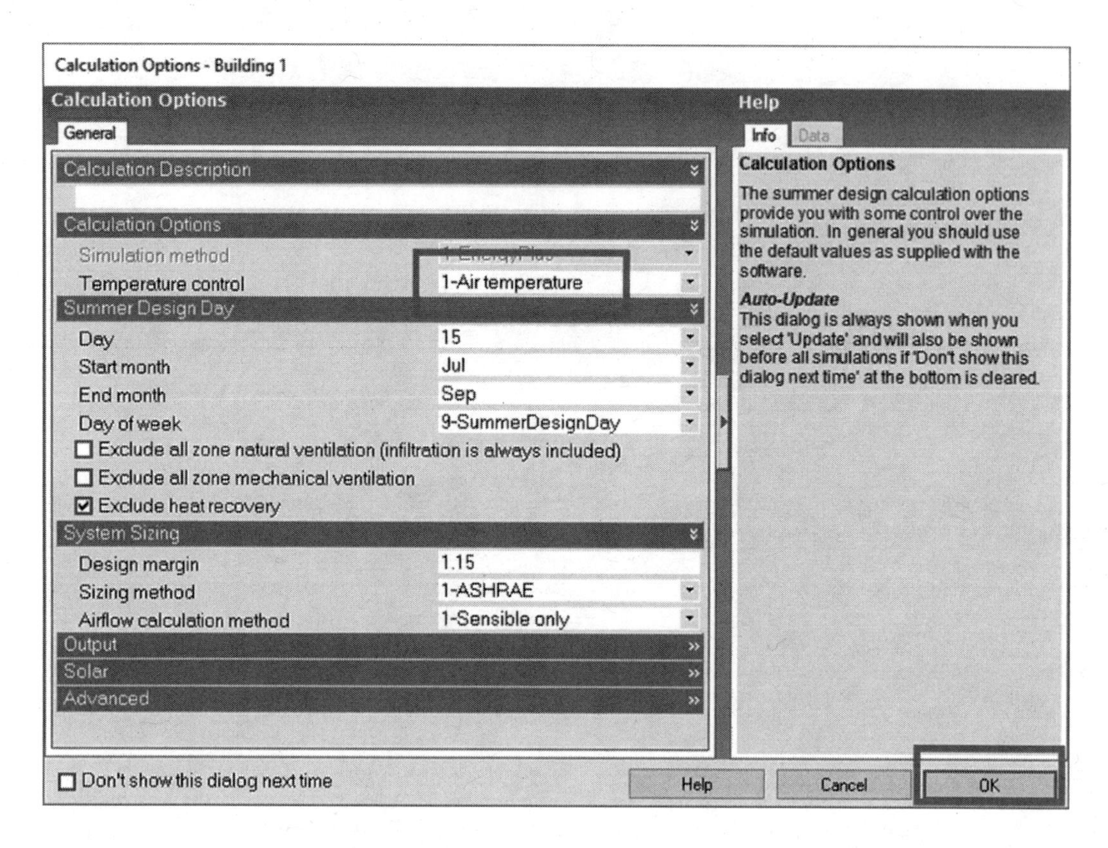

The **Day of week** field is used to identify the appropriate daily profile within the schedules to use for cooling design calculations. The **Day** type should be the day when the most extreme conditions and highest cooling loads are expected. You should normally keep the **9-SummerDesignDay** default option while running simulations.

Step 8: Select the **Summary** tab. Record the results.

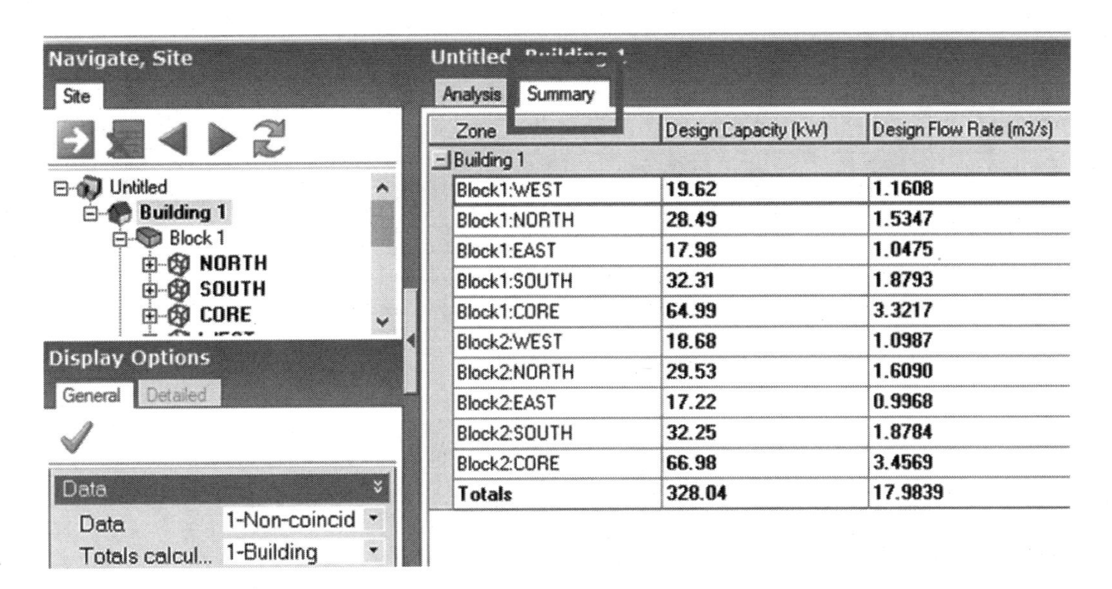

Step 9: Select the **Simulation** screen tab. In the **Edit Calculation Options** screen, select the **Options** tab. Ensure that **Air temperature** is selected. Ensure that **Hourly Simulation** is selected in **General** tab. Click **OK**.

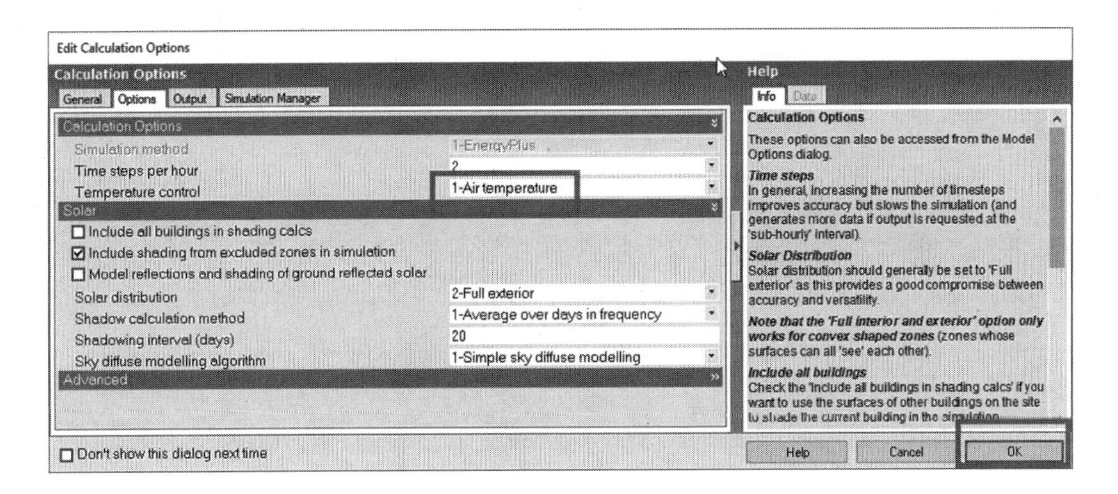

Step 10: Click **South** under **Ground Floor** in the navigation tree. Record the air temperature and operative temperature for 4 April. (Data displayed on the screen are for the whole year; you can scroll to get the data for 27 March.)

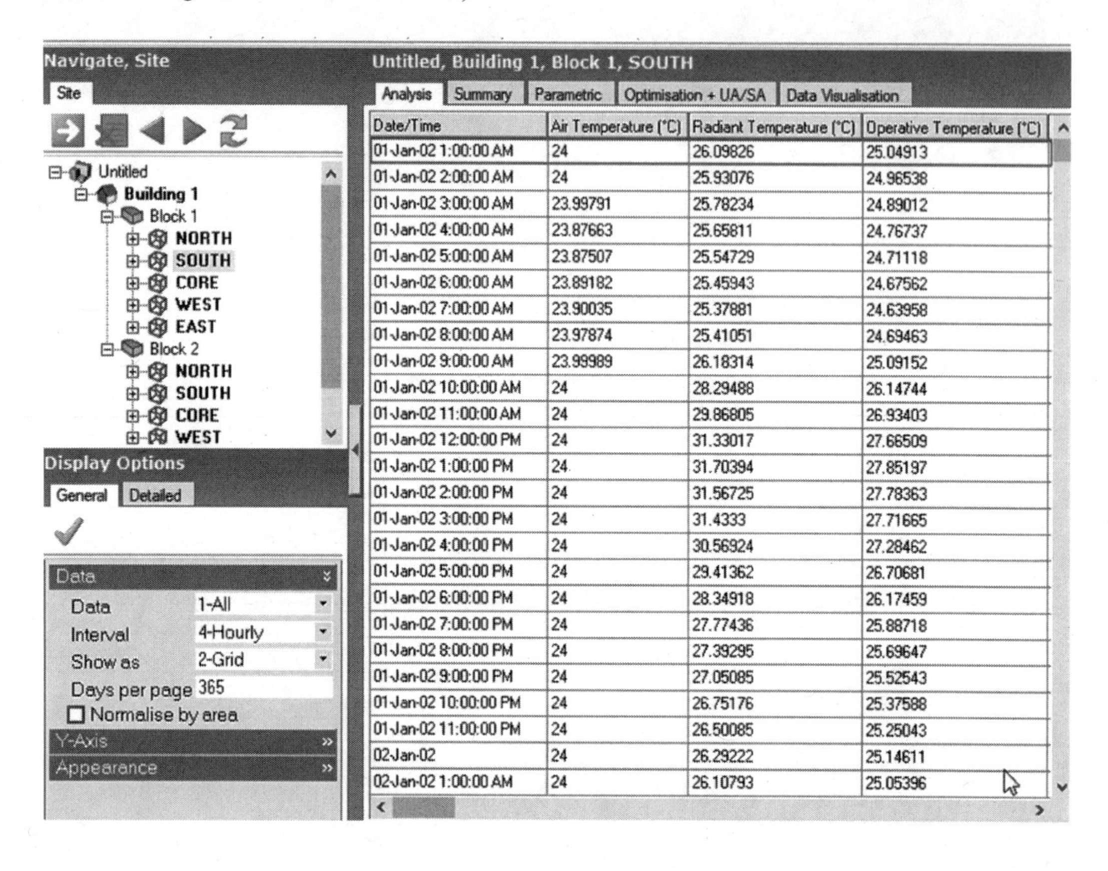

Date/Time	Air Temperature (°C)	Radiant Temperature (°C)	Operative Temperature (°C)
01-Jan-02 1:00:00 AM	24	26.09826	25.04913
01-Jan-02 2:00:00 AM	24	25.93076	24.96538
01-Jan-02 3:00:00 AM	23.99791	25.78234	24.89012
01-Jan-02 4:00:00 AM	23.87663	25.65811	24.76737
01-Jan-02 5:00:00 AM	23.87507	25.54729	24.71118
01-Jan-02 6:00:00 AM	23.89182	25.45943	24.67562
01-Jan-02 7:00:00 AM	23.90035	25.37881	24.63958
01-Jan-02 8:00:00 AM	23.97874	25.41051	24.69463
01-Jan-02 9:00:00 AM	23.99989	26.18314	25.09152
01-Jan-02 10:00:00 AM	24	28.29488	26.14744
01-Jan-02 11:00:00 AM	24	29.86805	26.93403
01-Jan-02 12:00:00 PM	24	31.33017	27.66509
01-Jan-02 1:00:00 PM	24	31.70394	27.85197
01-Jan-02 2:00:00 PM	24	31.56725	27.78363
01-Jan-02 3:00:00 PM	24	31.4333	27.71665
01-Jan-02 4:00:00 PM	24	30.56924	27.28462
01-Jan-02 5:00:00 PM	24	29.41362	26.70681
01-Jan-02 6:00:00 PM	24	28.34918	26.17459
01-Jan-02 7:00:00 PM	24	27.77436	25.88718
01-Jan-02 8:00:00 PM	24	27.39295	25.69647
01-Jan-02 9:00:00 PM	24	27.05085	25.52543
01-Jan-02 10:00:00 PM	24	26.75176	25.37588
01-Jan-02 11:00:00 PM	24	26.50085	25.25043
02-Jan-02	24	26.29222	25.14611
02-Jan-02 1:00:00 AM	24	26.10793	25.05396

Step 11: Repeat the preceding steps (the cooling design and simulation) by selecting the operative temperature control in place of the air temperature control.

Operative temperature is defined as a uniform temperature of a radiantly black enclosure in which an occupant would exchange the same amount of heat by radiation plus convection as in the actual non-uniform environment. It is the combined effect of the mean radiant temperature and air temperature calculated as the average of the two. It is also known as the *dry resultant temperature* or *resultant temperature*. It can be calculated as follows:

$$t_0 = \frac{\left[\left(t_a \times \sqrt{10v}\right) + t_{mr}\right]}{1 + \sqrt{10v}}$$

where v = air velocity

t_a = air temperature

t_{mr} = mean radiant temperature

The *mean radiant temperature* of an environment is defined as the uniform temperature of an imaginary black enclosure that would result in the same heat loss by radiation from the person as the actual enclosure.

Step 12: Select **Operative temperature** from the **Temperature control** drop-down list. Click **OK**, and get the summary of analysis.

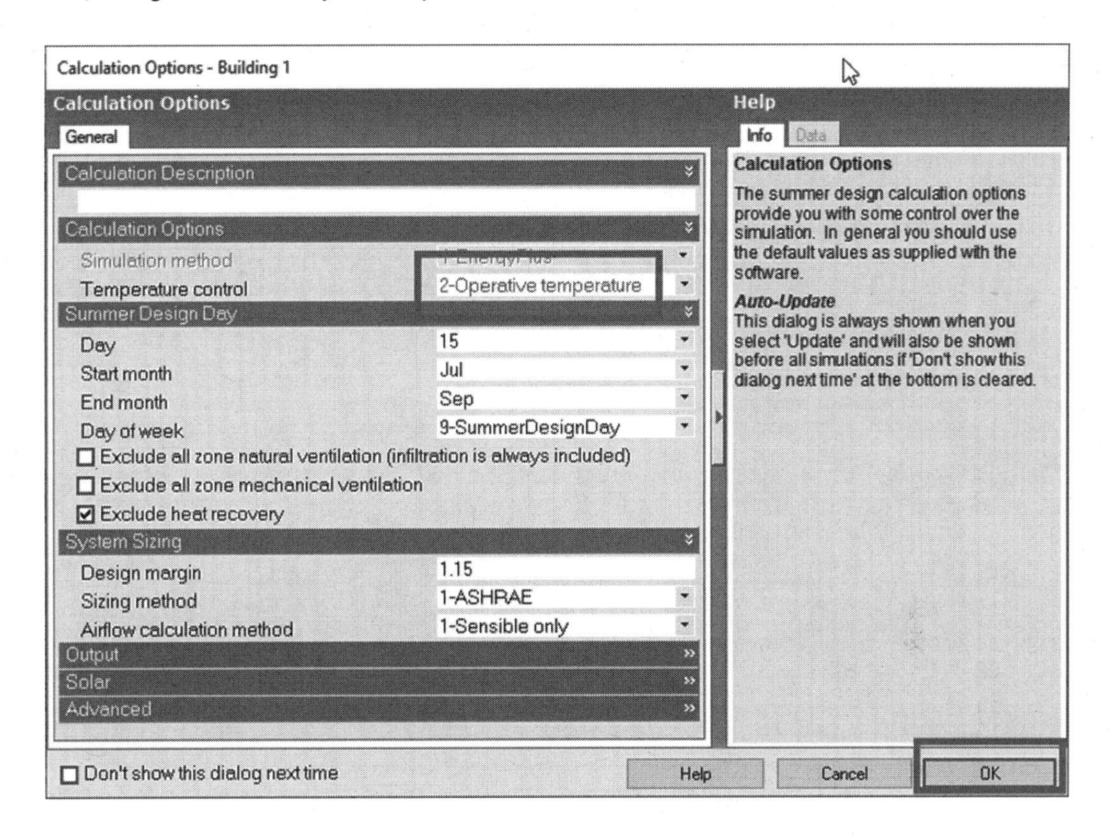

Step 13: Select the **Summary** tab. Record the results in a separate spreadsheet (hourly values for AT and OT on 27 March for all zones on the ground floor, cooling design data and annual energy consumption results). Now compare the results for both cases (Table 6.1).

Step 14: Plot AT and OT for the **Ground Floor: South** zone with the air temperature control for 27 March.

TABLE 6.1
Design capacity of cooling equipment

	Design capacity of cooling equipment (kW)	
Zone	**Air temperature control**	**Operative temperature control**
Ground: East	17.98	22.88
Ground: West	19.62	24.75
Ground: North	28.49	31.6
Ground: South	32.31	38.78
Ground: Core	64.99	67.18

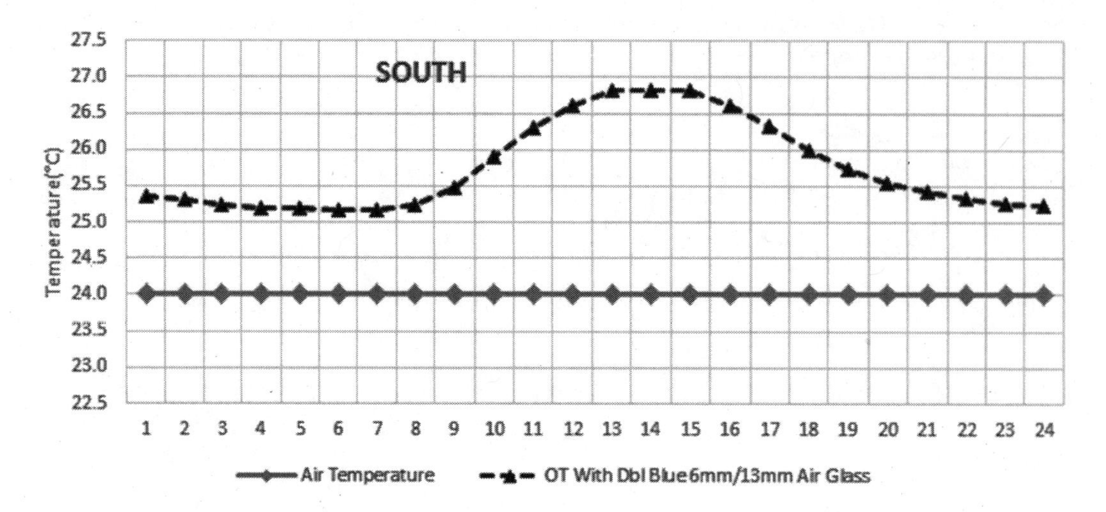

You can see that the zone air temperature is maintained at 24°C. However, the operative temperature of the zone is not constant over the day. In this case, the operative temperature of the zone is higher than the zone air temperature because of the higher temperature of the exposed walls and windows. It can also be noted that this difference is highest during the afternoon because the exposed surfaces of the south zone absorb solar radiation, resulting in a higher surface temperature.

A higher operative temperature can cause discomfort despite the air temperature being maintained at 24°C. Hence, to achieve comfort in the zone, there is a need to set the thermostat based on the operative temperature.

Step 15: Now plot temperatures (AT and OT) for the **Ground Floor: South** zone with the operative temperature control.

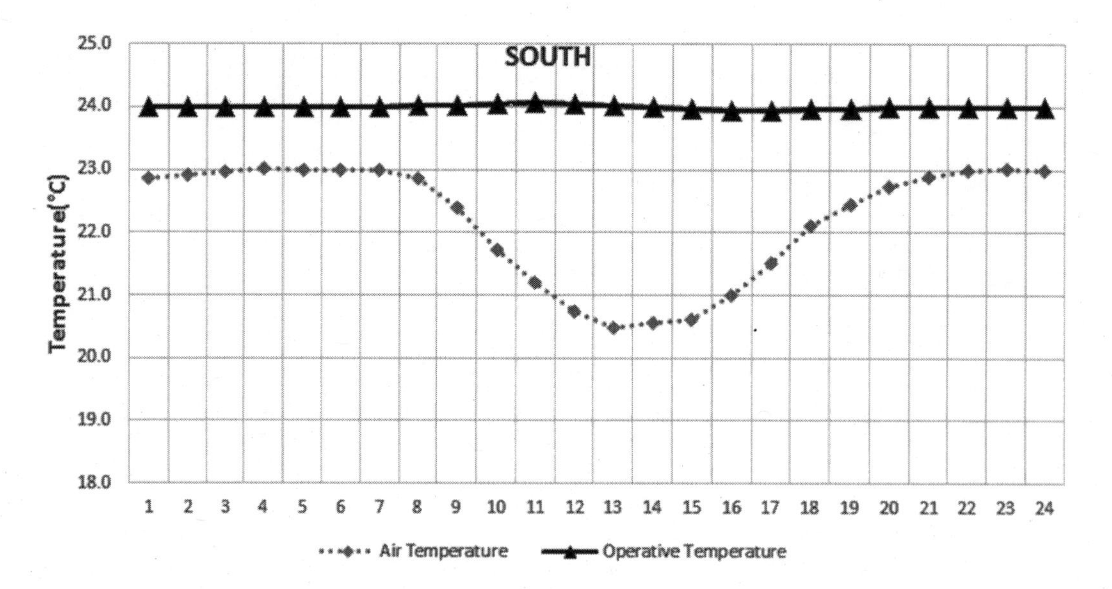

It can be seen that for maintaining a constant operative temperature during the afternoon, the air temperature was reduced significantly to compensate for the higher surface temperature in the zones.

Step 16: Plot a chart for the annual cooling energy consumption for the air and operative temperature controls.

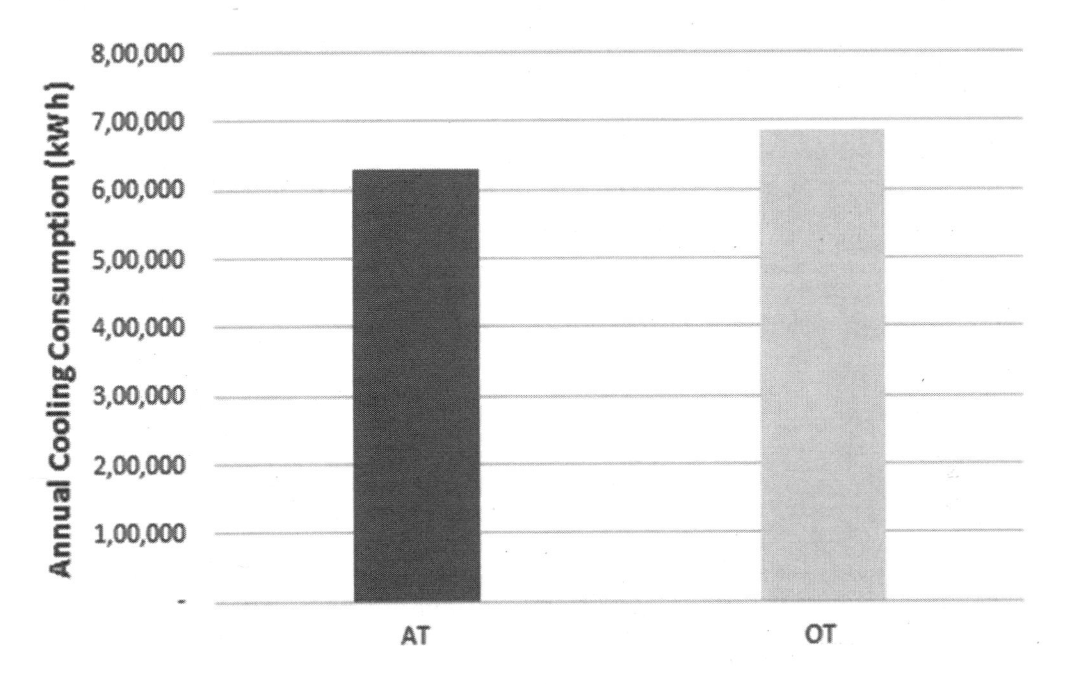

As shown in Step 15, because the air temperature was reduced below 24°C to compensate for the higher surface temperature, the energy consumption in the operative temperature control mode is higher than that in the air temperature control mode. Now repeat the preceding steps for the following two cases:

- Add a local shading 0.5-m overhang.
- Use **Vertical glazing, 0%–40% of wall, U-1.20 (6.81), SHGC-0.25** glass with local shading as a 0.5-m overhang

Simulate both models for the air temperature–based thermostat control. Record the results, and plot the temperatures for all the zones on the ground floor. Also run annual simulations and observe the energy consumption.

Step 17: Plot the temperatures for all zones on the ground floor with the air temperature for 4 April.

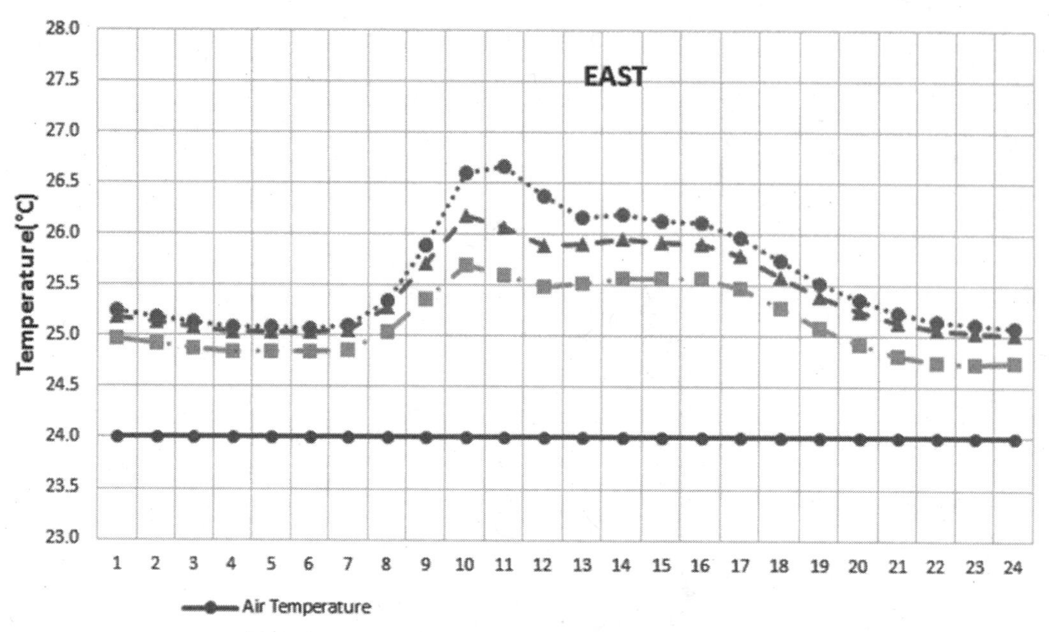

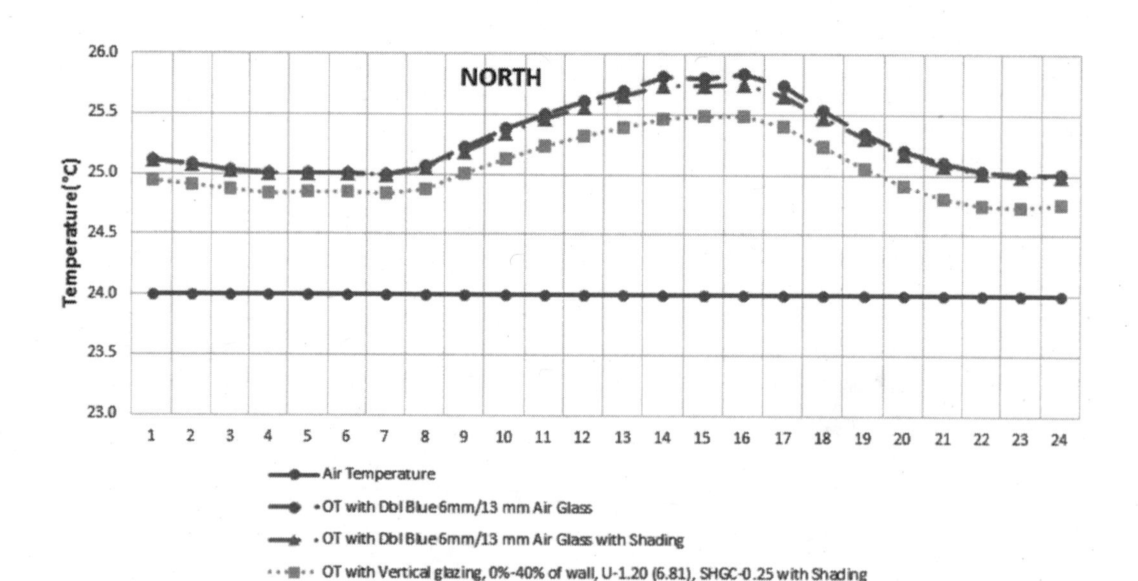

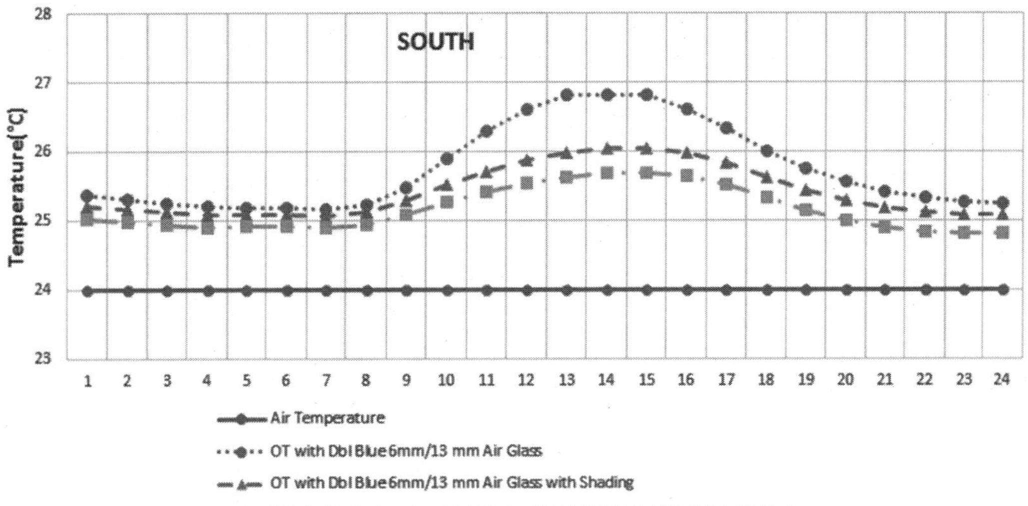

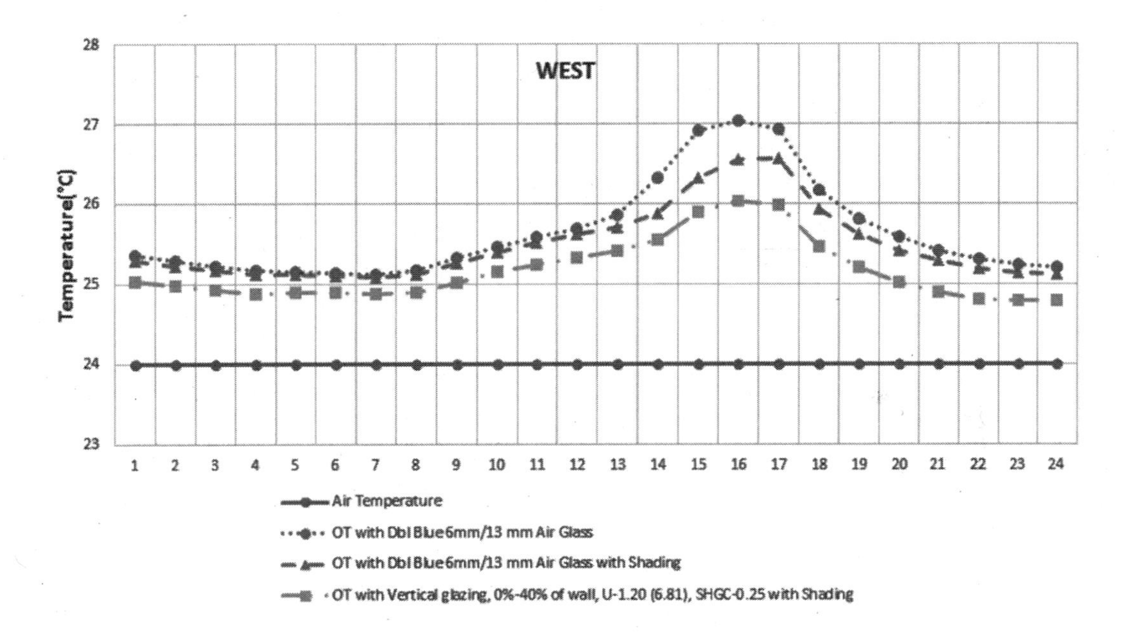

The results show that by putting a shade above the opening reduces the operative temperature. Also, use of high-performance glass also reduces the operative temperature as the high-performance glass surface tends to remain cooler than low-performance glasses. It can also be seen that the pattern of operative temperature is different for each zone. It is governed by the time of day when the zone receives solar radiation.

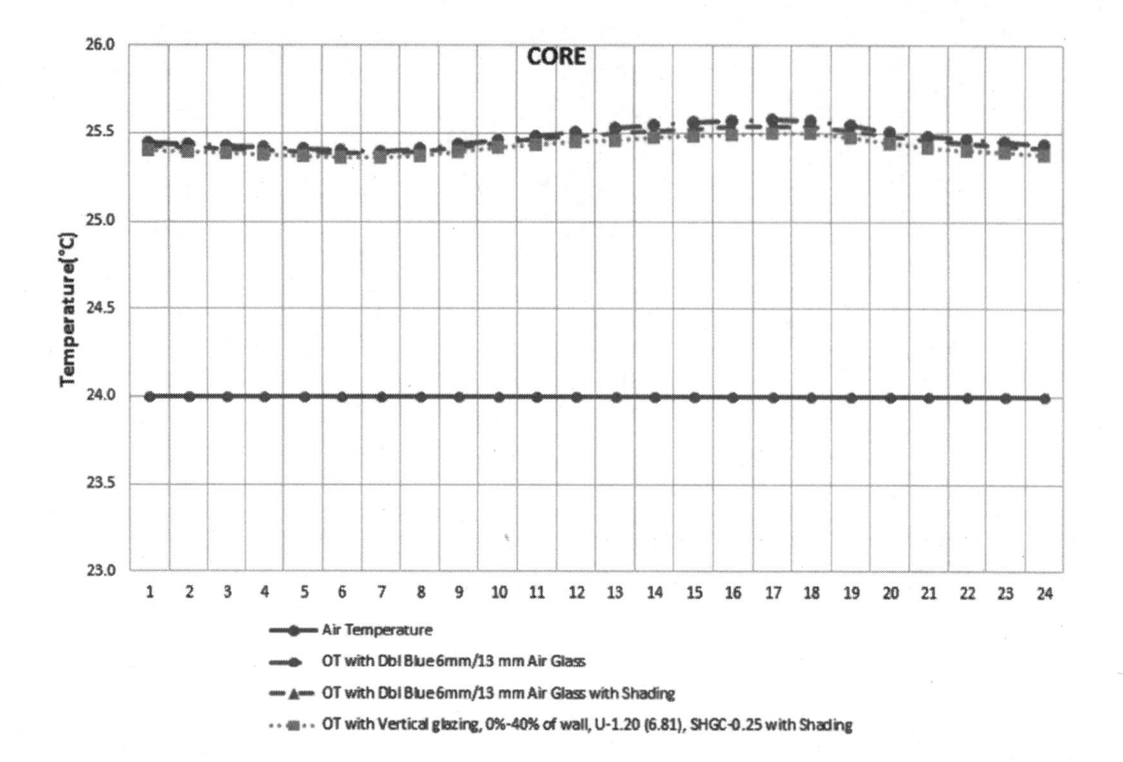

For the core zone, it can be seen that the variation in operative temperature is minimal because the walls are not receiving direct solar radiation and thus not getting heated up to the extent of the perimeter zones.

Step 18: Now plot temperatures for the **Ground Floor: South** zone with the operative temperature control.

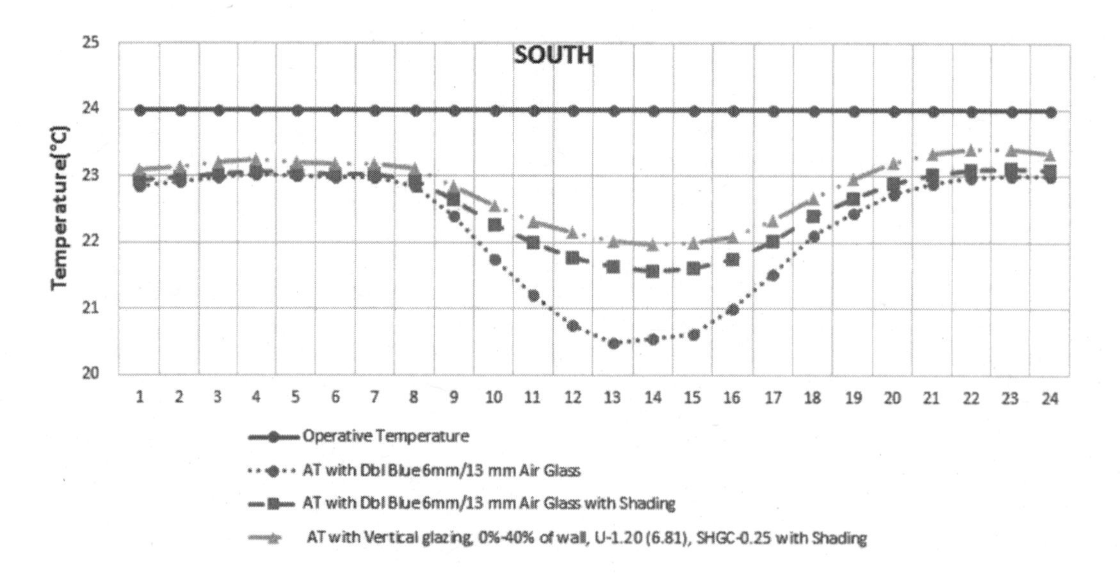

Step 19: Plot a chart for the annual cooling energy consumption for the air and operative temperature controls.

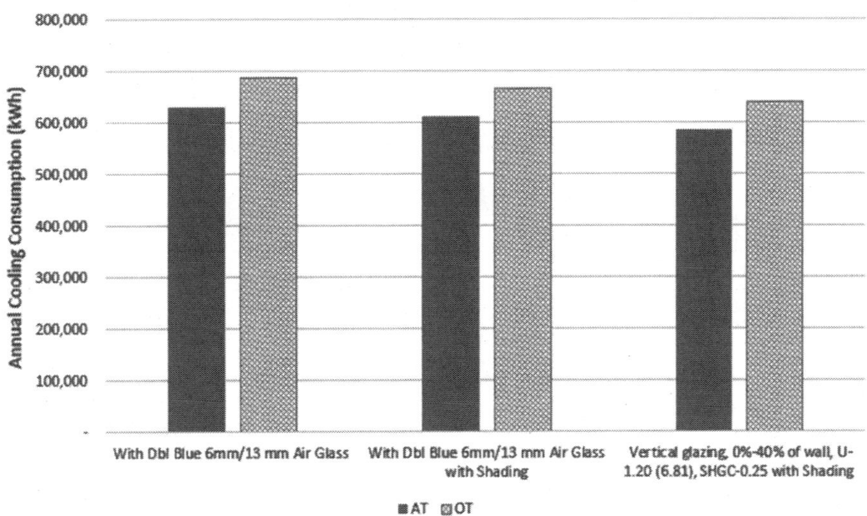

Analysis of Results

The thermostat control in conventional HVAC systems is activated based on the air temperature; however, human comfort is a function of operative temperature. Operative temperatures include the combined effect of air temperature and mean radiant temperature. During summer, perimeter spaces are uncomfortable because of higher operative temperatures. Therefore, there is an increase in the cooling equipment sizing of the system because the operative temperature setpoint continues to condition the building until comfort conditions are met.

You can see that the effect of the operative temperature setpoint is more noticeable in perimeter zones of the building than in the core zone. This can be explained by the fact that the perimeter zones have more surfaces that are connected with the outdoor environment, whereas for the core zone, only the roof is connected to the outdoor environment.

EXERCISE 6.1

Repeat this tutorial for 20% WWR (Table 6.2).

TABLE 6.2
Comparison of the design capacity of cooling equipment for air temperature and operative temperature controls

	Design capacity of cooling equipment (kW)	
Zone	**Air temperature control**	**Operative temperature control**
Ground: East		
Ground: West		
Ground: North		
Ground: South		
Ground: Core		

Tutorial 6.2

Evaluating the Impact of Design Day Selection

GOAL

To find the cooling equipment capacity using the design day and annual energy simulation approaches

WHAT ARE YOU GOING TO LEARN?

- Show to size using two methods: design day and annual energy simulation methods

PROBLEM STATEMENT

In this tutorial, you are going to use a **50- × 25-m** model with a **5-m** perimeter depth. Determine the peak cooling load

1. With the design day 15 July (as explained in Tutorial 6.1)
2. Using annual energy simulations

Use the **New Delhi/Palam, India** weather file. Find the total cooling load for both options. Also note the time of occurrence of the maximum cooling load in each case.

Cooling design calculations are carried out to determine the capacity of the mechanical cooling equipment required to meet the hottest summer design weather conditions likely to be encountered at the site location.

Cooling design simulations using EnergyPlus have the following characteristics:

- Periodic steady-state external temperatures calculated using the maximum and minimum design summer weather conditions
- No wind
- Includes solar gains through windows and scheduled natural ventilation
- Includes internal gains from occupants, lighting and other equipment
- Includes consideration of heat conduction and convection between zones of different temperatures

For buildings situated in the northern hemisphere, cooling design calculations are made for the month of July, and for buildings in the southern hemisphere, they are made for the month of January.

Source: http://www.designbuilder.co.Uk/helpv6.0/Content/_Cooling_design_simulation.htm.

SOLUTION

Step 1: Open a new model, and create a **50- × 25-m** building with a **5-m** perimeter depth. Select the **Cooling design** screen tab. The **Calculation Options** screen appears.

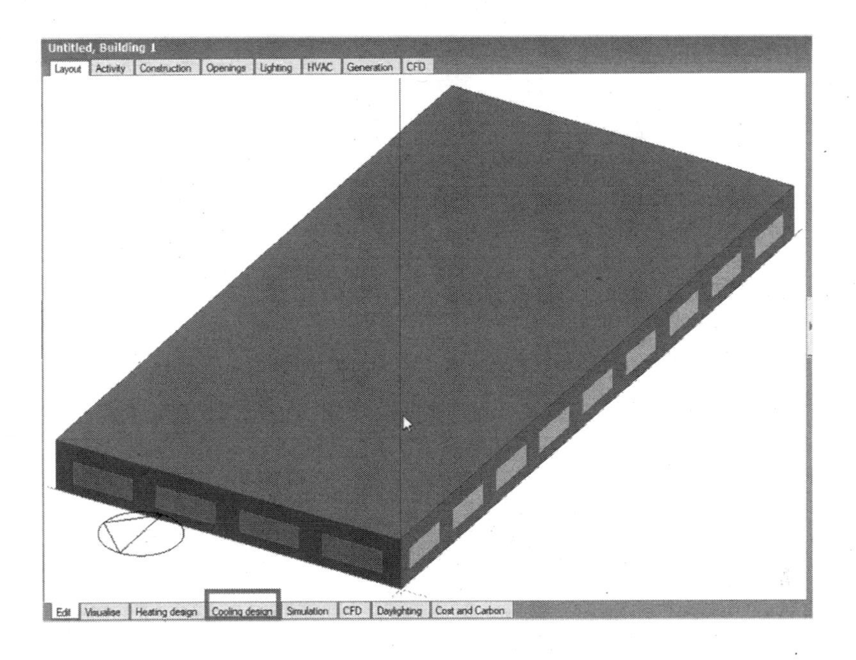

Step 2: Click **OK**. The results are displayed in the **Analysis** tab.

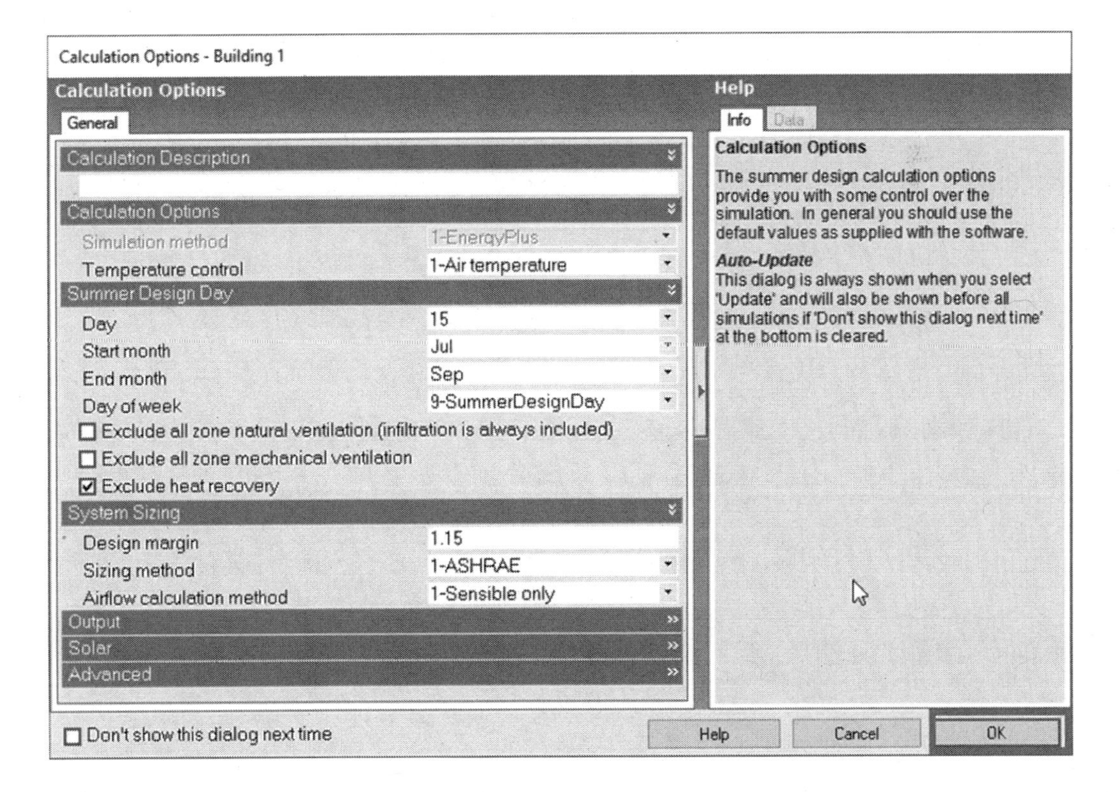

Step 3: Select the **Summary** tab. It shows the design capacity for 15 July. Select **Coincident**, and record the results.

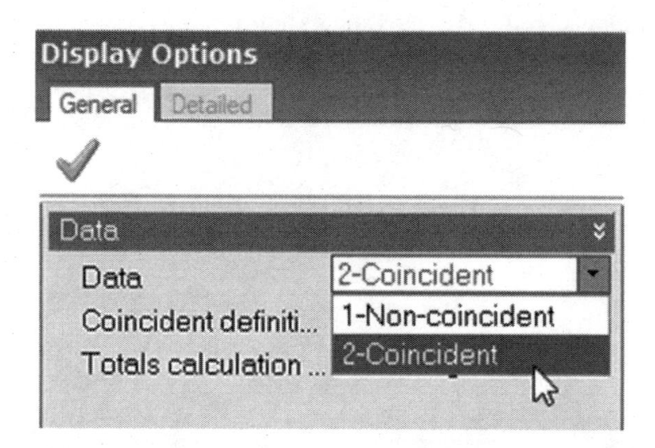

Untitled, Building 1

Analysis | Summary

Zone	Design Capacity (kW)	Design Flow Rate (m3/s)	Total Cooling Load (kW)	Sensible (kW)	Latent (kW)	Air Temperature (°C)	Humidity (%)	Time of Max Cooling	Max Op Temp in Day (°C)	Floor Area (m2)	Volume (m3)
Building 1											
Block1:WEST	20.02	1.1629	17.41	14.30	3.11	24.0	49.8	Jul 15:30	28.7	90.2	315.6
Block1:NORTH	32.86	1.7955	28.57	22.08	6.50	24.0	51.2	Jul 15:30	27.0	203.5	712.2
Block1:EAST	16.15	0.9023	14.05	11.09	2.95	24.0	50.7	Jul 15:30	27.4	91.3	319.7
Block1:CORE	78.99	4.1208	68.69	50.66	18.02	24.0	52.5	Jul 15:30	26.7	588.1	2058.5
Block1:SOUTH	[illegible]	1.8102	28.82	22.26	6.56	24.0	51.2	Jul 15:30	27.0	206.1	721.4
Totals	**181.17**	**9.7917**	**157.54**	**120.39**	**37.15**	**24.0**	**51.7**	**Jul 15:30**	**28.7**	**1179.3**	**4127.4**

Cooling design does not require the weather data file for the calculation. DesignBuilder has all the required information in the **ASHRAE_2005_Yearly_DesignConditions.xls** file. This file can be accessed at **C:\ProgramData\DesignBuilder\Weather Data**. These details automatically get loaded in DesignBuilder when the location is selected.

In the following steps, you are going to determine the cooling load using annual energy simulations.

Step 4: Select the **Simulation with Hourly** tab. After completion of the simulation, select **System loads** from the **Data** drop-down list.

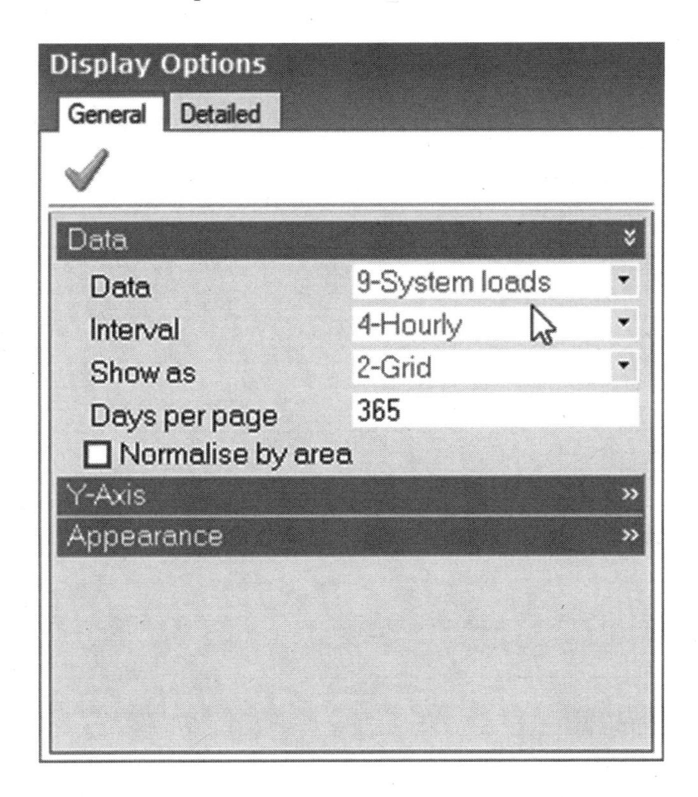

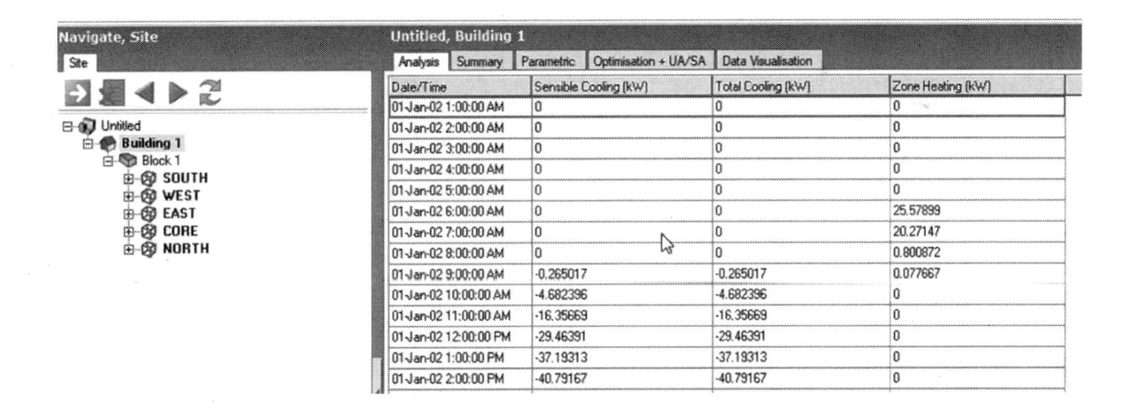

Date/Time	Sensible Cooling (kW)	Total Cooling (kW)	Zone Heating (kW)
01-Jan-02 1:00:00 AM	0	0	0
01-Jan-02 2:00:00 AM	0	0	0
01-Jan-02 3:00:00 AM	0	0	0
01-Jan-02 4:00:00 AM	0	0	0
01-Jan-02 5:00:00 AM	0	0	0
01-Jan-02 6:00:00 AM	0	0	25.57899
01-Jan-02 7:00:00 AM	0	0	20.27147
01-Jan-02 8:00:00 AM	0	0	0.800872
01-Jan-02 9:00:00 AM	-0.265017	-0.265017	0.077667
01-Jan-02 10:00:00 AM	-4.682396	-4.682396	0
01-Jan-02 11:00:00 AM	-16.35669	-16.35669	0
01-Jan-02 12:00:00 PM	-29.46391	-29.46391	0
01-Jan-02 1:00:00 PM	-37.19313	-37.19313	0
01-Jan-02 2:00:00 PM	-40.79167	-40.79167	0

Step 5: Click the **Export** icon to export the results to the spreadsheet (Table 6.3).

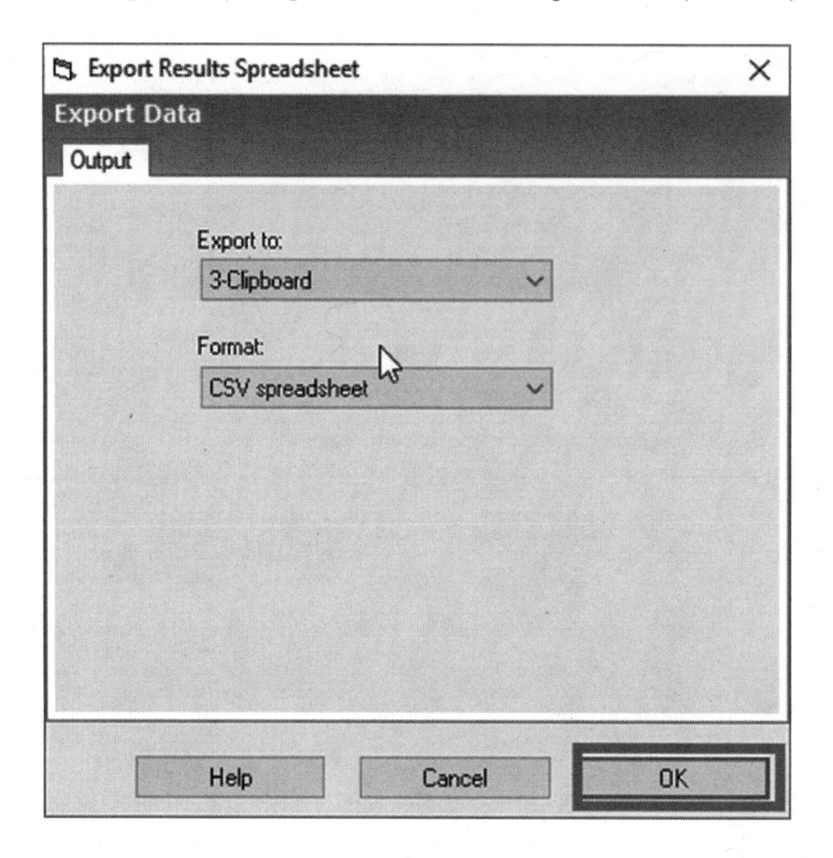

TABLE 6.3
Total cooling load

Day and month	Time	Total cooling (kW)
19 June	4:00:00 p.m.	−184.7126
19 June	3:00:00 p.m.	−184.1557
17 June	3:00:00 p.m.	−183.761
17 June	12:00:00 p.m.	−183.3166
17 June	11:00:00 p.m.	−183.2973
19 June	12:00:00 p.m.	−182.8737
17 June	4:00:00 p.m.	−181.0183
17 June	10:00:00 p.m.	−180.8233
18 June	3:00:00 p.m.	−179.1169

Step 6: Open the spreadsheet and sort the results for total cooling (kW) in decreasing order. This provides the peak total cooling (kW) of the building. Compare the results for the building (Table 6.4). The results show that there is a difference between the peak total cooling load of the building with the design day and without explicitly defining the design day. One should always be cautious while selecting the design day for cooling. Sometimes the design day might not represent the day of the maximum total cooling load. Save the simulation model for use in next tutorial.

TABLE 6.4
Total cooling load for design day and annual energy simulation

With design day on 15 July		With annual energy simulation	
Total cooling load (kW)	Date and time of peak	Total cooling load (kW)	Date and time of peak (building)
181.17	Jul 15 at 15:00 h	184.71	June 19 at 16:00 h

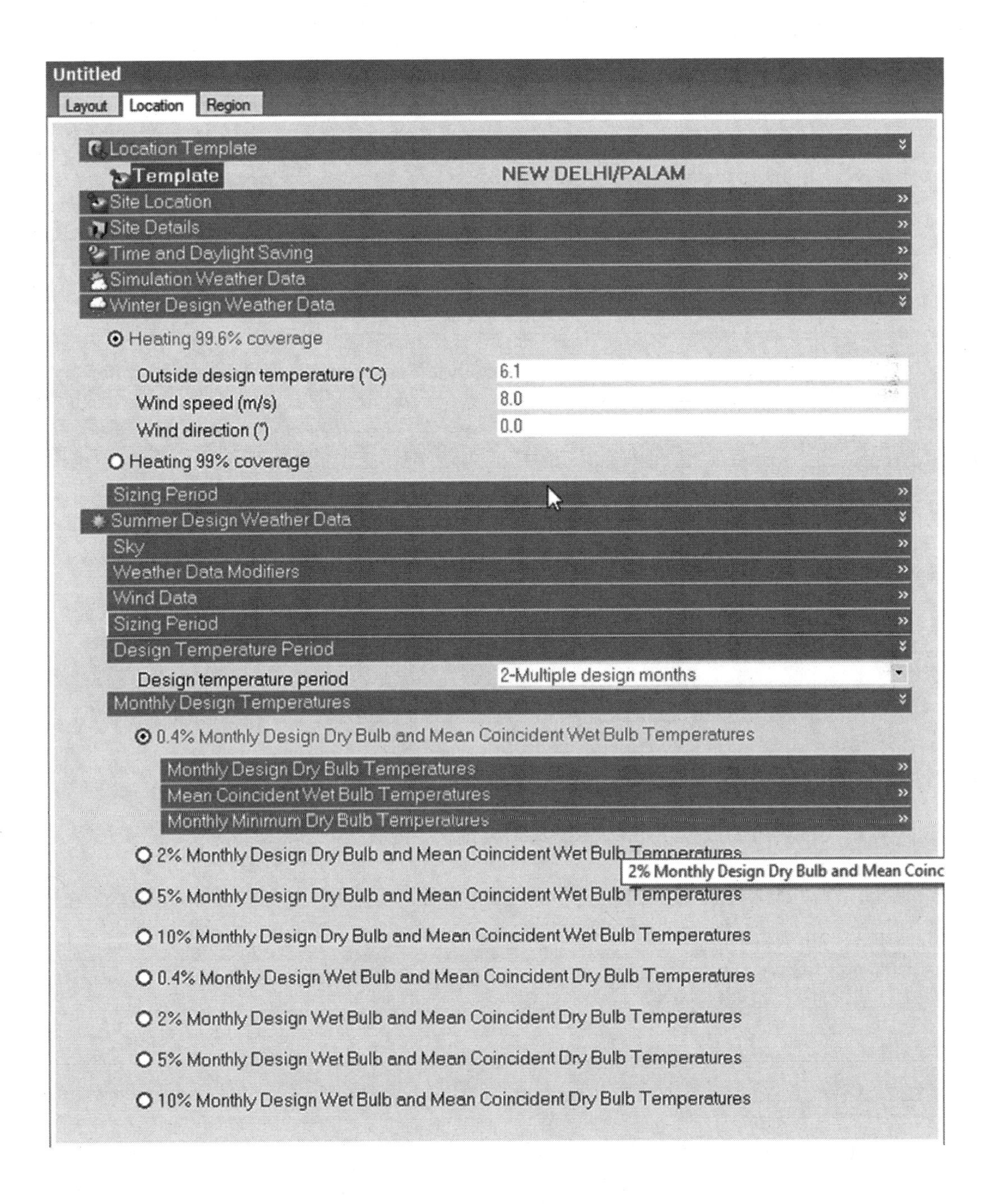

In some cases, design conditions, namely maximum dry bulb, concurrent wet bulb and minimum dry bulb temperature, are known to HVAC designers. As a third alternate, these conditions can be directly filled into DesignBuilder.

EXERCISE 6.2

Repeat the preceding steps for London to find the heating design capacity. You can use the **Heating Sizing** tab for this (Table 6.5).

TABLE 6.5
Total heating design capacity for design day and annual energy simulation

With design day on 15 Jan		With annual energy simulation	
Total heating design capacity (kW)	Date and time of peak heating load	Total heating design capacity (kW)	Date and time of peak heating load (building)

Heating design calculations are carried out to determine the size of the heating equipment required to meet the coldest winter design weather conditions likely to be encountered at the site location. The simulation calculates the heating capacities required to maintain the temperature setpoints in each zone and displays the total heat loss broken down as follows:

- Glazing
- Walls
- Partitions
- Floors
- Roofs
- External infiltration
- Internal natural ventilation (i.e. the heat lost to other cooler adjacent spaces through windows, vents, doors and holes)

Tutorial 6.3

Evaluating the Impact of the Airflow Calculation Method

GOAL

To evaluate the impact of the airflow calculation method on HVAC equipment sizing

WHAT ARE YOU GOING TO LEARN?

- Chow to change the airflow calculation method and find the design air flow

PROBLEM STATEMENT

In this tutorial, you are going to use the simulation model used for Tutorial 6.2 (50- × 25-m model with a 5-m perimeter depth) with building usage as a classroom. Find the cooling equipment sizing and design flow rates for the following airflow calculation methods:

1. Sensible
2. Sensible + latent

Use the **AZ – PHOENIX/SKY HARBOR, USA** weather file.

SOLUTION

Step 1: Open the model saved in Tutorial 6.2 (a **50- × 25-m** building with a **5-m** perimeter depth).

Step 2: Select the **Activity** tab. Select **Classroom** in the **Universities and colleges** section.

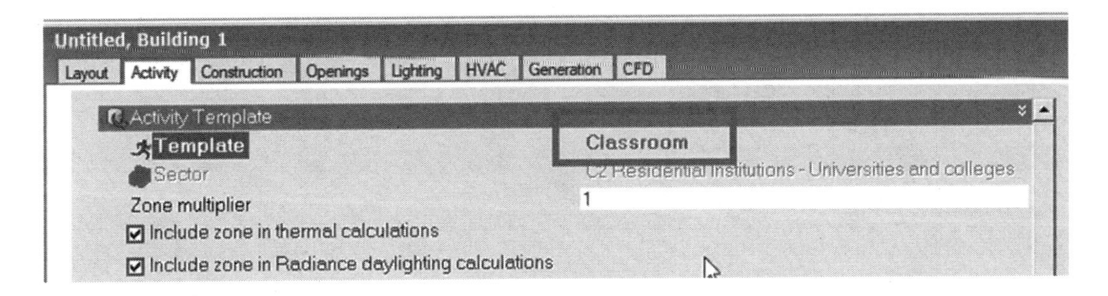

Step 3: Select the **Cooling design** screen tab. The **Calculation Options** screen appears.

Step 4: Select **Sensible only** from the **Airflow calculation method** drop-down list. Click **OK**. The results are displayed in the **Analysis** tab.

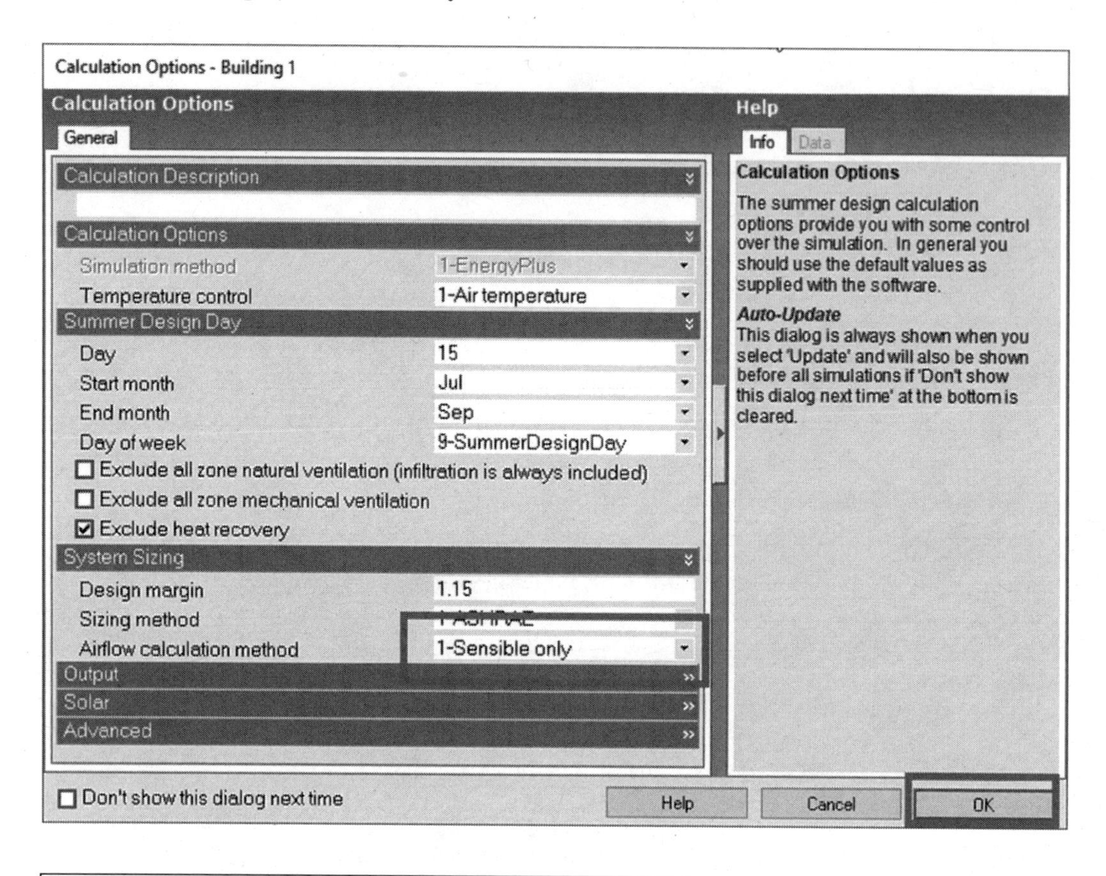

The supply air for cooling is calculated as

$$Q_S = \frac{q_S}{C_1(t_R - t_S)}$$

The supply air for dehumidification is calculated as

$$Q_L = \frac{q_L}{C_2(W_R - W_S)}$$

where Q_S = supply air volume required to satisfy the peak sensible load
Q_L = supply air volume required to satisfy the peak latent load
q_S = peak sensible load
q_L = peak latent load
t_R = room air temperature
t_S = supply air temperature
W_R = room humidity ratio
W_S = humidity ratio of the dehumidified supply air
C_1 = 1.23 (for calculation in SI units)
C_2 = 3,010 (for calculation in SI units)

Step 5: Select the **Summary** tab. Record the results.

Untitled, Building 1

Analysis | Summary

Zone	Design Capacity (kW)	Design Flow Rate (m3/s)	Total Cooling Load (kW)	Sensible (kW)	Latent (kW)	Air Temperature (°C)	Humidity (%
Building 1							
Block1:EAST	**14.37**	**1.1229**	12.50	12.50	0.00	23.0	45.6
Block1:WEST	**16.99**	**1.3270**	14.77	14.77	0.00	23.0	46.0
Block1:NORTH	**28.27**	**2.2081**	24.58	24.58	0.00	23.0	46.5
Block1:SOUTH	**29.46**	**2.3[illegible]17**	25.62	25.62	0.00	23.0	46.4
Block1:CORE	**63.24**	**4.9400**	54.99	54.99	0.00	23.0	47.1
Totals	152.33	11.8996	132.46	132.46	0.00	23.0	46.7

Step 6: Repeat the preceding steps to select **Sensible+latent** from the **Airflow calculation method** drop-down list. Click **OK**.

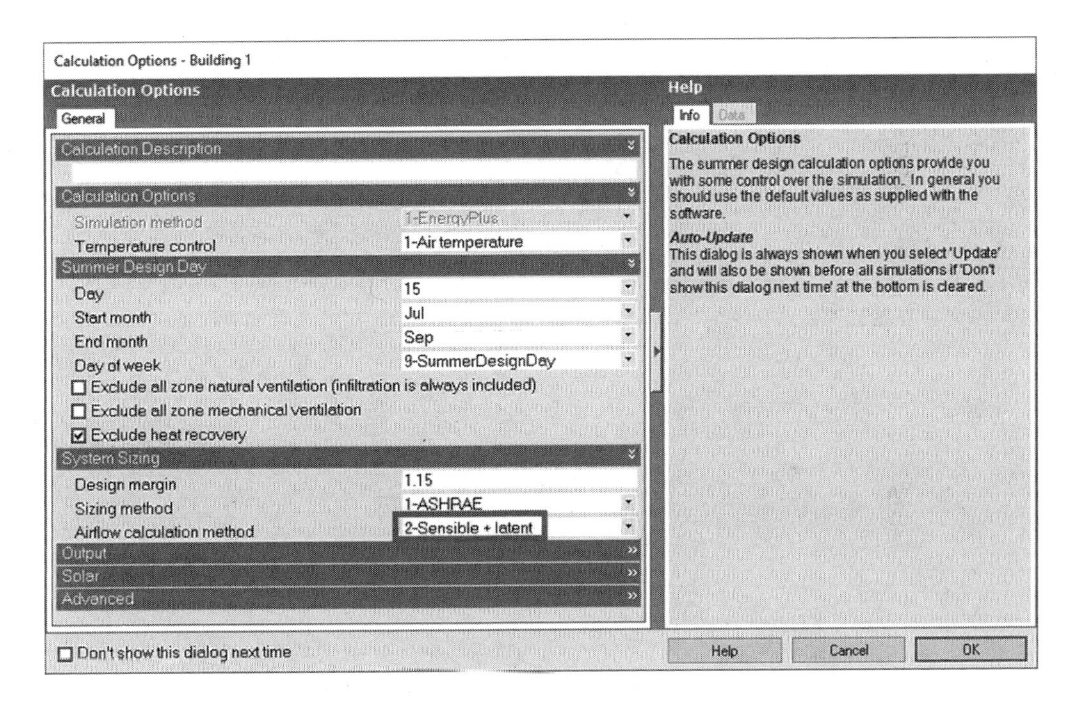

Step 7: Select the **Summary** tab. Record the results.

Untitled, Building 1

Analysis | Summary

Zone	Design Capacity (kW)	Design Flow Rate (m3/s)	Total Cooling Load (kW)	Sensible (kW)	Latent (kW)	Ai
Building 1						
Block1:EAST	**14.37**	**0.9750**	12.50	12.50	0.00	2[illegible]
Block1:WEST	**16.99**	**1.1341**	14.77	14.77	0.00	2[illegible]
Block1:NORTH	**28.27**	**1.8560**	24.58	24.58	0.00	2[illegible]
Block1:SOUTH	**29.46**	**1.9395**	25.62	25.62	0.00	2[illegible]
Block1:CORE	**63.24**	**4.0589**	54.99	54.99	0.00	2[illegible]
Totals	152.33	9.9635	132.46	132.46	0.00	2[illegible]

Compare both cases (Table 6.6).

TABLE 6.6
Design capacity and design flow rate for each zone

Zone	Design capacity (kW)	Design flow rate (m^3/s)	
		Sensible only	Sensible + latent
Core	63.24	4.94	4.06
East	14.37	1.12	0.98
West	16.99	1.33	1.13
North	28.27	2.21	1.86
South	29.46	2.30	1.94
Total	152.33	11.90	9.96

Factors that influence the sensible cooling load

Glass, windows or doors:

- Solar radiation striking windows, skylights or glass doors
- Thermal resistance of exterior walls
- Partitions (that separate spaces of different temperatures)
- Ceilings under an attic
- Thermal resistance of roofs and floors
- Air infiltration through cracks/gaps in the building, doors and windows
- Building occupants
- Equipment and appliances operated in summer
- Light

Factors that influence the latent cooling load

Moisture is introduced into a structure through

- Building occupants
- Equipment and appliances that release vapour in space, such as a teakettle
- Air infiltration through cracks/gaps in the building, doors and windows and frequent door opening to the ambience

EXERCISE 6.3

Repeat this tutorial for the London Getwick Airport, UK (Table 6.7).

TABLE 6.7
Design capacity and design flow rate for each zone

		Design flow rate (m^3/s)	
Zone	**Design capacity (kW)**	**Sensible only**	**Sensible + latent**
Core			
East			
West			
North			
South			
Total			

7 Unitary HVAC Systems

In most unitary heating, ventilation and air-conditioning (HVAC) systems, coefficient of performance (COP) and fan (condenser and evaporator) properties are the two key aspects that govern their energy consumption. A higher COP is associated with a reduction in energy consumption. Unitary systems require a fan for blowing air over the condenser as well as evaporator tubes. A fan is characterized by its static pressure and volumetric airflow rate. High static pressure and flow rate help in achieving the setpoint faster; however, they are also associated with higher energy consumption. If these parameters are not specified properly, results of the energy model could deviate significantly from the actual performance. This chapter explains the method of modelling unitary systems by specifying COP and fan properties.

Tutorial 7.1

Evaluating the Impact of Unitary Air-Conditioner Coefficient of Performance (COP)

GOAL

To evaluate the impact of unitary air-conditioner COP on building energy performance

WHAT ARE YOU GOING TO LEARN?

- How to model a unitary HVAC system
- How to change the cooling system COP

PROBLEM STATEMENT

In this tutorial, you are going to create a 50- × 25-m five-zone model with 5-m perimeter depth. You will model the unitary HVAC system and then change the COP of the system. You will change the cooling COP from 1.5 to 3.5 with increments of 0.5. Use **FL – MIAMI, USA** weather file. Find energy consumption for all COP values.

COP = desired effect/work input

For cooling:

$$\text{COP}_{\text{cooling}} = \frac{Q_o}{W} \tag{7.1}$$

For heating:

$$\text{COP}_{\text{Heating}} = \frac{Q_k}{W} \tag{7.2}$$

where Q_o = heat absorbed in the evaporator
Q_k = heat rejected in the condenser
W = compressor work

The COP of the refrigeration cycle is a dimensionless index used to indicate the performance of a thermodynamic cycle or thermal system. The magnitude of COP is usually greater than one.

SOLUTION

Step 1: Open a new project, and create a **50- × 25-m** building with five zones and a **5-m** perimeter depth. Select the **HVAC** tab.

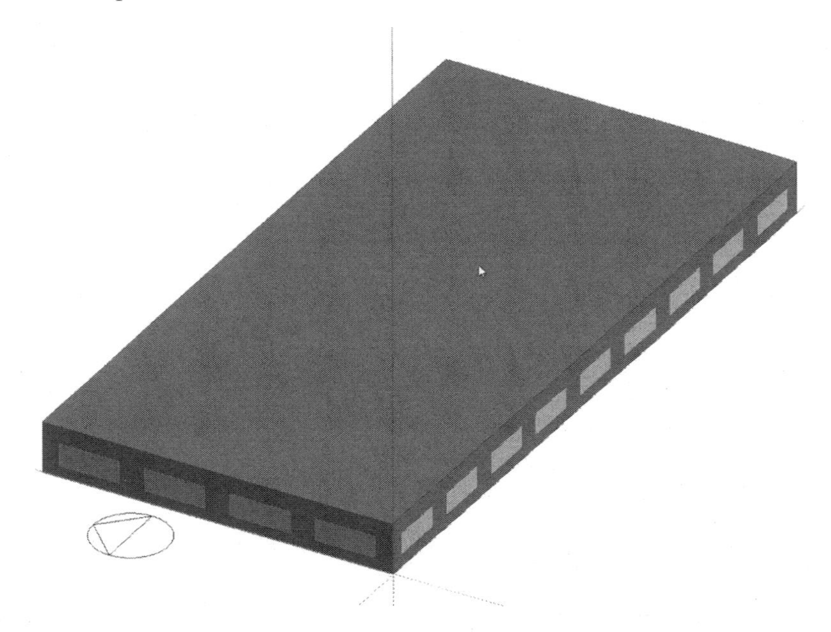

Step 2: Make sure that **Simple HVAC detail** is displayed under the **Help** tab.

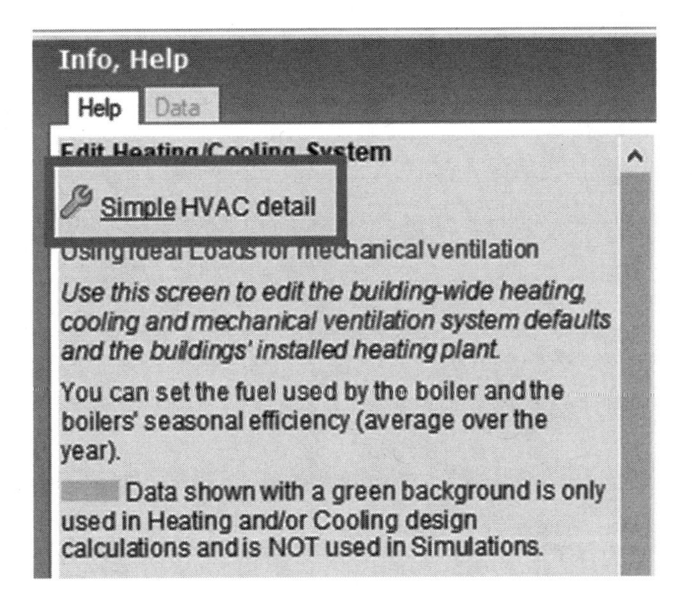

Two HVAC model options are available in DesignBuilder. **Simple HVAC** is suitable for use at early design stages. The heating/cooling system is modelled using basic loads calculation algorithms. **Detailed HVAC** models the HVAC system in full detail using EnergyPlus air- and water-side components linked together on a schematic layout drawing. This option will usually involve more work in setting up the setpoints, especially for large models, but there is increased flexibility.

Source: https://designbuilder.co.uk/helpv6.0/Content/_HVAC_model_detail.htm.

Step 3: Click **Template** under the **HVAC Template** section. It displays three dots. Click the three dots. The **Select the HVAC template** screen appears.

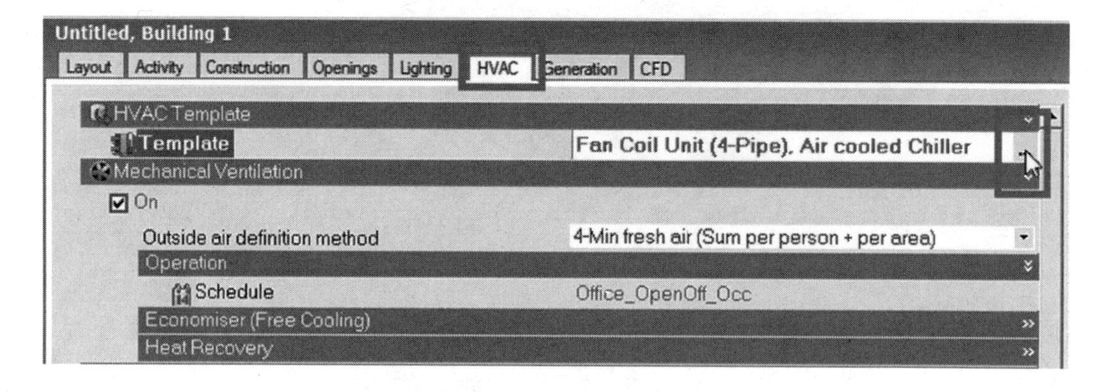

Step 4: Select **Packaged DX**. Click **OK**.

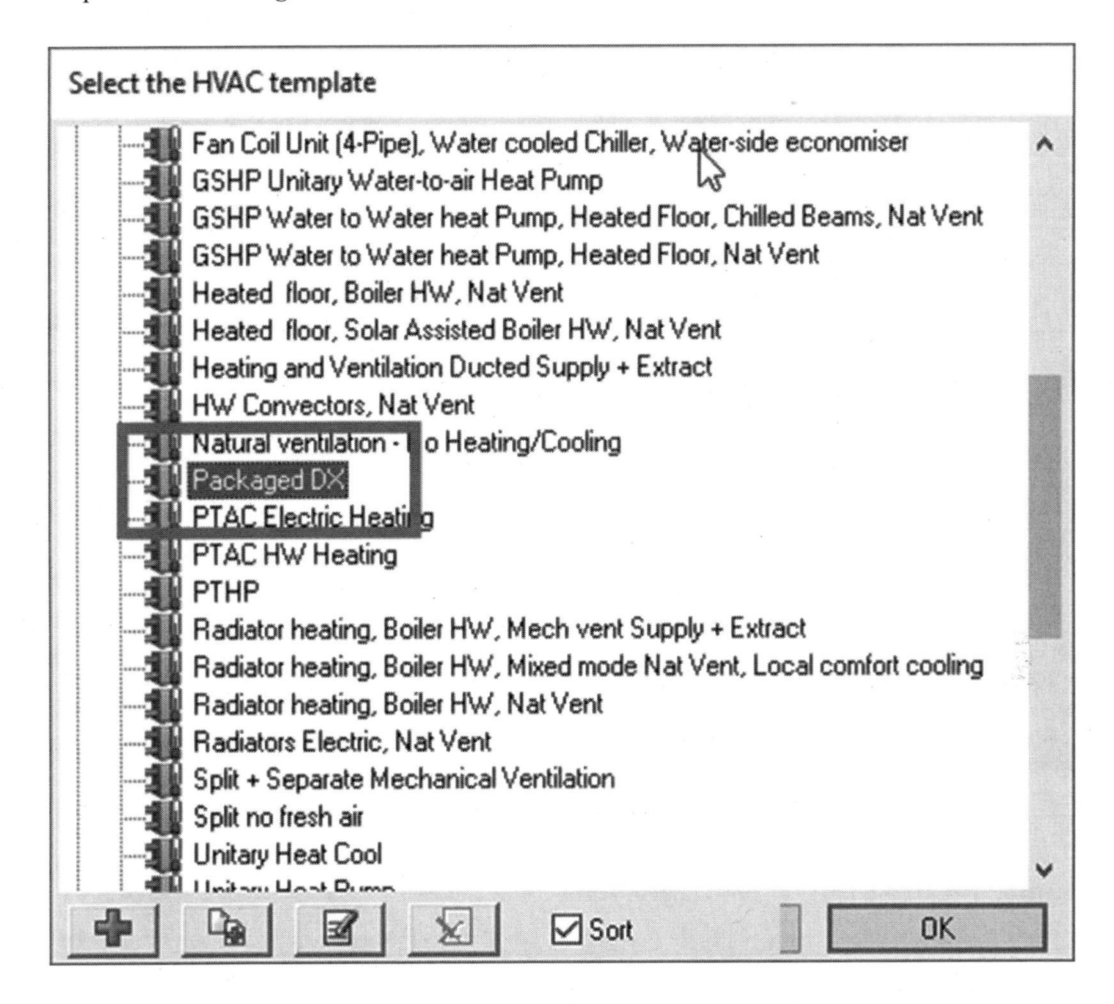

A screen appears showing the selected template as **Packaged DX**.

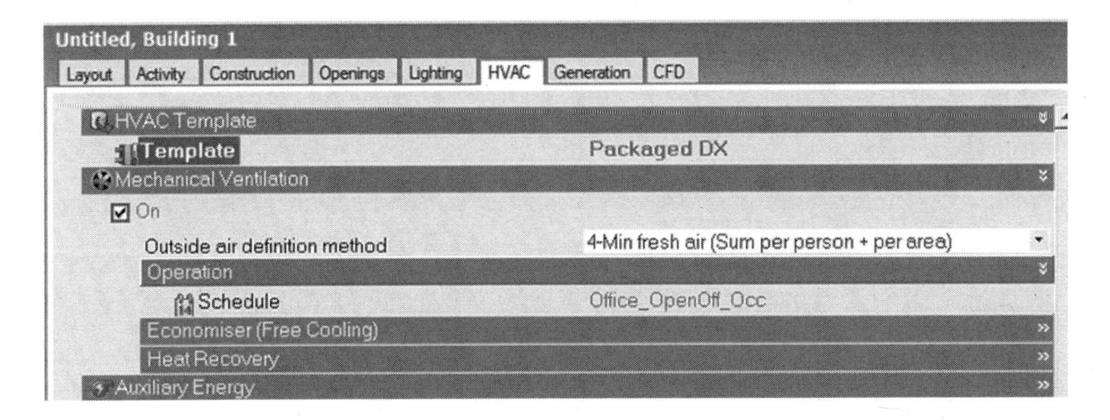

A *unitary system* combines heating, cooling and fan sections all in one or a few assemblies for simplified application and installation and is used in most classes of buildings, particularly where low initial cost and simplified installation are important. In DesignBuilder, the **Unitary single zone** option allows you to model simple single constant volume direct expansion (DX)–based HVAC configurations with several heating options. Direct expansion includes single-packaged rooftop systems commonly seen in commercial buildings and split systems commonly seen in residential buildings.

Source: https://designbuilder.co.uk/helpv6.0/#Cooling_Coil_-_DX.htm.

Step 5: Enter **1.5** as the **Cooling system seasonal CoP** in the **Cooling** section.

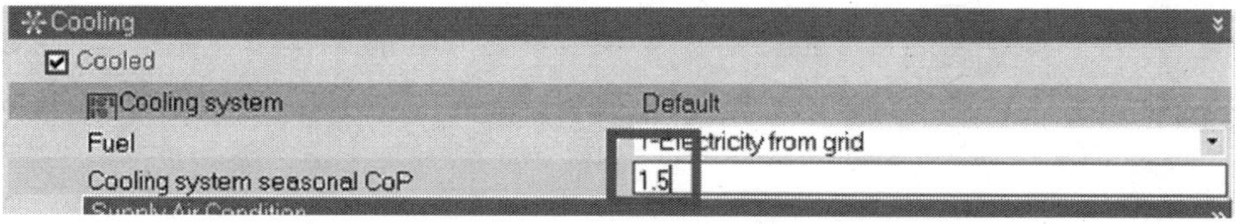

Step 6: Perform an annual simulation, and record the results.

Step 7: Repeat the preceding steps for all other COP values, and record the results (Table 7.1). Compare the results.

TABLE 7.1
Variation in annual energy consumption (kWh) with COP

	COP				
End-use category	**1.5**	**2**	**2.5**	**3**	**3.5**
Room electricity	52,291.88	52,291.88	52,291.88	52,291.88	52,291.88
Lighting	75,573.10	75,573.10	75,573.10	75,573.10	75,573.10
Cooling electricity)	186,008.00	139,506.00	111,604.80	93,004.01	79,717.73
Domestic hot water (DHW, electricity)	4,731.55	4,731.55	4,731.55	4,731.55	4,731.55

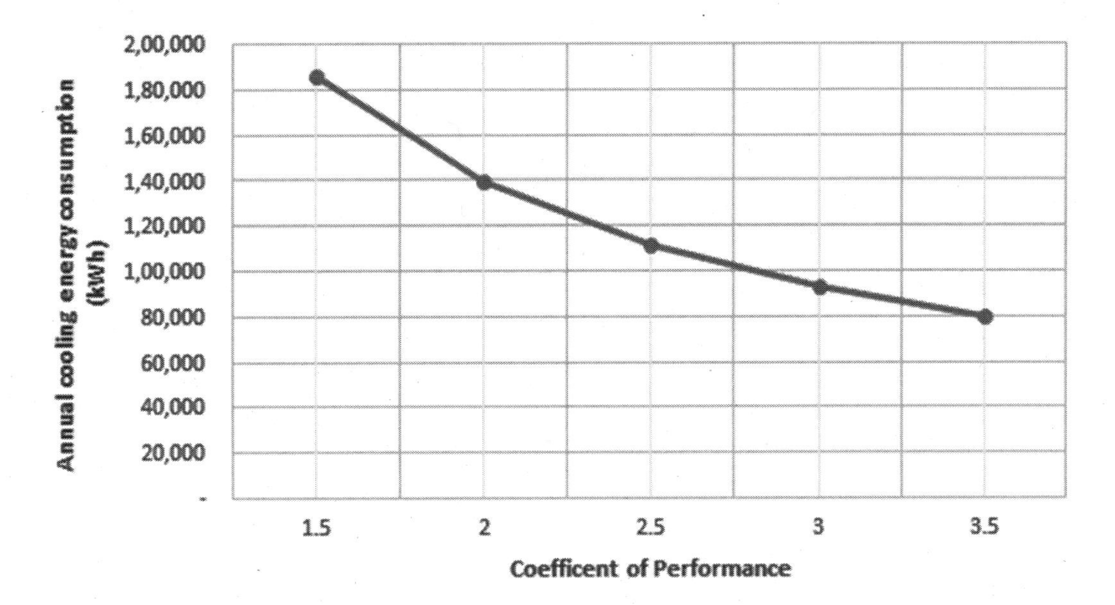

Results show that there is a decrease in energy consumption with an increase in the COP of the system. Systems with higher COP values require lesser energy input to remove the same amount of heat from the thermal zone. Save the model to use in the next tutorial.

EXERCISE 7.1

Repeat this tutorial for the heating system COP. Use the **CA-SAN FRANCISCO INTL, USA** weather data.

	Annual energy consumption (kWh)				
	COP				
End-use category	**1.5**	**2.0**	**2.5**	**3.0**	**3.5**
Room electricity					
Lighting					
System fans					
Heating (electricity)					
Cooling (electricity)					
DHW (electricity)					

Tutorial 7.2

Evaluating the Impact of Fan Efficiency of a Unitary Air-Conditioning System

GOAL

To evaluate the impact of fan efficiency of a unitary air-conditioning system on energy performance

WHAT ARE YOU GOING TO LEARN?

- How to change the fan efficiency

PROBLEM STATEMENT

In this tutorial, you are going to use a 50- × 25-m five-zone model with a 5-m perimeter depth. Model the unitary HVAC system with the following three fan efficiencies:

1 0.600
2 0.700
3 0.800

Find the change in energy consumption for all cases. Use **Rio de Janeiro (AERO), Brazil** weather location.

SOLUTION

Step 1: Open a new project, and create a **50- × 25-m** building with a **5-m** perimeter depth.

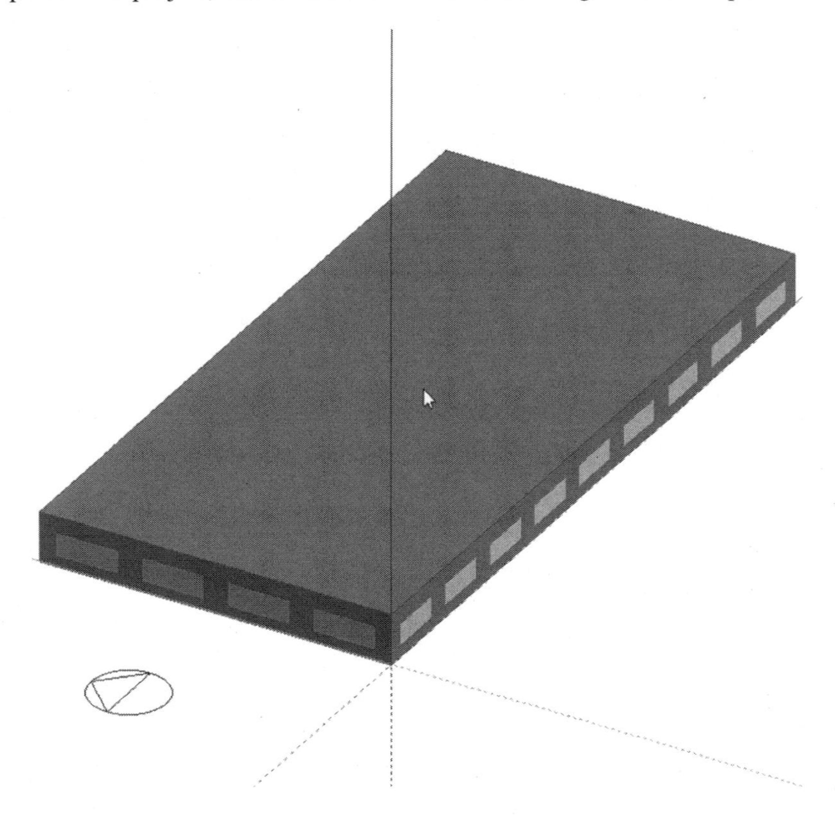

Step 2: Select the **HVAC** tab. Click the **Simple** link in the **Help** tab. The **Model Options – Building and Block** screen appears.

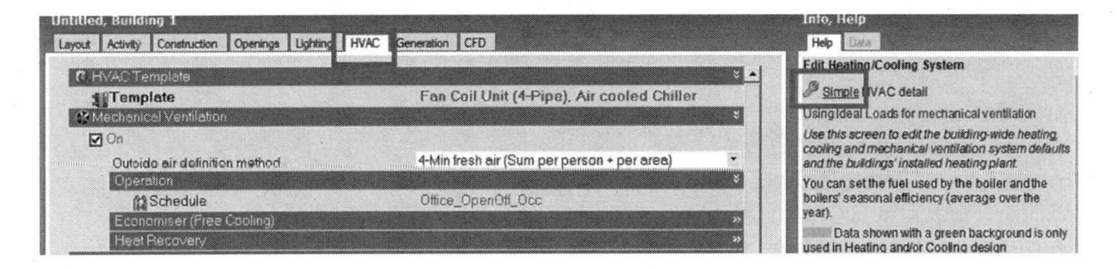

Step 3: Click **Detailed** under the **HVAC** slider. Select **Detailed HVAC Data** from the **Detailed HVAC** drop-down list. Click **OK**. This displays the **<HVAC system>** option under **Building 1**.

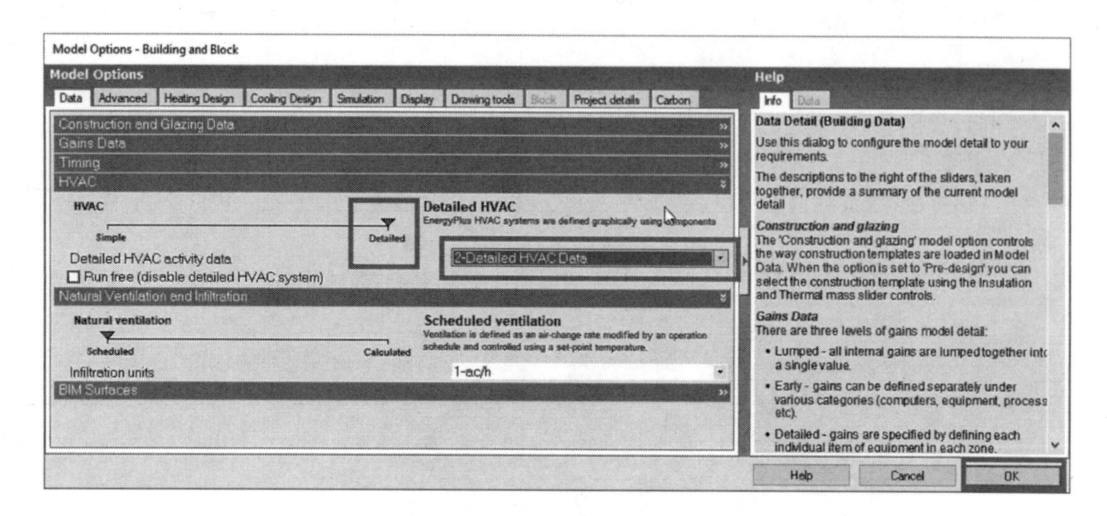

Step 4: Click **<HVAC System>**. This displays an initializing HVAC progress bar, and subsequently the **Load HVAC template** screen appears.

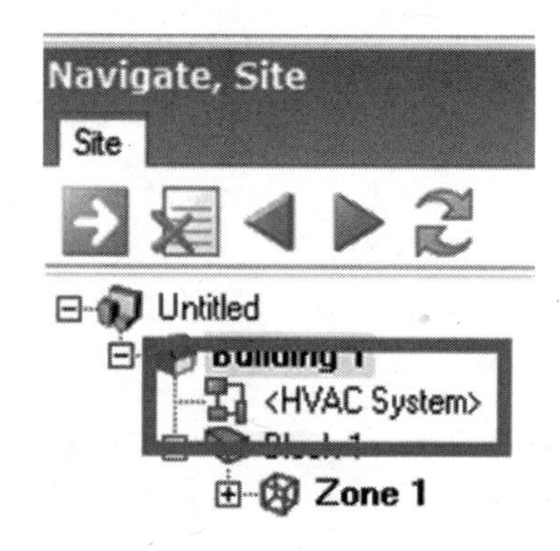

Step 5: Click the **Detailed HVAC** template. It displays three dots. Click the three dots. The **Select the Detailed HVAC Template** screen appears.

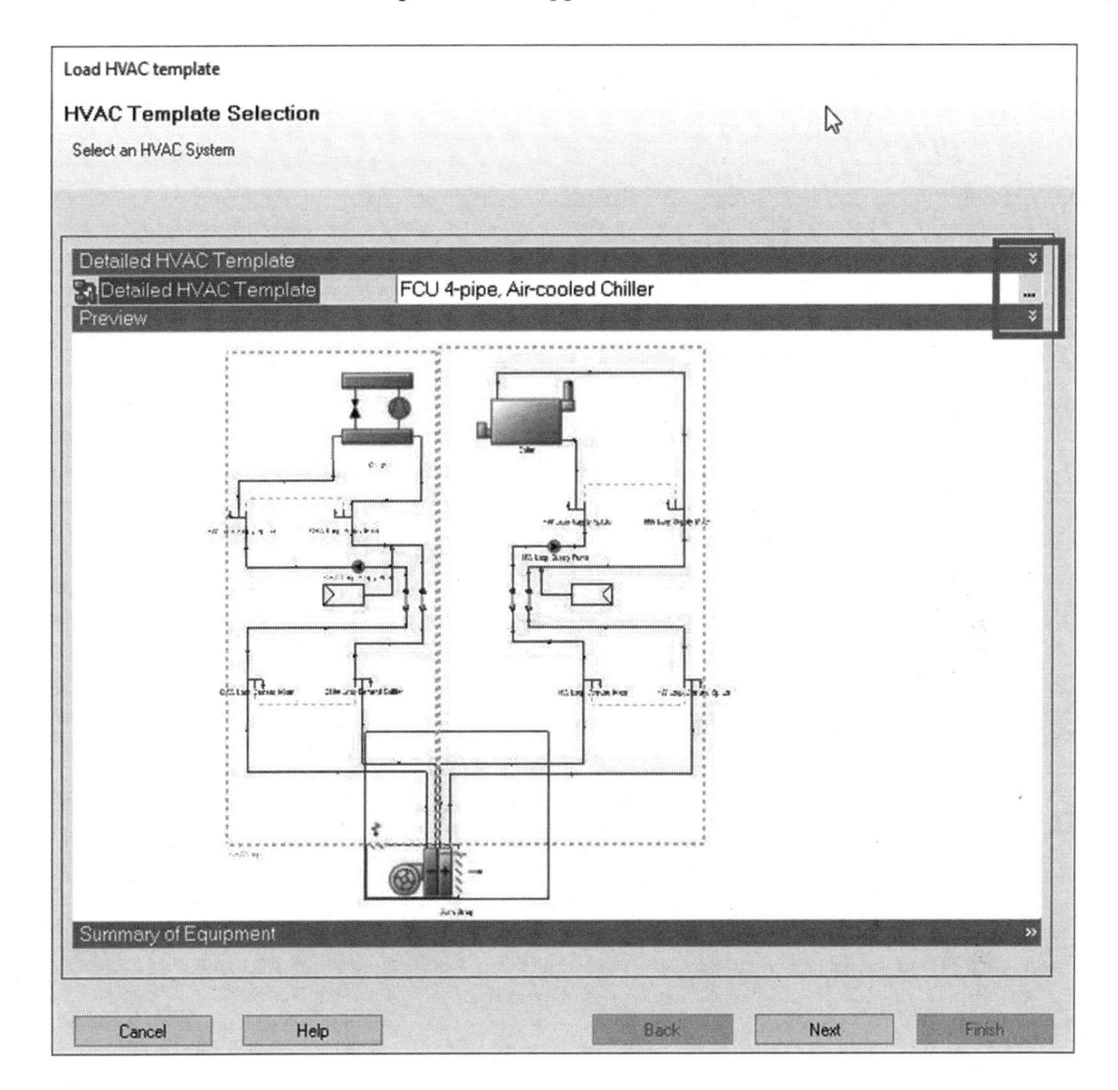

Step 6: Select **Unitary Heat Cool** under **Select Detailed HVAC** template. The **Load HVAC template** screen appears.

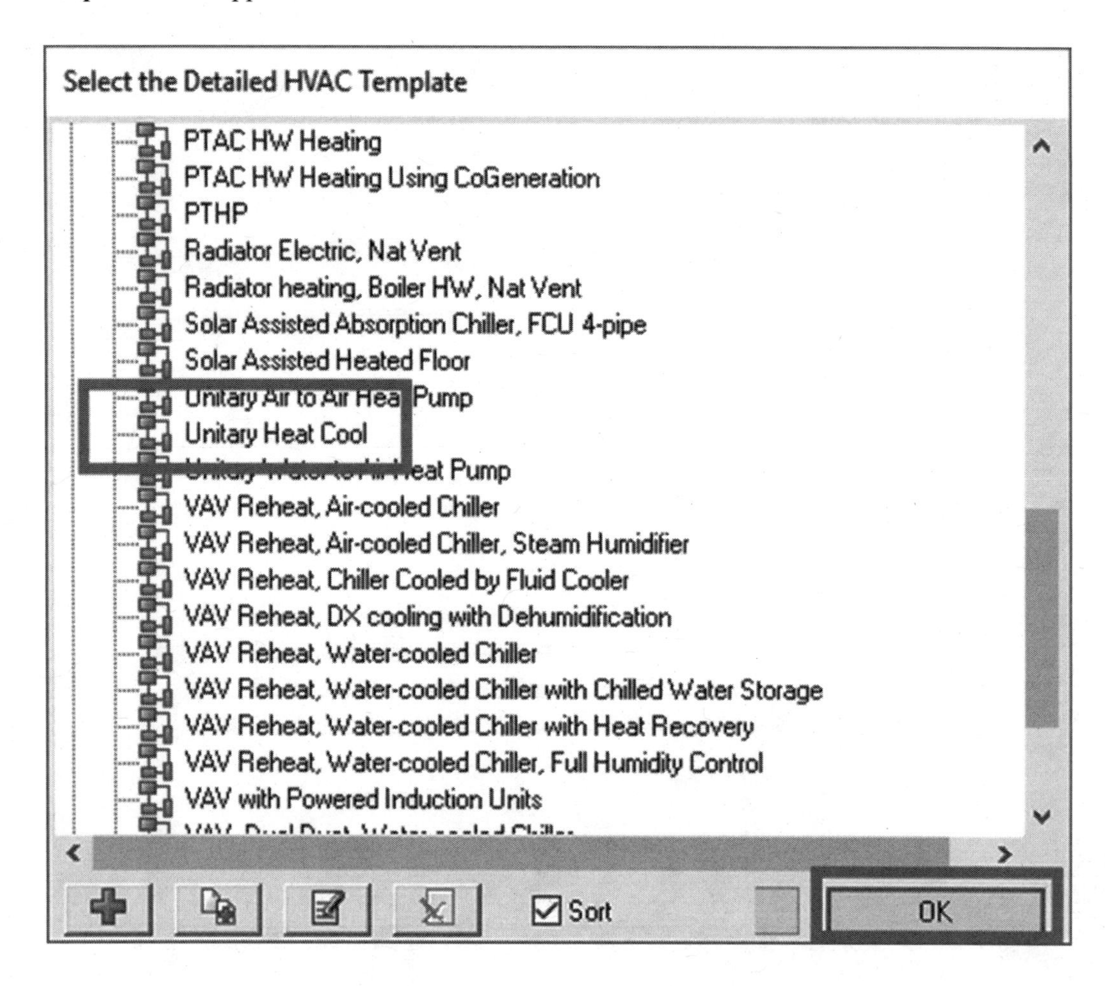

Step 7: Click **Next**. This displays all the zones.

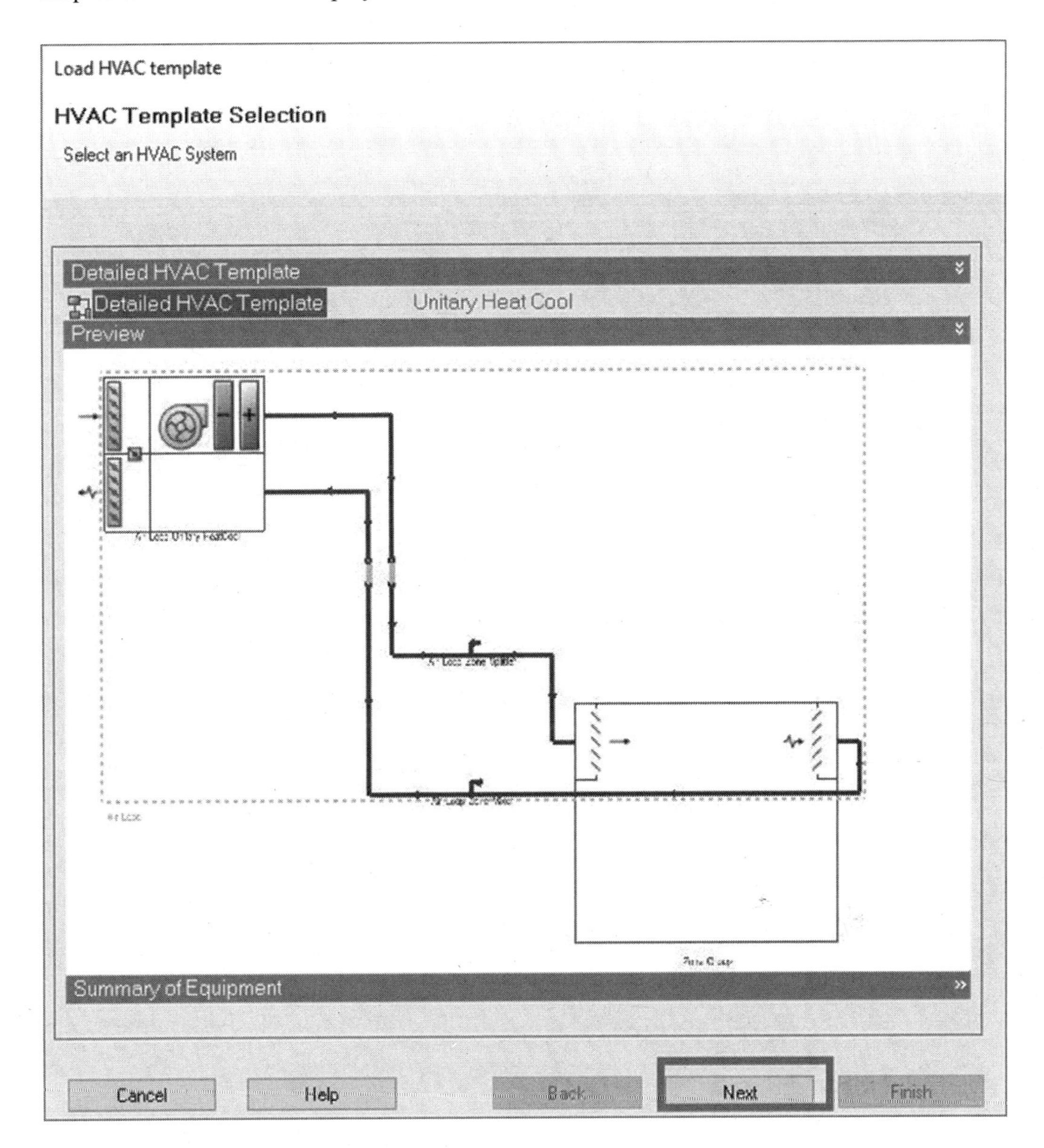

Step 8: Select the **Building 1** checkbox, and select **south zone** as contol zone for unitary system. Click **Finish**.

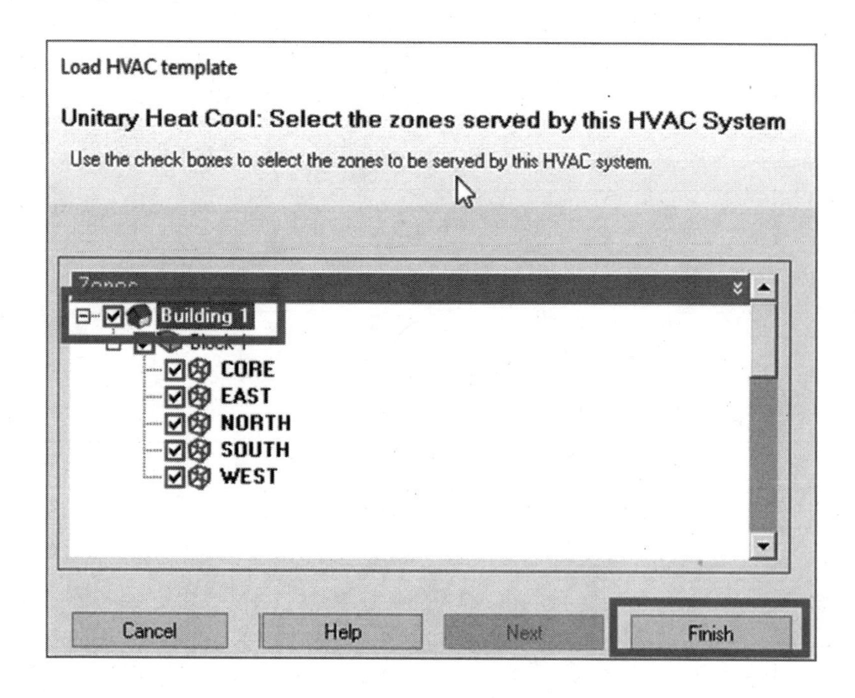

Step 9: In the navigation tree, select **Air Loop Unitary HeatCool Supply Fan** from the **<HVAC System>** tree. You need to click plus (+) to expand the tree.

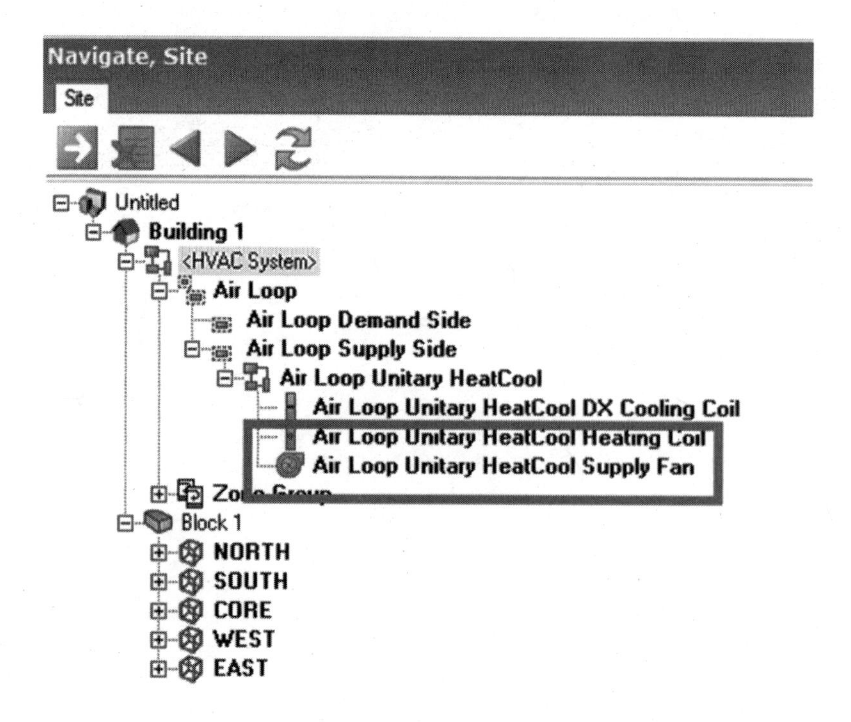

Step 10: Click **Edit component**.

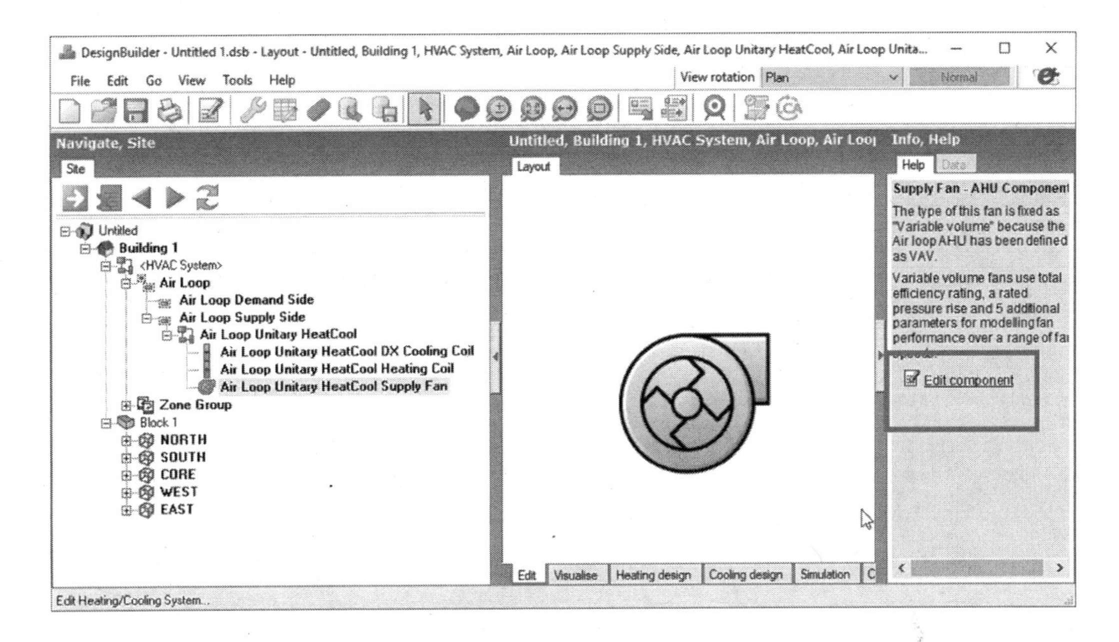

Step 11: Enter **0.600** for **Fan total efficiency** (%). Click **OK**.

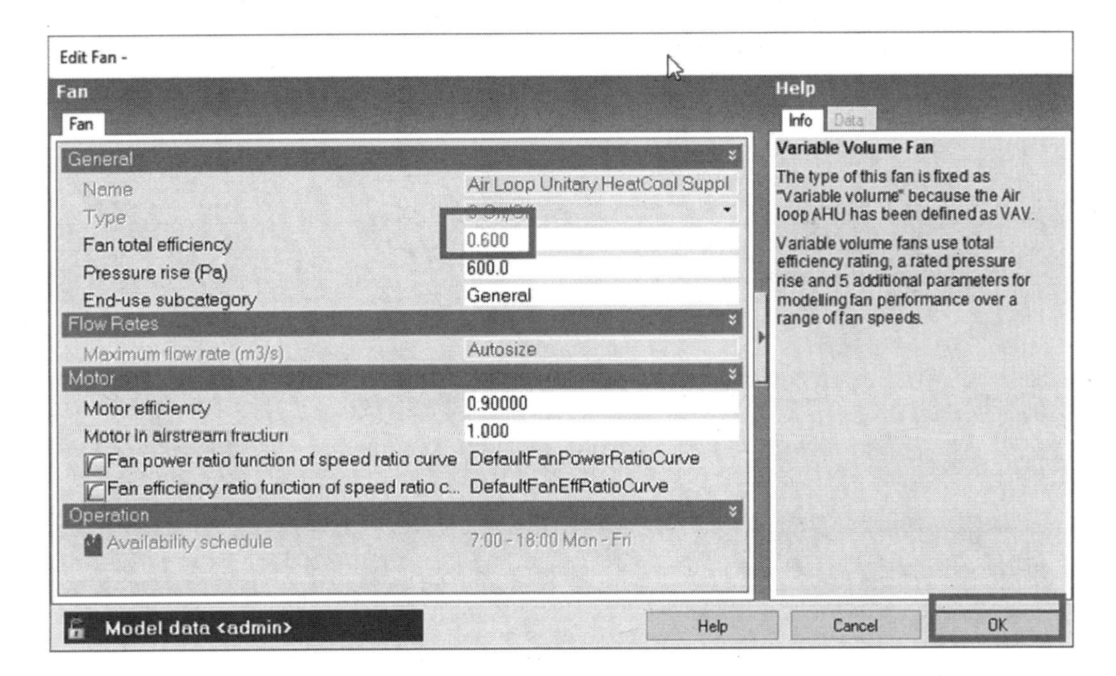

A ***fan*** is a power-driven rotary machine that causes a continuous flow of air. There are two broad categories of fans: axial and centrifugal.

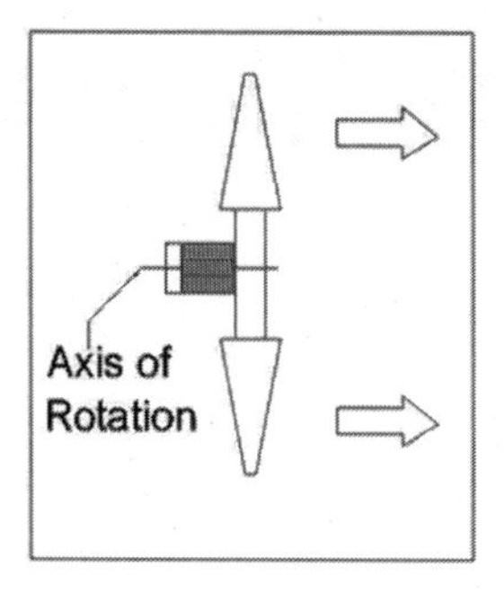

Axial Flow Fan

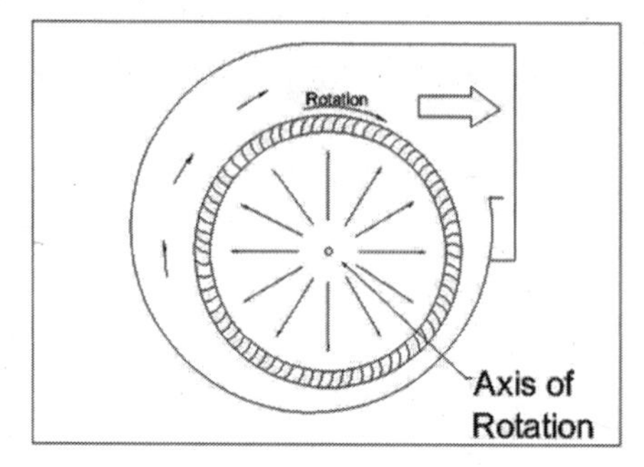

Centrifugal Fan

In axial-flow fans, air moves parallel to the shaft. This type of fan is used in many applications such as a cooling fan for electronics and wind tunnels. A centrifugal fan blows air at 90 degrees to the intake of the fan. Mostly, it is used for HVAC applications.

Fan Efficiency

The *fan efficiency* is the ratio between the power transferred to the airflow and the power used by the fan:

$$\eta = \frac{\Delta PQ}{W_{\text{fan}}} \tag{7.3}$$

where η = fan efficiency (values between zero and one)
ΔP = total pressure (Pa)
Q = air volume delivered by the fan (m^3/s)
W_{fan} = power used by the fan (W)

Step 12: In the navigation tree, click **Building 1**. Perform an annual simulation, and record the results.

Untitled, Building 1

Analysis | Summary | Parametric | Optimisation + UA/SA | Data Visualisation

Date/Time	Room Electricity (kWh)	Lighting (kWh)	System Fans (kWh)	Cooling (Electricity) (kWh)
12:00:00 AM	51112.08	73868.04	24506.64	93067.18

TABLE 7.2
Variation of annual energy consumption with fan efficiency

End use	Annual energy consumption (kWh)		
	0.60	**0.70**	**0.80**
Room electricity	51,112.08	51,112.08	51,112.08
Lighting	73,868.04	73,868.04	73,868.04
System fans	24,506.64	21,005.69	18,379.98
Cooling (electricity)	93,067.18	92,085.14	91,348.52

Step 13: Repeat the preceding steps for 0.700 and 0.800 fan efficiency. Compare the results (Table 7.2). It can be seen that in addition to the reduction in energy consumption under system fans, there is a decrease in cooling (electricity) with the increase in fan efficiency. This is due to the fact that with a high-efficiency fan, less heat is added to the air while it passes over the fan motor, thus requiring lesser cooling effect to be delivered by the system. Save the simulation model with 0.7 fan efficiency to use in the next tutorial.

EXERCISE 7.2

Repeat this tutorial for fan efficiency varying from 0.3 to 0.9 with steps of 0.1 for the climate of Melbourne, Australia.

Tutorial 7.3

Evaluating the Impact of Fan Pressure Rise (Static Pressure)

GOAL

To evaluate the impact of fan pressure on building energy performance

WHAT ARE YOU GOING TO LEARN?

- Changing fan pressure rise

PROBLEM STATEMENT

In this tutorial, you are going to use the simulation model created in Tutorial 7.2 (50- × 25-m model with a 5-m perimeter depth). Model the unitary HVAC system with the following two fan pressures:

1. 500 Pa
2. 750 Pa

Find the change in energy consumption for both cases. Use the **Rio de Janeiro (AERO), Brazil** weather location.

SOLUTION

Step 1: Open the project file used in Tutorial 7.2. In the navigation tree, select **Air Loop Unitary HeatCool Supply Fan**. Click **Edit component**.

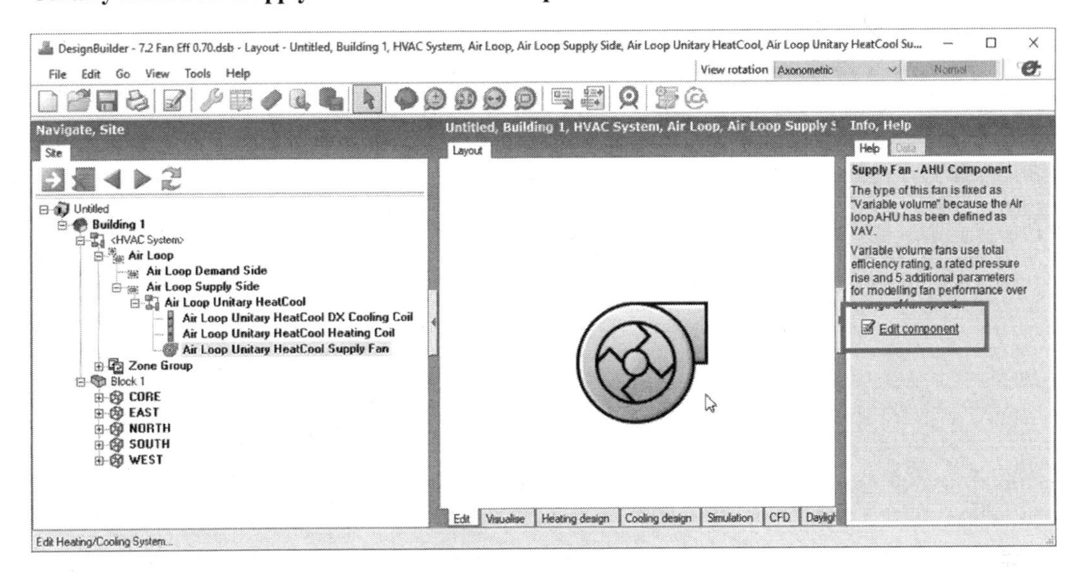

Step 2: Enter **500** for **Pressure rise (Pa)**. Click **OK**.

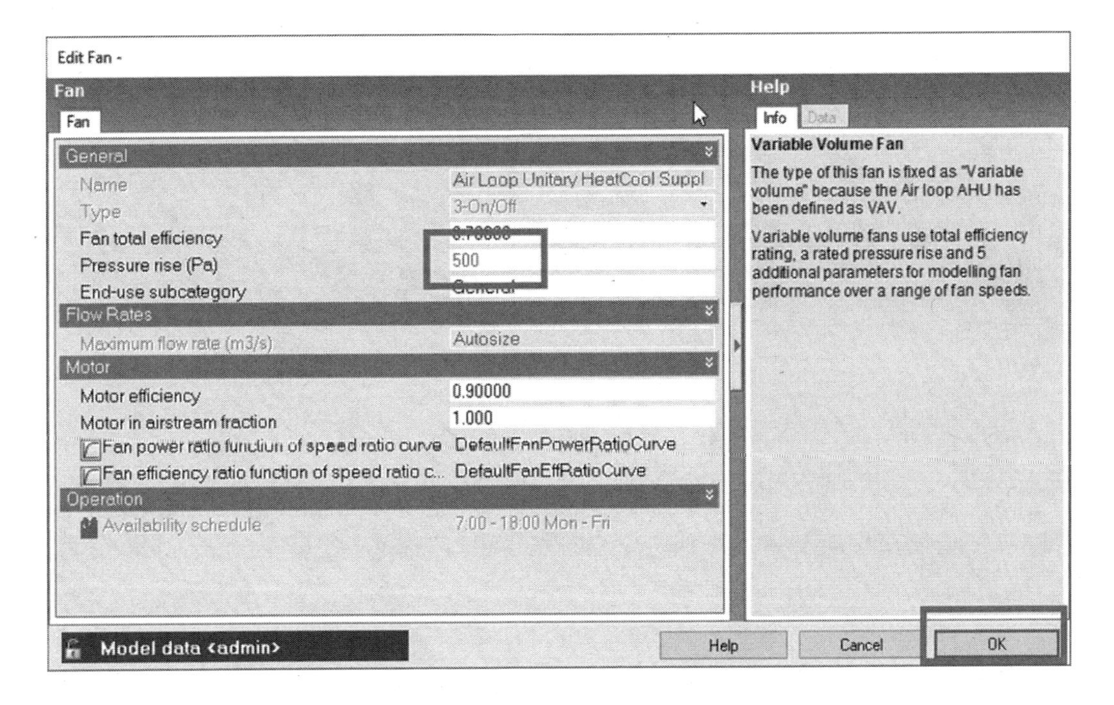

Fans provide energy to air that helps the air move through the ducts and other parts of the air side of an HVAC system, such as grills, diffusers, air filters, humidifiers and dampers. All these components impose resistance to the flow of air. To overcome such resistance, an increase in pressure is required.

As seen in Equation (7.3), fan power is proportional to pressure rise:

- A typical fan running at a fixed speed can provide a greater volumetric flow rate for systems with smaller total pressure drops (if we are to the right of the peak in the fan curve).
- Static pressure is used to overcome the pressure drop due to various ventilation system components on the airflow path within a given system. For mechanical ventilation systems, the fans create positive static pressure to move air through the system.
- The positive static pressure created by the fan equals the negative static pressure created by resistance as air navigates obstacles in the ventilation path plus the head required to impart sufficient kinetic energy to the air so as to diffuse it into the space.

Proper air filtration results in better conditions and air quality for occupants of the building and also increases the longevity of the HVAC system. However, because of the high negative pressure or resistance to flow with filtration devices, their use results in a rise in the static pressure requirement of the fan for maintaining the same flow rate.

Step 3: In the navigation tree, click **Building 1**. Perform an annual simulation, and record the results.

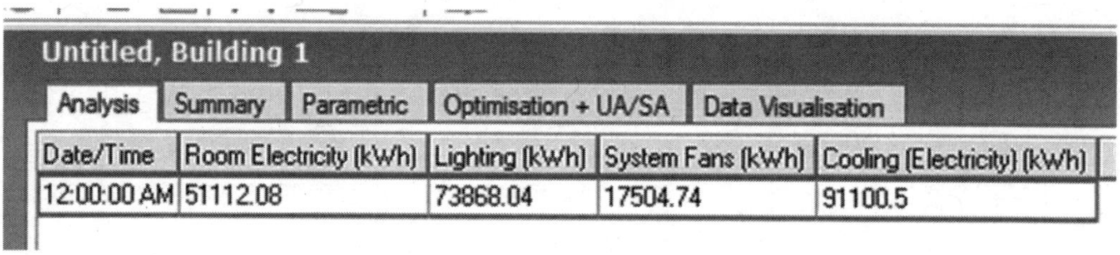

Untitled, Building 1

Analysis | Summary | Parametric | Optimisation + UA/SA | Data Visualisation

Date/Time	Room Electricity (kWh)	Lighting (kWh)	System Fans (kWh)	Cooling (Electricity) (kWh)
12:00:00 AM	51112.08	73868.04	17504.74	91100.5

Step 4: Repeat the preceding steps for a fan pressure of **750 Pa**. Compare the results (Table 7.3). From the results, you can see that there is an increase in fan energy consumption due to the rise in fan pressure. Also, you can see that there is an increase in cooling energy consumption with the increase in pressure rise due to the addition of more heat to the air passing over a fan motor of higher power rating.

TABLE 7.3
Variations in annual energy consumption with a fan pressure rise

	Annual energy consumption (kWh)	
End-use category	**500 Pa**	**750 Pa**
Room electricity	51,112.08	51,112.08
Lighting	73,868.04	73,868.04
System fans	17,504.74	26,257.12
Cooling (electricity)	91,100.50	93,555.52

EXERCISE 7.3

In the tutorial, make **Motor** in **Airstream fraction** zero, and observe the effect on the cooling and fan energy consumption. Use the same **Rio de Janeiro (AERO), Brazil** weather file (Table 7.4).

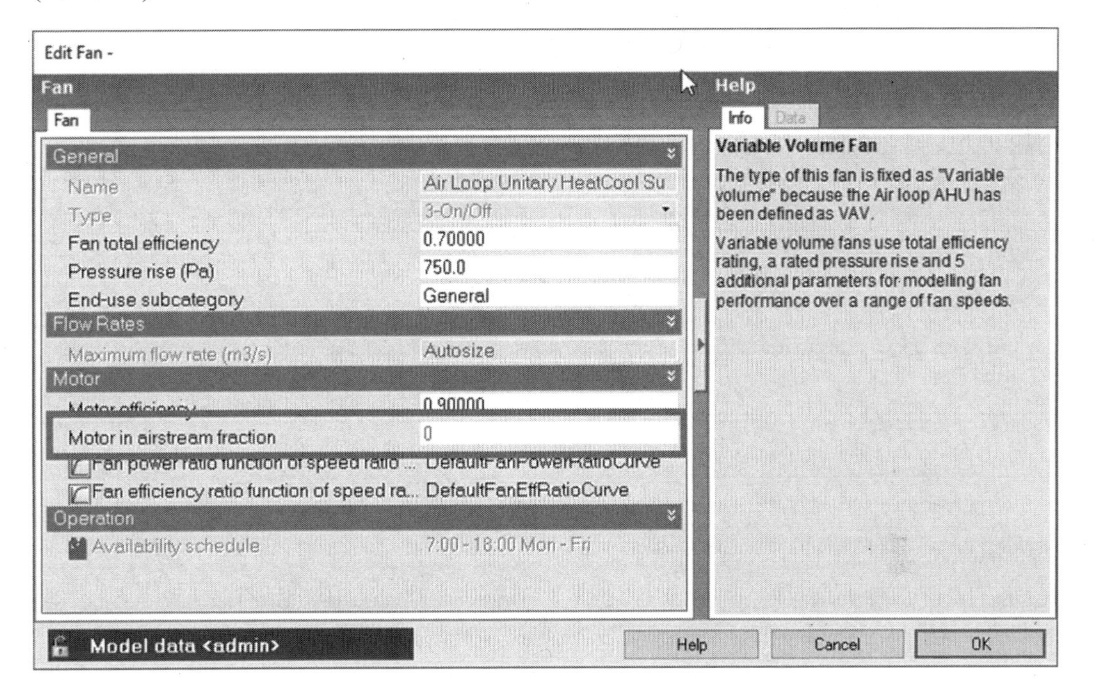

Motor in Airstream Fraction

A fraction of the motor heat is added to the airstream. A value of zero means that the motor is completely outside the airstream. A value of one means that all the motor heat loss will go into the airstream and act to cause a temperature rise. This value must be between zero and one.

Source: https://designbuilder.co.uk/helpv6.0/Content/Fan_-_Constant_Volume.htm.

TABLE 7.4
Variation in annual energy consumption with fan efficiency

	Annual energy consumption (kWh)	
End-use category	**500 Pa**	**750 Pa**
Room electricity		
Lighting		
System fans		
Cooling (electricity)		

Tutorial 7.4

Evaluating the Impact of Heat Pumps on Heating Energy Consumption

GOAL

To evaluate the impact of heat pumps on heating energy consumption

WHAT ARE YOU GOING TO LEARN?

- How to model a heat pump system

PROBLEM STATEMENT

In this tutorial, you are going to use 10- × 10-m model building. You need to model the following HVAC systems for heating:

1. Electrical resistance
2. Heat pump

Use the **New Delhi – Palam, India** weather location. Perform simulations for all winter months.

SOLUTION

Step 1: Open a new project, and create a **10- × 10-m** building.

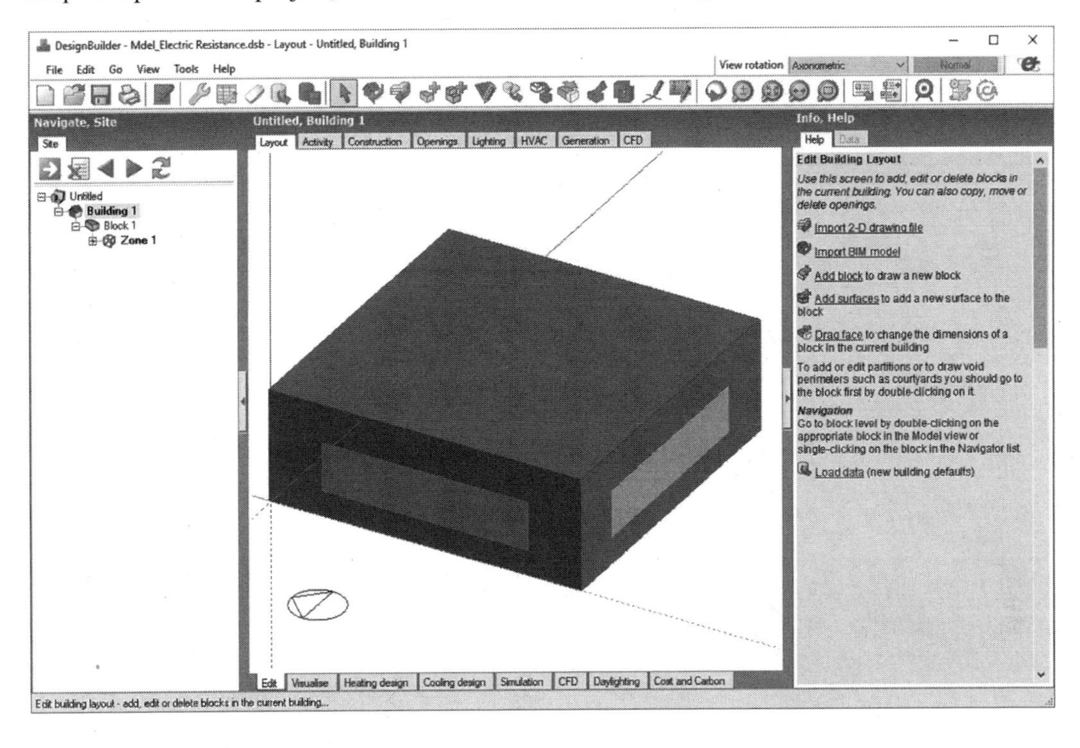

Step 2: Click the **Edit** menu, and select **Model Options**. Switch to **Detailed HVAC**. Click **<HVAC System>**. Click the **Load HVAC template** link. Click **Next**.

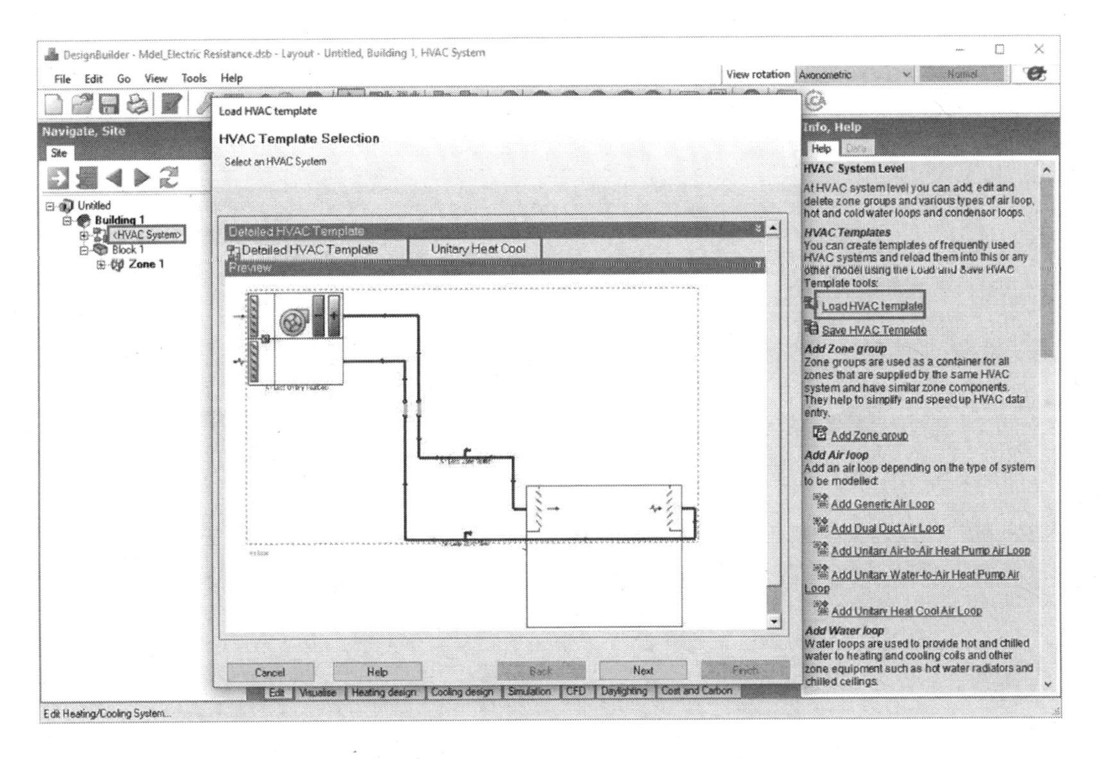

Step 3: Click the **<Select zone>** link.

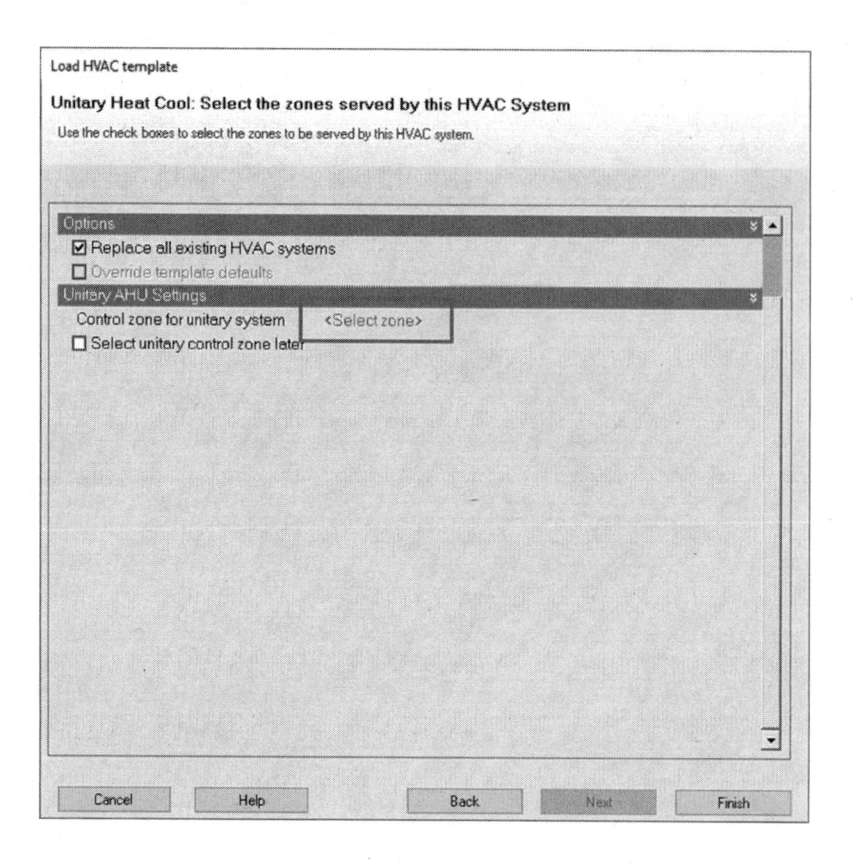

Step 4: Select **Block1:Zone1** for **Control zone for unitary system**.

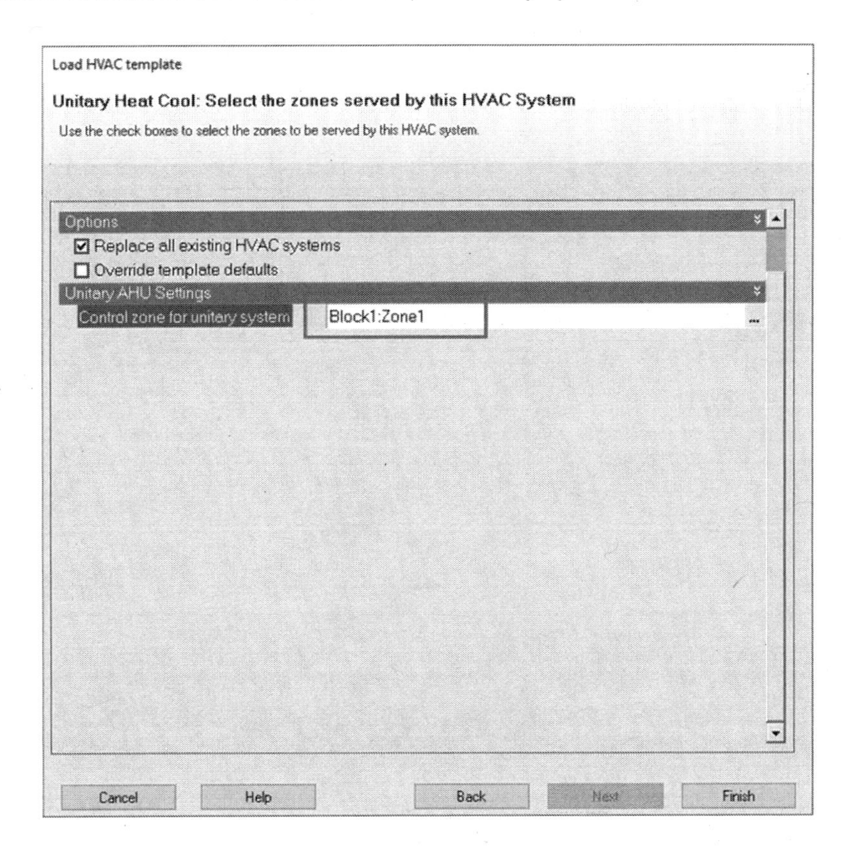

Step 5: Expand the **Air Loop**. Select **Air Loop Unitary HeatCool Heating Coil**. Click **Edit component** under the **Help** tab.

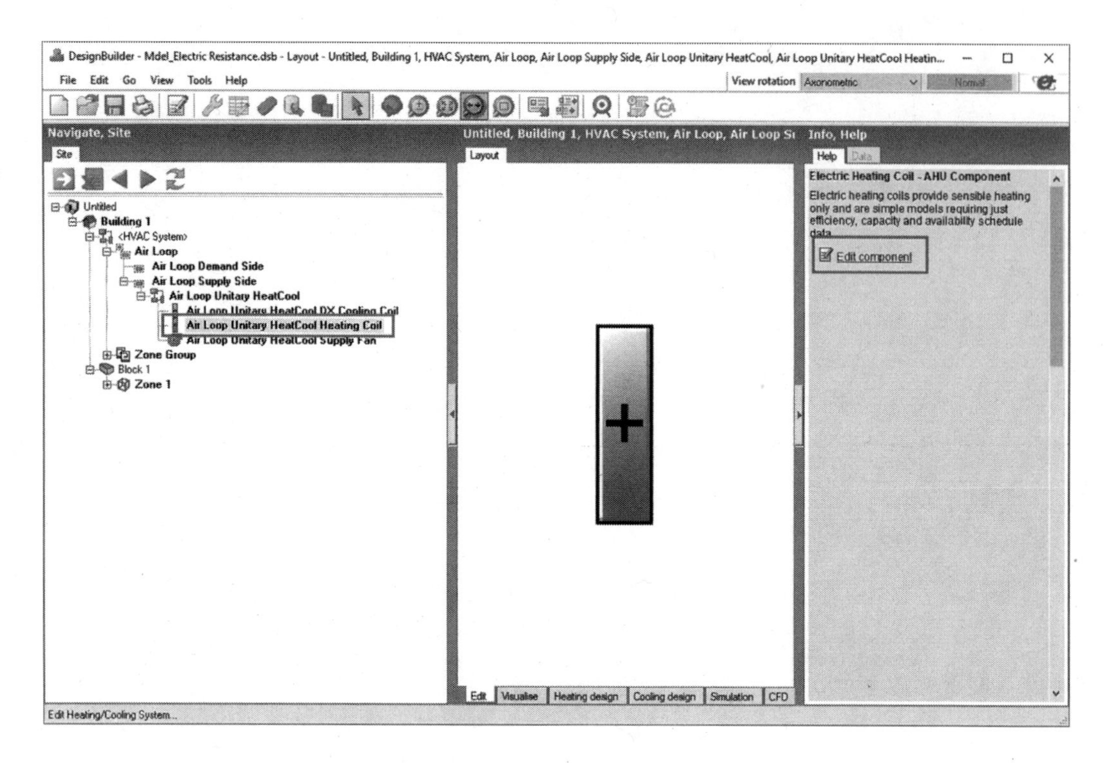

Step 6: You can see that **2-Electric** is selected in the **Type** drop-down list.

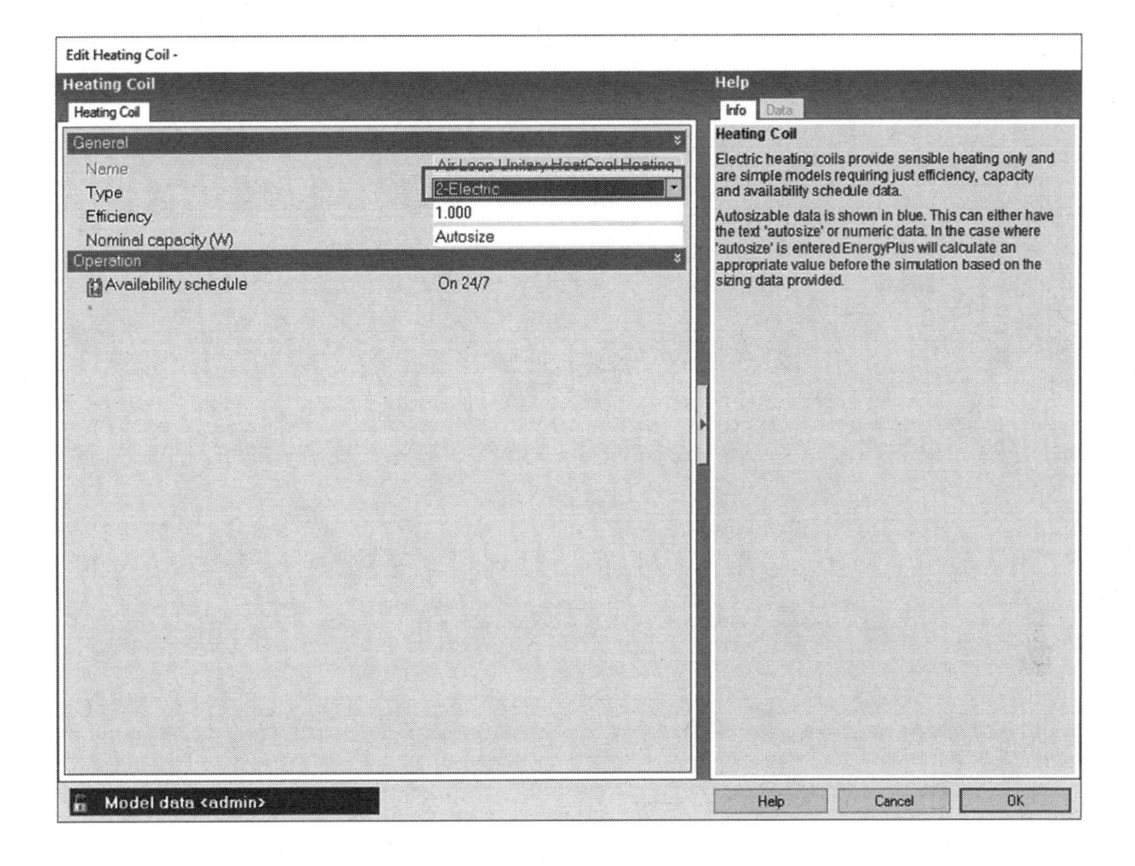

Efficiency of an Electric Heater

Electric resistance heating is 100% energy efficient in the sense that all the incoming electric energy is converted to heat. However, most electricity is produced from coal, gas or oil generators that convert only about 30% of the fuel's energy into electricity. Because of electricity generation and transmission losses, electric heat is often more expensive than heat produced in homes or businesses that use combustion appliances such as natural gas, propane and oil furnaces.

An air-source heat pump can provide efficient heating and cooling for your home. When installed properly, an air-source heat pump can deliver one-and-a-half to three times more heat energy to a home than the electrical energy it consumes. This is possible because a heat pump moves heat rather than converting it from a fuel like combustion heating systems do.

Source: www.energy.gov/energysaver/home-heating-systems/electric-resistance-heating and www.energy.gov/energysaver/heat-pump-systems/air-source-heat-pumps.

Step 7: Click the **Activity** tab. Select **24 × 7 Generic Office Area**.

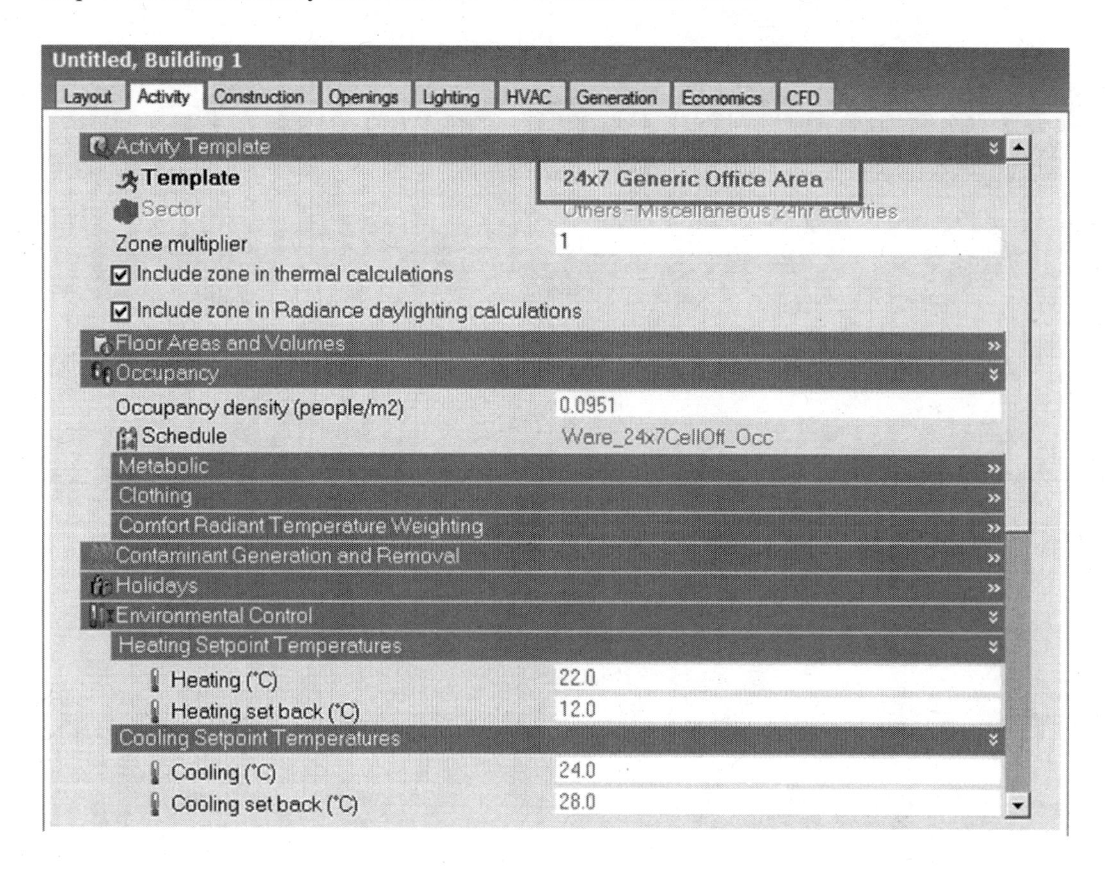

Step 8: Enter **15** in the **Fresh air (l/s-person)** text box and **0.120** in the **Mech vent per area (l/s-m²)** text box. Enter **7** in the **Power density (W/m²)** text box.

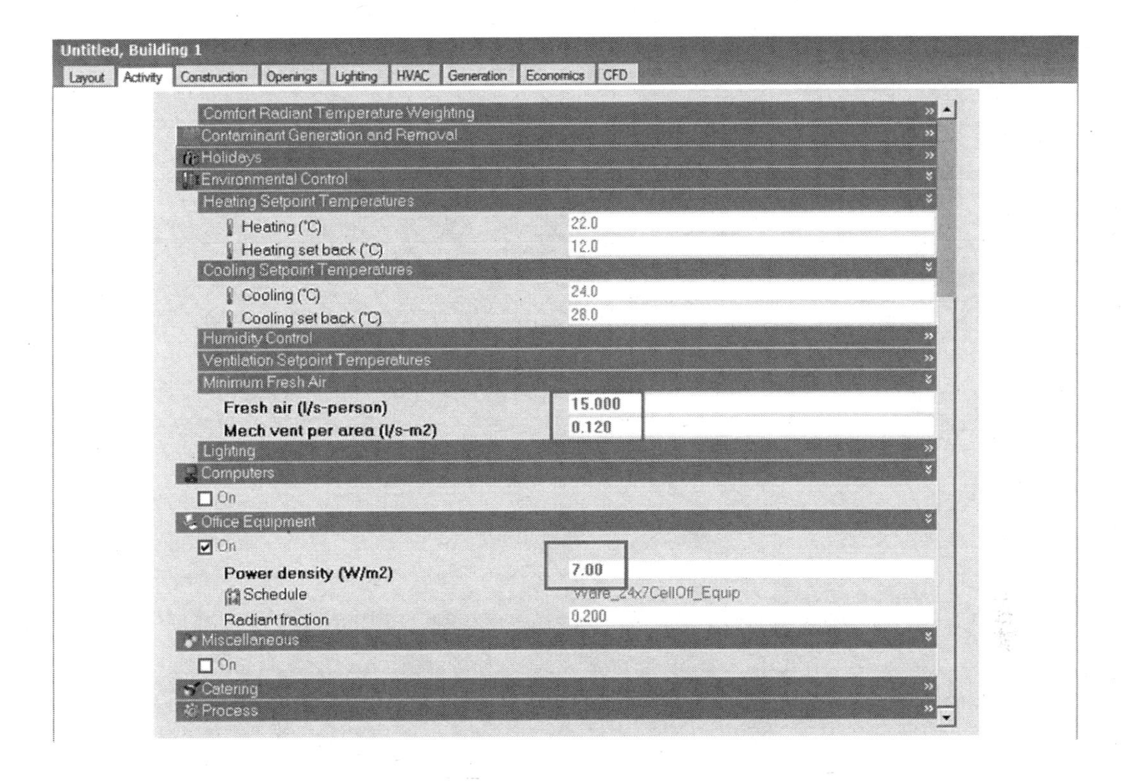

Step 9: Click the **Lighting** tab. Enter **6.0** in the **Power density (W/m²)** text box.

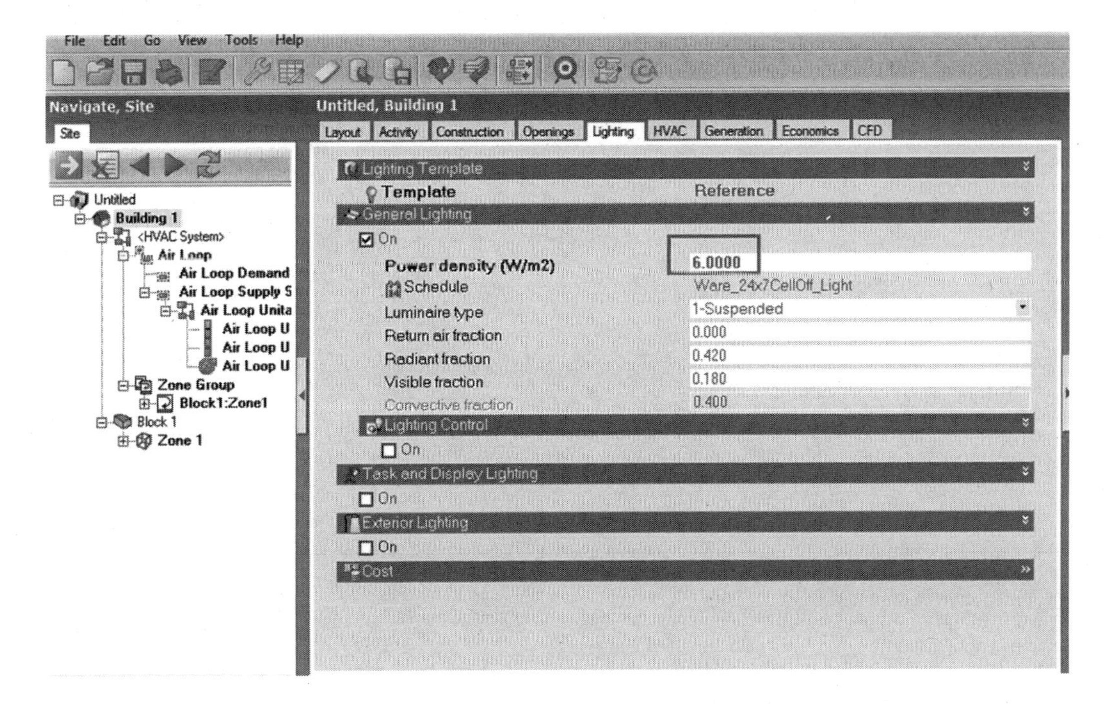

Step 10: Click the **Construction** tab. Select **Brick cavity with desnse plaster** as external wall and **100 concrete slab** as the flat roof.

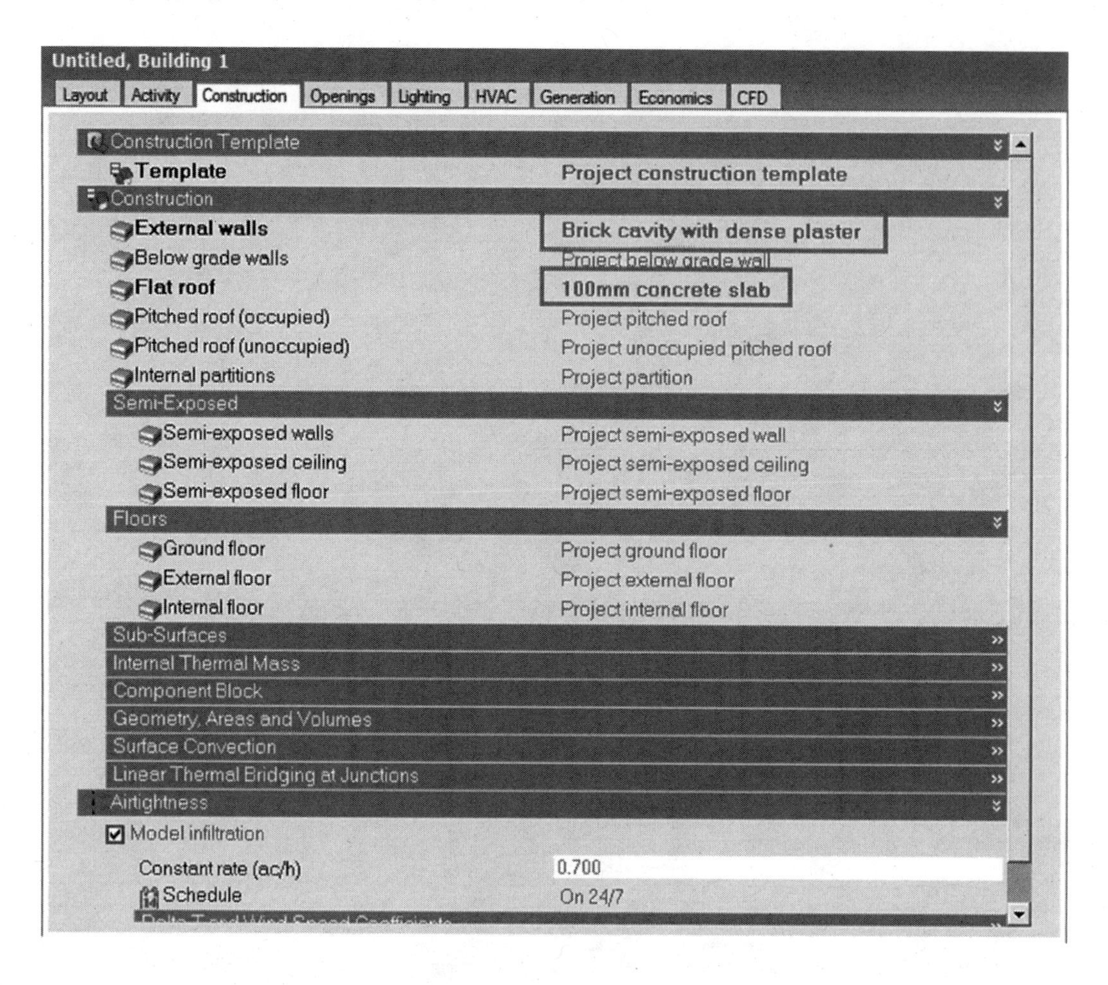

Step 11: Simulate the model for all winter months.

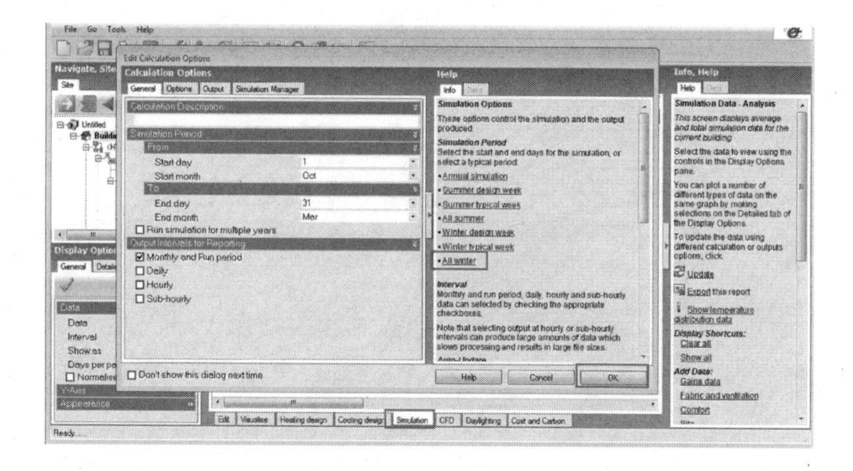

Step 12: Record the results.

End use annual energy consuption with electric resistance heating				
Room electricity (kWh)	**Lighting (kWh)**	**System fans (kWh)**	**Heating (electricity) (kWh)**	**Cooling (electricity) (kWh)**
1,909.26	2,244.36	2,124.80	285.18	2,186.29

Step 13: Repeat Step 12 to select **Unitary Air to Air Heat Pump**.

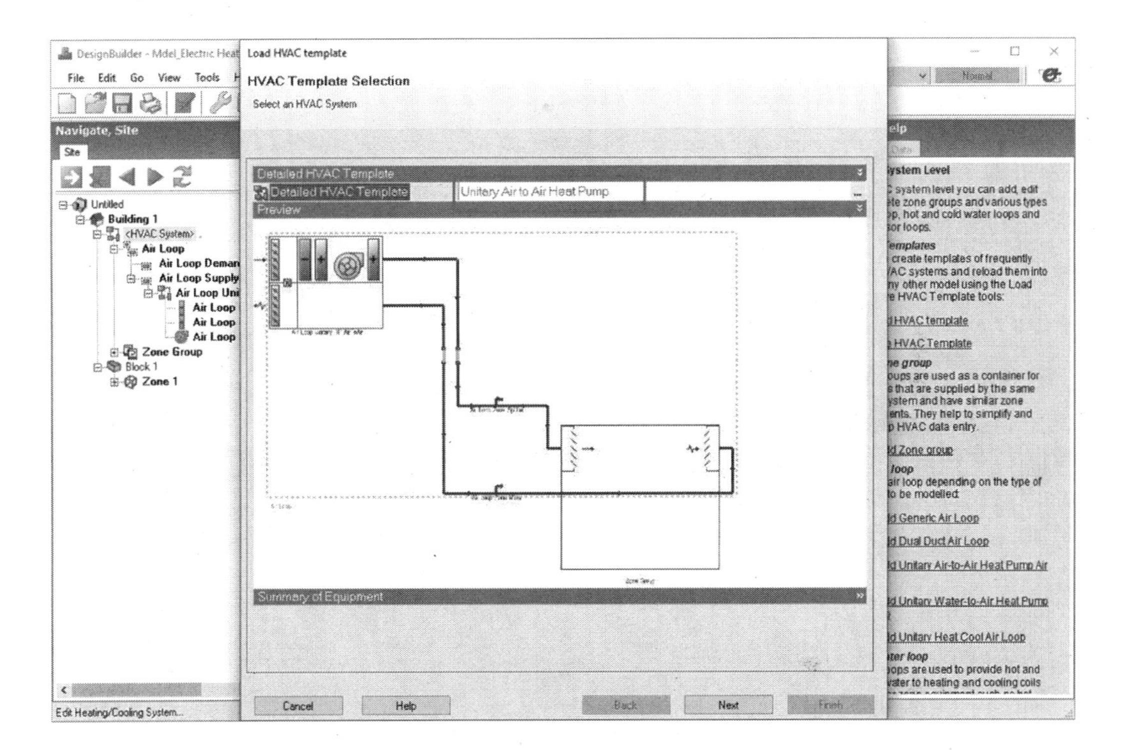

Step 14: Select **Block1:Zone1** as **the Control zone for unitary system.** Click **Finish.**

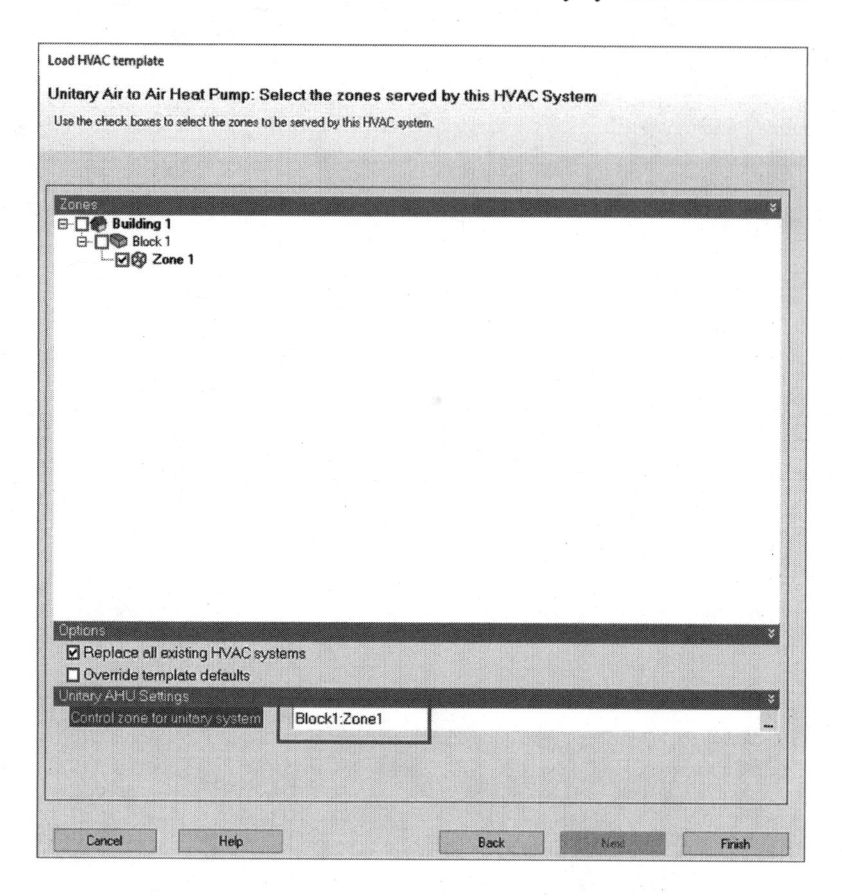

Step 15: Expand the **Air Loop**. Select **Air Loop Unitary HP AirToAir Heating Coil**.

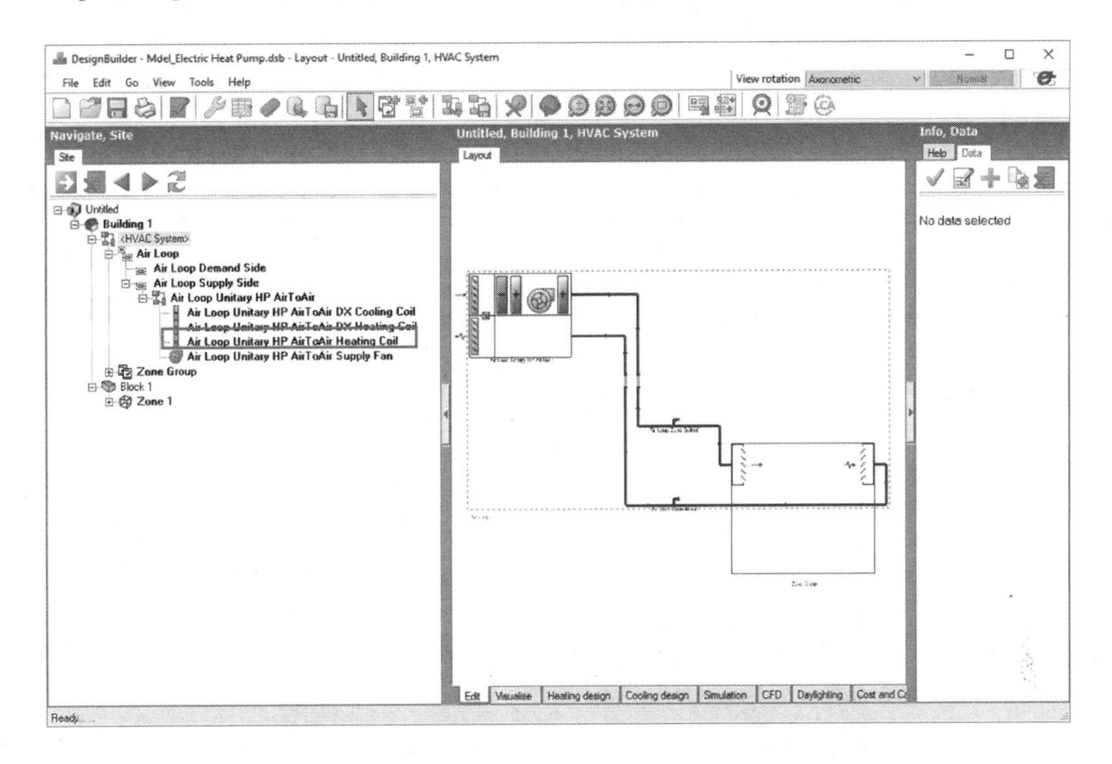

The *unitary air-to-air heat pump* (AHU) incorporates an EnergyPlus unitary air-to-air heat pump component. The unitary air-to-air heat pump component itself comprises a supply fan, a DX cooling coil component, a DX heating coil component, and an electric, hot-water or fuel supplementary heating coil component.

Source: https://designbuilder.co.uk/helpv6.0/Content/Unitary_Heat_Pump_AHU.htm.

Step 16: Click **Edit component** in the **Help** tab. Enter **3.0** in the **Gross rated COP** text box.

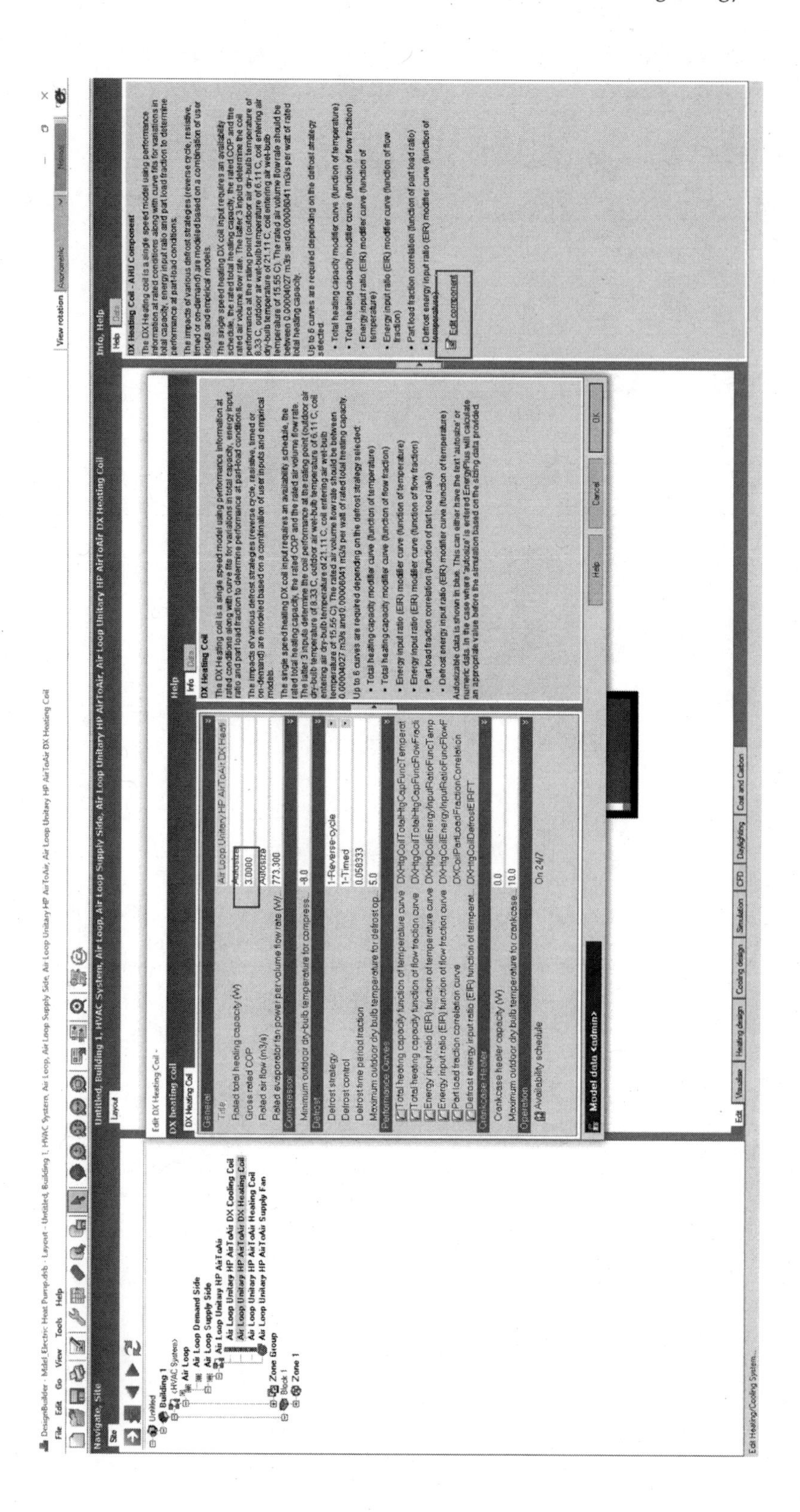
Untitled, Building 1, HVAC System, Air Loop, Air Loop Supply Side, Air Loop Unitary HP AirToAir, Air Loop Unitary HP AirToAir DX Heating Coil
Navigate, Site
Edit DX Heating Coil -
DX heating coil
General
Title
Air Loop Unitary HP AirToAir DX Heati
Rated total heating capacity (W)
Autosize
Gross rated COP
3.0000
Rated air flow (m3/s)
Autosize
Rated evaporator fan power per volume flow rate (W/
773.300
Compressor
Minimum outdoor dry-bulb temperature for compress
-8.0
Defrost
Defrost strategy
1-Reverse-cycle
Defrost control
1-Timed
Defrost time period fraction
0.058333
Maximum outdoor dry-bulb temperature for defrost op
5.0
Performance Curves
Crankcase Heater
Crankcase heater capacity (W)
0.0
Maximum outdoor dry-bulb temperature for crankcase
10.0
Operation
Availability schedule
On 24/7
Help
DX Heating Coil
Info, Help
DX Heating Coil - AHU Component
Edit component
Help
Cancel
OK
Model data <admin>

Step 17: Simulate the model, and record the results.

End use annual energy consuption with electric heat pump				
Room electricity (kWh)	Lighting (kWh)	System fans (kWh)	Heating (electricity) (kWh)	Cooling (electricity) (kWh)
1,909.26	2,244.36	2,124.80	105.15	2,216.42

Step 18: Compare the results.

TABLE 7.5
Comparison of end-use energy consumption

Room electricity (kWh)	Lighting (kWh)	System fans (kWh)	Heating (electricity) (kWh)	Cooling (electricity) (KWh)
1,909.26	2,244.36	2,124.80	285.18	2,186.29
1,909.26	2,244.36	2,124.80	105.15	2,216.42
0.00%	0.00%	0.00%	63.13%	−1.38%

If electricity is the only choice, heat pumps are preferable in most climates, as they easily reduce electricity use when compared with electric resistance heaters. Table 7.5 shows the energy savings by use of heat pumps over electric resistance heaters.

8 Heating, Ventilation and Air Conditioning

Central Water Side

Energy consumption of a centralized heating, ventilating and air-conditioning (HVAC) system depends on the selection of individual components and their integration in the entire system. Major variations include the type of condenser for heat rejection, chilled-water pumping scheme and efficiency of the chiller and boiler. This chapter, through its tutorials, explains the method of modelling central HVAC systems. Different variations of systems are explained, such as types of chillers, cooling towers and the use of variable-speed drives in HVAC components. The impact of each variation on energy consumption is also explained.

Tutorial 8.1

Evaluating the Impact of Air- and Water-Cooled Chillers

GOAL

To evaluate the impact of air- and water-cooled chillers on building energy consumption

WHAT ARE YOU GOING TO LEARN?

- How to model a central HVAC system
- How to model an air-cooled chiller
- How to model a water-cooled chiller

PROBLEM STATEMENT

In this tutorial, you are going to use a G + 5–floor building model. Each floor has a 50- × 25-m area and five zones with a 5-m perimeter depth. You need to make use of the floor/zone multiplier option to model the building. Model HVAC systems having variable air volume (VAV) with reheat.

Use the following two options for a chiller in the HVAC system:

1. Air cooled
2. Water cooled

Compare the energy consumption of the two chillers for the **AZ-PHOENIX/SKY HARBOR, USA** weather location. Phoenix has long, very hot summers and short, mild winters. The climate is arid, with plenty of sunshine and clear skies.

SOLUTION

Step 1: Open a new project, and create a building with a ground floor having a **50- × 25-m** area and five zones with a **5-m** perimeter depth.

Step 2: Copy the ground floor, and create two other floors. (You can refer to Tutorial 6.1 to copy/clone floors.)

Step 3: Select **Plan** from **View rotation** drop-down list. Click the **Add block** icon.

Before drafting, select **Block type** as **Component block** and **Adiabatic** in the **Component block type**, and enter **10.50** in **Height (m)**.

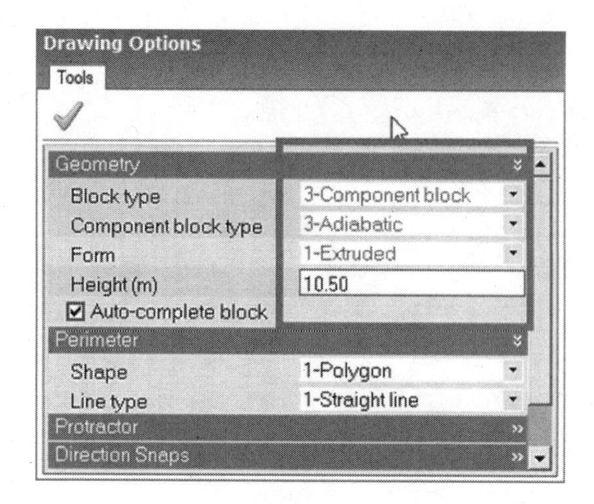

Go to the **Axonometric** view, click in the corner and make sure that the roof is selected when you click the point.

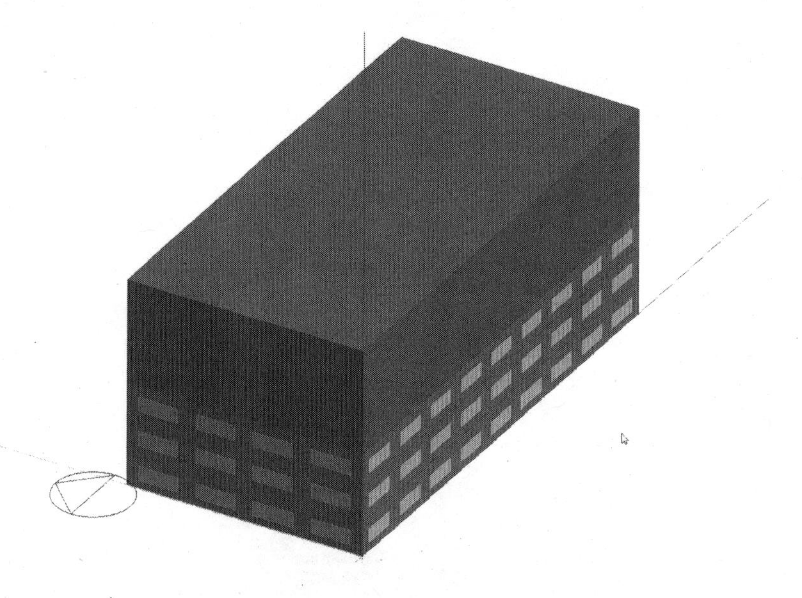

With the help of the **Move Selected Object** icon, the roof can be adjusted.

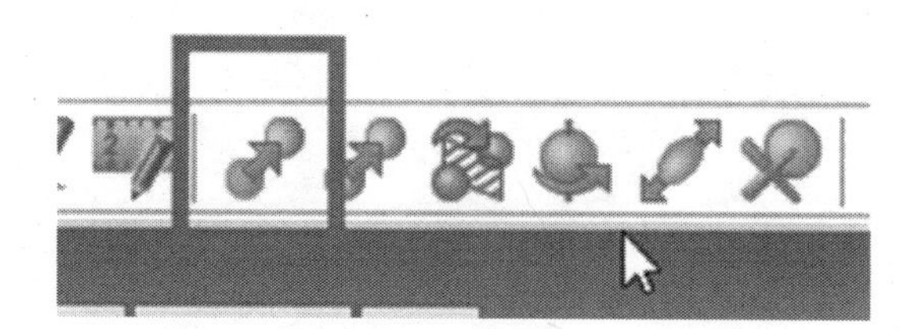

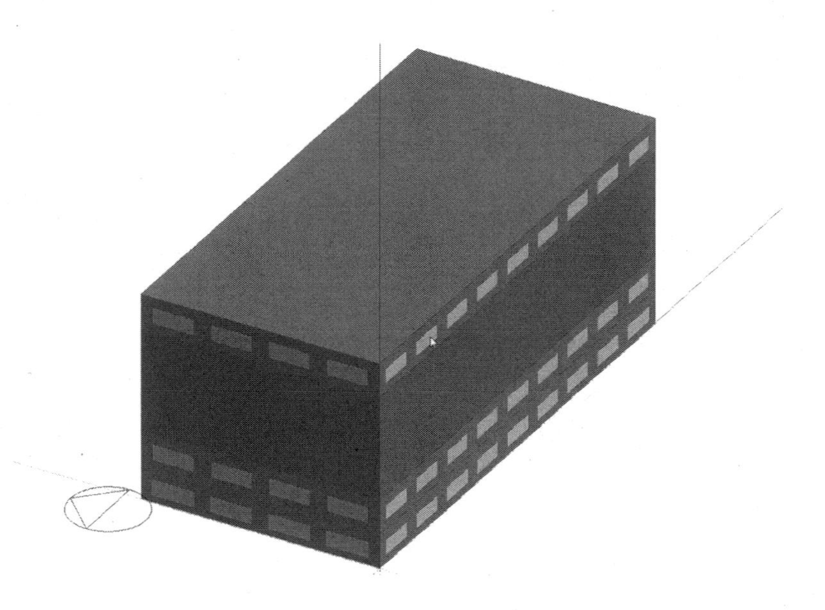

Step 4: Click **MIDDLE floor**. Click **the Activity** tab. Enter **4** in **Zone multiplier** text box. (You can also refer to Chapter 2 for floor/zone multiplier.)

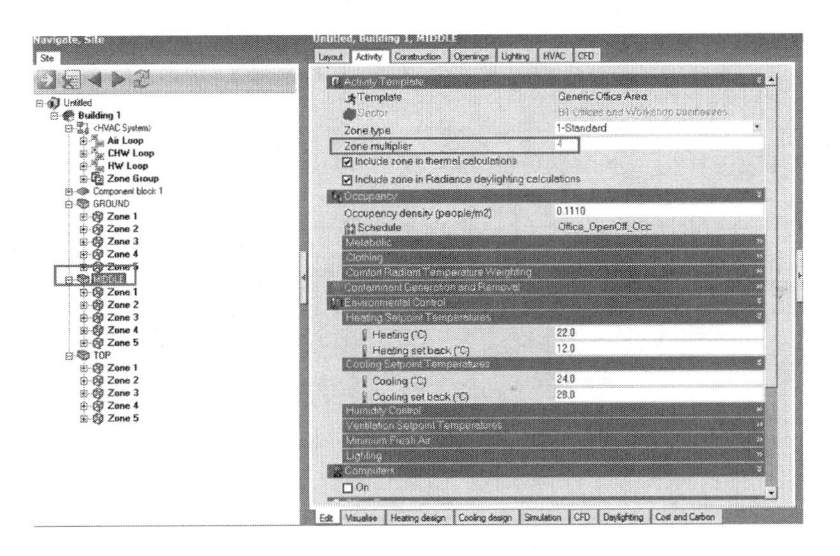

Step 5: Select the **HVAC** tab. Click **Simple link** under **Help** tab. The **Model Options – Building and Block** screen appears.

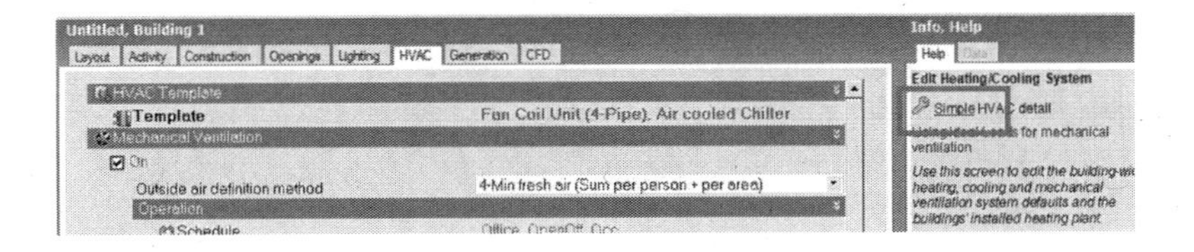

Step 6: Click **Detailed** under the **HVAC** slider. Select **Detailed HVAC Data** from the **Detailed HVAC** drop-down list. Click **OK**. It displays the **<HVAC system>** option under **Building 1**.

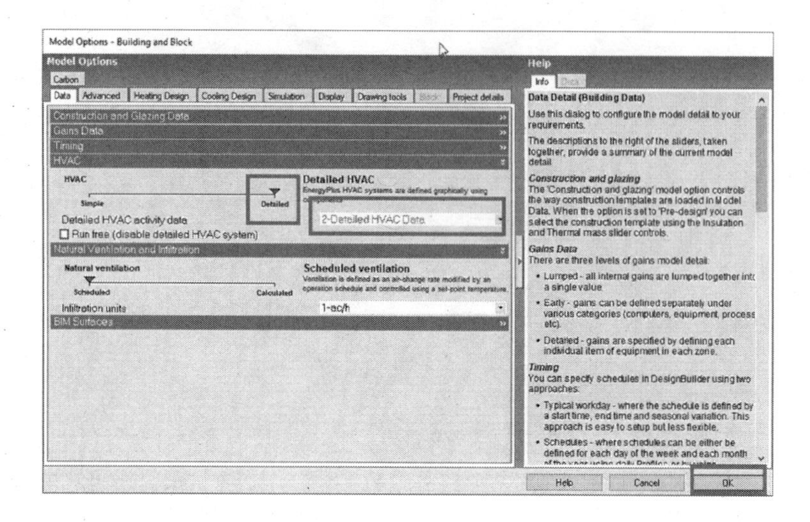

The HVAC system is modelled in detail using EnergyPlus air- and water-side components linked together on a schematic layout drawing. In this case, HVAC data are accessed by clicking on the **<HVAC System** > navigator node.

Step 7: Click **<HVAC System>**. It displays the initializing HVAC progress bar, and subsequently, the **Load HVAC template** screen appears.

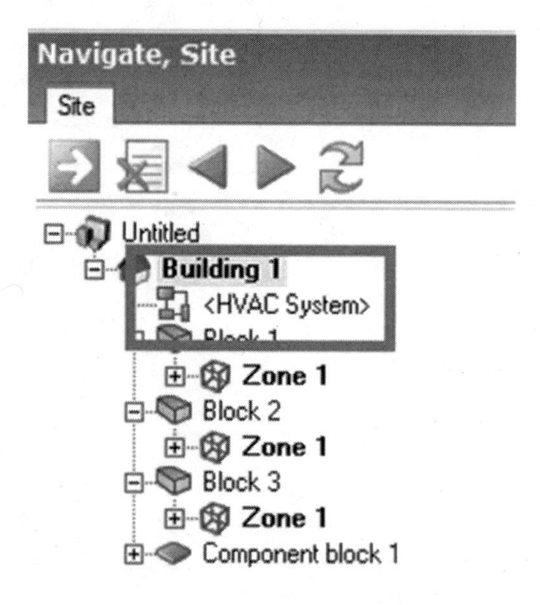

Step 8: Select **VAV Reheat, Air-cooled Chiller** from the **Detailed HVAC** template. Click **Next.**

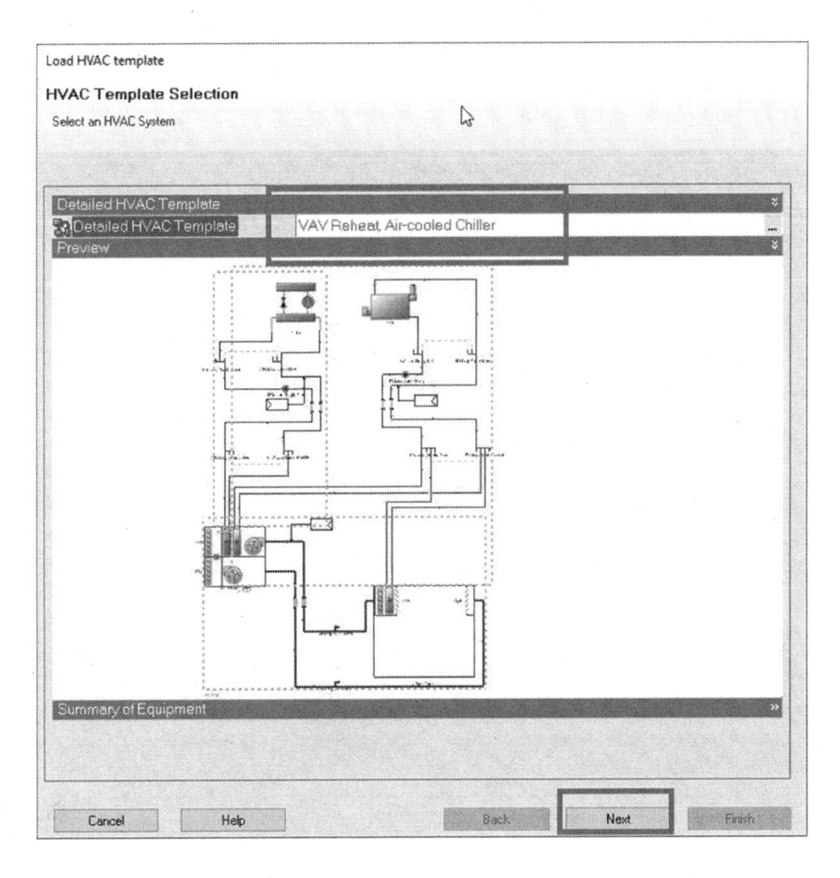

In medium to large air-conditioning systems, chilled water from the central plant is used to cool the air at the coils in an air-handling unit (AHU). Based on the type of condenser cooling, chillers are of two types: air cooled and water cooled.

- An air-cooled chiller has a condenser that is cooled by ambient air. Air-cooled chillers are preferred for small or medium-sized installations and are preferred in cases where there is not enough water.
- A water-cooled chiller has a condenser connected with a cooling tower. Cold water is obtained through partial evaporation of water via the ambient air that is used to facilitate heat rejection from the condenser. The use of a cooling tower to cool the condenser increases the efficiency of water-cooled chillers over air-cooled systems in which ambient air cools the condenser.

Factor	Air cooled	Water cooled
Efficiency	Less	High
Cost	Less	High
Maintenance	Less	High

Step 9: Select all the checkboxes from the **Zones** section. Click **Finish**. This displays the system layout.

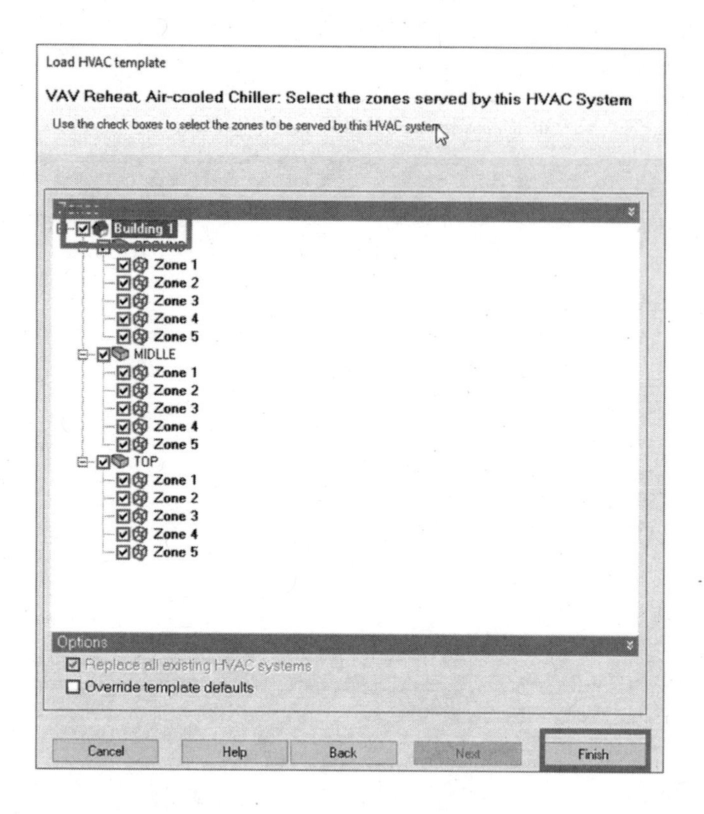

Step 10: In the navigation tree, expand the **CHW Loop**.

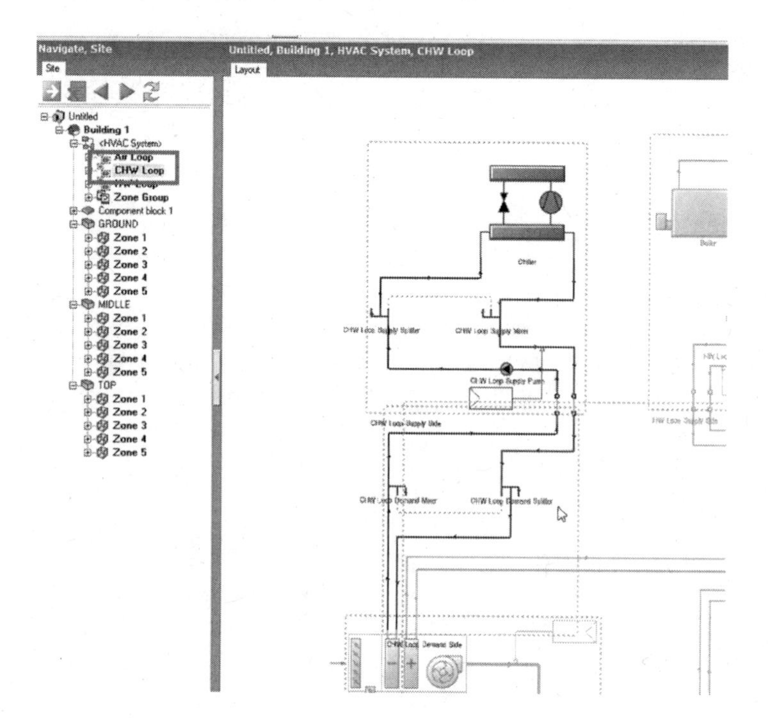

Step 11: Expand the **CHW Loop Supply Side**, and click **Chiller**. The **Chiller** layout appears. Click **Edit component** under the **Help** tab. The **Chiller** data appear.

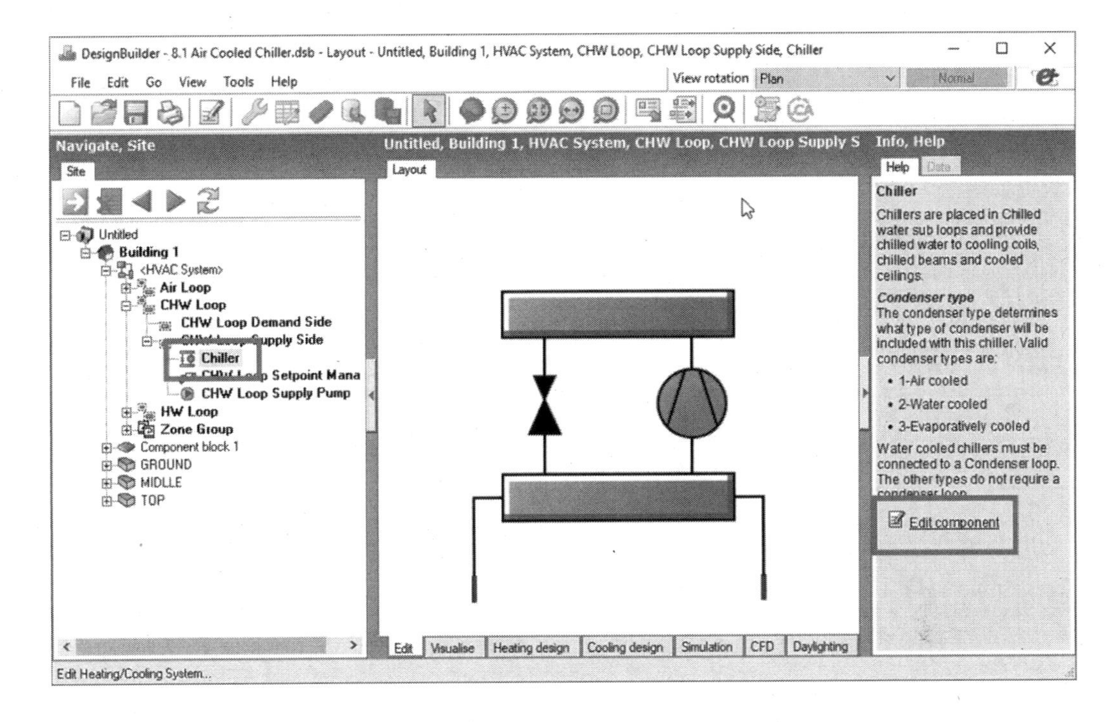

Step 12: Make sure that **Air Cooled Default** is selected for the **Chiller template**.

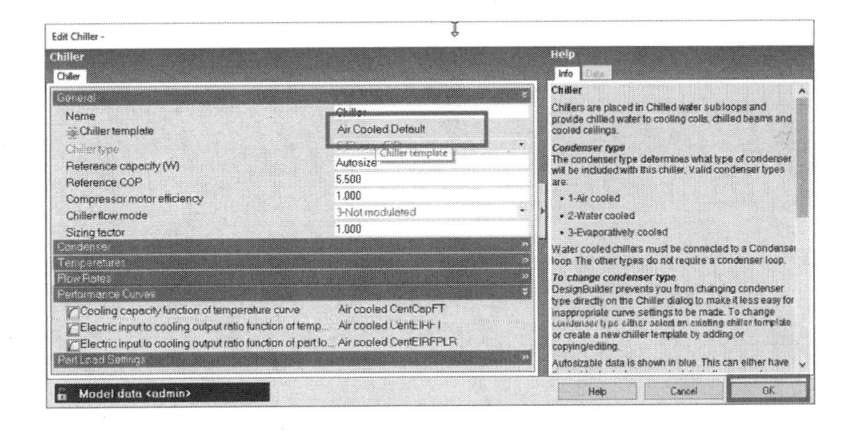

Step 13: Simulate the model, and record annual fuel breakdown results.

Step 14: Select the **Edit** tab. In the navigation tree, click **HVAC System**, and select **Load HVAC template**.

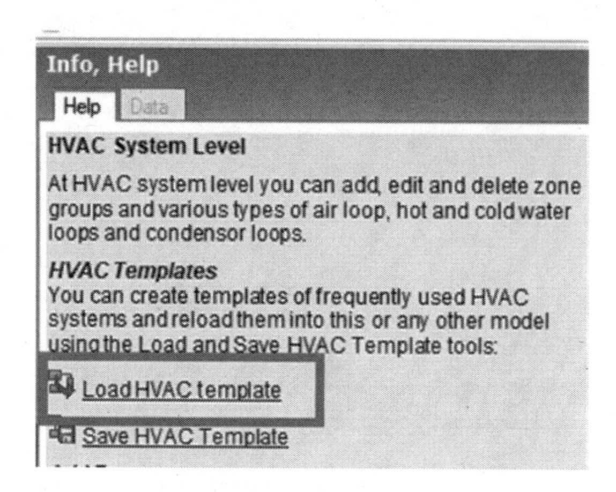

Step 15: Repeat the preceding steps to select **VAV Reheat, Water-cooled Chiller** from **Select the Detailed HVAC template.**

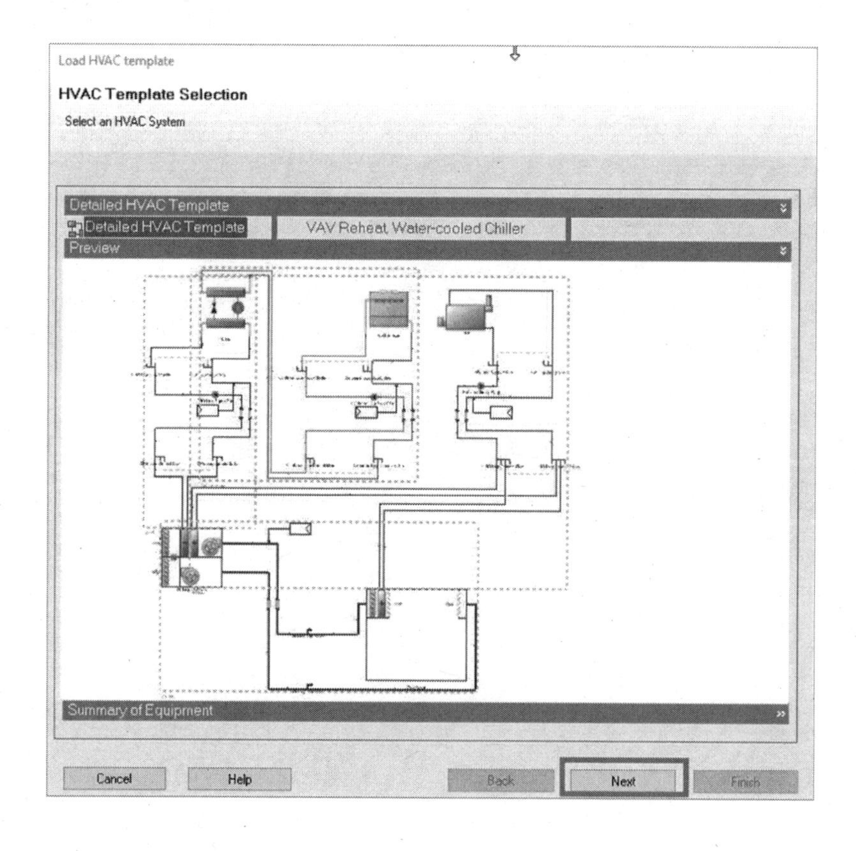

Step 16: Select **DOE-2 Centrifugal/5.50COP** in the **Chiller template** field.

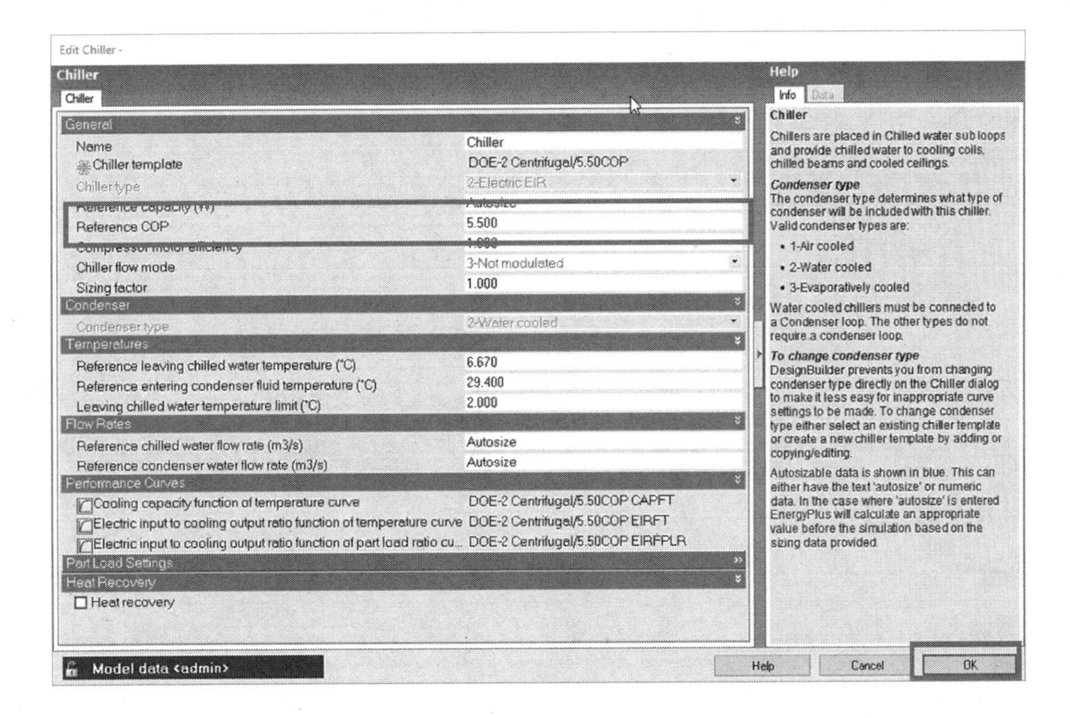

Step 17: Perform an annual energy simulation, and compare the results for both cases (Table 8.1).

Air-cooled chillers do not have a cooling tower for condenser cooling. In DesignBuilder, a heat-rejection term is used for reporting cooling tower energy consumption. Hence, for the air-cooled chiller in the end-use category **Heat Rejection**, there is no value.

You can also see that the cooling energy consumption with the water-cooled chiller is less than with the air-cooled chiller. The use of water instead of air increases the efficiency of heat transfer from the condenser, thereby increasing the efficiency of the chiller. However, because of the addition of a water loop, a circulation pump called the *condenser water pump* is required, which increases the energy consumption with system pumps. Save the simulation model with the water-cooled chiller to use in subsequent tutorials.

TABLE 8.1
Energy consumption for air- and water-cooled chillers

End-use category	Annual energy consumption (kWh)	
	Air-cooled chiller	Water-cooled chiller
Room electricity	306,674.40	306,674.40
Lighting	443,211.10	443,211.10
System fans	335,040.80	335,040.80
System pumps	2538.20	17,309.70
Heating (gas)	199,962.60	199,962.60
Cooling (electricity)	992,575.40	510,120.10
Heat rejection	Not applicable	226,757.1

Tutorial 8.2

Evaluating the Impact of a Variable-Speed Drive (VSD) on a Chiller

GOAL

To evaluate the impact of a VSD on a centrifugal chiller on building energy consumption

WHAT ARE YOU GOING TO LEARN?

- How to model a chiller with a VSD

PROBLEM STATEMENT

In this tutorial, you are going to use the water-cooled chiller model saved in Tutorial 8.1 (a 50- × 25-m six-floor model with a 5-m perimeter depth). You need to select the following two chillers and find out energy consumption in both cases:

1. **EIRchiller Centrifugal Carrier 19XR 1213kW/7.78COP/Vanes**
2. **EIRchiller Centrifugal Carrier 19XR 1143kW/6.57COP/VSD**

Use the AZ-PHOENIX/SKY HARBOR, USA weather location.

SOLUTION

Step 1: Open the simulation model saved in Tutorial 8.1. Click **<HVAC System>** in the navigation pane.

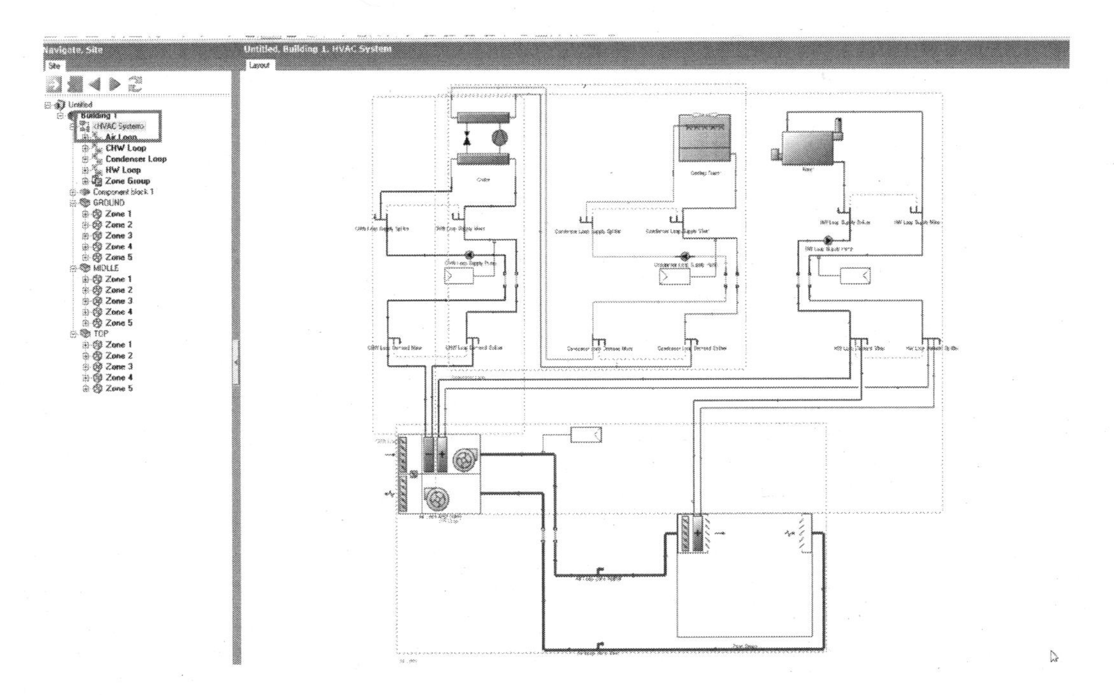

Step 2: Click **CHW Loop**. The components under the **CHW loop** appear.

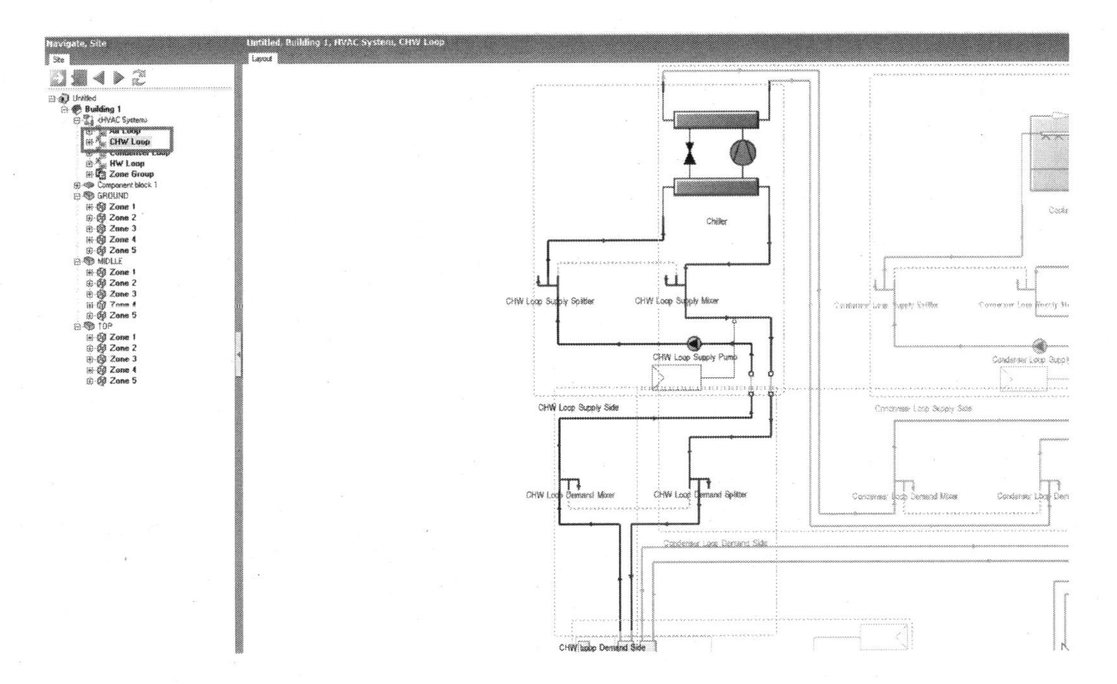

Step 3: Click **CHW Loop Supply Side.**

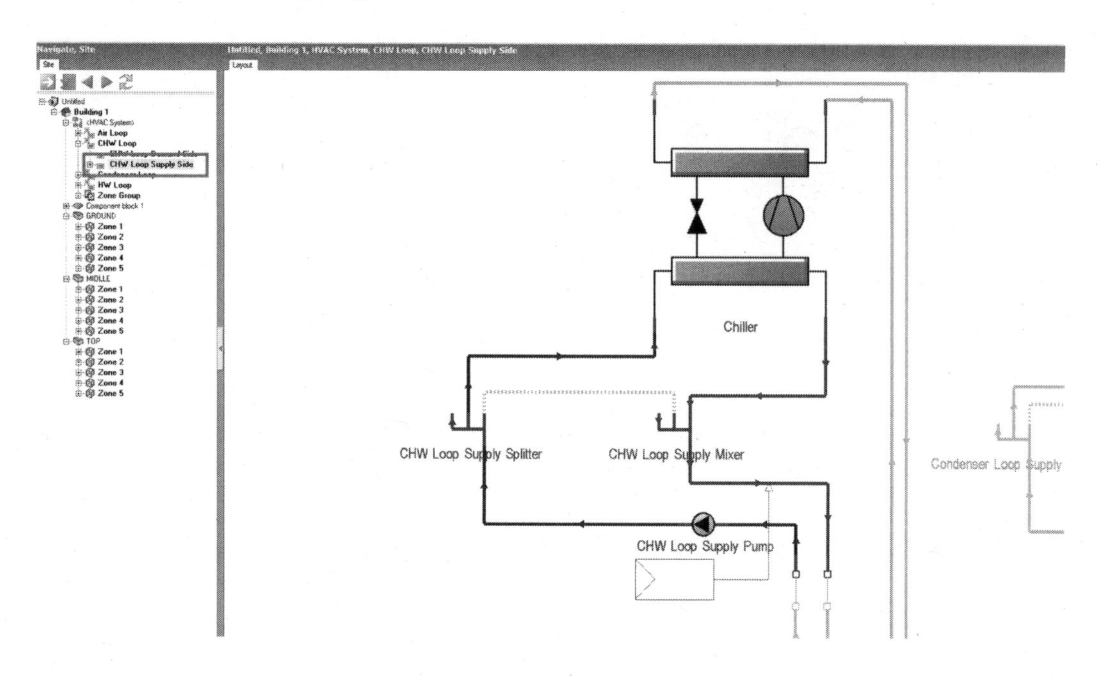

Step 4: Click **Chiller**. The chiller layout appears on the **Layout** tab. Click **Edit component** under the **Help** tab. The **Edit Chiller** screen appears.

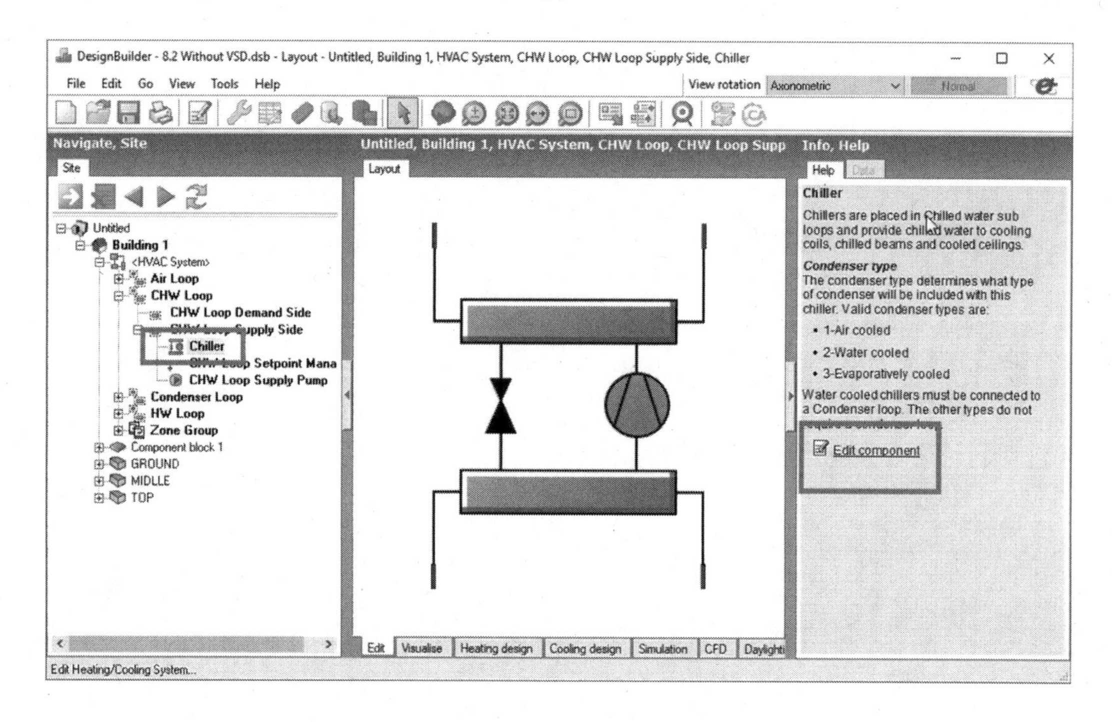

Step 5: Click **DOE-2 Centrifugal/5.5COP**. Three dots appear. Click the three dots. The **Select the Chiller** screen appears.

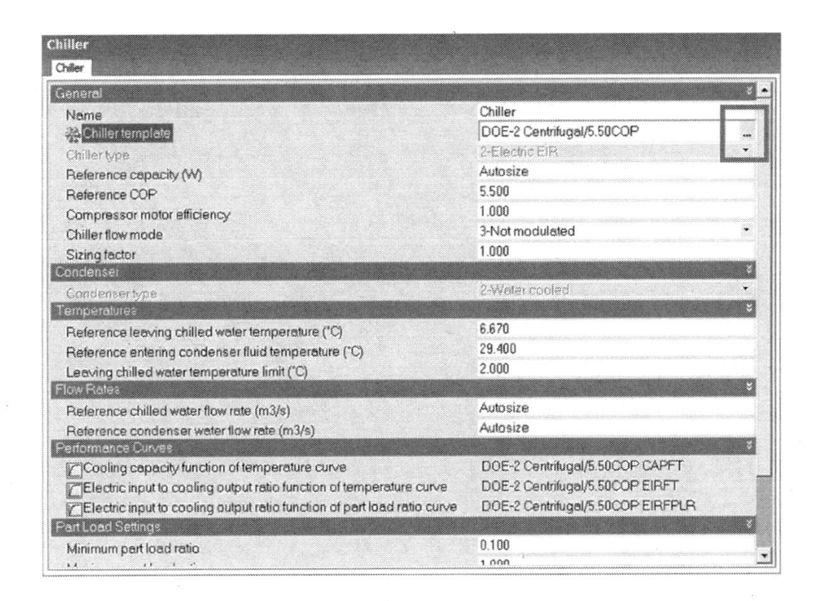

Step 6: Click **ElectricEIRChiller Centrifugal Carrier 19XR 1213kW/7.78COP/Vanes.**

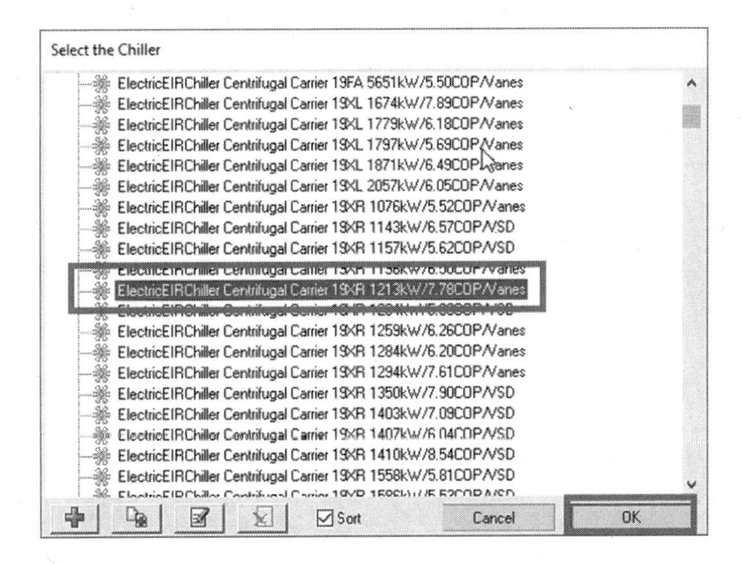

Step 7: Click Electric input to cooling output ratio function of part load ratio curve. The EIR versus part-load curve is displayed on the right side.

> Numerically, *energy input ratio* (EIR) is the inverse of COP. The part-load ratio is the ratio of actual cooling load delivered at any point in time compared with the chiller's cooling capacity.

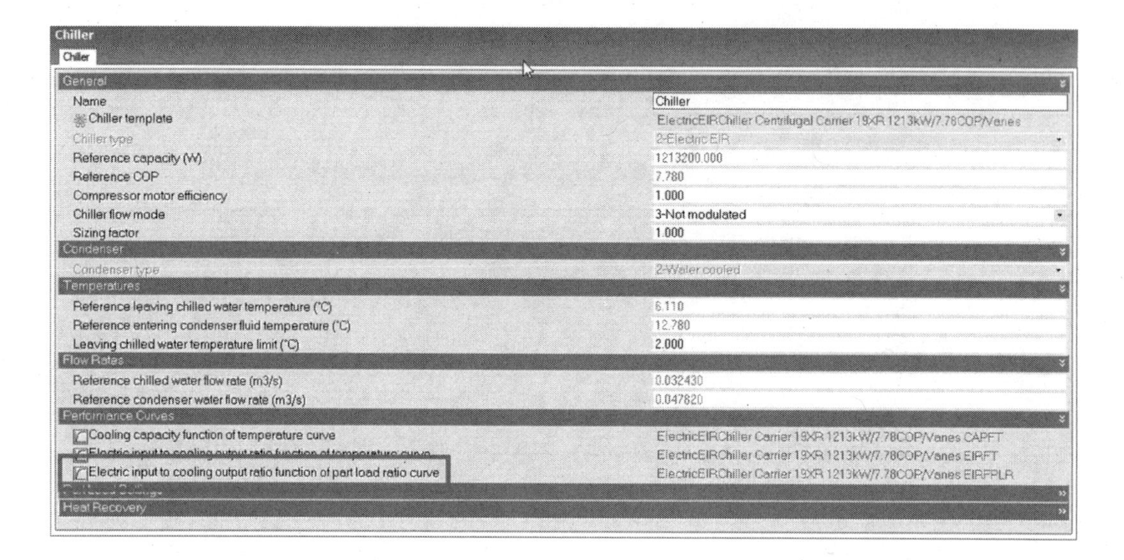

You can also view the curve coefficients.

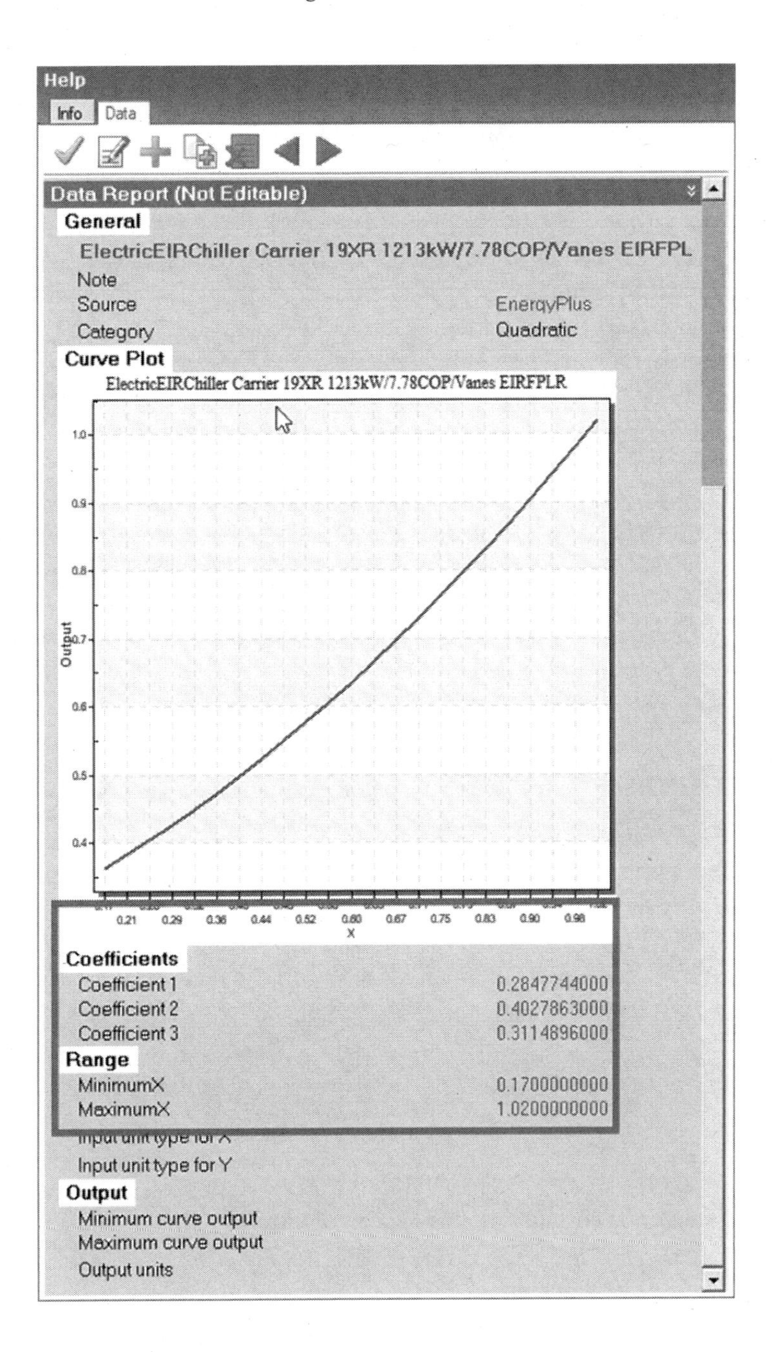
Help
Info
Data
Data Report (Not Editable)
General
ElectricEIRChiller Carrier 19XR 1213kW/7.78COP/Vanes EIRFPL
Note
Source
EnergyPlus
Category
Quadratic
Curve Plot
ElectricEIRChiller Carrier 19XR 1213kW/7.78COP/Vanes EIRFPLR
Output
X
Coefficients
Coefficient 1
0.2847744000
Coefficient 2
0.4027863000
Coefficient 3
0.3114896000
Range
MinimumX
0.1700000000
MaximumX
1.0200000000
Input unit type for Y
Output
Minimum curve output
Maximum curve output
Output units

Performance Curves

Cooling Capacity Function of Temperature Curve

This biquadratic performance curve parameterizes the variation in the cooling capacity as a function of the leaving chilled-water temperature and the entering condenser-fluid temperature. The output of this curve is multiplied by the reference capacity to give the cooling capacity at specific temperature operating conditions (i.e. at temperatures different from the reference temperatures). The curve should have a value of 1.0 at the reference temperatures and the flow rates specified earlier. This biquadratic curve should be valid for the range of water temperatures anticipated for the simulation.

Electric-Input-to-Cooling-Output Ratio Function of Temperature Curve

This biquadratic performance curve parameterizes the variation in energy input to cooling output ratio (EIR) as a function of the leaving chilled-water temperature and the entering condenser-fluid temperature. The output of this curve is multiplied by the reference EIR (inverse of the reference COP) to give the EIR at specific temperature operating conditions (i.e. at temperatures different from the reference temperatures). The curve should have a value of 1.0 at the reference temperatures and the flow rates specified earlier. This biquadratic curve should be valid for the range of water temperatures anticipated for the simulation.

Electric-Input-to-Cooling-Output Ratio Function of Part-Load Ratio Curve

This quadratic performance curve parameterizes the variation in the EIR as a function of the part-load ratio. The output of this curve is multiplied by the reference EIR (inverse of the reference COP) and the energy-input-to-cooling-output ratio function of the temperature curve to give the EIR at the specific temperatures and part-load ratio at which the chiller is operating. This curve should have a value of 1.0 when the part-load ratio equals 1.0. This quadratic curve should be valid for the range of part-load ratios anticipated for the simulation.

Source: http://designbuilder.co.uk/helpv6.0/Content/Performance_Curves.htm.

Step 8: Perform an annual simulation, and record the results.

Step 9: Repeat the preceding steps to select **ElectricEIRChiller Centrifugal Carrier 19XR 1143kW/6.57COP/VSD**.

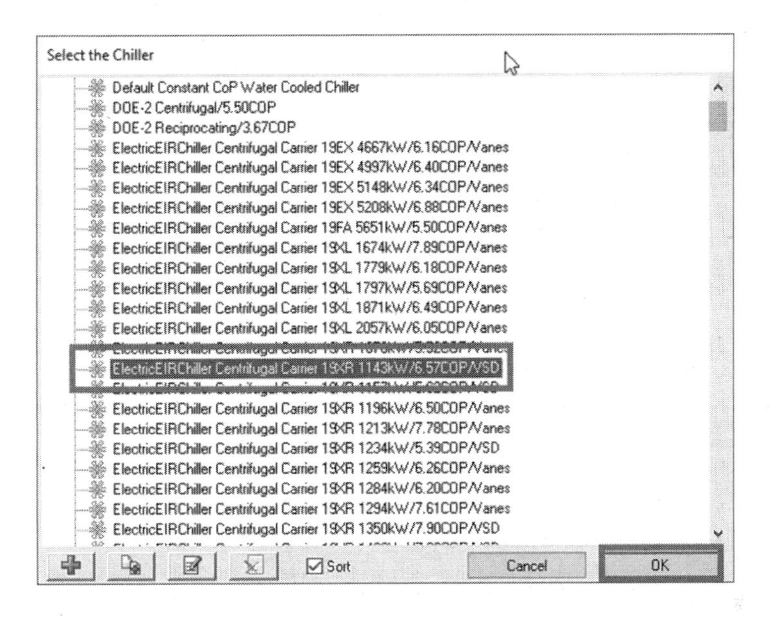

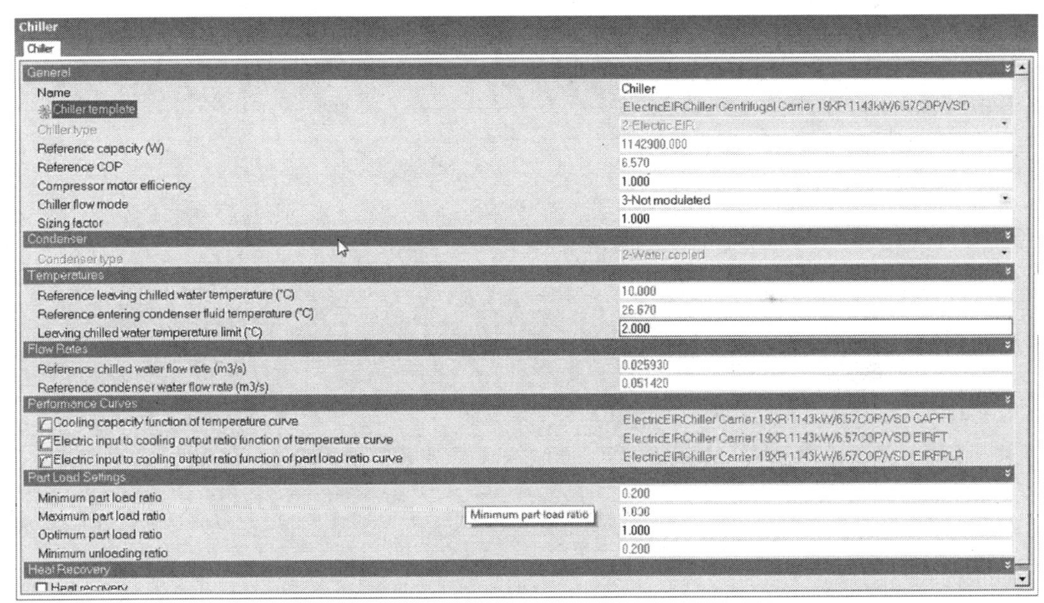

VSD on a chiller provides you with better efficiency while operating in part-load conditions.

You can also view the curve coefficients.

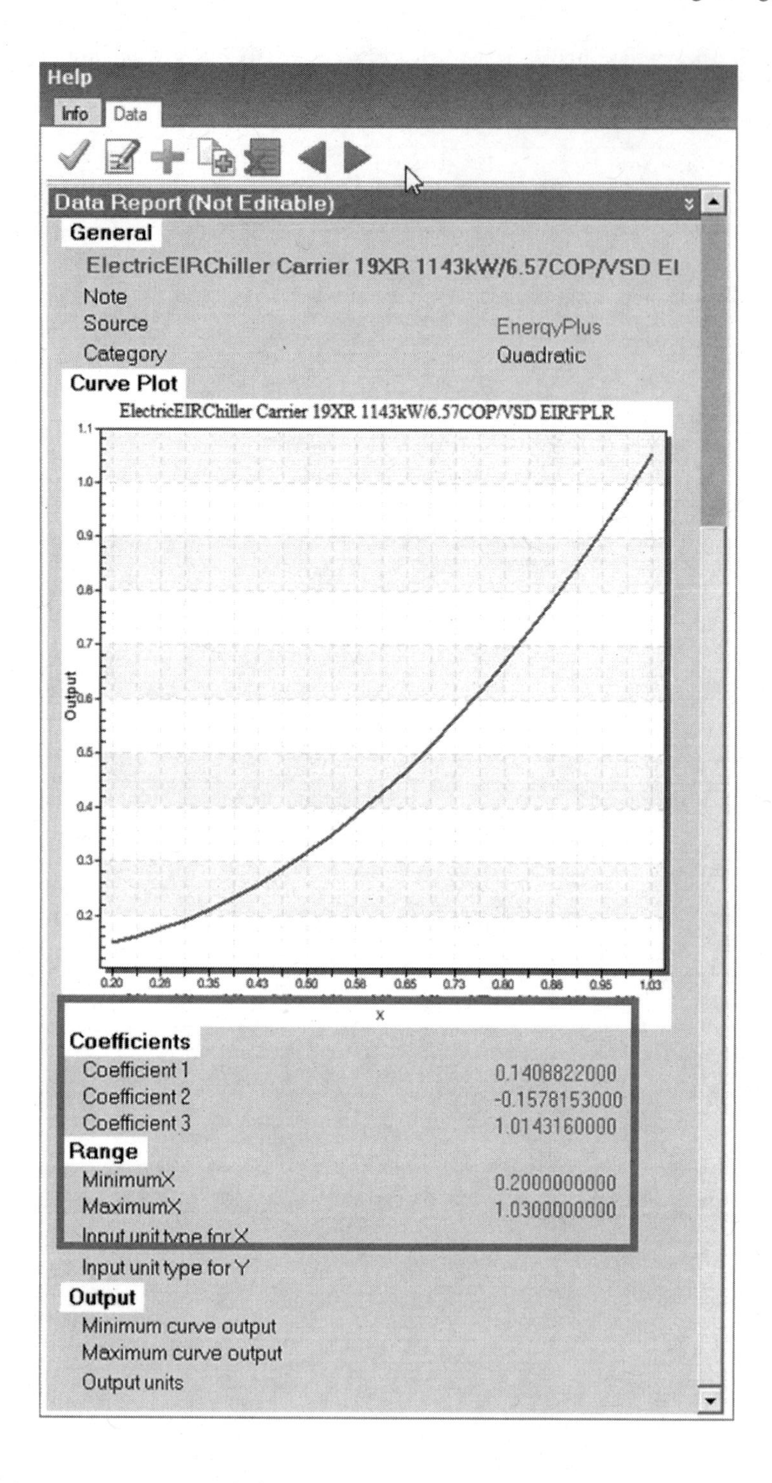
Help
Info
Data
Data Report (Not Editable)
General
ElectricEIRChiller Carrier 19XR 1143kW/6.57COP/VSD EI
Note
Source
EnergyPlus
Category
Quadratic
Curve Plot
ElectricEIRChiller Carrier 19XR 1143kW/6.57COP/VSD EIRFPLR
Output
x
Coefficients
Coefficient 1
0.1408822000
Coefficient 2
-0.1578153000
Coefficient 3
1.0143160000
Range
MinimumX
0.2000000000
MaximumX
1.0300000000
Input unit type for X
Input unit type for Y
Output
Minimum curve output
Maximum curve output
Output units

Step 10: Simulate the model, and record the results. Compare the energy consumption for both cases (Table 8.2).

TABLE 8.2
Effect of a VSD on a chiller

End-use category	Annual energy consumption (kWh)	
	Chiller without VSD	Chiller with VSD
Room electricity	306,674.40	306,674.40
Lighting	443,211.10	443,211.10
System fans	335,040.80	335,040.80
System pumps	14,426.80	15,299.10
Heating (gas)	199,962.60	199,962.60
Cooling (electricity)	491,727.70	224,234.40
Heat rejection	177,262.40	186,465.20

There is a significant reduction in cooling energy consumption with a VSD chiller. If you want to analyse the results, you need to look at the cooling-load profile of the building. Installing a VSD on a chiller provides you with better efficiency while operating in part-load conditions.

Tutorial 8.3

Evaluating the Impact of a VSD on a Chilled-Water Pump

GOAL

To evaluate the impact of a VSD on a chilled-water pump on building energy consumption

WHAT ARE YOU GOING TO LEARN?

- How to model a variable-speed chilled-water pump

PROBLEM STATEMENT

In this tutorial, you are going to use the water-cooled chiller model saved in Tutorial 8.1 (a 50- × 25-m six-floor model with a 5-m perimeter depth). You need to select the following two configurations of a chilled-water pump and compare energy consumption in both cases:

- Constant flow
- Variable flow

Use the **FL – MIAMI, USA** weather location. Miami has a tropical monsoon climate with hot, humid summers and short, warm winters, with a marked drier season in the winter.

SOLUTION

Step 1: Open the simulation model saved in Tutorial 8.1. Click **<HVAC System>** in the navigation pane. Click the **Load HVAC template**.

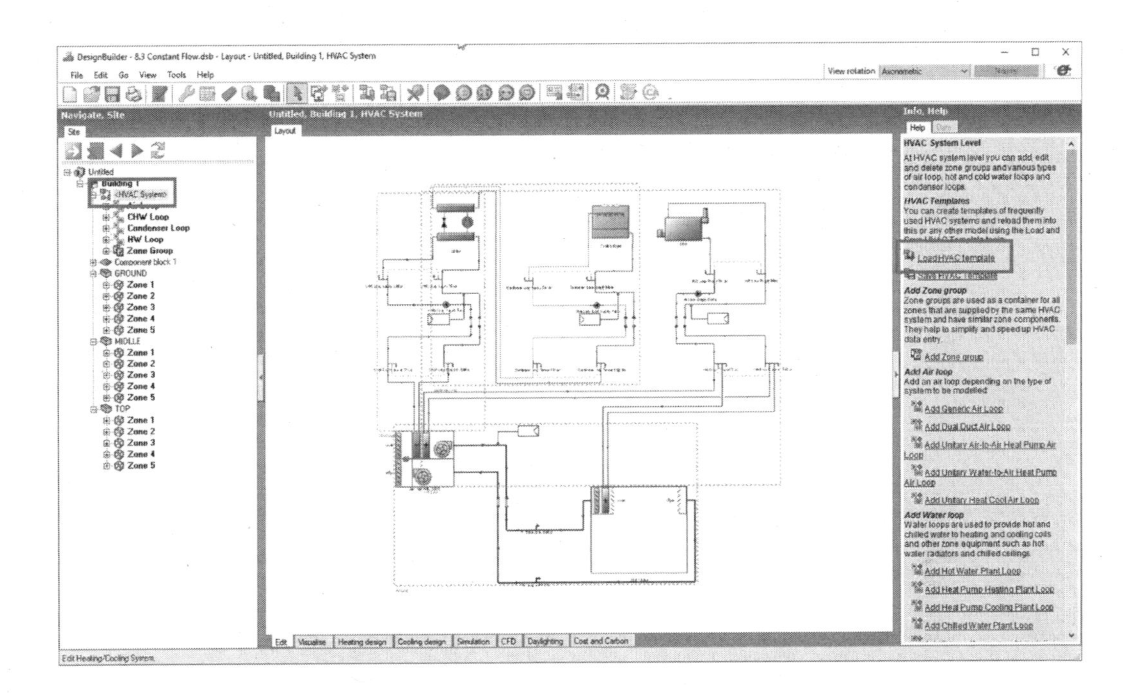

Step 2: Select **System No. 8 VAV with PFP Boxes** from **ASHRAE 90.1 Appendix G baseline**. Click **OK**. The **Load HVAC template** appears.

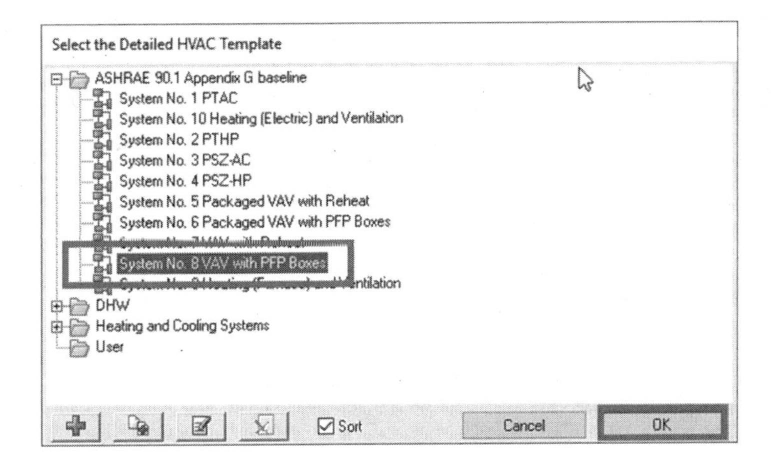

Step 3: Click **Next**. Select all the checkboxes from the **Zones** section. Click **Finish**. This displays the system layout.

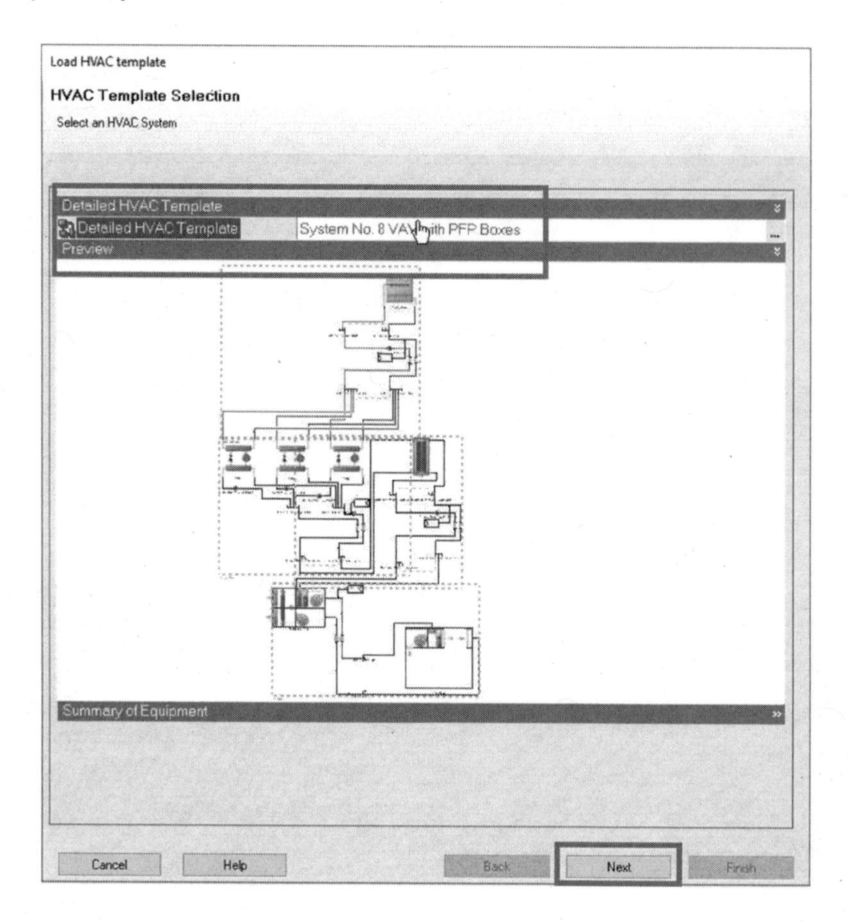

Step 4: Select **CHW Loop**. Click **Edit loop data** under the **Help** tab. The **Edit Plant loop** screen appears.

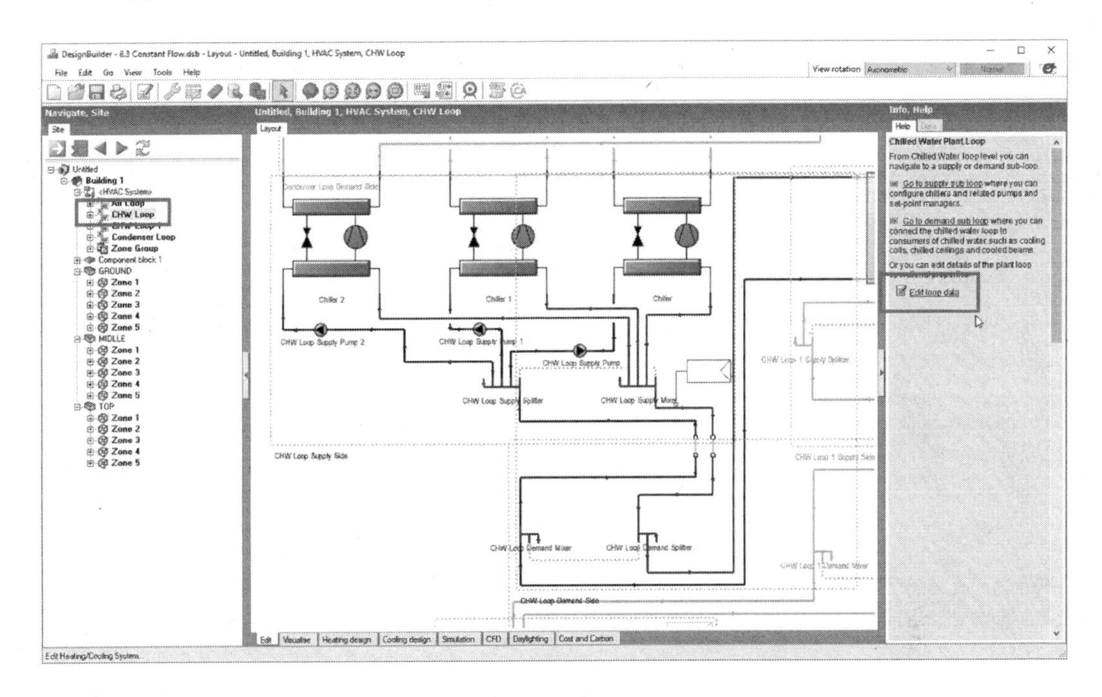

Step 5: Select **Constant flow** from the **Plant loop flow** type.

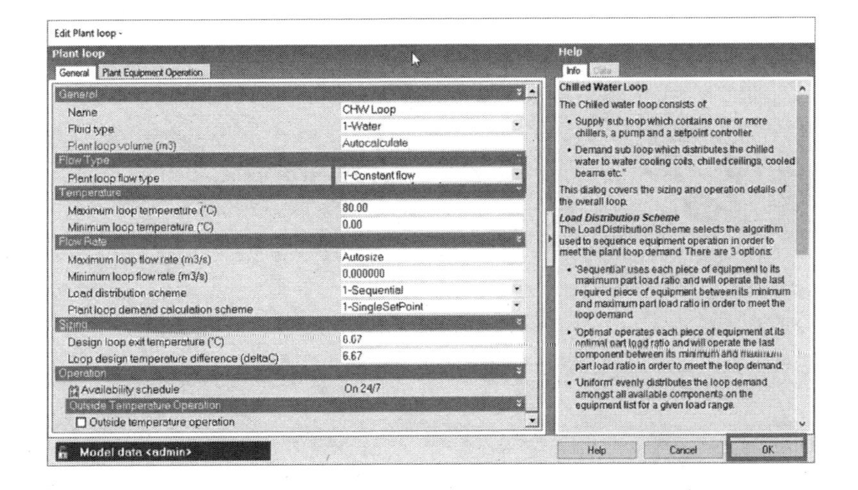

Step 6: Perform an annual energy simulation, and record the results.

Step 7: Repeat the preceding steps for a **Variable flow** CHW loop.

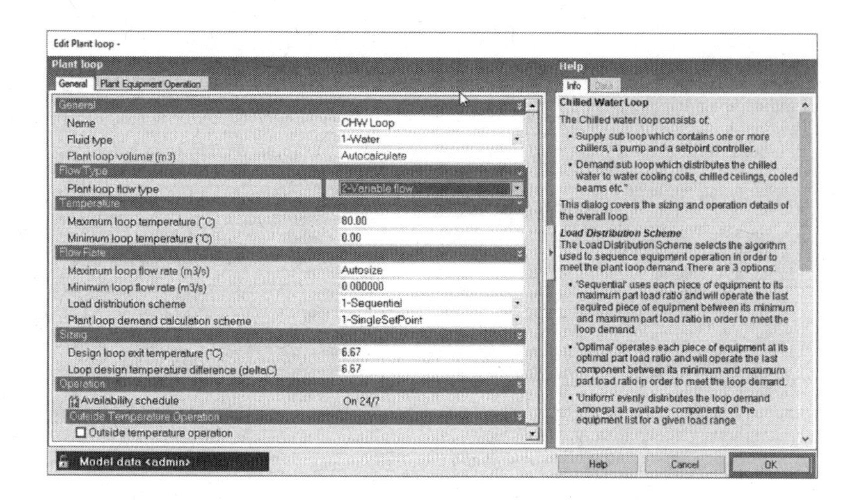

Step 8: Perform an annual energy simulation, and record the results. Save the simulation model to use in subsequent tutorials. Compare the energy consumption in both cases (Table 8.3).

You can see that there is a reduction in pump energy, cooling energy and energy consumption for heat rejection. Where variable flow is used, it can be clearly seen that there is less energy consumption for system pumps. Also, there is a small decrease in cooling energy consumption of the chiller and heat rejection owing to the change in operating conditions of the chiller.

TABLE 8.3

Energy consumption for constant and variable flow on a chilled-water pump

End-use category	Annual energy consumption (kWh)	
	Constant flow	Variable flow
Room electricity	306,674.40	306,674.40
Lighting	443,211.10	443,211.10
System fans	241,799.00	241,799.30
System pumps	513,022.60	474,218.10
Heating (electricity)	11,109.90	11,109.90
Cooling (electricity)	549,794.30	509,898.10
Heat rejection	478,423.90	477,866.70

Tutorial 8.4

Evaluating the Impact of a Cooling Tower Fan Type

GOAL

To evaluate the impact of a cooling tower fan type on building energy consumption

WHAT ARE YOU GOING TO LEARN?

- How to model a cooling tower with a single-speed fan
- How to model a cooling tower with a double-speed fan

PROBLEM STATEMENT

In this tutorial, you are going to use the variable-flow model saved in Tutorial 8.3 (a 50- × 25-m six-floor model with a 5-m perimeter depth). You need to select the following two configurations of the chilled-water pump and determine the energy consumption in both cases:

1. Single-speed fan cooling tower
2. Double-speed fan cooling tower

Use the AZ-PHOENIX/SKY HARBOR, USA weather location.

A *cooling tower* rejects heat extracted from a building to the atmosphere by the evaporation of water. Cooling tower fans help in governing the airflow and rate of evaporation.

SOLUTION

Step 1: Open the variable-flow model saved in Tutorial 8.3. Expand **<HVAC System>** in the navigation pane.

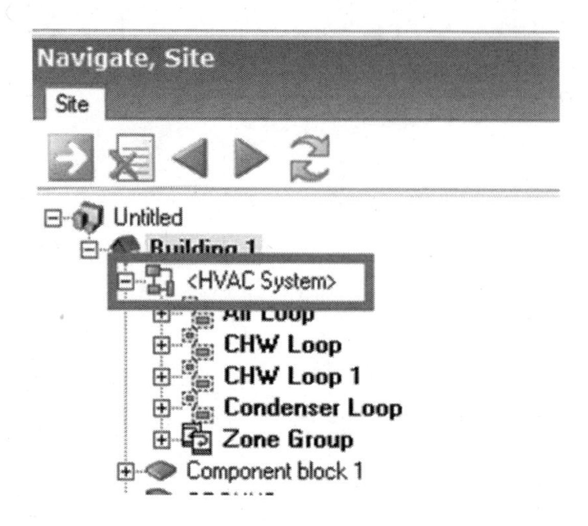

Step 2: Select **Cooling Tower** under **Condenser Loop Supply Side**. Click **Edit component** under the **Help** tab. The **Edit Cooling Tower** screen appears.

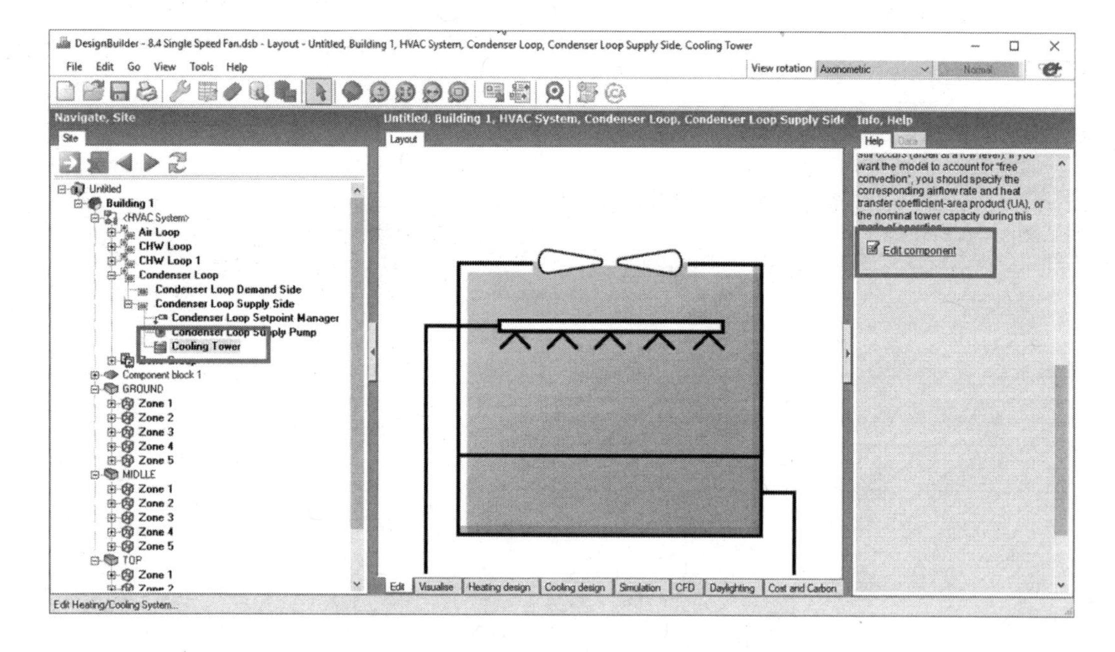

Step 3: Select **Single speed** for cooling tower type.

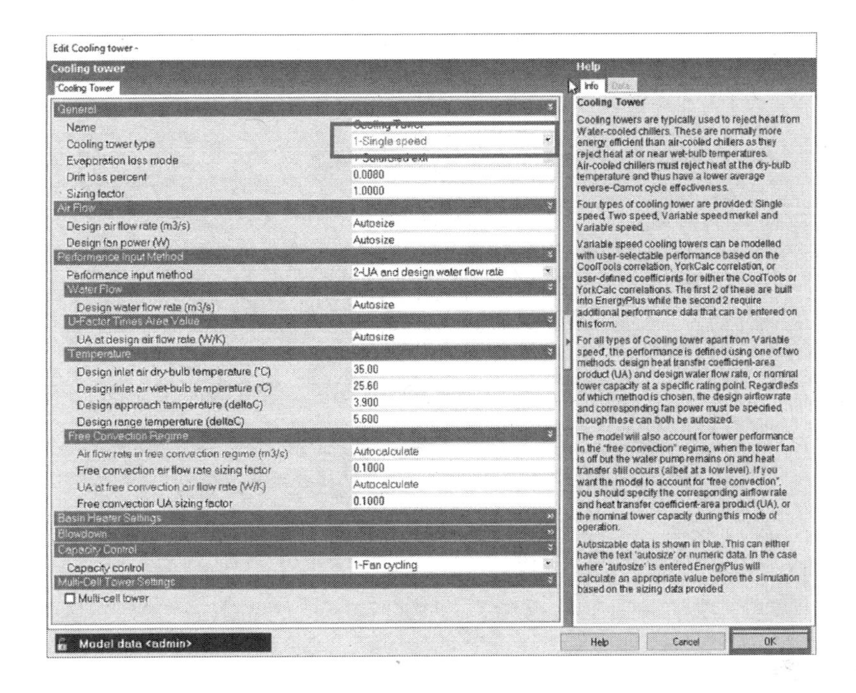

Step 4: Simulate the model, and record the results.

Step 5: Repeat the preceding steps to select **Two speed** for cooling tower type.

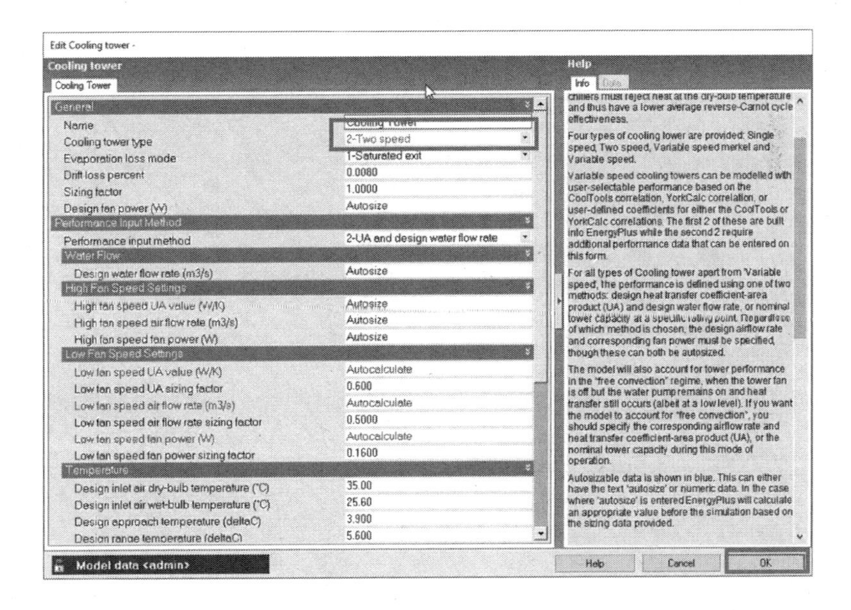

A multispeed/variable-speed-drive (VSD) fan is the preferred method of capacity control for cooling towers. By matching the fan motor speeds to the required heat rejection, multispeed/VSD cooling towers can significantly reduce energy consumption for heat rejection.

Step 6: Perform an annual energy simulation, and record the results. Save the simulation model to be used in the next tutorial. Compare results for both cases (Table 8.4).

TABLE 8.4
Energy consumption for single-speed and double-speed cooling tower fans

End-use components	Annual energy consumption (kWh)	
	Single-speed cooling tower fan	Two-speed cooling tower fan
Room electricity	306,674.40	306,674.40
Lighting	443,211.10	443,211.10
System fans	300,825.20	300,825.20
System pumps	326,676.70	326,676.70
Heating (electricity)	78,699.32	78,699.32
Cooling (electricity)	355,371.90	355,371.90
Heat rejection	148,584.70	114,571.60

You can see from the results that there is a significant reduction in energy consumption under heat rejection.

Tutorial 8.5

Evaluating the Impact of a Condenser Water Pump with a VSD

GOAL

To evaluate the impact of using a VSD with a condenser water pump on building energy consumption

WHAT ARE YOU GOING TO LEARN?

- How to model a VSD on a condenser water pump

PROBLEM STATEMENT

In this tutorial, you are going to use the double-speed model saved in Tutorial 8.4 (a 50- × 25-m six-floor model with a 5-m perimeter depth). You need to select the following two configurations of condenser water pump and determine the energy consumption in both cases:

- One-speed condenser water pump
- Variable-speed condenser water pump

Use the **FL – MIAMI, USA** weather location.

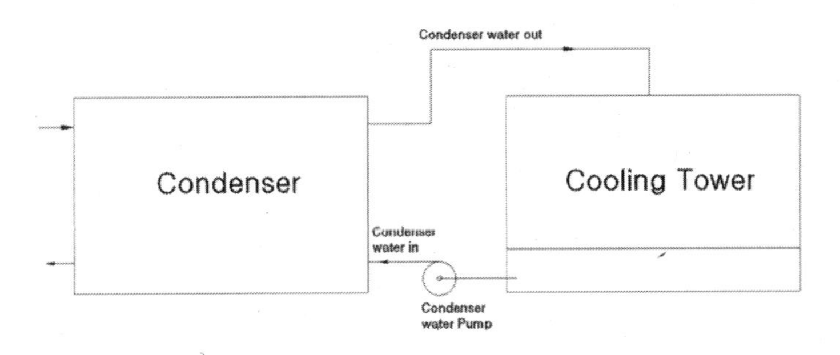

SOLUTION

Step 1: Open the double-speed model saved in Tutorial 8.4. Expand **<HVAC System>** in the navigation pane.

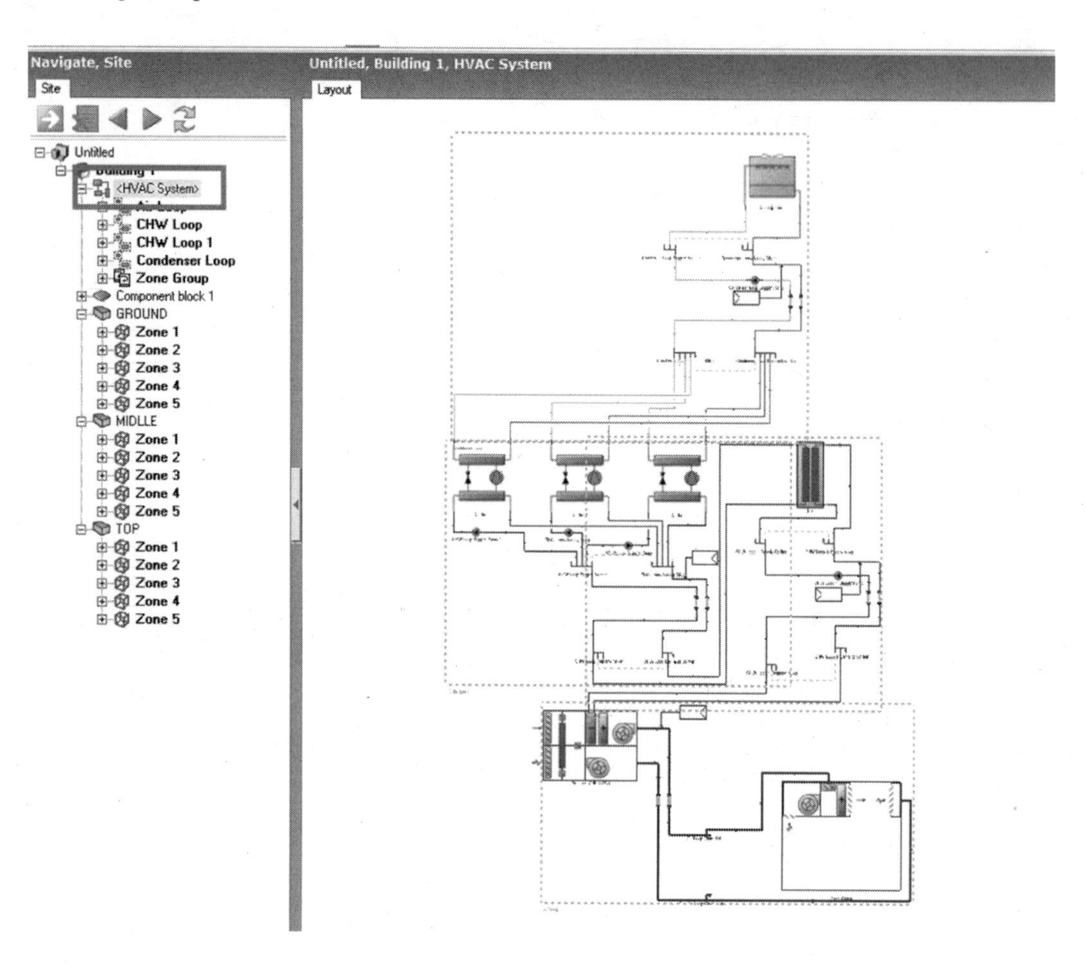

Step 2: Click **Condenser Loop**. This displays a condenser loop diagram in the **Layout** tab.

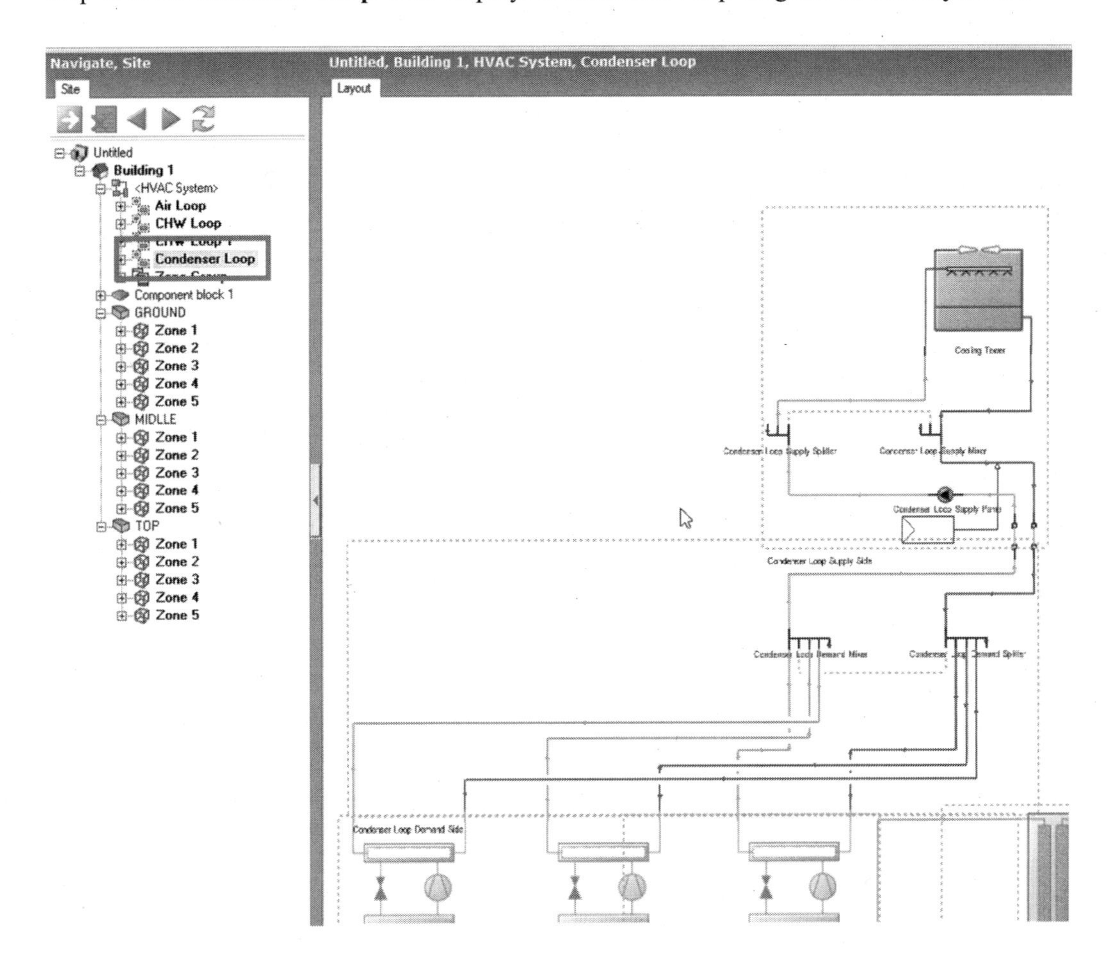

Step 3: Click **Condenser Loop Supply Pump**. Click **Edit component** under the **Help** tab. The **Edit Pump** screen appears.

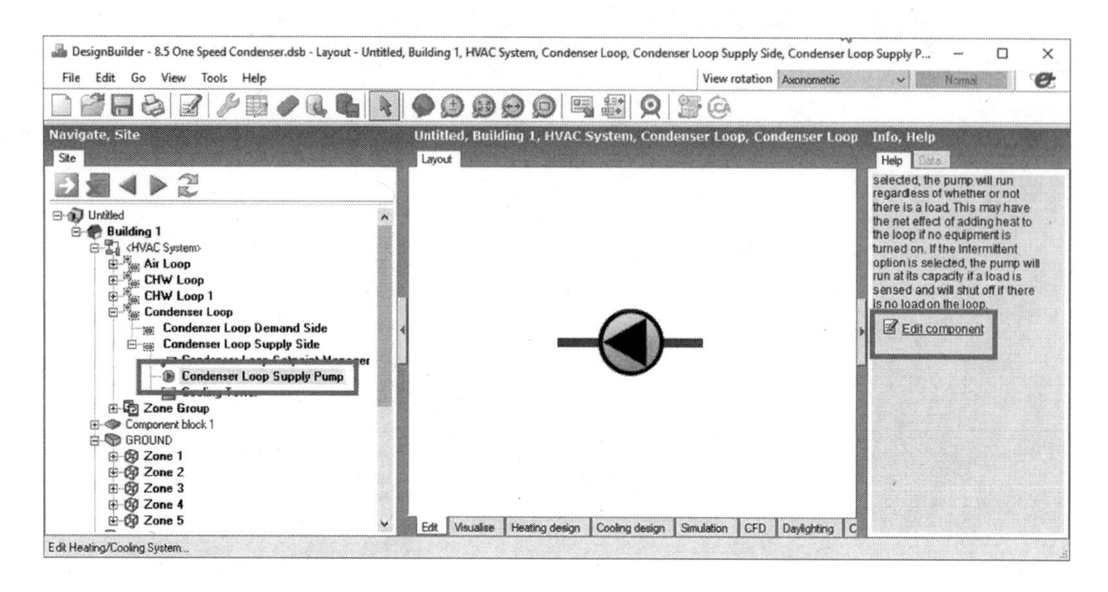

Step 4: Select **Constant speed** from the **Type** drop-down list. Click **OK**.

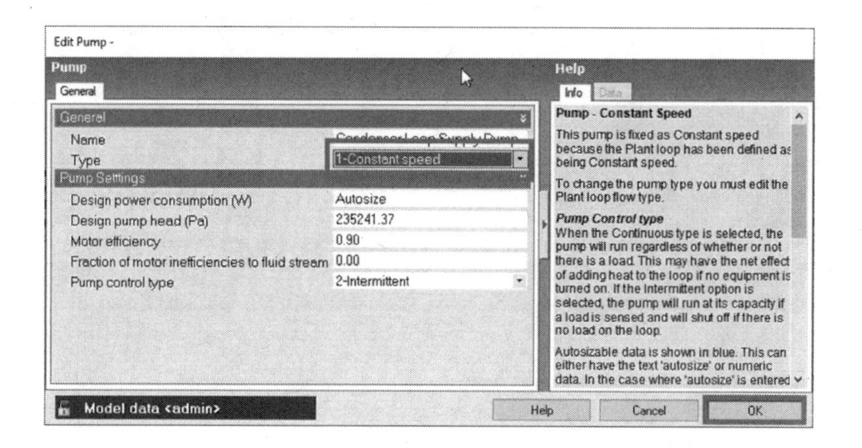

Step 5: Simulate the model, and record the results.

Step 6: Repeat the preceding steps to select **Variable speed** from the **Type** drop-down list.

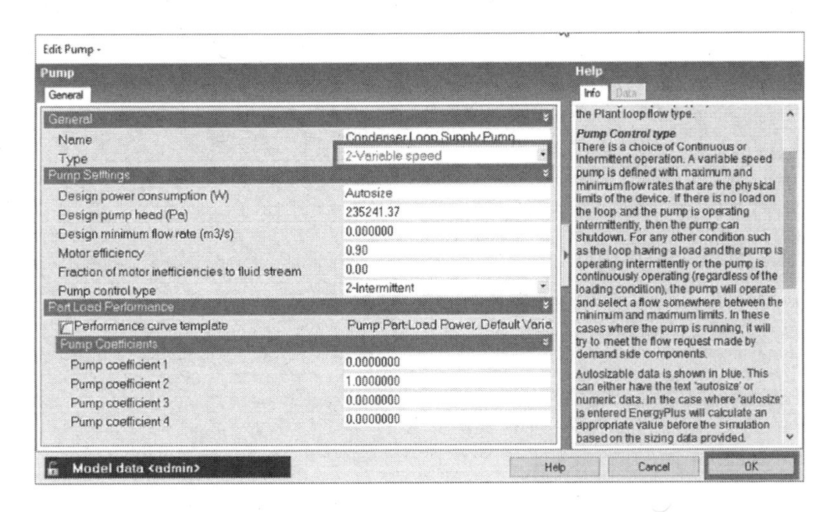

Step 7: Simulate the model, and record the results. Compare energy consumption for both cases (Table 8.5).

TABLE 8.5
Energy consumption for single- and variable-speed condenser water pumps

End-use category	Annual energy consumption (kWh)	
	Single-speed condenser water pump	**Variable-speed condenser water pump**
Room electricity	306,674.40	306,674.40
Lighting	443,211.10	443,211.10
System fans	241,799.30	241,799.30
System pumps	474,218.10	218,124.50
Heating (electricity)	11,109.90	11,109.90
Cooling (electricity)	509,898.10	508,458.60
Heat rejection	477,866.70	455,361.60

A reduction in energy consumption under system pumps can be seen in the case of a variable-speed condenser water pump. A small reduction in energy consumption for heat rejection can also be seen.

Tutorial 8.6

Evaluating the impact of boiler nominal thermal efficiency

GOAL

To evaluate the impact of boiler efficiency on building energy performance

WHAT ARE YOU GOING TO LEARN?

- How to change boiler efficiency

PROBLEM STATEMENT

In this tutorial, you are going to use the air-cooled chiller model saved in Tutorial 8.1 (a 50- × 25-m six-floor model with a 5-m perimeter depth). You need to simulate the model with the boiler efficiencies ranging from 0.89 to 0.95 in increments of 0.02. Determine the energy consumption for all cases. Use the **PARIS-AEROPORT CHAR, France** weather location.

Nominal Thermal Efficiency

This is the heating efficiency (as a fraction between zero and one) of a boiler's burner relative to the higher heating value (HHV) of fuel at a part-load ratio of 1.0. Manufacturers typically specify the efficiency of a boiler using the HHV of the fuel. For the rare occurrence where a manufacturer's (or a particular data set's) thermal efficiency is based on the lower heating value (LHV) of the fuel, multiply the thermal efficiency by the lower-to-higher-heating-value ratio. For example, assume that a fuel's LHV and HHV are approximately 45,450 and 50,000 kJ/kg, respectively. For a manufacturer's thermal efficiency rating of 0.90 (based on the LHV), the nominal thermal efficiency entered here is 0.82 (i.e. 0.9 × (45,450/50,000)).

Heating value is the amount of heat produced by a complete combustion of fuel, and it is measured as a unit of energy per unit mass or volume of substance (e.g. kcal/kg, kJ/kg, J/mol and Btu/m^3). HHV is the gross calorific value, defined as the amount of heat released when fuel is combusted and the products have returned to a temperature of 25°C. The heat of condensation of the water is included in the total measured heat. The LHV is the net calorific value and is determined by subtracting the heat of vaporization of water vapour (generated during combustion of fuel) from the HHV.

Source: http://designbuilder.co.uk/helpv6.0/#Boilers.htm.

SOLUTION

Step 1: Open the simulation model created in Tutorial 8.1 with an air-cooled chiller. Click **HW Loop.**

Step 2: Click **HW Loop Supply Side.**

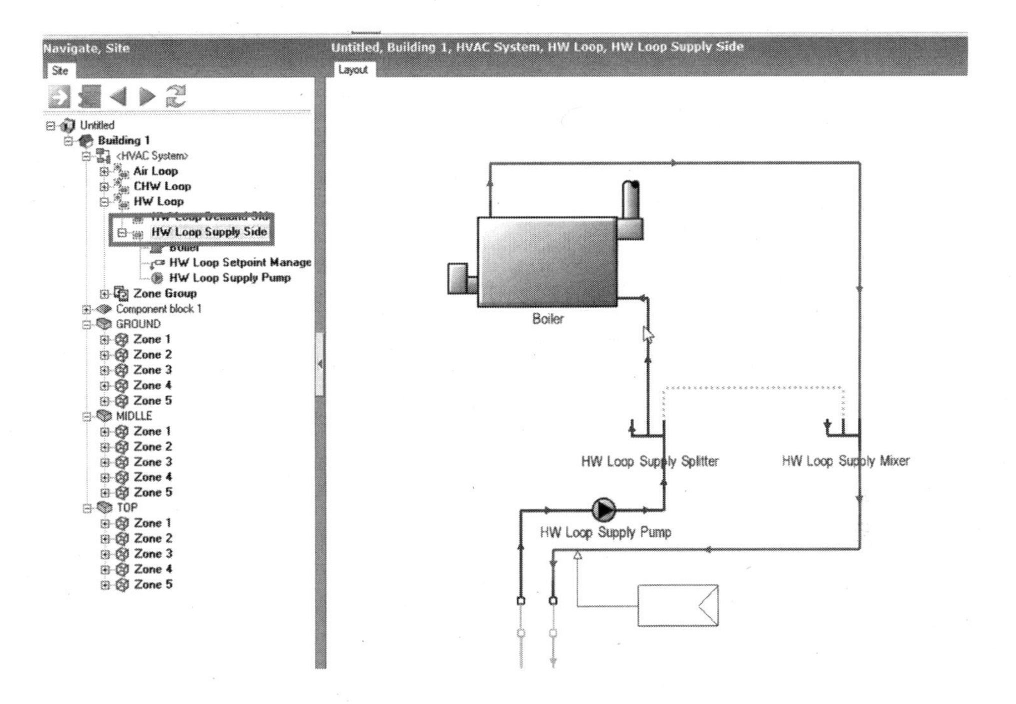

Step 3: Click **Boiler**. Click **Edit component** under the **Help** tab. The **Edit Hot Water Boiler** screen appears.

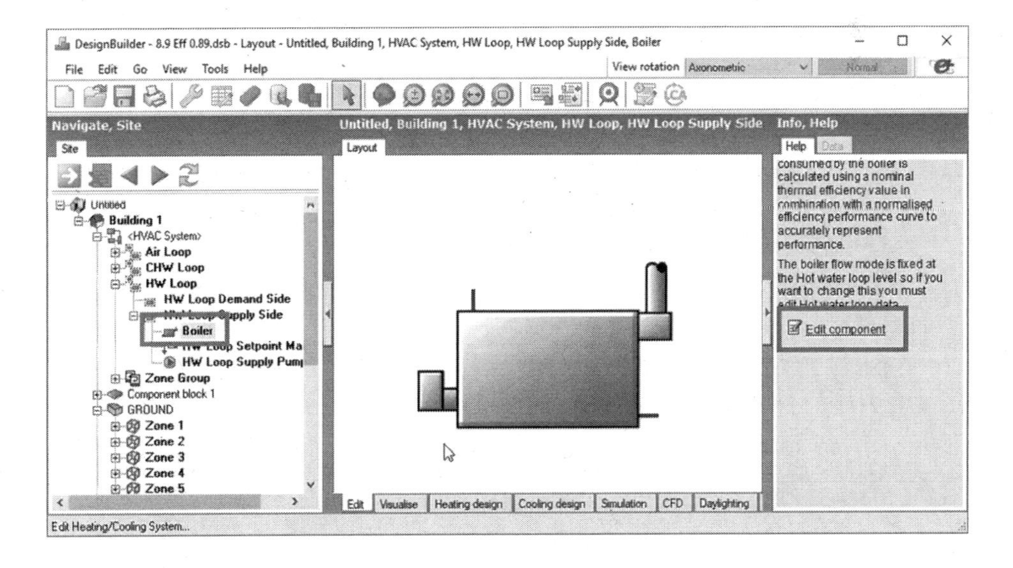

Step 4: Enter **0.890** in the **Nominal thermal efficiency** text box under the **Efficiency** subtab.

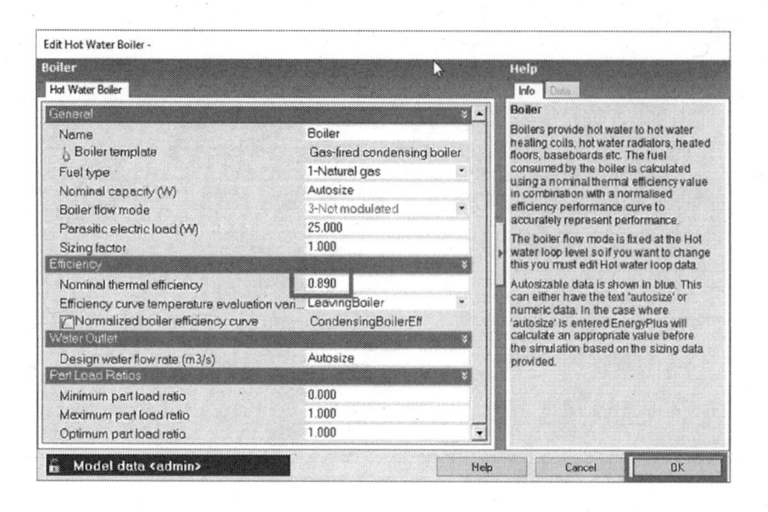

Step 5: Simulate the model, and record the results.

Step 6: Repeat the preceding steps to simulate the model with **0.91**, **0.93** and **0.95** nominal thermal efficiencies (Table 8.6).

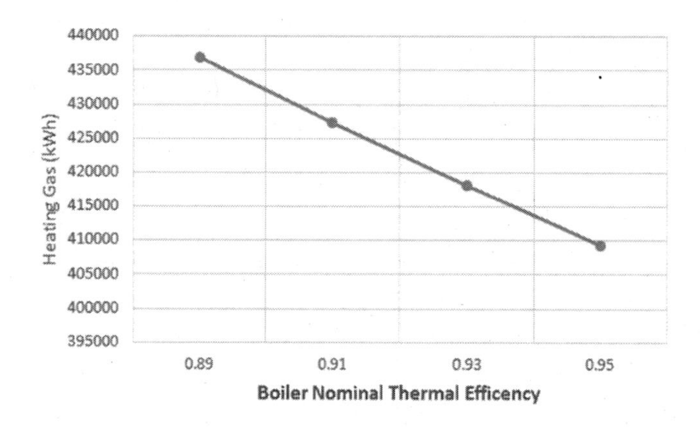

TABLE 8.6

Energy consumption with changes in nominal thermal efficiency

	Annual energy consumption (kWh)			
End-use category	**0.89**	**0.91**	**0.93**	**0.95**
Room electricity	306,674.40	306,674.40	306,674.40	306,674.40
Lighting	443,211.10	443,211.10	443,211.10	443,211.10
System fans	271,578.00	271,578.00	271,578.00	271,578.00
System pumps	880.00	880.00	880.00	880.00
Heating (gas)	824,718.00	806,592.30	789,246.30	772,630.60
Cooling (electricity)	582,605.70	582,605.70	582,605.70	582,605.70

Tutorial 8.7

Evaluating the Impact of Chiller Sequencing

GOAL

To evaluate the impact of chiller sequencing on energy consumption

WHAT ARE YOU GOING TO LEARN?

- How to model a plant manager

PROBLEM STATEMENT

In this tutorial, you are going to use a 100- × 50-m model (five zones with a 10-m perimeter depth) with four identical floors. You need to model a **VAV reheat water cooled chiller system.** Add one more chiller and perform a comparison for the following cases:

1. With chiller plant manager
2. Without chiller plant manager

You need to simulate the model with the data shown in Table 8.7.

TABLE 8.7
Input parameters

Parameter	Value
Building operation schedule	Daytime office
Chiller 1 cooling capacity	1,200 kW
Chiller 2 cooling capacity	1,200 kW
Plant operation to meet building cooling load	Up to 1,200 kW: Chiller 1 operational and chiller 2 off Above 1,200 kW: Equal load on chiller 1 and chiller 2

Perform a simulation for a summer design week, and compare the energy consumption for both cases. Use the **New Delhi Palam, India** weather location.

Five load distribution schemes are employed in EnergyPlus. The following figure illustrates the plant load distribution algorithm. The total loop demand is calculated and used in the **ManagePlantLoopOperation** routine to determine which equipment is available based on the supervisory control scheme specified by the user. Once all available components have been identified, the loop demand is distributed to the available components based on the user-specified load distribution scheme.

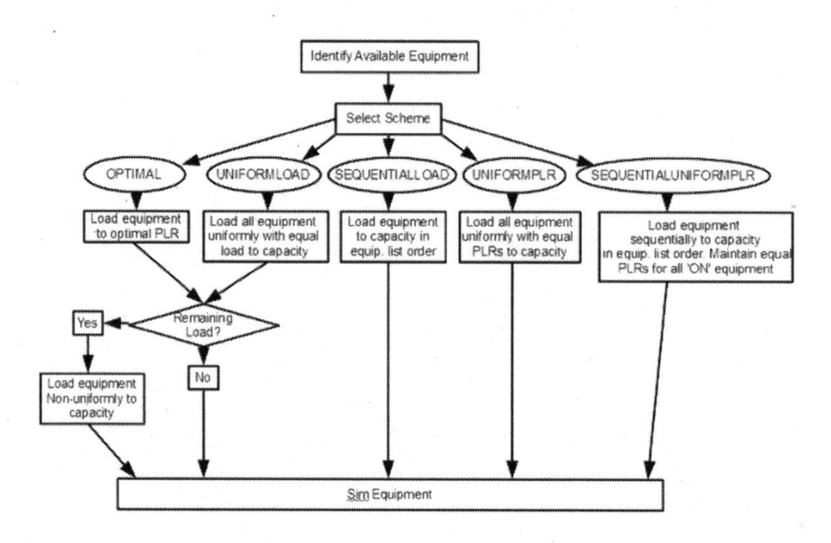

Source: EnergyPlus Input-Output Guide.

SOLUTION

Step 1: Open a new project, and create a **100- × 50-m** four-floor building with the specifications provided in the problem statement.

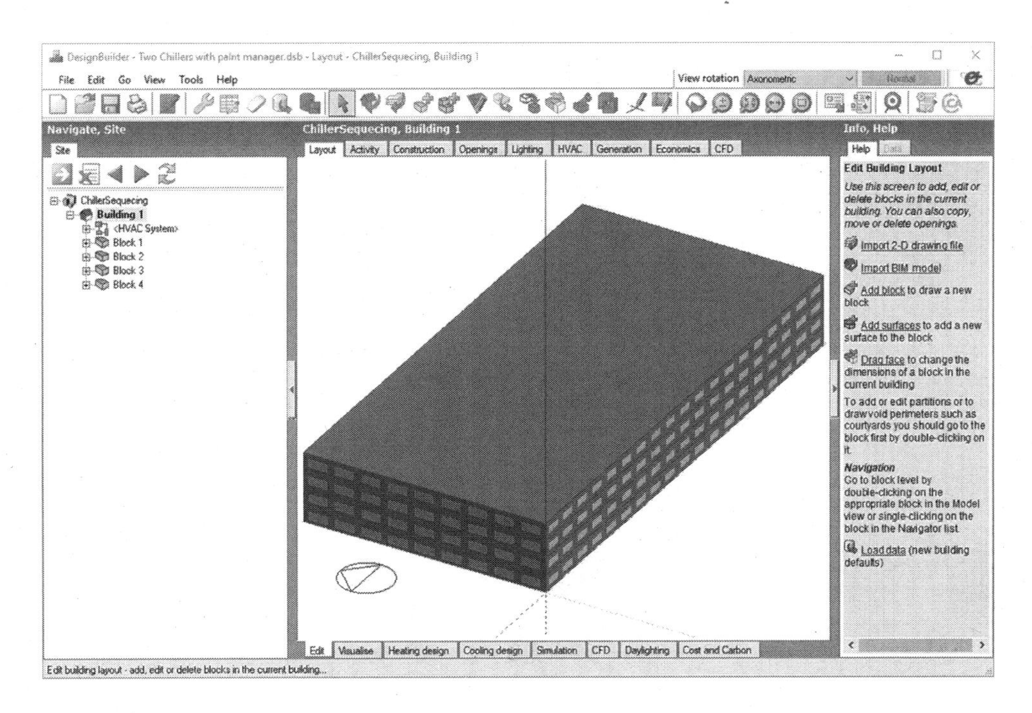

Step 2: Click the **Edit** menu, and select the **Model Options** switch to detailed HAVC. Click **HVAC System**. Click the **Load HVAC template** link. Select **VAV Reheat Water-cooled Chiller**.

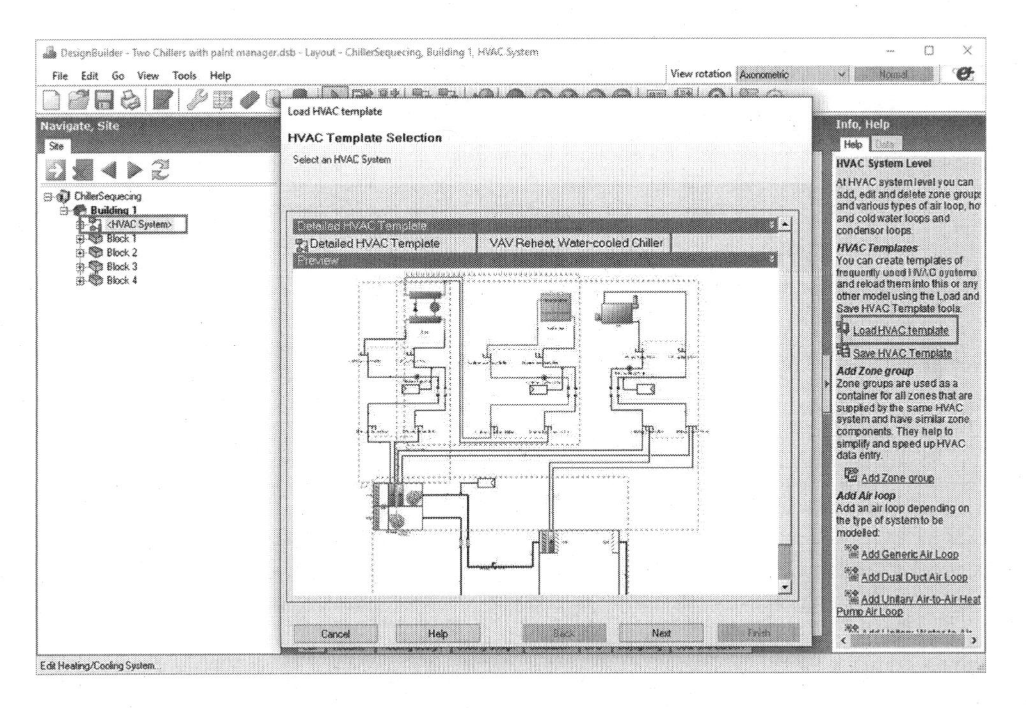

Step 3: Expand the **CHW Loop**.

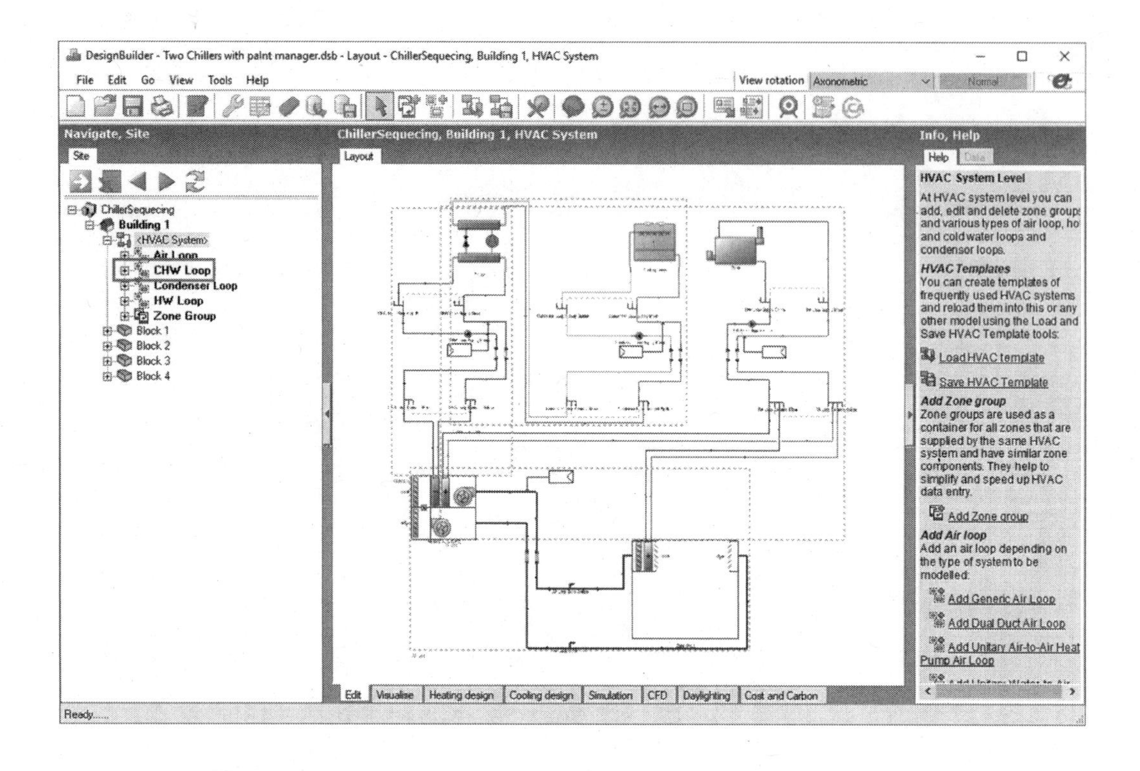

Step 4: Click **CHW Loop Supply Side**. Click **Add Chiller**.

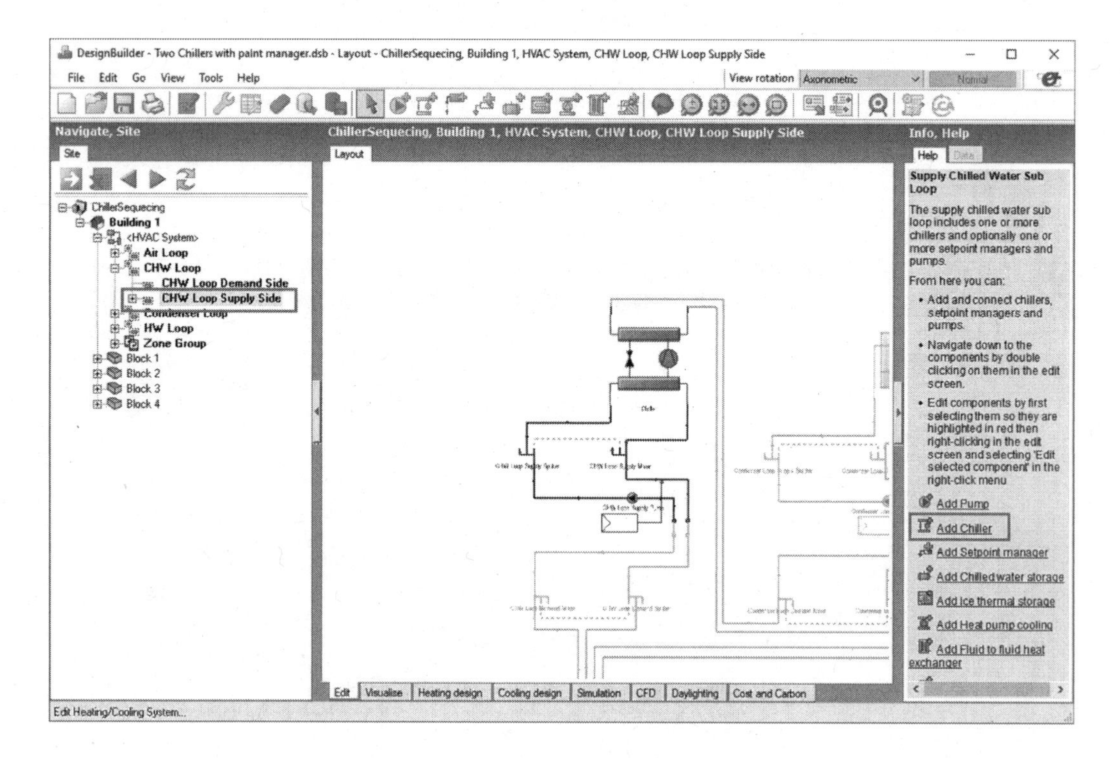

Step 5: Select **Add Chiller – Electric EIR**.

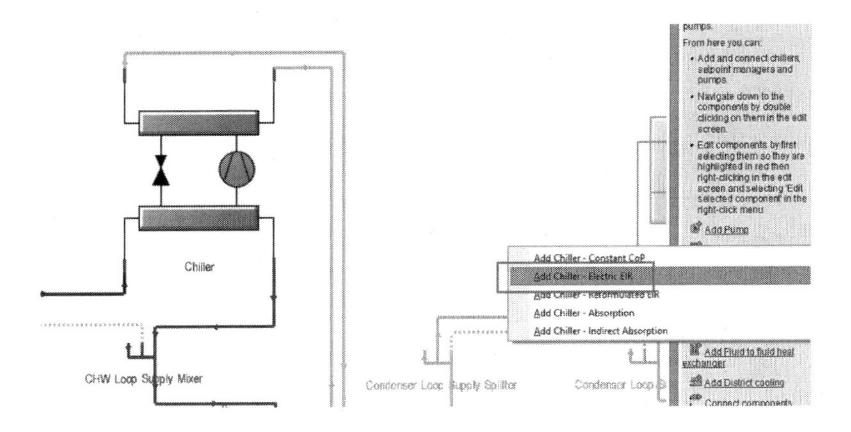

Step 6: Click **Connect components**.

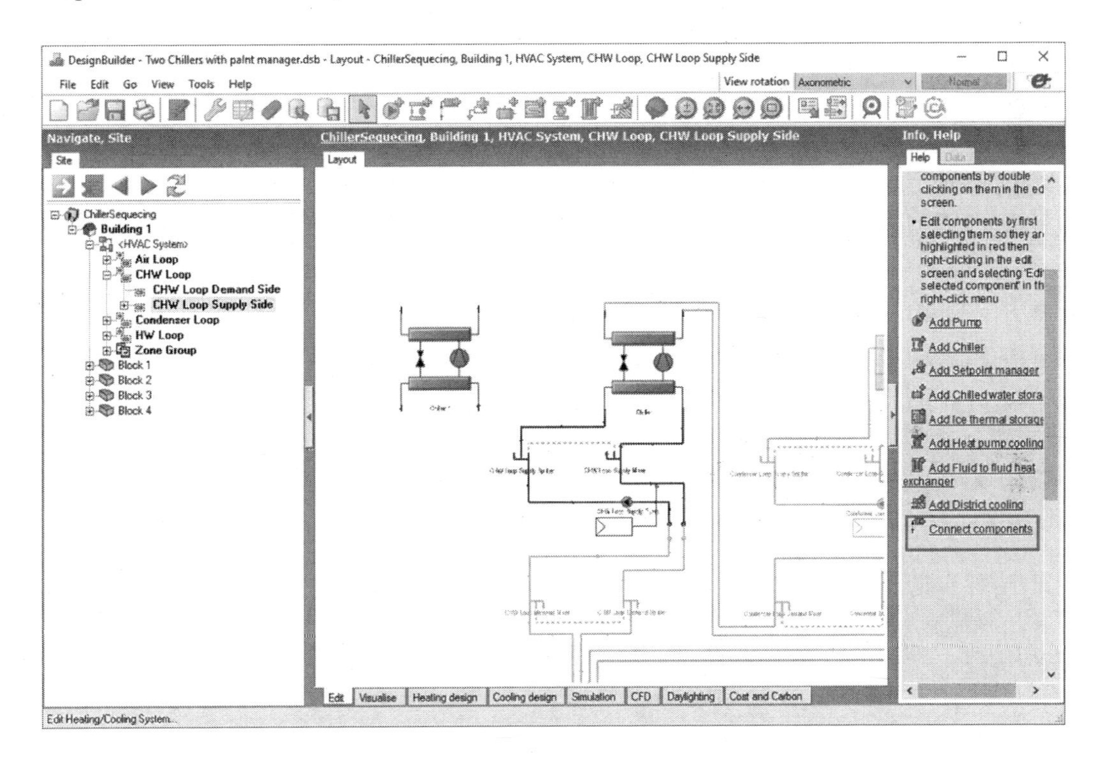

Step 7: Complete the **Evaporator Loop**.

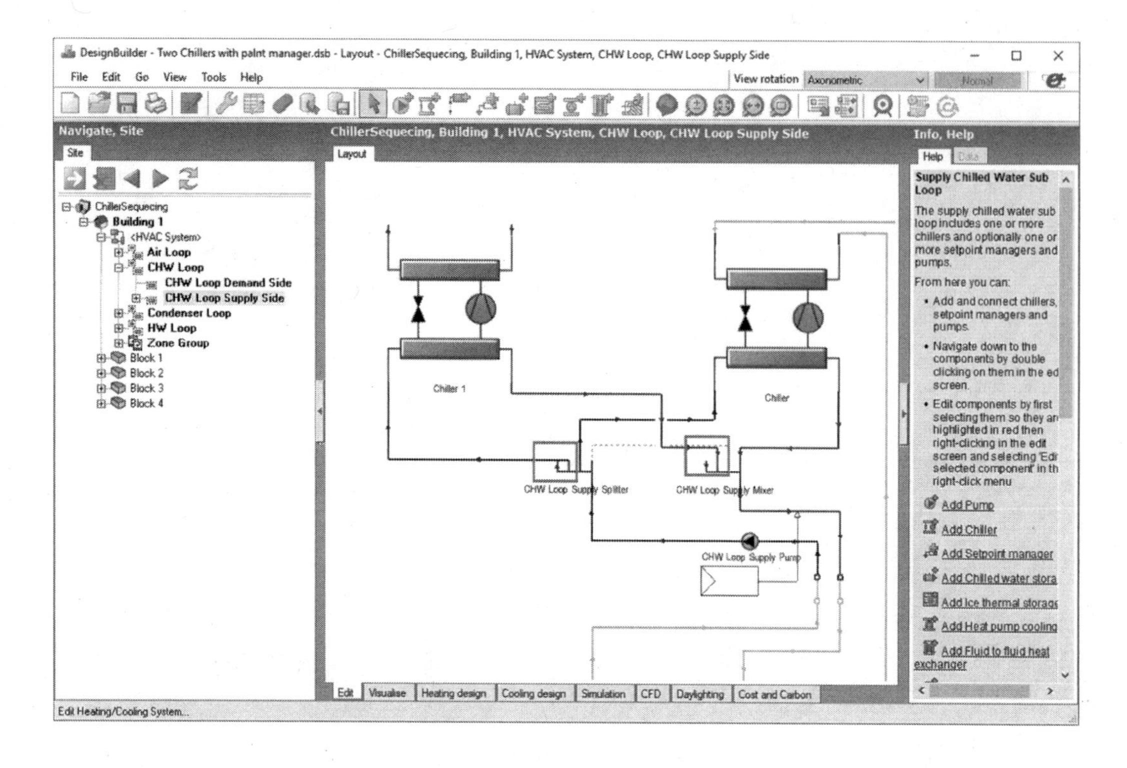

Step 8: Click **Condensor Loop Demand Side**.

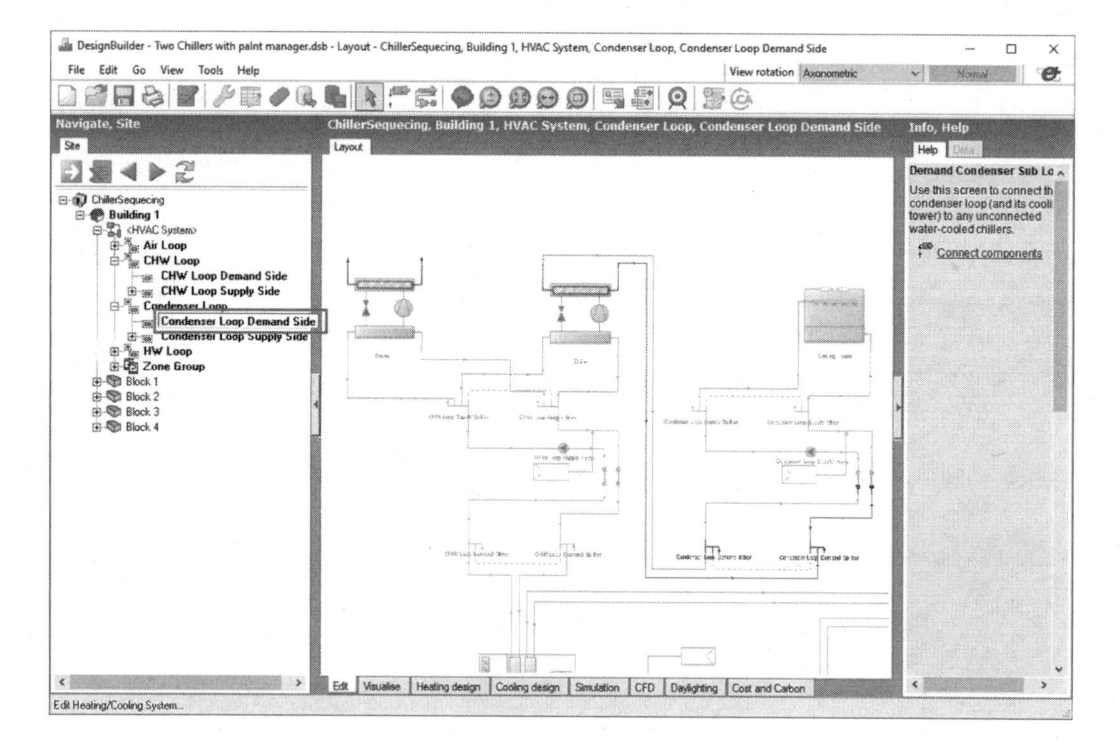

Step 9: Connect the condenser loop.

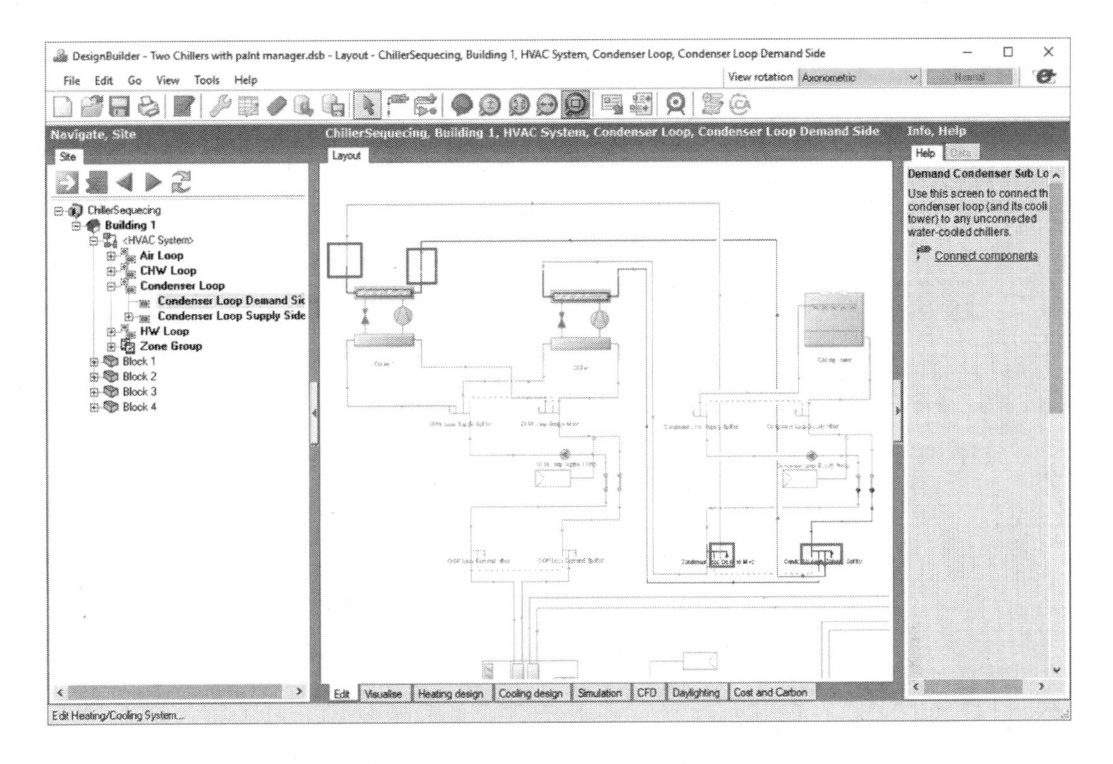

Step 10: Rename the chillers as **Chiller_1** and **Chiller_2**.

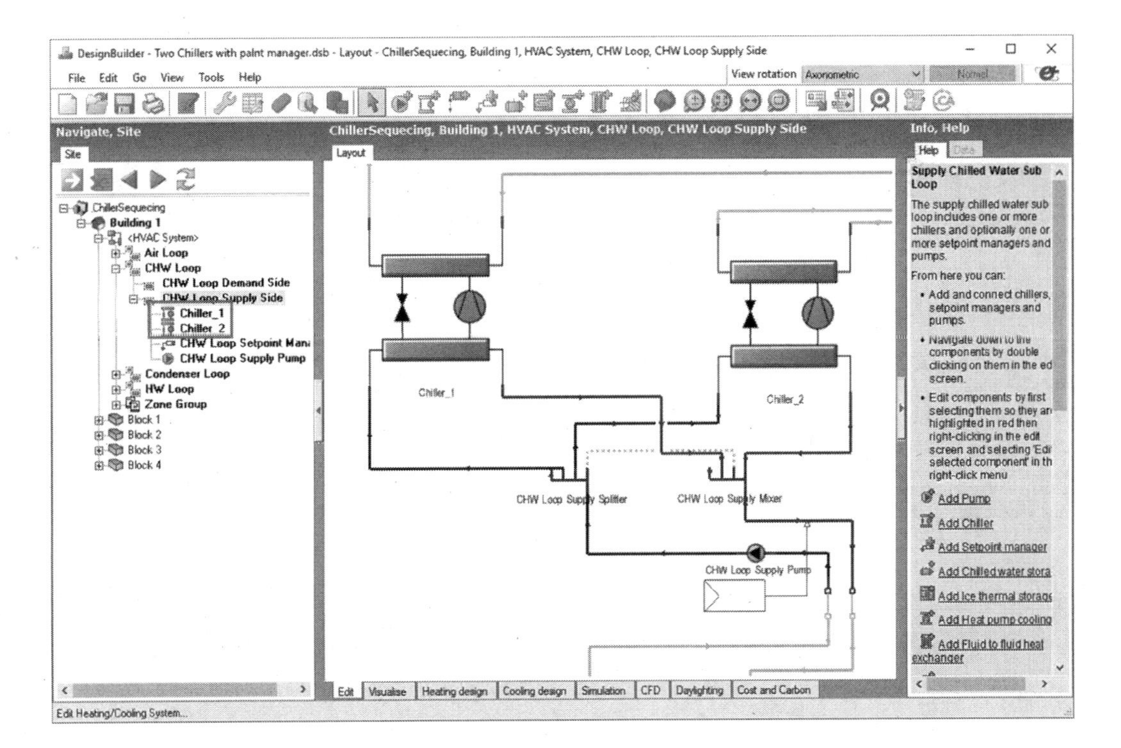

Step 11: Click **Chiller_1**. Click **Edit component**.

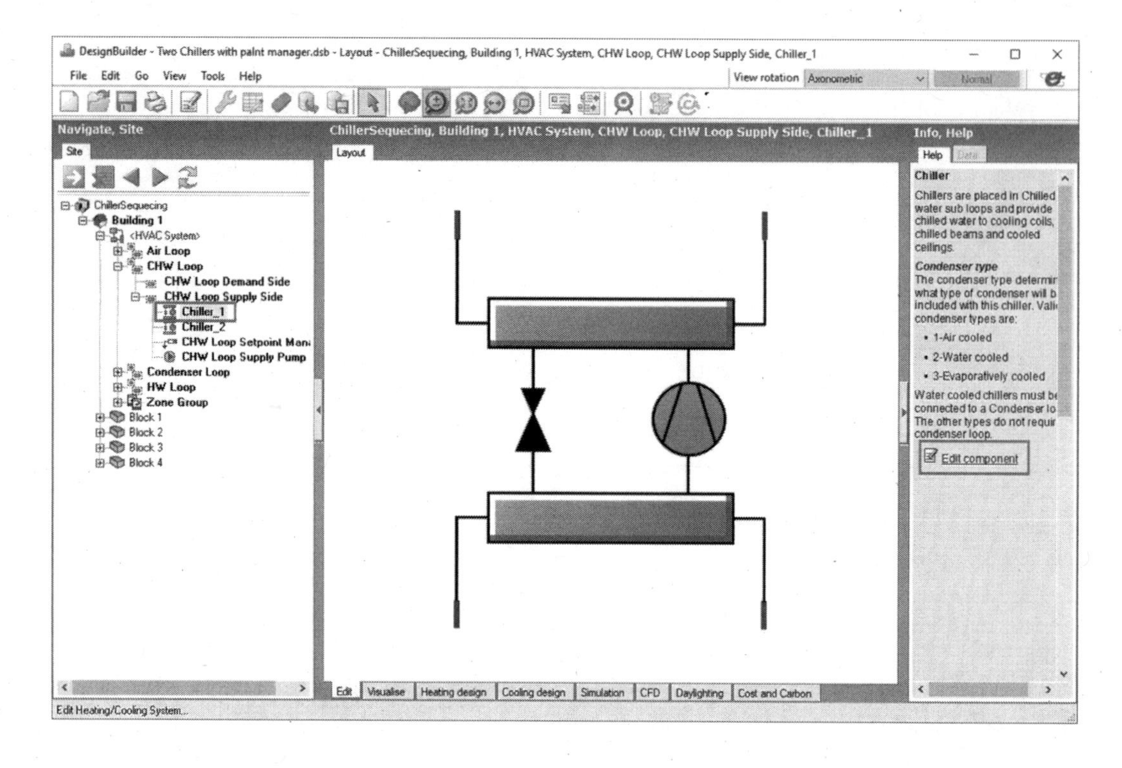

Step 12: Select **EnergyPlus Screw Chiller** for **Chiller template**. Enter **1,200,000** in the **Reference capacity (W)** text box.

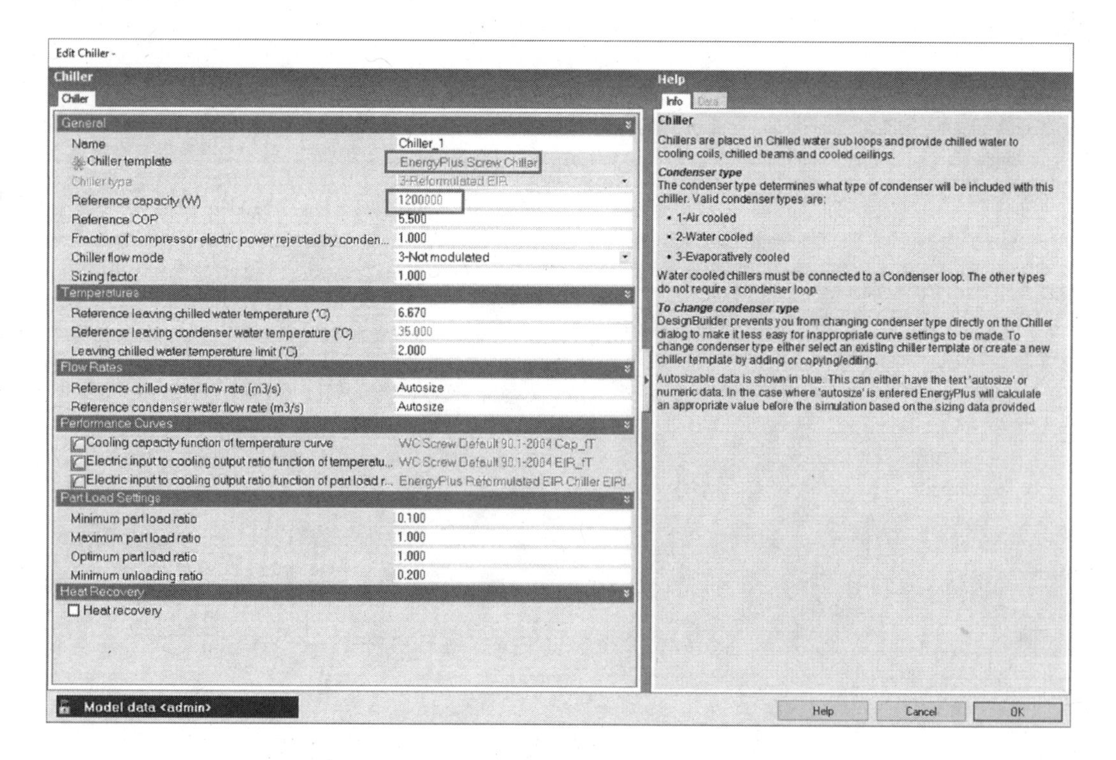

Step 13: Click **CHW Loop.** Click **Edit loop data.** Select **1-Sequential** from **Load distribution scheme** drop-down list.

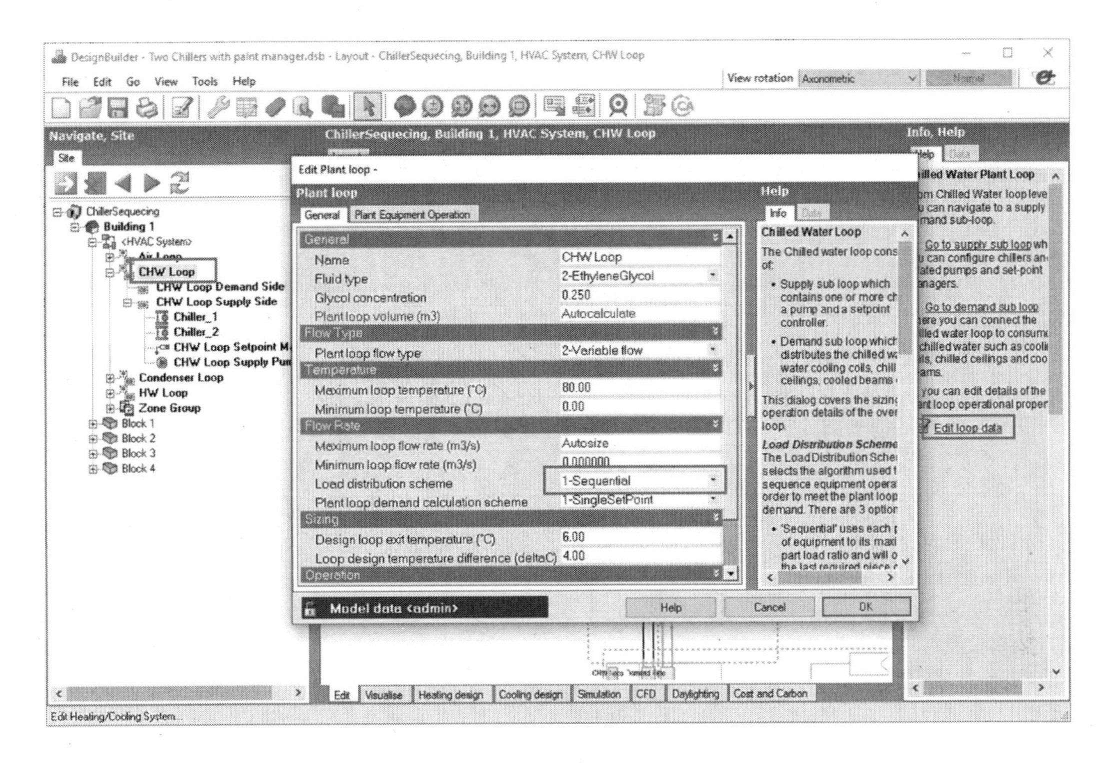

Step 14: Simulate the model, and record the results.

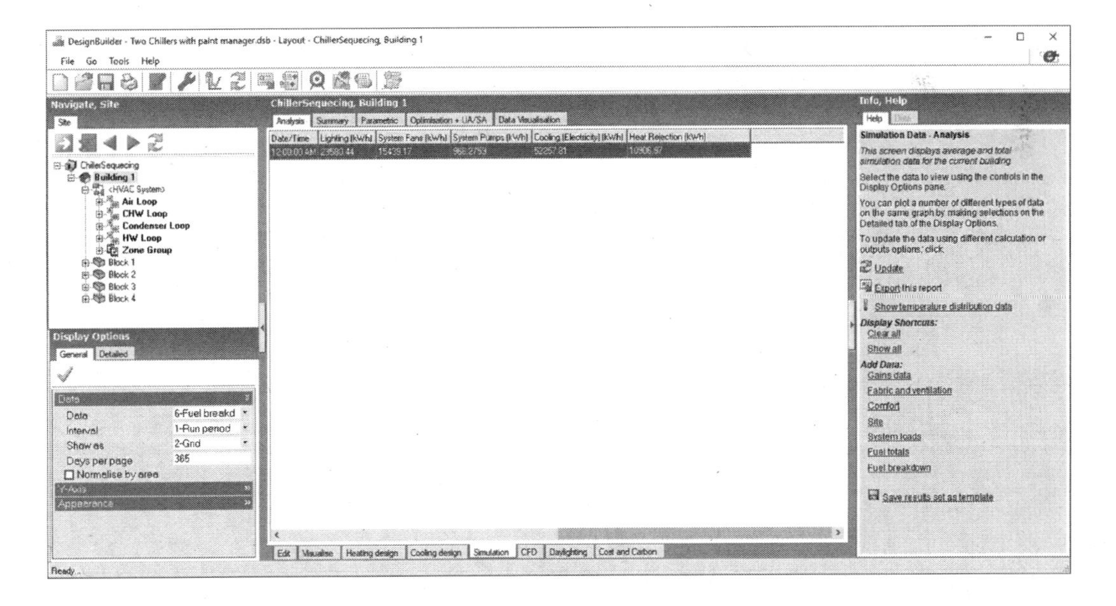

Step 15: Use **Result Viewer** to view the load on both chillers.

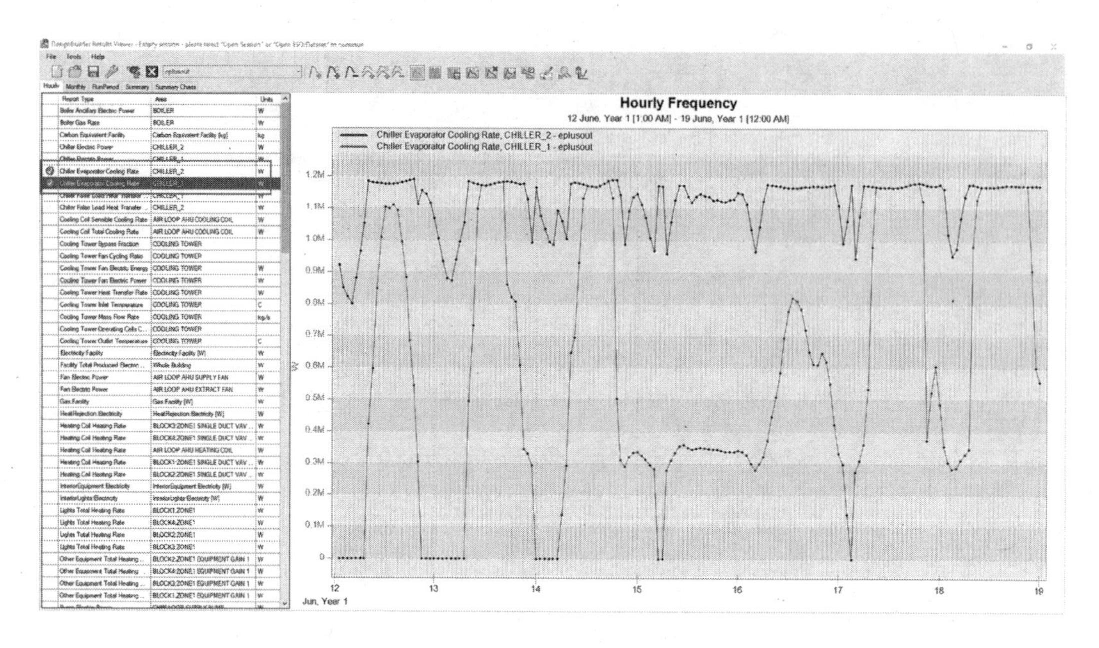

Step 16: Click the **Save Grid to CSV file** icon to save the results. Table 8.8 shows the results.

TABLE 8.8
Chiller evaporator cooling rate for both chillers

Date/time	Chiller evaporator cooling rate (W)		Total
	Chiller_1	Chiller_2	
12/06/2002 0:30	9,66,069	–	9,66,069
12/06/2002 1:00	8,79,504	–	8,79,504
12/06/2002 1:30	8,60,300	–	8,60,300
12/06/2002 2:00	8,43,395	–	8,43,395
12/06/2002 2:30	8,27,572	–	8,27,572
12/06/2002 3:00	8,19,078	–	8,19,078
12/06/2002 3:30	8,07,669	–	8,07,669
12/06/2002 4:00	7,94,614	–	7,94,614
12/06/2002 4:30	8,49,661	–	8,49,661
12/06/2002 5:00	9,10,286	–	9,10,286
12/06/2002 5:30	9,47,061	–	9,47,061

(Continued)

TABLE 8.8 (Cont.)

	Chiller evaporator cooling rate (W)		
Date/time	Chiller_1	Chiller_2	Total
12/06/2002 6:00	9,84,531	–	9,84,531
12/06/2002 6:30	10,29,065	–	10,29,065
12/06/2002 7:00	10,74,372	–	10,74,372
12/06/2002 7:30	11,86,805	3,84,224	15,71,029
12/06/2002 8:00	11,84,487	4,61,090	16,45,577
12/06/2002 8:30	11,82,934	5,53,666	17,36,600
12/06/2002 9:00	11,81,297	6,40,962	18,22,259
12/06/2002 9:30	11,80,307	8,05,852	19,86,159
12/06/2002 10:00	11,79,226	8,93,114	20,72,340
12/06/2002 10:30	11,78,467	9,76,362	21,54,829
12/06/2002 11:00	11,77,776	10,39,434	22,17,210
12/06/2002 11:30	11,77,148	10,83,513	22,60,661
12/06/2002 12:00	11,76,591	11,22,089	22,98,679
12/06/2002 12:30	11,76,495	10,95,834	22,72,329
12/06/2002 13:00	11,76,693	11,02,256	22,78,949
12/06/2002 13:30	11,76,611	11,06,763	22,83,374
12/06/2002 14:00	11,76,524	11,13,991	22,90,515
12/06/2002 14:30	11,77,629	11,15,051	22,92,681
12/06/2002 15:00	11,78,986	10,70,459	22,49,445
12/06/2002 15:30	11,80,101	10,22,079	22,02,179
12/06/2002 16:00	11,81,152	9,67,013	21,48,165
12/06/2002 16:30	11,83,343	9,08,824	20,92,167
12/06/2002 17:00	11,85,776	8,58,326	20,44,102
12/06/2002 17:30	11,87,968	7,24,099	19,12,067
12/06/2002 18:00	11,90,335	6,47,290	18,37,625
12/06/2002 18:30	11,91,408	5,61,752	17,53,160
12/06/2002 19:00	11,92,282	5,23,842	17,16,124
12/06/2002 19:30	11,26,625	3,05,990	14,32,614
12/06/2002 20:00	10,97,107	3,02,950	14,00,057
12/06/2002 20:30	11,55,003	–	11,55,003
12/06/2002 21:00	11,56,781	–	11,56,781
12/06/2002 21:30	11,48,867	–	11,48,867
12/06/2002 22:00	11,40,605	–	11,40,605
12/06/2002 22:30	11,24,087	–	11,24,087
12/06/2002 23:00	11,07,428	–	11,07,428
12/06/2002 23:30	10,81,754	–	10,81,754

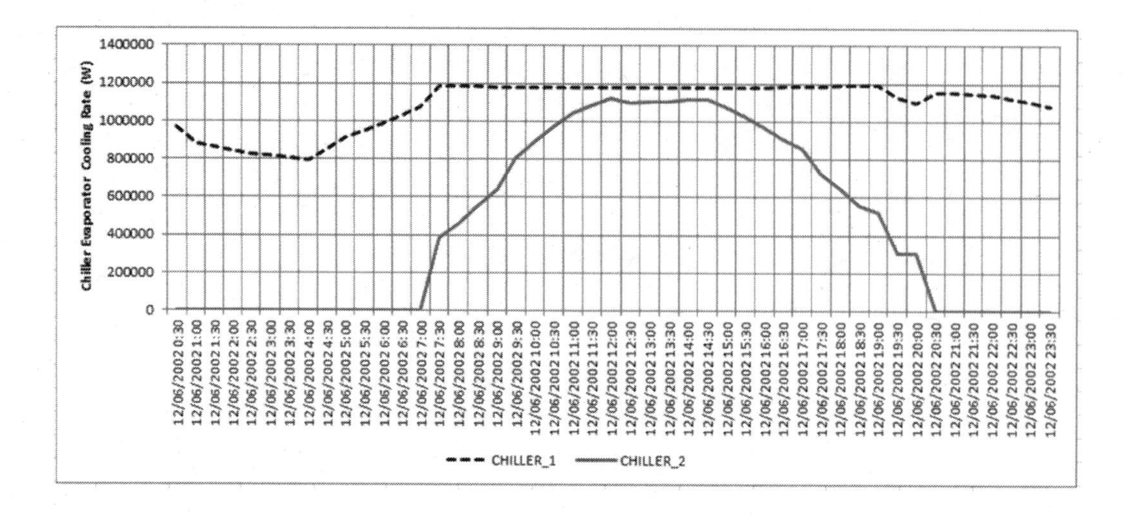

It can be seen that once the chiller 2 reaches its peak, chiller 1 starts. This shows that both are working in sequence.

Step 17: Rename and save the file for modelling the chillers with plant manager.

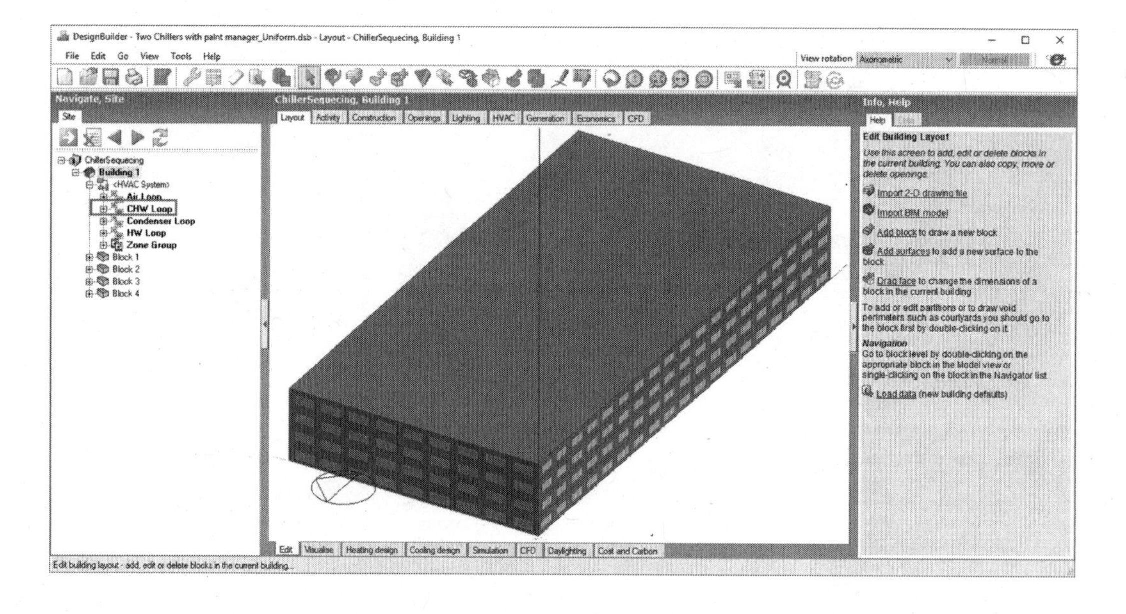

Step 18: Click **CHW Loop.** Click **Edit loop data.** Select **3-Uniform** from the **Load Distribution scheme** drop-down list.

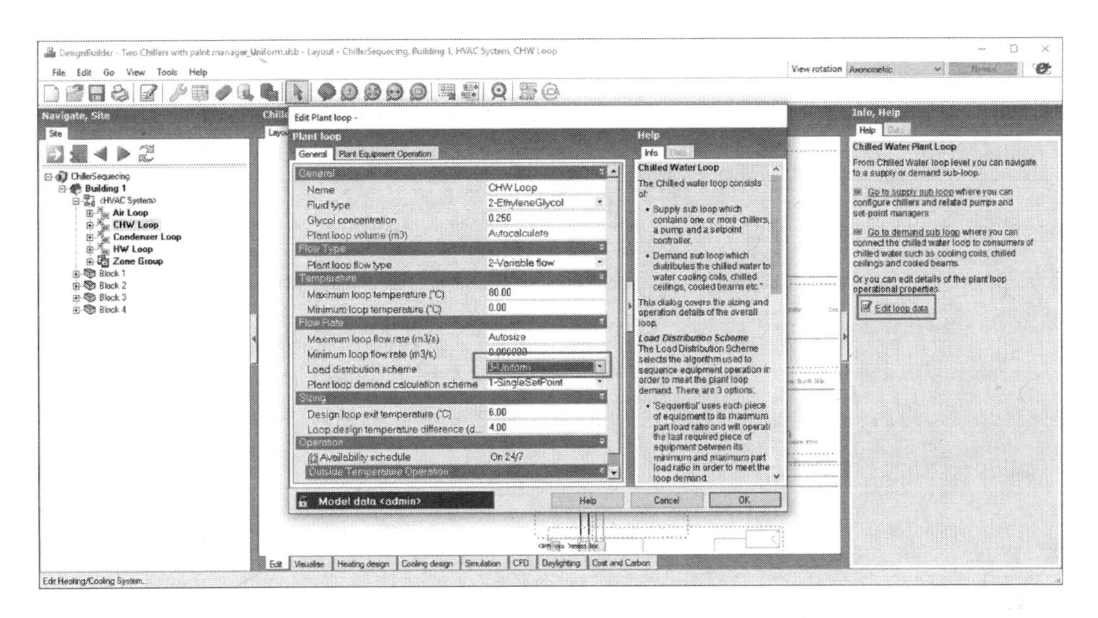

Step 19: Click the **Plant Equipment Operation** tab. Enter **2** in **Number of ranges** text box.

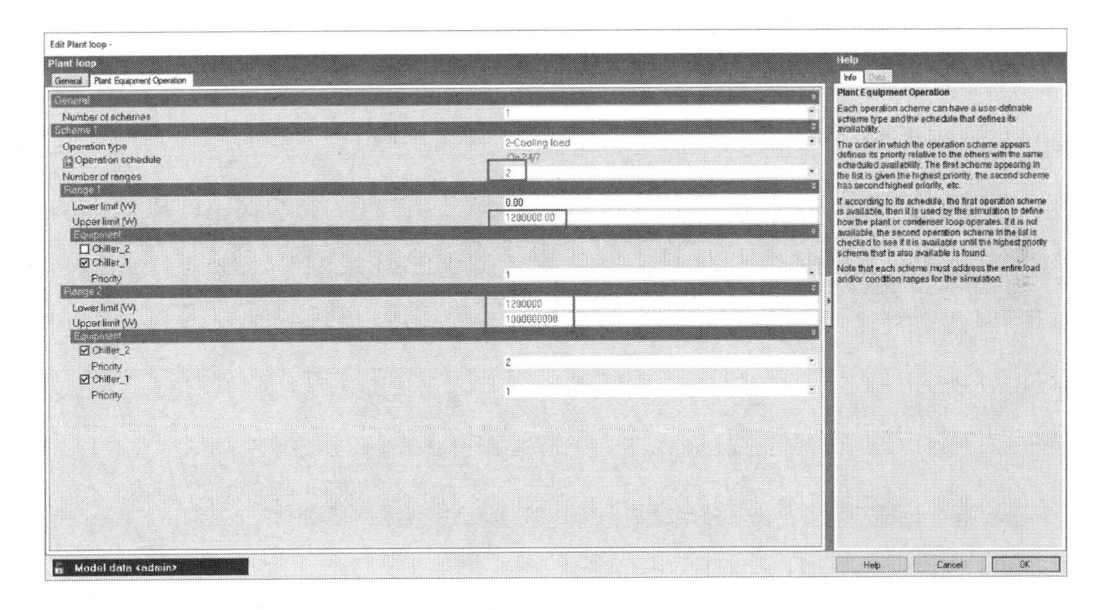

Step 20: Simulate the model, and record the results.

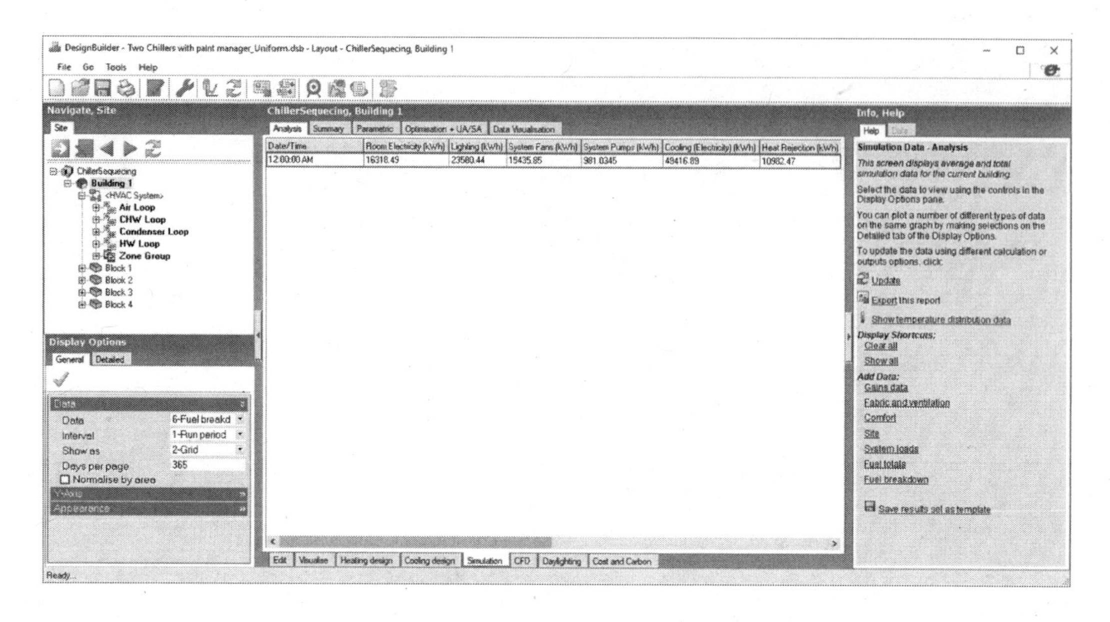

Step 21: Use **Result Viewer** to view the load on both the chillers. Table 8.9 shows the results.

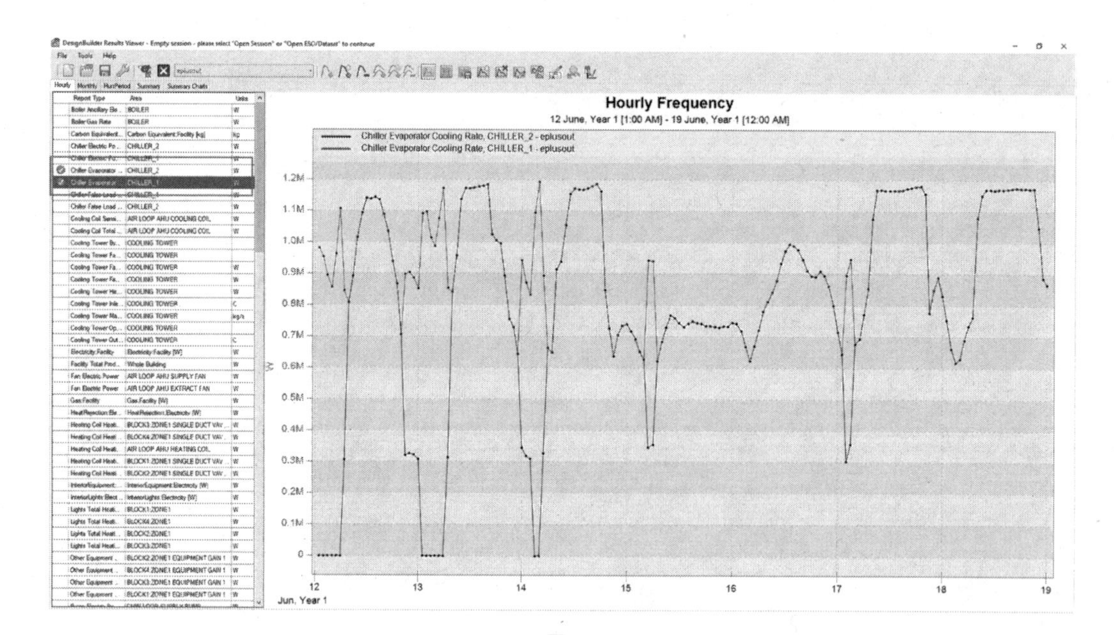

TABLE 8.9
Chiller evaporator cooling rate for both chillers with plant manager

	Chiller evaporator cooling rate (W)		
Date/time	**Chiller 1**	**Chiller 2**	**Total**
12/06/2002 0:30	9,31,590	–	9,31,590
12/06/2002 1:00	8,97,441	–	8,97,441
12/06/2002 1:30	8,64,794	–	8,64,794
12/06/2002 2:00	8,50,797	–	8,50,797
12/06/2002 2:30	8,37,169	–	8,37,169
12/06/2002 3:00	8,25,866	–	8,25,866
12/06/2002 3:30	8,18,924	–	8,18,924
12/06/2002 4:00	8,06,674	–	8,06,674
12/06/2002 4:30	7,93,597	–	7,93,597
12/06/2002 5:00	8,48,697	–	8,48,697
12/06/2002 5:30	9,09,380	–	9,09,380
12/06/2002 6:00	9,46,177	–	9,46,177
12/06/2002 6:30	9,83,650	–	9,83,650
12/06/2002 7:00	10,28,184	–	10,28,184
12/06/2002 7:30	10,73,494	–	10,73,494
12/06/2002 8:00	7,80,484	7,80,484	15,60,968
12/06/2002 8:30	8,19,639	8,19,639	16,39,278
12/06/2002 9:00	8,63,521	8,63,521	17,27,042
12/06/2002 9:30	9,07,125	9,07,125	18,14,251
12/06/2002 10:00	9,88,365	9,88,365	19,76,730
12/06/2002 10:30	10,31,692	10,31,692	20,63,384
12/06/2002 11:00	10,72,860	10,72,860	21,45,719
12/06/2002 11:30	11,04,076	11,04,076	22,08,152
12/06/2002 12:00	11,25,872	11,25,872	22,51,744
12/06/2002 12:30	11,44,884	11,44,884	22,89,768
12/06/2002 13:00	11,31,880	11,31,880	22,63,760
12/06/2002 13:30	11,34,921	11,34,921	22,69,843
12/06/2002 14:00	11,37,286	11,37,286	22,74,572
12/06/2002 14:30	11,40,911	11,40,911	22,81,823
12/06/2002 15:00	11,41,456	11,41,456	22,82,912
12/06/2002 15:30	11,19,823	11,19,823	22,39,645
12/06/2002 16:00	10,96,427	10,96,427	21,92,855
12/06/2002 16:30	10,69,664	10,69,664	21,39,329
12/06/2002 17:00	10,42,380	10,42,380	20,84,759
12/06/2002 17:30	10,18,288	10,18,288	20,36,576
12/06/2002 18:00	9,52,719	9,52,719	19,05,439
12/06/2002 18:30	9,15,681	9,15,681	18,31,362
12/06/2002 19:00	8,74,198	8,74,198	17,48,397
12/06/2002 19:30	8,55,876	8,55,876	17,11,753
12/06/2002 20:00	7,11,294	7,11,294	14,22,589
12/06/2002 20:30	6,96,067	6,96,067	13,92,134
12/06/2002 21:00	11,48,047	–	11,48,047

(*Continued*)

TABLE 8.9 (Cont.)

Date/time	Chiller evaporator cooling rate (W)		Total
	Chiller 1	Chiller 2	
12/06/2002 21:30	11,51,439	–	11,51,439
12/06/2002 22:00	11,43,588	–	11,43,588
12/06/2002 22:30	11,35,426	-	11,35,426
12/06/2002 23:00	11,19,014	-	11,19,014
12/06/2002 23:30	11,02,468	-	11,02,468

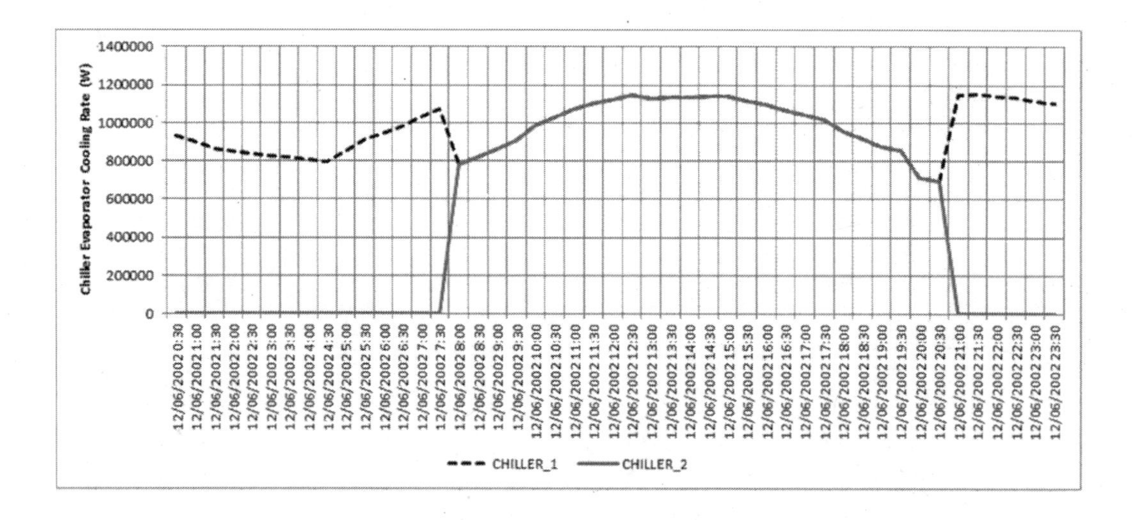

Step 22: Compare the results (see Table 8.10).

TABLE 8.10
Annual energy consumption with sequential and uniform operation

Operating mode	Room electricity (kWh)	Lighting (kWh)	System fans (kWh)	System pumps (kWh)	Cooling (electricity) (kWh)	Heat rejection (kWh)
Sequential	16,318.49	23,580.44	15,439.17	966.28	52,257.81	10,906.97
Uniform	16,318.49	23,580.44	15,435.85	981.03	49,416.89	10,982.40

You can see that because of different sequencing strategies, chiller energy consumption varies.

Tutorial 8.8

Evaluating the Impact of Thermal Storage and Time-of-Use Tariffs

GOAL

To evaluate the impact of thermal storage and time-of-use tariffs on energy cost

WHAT ARE YOU GOING TO LEARN?

- How to model a chilled-water thermal storage system
- How to model an electricity tariff with time-of-use (TOU) rates

PROBLEM STATEMENT

In this tutorial, you are going to use a 100- × 50-m model (five zones with a 10-m perimeter depth) with two identical floors. You need to model **VAV reheat water cooled chiller system** and perform comparisons for the following cases:

1. With chilled-water thermal storage
2. Without chilled-water thermal storage

You need to simulate the model with the data provided in Table 8.11.

TABLE 8.11
Input parameters

Parameter	Value
Building operating schedule	Daytime office
Thermal storage charging schedule	12–6 a.m.
Electricity tariff	7 a.m. to 8 p.m.: $0.18 per kWh 8 p.m. to 7 a.m.: $0.09 per kWh
Plant operation to meet building cooling load	Priority 1: Thermal storage Priority 2: Chiller

Perform a simulation for a summer design week, and compare the energy cost for both cases. Use the **New Delhi Palam, India** weather location.

Benefits of a Chilled-Water Storage System

Chilled water is normally generated using the off-peak energy supply, stored in chilled-water storage tanks and then distributed for use during peak hours. The economic benefits of chilled-water storage systems therefore generally rely on lower off-peak electrical rates.

Chillers running at lower ambient temperatures, which are typical of off-peak use, tend to have higher efficiencies. It is usually possible to install smaller chillers, heat rejection equipment and pumps when stored chilled water is used to meet peak cooling loads.

Chilled-water storage tanks are typically placed on the supply side of a primary chilled-water loop in parallel with one or more chillers.

Source: https://designbuilder.co.uk/helpv6.0/#ChilledWaterStorage.html.

SOLUTION

Step 1: Open a new project, and create a **100- × 50-m** two-floor building with the specifications provided in the problem statement.

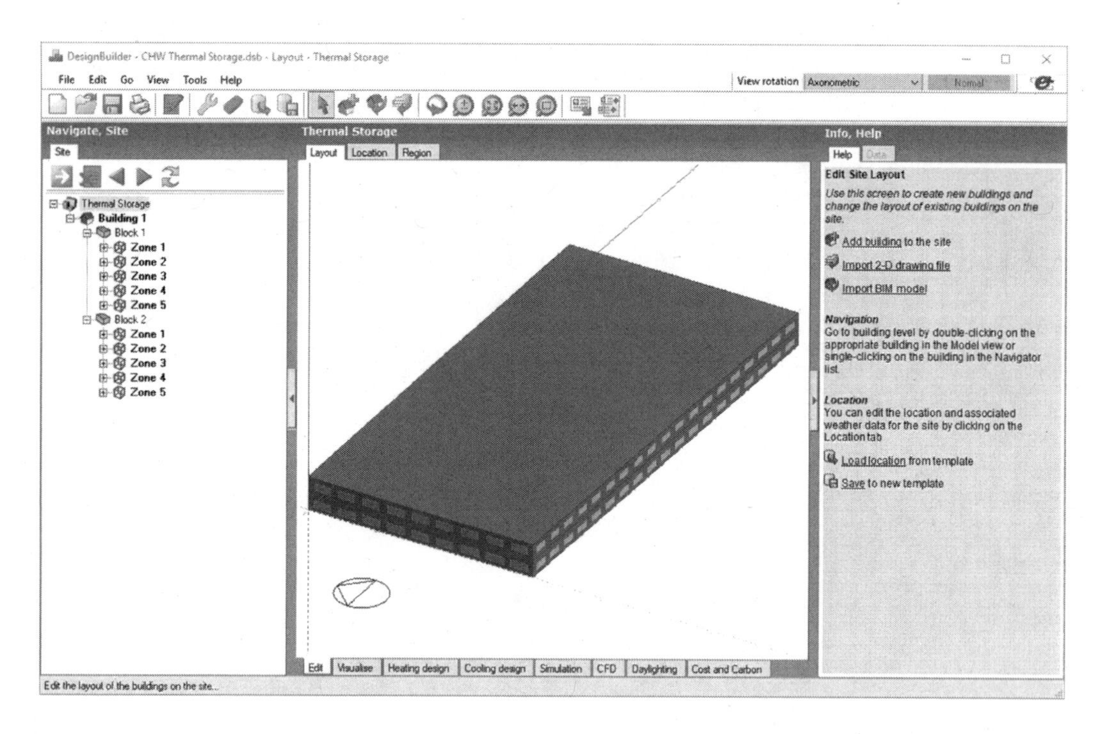

Step 2: Click **Building 1**. Click the **Activity** tab, and make sure that the **Generic Office Area** template is selected.

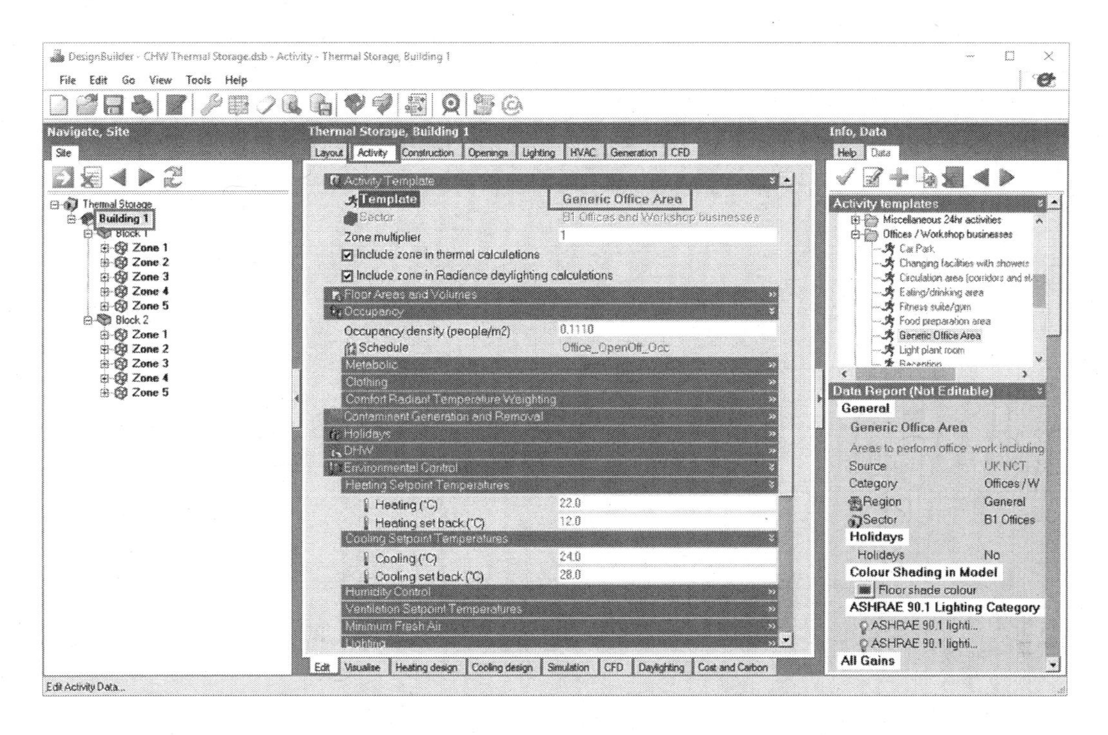

Step 3: Click the **Edit** menu, and select **Model Options** to switch to detailed HAVC. Click **HVAC System**. Click the **Load HVAC template** link. Select **VAV Reheat Water-cooled Chiller**.

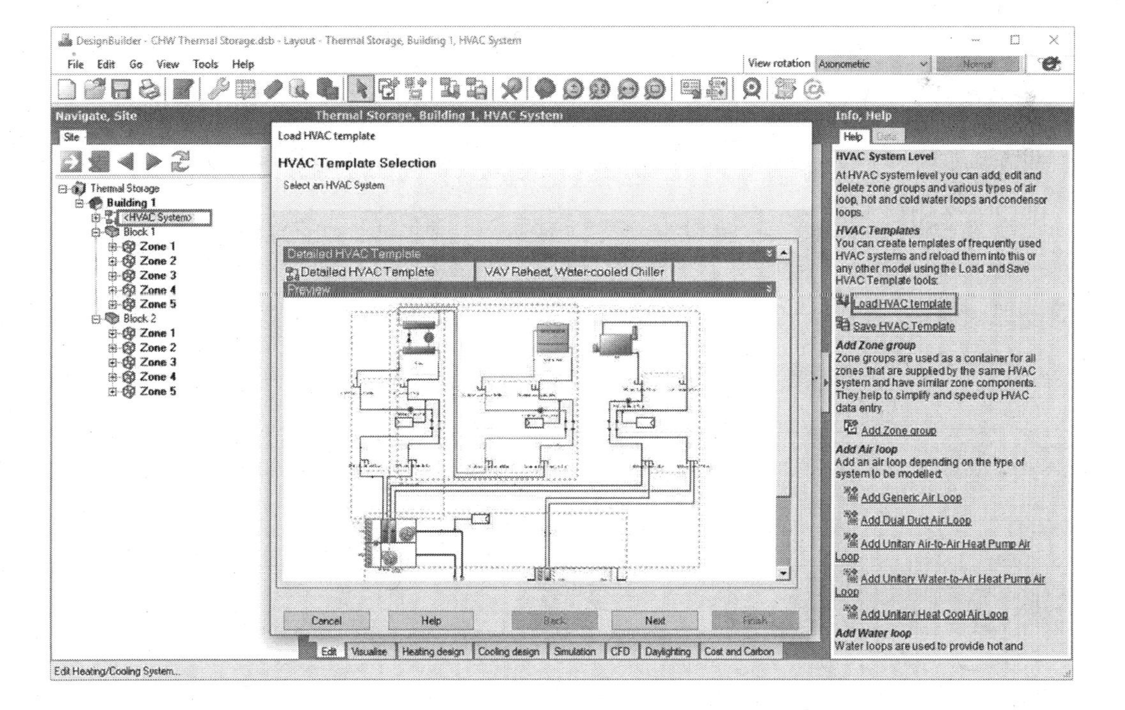

Step 4: Expand the **CHW Loop**. Click **CHW Loop Supply Side**.

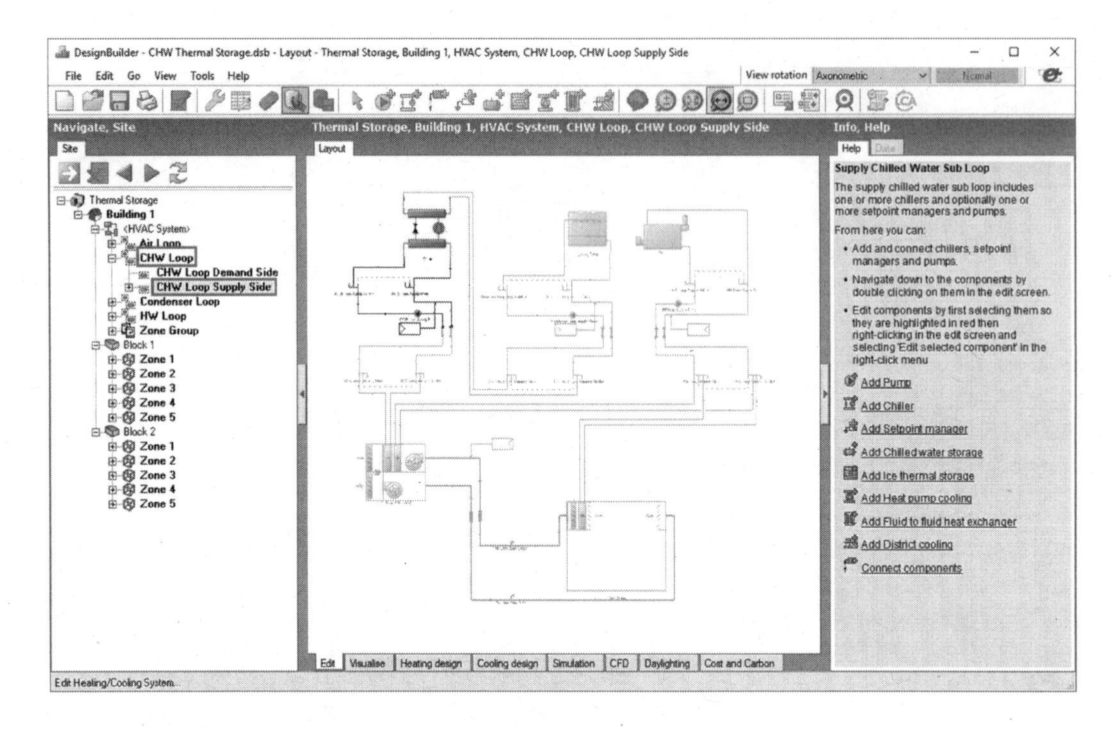

Step 5: Click **Add Chilled Water Storage** under the **Help** tab.

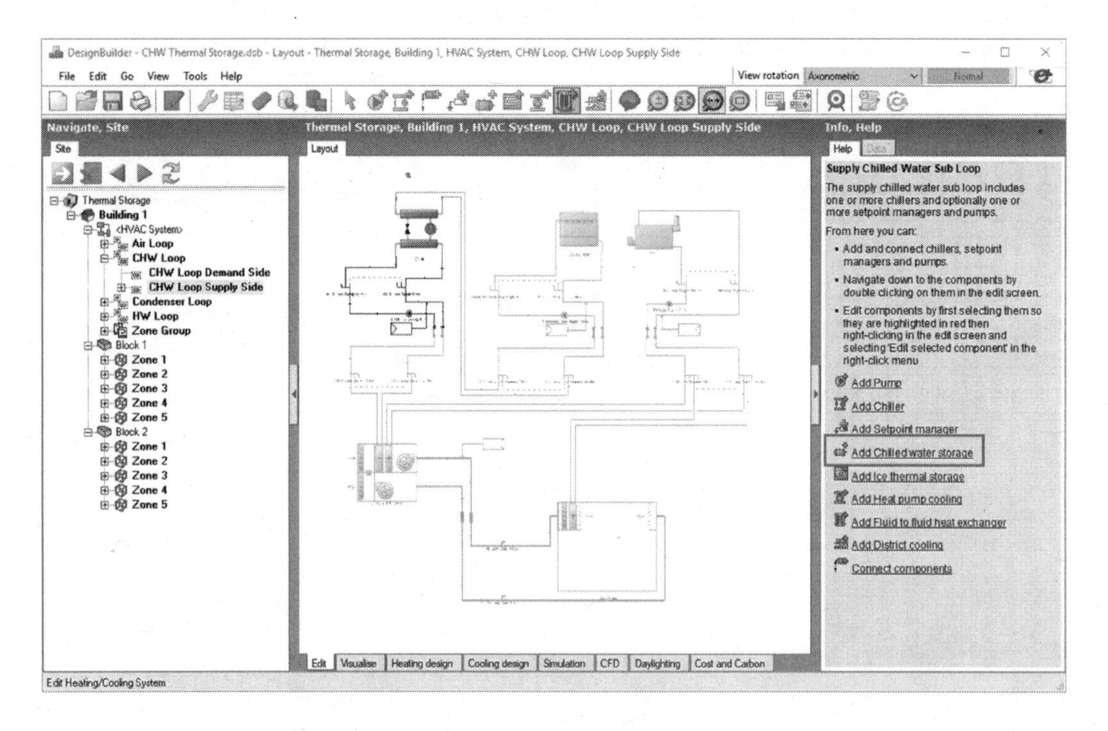

Step 6: Place **Chilled Water Storage** on the **Layout**.

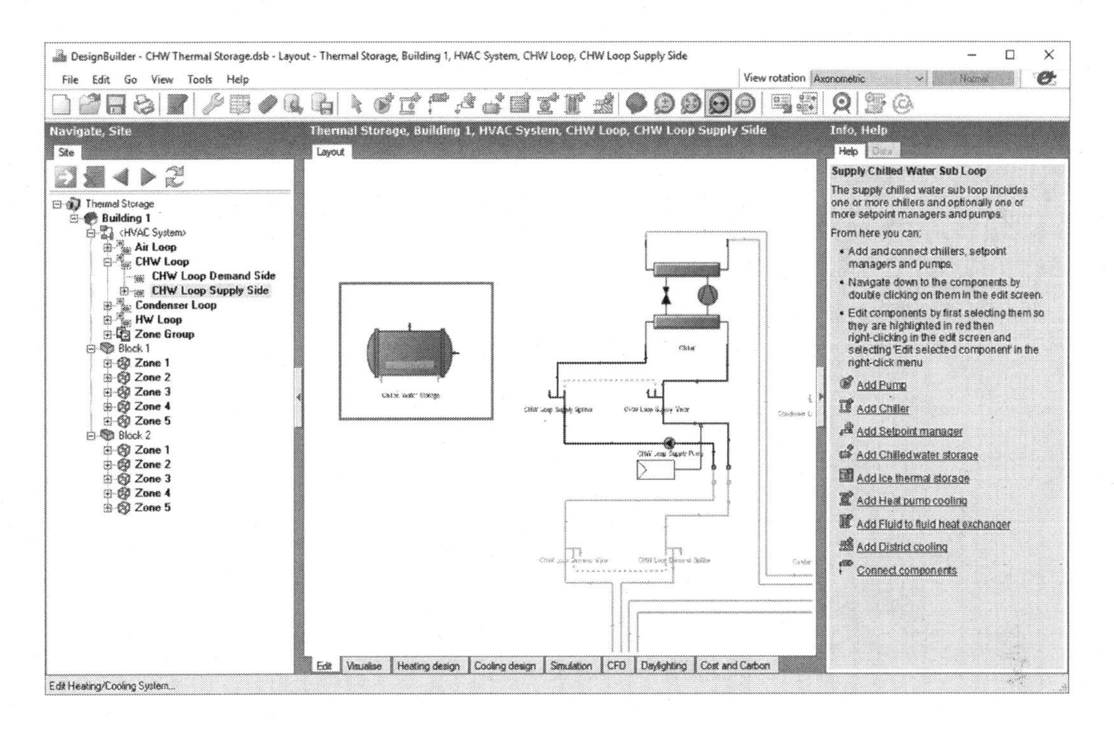

Step 7: Click **Connect components**.

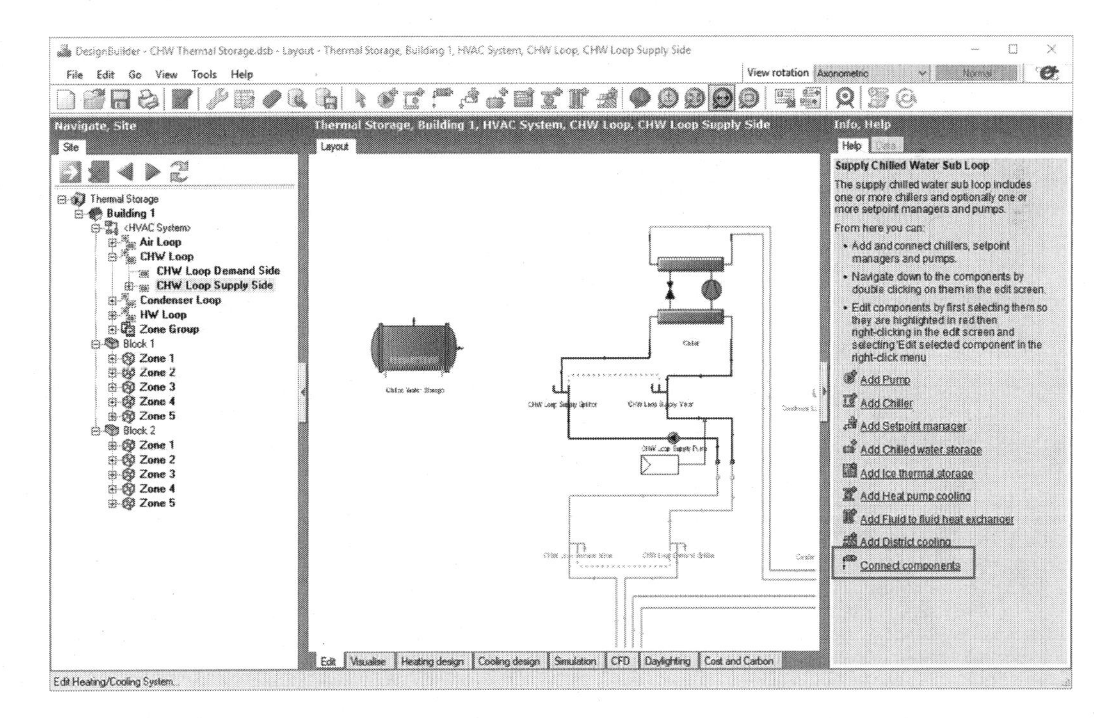

Step 8: Click the inlet of the **Chilled Water Storage**.

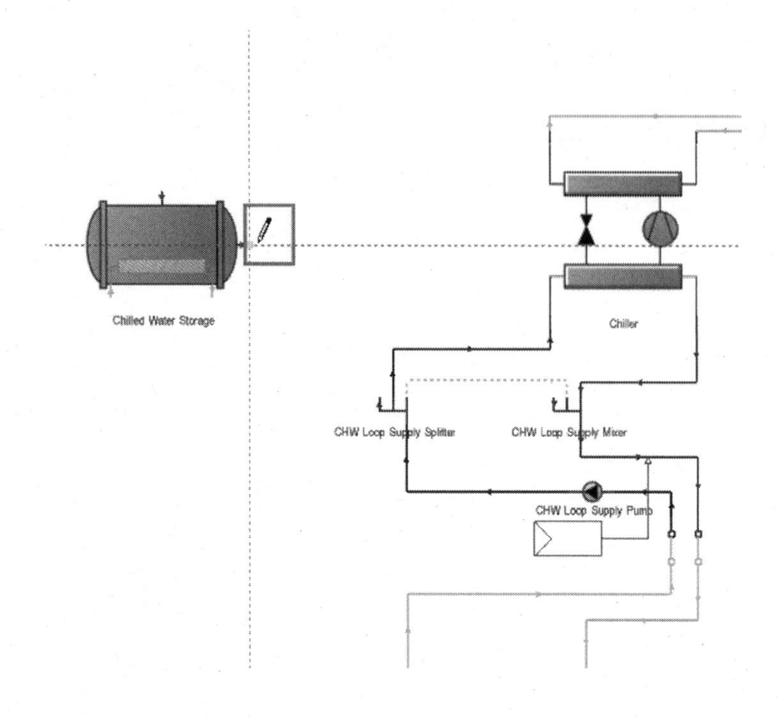

Step 9: Drag the cursor and connect it to **CHW Loop Supply mixer**.

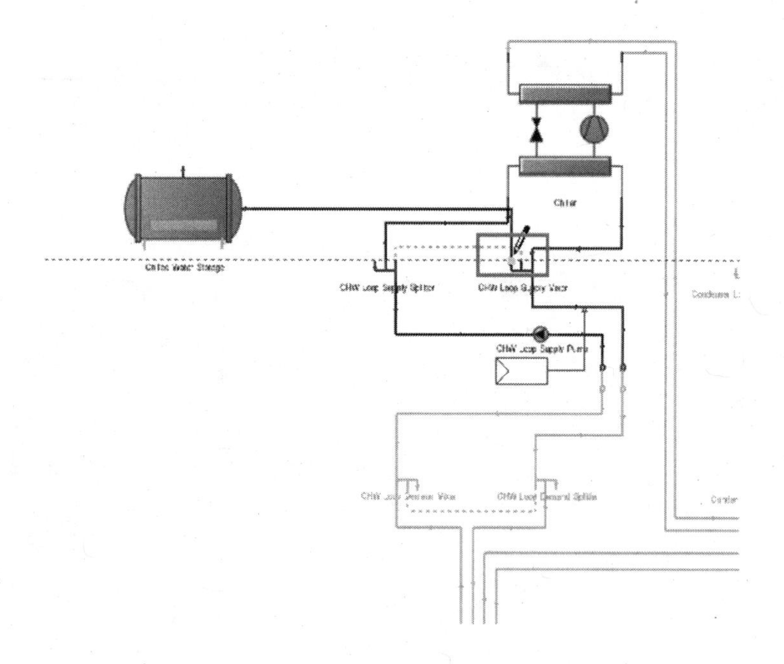

Step 10: Connect the outlet of the **Chilled Water Storage** to **CHW Loop Supply Splitter**.

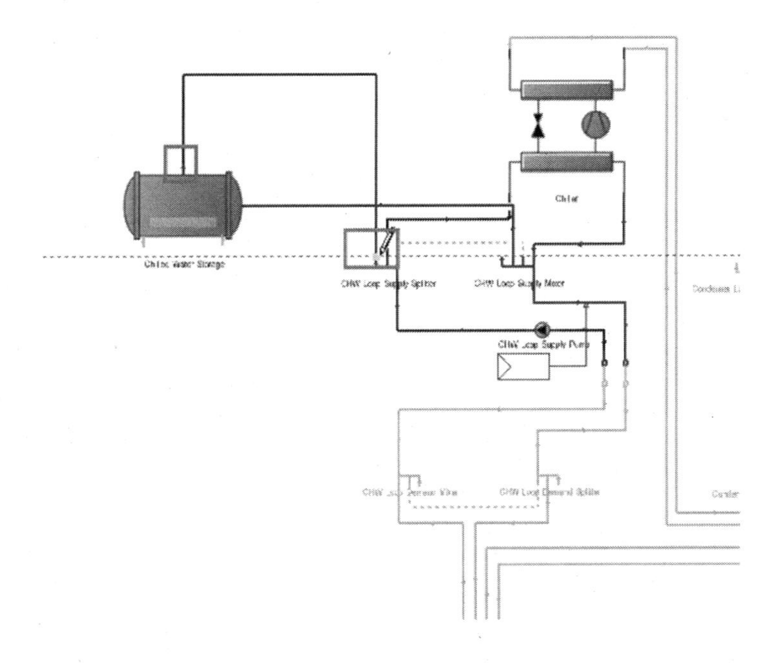

Step 11: Click **CHW Loop Demand Side**. Click **Connect components**.

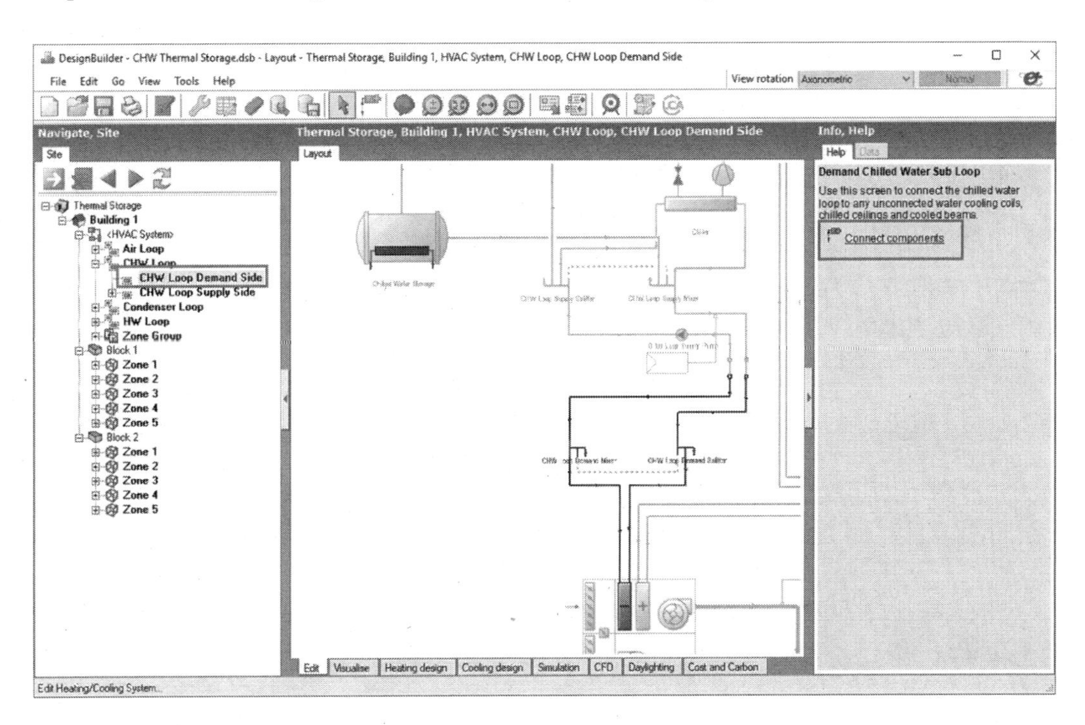

Step 12: Connect the **Chilled Water Storage** to demand-side loop.

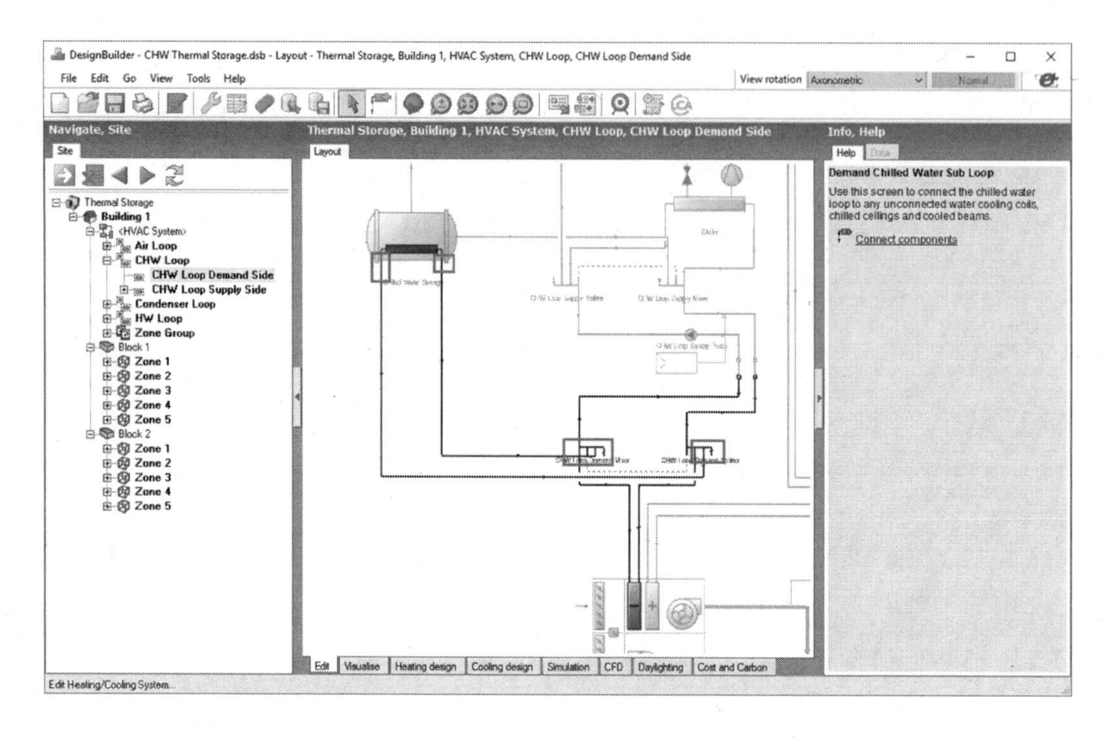

Step 13: Click **CHW Loop**. Click **Edit loop data**.

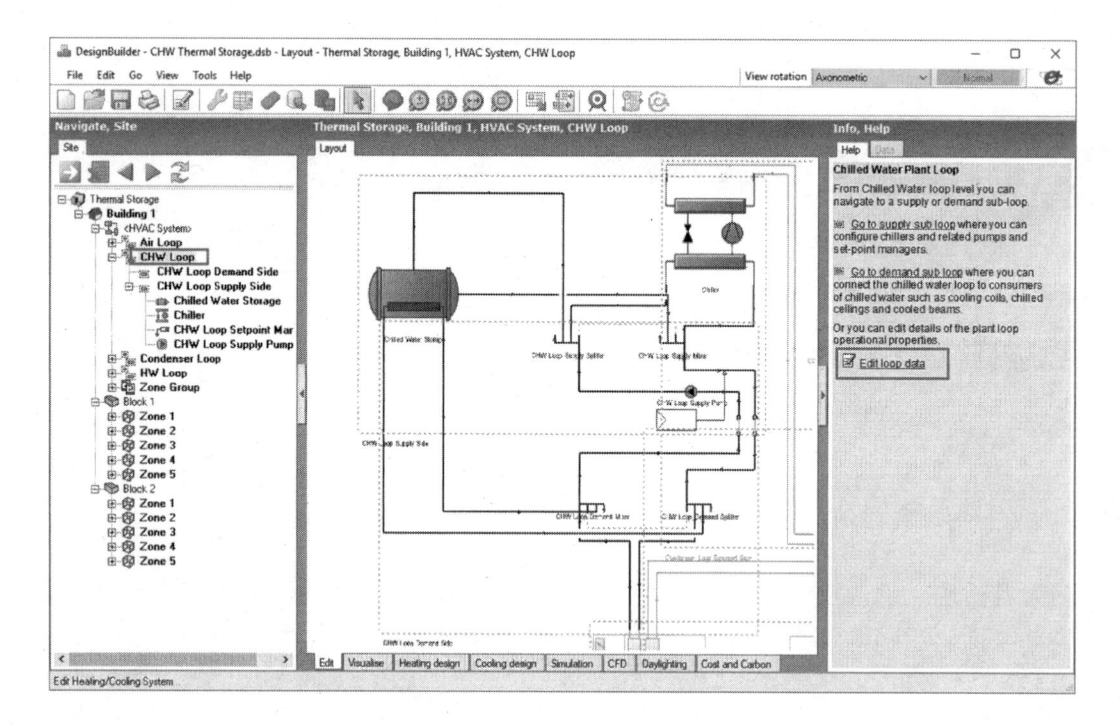

Step 14: Click the **Plant Equipment Operation** tab.

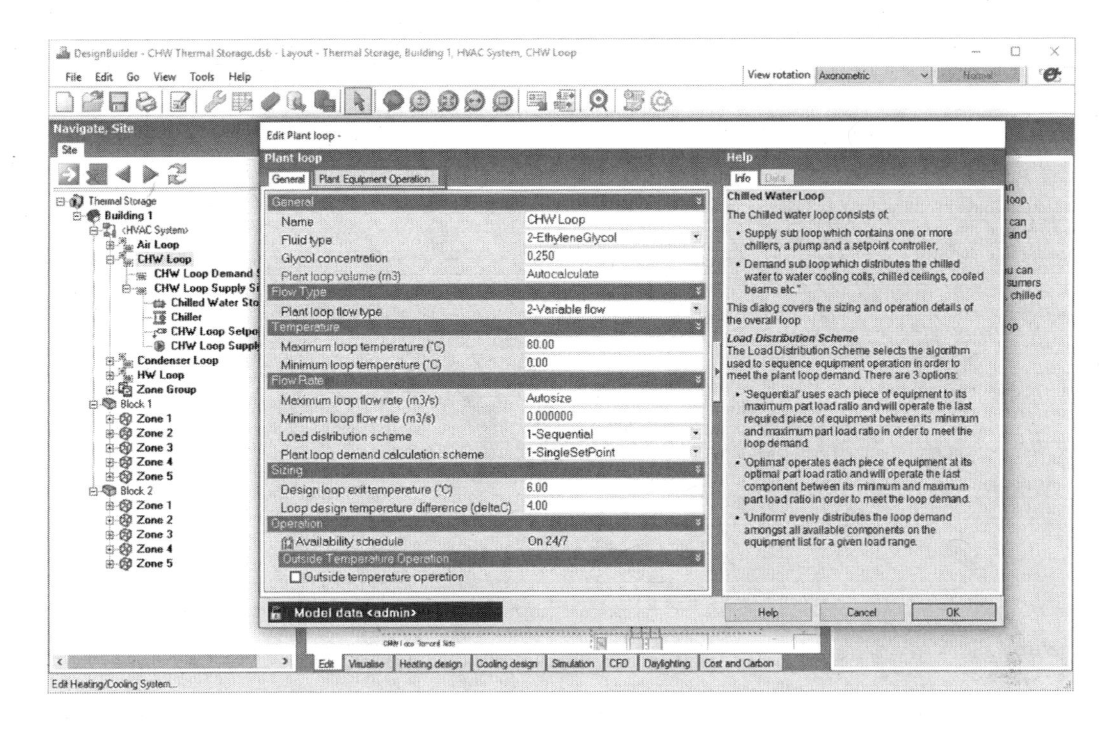

Step 15: Select the **Chilled Water Storage Priority** checkbox, and select **1** from the drop-down list.

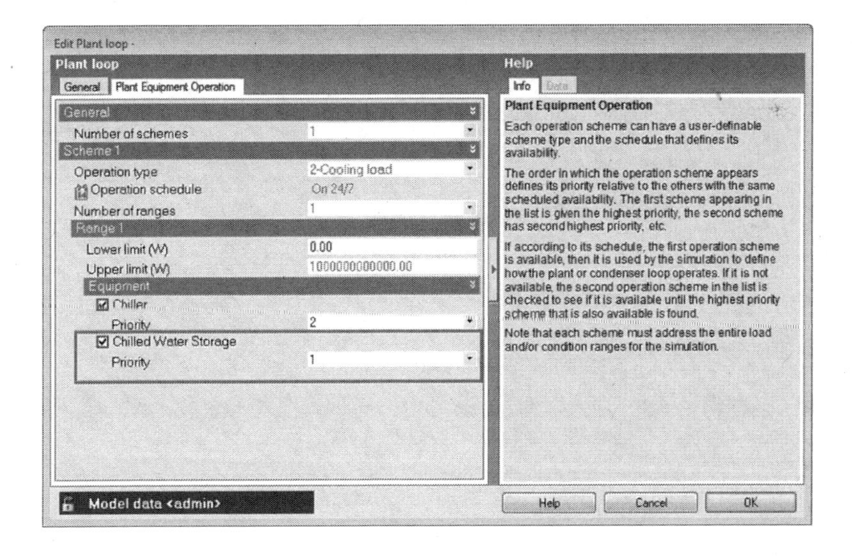

Step 16: Click **Chilled Water Storage**. Click **Edit component**.

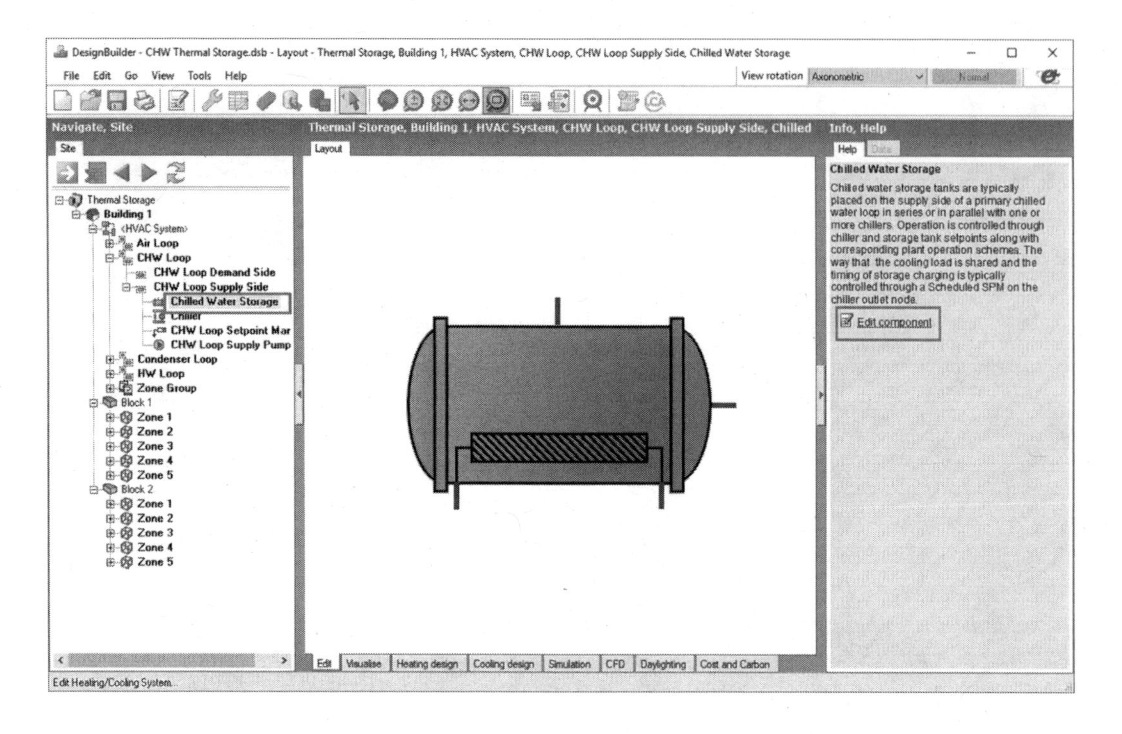

Step 17: Click the **On 24/7 Source side availability schedule**.

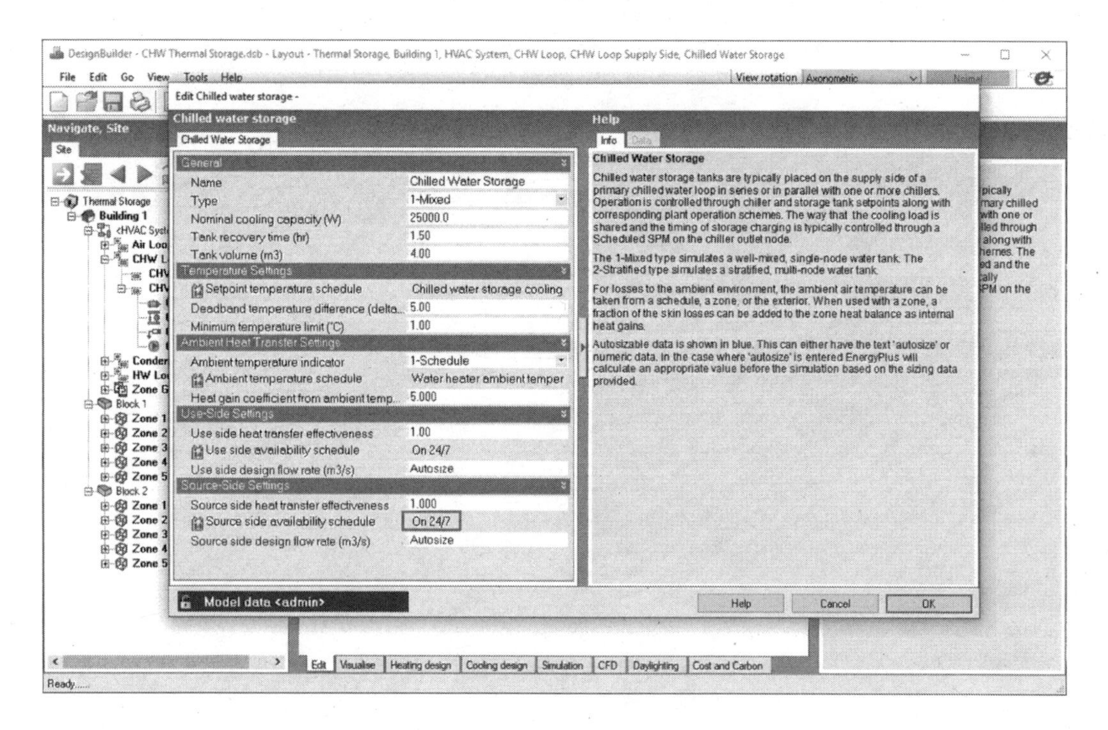

Step 18: Create a new schedule named **Night time schedule 11 PM to 6 AM.**

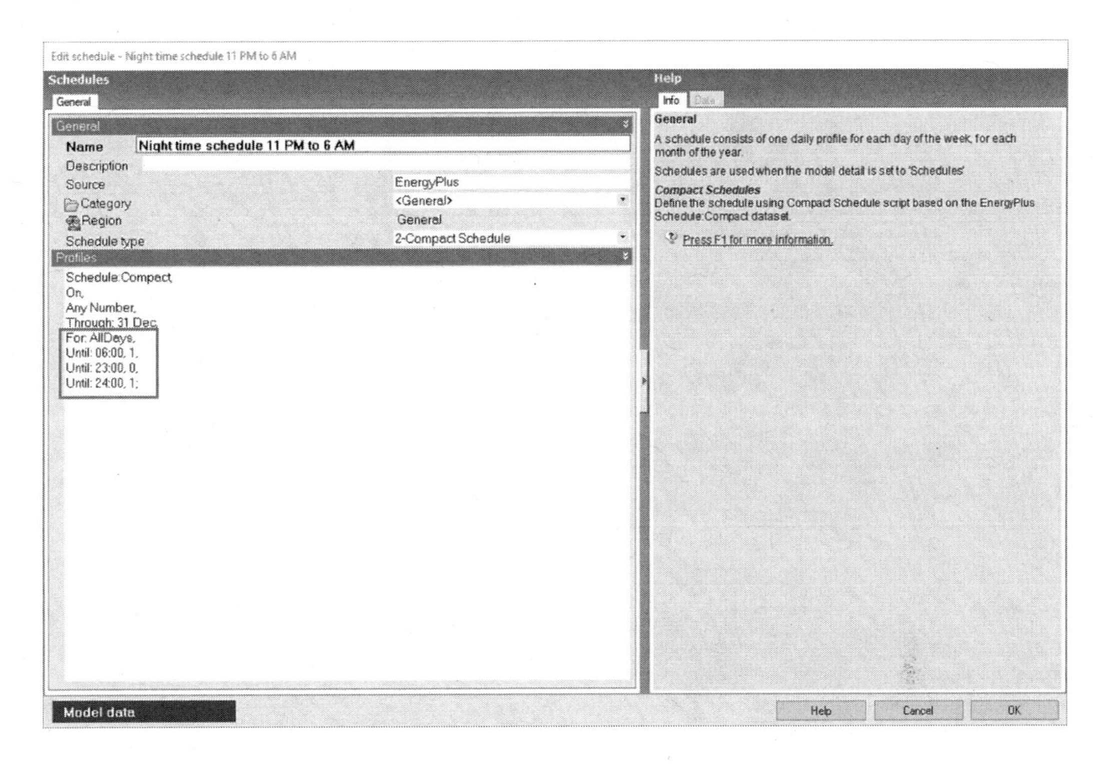

Step 19: Apply this schedule to the **Source side availability schedule**.

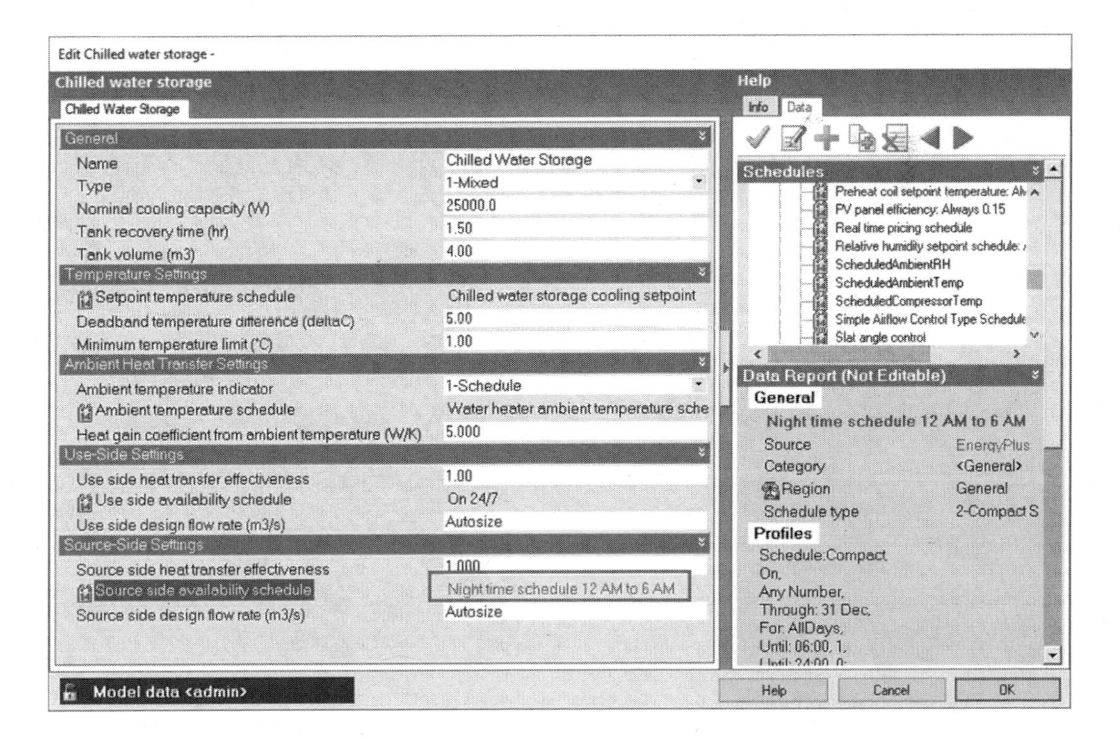

Step 20: Enter **1,000,000** in the **Nominal cooling capacity (W)** text box. Enter **7.00** in the **Tank recovery time (hr)** text box. Enter **1500** in the **Tank volume (m^3)** text box.

Edit Chilled water storage -

Chilled water storage

Chilled Water Storage

General	
Name	Chilled Water Storage
Type	1-Mixed
Nominal cooling capacity (W)	1000000.0
Tank recovery time (hr)	7.00
Tank volume (m3)	1500
Temperature Settings	
Setpoint temperature schedule	Chilled water storage cooling setpoint
Deadband temperature difference (deltaC)	0.50
Minimum temperature limit (°C)	1.00
Ambient Heat Transfer Settings	
Ambient temperature indicator	1-Schedule
Ambient temperature schedule	Water heater ambient temperature schedule: Always 20.00
Heat gain coefficient from ambient temperature (W/K)	5.000
Use-Side Settings	
Use side heat transfer effectiveness	1.00
Use side availability schedule	On 24/7
Use side design flow rate (m3/s)	Autosize
Source-Side Settings	
Source side heat transfer effectiveness	1.000
Source side availability schedule	Night time schedule 11 PM to 6 AM
Source side design flow rate (m3/s)	Autosize

Step 21: Click **Air Loop AHU**. Click **Edit component**.

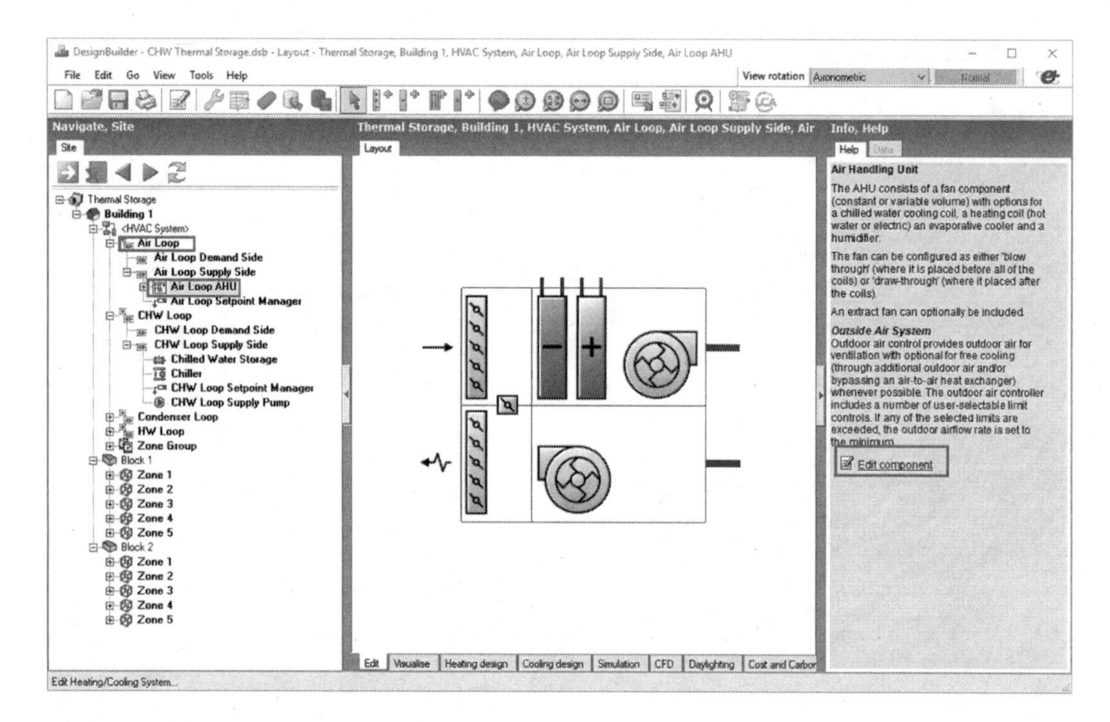

Step 22: Click **On 24/7**.

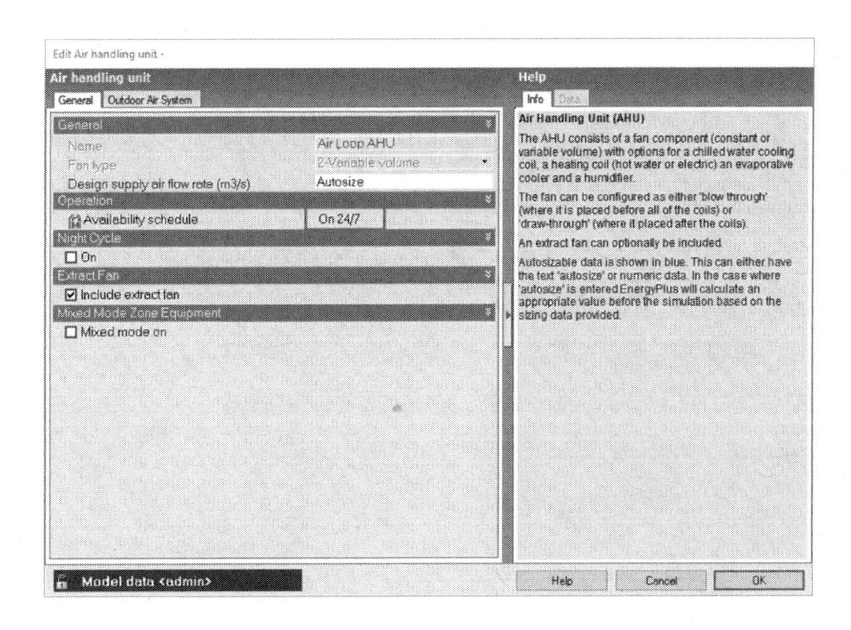

Step 23: Create a new schedule **AHU Operation 7 AM to 7 PM**. Click **OK**.

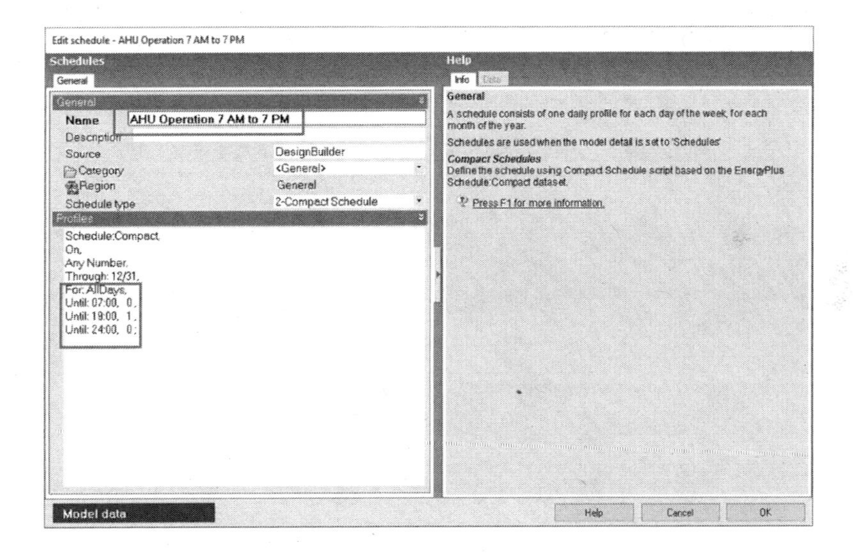

Step 24: Apply the schedule.

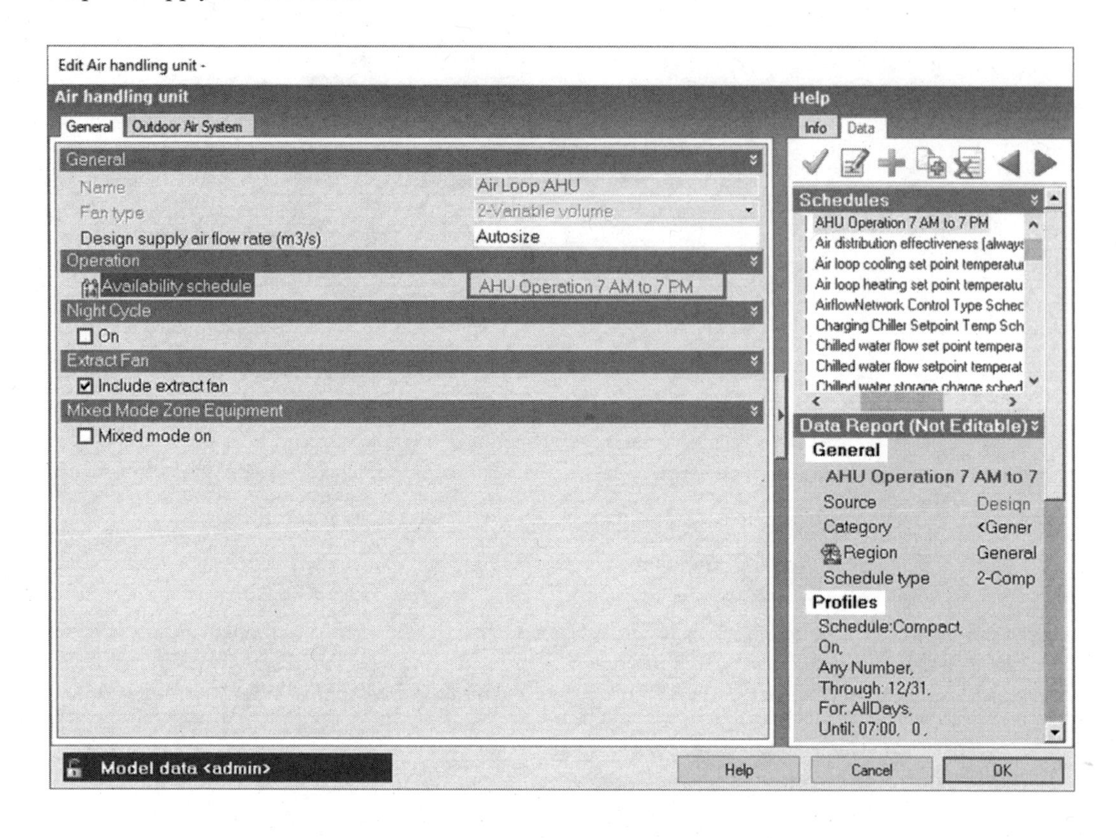

Step 25: Click the **Simulation** screen. Click **Summer design week**.

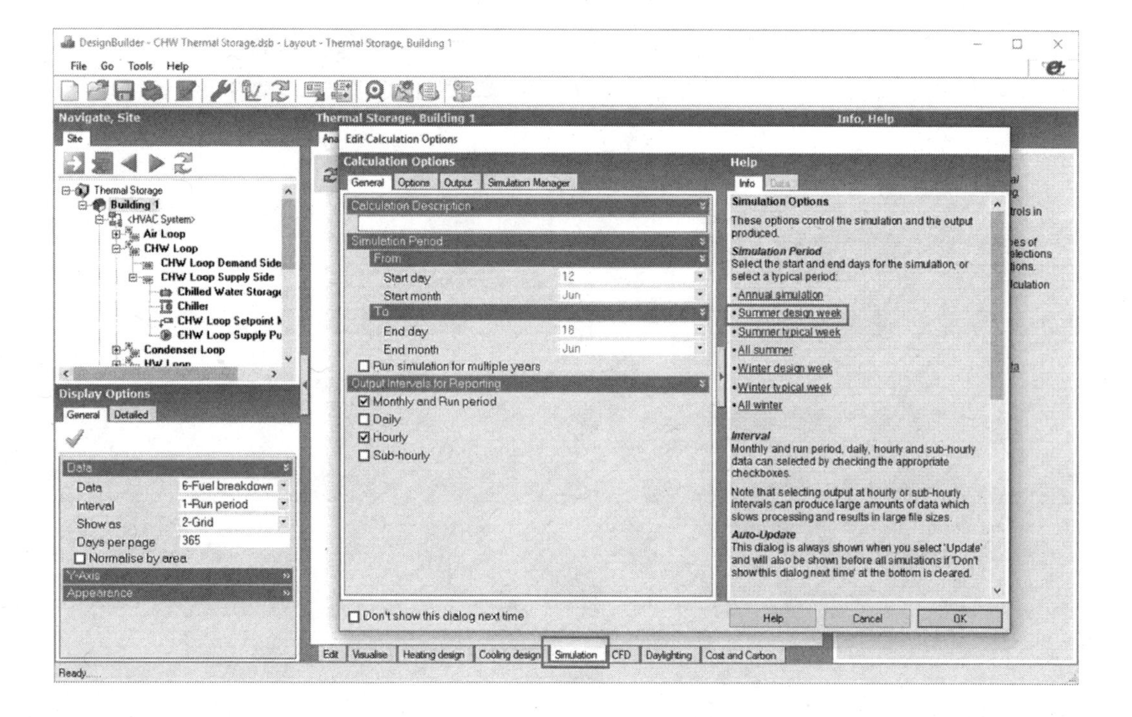

Step 26: After the simulation is over, click the **View EnergyPlus results** icon.

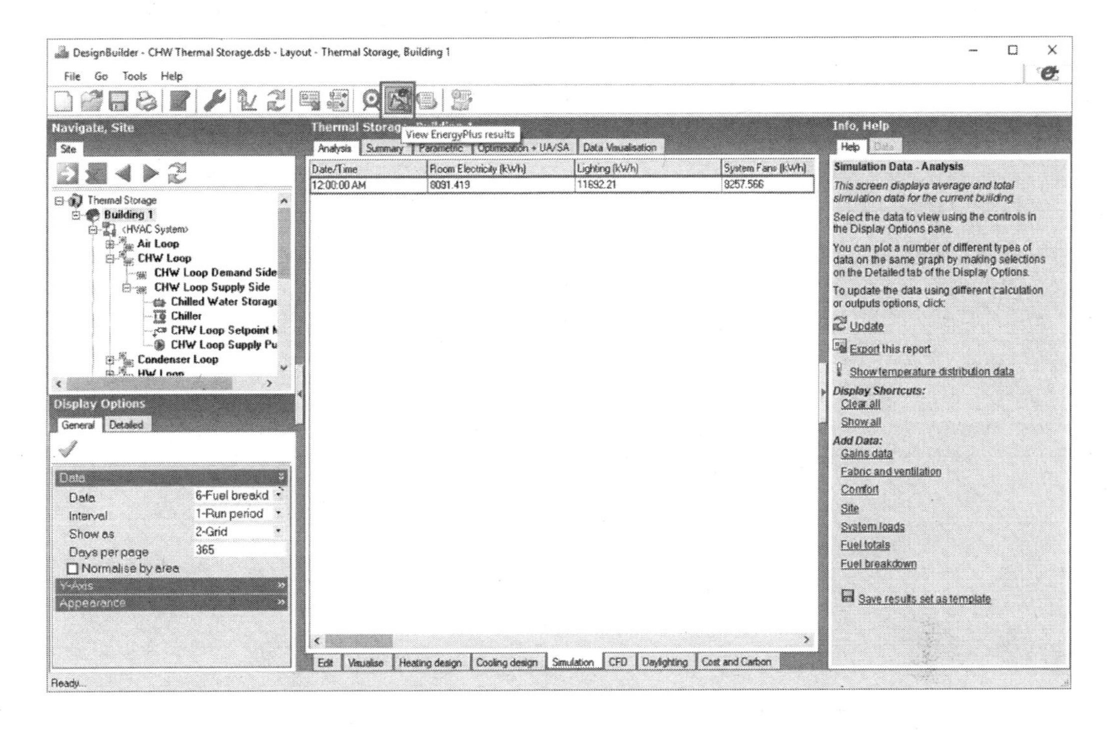

Step 27: Click the **Hourly** tab. Select **Chiller Evaporator Cooling Rate**.

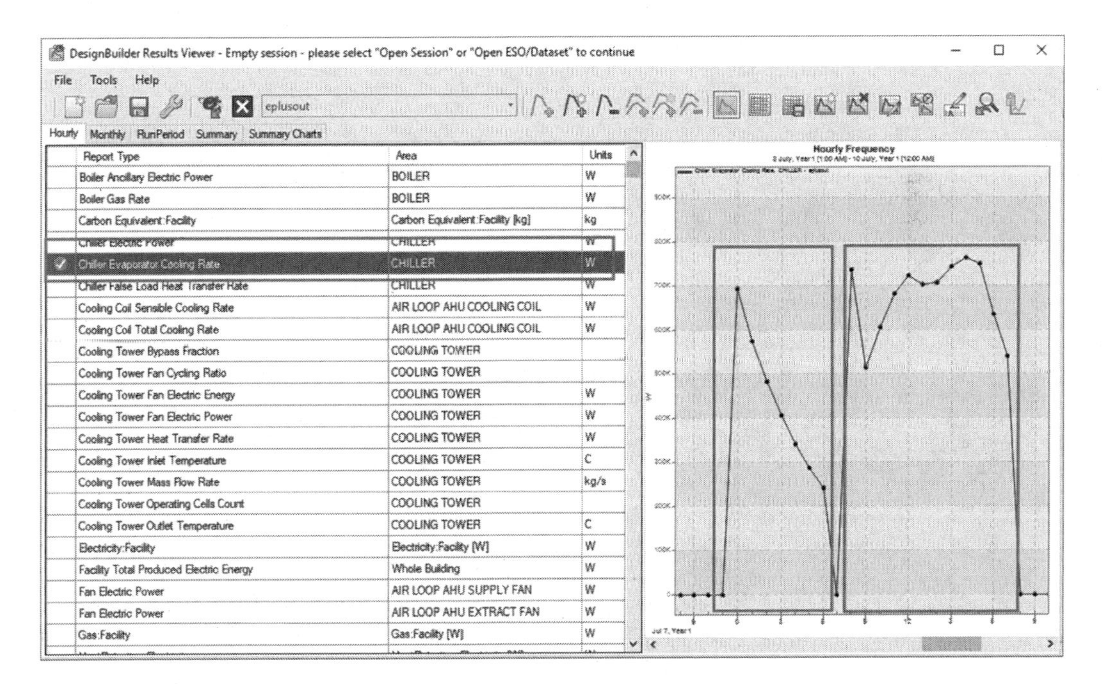

You can see that the chiller is operational at night. It is charging the chilled-water storage. It can also be seen that the chiller evaporator cooling rate decreases and then increases at the start of the occupancy.

Step 28: Click the **Edit** tab.

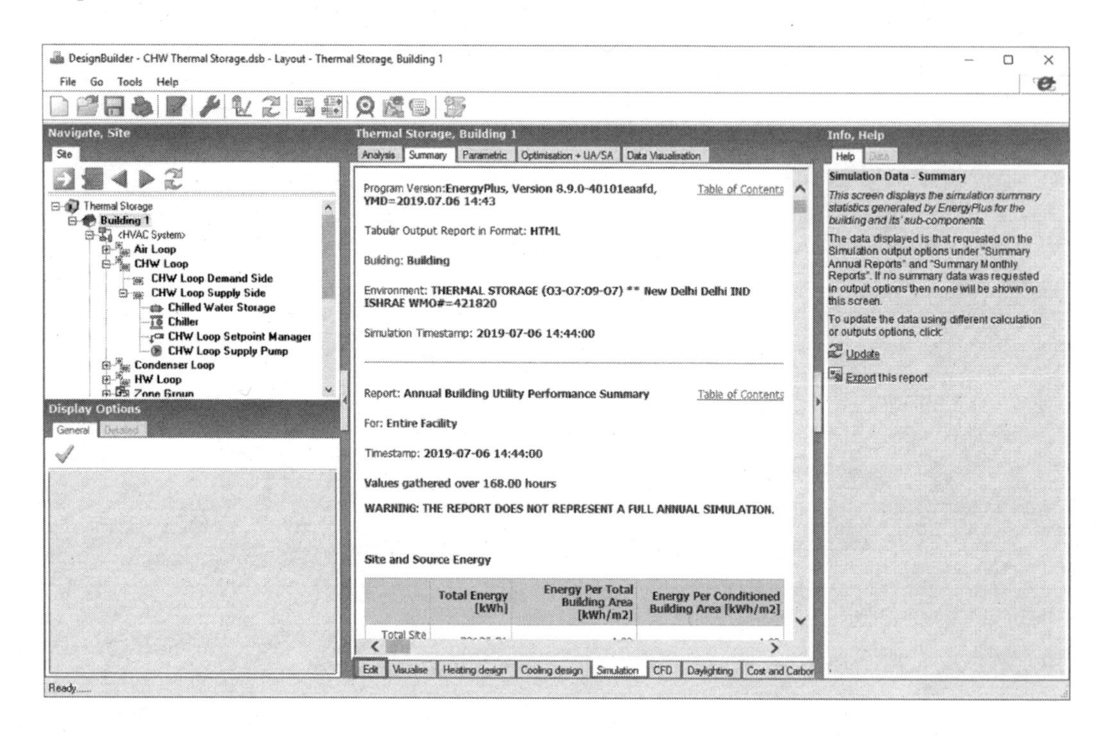

Step 29: Click the **Economics** tab.

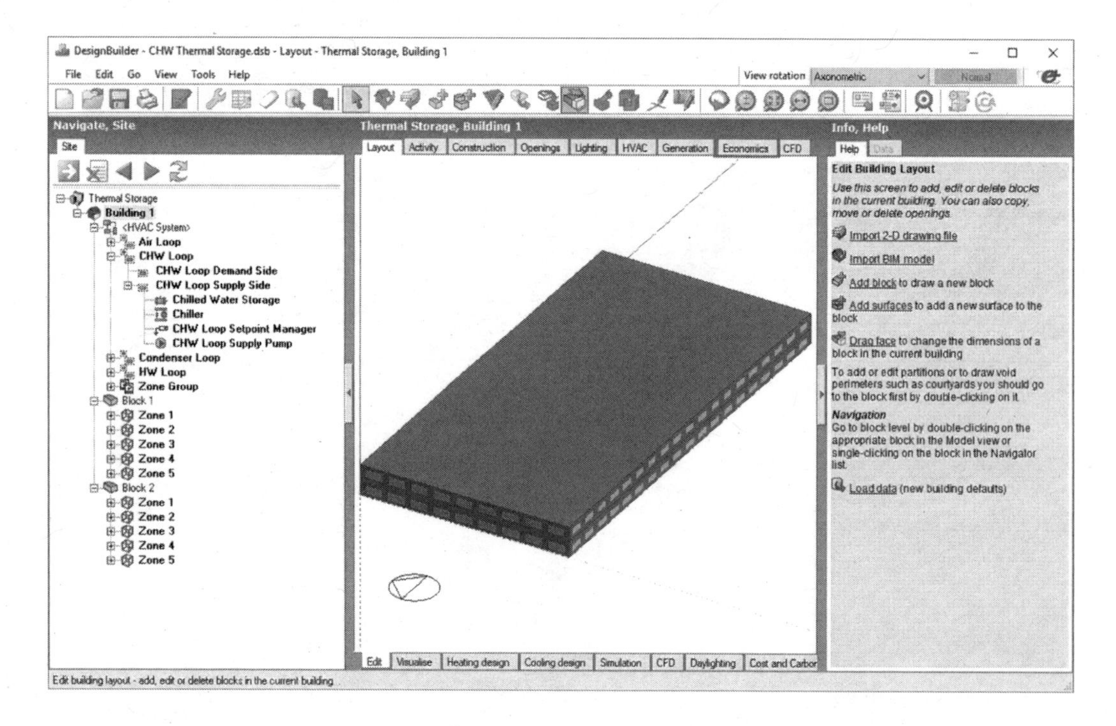

Step 30: Click on **Flat electricity charge**. Create a copy of the existing schedule.

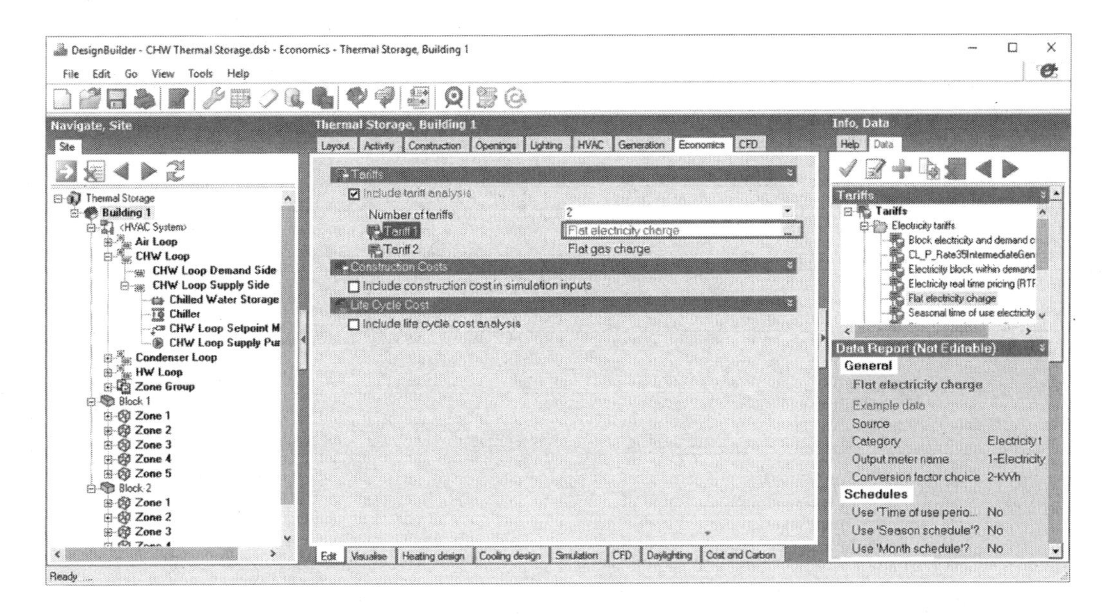

Step 31: Edit the schedule. Enter **Time of use electricity_Project** in the **Tariff name** text box. Select the **Use 'Time of use period schedule'?** checkbox.

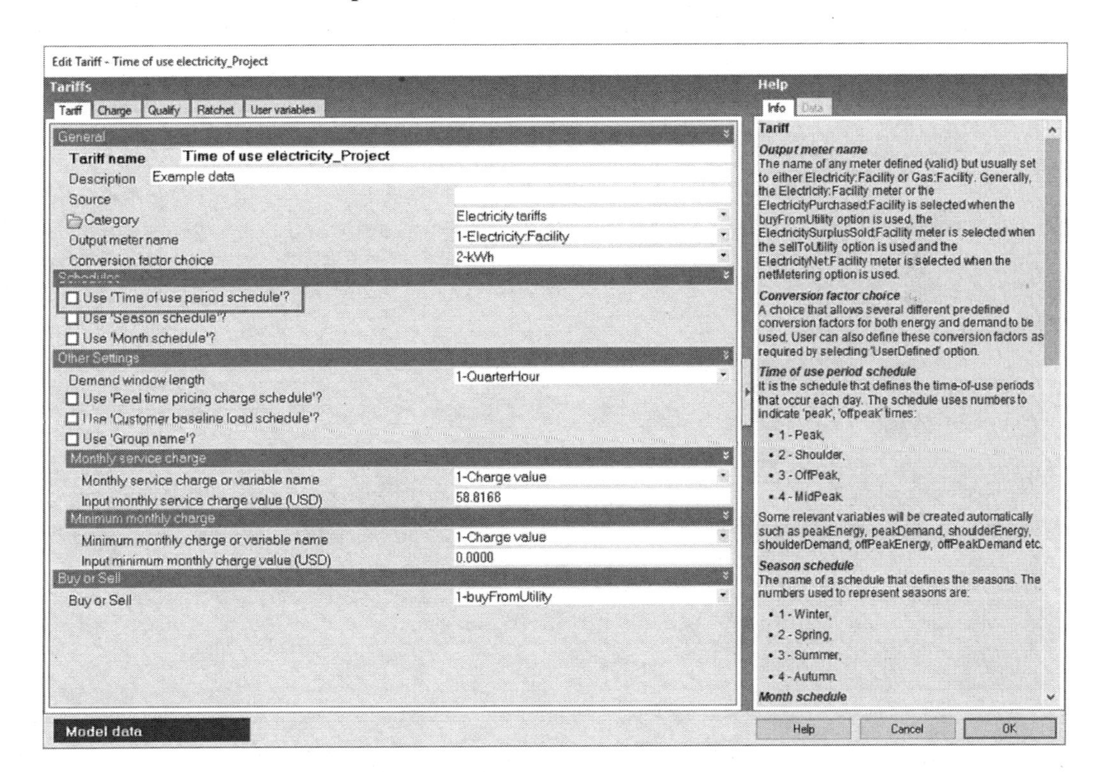

Step 32: Create a new schedule named **Time of day schedule_Project** by editing the **Time of use period schedule**.

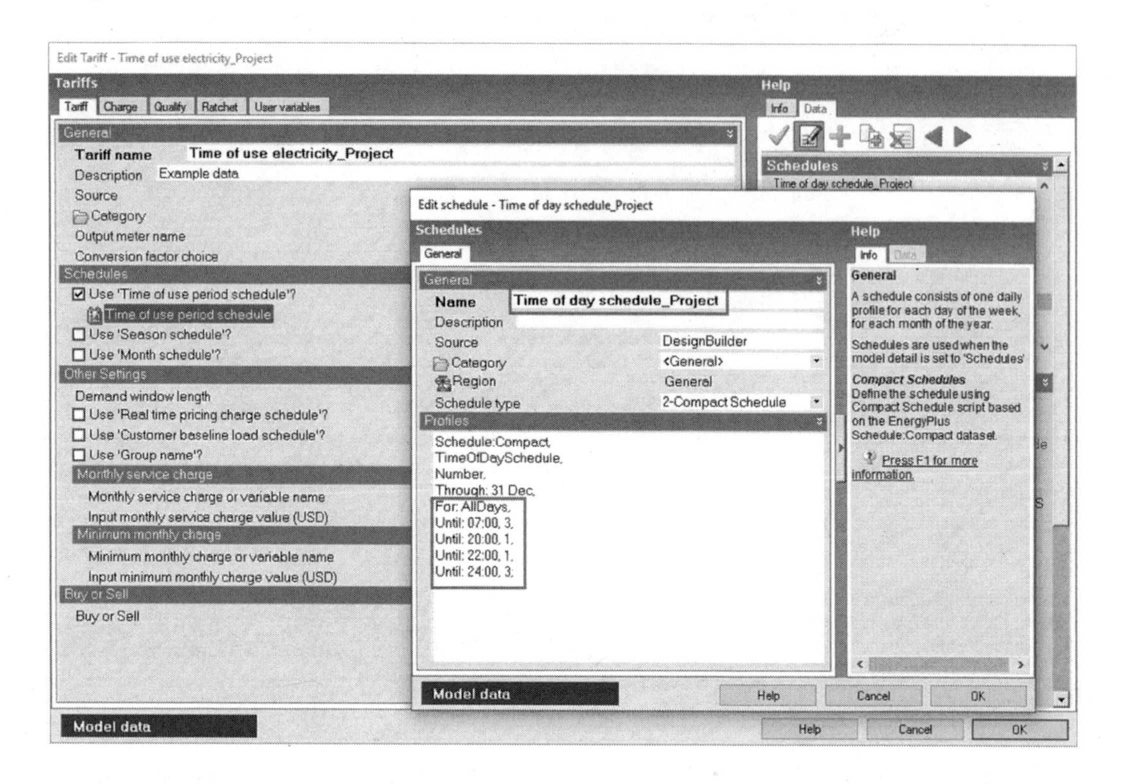

Time-of-Use Period Schedule

Select the schedule that defines the time-of-use periods that occur each day. The period schedule is used to determine which variables are defined. The values for the different variables are

1 = **Peak**
2 = **Shoulder**
3 = **OffPeak**
4 = **MidPeak**

Source: https://designbuilder.co.uk/helpv6.0/Content/TariffAnalysisTariff.html.

Step 33: Apply **Time of day schedule_Project**. Click the **Charge** tab.

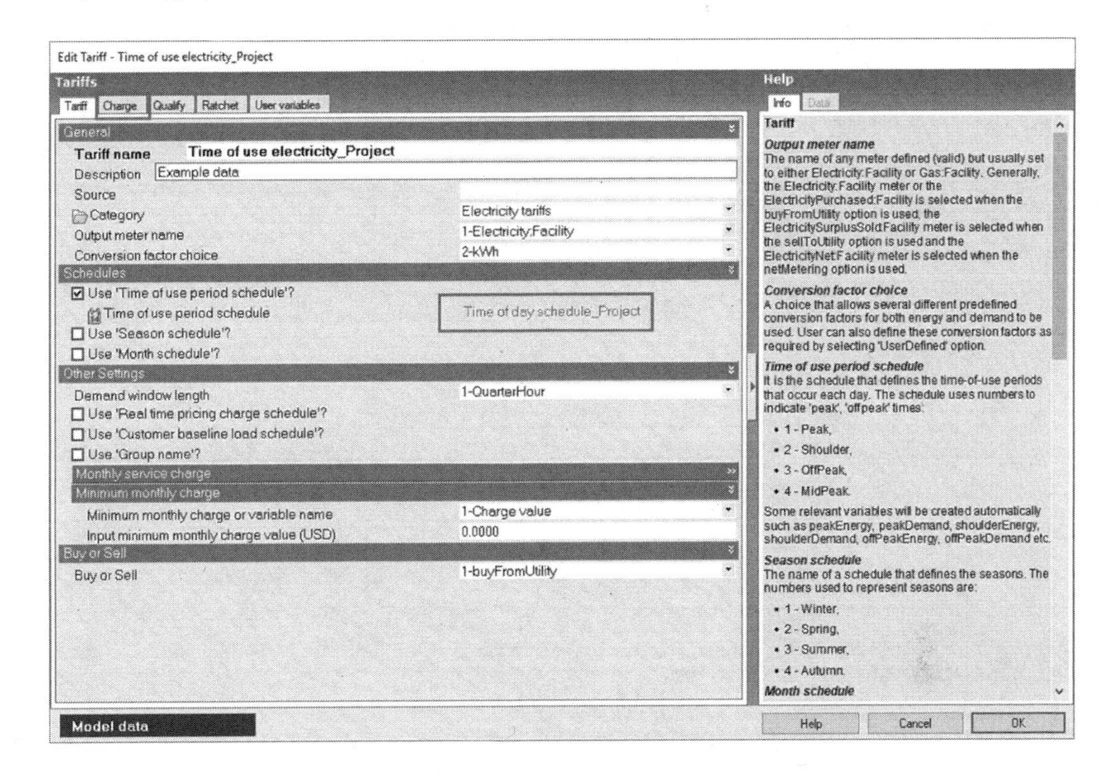

Step 34: Select **2** from **Number of charges** drop-down list. Under **Charge 1 settings**, enter **OnPeak** in the **Charge variable name** text box. Enter **0.18** in the **Input cost per unit value (USD)** text box. Under **Charge 2 settings**, enter **OffPeak** in the **Charge variable name** text box. Enter **0.09** in the **Input cost per unit value (USD)** text box.

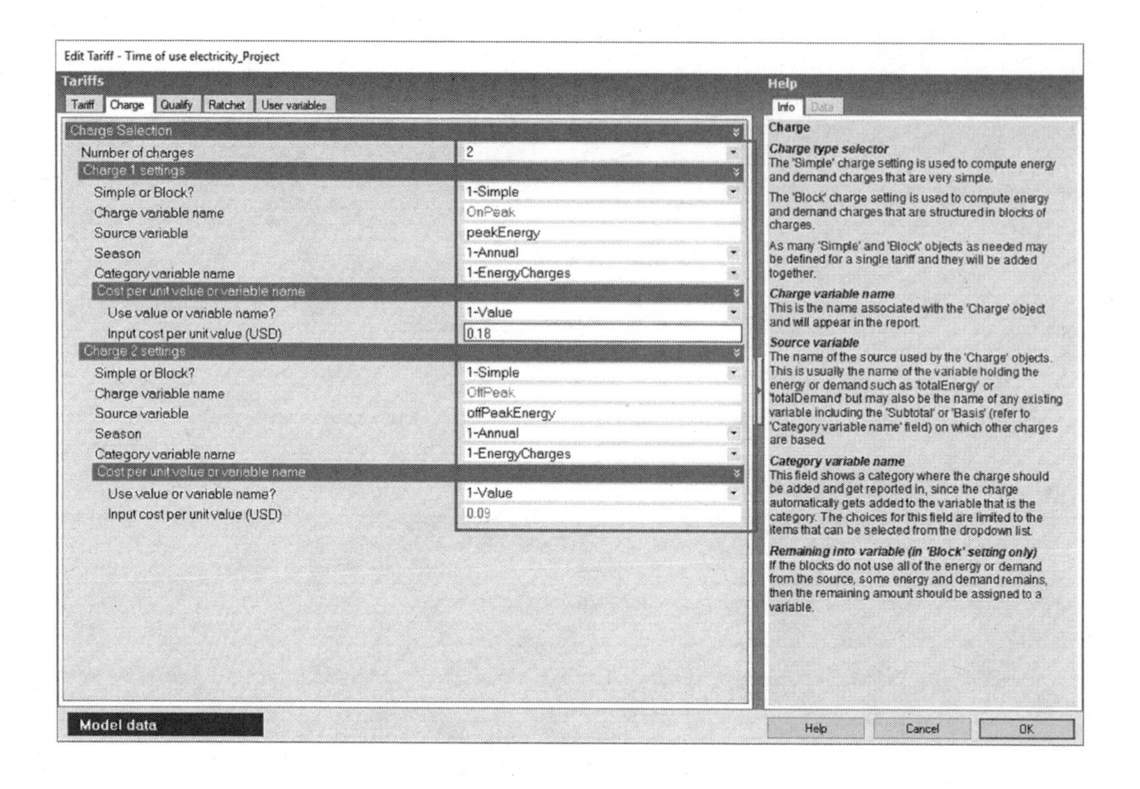

Step 35: Apply **Time of use electricity_Project**.

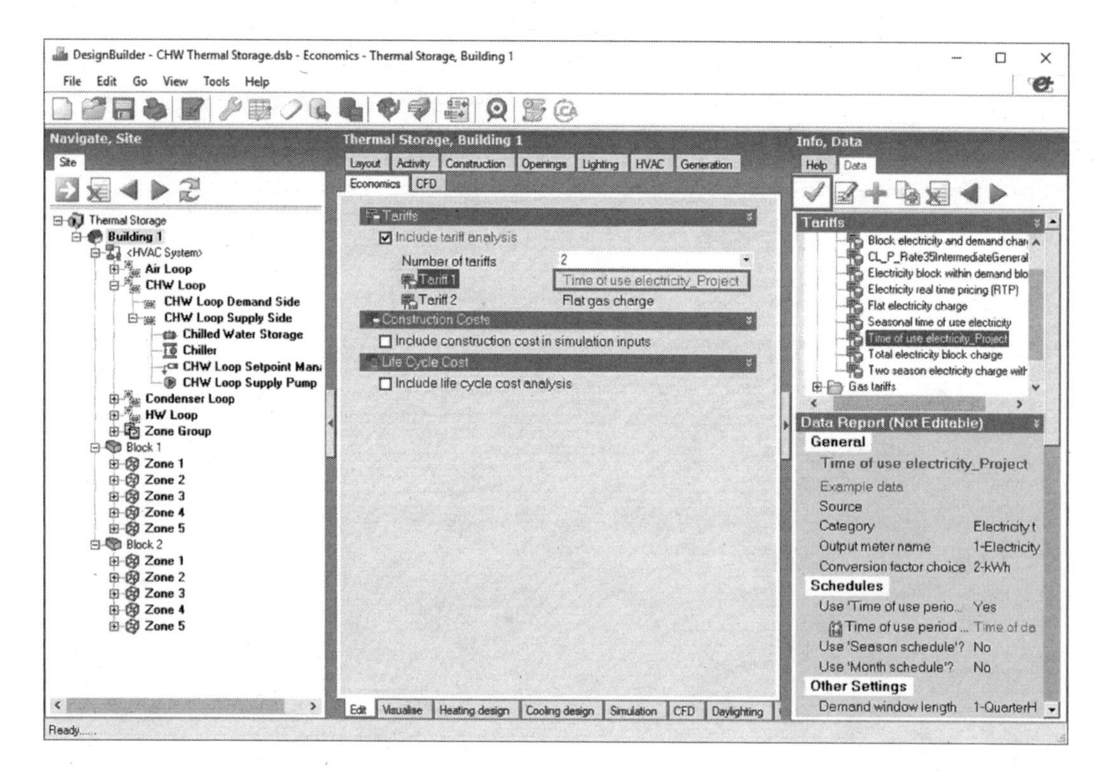

Step 36: Simulate the model, and record the **Total Energy Cost [$]** for **Electricity**.

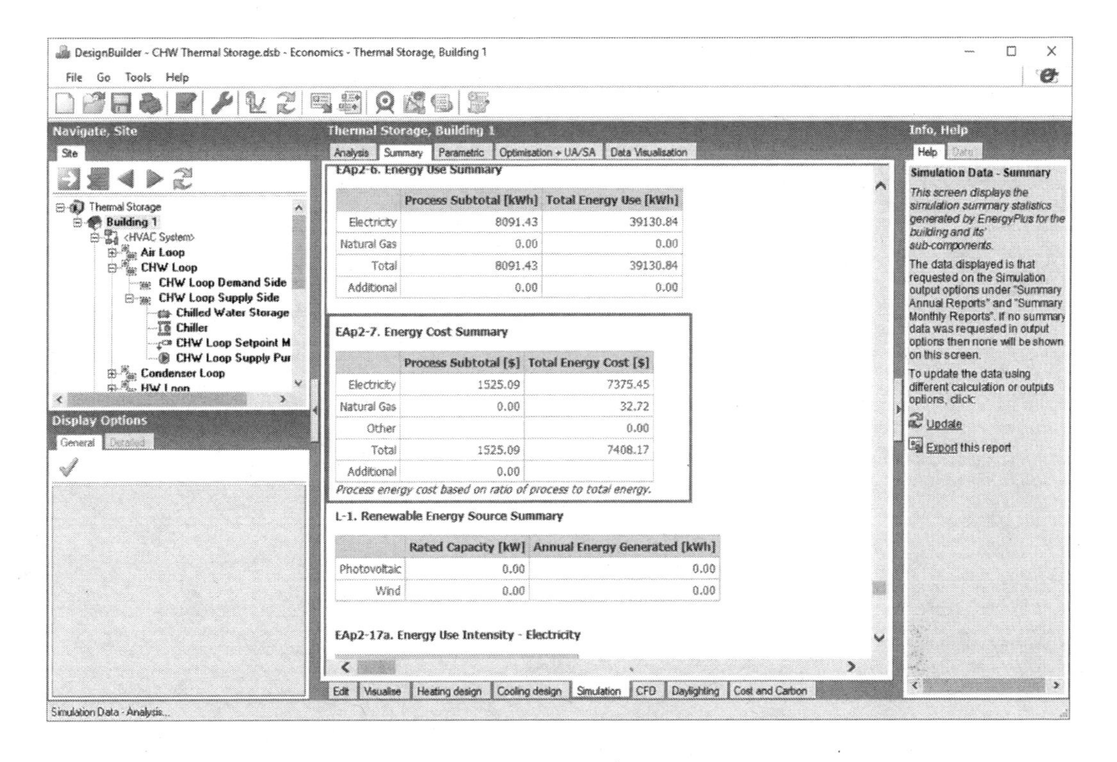

Step 37: Save as the model to simulate it without thermal storage.

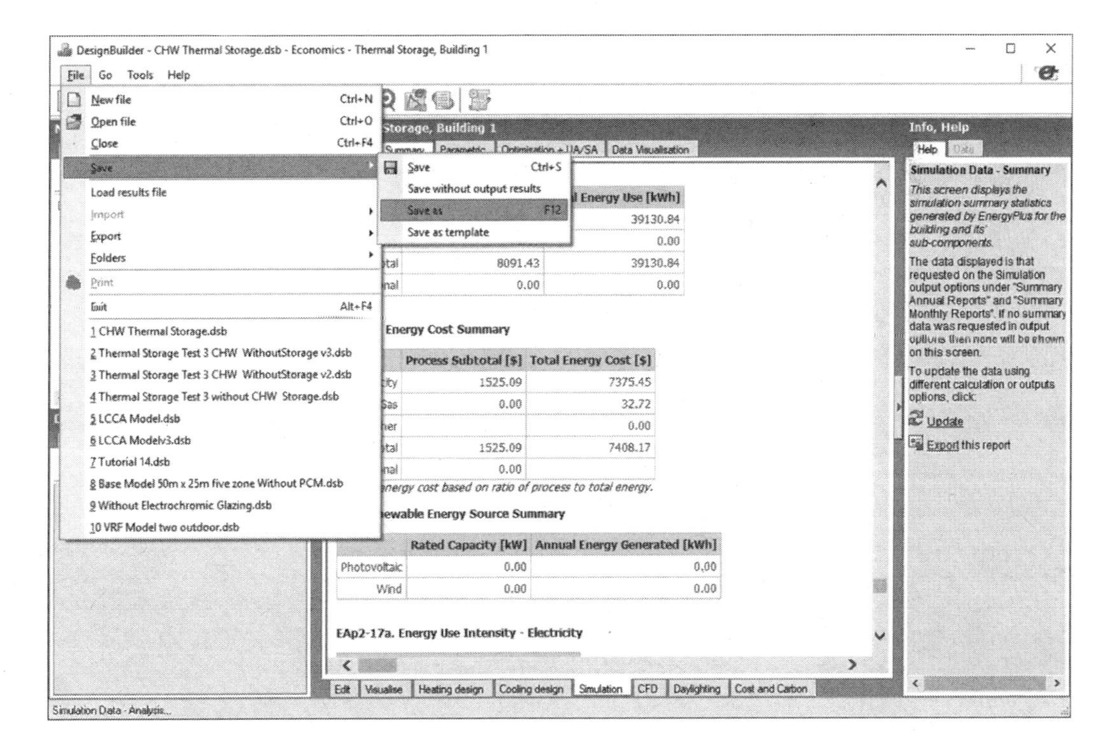

Step 38: Click **Chilled Water Storage**, and edit the storage properties. Enter **10.0** in the **Nominal cooling capacity (W)** text box. Enter **1.00** in the **Tank volume (m^3)** text box. This has been done to virtually remove the thermal storage from the system.

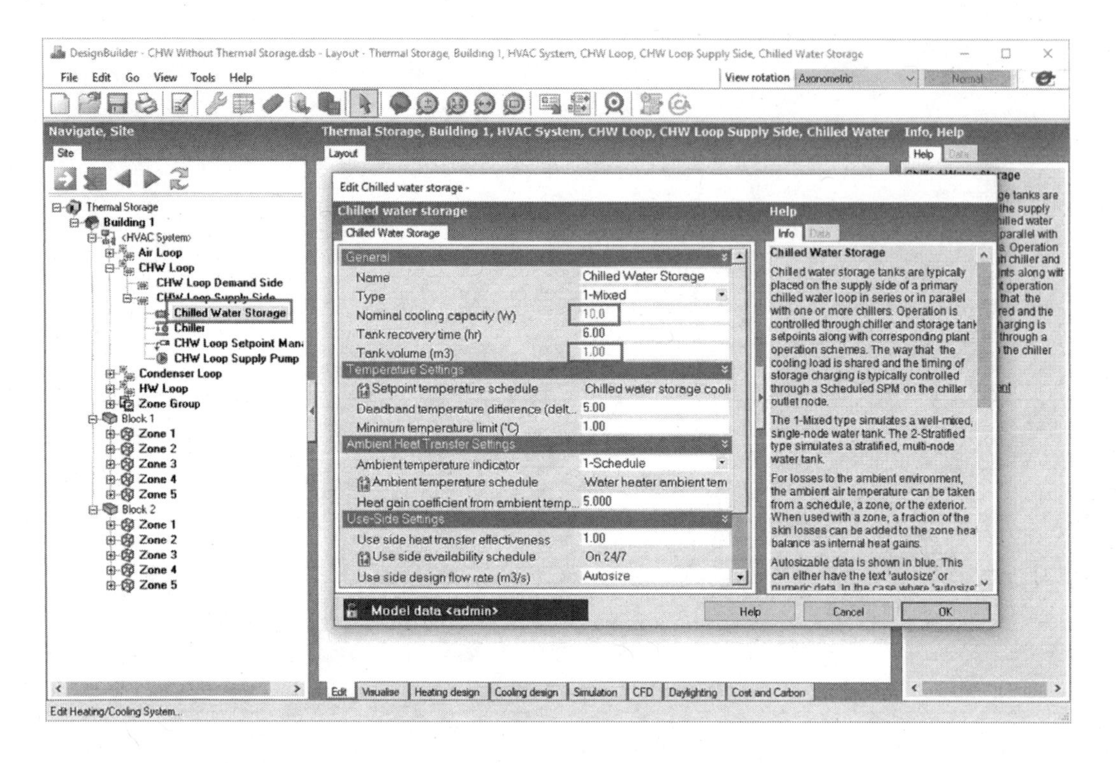

This will reduce the size of the thermal storage tank, thereby reducing the thermal storage to negligible.

Step 39: Simulate the model, and record the results.

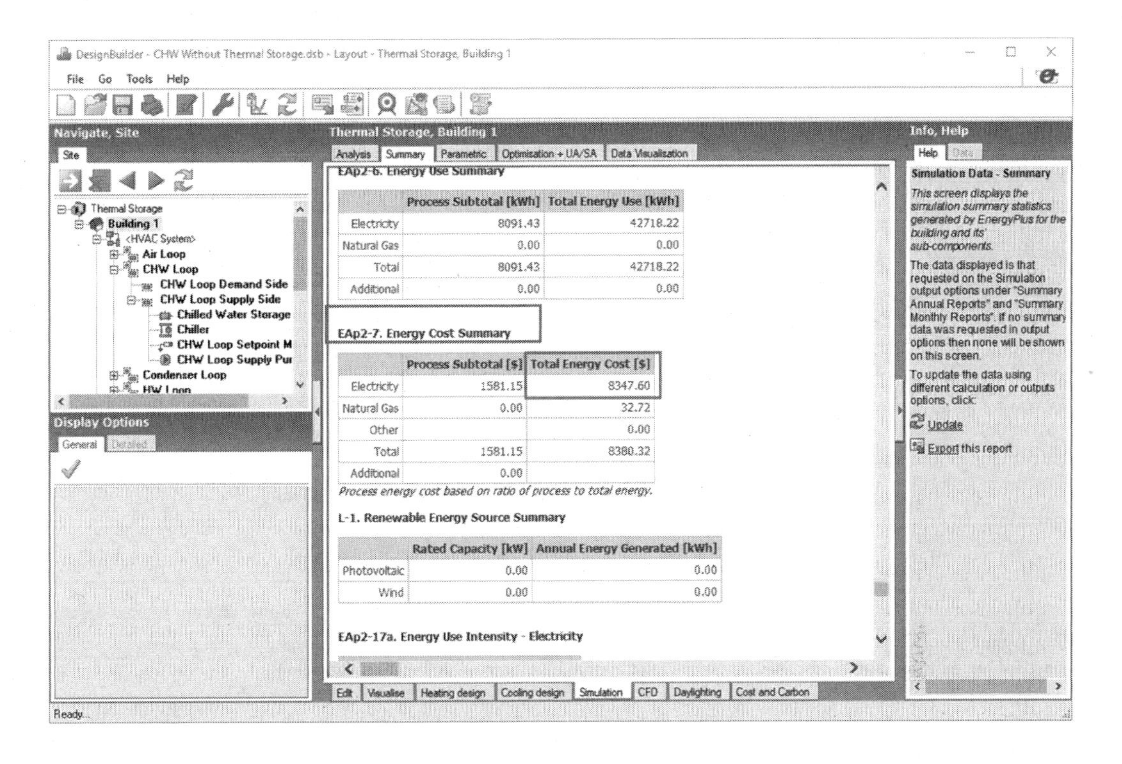

Step 40: Record the chiller cooling load using the **Results Viewer**.

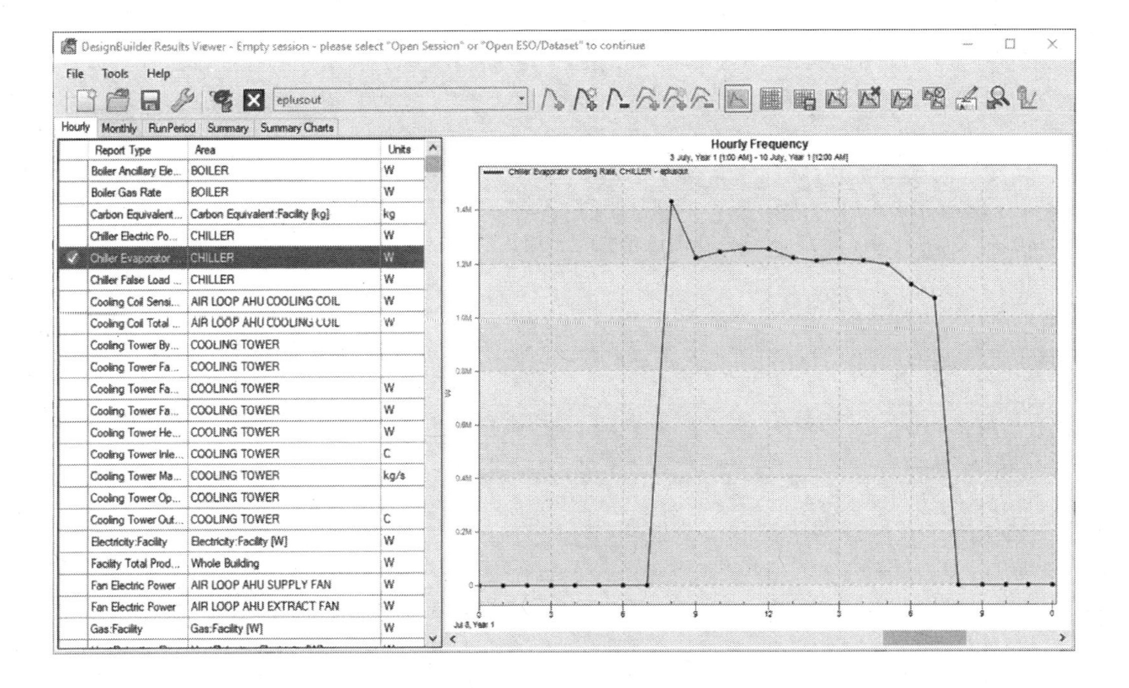

Step 41: Compare the cost results and cooling load profile in both cases (see Table 8.12).

TABLE 8.12
Cooling Energy Cost with and without thermal energy storage

Cooling energy cost	USD ($)
With chilled-water thermal storage	7,375.45
Without chilled-water thermal storage	8,347.60

You can see that with chilled-water thermal storage, the chiller works at night to store chilled water. Night-time energy cost is off-peak, so there is a reduction in the total cost for cooling energy.

9 Heating, Ventilation and Air Conditioning

Central Air Side

In a centralized heating, ventilation and air-conditioning (HVAC) system, low-side components such as fans, ducts, heat recovery and an air-side economiser are used to deliver and managing cool air to spaces. Energy savings potential from an air-side economizer and heat recovery based on dry bulb temperature and enthalpy for a particular location are discussed in this chapter. This chapter also covers the use of demand control ventilation to limit outside air quantities based on zone air carbon dioxide (CO_2) concentration to save energy.

Tutorial 9.1

Evaluating the Impact of an Air-Side Economiser

GOAL

To evaluate the impact of an air-side economiser on building energy performance

WHAT ARE YOU GOING TO LEARN?

- How to model an air-side economiser (free cooling system)

PROBLEM STATEMENT

In this tutorial, you are going to use the variable-flow model saved in Tutorial 8.3 (a 50- × 25-m six-level model with a 5-m perimeter depth). Model the unitary HVAC system with the following options for the air-side economiser:

1. None
2. Fixed dry bulb temperature based
3. Fixed enthalpy based

Determine the change in energy consumption for all three cases. Use the **New Delhi/Palam, India** weather location. New Delhi is located in a monsoon-influenced humid subtropical climate with high variation between summer and winter in terms of both temperature and rainfall. The temperature varies from 46°C in summer to around 4°C in winter.

> An *economiser* is an adjustable fresh air intake unit that can draw up to 100% outside air when the outside air is cooler than the temperature inside a building and not humid, thereby providing free cooling. Air-side economisers in HVAC systems can save energy in buildings by using cool outside air to cool the indoor space. When the temperature and/or enthalpy of the outside air is less than the temperature/enthalpy of the re-circulated air, conditioning the outside air is more energy efficient than conditioning re-circulated air. When the outside air is both sufficiently cool and dry (depending on the climate), no additional conditioning is required; this portion of air-side economiser control scheme is called *free cooling*.

SOLUTION

Step 1: Open the variable-flow model saved in Tutorial 8.3.

Step 2: Click **Air Loop AHU**. Click **Edit component** under the **Help** tab. The **Air handling unit** screen appears.

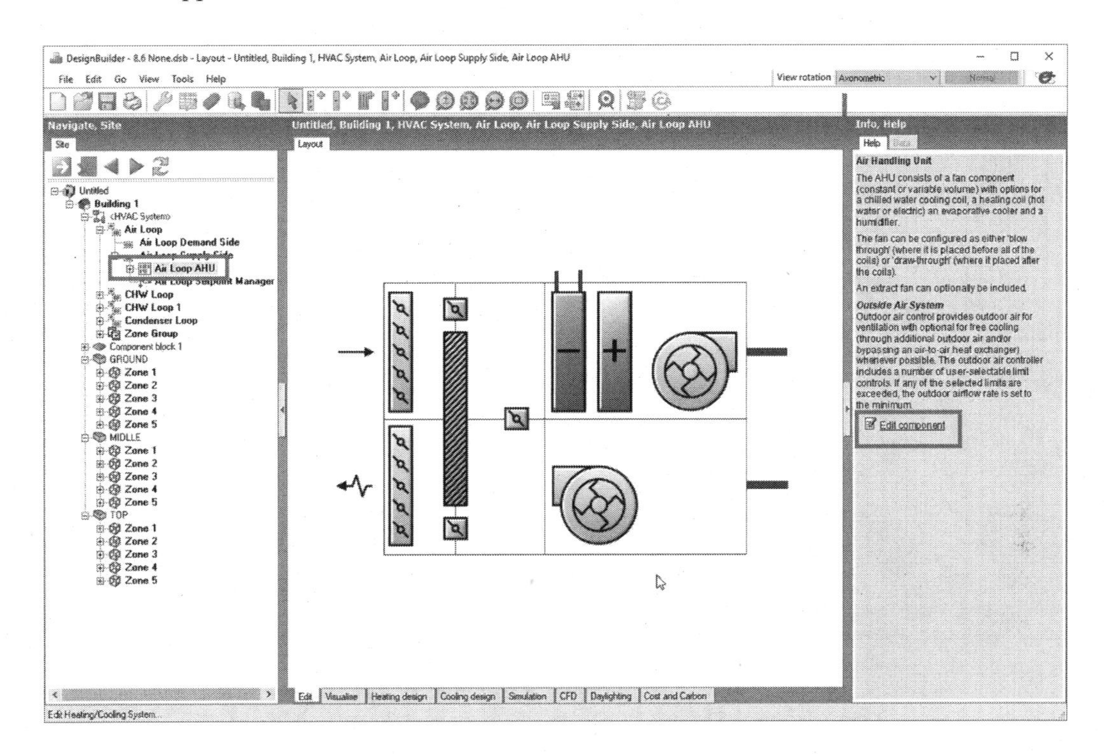

Step 3: Select the **Outdoor Air System** tab.

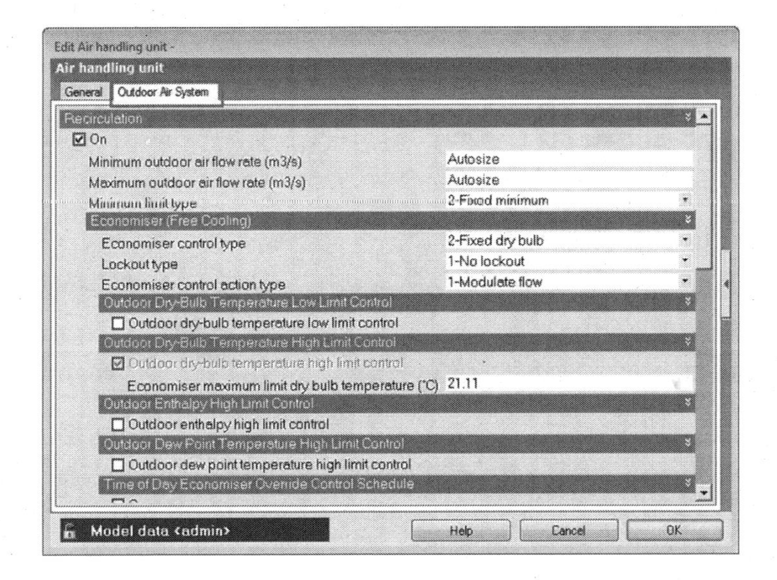

Step 4: Select **No economiser** from the **Economiser control** type drop-down list under **Economiser (Free Cooling).**

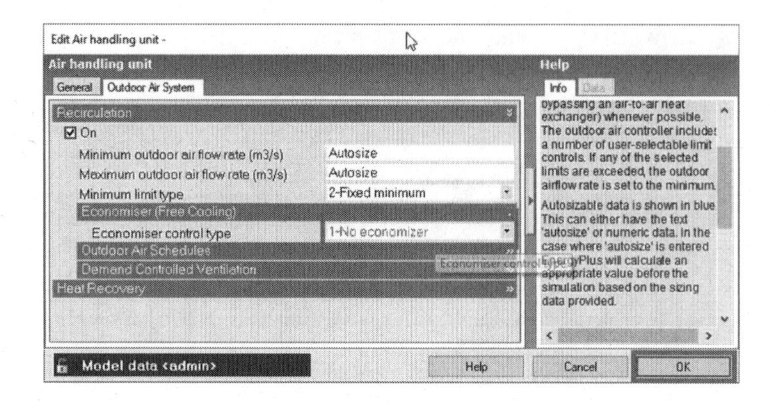

Step 5: Simulate the model, and record the results. Repeat the preceding steps to select the fixed dry bulb air-side economiser control type.

Step 6: Select **Fixed dry bulb** from the **Economiser control type** drop-down list under **Economiser (Free Cooling).**

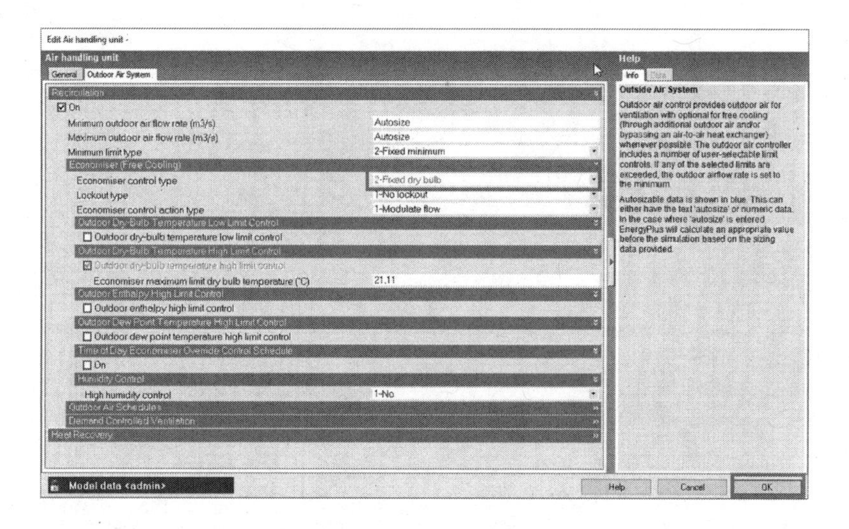

Step 7: Simulate the model, and record the results. Repeat the preceding steps to select fixed enthalpy air-side economiser control type.

Step 8: Select **Fixed enthalpy** from the **Economiser control type** drop-down list under **Economiser (Free Cooling).**

Step 9: Simulate the model, and record the results. Compare the results (Table 9.1). Save the model with no economiser for use in later tutorials.

> *Enthalpy* is the total heat content of the air. This covers the combined effect of temperature and humidity.

TABLE 9.1

Energy consumption for no, fixed dry bulb temperature (DBT) and fixed enthalpy-based economisers

	Annual energy consumption (kWh)		
End-use category	**No economiser**	**Fixed dry bulb temperature–based economiser**	**Fixed enthalpy–based economiser**
Room electricity	306,674.40	306,674.40	306,674.40
Lighting	443,211.10	443,211.10	443,211.10
System fans	281,229.10	281,341.80	281,340.80
System pumps	539,415.00	428,364.30	428,367.40
Heating (electricity)	12,366.60	45,697.30	45,697.30
Cooling (electricity)	528,084.10	484,802.40	485,059.80
Heat rejection	364,445.60	364,143.40	364,158.40

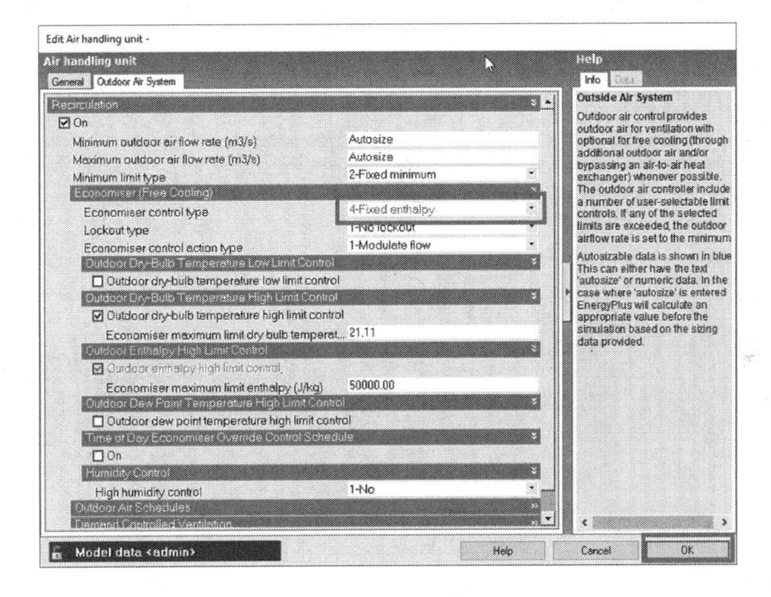

You can see that application of free cooling leads to a reduction in annual cooling energy consumption. Enthalpy type air-side economisers result in higher energy savings compared with dry bulb type economisers. There is a reduction in energy consumption of all the HVAC components, namely cooling, pumps and heat rejection.

EXERCISE

Compare the energy savings in Miami, Florida, and London Gatwick Airport using both types of air-side economisers.

Tutorial 9.2

Evaluating the Impact of Supply Air Fan Operating Mode during Unoccupied Hours

GOAL

To evaluate the impact of fan operating mode during unoccupied hours on energy consumption

WHAT ARE YOU GOING TO LEARN?

- How to change fan operating mode during unoccupied hours

PROBLEM STATEMENT

In this tutorial, you are going to use the no-economiser model saved in Tutorial 9.1 (a 50- × 25-m six-level model with a 5-m perimeter depth). You need to simulate the model with the following two options for the New Delhi/Palam, India weather location for supply air fan operation:

1. Stay off
2. Cycle on any

Determine the energy consumption in both cases.

Applicability Schedule (Night Cycle)

This schedule determines whether or not for a given time period this mechanism is to be applied. Schedule values greater than zero (usually one is used) indicate that the night-cycle mechanism is to be applied, whereas schedule values less than or equal to zero (usually zero is used) indicate that the mechanism is not used for this time period.

Control Type

The possible inputs are as follows:

- *Stay off* means that the night-cycle mechanism will have no effect – air handling unit (AHU) on/off will be determined by the fan schedule.
- *Cycle on any* means that if any zone served by the air loop incorporating this AHU has an air temperature outside the cooling or heating setpoints, the central fan will turn on even though the fan schedule indicates that the fan is off.

This setting is used to enable cycling of an air system when one or more zones become too hot or too cold. A common requirement for this mechanism is where the AHU is turned off at night. However, if the building gets too cold, there might be condensation on the walls and other damage. Thus the control system is usually programmed to turn the system on if either a specified control thermostat or any thermostat shows a zone temperature of less than a night-time setpoint. Similarly, there might be a concern about a building getting too hot. Again, the control system is programmed to turn the AHU back on if one or any zone temperature exceeds a night-time cooling setpoint.

This mechanism offers considerable flexibility in determining how the night-time on/off decision will be made. The temperature in one specific zone may be used or the temperatures in all the zones connected to the AHU may be sampled. You can specify a temperature tolerance and a run time for the system once it switches on. There is also an applicability schedule for scheduling when this mechanism may be applied.

Source: www.designbuilder.co.uk/helpv6.0/#Generic_AHU.htm?Highlight=Generic.

Heating Setback Setpoint Temperature

Some buildings require a low level of heating during unoccupied periods to avoid condensation/frost damage or to prevent the building from becoming too cold and to reduce peak heating requirements at start-up. Enter the setpoint temperature to be used at night, on weekends and on holidays during the heating season.

Cooling Setback Setpoint Temperature

Some buildings require a low level of cooling during unoccupied periods to prevent the building from becoming too hot and to reduce the start-up cooling load the next morning. Enter the setpoint temperature to be used at night, on weekends and on holidays during the cooling season.

Source: www.designbuilder.co.uk/helpv6.0/Content/_Environmental_comfort.htm.

SOLUTION

Step 1: Open the model saved in Tutorial 9.1. Enter **8.0** in the **Heating set back (°C)** and **35.0** in the **Cooling set back (°C)** text boxes in the Activity tab.

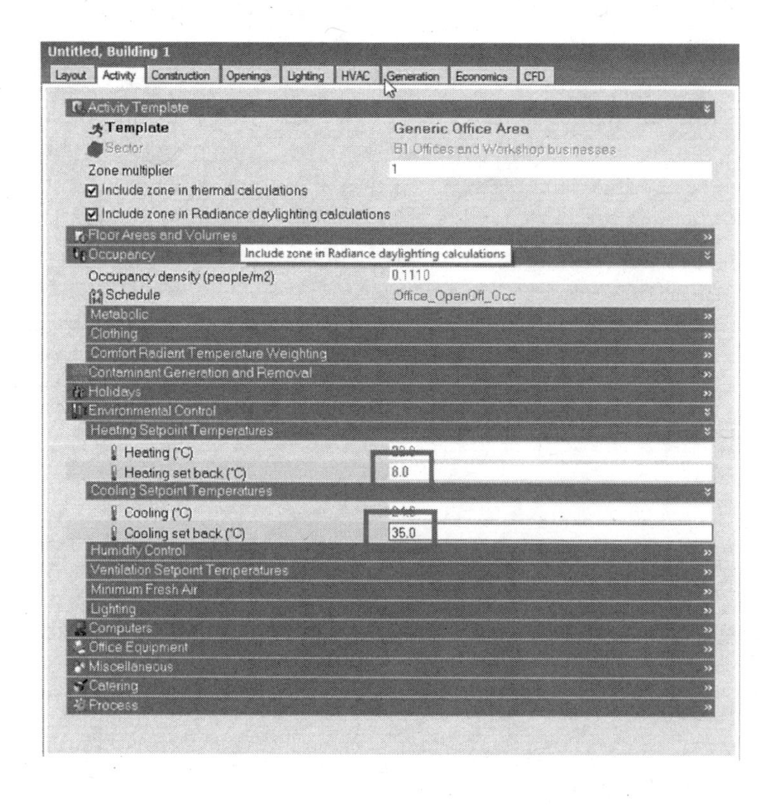

Step 2: Click **Air Loop AHU** in the navigation tree. Click **Edit component** under the **Help** tab. The **Edit Air handling unit** screen appears.

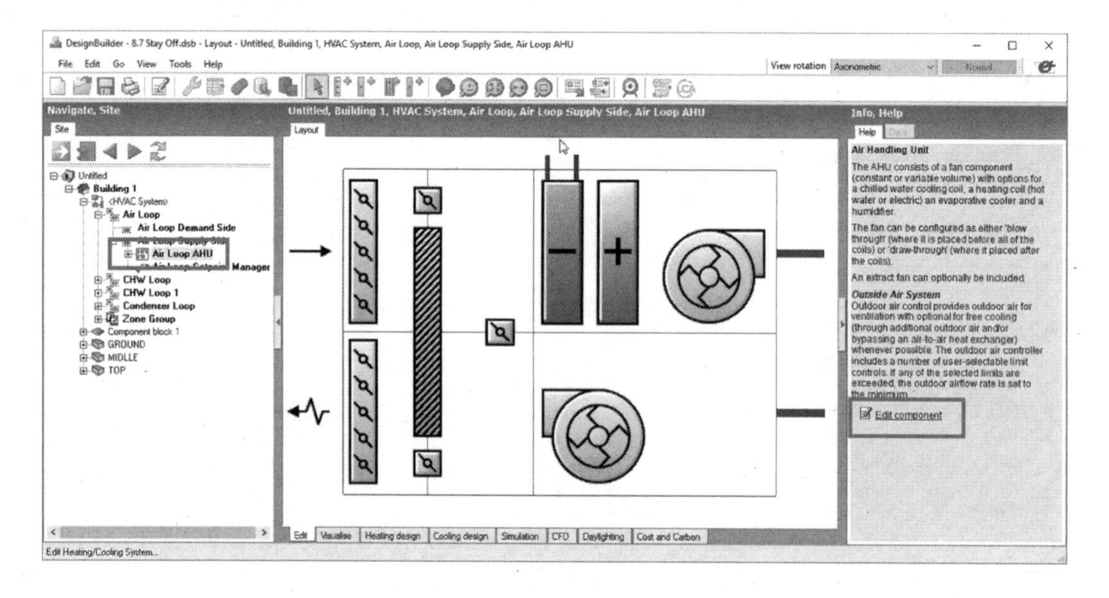

Step 3: Under the **Night Cycle** section, select **Stay off** from the **Control type** drop-down list, and select **8:00–18:00 Mon–Fri** under the **Operation** section and make sure that the **Heat Recovery** checkbox is clear under the **Outdoor Air System** tab.

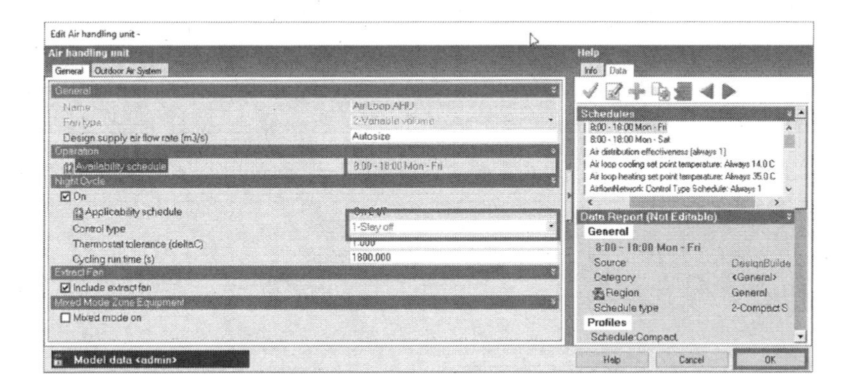

Step 4: Perform an hourly simulation, and record the results.

Step 5: Repeat the preceding steps to select **Cycle on any** from the **Control type** drop-down list, and select **8:00–18:00 Mon–Fri** under the **Operation** section.

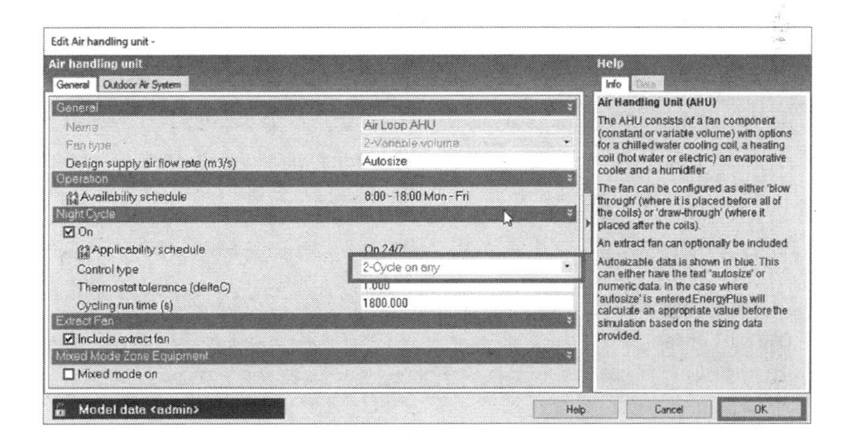

Step 6: Perform an hourly simulation, and record the results. Compare the results (Table 9.2).

TABLE 9.2
Energy consumption with changes in the supply air fan operating mode

	Annual energy consumption (kWh)	
End-use category	**Stay off**	**Cycle on any**
Room electricity	306,674.40	306,674.40
Lighting	443,211.10	443,211.10
System fans	175,849.00	221,768.80
System pumps	171,100.30	330,536.00
Heating (electricity)	1,036.80	1,940.30
Cooling (electricity)	275,730.30	404,901.70
Heat rejection	123,507.10	244,412.40

You can see that there is an increase in system fan cooling and heating energy consumption. This occurs because when the Cycle on any option is selected, fans run longer during unoccupied hours.

Codes such as ASHRAE 90.1–2010, Appendix G, require that schedules for HVAC fans that provide outdoor air for ventilation should run continuously whenever spaces are occupied and should be cycled on and off to meet heating and cooling loads during unoccupied hours.

Compare hourly cooling energy consumption for 21 June (Table 9.3).

TABLE 9.3
Cooling energy consumption on 21 June

	Cooling (electricity) (kW)	
Date/time	**Stay off**	**Cycle on any**
21-Jun-02	0.00	50.17
21-06-02 1:00	0.00	48.72
21-06-02 2:00	0.00	46.45
21-06-02 3:00	0.00	45.89
21-06-02 4:00	0.00	44.90
21-06-02 5:00	0.00	46.70
21-06-02 6:00	0.00	42.14
21-06-02 7:00	0.00	42.28
21-06-02 8:00	0.00	61.53
21-06-02 9:00	236.53	157.61
21-06-02 10:00	196.50	169.37
21-06-02 11:00	203.69	118.54
21-06-02 12:00	116.84	173.35
21-06-02 13:00	174.06	172.47
21-06-02 14:00	177.85	176.03
21-06-02 15:00	117.52	176.80
21-06-02 16:00	224.78	174.27
21-06-02 17:00	182.00	170.17
21-06-02 18:00	178.08	167.77
21-06-02 19:00	0.00	58.77
21-06-02 20:00	0.00	71.17
21-06-02 21:00	0.00	60.78
21-06-02 22:00	0.00	57.80
21-06-02 23:00	0.00	56.47

You can see that with the **Stay off** option, there is a high cooling load at the start of the HVAC system, and with **Cycle on any** option, the HVAC system does not get a starting spike.

Tutorial 9.3

Evaluating the Impact of Heat Recovery between Fresh and Exhaust Air

GOAL

To evaluate the impact of recovering heat between fresh air intake and exhaust on building energy performance

WHAT ARE YOU GOING TO LEARN?

- How to model a heat recovery system

PROBLEM STATEMENT

In this tutorial, you are going to use the air-cooled chiller model saved in Tutorial 8.1 (a 50- × 25-m six-level model with a 5-m perimeter depth). You need to simulate the model with the following three options:

1. No heat recovery
2. Sensible heat recovery
3. Enthalpy-based heat recovery

Determine the energy consumption for all cases. Use the New Delhi/Palam, India weather location.

> *Energy recovery ventilation* is the energy recovery process of exchanging the energy contained in air exhausted from a building or space air and using it to treat (precondition) the incoming outdoor ventilation air in an HVAC system. Air-to-air energy recovery reduces energy use and can significantly reduce heating and cooling system size. The driving force behind the exchange is the difference in temperature between the opposing airstreams, which is also called the *thermal gradient*.

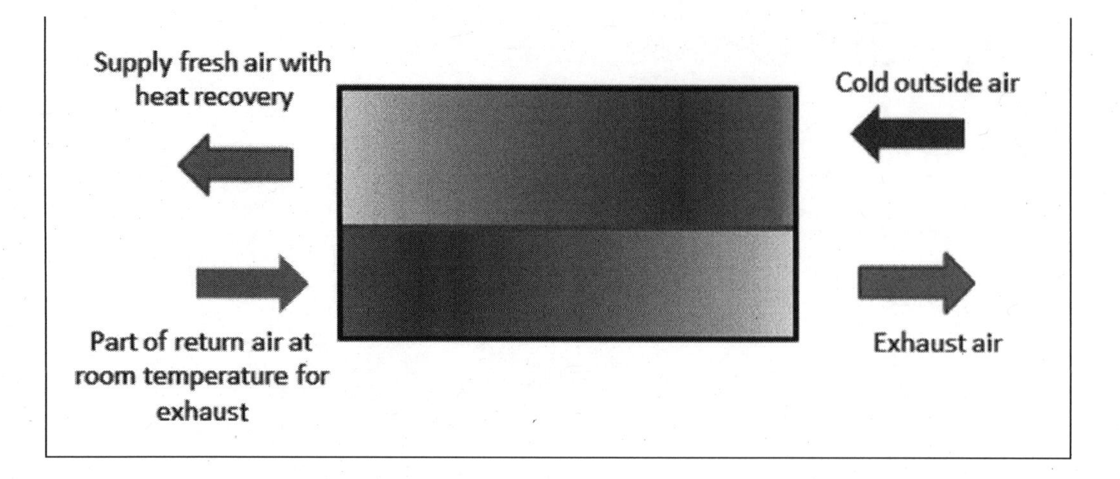

There are two types of heat recovery:

1. Sensible
2. Enthalpy

Sensible heat recovery is made possible by the use of fixed-plate heat exchangers. A fixed-plate heat exchanger has no moving parts and consists of alternating layers of plates that are separated and sealed. Typical flow is cross-current, and because most of the plates are solid and non-permeable, sensible-only transfer is the result. Sensible heat recovery is also possible via a rotating-wheel heat exchanger.

Enthalpy heat recovery is made possible by the use of a rotating-wheel heat exchanger. Rotating-wheel heat exchangers consist of a rotating cylinder filled with an air-permeable material, resulting in a large surface area. The surface area is the medium for the sensible energy transfer. As the wheel rotates between the ventilation and exhaust airstreams, it picks up heat energy and releases it into the colder airstream.

The enthalpy exchange is accomplished through the use of desiccants. Desiccants transfer moisture via the process of adsorption, which is predominantly driven by the difference in the partial pressures of vapour within the opposing airstreams. Typical desiccants are silica gel and molecular sieves.

SOLUTION

Step 1: Open the air-cooled chiller model saved in Tutorial 8.1. Select the **Activity** tab. Enter **7.100** in the **Fresh air (l/s-person)** text box in the **Minimum Fresh Air** tab, and select **24 × 7 Generic Office Area** as the template.

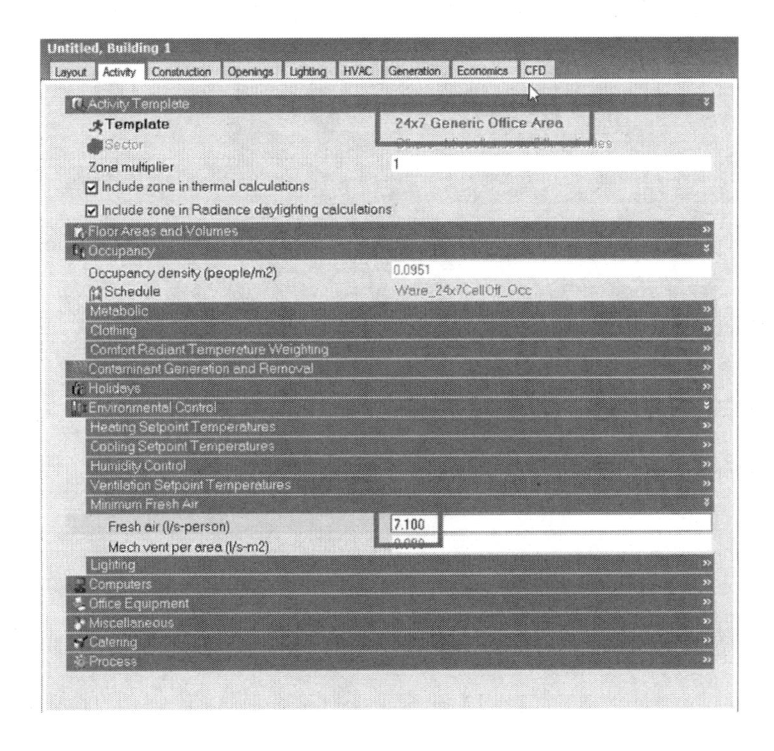

Step 2: Click **Air Loop AHU** in the navigation tree. Click **Edit component** under the **Help** tab. The **Edit Air Handling Unit Data** screen appears.

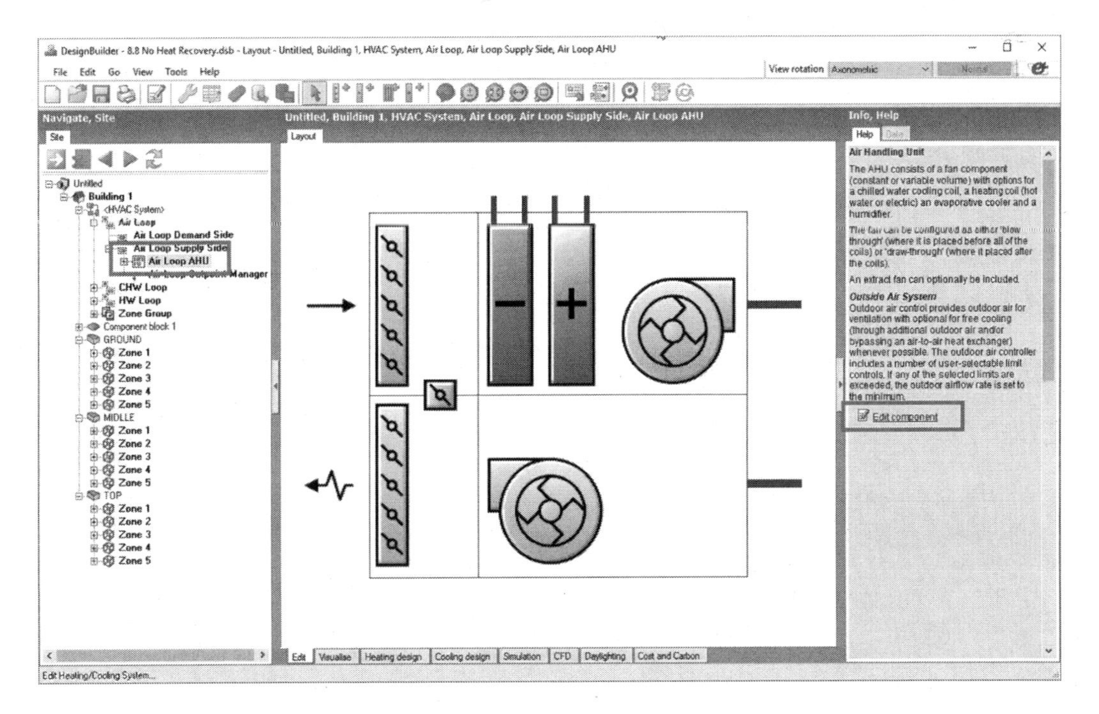

Step 3: Select the **Outdoor Air System** tab. Select the **On** checkbox under **Heat Recovery**. Make sure that **Plate** is selected from the **Heat exchanger type** drop-down list.

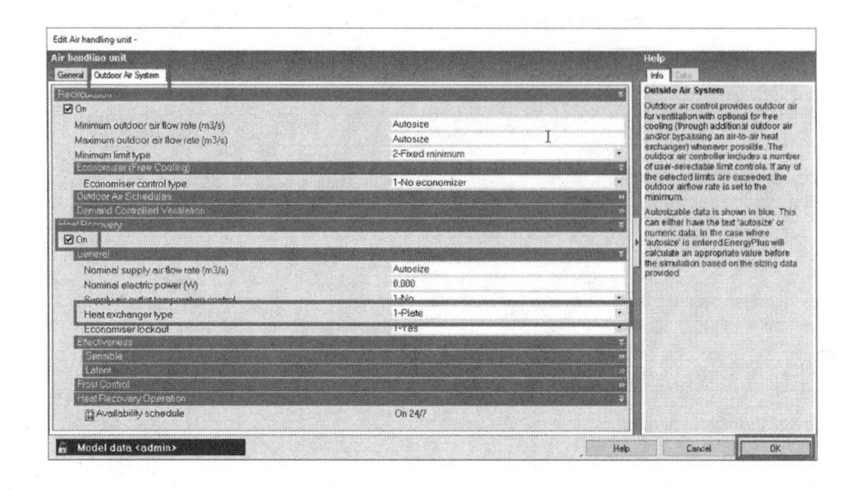

Step 4: Click **HW Loop** in the navigation tree. Click **Edit loop data** under the **Help** tab. The **Edit Plant loop** screen appears.

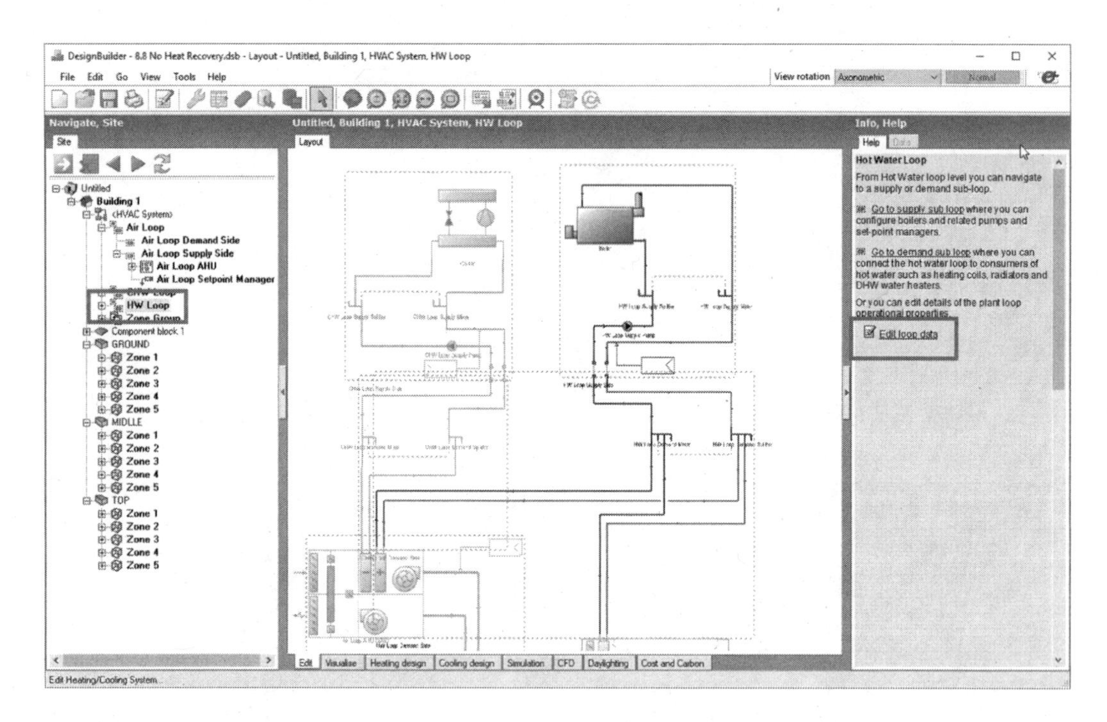

Step 5: Select **Off 24/7** in the **Availability schedule**. Click **OK.**

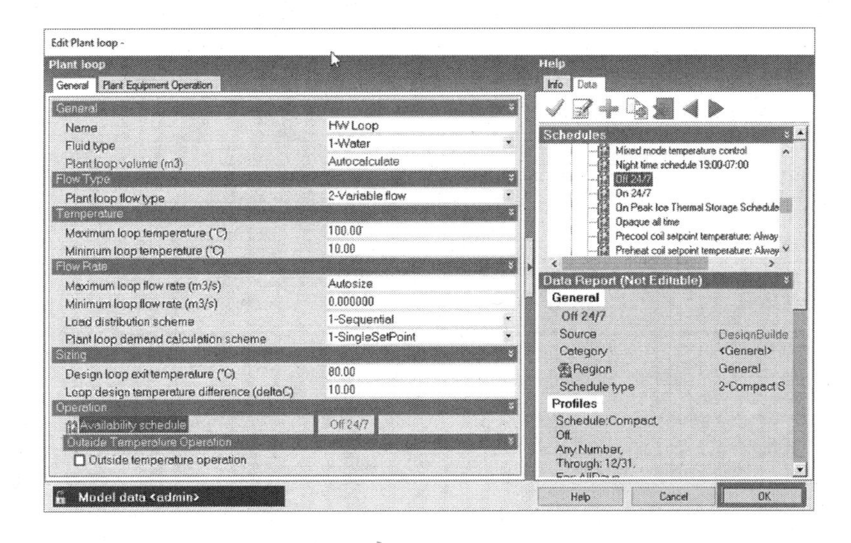

Step 6: Simulate the model, and record the results.

Step 7: Select **Rotary** from the **Heat exchanger type** drop-down list. Enter **0.70** under **Latent effectiveness**. Click **OK.**

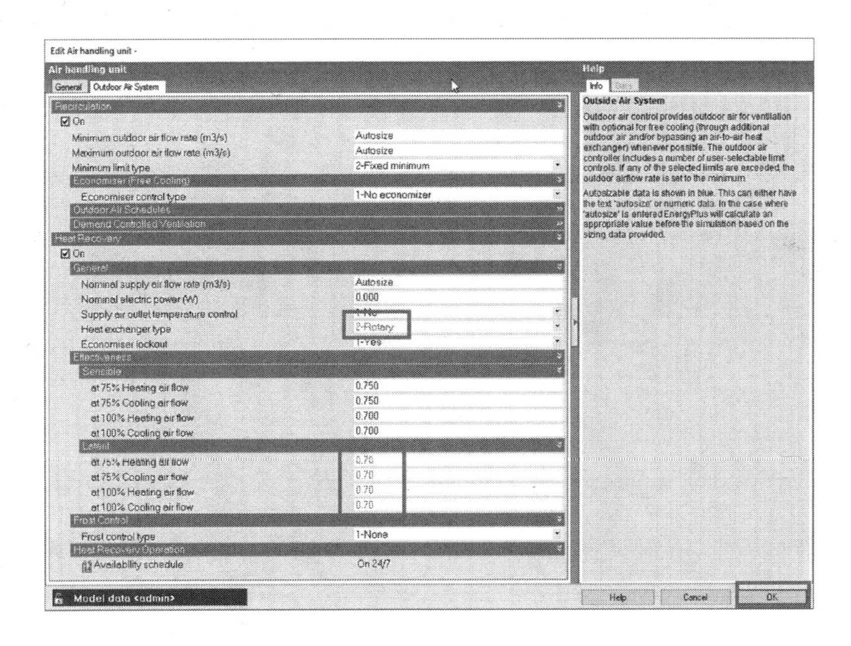

Step 8: Simulate the model, and record the results.

Step 9: Simulate the model and record the results without heat recovery. You can do this by clearing the **On** checkbox under the **Heat Recovery** section. Click **OK**. Compare the results (Table 9.4).

TABLE 9.4
Effect of exhaust air heat recovery in a HVAC system

	Annual energy consumption (kWh)		
End-use category	**None**	**Sensible heat recovery**	**Enthalpy heat recovery**
Room electricity	542,791.10	542,791.10	542,791.10
Lighting	1,239,632.00	1,239,632.00	1,239,632.00
System fans	361,071.90	361,062.80	361,040.50
System pumps	4,364.90	4,346.90	3,872.30
Cooling (electricity)	1,305,475.00	1,276,526.00	1,221,968.00

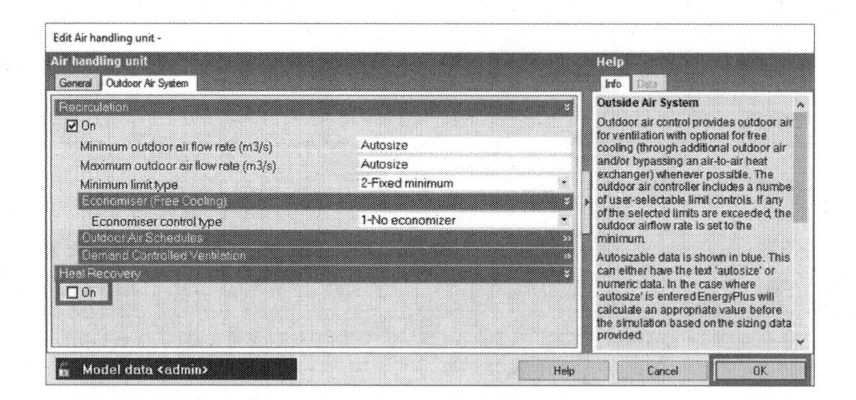

You can see that there is a decrease in cooling energy consumption with sensible heat recovery. Use of enthalpy heat recovery gives even higher savings.

Generally, there is an increase in fan energy consumption with a heat recovery system. This is due to the increase in static pressure of the fan for the heat recovery wheel. To model this effect, you need to change the fan pressure rise.

> Building an air-conditioning system requires a fresh air supply to maintain indoor air quality. There are different standards such as ASHRAE Standard 60.1–2010 that specify minimum fresh air requirements in different space types. In a cold climate, the temperature of fresh air is lower than that of the exhaust air (which is nearly at room temperature). Bringing fresh air from low to room temperature requires heating energy. Similarly, in hot climates, bringing fresh air from hot to room temperature requires cooling energy. Recovering heat or coolness from outgoing air offers an energy savings opportunity.

EXERCISE

Repeat this tutorial for the Montreal, Canada, climate.

Tutorial 9.4

Evaluating the Impact of a Variable-Refrigerant-Flow (VRF) System

GOAL

To demonstrate the use of VRF system sizing incorporating load diversity

WHAT ARE YOU GOING TO LEARN?

- How to model a VRF system

PROBLEM STATEMENT

In this tutorial, you are going to use a 50- × 30-m model with four identical floors. You need to model a VRF HVAC system. simulate the model with the operation indicated in Table 9.5. Use the **GUANGZHOU, China** weather location.

> A VRF system is an air-conditioning system that varies the refrigerant flow rate using variable-speed compressor(s) in the outdoor unit and the electronic expansion valves (EEVs) located in each indoor unit. The system meets the space cooling or heating load requirements by maintaining the zone air temperature at the setpoint. The ability to control the refrigerant mass flow rate according to the cooling and/or heating load enables the use of as many as 60 or more indoor units with differing capacities in conjunction with one single outdoor unit. This unlocks the possibility of having individualized comfort control, simultaneous heating and cooling in different zones and heat recovery from one zone to another. It also may lead to more efficient operations during part-load conditions.
>
> *Source:* https://designbuilder.co.uk/helpv6.0/Content/VRFLoops.htm.

TABLE 9.5
Activity schedule and outdoor VRF unit for building floor levels

Level	Activity	Outdoor VRF unit
First	**24X7_Circulation_Occ**	**VRF_1**
Second	**24X7_Circulation_Occ**	**VRF_1**
Third	**7 AM to 1 PM**	**VRF_2**
Fourth	**2 PM to 6 PM**	**VRF_2**

SOLUTION

Step 1: Open a new project, and create a 5**0- × 30-m** building.

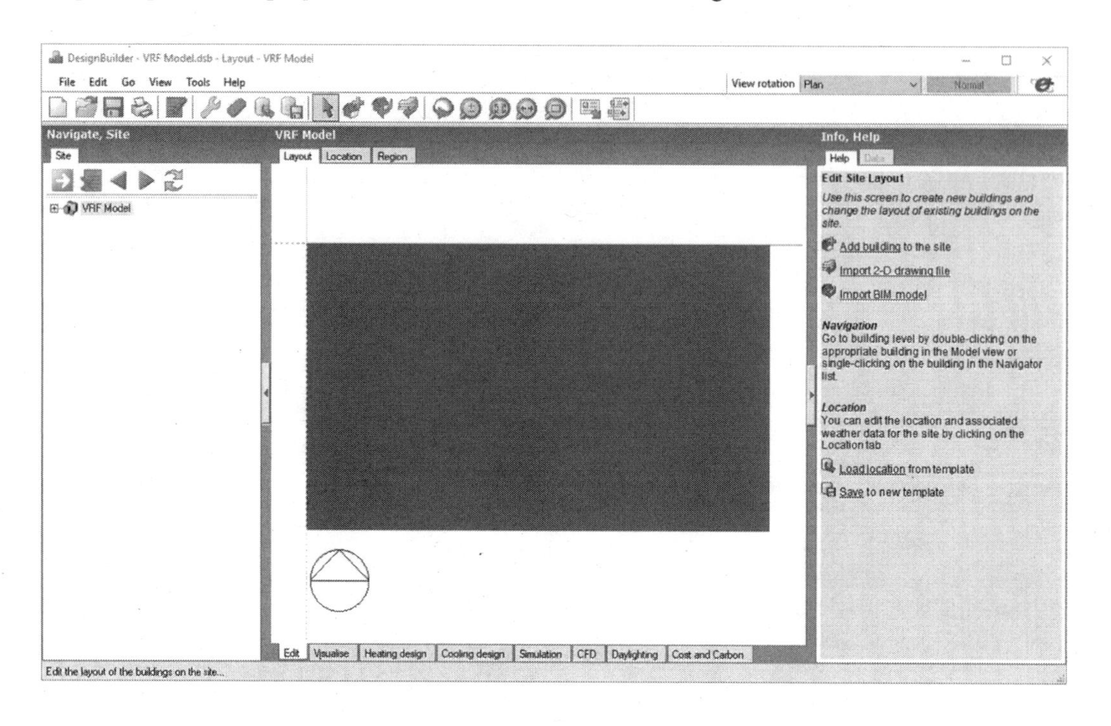

Step 2: Clone the floor, and add three more floors.

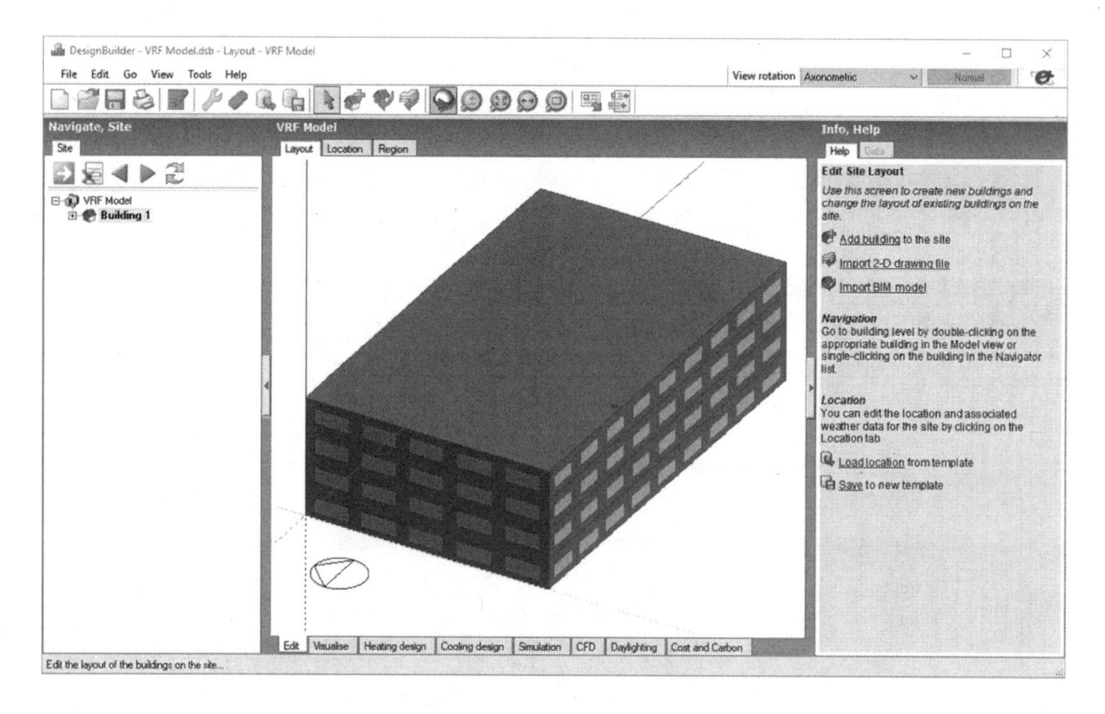

Step 3: Rename the floors as **LEVEL First**, **LEVEL Second**, **LEVEL Third** and **LEVEL Fourth**.

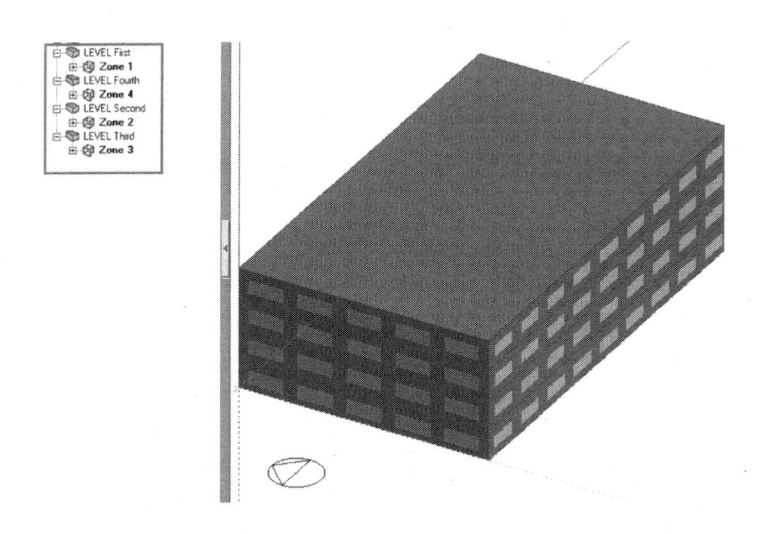

Step 4: Click the **Edit** menu, and select **Model Options**.

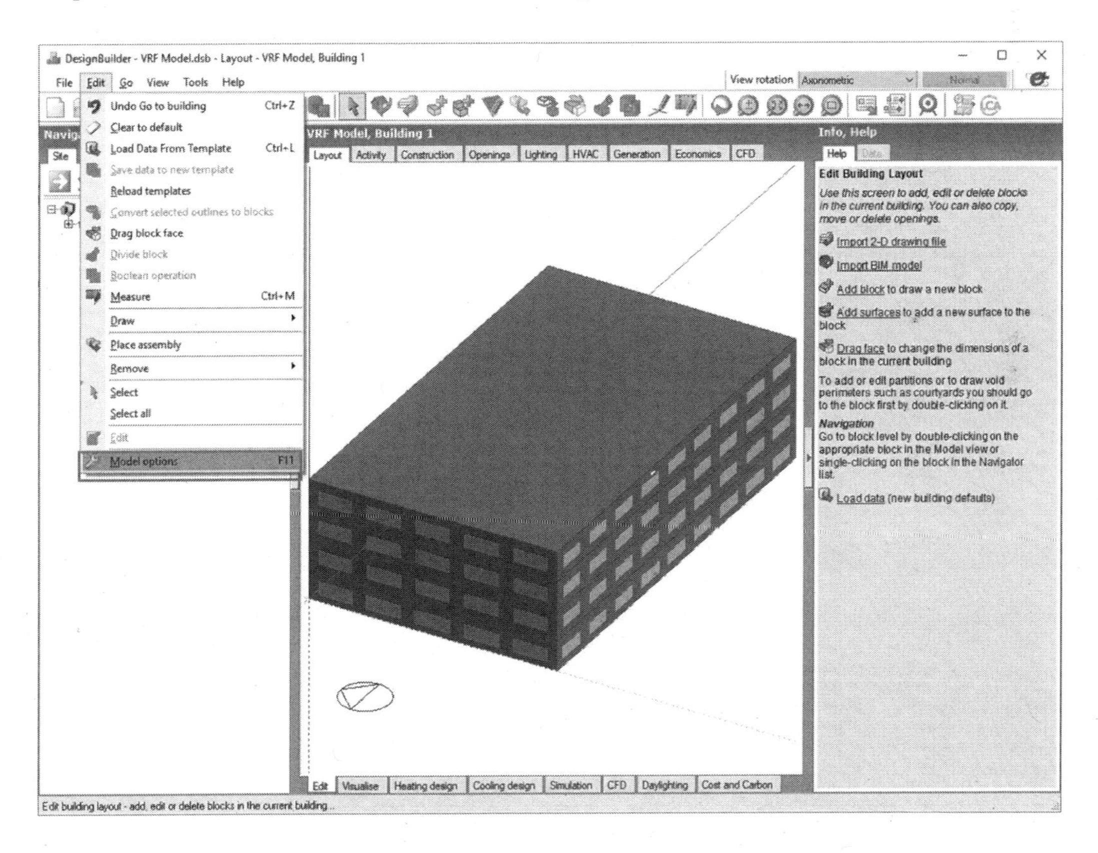

Step 5: Move the pointer to **Detailed** under **HVAC**. Select **2-Detailed HVAC Data** from the **Detailed HVAC data active data** drop-down list.

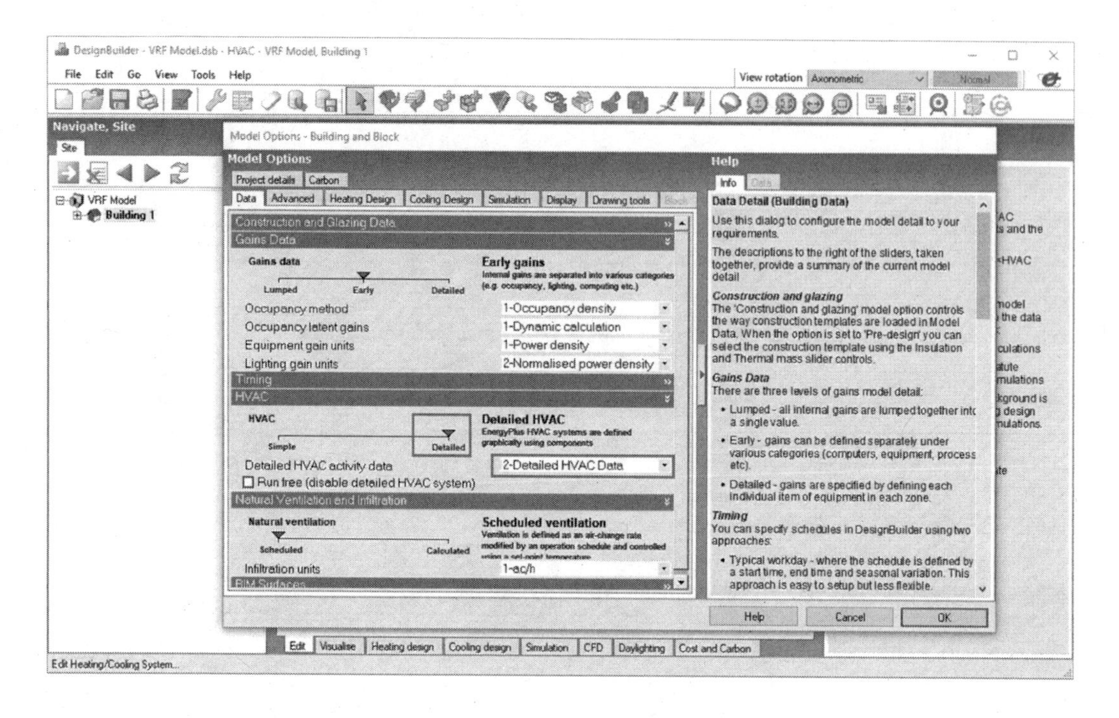

Step 6: Click **HVAC System** in the navigation panel.

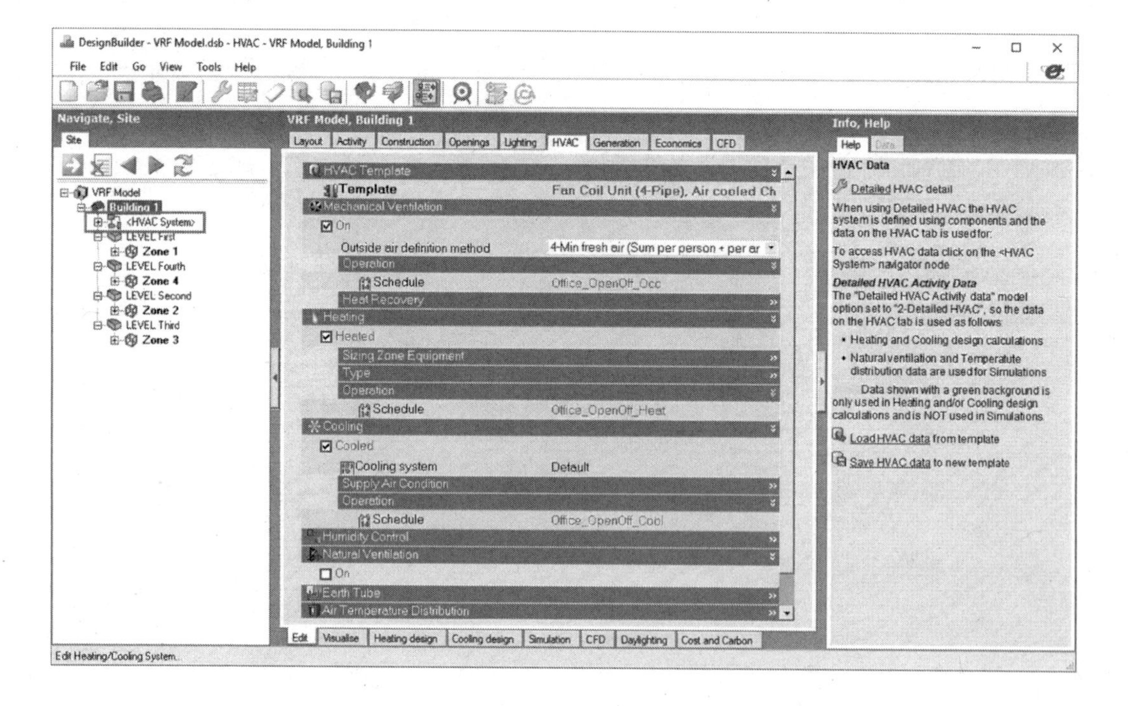

Step 7: Click **FCU 4-pipe, Air-cooled Chiller**. Three dots appear.

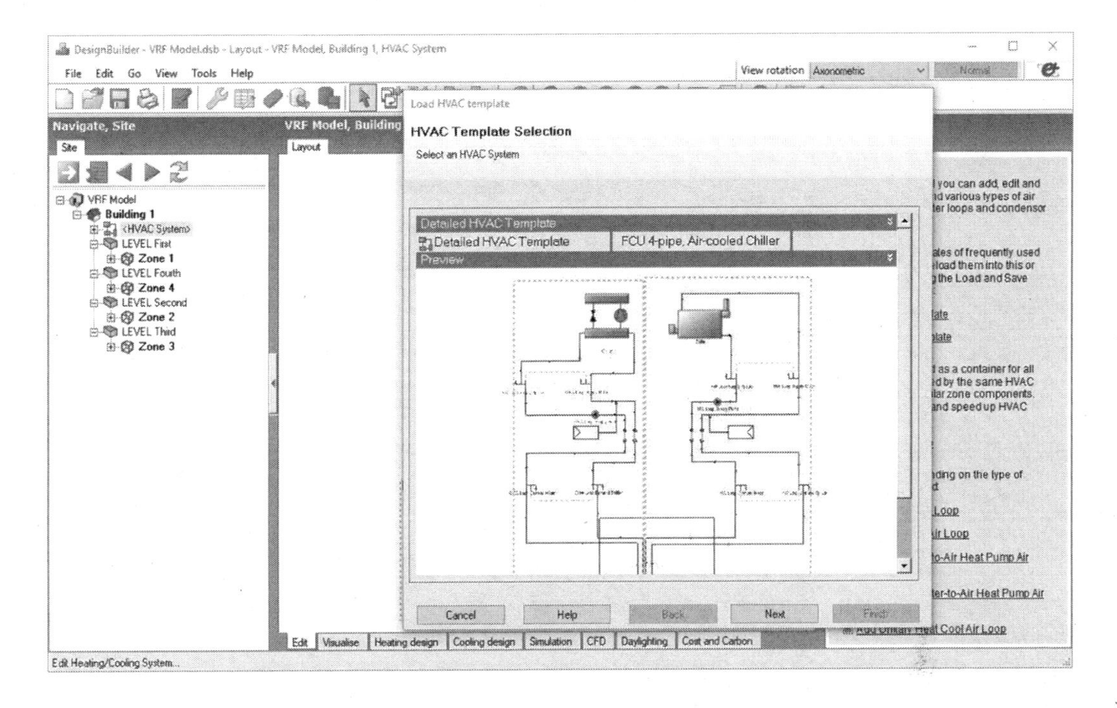

Step 8: Click the three dots, and select **VRF with HR and DOAS**.

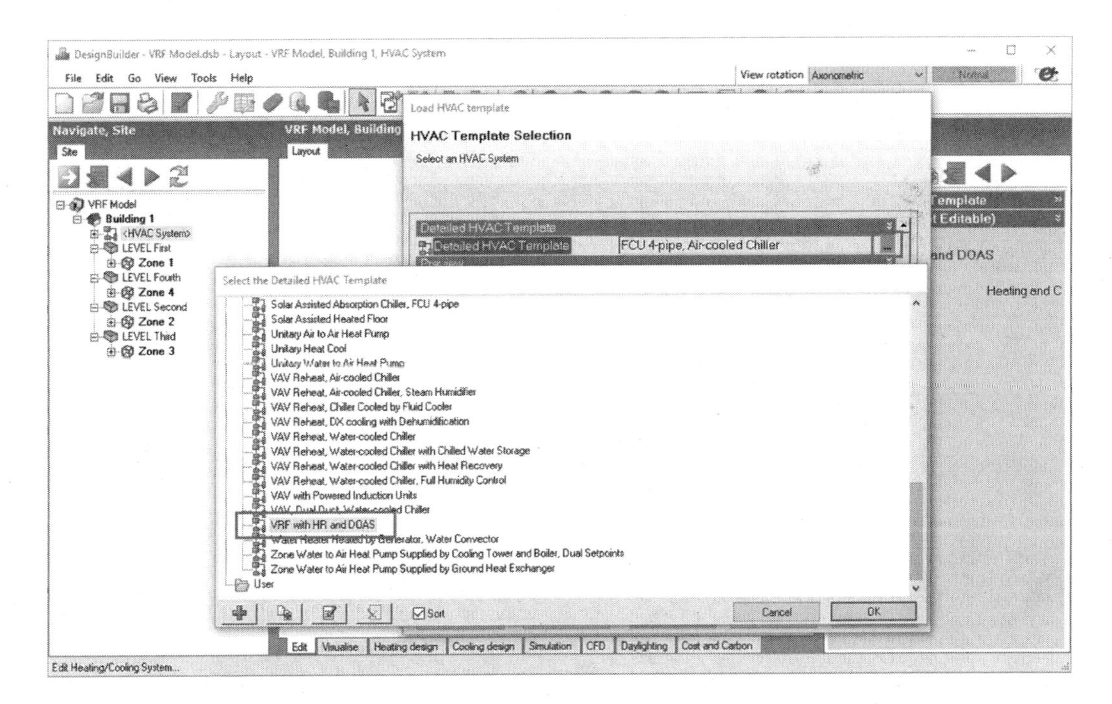

Step 9: Click **Next**.

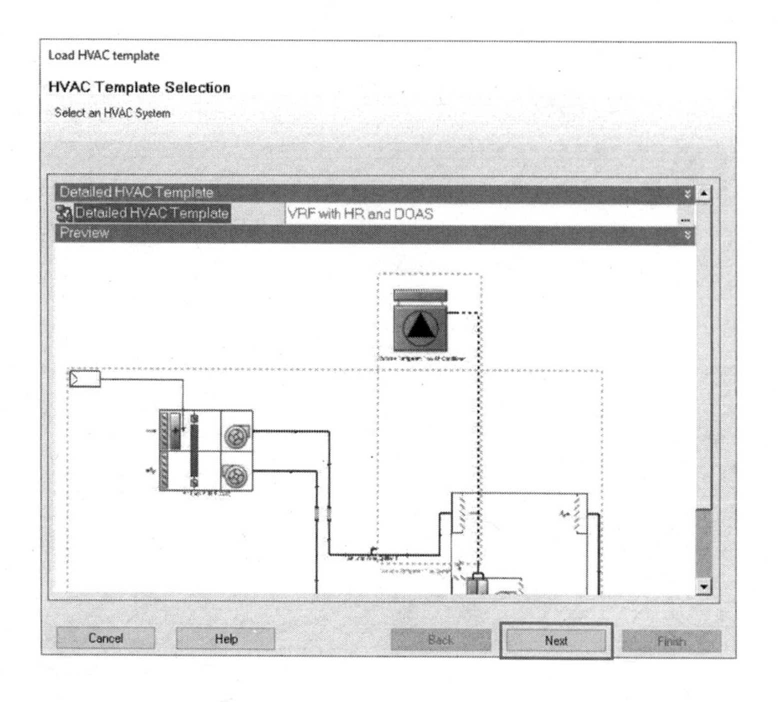

Step 10: Select the **Building 1** checkbox. Click **Finish**.

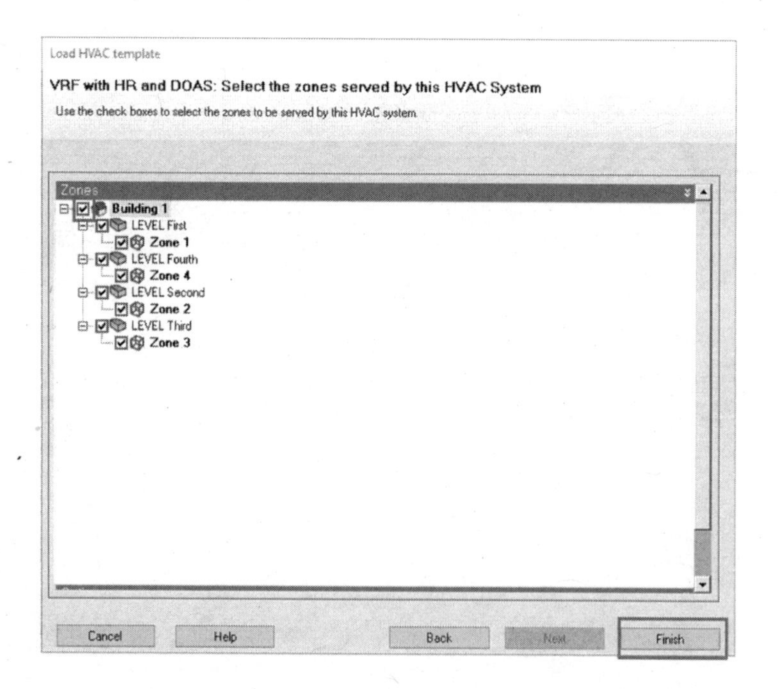

Step 11: Click **Building 1**.

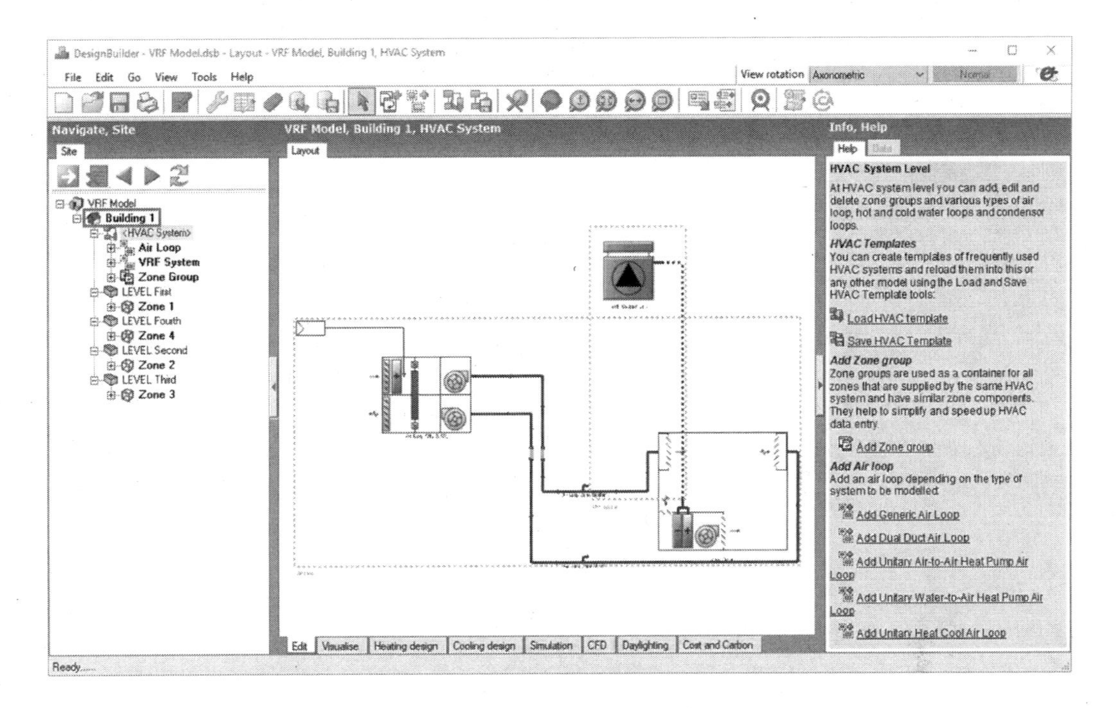

Step 12: Click the **Activity** tab.

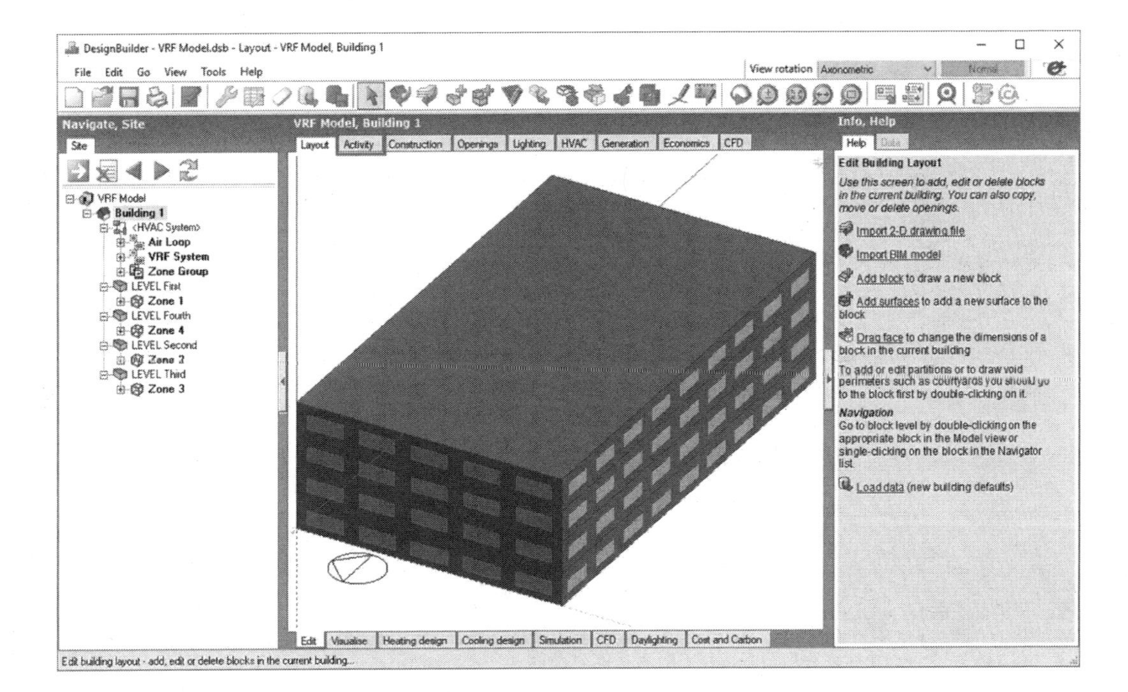

Step 13: Click **LEVEL First**. Select **24X7_Circulation_Occ schedule** under **the Occupancy** and **Office Equipment** schedules.

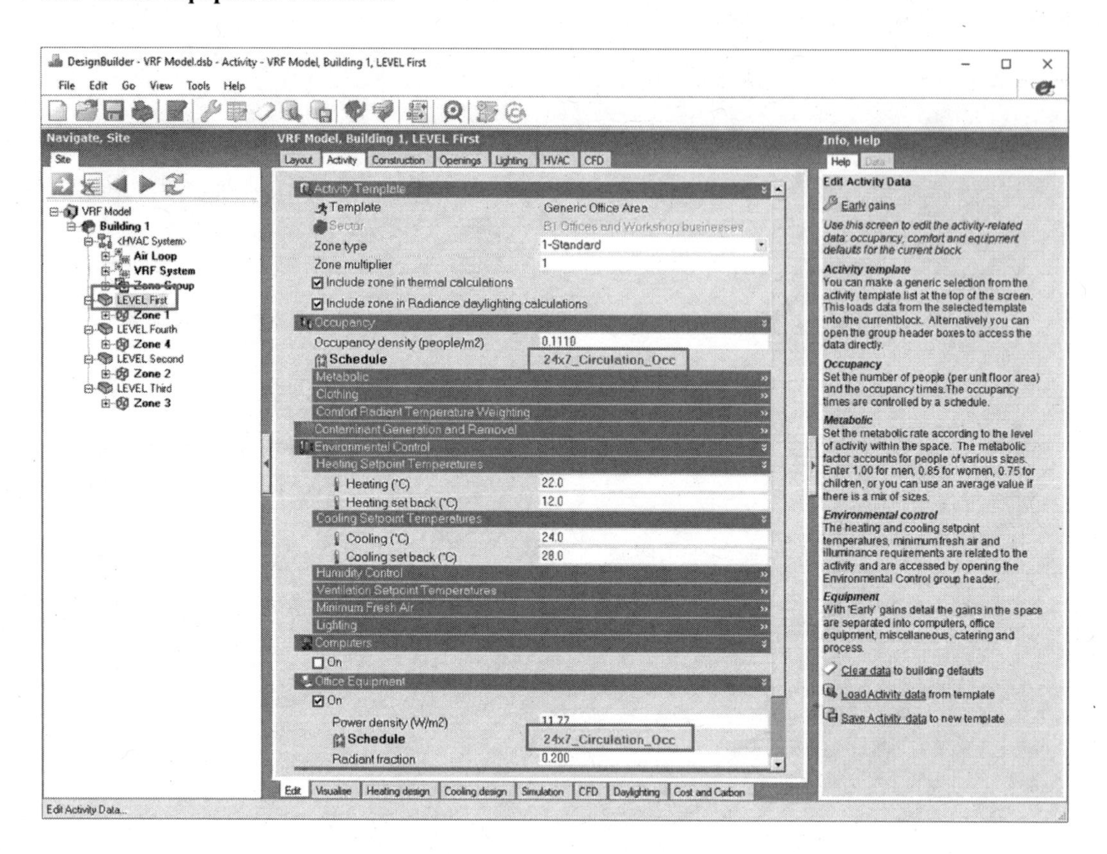

Step 14: Click the **Lighting** tab. Select **24X7_Circulation_Occ schedule**.

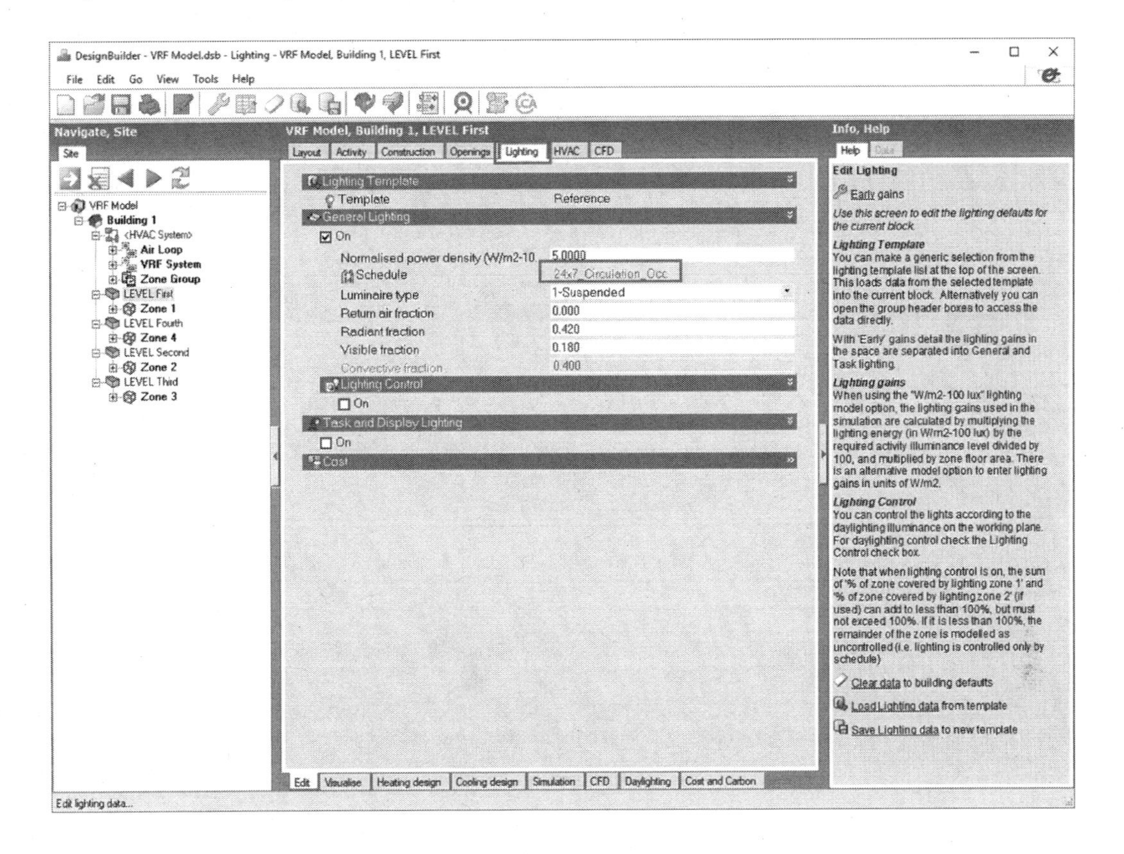

Step 15: Repeat the preceding three steps to apply **24X7_Circulation_Occ schedule** to **LEVEL Second**.

Step 16: Create a schedule named **7 AM to 1 PM**.

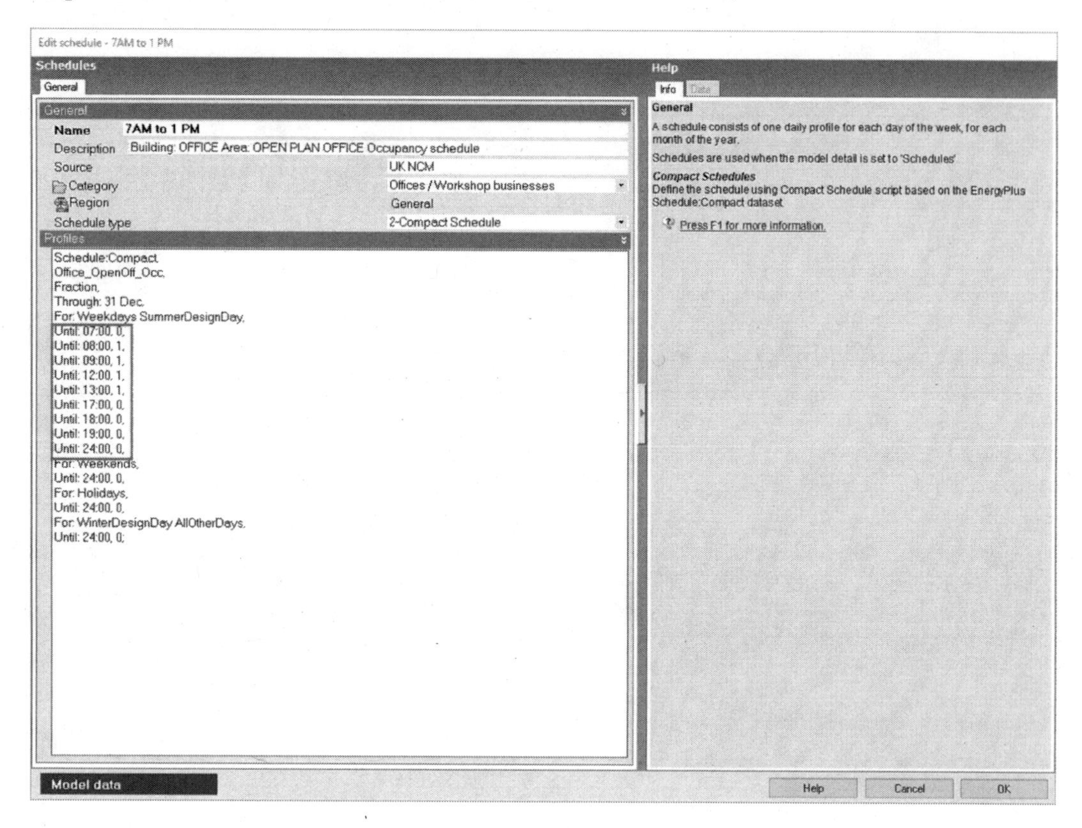

Step 17: Click **LEVEL Third**. Select the **7AM to 1 PM schedule** under **the Occupancy** and **Office Equipment** schedules.

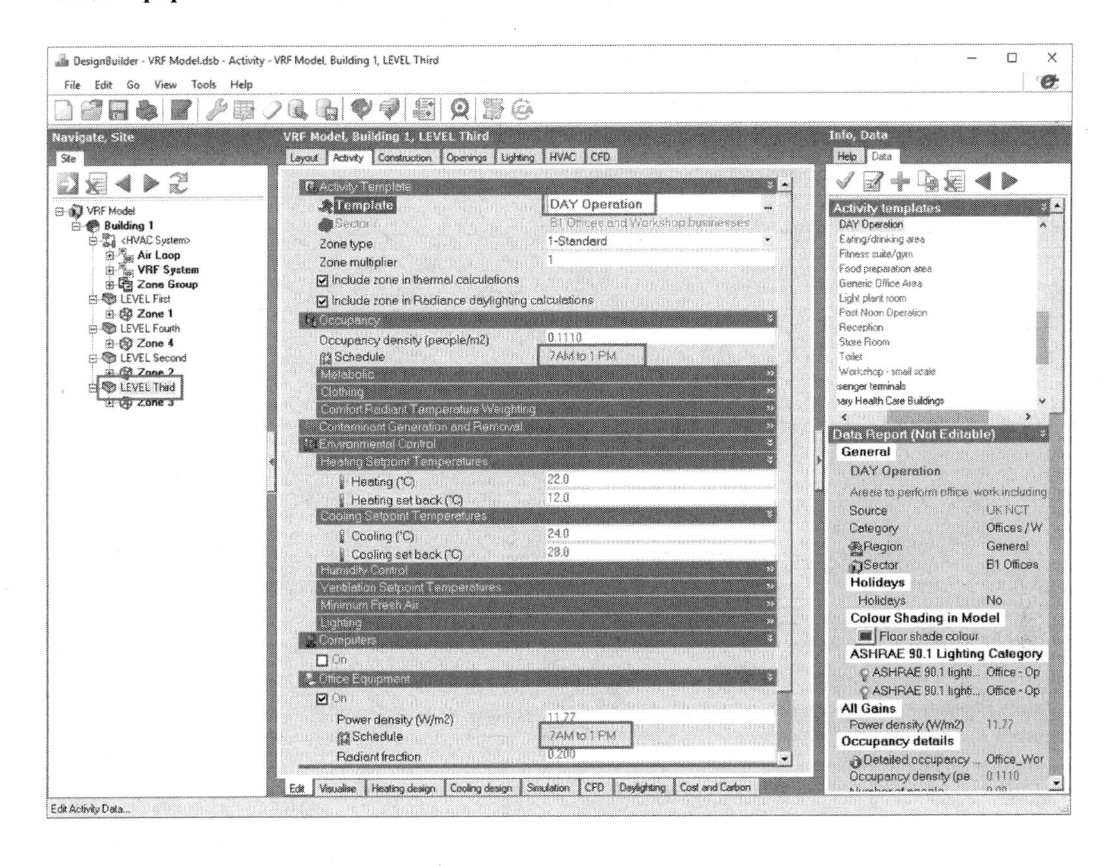

Step 18: Click the **Lighting** tab. Select the **7AM to 1 PM** schedule.

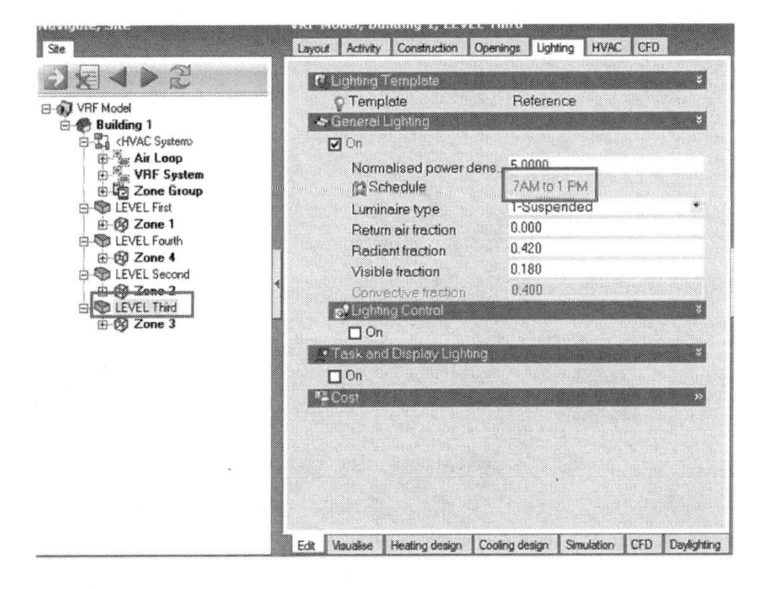

Step 19: Create and edit an operation schedule named 2 **PM to 6 PM**.

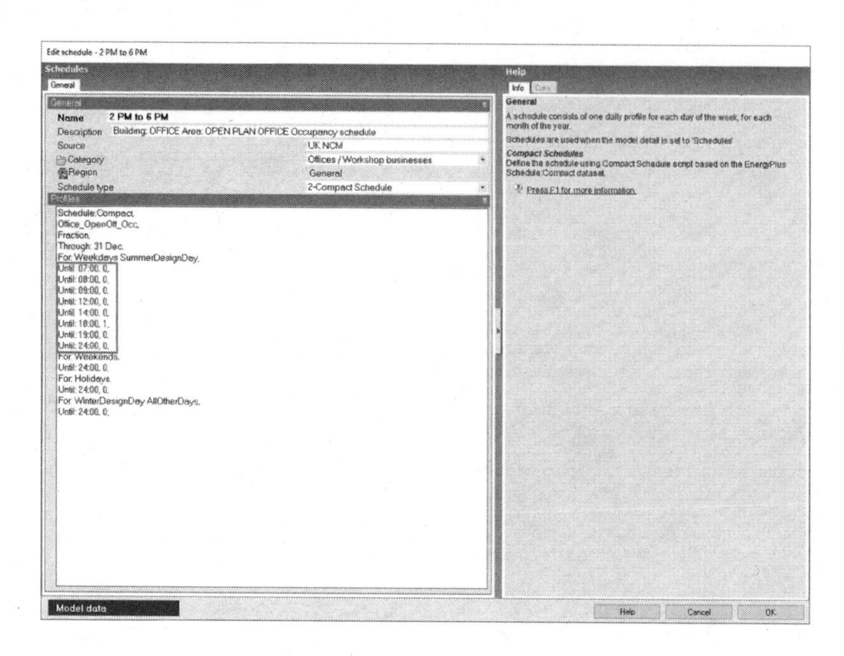

Step 20: Click **LEVEL Fourth**. Select the **2 PM to 6 PM schedule** under **the Occupancy** and **Office Equipment** schedules.

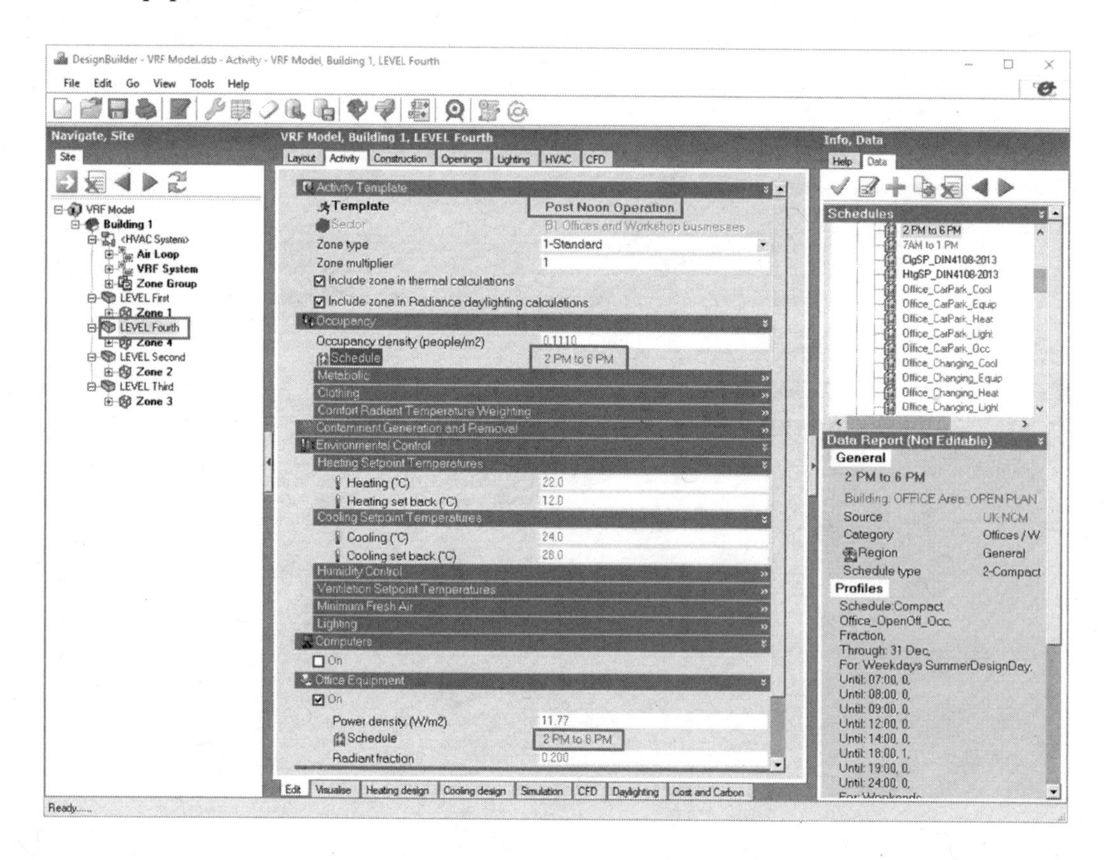

Step 21: Click the **Lighting** tab. Select the **2 PM to 6 PM** schedule.

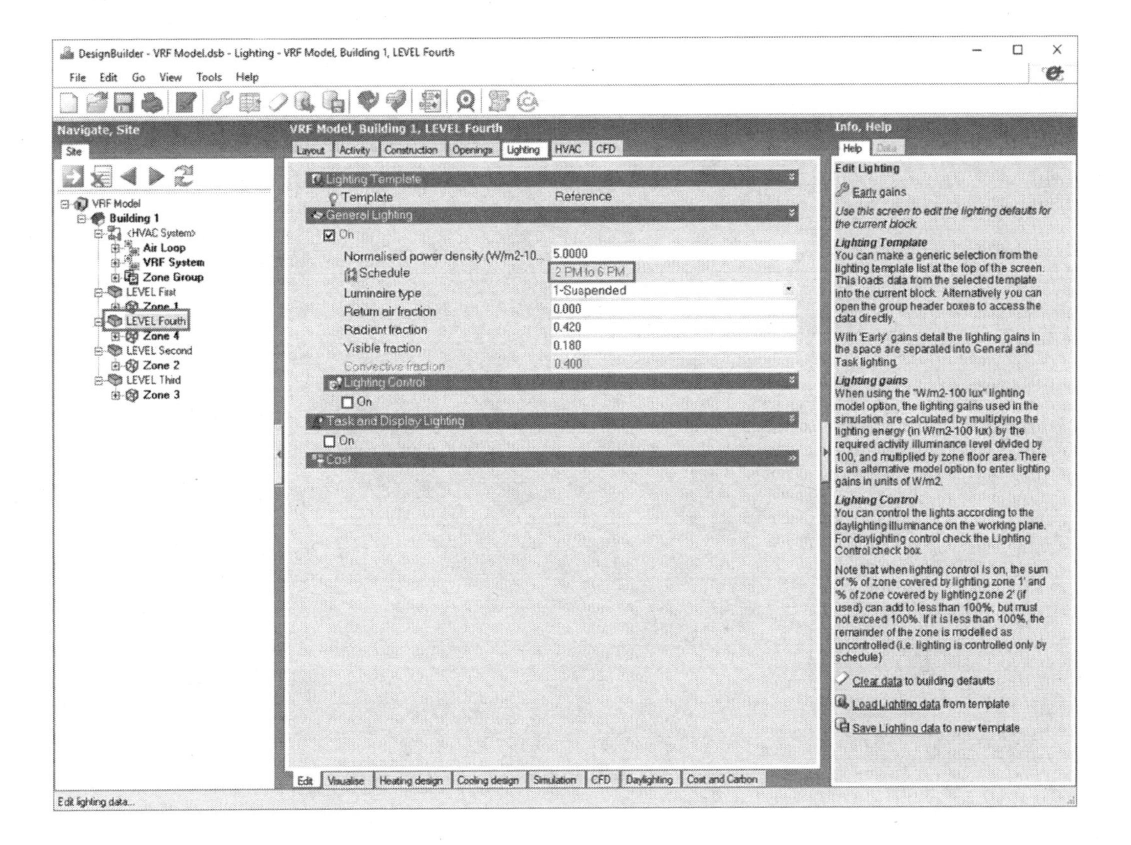

Step 22: Click **HVAC System**.

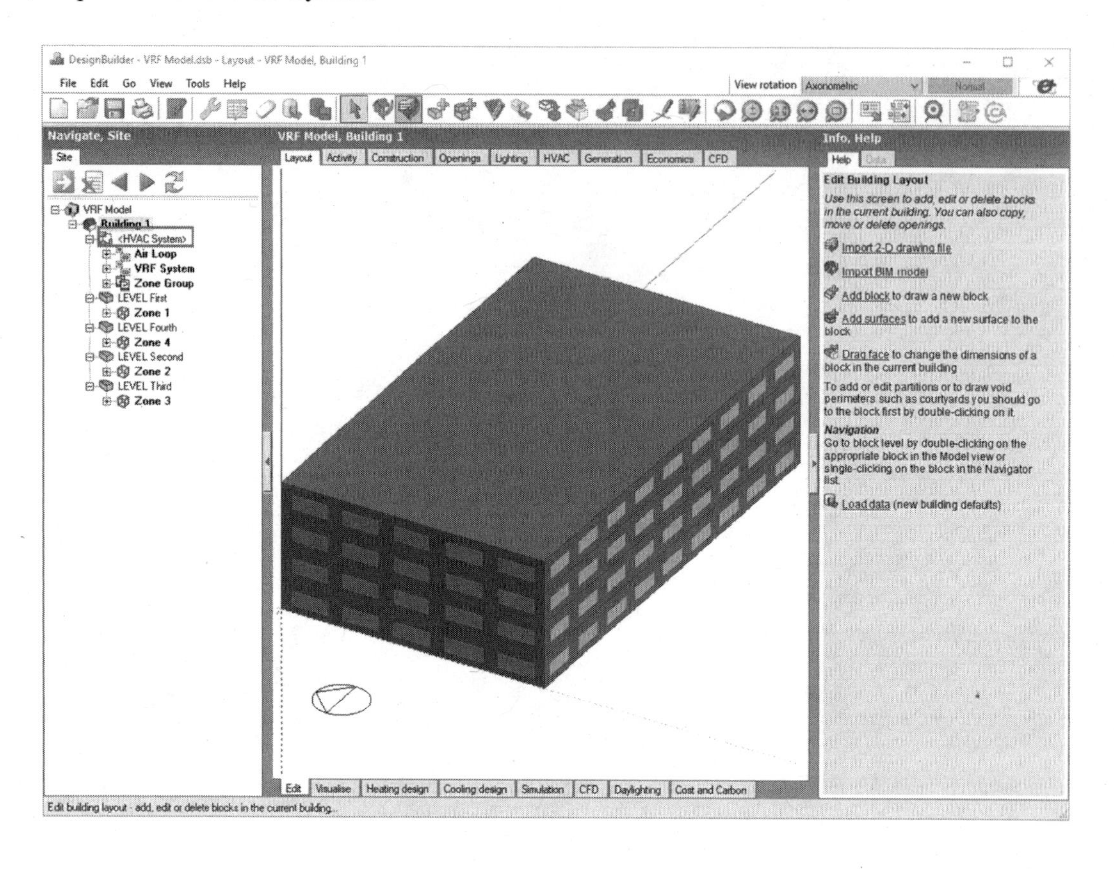

Step 23: Expand the **Zone Group** tree. Select **LEVEL:Third:Zone3**. Click **Edit HVAC zone**.

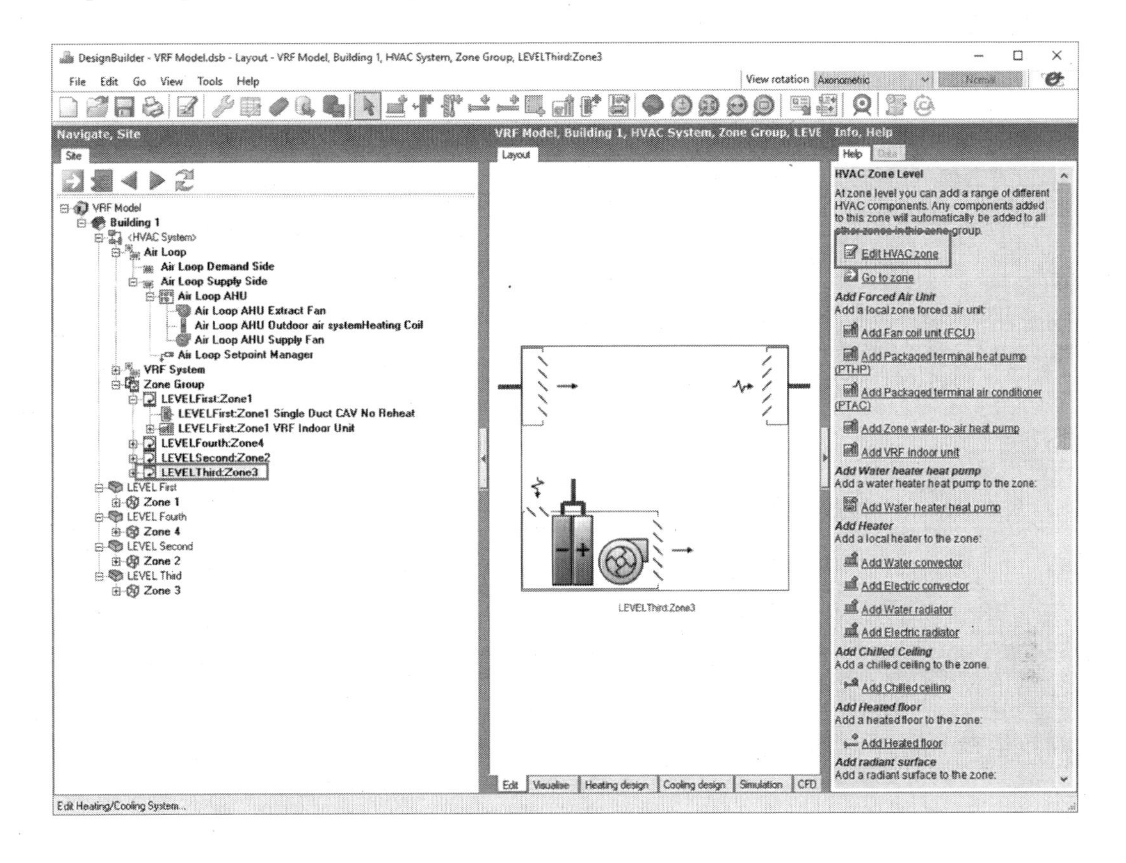

Step 24: Click **Cooling set point schedule**. Three dots appear.

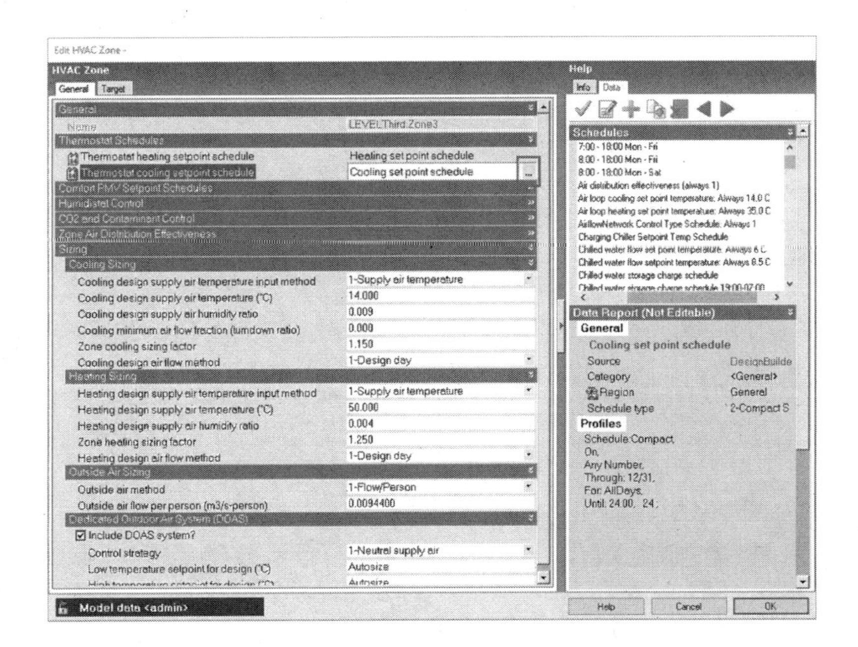

Step 25: Copy the existing schedule, and edit it with assign the name **Cooling set point schedule DAY**.

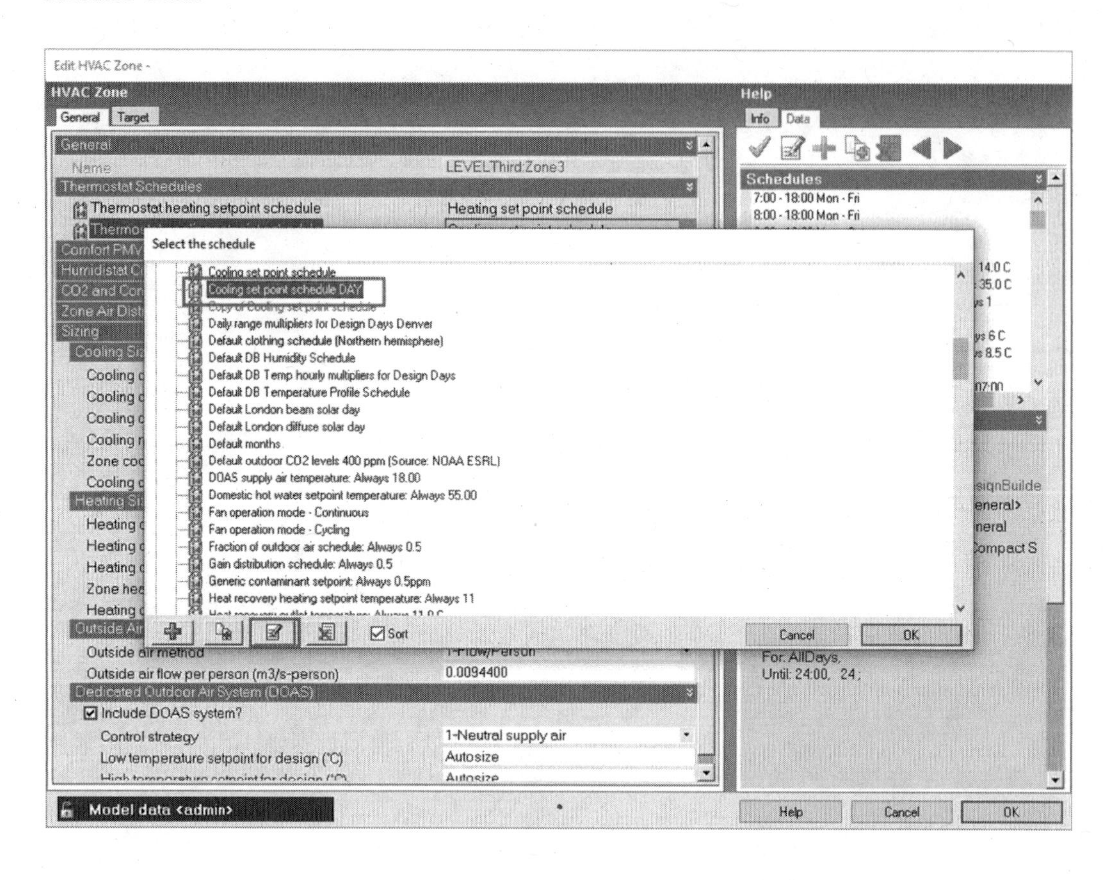

Step 26: Edit the schedule as shown in the following figure.

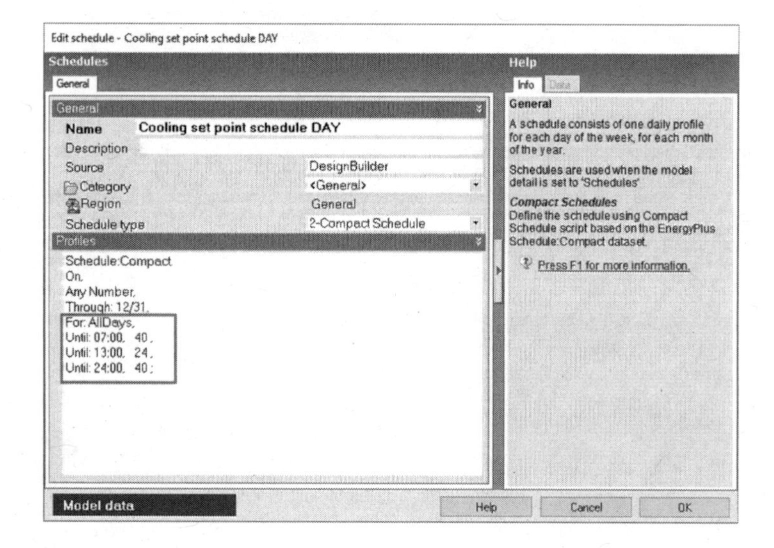

Step 27: Click the **Select this data** icon to apply **Cooling set point schedule DAY**.

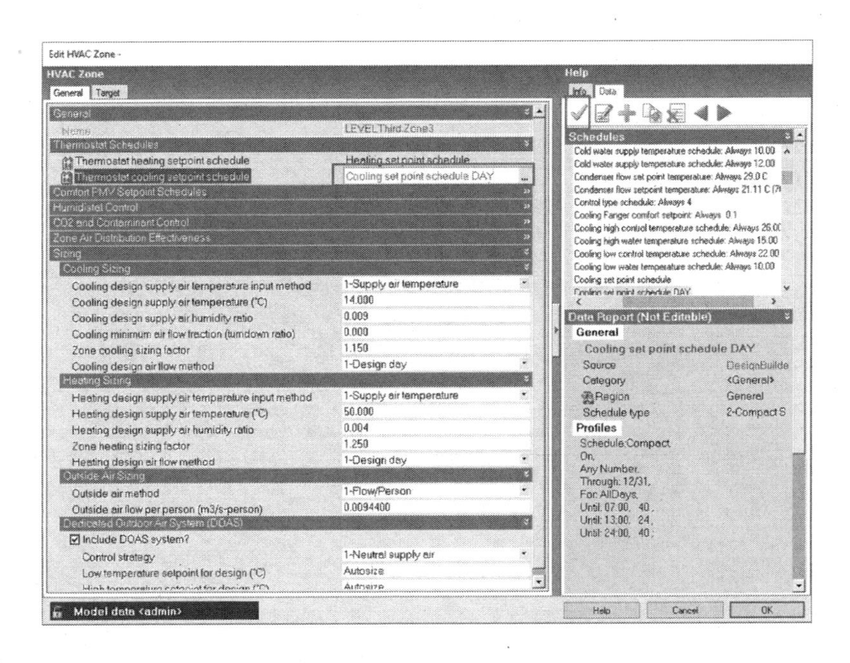

Step 28: Create a cooling setpoint schedule **Cooling set point schedule After Noon**.

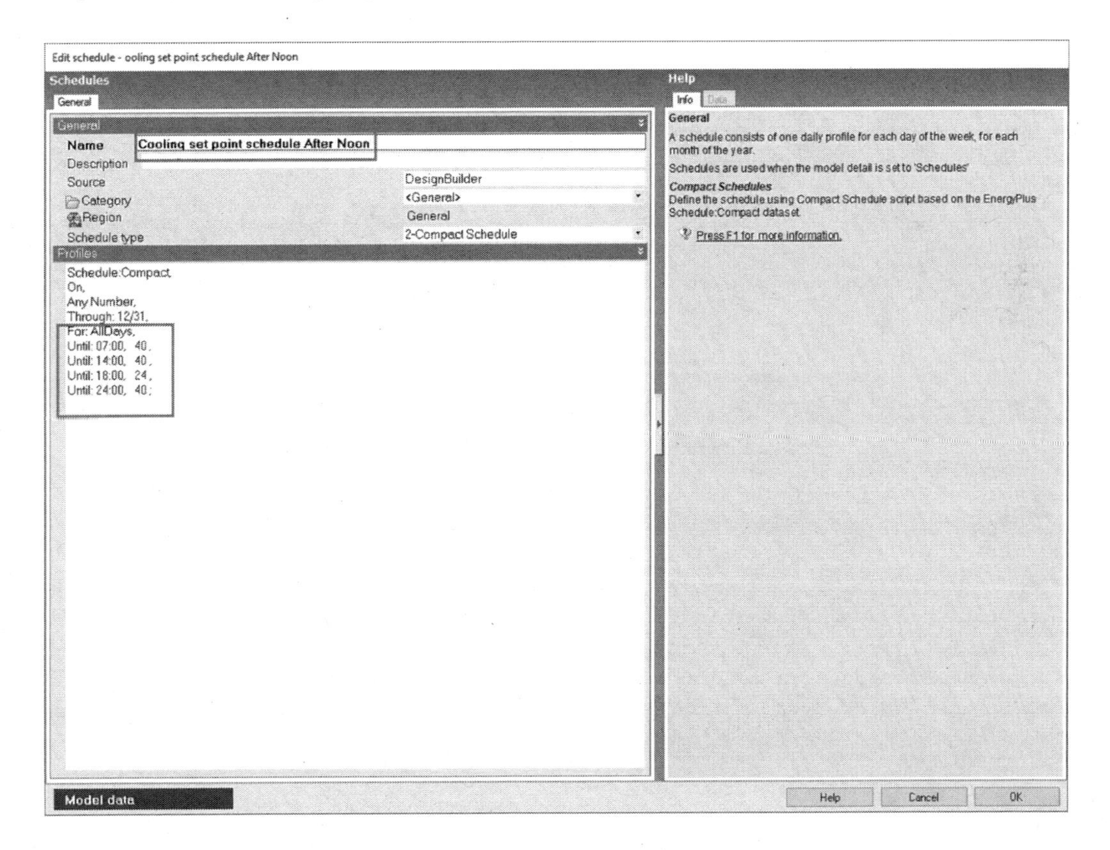

Step 29: Click **LEVEL:Fourth:Zone 4**. Click the **Select this data** icon to apply **Cooling set point schedule After Noon**. Click **OK**.

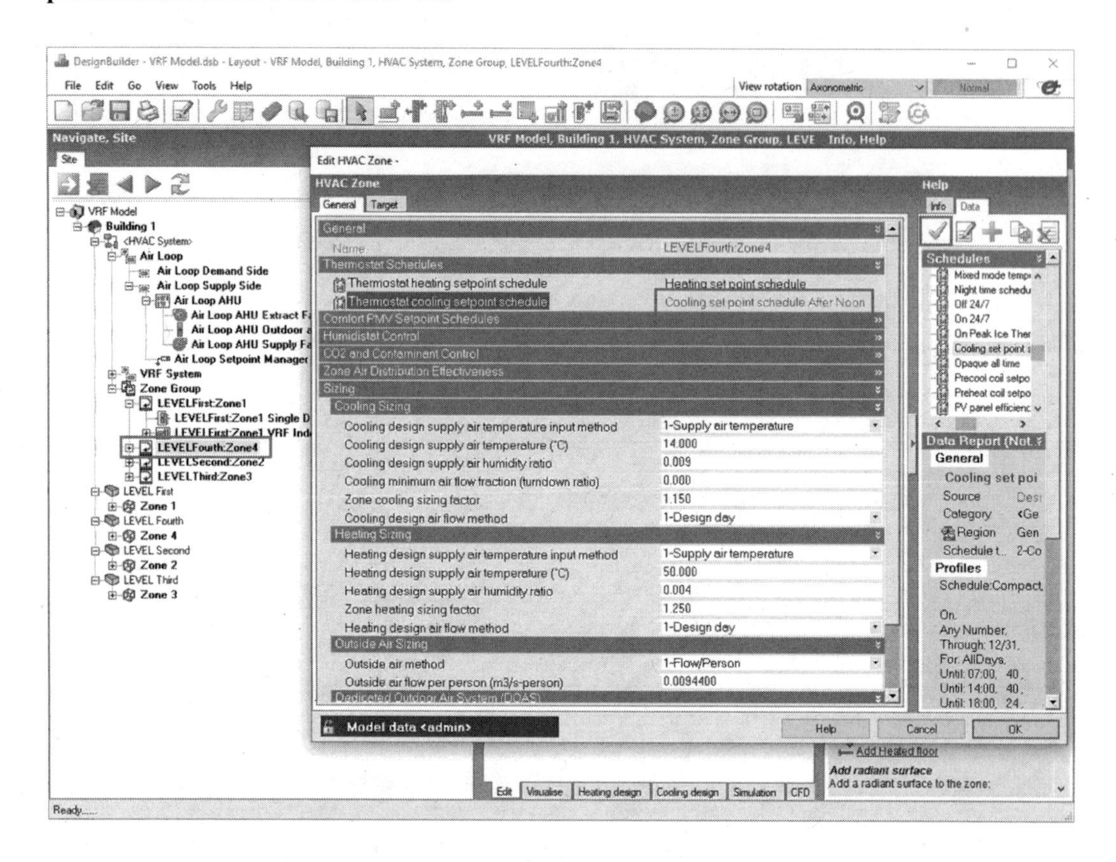

Step 30: Simulate the model for a **Summer design week**.

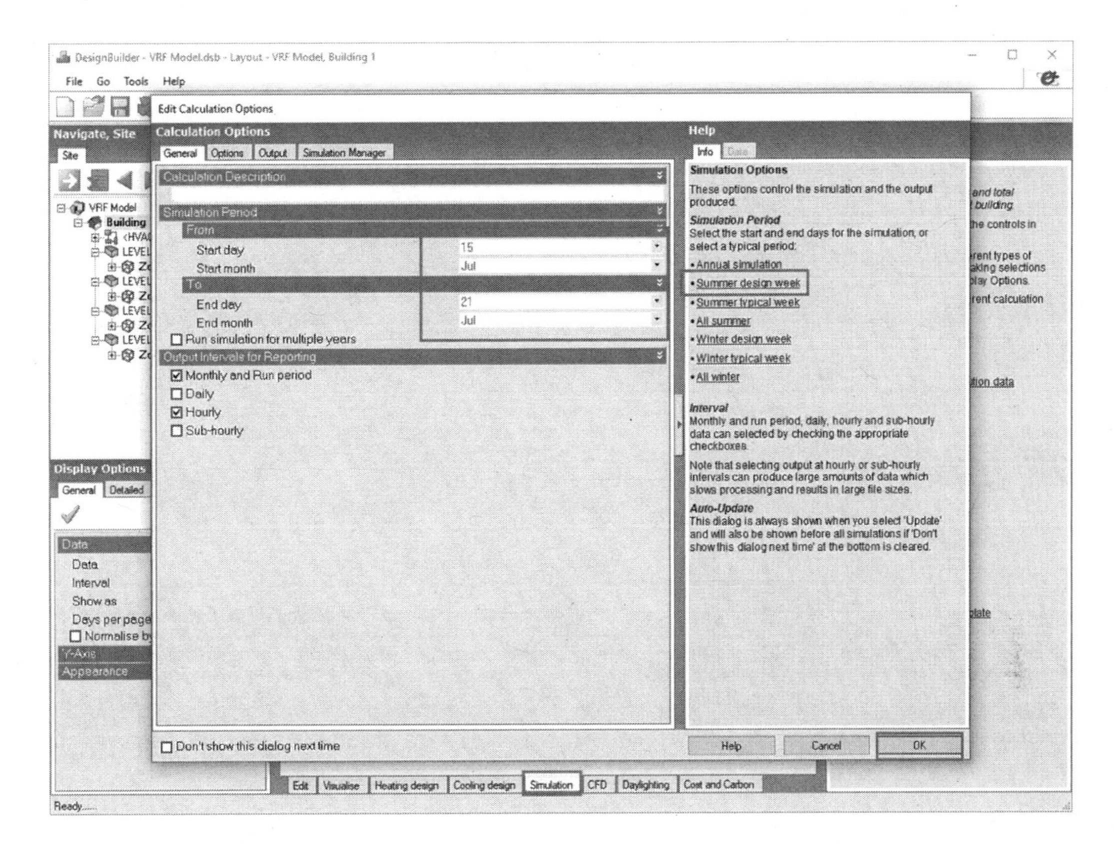

Step 31: After the simulation is finished, click the **View EnergyPlus Results** icon.

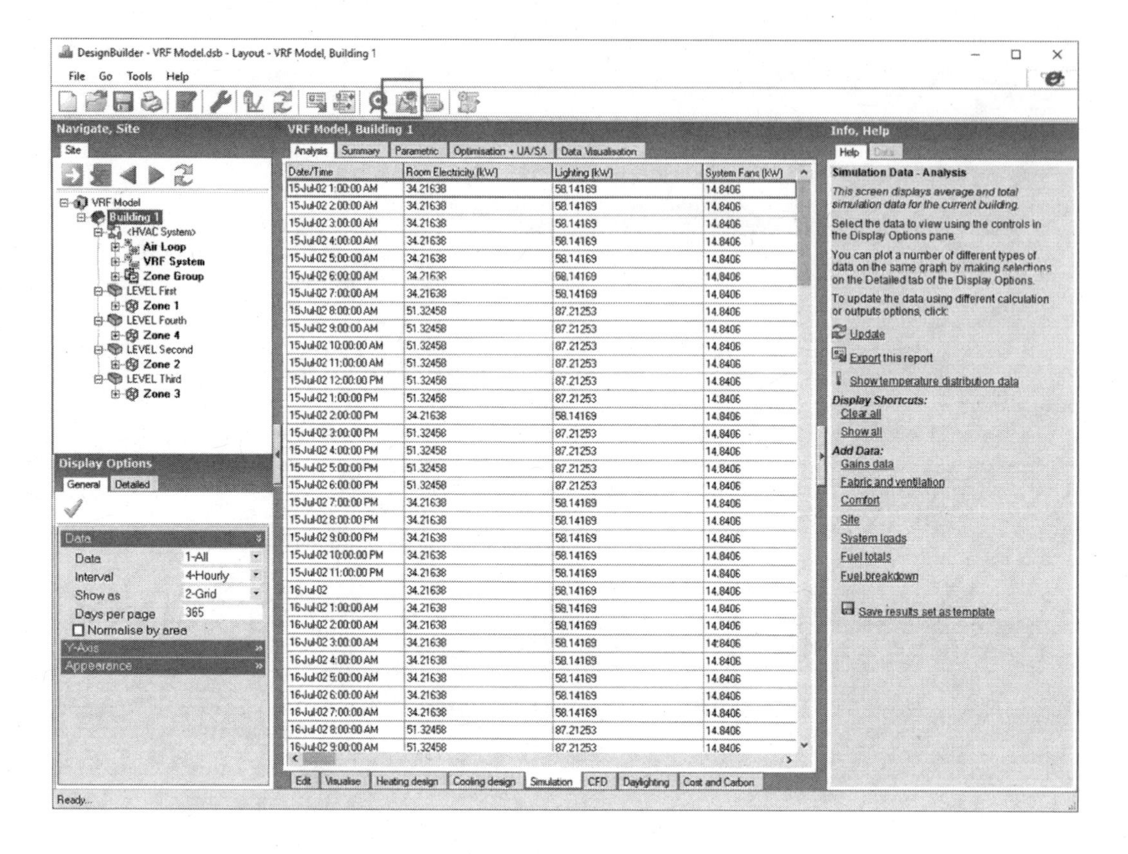

Step 32: Click the **Hourly** tab.

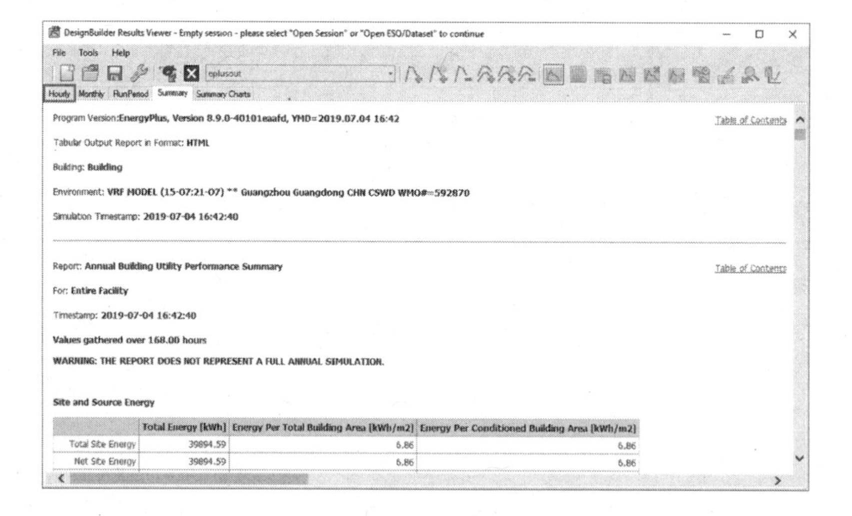

Step 33: Select **Zone VRF Air Terminal Total Cooling Rate** for all VRF indoor units. Zoom in to view the results for one day.

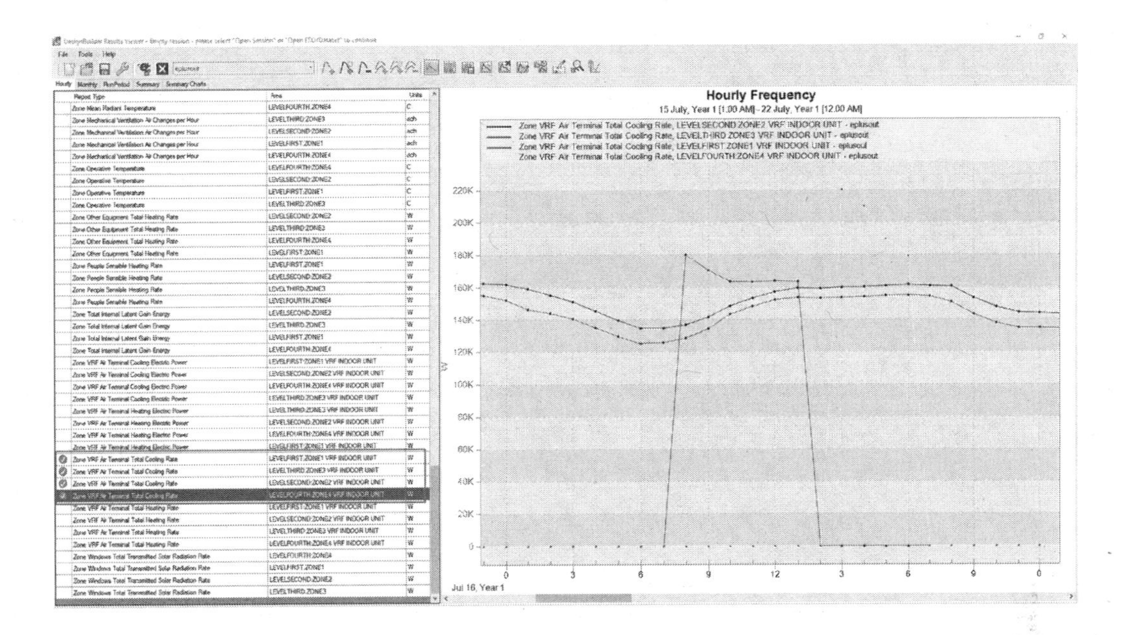

You can see that for levels 1 and 2, the total cooling rate is distributed to 24 hours. For level 3, cooling load is from 7 a.m. to 1 p.m., and for level 4, the cooling load is from 2 to 6 p.m. As of now, you have modelled one outdoor VRF unit for all indoor units. Now you can add one more outdoor VRF unit and attach level 3 and 4 systems to it.

Step 34: Select **HVAC system**. Click **Clone selected object(s)**.

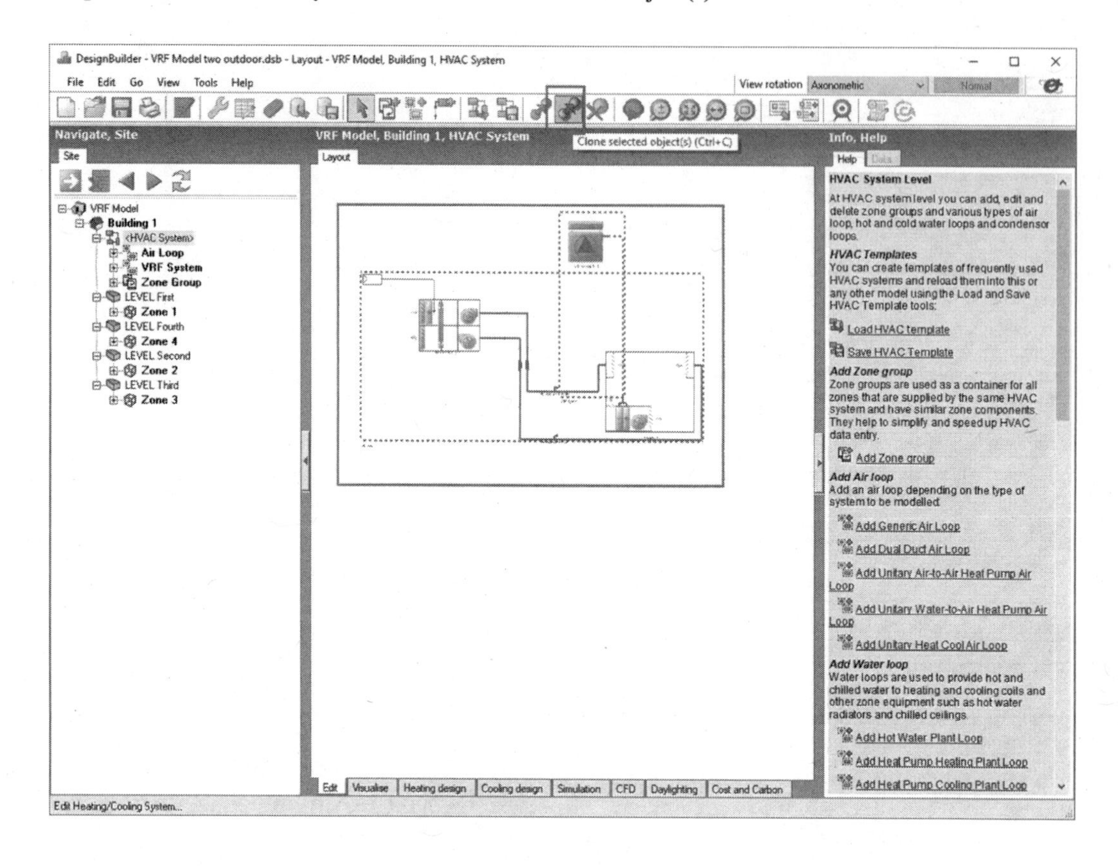

Step 35: Drag the system to clone it.

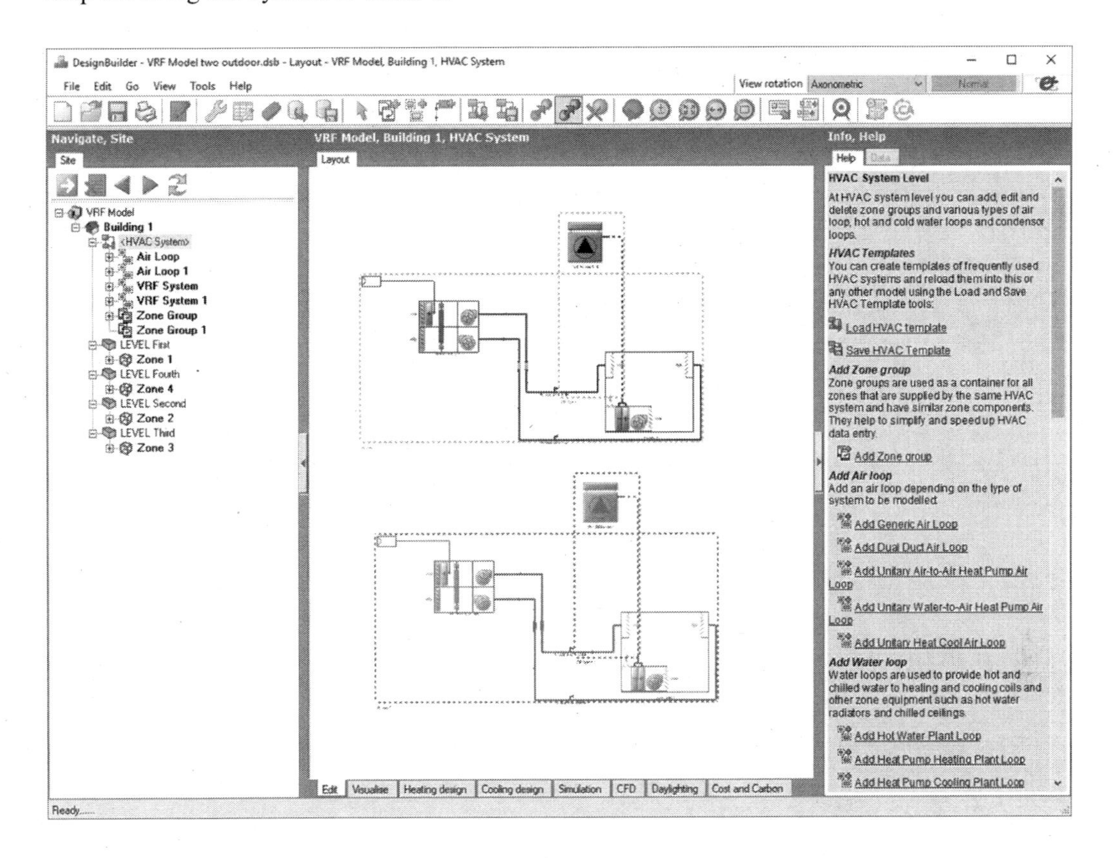

Step 36: Click **Zone Group**. Click **Edit zone Group**.

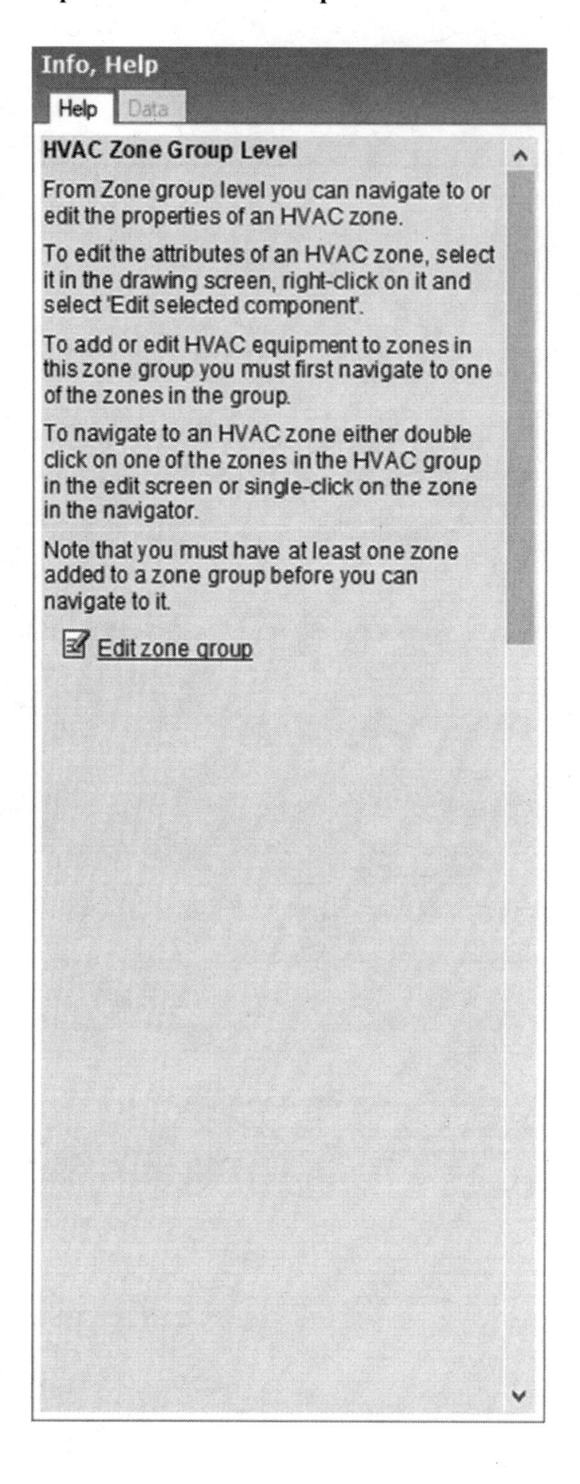

Step 37: Clear the **Zone 3** and **Zone 4** checkboxes.

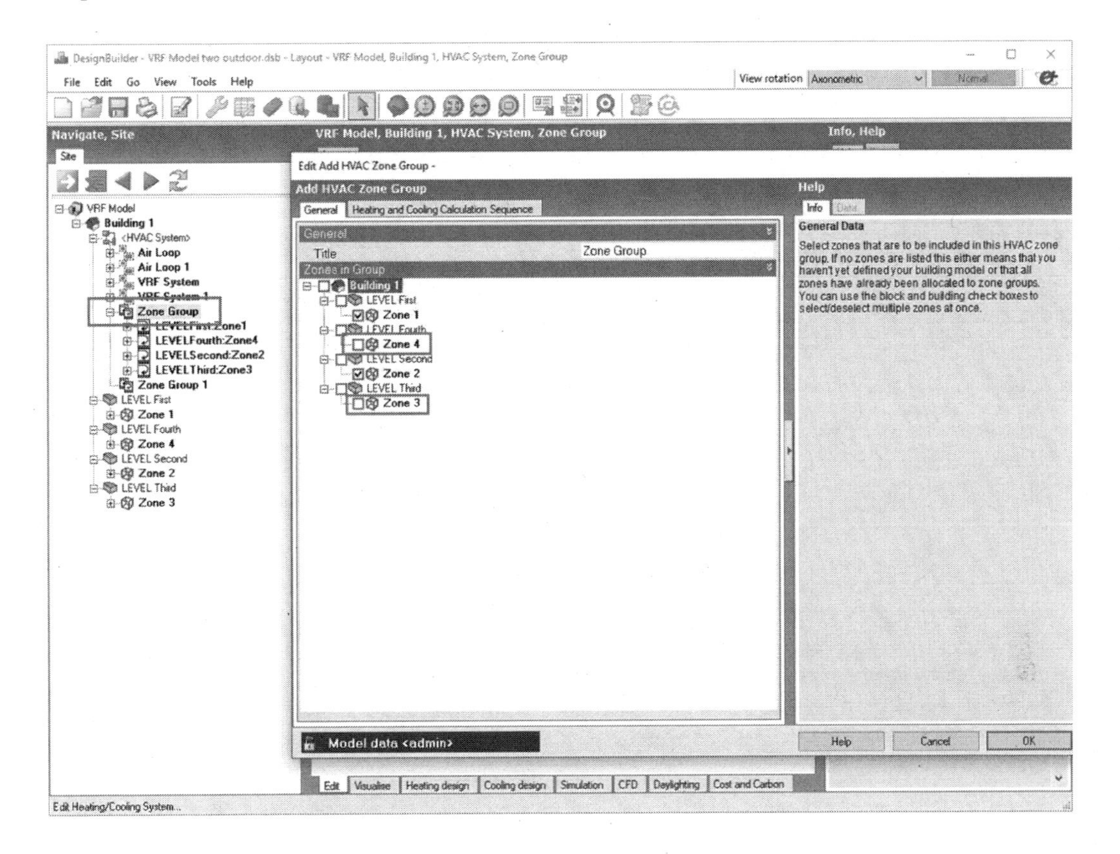

Step 38: Click **Add Zone group** from **Info, Help**.

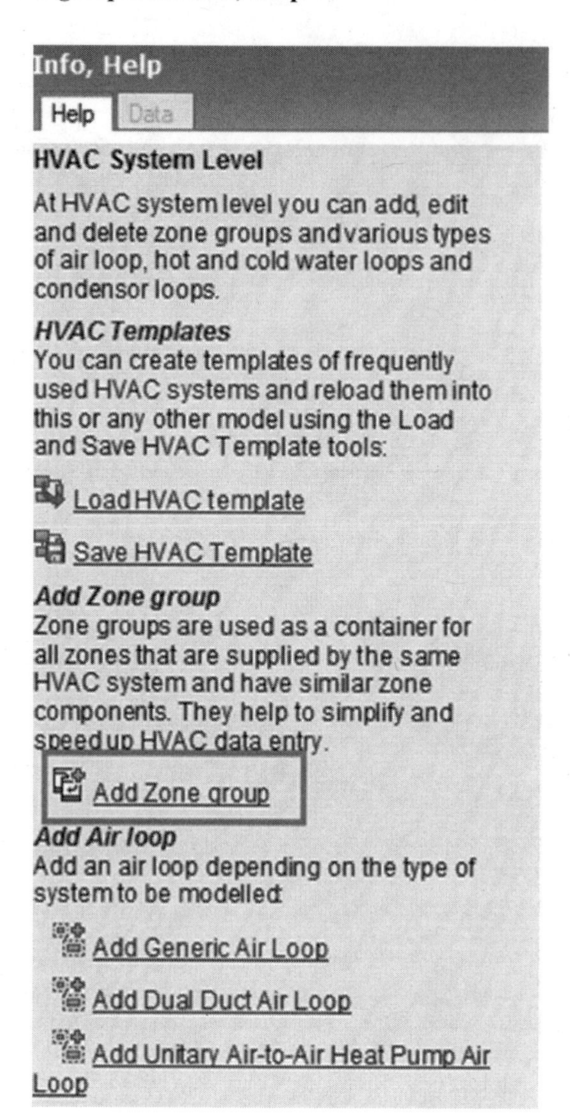

Step 39: Click **Zone Group 1**.

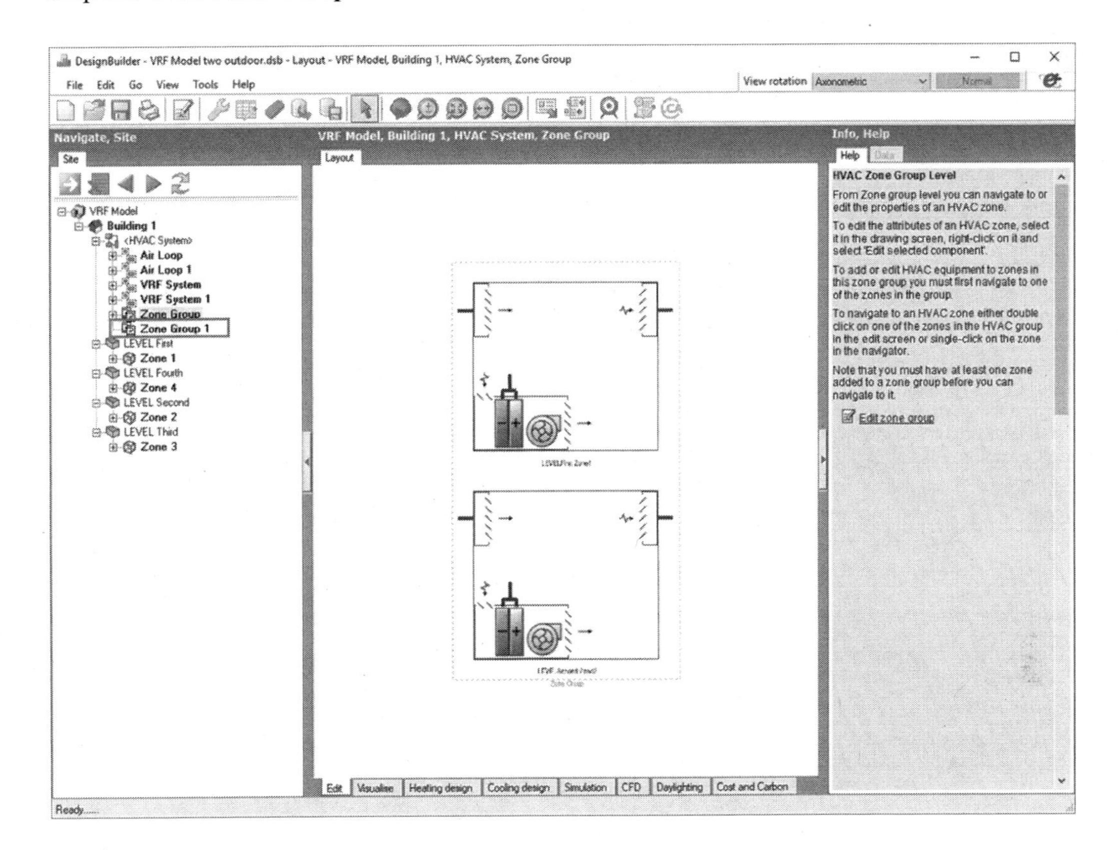

Step 40: Click **Yes**. The **Edit Add HVAC Zone Group** screen appears.

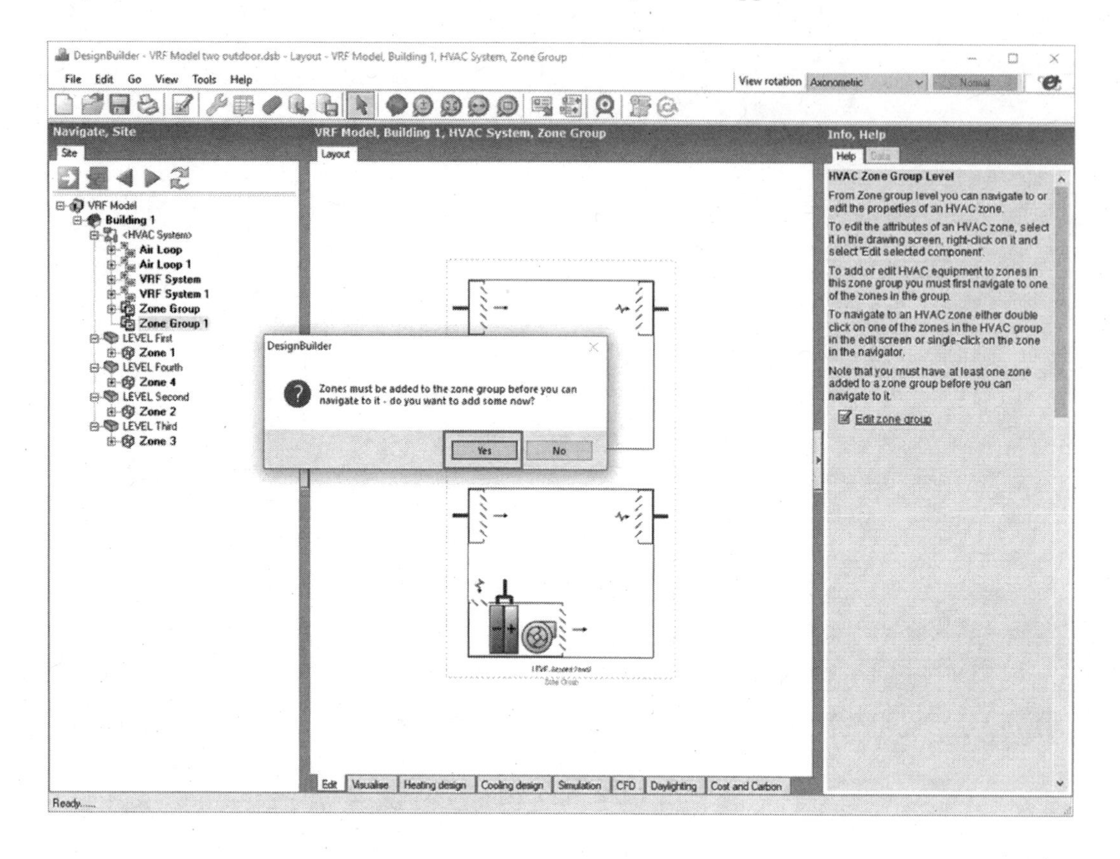

Step 41: Select **Zone 3** and **Zone 4** checkboxes. Click **OK**.

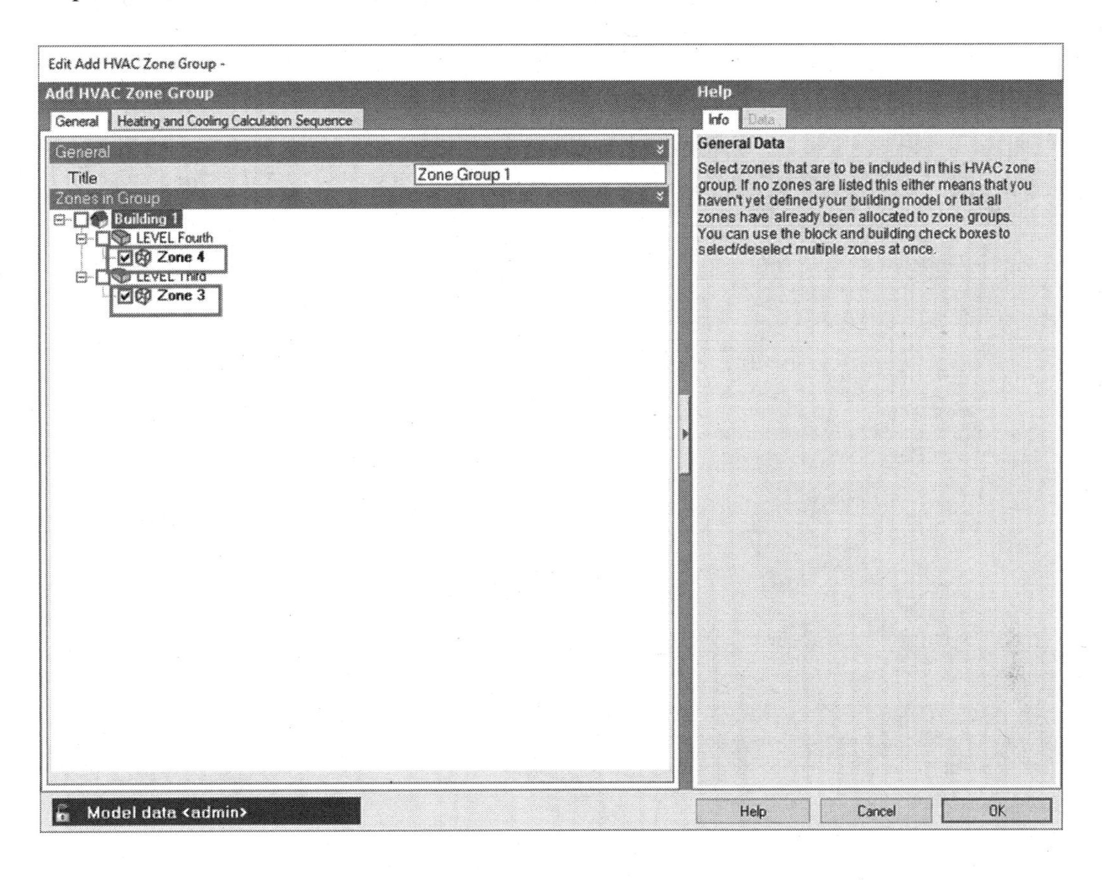

Step 42: Click the **Simulation screen** tab.

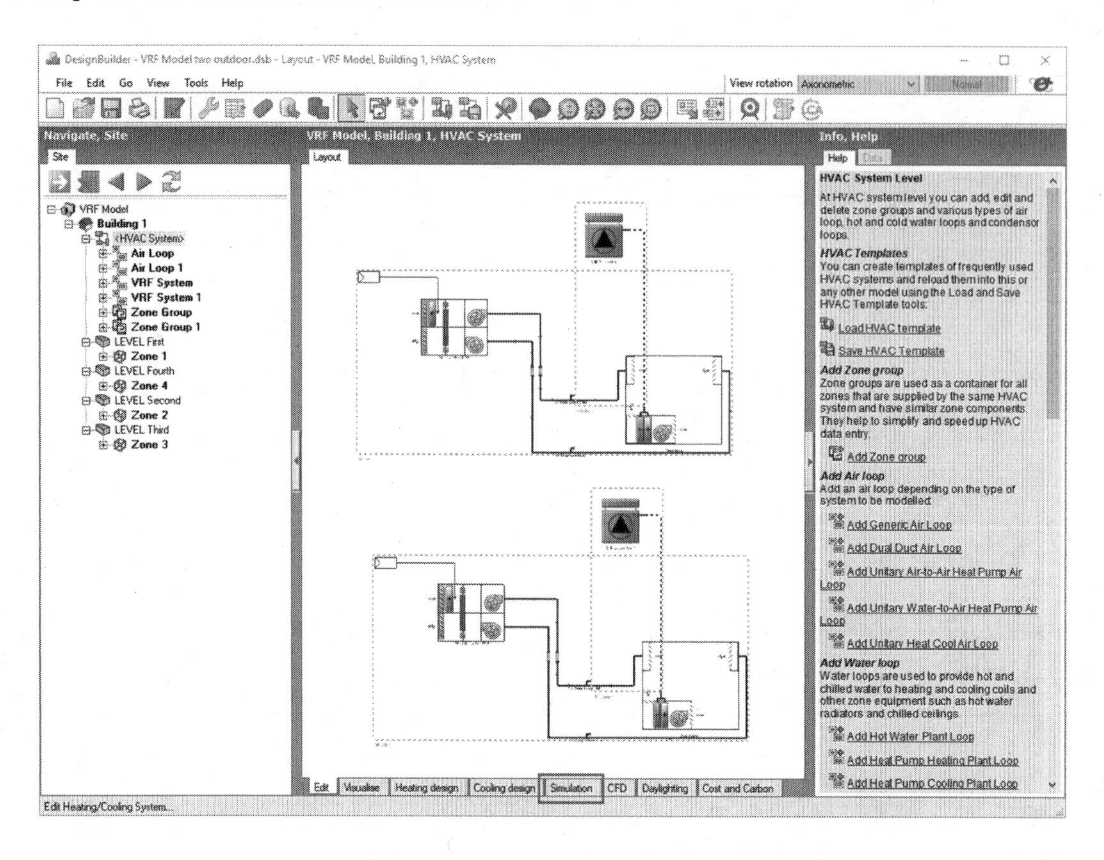

Step 43: After the simulation, open **Results Viewer** and select **VRF Heat Pump Terminal Unit Cooling Load Rate** for **VRF OUTDOOR UNIT** and **VRF OUTDOOR UNIT 1**.

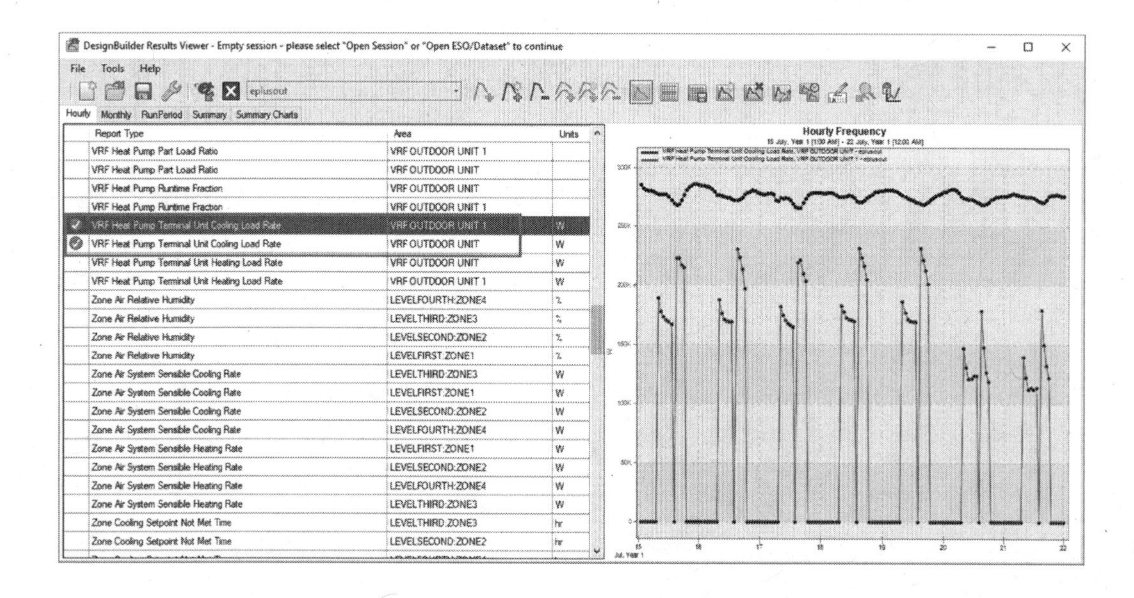

Step 44: Zoom the results for a single day.

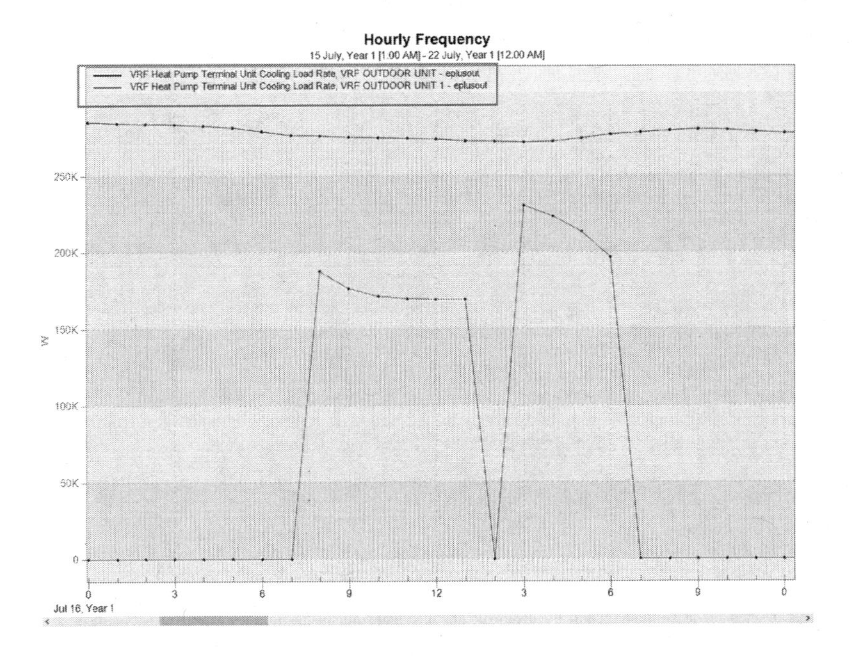

You can see that the VRF outdoor unit has an almost constant load for 24 hours with a peak of around 280 kW. Also, VRF outdoor unit 1 has a peak cooling load of 230 kW. The benefit of using two outdoor units is that cooling load is distributed over two units, and outdoor unit need not run on very low part loads.

Tutorial 9.5

Evaluating the Impact of Demand Control Ventilation

GOAL

To evaluate the impact of demand control ventilation

WHAT ARE YOU GOING TO LEARN?

- How to model demand control ventilation

PROBLEM STATEMENT

In this tutorial, you are going to use a 10- × 10-m model building. You need to model a **VAV Reheat DX cooling with Dehumidificaiton** HVAC system

1. Without demand control ventilation
2. With demand control ventilation

Use the **New Delhi Palam, India** weather location.

SOLUTION

Step 1: Create a **10- × 10-m** model.

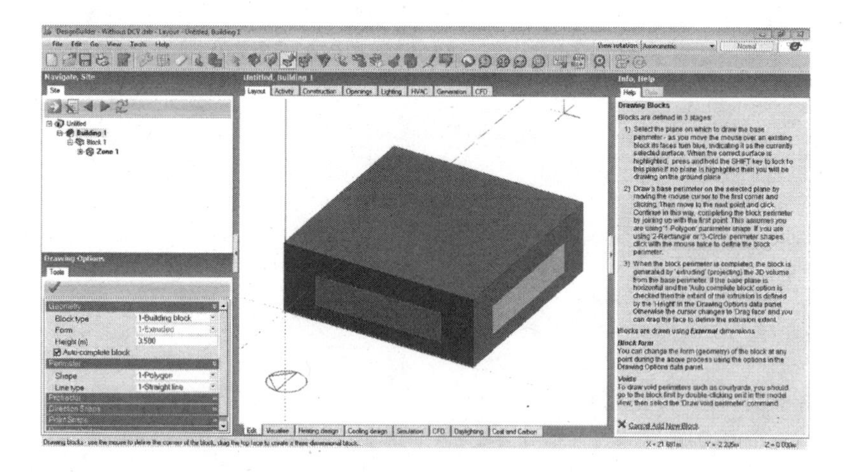

Step 2: Click **Untitled** under **Site**. Click the **Location** tab. Expand **Site Details**. Expand **Outdoor Air CO2 and Contaminants**.

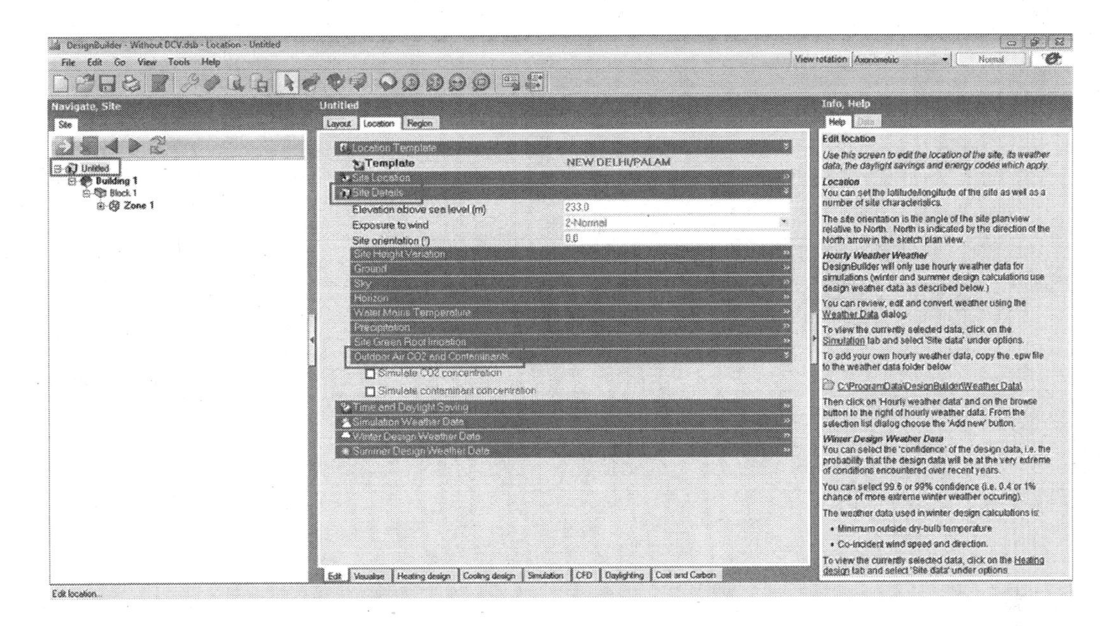

Step 3: Select the **Simulate CO2 concentration** checkbox.

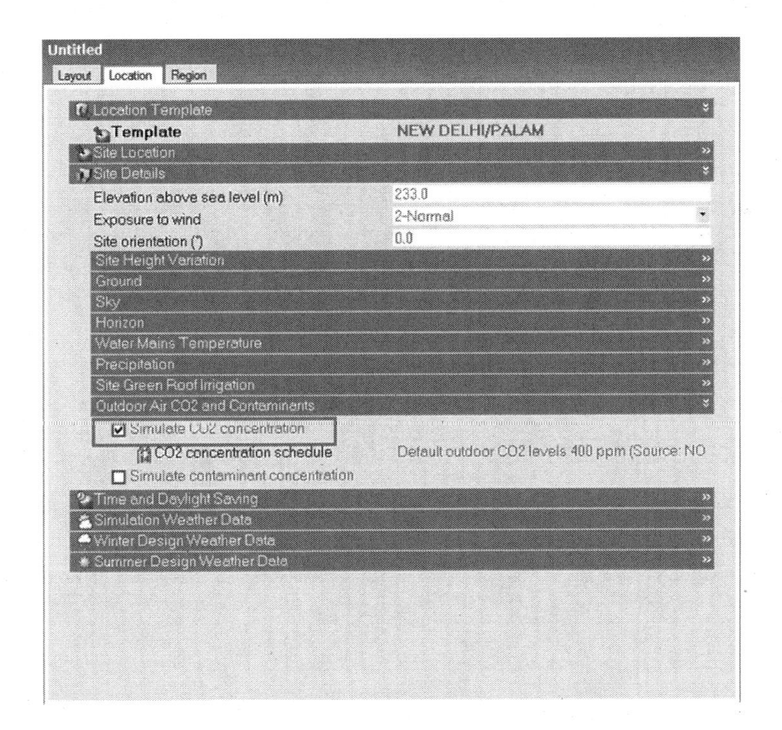

Step 4: Click the **Edit** menu, and select **Model Options.** Switch to **Detailed HAVC**. Click **HVAC System**. Click the **Load HVAC template** link. Select **VAV Reheat DX cooling with Dehumidificaiton**.

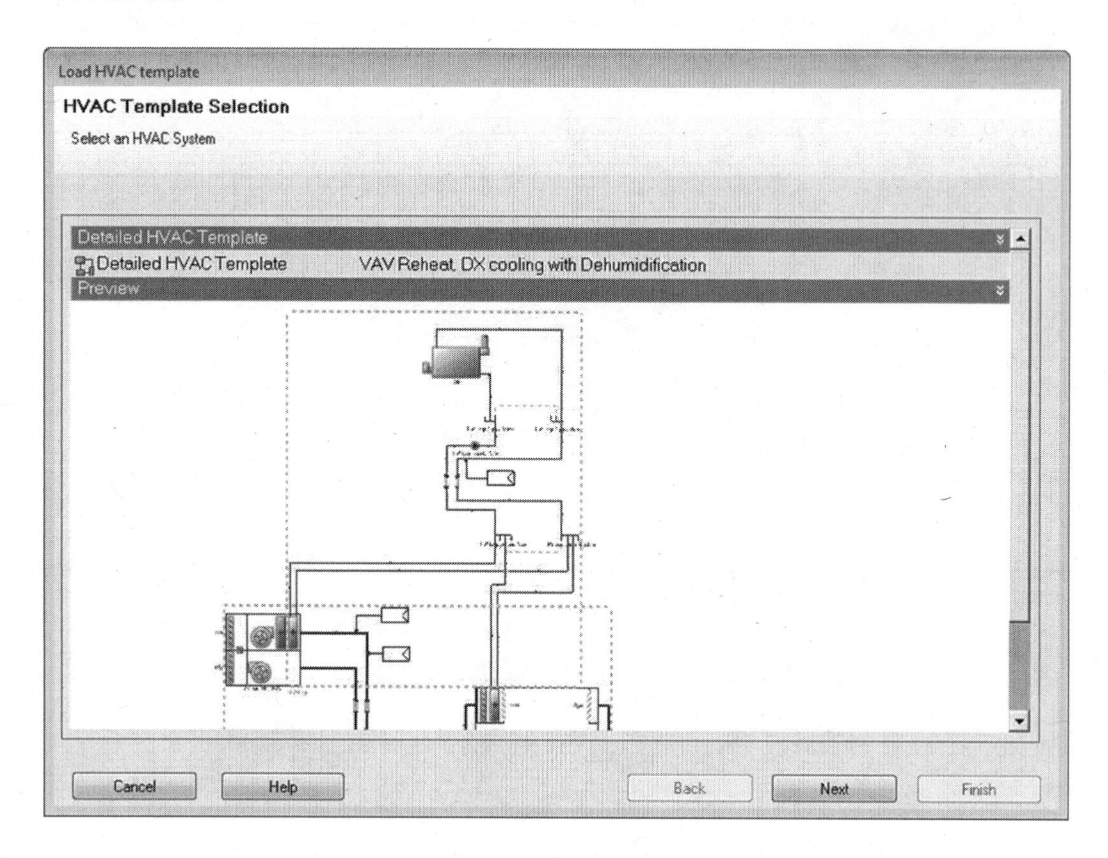

Step 5: Click **Air Loop**.

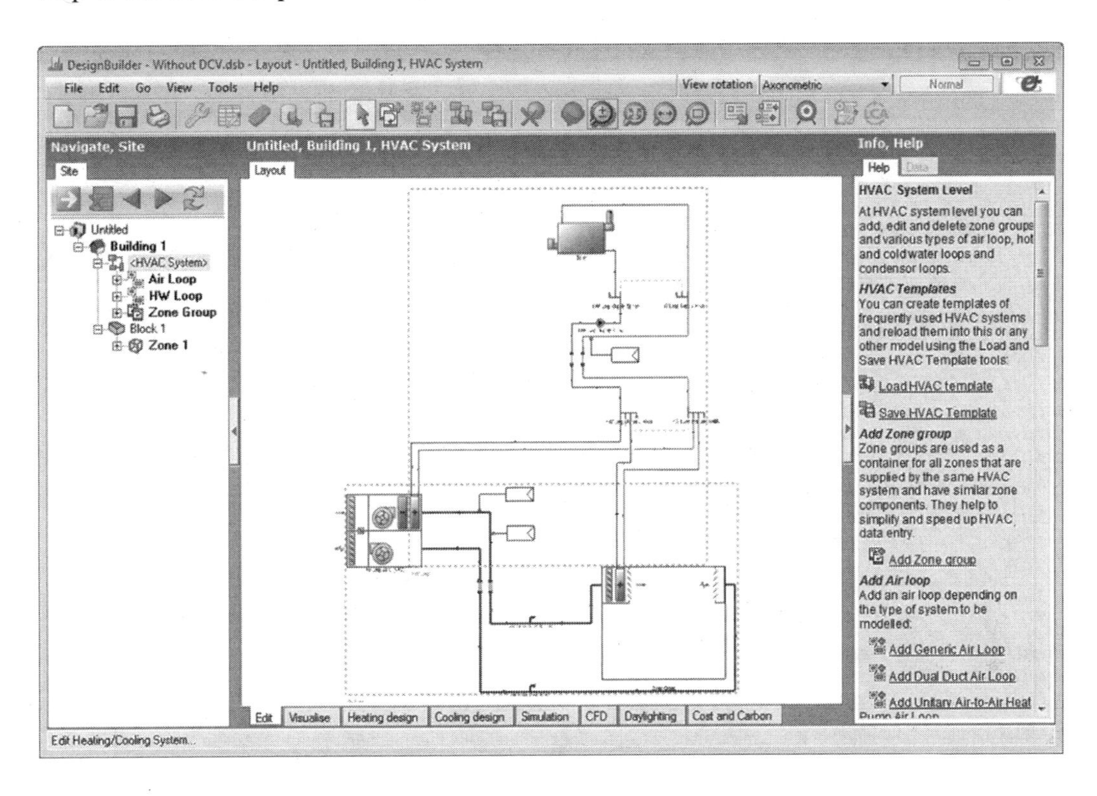

Step 6: Expand **Air Loop**. Select **Air Loop AHU**. Click **Edit component** under the **Help** tab.

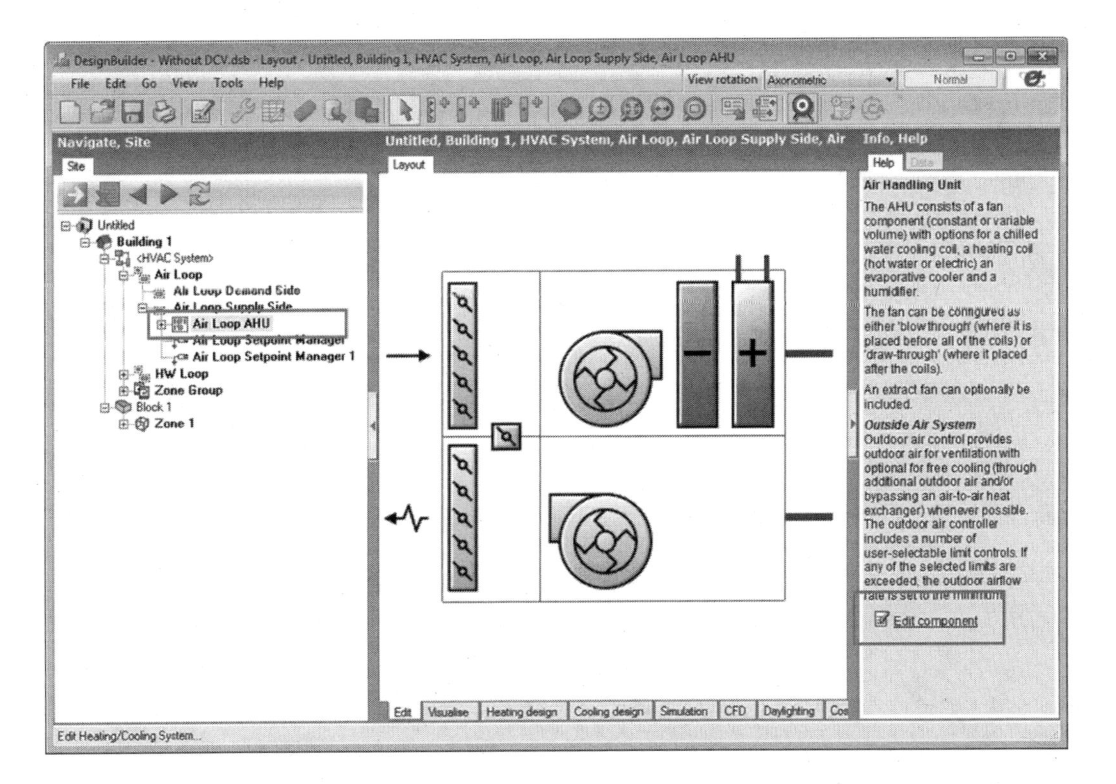

Step 7: Make sure that the **Demand controlled ventilation** checkbox is clear. Click **OK**.

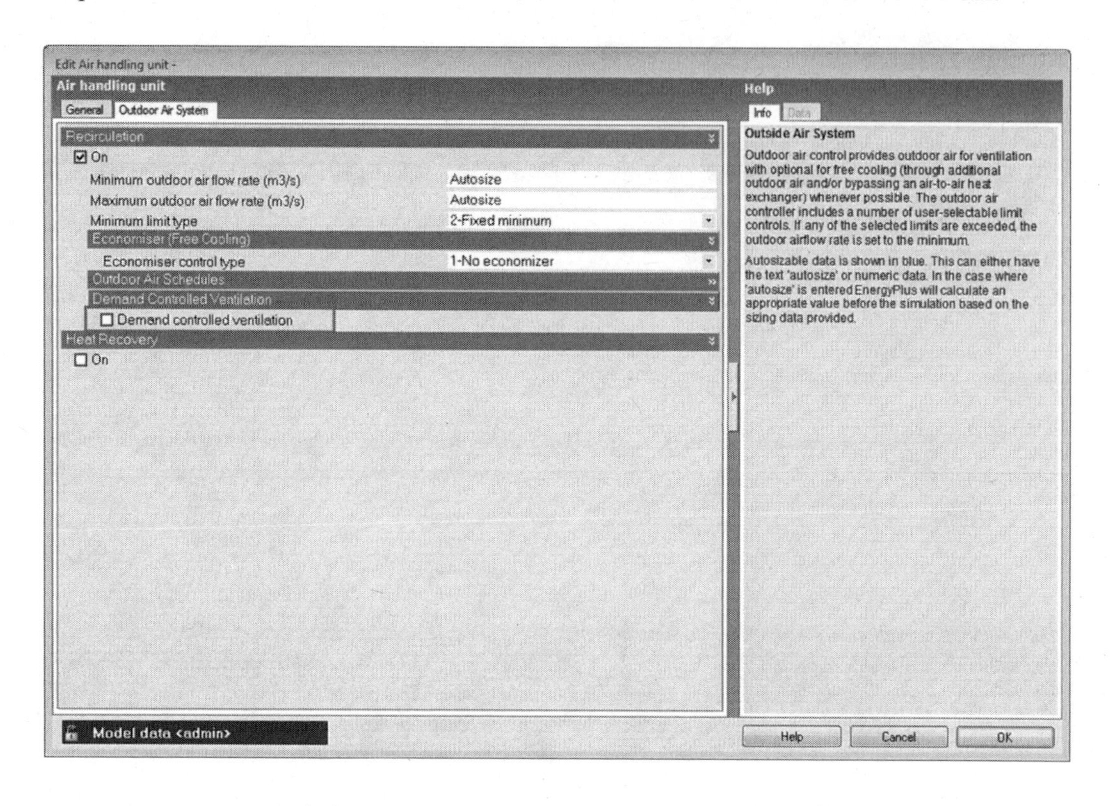

Step 8: Click the **Simulation** screen tab. Click **Update data**.

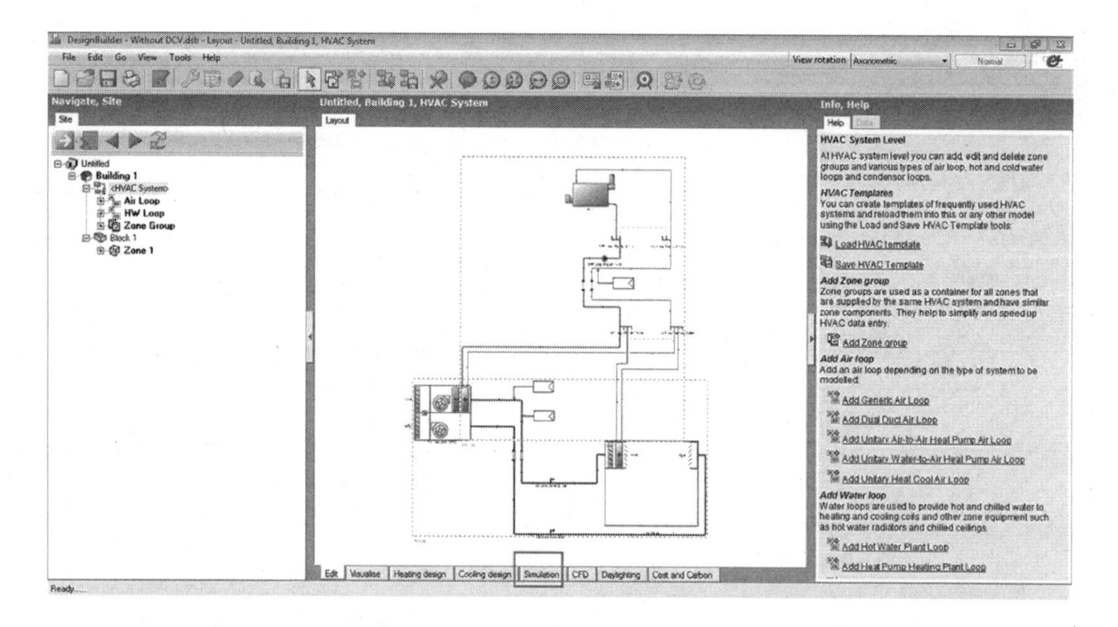

Step 9: Click the **Output** tab. Select the **HVAC system mass flow rates** checkbox. Click **OK**.

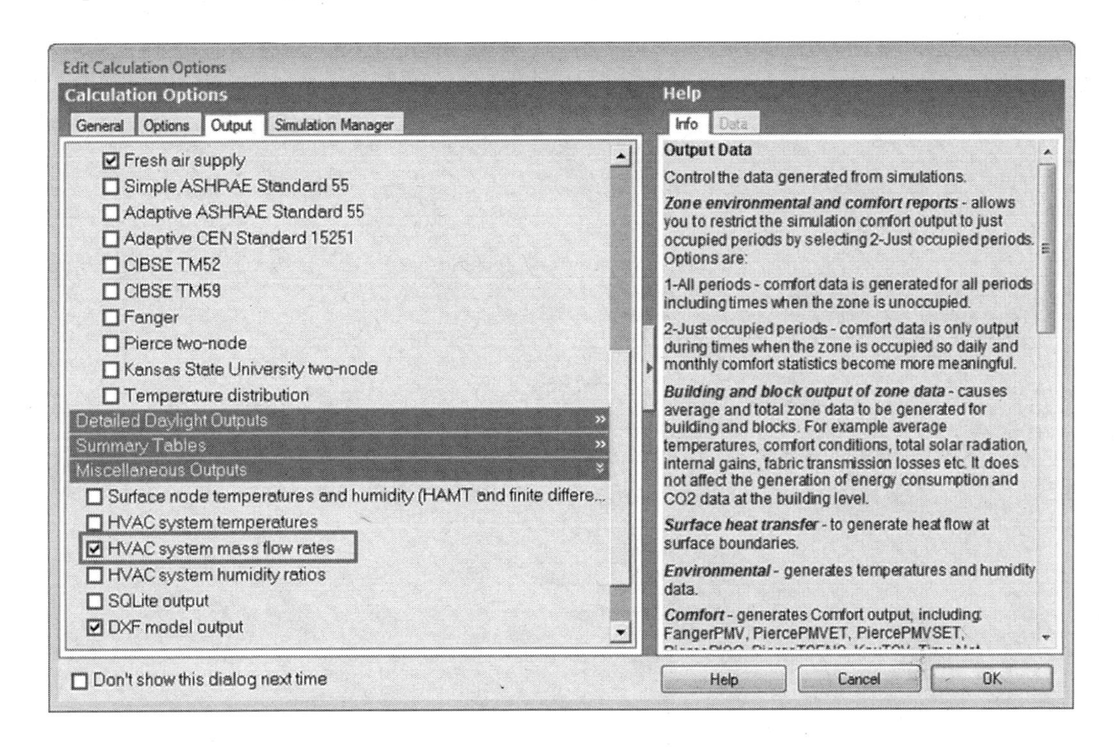

Step 10: After the simulation is over, click the **Results Viewer** tab.

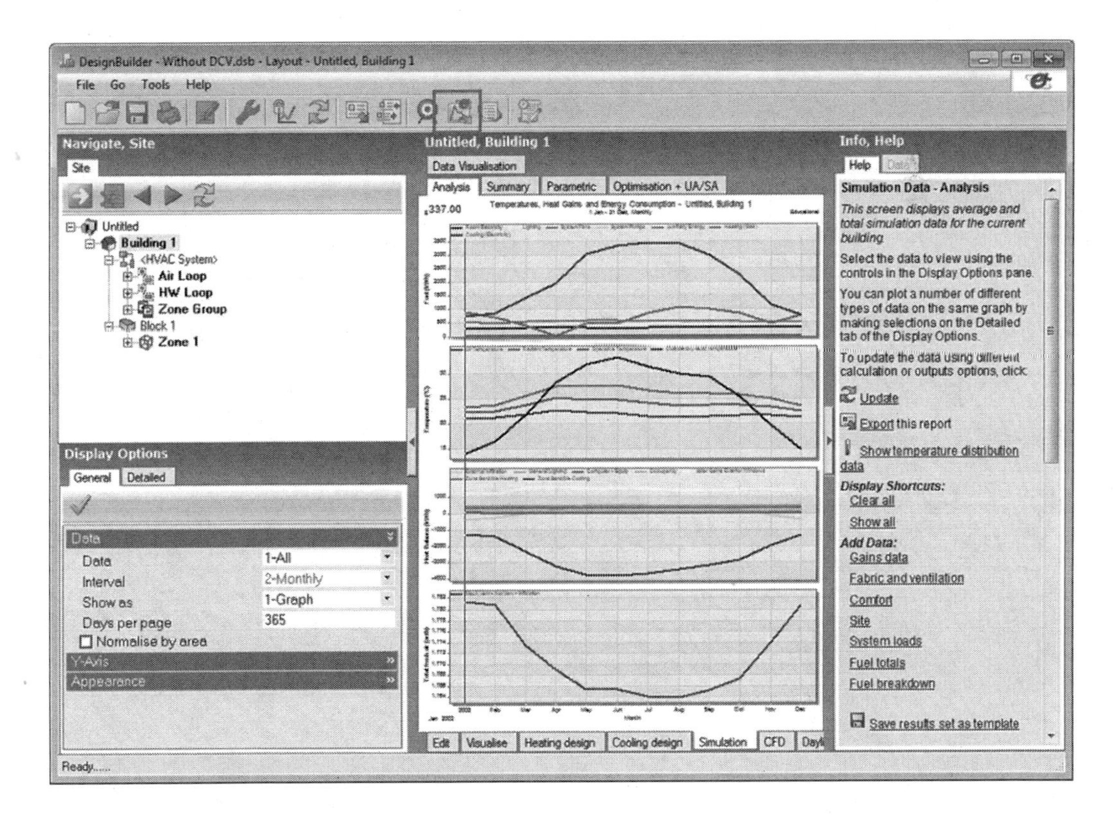

Step 11: Select **System Node Mass Flow Rate AIR LOOP AHU OUTDOOR AIR INLET**. You can see that the outdoor air mass flow rate is constant.

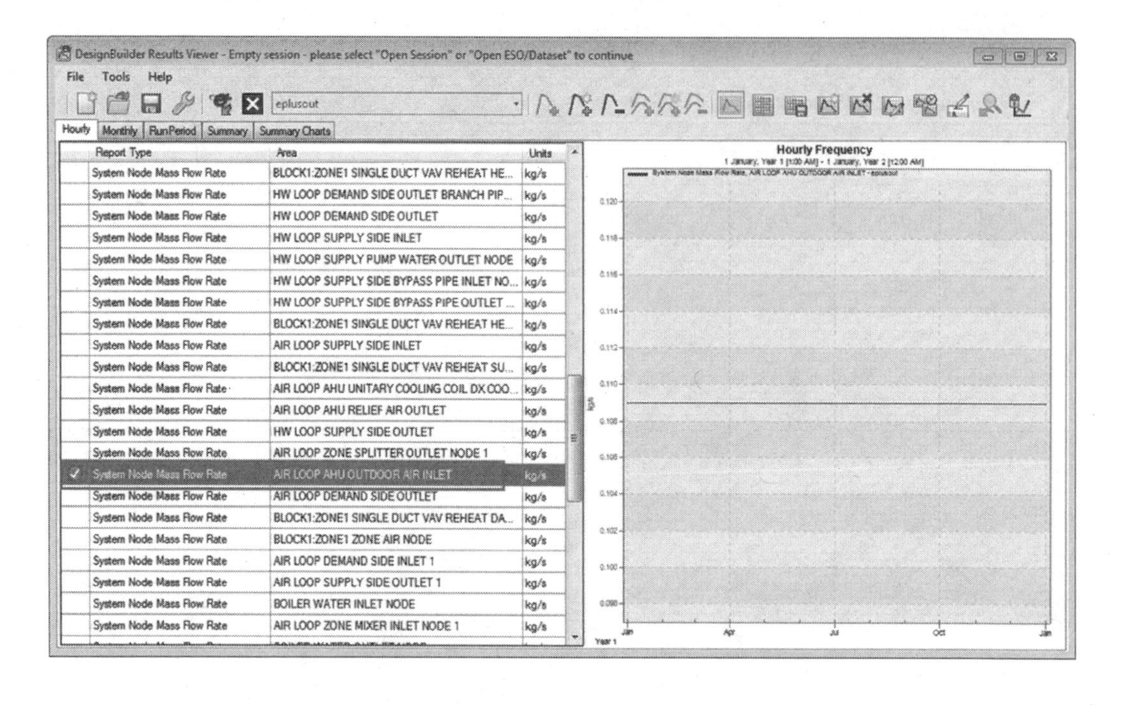

Step 12: Click **Zone Air CO2 Concentration Block1:ZONE1**.

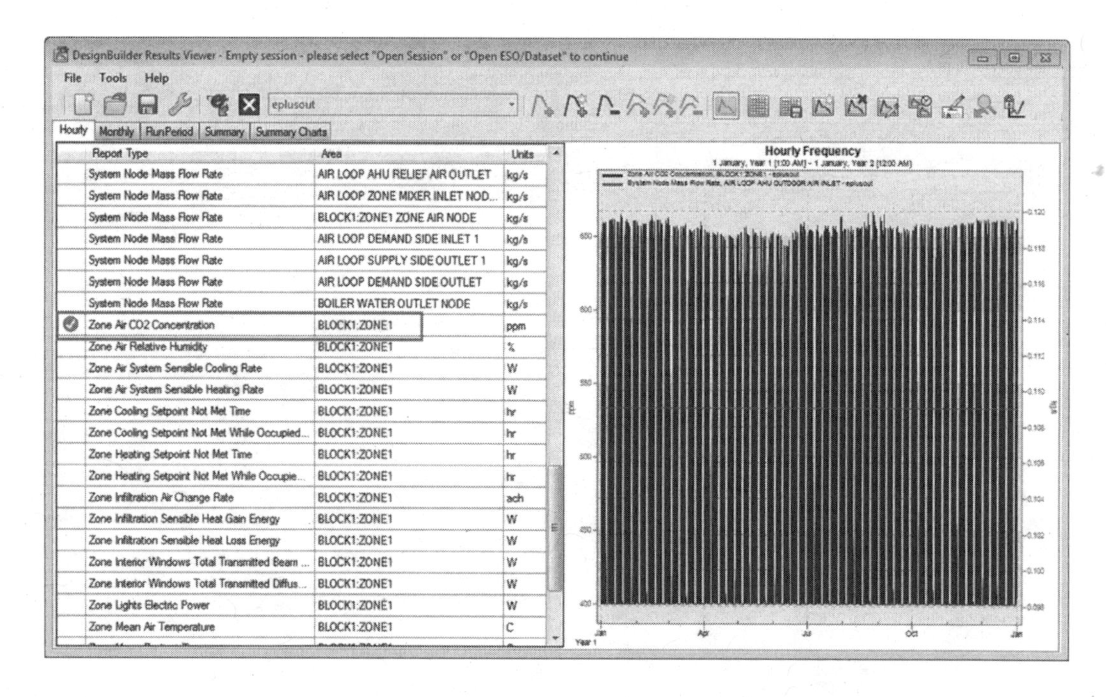

Step 13: Zoom in on the view. You can see that because of the constant airflow and change in occupancy, there is variation in **Zone air CO2** concentration.

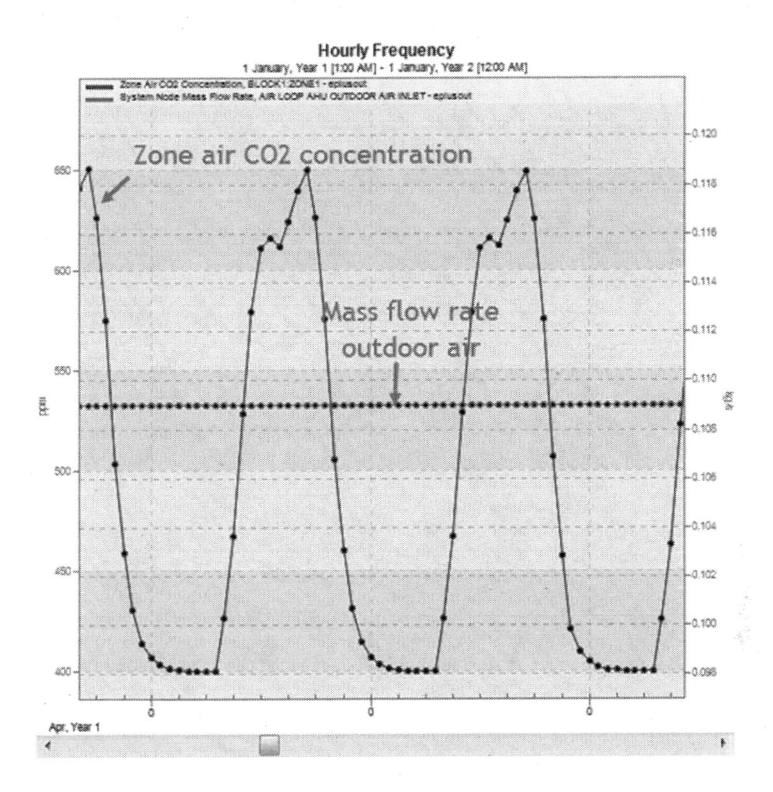

Also record the energy consumption results.

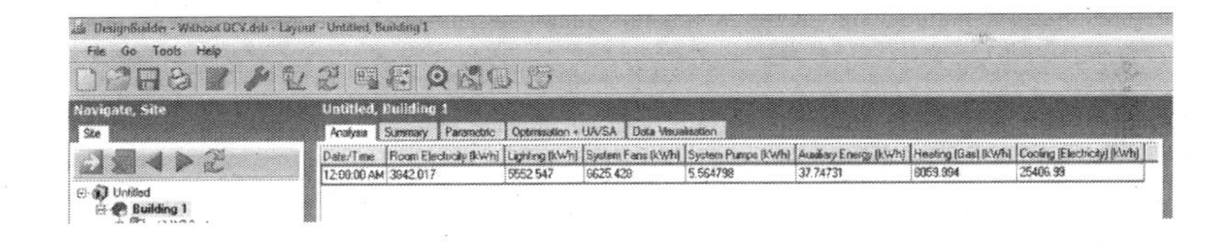

Date/Time	Room Electricity (kWh)	Lighting (kWh)	System Fans (kWh)	System Pumps (kWh)	Auxiliary Energy (kWh)	Heating (Gas) (kWh)	Cooling (Electricity) (kWh)
12:00:00 AM	3942.017	5552.547	6625.429	5.564798	37.74731	8059.894	25406.99

Step 14: Rename the file and add **Demand Control Ventilation** in the model.

Step 15: Select **Air Loop**. Click **Edit component** under the **Help** tab. The **Edit air loop** screen appears. Select **2-Ventialtion** from the **System outdoor air method** drop-down list.

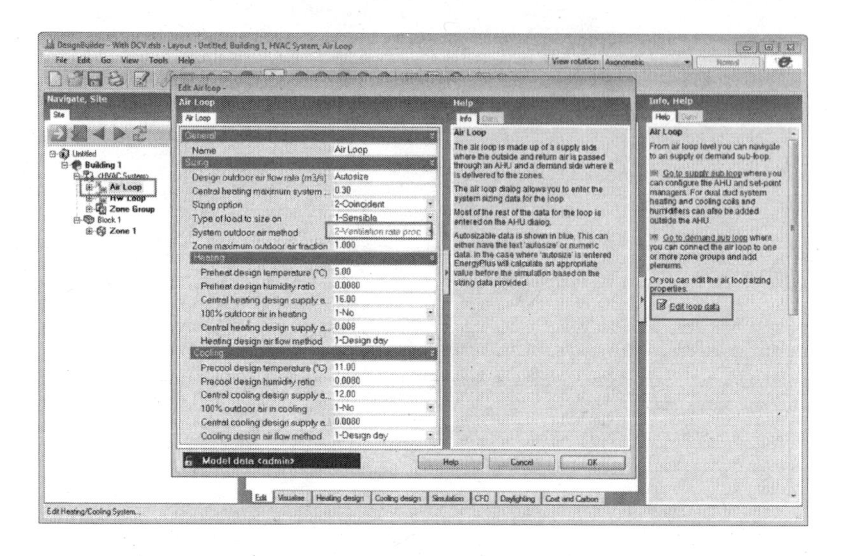

Step 16: Expand **Air Loop**. Select **Air Loop AHU**. Click **Edit component** under the **Help** tab.

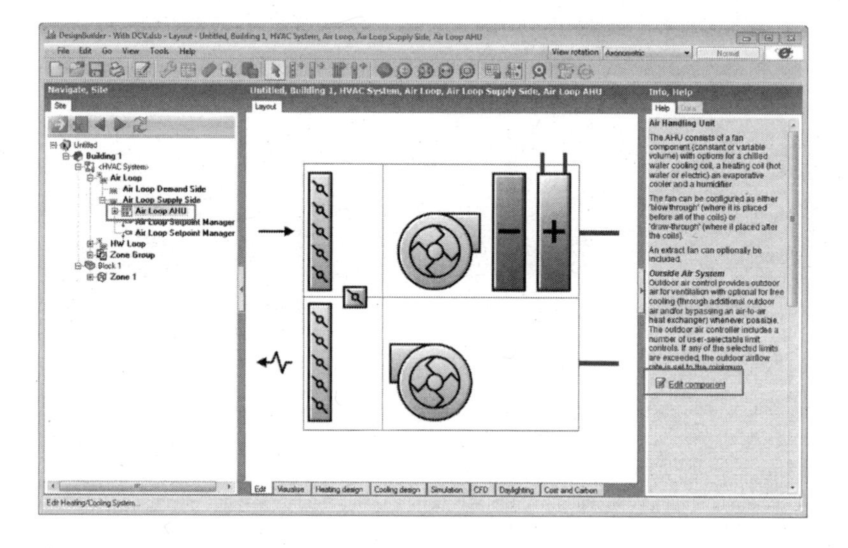

Step 17: Enter **0** in the **Minimum outdoor air flow rate (m^3/s)** text box. Select the **Demand Controlled Ventilation** checkbox. Select **3-Indoor Air Quality Procedure (IAQP)** from the **System outdoor air method** drop-down list. Click **OK**.

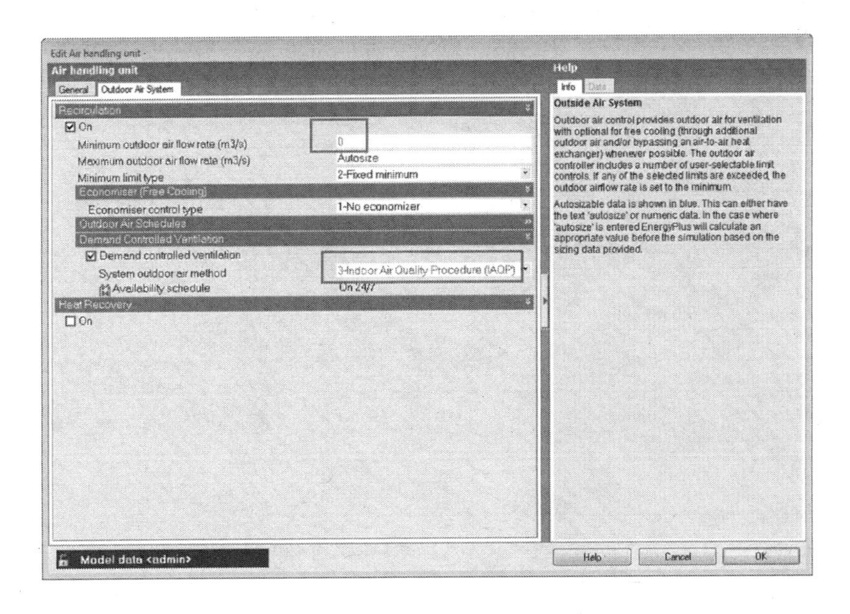

Step 18: Click **Block1:Zone1** under **Zone Group**. Click **Edit HVAC Zone**.

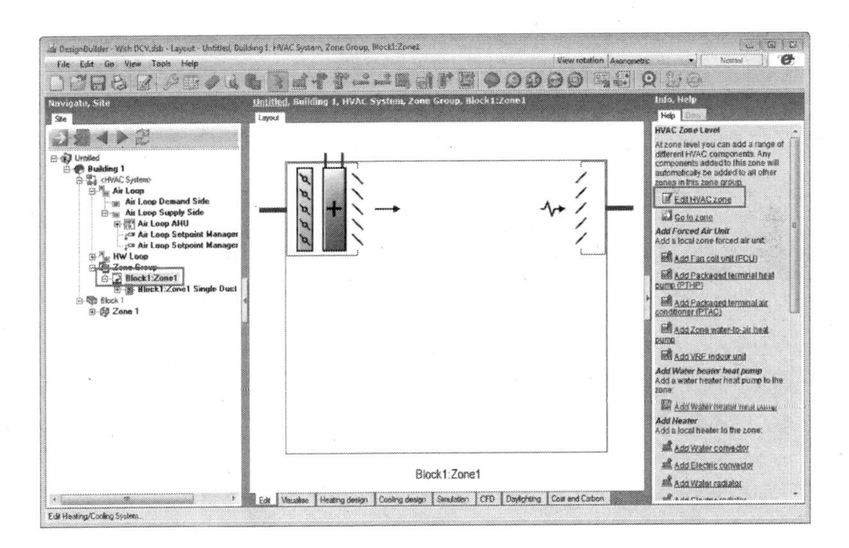

Step 19: Select the **CO2 and concentration control** checkbox. Click **OK**.

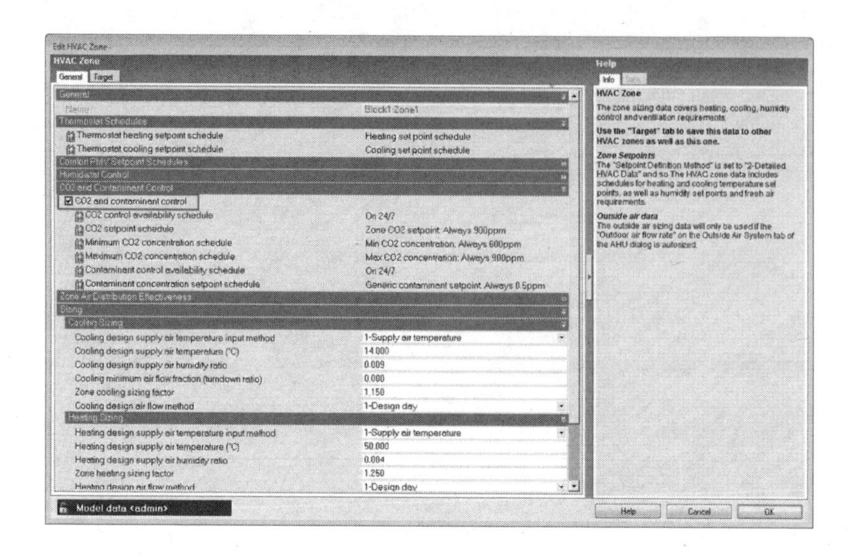

Step 20: Select **block1:Zone1 Single Duct VAV Reheat**. Click **Edit component**.

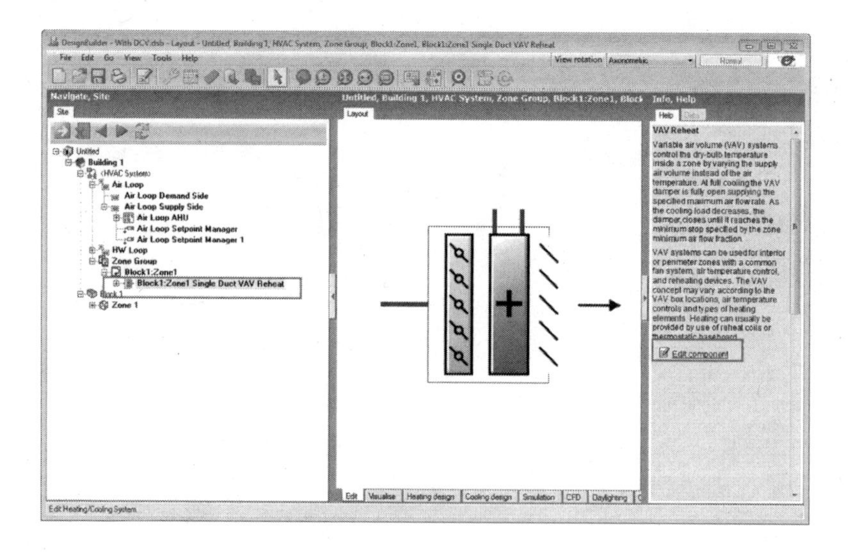

Step 21: Expand **Outside Air**. Select the **Control on outdoor air flow** checkbox. Click **OK**.

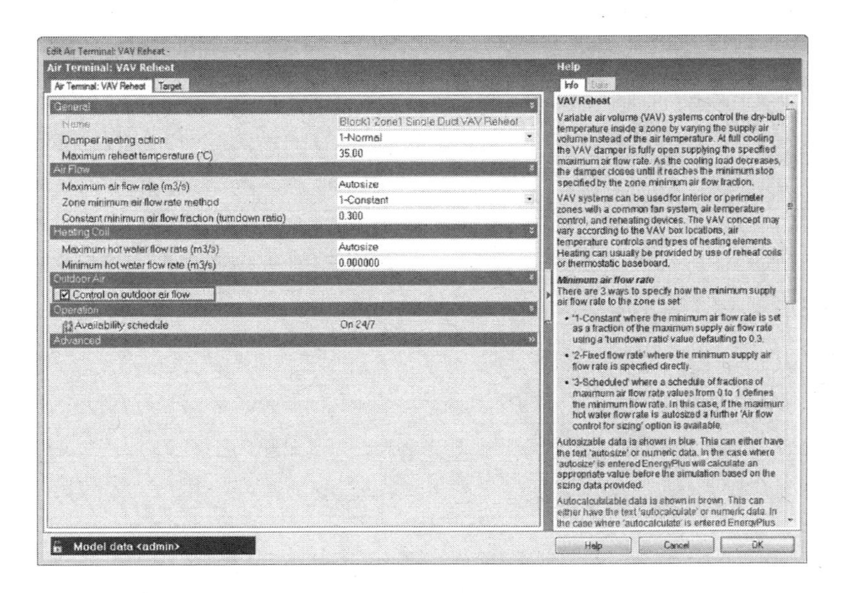

Step 22: Simulate the model, and record the results.

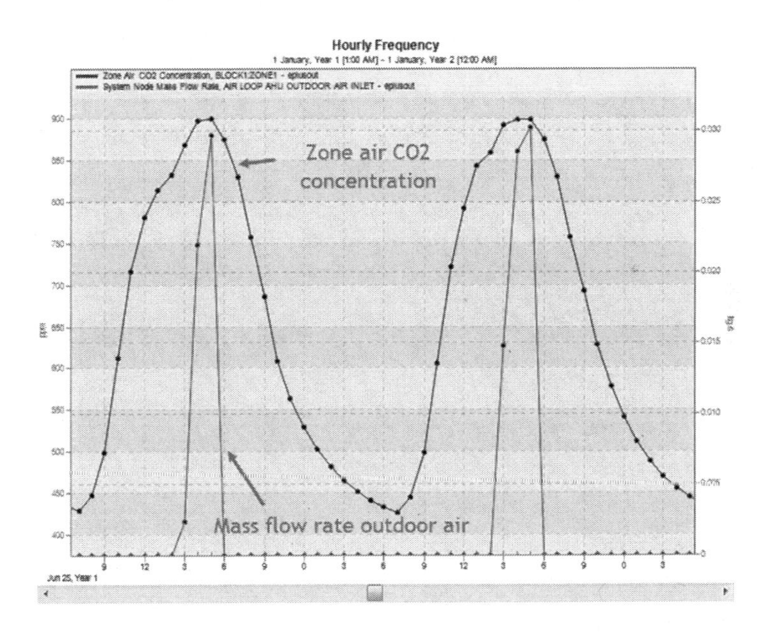

You can see that with demand control ventilation, the outdoor air varies based on the requirements. Also record the annual energy consumption. Compare the annual energy consumption for both cases (see Table 9.6).

TABLE 9.6
Annual energy consumption with and without demand control ventilation (DCV)

Room electricity (kWh)	Lighting (kWh)	System fans (kWh)	System pumps (kWh)	Auxiliary energy (kWh)	Heating (gas) (kWh)	Cooling (electricity) (kWh)
3,842	5,553	6,625	5.6	38	8,060	25,407
3,842	5,553	7,026	1.3	25	5,370	19,772
0%	0%	−6%	76%	33%	33%	22%

Energy savings in the DCV case appears to be the result of a reduction in the mass flow rate of outdoor air.

10 Natural Ventilation

The method of simulating non-air-conditioned buildings is somewhat different from that of simulating heating and cooling systems. Ventilation in non-air-conditioned buildings is often achieved using either windows or ventilation fans. Simulation tools can model both cases and can predict the thermal conditions of indoors. Mixed-mode buildings, which use both natural ventilation when ambient conditions are moderate and HVAC system when conditions are harsh, can also be modelled by defining the openings of windows and operation of the HVAC system in the simulation model. In this chapter, the method of modelling non-air-conditioned buildings is explained via six tutorials. Various aspects of modelling of non-air-conditioned buildings, such as how to define window openings, how to define scheduled opening and closing of windows and the impact of window openings on indoor conditions, are explained. Design issues such as the sizes of the openable windows and the temperature-controlled automatic opening of windows are also discussed.

Tutorial 10.1

Evaluating the Impact of Wind Speed on Natural Ventilation

GOAL

To evaluate the impact of a change in wind speed on the ventilation rate

WHAT ARE YOU GOING TO LEARN?

- How to model natural ventilation
- How to define glazing area openings for natural ventilation

There are two general approaches to natural ventilation and infiltration modelling in DesignBuilder depending on the setting of the natural ventilation model option:

- Scheduled, in which the natural ventilation change rate is explicitly defined for each zone in terms of a maximum air-changes-per-hour (ACH) value and a schedule, and infiltration air change rate is defined by a constant ACH value. A range of control options is provided.
- Calculated, where natural ventilation and infiltration are calculated based on window openings, cracks, buoyancy and wind-driven pressure differences, crack dimensions and so on. Control options are provided.

Source: www.designbuilder.co.uk/helpv6.0/Content/CalculatedNatVent.htm.

PROBLEM STATEMENT

In this tutorial, you are going to use a 10- × 10-m single-zone model. Find the zone air changes for the model with natural ventilation. Use the **New Delhi/Safdarjung, India** weather location.

SOLUTION

Step 1: Open a new blank project file, and create a **10- × 10-m** single-zone building. Select the **Activity** tab.

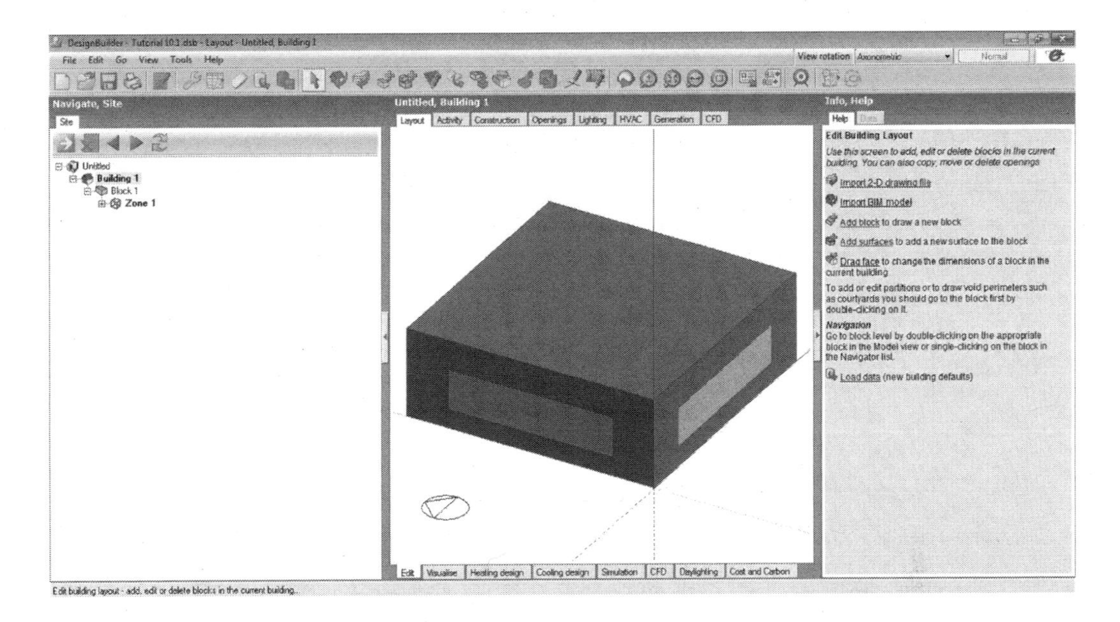

Step 2: Select the **24×7 Generic Office Area** template. Set **Fresh air (l/s-person)** and **Mech vent per area (l/s-m^2)** to **0**.

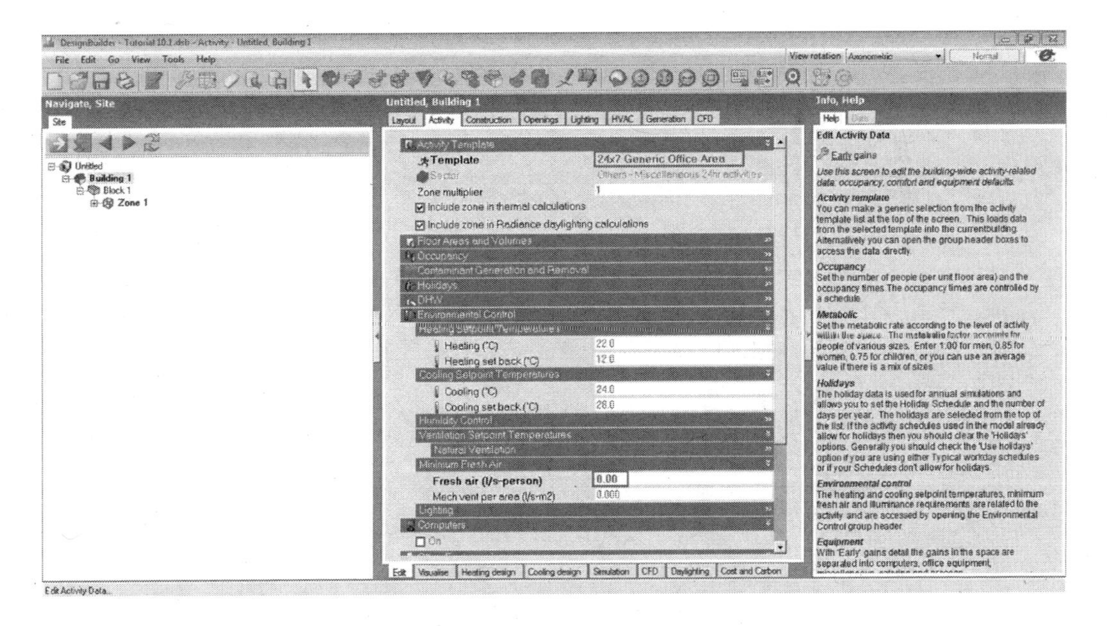

Step 3: Select the **HVAC** tab. Select the **Natural ventilation – No Heating/Cooling** template.

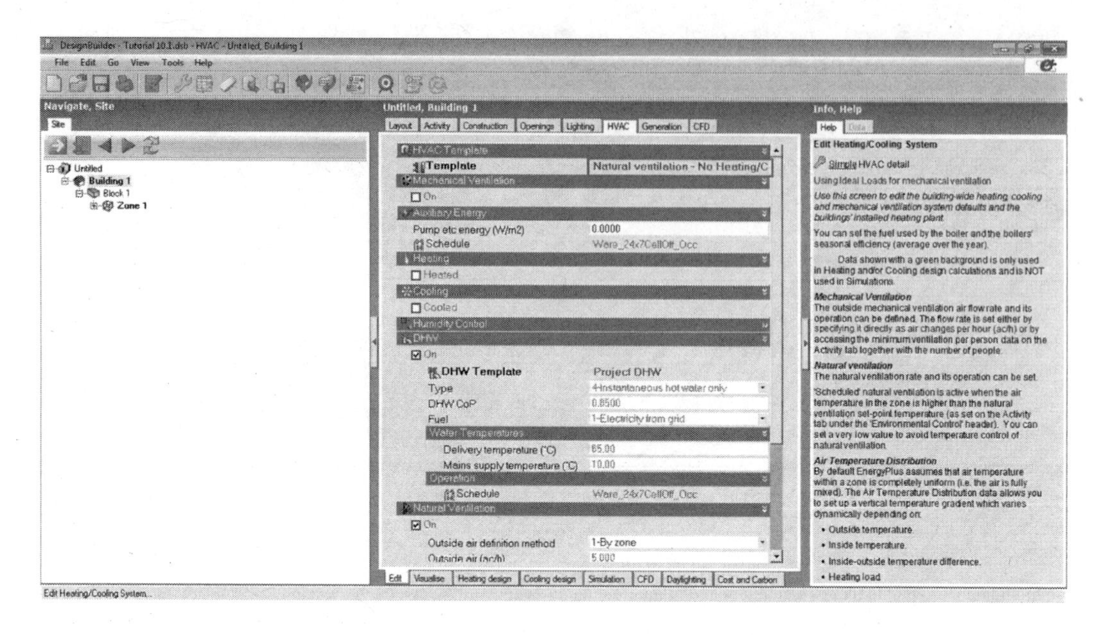

Step 4: Click **Simple** under the **Help** tab.

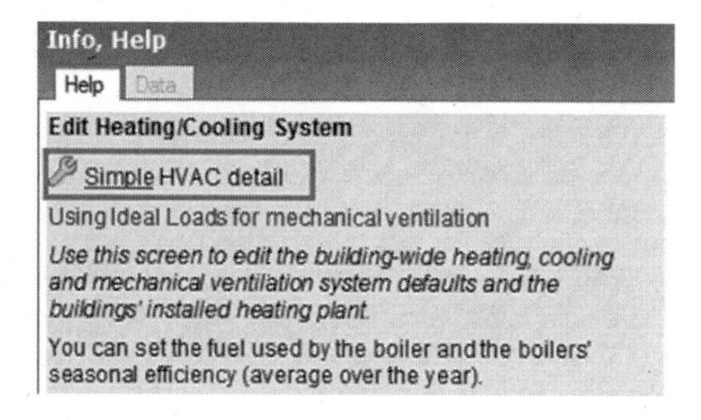

Step 5: Click **Calculated** under **Natural ventilation**. Click **OK**.

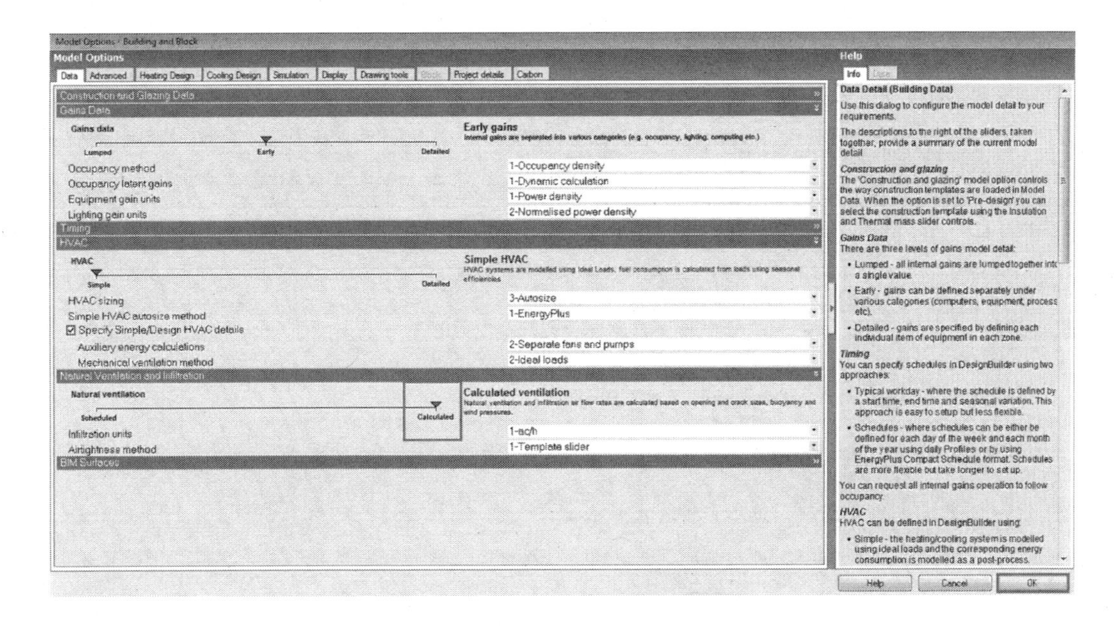

Step 6: Click the **Options** subsection in the **Natural Ventilation** section, and select **Constant** for the **External control mode** and **Internal control mode** drop-down lists.

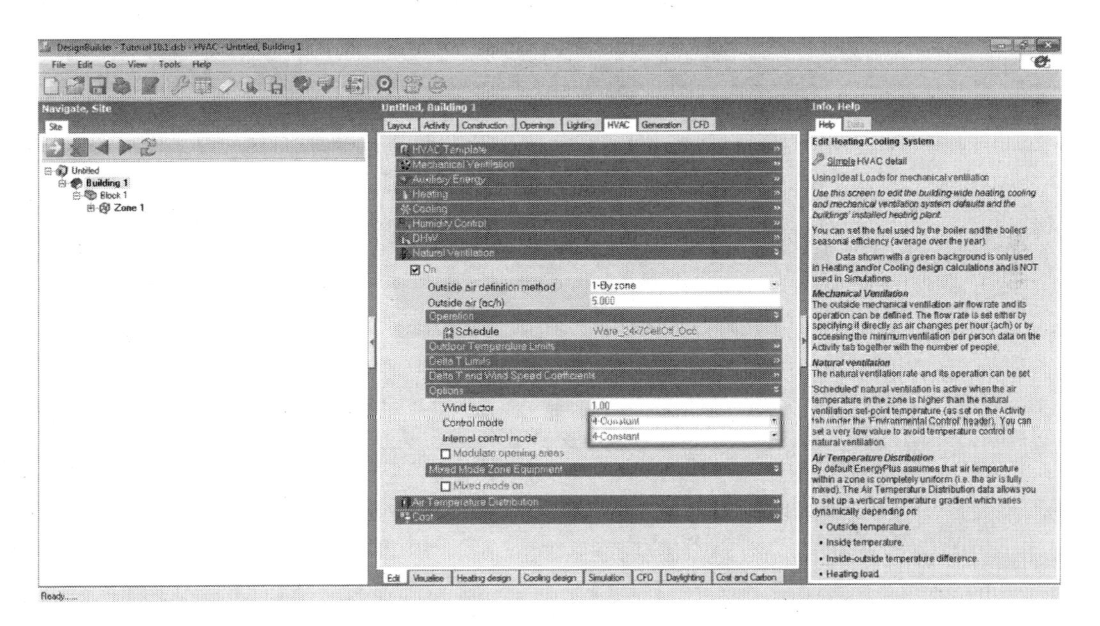

Constant: Whenever an opening's operating schedule allows venting, all the zone's openable windows and doors are open, independent of indoor or outdoor conditions. Note that *constant* here means that the size of each opening is fixed while venting; the airflow through each opening can, of course, vary from time step to time step. This option allows modelling of a window that is opened for fresh air regardless of inside/outside temperature/enthalpy.

Source: www.designbuilder.co.uk/helpv6.0/Content/CalculatedNatVent.htm.

Step 7: Select the **Opening** tab. Set **Window to wall %** to **20.00**. Click the **Free Aperture** section. Set **% Glazing area opens** to **100.0**, and select **On 24/7** in the **Operation schedule**. Select **Opening Position** as **4-Left** in the **Free Aperture** section.

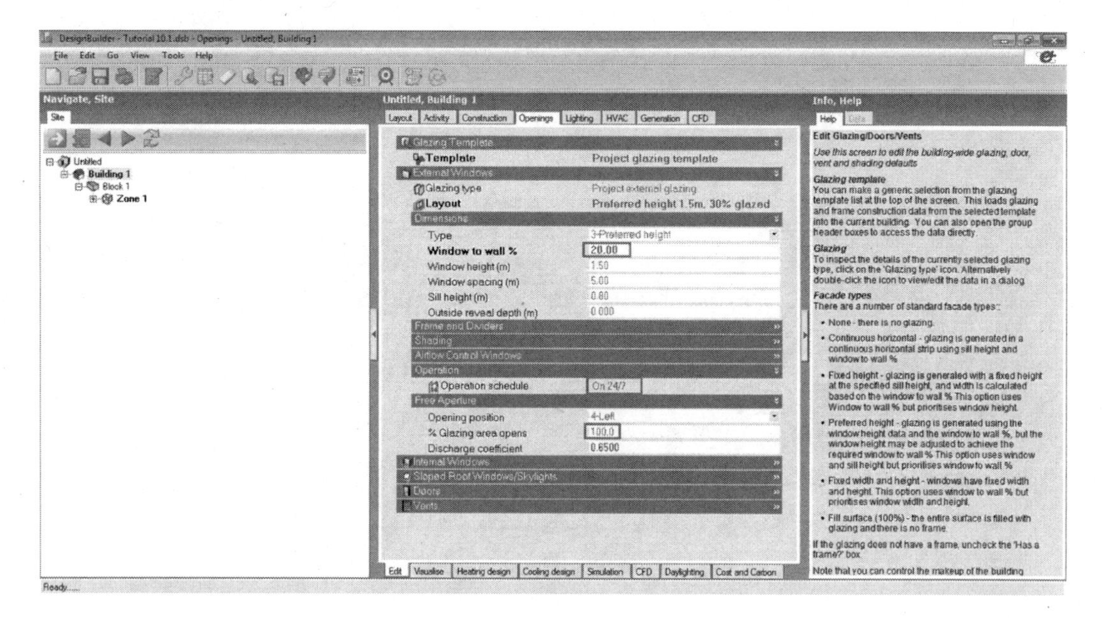

Step 8: Perform an hourly simulation.

Step 9: Click the links **Fabric and ventilation** and **Site** from **Add Data** to record the results for **Wind speed** and **Mechanical ventilation**. Select **Grid** in the **Show as** drop-down list.

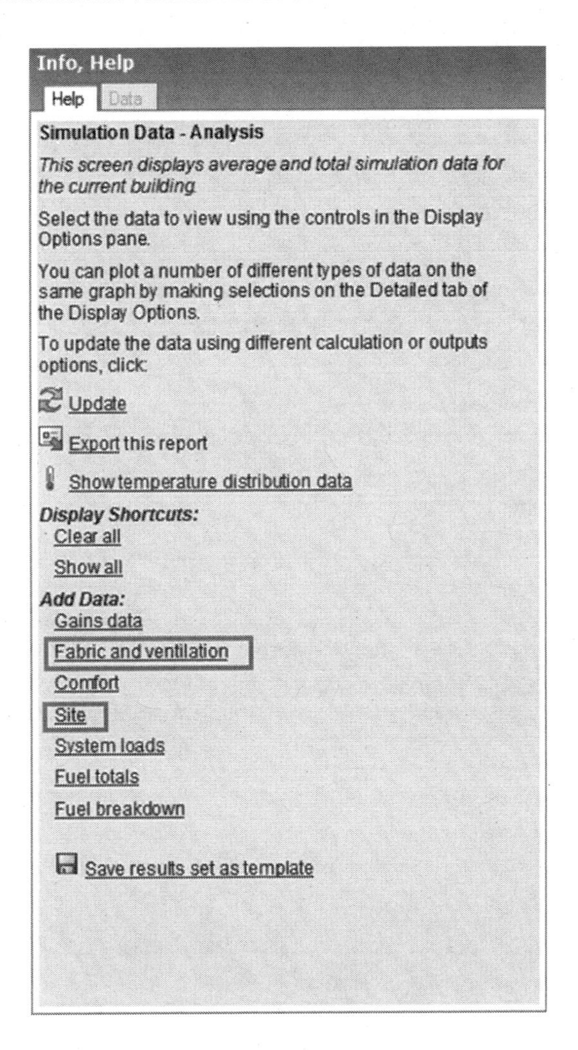

Untitled, Building 1

Analysis | Summary | Parametric | Optimisation + UA/SA | Data Visualisation

Date/Time	Outside Dry-Bulb Temperature (°C)	Outside Dew-Point Temperature...	External Air (kW)	Mech Vent + Nat Vent + Infiltration (ac/h)	Wind Speed (m/s)	Wind D...
01/01/2002 1:00:00 AM	7.55	4.5	-6.199422	31.40495	1.725	290.25
07/06/2002 5:00:00 PM	40.1	16.05	-2.356845	31.41512	1.8	339
08/06/2002 11:00:00 PM	31.225	14.25	-4.784535	31.43566	1.75	268.25
26/06/2002 6:00:00 PM	31.95	24.55	-3.619122	31.43626	1.675	201
10/09/2002 1:00:00 PM	32.55	23.2	-2.759041	31.44024	1.875	351
09/06/2002 10:00:00 PM	34.125	15.425	-4.109697	31.4428	1.75	293.25
05/05/2002 1:00:00 PM	31.35	17.725	-2.514278	31.45139	1.925	101.75
10/09/2002 2:00:00 AM	26.9	23.55	-4.505457	31.45476	1.8	274.75
16/07/2002 3:00:00 PM	34.9	24.725	-2.439424	31.46229	1.625	58
19/08/2002 3:00:00 PM	34.25	25.5	-3.015549	31.4711	1.7	205.75
06/04/2002 2:00:00 AM	23.625	12.775	-4.597131	31.47716	1.6	295.5
03/01/2002 2:00:00 AM	8.325	3.875	-5.619163	31.50995	1.575	296.25
24/08/2002 2:00:00 PM	27.975	25.425	-4.004414	31.52348	1.85	84.25
04/06/2002 5:00:00 PM	38.7	19.825	-2.676942	31.52758	1.85	5.5
10/07/2002 2:00:00 PM	39.125	22.725	-1.743928	31.5577	1.825	339
12/06/2002 8:00:00 PM	38.675	16.7	-3.272416	31.55843	1.8	272
20/04/2002 8:00:00 AM	27.85	11.975	-4.016714	31.56822	1.65	234
29/05/2002 4:00:00 AM	31.45	16.825	-4.770756	31.56866	1.625	304.75
05/11/2002	20.575	13.225	-4.825194	31.5694	1.825	96.25
19/12/2002 4:00:00 PM	21.225	13.625	-3.148634	31.57692	1.875	187.25
28/04/2002 6:00:00 PM	37.675	13.875	-2.682615	31.59036	1.675	246
12/09/2002 4:00:00 AM	26.3	22.4	-4.436107	31.60056	1.825	254.5
16/01/2002 3:00:00 PM	13.05	12.575	-4.098061	31.60768	1.85	108.25
05/11/2002 5:00:00 PM	26.325	13.025	-4.044408	31.6408	1.85	343.25
11/10/2002 2:00:00 AM	25.025	20.9	-4.533219	31.64738	1.725	154.25
04/01/2002 8:00:00 PM	14.025	9.025	-5.266687	31.66051	1.725	115.5
10/10/2002 1:00:00 AM	25.75	21.85	-4.101714	31.66289	1.625	145.75

Export the results to a spreadsheet, and plot a scatter graph between **Mech Vent + Nat Vent + Infiltration (ac/h)** and **Wind speed (m/s)**.

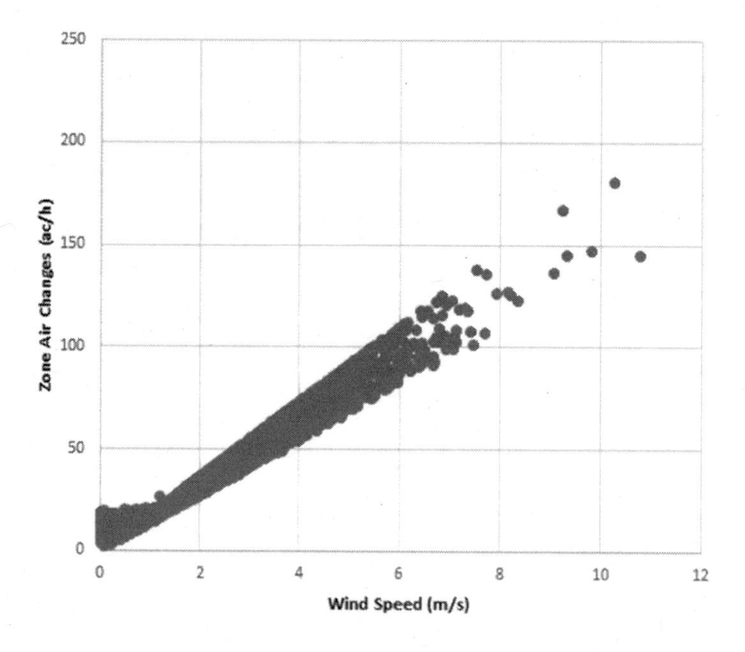

You can see that with an increase in wind speed, there is an increase in zone air changes.

Step 10: Repeat this tutorial with **Scheduled** under **Natural ventilation**.

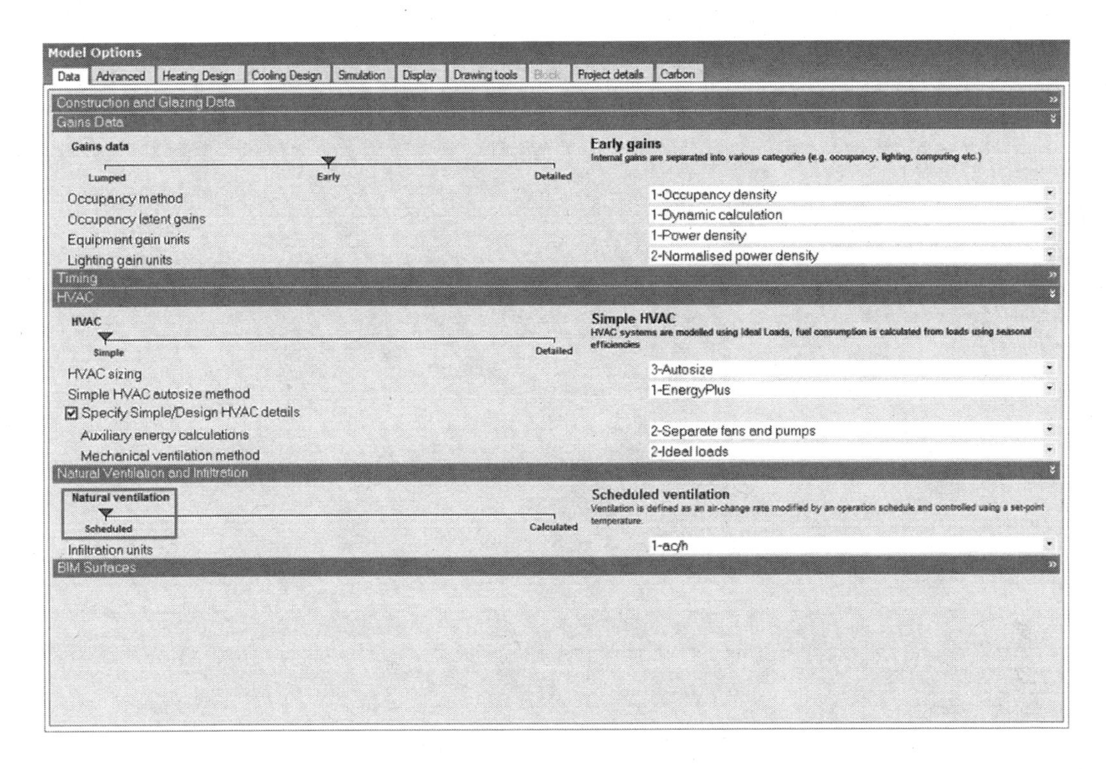

Scheduled: The ventilation rates are predefined using a maximum air change rate modified by operation schedules.

Source: www.designbuilder.co.uk/helpv6.0/Content/_Ventilation_model_detail.htm.

Step 11: Select the **Activity** tab. Clear both the **Indoor min and max temperature control** checkboxes in the **Natural Ventilation** section.

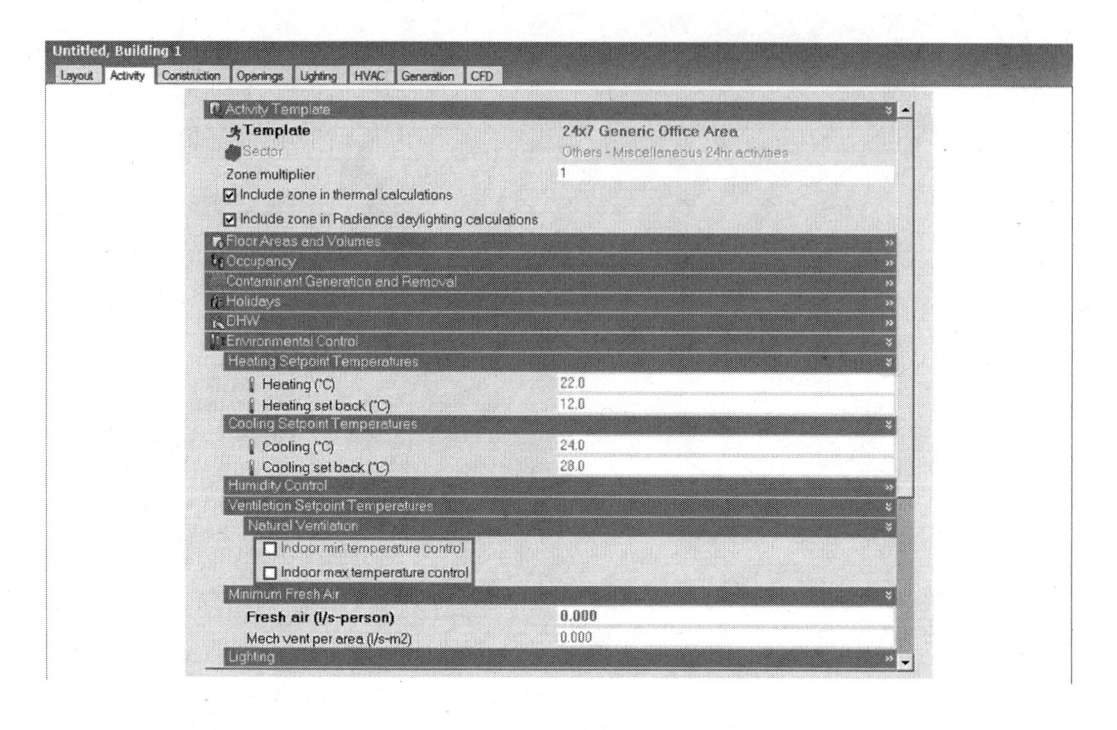

Step 12: Select the **Construction** tab. Clear **Model infiltration** in the **Airtightness** section.

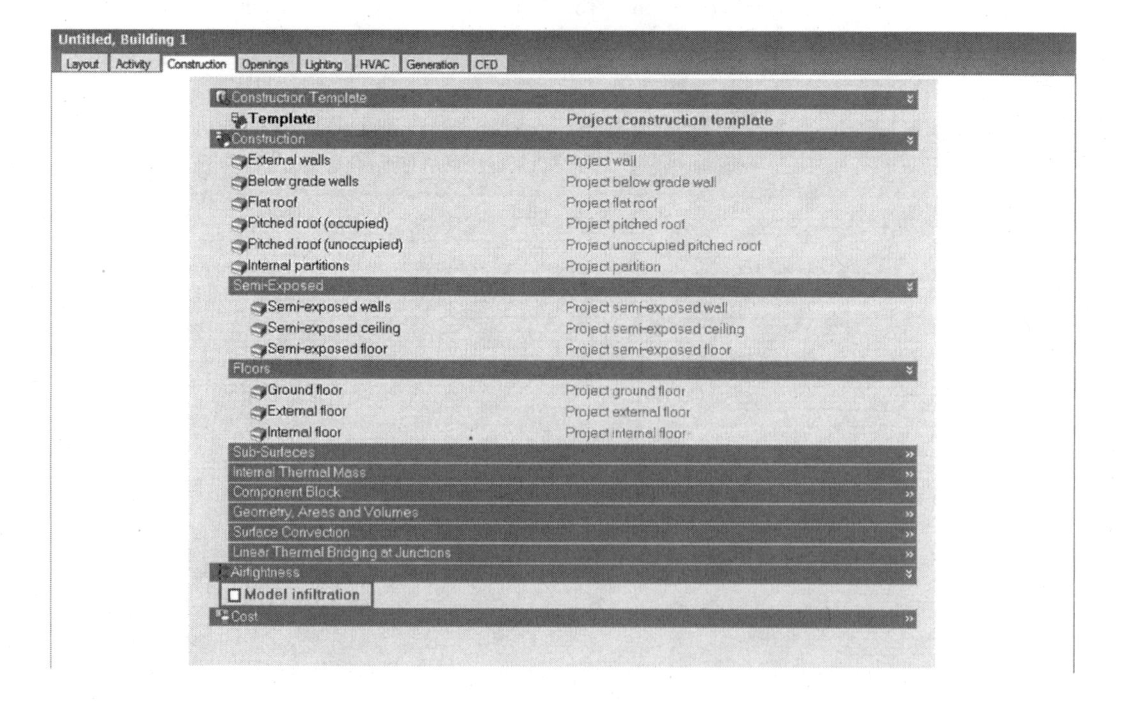

Step 13: Enter **3.000** for **Outside air (ac/h)** in the **Natural Ventilation** section, and select **On 24/7** in the **Operation** section.

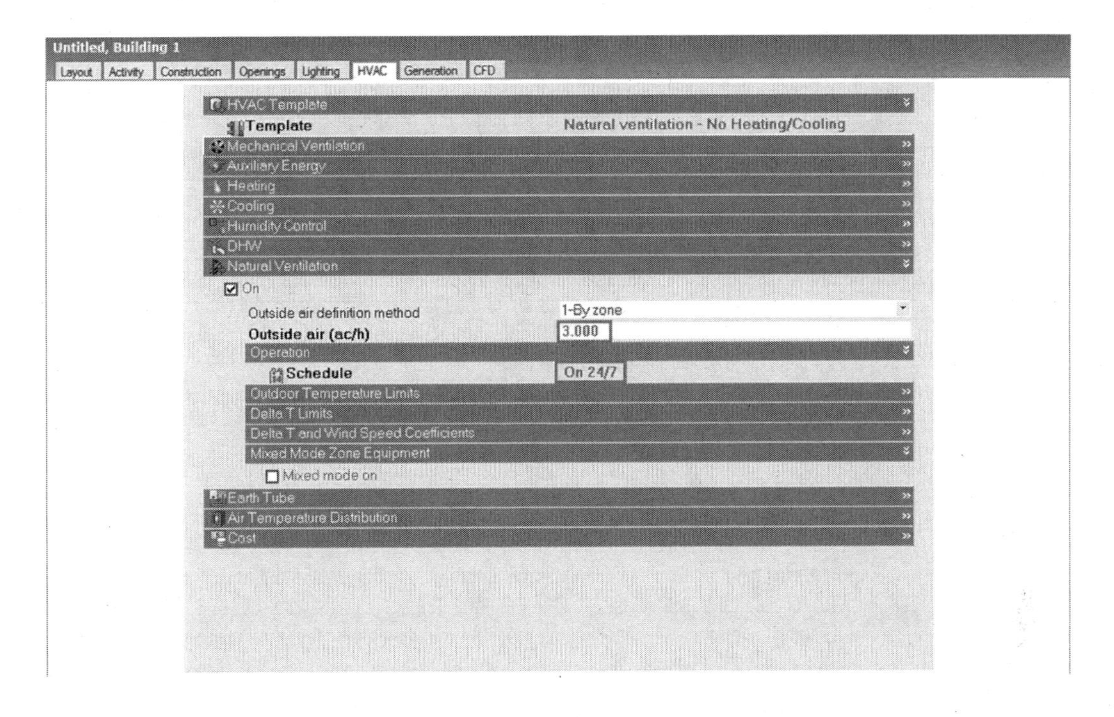

Perform an hourly simulation, and record the results.

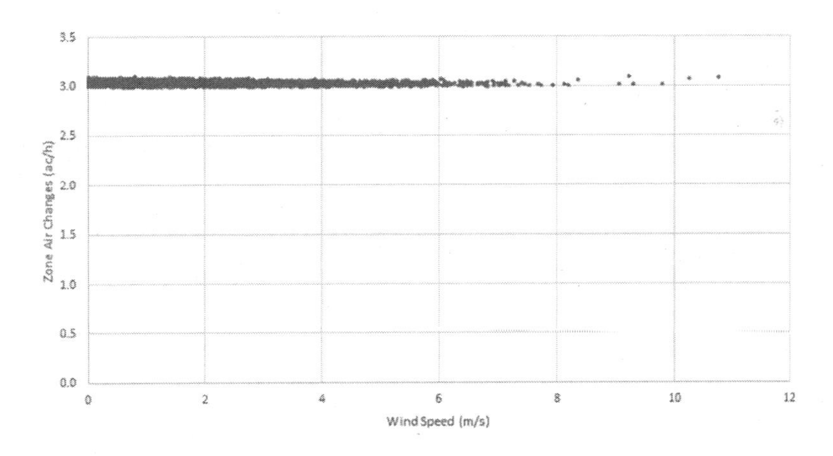

You may note that there is no change in ventilation rate with wind speed because in this option, the ventilation is fixed.

EXERCISE 10.1

Repeat this tutorial for **% Glazing area opens to 50**. Plot the chart between **Zone air changes (ac/h)** and **Wind speed (m/s)**. Observe the change in **Air changes per hour**.

Tutorial 10.2

Evaluating the Impact of Natural Ventilation with Constant Wind Speed and Direction

GOAL

To understand the impact of natural ventilation with constant wind and direction

WHAT ARE YOU GOING TO LEARN?

- How to change the weather data file for wind speed and direction

PROBLEM STATEMENT

In this tutorial, you are going to use a 10- × 10-m single-zone model with a window on the north and south facades. You are going to change wind velocity and direction in the weather file as follows:

- Wind velocity 1 and 2 m/s
- Wind direction 0 degrees (wind from the north to south)

Then simulate the weather file and observe the effect on air changes per hour. Use the **New Delhi/Safdarjung, India** weather location.

SOLUTION

First, you are going to modify the weather data file for wind speed and direction.

Step 1: Download the weather data for New Delhi from the following link: https://energyplus.net/weather-location/asia_wmo_region_2/IND//IND_New.Delhi.421820_ISHRAE. Then download the .epw file or copy the file from the **Weather data** folder.

Step 2: Download the following **CSV editor** tool: http://csved.sjfrancke.nl (by Sam Francke). (You can use any CSV editor; for this tutorial, we used CSVed 2.4.)

Step 3: Open the **CSV editor** tool, and select the downloaded .epw file of New Delhi weather location.

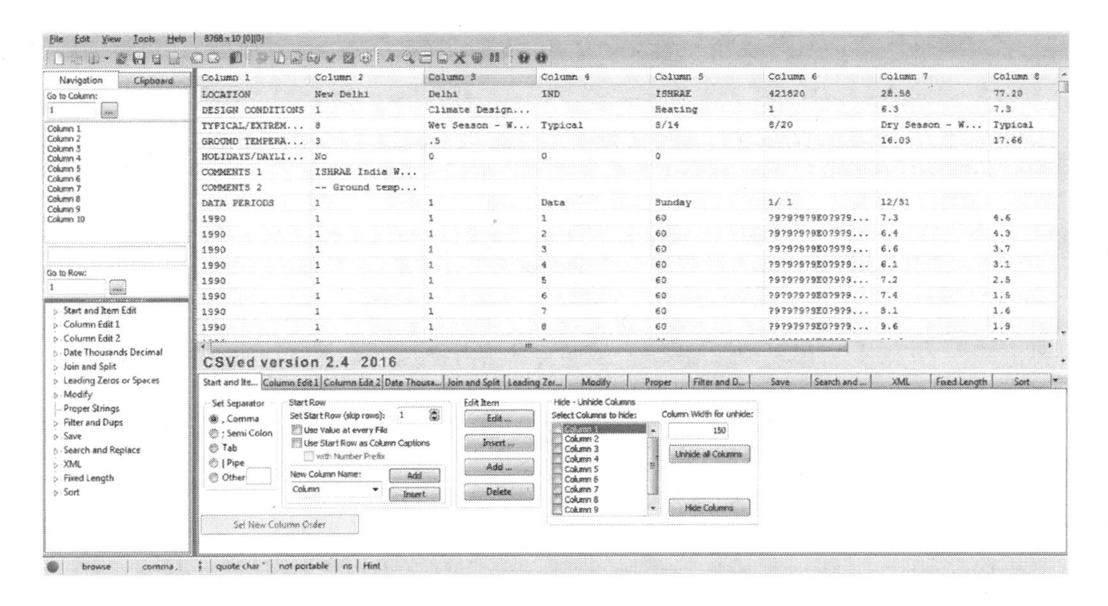

Source: From http://csved.sjfrancke.nl/.

Step 4: After opening the file, go to **Set Start Row** in the **Start and Item Edit** tab. Set **Start Row** to **9**.

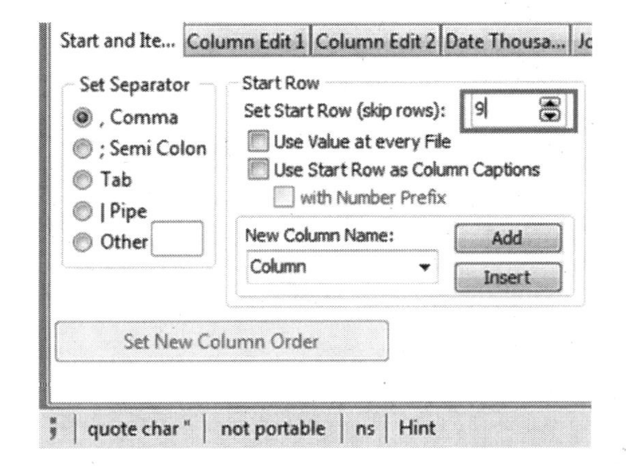

In the .epw file, columns 21 and 22 represent the wind direction and speed, respectively.

Column 19	Column 20	Column 21	Column 22	Column 23	Column 24	Column 25
0	0	289	1.4	0	5	1.3
0	0	270	1.7	0	5	1.3
0	0	265	2.0	0	1	1.3
0	0	248	1.9	0	7	1.3
0	0	270	1.7	0	5	1.3
0	0	276	2.6	1	7	1.3
0	0	280	3.3	2	8	1.3
0	0	292	3.9	3	4	1.3
0	0	293	3.9	5	9	1.3
0	0	297	3.9	6	0	1.3
0	0	292	3.9	6	8	1.3
0	0	294	4.4	6	4	1.3
0	0	290	4.7	6	0	1.3
0	0	292	5.0	5	8	1.3
0	0	287	4.5	3	2	1.3
0	0	291	3.9	2	4	1.3

Source: From http://csved.sjfrancke.nl/.

Step 5: For changing the wind direction to 0 degrees, select the **Column Edit 1** tab, select **Column 21** from the drop-down list and enter **0** in the **New Value** text box. Click **Edit Column**.

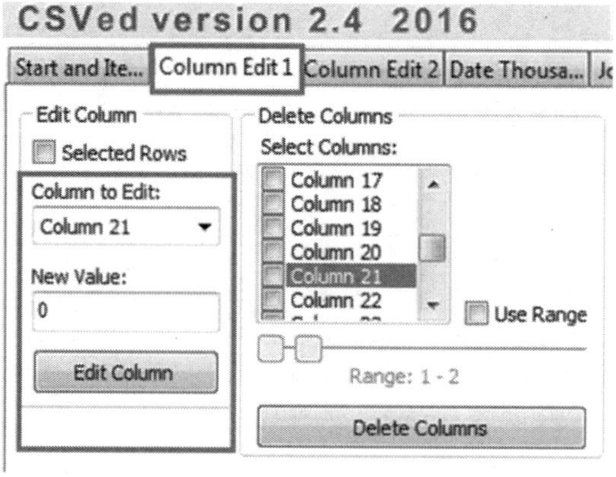

Step 6: Similarly, for changing the wind speed to 1 m/s, select **Column 22**, and enter **1** in the **New Value** text box. Click **Edit Column**.

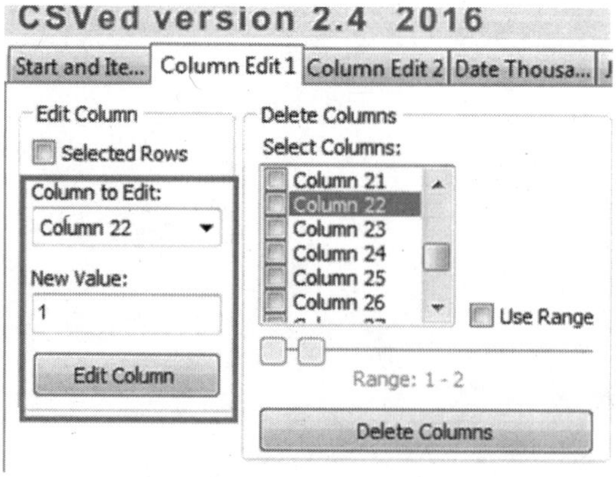

Step 7: Save the file with the name **Delhi Constant Wind.epw** on your desktop. Open **DesignBuilder**. Select the **File > Folders > Weather data** folder and paste it as the **Delhi Constant Wind.epw** weather file.

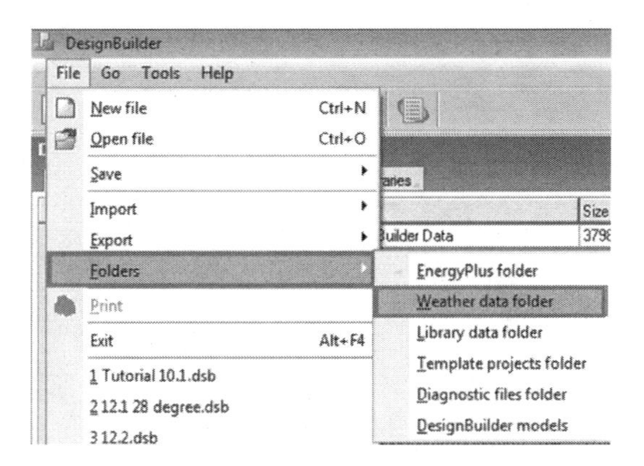

Step 8: Open the project file saved in Tutorial 10.1 with calculated natural ventilation and delete both the east and west facades.

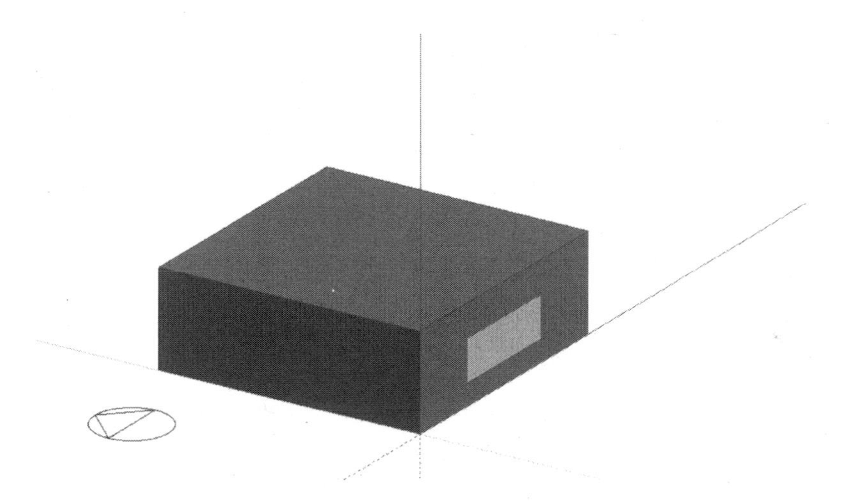

Step 9: Select the **Construction** tab. Select **Outer** for the **Zone volume calculation method** and **Zone floor area calculation method** in the **Geometry, Areas and Volumes** section.

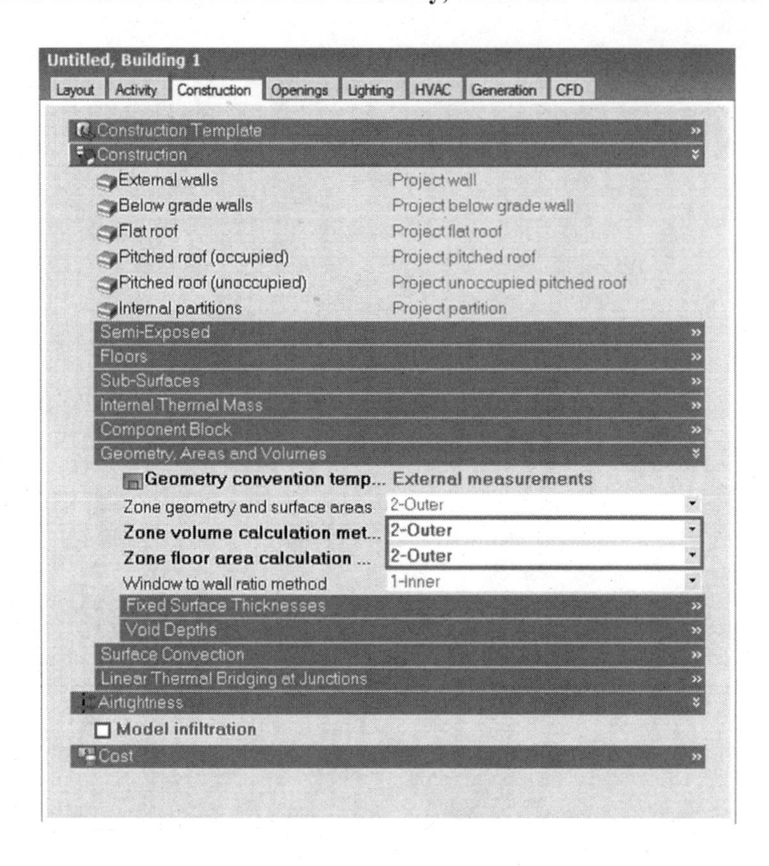

Step 10: Select the **Location** tab, and select **Hourly weather data** under the **Simulation Weather Data** section.

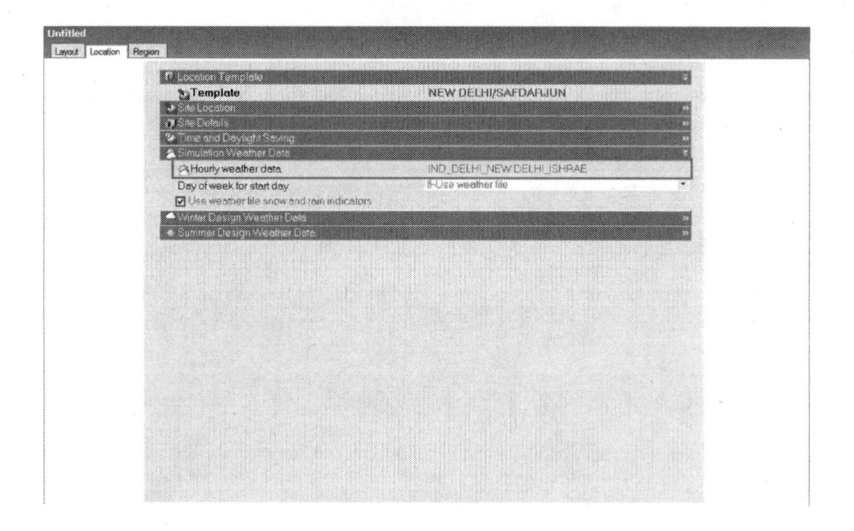

Step 11: Click the **Add new item** icon under the **Help** tab.

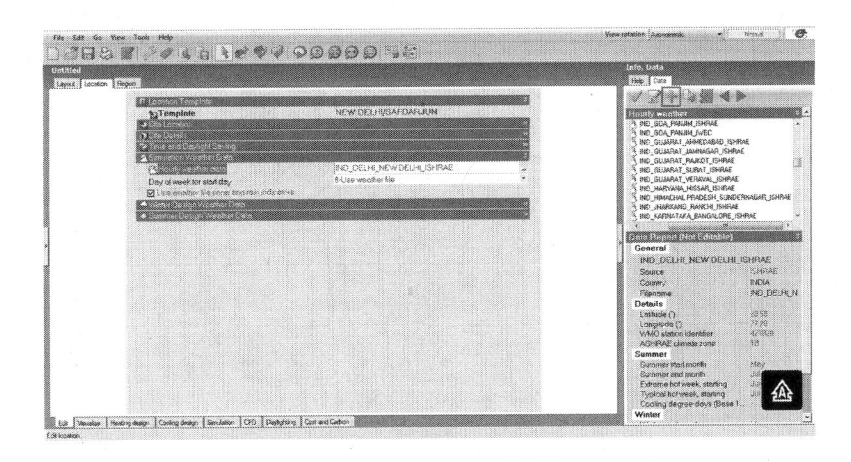

Step 12: In the **Edit hourly weather – Delhi Constant Wind** screen, rename with your custom weather file, and search for the file by clicking **Filename**. Click **OK** to successfully select the file.

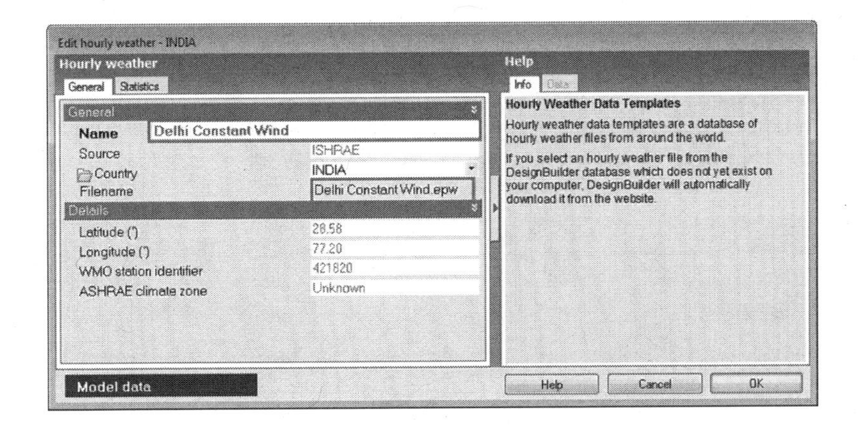

Step 13: Select **Exposed** from the **Exposure to wind** drop-down list in the **Site Details** section.

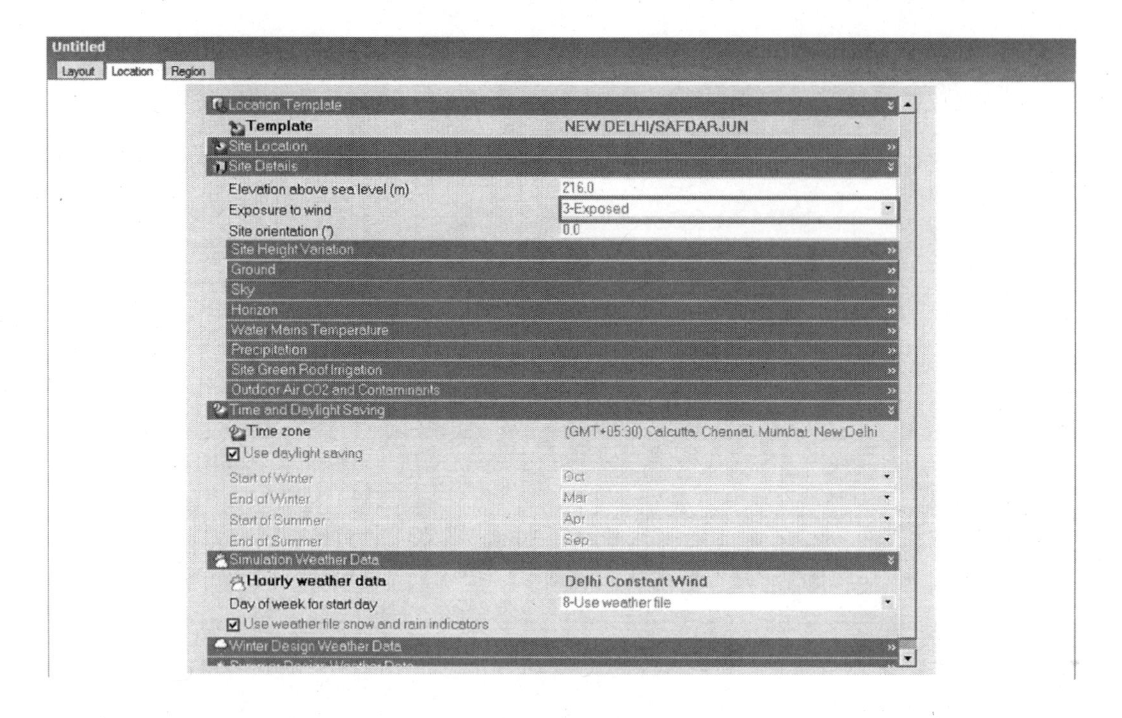

Step 14: Simulate the model, and record the results. Observe the results for **Wind Speed**, **Wind Direction** and **Mech Vent + Nat Vent + Infiltration (ac/h)**.

Untitled, Building 1

Analysis | Summary | Parametric | Optimisation + UA/SA | Data Visualisation

Date/Time	Outside Dry-Bulb Temperature (°C)	Outside Dew-Point Temperature...	External Air (kW)	Wind Speed (m/s)	Wind Direction (°)
01/01/2002 1:00:00 AM	7.55	4.5	-6.766246	1	0
01/01/2002 2:00:00 AM	6.625	4.375	-6.988322	1	0
01/01/2002 3:00:00 AM	6.55	3.85	-6.605448	1	0
01/01/2002 4:00:00 AM	6.225	3.25	-6.579642	1	0
01/01/2002 5:00:00 AM	6.925	2.65	-5.811839	1	0
01/01/2002 6:00:00 AM	7.35	1.75	-5.432754	1	0
01/01/2002 7:00:00 AM	7.925	1.575	-4.931956	1	0
01/01/2002 8:00:00 AM	9.225	1.825	-4.407452	1	0
01/01/2002 9:00:00 AM	11.25	2.725	-3.761882	1	0
01/01/2002 10:00:00 AM	13.9	3.525	-2.975247	1	0
01/01/2002 11:00:00 AM	16.7	3.775	-2.416214	1	0
01/01/2002 12:00:00 PM	19.05	3.125	-2.024568	1	0
01/01/2002 1:00:00 PM	20.425	2.3	-1.865415	1	0
01/01/2002 2:00:00 PM	20.775	1.125	-2.09947	1	0
01/01/2002 3:00:00 PM	20.2	0.65	-2.656234	1	0
01/01/2002 4:00:00 PM	19.1	0.9	-3.180876	1	0
01/01/2002 5:00:00 PM	17.9	1.75	-3.512373	1	0
01/01/2002 6:00:00 PM	16.85	2.3	-3.779998	1	0
01/01/2002 7:00:00 PM	15.775	2.625	-3.827676	1	0
01/01/2002 8:00:00 PM	14.675	2.25	-3.910826	1	0
01/01/2002 9:00:00 PM	13.275	2.175	-4.504936	1	0
01/01/2002 10:00:00 PM	11.7	2.35	-5.190884	1	0
01/01/2002 11:00:00 PM	10.025	3.15	-5.931063	1	0
02/01/2002	8.625	4	-6.473688	1	0
02/01/2002 1:00:00 AM	7.7	4.725	-6.65128	1	0
02/01/2002 2:00:00 AM	7.275	4.825	-6.533073	1	0
02/01/2002 3:00:00 AM	7.125	4.5	-6.320604	1	0

Untitled, Building 1

Analysis | Summary | Parametric | Optimisation + UA/SA | Data Visualisation

Date/Time	Solar Azimuth (°)	Atmospheric Pressure (Pa)	Direct Normal Solar (kW)	Diffuse Horizontal Solar (kW)	Mech Vent + Nat Vent + Infiltrati...
01/01/2002 9:00:00 AM	128.517	101000	0.20125	0.04475	21.23456
05/02/2002 6:00:00 PM	249.8215	101000	0.0435	0.0325	21.23456
01/03/2002 9:00:00 PM	279.3917	101000	0	0	21.23457
13/04/2002 10:00:00 AM	110.5309	101000	0.91575	0.0805	21.23458
25/07/2002 8:00:00 PM	296.763	99675	0	0	21.2346
07/05/2002 12:00:00 PM	147.5718	100250	0.49625	0.327	21.2346
12/02/2002 1:00:00 PM	183.383	101000	0.06675	0.26775	21.23461
20/01/2002 9:00:00 AM	125.1164	102000	0.16875	0.05075	21.23462
10/10/2002 8:00:00 PM	276.1412	100000	0	0	21.23462
16/10/2002 9:00:00 AM	119.6413	100000	0.41475	0.1205	21.23464
08/09/2002 6:00:00 PM	271.2	100000	0.0075	0.048	21.23464
06/01/2002 1:00:00 PM	185.3497	101000	0.72775	0.12525	21.23464
29/11/2002 11:00:00 AM	155.6138	102000	0.6505	0.12925	21.23464
27/05/2002 12:00:00 PM	135.7975	99825	0.79175	0.2	21.23464
20/04/2002 10:00:00 AM	107.6766	100000	0.96675	0.08225	21.23465
23/01/2002 12:00:00 PM	165.1909	102000	1.032	0.0465	21.23466
17/12/2002 5:00:00 AM	101.8248	101000	0	0	21.23467
03/11/2002 8:00:00 AM	115.7013	101000	0.1025	0.03325	21.2347
20/01/2002 1:00:00 PM	184.0056	101250	0.74375	0.1285	21.2347
04/06/2002 12:00:00 PM	130.365	100000	0.9195	0.1425	21.2347
02/01/2002 11:00:00 AM	152.4271	101000	0.97325	0.04325	21.2347
15/06/2002 3:00:00 PM	268.9895	99325	0.2975	0.43125	21.23472
19/02/2002 11:00:00 AM	142.6681	101000	0.21125	0.255	21.23472
07/02/2002 7:00:00 AM	104.8267	101750	0	0	21.23472
02/03/2002 9:00:00 AM	114.1038	101000	0.779	0.046	21.23473
13/03/2002 11:00:00 AM	137.1409	101000	1.03525	0.06825	21.23474
16/02/2002 9:00:00 AM	118.2401	101000	0.327	0.07575	21.23475

You can see that air changes per hour are almost constant throughout the year.

Step 15: Modify the weather data with **Wind speed** as **2 m/s** and **Wind direction** as **0 degrees** with the help of previous steps.

Step 16: Simulate the model, and record the results. Observe the results in **Wind Speed**, **Wind Direction** and **Mech Vent + Nat Vent + Infiltration (ac/h).**

Untitled, Building 1

Analysis | Summary | Parametric | Optimisation + UA/SA | Data Visualisation

Date/Time	Outside Dry-Bulb Temperature (°C)	Outside Dew-Point Temperature...	External Air (kW)	Wind Speed (m/s)	Wind Direction (°)
01/01/2002 1:00:00 AM	7.55	4.5	-7.149524	2	0
01/01/2002 2:00:00 AM	6.625	4.375	-7.356702	2	0
01/01/2002 3:00:00 AM	6.55	3.85	-6.873567	2	0
01/01/2002 4:00:00 AM	6.225	3.25	-6.844374	2	0
01/01/2002 5:00:00 AM	6.925	2.65	-5.937839	2	0
01/01/2002 6:00:00 AM	7.35	1.75	-5.520654	2	0
01/01/2002 7:00:00 AM	7.925	1.575	-4.950895	2	0
01/01/2002 8:00:00 AM	9.225	1.825	-4.391019	2	0
01/01/2002 9:00:00 AM	11.25	2.725	-3.688633	2	0
01/01/2002 10:00:00 AM	13.9	3.525	-2.849156	2	0
01/01/2002 11:00:00 AM	16.7	3.775	-2.278052	2	0
01/01/2002 12:00:00 PM	19.05	3.125	-1.785168	2	0
01/01/2002 1:00:00 PM	20.425	2.3	-1.631165	2	0
01/01/2002 2:00:00 PM	20.775	1.125	-1.928657	2	0
01/01/2002 3:00:00 PM	20.2	0.65	-2.499223	2	0
01/01/2002 4:00:00 PM	19.1	0.9	-3.114091	2	0
01/01/2002 5:00:00 PM	17.9	1.75	-3.477448	2	0
01/01/2002 6:00:00 PM	16.85	2.3	-3.774807	2	0
01/01/2002 7:00:00 PM	15.775	2.625	-3.82631	2	0
01/01/2002 8:00:00 PM	14.675	2.25	-3.923989	2	0
01/01/2002 9:00:00 PM	13.275	2.175	-4.6044	2	0
01/01/2002 10:00:00 PM	11.7	2.35	-5.390097	2	0
01/01/2002 11:00:00 PM	10.025	3.15	-6.227318	2	0
02/01/2002	8.625	4	-6.805517	2	0
02/01/2002 1:00:00 AM	7.7	4.725	-7.016041	2	0
02/01/2002 2:00:00 AM	7.275	4.825	-6.83005	2	0
02/01/2002 3:00:00 AM	7.125	4.5	-6.563624	2	0

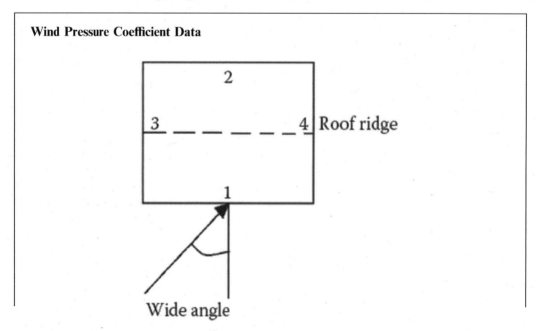

Untitled, Building 1

Analysis | Summary | Parametric | Optimisation + UA/SA | Data Visualisation

Date/Time	Solar Azimuth (°)	Atmospheric Pressure (Pa)	Direct Normal Solar (kW)	Diffuse Horizontal Solar (kW)	Mech Vent + Nat Vent + Infiltrati...
01/01/2002 1:00:00 AM	35.17849	101000	0	0	42.56889
01/01/2002 2:00:00 AM	77.21034	101000	0	0	42.57803
01/01/2002 3:00:00 AM	88.214	101000	0	0	42.55647
01/01/2002 4:00:00 AM	95.01023	101000	0	0	42.54217
01/01/2002 5:00:00 AM	100.8018	101000	0	0	42.55381
01/01/2002 6:00:00 AM	106.5388	101000	0	0	42.52799
01/01/2002 7:00:00 AM	112.7482	101000	0	0	42.5321
01/01/2002 8:00:00 AM	119.8971	101000	0.013	0.0095	42.52536
01/01/2002 9:00:00 AM	128.517	101000	0.20125	0.04475	42.51382
01/01/2002 10:00:00 AM	139.2282	101000	0.687	0.06675	42.49766
01/01/2002 11:00:00 AM	152.5845	101000	0.903	0.05675	42.4804
01/01/2002 12:00:00 PM	168.5691	101000	1.13725	0.0265	42.47214
01/01/2002 1:00:00 PM	185.9516	101000	0.85225	0.09125	42.46245
01/01/2002 2:00:00 PM	202.5799	101000	0.6505	0.1395	42.45009
01/01/2002 3:00:00 PM	216.8228	101000	0.6195	0.123	42.47005
01/01/2002 4:00:00 PM	228.3296	101000	0.38725	0.12425	42.48653
01/01/2002 5:00:00 PM	237.5457	101000	0.13125	0.08	42.49532
01/01/2002 6:00:00 PM	245.1006	101000	0.0045	0.0085	42.50195
01/01/2002 7:00:00 PM	251.5569	101000	0	0	42.50348
01/01/2002 8:00:00 PM	257.3938	101000	0	0	42.50496
01/01/2002 9:00:00 PM	263.0977	101000	0	0	42.51886
01/01/2002 10:00:00 PM	269.4093	101000	0	0	42.54028
01/01/2002 11:00:00 PM	278.3372	101000	0	0	42.55868
02/01/2002	302.2123	101000	0	0	42.57159
02/01/2002 1:00:00 AM	34.1753	101000	0	0	42.56785
02/01/2002 2:00:00 AM	76.83969	101000	0	0	42.54734
02/01/2002 3:00:00 AM	88.00227	101000	0	0	42.54525

You can calculate the air changes per hour for two windows with the following equation (from *ASHRAE Handbook of Fundamentals*, 2015. © 2015 ASHRAE, www.ashrae.org):

$$Q = U_w \times \sqrt{\frac{C_{p1} - C_{p2}}{\left(\frac{1}{A_1 \times C_1}\right)^2 + \left(\frac{1}{A_2 \times C_2}\right)^2}}$$

where C_p is the pressure drag coefficient, C is the discharge coefficient, U is the wind speed and A is the area of the opening.

Wind Pressure Coefficient Data

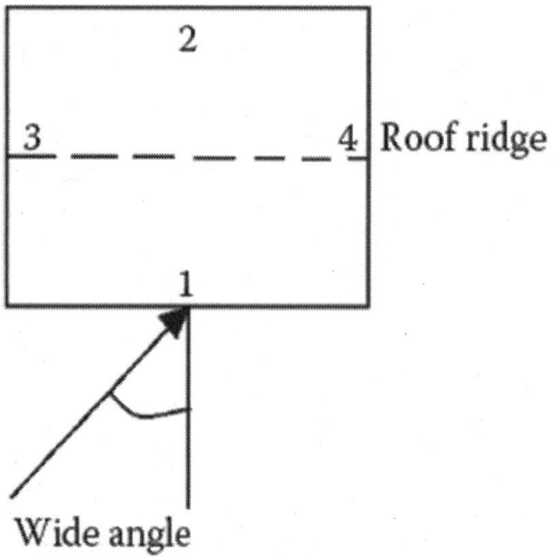

Low-rise buildings (up to three storeys)
Length-to-width ratio: 1:1
Shielding condition: exposed
Wind speed reference level: building height
Wind angle (Table 10.1)

TABLE 10.1
Wind pressure coefficient data

Location	0	45	90	135	180	225	270	315
Face 1	0.70	0.35	−0.50	−0.40	−0.20	−0.40	−0.50	0.35
Face 2	−0.20	−0.40	−0.50	0.35	0.70	0.35	−0.50	−0.40
Face 3	−0.50	0.35	0.70	0.35	−0.50	−0.40	−0.20	−0.40
Face 4	−0.50	−0.40	−0.20	−0.40	−0.50	0.35	0.70	0.35

Sources: www.designbuilder.co.uk/downloads/AIVCWindPressureCoefficientData.pdf and www.designbuilder.co.uk/helpv6.0/Content/Pressure_Coefficients_Data.htm.

Wind Pressure Coefficient Templates

Wind pressure coefficients are used when the **Natural ventilation model** option is set to **Calculated**. The EnergyPlus **Airflow network** calculations use pressure coefficients when calculating wind-induced pressure on each surface during simulations when the **Calculated Natural ventilation** option is selected.

DesignBuilder is supplied with a database of wind pressure coefficients based on the data from Martin Liddament, *Air Infiltration Calculation Techniques: An Applications Guide* (Paris: Air infiltration and Ventilation Centre, International Energy Agency, 1986). The C_p data are buildings of three storeys or fewer with square surfaces and for three levels of site exposure. The data are given in 45-degree increments.

Sources: https://designbuilder.co.uk/downloads/AIVCWindPressureCoefficientData.pdf and https://designbuilder.co.uk/helpv6.0/Content/Pressure_Coefficients_Data.htm.

Discharge Coefficient

The *discharge coefficient*, which ranges from zero to one, is a measure of the amount of pressure that the air loses as it goes through every opening of flow resistance. The default of 0.65 is a good general-purpose setting.

Source: https://designbuilder.co.uk/helpv6.0/Content/Calculated_Natural_Ventilation_Options.htm.

Wind speed provided in the weather file is for 10-m height. You can use the following formula to calculate the wind speed at 2-m height:

$$u = u_r \left(\frac{z}{z_r}\right)^{\alpha}$$

where α is approximately 1/7, or 0.143 (it varies with local topography). Tables 10.2 and 10.3 give the calculation for natural ventilation in the zone.

TABLE 10.2
Calculation of zone air changes per hour at a wind speed of 1 m/s at 10-m height

Wind speed at 2 m	0.72 m/s
Area of window	6.59 m^2
C_{p1}	0.70
C_{p2}	−0.20
C_1	0.65
C_2	0.65
Q	2.07 m^3/s
Q	7,442 m^3/h
Zone volume	350 m^3
Air changes	21.26 ac/h

You need to refer Table 10.1 for wind pressure coefficients.

TABLE 10.3
Calculation of zone air changes per hour at a wind speed of 2 m/s at 10 m-height

Wind speed at 2 m	1.44 m/s
Area of window	6.59 m^2
C_{p1}	0.70
C_{p2}	−0.20
C_1	0.65
C_2	0.65
Q	4.13 m^3/s
Q	14,885 m^3/h
Zone volume	350 m^3
Air changes	42.53 ac/h

Air changes values calculated from the formula are matched with DesignBuilder simulated values.

EXERCISE 10.2

Repeat this tutorial for a window area of 16 m^2.

Tutorial 10.3

Evaluating the Impact of a Window Opening and Closing Schedule

GOAL

To evaluate the impact of a window opening and closing schedule

WHAT ARE YOU GOING TO LEARN?

- How to define a window opening and closing schedule for natural ventilation

PROBLEM STATEMENT

In this tutorial, you are going to use the model saved in Tutorial 10.2 and add windows in the remaining two directions. You need to simulate the model for the following two window operation schedules:

1. On (windows are always open)
2. 00–18:00 Mon–Sat

Use the **New Delhi/Safdarjung, India weather location.**

SOLUTION

Step 1: Open the project saved in Tutorial 10.2.

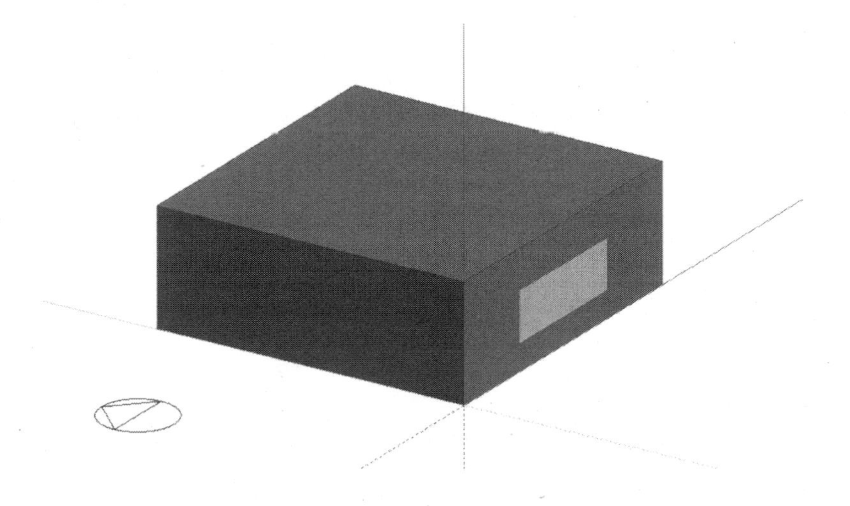

Step 2: In the navigation tree, select **East facing wall** by selecting **Wall** – **35.000 m^2 – 90.0°**.

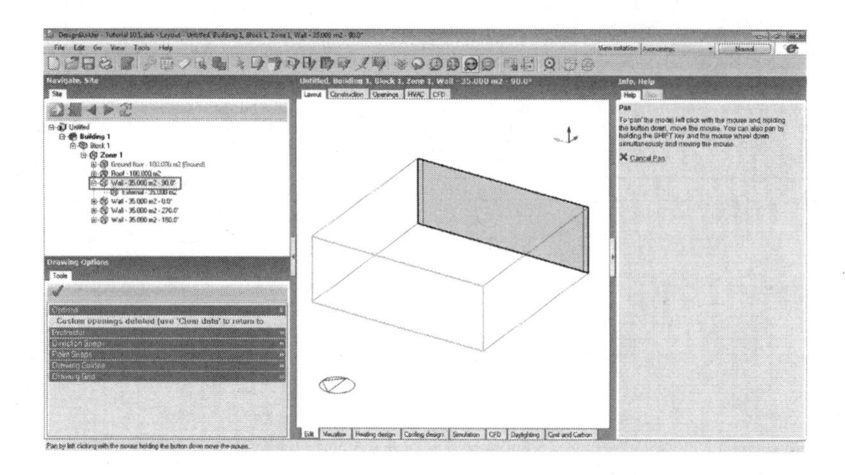

Step 3: Select **Left** in the **View rotation** drop-down list, and then select **Draw window** in the **Help** tab to draw a window with the same dimensions as that on the north or south side.

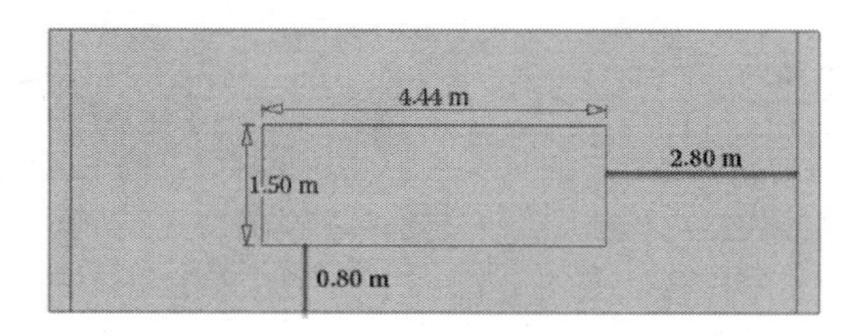

Step 4: Similarly, draw the west window.

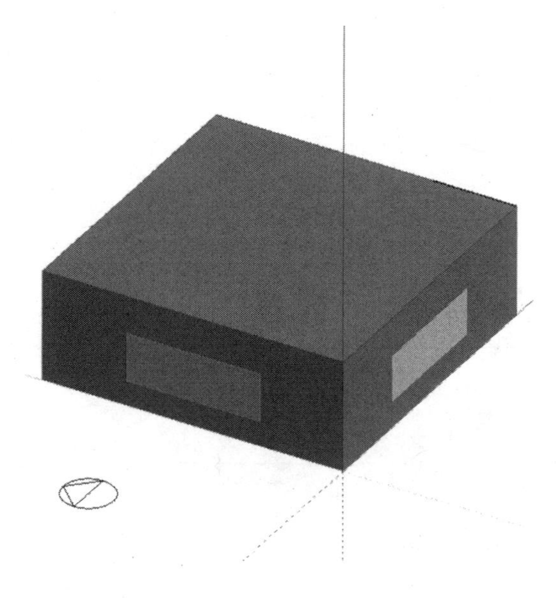

Step 5: Select the **Openings** tab. Click the **Operation** section. Make sure that **Operation schedule** is **On 24/7**.

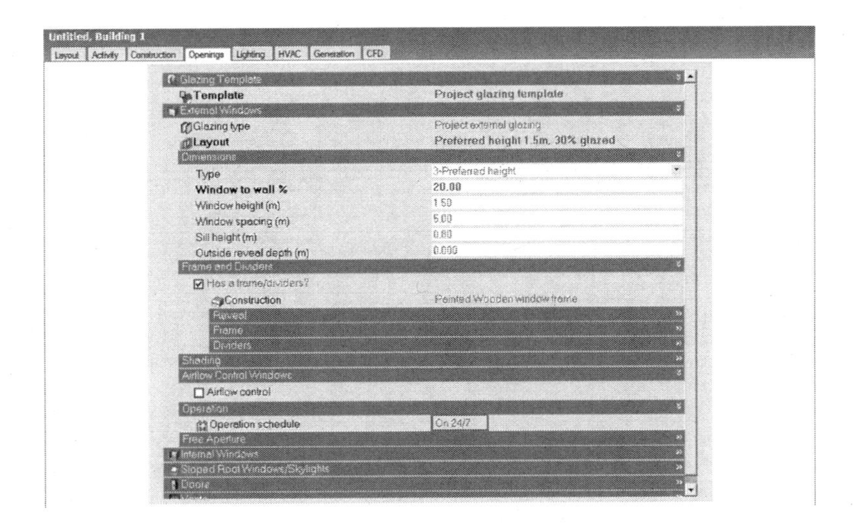

Simulate the model, and record the results by exporting them to a spreadsheet.

Step 6: Repeat the preceding steps for an **Operation schedule** of **8:00–18:00 Mon–Sat**.

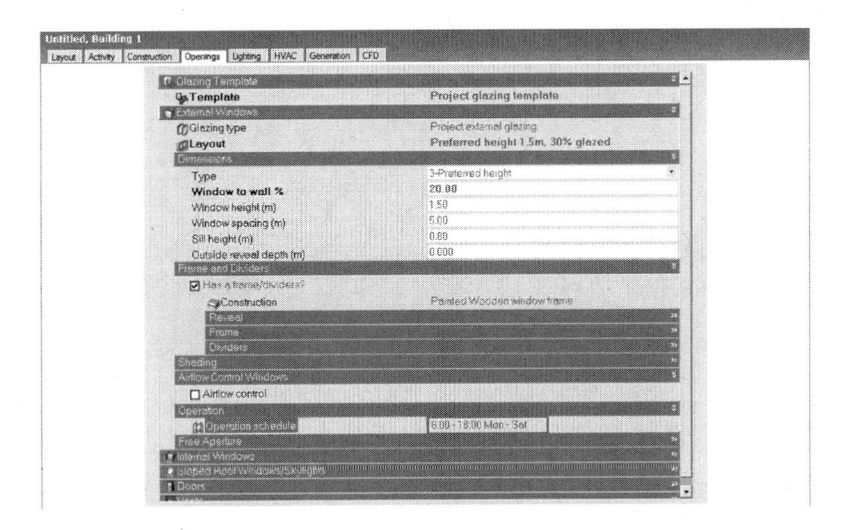

Simulate the model, and record the results. Plot the graph for 1 and 2 January for **Mech Vent + Nat Vent + Infiltration (Air changes per hour, ac/h)**.

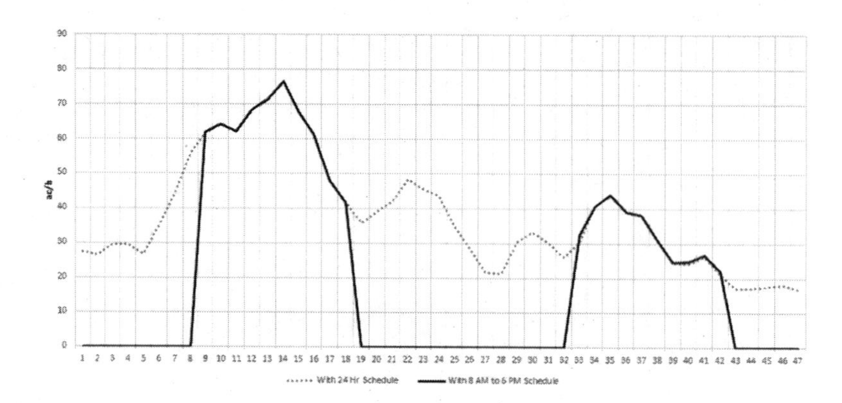

You can see the impact of window opening and closing on air change rate.

Tutorial 10.4

Evaluating the Impact of Window Opening Control Based on Temperature

GOAL

To evaluate the impact of window opening control based on zone and outdoor air temperature

WHAT ARE YOU GOING TO LEARN?

- How to define window operation control for natural ventilation

PROBLEM STATEMENT

In this tutorial, you are going to use the calculated model saved in Tutorial 10.1. You are going to simulate the model for **Temperature control mode**. Use the **New Delhi/Safdarjung, India** weather location.

SOLUTION

Step 1: Open the model saved in Tutorial 10.1 with **Natural ventilation** set to **calculate**.

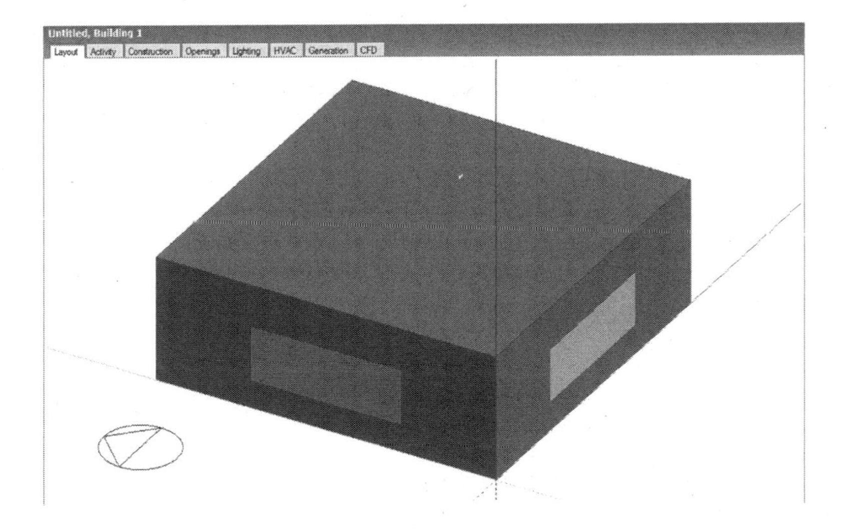

Step 2: Select the **Activity** tab. Set **Min temperature** to **20.0**.

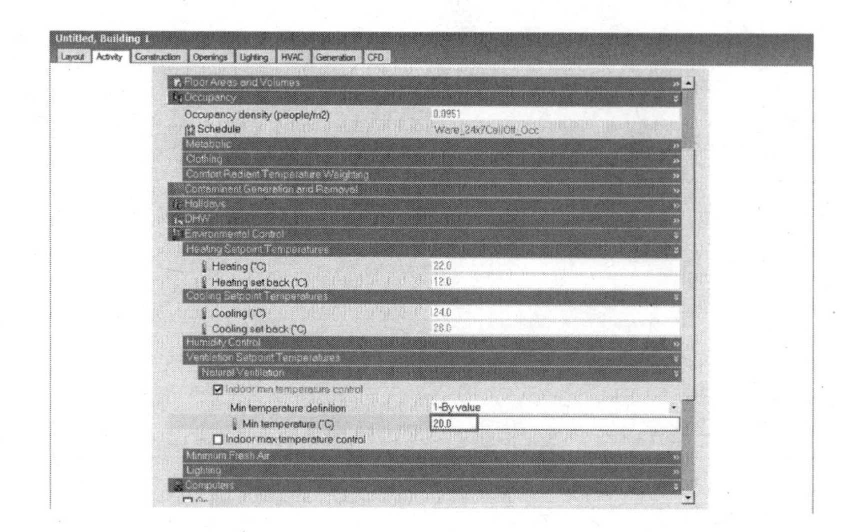

Natural Ventilation Minimum Temperature

This is the fixed indoor temperature below which ventilation is shut off. The control is visible when the **1-By** value option is selected for **Min temperature** definition. It can be thought of as the cooling setpoint temperature which controls the activation of natural ventilation. If the inside air temperature is greater than this setpoint temperature (and the **Natural ventilation** operation schedule is on), then natural ventilation can take place.

Source: www.designbuilder.co.uk/helpv6.0/Content/_Environmental_comfort.htm.

Step 3: Select the **Openings** tab. Click the **Operation** section. Make sure that **Operation schedule** is set to **On 24/7**.

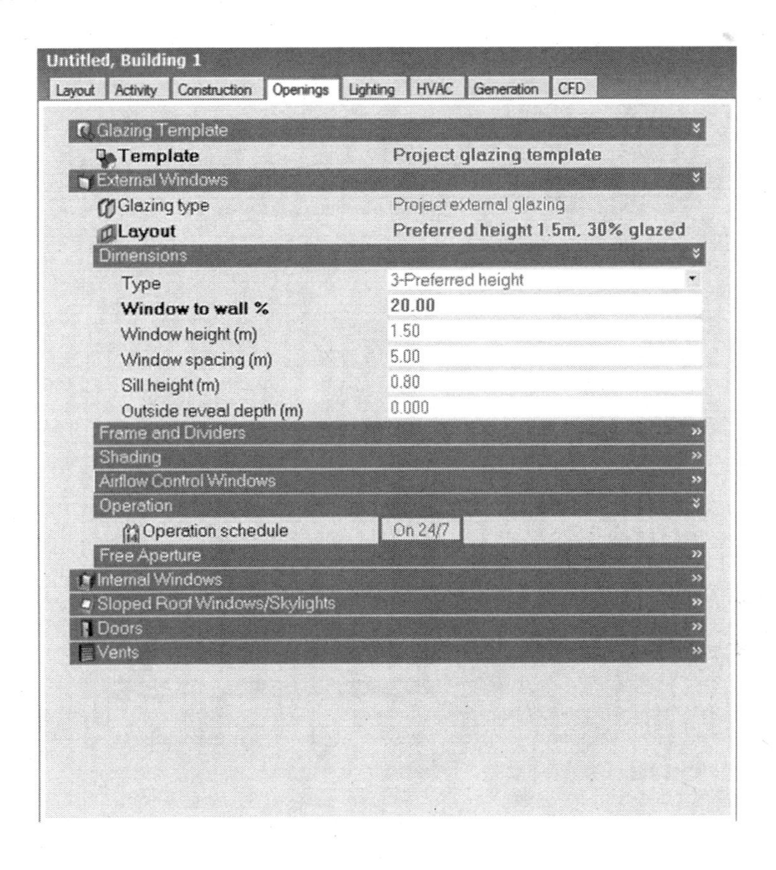

Step 4: Select the **HVAC** tab. Enter **Outside air (ac/h)** as **3.000**. Select **Temperature** from the **External** and **Internal Control mode** drop-down lists.

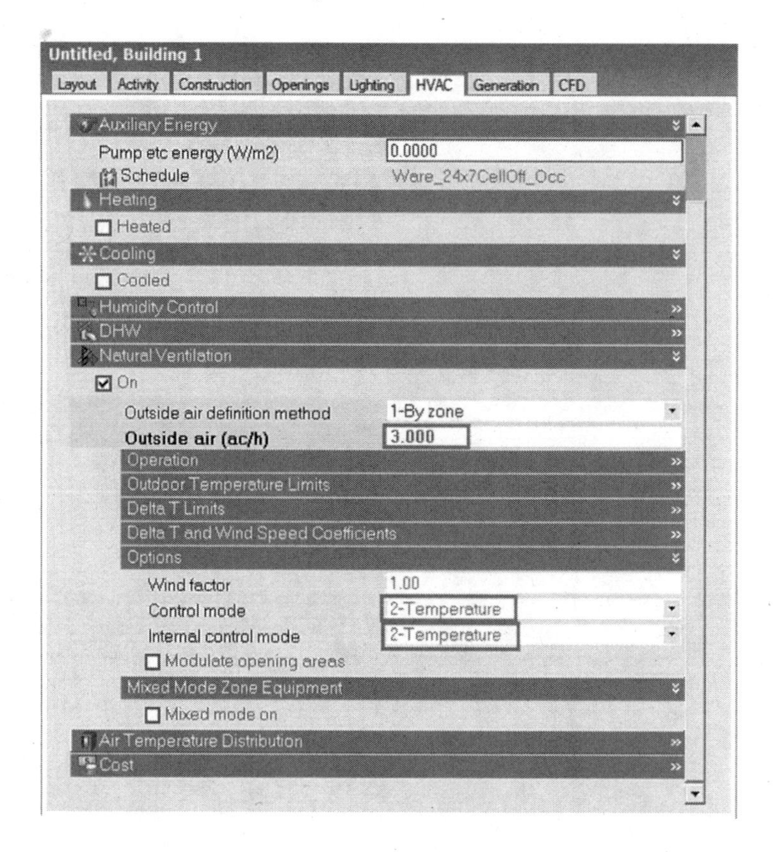

To see the effect of window opening controls, we need the following:

- Generate and view the results at system time steps instead of fixed user-defined time steps. This can be achieved by selecting the **detailed** frequency option on an HVAC output variable (e.g. **Zone Air Temperature**).
- Generate more output variables such as window opening factor, zone infiltration and zone air temperature.

Field: Ventilation Control Mode

This specifies the type of zone-level natural ventilation control. The windows are 'open' when $T_{zone\_air} > T_{setpoint}$ and $t_{zone\_air} > T_{outside\_air}$ and the schedule value = 1, where $T_{outside\_air}$ is equal to the outdoor air temperature, T_{zone_air} is equal to the previous time step's zone air temperature and $T_{setpoint}$ is equal to the **Vent Temperature Schedule** value.

Source: www.designbuilder.co.uk/helpv6.0/Content/_Operation2.htm.

The *system time step* is a variable-length time step that governs the driving time step for HVAC and plant system modelling. The user cannot directly control the system time step (except by use of the **ConvergenceLimits** object). When the HVAC portion of the simulation begins its solution for the current zone time step, it uses the zone time step as its maximum length but can then reduce the time step, as necessary, to improve the solution. Users can see the system time step used if they select the **detailed** frequency option on an HVAC output variable (e.g. **Zone Air Temperature**). In contrast, the **Zone** variables will only be reported on the zone time step (e.g. **Zone Mean Air Temperature**). Note that hourly data (such as outdoor conditions expressed by **Design Days** or **Weather data**) are interpolated to the **Zone Time** step.

Source: U.S. Department of Energy, *EnergyPlus Version 8.9.0 Documentation, Input-Output Reference*, March 2018; available at https://energyplus.net/documentation.

Zone Mean Air Temperature (°C)

Zone mean air temperature is the average temperature of the air temperatures at the system time step. Remember that the zone heat balance represents a 'well-stirred' model for a zone; therefore, there is only one mean air temperature to represent the air temperature for the zone. This is reported only at the zone time step.

Zone Air Temperature (°C)

This is very similar to the mean air temperature. The 'well-stirred' model for the zone is the basis, but this temperature is also available at the **detailed** system time step.

Source: U.S. Department of Energy, *EnergyPlus Version 8.9.0 Documentation, Input-Output Reference*, March 2018; available at https://energyplus.net/documentation.

Step 5: Minimize the **DesignBuilder** window. Create a new file named **output.idf**. You can use any text editor such as **Notepad++** or **Editplus**. Write additional output variables as follows:

Output:Variable, *, Zone Air Temperature, Detailed;
Output:Variable, *, AFN Surface Venting Window or Door Opening Factor, Detailed;
Output:Variable, *, AFN Zone Infiltration, Detailed;

```
Output:Variable, *, Zone Air Temperature, Detailed;
Output:Variable, *, AFN Surface Venting Window or door Opening Factor,
Detailed;
Output:Variable, *, AFN Zone Infiltration, Detailed;
```

This .idf file needs to be placed in the path shown in the following figure.

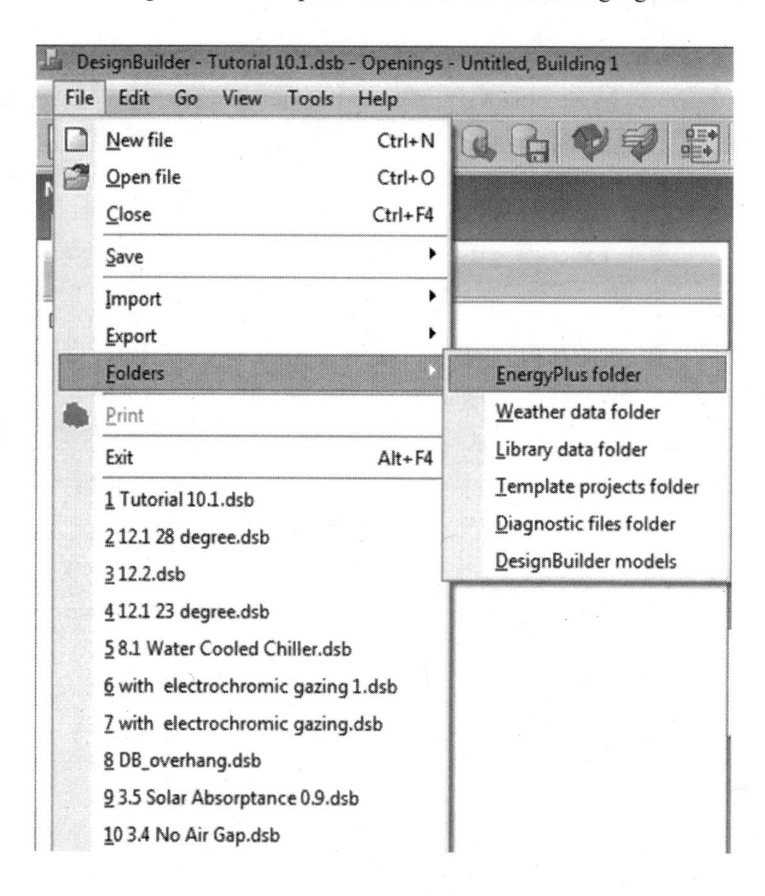

Step 6: Maximize the **DesignBuilder** window, and select the **HVAC** tab. Click **Simple** in the **Help** tab. The **Model Options – Building and Block** screen appears.

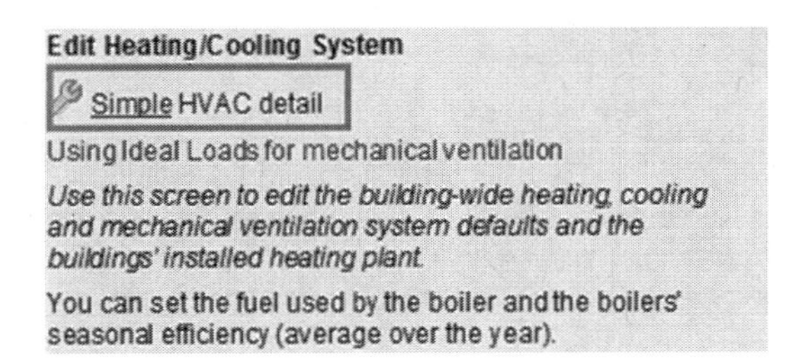

Step 7: Select the **Simulation** screen tab. Click the **Advanced** section.

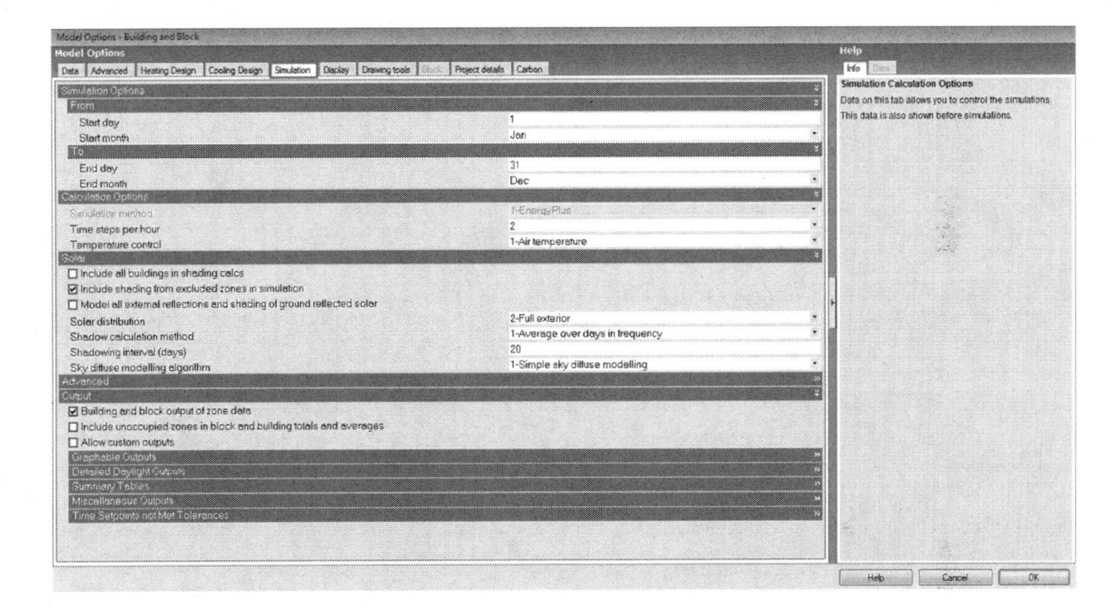

Step 8: Select the **IDF File 1** checkbox, and select the **Output.idf** file from the **EnergyPlus** folder. Click **OK**.

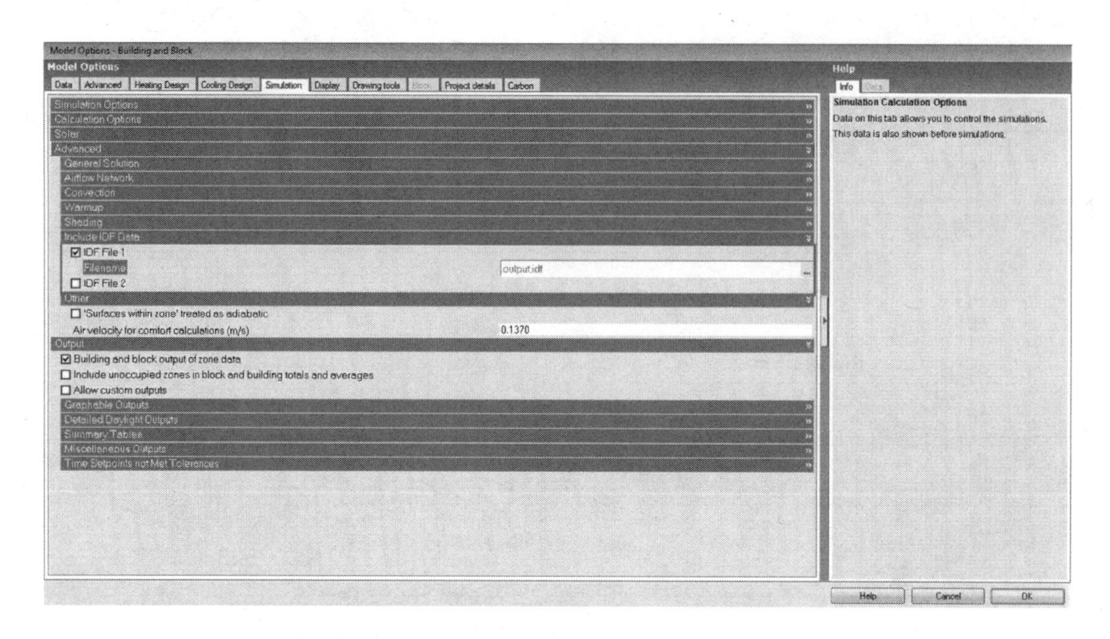

Step 9: Select the **Simulation** tab. The **Edit Calculation Options** screen appears. Click **OK**.

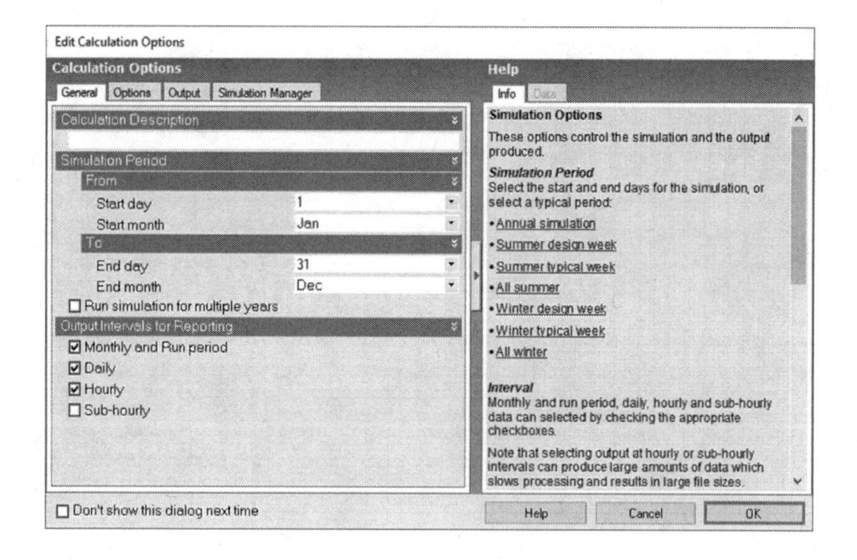

After the simulation, the results appear on the screen, but DesignBuilder does not support viewing 'Detailed' reporting frequency. After the simulation, an ESO file is generated in the **EnergyPlus** folder which contains results at detailed reporting frequency. You can view this in the **xESOView** tool.

xESOView can be downloaded from http://xesoview.sourceforge.net/.

Step 10: Click the **View EnergyPlus results** icon. From the **EnergyPlus** folder, select the **eplusout.eso** file. The **xESOView** tool will open.

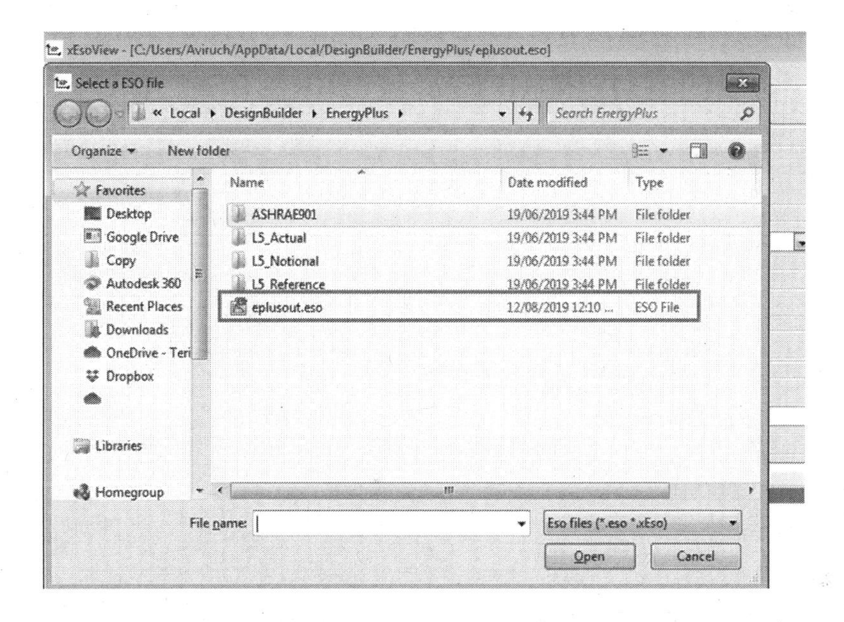

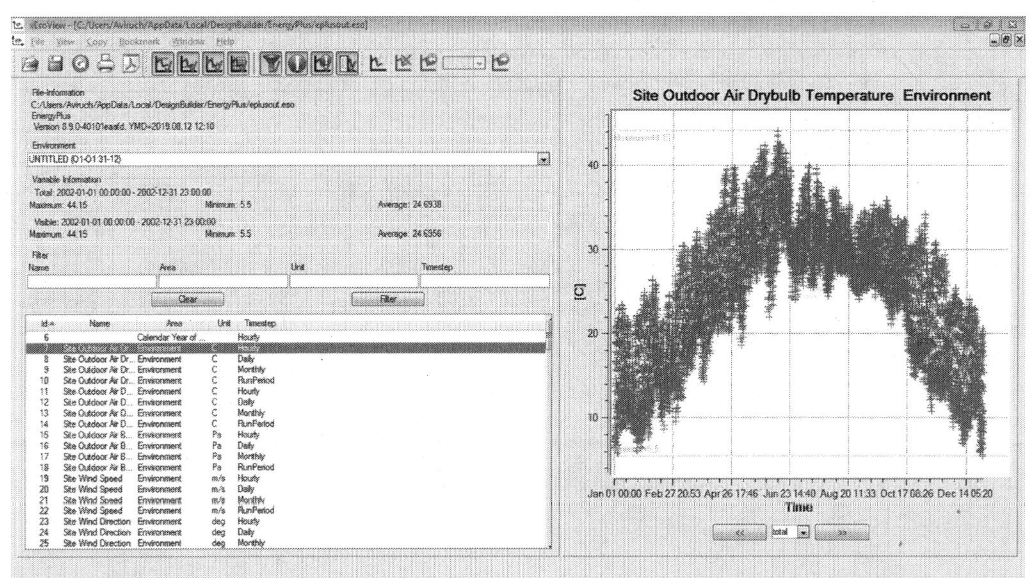

Step 11: Enter the **Site Outdoor Air Drybulb Temperature** in the **Name** field of the **Filter** section to view the variable. From the filtered list, and select the variable with hourly **Timestep**.

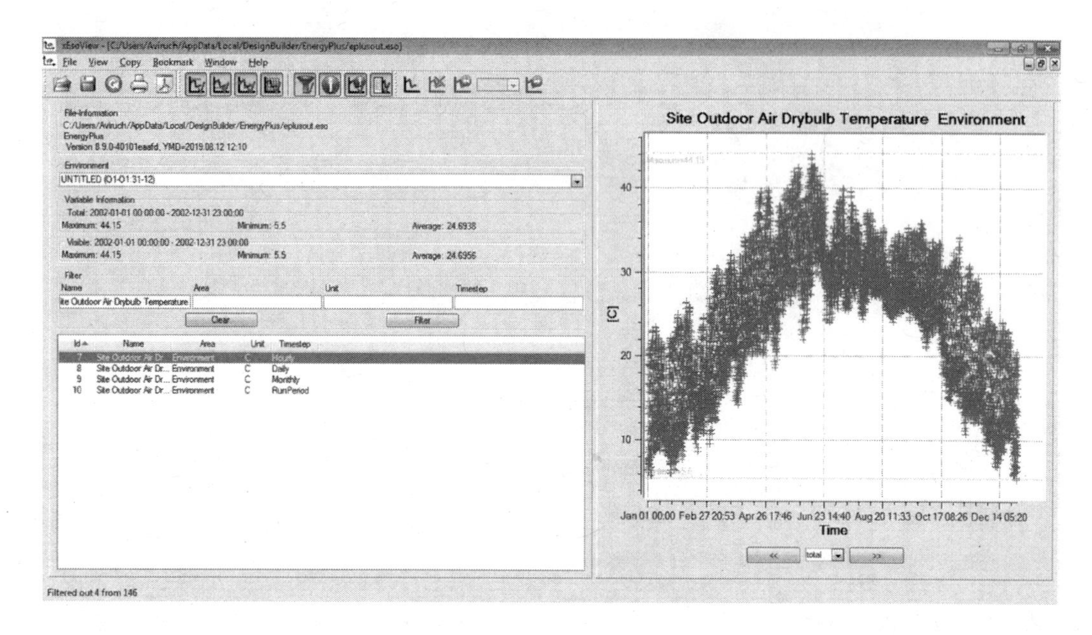

You can copy the variable data to view in a spreadsheet.

Step 12: Click **Copy**, and select **Copy variable data**.

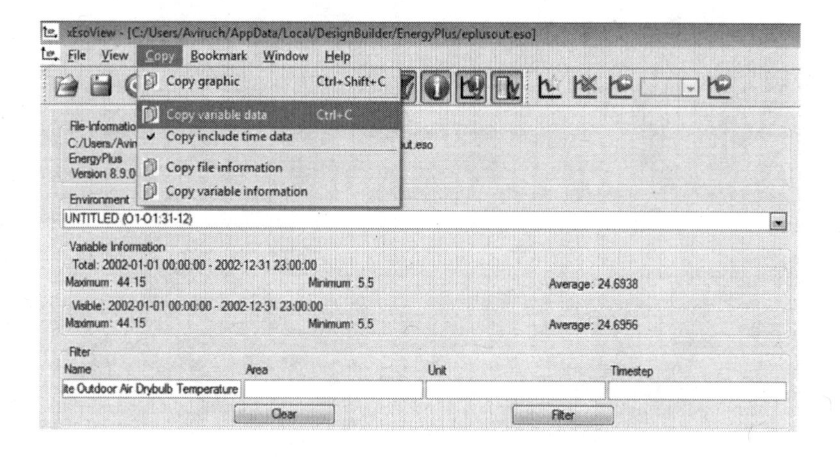

You can copy only one variable at a time.

Step 13: Open a spreadsheet program, and paste the data as shown in the following figure.

	A	B	C
1	Date/Time	Site Outdoor Air Drybulb Temperature Environment [C]	
65	01/01/2002 1:29		
66	01/01/2002 1:30		
67	01/01/2002 1:40		
68	01/01/2002 1:50		
69	01/01/2002 1:00	6.625	
70	01/01/2002 2:00		
71	01/01/2002 2:06		
72	01/01/2002 2:12		
73	01/01/2002 2:18		
74	01/01/2002 2:24		
75	01/01/2002 2:30		
76	01/01/2002 2:00	6.55	
77	01/01/2002 3:00		
78	01/01/2002 3:07		
79	01/01/2002 3:15		
80	01/01/2002 3:22		
81	01/01/2002 3:30		
82	01/01/2002 3:40		
83	01/01/2002 3:50		
84	01/01/2002 3:00	6.225	
85	01/01/2002 4:00		
86	01/01/2002 4:30		
87	01/01/2002 4:00	6.925	
88	01/01/2002 5:00		
89	01/01/2002 5:30		
90	01/01/2002 5:00	7.35	
91	01/01/2002 6:00		
92	01/01/2002 6:01		
93	01/01/2002 6:02		
94	01/01/2002 6:03		
95	01/01/2002 6:04		
96	01/01/2002 6:05		
97	01/01/2002 6:06		
98	01/01/2002 6:07		
99	01/01/2002 6:08		
100	01/01/2002 6:09		

Similarly, copy and paste the following variables with **Timestep** as **hourly:**

- **AFN Surface Venting Window or Door Opening Factor BLOCK1:ZONE1_WALL_2_0_0_0_0_0_WIN []**
- **AFN Surface Venting Window or Door Opening Factor BLOCK1:ZONE1_WALL_3_0_0_0_0_0_WIN []**
- **AFN Surface Venting Window or Door Opening Factor BLOCK1:ZONE1_WALL_4_0_0_0_0_0_WIN []**
- **AFN Surface Venting Window or Door Opening Factor BLOCK1:ZONE1_WALL_5_0_0_0_0_0_WIN []**
- **AFN Zone Infiltration Air Change Rate BLOCK1: ZONE1 [ach]**
- **Zone Air Temperature BLOCK1:ZONE1 [C]**

H159 =IF(AND(G159>20,G159>B159),TRUE,FALSE)

	A	B	C	D	E	F	G	H
1	Date/Time	Site Outdoor Air Drybulb Temperature Environment [C]	AFN Surface Venting Window or Door Opening Factor BLOCK1:MYZONE_WALL_2_0_0_0_0_WIN []	AFN Surface Venting Window or Door Opening Factor BLOCK1:MYZONE_WALL_3_0_0_0_0_WIN []	AFN Surface Venting Window or Door Opening Factor BLOCK1:MYZONE_WALL_4_0_0_0_0_WIN []	AFN Surface Venting Window or Door Opening Factor BLOCK1:MYZONE_WALL_5_0_0_0_0_WIN []	Zone Air Temperature BLOCK1: MYZONE [C]	Check if Tzone>Tout and Tzone>20
153	01/01/2002 9:11		-	-	-	-	19.0465	FALSE
154	01/01/2002 9:13		-	-	-	-	19.2503	FALSE
155	01/01/2002 9:15		-	-	-	-	19.4357	FALSE
156	01/01/2002 9:16		-	-	-	-	19.6045	FALSE
157	01/01/2002 9:18		-	-	-	-	19.7579	FALSE
158	01/01/2002 9:20		-	-	-	-	19.8975	FALSE
159	01/01/2002 9:22	13.90	-	-	-	-	20.0244	TRUE
160	01/01/2002 9:24		0.93	0.93	0.93	0.93	16.7358	FALSE
161	01/01/2002 9:26		-	-	-	-	14.9469	FALSE
162	01/01/2002 9:28		-	-	-	-	14.742	FALSE
163	01/01/2002 9:30		-	-	-	-	18.8156	FALSE
164	01/01/2002 9:35		-	-	-	-	19.6548	FALSE
165	01/01/2002 9:40		-	-	-	-	20.2963	TRUE
166	01/01/2002 9:45	13.90	0.94	0.94	0.94	0.94	16.5165	FALSE
167	01/01/2002 9:50		-	-	-	-	15.5298	FALSE

You can see from the preceding that if the following conditions are true, then there is natural ventilation:

$$T_{zone\_air} > T_{setpoint} \quad \text{and} \quad T_{zone\_air} > T_{outside\_air}$$

Please note that the zone setpoint is 20°C. You can see that the windows open in the succeeding timestep after the above-mentioned condition is met. Please note that the zone infiltration air change rate is reported hourly and not on the system time step. Hence, even when the windows are open, the zone ventilation will not be reported in that time step; it will be reported in the hourly time step.

USE OF ENERGY MANAGEMENT SYSTEM (EMS) IN DESIGNBUILDER TO CHANGE THE CONSTRUCTION STATE OF WINDOW TO ACCOUNT FOR CHANGE IN THE SOLAR HEAT GAIN COEFFICIENT (SHGC)

EMS is an advanced feature in EnergyPlus which provides a way to develop custom control and modelling routines for EnergyPlus models. EMS provides high-level supervisory control to over-ride selected aspects of EnergyPlus modelling. A small programming language called the *EnergyPlus Runtime Language* (Erl) is used to describe the control algorithms. EnergyPlus interprets and executes the Erl program as the model is being run.

Source: U.S. Department of Energy, *EnergyPlus Version 8.9.0 Documentation, Application Guide for EMS*, March 2018; available at https://energyplus.net/documentation.

Step (a): Click on the **Scripts** icon. The **Script Manager** screen appears.

Step (b): Select the **Enable scripts** checkbox.

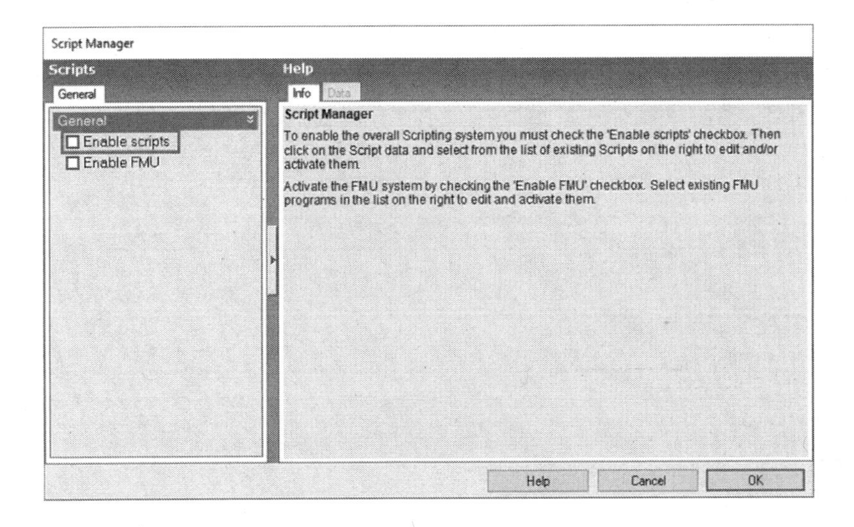

Step (c): Click the **Add new item** button. The **Edit Script – EMS** screen appears.

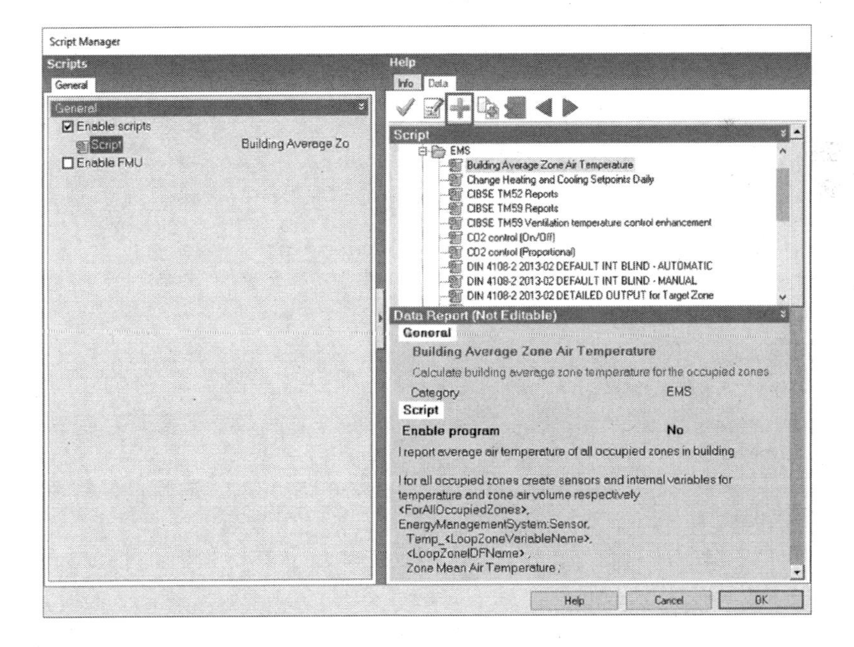

Step (d): Enter **WindowConstruct** in the **Name** text box. Select the **Enable program** checkbox in the **Script** subtab.

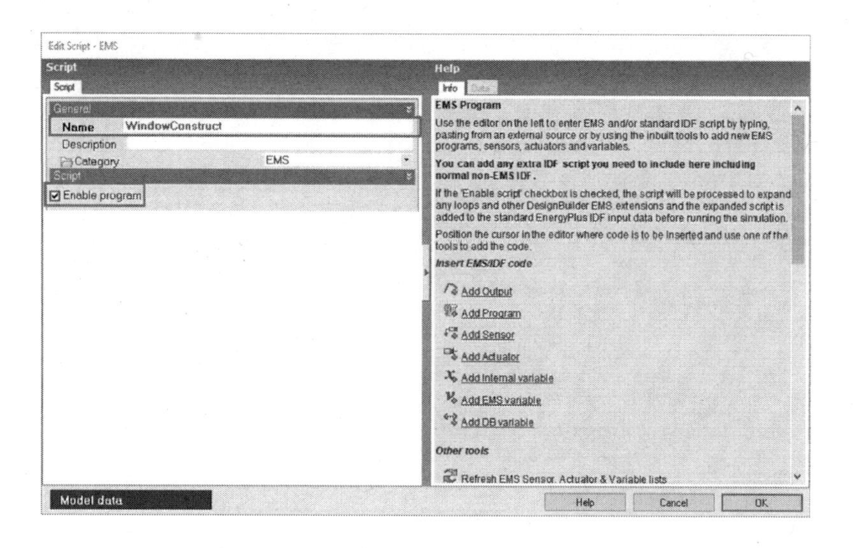

Step (e): In the **Script** area, write construction definitions for windows as follows:

```
WindowMaterial:SimpleGlazingSystem,    ! – Simple glazing material used to define an
                                           entire glazing system
   Glazing 1,                          ! – Name
   3,                                  ! – U-value W/m2/K
   .27,                                ! – Solar Heat Gain Coefficient
   .744;                               ! – Visible Transmittance

WindowMaterial:SimpleGlazingSystem,    ! – Simple glazing material used to define an
                                           entire glazing system
   Glazing 2,                          ! – Name
   5.6,                                ! – U-value W/m2/K
   .8,                                 ! – Solar Heat Gain Coefficient
   .9;                                 ! – Visible Transmittance

Construction,
   Glazing state – closed,
   Glazing 1;

Construction,
   Glazing state – open,
   Glazing 2;
```

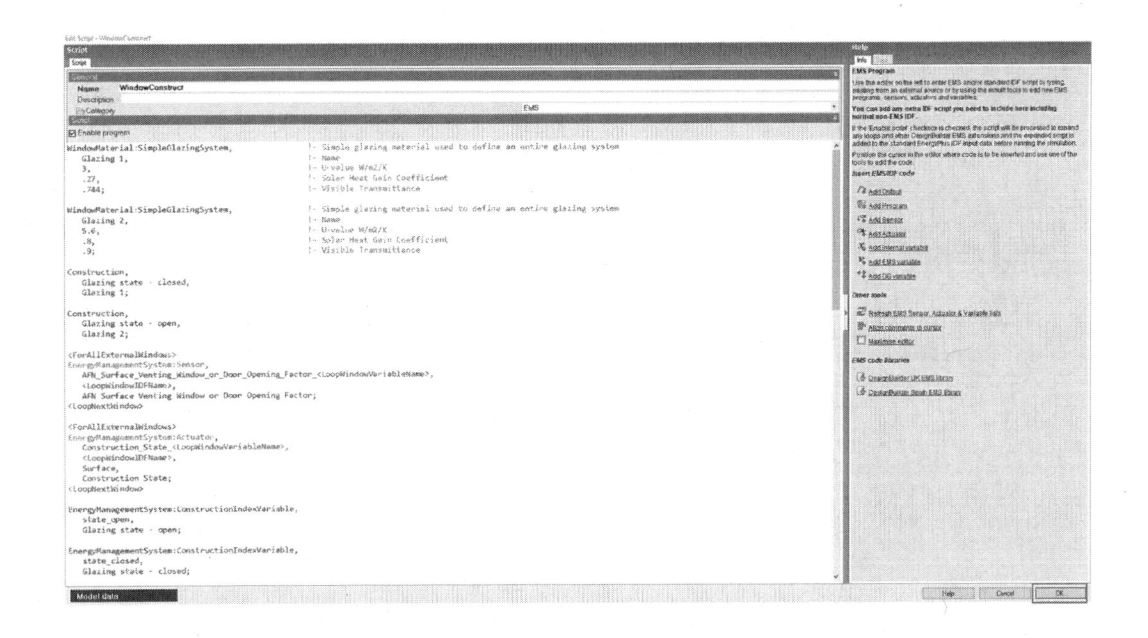

> The **WindowMaterial:SimpleGlazingSystem** object defines the construction parameters of windows in EnergyPlus. The construction object in EnergyPlus defines each layer of the window construction listed in order from 'outside' to 'inside'.
>
> *Source*: U.S. Department of Energy, *EnergyPlus Version 8.9.0 Documentation, Input-Output Reference*, March 2018; available at https://energyplus.net/documentation.

Step (f): Write the sensor definitions for EMS as follows:

```
<ForAllExternalWindows>
EnergyManagementSystem:Sensor,
  AFN_Surface_Venting_Window_or_Door_Opening_Factor_<LoopWindowVariableName>,
  <LoopWindowIDFName>,
  AFN Surface Venting Window or Door Opening Factor;
<LoopNextWindow>
```

In this script **AFN_Surface_Venting_Window_or_Door_Opening_Factor_** is the sensor which gives information about the current time step value of the venting opening factor for a particular window or door.

EMS sensors are normal EnergyPlus output variables to provide a general way of obtaining a wide variety of input data with minimal complications. The RDD file is an important resource for EMS users. The RDD file is an output from running EnergyPlus and is called *eplusout. rdd.* This output file is often needed to develop EMS input, so we need to do an initial run of the model with traditional controls to obtain an RDD file. Various types of output variables in EnergyPlus can be used as sensors in the EMS.

Source: U.S. Department of Energy, *EnergyPlus Version 8.9.0 Documentation, Application Guide for EMS*, March 2018; available at https://energyplus.net/documentation.

Step (g): Write the actuator definitions for EMS as follows:

```
<ForAllExternalWindows>
EnergyManagementSystem:Actuator,
  Construction_State_<LoopWindowVariableName>,
  <LoopWindowIDFName>,
  Surface,
  Construction State;
<LoopNextWindow>
```

In this script, **Construction_State_** is the actuator which will change the construction parameters of window based on the venting opening factor.

EMS actuators are the conduits by which Erl programs control EnergyPlus simulations. They actuate selected features inside EnergyPlus. Rather than add a new set of controls and component models that have EMS awareness, they generally override established features. Similar to how EnergyPlus reports the available output variables to the RDD file, a list of available actuators is written to the EDD file (depending on the settings in **Output:EnergyManagementSystem**). Note that the EDD file is *only* produced if you have set up EMS/Erl programs. To use an actuator in EMS, you need to enter an **EnergyManagementSystem:Actuator** input object.

Source: U.S. Department of Energy, *EnergyPlus Version 8.9.0 Documentation, Application Guide for EMS*, March 2018; available at https://energyplus.net/documentation.

Step (h): Write the **Construction Index Variables** as follows:

```
EnergyManagementSystem:ConstructionIndexVariable,
  state_open,
  Glazing state - open;

EnergyManagementSystem:ConstructionIndexVariable,
  state_closed,
  Glazing state - closed;
```

In this script, the **state_open** variable will be used to signify the construction state of windows in the opened state, and the **state_closed** variable will be used to signify the construction state of windows in the closed state.

Step (i): Write the **ProgramCallingManager** definition for EMS:

```
EnergyManagementSystem:ProgramCallingManager,
   Window Construct,
   BeginTimestepBeforePredictor,
   WindowConstruct;
```

In this script, **BeginTimestepBeforePredictor** is used as the calling point when the EMS will be executed.

> **ProgramCallingManager** requires the user to describe the timing for when the Erl programs are run. These EMS calling points correspond to places inside the EnergyPlus program where and when the EMS can be called to do something. The EMS offers a wide range of calling points.
>
> *Source*: U.S. Department of Energy, *EnergyPlus Version 8.9.0 Documentation, Application Guide for EMS*, March 2018; available at https://energyplus.net/documentation.

Step (j): Write the **Program definition** for EMS:

```
EnergyManagementSystem:Program,
   WindowConstruct,

<ForAllExternalWindows>
If AFN_Surface_Venting_Window_or_Door_Opening_Factor_<LoopWindowVariable
    Name> > 0,
   Set Construction_State_<LoopWindowVariableName> = state_open>,
Elseif AFN_Surface_Venting_Window_or_Door_Opening_Factor_<LoopWindowVariable
    Name> <= 0,
   Set Construction_State_<LoopWindowVariableName> = state_closed,
Else,
   Set Construction_State_<LoopWindowVariableName> = state_closed,
Endif,
<LoopNextWindow>
;
```

In this script, the custom program is written which changes the construction state of windows based on the window opening factor of the current time step.

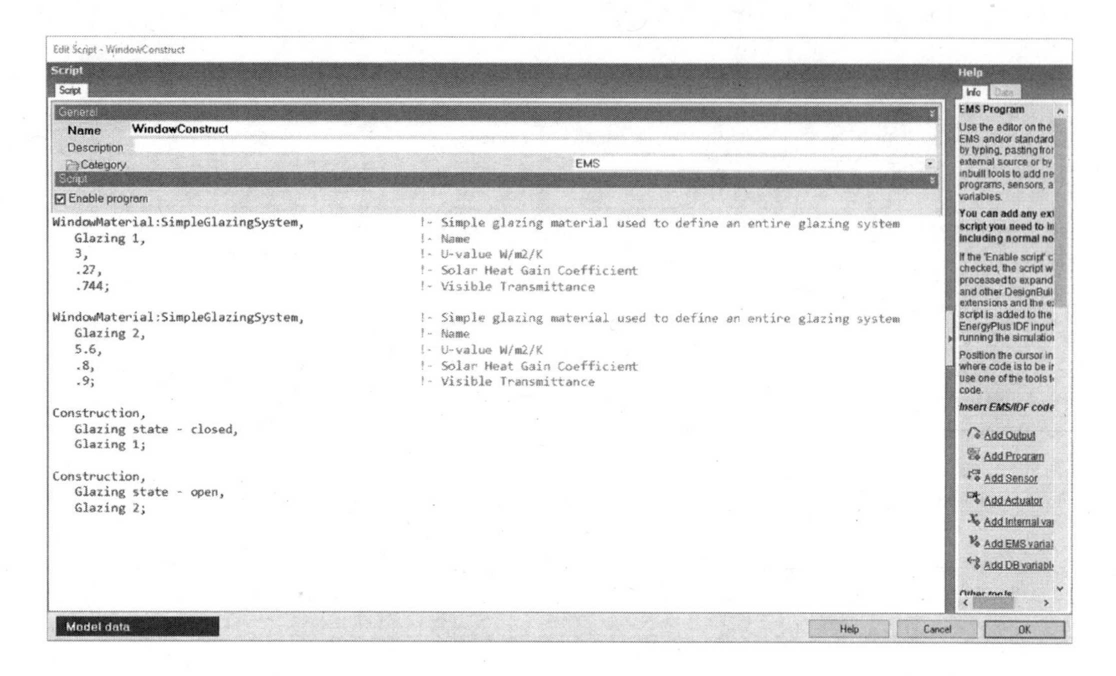

Step (k): Click the **Simulation screen** tab. The **Edit Calculation Options** screen appears.

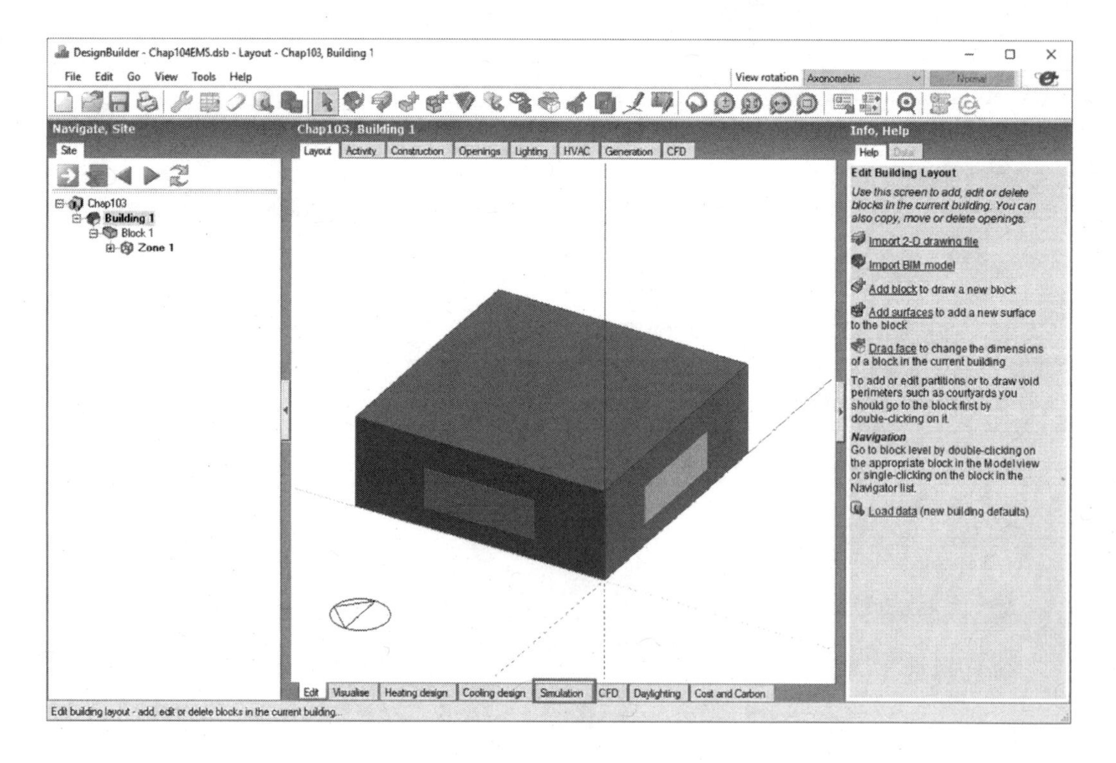

Step (l): Click the **Options** tab. Select **2-Timestep frequency** from the **Shadow calculation method** drop-down list. Click **OK**.

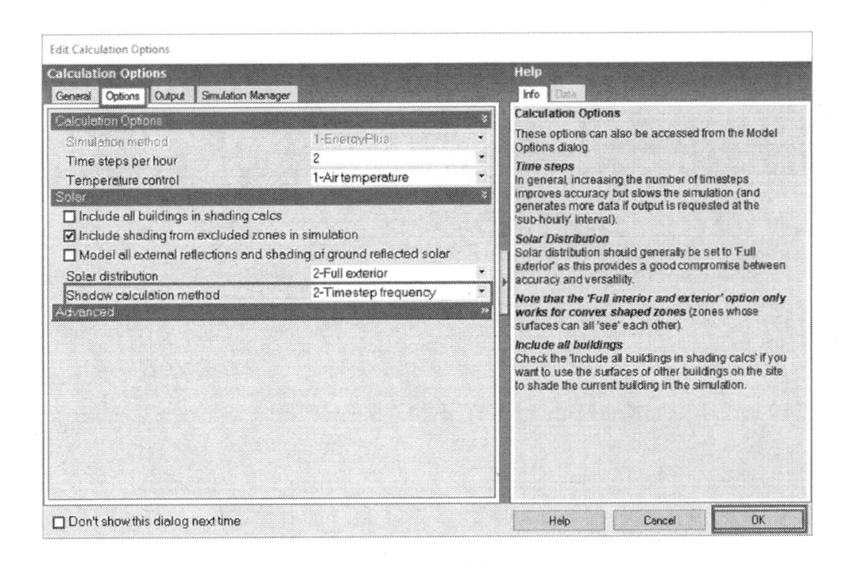

Step (m): Record the model results.

Tutorial 10.5

Evaluating the Impact of Window Opening Area Modulation on Natural Ventilation

GOAL

To evaluate the impact of window opening area modulation based on zone and outdoor air temperature

WHAT ARE YOU GOING TO LEARN?

- How to define window opening area modulation

PROBLEM STATEMENT

In this tutorial, you are going to use the model saved in Tutorial 10.4. You are going to simulate the model for **Temperature control mode**. Use the **New Delhi/Safdarjung, India** weather location.

The modulation takes the following form when **Ventilation Control Mode = Temperature**:

$T_{zone} - T_{out}$ = (lower value on inside/outside temperature difference for modulating the venting open factor)

Then the multiplication factor = 1.0.

(Lower value on inside/outside temperature difference for modulating the venting open factor) < $T_{zone} - T_{out}$ < (upper value on inside/outside temperature difference for modulating the venting open factor)

Then the multiplication factor varies linearly from 1.0 to (limit value on multiplier for modulating the venting open factor).

$T_{zone} - T_{out}$ = (upper value on inside/outside temperature difference for modulating the venting open factor)

Then the multiplication factor = (limit value on multiplier for modulating the venting open factor).

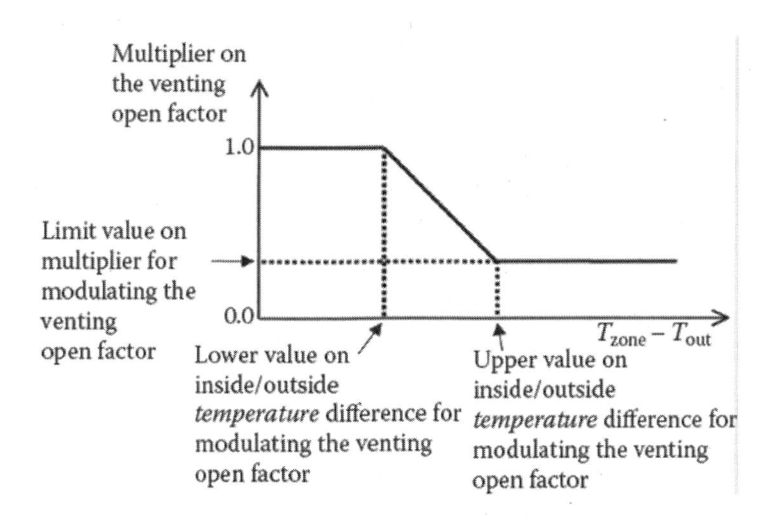

Source: U.S. Department of Energy, *EnergyPlus Version 8.9.0 Documentation, Input-Output Reference*, March 2018; available at https://energyplus.net/documentation.

SOLUTION

Step 1: Open the project saved in Tutorial 10.4.

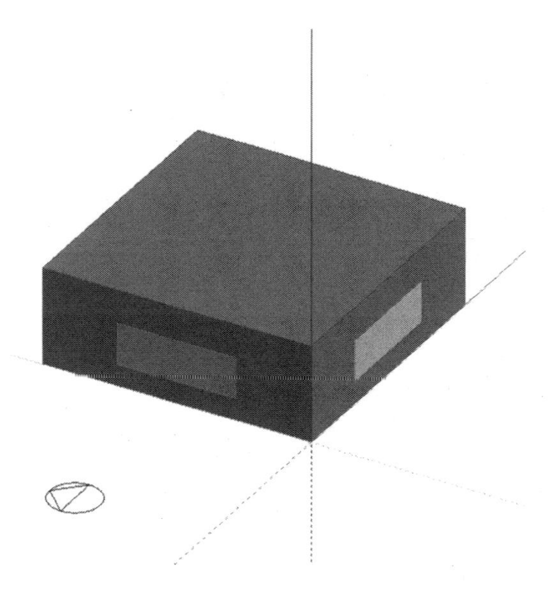

Step 2: Select the **HVAC** tab. Expand the **Natural Ventilation** section.

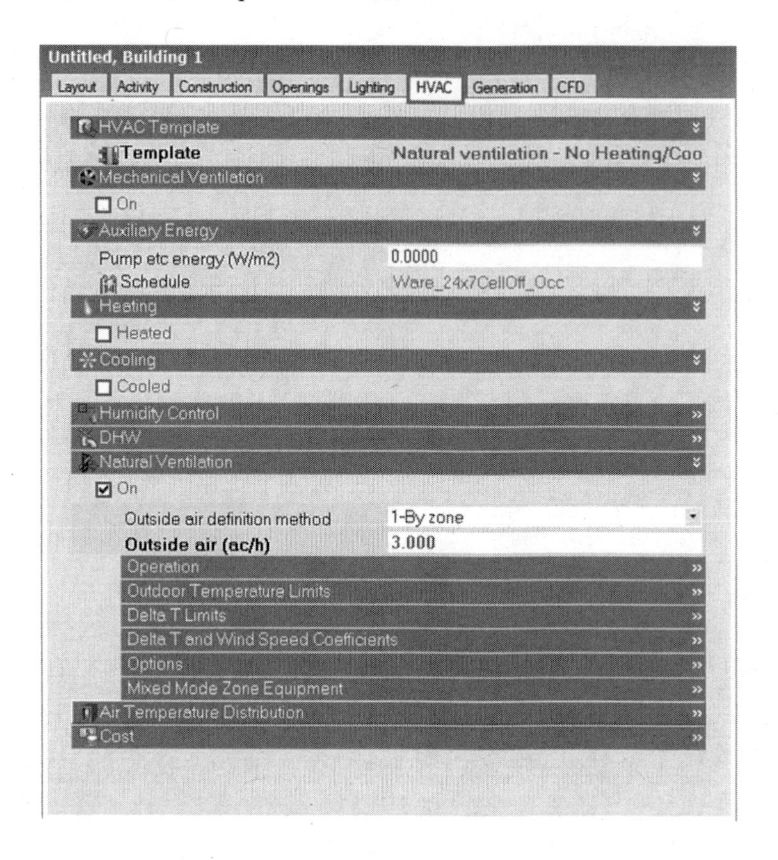

Step 3: Select the **Modulate opening areas** checkbox. Enter the lower value of $T_{in} - T_{out}$ as **5.0** and the upper value as **15**.

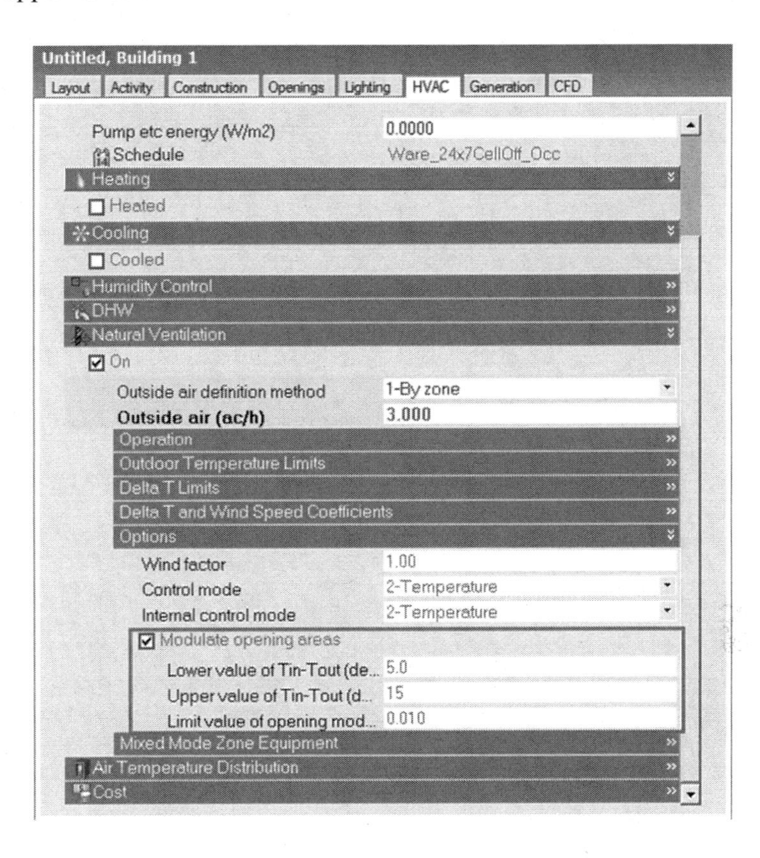

Step 4: Minimize the **DesignBuilder** window. Create an **output.idf** file. Write additional output variables:

Output: Variable, *, Zone Air Temperature, Detailed;
Output:Variable, *, AFN Surface Venting Window or Door Opening Factor, Detailed;
Output: Variable, *, AFN Zone Infiltration, Detailed;

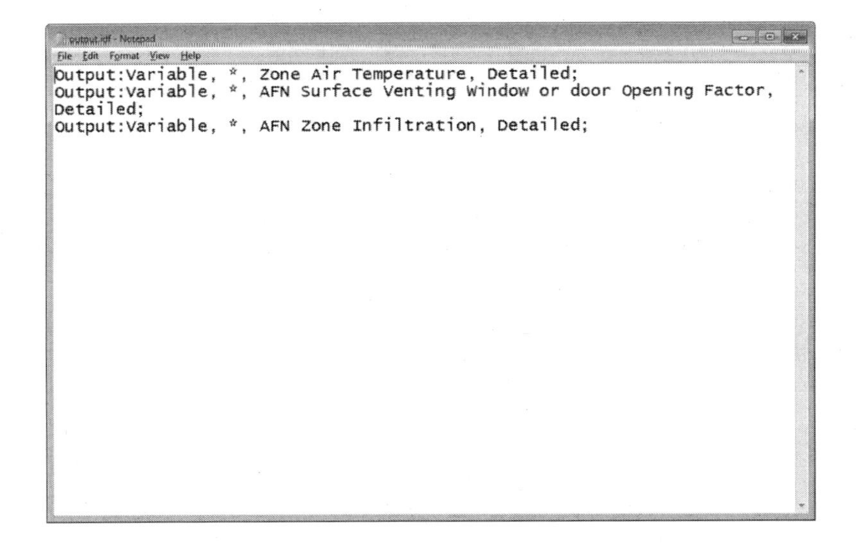

This .idf file needs to be placed in the path shown in the following figure.

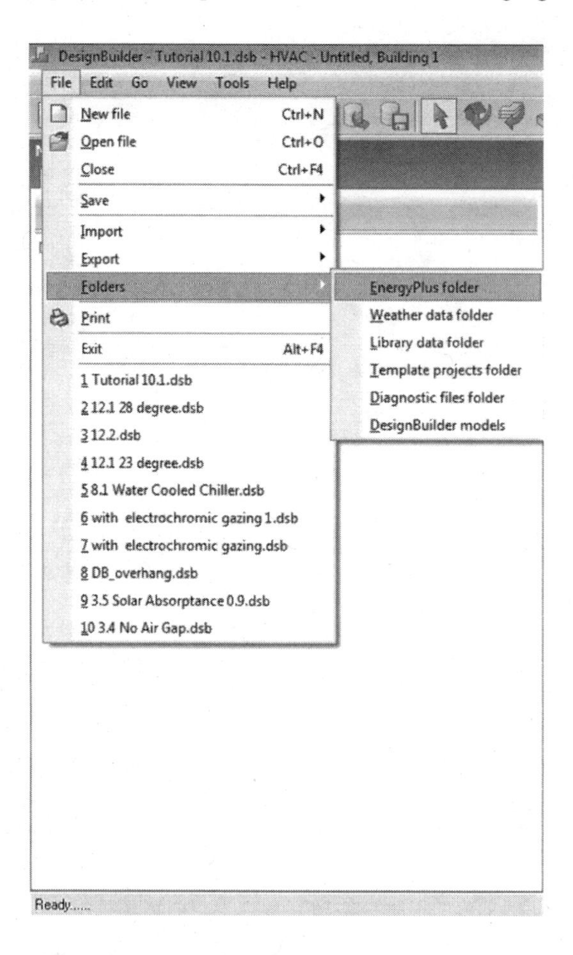

Step 5: Maximize the **DesignBuilder** window, and select the **HVAC** tab. Click **Simple** under the **Help** tab. The **Model Options – Building and Block** screen appears.

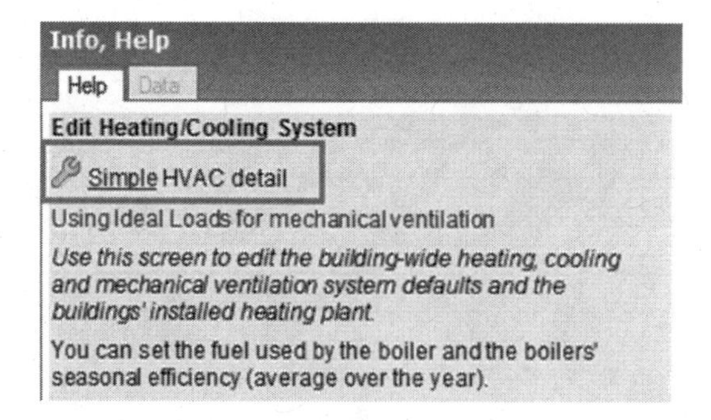

Step 6: Select the **Simulation** tab. Click the **Advanced** section.

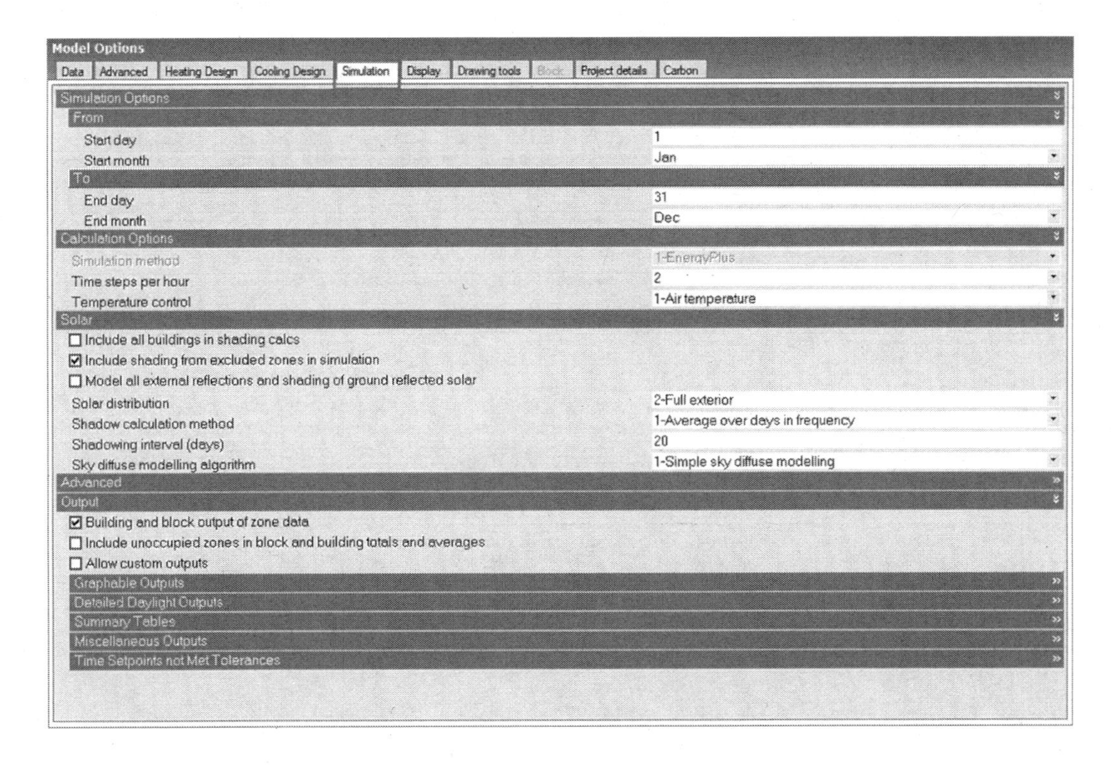

Step 7: Select the **IDF File 1** checkbox, and select the **Output.idf** file from the **EnergyPlus** folder. Click **OK**.

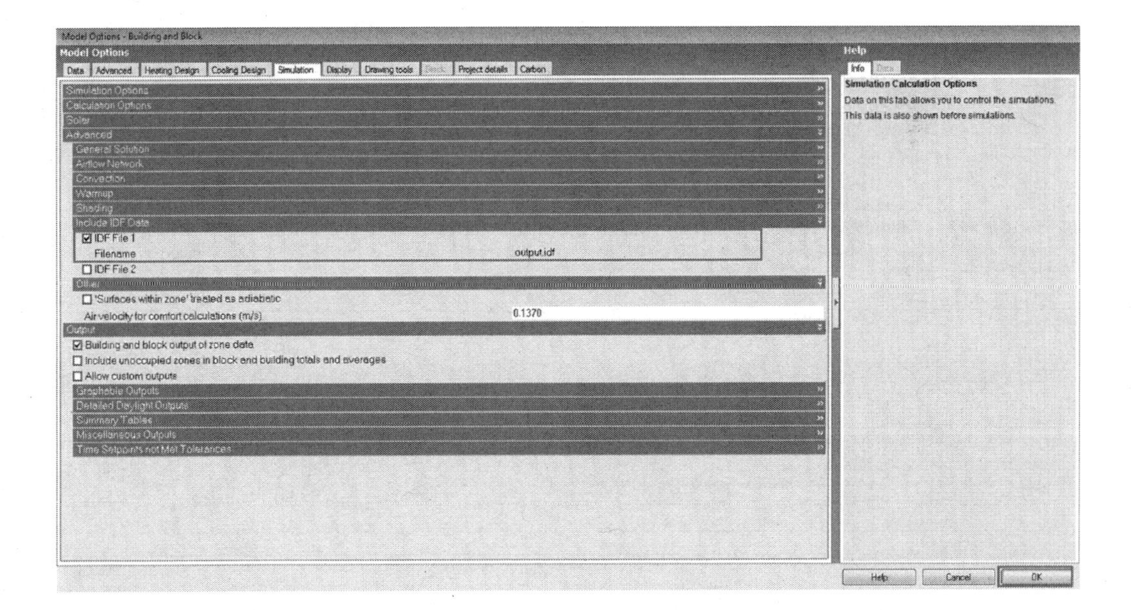

Step 8: Select the **Simulation** screen tab. The **Edit Calculation Options** screen appears. Click **OK**.

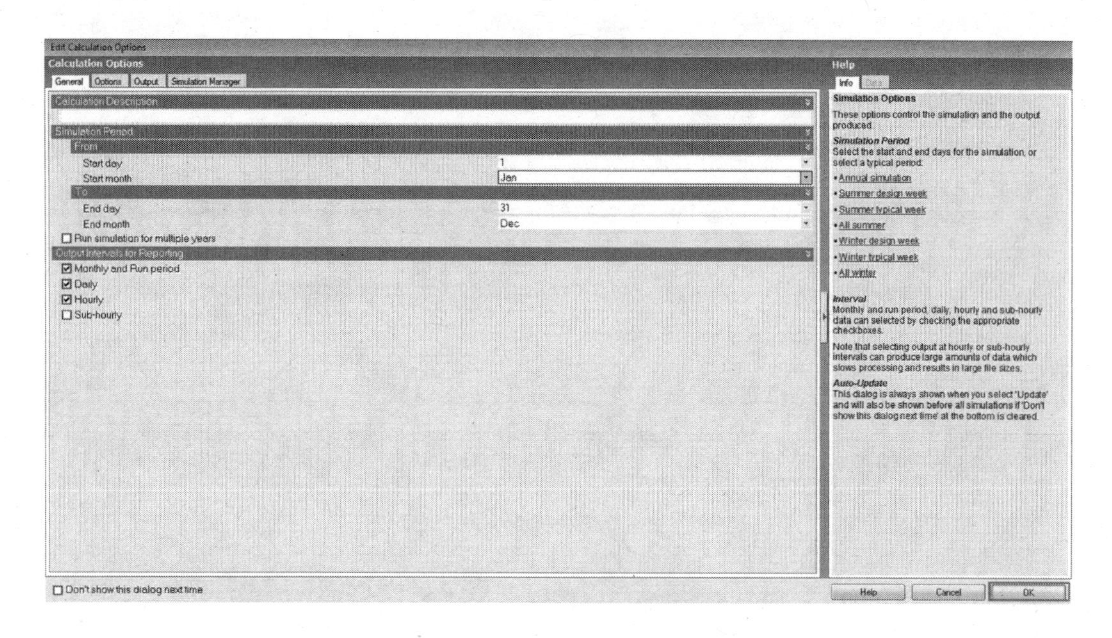

Step 9: Select the **EnergyPlus** results. Select the **eplusout** file so that the file will automatically open in the **xESOView** tool.

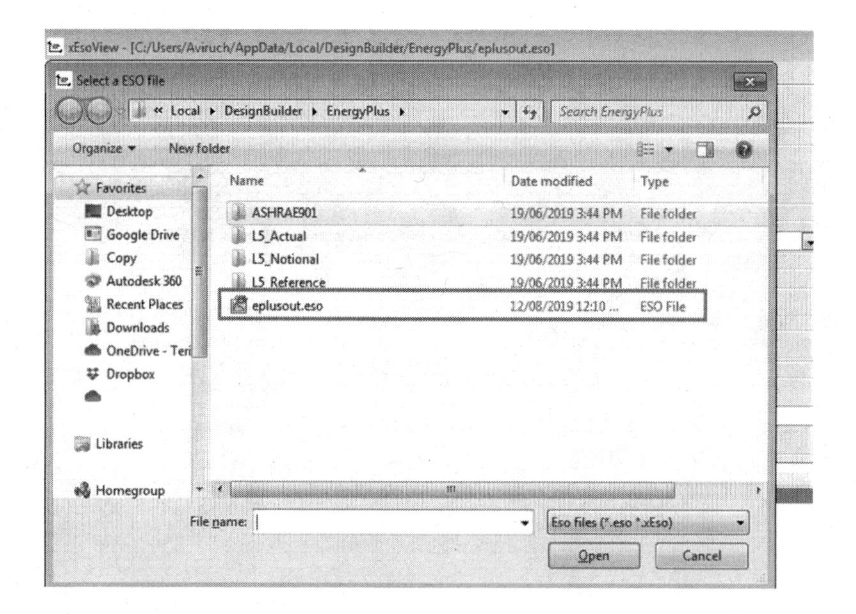

The **xEsoView** screen appears:

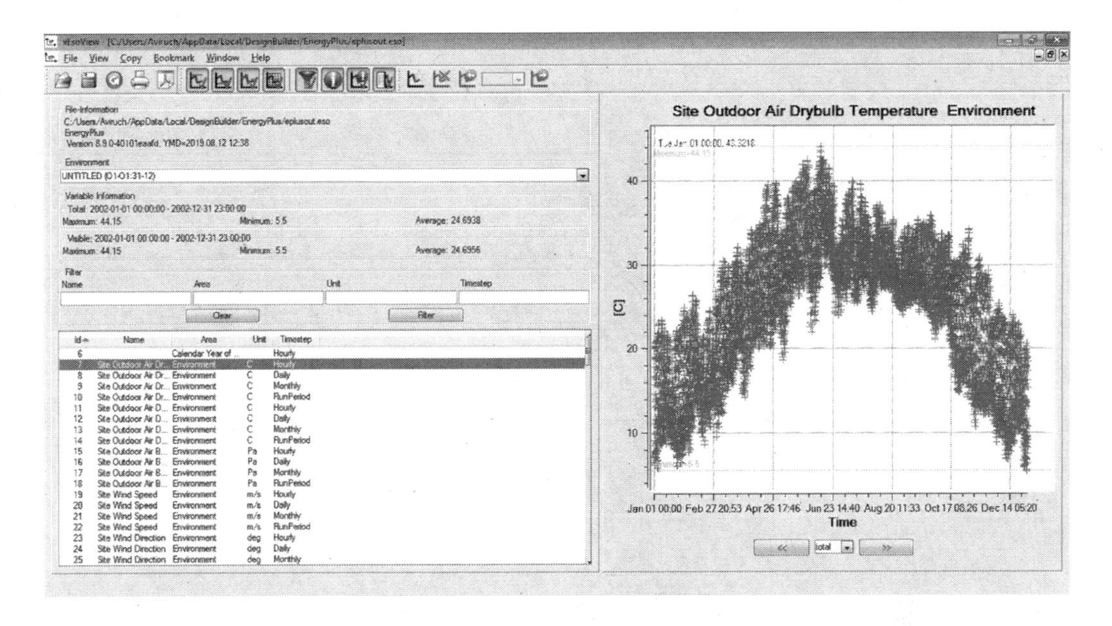

Step 10: Enter **Site Outdoor Air Drybulb Temperature** in the **Name** field of the **Filter** section to view the variable.

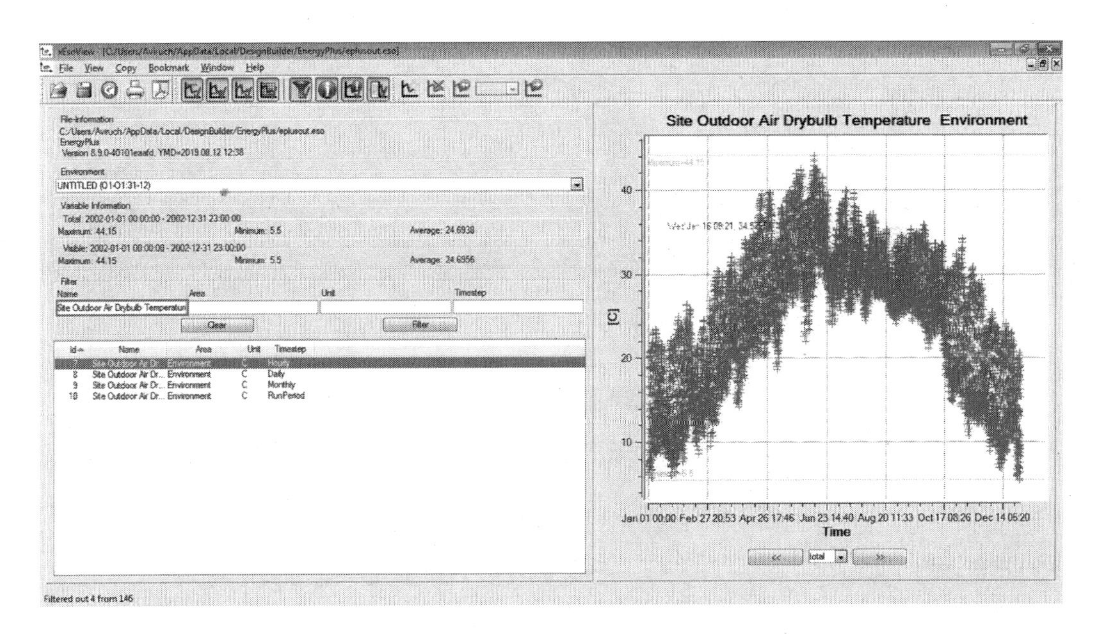

You can copy the variable data to view in a spreadsheet.

Step 11: Click **Copy**, and select **Copy variable data**.

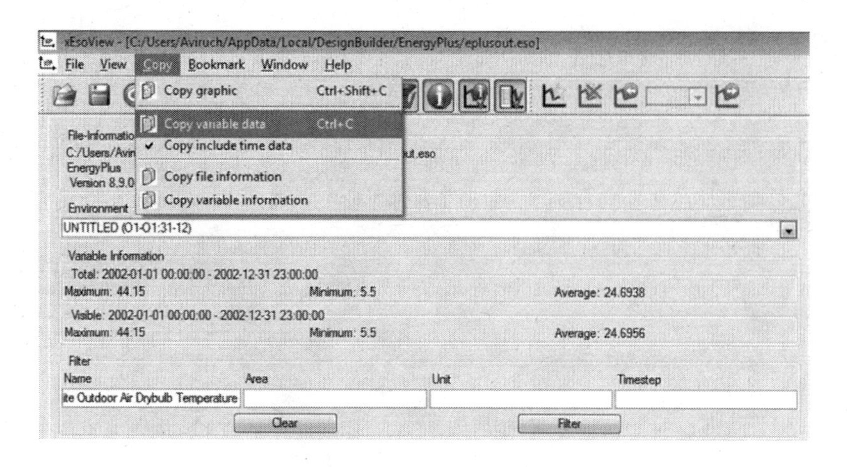

Step 12: Open a spreadsheet, and paste the data.

	A	B	C
1	Date/Time	Site Outdoor Air Drybulb Temperature Environment [C]	
65	01/01/2002 1:29		
66	01/01/2002 1:30		
67	01/01/2002 1:40		
68	01/01/2002 1:50		
69	01/01/2002 1:00	6.625	
70	01/01/2002 2:00		
71	01/01/2002 2:06		
72	01/01/2002 2:12		
73	01/01/2002 2:18		
74	01/01/2002 2:24		
75	01/01/2002 2:30		
76	01/01/2002 2:00	6.55	
77	01/01/2002 3:00		
78	01/01/2002 3:07		
79	01/01/2002 3:15		
80	01/01/2002 3:22		
81	01/01/2002 3:30		
82	01/01/2002 3:40		
83	01/01/2002 3:50		
84	01/01/2002 3:00	6.225	
85	01/01/2002 4:00		
86	01/01/2002 4:30		
87	01/01/2002 4:00	6.925	
88	01/01/2002 5:00		
89	01/01/2002 5:30		
90	01/01/2002 5:00	7.35	
91	01/01/2002 6:00		
92	01/01/2002 6:01		
93	01/01/2002 6:02		
94	01/01/2002 6:03		
95	01/01/2002 6:04		
96	01/01/2002 6:05		
97	01/01/2002 6:06		
98	01/01/2002 6:07		
99	01/01/2002 6:08		
100	01/01/2002 6:09		

You can check if natural ventilation is working as per the controls applied.

H173 =IF(AND(G173>20,G173>B173),TRUE,FALSE)

	A	B	C	D	E	F	G	H
1	Date/Time	Site Outdoor Air Drybulb Temperature Environment [C]	AFN Surface Venting Window or Door Opening Factor BLOCK1:MYZONE_WALL_2_0_0_0_0_WIN []	AFN Surface Venting Window or Door Opening Factor BLOCK1:MYZONE_WALL_3_0_0_0_0_WIN []	AFN Surface Venting Window or Door Opening Factor BLOCK1:MYZONE_WALL_4_0_0_0_0_WIN []	AFN Surface Venting Window or Door Opening Factor BLOCK1:MYZONE_WALL_5_0_0_0_0_WIN []	Zone Air Temperature BLOCK1: MYZONE [C]	Check if Tzone>20 and Tzone>Tout
70	01/01/2002 8:28		0	0	0	0	19.2566	FALSE
71	01/01/2002 8:30		0	0	0	0	18.8952	FALSE
72	01/01/2002 8:37		0	0	0	0	19.6496	FALSE
73	01/01/2002 8:45	11.25	0	0	0	0	20.3496	TRUE
74	01/01/2002 8:52	11.25	0.648556	0.648556	0.648556	0.648556	14.5185	FALSE
75	01/01/2002 8:00	11.25	-1.11E+06	-1.11E+06	-1.11E+06	-1.11E+06	-1.11E+06	FALSE
76	01/01/2002 9:00		0	0	0	0	18.8664	FALSE
77	01/01/2002 9:04		0	0	0	0	19.5796	FALSE
78	01/01/2002 9:08		0	0	0	0	20.1608	TRUE
79	01/01/2002 9:12		0.805851	0.805851	0.805851	0.805851	16.0044	FALSE
80	01/01/2002 9:17		0	0	0	0	14.7596	FALSE
81	01/01/2002 9:21		0	0	0	0	15.9434	FALSE
82	01/01/2002 9:25		0	0	0	0	17.7259	FALSE

You can see from the preceding that if the following conditions are true, then there is natural ventilation:

$$T_{\text{zone\_air}} > T_{\text{setpoint}} \quad \text{and} \quad T_{\text{zone\_air}} > T_{\text{outside\_air}}$$

and **Window opening fraction** depends on the $T_{\text{zone}} - T_{\text{out}}$.

Tutorial 10.6

Evaluating the Impact of Mixed-Mode Operation

GOAL

To evaluate the impact of mixed-mode operation on energy performance

WHAT ARE YOU GOING TO LEARN?

- How to model mixed-mode ventilation

PROBLEM STATEMENT

In this tutorial, you are going to use the model saved in Tutorial 10.4. Determine the energy consumption for the following two cases:

1. With mixed mode
2. Fully air conditioned

Use the **New Delhi/Safdarjung – India** weather location.

Mixed-Mode Cooling

In mixed-mode buildings, natural ventilation is used as the primary means of providing cooling, and active cooling is introduced when this is inadequate to provide comfort conditions.

Mixed-Mode Building

Mixed mode refers to a hybrid approach to space conditioning that uses a combination of natural ventilation from operable windows (either manually or automatically controlled) and mechanical systems that include air distribution equipment and refrigeration equipment for cooling. A well-designed mixed-mode building begins with an intelligent facade design to minimize cooling loads. It then integrates the use of air conditioning when and where necessary, with the use of natural ventilation whenever it is feasible or desirable, to maximize comfort while avoiding the significant energy use and operating costs of year-round air conditioning.

Source: http://www.cbe.berkeley.edu/mixedmode.

SOLUTION

Step 1: Open the file saved in Tutorial 10.4.

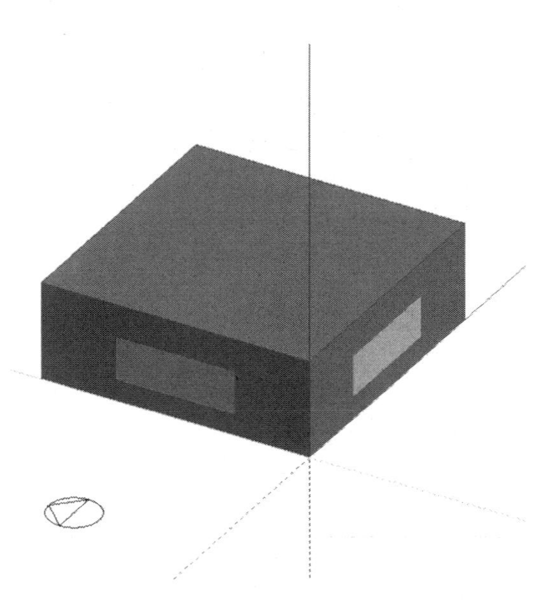

Step 2: Select the **HVAC** tab. Select the **HVAC template** as **Radiator heating, Boiler HW, Mixed mode Nat Vent, Local comfort cooling**.

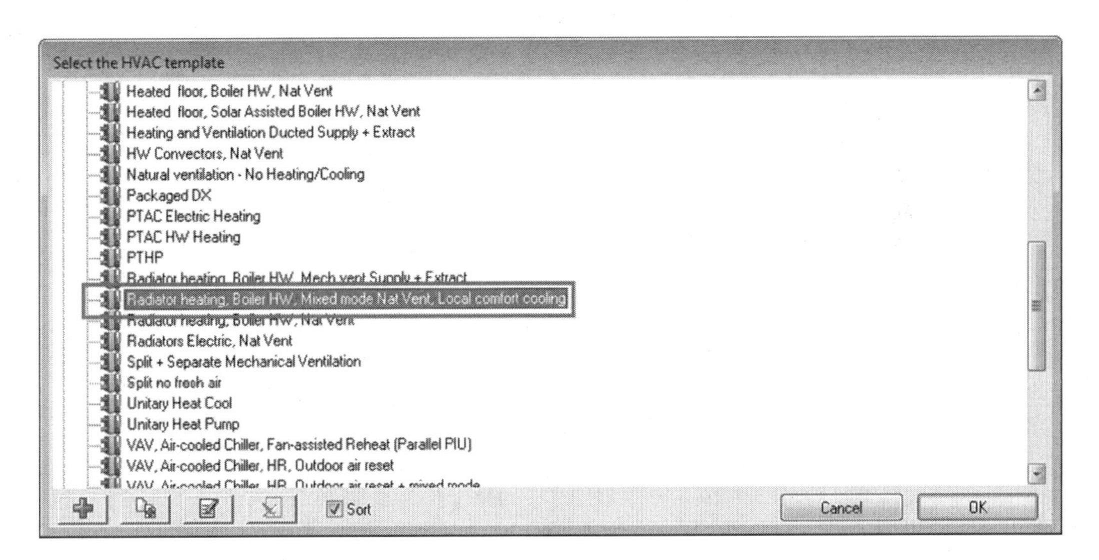

Step 3: Clear the **Heated** checkbox in the **Heating** section, and make sure that **Outside air (ac/h)** is **3.000**.

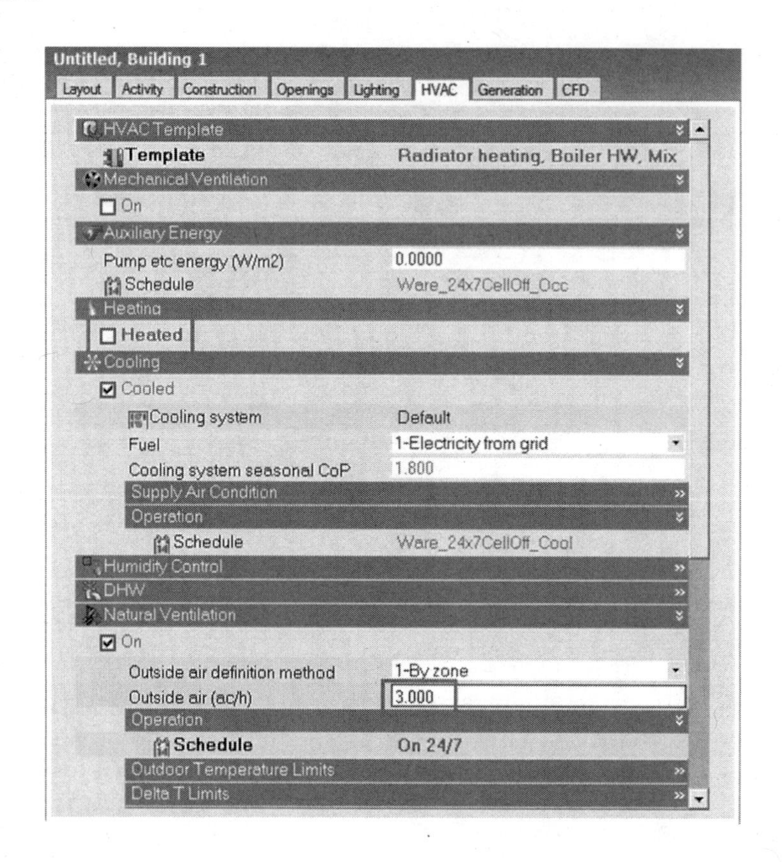

Step 4: Ensure that the **Mixed mode on** checkbox is selected.

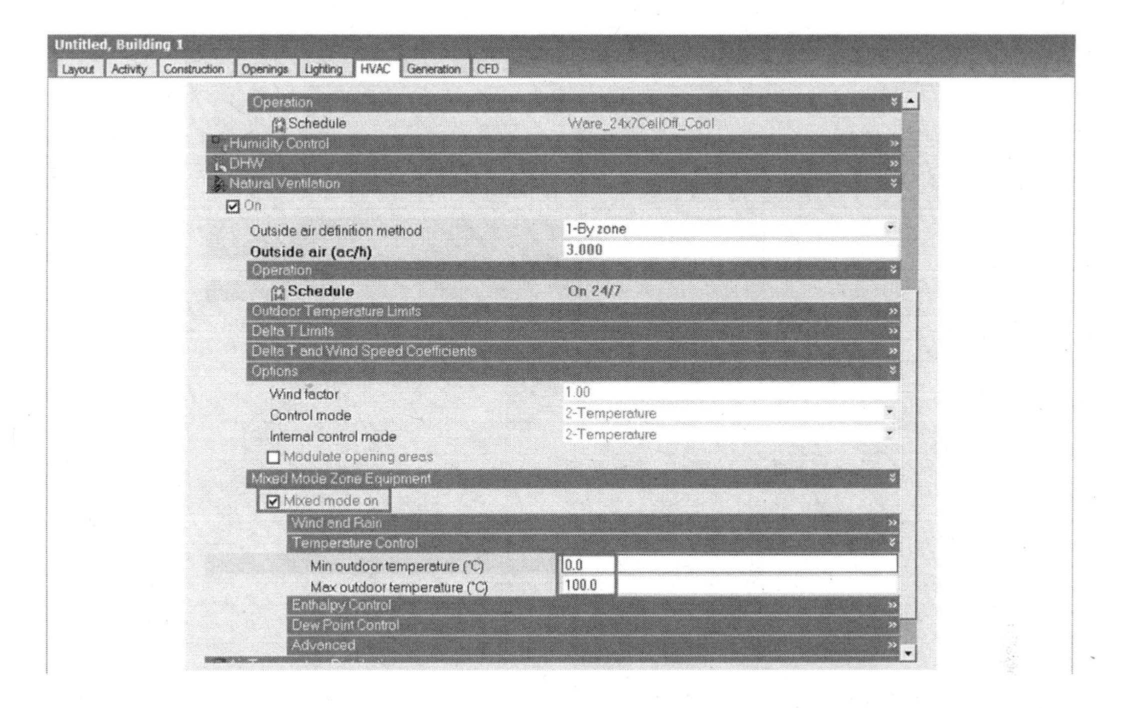

Step 5: Minimize the **DesignBuilder** window. Create an **output.idf** file. Write additional output variables as follows:

Output: Variable, *, Zone Air Temperature, Detailed;
Output: Variable, *, AFN Surface Venting Window or Door Opening Factor, Detailed;
Output: Variable, *, AFN Zone Infiltration Air Change Rate, Detailed;
Output: Variable, *, Zone Ideal Loads Supply Air Total Cooling Rate, Detailed;

This .idf file needs to be placed in the path shown in the following figures.

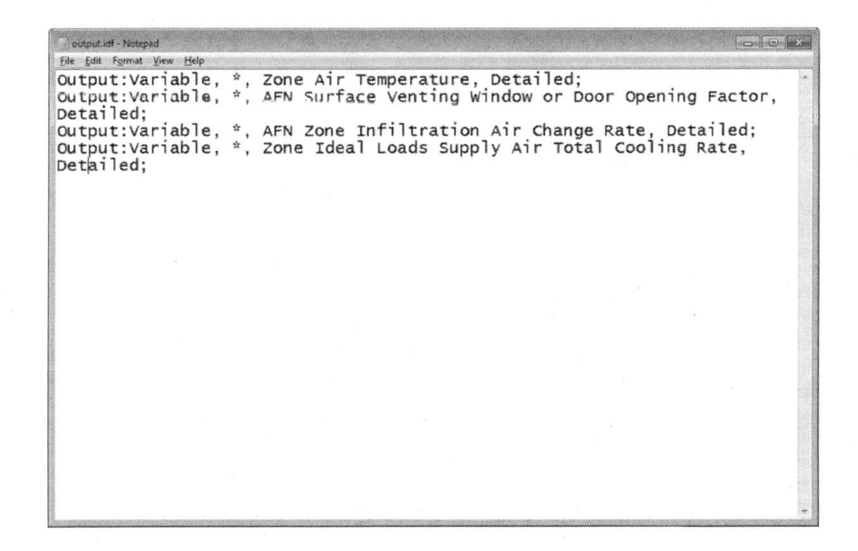


```
Output:Variable, *, Zone Air Temperature, Detailed;
Output:Variable, *, AFN Surface Venting Window or Door Opening Factor,
Detailed;
Output:Variable, *, AFN Zone Infiltration Air Change Rate, Detailed;
Output:Variable, *, Zone Ideal Loads Supply Air Total Cooling Rate,
Detailed;
```

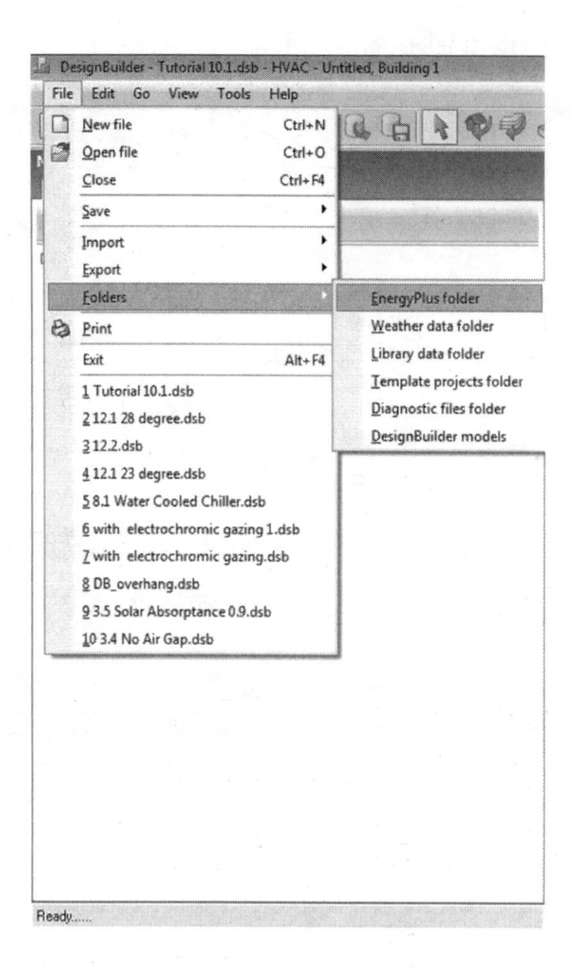

Step 6: Maximize the **DesignBuilder** window, and select the **HVAC** tab. Click **Simple** under the **Help** tab. The **Model Options – Building and Block** screen appears.

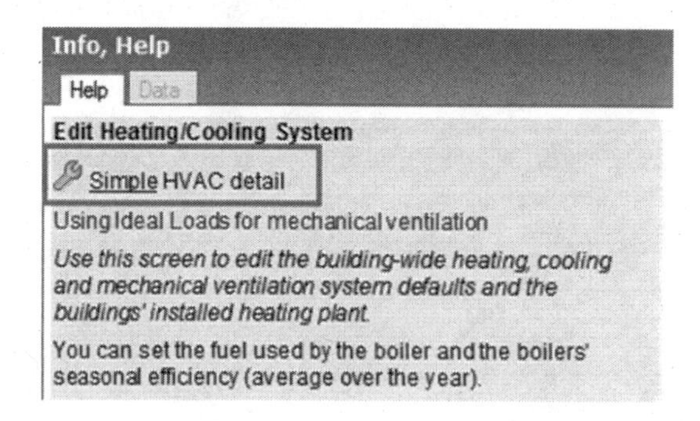

Step 7: Select the **Simulation** tab. Click the **Advanced** section.

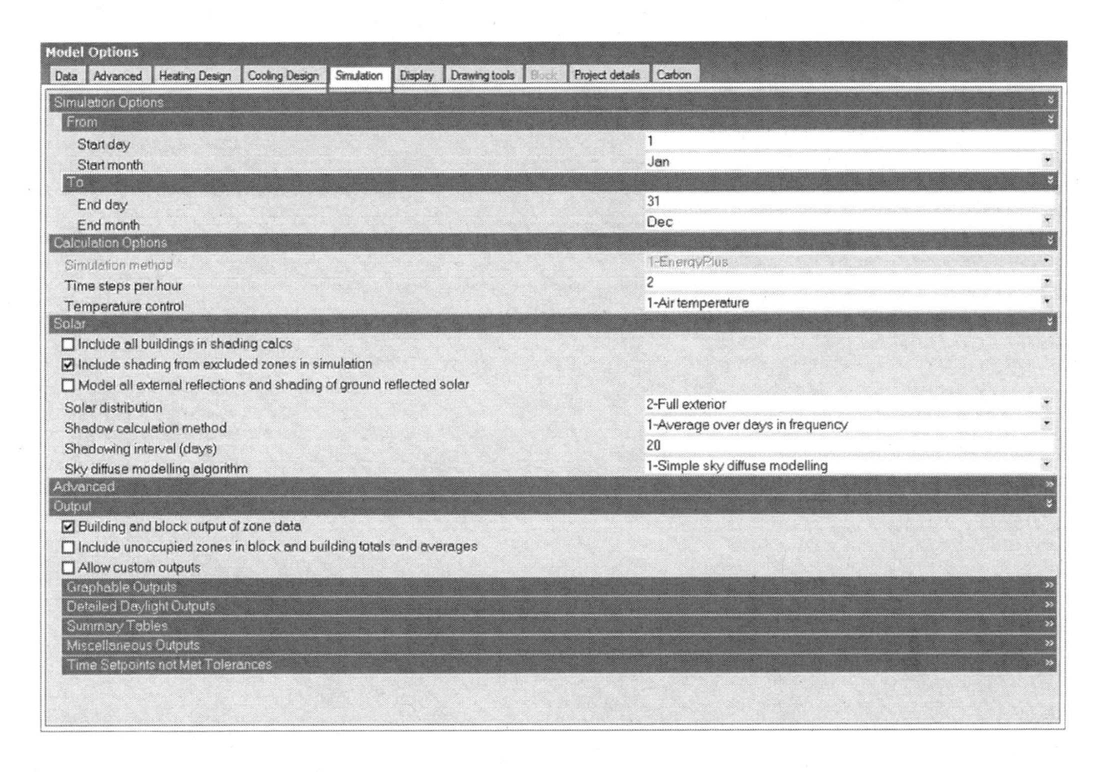

Step 8: Select the **IDF File 1** checkbox, and select the **output.idf** file from the **EnergyPlus** folder. Click **OK**.

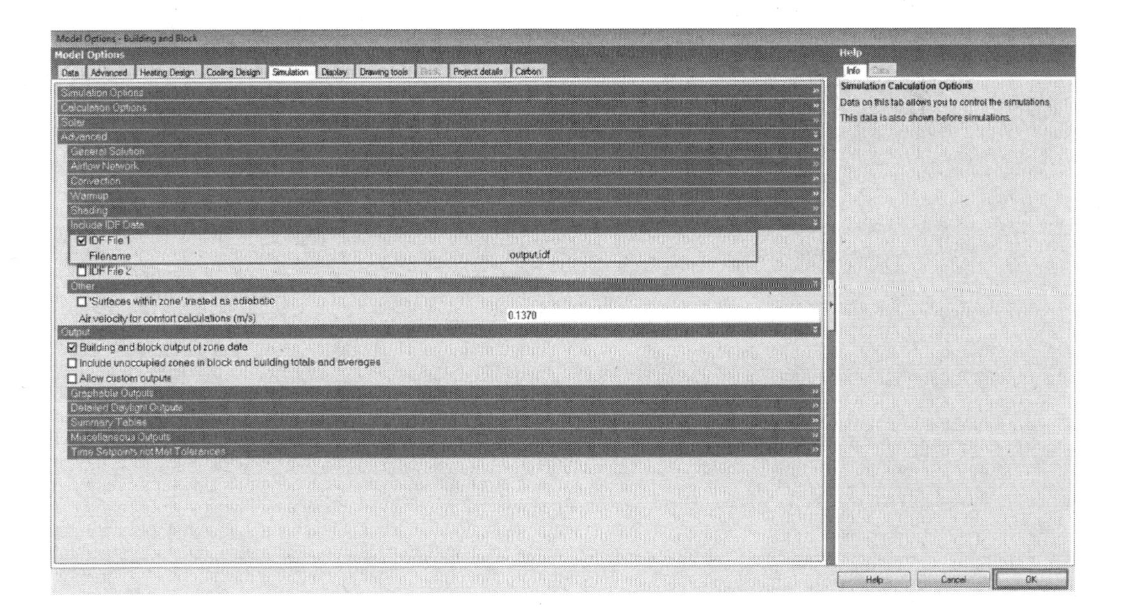

Step 9: Select the **Simulation** screen tab. The **Edit Calculation Options** screen appears. Click **OK**.

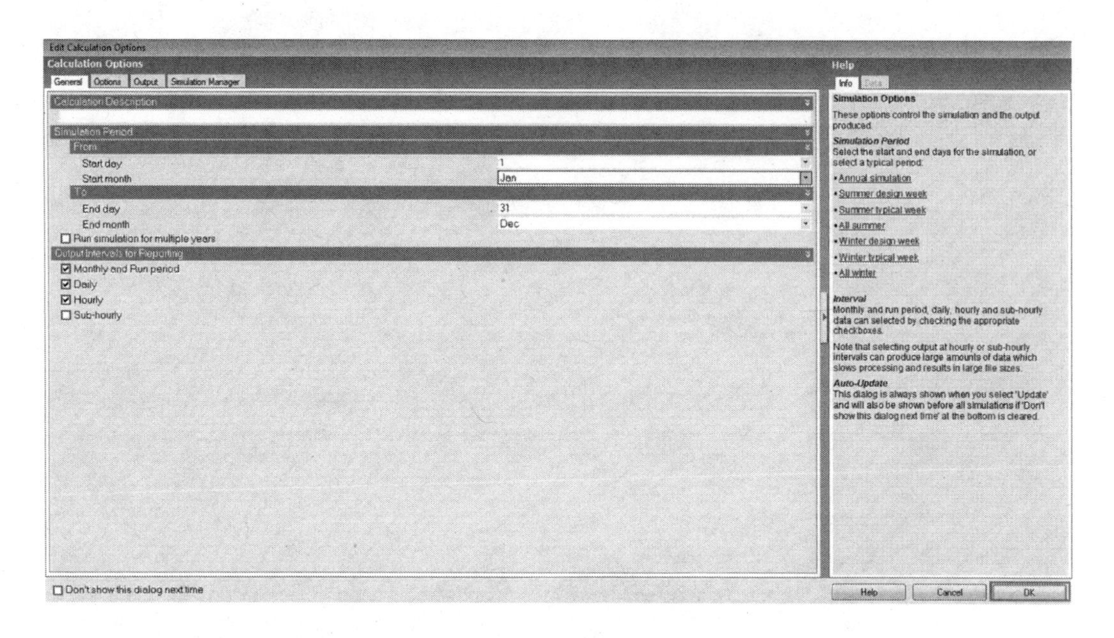

Step 10: Select the **EnergyPlus** results. Select the **eplusout** file so that the file will automatically open in the **xESOView** tool.

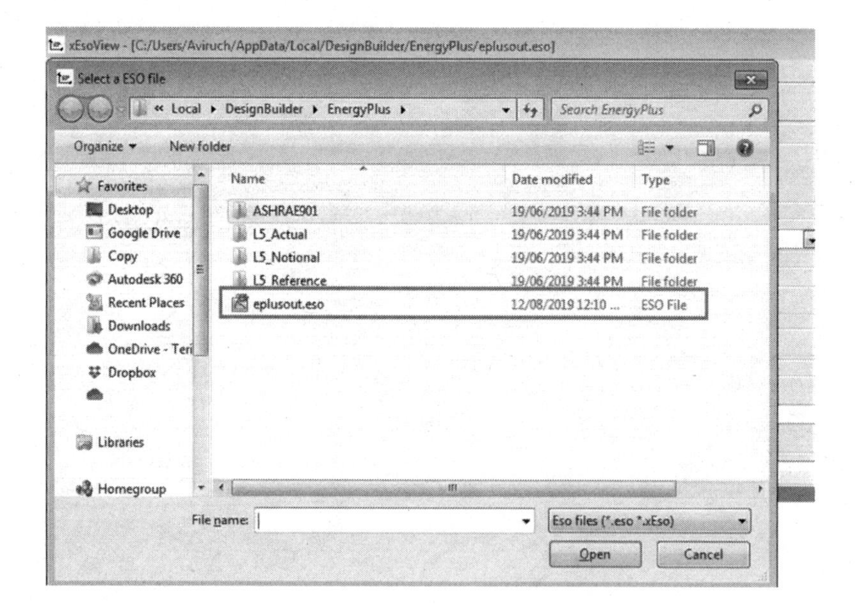

The **xEsoView** screen appears:

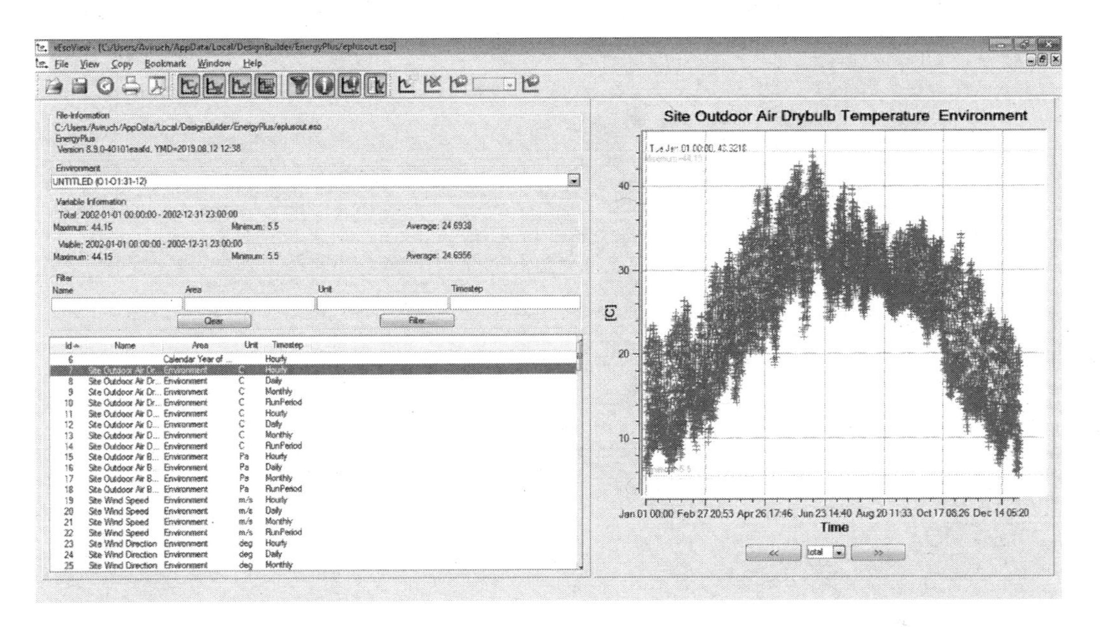

Step 11: Enter **Site Outdoor Air Drybulb Temperature** in the **Name** field of the **Filter** section to view the variable.

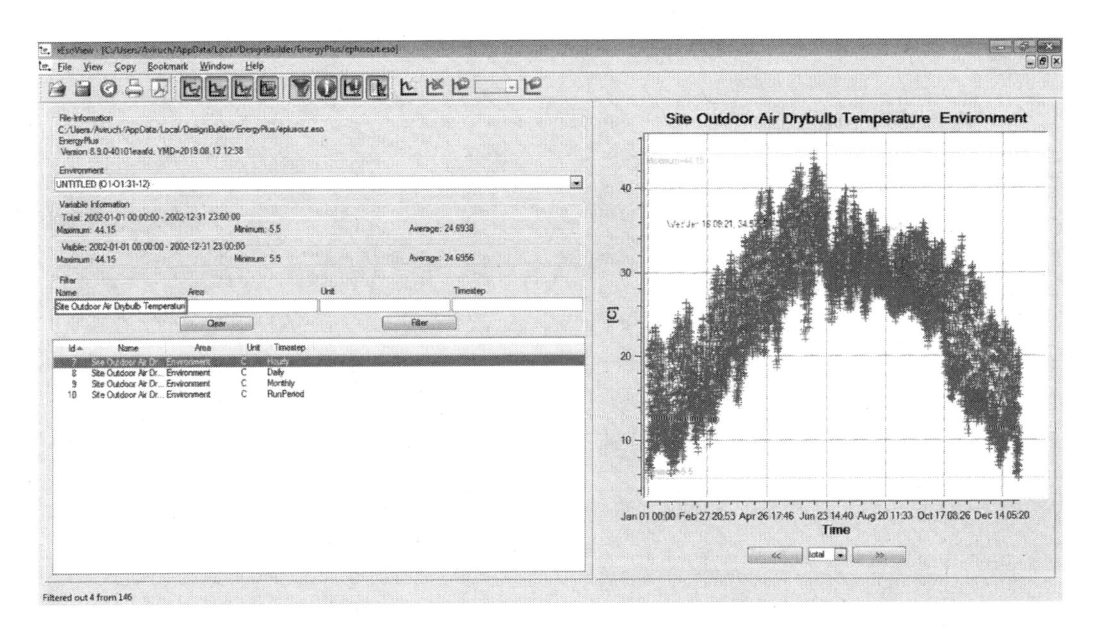

You can copy the variable data to view it in a spreadsheet.

Step 12: Click **Copy**; then select **Copy variable data**.

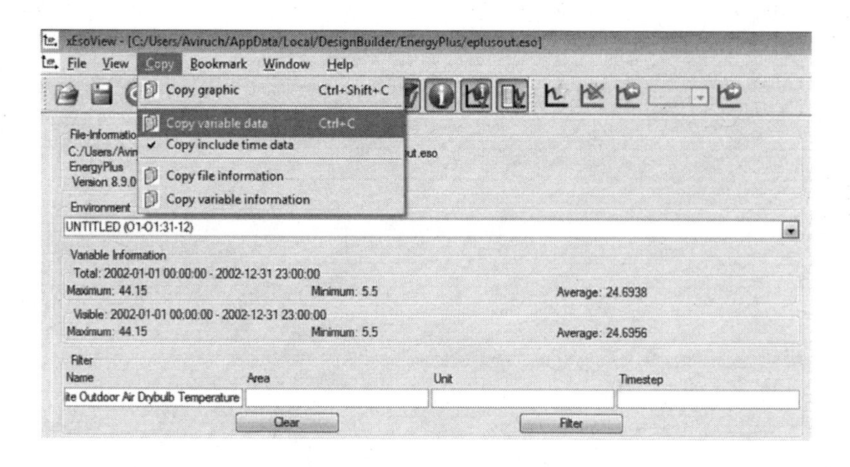

Step 13: Open a spreadsheet, and paste the data.

Date/Time	Site Outdoor Air Drybulb Temperature Environment [C]	AFN Surface Venting Window or Door Opening Factor BLOCK1:MYZONE_WALL_2_0_0_0_0_WIN []	AFN Surface Venting Window or Door Opening Factor BLOCK1:MYZONE_WALL_3_0_0_0_0_WIN []	AFN Surface Venting Window or Door Opening Factor BLOCK1:MYZONE_WALL_4_0_0_0_0_WIN []	AFN Surface Venting Window or Door Opening Factor BLOCK1:MYZONE_WALL_5_0_0_0_0_WIN []	AFN Zone Infiltration Air Change Rate BLOCK1:MYZONE [ach]	Zone Air Temperature BLOCK1: MYZONE [C]	Zone Ideal Loads Supply Air Total Cooling Rate BLOCK1:MYZONE IDEAL LOADS AIR [W]
17/11/2002 9:46		1.00	1.00	1.00	1.00		23.3292	
17/11/2002 9:49		1.00	1.00	1.00	1.00		23.3389	
17/11/2002 9:51		1.00	1.00	1.00	1.00		23.3397	
17/11/2002 9:54		1.00	1.00	1.00	1.00		23.3377	
17/11/2002 9:57		1.00	1.00	1.00	1.00		23.3367	
17/11/2002 9:00	22.23					31.00		0.00
17/11/2002 10:00		0.00	0.00	0.00	0.00		23.6863	
17/11/2002 10:03		0.00	0.00	0.00	0.00		24.3996	
17/11/2002 10:06		0.00	0.00	0.00	0.00		24.0324	
17/11/2002 10:09		0.00	0.00	0.00	0.00		24.0003	
17/11/2002 10:12		0.00	0.00	0.00	0.00		24	
17/11/2002 10:15		0.00	0.00	0.00	0.00		24	
17/11/2002 10:18		0.00	0.00	0.00	0.00		24	
17/11/2002 10:21		0.00	0.00	0.00	0.00		24	
17/11/2002 10:24		0.00	0.00	0.00	0.00		24	
17/11/2002 10:27		0.00	0.00	0.00	0.00		24	
17/11/2002 10:30		0.00	0.00	0.00	0.00		24	
17/11/2002 10:00	24.93					0.00		1918.43
17/11/2002 11:00		0.00	0.00	0.00	0.00		24	
17/11/2002 11:30		0.00	0.00	0.00	0.00		24	
17/11/2002 11:00	26.65					0.00		2236.20
17/11/2002 12:00		0.00	0.00	0.00	0.00		24	
17/11/2002 12:30		0.00	0.00	0.00	0.00		24	
17/11/2002 12:00	27.83					0.00		2618.66
17/11/2002 13:00		0.00	0.00	0.00	0.00		24	
17/11/2002 13:30		0.00	0.00	0.00	0.00		24	
17/11/2002 13:33		0.00	0.00	0.00	0.00		24.6052	
17/11/2002 13:36		0.00	0.00	0.00	0.00		24.214	
17/11/2002 13:39		0.00	0.00	0.00	0.00		24.0081	
17/11/2002 13:42		0.00	0.00	0.00	0.00		24.0004	
17/11/2002 13:45		0.00	0.00	0.00	0.00		24	
17/11/2002 13:48		0.00	0.00	0.00	0.00		24	
17/11/2002 13:51		0.00	0.00	0.00	0.00		24	

You can see from the preceding that if the following conditions are true, then there is natural ventilation:

$$T_{\text{zone\_air}} > T_{\text{setpoint}} \quad \text{and} \quad T_{\text{zone\_air}} > T_{\text{outside\_air}}$$

When the outside temperature is more than the zone temperature, natural ventilation in OFF, and mechanical cooling takes place.

Step 14: Clear the **On** checkbox to switch off the natural ventilation so that the zone runs in the **Air condition** mode.

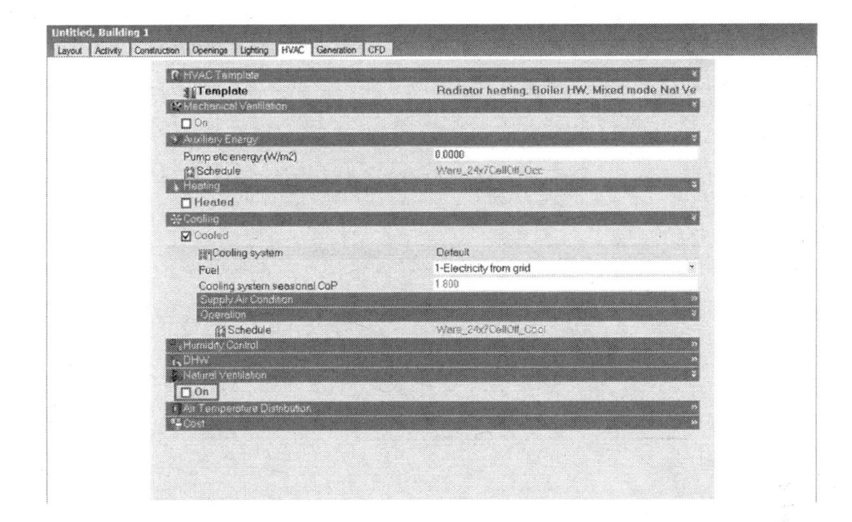

Simulate the model, and record the monthly energy consumption results. Compare the results in both cases.

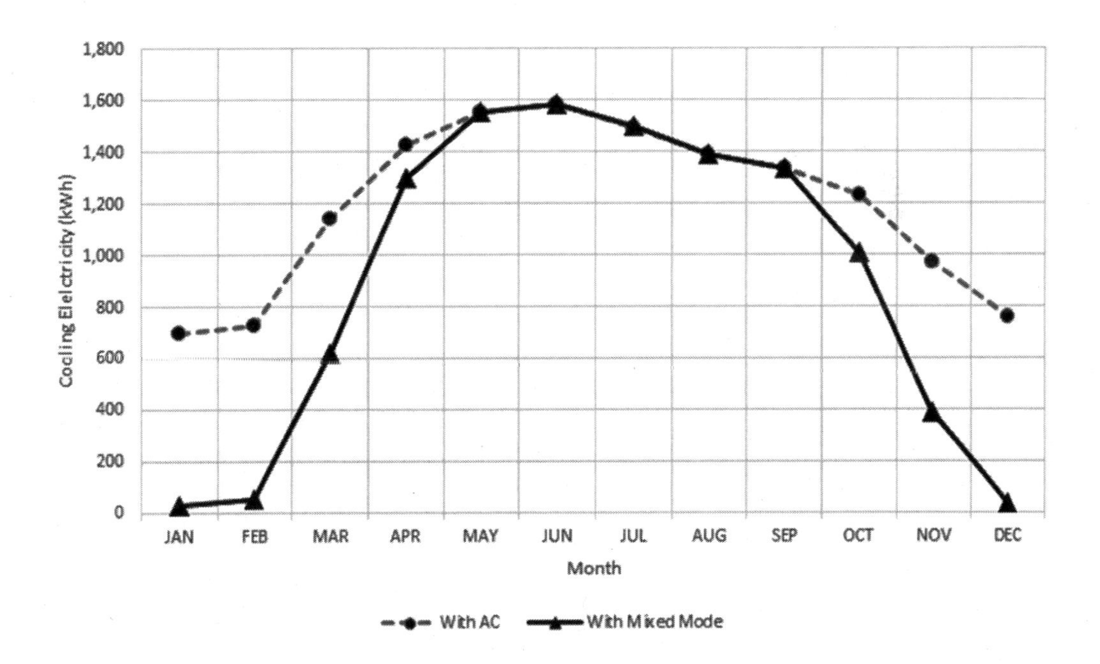

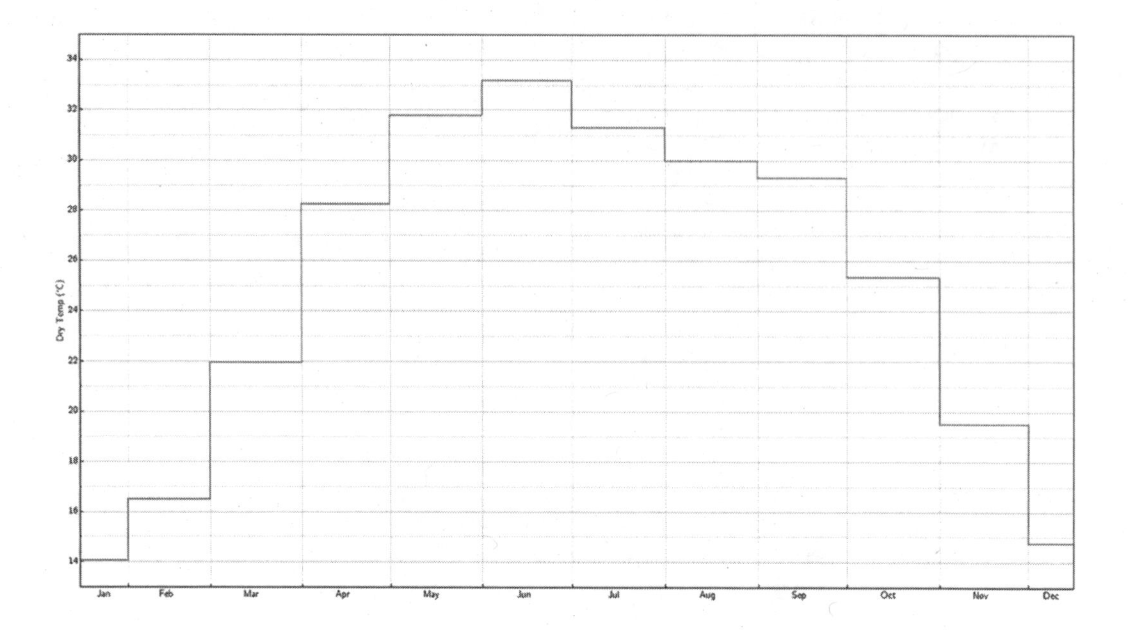

You can see that natural ventilation is effective in climate of New Delhi from January to March and October to December.

EXERCISE 10.6

Repeat this tutorial for window area modulation based on zone and outdoor air temperature.

11 Simulation Parameters

This chapter will help you to understand the nuances of simulation engine settings that not only affect the accuracy of calculations but also affect the run time of a model. This becomes very important especially for large building models. Three concepts covered in this chapter are time step, which may be treated analogous to the least count of the model; method of calculation for energy balance; and the algorithm for convective heat transfer in various building components. Simulation tools offer freedom to choose a smaller time step at the cost of a significant increase in run time. Similarly, the calculation method and convection algorithm are also associated with the accuracy of calculation at the cost of run time. These tutorials help in understanding the methods as well as the extent of difference that is obtained in the results when using different approaches. This information can be useful to the simulator to decide the appropriate simulation setting as per the availability of computing power, time and requirement of accuracy.

Tutorial 11.1

Evaluating the Impact of Time Steps per Hour on Run Time

GOAL

To evaluate the change in simulation run time with a change in time steps per hour

WHAT ARE YOU GOING TO LEARN?

- How to change time steps per hour

PROBLEM STATEMENT

In this tutorial, you are going to use the water-cooled chiller model saved in Tutorial 8.1 (a 50- × 25-m six-level model with a 5-m perimeter depth). You are going to use the following time steps per hour: 2, 10, 30 and 60. Determine the change in energy consumption and run time for all cases. Use the **Brisbane Aero, Australia** weather location.

Simulation time steps define the interval at which the heat transfer calculations are performed. In EnergyPlus (which is the simulation engine of DesignBuilder), this minimum time step is one, which means that the heat transfer and load calculation are performed on an hourly basis. The maximum number of time steps that can be assigned is 60, which means that the calculations are performed for every minute. The allowed options for time steps are 1–6, 10, 15, 20, 30 and 60. The higher the number of time steps, the more precise are the results.

Source: http://www.designbuilder.co.uk/helpv6.0/Content/Calculation_Options.htm.

Caution: Note the difference between simulation time steps, simulation period (also called *run period*) and run time. A run period is the time of the year for which the calculation should be performed, whereas a time step is the frequency at which these calculations are performed. Further, run time is the time taken for performing energy simulation. Run time depends on several factors, such as the complexity of the model, the speed of the computer hardware, the run period and the time step.

SOLUTION

Step 1: Open the simulation model saved in Tutorial 8.1. Select the **Simulation** screen tab. The **Edit Calculation Options** screen appears.

Step 2: Select the **Options** tab.

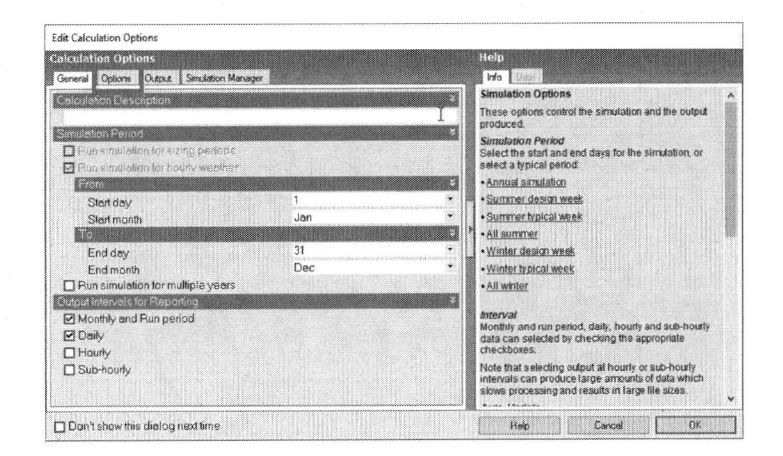

Step 3: Select **2** from the **Time steps per hour** drop-down list, and click **OK**.

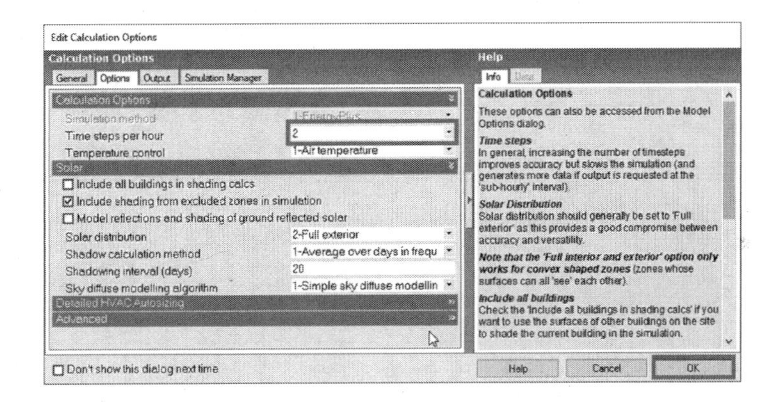

Step 4: Perform an annual simulation, and record the results for energy and run time.

How to Record Run Time

After the simulation is complete, open the **eplusout.err** file from the **EnergyPlus** folder. You can use any text editor to view this file.

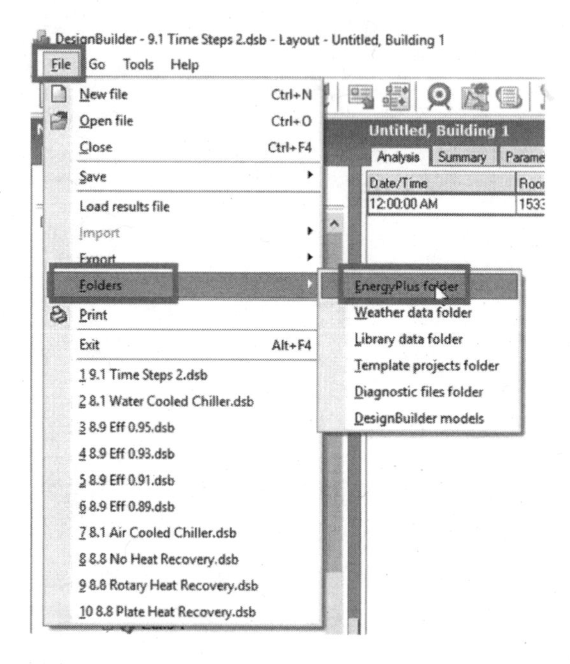

At the end of the file, you can find **Elapsed Time**. You need to record the elapsed time.

Source: www.designbuilder.co.uk/helpv6.0/Content/_DesignBuilder_files_location_and_extensions.htm.

Step 5: Repeat the preceding steps for the time steps 10, 30 and 60. Compare the run times for all the cases (Table 11.1).

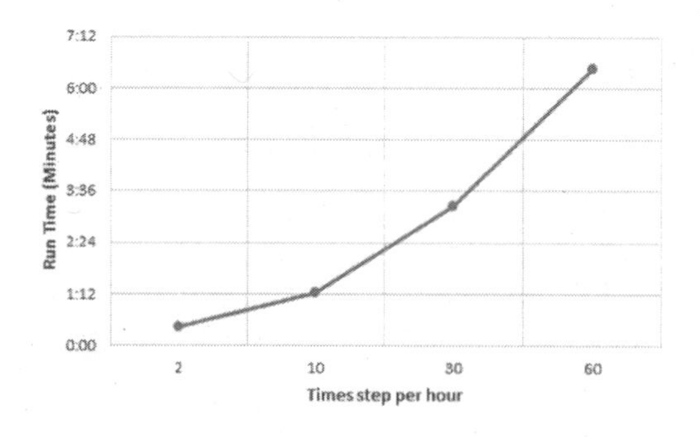

TABLE 11.1
Variation in annual energy consumption with variations in time steps

Factor	Two time steps per hour	Ten time steps per hour	Thirty time steps per hour	Sixty time steps per hour
Run time (min)	0:26	1:13	3:14	6:28
Room electricity	306,674.40	306,674.40	306,674.40	306,674.40
Lighting	443,211.10	443,211.10	443,211.10	443,211.10
System fans	240,545.30	243,184.10	243,659.60	243,736.30
System pumps	11,109.60	11,228.00	11,219.70	11,186.50
Heating (gas)	124,628.80	134,141.80	136,112.00	136,594.00
Cooling (electricity)	380,597.20	383,276.50	380,718.70	376,632.10
Heat rejection	149,786.90	149,971.80	148,558.10	148,492.00

You can see from the results that as the number of time steps per hour increases, the simulation run time also increases. You can also see a slight change in the HVAC energy consumption resulting from the change in the resolution of energy calculations. Please note that because the above-mentioned run times depend on the system configuration, your results might differ from the ones shown here. However, the trend would remain the same. In most cases, the difference in results is very less, of the order of 1%; hence, unless necessary, use of smaller time steps is not recommended. Save the simulation model with two time steps per hour to use in later tutorials.

Tutorial 11.2

Evaluating the Impact of the Solar Distribution Algorithm

GOAL

To evaluate the impact of the solar distribution algorithm on energy consumption and simulation run time

WHAT ARE YOU GOING TO LEARN?

- How to change the solar distribution algorithm

PROBLEM STATEMENT

In this tutorial, you are going to use the two time steps per hour model saved in Tutorial 11.1 (a 50- × 25-m six-level model with a 5-m perimeter depth). Add a 1-m overhang on all windows. You need to select the following solar distribution:

1. Full exterior
2. Minimal shadowing

Determine the change in energy consumption for both cases. Use the **AZ – PHOENIX/ SKY HARBOR, Arizona, USA** weather location.

This option determines how EnergyPlus treats beam solar radiation and reflectance from exterior surfaces that strike the building and, ultimately, enter the zone.

1. *Minimal shadowing*. In this case, there is no exterior shadowing except from window and door reveals. All beam solar radiation entering the zone is assumed to fall on the floor, where it is absorbed according to the floor's solar absorptance. Any radiation reflected by the floor is added to the transmitted diffuse radiation, which is assumed to be uniformly distributed on all interior surfaces. If no floor is present in the zone, the incident beam solar radiation is absorbed on all interior surfaces according to their absorptances. The zone heat balance is then applied at each surface and on the zone's air, with the absorbed radiation being treated as a flux on the surface.
2. *Full exterior*. In this case, shadow patterns on exterior surfaces caused by detached shading, wings, overhangs and exterior surfaces of all zones are computed. As for minimal shadowing, shadowing by window and door reveals is also calculated. Beam solar radiation entering the zone is treated as for minimal shadowing—all beam solar radiation entering the zone is assumed to fall on the

floor, where it is absorbed according to the floor's solar absorptance. Any radiation reflected by the floor is added to the transmitted diffuse radiation, which is distributed among interior surfaces according to view factors. If no floor is present in the zone, the incident beam solar radiation is absorbed on all interior surfaces according to their absorptances.

3. *Full interior and exterior.* This is the same as full exterior except that instead of assuming that all transmitted beam solar radiation falls on the floor, the program calculates the amount of beam radiation falling on each surface in the zone, including floor, walls and windows, by projecting the sun's rays through the exterior windows, taking into account the effect of exterior shadowing surfaces and window shading devices. If this option is used, you should be sure that the surfaces of the zone totally enclose a space. This can be determined by viewing the **eplusout.dxf** file with an external DXF viewer program.

Source: www.designbuilder.co.uk/helpv6.0#Solar_Options.htm.

SOLUTION

Step 1: Open the simulation model saved in Tutorial 11.1 with two time steps per hour.

Step 2: Select the **Openings** tab. Select the **Local Shading** checkbox in the **Shading** section. Select **1.0 m Overhang** as **Type**.

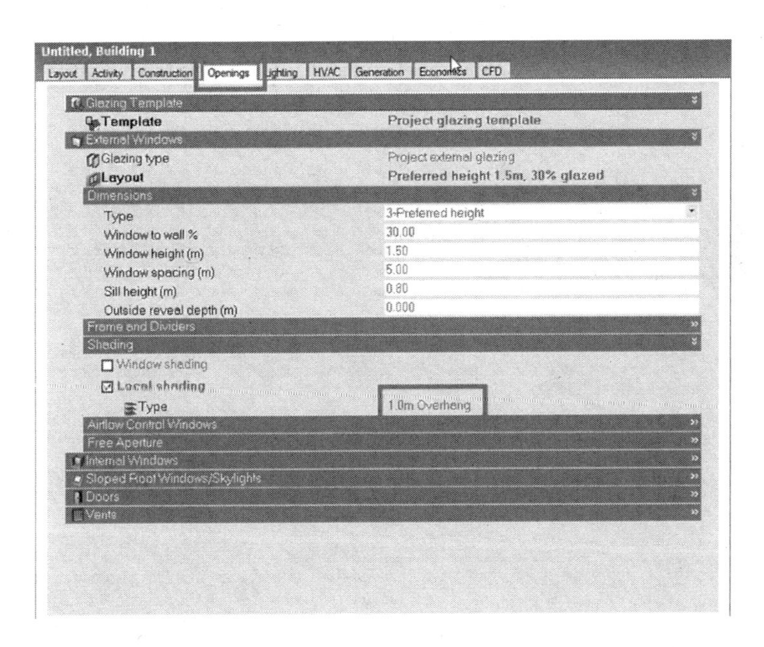

Step 3: Select the **HVAC** tab, and select **Detailed HVAC detail** under **Info, Help**.

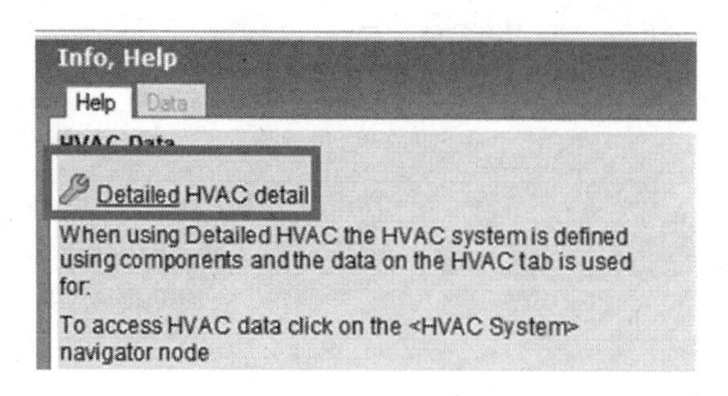

Step 4: Click **Simple** under the **HVAC** slider.

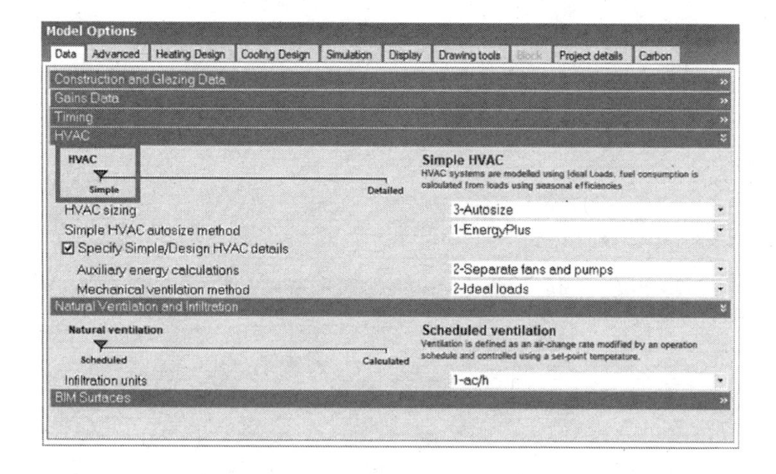

Step 5: Select the **Simulation** screen tab. The **Edit Calculation Options** screen appears.

Step 6: Select the **Options** tab. Expand the **Solar** section. Select **Full exterior** from the **Solar distribution** drop-down list. Click **OK**.

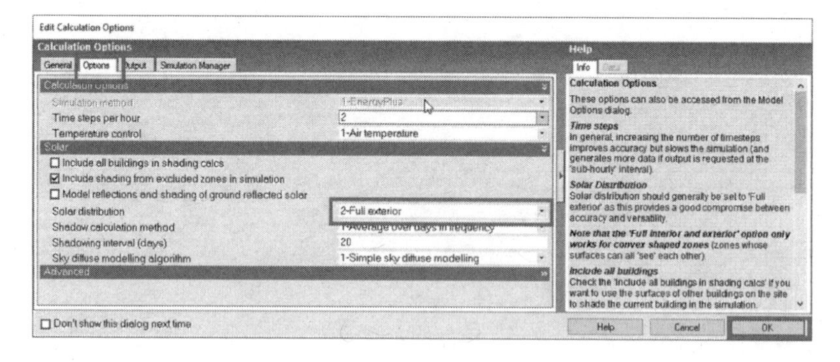

Step 7: Simulate the model, and record the results for energy and simulation run time.

Step 8: Select the **Summary** tab, and click **Table of Contents**.

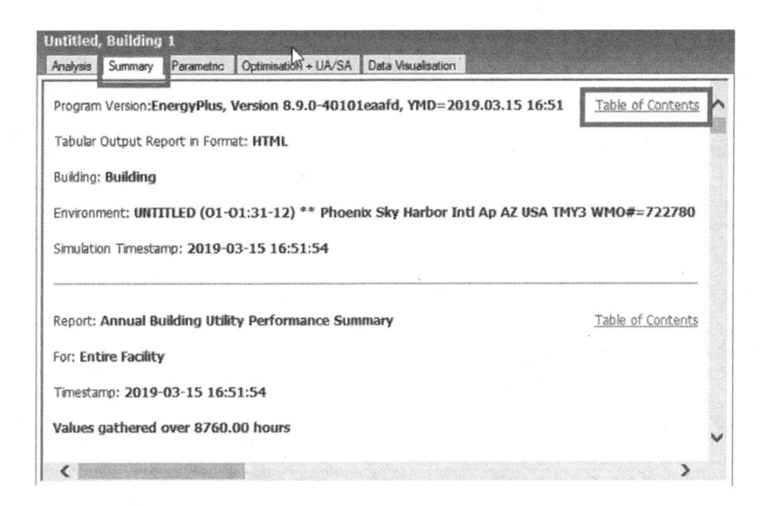

Step 9: Click the **Sensible Heat Gain Summary** link.

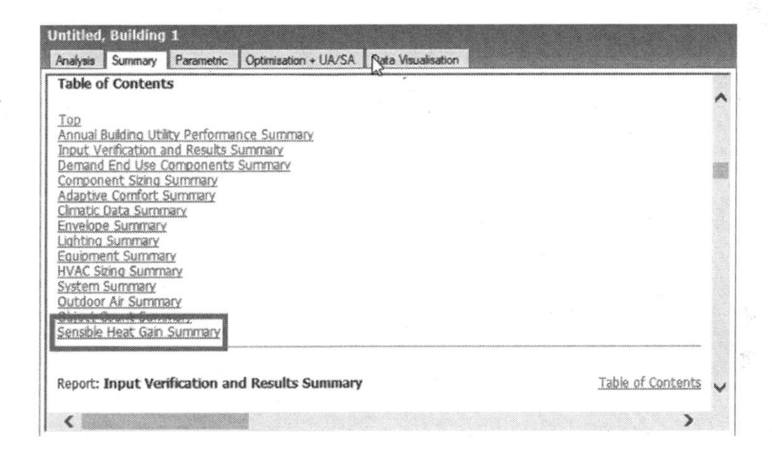

Step 10: Copy the table **Annual Building Sensible Heat Gain Components** to a spreadsheet program. (You need to select the table and right-click and select **Copy and Paste** in the spreadsheet.)

Untitled, Building 1, Top, West

Analysis | Summary | Parametric | Optimisation + UA/SA | Data Visualisation

Report: **Sensible Heat Gain Summary** Table of Contents

For: **Entire Facility**

Timestamp: **2019-12-03 17:23:32**

Annual Building Sensible Heat Gain Components

	HVAC Zone Eq & Other Sensible Air Heating [kWh]	HVAC Zone Eq & Other Sensible Air Cooling [kWh]	HVAC Terminal Unit Sensible Air Heating [kWh]	HVAC Terminal Unit Sensible Air Cooling [kWh]	HVAC Input Heated Surface Heating [kWh]	HVAC Input Cooled Surface Cooling [kWh]	People Sensible Heat Addition [kWh]	Lights Sensible Heat Addition [kWh]	Equipment Sensible Heat Addition [kWh]	Window Heat Addition [kWh]	Interzone Air Transfer Heat Addition [kWh]	Infiltration Heat Addition [kWh]	Opaque Surface Conduction and Other Heat Addition [kWh]	Equipment Sensible Heat Removal [kWh]	Window Heat Removal [kWh]	Interzone Air Transfer Heat Removal [kWh]	Infiltration Heat Removal [kWh]
GROUNDFOOR:WEST	259.253	-17358.55	0.000	0.000	0.000	0.000	1621.661	5635.093	3899.133	13466.820	0.000	1402.079	0.100	0.000	-3365.45	0.000	-2784.59
GROUNDFOOR:NORTH	882.002	-29482.90	0.000	0.000	0.000	0.000	3710.680	12850.443	8891.706	14246.342	0.000	3489.633	1.867	0.000	-5593.69	0.000	-5377.70
GROUNDFOOR:EAST	276.791	-18019.99	0.000	0.000	0.000	0.000	1615.671	5635.093	3899.133	13268.333	0.000	1427.365	0.228	0.000	-2929.65	0.000	-2734.12
GROUNDFOOR:CORE	969.187	-64497.17	0.000	0.000	0.000	0.000	10580.311	36897.495	25530.768	0.000	0.000	10969.276	1.297	0.000	0.000	0.000	-16387.10
GROUNDFOOR:SOUTH	325.625	-37013.42	0.000	0.000	0.000	0.000	3676.343	12850.443	8891.706	26826.471	0.000	3459.699	0.189	0.000	-6498.78	0.000	-6576.59
MIDDLE:WEST	861.780	-54504.14	0.000	0.000	0.000	0.000	5865.174	22540.371	15596.532	50680.189	0.000	4058.391	0.450	0.000	-14354.47	0.000	-11431.26
MIDDLE:NORTH	3080.106	-106240.24	0.000	0.000	0.000	0.000	13948.885	51401.774	35566.825	52946.658	0.000	11350.708	3.131	0.000	-23991.29	0.000	-22137.33
MIDDLE:EAST	962.152	-56051.43	0.000	0.000	0.000	0.000	5733.083	22540.371	15596.532	49732.373	0.000	3994.761	0.480	0.000	-12567.62	0.000	-11104.24
MIDDLE:CORE	2928.816	-271582.43	0.000	0.000	0.000	0.000	41954.321	147589.980	102123.072	0.000	0.000	40702.860	3252.990	0.000	0.000	0.000	-66969.61
MIDDLE:SOTUH	1015.677	-127111.47	0.000	0.000	0.000	0.000	13644.714	51401.774	35566.825	101820.325	0.000	11030.592	7.757	0.000	-27755.19	0.000	-26713.01
TOP:WEST	354.959	-17825.89	0.000	0.000	0.000	0.000	1624.867	5635.093	3899.133	12808.932	0.000	1300.587	0.286	0.000	-3619.57	0.000	-2848.94
TOP:NORTH	1139.972	-32745.54	0.000	0.000	0.000	0.000	3715.753	12850.443	8891.706	13308.488	0.000	3137.706	1320.915	0.000	-6099.10	0.000	-5520.35
TOP:EAST	382.640	-18591.56	0.000	0.000	0.000	0.000	1620.213	5635.093	3899.133	12580.886	0.000	1317.489	0.841	0.000	-3132.86	0.000	-2743.41
TOP:CORE	1263.879	-74821.08	0.000	0.000	0.000	0.000	10599.074	36897.495	25530.768	0.000	0.000	9412.914	7999.001	0.000	0.000	0.000	-16882.06
TOP:SOUTH	498.007	-38908.29	0.000	0.000	0.000	0.000	3687.122	12850.443	8891.706	25449.123	0.000	3116.413	1.758	0.000	-7014.81	0.000	-6609.37
Total Facility	15210.847	-964754.10	0.000	0.000	0.000	0.000	123597.872	443211.403	306674.680	387134.940	0.000	110170.473	12591.291	0.000	-116922.47	0.000	-206819.67

Step 11: Repeat the preceding steps to select **Minimal shadowing** from the **Solar distribution** drop-down list.

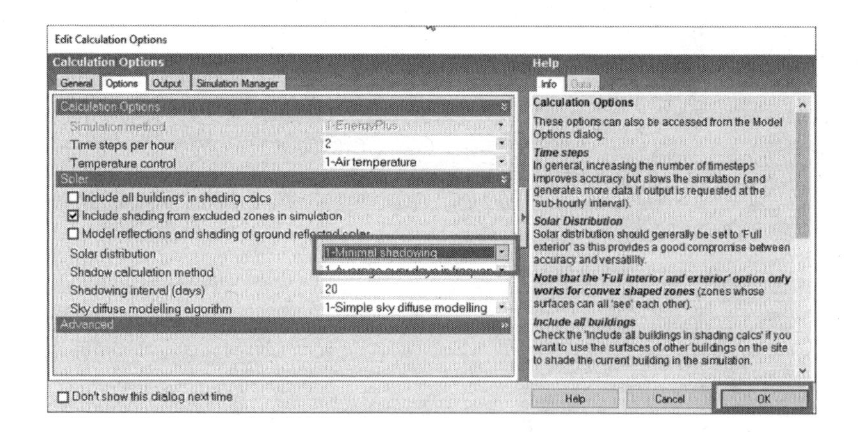

Step 12: Click **OK**. Record the results. Compare results for both simulations (Tables 11.2 and 11.3).

It can be seen from the results that with the **Minimal shadowing** option, there is an increase in HVAC energy consumption, as there is no exterior shadowing considered in the calculations except from window and door reveals. The results also show that the window heat gain is lesser with **Full exterior**. This is because the shadow patterns on the exterior surface caused by overhangs and exterior surfaces are taken into account for calculations. Save the model to use in the next tutorial.

TABLE 11.2
Variation of annual energy consumption with variation in local shading

	Annual fuel breakdown consumption (kWh)	
Factor	**Full exterior**	**Minimal shadowing**
Run time (min)	0:19	0:18
Room electricity	306,674.40	306,674.40
Lighting	443,211.10	443,211.10
Heating (gas)	18,867.30	15,329.40
Cooling (electricity)	497,957.70	543,281.20
DHW (electricity)	27,748.90	27,748.90

TABLE 11.3
Heat gains from window

	Window heat addition (GJ)	
Location	**Full exterior**	**Minimal shadowing**
Ground: west	13,466.80	20,132.20
Ground: north	14,246.30	15,949.00
Ground: east	13,268.30	19,886.30
Ground: south	0.00	0.00
Ground: core	26,826.50	47,346.70
Middle: west	50,680.20	75,945.20
Middle: north	52,946.70	59,331.20
Middle: east	49,732.40	74,642.80
Middle: south	0.00	0.00
Middle: core	101,820.00	179,577.60
Top: west	12,808.90	19,231.40
Top: north	13,308.50	15,015.70
Top: east	12,580.90	18,957.70
Top: south	0.00	0.00
Top: core	25,449.10	45,158.00
Total facility	387,135.00	591,173.80

EXERCISE 11.1

Repeat this tutorial for the full interior and exterior solar distribution algorithm.

Tutorial 11.3

Evaluating the Impact of the Solution Algorithm

GOAL

To evaluate the building energy performance and run time with a change in the solution algorithm

WHAT ARE YOU GOING TO LEARN?

- How to change the solution algorithm

PROBLEM STATEMENT

In this tutorial, you are going to use the simulation model saved in Tutorial 11.1 with two time steps per hour. You need to select the following algorithms:

1. Conduction transfer function (CTF)
2. Finite difference

Determine the change in energy consumption with both cases. Use the **AZ – PHOENIX/ SKY HARBOR, Arizona, USA** weather location.

CTF: the default method used in EnergyPlus for CTF calculations is known as the *state space method*. CTF is a sensible-heat-only solution not taking into account moisture storage or diffusion in the construction elements.

Finite difference: this solution technique uses a one-dimensional finite-difference solution in the construction elements. It is a sensible-heat-only solution and does not take into account moisture storage or diffusion in the construction elements.

Finite-difference settings: The settings listed below are required when the general solution algorithm is set to **2–Finite Difference** or if any constructions used in the simulation override the general setting to use the finite-difference algorithm.

Difference scheme: This field determines the solution scheme used by the conduction finite-difference model. There are two options:

1. *Fully implicit first-order*, which is first order in time and is more stable over time, but it may be slower than option 2.

2. *Crank Nicholson second order*, which is second order in time and may be faster than option 1, but it can be unstable over time when boundary conditions change abruptly and severely.

Source: http://www.designbuilder.co.uk/helpv6.0/Content/Advanced_Calculation_Options.htm.

SOLUTION

Step 1: Open the simulation model saved in Tutorial 11.1 with two time steps per hour. Select the **Simulation** screen tab.

Step 2: Select the **Options** tab, and expand the **Advanced** section.

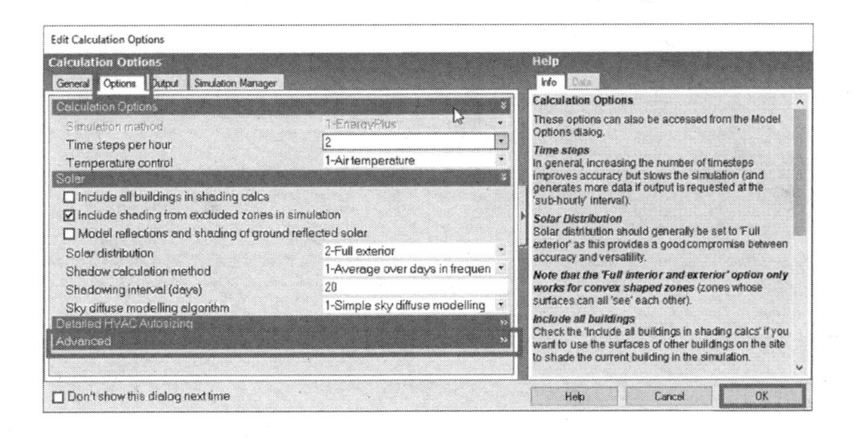

Step 3: Select **Conduction Transfer Function** from the **Solution algorithm** drop-down list.

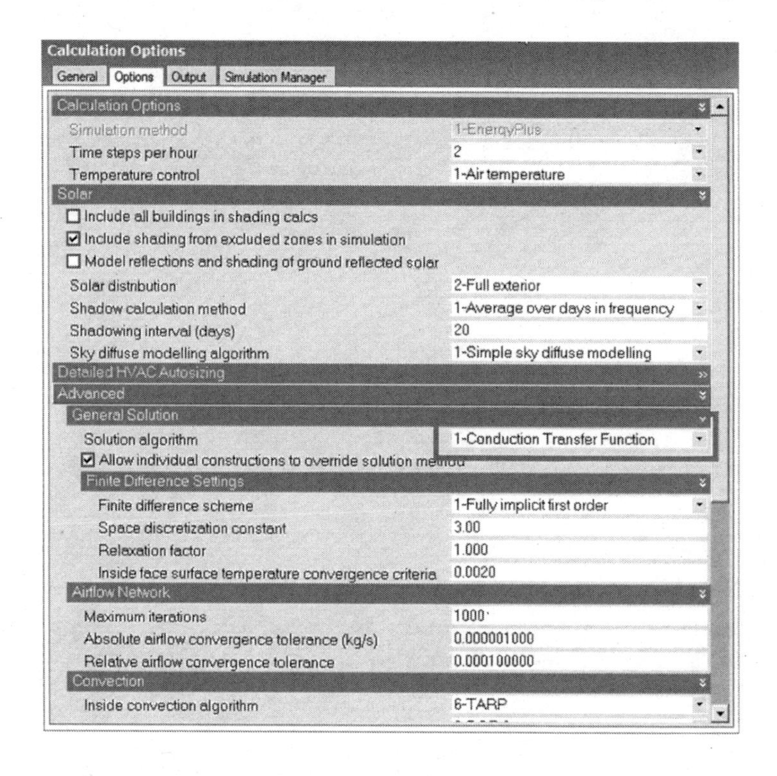

Step 4: Simulate the model, and record the results.

Step 5: Repeat the preceding steps, and select **Finite Difference** from the **Solution algorithm** drop-down list.

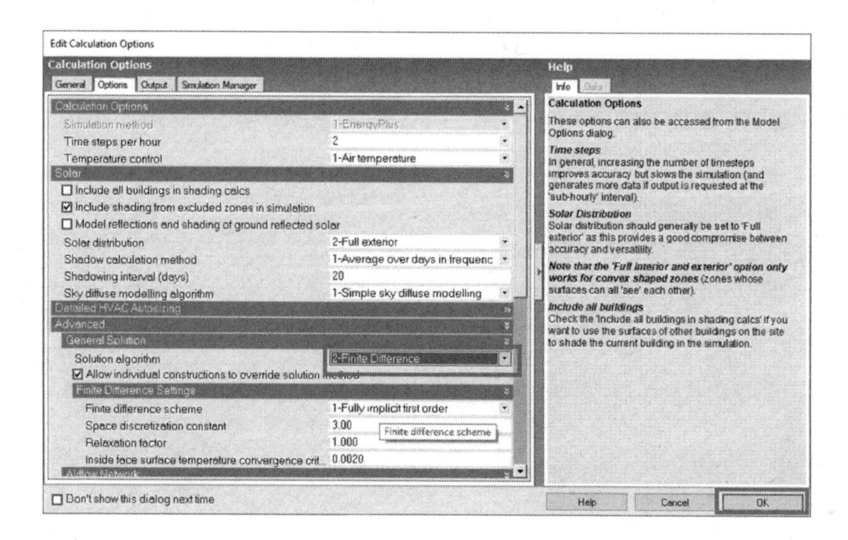

Step 6: Simulate the model, and record the results. Compare the results for both simulations (Table 11.4).

TABLE 11.4
Variation in simulation run time with variation in the solution algorithm

	Annual fuel breakdown consumption (kWh)	
Factor	**Conduction transfer function**	**Finite difference (fully implicit first order)**
Run time (min)	0:27	4:01
Room electricity	306,674.40	306,674.40
Lighting	443,211.10	443,211.10
System fans	335,040.80	338,411.50
System pumps	17,309.70	17,492.70
Heating (gas)	199,962.60	211,392.50
Cooling (electricity)	510,120.10	514,833.60
Heat rejection	226,757.10	228,356.20

You can see from the results that with the finite-difference method there is an increase in the simulation run time.

Tutorial 11.4

Evaluating the Effect of the Inside Convection Algorithm

GOAL

To evaluate building energy performance with a change in the inside convection algorithm

WHAT ARE YOU GOING TO LEARN?

- How to change the inside convection algorithm

PROBLEM STATEMENT

In this tutorial, you are going to use the simulation model saved in Tutorial 11.1 with two time steps per hour. You are going to select the following algorithms:

1. Adaptive convection
2. Simple
3. Chartered Institution of Building Services Engineers (CIBSE)
4. Thermal Analysis Research Program (TARP)

Find the change in energy consumption for all cases.
Use **PARIS-AEROPORT CHAR, France** weather location.

Inside Convection Algorithm

You can select from six main EnergyPlus inside convection algorithms for calculating the convection between internal zone surfaces and the rest of the zone air in the simulation calculations. Unless you have a good reason to do so, you are advised to use the default TARP convection algorithm.

1. *Adaptive convection algorithm.* This advanced option provides a dynamic selection of convection models based on conditions. Beausoleil-Morrison (Beausoleil-Morrison, I., 'The adaptive coupling of heat and air flow modeling within dynamic whole-building simulations', PhD Thesis, University of Strathclyde, Glasgow, UK, 2000) developed a methodology for dynamically managing the selection of h_c equations, called the *adaptive convection algorithm*. The algorithm is used to select among the available h_c equations for the one that is most appropriate for a given surface at a given time. As Beausoleil-Morrison notes, the adaptive convection algorithm is intended to be expanded and altered to reflect different classification schemes and/or new h_c equations. The

adaptive convection algorithm implemented in EnergyPlus for the inside face has a total of 45 different categories for surfaces and 29 different options for h_c equation selections. The tables provided in the engineering document summarize the categories and default assignments for h_c equations.

2. *Simple.* The simple convection model uses constant coefficients for different heat transfer configurations, employing the criteria to determine reduced and enhanced convections. The coefficients are taken directly from Walton (Walton, G. N., *Thermal Analysis Research Program Reference Manual* (NBSSIR 83-2655). Washington, DC: National Bureau of Standards (now NIST), 1983. This is the documentation for TARP.). Walton derived his coefficients from the surface conductance for $\varepsilon = 0.90$ found in the *ASHRAE Handbook* (ASHRAE, *ASHRAE Handbook: Fundamentals.* Atlanta, GA: American Society of Heating, Refrigerating, and Air-Conditioning Engineers, 1983, table 1). The radiative heat transfer component was estimated at $1.02 \times 0.9 = 0.918$ Btu/h-ft^2-°F and then subtracted off. Finally, the coefficients were converted to SI units to yield the following values: for a vertical surface, $h_c = 3.076$; for a horizontal surface with reduced convection, $h_c = 0.948$; for a horizontal surface with enhanced convection, $h_c = 4.040$; for a tilted surface with reduced convection, $h_c = 2.281$; and for a tilted surface with enhanced convection, $h_c = 3.870$.
3. *CIBSE.* This algorithm applies constant heat transfer coefficient derived from traditional CIBSE values.
4. *Ceiling diffuser.* This is a mixed and forced convection model for ceiling diffuser configurations. The model correlates the heat transfer coefficient to the air change rate for ceilings, walls and floors. The ceiling diffuser algorithm is based on empirical correlations developed by Fisher and Pedersen (Fisher, D. E., and Pedersen, C. O., 'Convective heat transfer in building energy and thermal load calculations', *ASHRAE Transactions* 103(Pt. 2), 1997). The correlation was reformulated to use the room outlet temperature as the reference temperature. The correlations are as follows: for floors, $h_c = 3.873 + 0.082 \times ACH^{0.98}$; for ceilings, $h_c = 2.234 + 4.099 \times ACH^{0.503}$; and for walls, $h_c = 1.208 + 1.012 \times ACH^{0.604}$.
5. *Cavity.* This algorithm was developed to model convection in a *Trombe wall zone*, that is, the airspace between the storage wall surface and the exterior glazing. (See the later sections on passive and active Trombe walls for more information about Trombe walls.) The algorithm is identical to the convection model (based on ISO 15,099; *Thermal Performance of Windows, Doors, and Shading Devices: Detailed Calculations.* Geneva: International Organization for Standardization, 2003) used in Window5 (Window is a software developed by the Lawrence Berkeley National Laboratory. More details can be found at https://windows.lbl.gov/software/window for convection between glazing layers in multipane window systems. Use of the algorithm for modelling an unvented Trombe wall has been validated against experimental data by Ellis (Ellis, Peter G., 'Development and validation of the unvented Trombe wall model in EnergyPlus'. Master's thesis, University of Illinois at Urbana-Champaign, 2003). This algorithm gives the convection coefficients for air in a narrow vertical cavity that is sealed and not ventilated. This applies both to the air gap in

between panes of a window and to the air gap between the Trombe wall glazing and the inner surface (often a selective surface). These convection coefficients are really the only difference between a normal zone and a Trombe zone.

6. *TARP.* This algorithm is based on variable natural convection based on the temperature difference from ASHRAE algorithms. This is the same as the old 'detailed' inside convection algorithm provided in earlier versions of DesignBuilder.

Source: www.designbuilder.co.uk/helpv6.0/Content/Surface_Convection.htm.

SOLUTION

Step 1: Open the simulation model saved in Tutorial 11.1 with two time steps per hour.

Step 2: Select the **Simulation** screen tab.

Step 3: Select the **Options** tab, and expand the **Advanced** section.

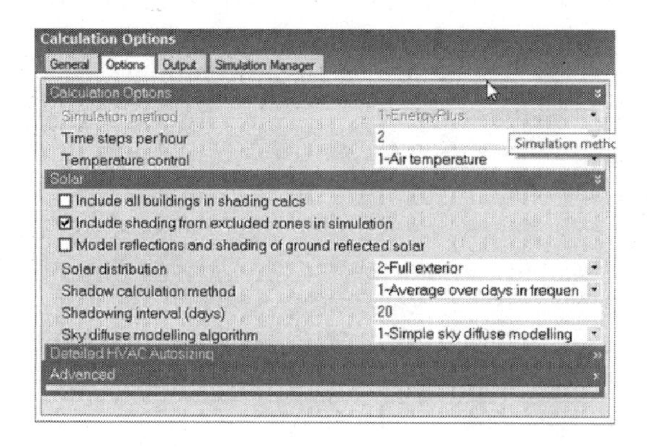

Step 4: Select **Adaptive Convection Algorithm** from the **Inside convection algorithm** drop-down list in the **Convection** section.

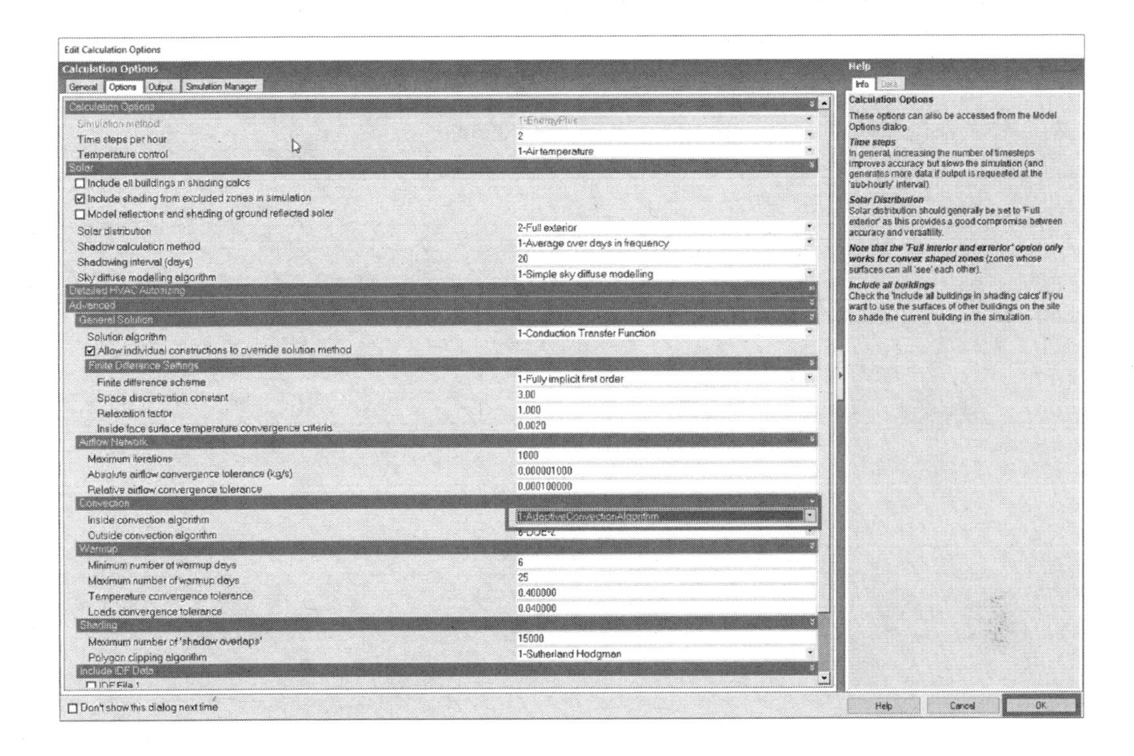

Step 5: Click **OK**, and note down the results.

Step 6: Repeat the preceding steps for the **Simple**, **CIBSE** and **TARP** algorithms from the **Inside convection algorithm** drop-down list. Compare the results (Table 11.5).

TABLE 11.5
Variation in simulation run time with variations in the inside convection algorithm

	Annual fuel breakdown consumption (kWh)			
Factor	**Adaptive convection algorithm**	**Simple**	**CIBSE**	**TARP**
Run time (min)	0:27	0.26	0.24	0.23
Room electricity	306,674.40	306,674.40	306,674.40	306,674.40
Lighting	443,211.10	443,211.10	443,211.10	443,211.10
System fans	260,918.30	279,606.70	281,272.40	271,578.00
System pumps	9,690.40	11,124.60	11,235.50	10,593.50
Heating (gas)	818,967.60	845,199.30	858,713.50	824,718.00
Cooling (electricity)	192,644.80	218,891.10	220,490.30	208,298.90
Heat rejection	121,383.10	135,161.40	136,291.80	129,901.80

Differences in the HVAC energy consumption and run time while using different algorithms can be seen in this table.

EXERCISE 11.2

Repeat this tutorial to evaluate the building energy performance with a change in the outside convection algorithm. You need to select the following algorithms:

1. Adaptive convection
2. Simple combined
3. TARP
4. DOE–2

Use the **PARIS-AEROPORT CHAR, France** weather location.

Tutorial 11.5

Evaluating the Impact of the Shadowing Interval

GOAL

To evaluate the impact of the shadowing interval on building energy consumption

WHAT ARE YOU GOING TO LEARN?

- How to change the shadowing interval

PROBLEM STATEMENT

In this tutorial, you are going to use the simulation model saved in Tutorial 11.2 with a full exterior. You are going to simulate the model for the following intervals: 5, 10, 20 and 30 days. Use the **New Delhi/Palam, India** weather location.

Shadowing interval is important for determining the amount of sun entering your building and, by inference, the amount of cooling or heating load needed for maintaining the building. Though termed *shadowing* calculations, the shadowing interval in effect determines the sun's position on a particular day in a weather file period simulation. (Each design day will use the date of the design day object.) Even though weather file data contain the amount of solar radiation, the internal calculation of the sun's position will govern how it affects various parts of the building.

By default, the calculations are done for every 20 days throughout a weather run period; an average solar position is chosen, and the solar factors (such as sunlit areas of surfaces) remain the same for that number of days. More integrated calculations are needed for controlling dynamic windows or shades.

Source: https://designbuilder.co.uk/helpv6.0/Content/Solar_Options.htm.

SOLUTION

Step 1: Open the simulation model saved in Tutorial 11.2.

Step 2: Select the **Simulation** screen tab.

Step 3: Select the **Options** tab.

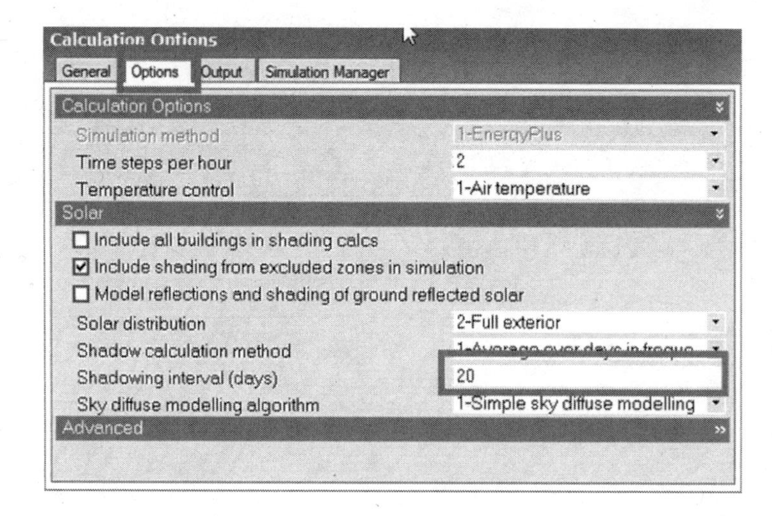

Step 4: Type **5** in the **Shadowing interval (days)**.

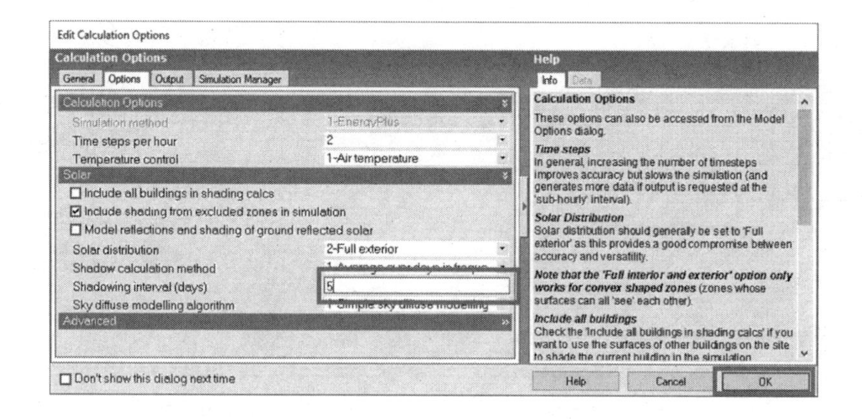

Step 5: Simulate the model, and record the results.

Step 6: Repeat the preceding steps for shading intervals of **10**, **20** and **30** days. Compare the data for all the cases (Table 11.6).

TABLE 11.6
Variation in simulation run time with variations in shadowing interval

	Annual fuel breakdown data			
	Shading interval (days)			
End-use component	**5**	**10**	**20**	**30**
Run time (min)	0:22	0:21	0:20	0:19
Room electricity	306,674.40	306,674.40	306,674.40	306,674.40
Lighting	443,211.10	443,211.10	443,211.10	443,211.10
Heating (gas)	10,135.40	10,143.70	10,179.70	10,177.20
Cooling (electricity)	748,183.60	748,074.30	748,097.10	747,715.40
DHW (electricity)	27,748.90	27,748.90	27,748.90	27,748.90

12 Renewable Energy System

This chapter discusses the modelling of photo-voltaic systems fitted on buildings for the generation of electricity. Modelling steps explained in this chapter include creating rooftop photo-voltaic systems, defining module properties and adding an electric load centre. Concepts covered in the tutorials include analyzing the impact of tilt angle on photo-voltaic plates, the impact of shading from roof top photo-voltaic systems and the impact of cell efficiency. Further, simulation of building-integrated photo-voltaic systems for glazing and opaque building components is also discussed.

Tutorial 12.1

Evaluating the Impact of Photo-Voltaic (PV) Panel Tilt Angle

GOAL

To evaluate the impact of the PV tilt angle on annual energy generation

WHAT ARE YOU GOING TO LEARN?

- How to create a rooftop PV system
- How to change the tilt angle of a PV panel
- How to add an electric load centre
- How to record annual energy generation

PROBLEM STATEMENT

In this tutorial, you are going to use a 10- × 10-m single-zone model. You are going to use the PV module described in Table 12.1 for the simulations.

TABLE 12.1
Properties of PV module

PV property	Value
Name	BPsolar 5170 (from DesignBuilder Library)
Cell type	Crystalline silicon
Active area	1.26 m^2
Rated electrical power	250 W_p

Use five modules in a series and two such series in parallel placed on the roof facing south. Determine annual electricity generation for following PV tilt angles from the horizontal roof surface:

1. Equal to the latitude angle of the location
2. Latitude angle + 5 degrees
3. Latitude angle − 5 degrees

Use the **New Delhi Palam India (latitude 28°N)** weather location.

PV Solar Collector Performance: Equivalent One Diode

This object describes the performance characteristics of PV modules to be modelled using an equivalent one-diode circuit.

Name: This field contains the unique name for the PV module performance data. The name is used as an identifier.

Cell type: This field is used to describe the type of technology used in the PV module. Two options are available:

1. Crystalline silicon
2. Amorphous silicon.

Cells in series: This is the number of individual cells wired in series to make up a single module. The typical number for a 12-volt crystalline silicon PV module is 36.

Active area: This field is the active area of the PV module in (m^2 or ft^2).

Rated electrical power: This numeric field contains the nominal electrical power output to be requested from the panel. It is normally equal to the rated power output of the generator (in W). This value is used only for supervisory control and generator dispatch; the actual power output for each time step is determined by the generator model. This value affects how much a generator is loaded (i.e. requested electrical power output) and can also impact the operation of an electrical storage unit if one is connected to the associated electric load centre.

Source: https://designbuilder.co.uk/helpv6.0/Content/PerformanceEquivalentOneDiodePV.htm.

SOLUTION

Step 1: Create a 10- × 10-m single-zone model.

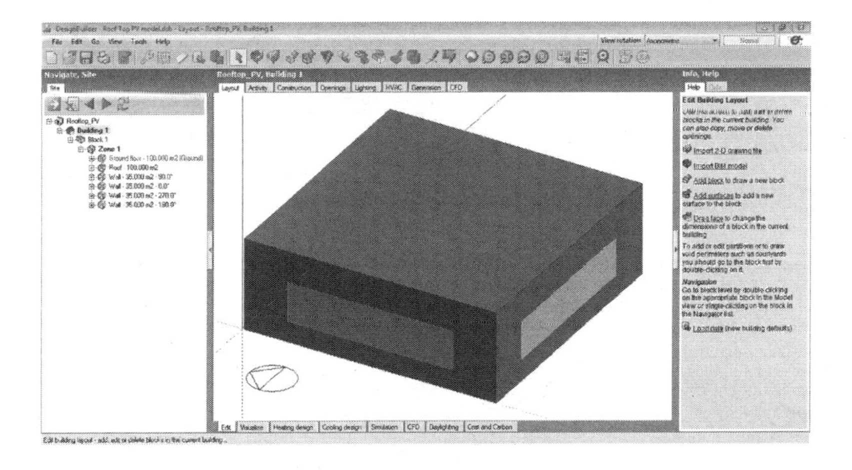

Step 2: Draw a construction line on the roof to mark placement of the solar PV collector (comprised of a total of 10 modules). Click the **Place construction line** icon.

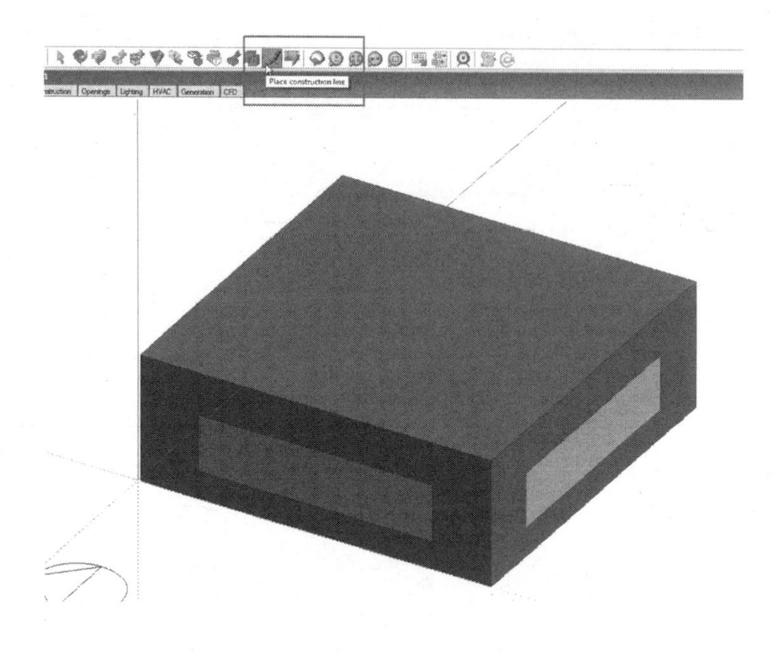

Step 3: Click on the south-west corner of the roof, and extend the line to the south-east corner.

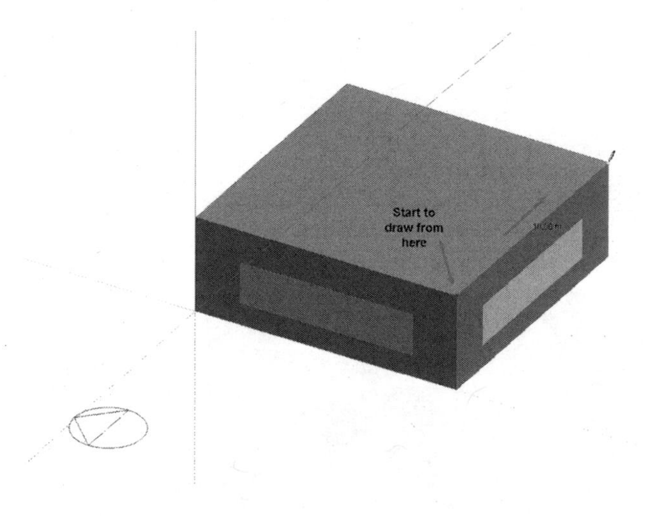

Step 4: Then click 1.26 m towards the north.

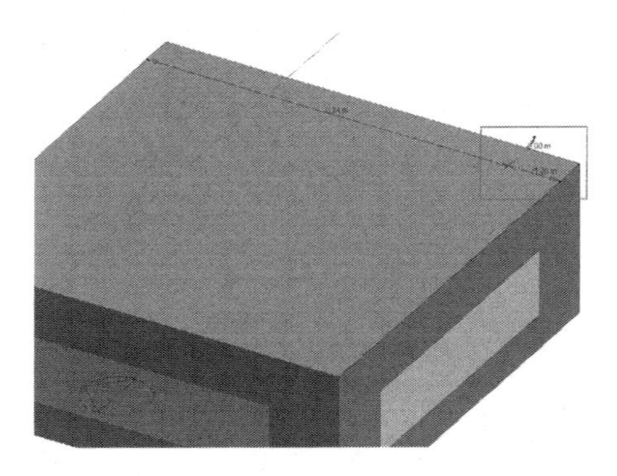

Step 5: Complete the rectangle using construction lines.

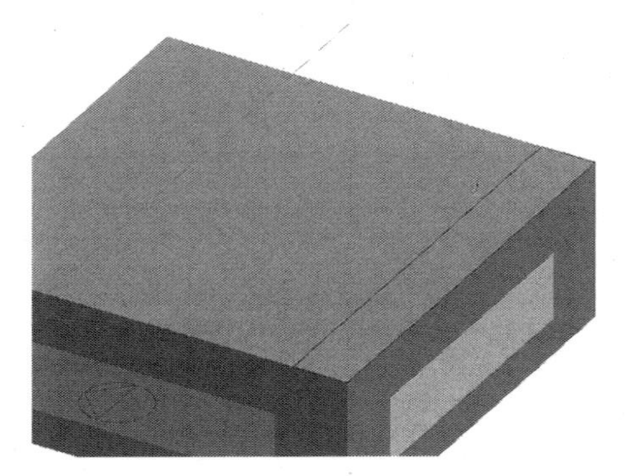

Step 6: Click the **Draw solar collector** icon.

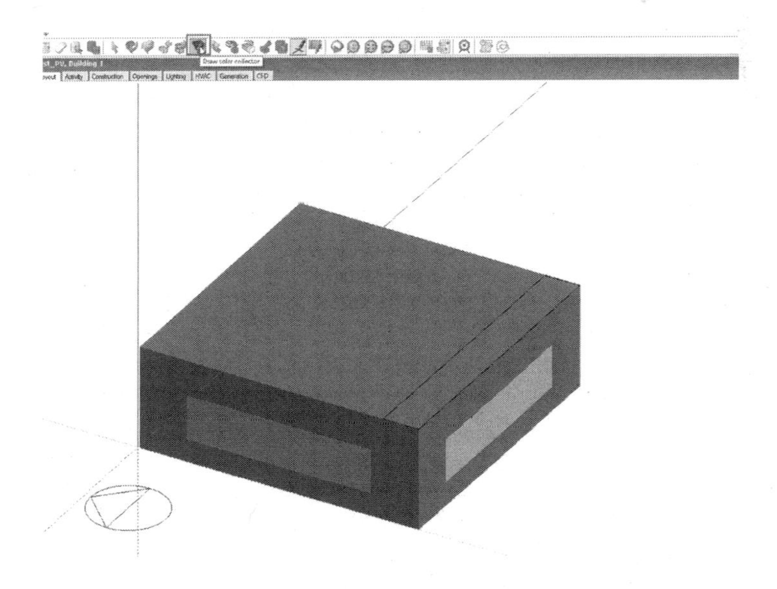

Step 7: Select **Add Solar Collector – Photovoltaic** from the submenu.

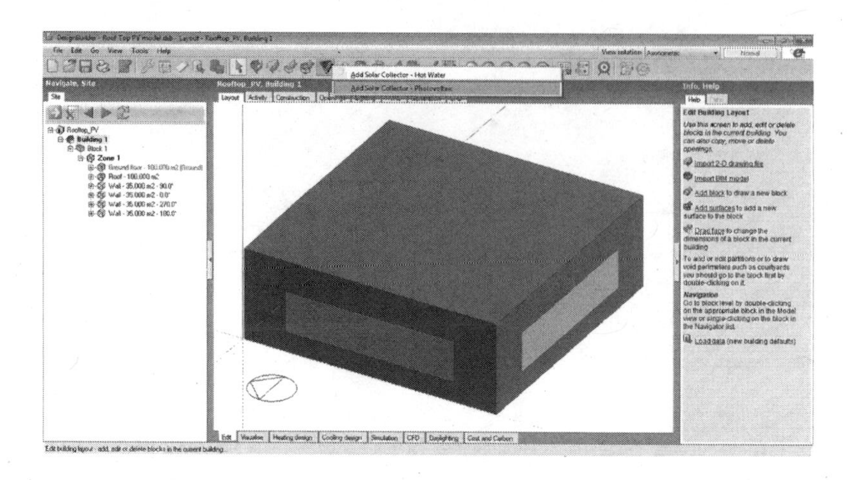

Step 8: Draw the solar PV collector on the construction line.

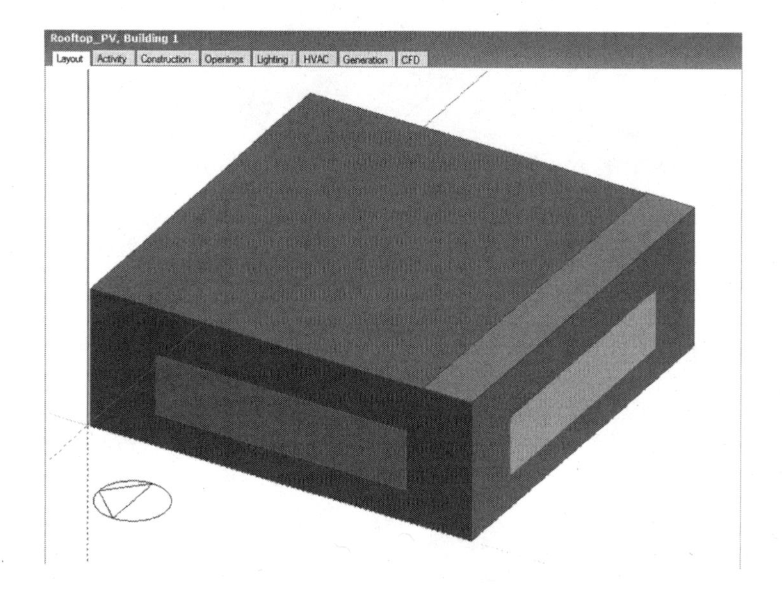

As active area of the solar cell on each module is 1.26 m^2, and the total 5 × 2 modules need to be connected; the required area of solar panel is 12.6 m^2.

Step 9: Select the PV panel drawn in Step 8. Select **Left** from the **View rotation** drop-down list. Click the **Rotate** icon.

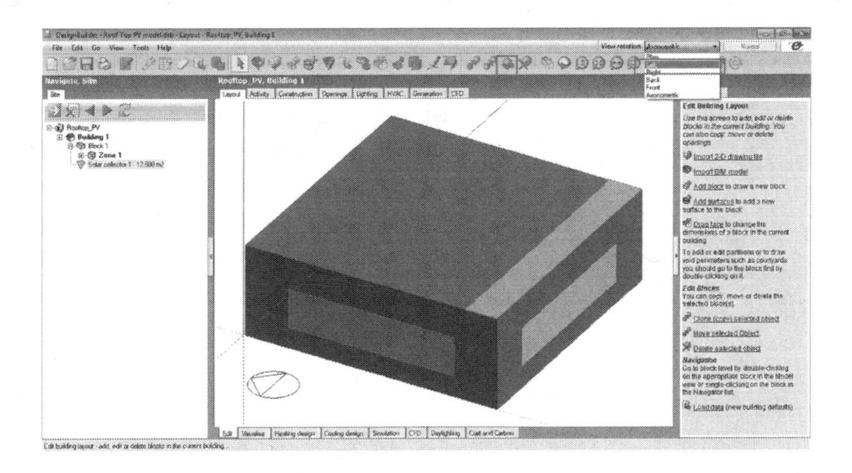

Step 10: Rotate the PV panel to 332 degree to set the tilt angle of the PV panel to 28 degrees from horizontal.

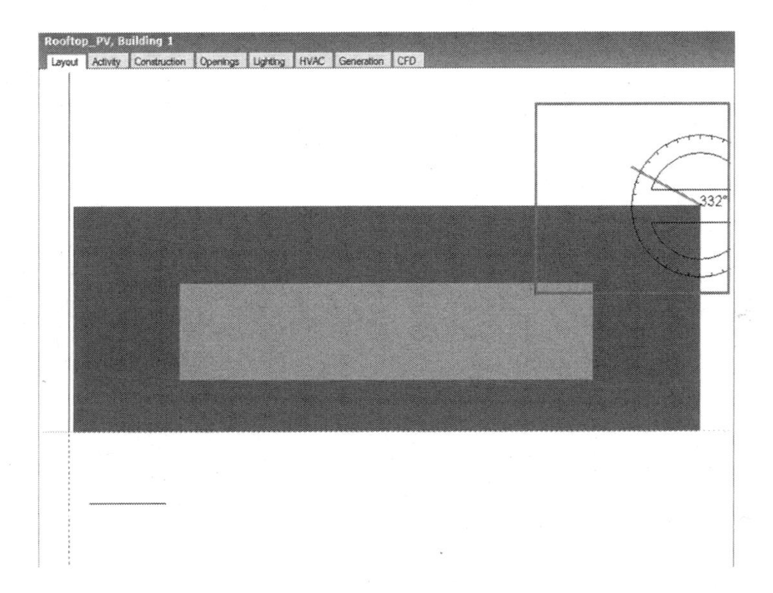

Orienting the PV panel in a direction and tilt to maximize its exposure to direct sunlight will optimize the collection efficiency. South facing in the northern hemisphere and north facing in southern hemisphere will optimize collection efficiency. The optimal tilt angle for PV modules at a site in the northern hemisphere is equal to the latitude.

Step 11: Click **Solar collector 1 – 12.600 m2** in the navigation tree.

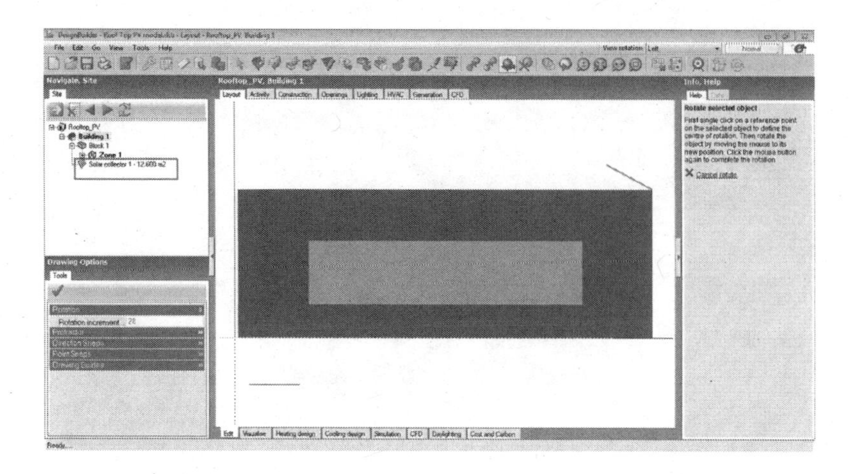

Step 12: Click the **Construction** tab. Select **2-Equivalent One-Diode** from the **Performance type** drop-down list. Select **BPsolar 5170** from **Performance model**. Enter **5** in the **Module in series** text box. Enter **2** in **Series strings in parallel** text box.

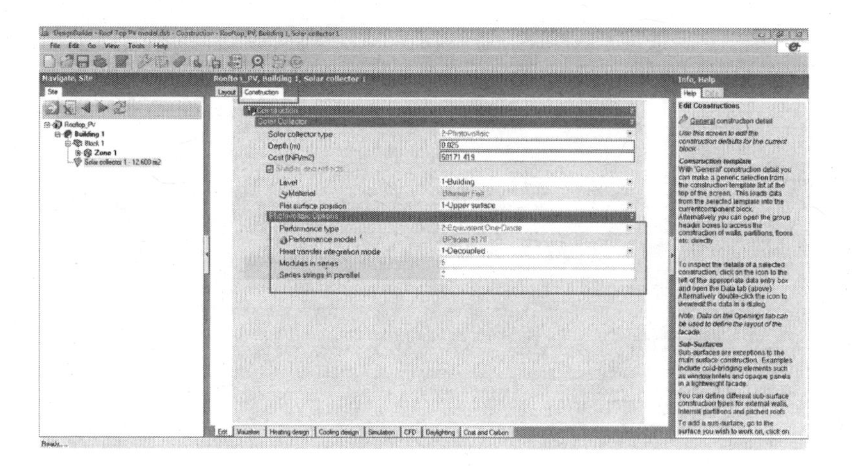

Step 13: Click **Building 1**, and then click the **Generation** tab. Select **Include electric load centres** checkbox.

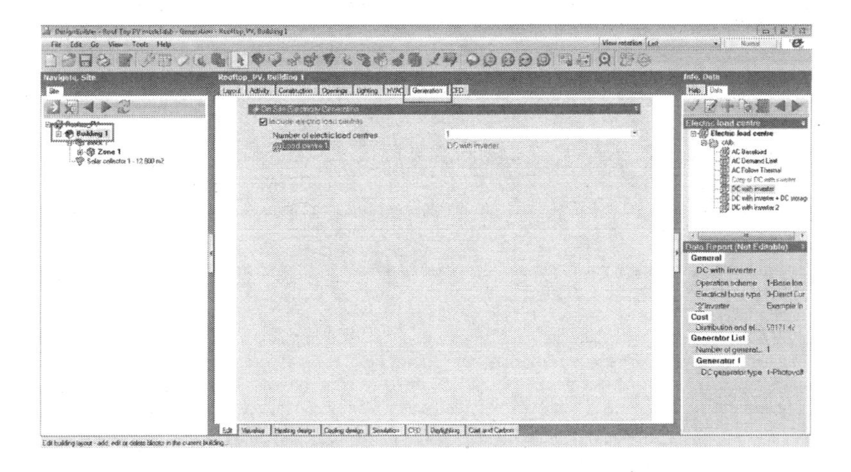

Direct Current with Inverter Load Centre

Direct current with inverter load centres collect direct-current (DC) power from various PV arrays, run the DC power through an inverter and produce alternating-current (AC) power. The PV arrays produce DC power based on the availability of sunshine and do not respond to load requests made by the electrical load centre. The AC output from the inverter is what is recorded as electricity production.

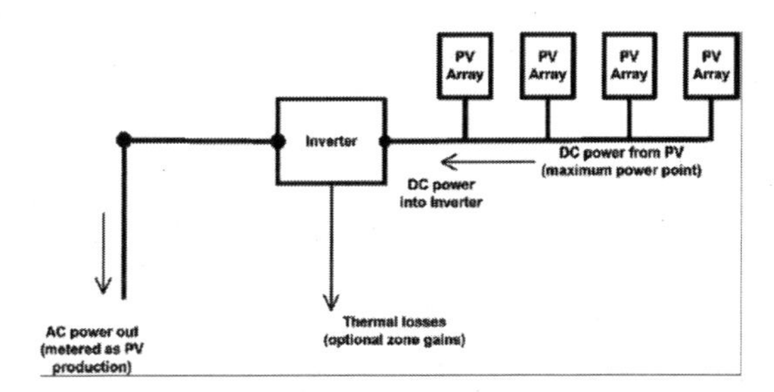

Source: https://designbuilder.co.uk/helpv6.0/Content/ElectricLoadDistribution.htm.

Step 14: Click on **Load centre 1**. Make a copy of **DC with inverter**. Edit **Copy of DC with inverter**.

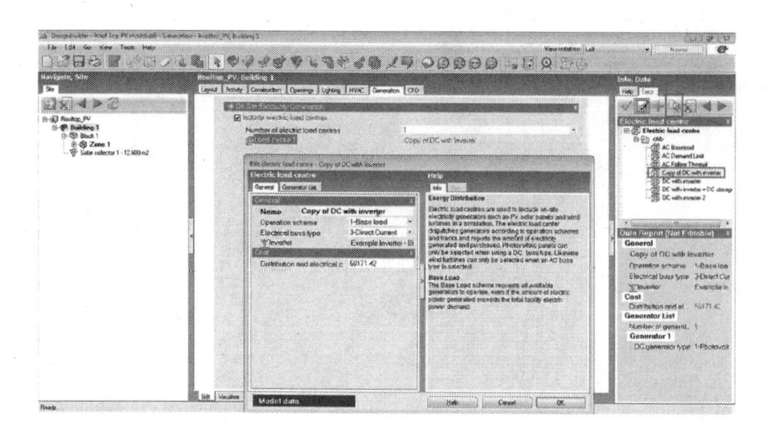

Step 15: Click the **Generator List** tab. Select **Solar collector 1** in **PV solar collector**. Click **OK**.

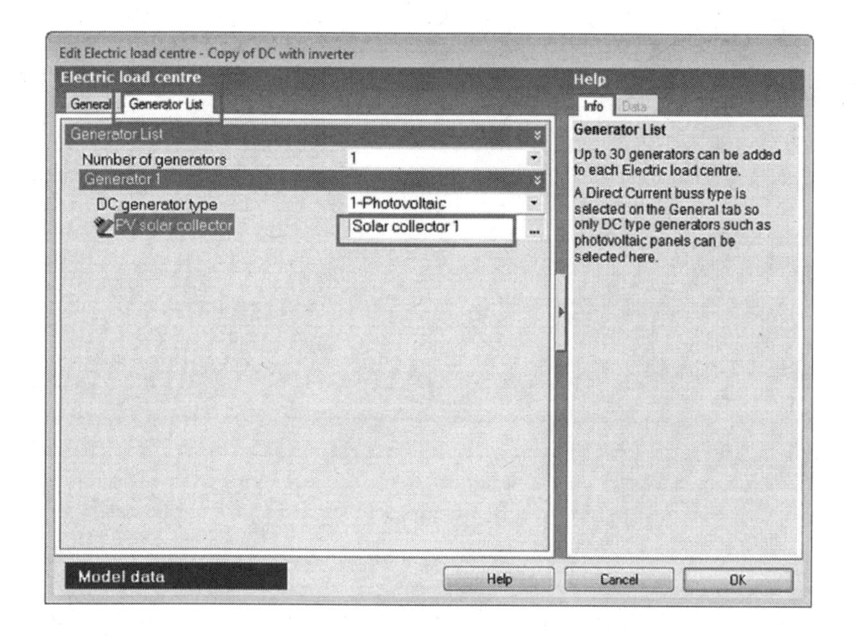

Step 16: Click the **Simulation** screen tab.

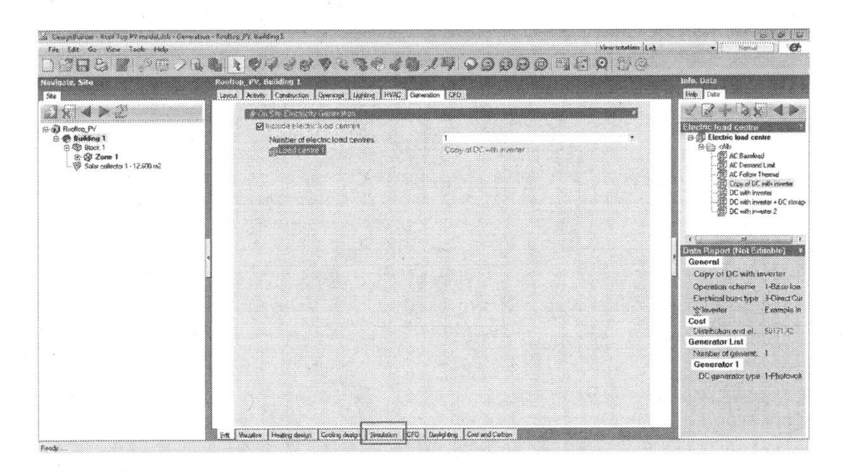

Step 17: Click **Update data**, and then click the **Output** tab. Select the **All Summary** check-box. Click **OK** to perform the simulation.

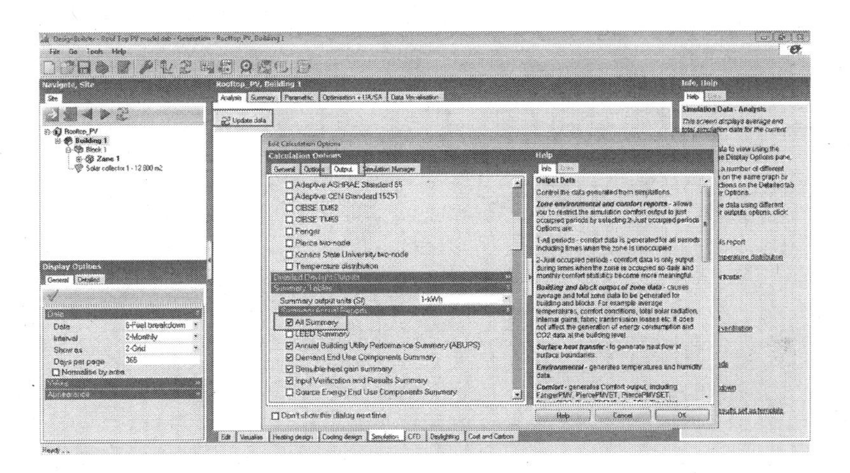

Step 18: Click the **Summary** tab. Scroll down to **Electric Load Satisfied** to note **Photovoltaic Power**.

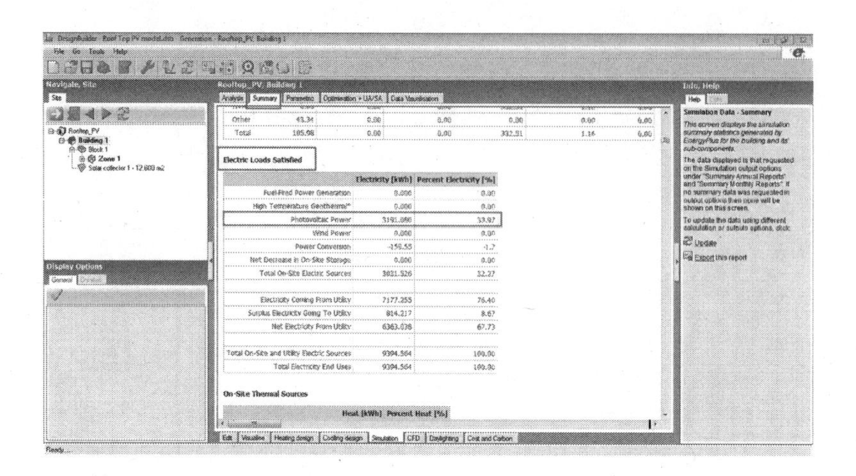

Step 19: Repeat the preceding steps for PV tilt angles of 23 and 33 degrees. Compare the annual electricity generation results (Table 12.2).

TABLE 12.2
Annual electricity generation with different PV tilt angles

Serial no.	Tilt angle (degees)	Annual electricity generation (kWh)
1	23	3,165
2	28	3,191
3	33	3,176

You can see that the maximum generation takes place when the PV collector is tilted at the latitude angle, which is 28 degrees for New Delhi. Save all the models to use in later tutorials.

Tutorial 12.2

Evaluating the Impact of Shading from Rooftop PV Panels

GOAL

To evaluate the impact of shading from rooftop PV panels on building energy consumption and cooling load

WHAT ARE YOU GOING TO LEARN?

- How to create a rooftop PV system with multiple collectors

PROBLEM STATEMENT

In this tutorial, you are going to use the model created in Tutorial 12.1. Add three more solar PV collectors (total area of PV collectors = 4×12.6 m^2). Determine building annual energy consumption and cooling load for the following:

1. With PV modules (roof is shaded with PV collectors' shadow)
2. Without PV modules (no shadow as there are no PV collectors)

Use the **New Delhi Palam India** weather location.

SOLUTION

Step 1: Open the model created in Tutorial 12.1 for 28 degrees of tilt.

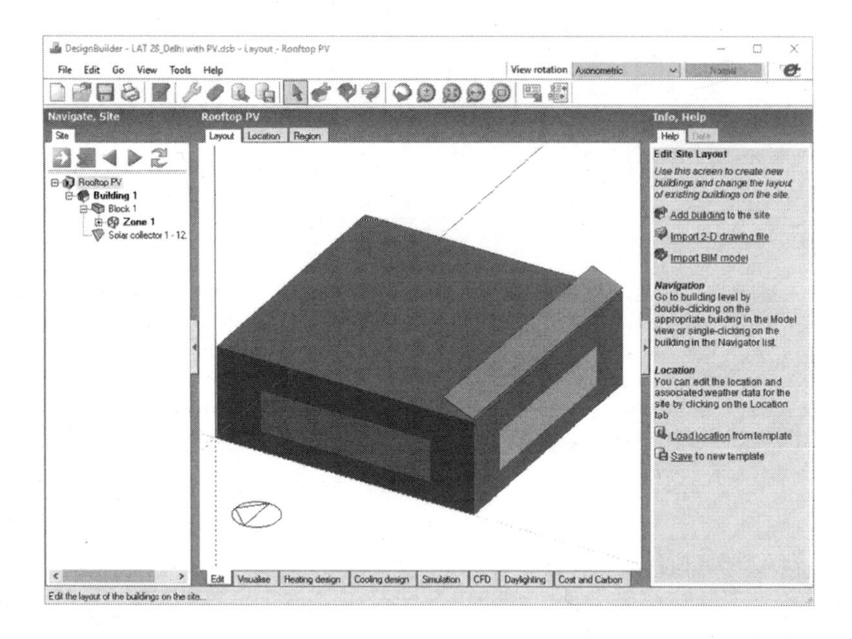

Step 2: Draw a construction line on the roof to mark the positions of the solar PV collectors. Click the **Place construction line** icon. Create a grid with construction lines dividing the roof into four equal strips.

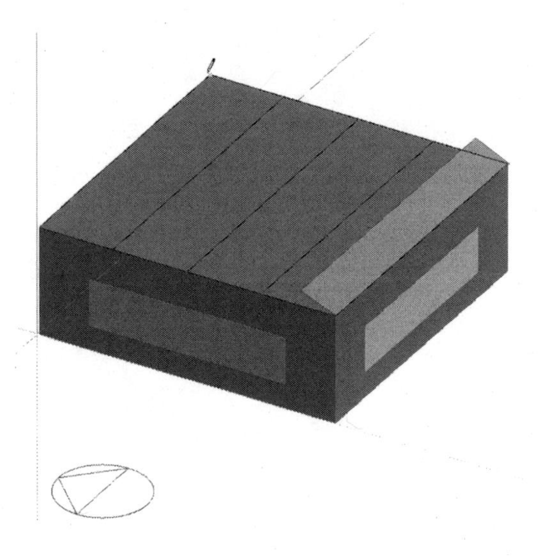

Step 3: Select **Left** from the **View rotation** drop-down list.

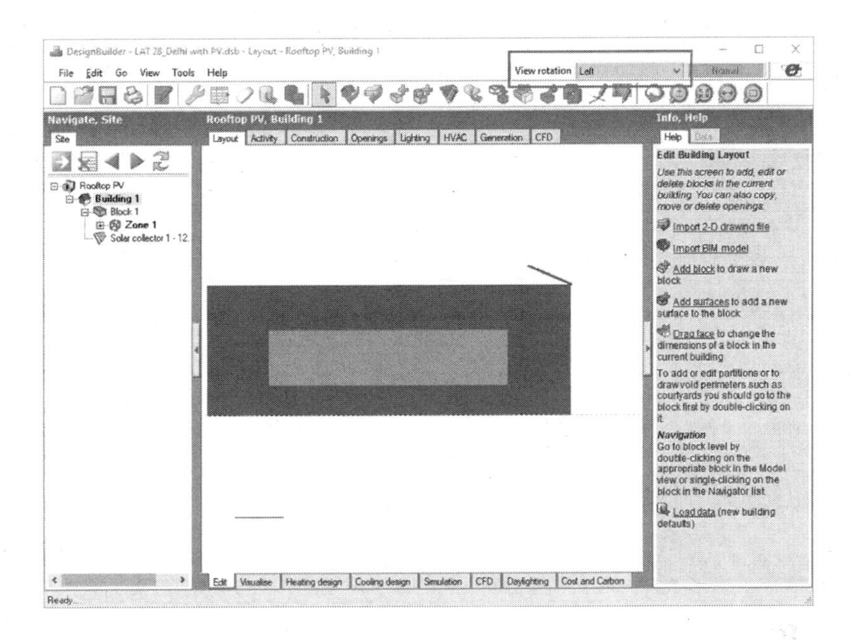

Step 4: Select **PV collector**, and copy the PV panel using the **Clone selected object** icon.

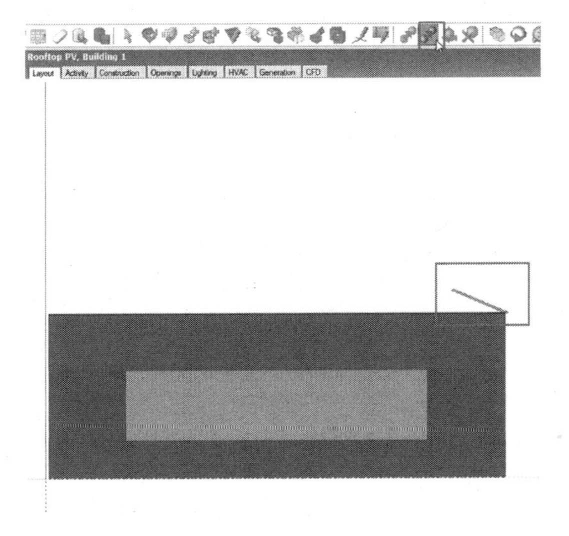

Step 5: Clone the solar collector, and place it on construction lines, as shown in the following figure.

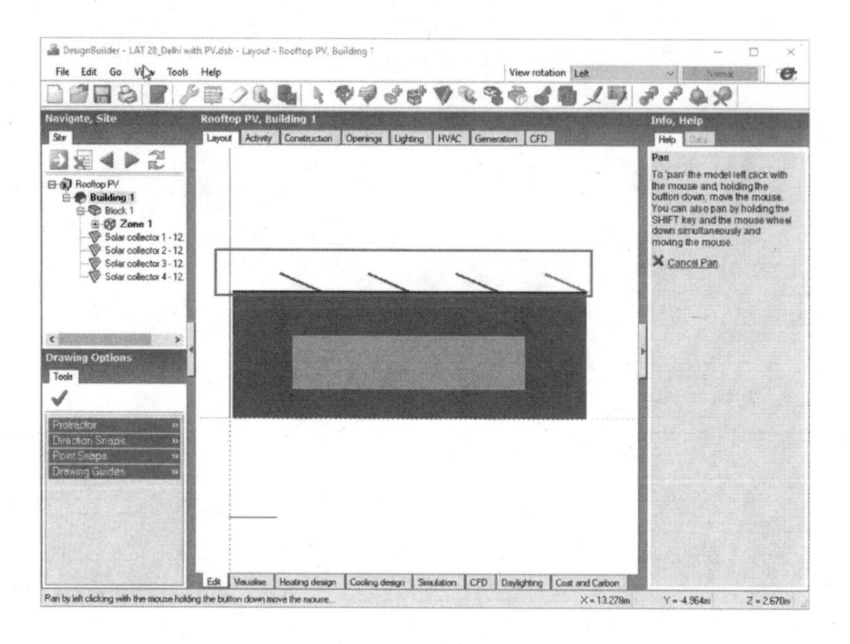

Step 6: Select **Axonometric** from the **View rotation** drop-down list.

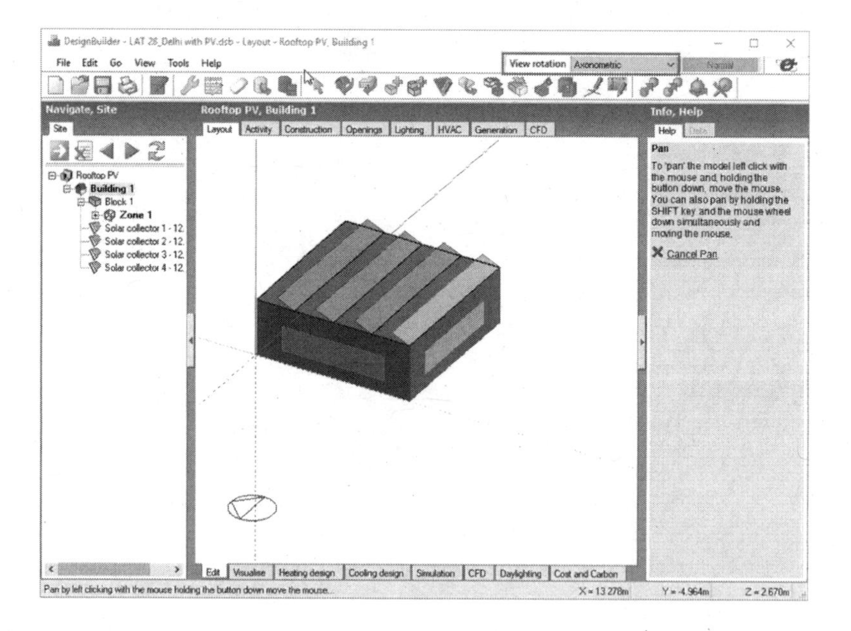

> Note that in cases where a solar collector is placed on a building surface (e.g. on a pitched-roof surface), it is not modelled as being fully coupled with the underlying surface; however, it does cast a shadow, as would any other block or shading surface.
>
> *Source*: https://designbuilder.co.uk/helpv6.0/Content/AddingSolarCollectors.htm.

Step 7: Click **Building 1**, and then click the **Generation** tab. Select the **Include electric load centre** checkbox. Create a copy of **DC with inverter** using **the Create copy of highlighted item** icon.

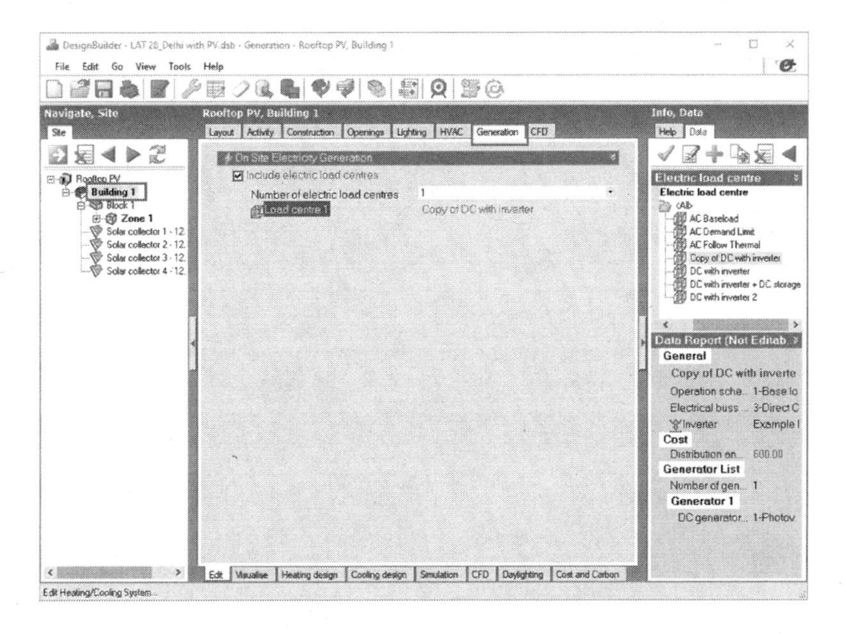

Step 8: Edit **Copy of DC with inverter**.

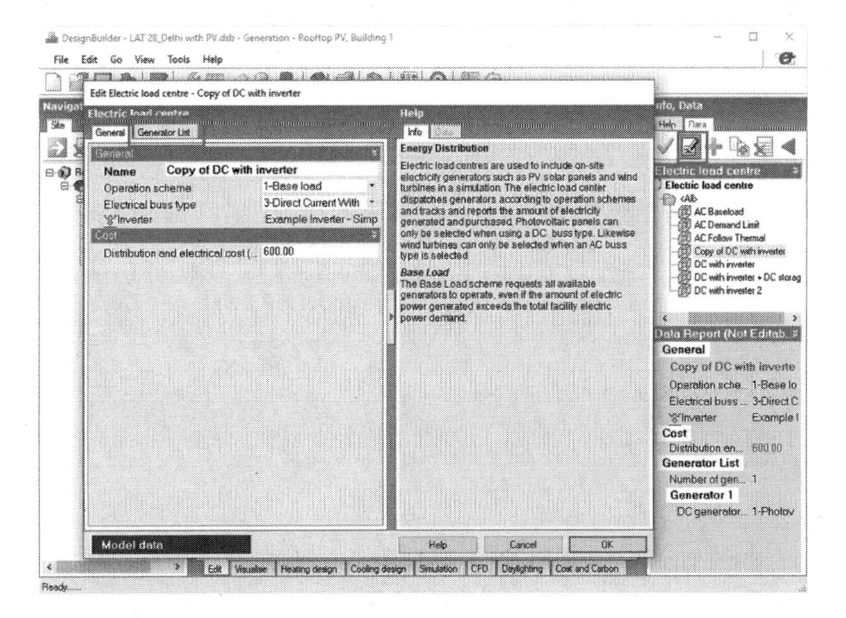

Step 9: Click the **Generator List** tab. Select **4** from **the Number of generators** drop-down list.

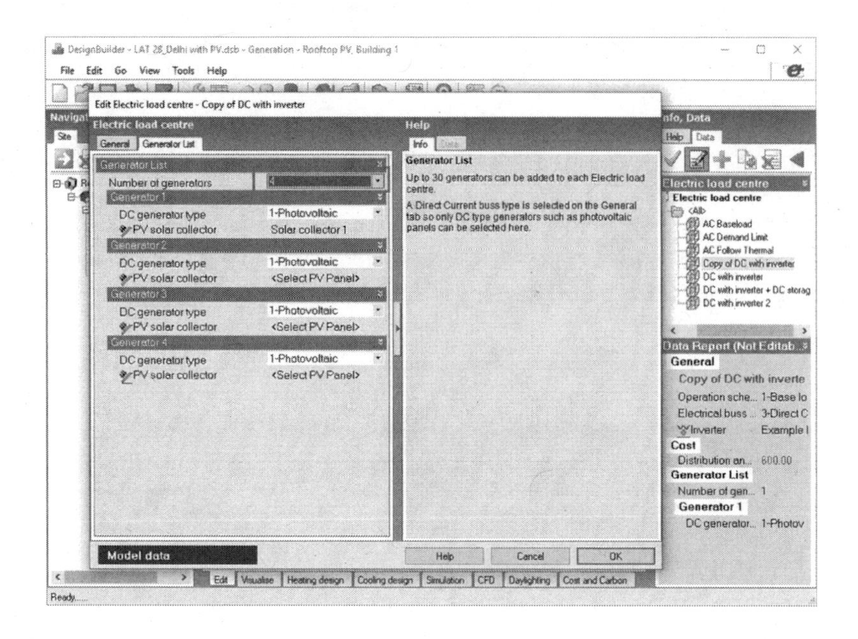

Step 10: Click **PV solar collector**, and select **Solar collector 2**.

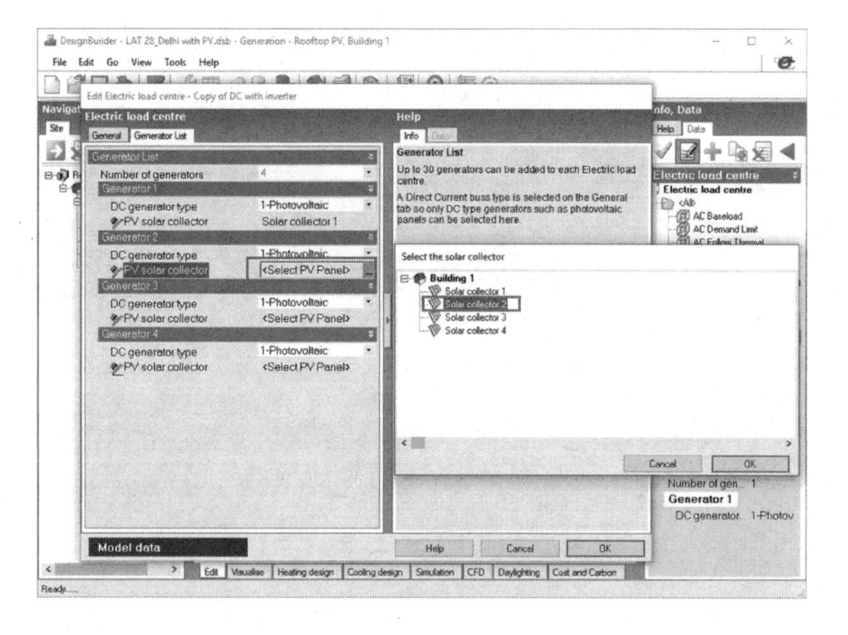

Step 11: Update other PV solar collectors.

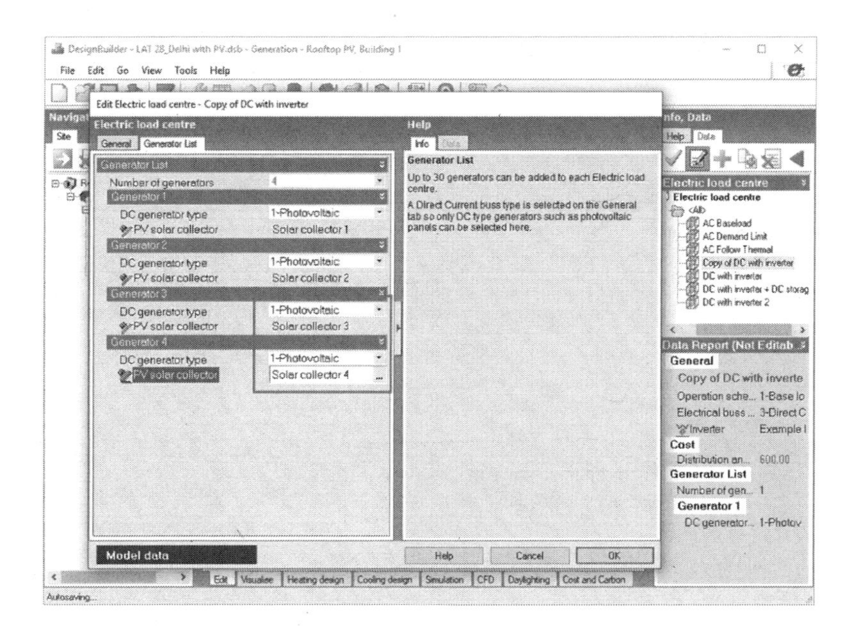

Step 12: Click the **Visualise** tab. Click the **Rendered View** tab. Select the **Show shadows** checkbox. Select **Apr** from the **Month** drop-down list.

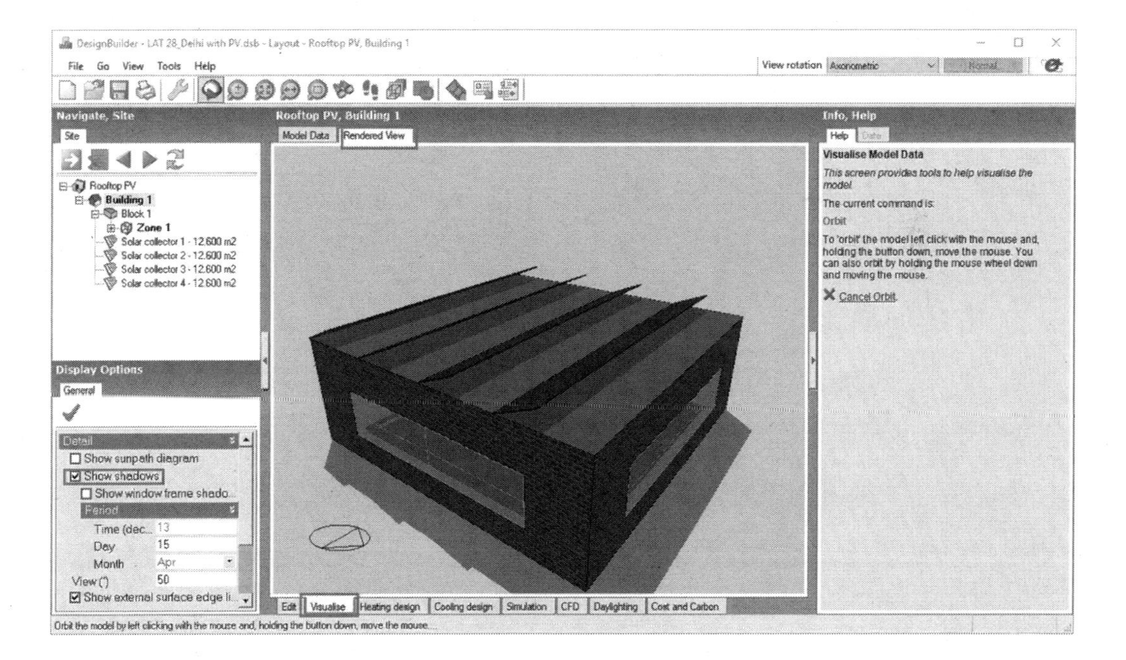

Step 13: Simulate the model, and record annual cooling energy consumption.

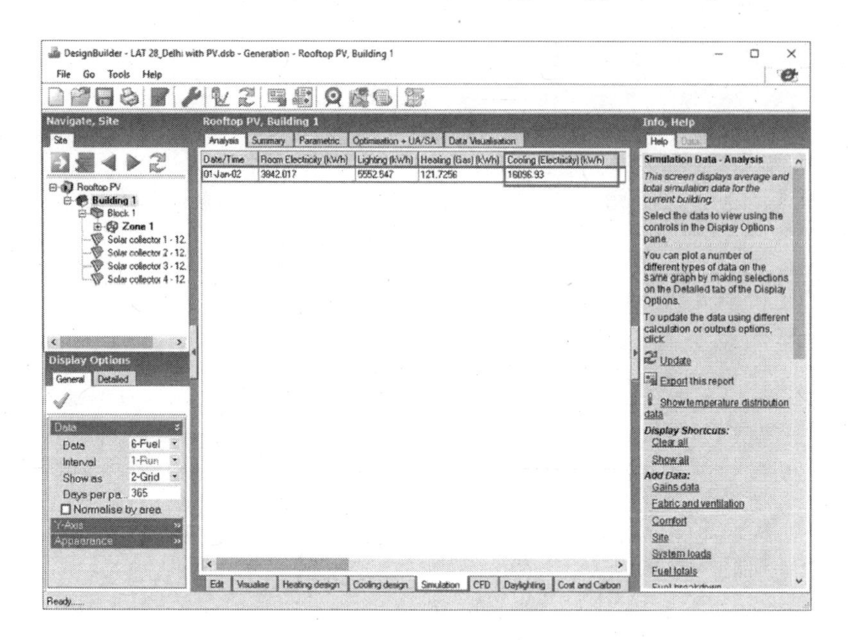

Step 14: Click **Cooling design** to calculate the cooling load.

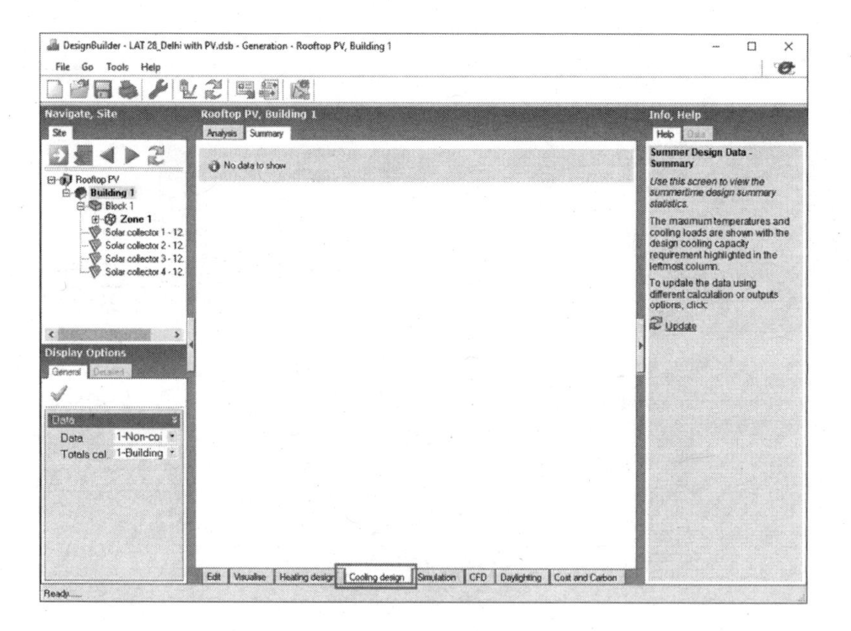

Step 15: Record the cooling load.

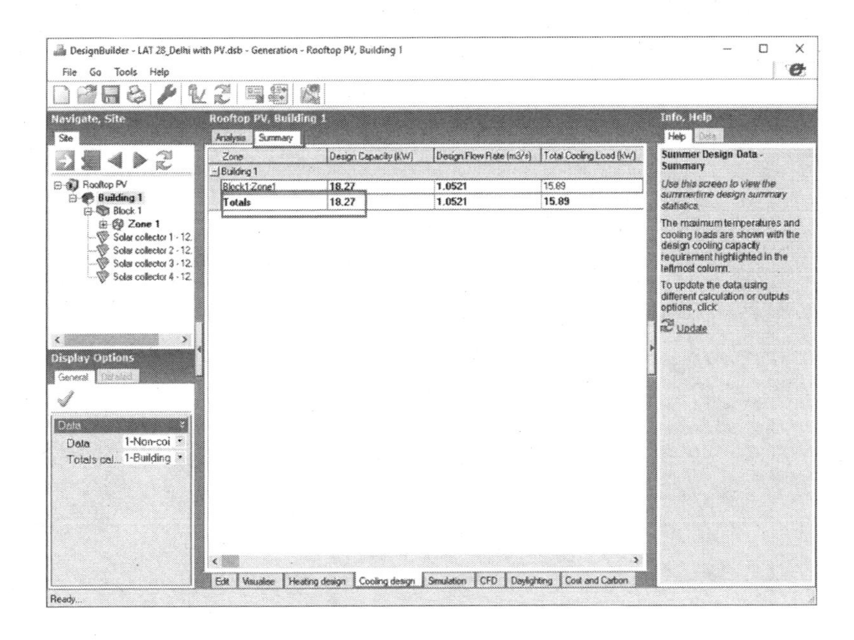

Step 16: Save the model. Select all the PV collectors. Delete them by clicking on the **Delete selected object(s)** icon.

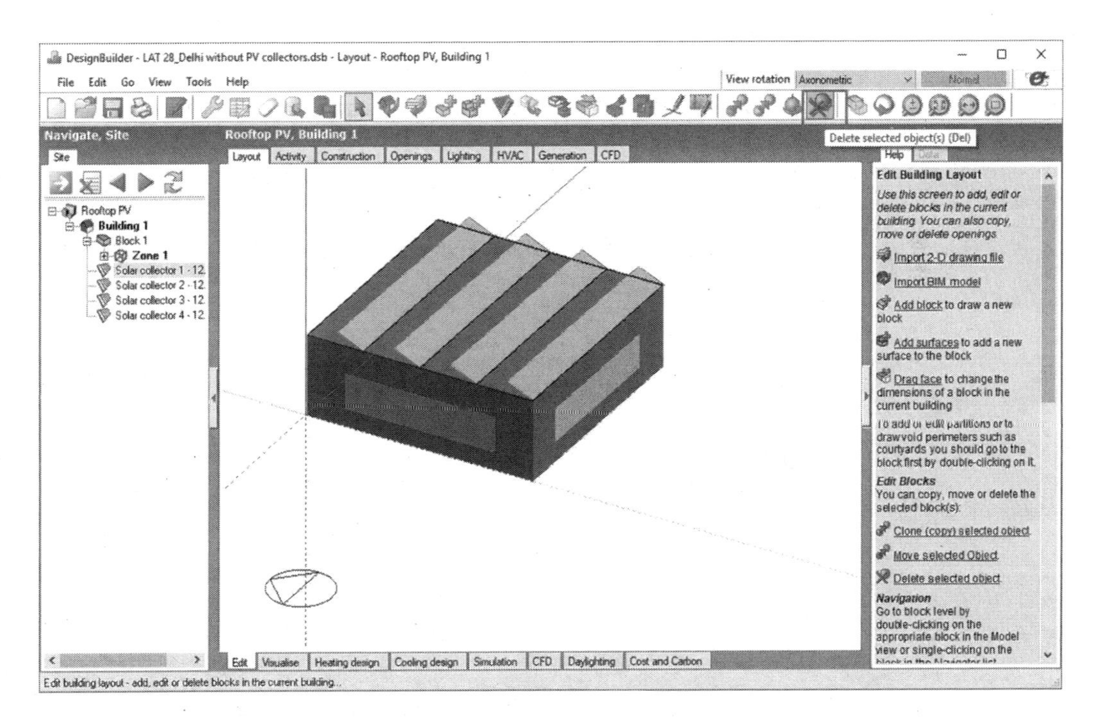

Step 17: Click **Yes** to **Delete**.

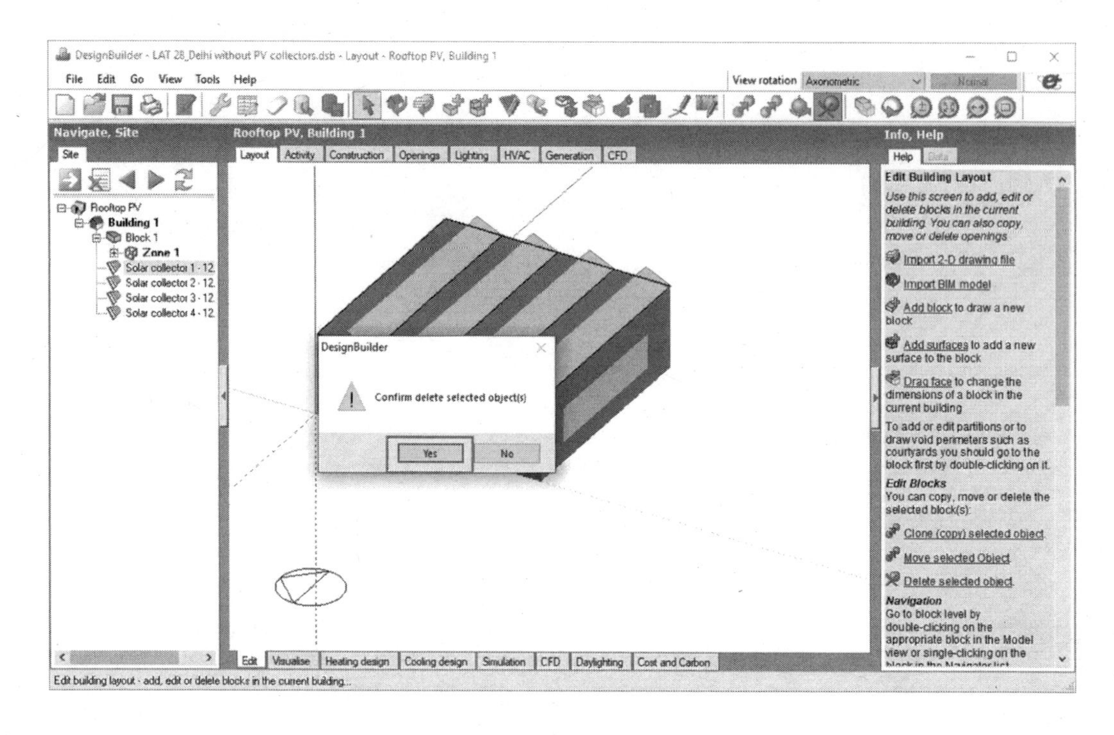

Step 18: Click the **Generation** tab.

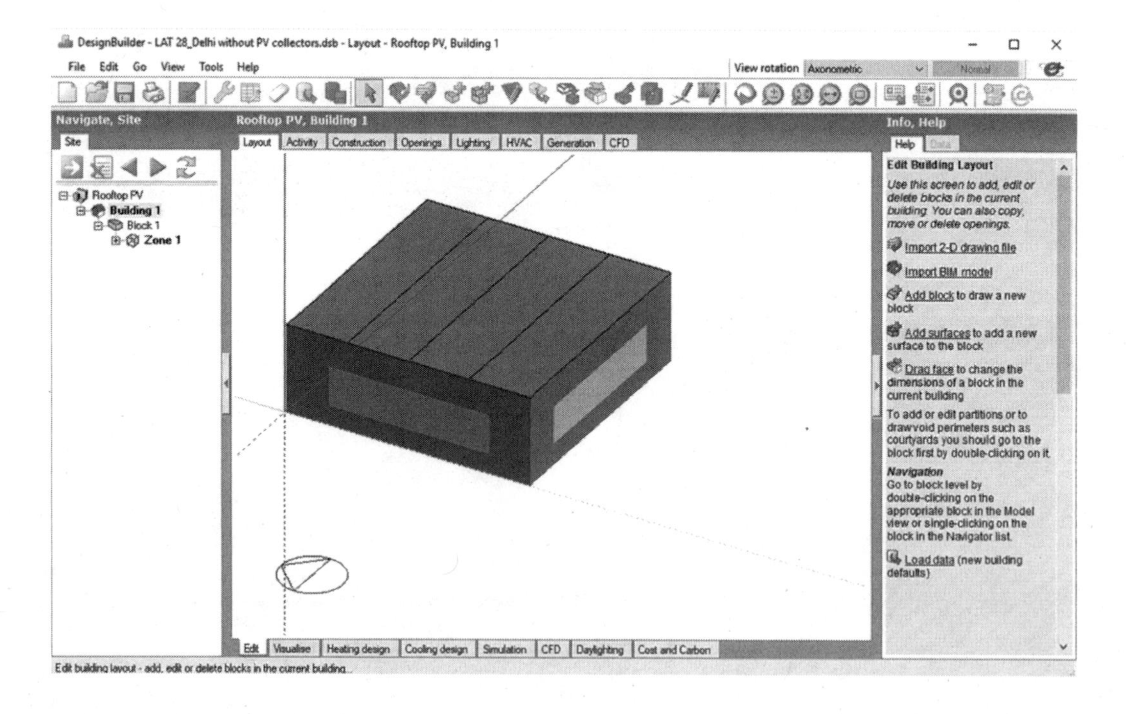

Step 19: Clear the **Include electric load centres** checkbox.

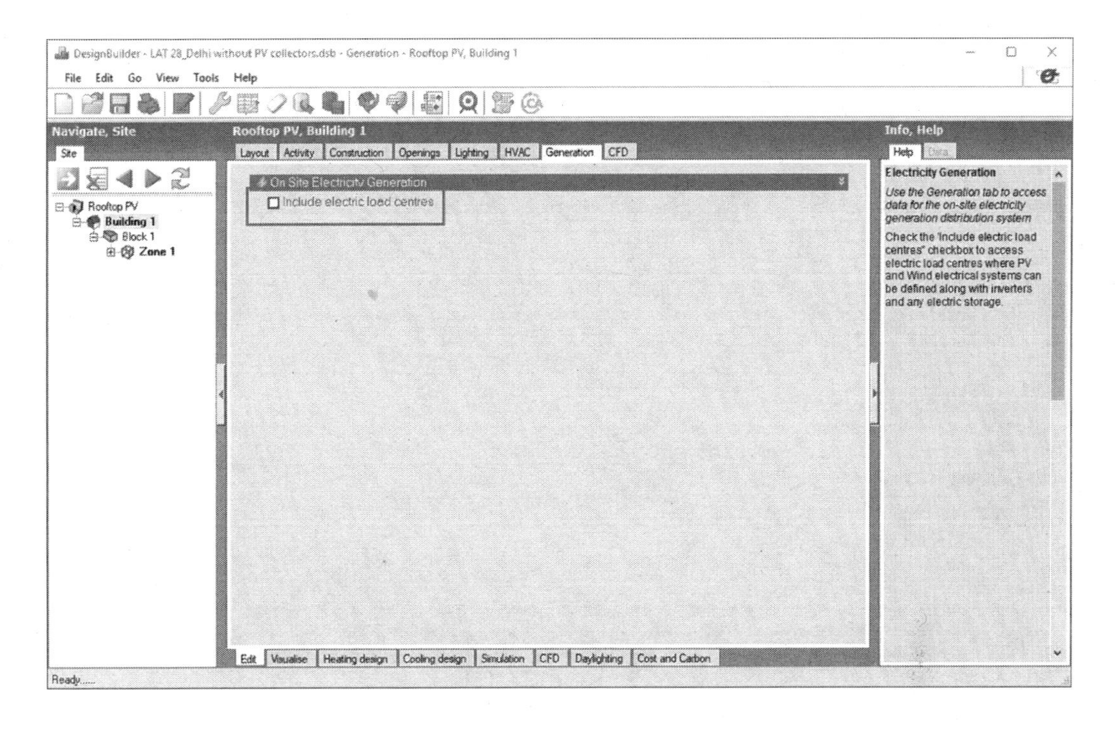

Step 20: Simulate the model, and record annual cooling energy consumption.

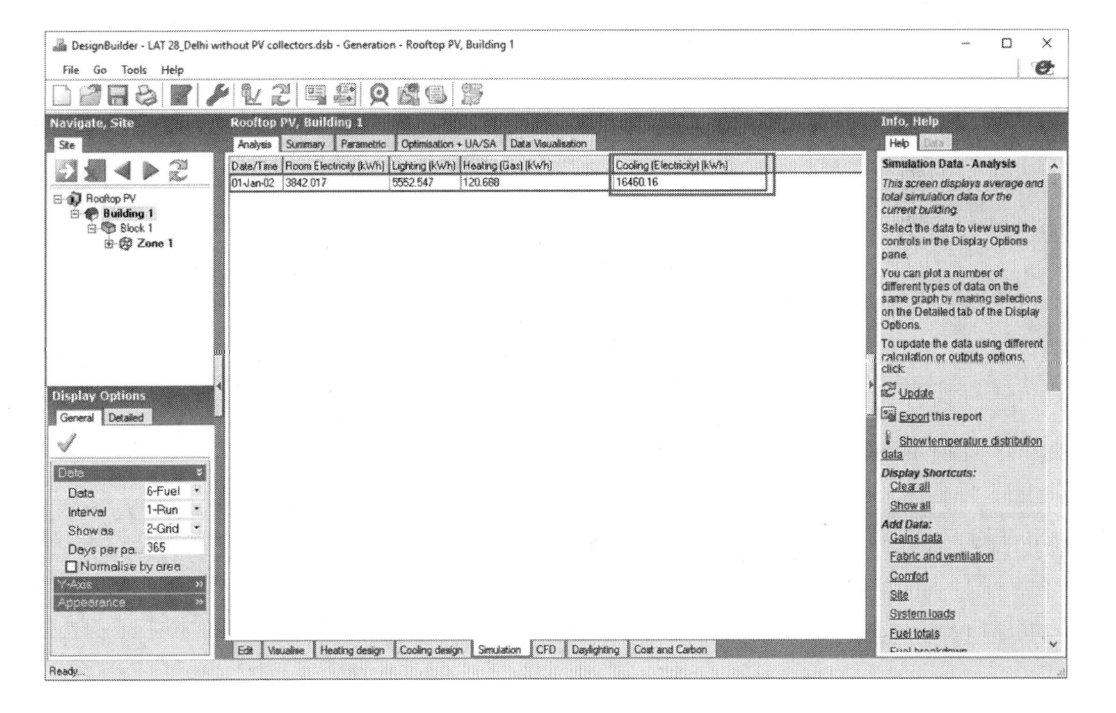

Step 21: Click **Cooling design** to calculate the cooling load.

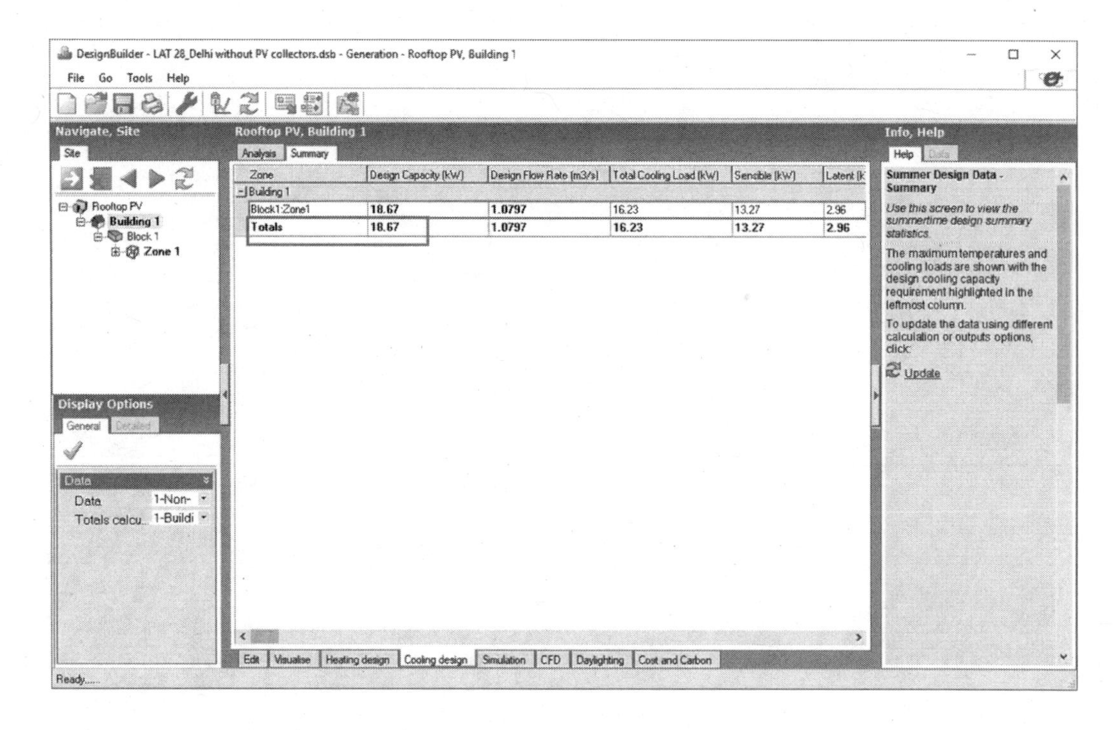

Step 22: Compare the results (Table 12.3).

TABLE 12.3
Annual energy consumption and cooling design capacity

Configuration	Annual cooling energy consumption (kWh)	Cooling design capacity (kW)
With rooftop PV	16,097	18.27
Without rooftop PV	16,460	18.67

You can see that because of the shadow from the PV collectors on the roof, there is a reduction in the annual energy consumption and cooling design capacity.

Tutorial 12.3

Evaluating the Impact of the Cell Efficiency of PV Panels

GOAL

To evaluate the impact of the cell efficiency of PV panels on annual energy generation

WHAT ARE YOU GOING TO LEARN?

- How to edit cell efficiency

PROBLEM STATEMENT

In this tutorial, you are going to use the model created in Tutorial 12.1. Determine building energy generation with following cell efficiencies:

1. 0.15
2. 0.20
3. 0.25

Use the **New Delhi Palam, India** weather location.

PV Cell Efficiency

This is the efficiency with which PV surfaces convert incident solar radiation into electricity. Solar PV materials are mainly classified into following categories:

- Monocrystalline silicon solar cells (efficiency 18%–25%)
- Polycrystalline silicon solar cells (efficiency 13%–20%)
- Thin-film solar cells (efficiency 9%–11%)
- Organic solar cells (efficiency 8%–12%)

Source: www.nrel.gov/pv/cell-efficiency.html.

SOLUTION

Step 1. Open the model created in Tutorial 12.1 for 28 degrees of PV tilt.

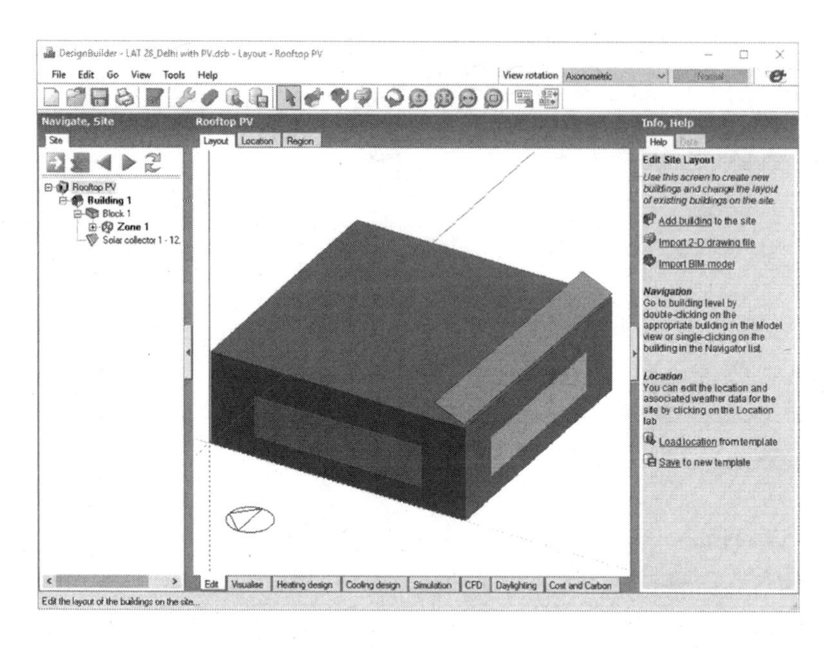

Step 2: Click **Solar collector 1 – 12.600 m2**. Click the **Construction** tab. Select **1-simple** from the **Performance type** drop-down list. Make sure that **PV constant Efficiency = 0.15** is selected for the **Performance model**.

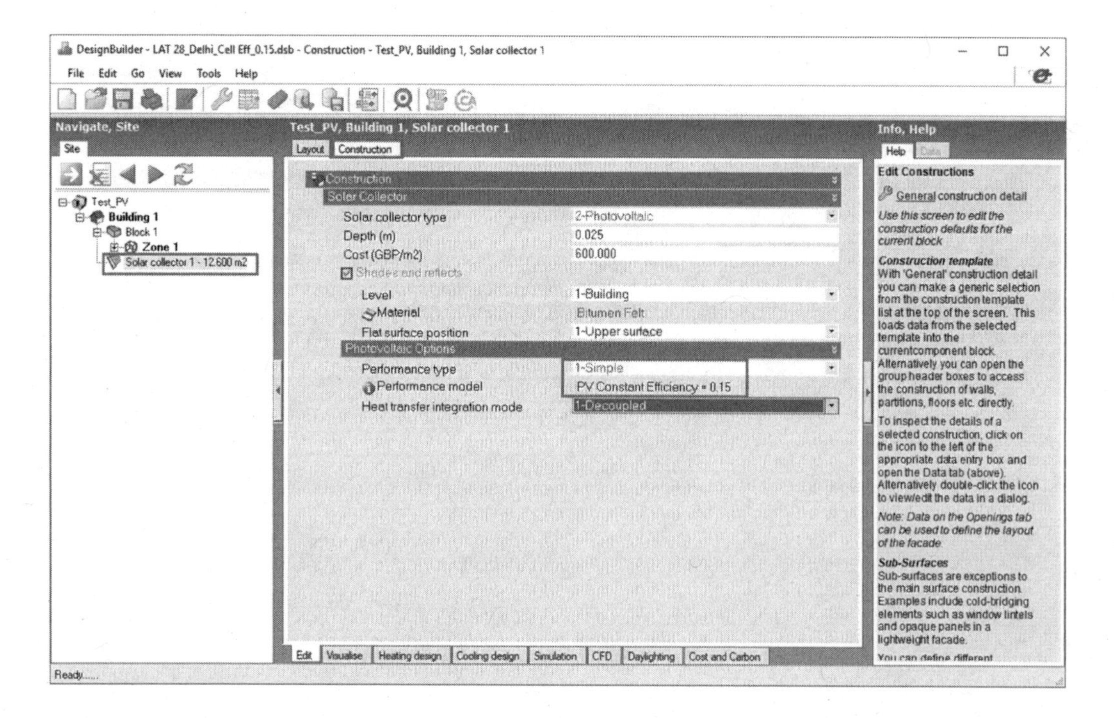

The **PV Generator Simple** component describes a simple model of PV panels that may be useful for early-phase design analysis. It provides direct access to the efficiency with which PV panels convert incident solar radiation to electricity and does not require arrays of specific modules to be defined. The full geometrical model for solar radiation is used, including shading and reflections, to determine the incident solar radiation on the panel. This model is intended to be useful for design purposes to quickly get an idea of the levels for annual production and peak power. The model can also accept arbitrary conversion efficiencies and does not require actual production units be tested to obtain performance coefficients.

Source: https://designbuilder.co.uk/helpv6.0/Content/PerformanceSimplePV.htm.

Step 3: Simulate the model, and record the annual energy generation.

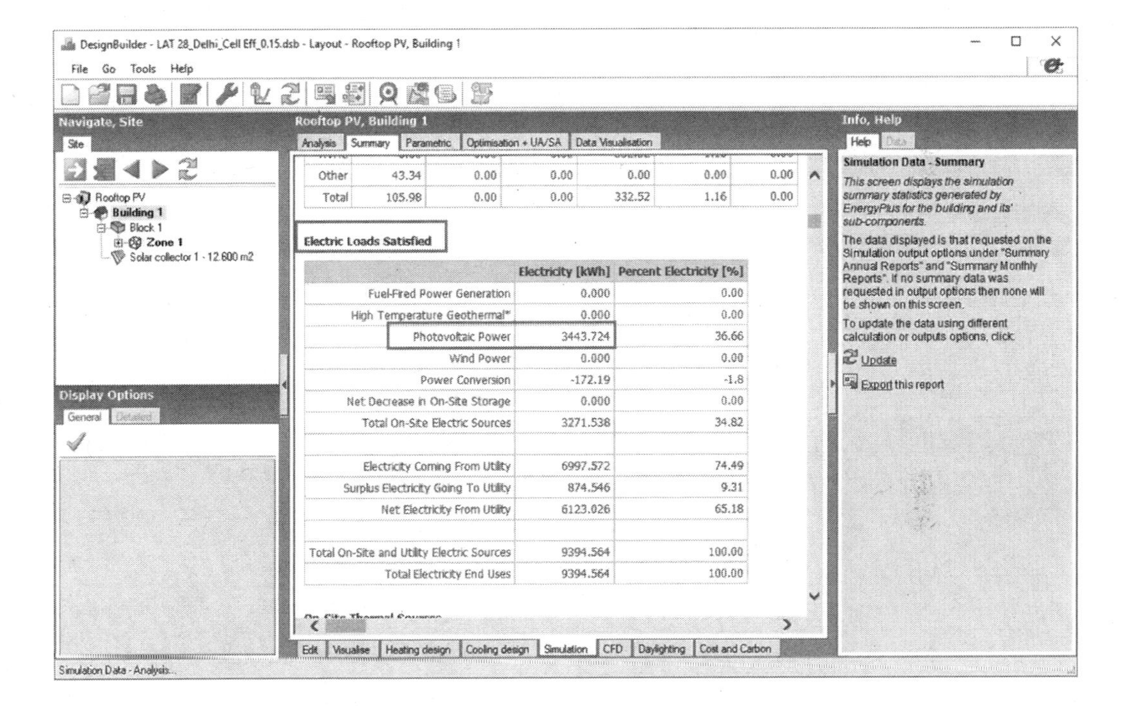

Step 4: Click the **Edit screen** tab.

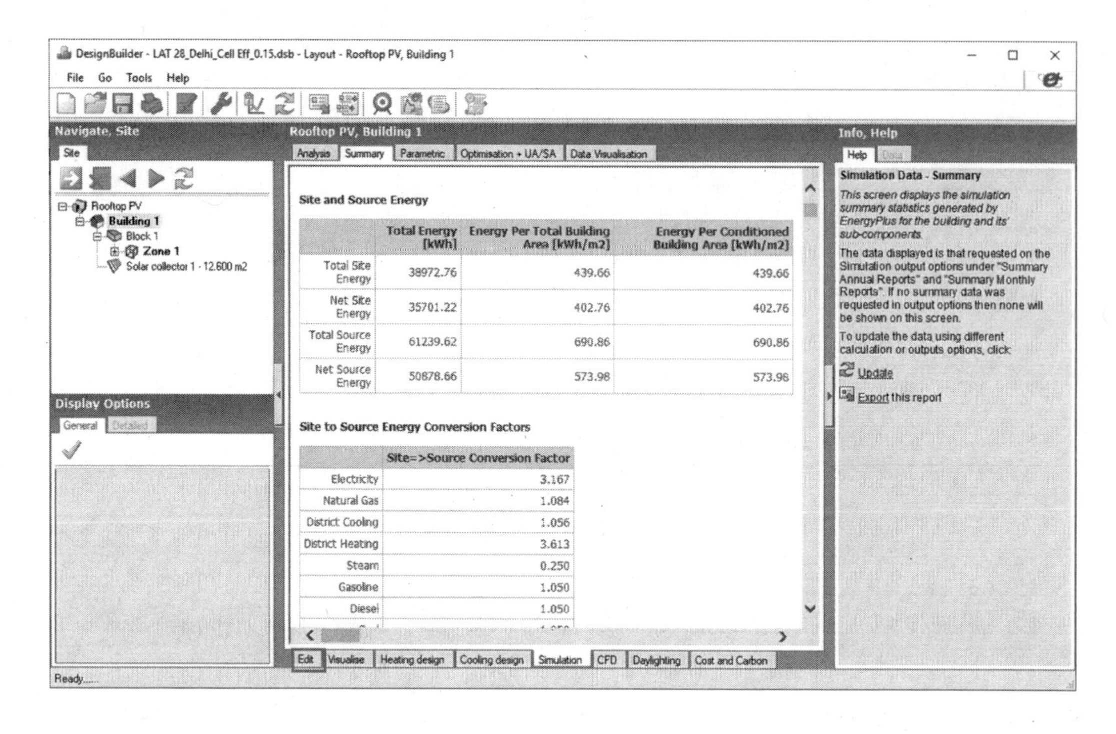

Step 5: Click **Solar collector 1 – 12.600 m2**. Click the **Construction** tab.

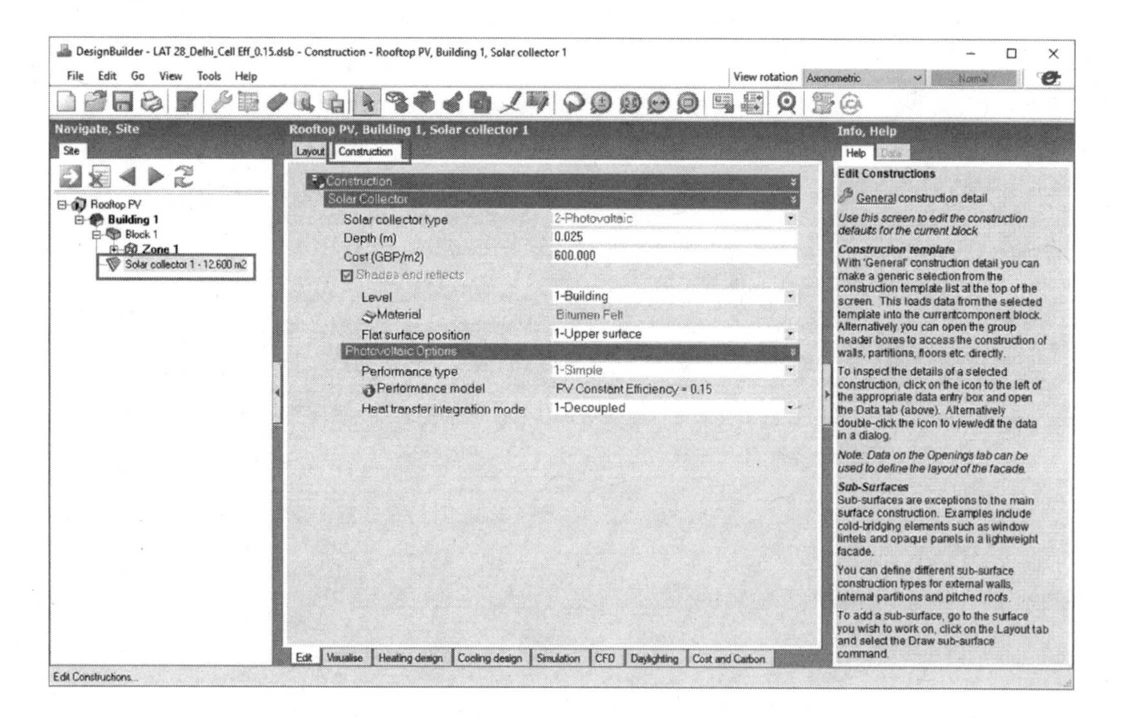

Step 6: Click **PV constant Efficiency = 0.15**. Click the **Create copy of highlighted item** icon.

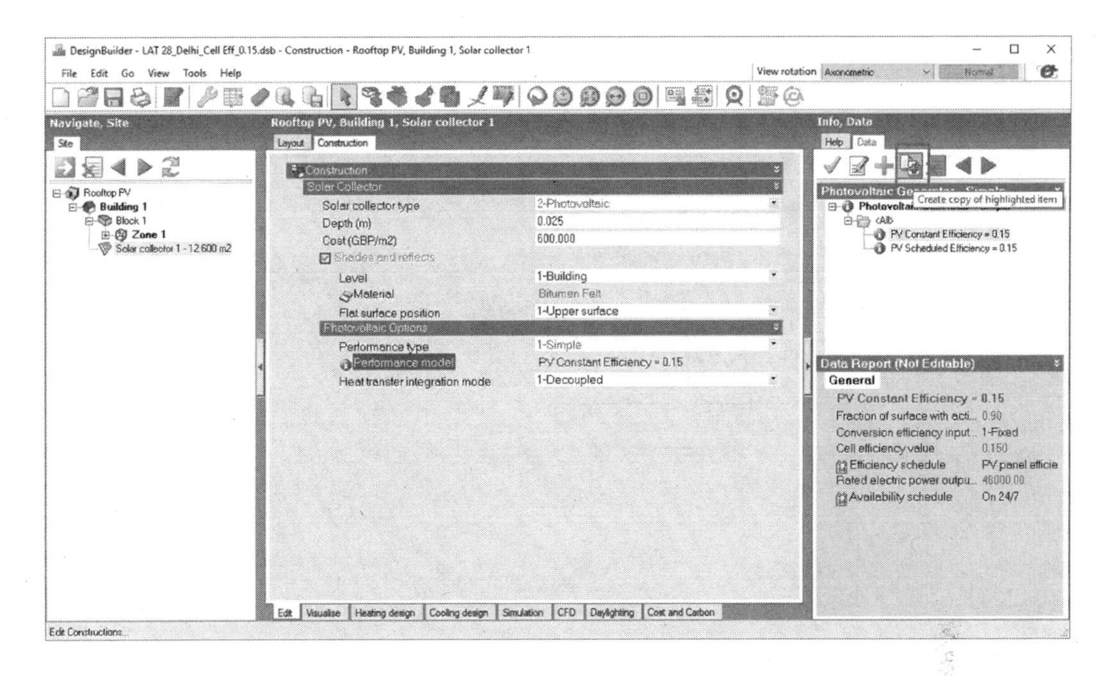

Step 7: Click the **Edit highlighted item** icon.

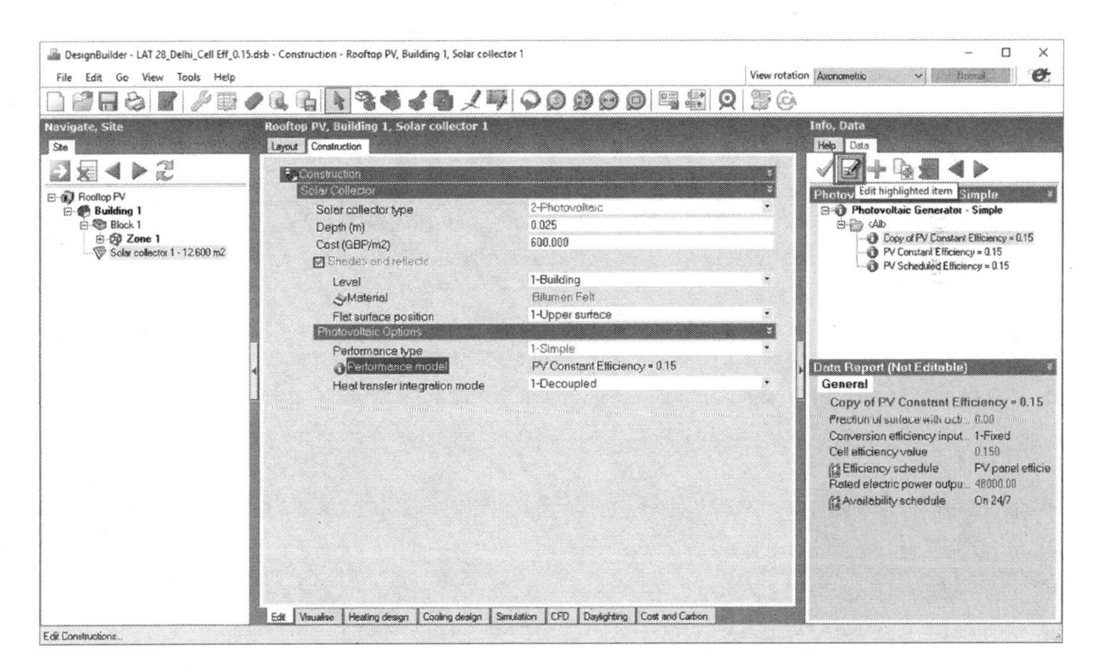

Step 8: Enter **PV constant Efficiency = 0.20** in the **Name** text box. Enter **0.20** in the **Cell efficiency value** text box. Click **OK**.

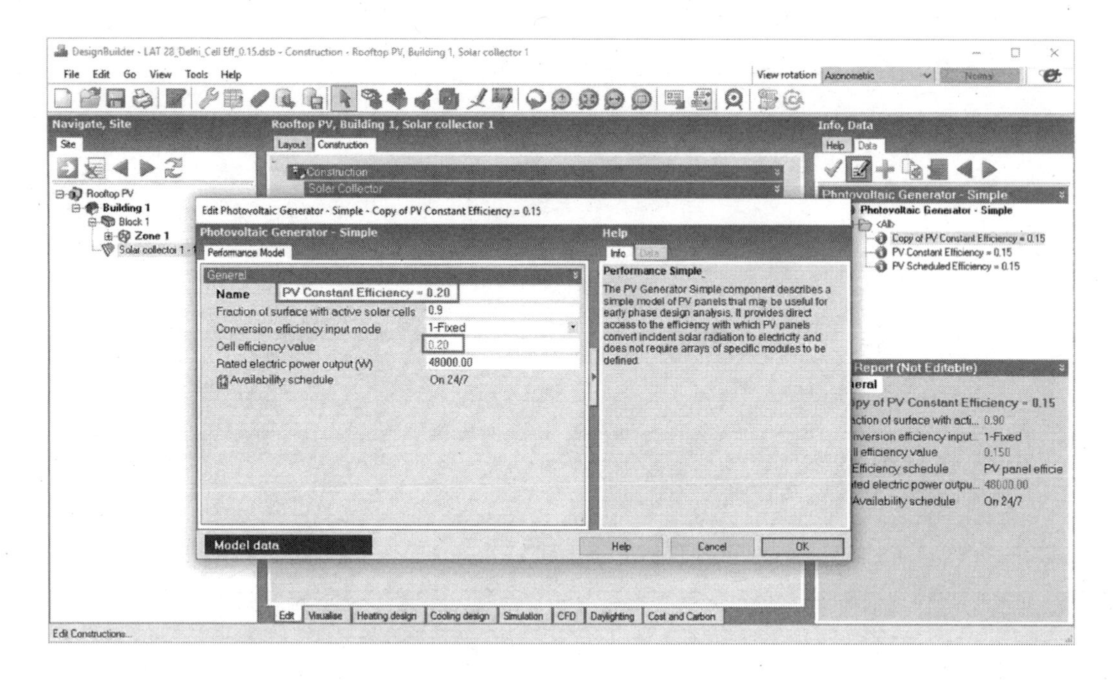

Step 9: Click the **Select this data** icon.

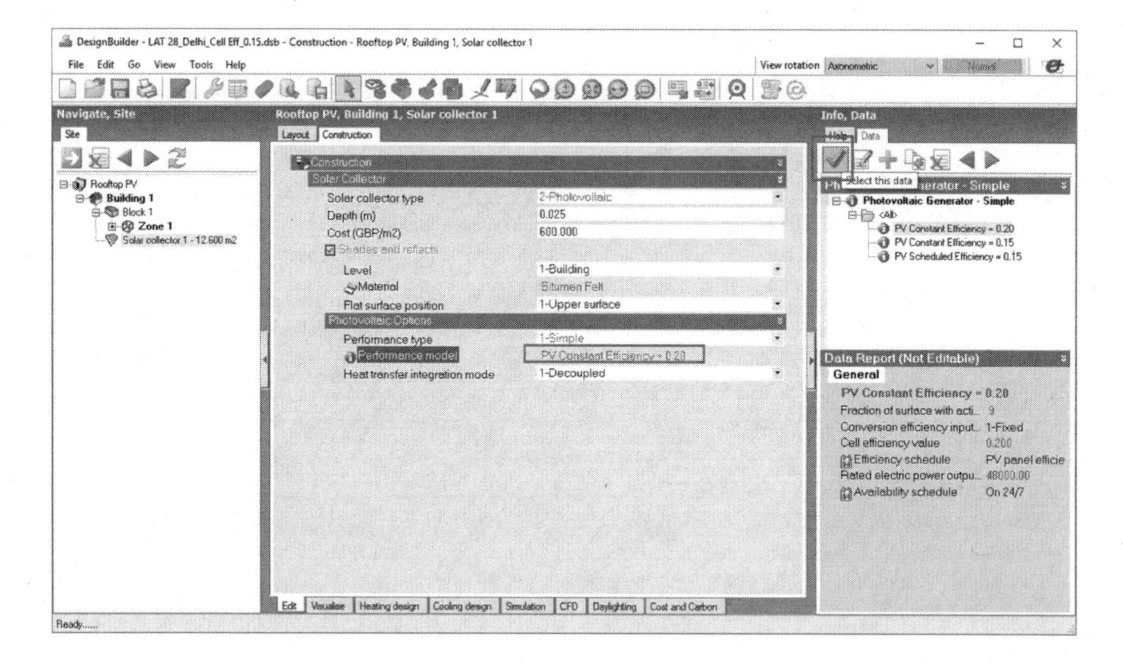

Step 10: Simulate the model, and record the annual energy generation.

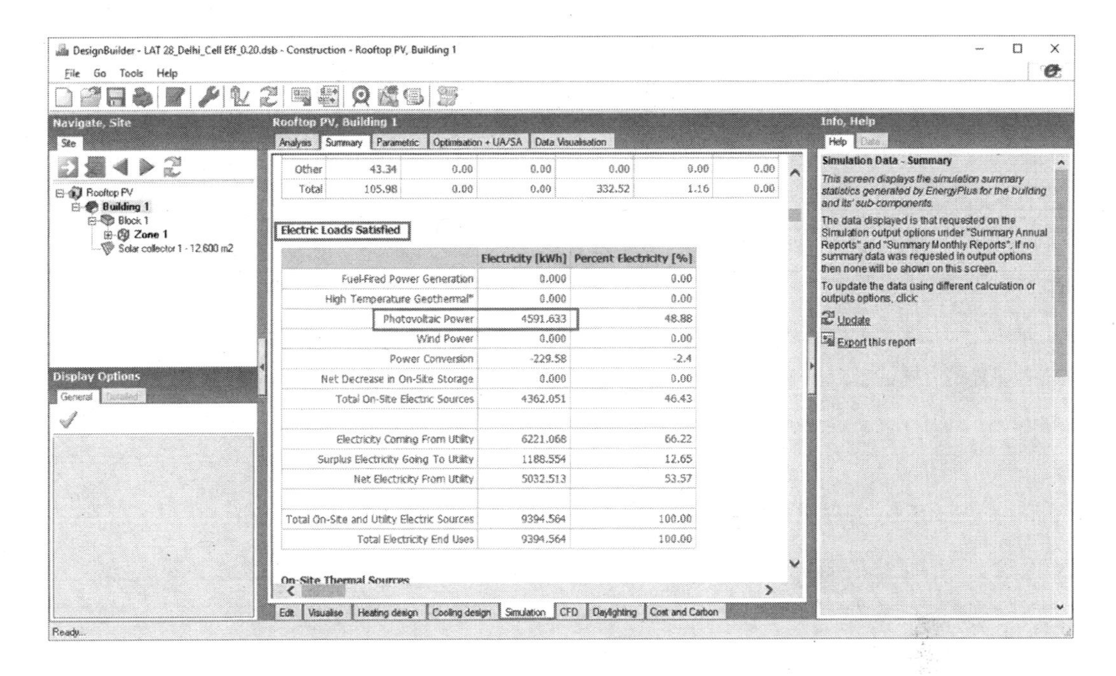

Step 11: Repeat the preceding steps for a cell efficiency of 0.25.

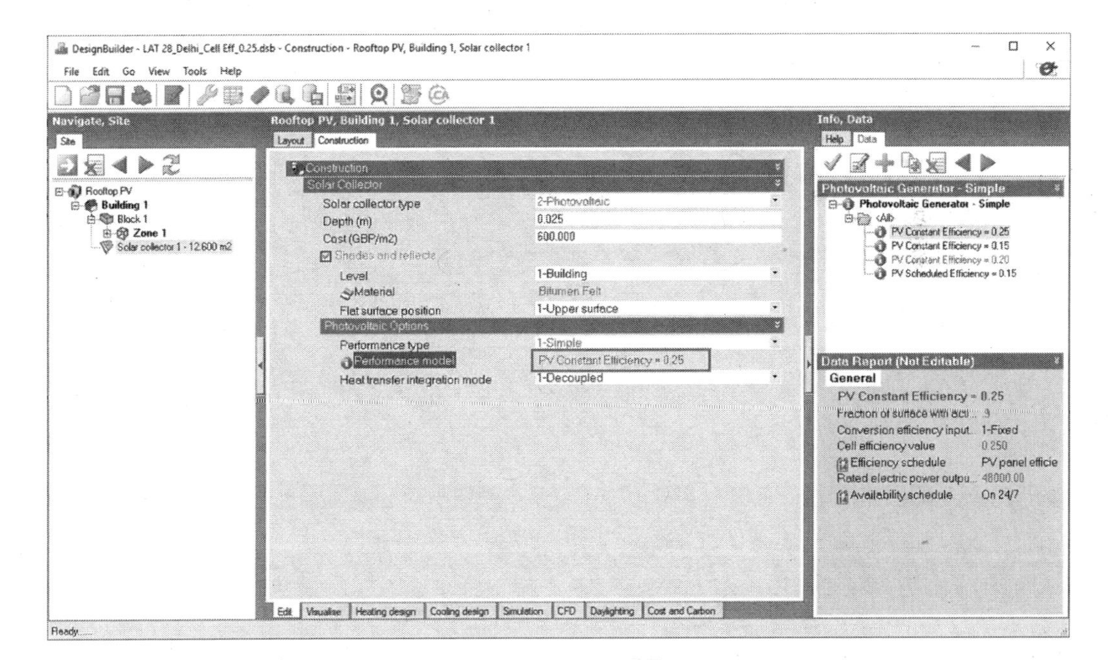

Step 12: Simulate the model, and record annual cooling energy consumption.

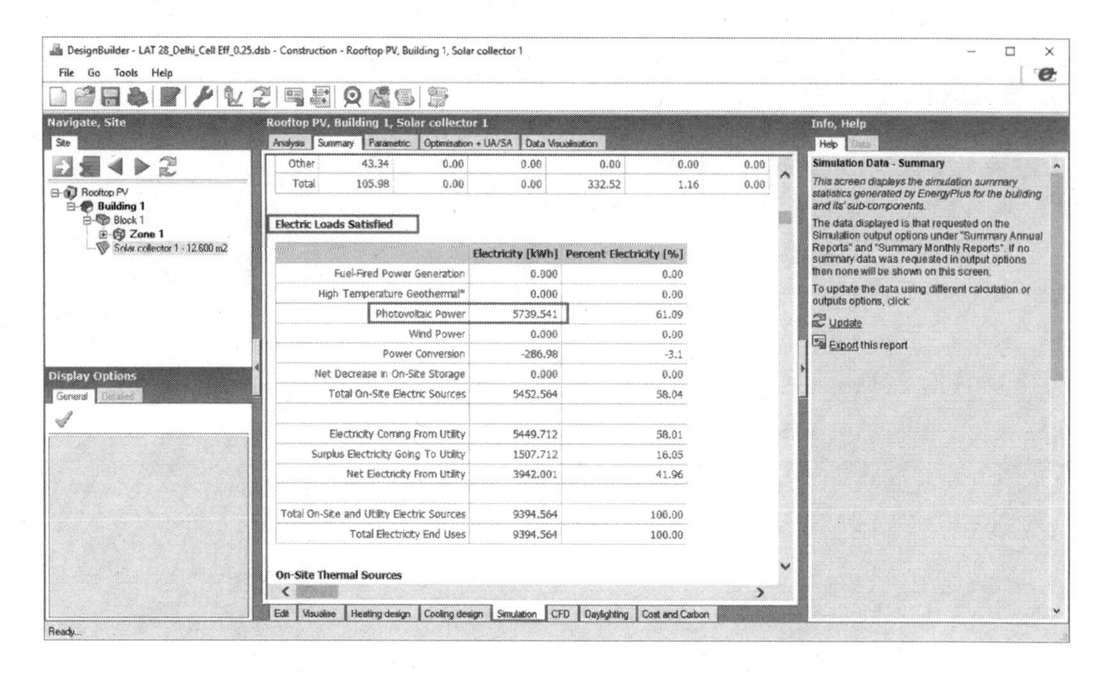

Step 13: Compare the results (Table 12.4).

TABLE 12.4
Annual energy generation with different PV cell efficiencies

PV cell efficiency	Annual energy generation (kWh)
0.15	3,443
0.20	4,591
0.25	5,739

Results shows that with an increase in PV cell efficiency, there is an increase in annual energy generation.

Tutorial 12.4

Evaluating the Performance of Glazing-Integrated PV Panels

GOAL

To evaluate the impact of glazing-integrated PV panels on the south facade on annual energy generation

WHAT ARE YOU GOING TO LEARN?

- How to model a glazing-integrated PV system

PROBLEM STATEMENT

In this tutorial, you are going to use a 50- × 30-m five-zone model. You are going to use the PV module described in Table 12.5 for the simulations.

TABLE 12.5
Building-integrated photovoltaic (BIPV) input parameters

PV Property	Value
Name	BPsolar 5170 (from DesignBuilder Library)
Cell type	Crystalline silicon
Active area	1.26 m^2
Rated electrical power	250 W_p

Model the glazing-integrated PV system on the south facade glazing of a building with an active area of 5 × 1.26 m^2. Determine annual electricity generation for the **Moskva, Russia** weather location.

SOLUTION

Step 1: Create a **50- × 30-m** five-zone model with a **5-m** perimeter depth.

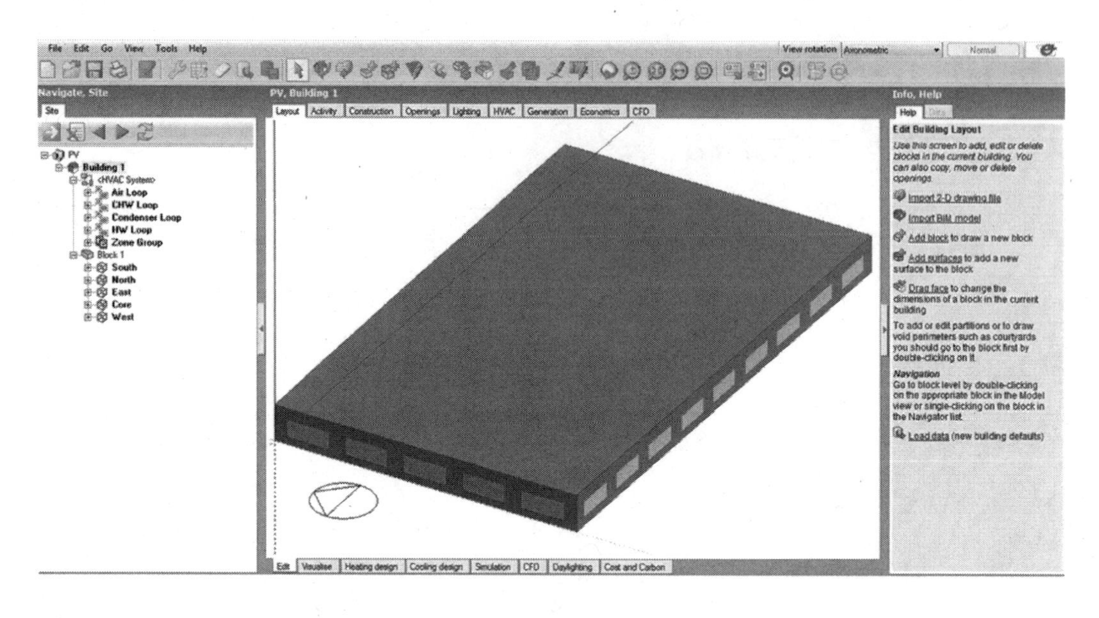

Step 2: Click the **Openings** tab.

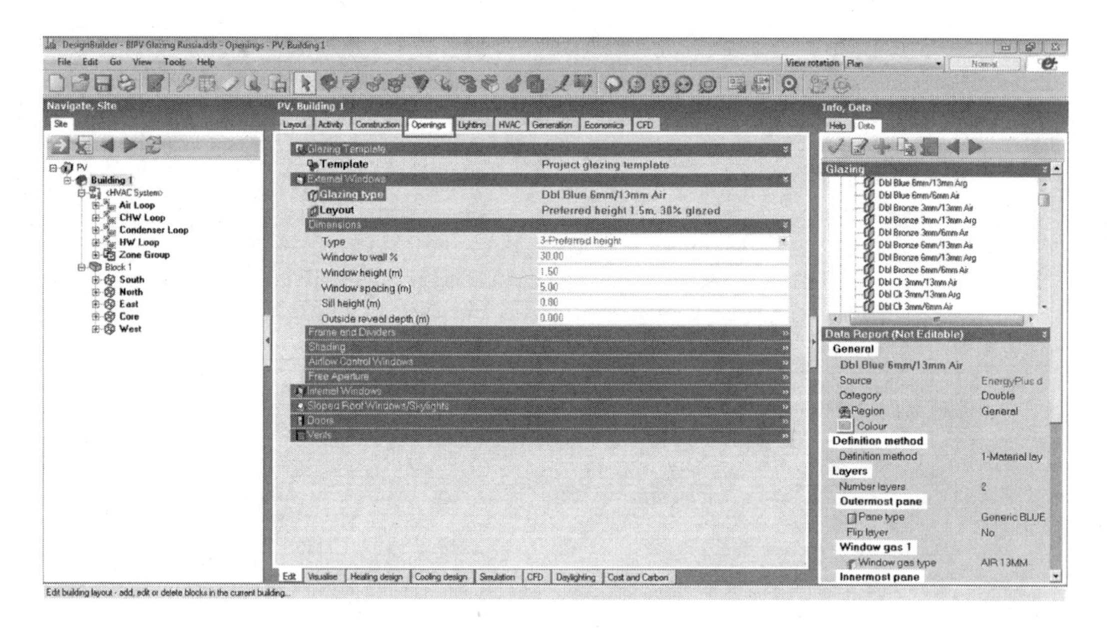

Step 3: Expand the **South** zone, and click on **Wall – 175.00 m2 – 180.** Click on **Glazing type.** Expand **Glazing Integrated Photovoltaics.** Select **Project BIPV Window**, and click **the Create copy of highlighted item** icon.

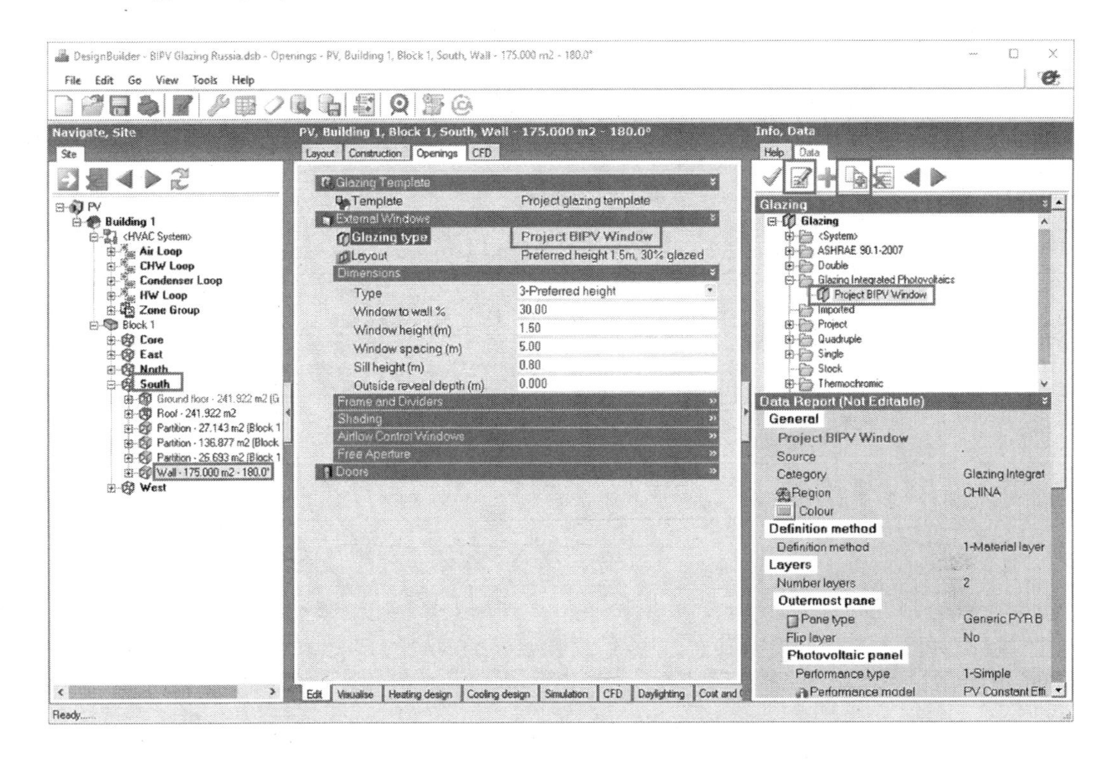

Step 4: Edit the **Copy of project BIPV Window**. Rename it **Glazing Integrated Photovoltaic.** Select **2-Equivalent One-Diode** from the **Performance type** drop-down list. Select **BPsolar 5170** from **Performance model**. Enter **5** in the **Module in series** text box. Enter **1** in the **Series strings in parallel** text box.

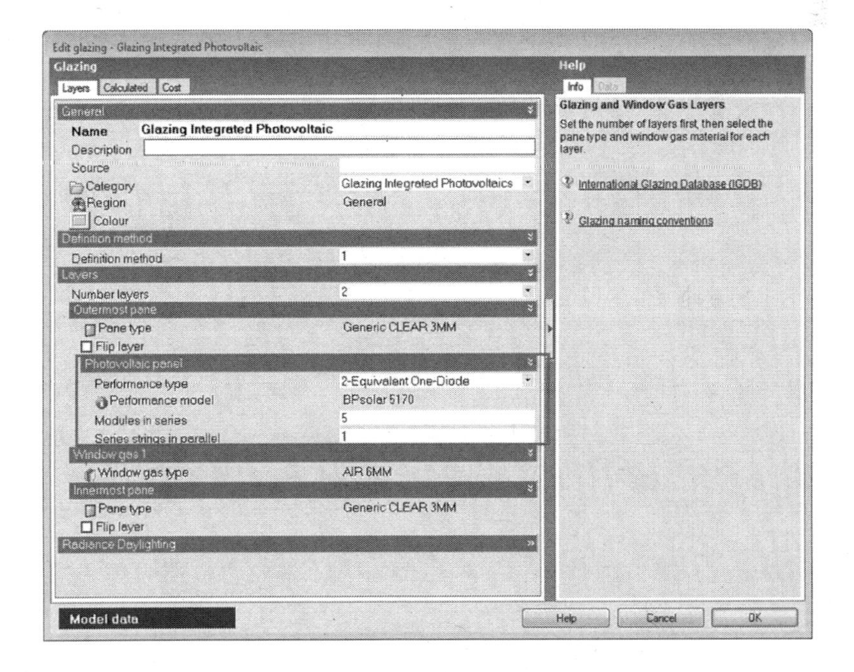

Available glazing area on a south wall is 51.5 m^2, and the active area of the solar cells on each module is 1.26 m^2. A total of 5 × 1 modules can easily be fitted on the available area.

Step 5: Click **Building 1**, and then click the **Generation** tab. Select the **Include electric load centres** checkbox. Click on **Load centre 1**.

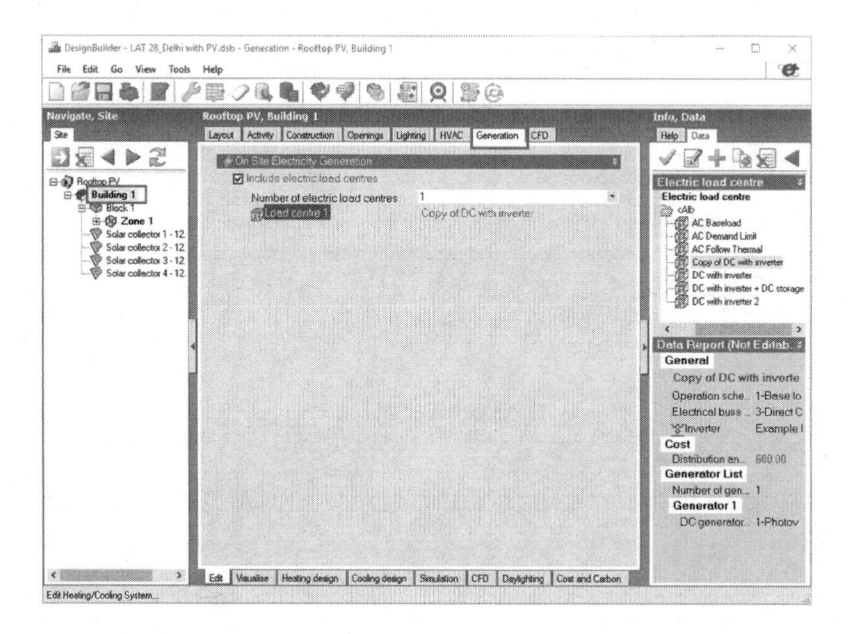

Step 6: Edit the **DC with inverter.** Enter **BIVP System Glazing** in the **Name** text box.

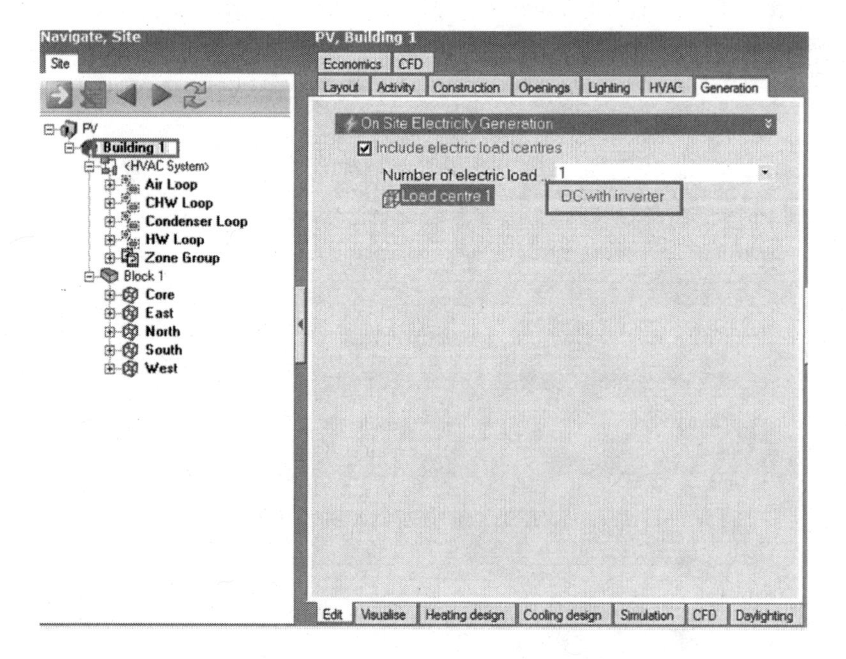

Step 7: Click the **Generator List** tab. Select **3-Glazing with BIPV** from the **DC generator type** drop-down list. Select **Glazing Integrated Photovoltic** in **Building integrated photovoltic glazing**.

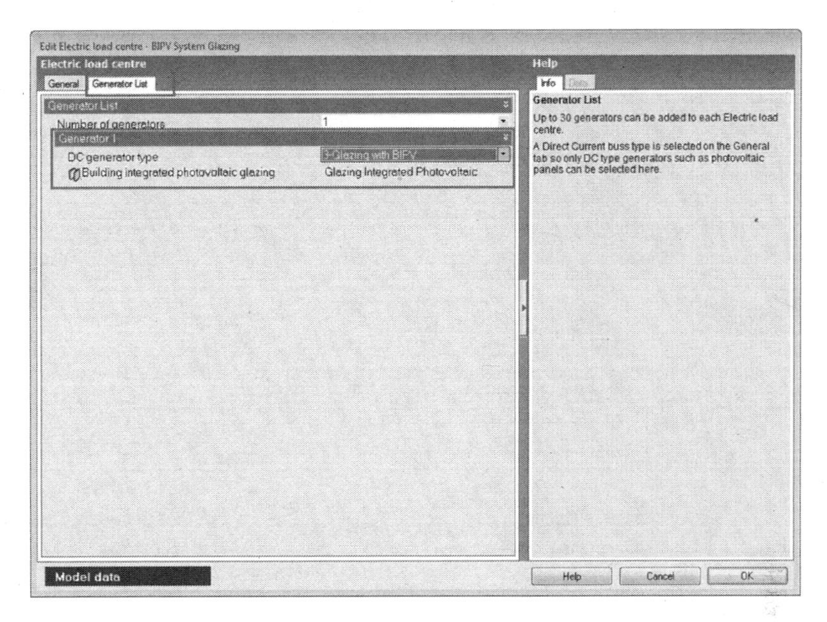

Step 8: Click the **Simulation screen** tab.

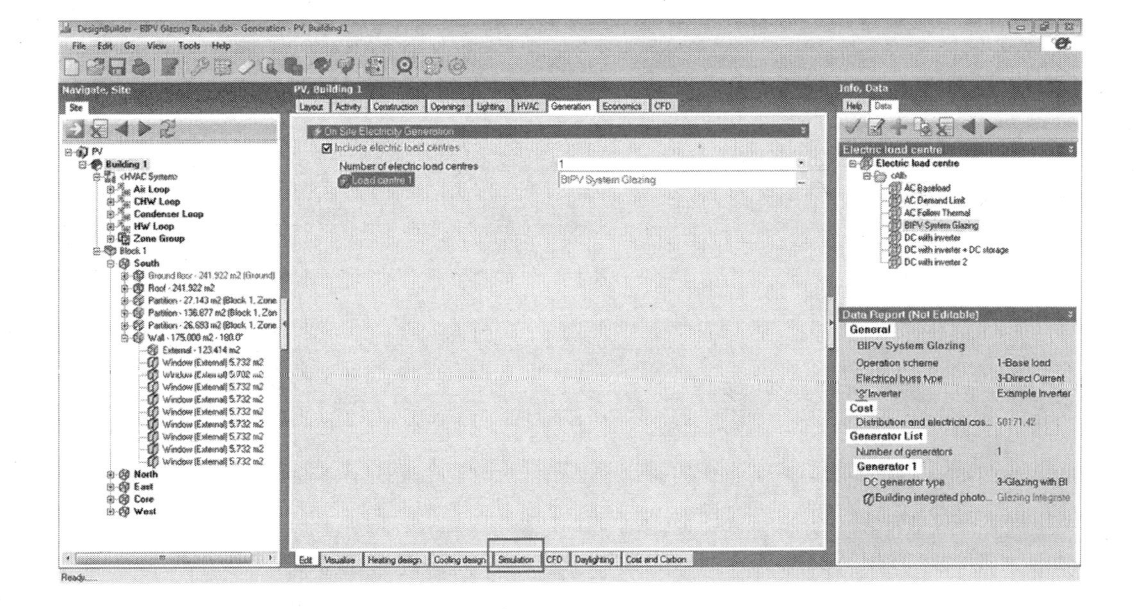

Step 9: Click **Update data**, and then click the **Output** tab. Select the **All Summary** checkbox. Click **OK** to perform the simulation.

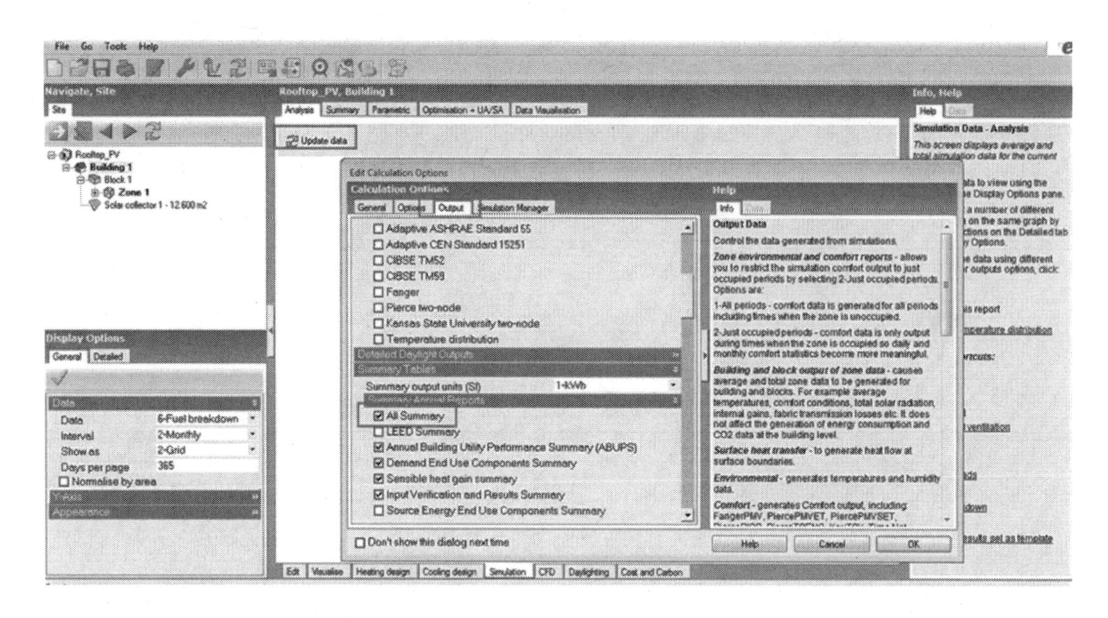

Step 10: Click the **Summary** tab. Scroll down to **Electric Load Satisfied** to note **Photovoltaic Power**.

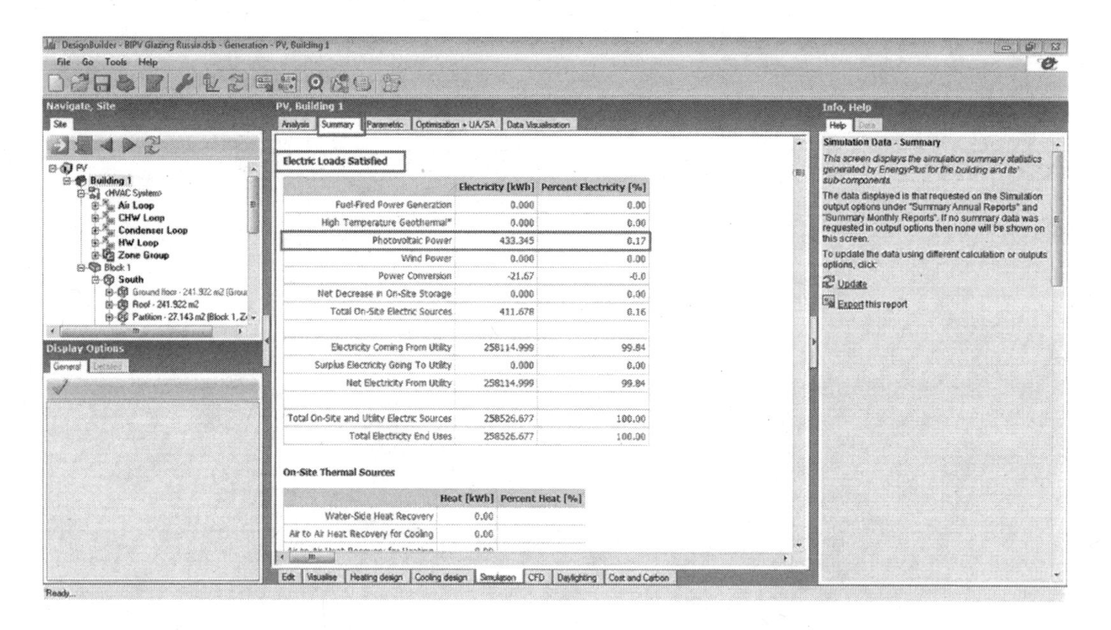

Record the result.

Tutorial 12.5

Evaluating the Performance of Opaque Building-Integrated PV Panels

GOAL

To evaluate the impact of BIPV panels on the south facade on annual energy generation

WHAT ARE YOU GOING TO LEARN?

- How to model a BIPV system
- How to size PV modules

PROBLEM STATEMENT

In this tutorial, you are going to use a 50- × 30-m five-zone model. You are going to use the PV module described in Table 12.6 for the simulations.

TABLE 12.6
BIPV input parameters

PV property	Value
Name	BPsolar 5170 (from DesignBuilder Library)
Cell type	Crystalline silicon
Active area	1.26 m^2
Rated electrical power	250 W_p

Model the BIPV system on the south facade of the building. Determine annual electricity generation for the **New Delhi Palam, India** weather location.

SOLUTION

Step 1: Create a **50- × 30-m** five-zone model with a **5-m** perimeter depth.

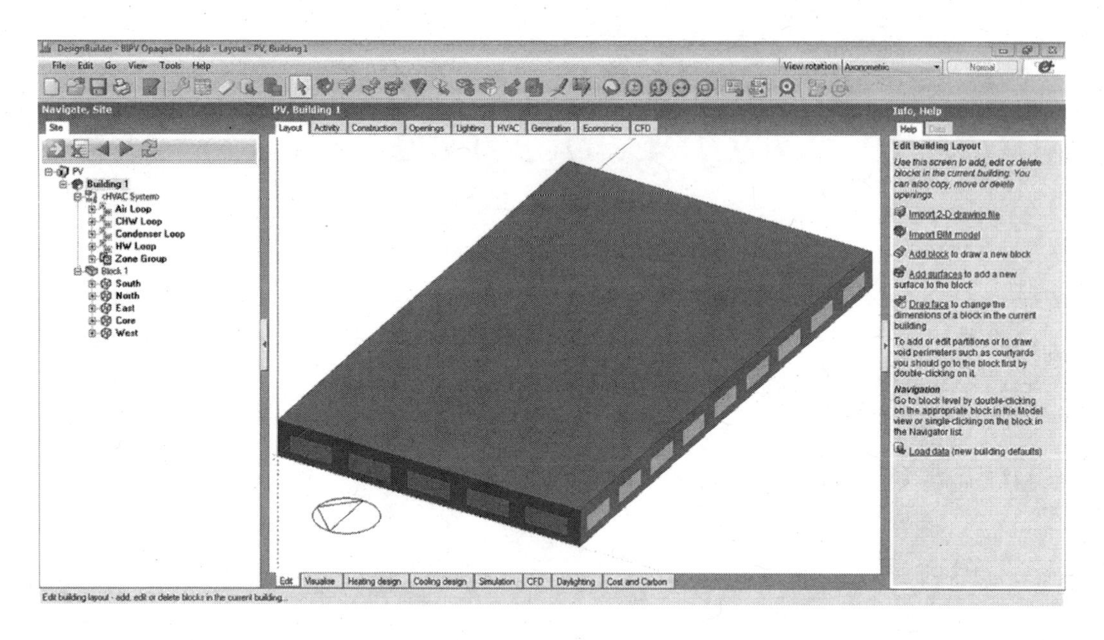

Step 2: Click the **Construction** tab.

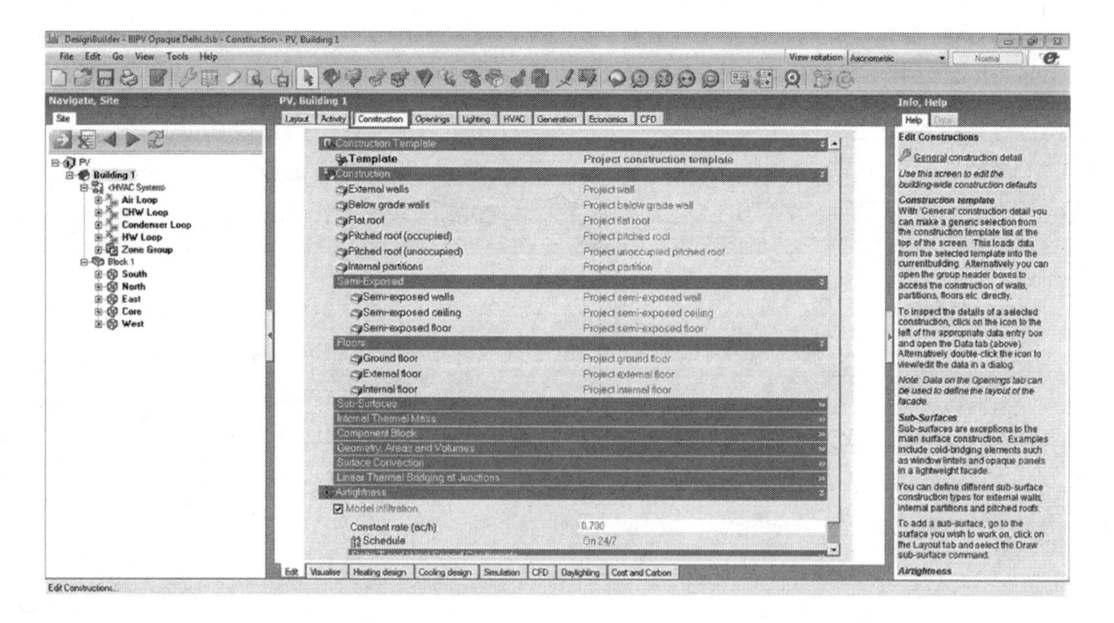

Step 3: Expand **South** zone, and click on **Wall – 175.00 m2 – 180.** Select **Project BIPV Wall** for **External walls**.

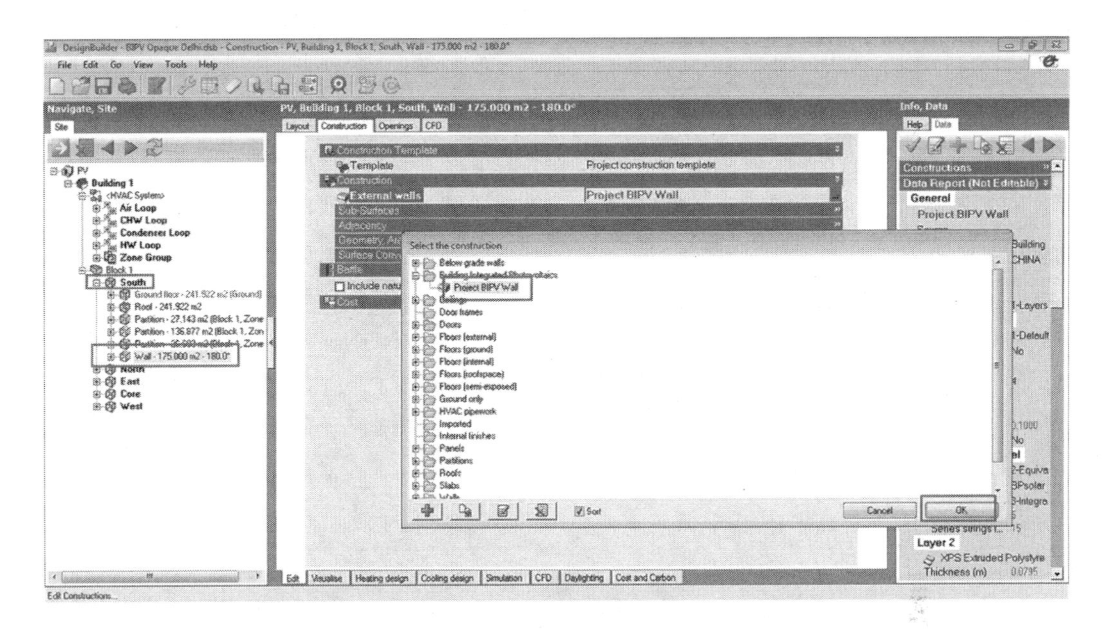

Step 4: Click the **Edit** icon. Select **2-Equivalent One-Diode** from the **Performance type** drop-down list. Select **BPsolar 5170** from **Performance model**. Enter **5** in the **Module in series** text box. Enter **15** in the **Series strings in parallel** text box.

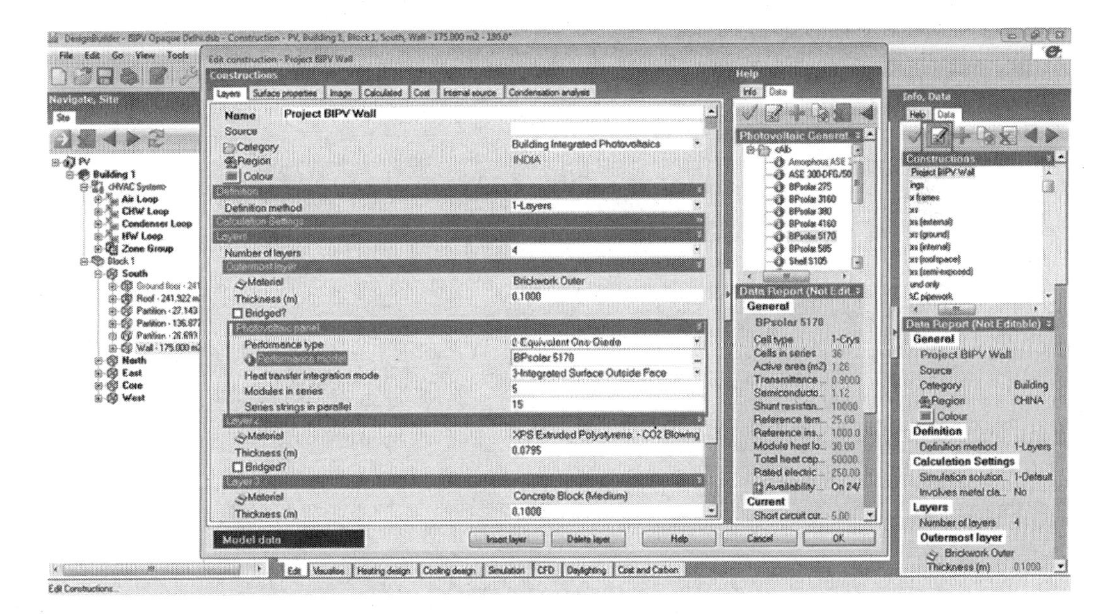

Available opaque area on the south wall is 123 m^2, and the active area of solar cells on each module is 1.26 m^2. A total of 15 × 5 modules can easily be fitted on the available area.

Step 5: Click **Building 1**, and then click the **Generation** tab. Select the**Include electric load centre** checkbox. Click **Load centre 1**.

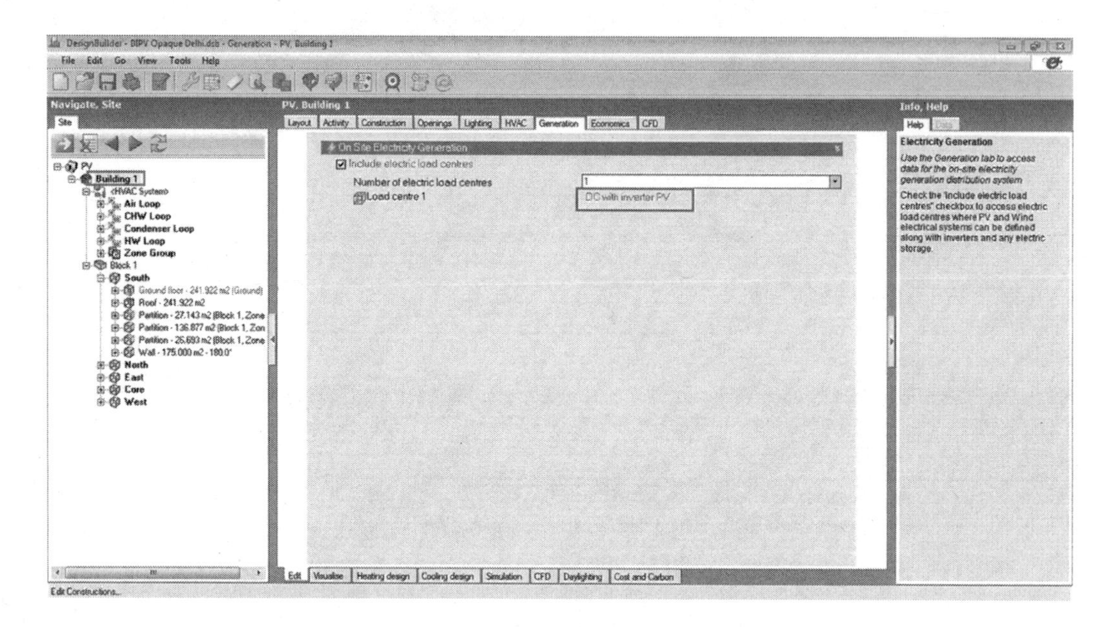

Step 6: Edit the **DC with inverter.** Enter **DC with inverter PV** in the **Name** text box.

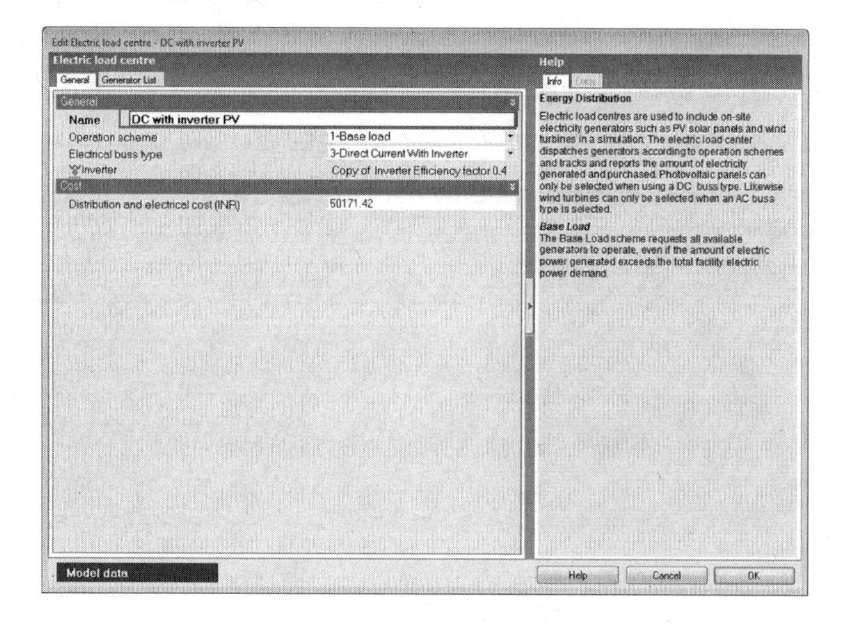

Step 7: Click the **Generator List** tab. Select **2-Construction with BIPV** from the **DC generator type** drop-down list. Select **Project BIPV Wall** for **Building integrated photovoltic constuction**.

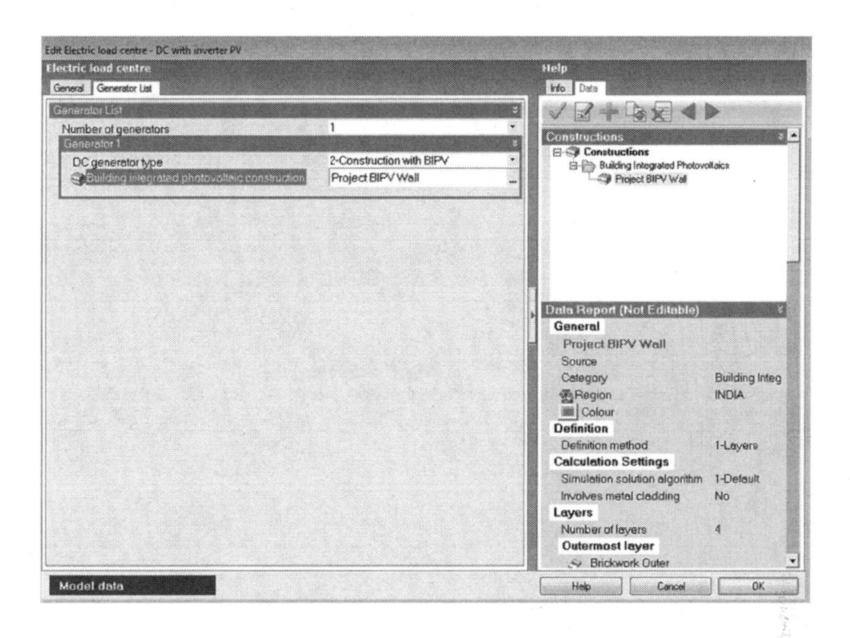

Step 8: Click the **Simulation screen** tab.

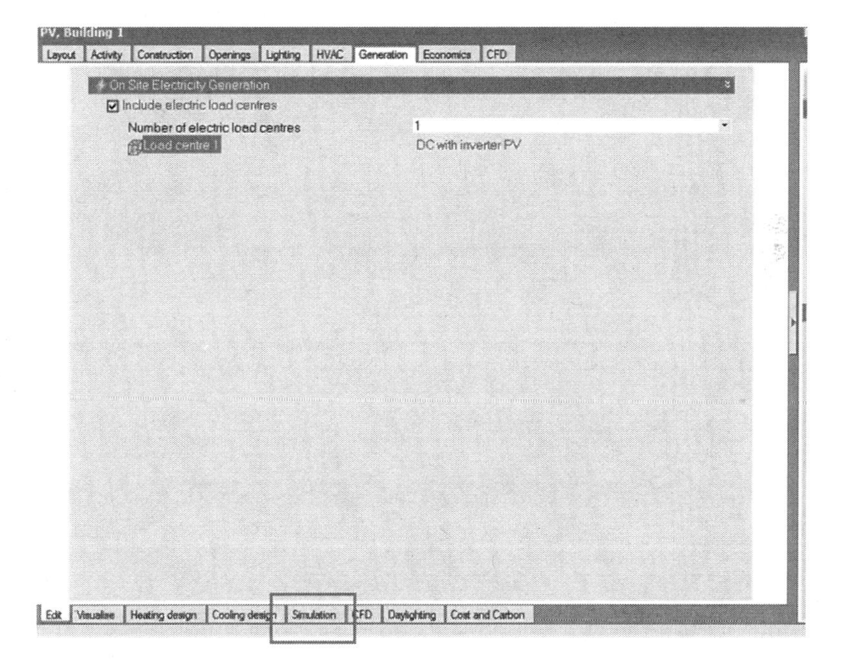

Step 9: Click **Update data**, and then click the **Output** tab. Select the **All Summary** checkbox. Click **OK** to perform the simulation.

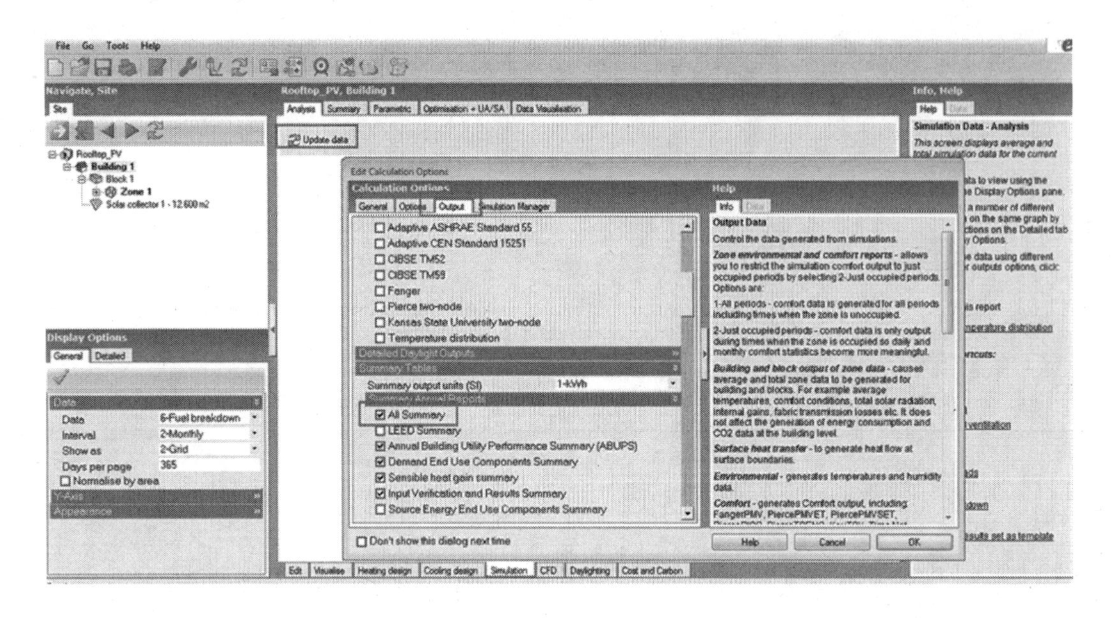

Step 10: Click the **Summary** tab. Scroll down to **Electric Load Satisfied** to note **Photovoltaic Power**.

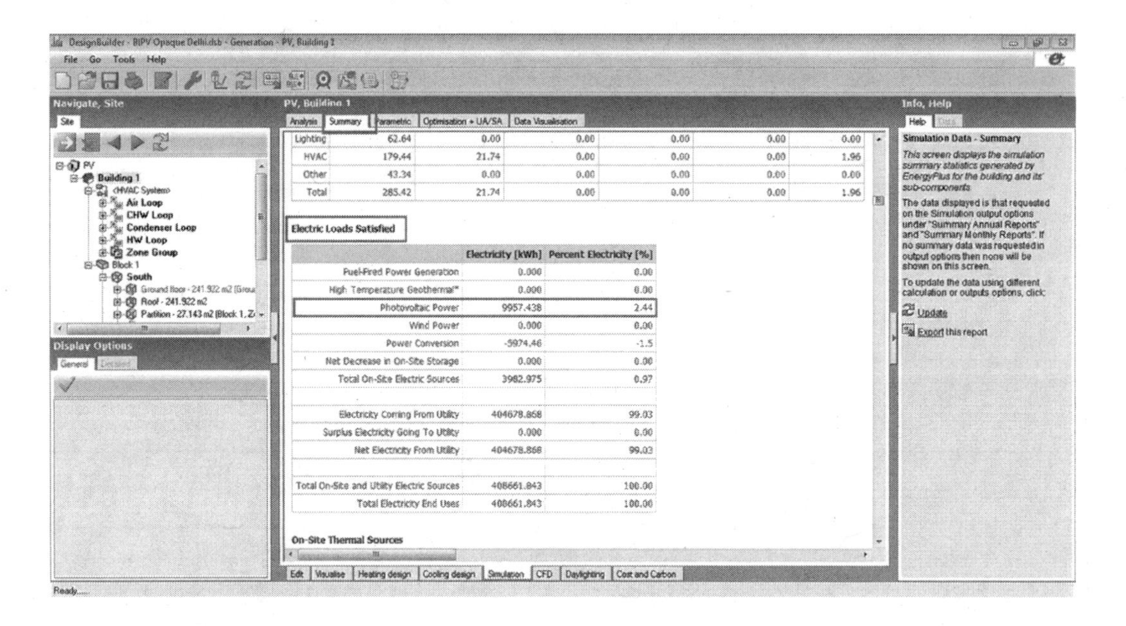

13 Costing, Sensitivity and Uncertainty Analysis

This chapter explains the cost-benefit analysis to compare alternatives for energy efficiency measures for building components. Generally, building components for energy efficiency have different initial and operating costs. In this situation, analysis of life-cycle cost parameters can be helpful in decision making. Concepts of life-cycle analysis are discussed through simple and discounted payback period and internal rate of return. This chapter also describes a method for performing sensitivity and uncertainty analysis.

Tutorial 13.1

Selecting Glazing Using Cost-Benefit Analysis

GOAL

To perform cost-benefit analysis for two glazing options

WHAT ARE YOU GOING TO LEARN?

- How to input material and energy costs in the desired currency
- How to input project life, inflation and discount rates
- How to view life-cycle cost output in terms of present value
- How to calculate simple and discounted payback period and internal rate of return

PROBLEM STATEMENT

In this tutorial, you are going to use a 50- × 25-m five-zone model with a 5-m perimeter depth. You are going to use the glass types shown in Table 13.1 for the simulations.

TABLE 13.1
Glazing input parameters

Glass	Properties	Cost (Indian rupee/m^2)
Single-glazing unit (SGU)	**Sgl Clr 6mm, SHGC-0.81, VLT-0.88**	800
Double-glazing unit (DGU)	**Dbl Ref-A-H Clr 6mm/6mm Air, SHGC-0.248, VLT-0.18**	3,000

Indian rupee (INR) is the official currency of India. Other parameters are as follows:

Window-to-wall ratio (WWR): 40%
Project life: 40 years
Electricity charges: INR10/kWh
Discount rate: 10%
Inflation rate: 4%

Determine net present value, simple and discounted payback periods and internal rate of return (IRR) to select glazing for the **New Delhi Palam, India** weather location.

Nominal Discount Rate

The *time value of money* is the concept that money available at the present time is worth more than the identical sum in the future because of its potential earning capacity. This core principle is that money can earn interest. The *discount rate* refers to the interest rate charged to banks. The *nominal discount rate* reflects the interest rates needed to make current and future expenditures have comparable equivalent values when general inflation is included.

Inflation Rate

Inflation is a phenomenon that results in a decrease in the purchasing power of money and an increase in the nominal value of revenue (i.e. cash inflows) and expenses (i.e. cash outflows).

Source: https://designbuilder.co.uk/helpv6.0/#LCCParameters.htm.

SOLUTION

Step 1: Create a **50- × 25-m** five-zone model with a **5-m** perimeter depth. Set the **WWR** to **40%**.

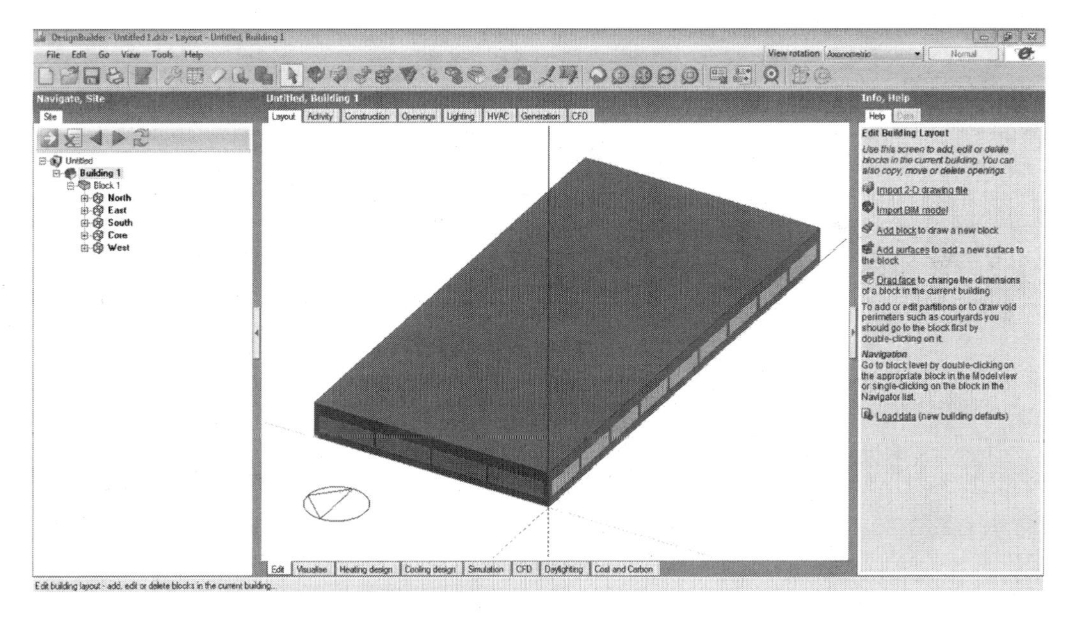

Step 2: Select the **PTHP template** in detail **HVAC system**. (Refer to Chapter 7 for modeling HVAC systems.)

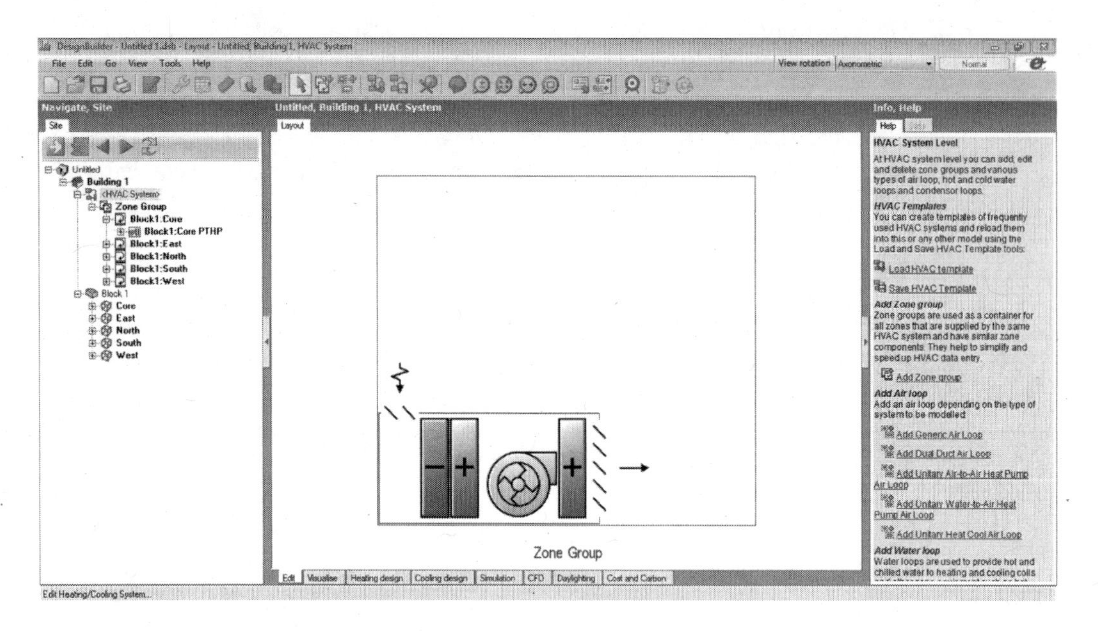

Step 3: Select **Program Options** from the **Tools** menu.

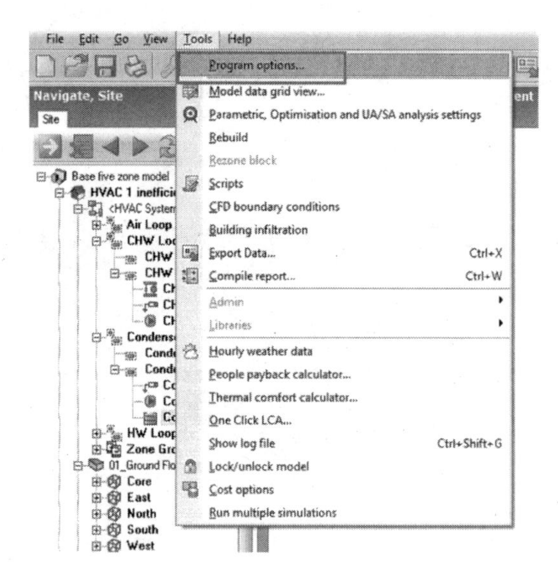

Step 4: Click the **International** tab. Select **Indian Rupee** from **Currency**. Click **OK**.

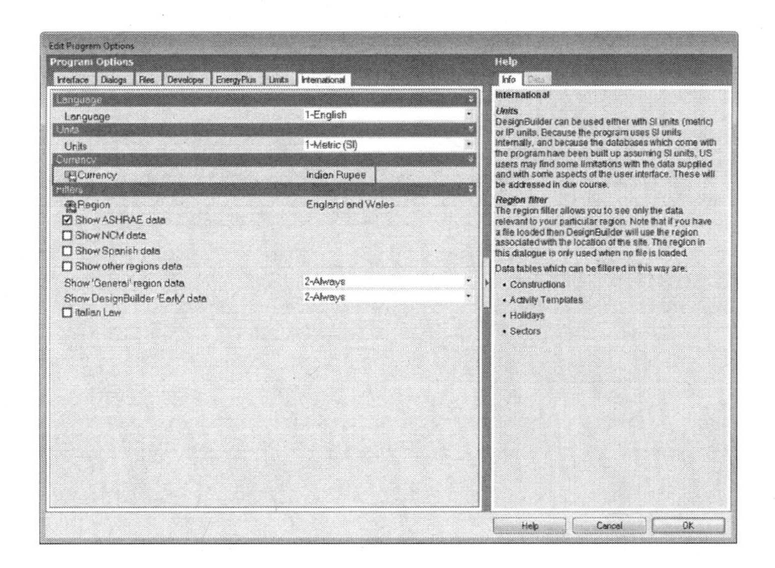

Step 5: Click the **Economics** tab.

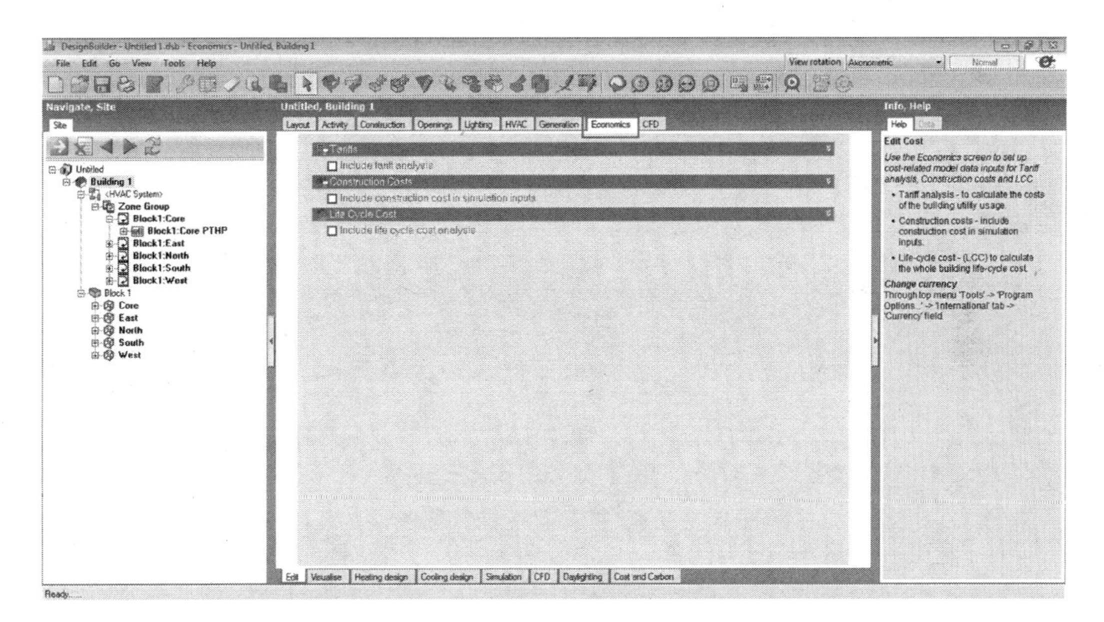

Step 6: Select the **Include tariff analysis** checkbox.

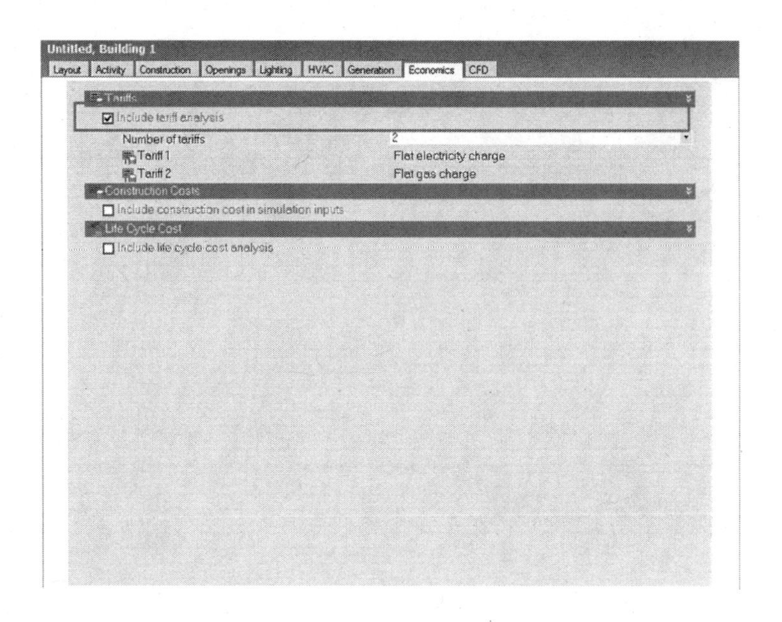

Step 7: Create a copy of the **Flat electricity charge**.

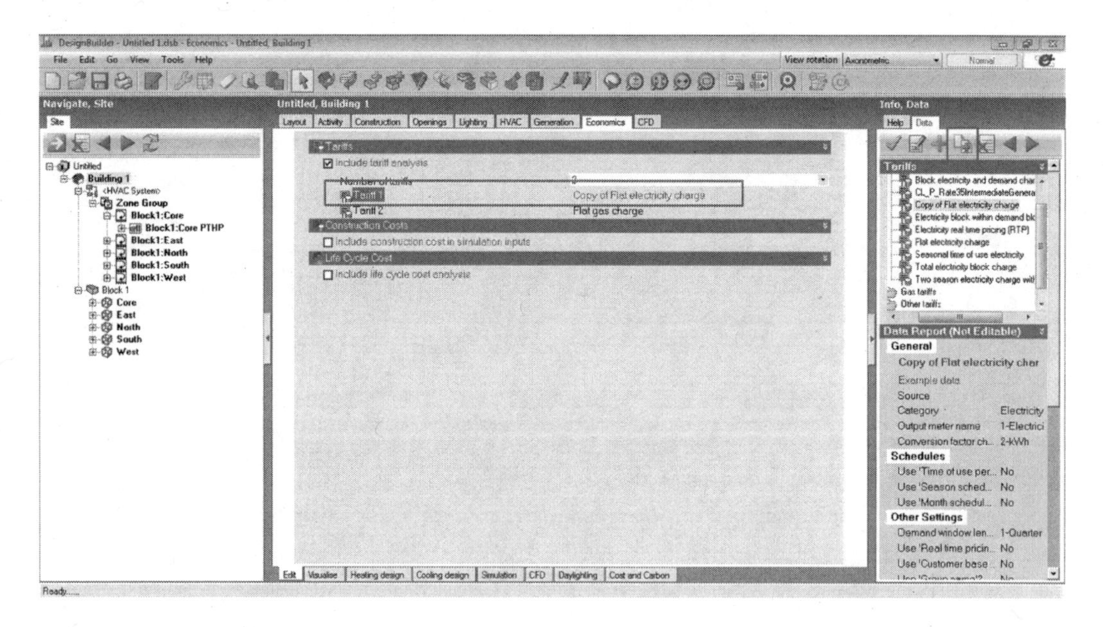

Step 8: Update **Tariff name** as **Flat electricity charge_Project**.

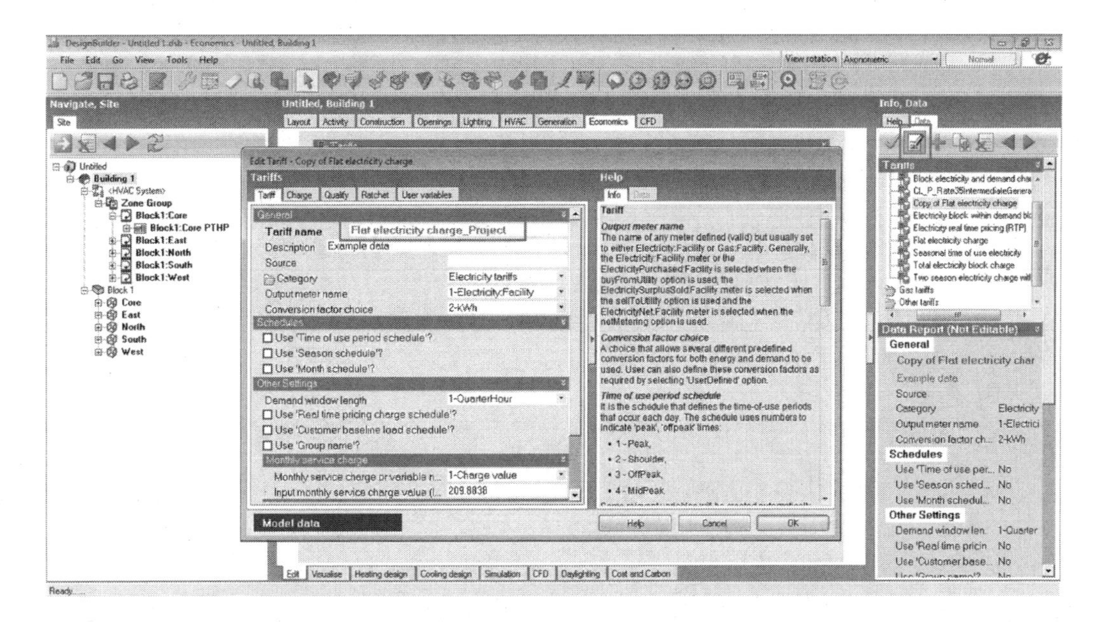

Step 9: Click on the **Charge** tab. Select **1** from the **Number of charges** drop-down list. Enter **10** in **Input cost per unit value (INR)**. Click **OK**.

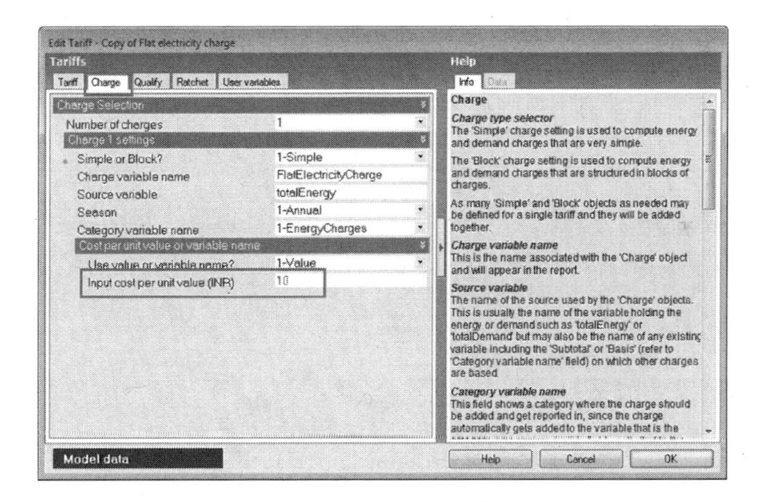

Monthly utility bills are often directly related to monthly energy consumption and monthly peak demand. The approach of tariff analysis allows you to model the individual component charges that make up the utility bill and report the results on a monthly basis. Any charges included in the **EnergyCharges** category are added together at the first step. The **EnergyCharges**, **DemandCharges** and **ServiceCharges** are added together to form the **Basis**. The **Basis**, **Adjustments** and **Surcharges** are added together to form the **Subtotal**. And finally, the **Total** is obtained from the summation of **Subtotal** and **Taxes**, where the **Total** represents the total monthly charges on that tariff for the energy source used. In addition, each category name is usually used as a source variable when setting charges and doing tariff calculations.

In detail, this module consists of the name of the tariff, the type of tariff, and other details about the overall tariff. The objects such as **Tariff**, **Charge** (either **Simple** or **Block**), **Qualify** and **Ratchet** stay in the same tariff package and will perform tariff calculations with the tariff name being given in the **Tariff** tab.

Source: https://designbuilder.co.uk/helpv6.0/#TariffAnalysis.htm.

Step 10: Select the **Include construction cost in simulation inputs** checkbox. Select the **Include life cycle cost analysis** checkbox.

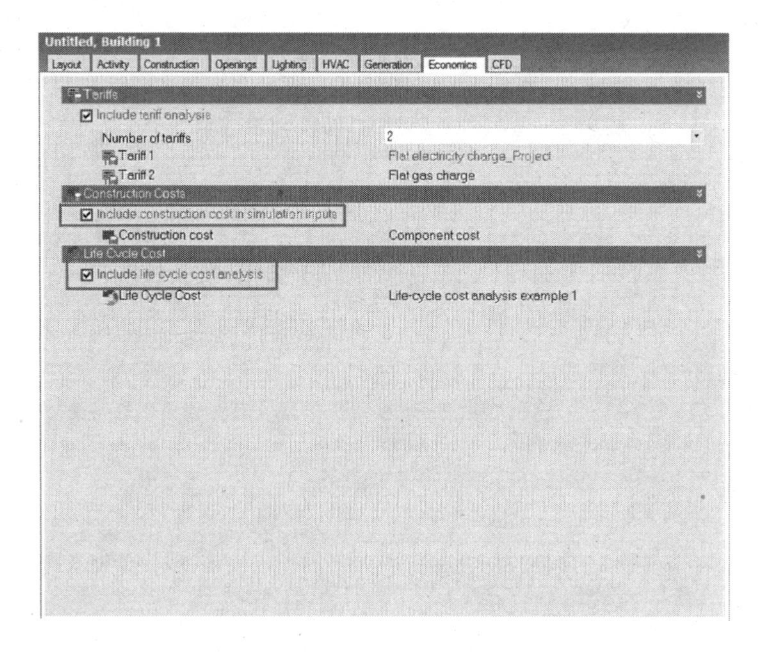

Construction Cost

Component cost modelling provides an early-design-stage estimate of the initial construction costs associated with the building and HVAC system being modelled. The construction costs input to the EnergyPlus costing calculations are derived from the cost outputs on the **Cost and Carbon** tab. Initially, the idea of EnergyPlus construction costs is to ensure that cost estimates are consistent with the EnergyPlus model and allow using results calculated during a simulation, such as equipment sizes and surface areas.

The DesignBuilder EnergyPlus cost estimating capabilities are intended for early-stage estimation and do not cover all types of analyses included in more detailed cost estimating software.

Source: https://designbuilder.co.uk/helpv6.0/#ConstructionCosts.htm.

Construction Costs: Adjustments

Data on the **Adjustments** tab can be used to perform various modifications to the construction costs to arrive at an estimate for total project costs. This allows the line-item model to be extended so that the overall costs of the project will reflect various profit and fees:

- Miscellaneous cost per conditioned area
- Regional adjustment factor
- Design and engineering fees
- Contractor fee
- Contingency
- Permits, bonding and insurance
- Commissioning fee

Source: https://designbuilder.co.uk/helpv6.0/Content/ConstructionCostsAdjustments.htm.

Step 11: Copy and edit **Life-cycle cost analysis example 1**.

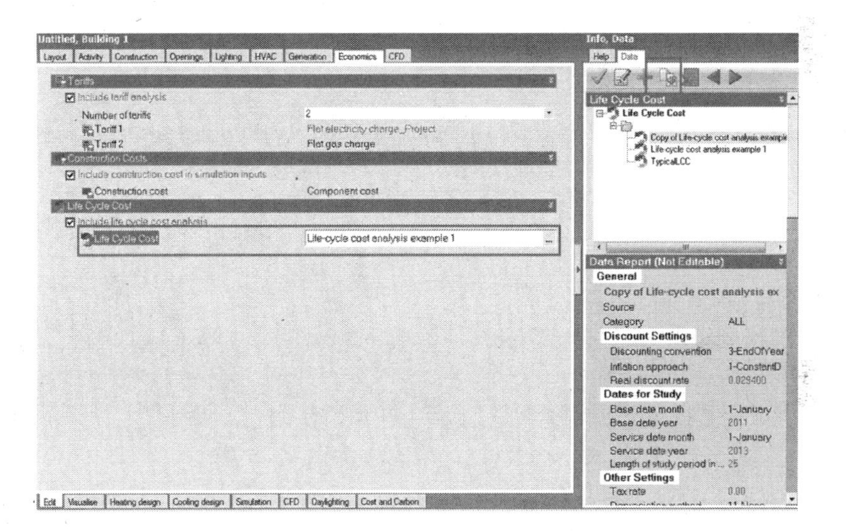

Step 12: Enter the **Life-cycle cost analysis_Project** in **Name** text box. Select **2-CurrentDollar** from the **Inflation approach** drop-down list. Enter **0.1** in the **Nominal discount rate** and **0.04** in the **Inflation rate** text boxes. Enter **40** in the **Length of study period in years** text box.

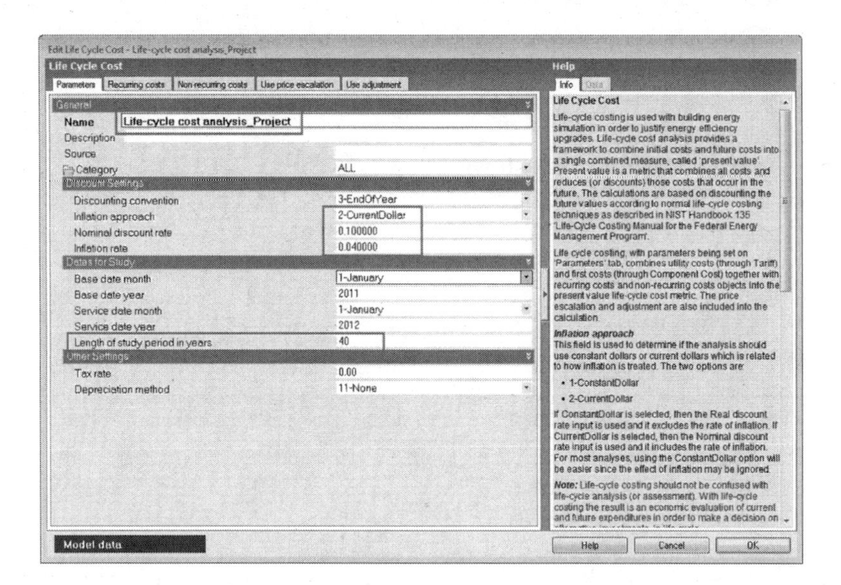

> *Life-cycle costing* (LCC) is used with building energy simulations to justify energy-efficiency upgrades. Many alternative building technologies that result in energy savings cost more initially or may cost more to maintain than traditional solutions. In order to justify selecting these energy-saving technologies, it is essential to combine both initial and future costs in the decision process. Using life-cycle costs provides a framework to combine initial costs and future costs into a single combined measure called the *present value.* Present value is a metric that combines all costs and reduces (or discounts) those that occur in the future. Discounting future costs is based on the principle of the time value of money.
>
> *Source:* https://designbuilder.co.uk/helpv6.0/#LifeCycleCost.htm.

Step 13: Select **0** from the **Number of recurring costs** drop-down list under the **Recurring costs** tab. To keep the tutorial simple, recurring cost is kept to zero.

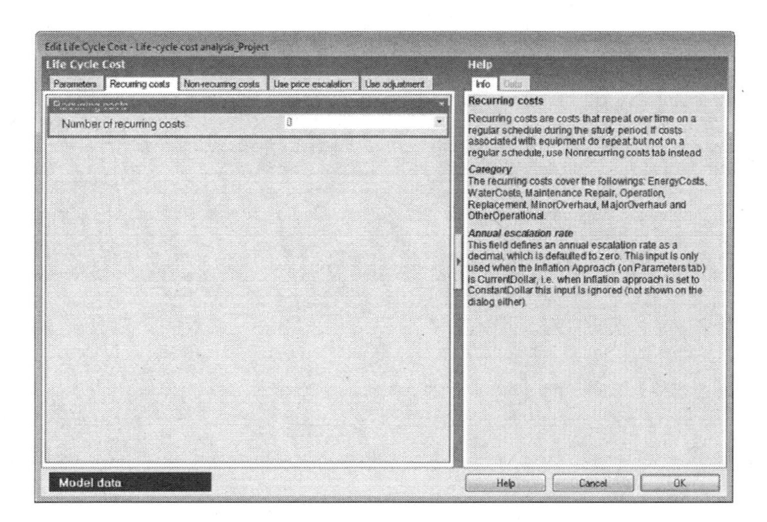

Recurring costs are costs that repeat over time on a regular schedule during the study period. Following are the categories of the recurring costs:

1. **EnergyCosts**
2. **WaterCosts**
3. **Maintenance**
4. **Repair**
5. **Operation**
6. **Replacement**
7. **MinorOverhaul**
8. **MajorOverhaul**
9. **OtherOperational**

Source: https://designbuilder.co.uk/helpv6.0/#LCCRecurringCosts.htm.

Step 14: Select **0** from the **Number of non-recurring costs** drop-down list under the **Non-recurring costs** tab. Click **OK**. To keep this tutorial simple, non-recurring cost is kept to zero.

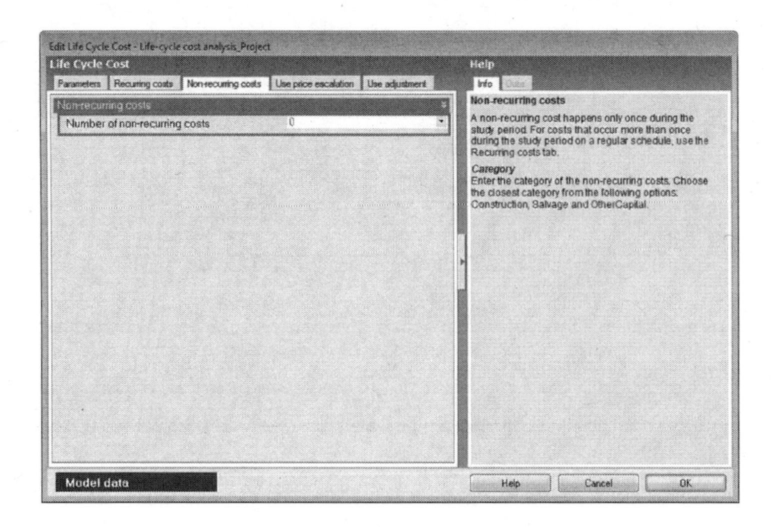

Non-recurring costs happen only once during a study period. Following are the categories of the non-recurring costs:

1. **Construction**
2. **Salvage**
3. **OtherCapital**

Source: https://designbuilder.co.uk/helpv6.0/#LCCNonRecurringCosts.htm.

Step 15: Select **Life-cycle cost analysis_Project**.

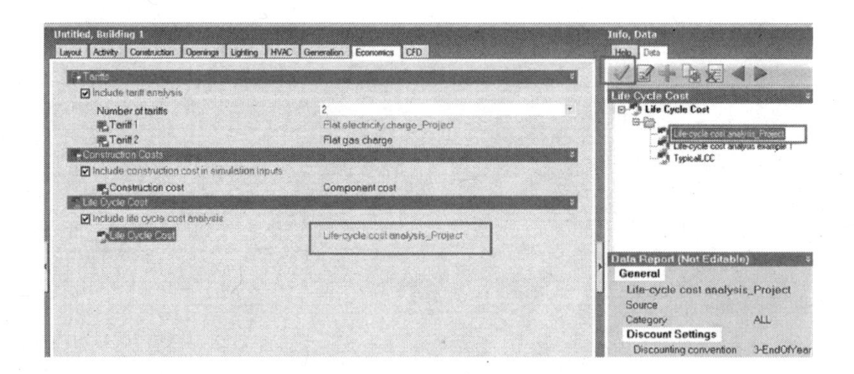

Step 16: Select the **Openings** tab. Copy and edit **Sgl Clr 6mm**. Enter **Sgl Clr 6mm_Cost 800** in the **Name** text box.

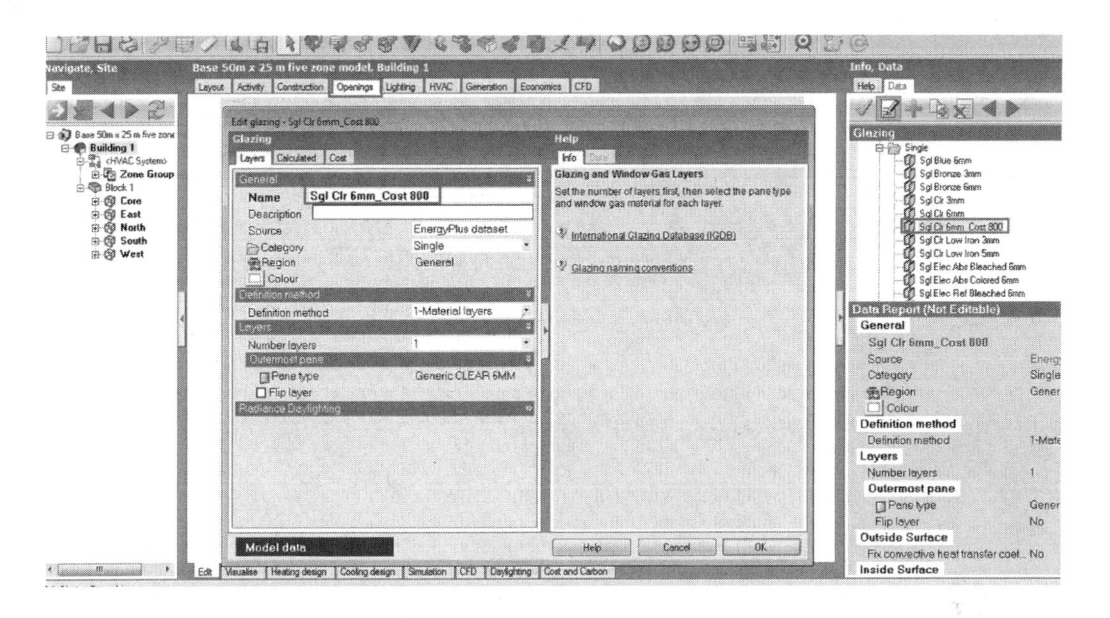

Step 17: Select the **Cost** tab. Enter **800** in the **Cost per area (INR/m2)** text box. Click **OK**.

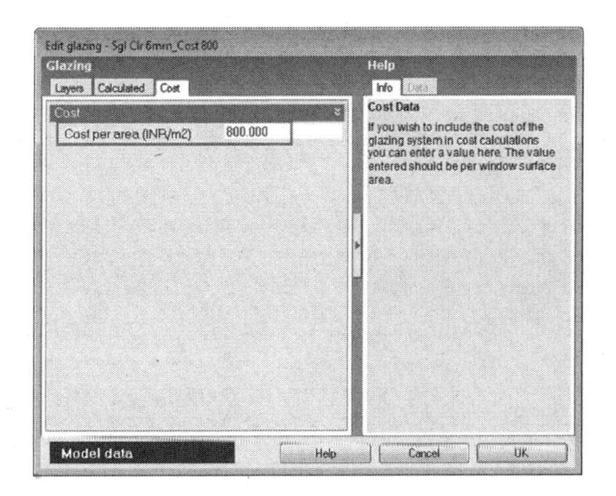

Step 18: Select **Sgl Clr 6mm_Cost 800** for **Glazing type**.

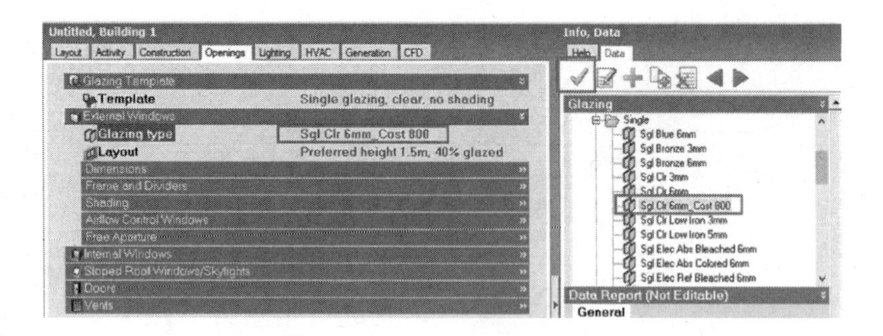

Cost Data

The capital cost of constructing a building is defined in a number of places in the model:

1. Within the building model on component dialogues:
   - Construction: cost of constructions
   - Material: cost of material layers (cost per area or volume)
   - Glazing: cost of glazing systems (cost per window area)
   - Window shading devices such as blinds and shades (cost per window area)
   - Local shading such as overhangs and louvres (cost per window area)
   - Wind turbines connected to load centres
   - Renewable generation distribution and electrical costs from load centres
2. On model data tabs:
   - Constructions: substructure, structure and surface finishes
   - Lighting: cost of lighting systems (cost per gross internal flooe area [GIFA])
   - HVAC: cost of HVAC systems (cost per GIFA or capacity)

Building cost summary data are found in the **Building Cost Summary** dialogue (accessed from the **Tools** menu). These data, together with the building geometry, allow the construction cost (aka capital expenditure or Capex) of the building to be calculated.

Source: https://designbuilder.co.uk/helpv6.0/#CostData.htm.

Step 19: Click on the **Simulation screen** tab. Click **Update data**.

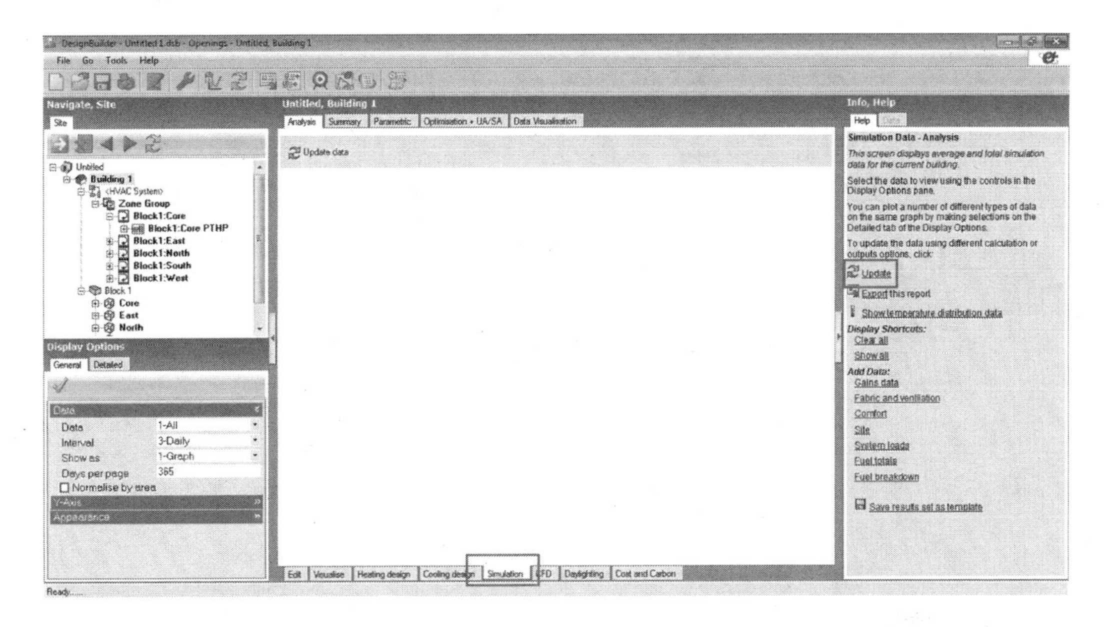

Step 20: Click the **Output** tab. Select the **Component Cost Economics Summary** checkbox. Click **OK**.

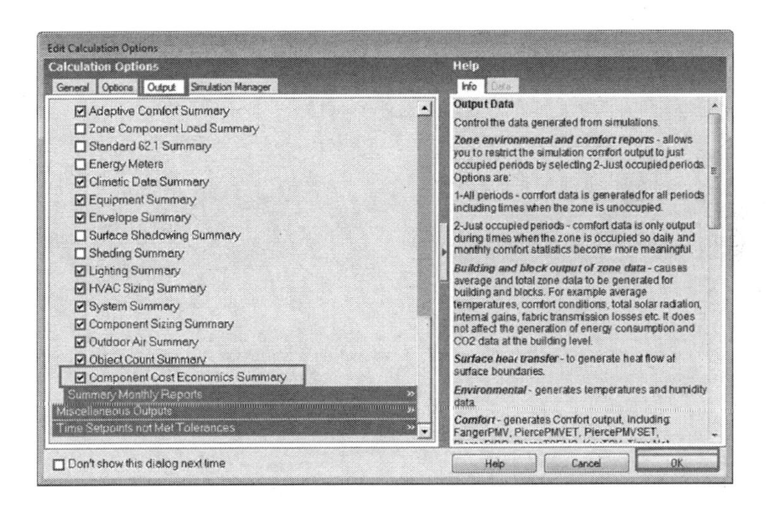

Step 21: After the simulation, click the **Summary** tab, and scroll down to read **Life-Cycle Cost Parameters**.

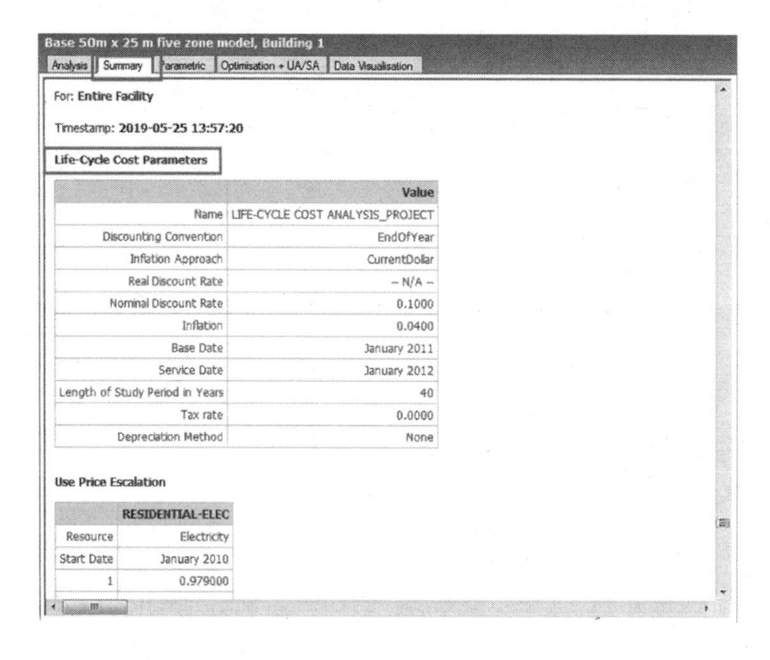

Step 22: Copy the **Present Value by Year** table to an external spreadsheet program.

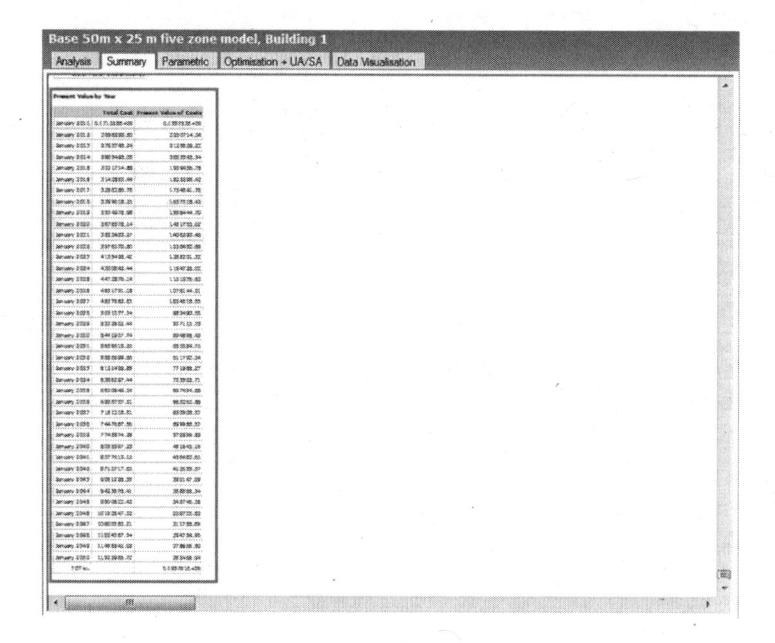

Step 23: Save the model. Repeat the preceding steps with **Dbl Ref-A-H Clr 6mm/6mm Air** glazing as given in the problem statement.

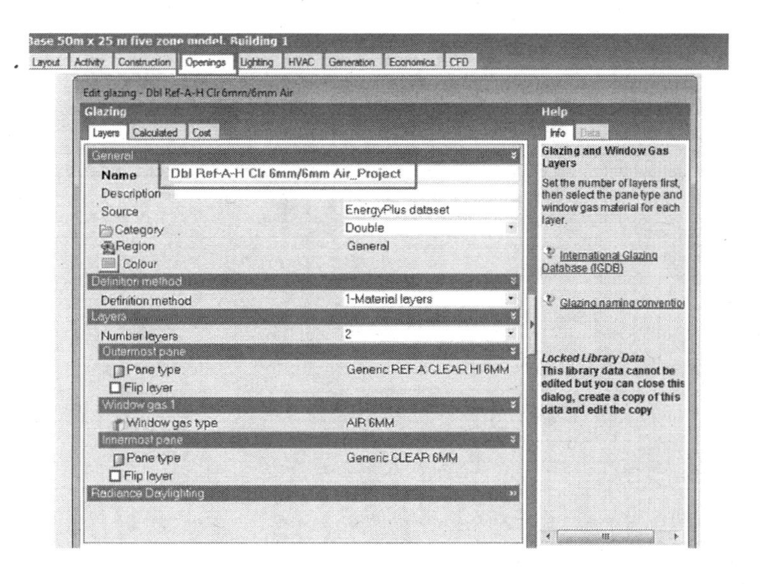

Step 24: Enter **3,000** in the **Cost per area (INR/m2)** text box.

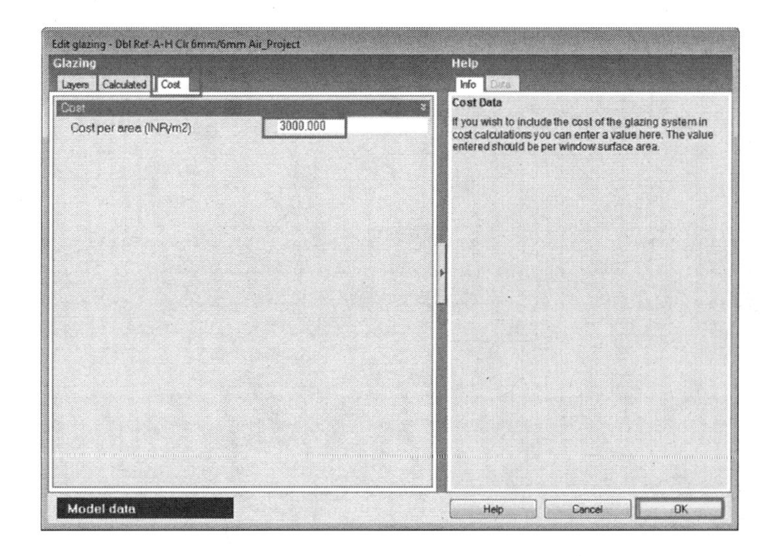

Step 25: Simulate the model, and record the results. Compare results in the spreadsheet program (Table 13.2).

TABLE 13.2
Spreadsheet for the costing calculation

	SGU		DGU					
Year	Total cost (INR) (A)	Present value of costs (B)	Total cost (INR) (C)	Present value of costs (D)	Difference in total costs (E = A − B)	Cumulative cash flow for total cost (F) (F(*n*) = E(*n*) + F(*n* − 1))	Difference in present value of costs (G = C −D)	Cumulative cash flow for discounted cost (H) (H(*n*) = G(*n*) + H(*n* − 1))
Jan-11	171,328,000	155,753,000	172,007,000	156,370,000	(−)679,000	(−)679,000	(−)617,000	(−)617,000
Jan-12	2,686,294	2,250,714	2,449,239	2,052,098	237,054	−441,946	198,616	−418,384
Jan-13	2,793,745	2,125,639	2,547,209	1,938,061	246,536	−195,410	187,579	−230,805
Jan-14	2,905,495	2,003,543	2,649,097	1,826,739	256,398	60,988	176,804	−54,001
Jan-15	3,021,715	1,909,457	2,755,061	1,740,955	266,654	327,641	168,501	114,500
Jan-16	3,142,583	1,823,398	2,865,264	1,662,491	277,320	604,961	160,907	275,408
Jan-17	3,268,287	1,734,842	2,979,874	1,581,749	288,412	893,374	153,092	428,500
Jan-18	3,399,018	1,637,518	3,099,069	1,493,014	299,949	1,193,323	144,504	573,004
Jan-19	3,534,979	1,556,445	3,223,032	1,419,095	311,947	1,505,270	137,350	710,354
Jan-20	3,676,378	1,481,753	3,351,953	1,350,995	324,425	1,829,694	130,758	841,112
Jan-21	3,823,433	1,406,290	3,486,031	1,282,191	337,402	2,167,096	124,099	965,211
Jan-22	3,976,371	1,338,453	3,625,473	1,220,340	350,898	2,517,994	118,113	1,083,324
Jan-23	4,135,425	1,268,201	3,770,492	1,156,288	364,934	2,882,928	111,913	1,195,238
Jan-24	4,300,842	1,194,723	3,921,311	1,089,294	379,531	3,262,459	105,429	1,300,667
Jan-25	4,472,876	1,131,377	4,078,164	1,031,537	394,712	3,657,171	99,839	1,400,506
Jan-26	4,651,791	1,076,144	4,241,290	981,179	410,501	4,067,672	94,965	1,495,471
Jan-27	4,837,863	1,024,816	4,410,942	934,380	426,921	4,494,593	90,436	1,585,907
Jan-28	5,031,377	982,491	4,587,380	895,790	443,998	4,938,591	86,701	1,672,608
Jan-29	5,232,632	937,114	4,770,875	854,417	461,758	5,400,349	82,696	1,755,304
Jan-30	5,441,938	894,896	4,961,710	815,926	480,228	5,880,577	78,971	1,834,275
Jan-31	5,659,615	853,885	5,160,178	778,533	499,437	–	75,352	–
Jan-32	5,886,000	811,792	5,366,585	740,155	519,415	–	71,637	–
Jan-33	6,121,440	771,956	5,581,249	703,834	540,191	–	68,122	–
Jan-34	6,366,297	733,534	5,804,499	668,802	561,799	–	64,731	–
Jan-35	6,620,949	697,495	6,036,679	635,944	584,271	–	61,551	–
Jan-36	6,885,787	663,263	6,278,146	604,733	607,642	–	58,530	–
Jan-37	7,161,219	630,909	6,529,272	575,234	631,947	–	55,675	–
Jan-38	7,447,668	599,956	6,790,442	547,012	657,225	–	52,944	–
Jan-39	7,745,574	570,600	7,062,060	520,247	683,514	–	50,353	–
Jan-40	8,055,397	461,643	7,344,543	420,905	710,855	–	40,738	–
Jan-41	8,377,613	436,463	7,638,324	397,947	739,289	–	38,516	–
Jan-42	8,712,718	412,656	7,943,857	376,240	768,860	–	36,415	–
Jan-43	9,061,226	390,147	8,261,611	355,718	799,615	–	34,429	–
Jan-44	9,423,675	368,866	8,592,076	336,315	831,600	–	32,551	–
Jan-45	9,800,622	348,746	8,935,759	317,971	864,863	–	30,775	–
Jan-46	10,192,647	329,724	9,293,189	300,627	899,458	–	29,097	–
Jan-47	10,600,353	311,739	9,664,917	284,229	935,436	–	27,510	–
Jan-48	11,024,367	294,735	10,051,514	268,726	972,854	–	26,009	–
Jan-49	11,465,342	278,659	10,453,574	254,068	1,011,768	–	24,590	–
Jan-50	11,923,956	263,459	10,871,717	240,210	1,052,239	–	23,249	–
Total		193,761,000		191,024,000				

Step 26: Compare the present value cost for both cases (Table 13.3).

TABLE 13.3
Present value cost for SGU and DGU

SGU	DGU
193,761,000	191,024,000

Computation of Present Value

PV formula for annually recurring uniform amounts

$$PV = A_0 \times UPV_{(n,d)}$$

The **Uniform Present Value (UPV) factor** is used to calculate the PV of a series of equal cash amounts, A_0, that recur annually over a period of n years, given d.

$$PV = A_0 \times \sum_{t=1}^{n} \frac{1}{(1+d)^t} = A_0 \times \frac{(1+d)^n - 1}{d(1+d)^n}$$

The **UPV factor** for d = 3% and n = 15 years is **11.94.**

PV formula for annually recurring non-uniform amounts

$$PV = A_0 \times UPV^*_{(n,d,e)}$$

$$PV = A_0 \times \sum_{t=1}^{n} \left(\frac{1+e}{1+d}\right)^t = A_0 \frac{(1+e)}{(d-e)}\left[1 - \left(\frac{1+e}{1+d}\right)^n\right]$$

PV: present value
d: discount rate
e: escalation rate
A: cash amount
n: period of years

Sources: Adapted from *NIST Handbook 135 LCC Manual*, available at https://www.wbdg.org/FFC/NIST/hdbk_135.pdf.

Computation of Internal Rate of Return

The internal rate of return (IRR) is a metric used in capital budgeting to estimate the profitability of potential investments. The IRR is a discount rate that makes the net present value (NPV) of all cash flows from a particular project equal to zero. IRR calculations rely on the same formula as does NPV.

$$IRR = NPV = \sum_{t=1}^{T} \frac{C_t}{(1+r)^t} - C_0 = 0$$

where C_t = net cash inflow during the period t

C_0 = total initial investment costs

r = the discout rate

t = the number of time periods

Simple and Discounted Payback

There are two payback measures that are often used for economic analysis of a capital investment: *simple payback* (SPB) and *discounted payback* (DPB). Both SPB and DPB measure the time required to recover initial investment costs. They are expressed as the number of years elapsed between the beginning of the service period and the time at which cumulative savings (net of any incremental investment costs incurred after the service date) are just sufficient to offset the incremental initial investment cost of the project. Both of these payback measures are relative measures; that is, they can only be computed with respect to a designated base case.

DPB is the preferred method of computing the payback period for a project because it requires that cash flows occurring each year be discounted to present value before accumulating them as savings and costs. If the DPB is less than the length of the service period used in the analysis, the project is generally cost-effective. This is consistent with the requirement that the LCC of the project alternative be lower than the LCC of the base case.

Source: https://www.wbdg.org/FFC/NIST/hdbk_135.pdf.

Step 27: Calculate the IRR using a spreadsheet program. Use the IRR function.

POWER =IRR(I3:I42)

	B	C	E	F	G	I	J	L	M	O	Q	R	S
1			Glass 2			Diff in normal value	Cumulative Cash Flow	Difference in PV		YEAR			
2	Total Cost (INR)	Present Value of Costs		Total Cost (INR)	Present Value of Costs								
3	17,13,28,000	15,57,53,000	Jan-11	########	15,63,70,000	-6,79,000		-6,17,000			IRR	=IRR(I3:I	
4	2686293.5	2250714.34	Jan-12	2449239.46	20,52,098	2,37,054	-4,41,946	1,98,616	-4,18,384	1	Simple Payback	2.86	Years
5	2793745.24	2125639.22	Jan-13	2547209.04	19,38,061	2,46,536	-1,95,410	1,87,579	-2,30,805	2	Discounted Payback	4	Years
6	2905495.05	2003543.34	Jan-14	2649097.4	18,26,739	2,56,398	60,988	1,76,804	-54,001	3			
7	3021714.85	1909456.76	Jan-15	2755061.29	17,40,955	2,66,654	3,27,641	1,68,501	1,14,500	4			
8	3142583.44	1823398.42	Jan-16	2865263.75	16,62,491	2,77,320	6,04,961	1,60,907	2,75,408				
9	3268286.78	1734841.78	Jan-17	2979874.3	15,81,749	2,88,412	8,93,374	1,53,092	4,28,500				
10	3399018.25	1637518.42	Jan-18	3099069.27	14,93,014	2,99,949	11,93,323	1,44,504	5,73,004				
11	3534978.98	1556444.7	Jan-19	3223032.04	14,19,095	3,11,947	15,05,270	1,37,350	7,10,354				
12	3676378.14	1481753.02	Jan-20	3351953.32	13,50,995	3,24,425	18,29,694	1,30,758	8,41,112				
13	3823433.27	1406290.48	Jan-21	3486031.45	12,82,191	3,37,402	21,67,096	1,24,099	9,65,211				
14	3976370.6	1338452.69	Jan-22	3625472.71	12,20,340	3,50,898	25,17,994	1,18,113	10,83,324				
15	4135425.42	1268201.32	Jan-23	3770491.62	11,56,288	3,64,934	28,82,928	1,11,913	11,95,238				
16	4300842.44	1194723.02	Jan-24	3921311.28	10,89,294	3,79,531	32,62,459	1,05,429	13,00,667				
17	4472876.14	1131376.62	Jan-25	4078163.74	10,31,537	3,94,712	36,57,171	99,839	14,00,506				
18	4651791.18	1076144.31	Jan-26	4241290.29	9,81,179	4,10,501	40,67,672	94,965	14,95,471				
19	4837862.83	1024815.55	Jan-27	4410941.9	9,34,380	4,26,921	44,94,593	90,436	15,85,907				
20	5031377.34	982490.58	Jan-28	4587379.57	8,95,790	4,43,996	49,38,591	86,701	16,72,608				
21	5232632.44	937113.73	Jan-29	4770874.76	8,54,417	4,61,758	54,00,349	82,696	17,55,304				
22	5441937.74	894896.43	Jan-30	4961709.75	8,15,926	4,80,228	58,80,577	78,971	18,34,275				
23	5659615.25	853884.73	Jan-31	5160178.14	7,78,533	4,99,437		75,352					
24	5885999.86	811792.24	Jan-32	5366585.26	7,40,155	5,19,415		71,637					
25	6121439.85	771956.27	Jan-33	5581248.67	7,03,834	5,40,191		68,122					
26	6366297.44	733533.71	Jan-34	5804498.62	6,68,802	5,61,799		64,731					
27	6620949.34	697494.85	Jan-35	6036678.56	6,35,944	5,84,271		61,551					
28	6885787.31	663262.86	Jan-36	6278145.71	6,04,733	6,07,642		58,530					
29	7161218.81	630908.57	Jan-37	6529271.53	5,75,234	6,31,947		55,675					

25052019 GLASS 1 28042019_OLd IRR Sheet5 03052019_old Sheet2 Sheet3

TABLE 13.4
Internal rate of return and payback calculation

IRR	39%
Simple payback	3 years
Discounted payback	4 years

In Table 13.4, payback period is calculated using Table 13.2. The cumulative cash flow is a positive number in the third year; that year is the simple payback year. Similarly, cumulated cash flow for discounted is positive in fourth year (refer to Table 13.2). You can see from cost-benefit analysis that double-glazed windows are a better option to choose because they have a lower net present value.

Tutorial 13.2

Selecting a HVAC System Using Cost-Benefit Analysis

GOAL

To perform cost-benefit analysis for two HVAC options

WHAT ARE YOU GOING TO LEARN?

- How to input the cost of HVAC and energy tariffs in a desired currency

PROBLEM STATEMENT

In this tutorial, you are going to use a G + 5 floor building model. Each floor has a 100- × 50-m area and five zones with a 10-m perimeter depth. You need to make use of floor/zone multiplier option to model the building. Model the HVAC systems as **VAV with reheat water-cooled chiller**. You are going to use the HVAC systems described in Table 13.5 for the simulations.

TABLE 13.5
Input HAVC system parameters

HVAC system	Properties	First cost (INR/m^2)	Maintenance cost (5% of first cost payable annually) (INR/year)
System 1	Chiller COP: 4.2 Cooling tower with single-speed fan	2,000	5% of INR 2,000 × 30,000 m^2 = 3,000,000
System 2	Chiller COP: 6.1 cooling tower with variable-speed fan	3,000	5% of INR 3,000 × 30,000 m^2 = 4,500,000

Other parameters are as follows:

WWR: 40%
Project life: 15 years
Maintenance cost: 5% of first cost payable annually
Electricity charges: Flat INR10/kWh
Discount rate: 10%
Inflation rate: 4%

Determine the net present value, simple and discounted payback periods and IRR to select HVAC system for the **New Delhi Palam India** weather location.

SOLUTION

Step 1: Open a new project, and create a floor having a **100- × 50-m** area and five zones with a **10-m** perimeter depth. Use the **Clone, floor multiplier and components** block to create other floors to make this model G + 5. (Refer to Tutorial 5.1 to see how to copy/clone floors.)

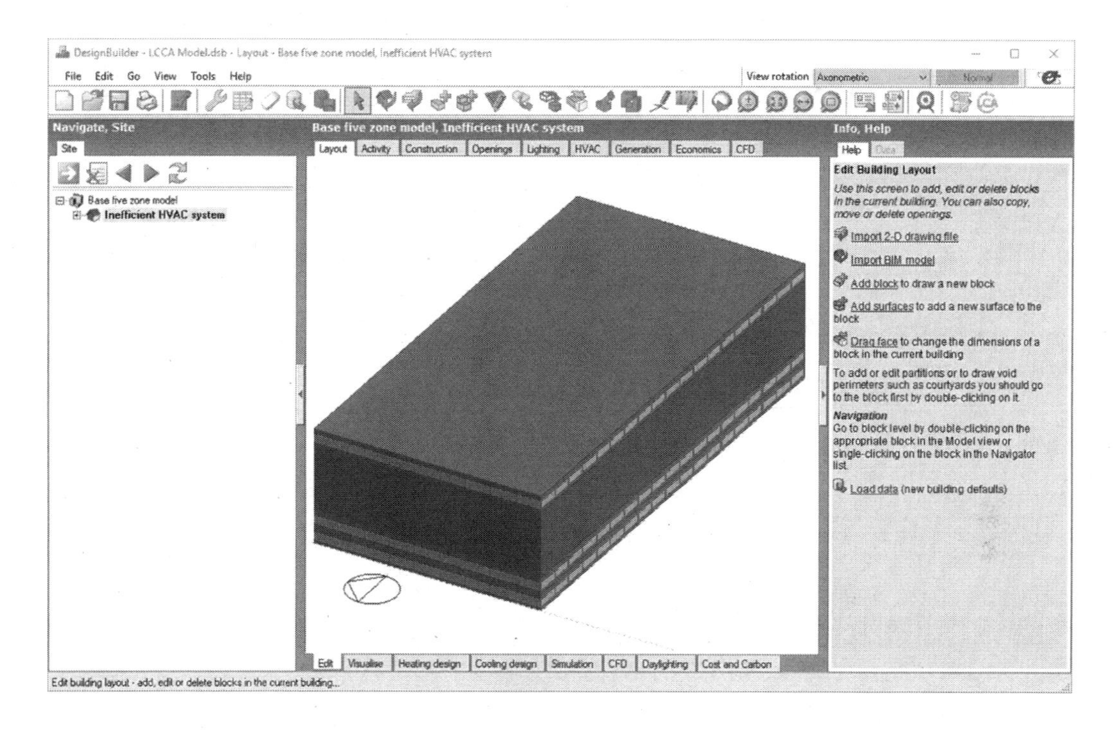

Step 2: Click the **Spanner** icon. The **Model Options – Building and Block** screen appears.

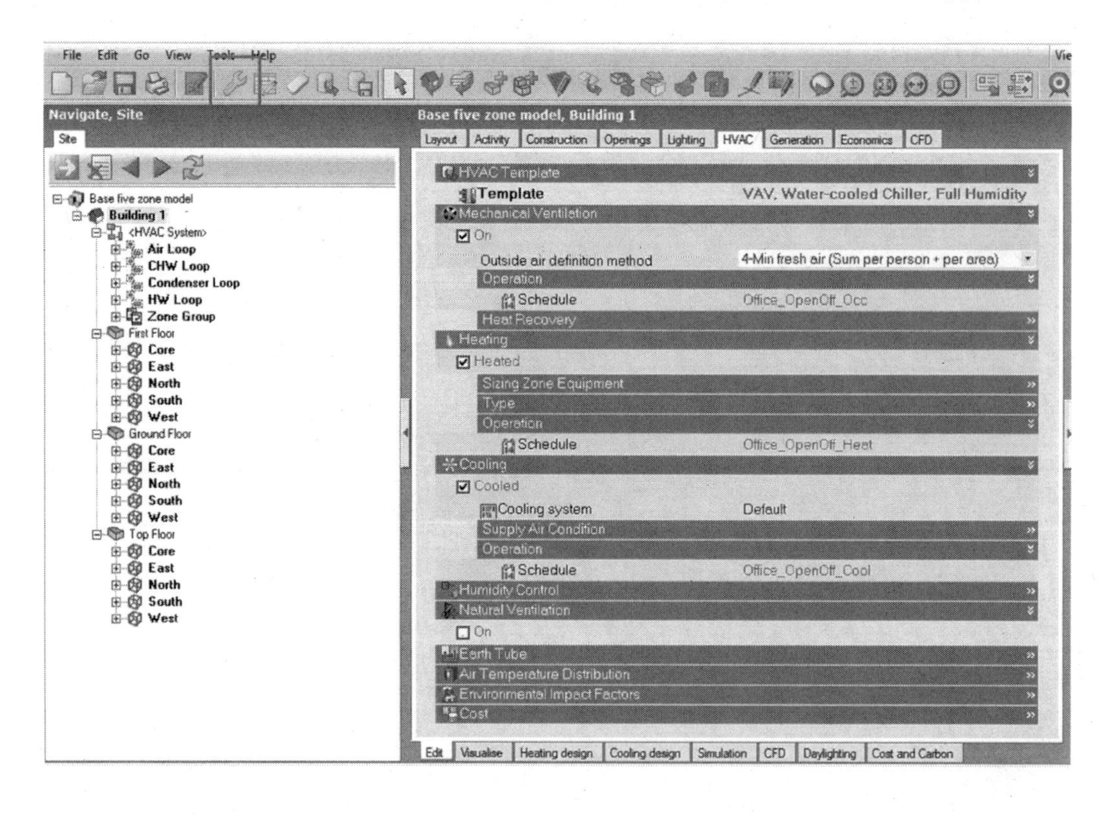

Step 3: Click **Detailed** under the **HVAC** slider. Select **Detailed HVAC Data** from the **Detailed HVAC** drop-down list. Click **OK**.

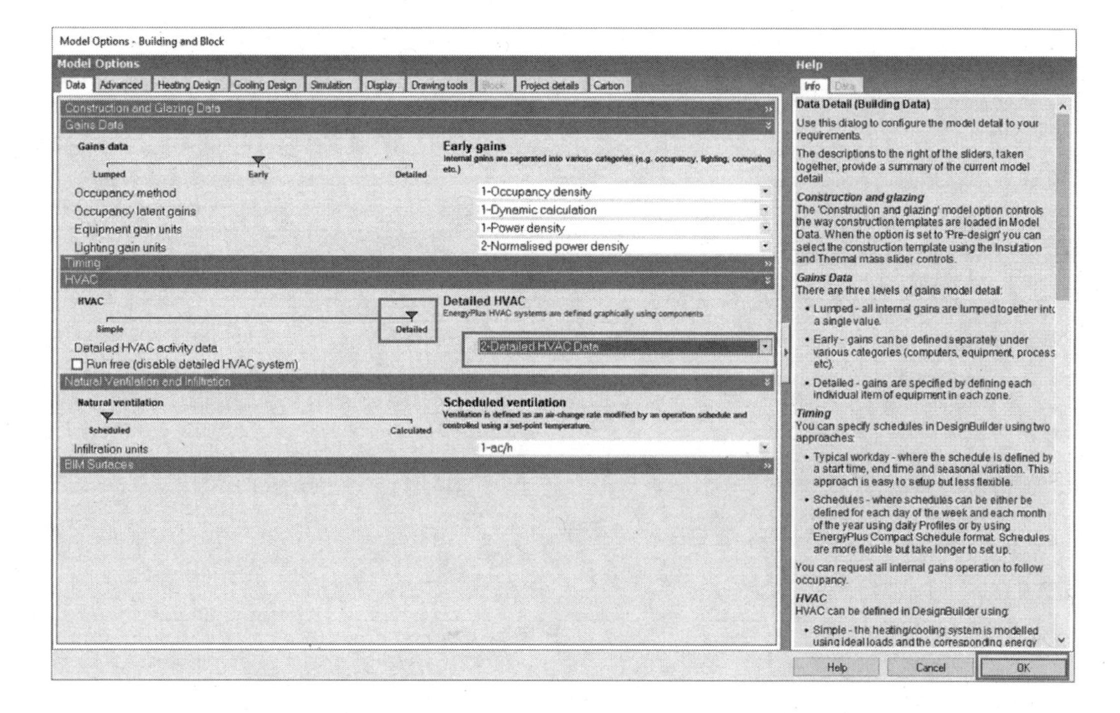

Step 4: Click **<HVAC System>** in the navigation pane. It displays initializing the **HVAC progress** bar, and subsequently, the **Load HVAC template** screen appears. Select **VAV Reheat, Water-cooled Chiller** from **Detailed HVAC Template**. Click **Next**. Then click **Finish**.

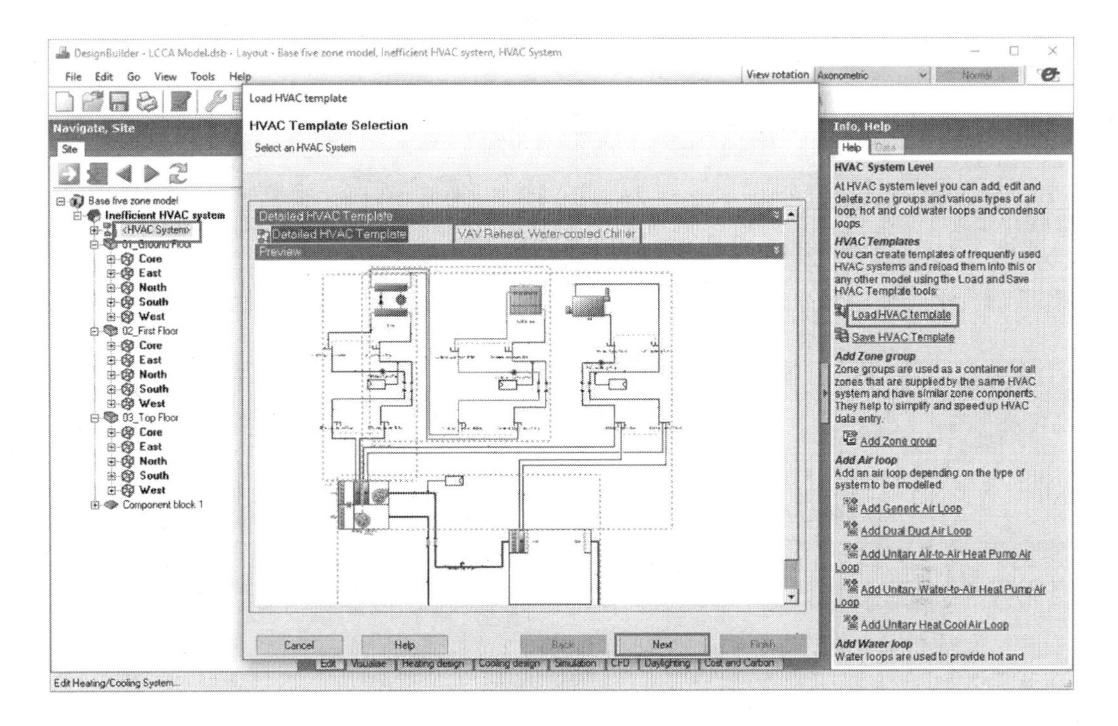

Step 5: Expand **CHW Loop Supply Side**, and click **Chiller**. The **Chiller** layout appears. Click **Edit component** under the **Info** panel on the right. The **Chiller** data appear.

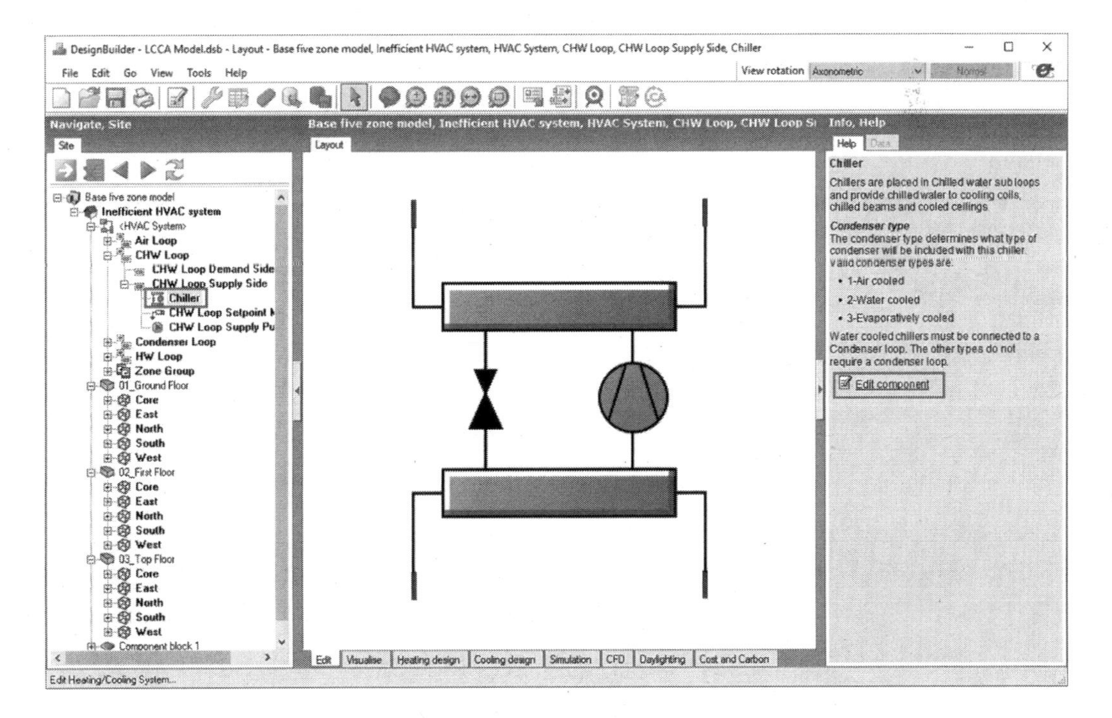

Step 6: Enter **4.2** in the **Reference COP** text box.

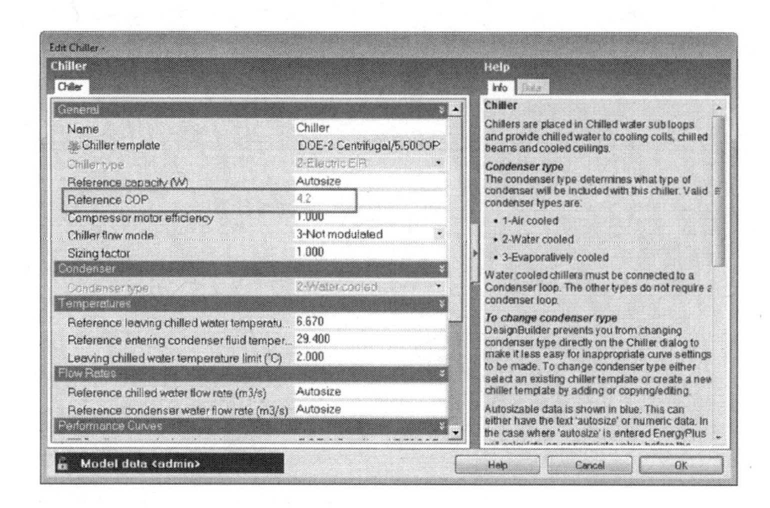

Step 7: Expand **Condensor loop**. Click on **Cooling Tower**. Select **1-Single speed** from the **Cooling tower type** drop-down list.

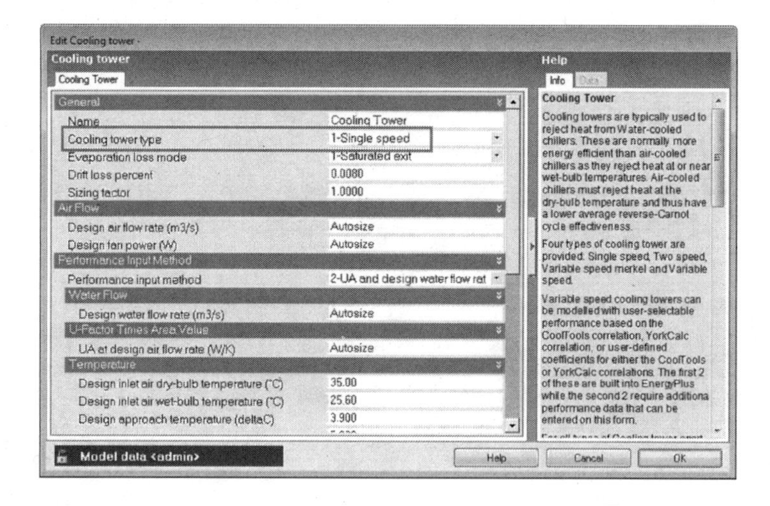

Step 8: Select **Program Options** from the **Tools** menu.

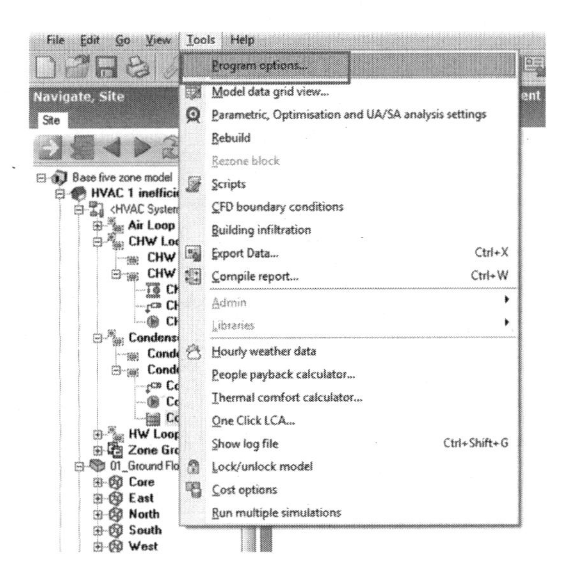

Step 9: Click the **International** tab. Select **Indian Rupee** from **Currency**. Click **OK**.

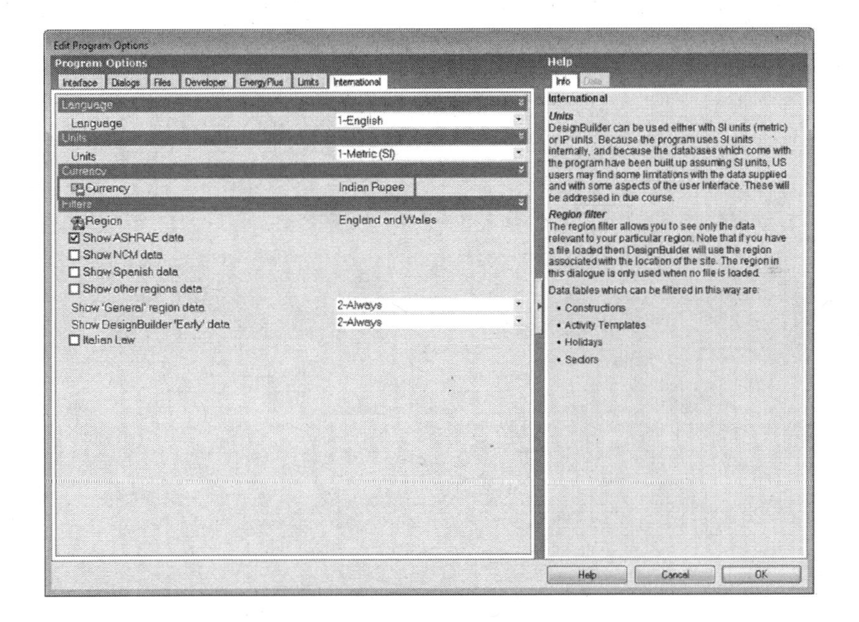

Step 10: Click the **HVAC** tab. Select the **Cost** section. Enter **2000** in the **HVAC cost per area (INR/m2 GIFA)** text box.

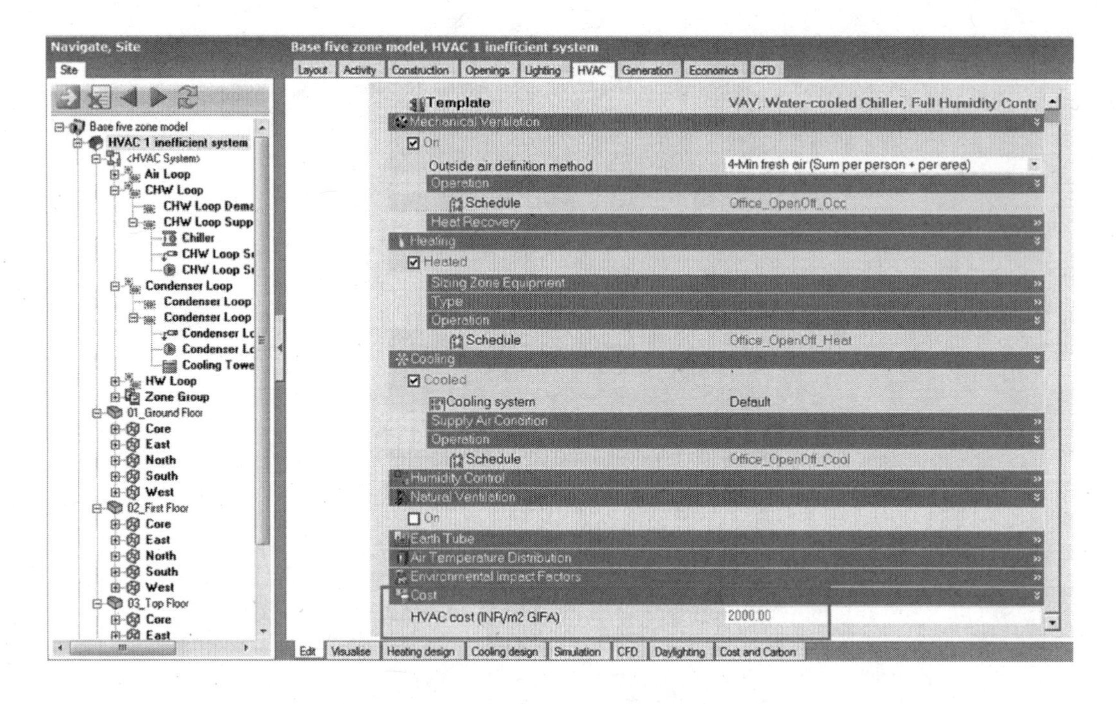

The HVAC cost is calculated zone by zone by multiplying the zone floor area by the cost per floor area (GIFA) for the zone on the **HVAC** tab. This data will normally be loaded from the **HVAC template** to ensure that lighting performance and cost data are consistent. This cost should include any contributions to the total cost of the building HVAC system from plant components and not only the components within the particular zone.

Source: https://designbuilder.co.uk/helpv6.0/Content/HVACCost.htm.

Step 11: Click the **Economics** tab.

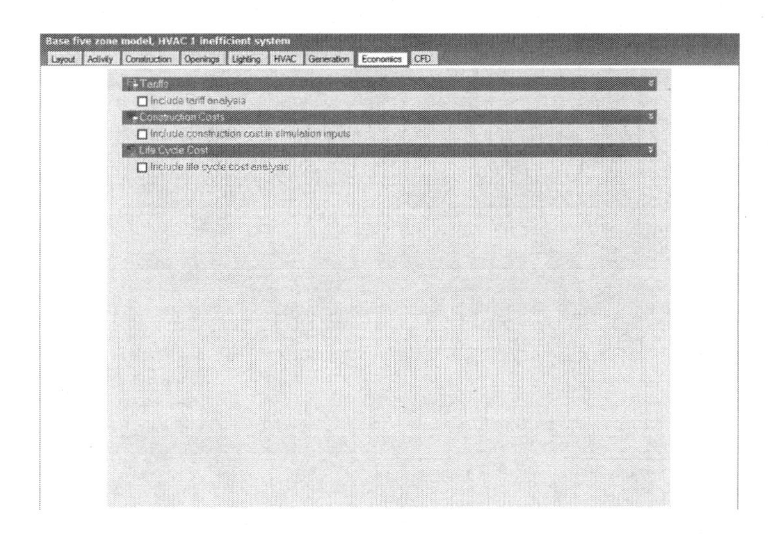

Step 12: Select the **Include tariff analysis** checkbox. Create a copy of the **Flat electricity charge**. Update **Tariff name** as **Flat electricity charge_Project**.

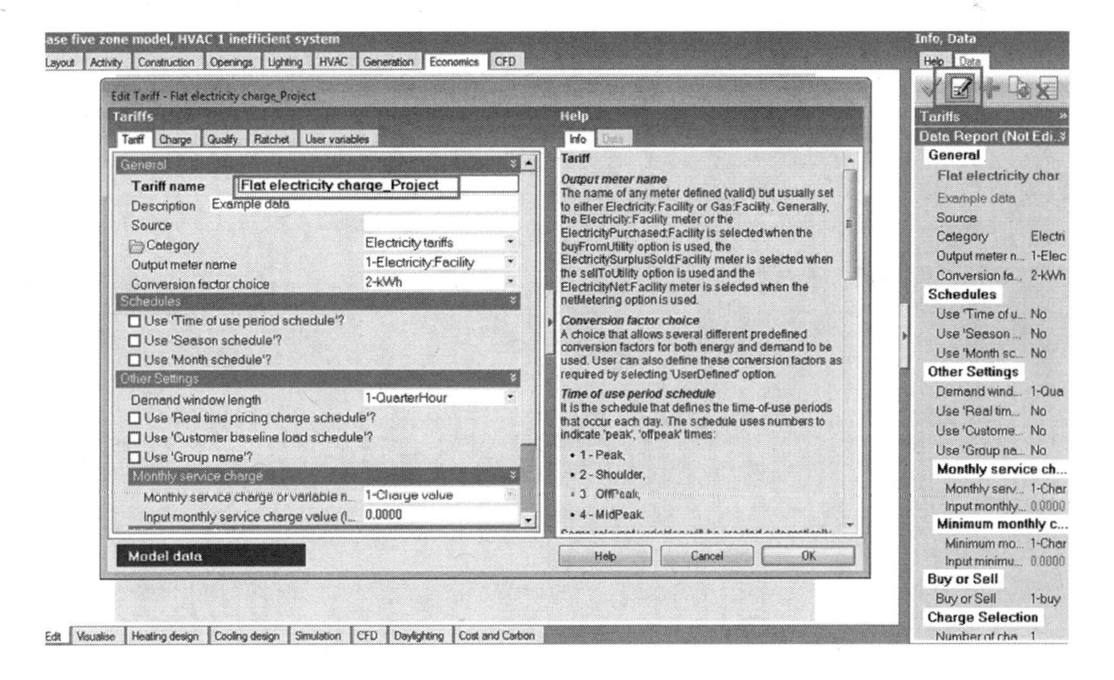

Step 13: Click on the **Charge** tab. Select **1** from the **Number of charges** drop-down list. Enter **10** in **Input cost per unit value (INR)**. Click **OK**.

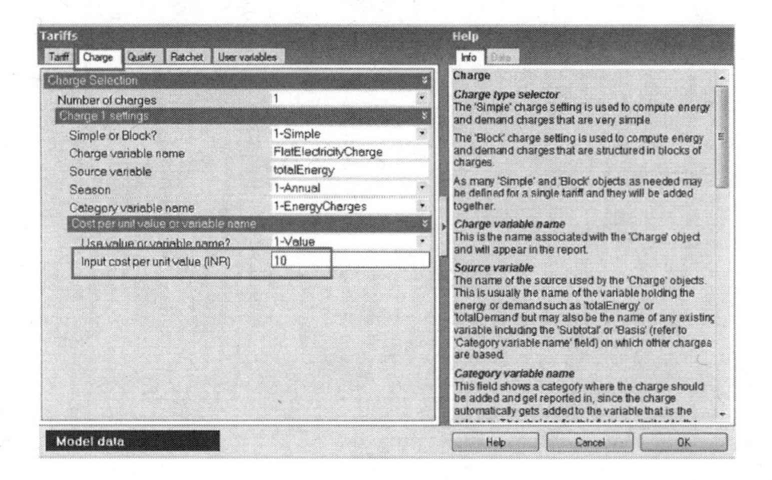

Step 14: Select the **Include construction cost in simulation inputs** checkbox. Select the **Include life cycle cost analysis** checkbox.

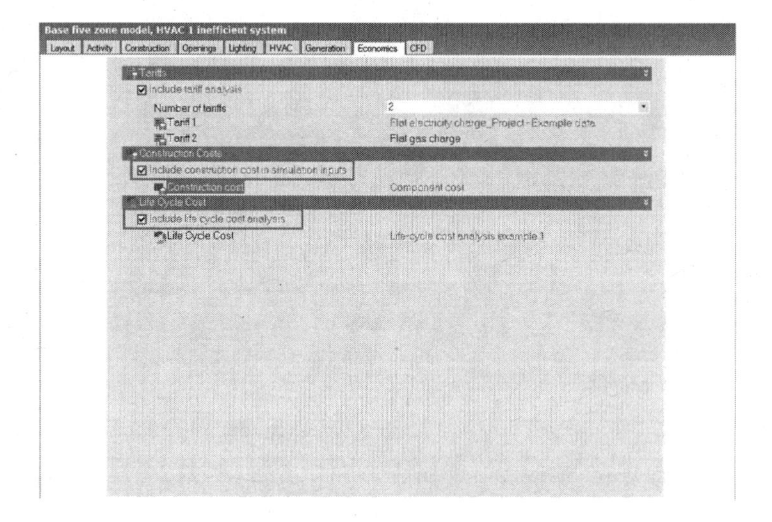

Step 15: Copy and edit **Life-cycle cost analysis example 1**.

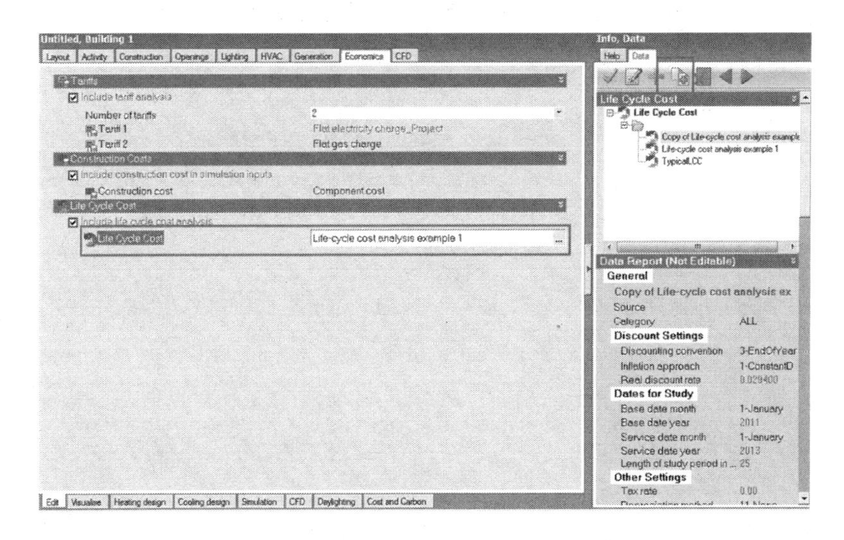

Step 16: Enter **Life-cycle cost analysis_Project** in the **Name** text box. Select **2-CurrentDollar** from the **Inflation approach** drop-down list. Enter **0.1** in the **Nominal discount rate** and **0.04** in the **Inflation rate** text boxes. Enter **15** in the **Length of study period in years** text box.

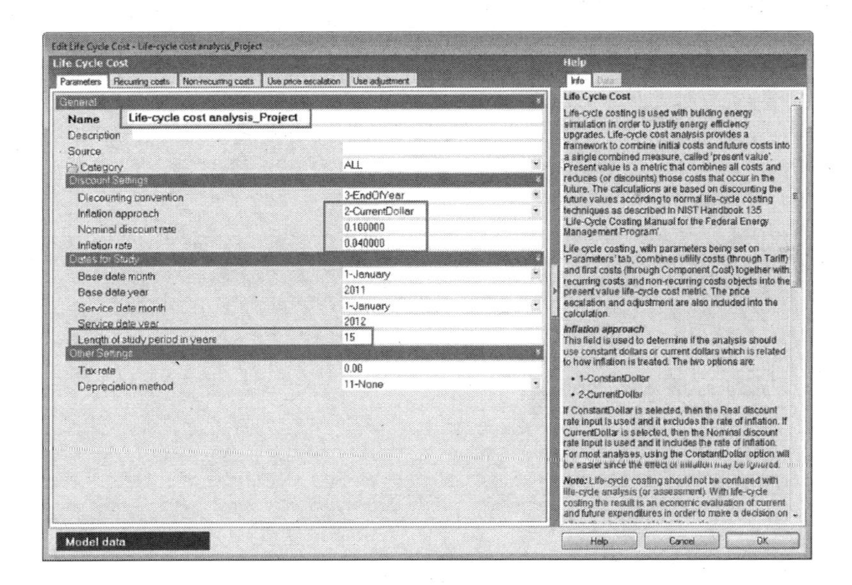

Step 17: Select **1** from the **Number of recurring costs** drop-down list in the **Recurring costs** tab. Enter **3,000,000** in the **Cost (INR)** text box.

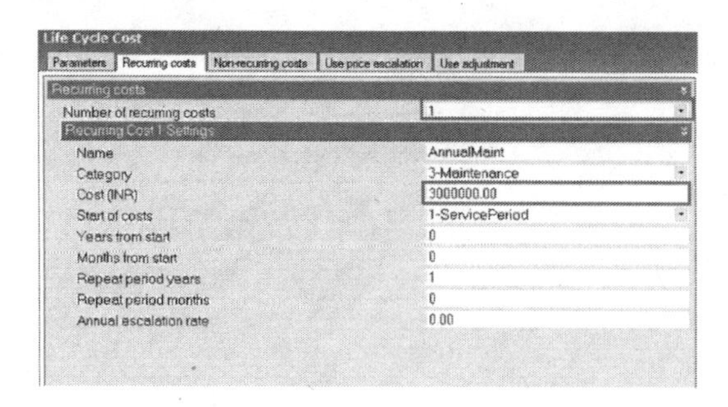

Step 18: Select **0** from the **Number of non-recurring costs** drop-down list in the **Non-recurring costs** tab. Click **OK**. To keep this problem simple, non-recurring cost is kept to zero.

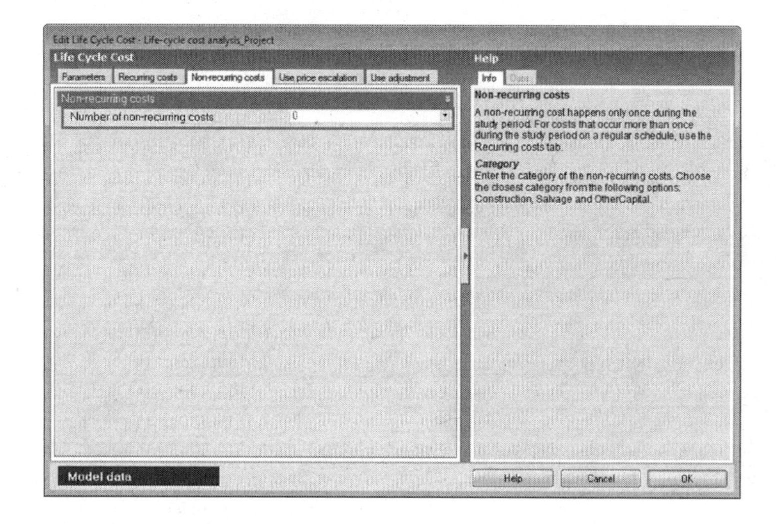

Step 19: Select **Life-cycle cost analysis_Project**.

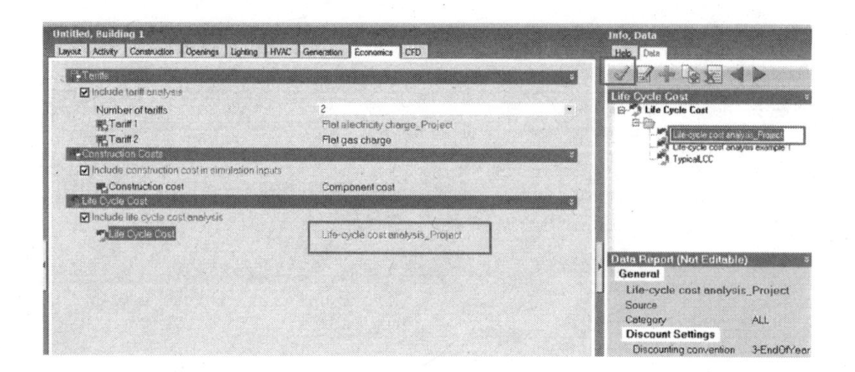

Step 20: Click **Component cost**. Create a copy and edit. Name it **Component cost_project**.

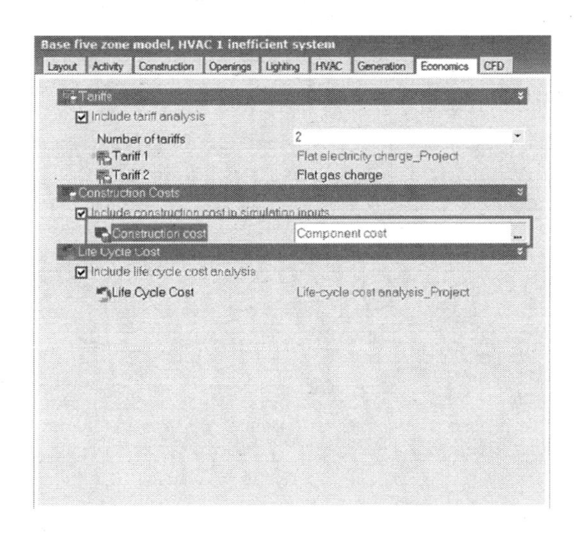

Step 21: Click the **Adjustments** tab. Enter **0.000** in the **Miscellaneous cost per conditioned area (INR/GIFA)** text box. Enter **1.000** in the **Regional adjustment factor** text box. Enter the other values as shown in the following figure. Click **OK**.

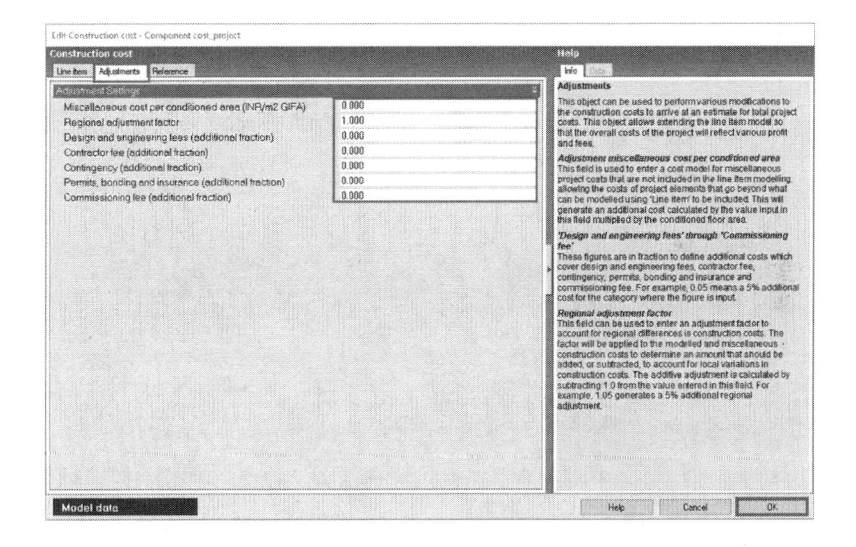

Step 22: Click on the **Simulation screen** tab. Click **Update data**.

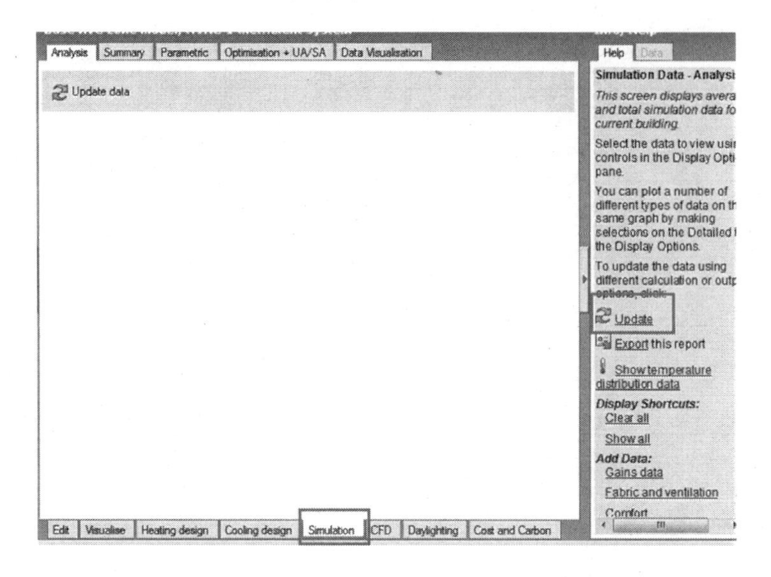

Step 23: Click the **Output** tab. Select the **Component Cost Economics Summary** checkbox. Click **OK**.

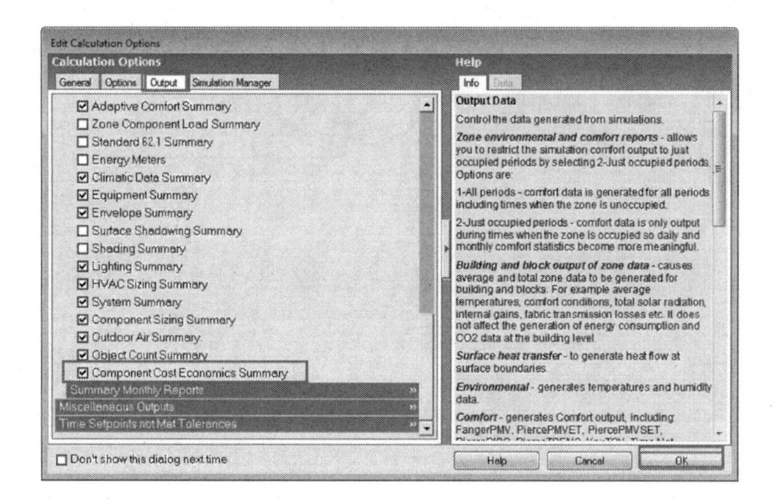

Step 24: After the simulations, click the **Summary** tab, and scroll down to read **Life-Cycle Cost Parameters**.

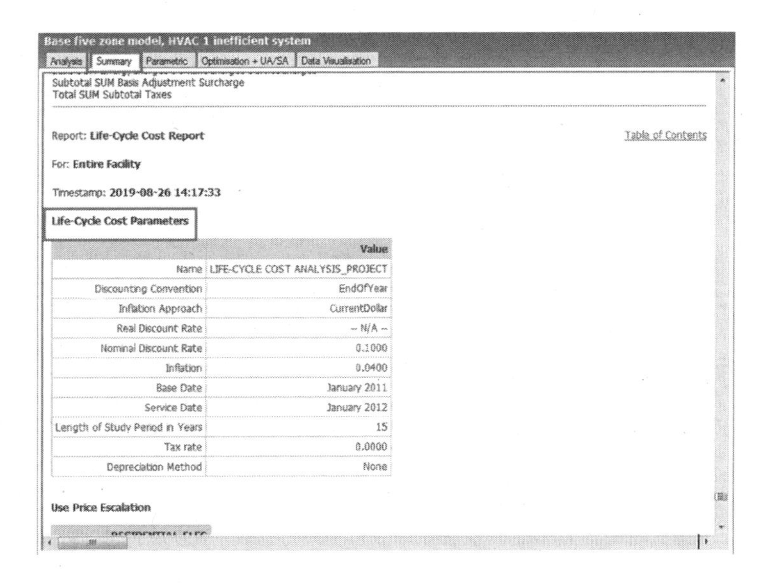

Base five zone model, HVAC 1 inefficient system

Analysis | Summary | Parametric | Optimisation + UA/SA | Data Visualisation

Subtotal SUM Basis Adjustment Surcharge
Total SUM Subtotal Taxes

Report: **Life-Cycle Cost Report**

Table of Contents

For: **Entire Facility**

Timestamp: **2019-08-26 14:17:33**

Life-Cycle Cost Parameters

	Value
Name	LIFE-CYCLE COST ANALYSIS_PROJECT
Discounting Convention	EndOfYear
Inflation Approach	CurrentDollar
Real Discount Rate	-- N/A --
Nominal Discount Rate	0.1000
Inflation	0.0400
Base Date	January 2011
Service Date	January 2012
Length of Study Period in Years	15
Tax rate	0.0000
Depreciation Method	None

Use Price Escalation

Step 25: Copy the **Present Value by Year** table to an external spreadsheet program.

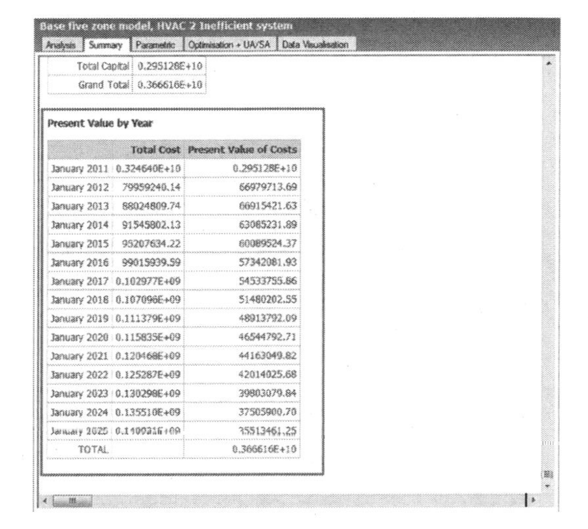

Base five zone model, HVAC 2 Inefficient system

Analysis | Summary | Parametric | Optimisation + UA/SA | Data Visualisation

Total Capital	0.295128E+10
Grand Total	0.366616E+10

Present Value by Year

	Total Cost	Present Value of Costs
January 2011	0.324640E+10	0.295128E+10
January 2012	79959240.14	66979713.69
January 2013	88024809.74	66915421.63
January 2014	91545802.13	63085231.89
January 2015	95207634.22	60089524.37
January 2016	99015939.59	57342081.93
January 2017	0.102977E+09	54533755.86
January 2018	0.107096E+09	51480202.55
January 2019	0.111379E+09	48913792.09
January 2020	0.115835E+09	46544792.71
January 2021	0.120468E+09	44163049.82
January 2022	0.125287E+09	42014025.68
January 2023	0.130298E+09	39803079.84
January 2024	0.135510E+09	37505900.70
January 2025	[illegible]	35513461.25
TOTAL		0.366616E+10

Step 26: Save the model. Repeat the preceding steps to model **HVAC System 2** as given in the problem statement.

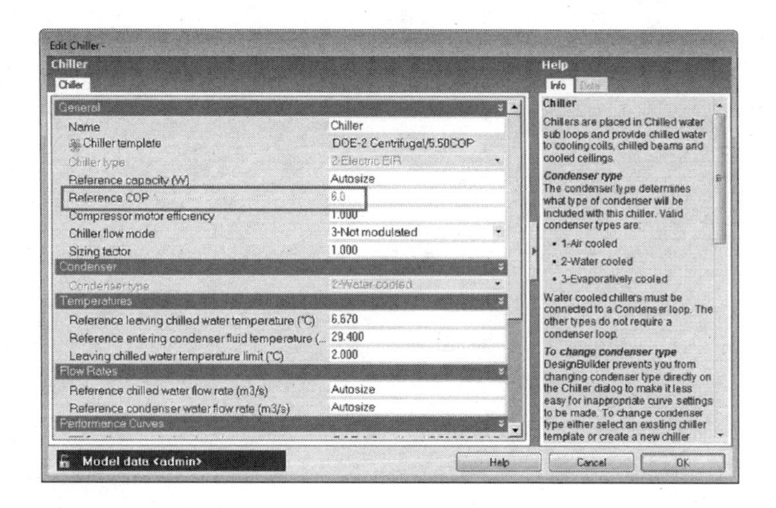

Step 27: Click on **Cooling Tower**. Click the **Edit component** link in the **Help** tab.

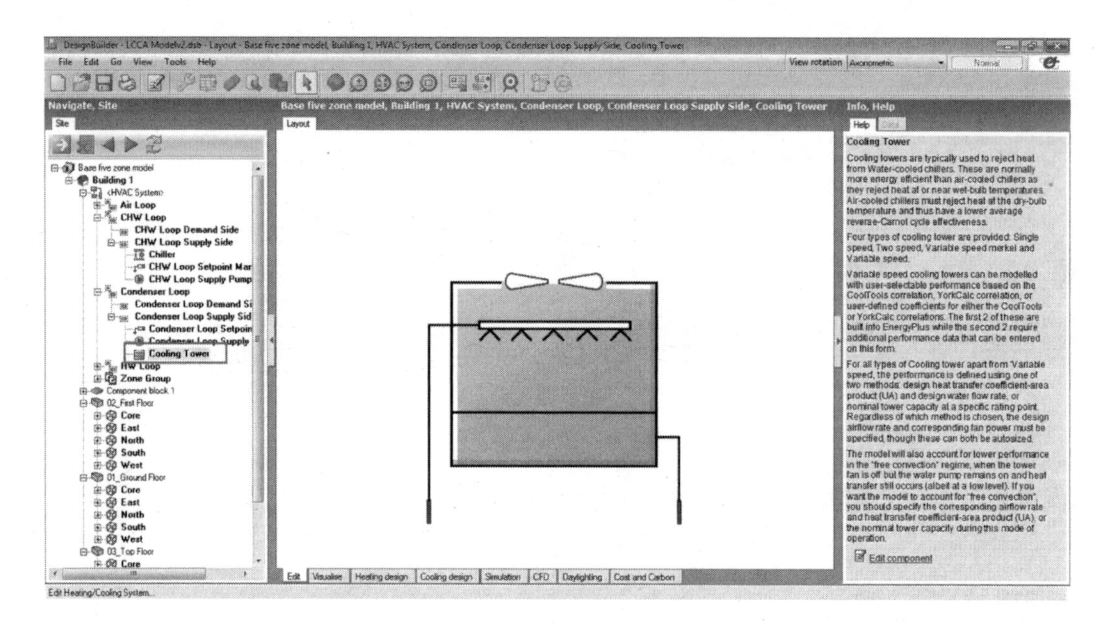

Step 28: Select **4-Variable Speed** from the **Cooling tower type** drop-down list.

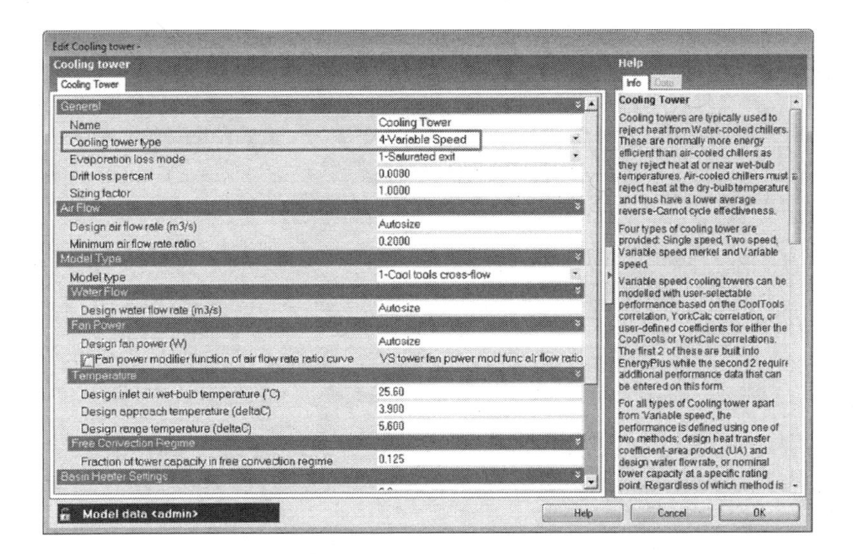

Step 29: Click the **HVAC** tab. Enter **3,000** in the **HVAC Cost (INR/m2 GIFA)** text box.

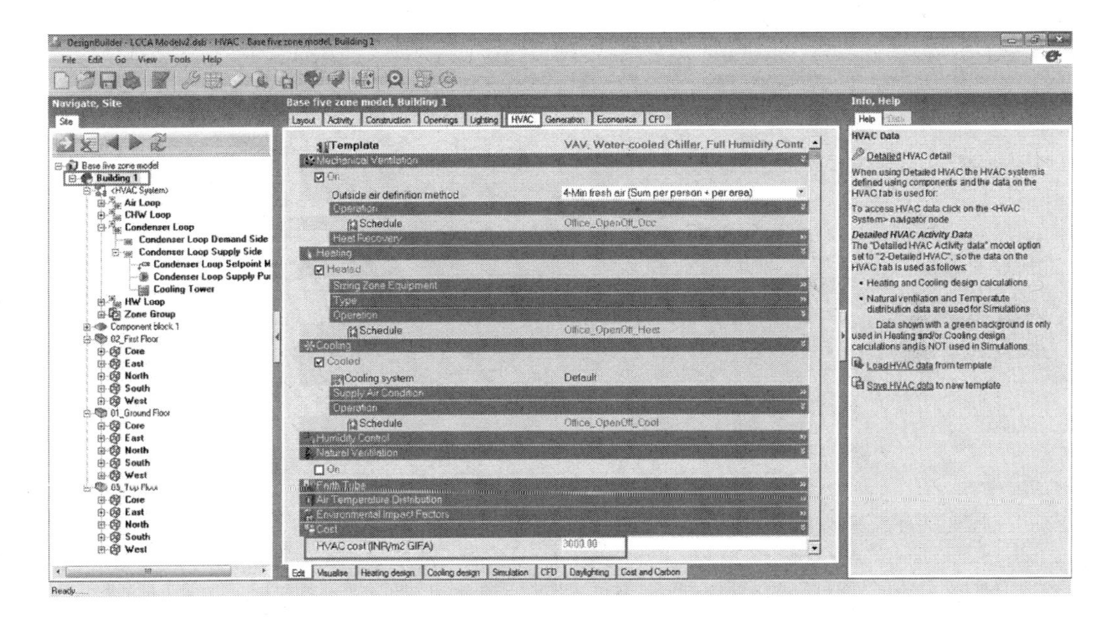

Step 30: Select the **Economics** tab. Edit **Life Cycle Cost**. Enter **4,500,000** in the **Cost (INR)** text box.

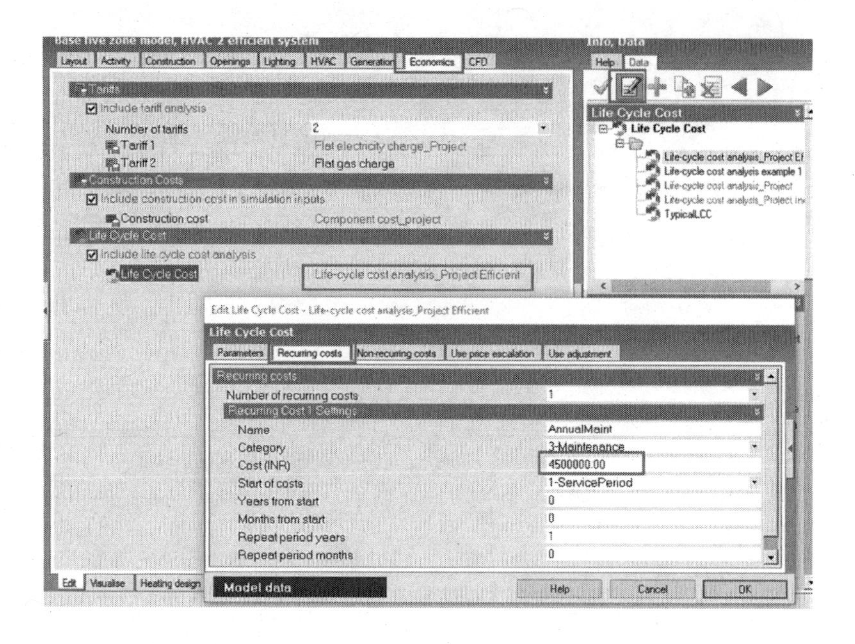

Step 31: Simulate the model, and record the results. Compare result in the spreadsheet program (Table 13.6).

TABLE 13.6
Results comparison in spreadsheet program

	HVAC System 1		HVAC System 2					
Year	Total cost (INR) (A)	Present value of costs (B)	Total cost (INR) (C)	Present value of costs (D)	Difference in total costs (E = A – B)	Cumulative cash flow for total cost (F) (F(n) = E(n) + F(n – 1))	Difference in present value of costs (G = C – D)	Cumulative cash flow for discounted cost (H) (H(n) = G(n) + H(n – 1))
Jan-11	2,154,490,000	1,958,630,000	2,184,490,000	1,985,900,000	(−)30,000,000	(−)30,000,000	(−)30,000,000	(−)30,000,000
Jan-12	83,079,240	69,558,226	74,862,997	62,656,445	8,216,243	(−)21,783,757	6,901,781	(−)23,098,219
Jan-13	86,402,410	65,696,488	77,857,517	59,179,571	8,544,893	(−)13,238,864	6,516,918	(−)16,581,301
Jan-14	89,858,506	61,932,786	80,971,817	55,793,725	8,886,689	(−)4,352,175	6,139,061	(−)10,442,240
Jan-15	93,452,846	58,999,939	84,210,690	53,140,427	9,242,156	4,889,981	5,859,512	(−)4,582,728
Jan-16	97,190,960	56,311,929	87,579,118	50,706,178	9,611,842	14,501,824	5,605,751	1,023,023
Jan-17	101,079,000	53,559,793	91,082,282	48,220,136	9,996,718	24,498,541	5,339,656	6,362,679
Jan-18	105,122,000	50,559,365	94,725,574	45,520,772	10,396,426	34,894,968	5,038,592	11,401,271
Jan-19	109,327,000	48,043,182	98,514,596	43,249,414	10,812,404	45,707,371	4,793,768	16,195,039
Jan-20	113,700,000	45,721,670	102,455,000	41,152,240	11,245,000	56,952,371	4,569,430	20,764,469
Jan-21	118,248,000	43,384,825	106,553,000	39,045,129	11,695,000	68,647,371	4,339,696	25,104,165
Jan-22	122,978,000	41,278,249	110,316,000	37,142,987	12,162,000	80,809,371	4,135,262	29,239,427
Jan-23	127,897,000	39,107,437	115,248,000	35,187,707	12,649,000	93,458,371	3,919,729	33,159,156
Jan-24	133,013,000	36,848,202	119,358,000	33,157,938	13,155,000	106,613,371	3,690,263	36,849,420
Jan-25	138,333,000	34,891,637	124,552,000	31,396,036	13,681,000	120,294,371	3,495,601	40,345,021
		2,664,520,000		2,621,450,000				

Step 32: Compare the present value cost for both cases (Table 13.7).

TABLE 13.7
Present value cost for the both types of HVAC systems

HVAC System 1	HVAC System 2
2,664,520,000	2,621,450,000

Step 33: Calculate the IRR using a spreadsheet program (Table 13.8).

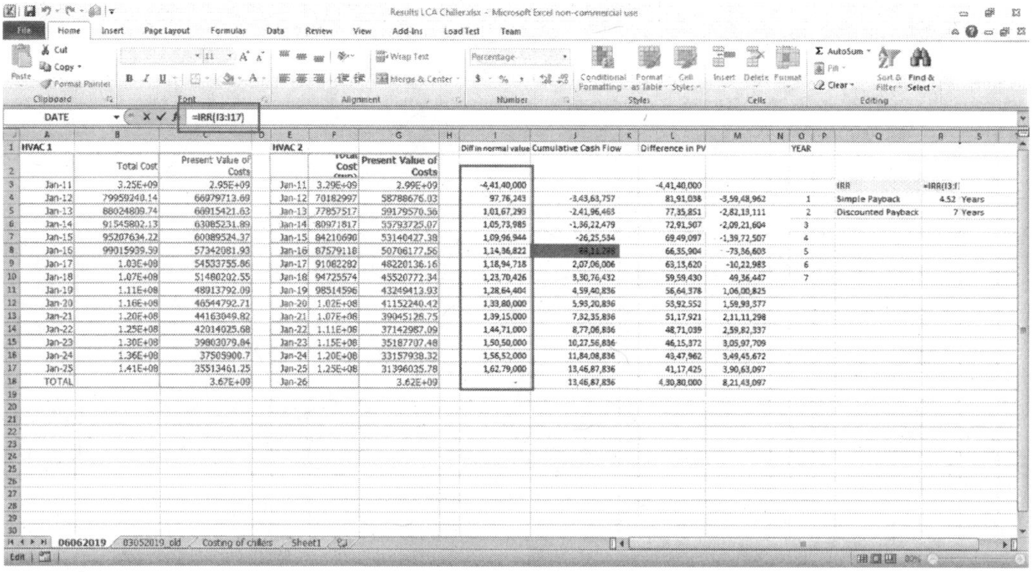

Results LCA Chillers.xlsx - Microsoft Excel non-commercial use

DATE =IRR(I3:I17)

	A	B	C	E	F	G	I	J	L	M	O	Q	R
1	HVAC 1			HVAC 2			Diff in normal value	Cumulative Cash Flow	Difference in PV		YEAR		
2		Total Cost	Present Value of Costs		Total Cost	Present Value of Costs							
3	Jan-11	3.25E+09	2.95E+09	Jan-11	3.29E+09	2.99E+09	-4,41,40,000		-4,41,40,000			IRR	=IRR(I3:I
4	Jan-12	79959240.14	66979713.69	Jan-12	70182997	58788676.03	97,76,243	-3,43,63,757	81,91,038	-3,59,48,962	1	Simple Payback	4.52 Years
5	Jan-13	88024809.74	66915421.63	Jan-13	77857517	59179570.56	1,01,67,293	-2,41,96,463	77,35,851	-2,82,13,111	2	Discounted Payback	7 Years
6	Jan-14	91545802.13	63085231.89	Jan-14	80971817	55793725.07	1,05,73,985	-1,36,22,479	72,91,507	-2,09,21,604	3		
7	Jan-15	95207634.22	60089524.37	Jan-15	84210690	53140427.38	1,09,96,944	-26,25,534	69,49,097	-1,39,72,507	4		
8	Jan-16	99015939.59	57342081.93	Jan-16	87579118	50706177.56	1,14,36,822	[illegible]	66,35,904	-73,36,603	5		
9	Jan-17	1.03E+08	54533755.86	Jan-17	91082282	48220136.16	1,18,94,718	2,07,06,006	63,13,620	-10,22,983	6		
10	Jan-18	1.07E+08	51480202.55	Jan-18	94725574	45520772.34	1,23,70,426	3,30,76,432	59,59,430	49,36,447	7		
11	Jan-19	1.11E+08	48913792.09	Jan-19	98514596	43249413.93	1,28,64,404	4,59,40,836	56,64,378	1,06,00,825			
12	Jan-20	1.16E+08	46544792.71	Jan-20	1.02E+08	41152240.42	1,33,80,000	5,93,20,836	53,92,552	1,59,93,377			
13	Jan-21	1.20E+08	44163049.82	Jan-21	1.07E+08	39045128.75	1,39,15,000	7,32,35,836	51,17,921	2,11,11,298			
14	Jan-22	1.25E+08	42014025.68	Jan-22	1.11E+08	37142987.09	1,44,71,000	8,77,06,836	48,71,039	2,59,82,337			
15	Jan-23	1.30E+08	39803079.84	Jan-23	1.15E+08	35187707.48	1,50,50,000	10,27,56,836	46,15,372	3,05,97,709			
16	Jan-24	1.36E+08	37505900.7	Jan-24	1.20E+08	33157938.32	1,56,52,000	11,84,08,836	43,47,962	3,49,45,672			
17	Jan-25	1.41E+08	35513461.25	Jan-25	1.25E+08	31396035.78	1,62,79,000	13,46,87,836	41,17,425	3,90,63,097			
18	TOTAL		3.67E+09	Jan-26		3.62E+09	-	13,46,87,836	4,30,80,000	8,21,43,097			

06062019 / 03052019_old / Costing of chillers / Sheet1

TABLE 13.8
Internal rate of return and payback calculation

IRR	**30%**
Simple payback	4
Discounted payback	5

In Table 13.8, payback period is calculated using Table 13.6. The cumulative cash flow is a positive number in the fourth year; that year is the simple payback year. Similarly cumulated cash flow for discounted is positive in the fifth year (refer to Table 13.6). You can see from cost-benefit analysis that HVAC System 2 is a better option to choose because it has a lower net present value.

Tutorial 13.3

Performing Sensitivity and Uncertainty Analysis

GOAL

To evaluate the impact of solar control design options of window sizes, glass solar heat gain coefficient (SHGC) and shading on net carbon emissions

WHAT ARE YOU GOING TO LEARN?

- How to set up design variables for sensitivity analysis and uncertainty analysis

PROBLEM STATEMENT

In this tutorial, you are going to use a 50- x 25-m five-zone model with a 5-m perimeter depth. You are going to use the design variables listed in Table 13.9.

TABLE 13.9
Glazing input parameters

Parameter	Distribution type	Properties
Glass SHGC	Uniform distribution (discrete)	0.2
		0.3
		0.4
		0.5
		0.6
WWR	Normal distribution (continuous)	Mean 40, standard deviation 10, lower limit 20 and upper limit 60
Local shading – overhang	Uniform distribution (discrete)	No shading
		0.5 m
		1.0 m
		1.5 m
		2.0 m

Perform sensitivity and uncertainty analysis on operational CO_2 emissions. Use the **New Delhi Palam, India** weather location.

The uncertainty in the inputs imposes a limit on confidence in the output of the model; that is, results of the simulation model are potentially erroneous. Such errors are quantified in uncertainty analysis, which treats the simulation output as 'probabilistic' rather than 'deterministic'. A related field to uncertainty analysis is sensitivity analysis. Whereas uncertainty analysis is a method to quantify the variability of a model output owing to uncertainty in the one or more input variables, sensitivity analysis is the study of how the uncertainty in each simulation output can be apportioned to various sources of uncertainty in its inputs.

Source: https://designbuilder.co.uk/helpv6.0/#UASA.htm.

SOLUTION

Step 1: Create a **50- × 25-m** five-zone model with a **5-m** perimeter depth.

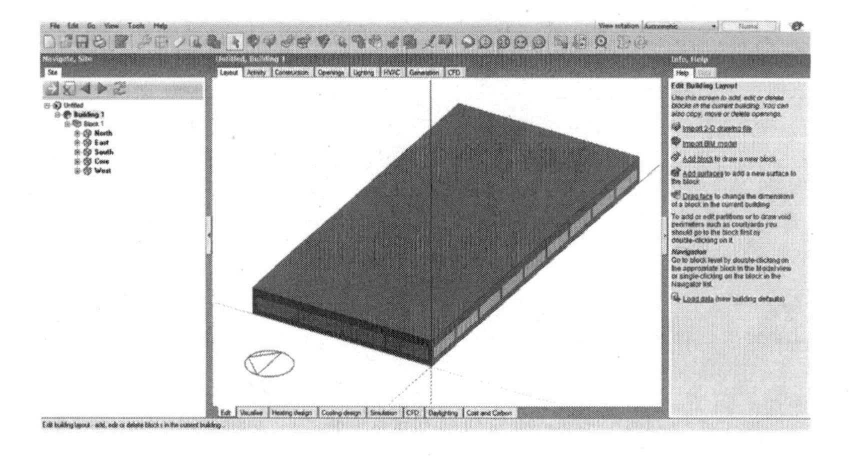

Step 2: Click the **Openings** tab. Click the **Add new item** icon.

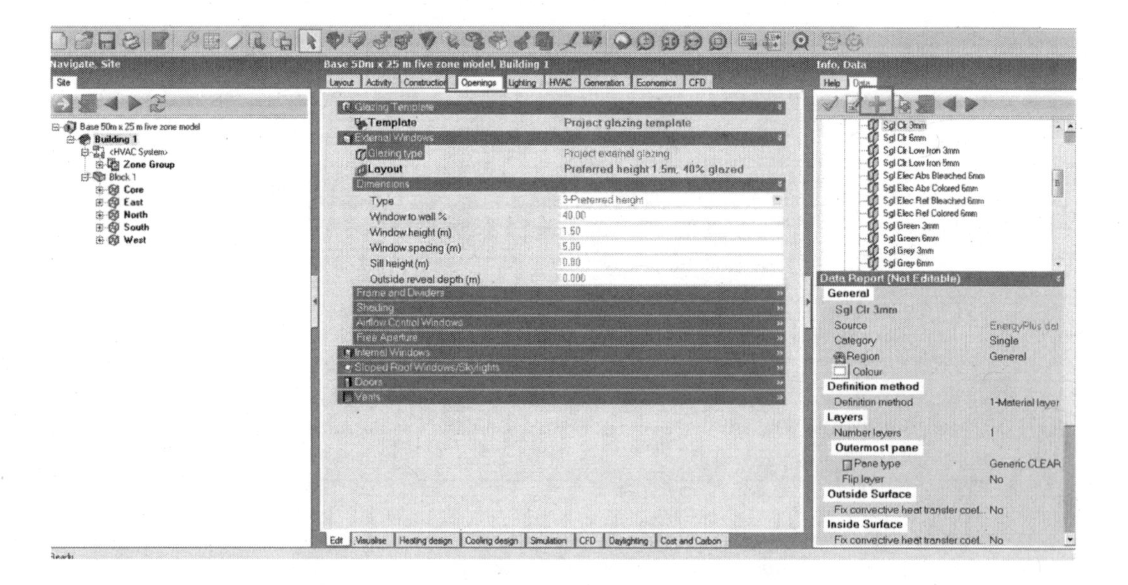

Step 2: Enter **Glass SHGC 0.2** in the **Name** text box. Select **2-Simple** from the **Definition method** drop-down list. Enter **0.2** in the **Total solar transmission (SHGC)** text box, **0.4** in the **Light transmission** text box and **5.500 in the U-value (W/m2-K)** text box. Click **OK**.

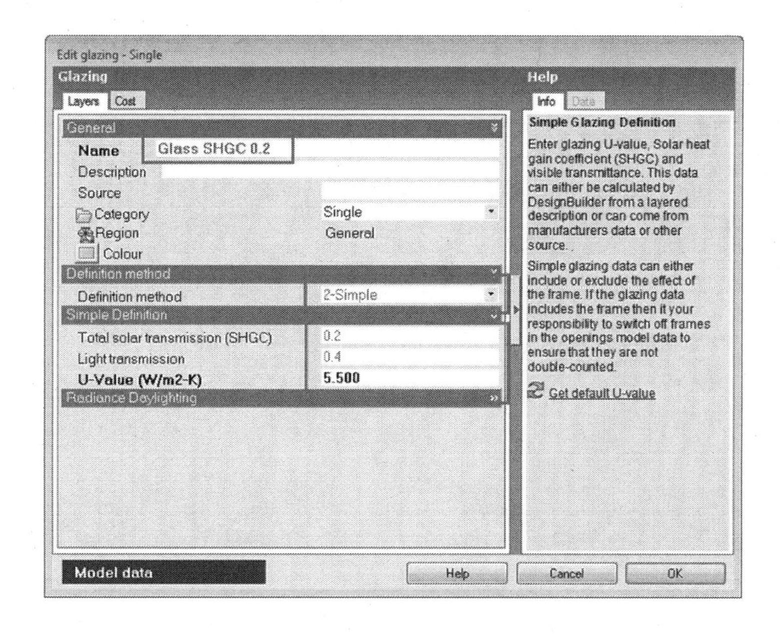

Step 3: Select **Glass SHGC 0.2**. Click the **Create copy of heighted item** icon.

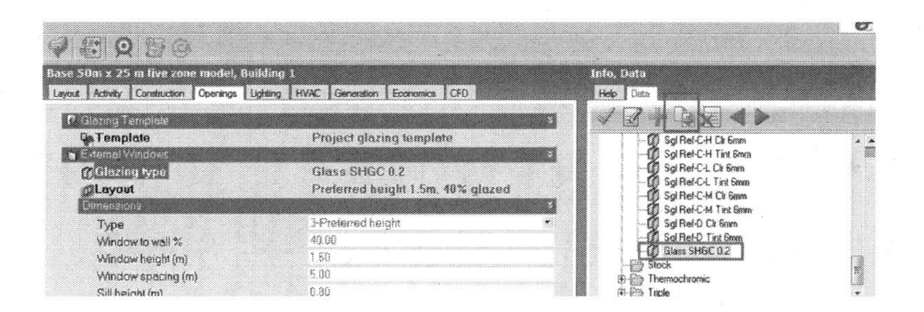

Step 4: Enter **Glass SHGC 0.3** in the **Name** text box. Enter **0.3** in the **Total solar transmission (SHGC)** text box. Click **OK**.

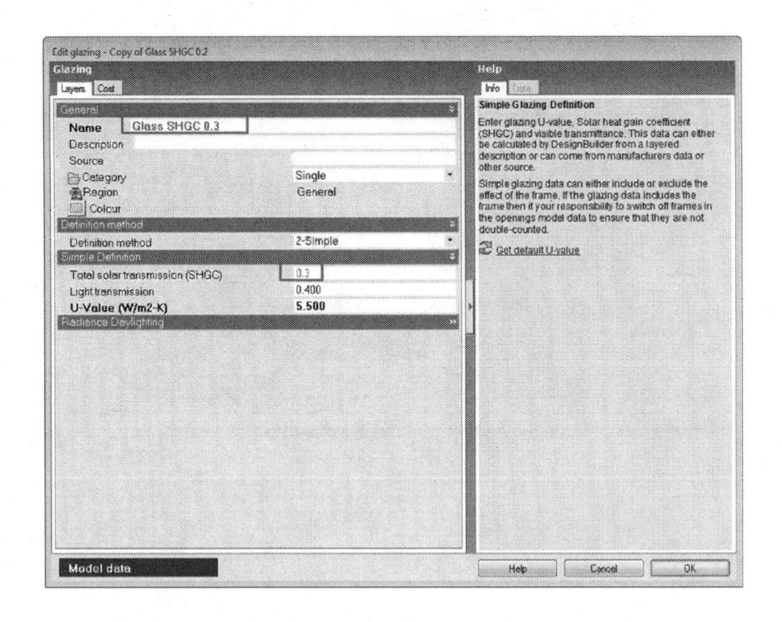

Similarly, create **Glass with SHGC 0.4**, **0.5** and **0.6**.

Step 5: Under the **Shading** subtab, select the **Local shading** checkbox. Select **1.0 m Overhang** in **Type**.

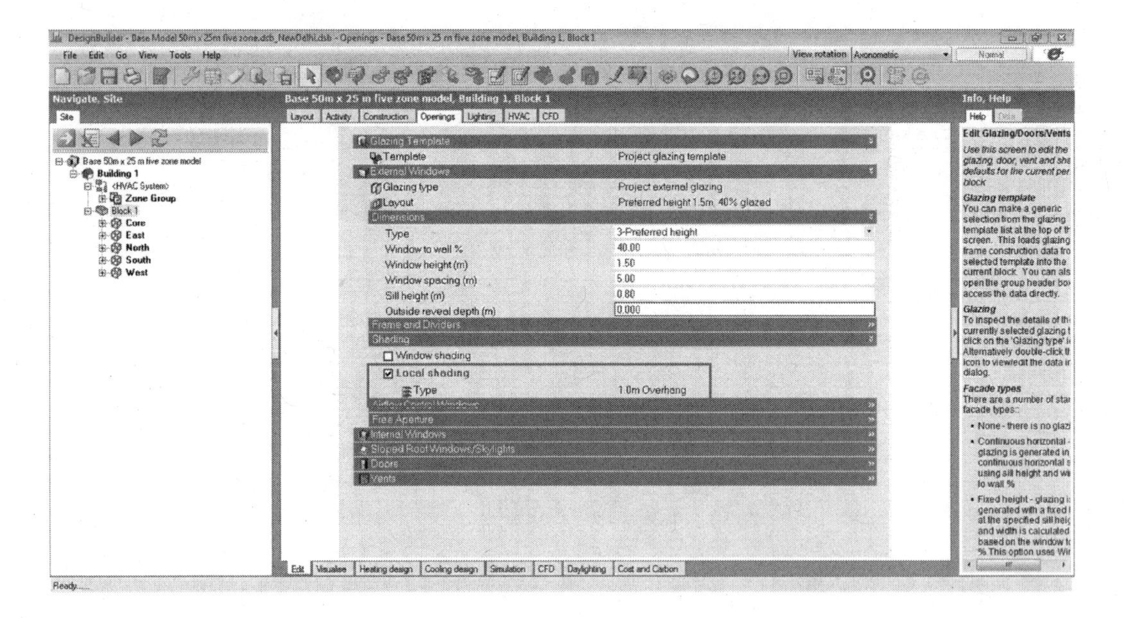

Step 6: Click the **Optimisation UA/SA** button.

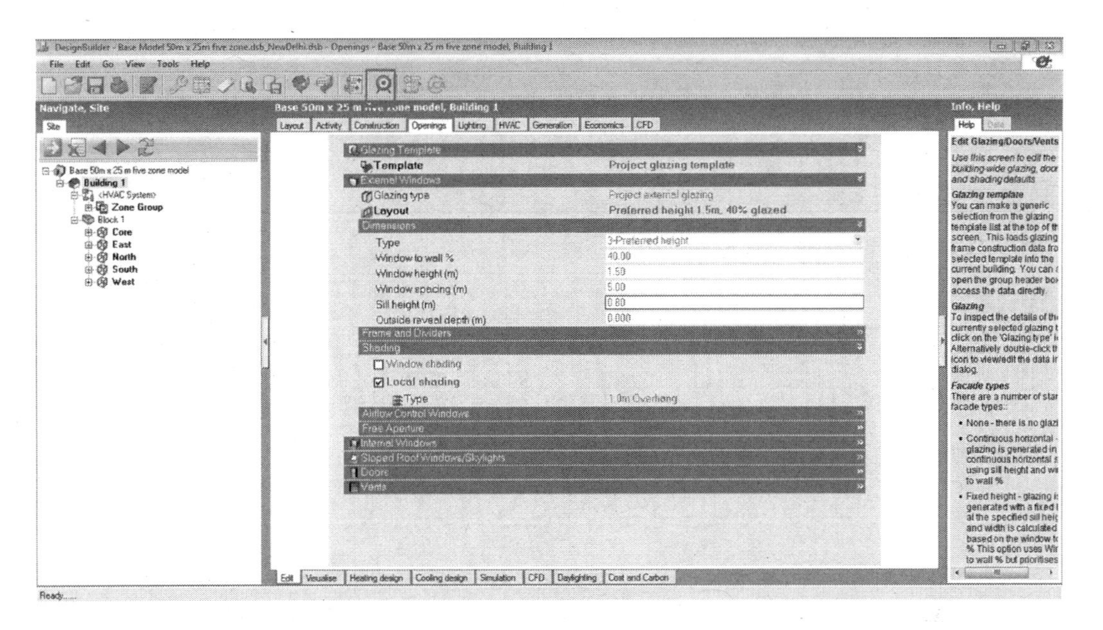

Step 7: Select **3-Uncertainty/Sensitivity** from the **Analysis** drop-down list. Select **2-Detailed** from the **Level** drop-down list.

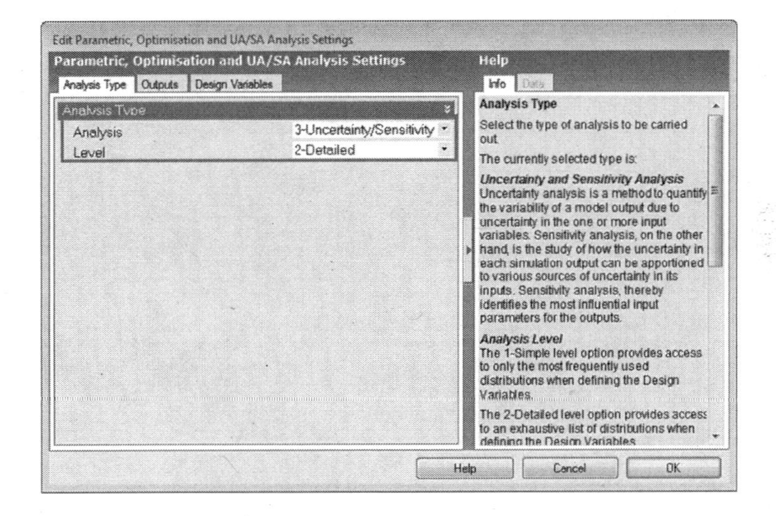

Step 8: Click the **Design Variables** tab. Click **Heating setpoint temperature**, and then click **Delete Variable** in the **Info** panel. Click **Cooling setpoint temperature**, and then click **Delete Variable** in the **Info** panel.

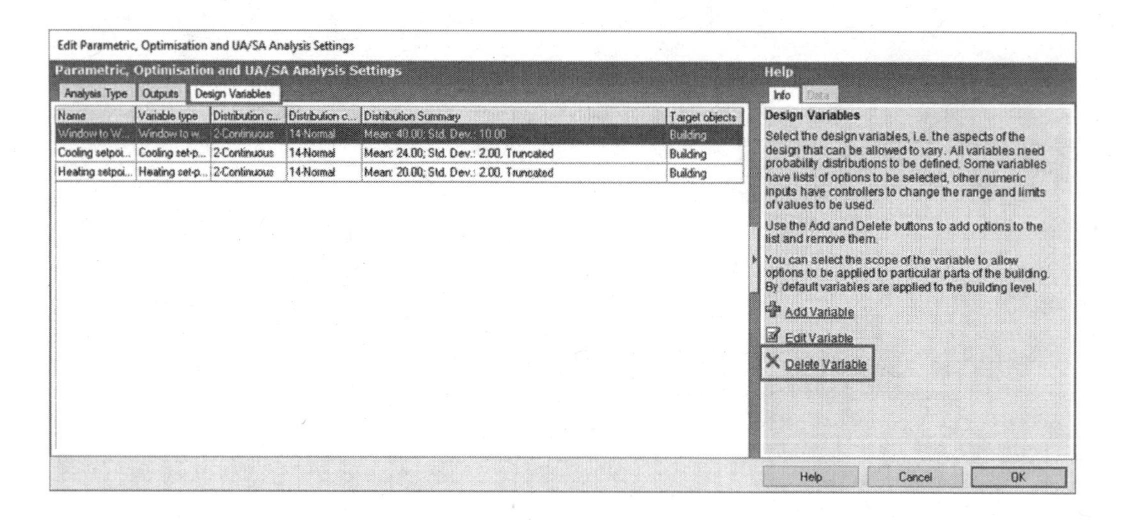

Step 9: Click **Add Variable** in the **Info** panel.

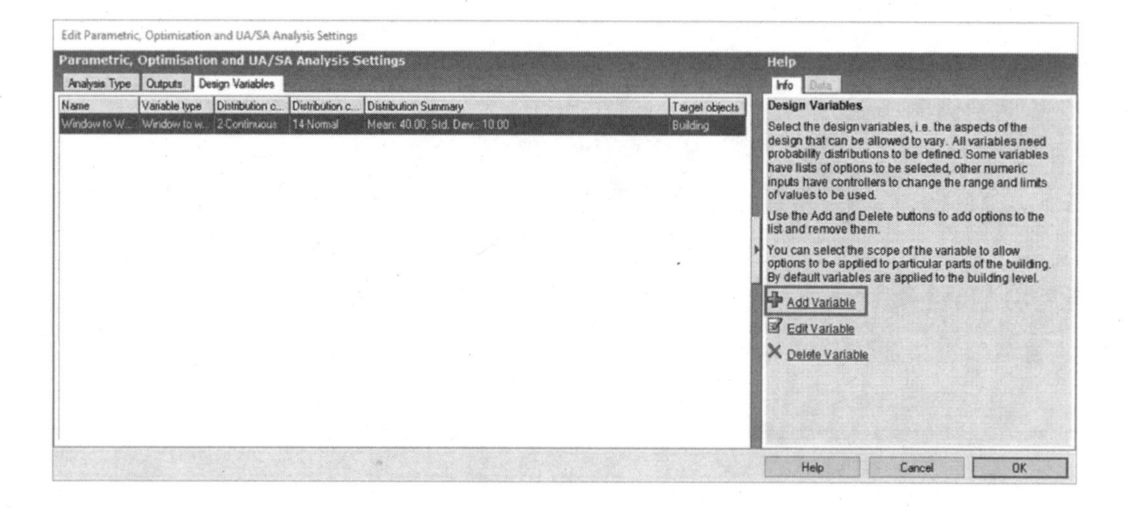

Step 10: Click **Variable type**. The **Variable types** tree appears. Select **Glazing type** under **Glazing/shading**.

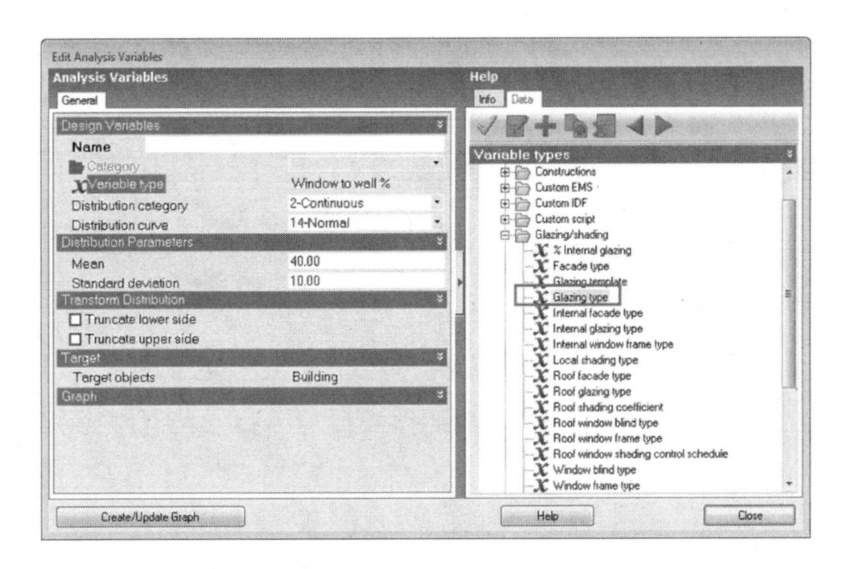

Step 11: Click the **Options list**. Click on the three dots to select the list.

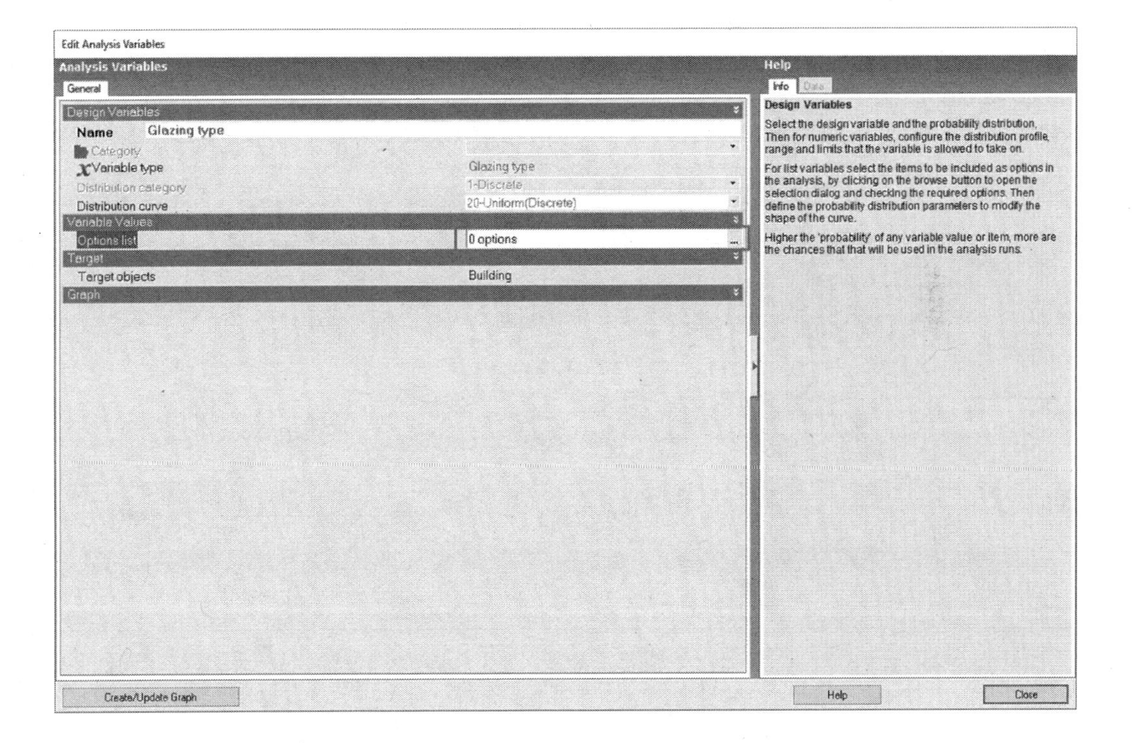

Step 12: Select the glazing types as provided in the problem statement. Click **OK**. The **Variable Option Order** screen appears.

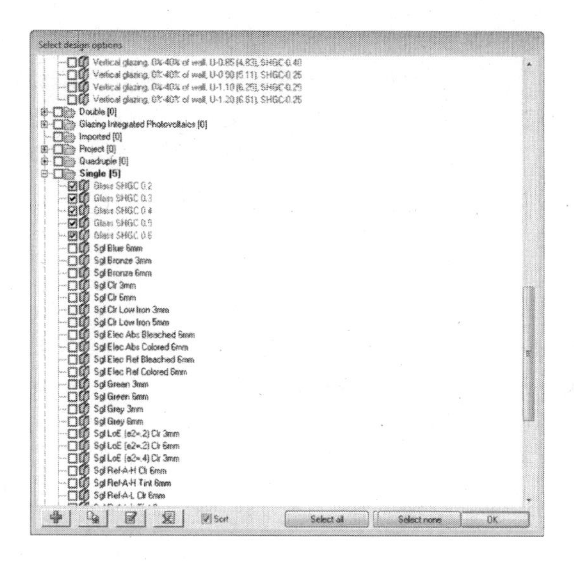

Step 13: Make sure that the variable options are in order. Click **OK**. When a variable is used to modify a single numeric setting within a set of components (e.g. SHGC of glazing here), then it is important that the design variable options with the only differing SHGCs are evenly spaced in each of the components in the list to ensure meaningful regression sensitivity analysis results. For example, here we analyse the effect of changing window SHGCs between 0.2 and 0.6; then we use **4 Options** with a fixed increment in SHGC value of 0.1.

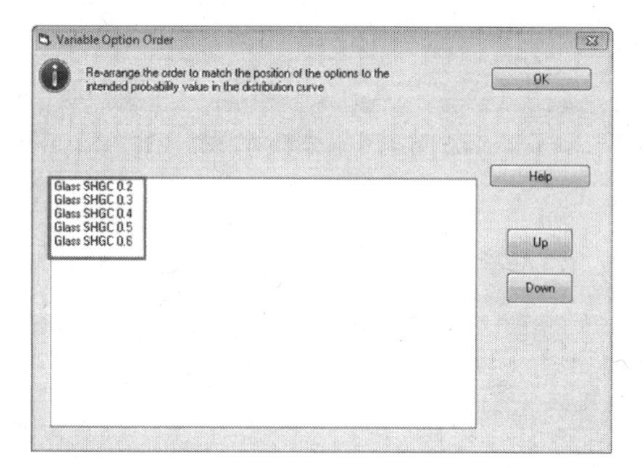

The **Options list** is only available for list-type variables. It allows you to select non-numeric options and create a list of various variables that will be sampled from during the UA/SA.

Source: https://designbuilder.co.uk//helpv6.0/Content/UASAVariables.htm.

Step 14: Make sure that **5 Options** appears in the **Option list**.

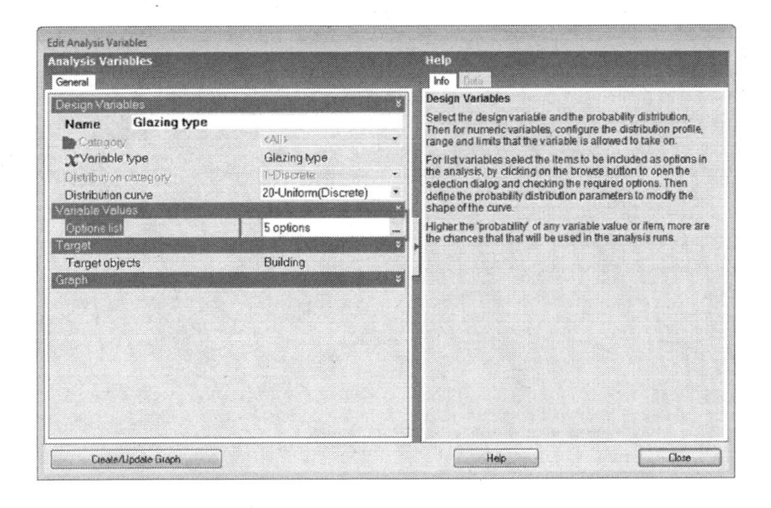

Step 15: Click **Create/Update Graph** to view the graphs. Click the **Close** button.

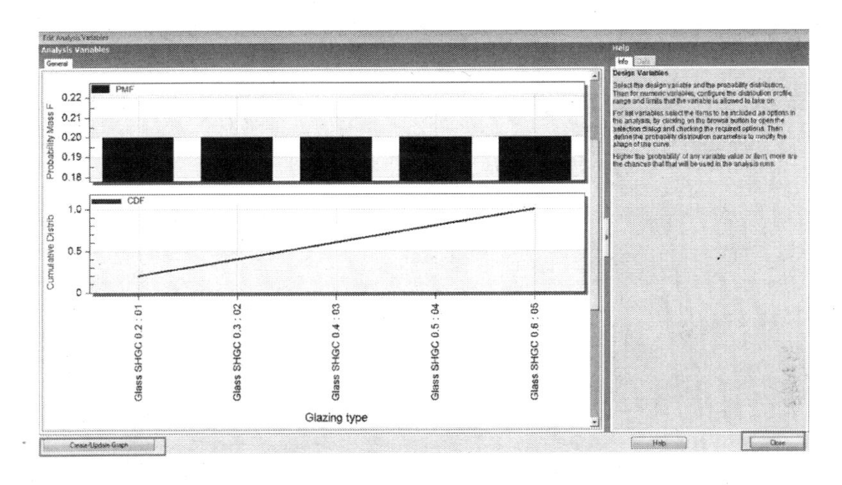

Step 16: Click on the **Add Variable** link in the **Info** panel.

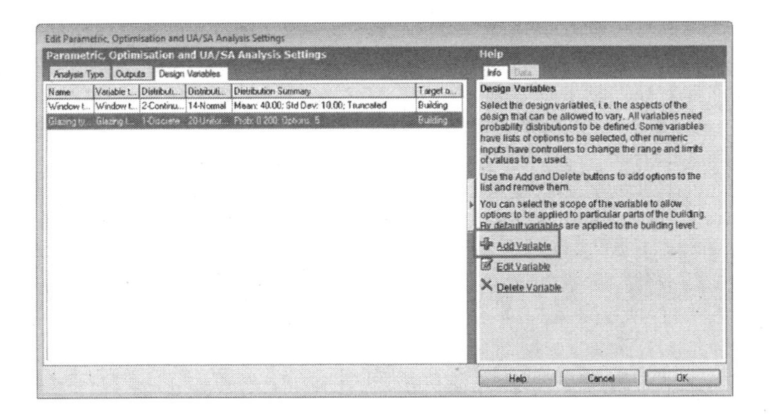

Step 17: Click **Variable type** under **Design Variables**. Select **Local shading type** under **Glazing/shading**.

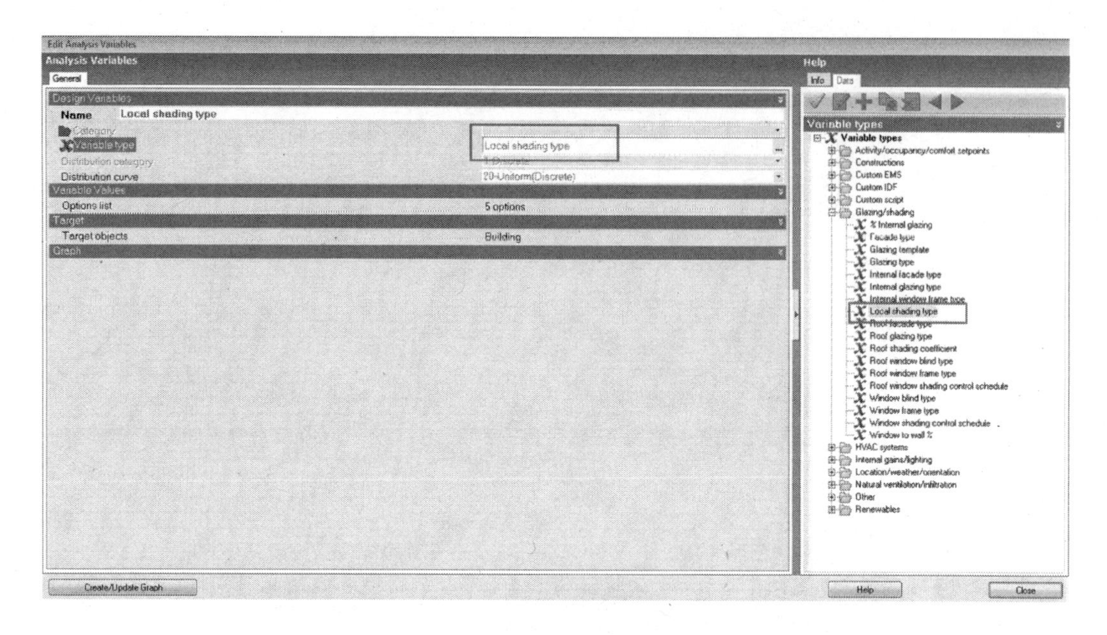

Step 18: Click on the **Options list** under **Variable values**, and select the overhangs as given in the problem statement.

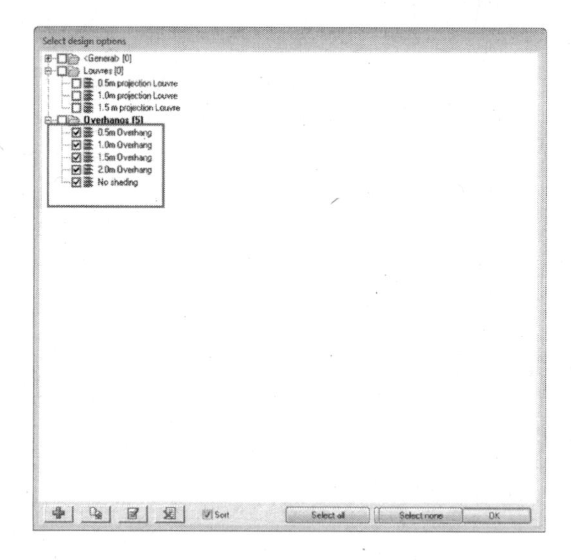

Step 19: Click **OK**.

Variable Option Order

Re-arrange the order to match the position of the options to the intended probability value in the distribution curve

OK

Help

No shading
0.5m Overhang
1.0m Overhang
1.5m Overhang
2.0m Overhang

Up

Down

Step 20: Click the **Close** button.

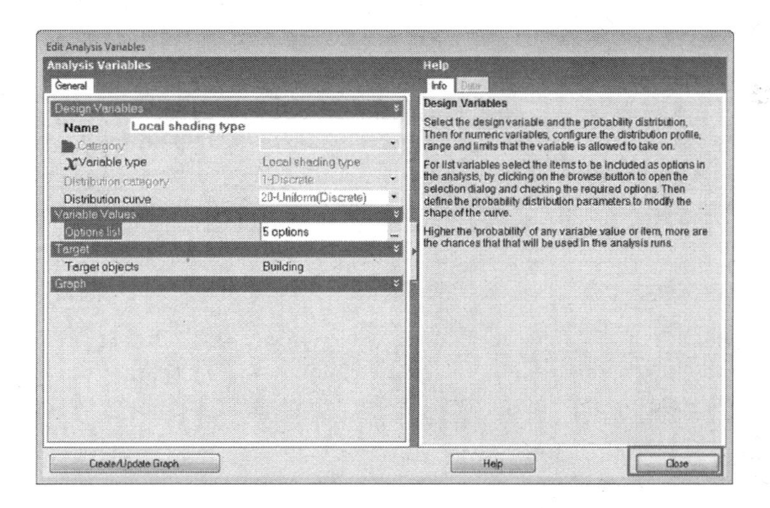

Step 21: Select **Window to wall ratio** under the **Design Variables** tab. Click **Edit Variable link** in the **Info** panel.

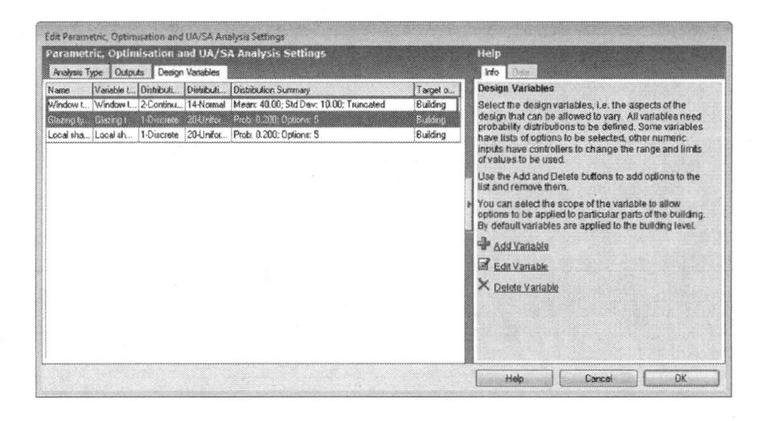

Step 22: Make sure that all variables values are as shown in the following figure.

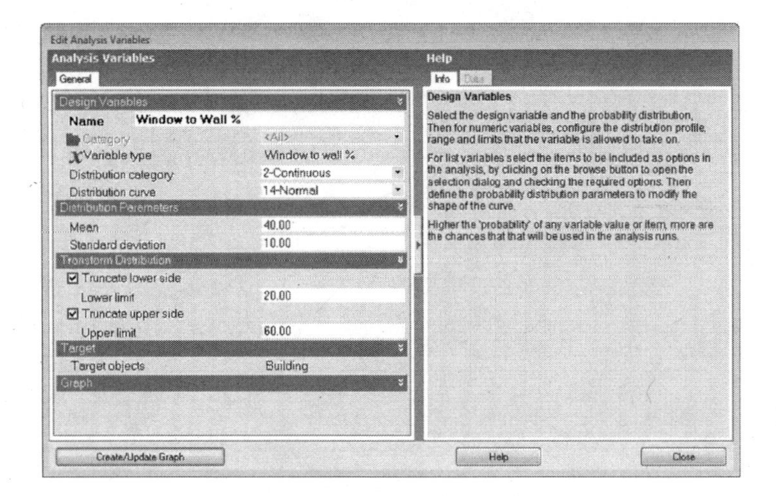

> **Truncate Lower Side**
>
> *Limit:* This value removes all the variable values below this number. In this case, the probability which is removed is redistributed across the remaining curve. The value should be more than the minimum value in the untruncated graph and less than truncation upper side limit.
>
> **Truncate Upper Side**
>
> *Limit:* This value removes all the variable values above this number. In this case, the probability which is removed is redistributed across the remaining curve. The value should be less than the maximum value in the untruncated graph and more than the truncation lower side limit.
>
> *Source:* https://designbuilder.co.uk//helpv6.0/Content/UASAVariables.htm.

Step 23: Click **Create/Update Graph** to view the graphs.

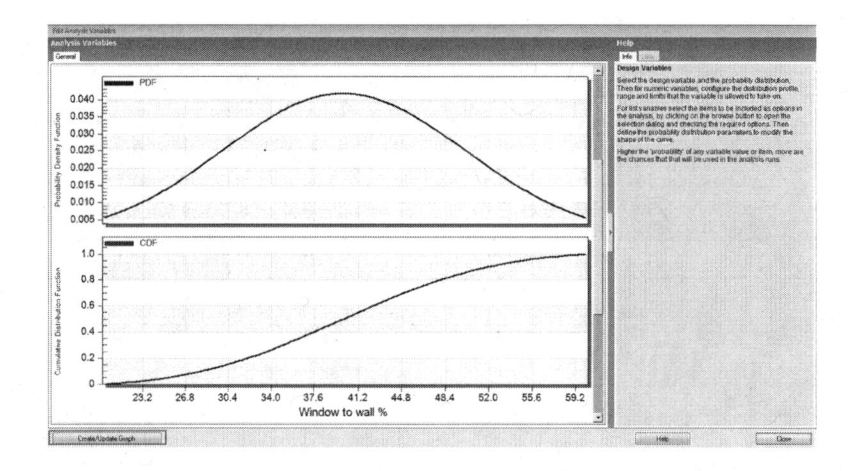

Step 24: Click **OK**.

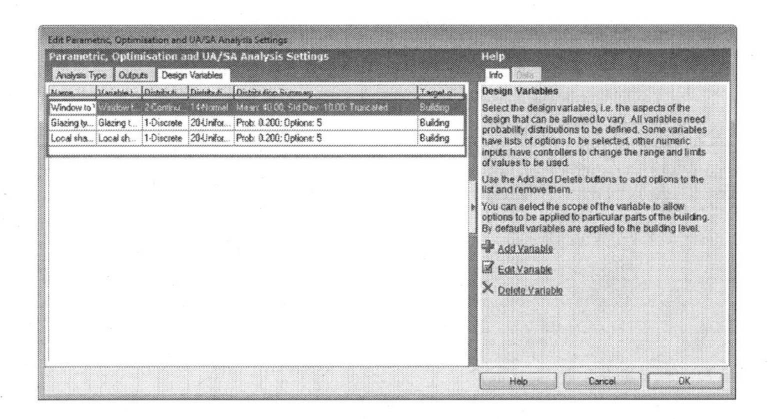

Step 25: Click the **Outputs** tab. Click the **Add Output** link.

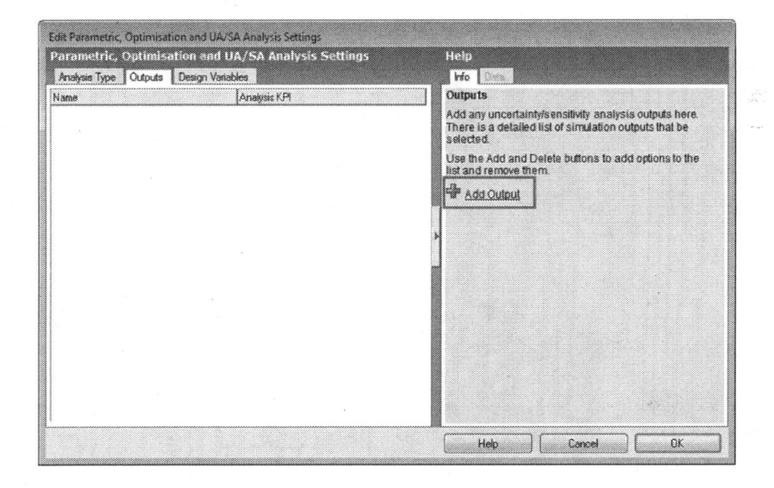

Step 26: Enter **CO2** in the **Name** text box. Click the **Close** button.

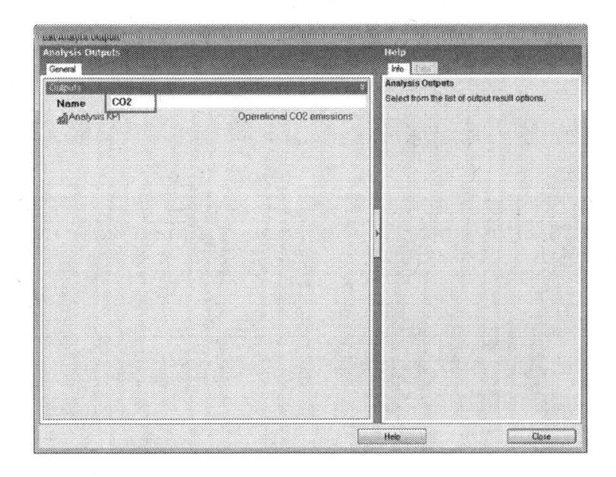

Step 27: Click **OK**.

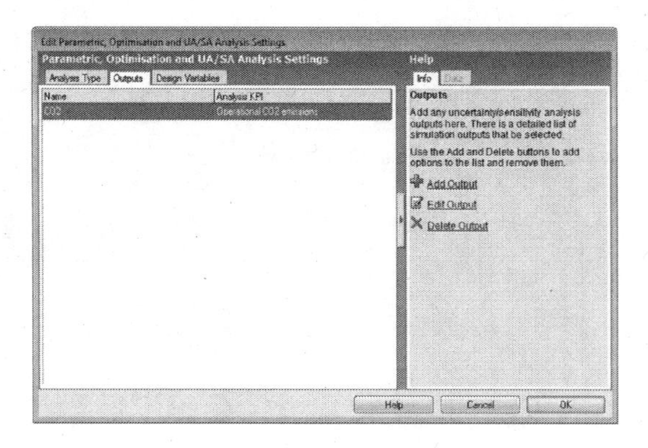

Step 28: Simulate the model.

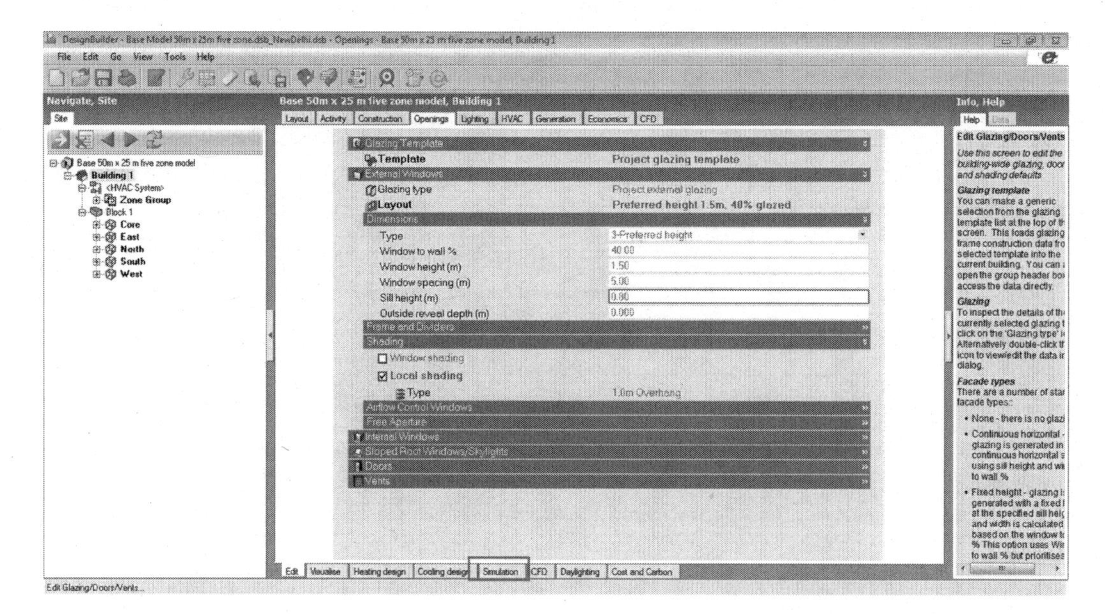

Step 29: Once the simulations are finished, click the **Optimisation + UA/SA** tab.

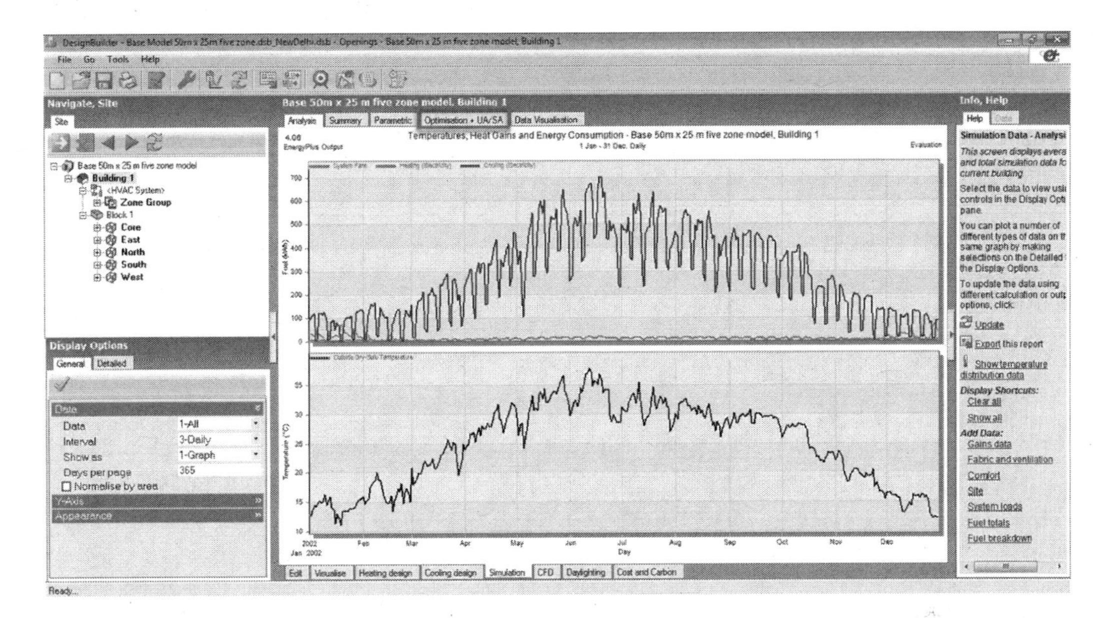

Step 30: Click the **Update** link.

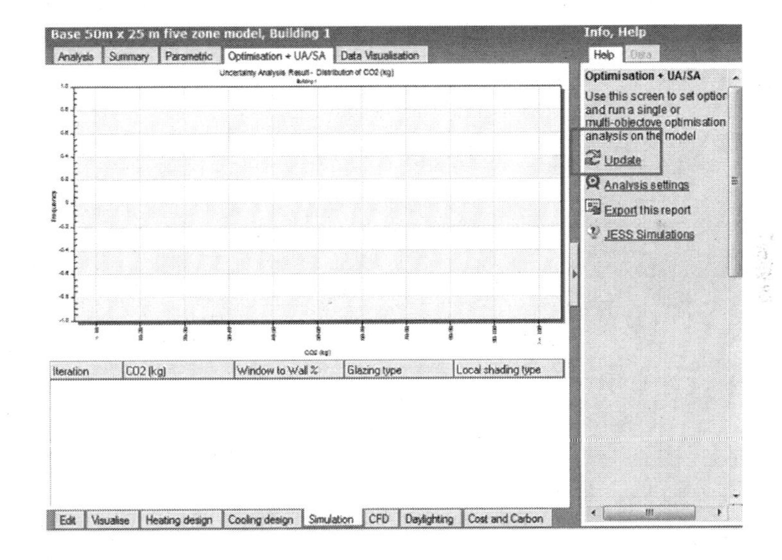

Step 31: Select **4-LHS** from the **Sampling option** drop-down list. Enter **150** in the **Number of runs** text box. Click the **Start** button. All simulations take around 15–30 minutes.

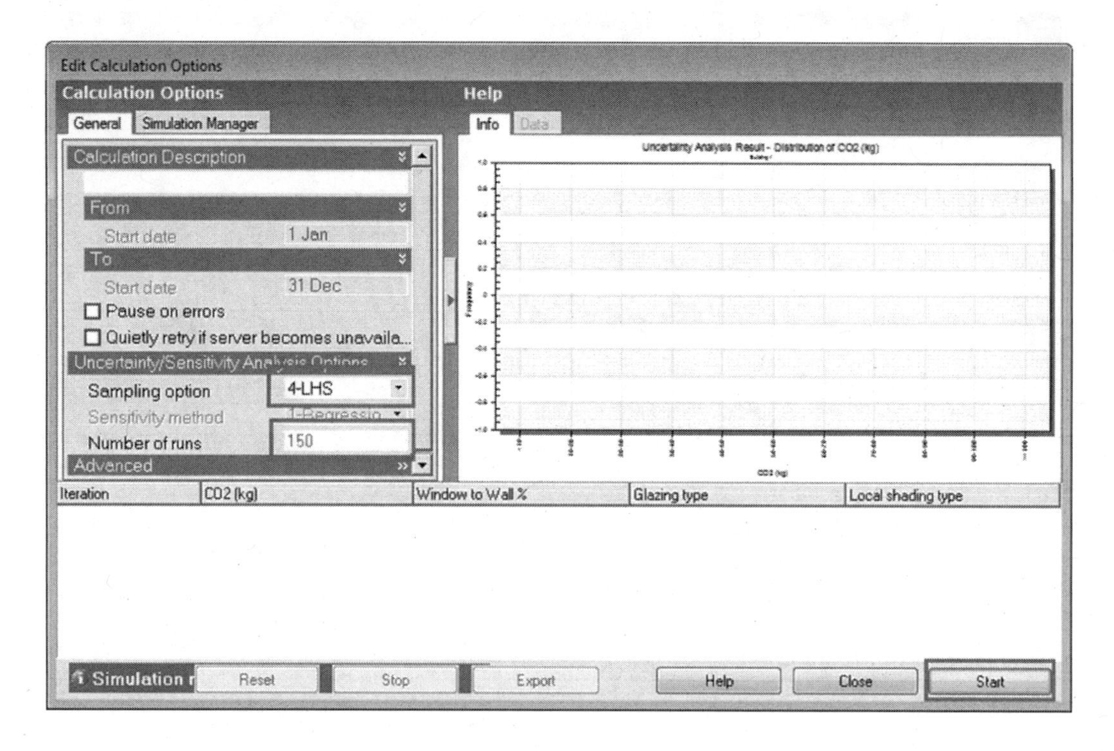

Number of Runs

The two most important factors to consider when selecting the value are the sampling option and the number of design variables. The following table can be used to select the minimum number of runs.

Sampling option	Number of runs rule of thumb
Random	20 times the number of variables
Random walk	Not used. Define number of trajectories instead.
Latin hypercube sampling (LHS)	10 times the number of variables
Sobol	15 times the number of variables
Halton	15 times the number of variables

For cases where any of the design variables are discrete and have more than 10 options, a higher value should be used to ensure that all options are simulated.

Source: https://designbuilder.co.uk/helpv6.0/#UASACalculationOptionsGeneral.htm.

Step 32: Once the analysis is finished, click **OK**.

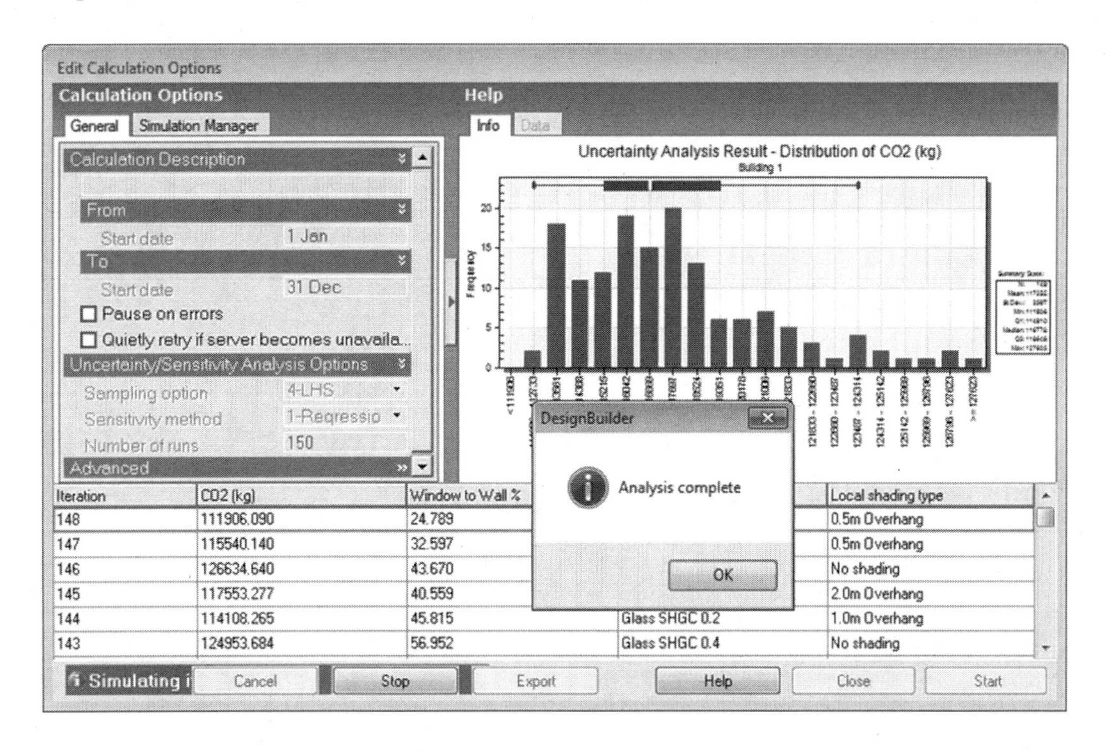

Step 33: Record the results. Click the **Close** button.

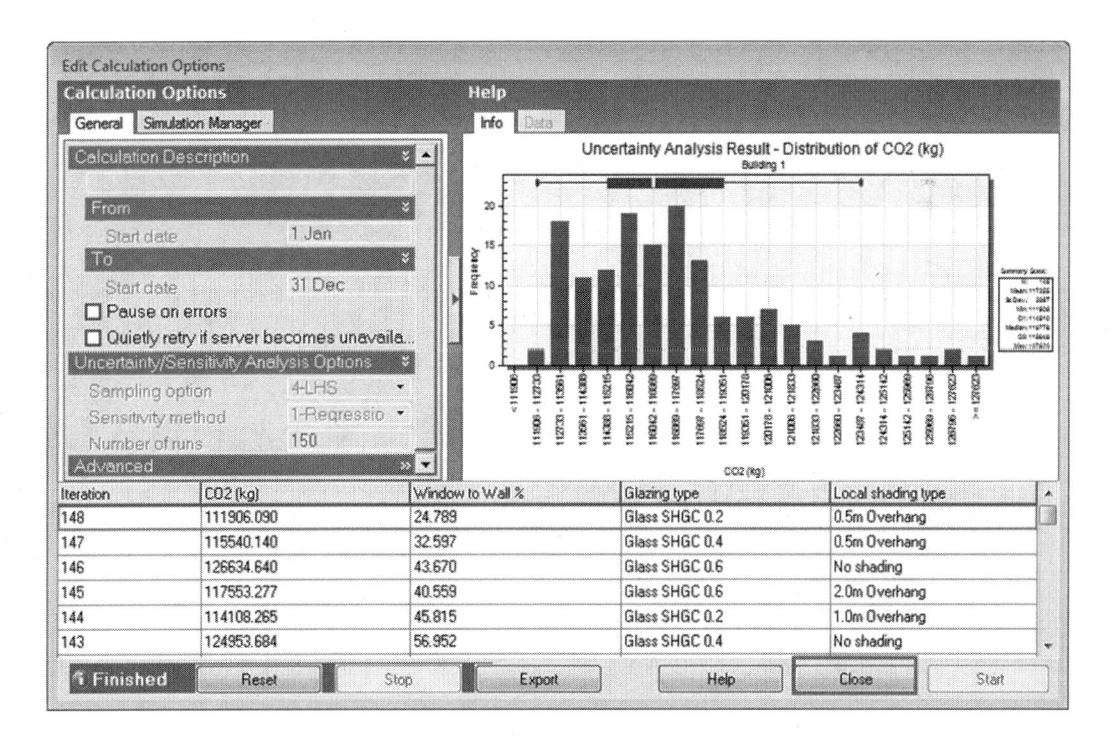

Step 34: Select the **1-Uncertainty analysis** from the **Analysis type** drop-down list. Record the results.

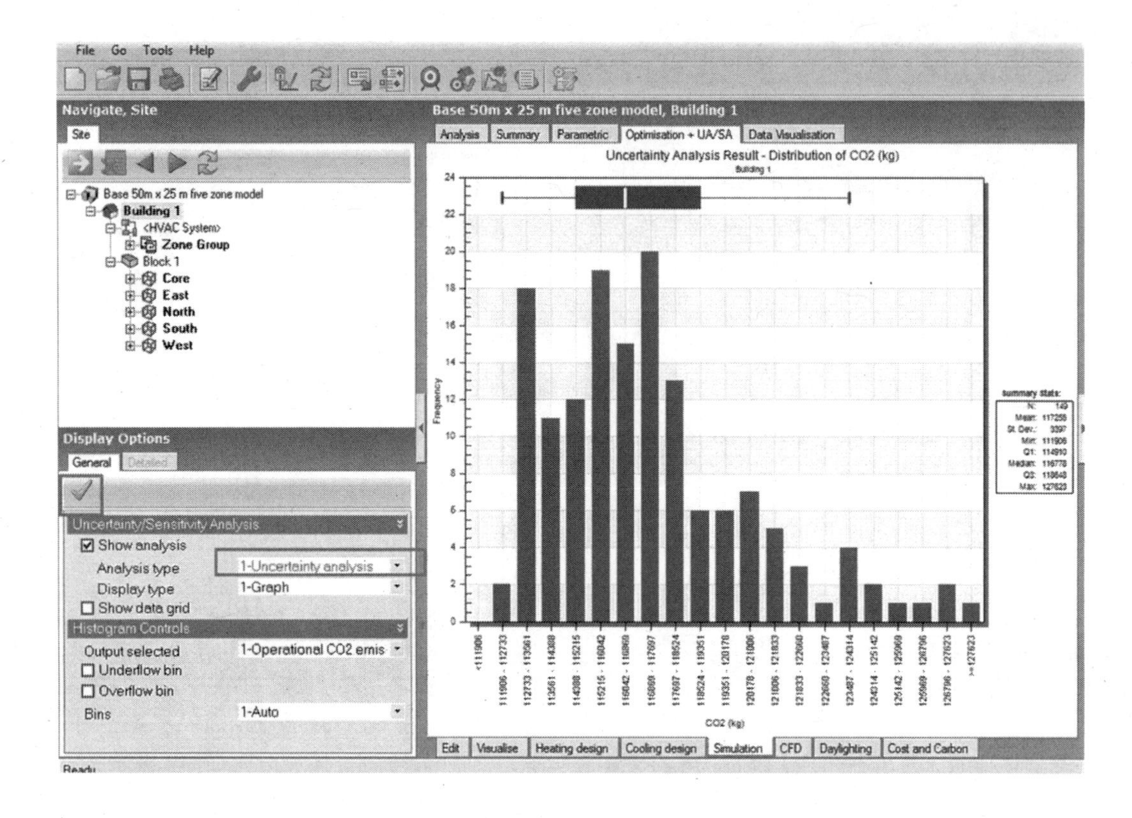

Step 35: Select **2-Sensitivity analysis** from the **Analysis type** drop-down list on left side of **Display** options. Record the results.

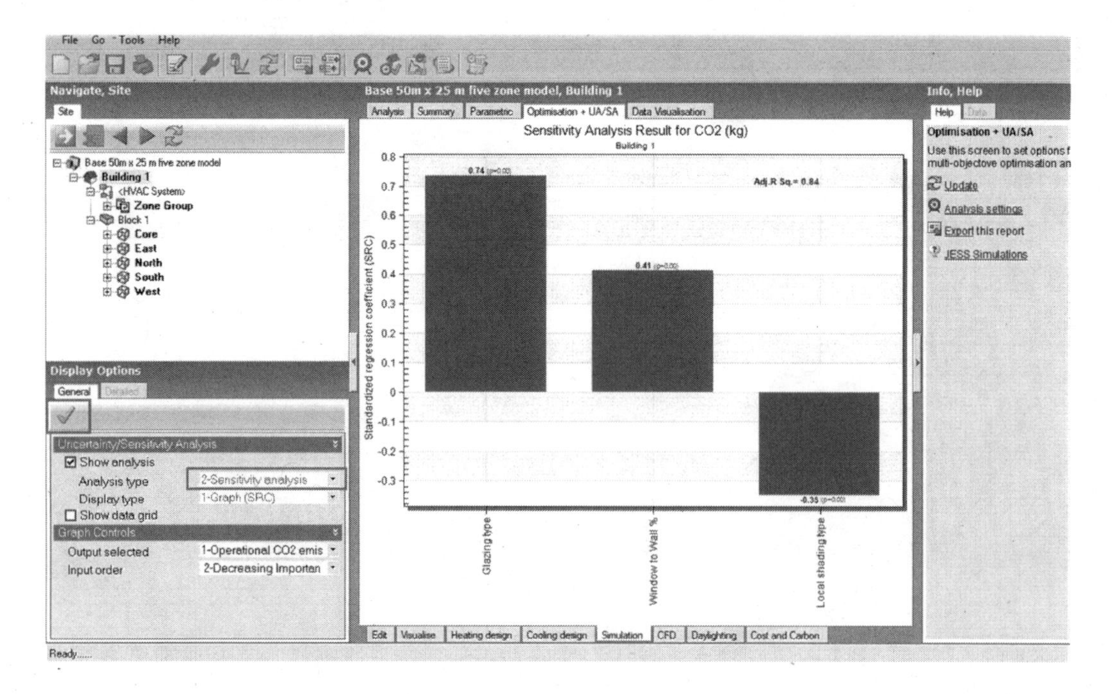

Uncertainty Analysis Outputs

After finishing the uncertainty and sensitivity analyses, uncertainty analysis results can be accessed from **Simulation** screen by setting the **Analysis type** to **1-Uncertainty Analysis** and the **Display type** to **1-Graph**. The following screenshot shows the main features of the uncertainty analysis graph.

Uncertainty Analysis Report

After finishing the uncertainty and sensitivity analyses, a report detailing the results of the uncertainty analysis can be accessed by setting **Analysis** to **1-Uncertainty Analysis** and **Display type** to **2-Report**. The uncertainty analysis report consists of the following sections:

Input variables: Details of all variables used in the analysis, including their distribution curves
Outputs (KPIs): Details of the outputs analysed
Analysis information: Analysis summary, details about the input samples
Uncertainty analysis results: Graphs and statistical details showing the variability in the output

Source: https://designbuilder.co.uk/helpv6.0/#UAGraph.htm.

Step 36: Select **6-Report** from the **Display type** drop-down list to view the sensitivity analysis report. This report helps in understanding the SA and UA results in plain English.

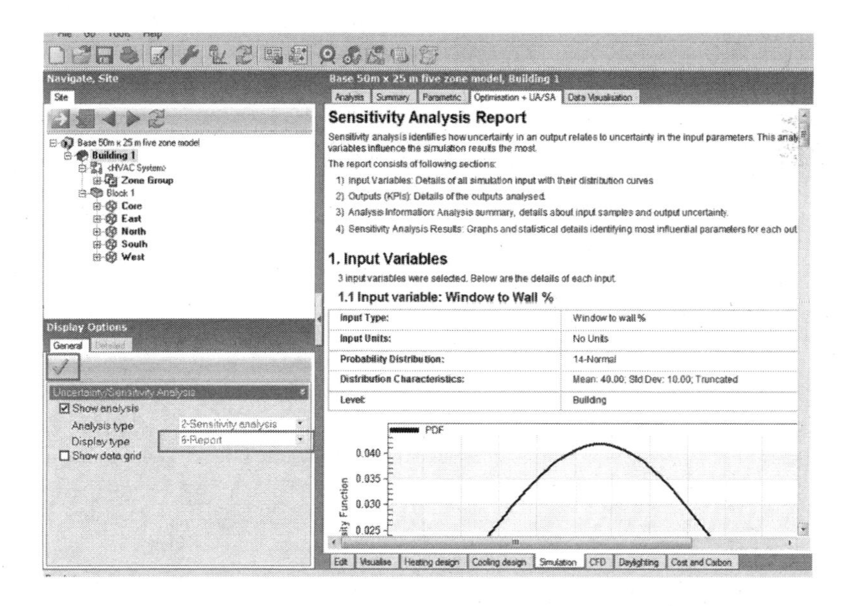

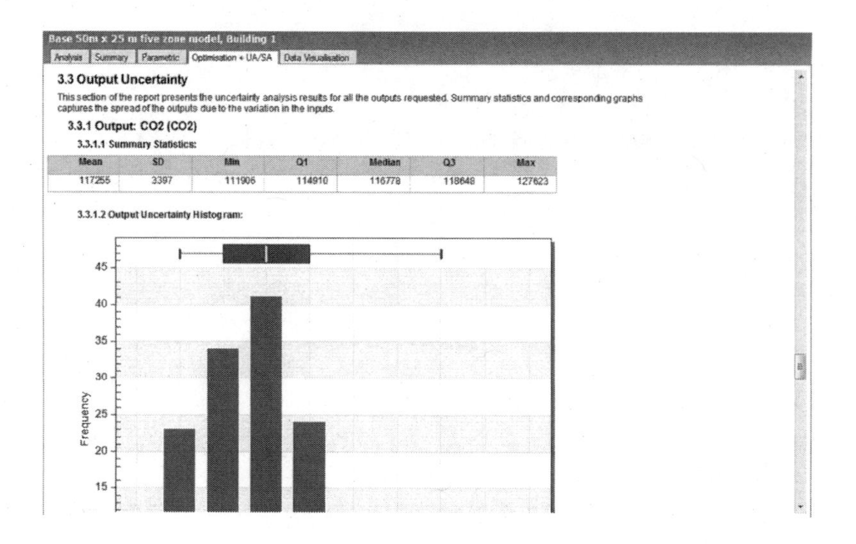

Sensitivity Analysis Outputs: Standardised Regression Coefficient

After finishing the uncertainty and sensitivity analyses, sensitivity analysis results can be accessed from the **Simulation** screen by setting the **Analysis type** to **1-Sensitivity Analysis** and the **Display type** to **1-Graph (SRC)**. The following screenshot shows the main features of the sensitivity analysis graph.

Sensitivity Analysis Report

After finishing the uncertainty and sensitivity analyses, the sensitivity analysis report can be accessed from the **Simulation** screen by setting the **Analysis type** to **1-Sensitivity Analysis** and the **Display type** to **6-Report**. The report consists of the following sections:

Input variables: Details of all design variables with their distribution curves
Outputs (KPIs): Details of all the outputs analysed
Analysis information: Analysis summary, details about input samples and output uncertainty
Sensitivity analysis results: Interpretation of the results, graphs and statistical details identifying most influential parameters for each output

Source: https://designbuilder.co.uk/helpv6.0/#SensitivityOutputsSRC.htm.

Through this tutorial, you have performed uncertainty analysis, which is a method to quantify the variability of a model output owing to uncertainty in one or more input variables. You have also performed sensitivity analysis to study how the uncertainty in each simulation output can be apportioned to various sources of uncertainty in its inputs.

14 Building Energy Code Compliance

This chapter explains building energy code compliance using a performance-based path which provides more design freedom and can lead to innovative designs. A performance-based path involves more complex energy simulations and trade-offs between systems. Larger commercial buildings that have multiple systems or varied uses and loads may find it more advantageous to follow a performance-based path for code compliance. In this chapter, in all tutorials (except Tutorial 14.19), no simulations are performed and only the modeling requirements are described. This chapter also describes a method for automating baseline building model creation for checking of compliance with ASHRAE Standard 90.1–2010.

Building energy code compliance in most countries has the following two paths:

1. Prescriptive
2. Performance

The prescriptive path gives specific requirements for all the building components, and all the requirements have to be met for compliance. This generally limits design freedom. The performance-based path provides more design freedom and can lead to innovative designs but involves more complex energy simulations and trade-offs between systems. Residential and smaller commercial buildings with singular heating, ventilation and air-conditioning (HVAC), service hot water and lighting systems are more likely to be designed by using a prescriptive approach. Larger commercial buildings that have multiple systems or varied uses and loads may find it more advantageous to follow a performance-based path for code compliance (see www.energycodes.gov/resource-center/ace/enforcement/step2).

The performance-based path requires the whole-building energy simulation. ASHRAE Standard 90.1–2010 has two methods for the whole-building simulation:

1. Energy cost budget (ECB) method
2. Performance-rating method given in ASHRAE 90.1, Appendix G

Both methods require energy simulation of the proposed building design and the budget building design.

Tutorial 14.1

Modelling Building Performance in Four Orientations

In most of the codes, the baseline building performance is generated by simulating the building with its actual orientation and again after rotating the entire building by 90, 180 and 270 degrees and then deriving the average from the results. This is illustrated by the following example. The field **Site orientation** can be used for achieving the desired orientation.

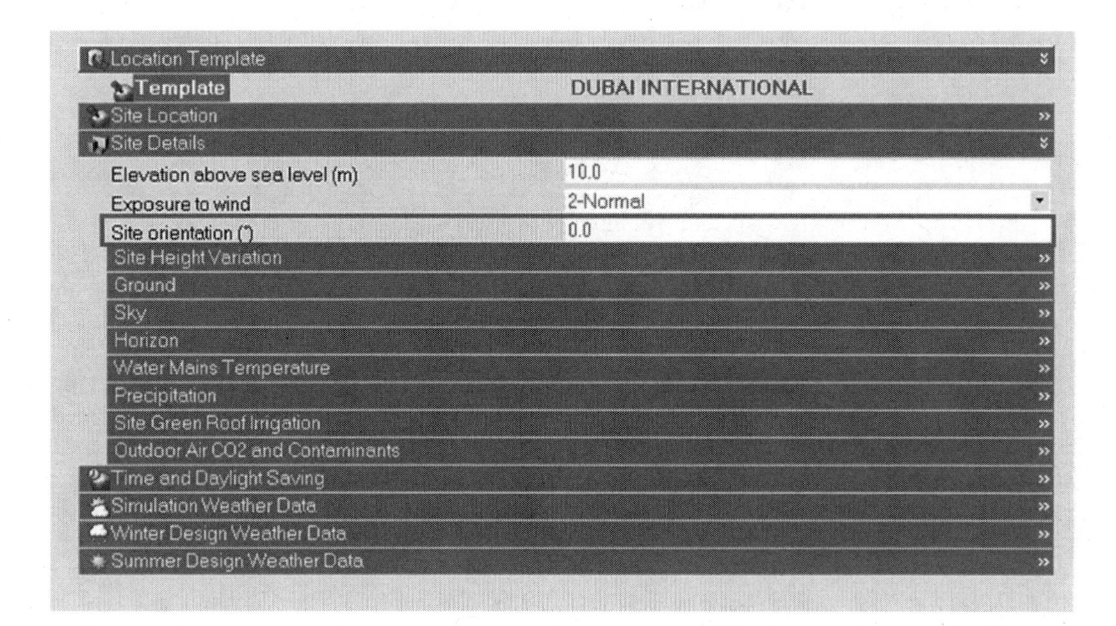

The building face is 150 degrees from the north. (The rectangle inside the circle is the building, and the face is shown with the bold line.)

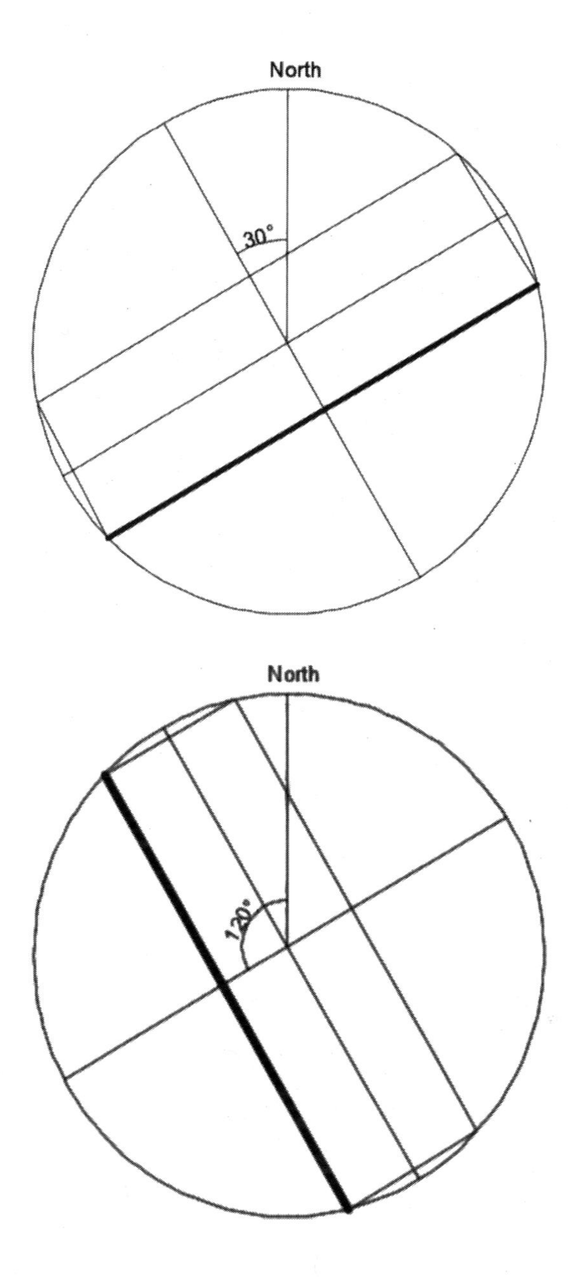

With 90 degrees of rotation from the building face.

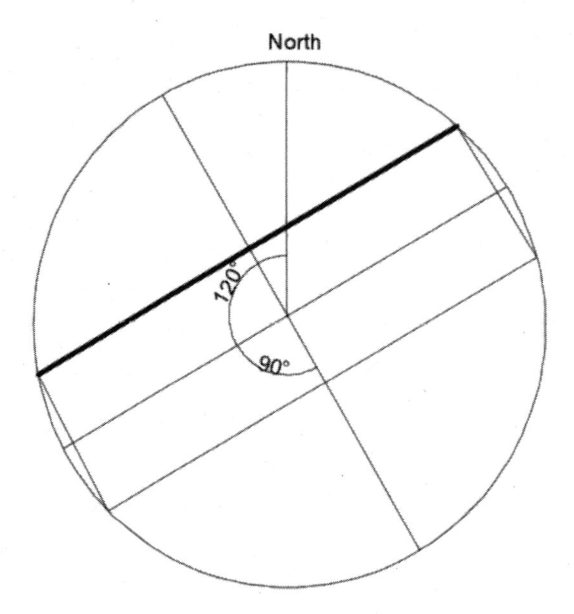

With 180 degrees of rotation from the building face.

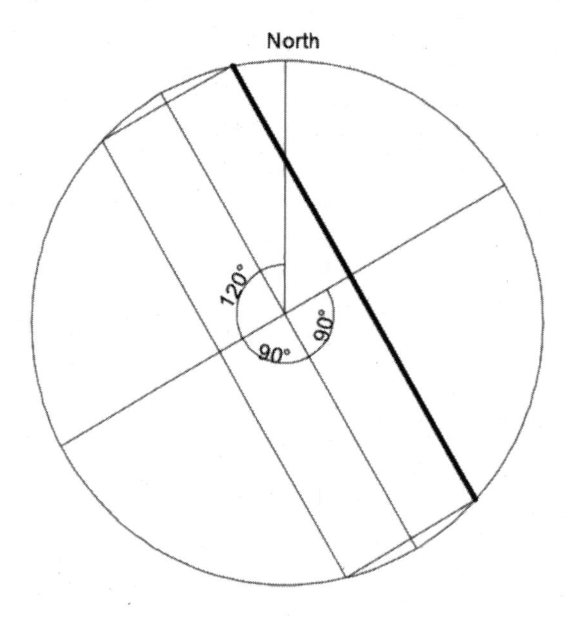

With 270 degrees of rotation from the building face.

Tutorial 14.2

Creating the Base-Case External Wall for ASHRAE Standard 90.1–2010, Appendix G

The energy cost budget (ECB) method requires modelling of opaque assemblies such as roofs and walls with the same heat capacity as in the proposed design, whereas in Appendix G of the standard, the performance-based method requires these to be modelled as lightweight/steel-framed.

ASHRAE Standard 90.1–2010, Appendix G, requires that the external wall be steel framed. You can refer to Table 5.5 of ASHRAE Standard 90.1–2010 for the *U*-value. In this tutorial you need to create a steel-framed external wall with a *U*-value of 0.705 W/m^2-K.

Step 1: Create a new external wall, and select **Wall, Steel-Framed, R-13 (2.3), U-0.124 (0.70) – 16 in. (400mm) On Centre, 3.5 in. (89mm) Depth Framing**.

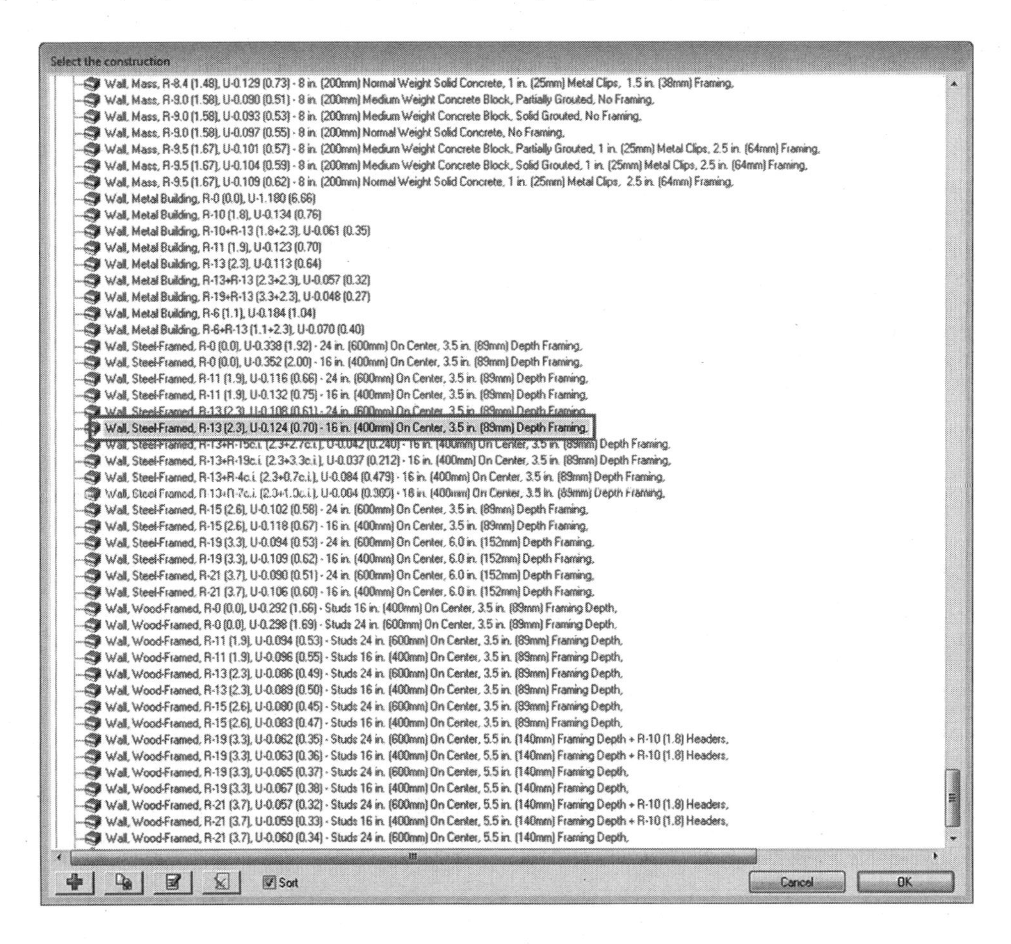

Step 2: Edit the construction to view properties.

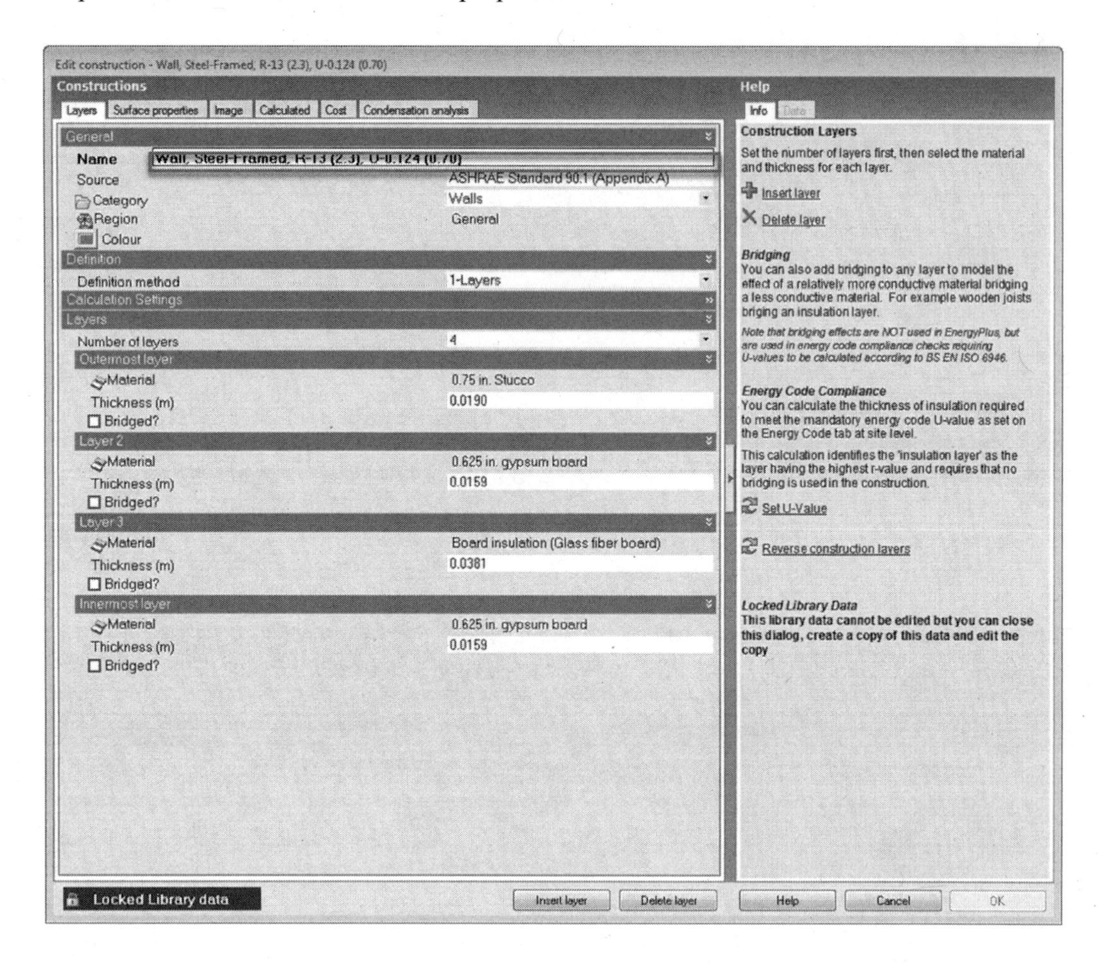

Step 3: Select the **Calculated** tab. It displays all construction properties.

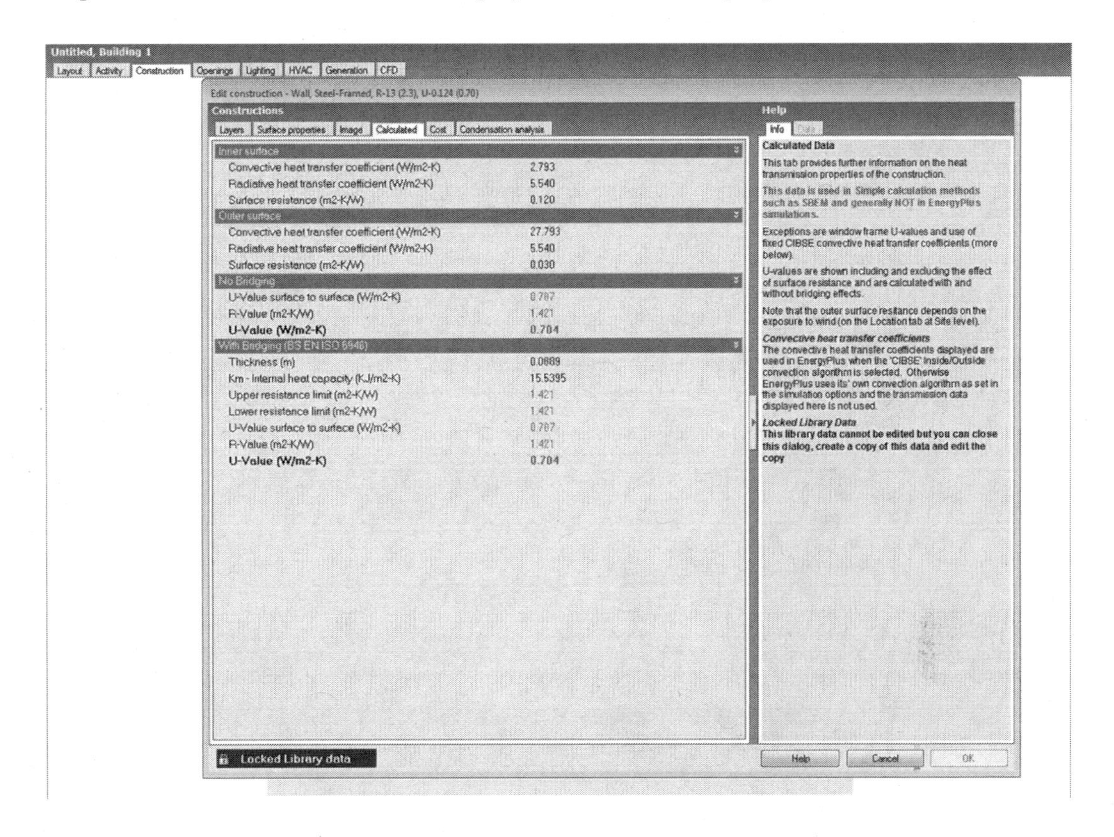

Please refer to Tutorial 3.1 to understand the effect of steel-framed and mass walls.

Tutorial 14.3

Modelling Flush Windows for the Base Case

As per ASHRAE Standard 90.1–2010, Appendix G, Table G3.1, you must not model shading projections on the fenestrations of the base-case building. The fenestrations need to be modelled as flush with the exterior wall or roof.

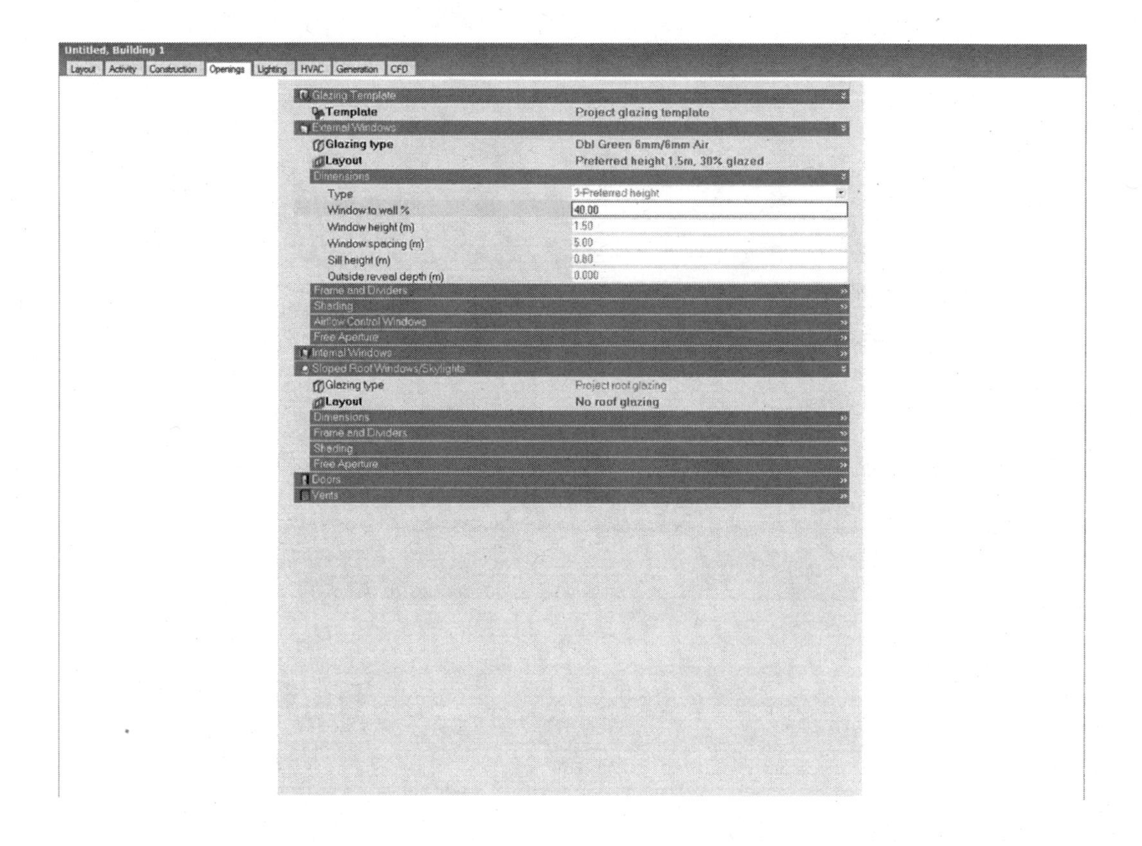

In the **Reveal** section, make sure that the **Outside reveal depth (m)** is zero for the base case. Window shading and local shading must not be modelled.

Tutorial 14.4

Selecting a HVAC System for the Base Case

As per ASHRAE Standard 90.1–2010, Section G3.1.3, HVAC type can be determined based on the conditioned floor area, number of floors and the fuel type. In DesignBuilder, you can select ASHRAE Standard 90.1, Appendix G, baseline systems from the template.

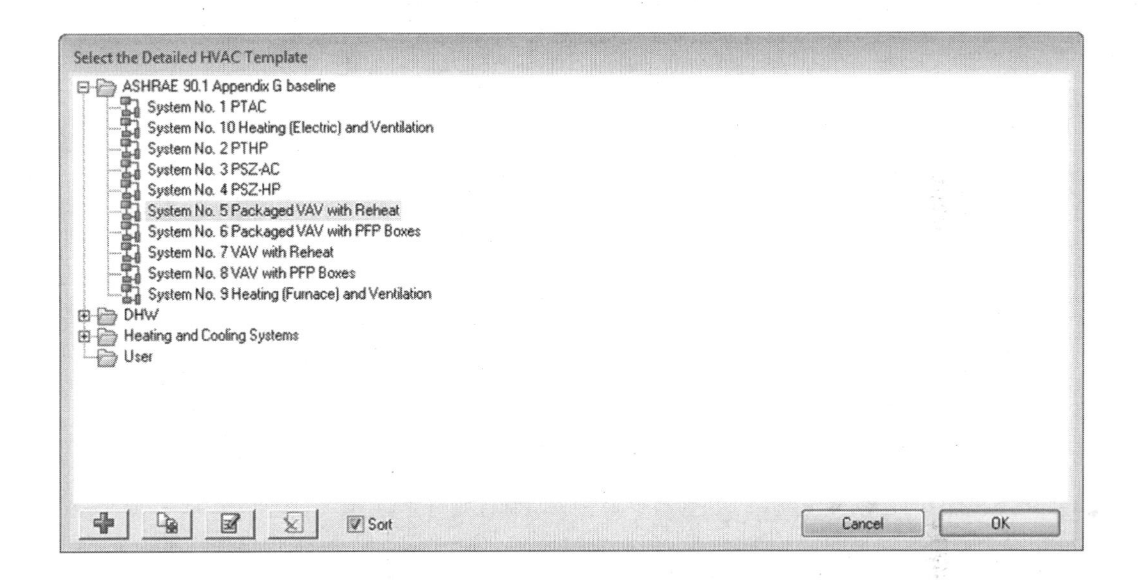

Tutorial 14.5

Calculating Fan Power for the Base Case

As per ASHRAE Standard 90.1–2010, Section G3.1.2.10, for the base case, system fan electrical power for supply, return, exhaust and relief (excluding power to fan-powered VAV boxes) needs to be calculated with the following formulas:

For Systems 1 and 2,

$$P_{\text{fan}}\ (\text{W}) = \text{CFM}_s \times 0.3$$

For Systems 3 and 4,

$$\text{Baseline fan motor brake horsepower} = \text{CFM}_s \times 0.00094 + A$$
$$\text{Baseline fan motor brake horsepower} = \text{CFM}_s \times 0.0013 + A$$

where A = sum of (PD × CFM_D/4,131)
CFM_s = maximum design supply airflow rate to conditioned spaces served by the system in cubic feet per minute
CFM_D = design airflow through each applicable device in cubic feet per minute
PD = each applicable pressure prop adjustment

(Please refer to Table 6.5.3.1.1B of ASHRAE 90.1 for PD adjustment values.)

MODELLING IN DESIGNBUILDER

$$\Delta P = 1{,}000 \times \text{SFP} \times \text{fan total efficiency}$$

The specific fan power (SFP) is a function of the volume flow of the fan and the electrical power input and is quoted for a particular flow rate.

$$\text{SFP} = P_e\ (\text{W})/V\ (\text{l/s})$$

where V is volume flow (l/s), and P_e is electrical power input (W) to the fan system or complete air movement installation.

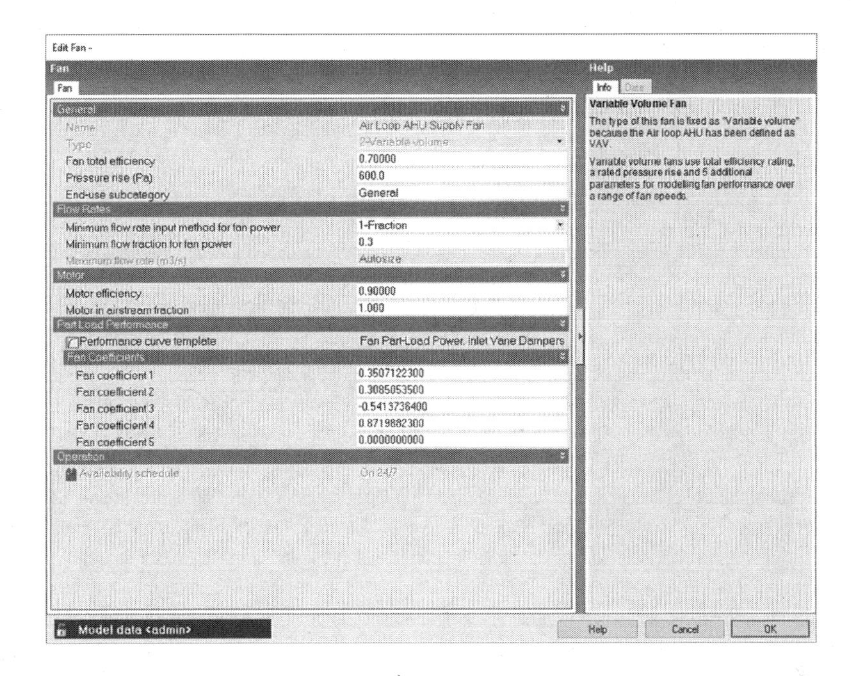

In this tutorial you will calculate the base-case power (kW) requirement for the air handling unit (AHU) having a supply air volume of 4,000 CFM (2.359 m^3/s; 1,888 l/s) for System 8. (Assume that the air filtering system's pressure drop is less than 1 inch WG when filters are clean.)

Variable-Volume Systems 5–8

$$\text{Baseline fan motor brake horsepower} = \text{CFMs} \times 0.0013 + A$$

where A = sum of (PD × CFM_D/4,131)

PD = each applicable pressure drop adjustment

$A = 0.97$

$$\text{Baseline fan motor brake horsepower} = 6.17 \text{ BHP} = 4{,}601.54 \text{ W}$$

$$\text{SFP} = P_e \text{ (W)}/V \text{ (l/s)}$$

$$\text{SFP} = 4{,}601.54/1{,}888 = 2.4 \text{ W/(l/s)}$$

Let fan efficiency be 0.6,

$$\Delta P = 1{,}000 \times \text{SFP} \times \text{fan total efficiency}$$

$$= 1{,}000 \times 2.4 \times 0.6$$

$$= 1{,}440 \text{ Pa}$$

Tutorial 14.6

Understanding Fan Cycling

As per ASHRAE Standard 90.1–2010, Section G3.1.2.5, Fan System Operation, supply and return fans shall operate continuously whenever spaces are occupied and shall be cycled to meet heating and cooling loads during unoccupied hours. Supply, return and/or exhaust fans will remain on during occupied and unoccupied hours in spaces that have health- and safety-mandated minimum ventilation requirements during unoccupied hours. Refer to Tutorial 9.2 for more details related to modelling.

Tutorial 14.7

Specifying Room-Air-to-Supply-Air Temperature Difference

As per ASHRAE Standard 90.1–2010, Section G3.1.2.9, System Types 1 through 8, a system's supply airflow rates for the baseline building design shall be based on a supply-air-to-room-air temperature difference of 11°C (20°F), or the minimum outdoor air flow rate, or on the airflow rate required to comply with applicable codes or accreditation standards, whichever is greater.

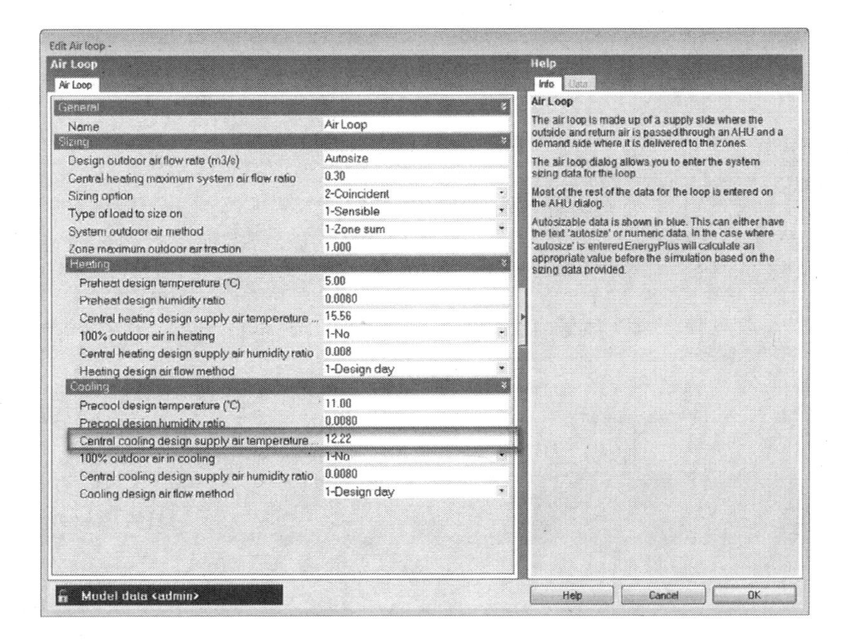

For example, if your room's air temperature setpoint is 23.22°C, then for the base case, you need to set the supply air temperature to 12.22°C.

Tutorial 14.8

Number of Chillers in the Base Case

As per ASHRAE Standard 90.1–2010, Section G3.1.3.7, Type and Number of Chillers (Systems 7 and 8), electric chillers shall be used in the baseline building design regardless of the cooling energy source. To choose the number of chillers in the base case, you need to determine the building peak cooling load. Based on the building cooling load, you can refer to Table G3.1.3.7 of ASHRAE Standard 90.1 to know about the numbers and type of chillers.

In DesignBuilder, you can add chillers with the following steps:

Step 1: Click **Chiller**. It shows the chiller layout.

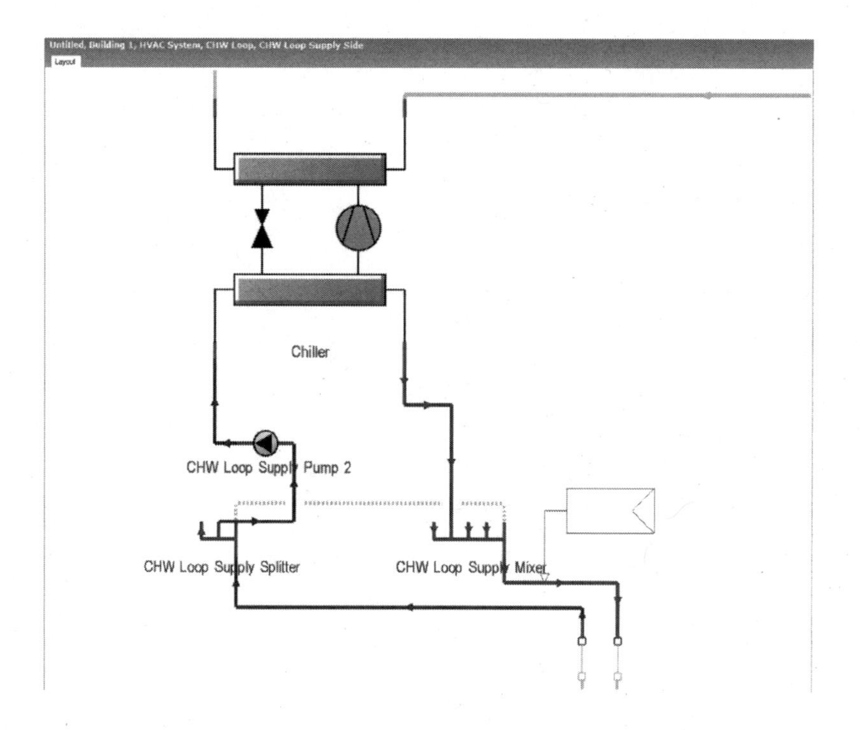

Step 2: Click the **Add Chiller** link. It displays one more chiller on the layout screen.

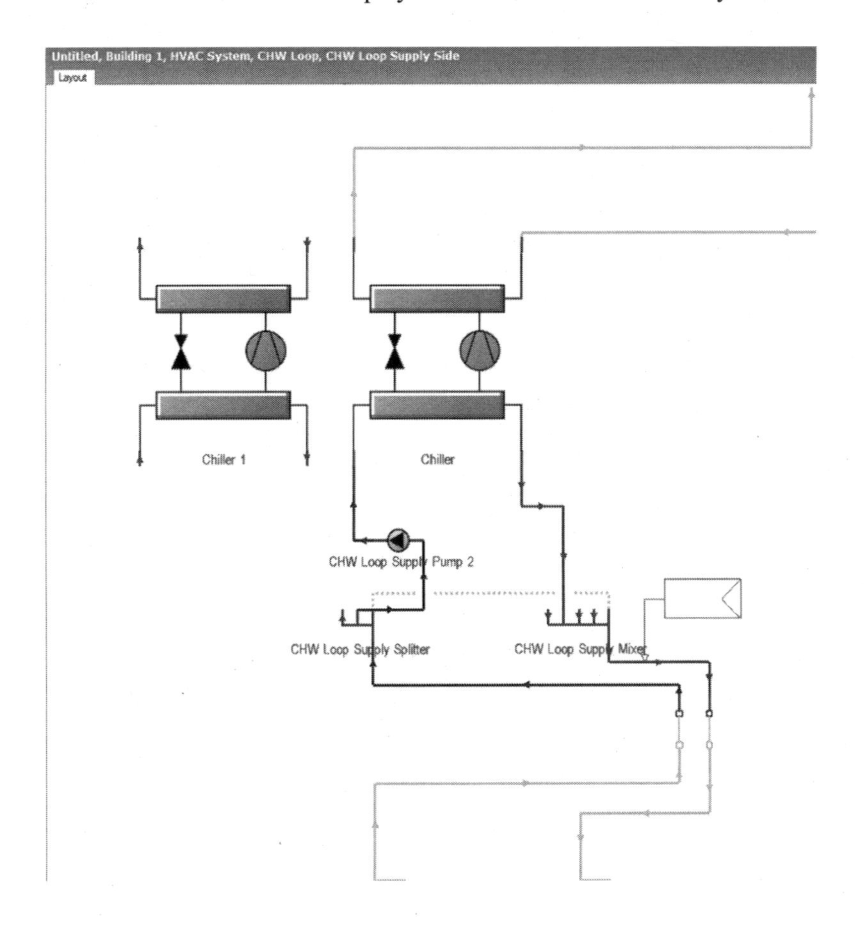

Step 3: Click **Connect components**. It shows a green dot when you click on the node of the chiller.

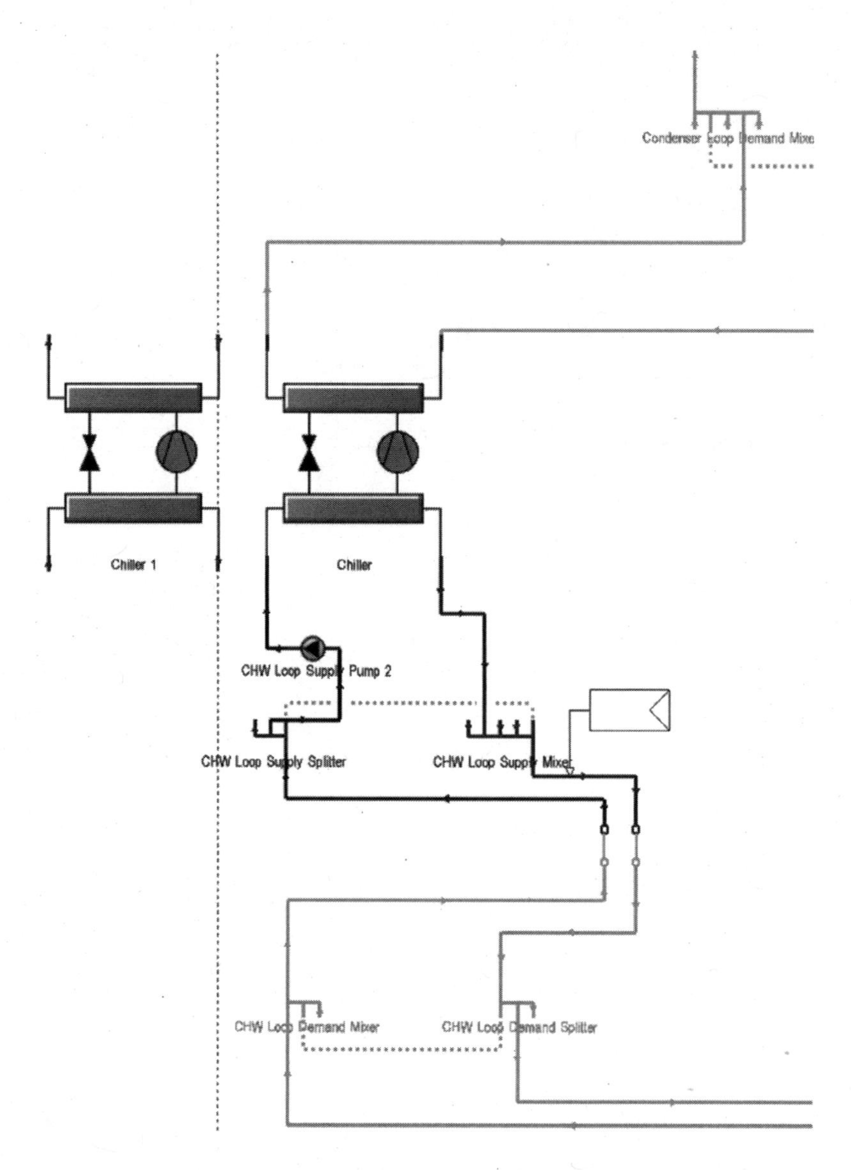

Step 4: Connect **Connector** to the chilled-water loop splitter.

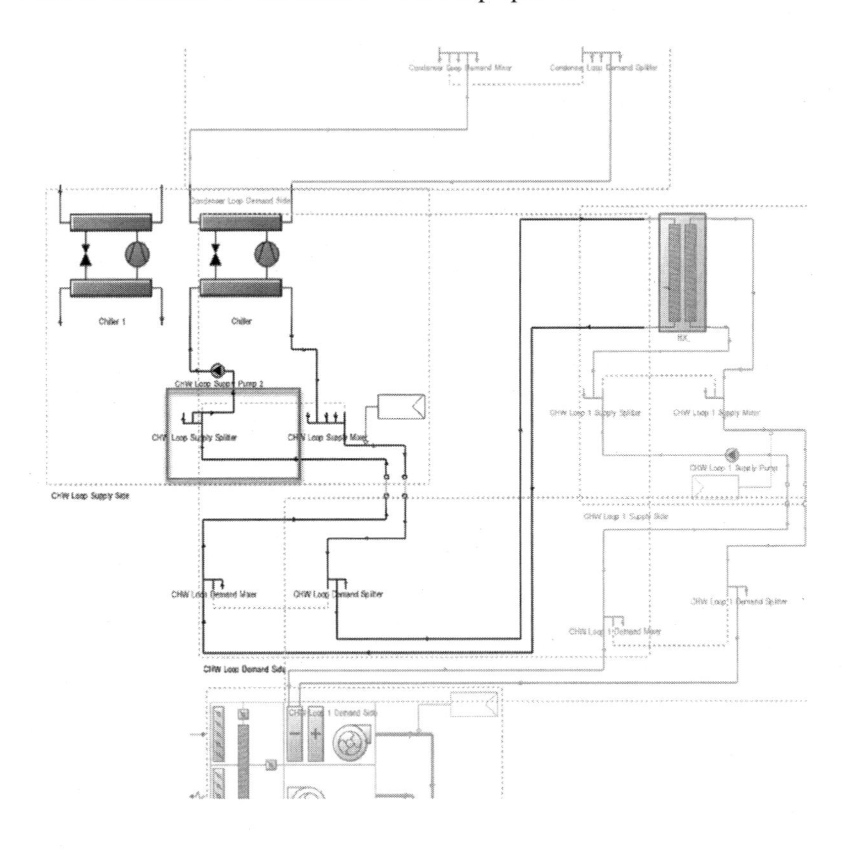

Step 5: Repeat Step 4 to connect the chilled-water output loop.

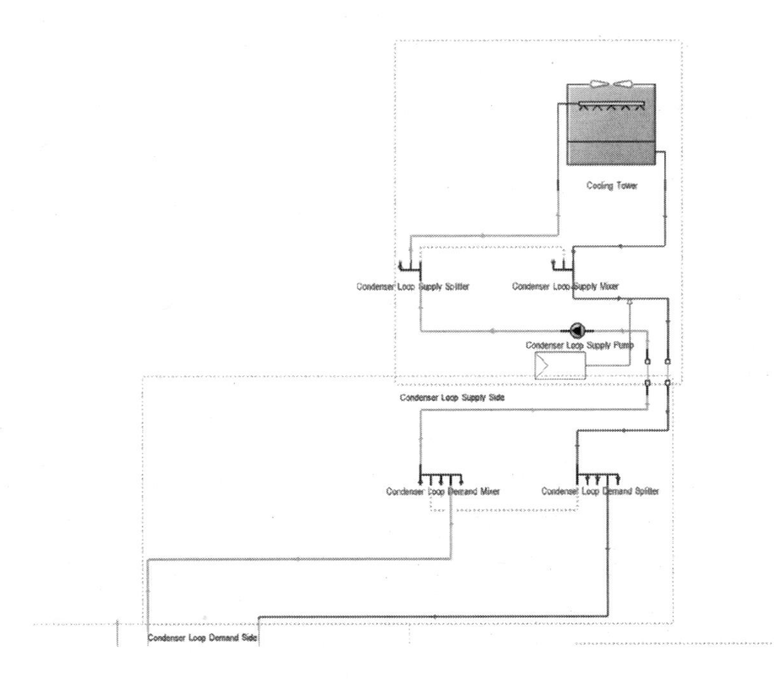

Step 6: Connect the **Condenser water loop** nodes.

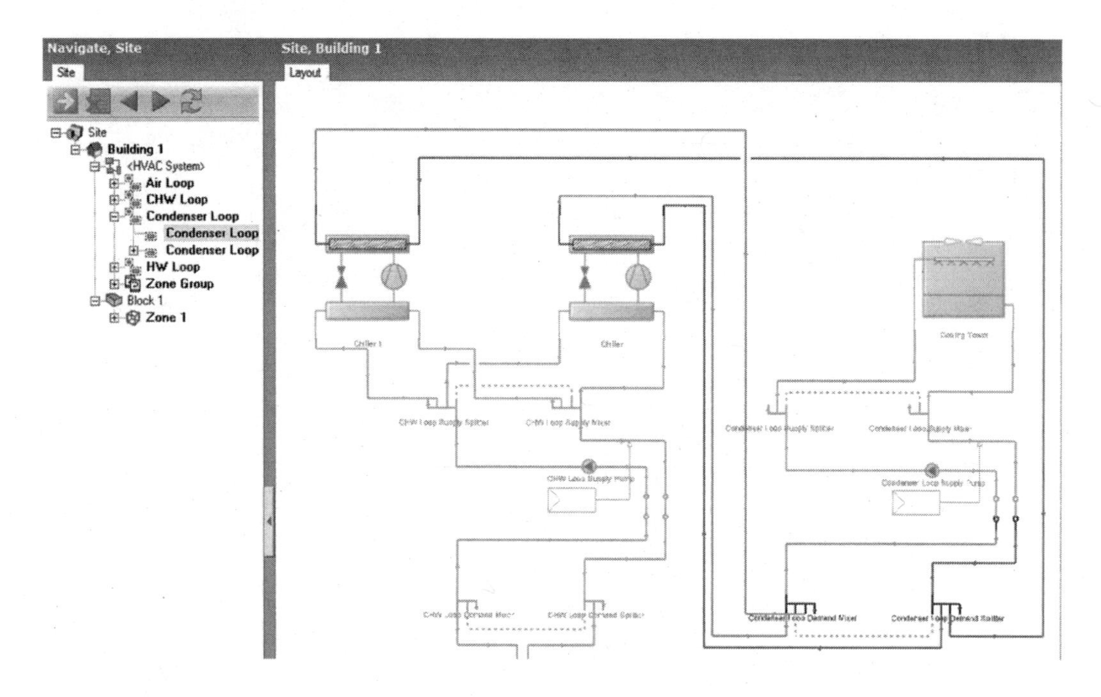

You can change the type of chiller by clicking on **Edit component.**

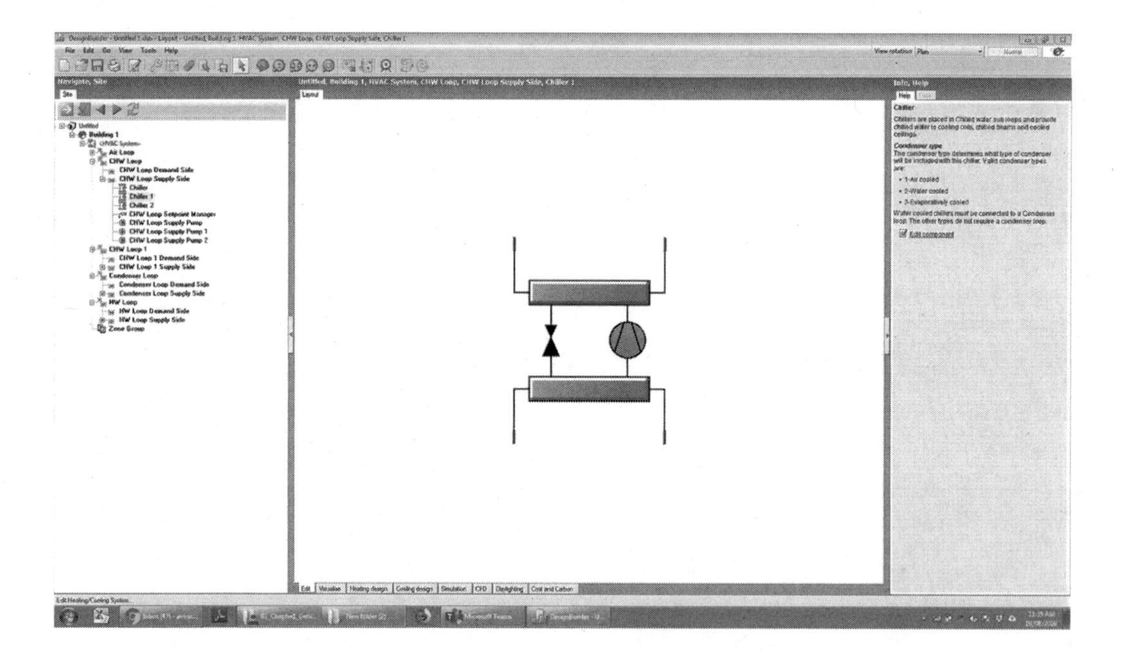

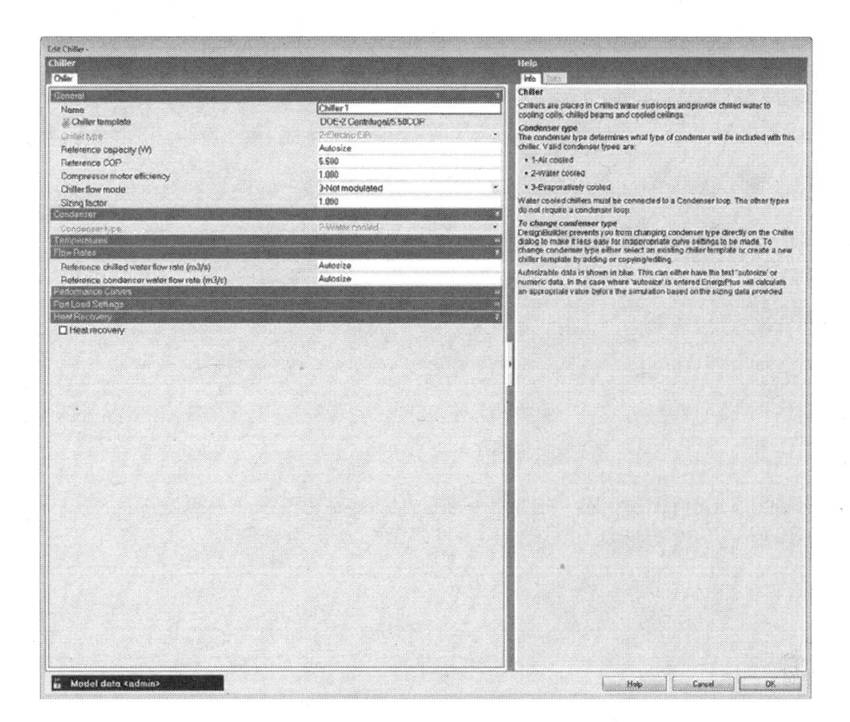
Chiller
Chiller
General
Name
Chiller 1
Chiller template
DOE-2 Centrifugal/5.58COP
Reference capacity (W)
Autosize
Reference COP
5.500
Compressor motor efficiency
1.000
Chiller flow mode
3-Not modulated
Sizing factor
1.000
Condenser
Flow Rates
Reference chilled water flow rate (m3/s)
Autosize
Reference condenser water flow rate (m3/s)
Autosize
Heat recovery
Help
Info
Chiller
Chillers are placed in Chilled water sub loops and provide chilled water to cooling coils, chilled beams and cooled ceilings.
Condenser type
The condenser type determines what type of condenser will be included with this chiller. Valid condenser types are:
• 1-Air cooled
• 2-Water cooled
• 3-Evaporatively cooled
Water cooled chillers must be connected to a Condenser loop. The other types do not require a condenser loop.
To change condenser type
DesignBuilder prevents you from changing condenser type directly on the Chiller dialog to make it less easy for inappropriate curve settings to be made. To change condenser type either select an existing chiller template or create a new chiller template by adding or copying/editing.
Autosizable data is shown in blue. This can either have the text 'autosize' or numeric data. In the case where 'autosize' is entered EnergyPlus will calculate an appropriate value before the simulation based on the sizing data provided
Model data <admin>
Help
Cancel
OK

Tutorial 14.9

Defining the Chilled-Water Supply Temperature Reset for the Base Case

As per ASHRAE Standard 90.1–2010, Section G3.1.3.9, Systems 7 and 8, the chilled-water supply temperature shall be reset based on the outdoor dry bulb temperature (DBT) by using the following schedule:

- Chilled-water temperature 7°C at DBT 27°C and above
- Chilled-water temperature 12°C at DBT 16°C and below
- Chilled-water temperature ramped linearly between 7 and 12°C at outdoor DBTs between 27 and 16°C

This can be modelled in **Setpoint Manager** by clicking on **CHW Loop Setpoint Manager**.

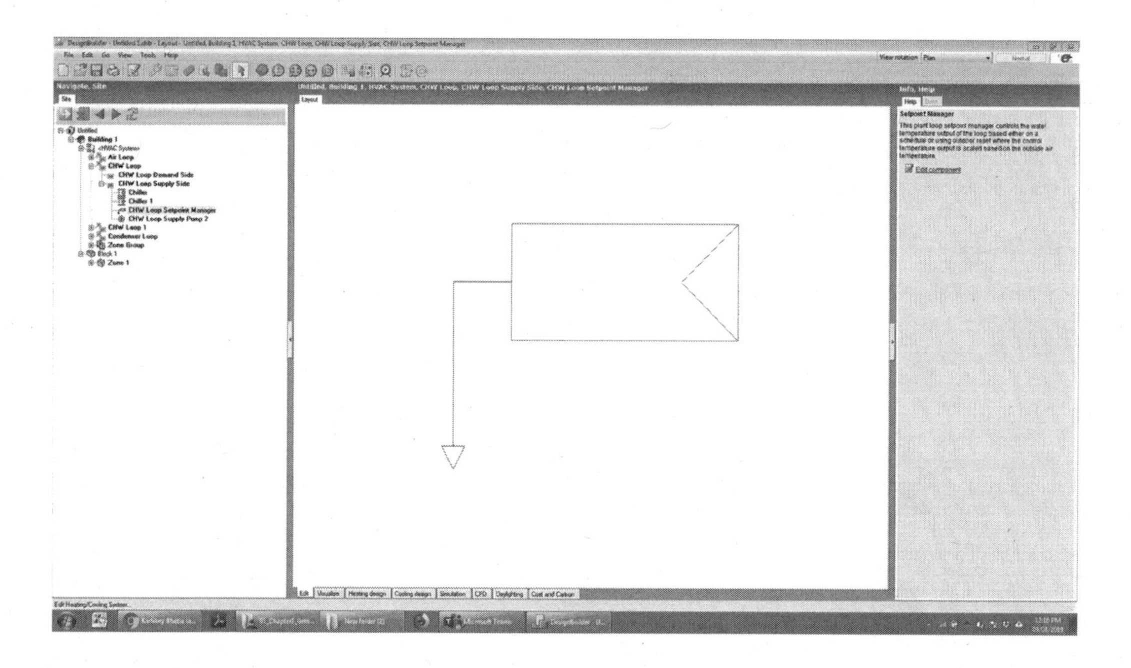

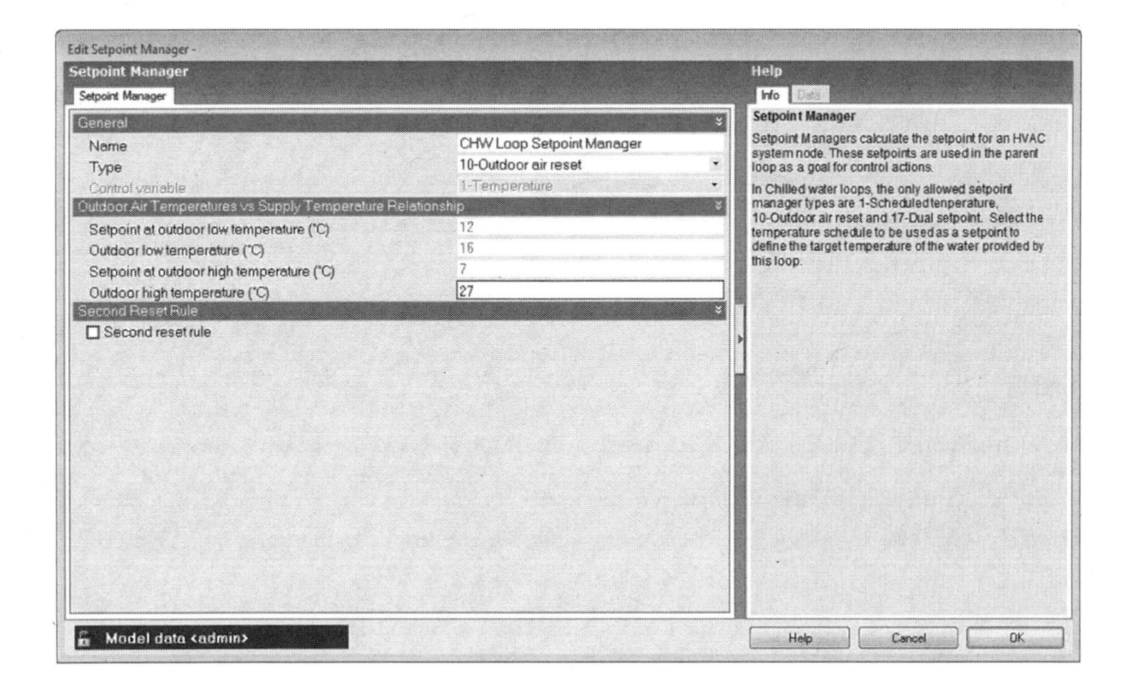
Edit Setpoint Manager -
Setpoint Manager
Setpoint Manager
General
Name
CHW Loop Setpoint Manager
Type
10-Outdoor air reset
Control variable
1-Temperature
Outdoor Air Temperatures vs Supply Temperature Relationship
Setpoint at outdoor low temperature (°C)
12
Outdoor low temperature (°C)
16
Setpoint at outdoor high temperature (°C)
7
Outdoor high temperature (°C)
27
Second Reset Rule
Second reset rule
Help
Info
Data
Setpoint Manager
Setpoint Managers calculate the setpoint for an HVAC system node. These setpoints are used in the parent loop as a goal for control actions.
In Chilled water loops, the only allowed setpoint manager types are 1-Scheduled tenperature, 10-Outdoor air reset and 17-Dual setpoint. Select the temperature schedule to be used as a setpoint to define the target temperature of the water provided by this loop.
Model data <admin>
Help
Cancel
OK

Tutorial 14.10

Type and Number of Boilers for the Base Case

As per ASHRAE Standard 90.1–2010, Section G3.1.3.2, for Systems 1, 5 and 7, the boiler plant shall use the same fuel as the proposed design and shall be a natural draft, except as noted in Section G3.1.1.1 (Purchase Heat). The baseline building design boiler plant shall be modelled as having a single boiler if the baseline building design plant serves a conditioned floor area of 1,400 m^2 or less and as having two equal-sized boilers for plants serving more than 1,400 m^2. Boilers shall be staged as required by the load.

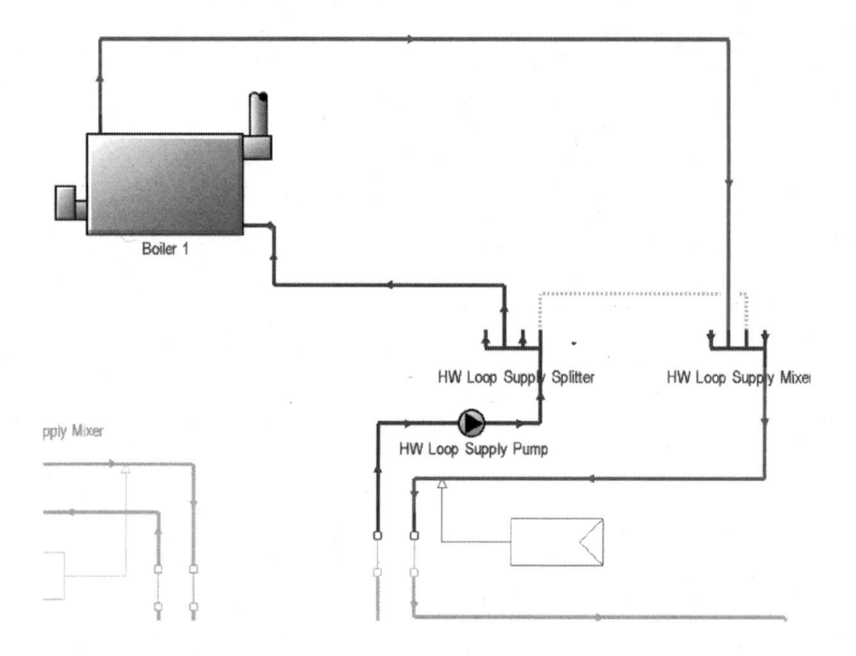

You can choose source **Fuel type** from the drop-down list.

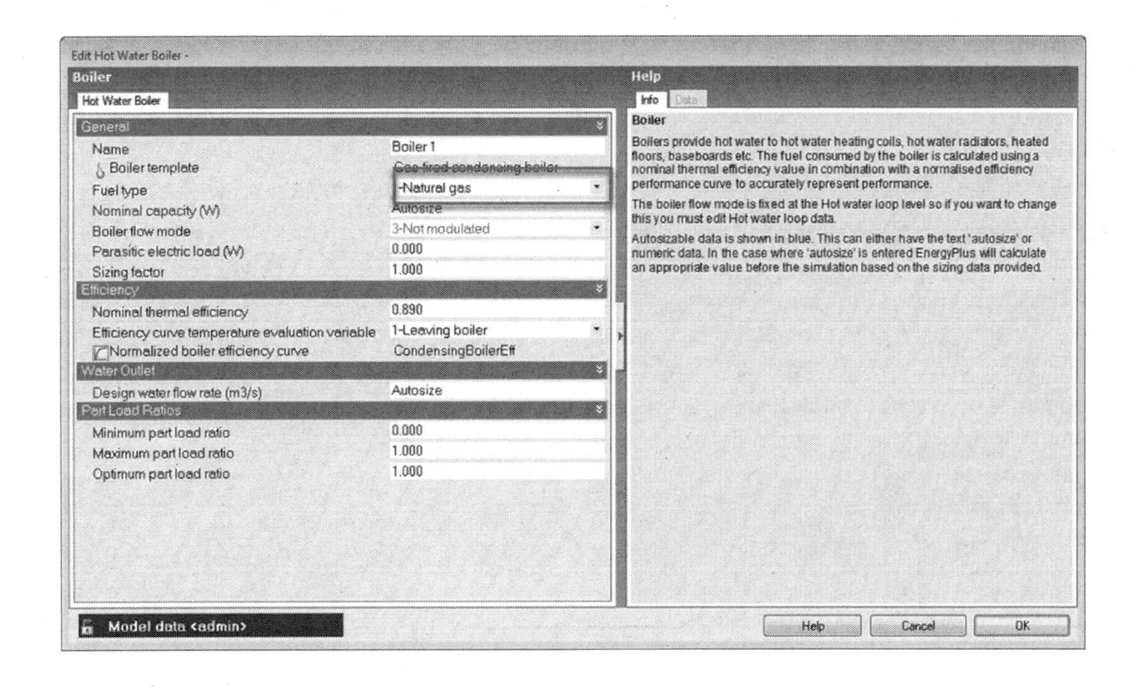

Fuel type should be the same as used in the proposed design.

Tutorial 14.11

Defining the Hot-Water Supply Temperature Reset

As per ASHRAE Standard 90.1–2010, Section G3.1.3.4, Hot-water Supply Temperature Reset (Systems 1, 5 and 7), the hot- water supply temperature shall be reset based on the outdoor dry bulb temperature by using the following schedule:

- 82°C at –7°C and below
- 66°C at 10°C and above
- Ramped linearly between 82 and 66°C at temperatures between –7 and 10°C

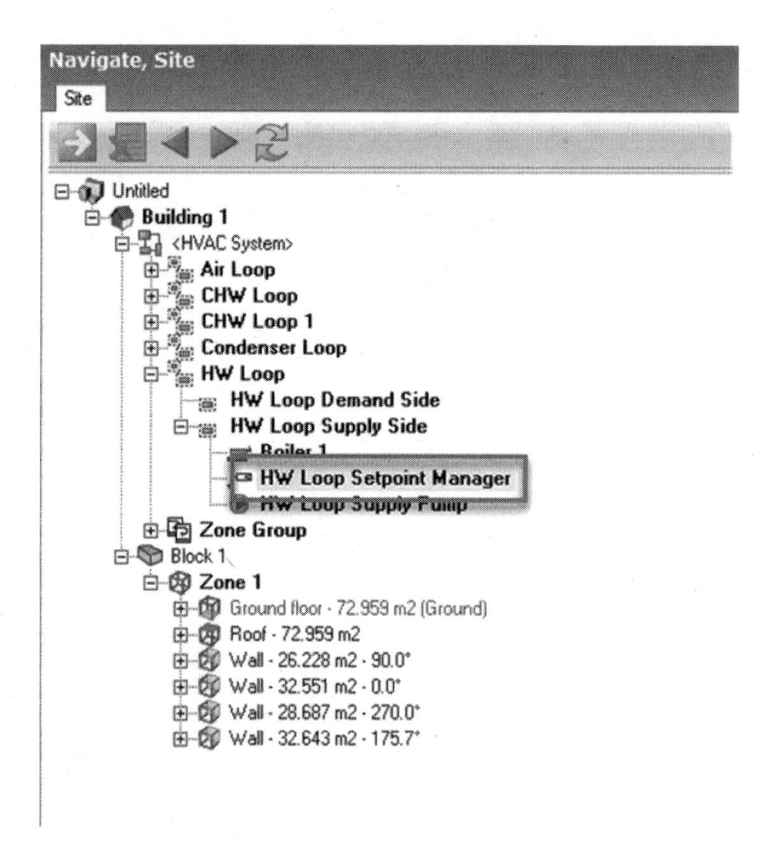

This can be modelled in **Setpoint Manager** by clicking on **HW Loop Setpoint Manager**.

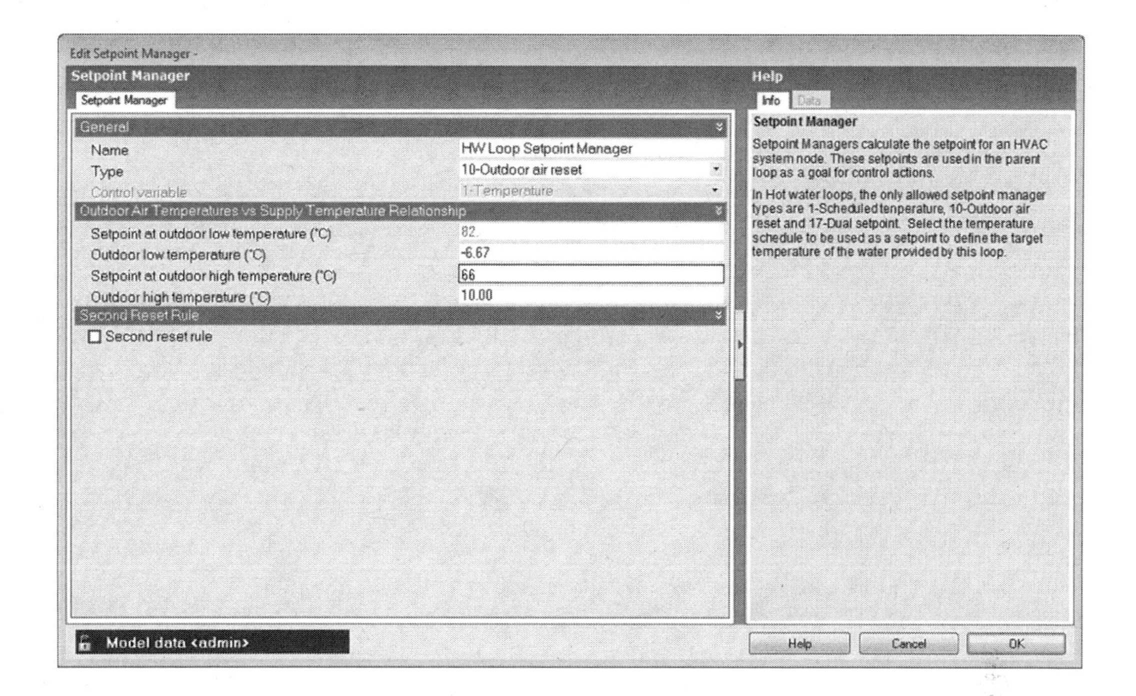
Edit Setpoint Manager -
Setpoint Manager
Setpoint Manager
General
Name
HW Loop Setpoint Manager
Type
10-Outdoor air reset
Control variable
1-Temperature
Outdoor Air Temperatures vs Supply Temperature Relationship
Setpoint at outdoor low temperature (°C)
82.
Outdoor low temperature (°C)
-6.67
Setpoint at outdoor high temperature (°C)
66
Outdoor high temperature (°C)
10.00
Second Reset Rule
Second reset rule
Help
Info
Data
Setpoint Manager
Setpoint Managers calculate the setpoint for an HVAC system node. These setpoints are used in the parent loop as a goal for control actions.
In Hot water loops, the only allowed setpoint manager types are 1-Scheduled tenperature, 10-Outdoor air reset and 17-Dual setpoint. Select the temperature schedule to be used as a setpoint to define the target temperature of the water provided by this loop.
Model data <admin>
Help
Cancel
OK

Tutorial 14.12

Hot-Water Pumps

As per ASHRAE Standard 90.1–2010, Section G3.1.3.5, Hot-water Pumps, the baseline building design hot-water pump power shall be 301 kW/1,000 l/s. The pumping system shall be modelled as primary only, with a continuous variable flow. Hot-water systems serving 11,148 m^2 or more shall be modelled with variable-speed drives, and systems serving less than 11,148 m^2 shall be modelled as riding the pump curve (see www.designbuilder.co.uk/helpv6.0/#Pump_-_Variable_Speed.htm). You can model this by clicking on **HW loop supply pump**.

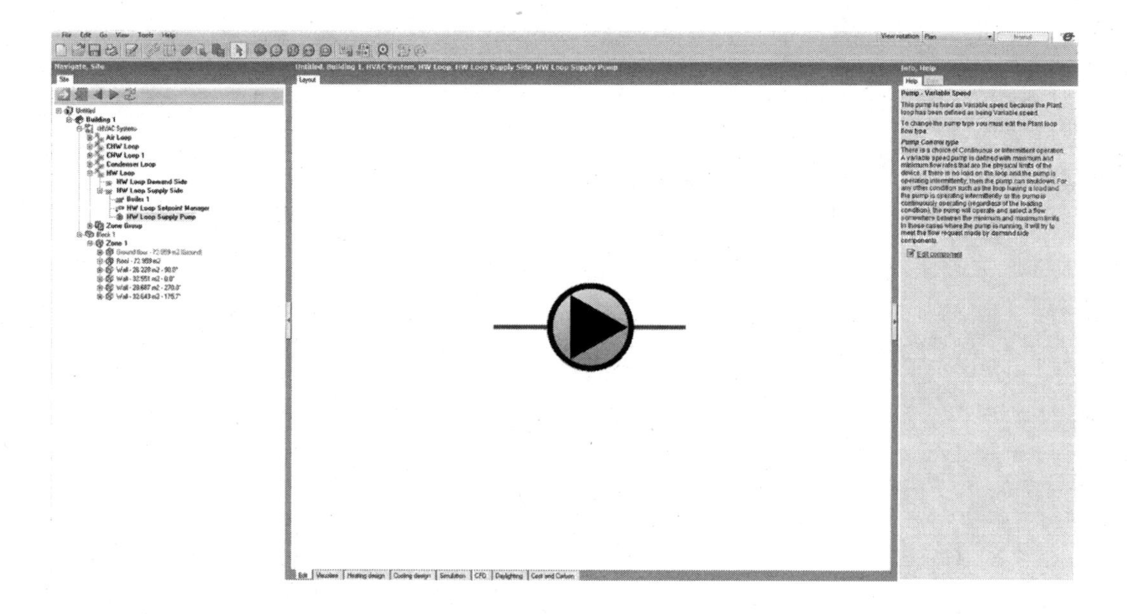

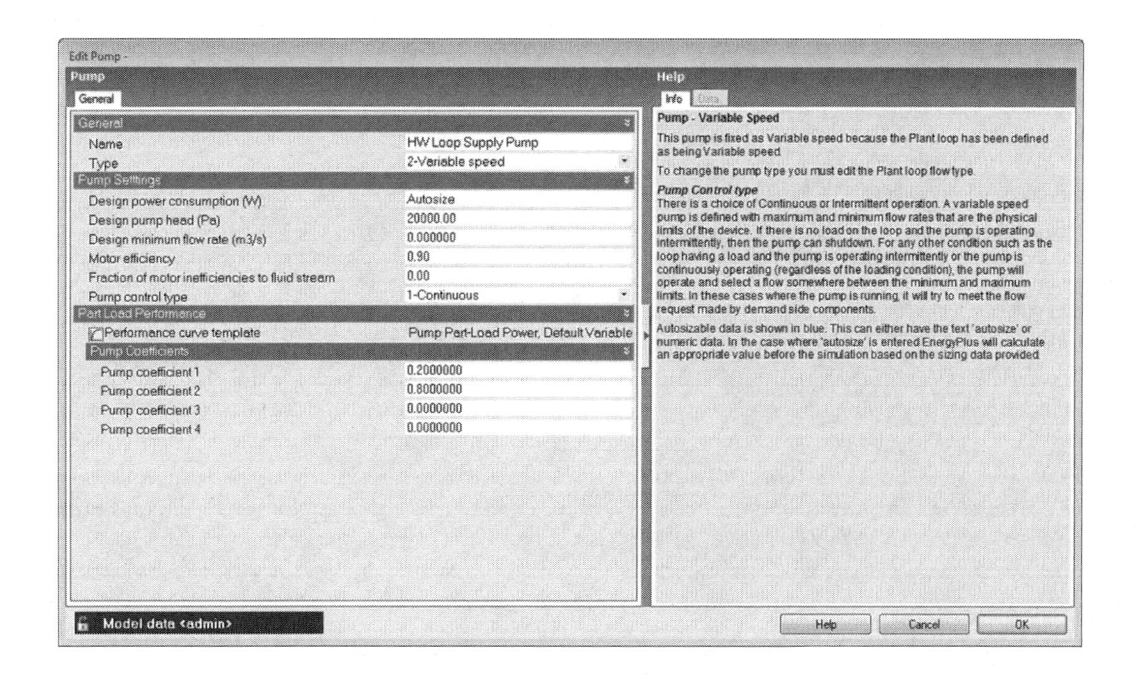
Edit Pump -
Pump
General
General
Name
HW Loop Supply Pump
Type
2-Variable speed
Pump Settings
Design power consumption (W)
Autosize
Design pump head (Pa)
20000.00
Design minimum flow rate (m3/s)
0.000000
Motor efficiency
0.90
Fraction of motor inefficiencies to fluid stream
0.00
Pump control type
1-Continuous
Part Load Performance
Performance curve template
Pump Part-Load Power, Default Variable
Pump Coefficients
Pump coefficient 1
0.2000000
Pump coefficient 2
0.8000000
Pump coefficient 3
0.0000000
Pump coefficient 4
0.0000000
Help
Info
Pump - Variable Speed
This pump is fixed as Variable speed because the Plant loop has been defined as being Variable speed.
To change the pump type you must edit the Plant loop flow type.
Pump Control type
There is a choice of Continuous or Intermittent operation. A variable speed pump is defined with maximum and minimum flow rates that are the physical limits of the device. If there is no load on the loop and the pump is operating intermittently, then the pump can shutdown. For any other condition such as the loop having a load and the pump is operating intermittently or the pump is continuously operating (regardless of the loading condition), the pump will operate and select a flow somewhere between the minimum and maximum limits. In these cases where the pump is running, it will try to meet the flow request made by demand side components.
Autosizable data is shown in blue. This can either have the text 'autosize' or numeric data. In the case where 'autosize' is entered EnergyPlus will calculate an appropriate value before the simulation based on the sizing data provided.
Model data <admin>
Help
Cancel
OK

Tutorial 14.13

Defining Exhaust Air Energy Recovery Parameters

As per ASHRAE Standard 90.1–2010, Section 6.5.6.1, each fan system shall have an energy recovery system when the system's supply airflow rate exceeds the value listed in Table 6.5.6.1, based on the climate zone and the percentage of outdoor airflow rate at design conditions. Energy recovery systems required by this section shall have at least 50% energy recovery effectiveness. Fifty percent energy recovery effectiveness shall mean a change in the enthalpy of the outdoor air supply equal to 50% of the difference between the outdoor air and return air enthalpies at design conditions. Provision shall be made to bypass or control the energy recovery system to permit air economiser operation, as required by Section 6.5.1.1. There are some exceptions to this, which you can find in Section 6.5.6.1. Refer to Tutorial 9.3 for modelling a heat recovery system.

Tutorial 14.14

Defining Economiser Parameters

As per ASHRAE 90.1–2010, Section G3.1.2.7, Economisers, outdoor air economisers shall not be included in baseline HVAC Systems 1, 2, 9 and 10. Outdoor air economisers shall be included in baseline HVAC Systems 3–8 based on climate, as specified in Table G3.1.2.6A. Exceptions can be found in the relevant section. The high-limit shutoff shall be a dry bulb switch with setpoint temperatures in accordance with the values in Table G3.1.2.6B. Refer to Tutorial 9.1 for more details.

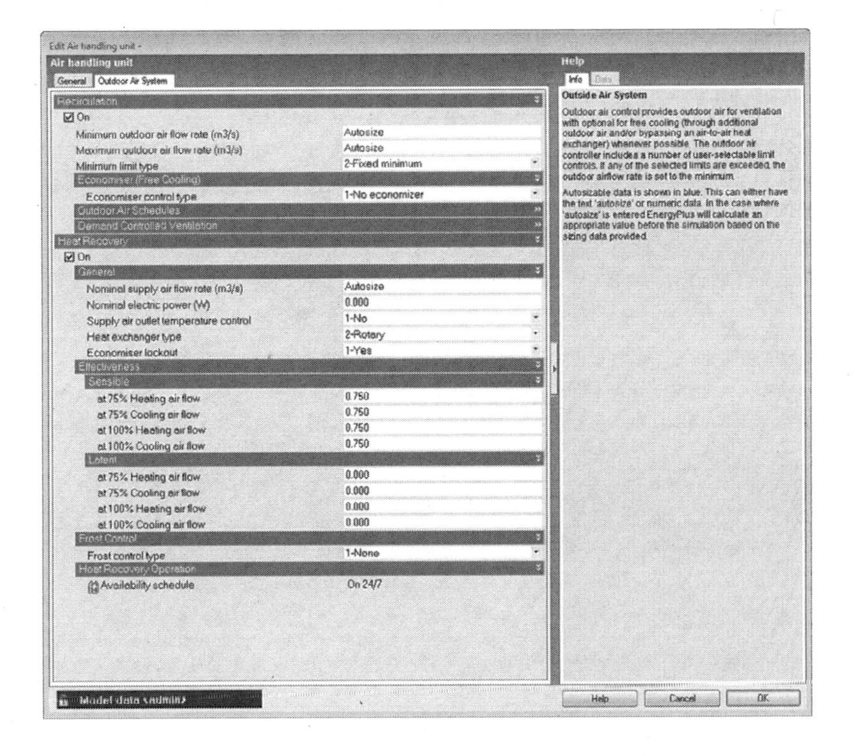

Tutorial 14.15

Finding Unmet Hours after Simulation

As per ASHRAE 90.1–2010, Section G3.1.2.3, unmet load hours for the proposed or baseline building designs should not exceed 300 h (of the 8,760 simulated hours). You can get unmet hours in the **Summary** tab after simulating the building.

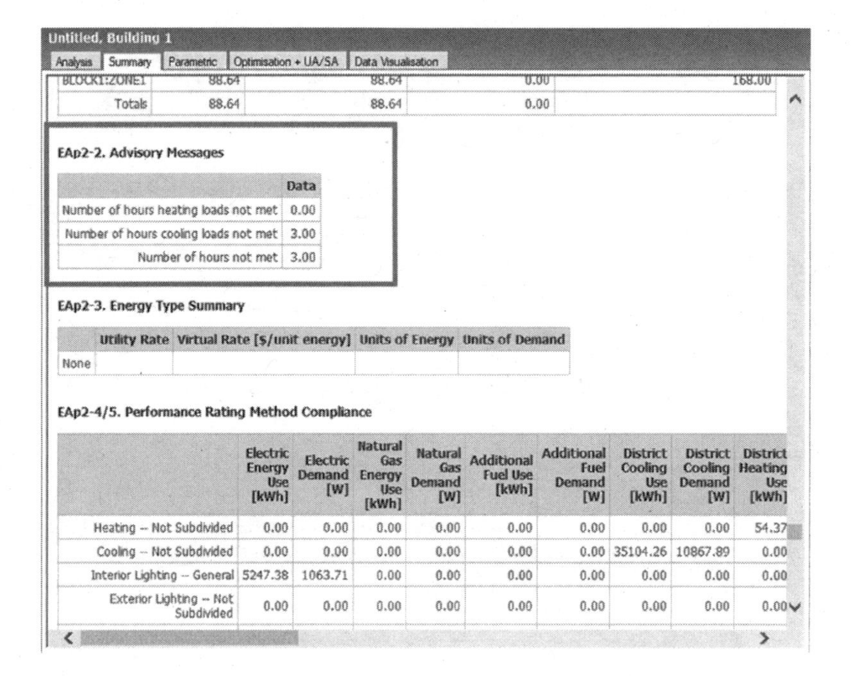

Tutorial 14.16

Generating the Performance-Rating Method Compliance Report in DesignBuilder

You can get the Leadership in Energy and Environmental Design (LEED) summary in DesignBuilder by selecting the **LEED Summary** checkbox.

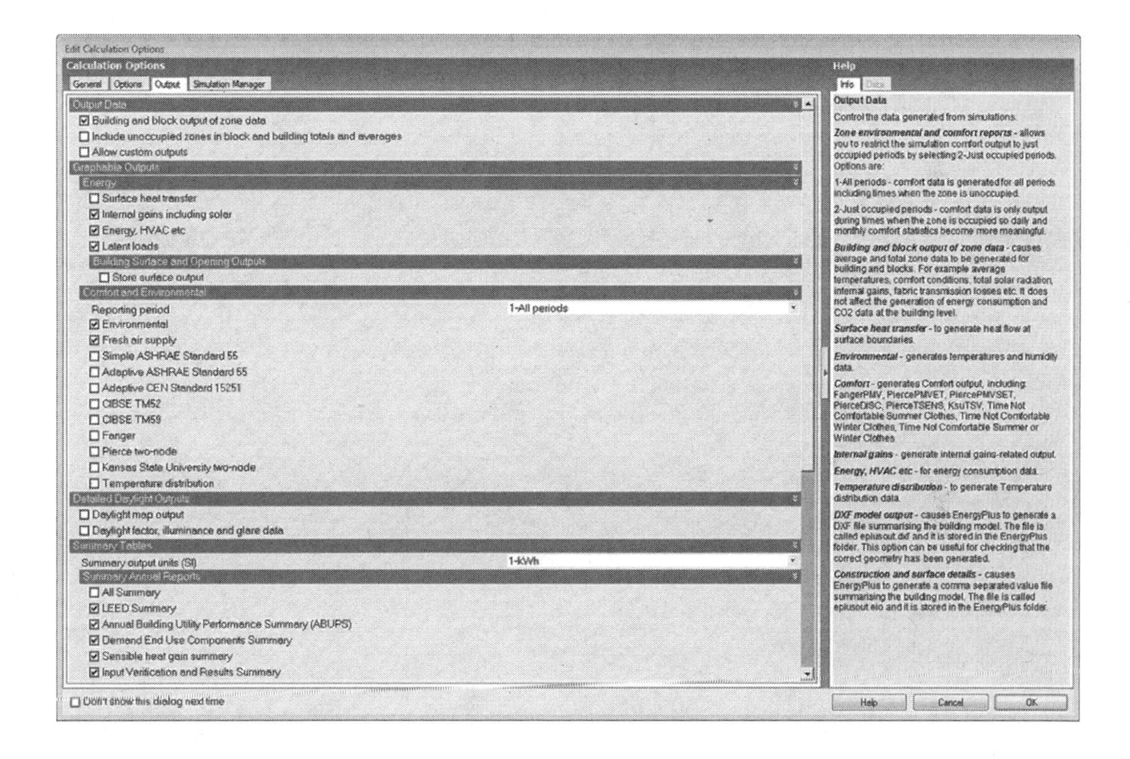

After simulation, you can get the LEED summary in the **Summary** tab.

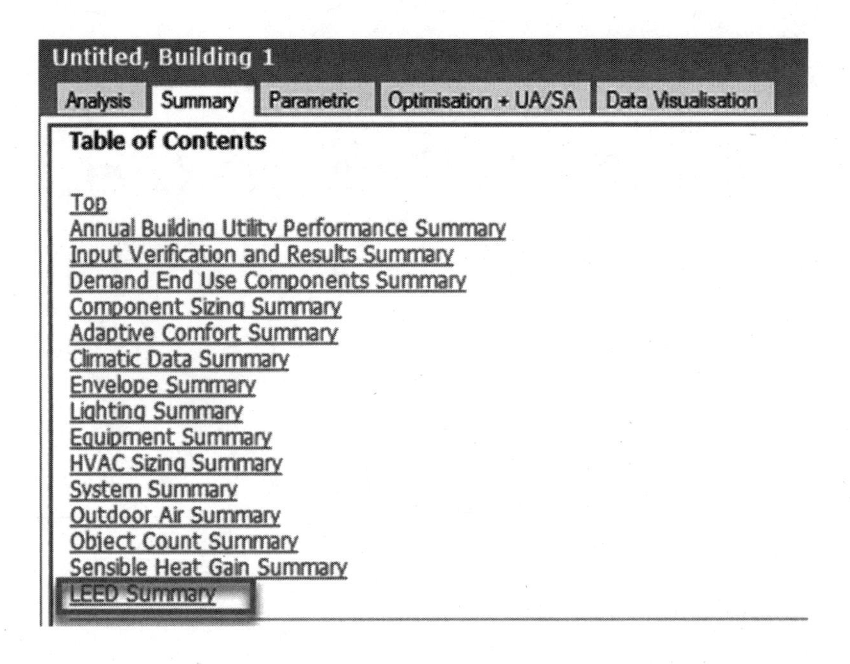

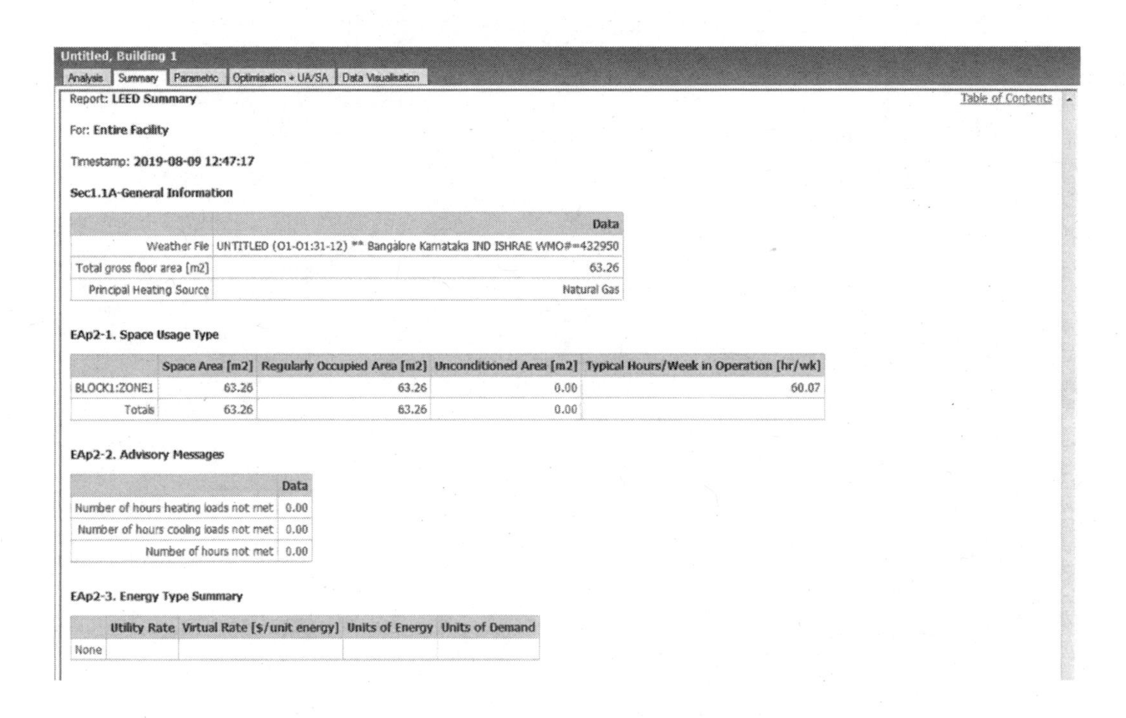

Untitled, Building 1

Analysis | Summary | Parametric | Optimisation + UA/SA | Data Visualisation

Report: **LEED Summary** Table of Contents

For: **Entire Facility**

Timestamp: **2019-08-09 12:47:17**

Sec1.1A-General Information

	Data
Weather File	UNTITLED (01-01:31-12) ** Bangalore Karnataka IND ISHRAE WMO#=432950
Total gross floor area [m2]	63.26
Principal Heating Source	Natural Gas

EAp2-1. Space Usage Type

	Space Area [m2]	Regularly Occupied Area [m2]	Unconditioned Area [m2]	Typical Hours/Week in Operation [hr/wk]
BLOCK1:ZONE1	63.26	63.26	0.00	60.07
Totals	63.26	63.26	0.00	

EAp2-2. Advisory Messages

	Data
Number of hours heating loads not met	0.00
Number of hours cooling loads not met	0.00
Number of hours not met	0.00

EAp2-3. Energy Type Summary

	Utility Rate	Virtual Rate [$/unit energy]	Units of Energy	Units of Demand
None				

Tutorial 14.17

Finding Process Load for the Base Case

Process loads must be identical for both the baseline building and the proposed building. However, project teams may follow the exceptional calculation method (ANSI/ASHRAE/IESNA Standard 90.1–2010, Section G 2.5) to document measures that reduce process loads. Documentation of process load energy savings must include a list of assumptions made for both the baseline and the proposed design and theoretical or empirical supporting information. You can search **EAp2-4/5. Performance Rating Method Compliance** in the **Summary** tab.

EAp2-4/5. Performance Rating Method Compliance

	Electric Energy Use [kWh]	Electric Demand [W]	Natural Gas Energy Use [kWh]	Natural Gas Demand [W]	Additional Fuel Use [kWh]	Additional Fuel Demand [W]	District Cooling Use [kWh]	District Cooling Demand [W]	District Heating Use [kWh]	District Heating Demand [W]
Heating -- Boiler	0.00	0.00	89.28	1363.83	0.00	0.00	0.00	0.00	0.00	0.00
Heating -- Boiler Parasitic	0.00	0.00	0.00	0.00	0.00	0.00	0.00	0.00	0.00	0.00
Cooling -- General	6880.62	2450.35	0.00	0.00	0.00	0.00	0.00	0.00	0.00	0.00
Interior Lighting -- General	3962.89	1265.29	0.00	0.00	0.00	0.00	0.00	0.00	0.00	0.00
Exterior Lighting -- Not Subdivided	0.00	0.00	0.00	0.00	0.00	0.00	0.00	0.00	0.00	0.00
Interior Equipment -- ELECTRIC EQUIPMENT#Block1:Zone1#05	2742.07	744.62	0.00	0.00	0.00	0.00	0.00	0.00	0.00	0.00
Exterior Equipment -- Not Subdivided	0.00	0.00	0.00	0.00	0.00	0.00	0.00	0.00	0.00	0.00
Fans -- General	3918.31	972.76	0.00	0.00	0.00	0.00	0.00	0.00	0.00	0.00
Pumps -- Not Subdivided	571.23	76.05	0.00	0.00	0.00	0.00	0.00	0.00	0.00	0.00
Heat Rejection -- General	1643.58	485.71	0.00	0.00	0.00	0.00	0.00	0.00	0.00	0.00
Heat Rejection -- Other	0.00	0.00	0.00	0.00	0.00	0.00	0.00	0.00	0.00	0.00
Humidification -- Not Subdivided	0.00	0.00	0.00	0.00	0.00	0.00	0.00	0.00	0.00	0.00
Heat Recovery -- Not Subdivided	0.00	0.00	0.00	0.00	0.00	0.00	0.00	0.00	0.00	0.00
Water Systems -- Not Subdivided	0.00	0.00	0.00	0.00	0.00	0.00	0.00	0.00	0.00	0.00
Refrigeration -- Not Subdivided	0.00	0.00	0.00	0.00	0.00	0.00	0.00	0.00	0.00	0.00
Generators -- Not Subdivided	0.00	0.00	0.00	0.00	0.00	0.00	0.00	0.00	0.00	0.00

EAp2-17b. Energy Use Intensity - Natural Gas

	Natural Gas [kWh/m2]
Space Heating	1.41
Service Water Heating	0.00
Miscellaneous (All)	1.41
Subtotal	1.41

EAp2-17c. Energy Use Intensity - Additional

	Additional [kWh/m2]
Subtotal	0.00
Miscellaneous	0.00

EAp2-18. End Use Percentage

	Percent [%]
Interior Lighting (All)	20.01
Space Heating	0.45
Space Cooling	34.74
Fans (All)	19.78
Service Water Heating	0.00
Receptacle Equipment	13.84
Miscellaneous	11.18

The total process energy cost must be equal to at least 25% of the baseline building performance. For buildings where the process energy cost is less than 25% of the baseline building energy cost, you should include documentation substantiating that the process energy inputs are appropriate.

Tutorial 14.18

Getting the ASHRAE 62.1 Standard Summary in DesignBuilder

Many rating systems require meeting the minimum requirements of Sections 4–7 of ASHRAE Standard 62.12007. You can get this summary from the DesignBuilder output.

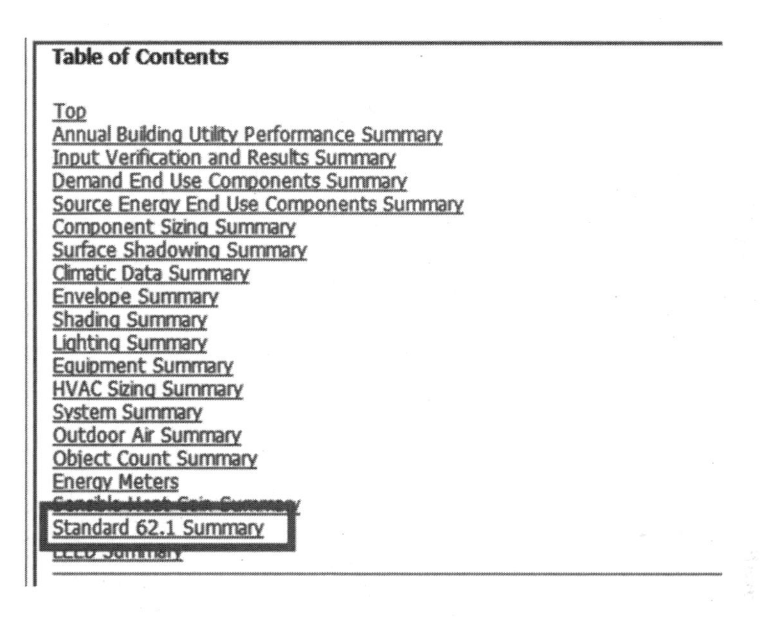

Zone Ventilation Calculations for Heating Design

	AirLoop Name	Box Type	Zone Primary Airflow - Vpz [m3/s]	Zone Discharge Airflow - Vdz [m3/s]	Minimum Zone Primary Airflow - Vpz-min [m3/s]	Zone Outdoor Airflow Heating - Voz htg [m3/s]	Primary Outdoor Air Fraction - Zpz	Primary Air Fraction - Ep	Secondary Recirculation Fraction- Er	Supply Air Fraction- Fa	Mixed Air Fraction - Fb	Outdoor Air Fraction - Fc	Zone Ventilation Efficiency - Evz
BLOCK1:ZONE4	AIR LOOP	AIRTERMINAL:SINGLEDUCT:VAV:REHEAT	0.0492	0.0492	0.0492	0.0112	0.3613	1.0000	0.0000	0.0000	0.0000	0.0000	1.0000
BLOCK1:ZONE1	AIR LOOP	AIRTERMINAL:SINGLEDUCT:VAV:REHEAT	0.0523	0.0523	0.0523	0.0159	0.4157	1.0000	0.0000	0.0000	0.0000	0.0000	1.0000
BLOCK1:ZONE3	AIR LOOP	AIRTERMINAL:SINGLEDUCT:VAV:REHEAT	0.0749	0.0749	0.0749	0.0227	0.4712	1.0000	0.0000	0.0000	0.0000	0.0000	1.0000
BLOCK1:ZONE2	AIR LOOP	AIRTERMINAL:SINGLEDUCT:VAV:REHEAT	0.0538	0.0538	0.0538	0.0170	0.4287	1.0000	0.0000	0.0000	0.0000	0.0000	1.0000

System Ventilation Calculations for Heating Design

Tutorial 14.19

Automating Baseline Building Model Creation

GOAL

To automate the process of ASHRAE 90.1–2010 baseline creation.

WHAT ARE YOU GOING TO LEARN?

- How to generate an ASHRAE 90.1–2010 baseline model
- How to view energy and cost output
- How to view unmet hours

PROBLEM STATEMENT

In this tutorial, you are going to use a G + 2 floor building model. Each floor has a 100- × 50-m area with five zones and a 10-m perimeter depth. Other modelling parameters are as follows:

S. no.	Model input parameter	Proposed case value
1	Exterior Wall Construction	Light Weight Metallic Cladding 50 mm + Foam – polyurethane 75 mm + Cement/plaster/mortar – gypsum plasterboard (10 mm)
2	Roof Construction	100 mm Concrete (1,800 kg/m^3) + Foam – polyurethane 75 mm + Cement/plaster/mortar – gypsum plasterboard (10 mm)
3	Glazing	Double glazed Unit – Dbl Blue 6 mm/13 mm Air U-2.662 W/m^2-K, SHGC 0.49 and VLT 50.5%
4	Window to Wall Ratio (%)	60
5	Fresh air CFM/Person	6 l/s-person and 0.305 l/s-m^2
6	Equipment Power Density (W/m^2)	8
7	Occupancy (m^2/person)	10
8	Lighting Power Density (W/m^2)	8
9	Shading Devices	Overhang 0.5 m on all windows
10	Daylight Sensors	Installed in all perimeter spaces
11	HVAC System Type	VAV system
12	Chiller Parameter	Water-cooled centrifugal chiller, COP 6.17
13	Winter Heating Source	Natural gas–fired boiler
14	Boiler efficiency	96%
15	Cooling Tower Fan	Variable speed
16	Energy Rates	Electricity 0.078 USD/kWh Gas 0.0545 USD/Therm

Compare the annual energy and annual energy cost for proposed and baseline models for **Florida Miami USA**.

> The DesignBuilder LEED module can generate ASHRAE 90.1, Appendix G, performance rating method buildings.
>
> *Source*: https://designbuilder.co.uk//leed.

SOLUTION

Step 1: Open DesignBuilder. Click the **File** menu, and select **New File**.

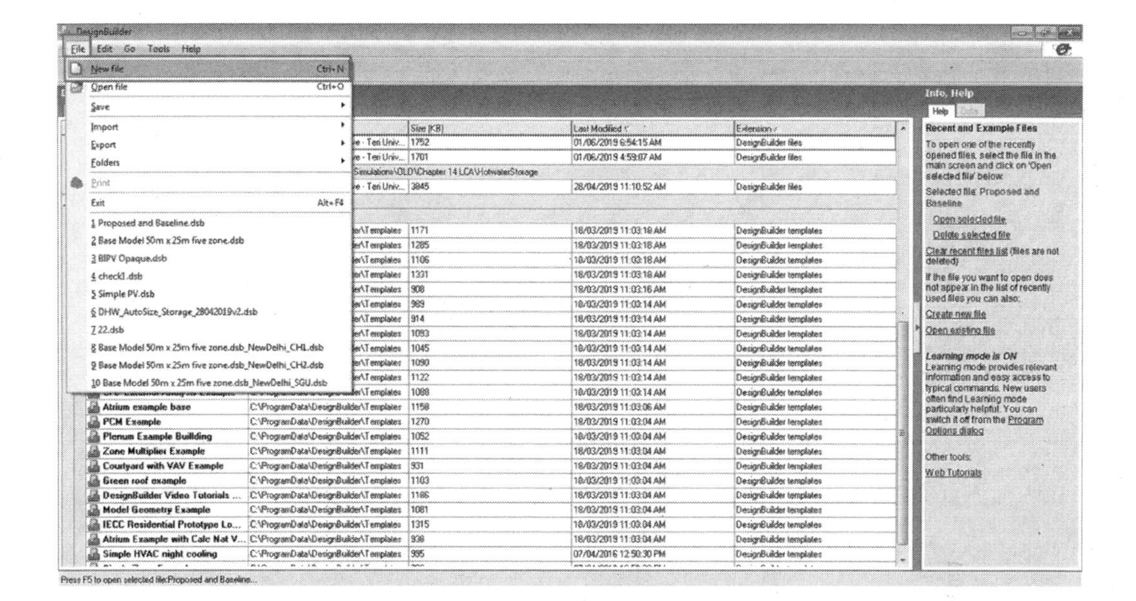

Step 2: Select **FL-MIAMI** as **Location**. Select the **ASHRAE 90.1 App G PRM** checkbox. Select **2-ASHRAE 90.1-2010** from the **ASHRAE 90.1 energy code** drop-down list.

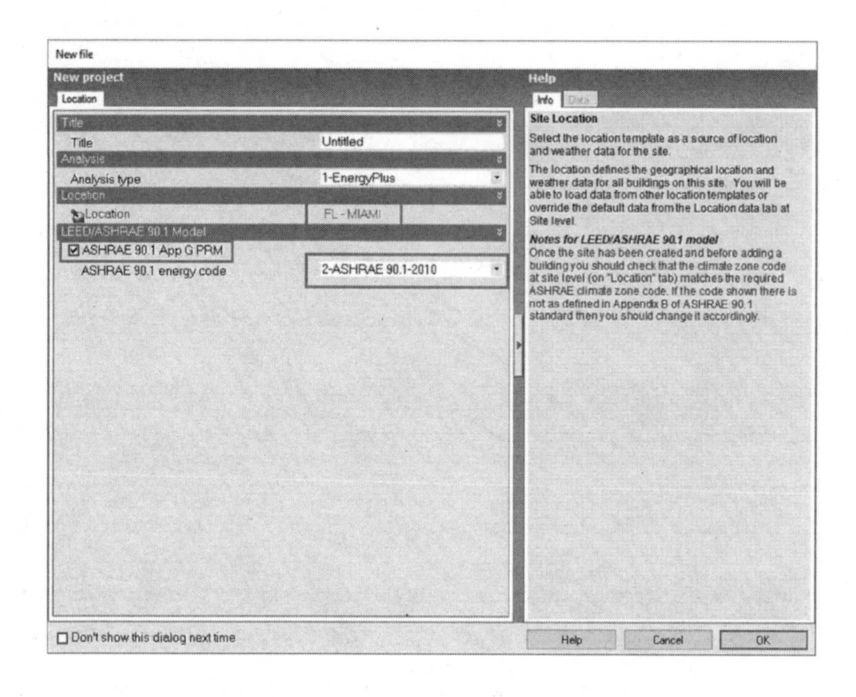

Step 3: Click on **Add new building**.

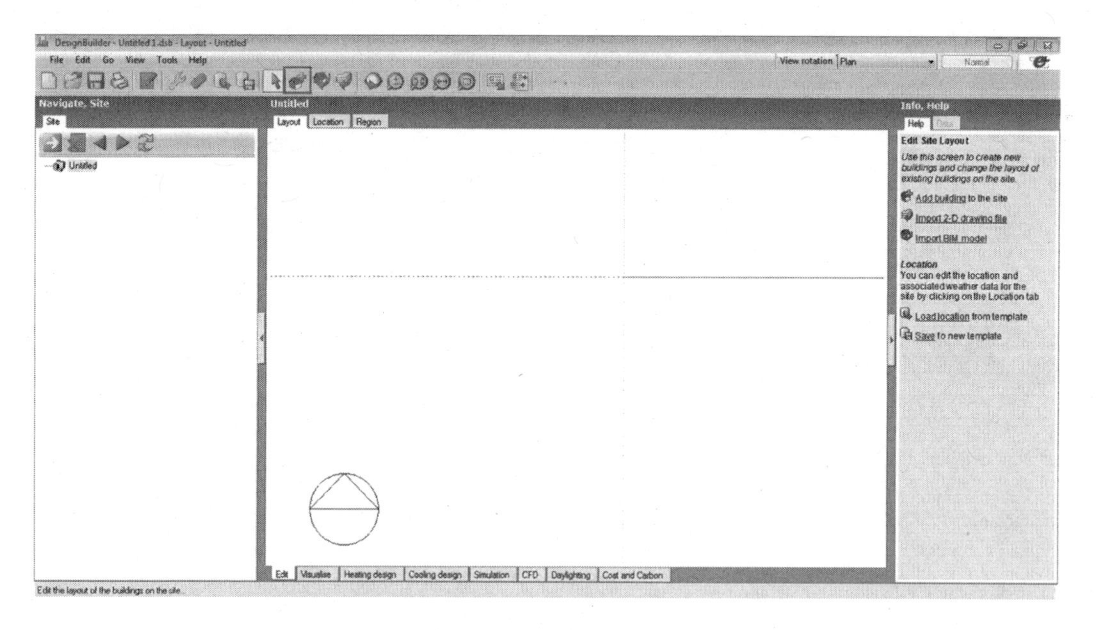

Step 4: Select **ASHRAE 90.1** for both **Type** and **Geometry convention template**. Select **1-Proposed** from **ASHRAE 90.1 building type**. Select **1-New building** from the **LEED modelling category** drop-down list. Select **1-Non-residential** from the **Primary building condition category** drop-down list.

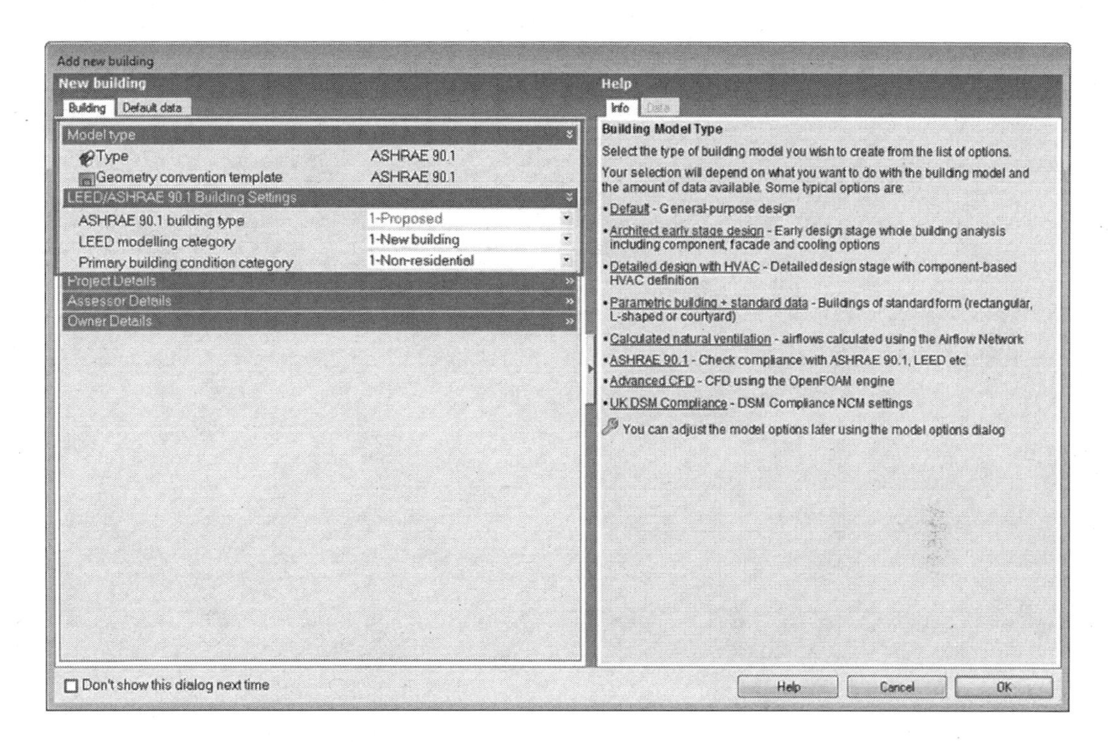

When the mandatory energy code is ASHRAE 90.1–2007 or 2010, then on the **Add new building** dialogue box, the **Model type** template shows ASHRAE 90.1. This loads some appropriate model data settings such as **Detailed HVAC** and **6 timesteps per hour** to the building model. On the **Add new building** dialogue box, you can select either **1-Proposed** or **2-Baseline** for the **ASHRAE 90.1 building type**. DesignBuilder does not currently generate different settings based on this selection, but it always generates baseline constructions, and construction templates in accordance with building envelope requirements defined in ASHRAE 90.1, Chapter 5 (detailed in Tables 5.5-1 through 5.5-8) can be selected as required for both proposed and baseline buildings. Following formal ASHRAE 90.1 PRM modelling procedure, you are strongly advised to create the proposed building first and create the baseline building from that.

Source: https://designbuilder.co.uk/helpv6.0/#ASHRAE90.1Modelling.htm.

Step 5: Create a **100- × 50-m** five-zone model with a **10-m** perimeter depth. Clone the floor to the other two floors.

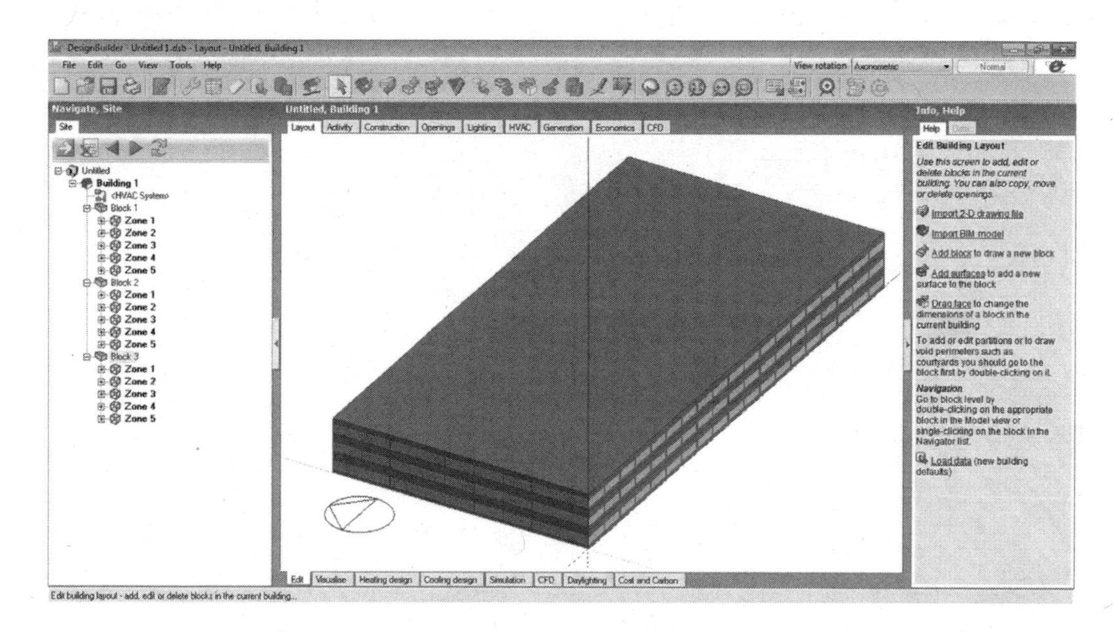

Step 6: Update the model with proposed case parameters as provided in the problem statement. (You need to make changes in parameters under the **Activity**, **Construction**, **Openings**, **Lighting**, **HVAC** and **Economics** tabs. Refer to previous chapters for details.)

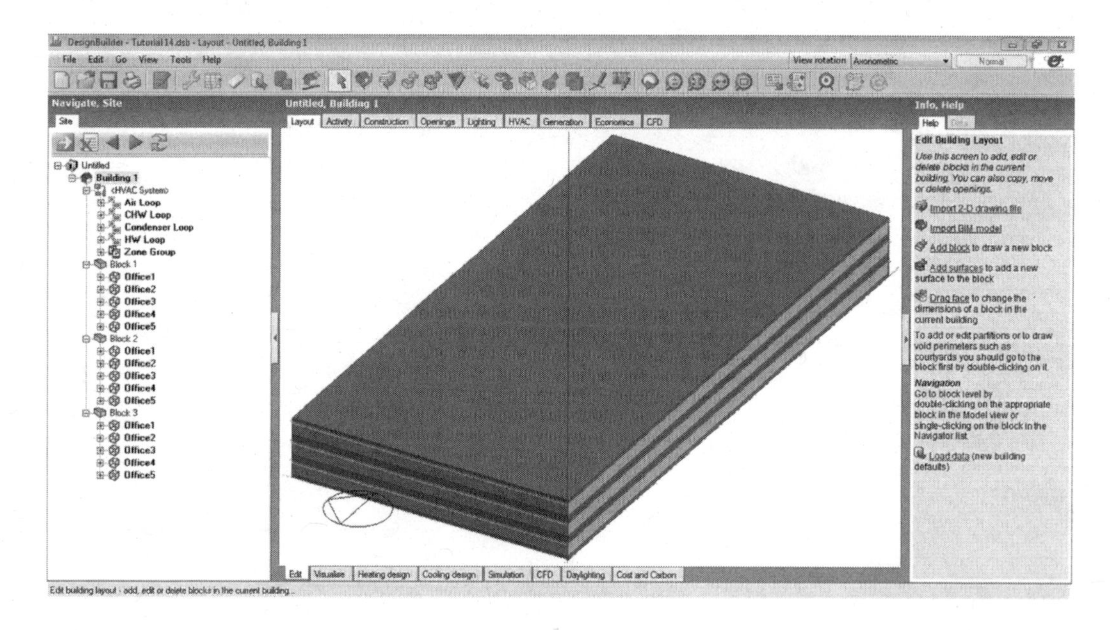

Step 7: Click **Generate baseline building**. The **Generate baseline building** screen appears.

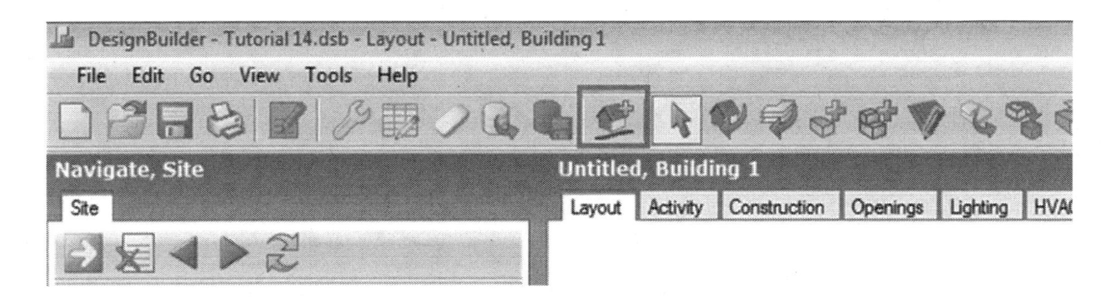

Step 8: Confirm that all the values are as shown in the following figure. Click **Next**.

Generate Baseline Building

Proposed Building Information

Building Use

ASHRAE 90.1 Jurisdiction	
ASHRAE 90.1 energy code	ASHRAE 90.1-2010
Building Details	
Primary building condition category	Non-residential
Model Details	
Site location	FL - MIAMI
Weather file	USA_FL_MIAMI INTL AP_TMY3
Climate zone	1A
Building Use	
Total floor area (m2)	15000.000
Total conditioned floor area (m2)	15000.000
Total primary conditioned floor area (m2)	15000.000

Cancel Help Back Next Finish

Step 8: Confirm that all the values are as shown in the following figure. Click **Next**.

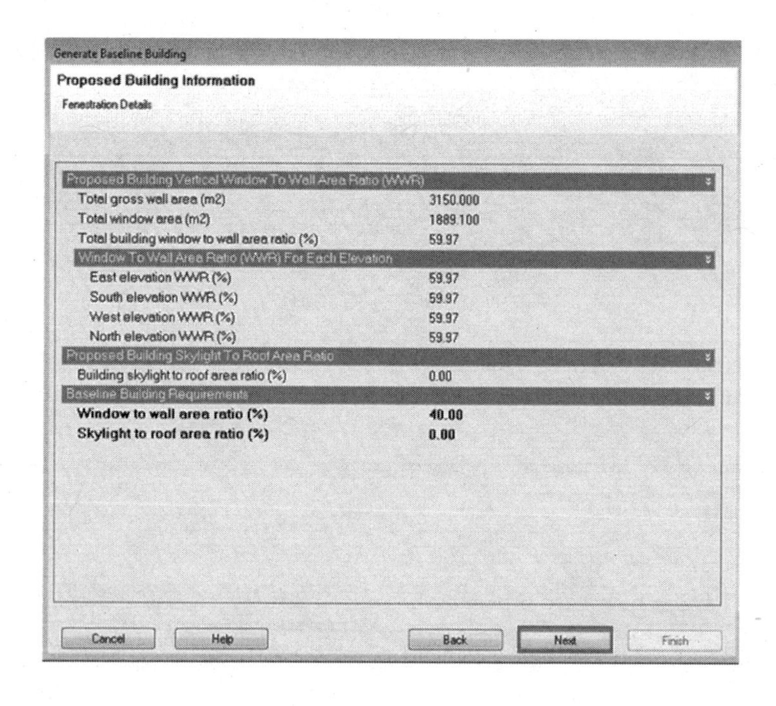

Step 9: Confirm that all the values are as shown in the following figure. Click **Finish**. The **ASHRAE 90.1 Baseline model creation progress** bar appears.

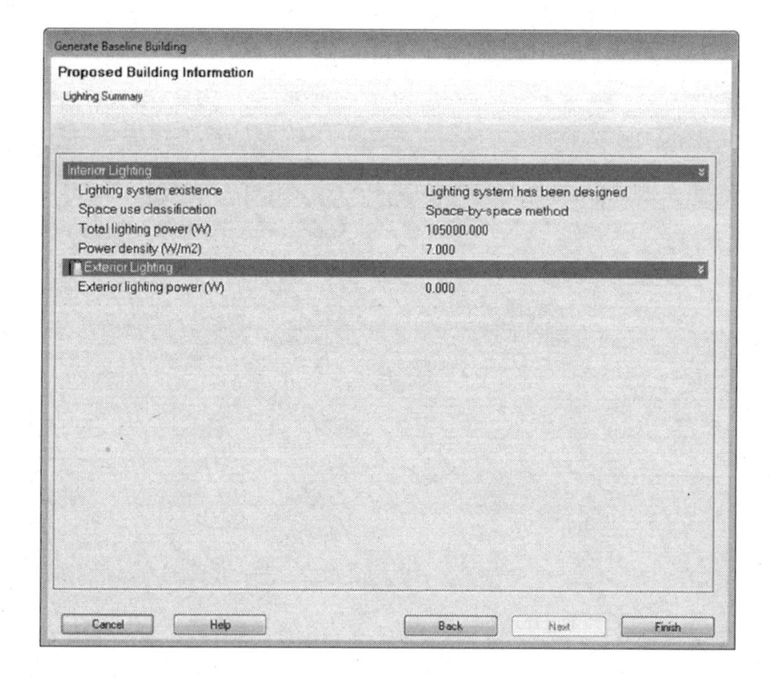

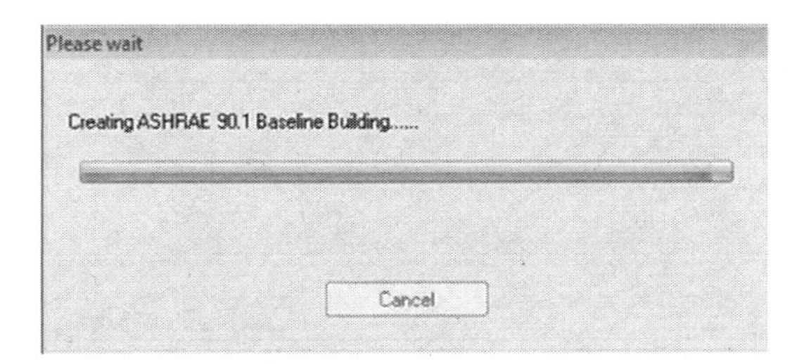

Once the models are created, you can see both the proposed building model (**Building 1**) and the baseline building model (**Building 1_Baseline building**) in the navigation tree.

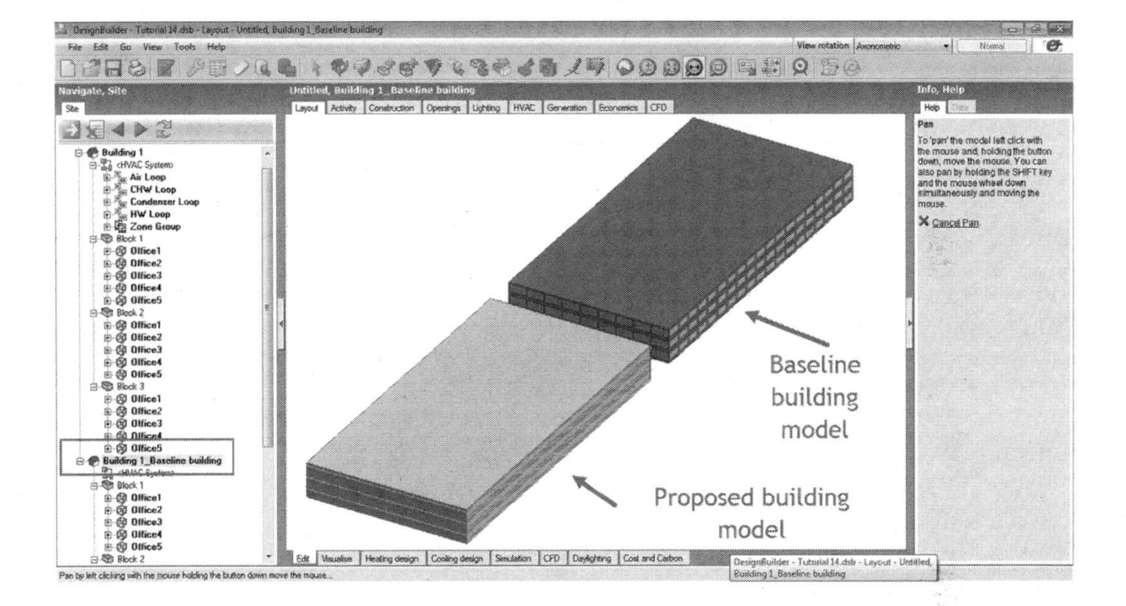

Step 10: Now you need to select the baseline HVAC system. Select **HVAC system** in the **Baseline building**.

A set of Appendix G HVAC systems for loading to baseline building models is provided in the **Detailed HVAC templates** list. The proposed building HVAC system should meet the requirements shown in Appendix G, Table G3.1-10. A modified copy of a baseline HVAC system can be used for the proposed building. For the proposed building design, where no heating or cooling systems exist or has been specified, the heating or cooling systems shall be identical to the systems modelled in the baseline building design.

Source: https://designbuilder.co.uk/helpv6.0/#ASHRAE90.1Modelling.htm.

Step 11: Select the **Do not want to run cooling design calculation or it is not needed** check-box. Click **Next**.

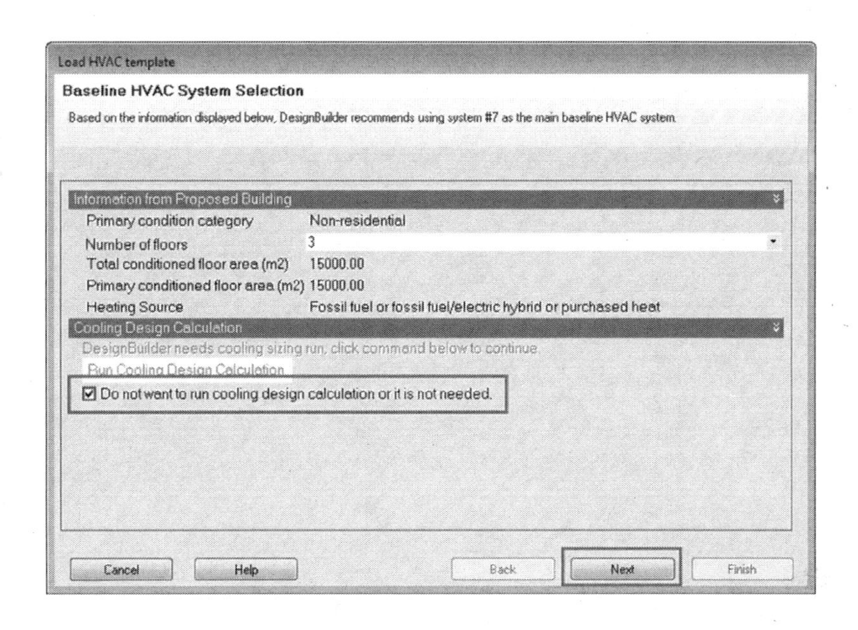

Step 12: Click **Next**.

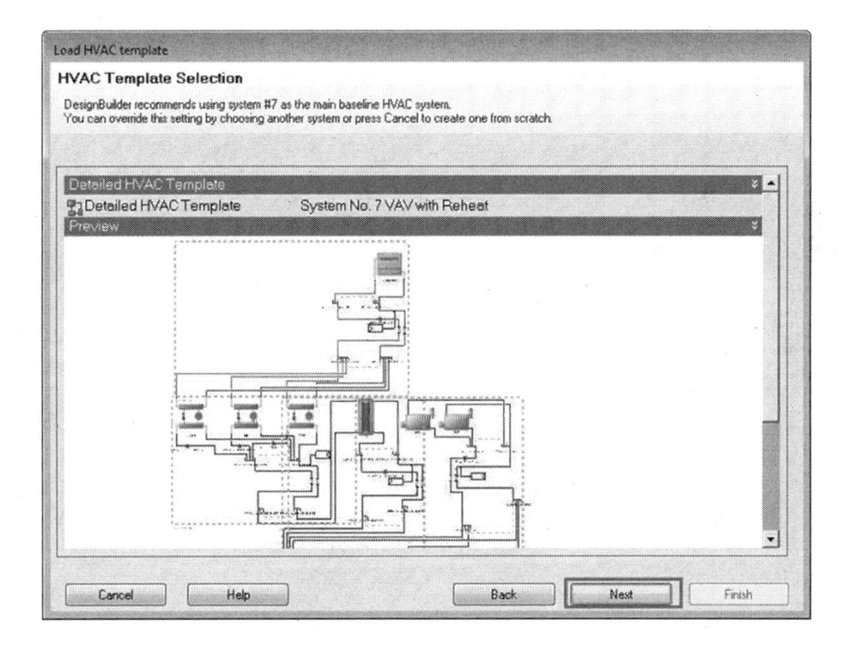

Step 13: Click **Finish**. The **Baseline HVAC system** appears in the layout.

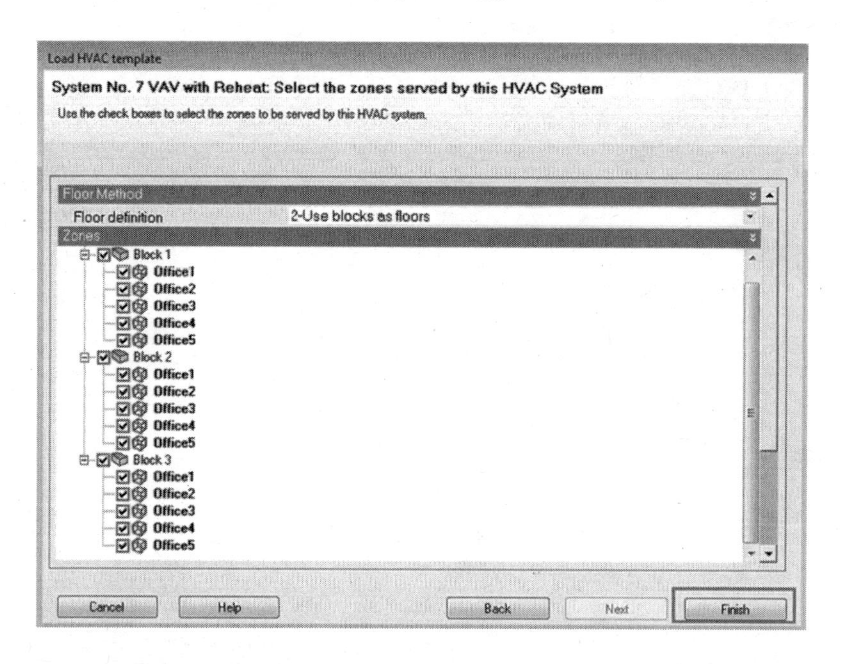

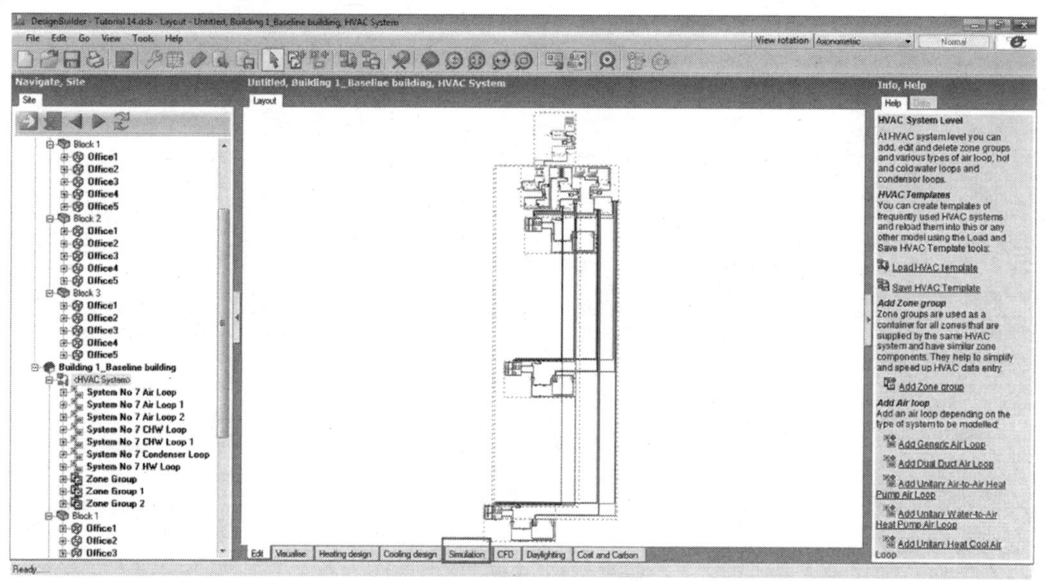

Step 14: Click **Building 1_Baseline building**. Click the **Openings** tab. Observe that WWR is set to 40% and shading is removed in the baseline model. Similarly, other parameters can be checked.

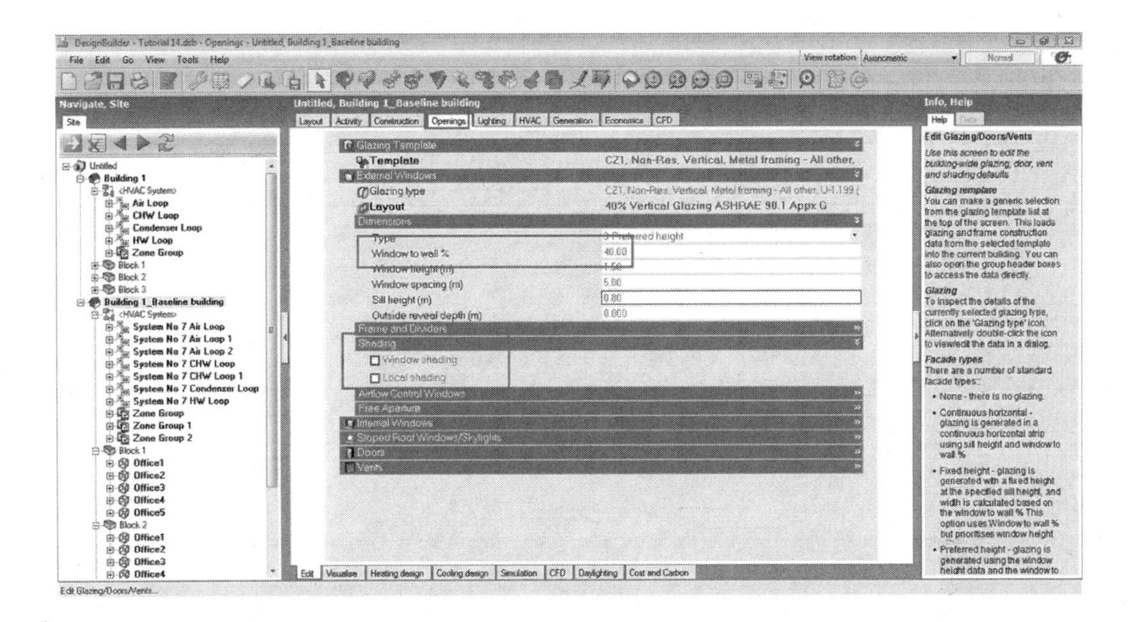

Step 15: Click the **Simulation screen** tab.

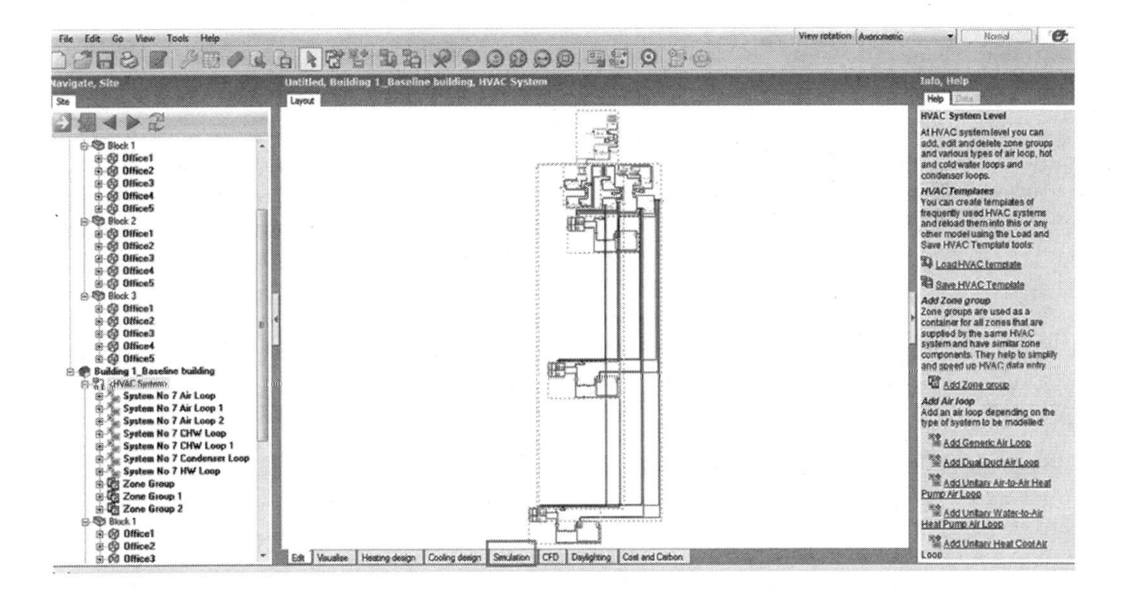

Step 16: Click the **ASHRAE 90.1** tab. Click **Update data**. The **Edit Calculation Options** screen appears.

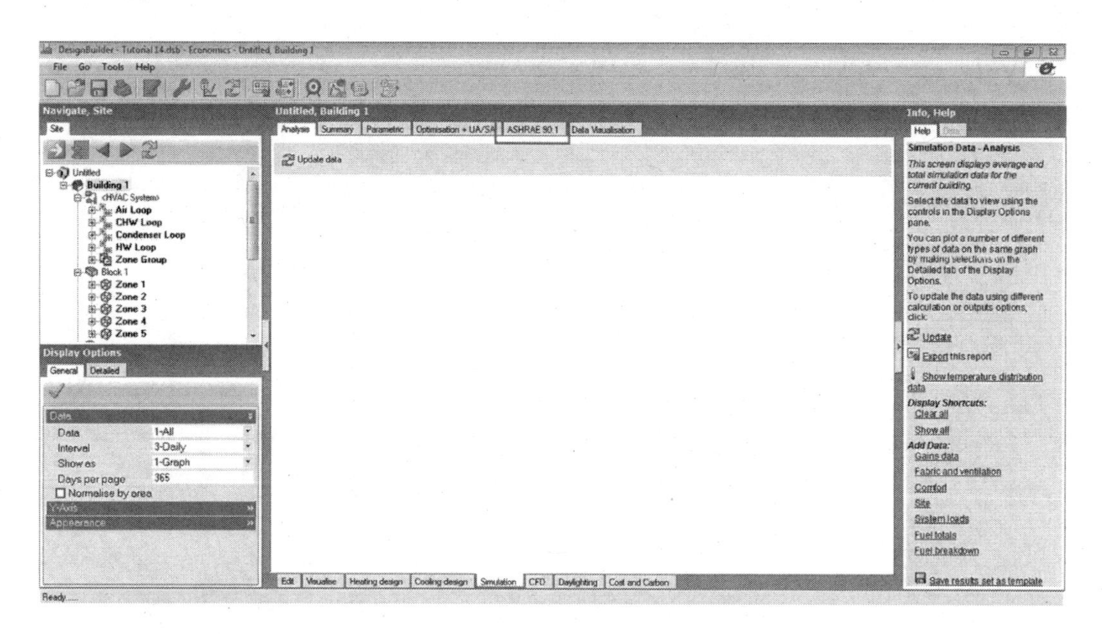

Step 17: Click **OK**. The baseline results appear in the output.

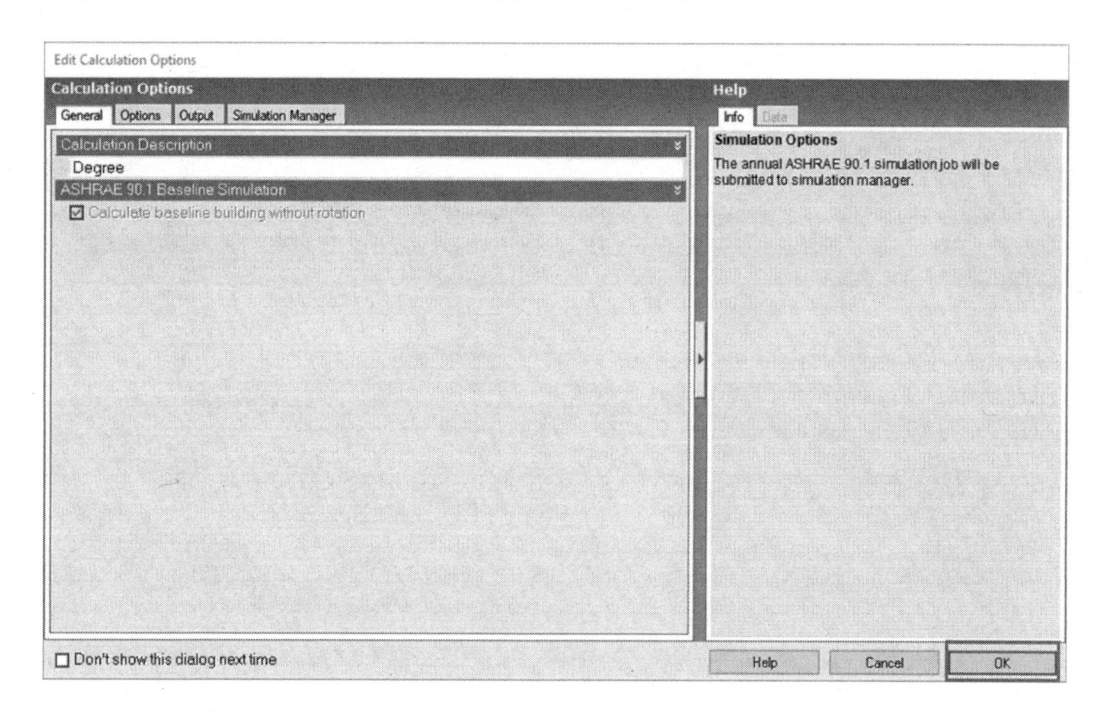

To view generated IDF files for all orientations, you need to open the **ASHRAE901** folder under **EnergyPlus** folder. The path of the files is **C:\Users\XXX\AppData\Local \DesignBuilder\EnergyPlus\ASHRAE901.**

BSRC-Aviruch ▸ AppData ▸ Local ▸ DesignBuilder ▸ EnergyPlus ▸ ASHRAE901

Name	Date modified	Type	Size
eplusout_B180.eio	01/06/2019 5:26 PM	EIO File	134 K
eplusout_B270.eio	01/06/2019 5:26 PM	EIO File	134 K
eplusout_P000.eio	01/06/2019 3:59 PM	EIO File	316 K
eplusout_B000.end	01/06/2019 5:26 PM	END File	1 K
eplusout_B090.end	01/06/2019 5:26 PM	END File	1 K
eplusout_B180.end	01/06/2019 5:26 PM	END File	1 K
eplusout_B270.end	01/06/2019 5:26 PM	END File	1 K
eplusout_P000.end	01/06/2019 3:59 PM	END File	1 K
in_B000.idf	01/06/2019 5:26 PM	EnergyPlus Input ...	545 K
in_B090.idf	01/06/2019 5:26 PM	EnergyPlus Input ...	545 K
in_B180.idf	01/06/2019 5:26 PM	EnergyPlus Input ...	545 K
in_B270.idf	01/06/2019 5:26 PM	EnergyPlus Input ...	545 K
in_P000.idf	01/06/2019 3:59 PM	EnergyPlus Input ...	515 K
eplusout_B000.err	01/06/2019 5:26 PM	Error log	22 K
eplusout_B090.err	01/06/2019 5:26 PM	Error log	22 K
eplusout_B180.err	01/06/2019 5:26 PM	Error log	22 K
eplusout_B270.err	01/06/2019 5:26 PM	Error log	22 K

eplusout_B270.dxf State: Shared Size: 35.4 KB
AutoCAD Drawing Interchange Date modified: 01/06/2019 5:26 PM Date created: 01/06/2019 5:26 PM

Step 18: Click **Building 1**, and update the results.

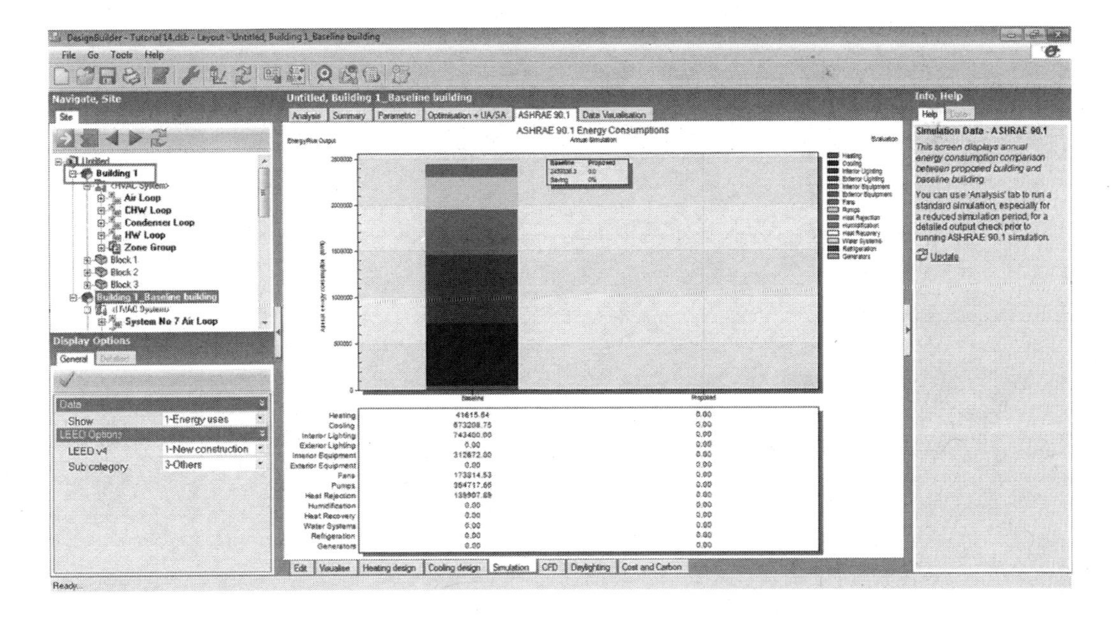

Step 19: Make sure that **1-Energy** uses is selected from **the Show** drop-down list. Record the savings.

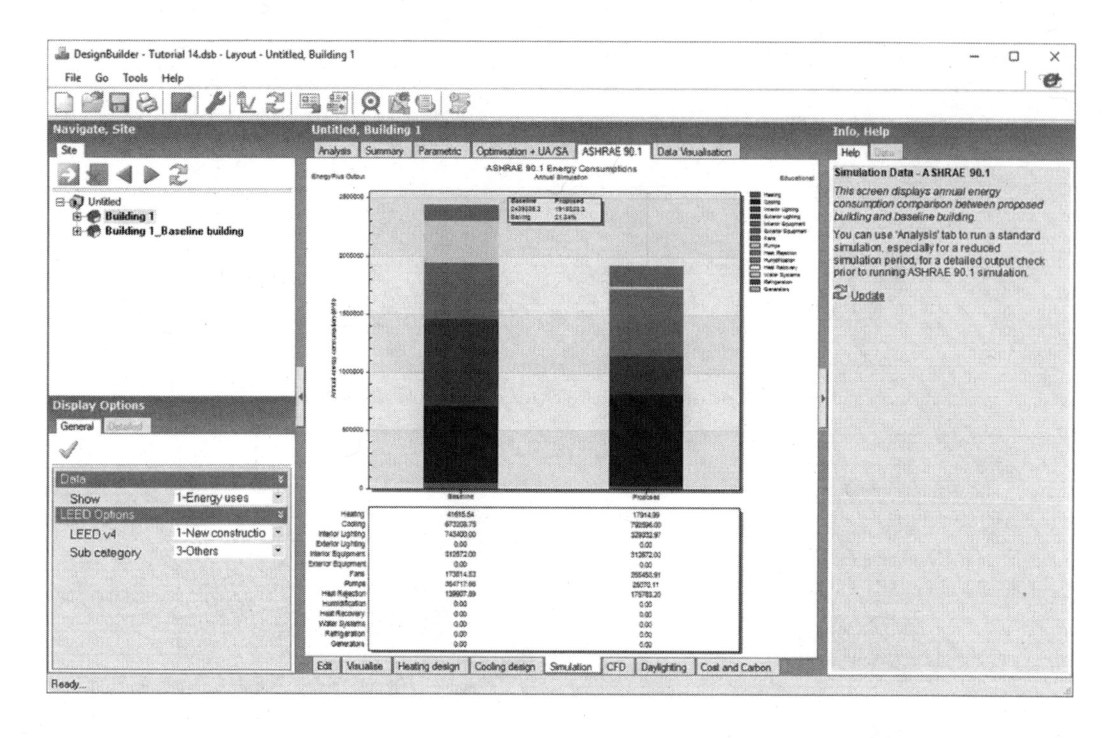

Step 20: Select 2-**Costs** from **the Show** drop-down list.

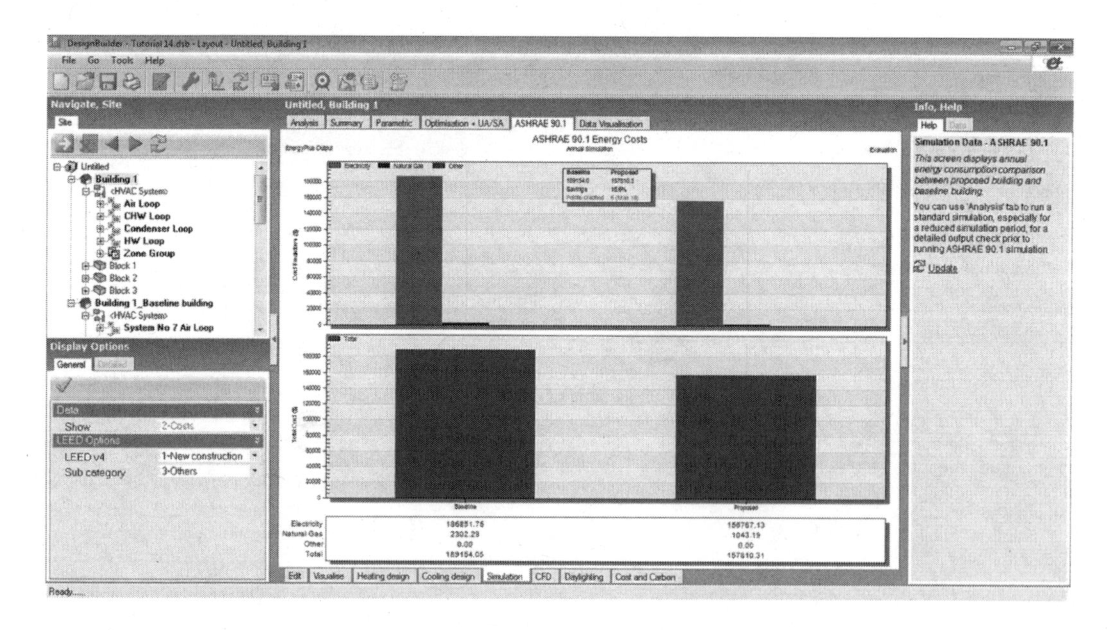

Step 21: Click the **Summary** tab to record the unmet hours.

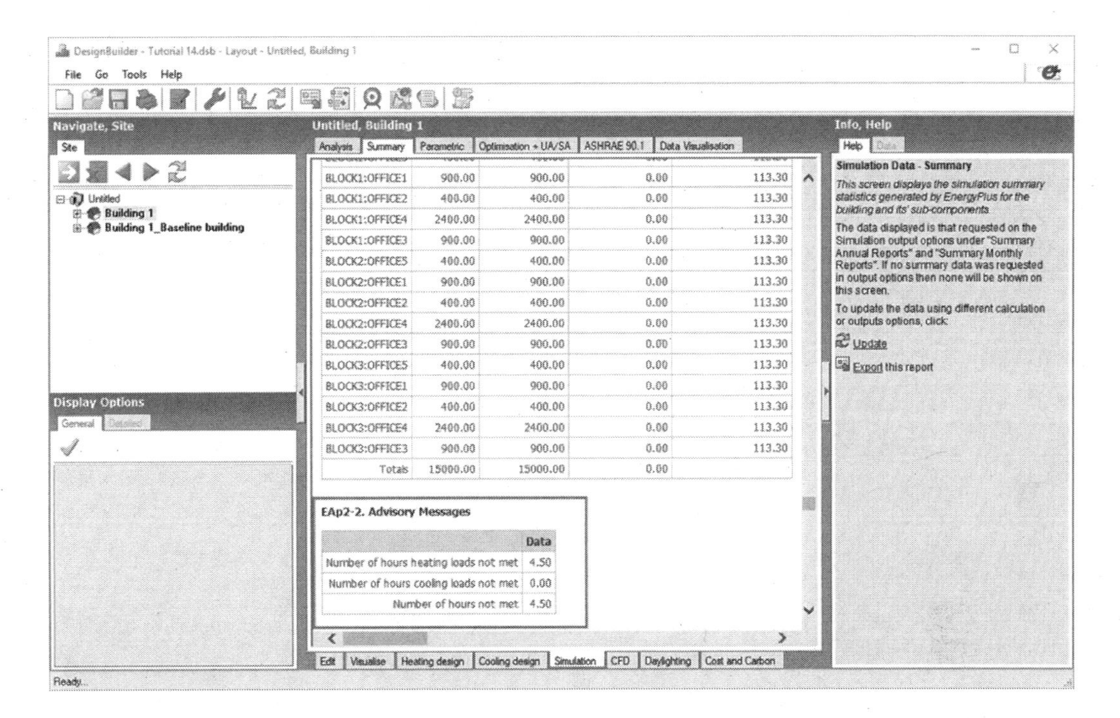

REFERENCE

American Society of Heating, Refrigerating and Air-Conditioning Engineers (ASHRAE), ASHRAE Standard 90.1–2010: *Energy Standard for Buildings Except Low-Rise Residential Buildings.* Atlanta, GA: ASHRAE, www.ashrae.org.